Zoology

Zoology

Steven A. Miller

College of the Ozarks

John P. Harley

Eastern Kentucky University

WCB　Wm. C. Brown Publishers

Book Team

Editor *Kevin Kane*
Developmental Editor *Margaret J. Manders*
Photo Editor *Mary Roussel*
Permissions Editor *Vicki Krug*
Visuals Processor *Joyce E. Watters*

 Wm. C. Brown Publishers

President *G. Franklin Lewis*
Vice President, Publisher *George Wm. Bergquist*
Vice President, Publisher *Thomas E. Doran*
Vice President, Operations and Production *Beverly Kolz*
National Sales Manager *Virginia S. Moffat*
Advertising Manager *Ann M. Knepper*
Marketing Manager *Craig S. Marty*
Editor in Chief *Edward G. Jaffe*
Managing Editor, Production *Colleen A. Yonda*
Production Editorial Manager *Julie A. Kennedy*
Production Editorial Manager *Ann Fuerste*
Publishing Services Manager *Karen J. Slaght*
Manager of Visuals and Design *Faye M. Schilling*

Cover photo: Atlantic puffins, copyright Dick Poe/Visuals Unlimited.

Photograph on page ii (Frontispiece): Arctic fox in summer pelage, copyright Bruce Cushing/Visuals Unlimited.

Photograph on page xvi (facing preface): Iguana, copyright R. DeGoursey/Visuals Unlimited.

Photo research by John Cunningham.

The credits section for this book begins on page 719, and is considered an extension of the copyright page.

Library of Congress Catalog Card Number: 90–80170

ISBN 0–697–03347–3

Printed in the United States of America by Wm. C. Brown Publishers, 2460 Kerper Boulevard, Dubuque, IA 52001

10 9 8 7 6 5 4 3 2 1

Contents

Part I: Introduction, Cells, Molecules, and Life

Part II: The Continuity of Life

Part III: Evolution

Part IV: Animallike Protists and Animalia

Part V: Form and Function

Part VI: Behavior and Ecology

Colorplates

A Guided Tour through Miller/Harley *Zoology*

This Guided Tour highlights the teaching aids provided in this textbook. The tools and aids noted here are found throughout the book and should be useful to the student in making the learning of zoology a stimulating and enjoyable experience.

Outline of the chapter tells what topics are included.

Concepts indicate the major lessons to be learned.

Have You Ever Wondered— questions to pique curiosity, with the answers keyed in text.

22

The Arthropods: Blueprint for Success

Outline

Evolutionary Perspective
  Classification and Relationships
    to Other Animals
Metamerism and Tagmatization
The Exoskeleton
Metamorphosis
Subphylum Trilobita
Subphylum Chelicerata
  Class Merostomata
  Class Arachnida
  Class Pycnogonida
Subphylum Crustacea
  Class Malacostraca
  Class Branchiopoda
  Class Copepoda
  Class Cirripedia
Further Phylogenetic
    Considerations

Concepts

1. Arthropods have been successful in almost all habitats on the earth. Some ancient arthropods were the first animals to live most of their lives in terrestrial environments.
2. Metamerism with tagmatization, an exoskeleton, and metamorphosis have contributed to the success of arthropods.
3. Members of the subphylum Trilobita are extinct arthropods that were a dominant form of life in the oceans between 345 and 600 million years ago.
4. Members of the subphylum Chelicerata have a body divided into two regions and have chelicerae. The class Merostomata contains the horseshoe crabs. The class Arachnida contains the spiders, mites, ticks, and scorpions. Some ancient arachnids were the first terrestrial arthropods, and modern arachnids have numerous adaptations for terrestrial life. The class Pycnogonida contains the sea spiders.
5. Animals in the subphylum Crustacea have biramous appendages and two pairs of antennae. The class Branchiopoda includes the fairy shrimp, brine shrimp, and water fleas. The class Malacostraca include the crabs, lobsters, crayfish, and shrimp, and the classes Copepoda and Cirrepedia include the copepods and barnacles, respectively.

Have You Ever Wondered:

[1] what the most abundant animal is?
[2] how an arthropod grows within the confines of a rigid exoskeleton?
[3] how arthropods were preadapted for terrestrialism?
[4] why some spiders go ballooning?
[5] what two spiders found in the United States are dangerous to humans?
[6] what causes the bite of a chigger to itch so badly?
[7] what mite lives in the hair follicles of most readers of this textbook?
[8] what crustaceans colonize the hulls of ships?

These and other useful questions will be answered in this chapter.

This chapter contains underlined evolutionary concepts.

342

Informative **Boxes** give vignettes covering a wide variety of topics.

Key Terms are set off in boldface type when first used. Also listed at end of chapter and in glossary.

Box 28.1

Poison Frogs of South America

A South American native stalks quietly through the jungle, peering into the tree branches overhead. A monkey's slight movements divulge its presence, and the hunter takes careful aim with what appears to be an almost toylike bow and arrow. The arrow sails true, and the monkey is hit. The arrow seems ineffectual at first, however, after a few moments, the monkey tumbles from the tree. Thousands of years of cultural evolution have taught these natives a deadly secret that makes effective hunting tools out of seemingly innocuous instruments.

All amphibians possess glandular secretions that are noxious or toxic to varying degrees. These glands are distributed throughout the skin and exude milky toxins designed to ward off potential predators. Toxic secretions are frequently accompanied by warning (aposematic) coloration that signals to predators the presence of noxious secretions.

Four genera of frogs (*Atopophrybnus, Colostethus, Dendrobates,* and *Phyllobates*) in the family Dendrobatidae live in tropical forests from Costa Rica to southern Brazil. The black and gold skin of *Dendrobates* (box figure 28.1) and *Phyllobates* signal their highly toxic secretions. South American natives use this toxin to tip their arrows. Frogs are killed with a stick and held over a fire. *Granular glands* in the skin release their venom, which is collected and allowed to ferment. Poisons collected in this manner are neurotoxins that prevent the transmission of nerve impulses between nerves and between nerves and muscles. Arrow tips dipped in this poison and allowed to dry contain sufficient toxin to paralyze a bird or small mammal.

Members of this family of frogs, in addition to being exploited by South American natives, have interesting reproductive habits. A female lays one to six large eggs in moist, terrestrial habitats. The female promptly abandons the eggs, but the male visits the clutch regularly and guards the eggs. The eggs hatch after approximately 2 weeks, and the tadpoles wiggle onto the male's back. The male then transports the tadpoles from the egg-laying site to water, where they are left to develop. The tadpoles metamorphose to the adult body form after approximately 6 weeks.

Box figure 28.1 A poison arrow frog (*Dendrobates pumilo*).

very dry deserts. Adults lack tails, and caudal (tail) vertebrae are fused into a rodlike structure called the *urostyle*. ...d end in webbed

...tilization is almost typically aquatic. ...l-developed tails. ...r the end of their ...ae are herbivores ...structure used in ...c and rapid meta...body form.

...toad" is more ver...refers to Anurans ...hey are more ter... These character...related taxa. True

toads belong to the family Bufonidae. Other familiar anurans include the leopard frog (*Rana pipiens*), the tree frog (*Hyla andersoni*), and the American toad (*Bufo americanus*) (colorplate 11g–i).

Stop and Ask Yourself:

1. What were two ancient lineages of amphibians? What groups of animals are the modern descendants of each lineage?
2. What animals are members of the order Caudata?
3. What order of amphibians is characterized by wormlike burrowing?
4. What order of amphibians is characterized by tail vertebrae fused into a urostyle?

The enterocoel hypothesis (Gr. *enteron*, gut + *koilos*, hollow) suggests that the eucoelom may have arisen as outpocketings of a primitive gut tract. This hypothesis is patterned after the method of eucoelom formation in deuterostomes (other than vertebrates; *see figure 20.1b*). The implication of this hypothesis is that mesoderm and the eucoelom formed from the gut of a diploblastic animal. If this is true, the triploblastic, acoelomate design would be secondarily derived by mesoderm filling the body cavity of a eucoelomate animal.

Unfortunately zoologists may never know which, if either, of these hypotheses is accurate. Some zoologists believe that the eucoelom may have arisen more than once in different evolutionary lineages, in which case, more than one explanation could be correct.

Molluscan Characteristics

Molluscs range in size and body form from the giant squid, measuring 18 m in length, to the smallest garden slug, less than 1 cm long. In spite of this diversity, the phylum Mollusca (mol-lus'kah) (L. *molluscus*, soft) is not difficult to characterize (table 20.1).

[1] The body of a mollusc is divided into two main regions—the head-foot and the visceral mass (figure 20.2). The **head-foot** is elongate with an anterior head, containing

the mouth and certain nervous and sensory structures, and an elongate foot, used for attachment and locomotion. A **visceral mass** contains the organs of digestion, circulation, reproduction, and excretion and is attached at the dorsal aspect of the head-foot.

The **mantle** of a mollusc usually attaches to the visceral mass, enfolds most of the body, and may secrete a shell that overlies the mantle. (Modifications of the mantle are described in discussions that follow.) The *shell* of a mollusc is secreted in three layers (figure 20.3). The outer layer of the shell is called the *periostracum*. This protein layer is secreted by mantle cells at the outer margin of the mantle. The middle layer of the shell, called the *prismatic layer*, is the thickest of the three layers and consists of calcium carbonate mixed with organic materials. It is also secreted by cells at the outer margin of the mantle. The inner layer of the shell, the *nacreous layer*, forms from

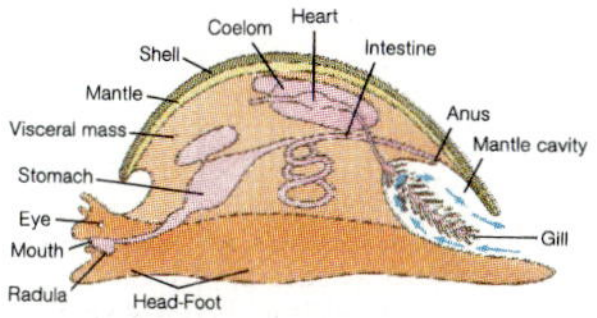

Table 20.1 Classification of the Mollusca

Phylum Mollusca (mol-lus'kah)
The eucoelomate animal phylum whose members possess a head-foot, visceral mass, mantle, and mantle cavity. Most molluscs also possess a radula and a shell.

Class Gastropoda (gas-trop'o-dah)
Shell, when present, usually coiled; body symmetry distorted by torsion; some monoecious species. *Nerita, Orthaliculus, Helix.*

Class Bivalvia (bi'val've-ah)
Body enclosed in a shell consisting of two valves, hinged dorsally; no head or radula; wedge-shaped foot. *Anodonta, Mytilus, Venus.*

Class Cephalopoda (sef'ah-lop'o-dah)
Foot modified into a circle of tentacles and a siphon; shell reduced or absent; head in line with the elongate visceral mass. *Octopus, Loligo, Sepia, Nautilus.*

Class Scaphopoda (ska-fop'o-dah)
Body enclosed in a tubular shell that is open at both ends; tentacles; no head. *Dentalium.*

Class Monoplacophora (mon'o-pla-kof'o-rah)
Molluscs with a single arched shell; foot broad and flat; certain structures serially repeated. *Neopilina.*

Class Aplacophora (a'pla-kof'o-rah)
Shell, mantle, and foot lacking; wormlike; head poorly developed; burrowing molluscs. *Neomenia.*

Class Polyplacophora (pol'e-pla-kof'o-rah)
Elongate, dorsoventrally flattened; head reduced in size; shell consisting of eight dorsal plates. *Chiton.*

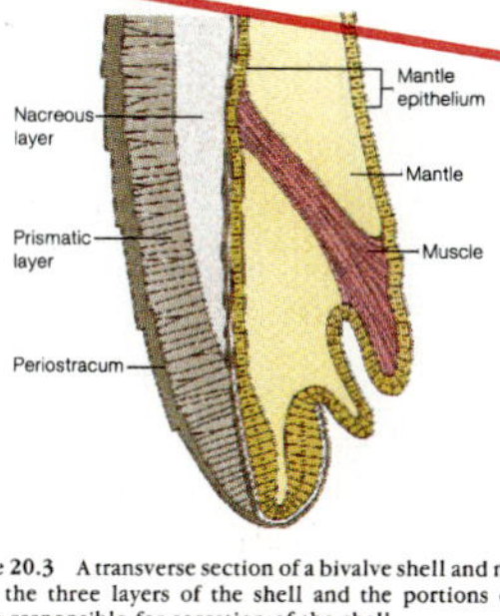

Figure 20.2 All molluscs possess three features unique to the phylum. The head-foot is a muscular structure usually used for locomotion and sensory perception. The visceral mass contains organs of digestion, circulation, reproduction, and excretion. The mantle is a sheet of tissue that enfolds the rest of the body and secretes the shell.

Figure 20.3 A transverse section of a bivalve shell and mantle shows the three layers of the shell and the portions of the mantle responsible for secretion of the shell.

Figure 20.3 redrawn, with permission, from *Living Invertebrates* by Pearse/Buchsbaum, copyright by The Boxwood Press.

Stop and Ask Yourself—a series of questions to use as a study aid—with answers in a separately available Study Guide.

Tables present concise summaries of information.

Colorful **Visuals** explain structures and processes clearly.

Arm
Calyx
Cirri

Figure 25.12 Crinoidea. A feather star (*Neometra*).

is resting on a substrate. Swimming is accomplished by raising and lowering the arms, and crawling results from using the tips of the arms to pull the animal over the substrate.

Maintenance Functions

Circulation, gas exchange, and excretion in crinoids are similar to other echinoderms. In feeding, however, crinoids differ from other living echinoderms. They use outstretched arms for suspension feeding. When a planktonic organism contacts a tube foot, it is trapped and carried to the mouth by cilia in ambulacral grooves. Although this method of feeding is different from the way other modern echinoderms feed, it probably reflects the original function of the water-vascular system

Figure 25.13 Photograph of a preserved sea daisy (*Xyloplax medusiformis*). This specimen is 3 mm in diameter.

Class Concentricycloidea

The class Concentricycloidea (kon-sen'tri-si-kloi"de-ah) (ME *consentrik*, having a common center + Gr. *kykloeides*, like a circle) contains a single described species, known as the sea daisy. Sea daisies have been recently discovered in deep oceans, on wood and other debris (figure 25.13). They lack arms and are less than one cm in diameter. The most distinctive feature of this species is two circular water-vascular rings that encircle the disclike body. The inner of the two rings probably corresponds to the ring canal of members of other classes because it has Polian vesicles attached. The outer ring contains tube feet and ampullae and probably corresponds to the radial canals of members of other classes. In addition, this animal lacks an internal digestive system. Instead, the surface of the animal that is applied to the substrate (decomposing wood) is covered by a thin membrane, called a *velum*, that digests and absorbs nutrients. Internally, there are five pairs of brood pouches where embryos are held during development. There are apparently no free-swimming larval stages. The mechanism for fertilization is unknown.

Further Phylogenetic Considerations

As described earlier, most zoologists believe that echinoderms evolved from bilaterally symmetrical ancestors. Radial symmetry probably evolved during the transition from active to more sedentary life styles; however, the oldest echinoderm fossils, about 600 million years old, give little direct evidence of how this transition occurred.

Ancient fossils do give clues regarding the origin of the water-vascular system and the calcareous endoskeleton. Of all living echinoderms, the crinoids most closely resemble

[7]

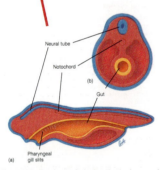

Neural tube
Notochord
(b)
Gut
(a)
Pharyngeal gill slits

Figure 10.7 The chordate body plan. The development of all chordates involves the formation of a neural tube, the notochord, gill slits, and a postanal tail. Derivatives of all three primary germ layers are present. (a) Side view. (b) Cross section.

dark pigment of the animal pole also absorbs heat from the sun, and the warming that results may promote development.

Initial Cleavages The first cleavage of the amphibian embryo, like that of echinoderms, is longitudinal. It begins at the animal pole, and divides the gray crescent in half. If the first cleavage is experimentally forced to pass to one side of the gray crescent, only the blastomere with the gray crescent will develop normally. Because of the large amount of yolk in the vegetal end of the egg, cleavages are slower than in the animal end. The amphibian morula, therefore, consists of many small cells at the animal end

of the embryo and fewer, larger cells at the vegetal end of the embryo (figures 10.9a-d and 10.10a-c). The amphibian blastula forms in much the same way as the echinoderm blastula, except that the yolky vegetal cells cause the blastocoel to form in the animal half of the embryo (figure 10.9e). Unlike that of echinoderms, the blastula wall of amphibians is composed of multiple cell layers.

Gastrulation The cells of the blastula that will develop into specific structures are grouped together on the surface of the blastula. Gastrulation involves the movement of some of these cells into the interior of the embryo. Embryologists have used dyes or carbon particles to mark the surface of the blastula, and then have followed the movements of these cells during gastrulation. *Fate maps* can be constructed to show what will happen to groups of cells on the surface of the blastula during gastrulation. Embryonic cells are designated according to their fate by "presumptive notochord," "presumptive endoderm," and so forth.

The first sign that gastrulation is beginning is the formation of a groove between the gray crescent and the vegetal region of the embryo. This groove is the slitlike blastopore. The animal-pole margin of the blastopore is called the *dorsal lip of the blastopore*. Cells at the bottom of the groove move to the interior of the embryo, and the groove spreads transversely (figures 10.9f and 10.10d). This groove is similar to that which occurs during echinoderm blastopore formation. In amphibians, however, superficial cells now begin to roll over the dorsal lip of the blastopore in a process called *involution*. Cells spread from the animal pole toward the blastopore, and replace those moving into the interior of the embryo. In the process, the ends of the slitlike blastopore continue to spread transversely and downward toward the vegetal pole, until one end of the slit meets and joins the opposite end of the slit. A ringlike blastopore now surrounds the protruding, yolk-filled cells near the vegetal end of the embryo (figure 10.10e). These

(a)
(b)

Figure 10.8 Fertilization of amphibian eggs. (a) Egg release and fertilization in frogs occurs when the male (above) mounts and grasps the female (below). This positioning is called amplexus. Eggs are fertilized by the male as they are released by the female. (b) The zygote of a frog shortly after fertilization. Each zygote has a jellylike coat. Note the pigmented animal pole.

Answers to the **Have You Ever Wondered** questions are keyed in the margin.

The large number of carefully chosen photographs emphasize important topics.

by sight and simply wait for prey to pass by. Olfaction plays an important role in prey detection by aquatic salamanders and caecilians.

Many salamanders are relatively unspecialized in their feeding methods, using only their jaws to capture prey. Anurans and plethodontid salamanders, however, use their tongue and jaws in a flip-and-grab feeding mechanism (figure 28.6). A true tongue is first seen in amphibians. (The "tongue" of fishes is simply a fleshy fold on the floor of the mouth. Fish food is swallowed whole and not manipulated by the "tongue.") The tongue of amphibians is attached at the anterior margin of the jaw and lies folded back over the floor of the mouth. Mucous and buccal glands on the tip of the tongue exude sticky secretions. When a prey comes within range, an amphibian lunges forward and flicks out its tongue. The tongue turns over, and the lower

[2]

jaw is depressed. The fact that the head can tilt on its single cervical vertebra aids in aiming the strike. The tip of the tongue entraps the prey, and the tongue and prey are flicked back inside the mouth. All of this may happen in 0.05 to 0.15 second! The prey is held by pressing it against teeth on the roof of the mouth, and the tongue and other muscles of the mouth push food toward the esophagus. The eyes sink downward during swallowing, and help force food toward the esophagus. Digestive processes are similar to those of other vertebrates (*see chapter 37*).

Circulation, Gas Exchange, and Temperature Regulation

The circulatory system of amphibians shows remarkable adaptations for a life that is divided between aquatic and terrestrial habitats. The separation of pulmonary and systemic circuits is less efficient in amphibians than in lungfishes (figure 28.7; *see figure 27.11b*). The atrium is partially divided in urodeles and completely divided in anurans. The ventricle has no septum. A *spiral valve* is present in the conus arteriosus or ventral aorta and helps direct blood into pulmonary and systemic circuits. As discussed later, gas exchange occurs across the skin of amphibians, as well as at the lungs. Therefore, blood entering the right side of the heart is nearly as well oxygenated as blood entering the heart from the lungs! When an amphibian is completely submerged, all gas exchange occurs across the skin and other moist surfaces; therefore, blood coming into the right atrium has a higher oxygen concentration than blood returning to the left atrium from the lungs. Under these circumstances, blood vessels leading to the lungs constrict, reducing blood flow to the lungs and conserving energy. This adaptation is especially valuable for those frogs and salamanders that overwinter in the mud at the bottom of a pond.

Fewer aortic arches are present in adult amphibians than in fishes. After leaving the conus arteriosus, blood may enter the *carotid artery* (aortic arch III), which takes blood to the head; the *systemic artery* (aortic arch IV), which takes blood to the body; or the *pulmonary artery* (aortic arch VI).

Figure 28.6 Flip-and-grab feeding of the toad, *Bufo americanus.*

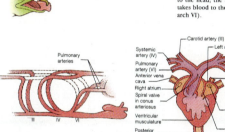

Pulmonary arteries
Systemic artery (IV)
Pulmonary artery (VI)
Anterior vena cava
Right atrium
Spiral valve in conus arteriosus
Ventricular musculature
Posterior vena cava
Carotid artery (III)
Left at[rium]
III IV VI

Margin annotations:

Evolutionary concepts are underscored throughout the text.

Numerous **Colorplates** occur at strategic places throughout the text to supplement the text illustrations.

End-of-Chapter sections provide an overview of the chapter and give opportunities for self-testing and for further study.

A list of **Key Terms**, with page numbers for easy reference and review. Also listed in Glossary.

■ Summary

1. Organic evolution is the change of a species over time.
2. Ideas of evolutionary change can be traced back to the ancient Greeks.
3. Jean Baptiste Lamark was an eighteenth century zoologist who advocated evolution, and proposed a mechanism—inheritance of acquired characteristics—to explain it.
4. Charles Darwin saw impressive evidence for evolutionary change while on a mapping expedition on the H.M.S. *Beagle.* The theory of uniformitarianism, South American fossils, and observations of tortoises and finches on the Galápagos Islands convinced Darwin that evolution occurs.
5. After returning from his voyage, Darwin began formulating his theory of evolution by natural selection. In addition to his experiences on his voyage, later observations of artificial selection and Malthus' theory of human population growth helped shape his theory.
6. Darwin's theory of natural selection includes the following elements: a. All organisms have a greater reproductive potential than is ever realized. b. Inherited variations exist. c. There is a constant struggle for existence in which those organisms that are best suited to their environment will survive. d. The adaptive traits present in the survivors will tend to be passed on to subsequent generations, and the nonadaptive traits will tend to be lost.
7. "Adaptation" may refer to a process of change or a result of change. In the latter sense, an adaptation is a structure or a process that increases an organism's potential to survive and reproduce in a given environment.
8. All evolutionary changes are not adaptive, nor do all evolutionary changes lead to perfect solutions to environmental problems.
9. Alfred Russel Wallace outlined a theory similar to Darwin's, but never accumulated significant evidence documenting his theory.
10. Modern evolutionary theorists apply principles of genetics, ecological theory, and geographic and morphological studies to solving evolutionary problems.

■ Key Terms

adaptation (p. 185)
continental drift (p. 187)
Galápagos Islands (p. 180)
natural selection (p. 184)
organic evolution (p. 177)

theory of evolution by natural selection (p. 177)
theory of inheritance of acquired characteristics (p. 177)
uniformitarianism (p. 180)

■ Critical Thinking Questions

1. In chapter 1 there is a discussion of the scientific method. Review that section and do the following: a. Outline a hypothesis and design a test of "inheritance of acquired characteristics." b. Define what is meant by the word "theory" in the theory of evolution by natural selection.

2. Assuming that you have already studied chapter 9, describe the implications of inheritance of acquired characteristics for the central dogma of molecular genetics.

3. Review the definition of "adaptation" in the sense of a result of evolutionary change. Imagine that two deer, A and B, are identical twins. Deer A is shot by a hunter before it had a chance to reproduce. Deer B is not shot and goes on to reproduce. According to our definition of adaptation is deer B more fit for its environment than deer A? Why or why not?

4. Why is the stipulation of "a specific environment" included in the definition of "adaptation?"

■ Suggested Readings

Books

Bowler, P. 1984. *Evolution: The History of an Idea.* Berkeley: University of California Press.
Darwin, C. 1894. *On the Origin of Species.* Reprint. 1975. Cambridge: Cambridge University Press.
Dodson, E. O. and Dodson, P. 1985. *Evolution: Process and Product.* Belmont: Wadsworth Publishing Company.
Eldredge, N. and Cracraft, J. 1980. *Phylogenetic Patterns and the Evolutionary Process: Method and Theory in Comparative Biology.* New York: Columbia University Press.
Futuyma, D. J. 1986. *Evolutionary Biology,* 2nd ed. Sutherland: Sinauer Associates, Inc.
Godfrey, L. R. 1985. *What Darwin Began.* Old Tappan: Allyn and Bacon, Inc.
Greene, J. C. 1959. *The Death of Adam: Evolution and its Impact on Western Thought.* Ames: Iowa State University Press.
Mayr, E. 1982. *The Growth of Biological Thought: Diversity, Evolution, and Inheritance.* Cambridge: Harvard University Press.
Smith, J. M. (ed.). 1982. *Evolution Now: A Century After Darwin.* New York: W. H. Freeman and Company.
Stebbins, G. L. 1982. *Darwin to DNA, Molecules to Humanity.* New York: W. H. Freeman and Company.
Volpe, E. P. 1985. *Understanding Evolution,* 5th ed. Dubuque: Wm. C. Brown Publishers.

Articles

Ayala, F. J. The mechanisms of evolution. *Scientific American* September, 1978.
Dickerson, R. E. Chemical evolution and the origin of life. *Scientific American* September, 1978.
Herbert, S. Darwin as a geologist. *Scientific American* May, 1986.
Lack, D. Darwin's finches. *Scientific American* April, 1953.
Lewontin, R. C. Adaptation. *Scientific American* September, 1978.
Mayr, E. Evolution. *Scientific American* September, 1978.

Preface

Because zoology is such a broad scientific discipline, teaching an introductory zoology course is as difficult as, or more difficult than, teaching a highly specialized, upper-division course. The problem is one of trying to decide what background information beginning zoology students need most. Part of the decision of what information to include in a general zoology course may be dictated by the background of the students and where the course fits into the various curricula at a particular college or university. As a result of such varied student backgrounds, all instructors must decide what can and cannot be covered, given their individual time constraints. In writing textbooks for a general zoology course, authors must also make similar decisions. It is difficult to write a comprehensive textbook that is also manageable and adaptable to a one or two semester (quarter) course. Authors include what they perceive others are teaching or should be teaching. In this textbook, we have attempted to present zoology in a way that is useful to most course formats and in the detail appropriate for a one or two semester (quarter) general zoology course. Throughout this textbook, zoology is presented as an exciting and dynamic scientific field that does not stand apart from the other life sciences.

The organization of this textbook is different from others currently available. Throughout, the authors have tried to emphasize the interrelationships of all life forms by covering common life processes early in the textbook and in more detail than is found in other zoology textbooks. These early topics include molecular biology, cell structure and function, genetics, development, and evolution. We have also tried to maintain an evolutionary theme throughout. The survey of the animal phyla emphasizes evolutionary relationships, aspects of animal organization that unite major animal phyla, and animal adaptations. This evolutionary theme continues with descriptions of evolutionary changes in the structure and function of selected organ systems.

The six part organization of this textbook should be adaptable to most course formats. Part I (Introduction, Cells, Molecules, and Life) covers the basics of cellular structure and function, including energy relationships and the functions of enzymes. Part II (The Continuity of Life) covers cellular and organismal processes responsible for the maintenance and propagation of life. Part III (Evolution) covers evolutionary processes. These are presented relatively early in the textbook to help students understand the forces behind evolution and the emphasis of evolution in parts IV and V. Evolution, along with cellular and molecular biology, are the foundations on which all zoology rests. It is imperative that the student understand these unifying processes and be able to understand their relationships to other disciplines in zoology. Part IV (Animallike Protists and Animalia) is a survey of the animal phyla and the animallike protists. Biological classification, animal organization, and animal phylogeny are introduced in chapter 15. Chapters 16 through 31 cover the animallike protists and most animal phyla. Part V (Form and Function) is an introduction to animal structure and function and is presented in a systems format. In this part of the textbook, comparative material is presented for invertebrate and vertebrate systems, with an emphasis on human structure and function. Part VI (Behavior and Ecology) is an introduction to behavior and ecology. In future decades, the public will be asked to make decisions regarding land use, preservation of natural areas, and preservation of threatened species. Appreciation of basic ecological principles is a prerequisite for making intelligent decisions concerning these issues. Helping students acquire this basic ecological knowledge is a goal of all zoology textbooks.

In addition to balance and flexibility in organization, this textbook has been designed to be an effective teaching tool. Readability has been enhanced by a relatively simple, direct, writing style; frequent headings; and an organized outline format in each chapter. The level of difficulty has been carefully set with the target audience in mind.

The many new terms encountered in studying general zoology can be a stumbling block for students. This textbook overcomes this potential difficulty by addressing and reinforcing a student's vocabulary development at three levels. (1) No new term or animal is used without it being clearly defined (derivations and phonetics are often given). A student does not have to be familiar with the terminology of zoology to use this textbook. (2) The most important terms are printed in boldface when they are first used and defined. (3) A complete and up-to-date glossary is included at the end of the textbook.

Because illustrations are so critical to a student's learning and enjoyment of general zoology, as many full-color photographs as possible have been used throughout the book. In addition, 32 pages of colorplates have been strategically placed in order to supplement the in-text photographs. All line art has been produced under the direct supervision of the authors and designed to illustrate and reinforce specific points in the textbook. Consequently, every illustration is directly related to the narrative and specifically cited where appropriate.

Supplementary Materials

Supplementary materials are available to assist instructors with their presentations and general course management.

1. An **Instructor's Manual** prepared by Barbara A. Taber, Southwest Missouri State University, provides examples of lecture reading schedules for courses with various emphases. In addition, each chapter contains a detailed outline, purpose, objectives, key terms, approximately 50 multiple choice test questions, summary, sources for audiovisual materials and computer software, and the answer key for the text's Stop and Ask Yourself questions.
2. A **Student Study Guide,** prepared by Jay M. Templin, Widener University, contains chapter summaries, outlines, key terms with phonetics, pretest assessment questions, various learning activities, mastery tests, crossword puzzles, and the answer key for the Stop and Ask Yourself questions.
3. A set of 60 acetate **transparencies** is available and may be used to supplement classroom lectures.
4. An **Answer Key** for the Stop and Ask Yourself questions that students can purchase as an abbreviated study guide.
5. **WCB TestPak** is a computerized testing service offered free upon request to adopters of this textbook. It provides a call-in/mail-in test preparation service. A complete test item file is also available on computer diskette for use with IBM compatible, Apple IIe or IIc, or Macintosh computers.
6. A laboratory manual, **General Zoology Laboratory Manual,** 2nd edition, by Stephen A. Miller, has been prepared to accompany the textbook. The class-tested exercises are modular and short so that an instructor can easily choose only those exercises that fit his or her particular course.
7. The **Customized Laboratory Manual** — each lab manual exercise is also available individually as offprints so students need buy only those exercises used in the laboratory. Contact your local sales rep for more details.

Acknowledgments

The authors wish to express their thanks to the reviewers who provided detailed criticism and analysis of the text-

book as it developed. Their suggestions greatly improved the final product. These reviewers, themselves zoology teachers, have taken time out from their busy teaching and research schedules to read our manuscript and to suggest numerous improvements.

Reviewers

- Ron Basmajian, Merced College
- Bayard H. Brattstrom, California State University—Fullerton
- William Brueske, Humboldt State University
- James E. Cole, Bloomsburg University of Pennsylvania
- Harry N. Cunningham, Jr., Pennsylvania State University at Erie—The Behrend College
- Opal H. Dakin, Hinds Community College
- J. William Dapper, Troy State University at Dothan
- Peggy Rae Dorris, Henderson State University
- DuWayne C. Englert, Southern Illinois University—Carbondale
- Stephen Ervin, California State University—Fresno
- William F. Evans, University of Arkansas
- Daniel R. Formanowicz, Jr., University of Texas—Arlington
- Mildred J. Galliher, Cochise College
- Judith Goodenough, University of Massachusetts—Amherst
- Earl A. Holmes, Friends University
- Ronald L. Hybertson, Mankato State University
- Thomas A. Leslie, Saddleback College
- Steele R. Lunt, University of Nebraska at Omaha
- Edward B. Lyke, California State University
- Richard N. Mariscal, Florida State University
- Neal F. McCord, Stephen F. Austin State University
- John C. McGrew, Colorado State University
- J. E. McPherson, Southern Illinois University
- Alex L. A. Middleton, University of Guelph
- Robert Powell, Avila College
- John D. Rickett, University of Arkansas—Little Rock
- Tim V. Roye, San Jacinto College
- Joseph L. Simon, University of South Florida
- Thomas P. Simon, Indiana University NW
- John Snyder, Furman University
- Dean Stevens, University of Vermont
- J. L. Sumich, Grossmont Community College
- Barbara A. Taber, Southwest Missouri State University
- Jay M. Templin, Widener University
- Olivia White, University of North Texas
- Richard L. Whitman, Indiana University NW
- Sr. Mary L. Wright, College of Our Lady of the Elms
- Harold L. Zimmack, Ball State University

We are grateful for special permission to redraw many original figures from the textbook of invertebrate zoology, *Living Invertebrates* by Vicki and John Pearse and Mildred and Ralph Buchsbaum (copyright by The Boxwood Press).

The production of a textbook requires the efforts of many people. We would like to express special appreciation to the editorial and production staffs of Wm. C. Brown Publishers and Science Tech Publishers. Marge Manders, our developmental editor at Wm. C. Brown, and her predecessors, John Stout and Mary Porter, showed extraordinary patience and commitment to the book. Kevin Kane and Renee Menee were always there with the needed support and direction to meet all scheduling demands. Tammy Ben kept all material moving in a plethora of directions. Much credit is also due to Tom and Kathie Brock and the staff at Science Tech, especially our editor, Katherine Noonan, for her meticulous scrutiny of every manuscript line, figure caption, table, and zoological concept.

Finally, but most importantly, we wish to extend appreciation to our families for their patience and encouragement. Our wives, Janice A. Miller and Jane R. Harley, have been supportive on a daily basis, for which we are grateful. To them we dedicate this book.

Stephen A. Miller
John P. Harley

Other Titles of Related Interest from Wm. C. Brown Publishers

Zoology

General Zoology Laboratory Guide, 11th edition (Complete and short versions) (1991)
by Charles F. Lytle and the late J. E. Wodsedalek

Anatomy of the Cat: Text and Dissection Guide (1990)
by Lionel J. Rosenzweig

Anatomy of the Necturus: Text and Dissection Guide (1988)
by Lionel J. Rosenzweig

Anatomy of the Vertebrates: A Laboratory Guide (1989)
by Donn D. Martin

Animal Variety: An Evolutionary Account, 4th edition (1980)
by Lawrence S. Dillon

Biology of the Invertebrates, 2nd edition (1991)
by Jan A. Pechenik

Animal Behavior

Animal Behavior: Mechanisms, Ecology, and Evolution, 3rd edition (1991)
by Lee C. Drickamer and Stephen H. Vessey

Laboratory Anatomy Guides

Laboratory Anatomy of the Human Body, 4th edition (1992)
by Bernard B. Butterworth

Laboratory Anatomy of the Perch, 4th edition (1992)
by Robert B. Chiasson and William J. Radke

Laboratory Anatomy of the Pigeon, 3rd edition (1984)
by Robert B. Chiasson

Laboratory Anatomy of the Fetal Pig, 8th edition (1988)
by Theron Odlaug

Laboratory Anatomy of the Shark, 5th edition (1988)
by Laurence M. Ashley and Robert B. Chiasson

Laboratory Anatomy of the Cat, 7th edition (1989)
by Robert B. Chiasson and Ernest S. Booth

Laboratory Anatomy of the White Rat, 5th edition (1988)
by Robert B. Chiasson

Laboratory Anatomy of the Frog, 5th edition (1988)
by Raymond A. Underhill

Laboratory Anatomy of the Rabbit, 3rd edition (1990)
by Charles A. McLaughlin and Robert B. Chiasson

Laboratory Anatomy of the Mink, 2nd edition (1979)
by David Klingener

Laboratory Anatomy of the Turtle (1962)
by Laurence M. Ashley

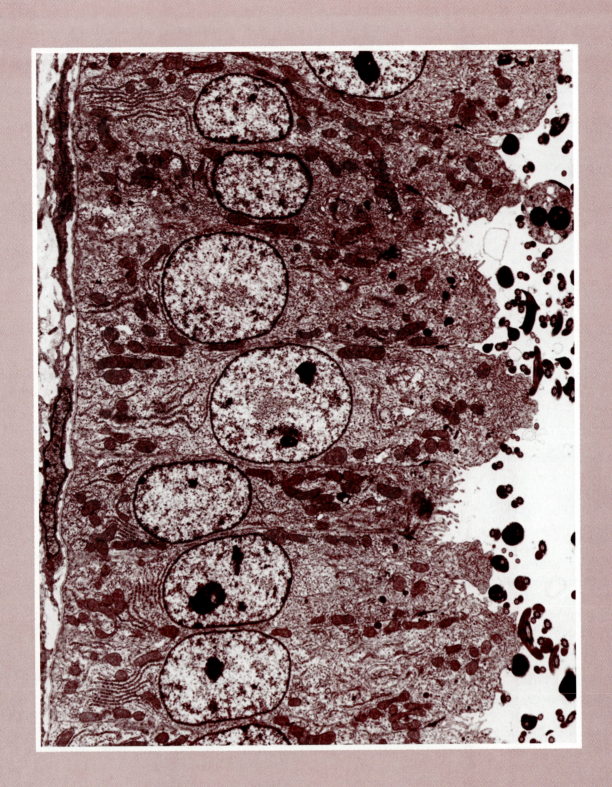

Introduction, Cells, Molecules, and Life

Part I

One of the fundamental principles of modern biology is that processes occurring in living things obey the laws of chemistry and physics, just as processes in nonliving systems do. The special properties of life arise because of the complex organization of living things and the chemical processes that occur in them. Biologists must understand certain principles of chemistry in order to better analyze the organization and function of life on earth.

Another key to understanding life on our planet is to recognize the role of cells. The cell is the fundamental organizational unit of life. As you learn more about cell function and structure, you will find that many cellular components and chemical processes are identical or very similar in cells from a variety of organisms. Recognition of these similarities at the cellular level provides an important unifying framework within which biologists approach the diversity of living things.

All living things require energy for maintenance, growth, and reproduction. Some organisms obtain energy by harnessing light energy from the sun and then producing materials they need (photosynthesis). Other organisms, however, are not capable of using light energy in this way. They must use photosynthesizing organisms as sources of energy and materials.

In chapter 1, we will examine some of the underlying principles of zoology. Chapters 2 through 5 cover the biologically important concepts of chemistry, some basic characteristics of the cells that constitute living things, and some of the energy relationships upon which animal life depends.

Zoology in Science and History

1

Concepts

1. The field of zoology, as other subdisciplines of biology, is a somewhat artificial category created by biologists to organize the study of life.
2. All life shares a common genetic molecule (DNA), unit of organization (the cell), evolutionary forces, and environment (the earth).
3. Zoology, the study of animals, is a very broad field and cannot be separated from either medicine or biology.
4. The scientific method is a procedure that allows an investigator to objectively analyze biological occurrences.

Have You Ever Wondered:

[1] what the science of life is called?
[2] what fundamental molecule carries the genetic code for life?
[3] how scientists know that the earth is about 5 billion years old?
[4] why the human appendix is evidence of evolution?
[5] who the father of medicine is?

These and other useful questions will be answered in this chapter.

This chapter contains underlined evolutionary concepts.

What is life? What are the processes that maintain living organisms? How did life arise? What are the natural laws that have been influencing and will continue to influence life on this planet? How can we explain life's diversity? These are some of the questions that inquiring minds have been asking for many centuries. Questions similar to these are asked in all fields of **biology** (Gr. *bios,* life + *logos,* to study), the science of life.

Since the seventeenth century, scientific observation and experimentation have expanded human knowledge and improved human health and welfare. At the same time, science has promoted technologies that destroy life. Technologies that give us nuclear medicine also give us nuclear wastes and threaten the world with nuclear destruction. Technologies that have made us highly mobile also pollute our environment. We, as scientists, are often applauded for our successes, but each of us must also be aware of the potential for misuse of the good that comes from our discipline. In the pursuit of answers to our "how" and "what" questions, we must also think about "why," "for whom," and "by whom." Science must never be pursued without regard for the ethical values of our society. It is, after all, the answers to these "why" questions that tell us our reason for being and establish the direction that science is to take in the world.

Biology, like all science, is limited only by the potential of the human mind to question and understand. In science, we assume that the mind is capable of understanding the natural processes that govern our universe, and that these natural processes are logical, orderly, and timeless. Using these assumptions, scientists have gone into the world, and beyond, seeking to understand nature. Regardless of whether you are a budding scientist or a student fulfilling basic science requirements, you have a natural curiosity and a desire to learn. One goal of this book is to bring out that curiosity, to help you see some of the questions that humans have asked in the past, and to help you develop a framework that can be used to answer questions in a systematic fashion.

A One-World View

In spite of life's diversity, there is a fundamental unity that embraces all of the biological disciplines. As modern biology probes deeper into the secrets of life, it becomes clearer that all life shares a common genetic blueprint (in DNA); a common organizational unit (the cell); common evolutionary forces that influence the form, function, and habitat of the animal; and a common environment (the earth).

Genetic Unity

All life is based upon a fundamental molecule that carries the genetic code. This molecule is *deoxyribonucleic acid (DNA)* and codes for all the proteins that make up the structural and functional components of life (figure 1.1). Understanding the structure of this molecule has led to tremendous advances in the field of biology called "molecular biology." These advances have left no area of biology untouched.

Units of inheritance, *genes,* were first described by Gregor Mendel (1822–1884). To Mendel, a gene was a "factor" known only by the effect it had on an organism's appearance and the pattern of inheritance it showed. By the early 1900s, knowledge of the gene was increasing and the description of the structure of DNA by James Watson and Francis Crick in 1953 was a turning point for biology. We now have a molecular basis for our concept of the gene, and a basis for understanding how actions are initiated within cells. We better understand how traits are passed from parents to their offspring. We understand what mutations are, and how mutations furnish variations that may allow a group of organisms to adapt to environmental changes. The fundamental importance of the DNA molecule in all aspects of biology should become clearer as you study later chapters.

The Fundamental Unit of Life

Knowledge of cell structure and function is a basic background upon which all disciplines in biology are built. At this level, biology transcends the somewhat artificial boundaries that we sometimes construct to organize life's diversity. Regardless of whether one is studying animal bi-

Figure 1.1 DNA is the genetic material of life. This model shows part of a DNA molecule.

ology, plant biology, or any other subdiscipline of biology, a strong background in cell biology is required.

The invention of simple microscopes in the mid 1600s allowed Robert Hooke (1635–1703) and Anton van Leeuwenhoek (1632–1723) to make early descriptions of cells. In the mid 1800s, the cell theory began to take form. The *Cell Theory* states that all living organisms are composed of cells or cell products. Cells are self-reproducing units that come from preexisting cells. Our modern understanding of the cell has changed tremendously since these early observations. Electron microscopy and other modern technologies have allowed biologists to probe deep into the interior of cells.

The study of the cell has wide-ranging implications for biology. The study of the structure (*anatomy*) and the function (*physiology*) of animal systems depends on a knowledge of the structure and function of the cells that make up the bodies of animals. For example, it is impossible to understand the function of a muscle without a basic understanding of the structure and function of the subunits of the muscle cell that promote contraction. Similarly, understanding the function of a nervous system requires that one appreciate the nerve cell, and have a basic knowledge of the structure and function of the membrane that surrounds that cell.

Just as one can never view an animal apart from its cells, one fails to perceive the animal by focusing only on the cell. It is the interaction of cells that determines what an animal will look like, how the animal will live, and frequently, when the animal will die.

Evolutionary Oneness and the Diversity of Life

Animals are united at all levels because of their common evolutionary origin and the shared forces that influenced their history. Evolutionary processes are remarkable for their relative simplicity, yet they are awesome because of the affects they have had on life forms.

The theory of organic evolution is the concept that organisms change over time. Charles Darwin (1809–1882) published convincing evidence for his "theory of evolution by natural selection" in *On the Origin of Species* in 1859 (figure 1.2). By 1900, most biologists were convinced that evolution occurs, and that natural selection is a reasonable explanation of how it occurs.

Evidence of Evolution   There is now a wealth of evidence that documents the basic premises of evolution. Nowhere is that evidence more convincing than in animal biology.

Biogeography   **Biogeography** is the study of the geographic distribution of plants and animals and attempts to explain why organisms are distributed the way they are. Biogeographic studies have shown that life forms in different parts of the world have had distinctive evolutionary histories. For example, large areas of the world often have

Figure 1.2   A bust of Charles Darwin in later life.

similar climatic and geographic factors. Each of these geographic regions usually has distinctive plants and animals that have particular roles in the environment. Compare the large meat-eating animals of North America (such as wolves and mountain lions) with those of Africa (wild dogs and African lions). Their similar design suggests a distant common ancestry and they have similar life-styles. Obvious differences, however, are the result of millions of years of independent evolution.

Recognizing that plants and animals in different parts of the world have distinctive evolutionary histories, biogeographers have divided the world into six major biogeographic regions (figure 1.3). Each region has a characteristic group of plants and animals, and even though one may move from one region to another and experience similar climates, the different plants and animals encountered may make it seem as if one has entered another world.

In some instances the same, or very similar, organisms exist in geographically separated places—with no representatives in intermediate territories. Darwin, for example, was surprised to find fossils of rhinoceroslike animals in South America. He knew of rhinos only from Africa. As we now know, even the barriers, such as oceans and mountain ranges, that now separate groups of animals were not always there. Movement of the earth's crust has changed, and continues to change, the geologic patterns of the earth, including the relationship of the land masses to each other.

Another important observation of biogeographers is that islands, even though they may be close to a continental mainland, have very unique groups of organisms. For example, on the Galàpagos Islands, off the coast of South America, Darwin found finches that have been found nowhere else, though they are obviously similar to finches on the mainland of South America. This pattern can most logically be explained by evolution. Groups of organisms have a common site of origin. Migration results in organisms spreading through a particular range. Isolation—as a result of geologic, climatic, genetic, or other events—fosters change in the isolated groups and they become noticeably distinct from the common ancestor.

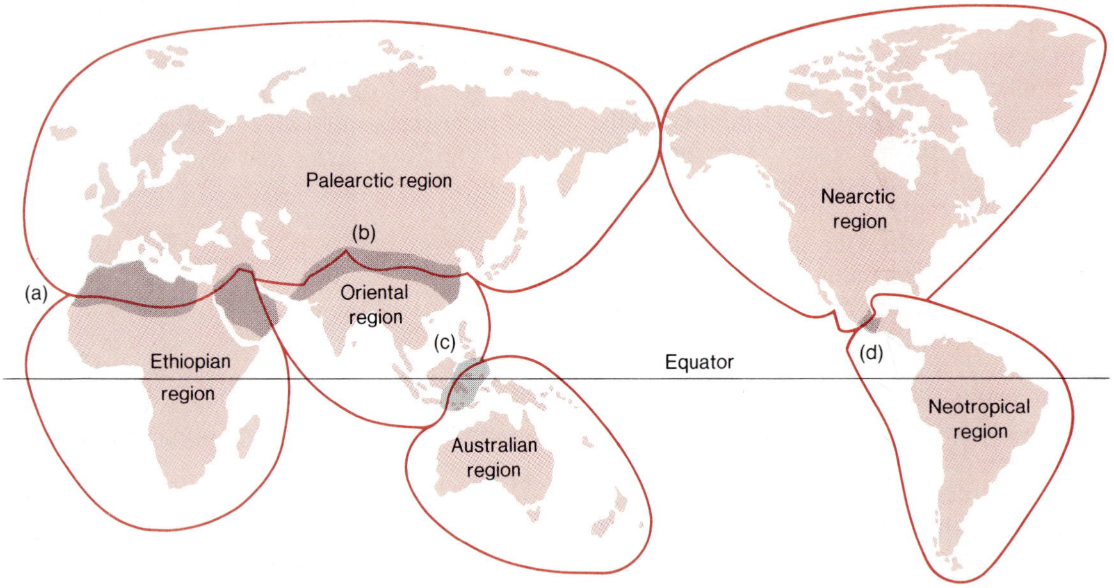

Palearctic region

(b)

(a)

Oriental region

(c)

Ethiopian region

Equator

Nearctic region

(d)

Neotropical region

Australian region

Figure 1.3 Biogeographic regions of the world are separated from one another by barriers, such as oceans, mountain ranges, and deserts. The Ethiopian region is separated from the Palearctic by the Sahara and Arabian Deserts (a), the Palearctic region is separated from the Oriental region by the Himalayan Mountains (b), the Oriental and Australian regions are separated by deep ocean channels (c), and the Nearctic and Neotropical regions are separated by the mountains of southern Mexico and the tropical lowlands of Mexico (d).

Paleontology Some of the most direct evidence for evolution comes from **paleontology** (Gr. *palaios,* old + *on,* existing + *logos,* to study), which is based on the study of the fossil record. **Fossils** are evidence of plants and animals that existed in the past and have become incorporated into the earth's crust (e.g., as rock or mineral) (figure 1.4). For fossilization to occur, an organism must be quickly covered by sediments to prevent scavenging, and in a way that seals out oxygen so as to slow decomposition. Fossilization is most likely to occur in aquatic or semiaquatic environments. Therefore, the fossil record is more complete for those groups of organisms living in or around water and for organisms with hard parts. This documentation provides some of the most convincing evidence for evolution and, at least for some groups of organisms, has resulted in nearly complete understanding of evolutionary lineages (figure 1.5).

Methods for dating fossils help with the reconstruction of evolutionary time scales. *Relative dating* techniques can be used to establish the time an animal lived in relation to animals that came before and after or the relative age of rock layers containing fossils. One technique is based on the principle that in an undisturbed sequence of sedimentary rocks, the oldest layer of rocks is bottommost, and the youngest layer of rocks is uppermost. Sedimentary rock is formed by the accumulation of small particles on the ocean floor. Newly deposited particles overlay older particles, and when sediments are converted to rock, the oldest layers and their fossils will be near the bottom. Another method of relative dating is called *fossil correlation* and involves

the relative aging of rock layers based on the fossils found in them.

Absolute or *radiometric dating* techniques are based on the fact that when nuclei of radioactive atoms break down (decay), they emit particles. The breakdown product is another kind of atom. One can compare the quantity of radioactive substance present in the rock with the quantity of breakdown product present. Knowing the rate of radioactive decay of the radioactive substance, one can calculate how long it has been since the original substance, and any

Figure 1.4 Fossils, such as this trilobite, are direct evidence of evolutionary change. Trilobites were in existence about 500 million years ago and became extinct about 250 million years ago. Fossils form when an animal dies and is covered with sediments. Water dissolves calcium from hard body parts and replaces calcium with another mineral, forming a hard replica of the original animal. This process is called mineralization.

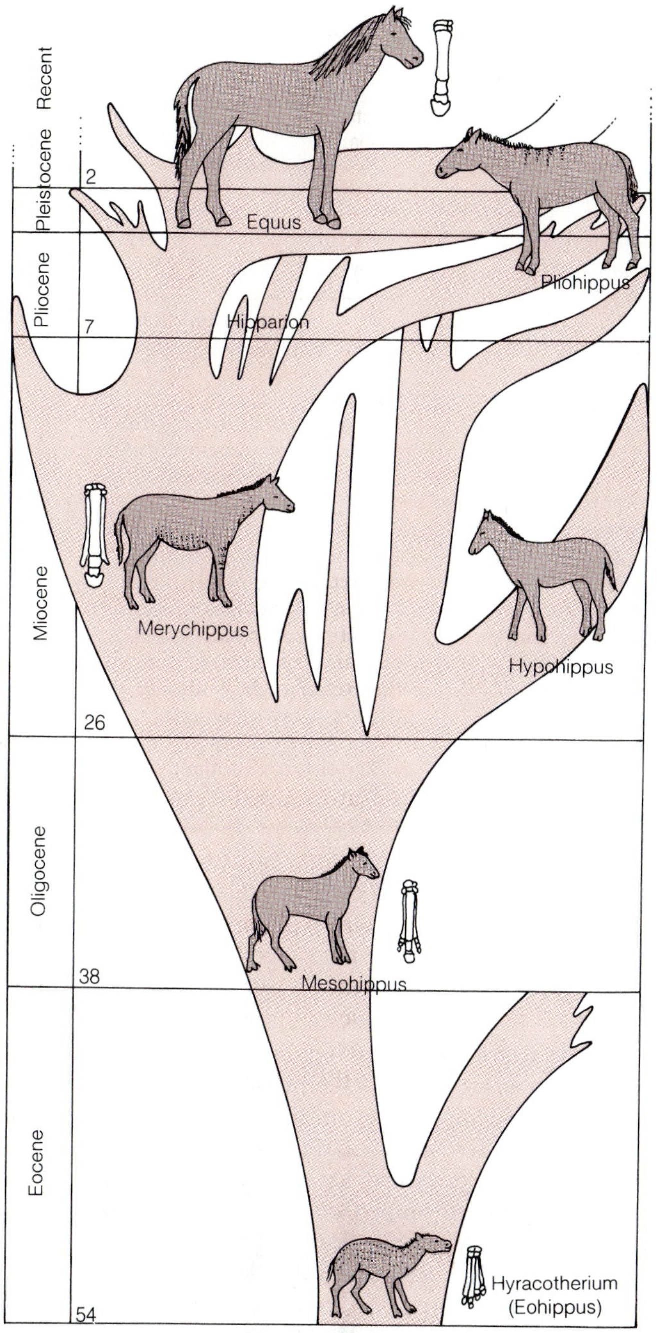

animals fossilized in that substance, where deposited. These methods are described further in chapter 2.

Using these methods, paleontologists have estimated the age of earth (about 5 billion years old), as well as the ages of many rocks and fossils. Table 1.1 shows major divisions of geologic time and some of the major evolutionary events associated with those time periods. [3]

Comparative Anatomy   A structure in one animal may resemble a structure in another animal because of a common evolutionary origin. This relationship is a fundamental premise upon which **comparative anatomy** is based. Comparative anatomists study the structure of fossilized and living animals, looking for similarities that could be indications of evolutionarily close relationships. Structures that are derived from common ancestry are said to be **homologous** (Gr. *homolog* + *os,* agreeing) (i.e., having the same or a similar relation). Some examples of homology are obvious. For example, the appendages of vertebrates have a common arrangement of similar bones, even though the function of the appendage may vary (figure 1.6). Along with other evidence, this similarity in appen-

Figure 1.5   Evolution of the horse has been traced back about 60 million years using the fossil record. *Hyracotherium (Eohippus)* was a dog-sized animal with four prominent toes on each foot. A single middle digit of the toe and vestigial digits on either side of that remain in modern horses. Note that evolutionary lineages are seldom simple ladders of change. Instead, numerous evolutionary side branches often meet with extinction.

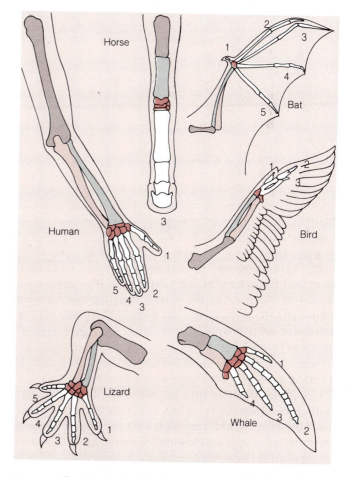

Figure 1.6   The forelimbs of vertebrates evolved from an ancestral pattern. Even in vertebrates as dissimilar as whales and bats, the same basic arrangement of bones can be observed. The digits (fingers) are numbered 1 (thumb) to 5 (little finger). The pattern of shading indicates homologous bones.

Table 1.1 Major Geological Eras

Era	Period	Millions of Years Ago	Major Biological Events	
			Plants	*Animals*
Cenozoic	Quaternary	2	Rise of herbaceous plants	Age of humans
	Tertiary	65	Dominance of the angiosperms	First hominids
				Rise of modern forms
				Mammals and insects dominate the land
Mesozioc	Cretaceous	130	Spread of angiosperms Decline of gymnosperms	Extinction of the dinosaurs
	Jurassic	180	First flowering plants (angiosperms)	Age of reptiles First mammals and first birds
	Triassic	230	Land plants dominated by gymnosperms	First appearance of the dinosaurs
Paleozoic	Permian	280	Land covered by forests of primitive vascular plants	Expansion of the reptiles Decline of the amphibians Extinction of the trilobites
	Carboniferous	350	Land covered by forests of coal-forming plants	Age of amphibians First appearance of reptiles
	Devonian	400	Expansion of primitive vascular plants over land	Fishes dominate the seas First insects First amphibians move onto land
	Silurian	435	First appearance of primitive vascular plants on land	Expansion of the fishes
	Ordovician	500	Marine algae	Invertebrates dominate the seas First fishes (jawless)
	Cambrian	600	Primitive marine algae	Age of invertebrates Trilobites abundant
Precambrian		4,500–5,000	*Aquatic Life Only* Origin of the invertebrates Origin of complex (eukaryotic) cells Origin of photosynthetic organisms Origin of primitive (prokaryotic) cells Origin of life	

dage structure indicates that the vertebrates evolved from a common ancestor.

Homologies are not always as obvious as the above example. Organisms may be only distantly related and have structures that have changed to the point that similarities are very difficult to recognize. For example, the bones of the middle ear of mammals are derived from some of the jaw bones of ancient fish. Comparing these evolutionary extremes by themselves may not reveal the homologies involved. Careful study of the fossil record, however, has revealed the important transitional stages.

Not all such similarities indicate homology. **Convergent evolution** occurs when two unrelated organisms adapt to similar conditions, resulting in superficial similarities in structure. For example, the wing of the bird and the wing of the insect are both adaptations for flight but are not homologous (figure 1.7). Any similarities existing between them are simply reflections of the fact that, to fly, an organism must have a broad, flat gliding surface. Instead of being homologous, these structures are said to be **analogous** (i.e., having a partial similarity). The comparative anatomist must, therefore, be careful of two fundamental

mistakes: describing homologies where none exist, and missing homologies that have been obscured by millions of years of independent evolution.

Structures are often retained in an organism, even though the structures may have lost their usefulness. They are often poorly developed, and are called **vestigial structures.** For example, the appendix of humans is the remnant of a larger, saclike structure that was involved in fermentation of plant materials. Such remnants of once useful structures are clear indications of change, hence evolution.

Molecular Biology In recent years, one scientific discipline that has yielded a wealth of information on evolutionary relationships is **molecular biology.** Just as animals can have homologous structures, animals may also have homologous processes. Ultimately, structure and function are based on the genetic blueprint found in all living animals, the DNA molecule. Related animals have DNA derived from their common ancestor. Because DNA carries the codes for proteins that make up each animal, related animals are expected to have similar proteins. With the modern laboratory technologies now available, biolo-

[4]

(a)

(b)

Figure 1.7 The wings of birds (a) and insects (b) are analogous. They are specialized for similar functions, and that similar function has coincidentally led to flat, planing surfaces required for flight. Both kinds of wings arose independently and, therefore, are not homologous.

gists can extract and analyze the structure of proteins from animal tissues, and compare the DNA of different animals. By looking for dissimilarities in the structure of related proteins and DNA, and by assuming relatively constant mutation rates, molecular biologists can estimate the elapsed time since divergence from a common ancestral molecule.

The above fields of study have generated impressive documentation of evolution since the initial studies of Darwin. Although investigators have unique problems with gathering and analyzing the data in each field of study, there is no doubt in the minds of the vast majority of scientists as to the reality of evolution, even though we may argue about the rates and the mechanisms involved.

Evolutionary theory has impacted biology like no other single theory. It has impressed scientists with the fundamental unity of all of biology. As you progress through this text, you will continually be reminded of the unity that exists within biology because of life's common origin.

Animals and Their Environment

In his foreword to *A Sand County Almanac* (1949), Aldo Leopold wrote:

> Conservation is getting nowhere because it is incompatible with our Abrahamic concept of land. We abuse land because we regard it as a commodity belonging to us. When we see land as a community to which we belong, we may begin to use it with love and respect. There is no other way for land to survive the impact of mechanized man, nor for us to reap from it the esthetic harvest it is capable, under science, of contributing to culture . . .

This passage is an attack on some of the misplaced values of our society, as well as a statement of hope for a better future. It has been over 40 years since Aldo Leopold wrote *A Sand County Almanac*. Have we come any closer to appreciating that humans are an integral part of the land? Have we even heard the challenges of Aldo Leopold, Rachel Carson (*Silent Spring*, 1962), and others?

The answer of course is not a simple one. In some parts of the world, the land is recovering from decades of abuse. One needs only to compare the polluted Lake Erie of the 1960s to the much cleaner lake of the 1990s. A lake once unfit for fishing and other recreational uses is now used extensively for these purposes. Tracts of land have been set aside for use as natural and wilderness areas. Many *nonbiodegradable* (not broken down by biological processes) substances have been taken out of the marketplace. These developments and others like them give reason for hope. Certainly some societies are beginning to realize that all life shares a dependence upon the earth's resources.

Even though there is hope, societies have not abandoned those values that treat land, water, and air as commodities that can be bought and sold, or neglected and abused. The problems are most acute in the Third-world countries. These countries look at industrialized nations of the world

and do all that is in their power to gain a similar wealth. In the process, the land suffers, population growth goes unchecked, toxic wastes are dumped, nuclear technologies become status symbols, plant and animal species become extinct, and deserts and famine expand.

Over the past 15 years both governmental and private agencies have undertaken large-scale evaluations of the environmental health of the world. The results of an early study commissioned by President Jimmy Carter in 1977 are published in *The Global 2000 Report to the President*. More recent studies have been made by organizations such as the Worldwatch Institute. The results of such studies deserve the serious attention of every concerned cititzen.

Population At the root of virtually all other environmental problems is global overpopulation. Population growth is expected to continue into the next century. Through the year 2000, most growth (92%) can be expected to occur in less-developed countries, where 5 billion out of a projected total of 6.35 billion will live. With a high proportion of the population at child-bearing age, even faster growth could occur in the next century. Problems with food distribution, health care, water, and poverty could drive people to the cities, and a proliferation of slums is certain to result. As the world population grows, it is expected that the disparity between the wealthiest and poorest nations will increase.

World Resources An obvious effect of overpopulation is the stress placed on world resources. Although new technologies continue to increase food production, most food is produced in industrialized countries that already have a high per-capita food consumption. Maximum oil production is expected during the 1990s. Although the quantity of other fuels is adequate for centuries of use, distribution is a problem, because the resources are not equally distributed. Continued use of fossil fuels adds more carbon dioxide to the atmosphere, contributing to the greenhouse effect and global warming. Deforestation of large areas of the world results from continued demand for forest products and fuel. This trend contributes to the greenhouse effect; causes severe, regional water shortages; and results in the extinction of many plant and animal species, especially in tropical forests. Preservation of forests would undoubtedly result in the identification of new species of plants and animals that could serve as important human resources: new foods, drugs, building materials, and predators of pests (figure 1.8).

Solutions There are no easy solutions to these problems. Unless we deal with the problem of overpopulation, however, it will be impossible to solve the other problems. What is certain is that there is no time for delay. We must forget about nationalistic boundaries and begin to work as a world community to prevent the spread of disease, famine, and other forms of suffering that go along with overpopulation. Bold and imaginative steps toward improved social and economic conditions and better resource man-

Figure 1.8 A tropical forest before (a) and after (b) clearcutting and burning to make way for agriculture. These soils will quickly become depleted, and will be abandoned for richer soils of adjacent forests. Loss of tropical forests results in extinction of many valuable forest species.

agement are needed. Essential to this work is the change in values that Aldo Leopold talked about in 1948. At the heart of this issue are the warnings that have been given to us over past decades. Again, from Aldo Leopold in *A Sand County Almanac*—

> Like winds and sunsets, wild things were taken for granted until progress began to do away with them. Now we face the question whether a still higher "standard of living" is worth its cost in things natural, wild, and free. For us the minority, the opportunity to see geese is more important than television, and the chance to find a pasque-flower is a right as inalienable as free speech.

Stop and Ask Yourself

1. What are four unifying themes of biology? Why is the study of each an important part of any biologist's training?
2. What is organic evolution?
3. How do each of the following contribute to the evidence for evolution? a) biogeography b) paleontology. c) molecular biology
4. What are the conclusions of studies on the environmental health of the world as regards: a) the earth's population in the year 2001, b) resources for the year 2001, and c) solutions for environmental problems?

What is Zoology?

Zoology shares with all of biology some common goals and concerns that have been described previously. As soon as zoologists begin to forget about these common elements they begin to lose the perspective that binds zoology to biology as a whole. **Zoology** (Gr. *zoon,* animal + *logos,* to study), the study of animals, is one of the broadest fields in all of science. Since the late 1800s, a wealth of information has accumulated, and it is virtually impossible for one person to keep abreast of all developments in zoology. The best that zoologists can hope to accomplish is to develop an expertise in a relatively restricted portion of the broader discipline of zoology. Then, by reading some of the less specialized journals, one can keep informed about developments in other areas of zoology, biology, and science as a whole.

Subdisciplines of Zoology

The diversity of the subdisciplines within zoology reflects the breadth of the field. These subdisciplines are based on particular functional, structural, or ecological interests that span many animal groups. It is not uncommon to specialize in the physiology, anatomy, or embryology of any one animal group or any one species (table 1.2). It is also possible to specialize in the structure or function of a particular organ system. For example, neurophysiologists specialize in the physiology of nervous systems, and cardiologists specialize in the functioning of the heart and its associated structures. Other zoologists specialize in the biology of particular animals or groups of animals (table 1.3). When one considers the sizes of these groups, it is no wonder that further specialization is also common. For example, there are approximately 300,000 described (and many undescribed) species of beetles! It is impossible for one person to be an expert in all areas of beetle biology. Therefore, one can specialize even further and become a beetle taxonomist, physiologist, or ecologist. It is obvious that the subdisciplines within zoology are not separated by sharply defined boundaries and much information is shared among them.

Historical Perspective

The early history of zoology, as all of biology, is essentially the same as the early history of philosophy and medicine. Ancient Egyptians and Greek biologists were also philos-

Table 1.2	Examples of Specializations in Zoology
Subdiscipline	**Description**
Anatomy	The study of the structure of entire organisms and their parts
Cytology	The study of the structure and function of cells
Ecology	The study of the interaction of organisms with their environment
Embryology	The study of the development of an animal from the fertilized egg to birth or hatching
Genetics	The study of the mechanisms of transmission of traits from parents to offspring
Histology	The study of tissues
Molecular biology	The study of subcellular details of animal structure and function
Parasitology	The study of animals that live in or on other organisms
Physiology	The study of the function of organisms and their parts
Systematics	The study of the classification of, and the evolutionary interrelationships between, animal groups

ophers. Some of the most important of these early biologists were Hippocrates (460 B.C.), who is called the "father of medicine" and Aristotle (384–322 B.C.). Aristotle was probably the greatest collector and organizer of knowledge in his time. He contributed valuable information on natural history and embryology to much later generations of scientists.

[5]

After a period with little scientific advancement, known as the "Dark Ages", a renaissance in interest in science began in Italy about 1100 A.D. The genius most often associated with this period is that of Leonardo da Vinci (1452–1519). He was an artist, an architect, a physicist, a biologist, and a philosopher. His interest in art led him to study the human body and he made detailed observations of the human eye and other aspects of anatomy.

Knowledge of physiology lagged somewhat. In the early 1600s, this situation changed when William Harvey (1578–1657) began his studies. His most significant contribution to human physiology was his demonstration that blood circulated through the body, moving away from the heart in the arteries and toward the heart in the veins. At this time advances were being made in the ability to ob-

Table 1.3	Examples of Specializations in Zoology by Taxonomic Categories
Entomology	The study of insects
Herpetology	The study of amphibians and reptiles
Ichthyology	The study of fish
Mammology	The study of mammals
Ornithology	The study of birds
Protozoology	The study of animallike protists

serve scientific materials with the development of microscopy by Anton van Leeuwenhoek. Along with other early microscopists, Robert Hooke, Marcello Malpighi, van Leeuwenhoek participated in the early development of the concept of the cell.

Although Carolus Linnaeus (1707–1778) is primarily remembered for his work with the collection and classification of plants, his work was also a milestone for zoology. His system of classification, **binomial nomenclature**, has been adopted for animals as well as plants. Each kind of organism is described with a two-part name. The first part indicates the *genus*, and the second part of the name indicates the *species* to which the organism belongs. Using these two-part names, each kind of organism can be recognized throughout the world. Above the species and genus level, organisms are grouped into families, orders, classes, phyla, and kingdoms, based on a hierarchy of relatedness (figure 1.9). Organisms in the same species are more closely related than organisms in the same genus, and organisms in the same genus are more closely related that organisms in the same order, and so on.

The eighteenth, nineteenth, and twentieth centuries saw a veritable explosion of all forms of scientific inquiry. The nineteenth century in particular has been called "The Scientific Age." The attitude of most scientists towards their work changed. Although the practical implications of scientific discoveries were by no means ignored, more attention was paid to the pursuit of knowledge, even when immediate rewards were not obvious. As has been the case ever since, the rewards of basic research generally have surfaced secondarily.

The Scientific Method

Over the years, scientific discoveries have been made as a result of several commonly recurring thought processes and activities that have become known as "the scientific method." This statement should not imply that there is a series of steps used by all scientists to study natural occurrences. Some problems lend themselves to solution by designing experiments; others are solved by detailed observations of processes or structures. **The scientific method** provides a frame of mind that allows the investigator to objectively analyze an occurrence. It often begins by observing natural phenomena with as little external interference as possible. Experiments may then be designed to answer questions that arise from such observations. The scientific method also involves recording and analyzing data as objectively as possible. The simple presence of the observer may interfere with what is being studied. In addition, it is very difficult for the observer to keep preconceptions from inadvertently influencing the results of a set of observations. For these reasons, science cannot depend on the results of one scientist. Rather, the scientific method depends upon repeated investigations by others for confirmation or rejection of experimental results.

Kingdom	Phylum	Class	Order	Family	Genus	Species

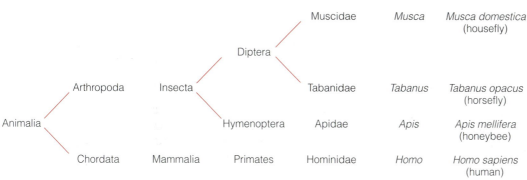

Figure 1.9 The classification of a housefly, horsefly, honeybee, and human.

Traditionally, the scientific method involves asking *questions* based on a set of *observations*. Being able to look at nature and ask relevant questions regarding biological (or other) processes is not as easy as it may seem. Knowing the proper questions to ask requires background in one's discipline as well as a perceptive mind. Asking the proper questions frequently determines whether the data gathered can be used to formulate reliable answers.

Following the initial questioning, the scientist attempts to formulate a reasonable explanation for the question. This explanation is called a **hypothesis.** In order to be of any use, a hypothesis must be testable. Frequently, a hypothesis is so broad that it cannot be directly tested. Rather, some preditions of the hypothesis are tested. The simplest explanations that seem to be consistent with the evidence available are usually the ones most easily tested.

The hypothesis is tested by collecting data from further observations or from an experiment that has been designed. It is essential to include a *control* in any experiment. A control group is treated the same as an experimental group, except that the variable being tested is omitted. A control serves as a basis for comparing the data from the experiment. Regardless of the outcome of an experiment, further testing is needed to back up any conclusions, and it is crucial that the results are communicated to other scientists. This communication may take the form of an article published in one of the hundreds of biological journals, or a paper presented orally at one of the many gatherings of scientists throughout the world. Regardless of the vehicle chosen, these reports must include discussions of the hypothesis, the methodology, the results, and the conclusions. The results of any biologist's work are thus available to the scrutiny of others.

After repeated testing, related hypotheses that have proven to be true may be assembled into a generalized statement that can be used in a predictive fashion and is called a **theory.** To a scientist, "theory" means something very different from the common use of the word as a conjecture or a guess. Scientists recognize that the details of what is known today will probably change with future research, and most realize that their knowledge is incomplete, as history has repeatedly shown. When scientists talk about the theory of evolution (or the theory of anything else), they are talking about a concept that has been proven to be true given years of research. What scientists can never do, however, is claim that all of the details of evolutionary (or other) mechanisms have been worked out.

Stop and Ask Yourself

5. What is zoology? Why do most zoologists choose to specialize in a field within zoology?
6. What is the scientific method?
7. What is a hypothesis?
8. What is a scientific theory? How is the word theory, used in this context, different from everyday usage?

Summary

1. Biology, the study of life, has four themes that unify all of its divergent disciplines. They are: (1) a common genetic molecule—DNA, (2) a common unit of organization—the cell, (3) a common evolutionary history, and (4) a common environment.

2. All life is based upon a common genetic molecule, DNA.

3. All animals are united by common processes occurring at a cellular level.

4. A common evolutionary origin and shared evolutionary forces influence the history of all animals.

5. Evidence for evolution comes from the fields of biogeography, paleontology, comparative anatomy, and molecular biology.

6. All animals share a common environment. Excessive population growth has placed undue pressure on all of the earth's resources. Human values must change to successfully check deterioration of the environment.

7. Zoology is the study of animals. It is a very broad field that requires zoologists to specialize.

8. Animals are classified into a hierarchical system of classification that uses a two-part name for every different kind of animal.

9. The scientific method provides a frame of mind that allows the investigator to objectively analyze an occurrence.

■ Key Terms

analogous (p. 8)
binomial nomenclature
   (p. 12)
biogeography (p. 5)
biology (p. 4)
comparative anatomy (p. 7)
convergent evolution (p. 8)
fossils (p. 6)
homologous (p. 7)

hypothesis (p. 13)
molecular biology (p. 8)
paleontology (p. 6)
the scientific method
   (p. 12)
theory (p. 13)
vestigial structures (p. 8)
zoology (p. 11)

■ Critical Thinking Questions

1. Zoologists need to consider the broader perspective that unites zoology with biology as a whole. Could this broader perspective also include the arts and humanities? Explain.

2. What is the difference between analogy and homology? In an evolutionary context, how could analogous structures arise?

3. Why is perfect objectivity impossible to attain?

4. Imagine that you have just gotten into your car and turned the key but, the engine did not start. Outline how you would go about solving this problem using an objective, scientific approach. (If you would rather substitute another everyday problem, go ahead.)

■ Suggested Readings

Books

Brown, L. R., Durning, A., Flavin, C., Heise, L., Jacobson, J., Postel, S., Renner, M., Shea, C. P., and Starke, L. 1989. *State of the World:* New York: W. W. Norton.

Carson, R. 1962. *The Silent Spring.* Burlington: Houghton Mifflin.

Danpier, W. C. 1966. *A History of Science and Its Relation to Philosophy and Religion.* New York: Cambridge University Press.

Gardner, E. J. 1972. *History of Biology.* Minneapolis: Burgess Publishing.

Gould, S. J. 1980. *The Panda's Thumb: More Reflections in Natural History.* New York: Norton.

Kuhn, T. S. 1970. *Structure of Scientific Revolutions.* Chicago: University of Chicago Press.

Leopold, A. 1949. *A Sand County Almanac.* New York: Sierra Club/Ballantine.

Sayre, A. 1975. *Rosalind Franklin and DNA.* New York: Norton.

Schrodinger, E. 1967. *What is Life?* and *Mind and Matter.* New York: Cambridge University Press.

The Global 2000 Report to the President: Entering the Twenty-First Century, Vol. 1–3. 1983. Washington D.C.: U.S. Government Printing Office.

Watson, J. D. 1968. *The Double Helix.* New York: Signet Books.

Articles

Bingham, R. 1982. On the life of Mr. Darwin. *Science '82* 3(3): 82.

Eiseley, L. C. Charles Darwin. *Scientific American* February, 1956.

Eiseley, L. C. Alfred Russel Wallace. *Scientific American* February, 1959.

Gould, S. J., 1980. Wallace's fatal flaw. *Natural History* 89:26.

Keele, K. D. 1978. The life and work of William Harvey. *Endeavour* 2:104.

The Chemistry of Life

2

Concepts

1. Animals are made up of molecules, which are collections of atoms bound to one another. The life processes within an animal are based, to a large degree, on the chemical properties of atoms, ions, and molecules.
2. The molecule most important to the evolution of life is water.
3. Carbohydrates and lipids serve as the principal source of energy for most animals.
4. Proteins and nucleic acids are large molecules composed of amino acids and nucleotides, respectively. These molecules provide the basis for structure, function, and genetic regulation of animals.

Have You Ever Wondered:

[1] what the four most common elements in most animals are?
[2] what radioactivity is?
[3] what holds water together?
[4] what type of chemical reaction repairs worn-out parts in animals?
[5] why oxidation-reduction reactions are important to animals?
[6] why water is called the "universal solvent?"
[7] why animals can't be "vaporized" as they are in science fiction movies?
[8] how it is possible that some insects can walk on water?
[9] how water moves from the roots of a tree to its uppermost branches?
[10] why ice cubes float?
[11] why a liquid fat is called an oil?

These and other useful questions will be answered in this chapter.

This chapter contains underlined evolutionary concepts.

Everything that goes on within the body of an animal depends on some kind of chemical activity. Without this chemical activity, life as we know it could not exist. This chapter presents the chemistry you will need to understand the basic chemical processes that enable animals to function in the environment as living entities.

Chemistry is the branch of science dealing with the composition of substances and changes that take place in their composition. A knowledge of chemistry is essential for understanding the structure (**anatomy** [Gr. *ana*, again + *temnein*, to cut]) and function (**physiology** [Gr. *physis*, nature]) of animals, because body functions involve chemical changes that occur in structural units, such as cells. As interest in the chemistry of living animals grew, and knowledge in this area expanded, a new subdivision of science called **biological chemistry**, or biochemistry ("the chemistry of life") emerged. Biochemistry is the study of the molecular basis of life.

Atoms and Elements: Building Blocks of All Matter

Matter is anything that occupies space and has mass. It includes all the solids, liquids, and gases in our environment, as well as those in bodies of all forms of life. **Mass** refers to the amount of matter in an object. Matter is composed of **elements** which are chemical substances that cannot be broken down into simpler units by ordinary chemical reactions. An element is designated by either a one- or two-letter abbreviation of its English or Latin name. For example, O is the symbol for the element oxygen, H stands for hydrogen, and Na for sodium. Currently, 106 elements are known, 92 of which occur naturally. About 15 elements are found in most animals, and four of these (carbon, hy-

[1]

drogen, oxygen, and nitrogen) account for the majority (97%) of an animal's body weight (table 2.1). The remaining 3% of an animal's weight consists primarily of sodium, calcium, phosphorus, potassium, and sulfur. Some elements present in trace amounts include manganese, magnesium, copper, iodine, iron, and chlorine.

Elements are composed of units of matter called atoms. An **atom** (Gr, *atomos*, indivisible) is the smallest part of an element that can enter into a chemical reaction. Although the atoms that make up each element are similar to each other, they differ from the atoms that make up the other elements. Atoms vary in size, weight, and the various ways they interact with each other. For example, some atoms are capable of combining with atoms like themselves or with other kinds of atoms, whereas others lack this ability.

Structure of Atoms

Atoms have two main parts—a central core called a *nucleus* and the surrounding *electron field* (figure 2.1). The nucleus contains two major particles, the positively charged **protons** (p^+) and the uncharged **neutrons** (n^0). Surrounding the nucleus are negatively charged particles called **electrons** (e^-) that exist in an electron field or "cloud." Any one electron is moving so rapidly around the nucleus that it cannot be found at any given point at any particular moment in time, therefore its location is given as an electron field or cloud.

Because the nucleus contains the protons, this part of the atom is always positively charged. The number of negatively charged electrons outside the nucleus is equal to the number of protons, thus, an atom is electrically uncharged or *neutral*.

The chemical and physical properties of an atom are determined by the number of protons and neutrons in its

Table 2.1	Naturally Occurring Elements in Animals			
Element	Symbol	Atomic Number	Atomic Mass	Percent Wet Weight of Body
Oxygen	O	8	16	65
Carbon	C	6	12	19
Hydrogen	H	1	1	10
Nitrogen	N	7	7	3
Calcium	Ca	20	40	2
Phosphorus	P	15	31	1
Potassium	K	19	39	0.5
Sulfur	S	16	32	0.3
Sodium	Na	11	23	0.2
Chlorine	Cl	17	35	0.2
Magnesium	Mg	12	24	0.1
Manganese	Mn	25	55	0.1
Iron	Fe	26	56	0.1
Copper	Cu	29	64	0.1
Iodine	I	53	127	0.1

(97% for the top four through Nitrogen; approx. 3% for Calcium through Iodine)

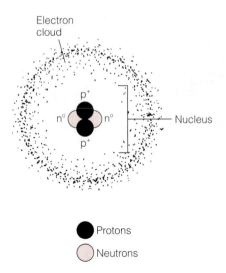

Figure 2.1 The nucleus of a helium (He) atom consists of two neutrons and two protons and is surrounded by two electrons that form a cloud of negative charges.

nucleus and by the number and arrangement of electrons in the electron field. The **atomic number** of an element is the number of protons in the nucleus of one of its atoms. The atomic number identifies an element. If an atom has one proton, for example, it is hydrogen; if it has six, it is carbon; and if an atom has eight protons, it is oxygen.

Another measure of an atom is its atomic mass. The **atomic mass** is equal to the number of neutrons *and* protons in its nucleus. For example, because carbon contains six protons and six neutrons, its atomic mass is 12 and is symbolized with a superscript preceding the element's symbol: ^{12}C (read ''carbon-12''). The atomic numbers and atomic masses of the more important elements in a typical animal are given in table 2.1.

Most naturally occurring elements are actually a mixture of slightly different forms of that element. All atoms of a given element have the same number of protons in the nucleus, but some have different numbers of neutrons, and thus different atomic masses. These different forms that have the same atomic number but different atomic masses are called **isotopes.** For example, the most common form of carbon atom has six protons and six neutrons in the

nucleus, and has a atomic mass of 12 (^{12}C). A carbon isotope with six protons and seven neutrons would have a atomic mass of 13 (^{13}C), and the carbon isotope with six protons and eight neutrons is 14 (^{14}C) (figure 2.2).

Some isotopes (e.g., ^{12}C and ^{13}C) are stable and do not break down. Other isotopes (e.g., ^{14}C) are unstable and tend to break down (decay or decompose) by periodically emitting small particles. These unstable isotopes are termed **radioactive.** Oxygen, iron, cobalt, iodine, and phosphorous are all elements that have radioactive isotopes.

When the nuclei of **radioisotopes** (radioactive isotopes) break down, they emit one or more of three major types of radioactivity: *alpha particles* (α), *beta particles* (β), or *gamma rays* (γ). An alpha particle consists of two neutrons and two protons. These particles do not travel very far and are not very penetrating. By contrast, beta particles are fast-moving electrons that have about 100 times more penetrating power than alpha particles. Gamma rays are not particles but a form of high-energy electromagnetic radiation. They travel at the speed of light and are so penetrating that they can only be stopped by very dense material, such as lead or concrete (box 2.1).

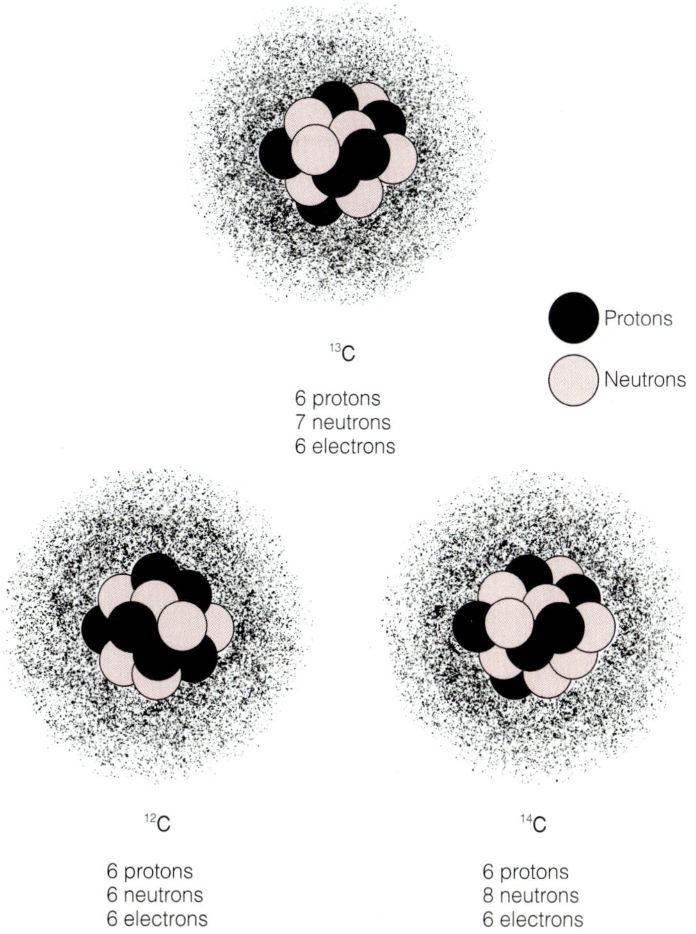

[2]

Energy-Level Shells

The electrons of an atom are distributed around its nucleus in orbitals called **energy-level shells** or **clouds of electronegativity** (figure 2.3). The location of these electrons in relation to the nucleus has a great influence on the way atoms react with each other. Seven energy-level shells are possible. Each shell can hold only a certain number of electrons. There are never more than two electrons in the shell nearest the nucleus; as many as eight electrons can be in each of the second and third shells; larger numbers fill the more distant shells. When an atom has a complete outer shell—that is, the shell holds the maximum number of electrons possible—the shell is complete and stable. An atom with an incomplete, or unstable, outer shell tends to gain, lose, or share electrons with another atom. For example, an atom of sodium has 11 electrons arranged as illustrated in figure 2.3: two in the first shell, eight in the second shell, and one in the third shell. As a result of this arrangement, this atom will tend to lose the single electron from its outer shell, which leaves the second shell filled and the atom stable.

^{13}C

6 protons
7 neutrons
6 electrons

● Protons

○ Neutrons

^{12}C

6 protons
6 neutrons
6 electrons

^{14}C

6 protons
8 neutrons
6 electrons

Figure 2.2 The nuclei of these carbon isotopes differ in their number of neutrons, although each has six protons. ^{12}C has six neutrons, ^{13}C has 7 neutrons, and ^{14}C has eight neutrons.

Stop and Ask Yourself

1. What is the relationship between matter and elements?
2. How are electrons, protons, and neutrons positioned within an atom?
3. What is an isotope?
4. What is the difference between atomic number and atomic mass?

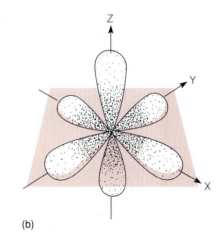

Outermost shell
(1 electron)

Second shell
(8 electrons)

Nucleus

Innermost shell
(2 electrons)

Na atom contains
11 electrons (e⁻)
11 protons (p⁺)
12 neutrons (n°)

Atomic number = 11
Atomic weight = 23

(a)

(b)

Figure 2.3 A diagram of a sodium atom showing its three energy-level shells (a). In reality, electrons are not found in a definite location, but travel rapidly in a three-dimensional space (x, y, and z planes) around the nucleus (b).

Molecules: Aggregates of Atoms

Atoms rarely exist in animals or the environment by themselves. Instead, they combine to form **molecules**. An example of a simple molecule is oxygen (O_2); the subscript 2 indicates that the oxygen molecule contains two oxygen atoms. Other molecules are made up of atoms of two or more elements, such as carbon dioxide (CO_2).

When atoms interact chemically to form molecules, the atoms are held together by electrical forces called *chemical bonds*. The kinds of bonds most important to the chemistry of animals are the covalent, hydrogen, and ionic bonds.

Covalent Bonds

When atoms *share* electrons with other atoms, the chemical bond that is formed is called a **covalent bond** (the prefix *co-*, indicates a shared condition) (figure 2.4). In covalent bonding, electrons are always shared in pairs. When a pair of electrons is shared (one from each molecule), a *single bond* is formed; when two pairs are shared,

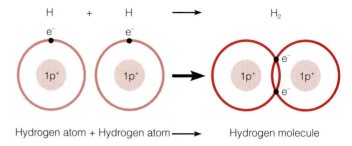

Figure 2.4 A single covalent bond is formed when two hydrogen atoms share a pair of electrons.

a *double bond* is formed; and when three are shared, a *triple bond* is formed.

In a molecule like H_2, the electrons spend as much time orbiting one nucleus as the other. Therefore, the distribution of charges is symmetrical, and the bond is called a **nonpolar covalent bond**. Because of this equal sharing, the molecule is electrically balanced, and the molecule as a whole is neutral.

In other molecules, such as H_2O, where two hydrogen atoms combine with one oxygen atom, the electrons spend more time orbiting the oxygen nucleus than the hydrogen nuclei. The electrical charge from the cloud of moving electrons is thus asymmetrical, and the bond is called a **polar covalent bond**. Such a bond leaves the oxygen atom with a slightly negative charge and the hydrogens with a slightly positive charge. The entire shape of the H_2O molecule reflects this polarity; rather than the linear arrangement H-O-H, the two hydrogens are at one end, a bit like the corners of a triangle. This shape and polarity can lead

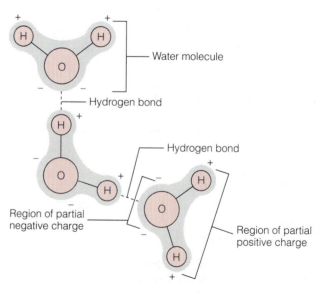

Figure 2.5 Hydrogen bonds are formed when oxygen atoms of different water molecules are weakly joined together by the attraction of the electronegative oxygen for the positively charged hydrogens. Because of the arrangement of electron orbitals and the bonding angles between oxygen and hydrogen, the molecule as a whole is polar (it carries a slight negative charge at one end and a slight positive charge at the other end). Many properties of water can be traced to this polarity.

Box 2.1

Isotopes: Estimations of Fossil Age

Radioactive isotopes can be used for dating rocks and fossil remains of animal life. As mentioned in the text, radioactive atoms undergo radioactive decay. When they decay, they become atoms of other elements. Carbon 14 (^{14}C) decays to nitrogen 14 (^{14}N), and the radioactive isotope uranium 238 (^{238}U) decays, through several steps, to the stable isotope lead 206 (^{206}Pb). Each radioactive isotope has a typical **half-life** (box table 2.1). Its half-life is the time it takes for one-half of the atoms of that isotope to undergo radioactive decay. After one half-life, only one-half the original atoms of the isotope are left; after two half-lives, only one-fourth are left, and so on.

Radioactive-dating is a technique in which isotopes can serve as "radioactive clocks" for measuring the passage of time (box figure 2.1). In order to apply radioactive-dating to a fossil or rock, three things must be known: (1) the half-life of the isotope being measured;

Box figure 2.1 Isotope-dating techniques can be used to determine the age of human bones, such as those found in this Neanderthal burial site.

Box Table 2.1	Radioisotopes Commonly Employed to Determine the Age of Rocks and Fossils	
Parent Element*	**Daughter Element***	**Half-Life (millions of years)**
Potassium 40	Argon 40	1,300
Rubidium 87	Strontium 87	47,000
Uranium 235	Lead 207	713
Uranium 238	Lead 206	4,510
Carbon 14	Nitrogen 14	0.0057

*The parent radioactive element emits particles (α or β) or rays (γ) as it decays into the daughter element. In 713,000,000 years, for example, 1 kg of uranium 235 will decay to form 0.5 kg of uranium 235 plus 0.5 kg of lead 207.

(2) how much of the isotope was originally present in the fossil or the rock containing the fossil; and (3) how much of the isotope is left. For example, by measuring the amounts of uranium 238 and lead 206 in a rock, and knowing that the half-life of uranium 238 is 4,510 million years, one can calculate when the rock was formed.

Carbon dating is one method of determining the age of the fossil remains of an organism because the carbon atom is the basic backbone of all living matter. Carbon 14 has such a short half-life (5,730 years), however, that it is only useful for fossils less than 30,000 years old. As a result, the more slowly decaying isotopes, shown in box table 2.1, have been used, along with a variety of geological methods, to establish the age of rocks. Fossils found in the rocks are assumed to be the same age as the rock in which they are embedded.

to the formation of another kind of chemical bond—the hydrogen bond.

another water molecule, and so forth, until many molecules are bonded together (figure 2.5).

Hydrogen Bonds

In those molecules where hydrogen is bonded to certain other atoms (e.g., C, N, or Fe), the hydrogen electron is drawn to another atom, leaving a proton behind. As a result, the hydrogen atom gains a slight positive charge. The remaining proton is attracted to negatively charged atoms of, for example, oxygen in nearby molecules. When this happens, a weak attraction, called a **hydrogen bond,** is formed. The hydrogen atom in one water molecule, for example, forms a hydrogen bond with the oxygen atom in

Ionic Bonds

If an atom *loses* one or more electrons, it becomes *positively* charged, because there are now more positively charged protons in the nucleus than negatively charged electrons surrounding the nucleus. This positive charge is shown as one or more "plus" signs. Conversely, if an atom *gains* one or more electrons, it becomes *negatively* charged, and this negative charge is shown as one or more "minus" signs. Once an atom either gains or loses electrons, it acquires an electrical charge and is called an **ion**

[3]

Figure 2.6 In the formation of sodium chloride, an electron from the outermost shell is transferred from the sodium atom to the chlorine atom, giving both atoms complete outermost shells. This transfer gives the sodium atom a net charge of $+1$ (sodium ion) and the chlorine atom a net charge of -1 (chloride ion). An ionic bond is formed between the oppositely charged ions.

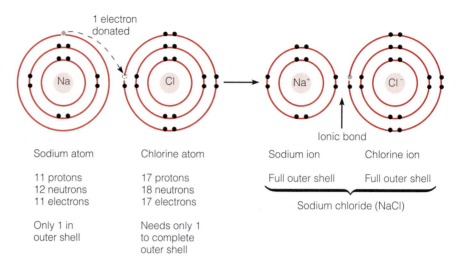

1 electron donated

Ionic bond

Sodium atom	Chlorine atom	Sodium ion	Chlorine ion
11 protons	17 protons	Full outer shell	Full outer shell
12 neutrons	18 neutrons		
11 electrons	17 electrons		

Sodium chloride (NaCl)

Only 1 in outer shell

Needs only 1 to complete outer shell

(Gr. *ion,* going). A positive ion is known as a *cation,* and a negative ion is an *anion.* Examples of cations are sodium (Na^+), potassium (K^+), hydrogen (H^+), calcium (Ca^{2+}) and iron (Fe^{3+}). Some anions are chloride (Cl^-), hydroxyl (OH^-), bicarbonate (HCO_3^-), sulfate (SO_4^{2-}), phosphate (PO_4^{3-}), and carboxyl (COO^-).

Ionic bonds are formed when an atom or group of atoms develops an electrical charge and becomes attracted to an atom or group of atoms with an opposite charge. Figure 2.6 shows how an ionic bond is formed between sodium and chlorine to produce sodium chloride. When a sodium atom and chlorine atom come together, an electron is donated from the sodium atom to the chlorine atom. This electron transfer changes the balance between the protons and electrons in each of the two atoms. The sodium atom ends up with one more proton than it has electrons, and the chlorine atom with one more electron than it has protons. The sodium atom is left with a net charge of $+1$ (Na^+), and the chlorine atoms's net charge is -1 (Cl^-). These unlike charges attract each other and form the ionic bond.

Stop and Ask Yourself

5. What is a chemical bond?
6. What is a double covalent bond?
7. How is a hydrogen bond formed?
8. What is an ion?

Chemical Reactions

In *chemical reactions,* bonds between atoms are broken or joined, and different combinations of atoms or molecules are formed. Four basic types of chemical reactions occur in animals: (1) combination reactions, (2) decomposition reactions, (3) hydrolysis and condensation reactions, and (4) oxidation-reduction reactions.

Combination Reactions

When two or more atoms, ions, or molecules combine to form a more complex substance, the process is called a **combination** (or synthesis) **reaction**: A + B → AB. The substances that are combined are the *reactants,* and the new molecule they form is the *product*:

$$C + O_2 \rightarrow CO_2$$

Carbon and oxygen are the reactants that combine to form carbon dioxide. The arrow means "to produce" or "yields" and illustrates the direction in which the reaction is moving. In animals, combination reactions are important in the growth of body parts and the repair of worn or damaged tissues. Combination reactions require an input of energy to occur.

Many chemical reactions are reversible, that is, after the reactants combine, the product can be decomposed in a later reaction to produce the original reactants. A *reversible reaction* can be symbolized using double arrows, as follows:

$$A + B \rightleftharpoons AB$$

Decomposition Reactions

Decomposition reactions are the opposite of combination reactions. They result in the breakage of chemical bonds to form two or more simple products (atoms, ions, or molecules): AB → A + B.

$$C_6H_{12}O_6 + 6O_2 \rightarrow 6CO_2 + 6H_2O$$

Hydrolysis and Condensation Reactions

When a **hydrolysis** (Gr. *hydro*, water + *lysis*, dissolution) **reaction** takes place, a molecule of water interacts with the reactant to break the reactant's bonds. The original reactant and water are then rearranged into different molecules:

$$C_{12}H_{22}O_{11} + H_2O \rightarrow C_6H_{12}O_6 + C_6H_{12}O_6$$

[4]

Sucrose and water combine to form glucose and fructose. In the above reaction, the bonds of sucrose are broken (hydrolyzed) by the water molecule to form the sugars, glucose and fructose. In general, most digestive and degradative processes in animals occur by hydrolysis.

Condensation reactions are often hydrolytic reactions in reverse, and involve molecules that unite to form larger molecules with the production of one or more molecules of water:

$$NH_3 + C_2H_3O_2H \rightarrow C_2H_5NO + H_2O$$

In this reaction, ammonia plus acetic acid produce acetamide and water.

Oxidation-Reduction Reactions

An **oxidation reaction** occurs when an atom or molecule *loses* electrons or hydrogen atoms:

$$Na \rightarrow e^- + Na^+$$

When a sodium ion loses a negatively charged electron, it is oxidized and produces a positively charged sodium ion. When an atom or molecule *gains* electrons or hydrogen atoms, the reaction is called **reduction:**

$$Cl_2 + 2e^- \rightarrow 2Cl^-$$

When a chlorine molecule gains two negatively charged electrons, it is reduced and produces two negatively charged chloride ions.

In the bodies of animals, oxidation and reduction always occur together. Therefore, when an *oxidation-reduction* reaction takes place, the oxidizing agent is *reduced* (it gains electrons), and the reducing agent is *oxidized* (it loses electrons). Oxidation-reduction reactions are especially important because they provide the energy that is used for most biological activities.

[5]

Acids, Bases, Salts, and Buffers

Any substance, such as sodium chloride (NaCl), that conducts electricity when in solution, is called an **electrolyte.** Many fluids contain strong electrolytes that break down (ionize) into ions. Most acids and bases are electrolytes.

An **acid** is a substance that releases hydrogen ions (H^+) when dissolved in water. Because a hydrogen atom without its electron (H^+) is only a proton, an acid can be described as a proton donor as follows:

$$HCl \rightarrow H^+ + Cl^-$$

One molecule of hydrochloric acid dissolves in water to give one hydrogen ion and one chloride ion.

In contrast, a **base** is a substance that releases hydroxyl ions (OH^-) when dissolved in water as follows:

$$NaOH \rightarrow Na^+ + OH^-$$

One molecule of sodium hydroxide dissolves in water to give one sodium ion and one hydroxyl ion.

When a base has dissolved in water, it acts to remove free protons from the water; therefore, it can be called a proton acceptor.

A **salt** is an ionic substance that contains an anion other than OH^- or O^{2-}. An acid can react with a base to produce a salt in what is called a *neutralization reaction* as follows:

$$HCl + NaOH \rightarrow NaCl + H_2O$$

Dissociation is the ability of some molecules to break up into charged ions in water. For example, hydrochloric acid is a *strong acid* because it dissociates completely in water into H^+ and Cl^- ions. Acetic acid, however, is a *weak acid* because it does not completely dissociate in water.

pH: Measuring Acidity and Alkalinity

The chemical reactions involved with life processes are often affected by the presence of hydrogen and hydroxyl ions; therefore, the concentrations of these ions in the body fluids are important. The higher the concentration of hydrogen ions (H^+), the *more* acidic a solution is, and the higher the concentration of hydroxyl (OH^-) ions, the more basic a solution is. A solution is *neutral* when the number of hydrogen ions equals the number of hydroxyl ions.

The numerical scale that measures acidity and alkalinity is called the **pH scale.** The pH scale runs from 0 to 14, with neutrality at 7. Table 2.2 shows the relationship between hydrogen ion concentration and pH. *Acidic solutions,* have a pH of less than 7, and *basic solutions* have a pH of between 7 and 14. Each whole number on the pH scale represents a tenfold change (logarithmic) in acidity, therefore, a solution with a pH of 3 is 10 times more acidic than a solution with a pH of 4. Actually, the pH value is equal to the negative logarithm of the hydrogen ion concentration

$$pH = -\log [H^+]$$

or

$$pH = \log(1/[H^+])$$

For example, a solution with a hydrogen ion concentration of 0.1 grams per liter has a pH value of 1.0; a concentration of 0.01 g H^+/l has a pH of 2.0; 0.001 g H^+/l has a pH of 3.0, and so forth.

pH: Control with Buffers

A stable internal environment in an animal can be maintained only if there is a relatively constant pH of the body

Table 2.2 The Relationship Between Hydrogen Ion (H⁺) Concentration, Hydroxyl Ion (OH⁻) Concentration, and pH

H$^+$ (Hydrogen Ion)		pH	OH$^-$ (Hydroxyl Ion)
$10^{0} = 1$		0	$10^{-14} = 0.00000000000001$
$10^{-1} = 0.1$		1	$10^{-13} = 0.0000000000001$
$10^{-2} = 0.01$		2	$10^{-12} = 0.000000000001$
$10^{-3} = 0.001$	Acidic	3	$10^{-11} = 0.00000000001$
$10^{-4} = 0.0001$		4	$10^{-10} = 0.0000000001$
$10^{-5} = 0.00001$		5	$10^{-9} = 0.000000001$
$10^{-6} = 0.000001$		6	$10^{-8} = 0.00000001$
$10^{-7} = 0.0000001$	Neutral	7	$10^{-7} = 0.0000001$
$10^{-8} = 0.00000001$		8	$10^{-6} = 0.000001$
$10^{-9} = 0.000000001$		9	$10^{-5} = 0.00001$
$10^{-10} = 0.0000000001$		10	$10^{-4} = 0.0001$
$10^{-11} = 0.00000000001$	Basic	11	$10^{-3} = 0.001$
$10^{-12} = 0.000000000001$		12	$10^{-2} = 0.01$
$10^{-13} = 0.0000000000001$		13	$10^{-1} = 0.1$
$10^{-14} = 0.00000000000001$		14	$10^{0} = 1$

fluids. Too much of a strong acid or base can destroy the stability of cells. Also, too sudden a change in pH may be destructive. The fluid systems of most animals contain chemical substances that help regulate the acid-base balance. These substances, called **buffers,** are combinations of weak acids or weak bases and their respective salts in solution. Buffers help body fluids resist changes in pH when small amounts of strong acids or bases are added. The most important buffers are the bicarbonates, phosphates, and organic molecules, such as amino acids and proteins.

An important buffer system is the carbonic acid-bicarbonate ion system, which is involved in the buffering of the blood of higher vertebrates:

$$H_2CO_3 \rightleftharpoons H^+ + HCO_3^-$$

Carbonic acid dissociates to form hydrogen ion and bicarbonate ion.

In this example, if H⁺ ions are added to the system, they combine with HCO_3^- to form H_2CO_3 (the reaction goes to the left). This reaction removes H⁺ and keeps the pH from changing. If excess OH⁻ are added, they react with the H⁺ to form water, and more H_2CO_3 will ionize and replace the H⁺ ions that were used (the reaction goes to the right). Again, pH stability is maintained.

Stop and Ask Yourself

9. What is an electrolyte? What is an acid? What is a base? What is a salt?
10. Why is HCl a strong acid?
11. What is the relationship between pH and H⁺ concentration?
12. How do buffers function? Why are buffers important to animals?

Water: Life's Most Precious Fluid

Water covers about 75% of the earth's surface and comprises from 50 to 99% of an animal's body. Life as it evolved on earth is inextricably tied to its principal ingredient, water. As a result, much of the chemistry of life is water chemistry. The way that life evolved was determined largely by the chemical properties of the liquid water in which its evolution occurred. For example, when life on earth was beginning, water provided a medium in which other molecules could move about and interact without being bound by strong ionic or covalent bonds. Life evolved as a result of these interactions.

Chemical Properties of Water

The unique properties of water depend on its bonding structure—two hydrogen atoms covalently bonded to one oxygen atom (see figure 2.5). The oxygen atom attracts electrons to itself and away from the hydrogen atoms. As a result, the oxygen atom obtains a slight negative charge and the hydrogens a slight positive charge. Because the two ends of a water molecule have different charges, like the two poles of a magnet, it is called a **polar molecule.** Polarity allows water molecules to form hydrogen bonds with other water molecules and with a large variety of other compounds, as well as to dissolve other polar substances.

1. Water is a "universal" solvent. A solvent is a liquid that dissolves something else. Because of its polar nature, more substances can dissolve in water than any other known liquid. Substances that dissolve in solvents are called *solutes.* A solute plus solvent equals a *solution.* When a polar substance dissolves, its individual molecules or ions separate and mingle with water molecules. For example, the Na⁺ and Cl⁻ ions of a salt crystal are attracted to one another because of their opposite elec-

[6]

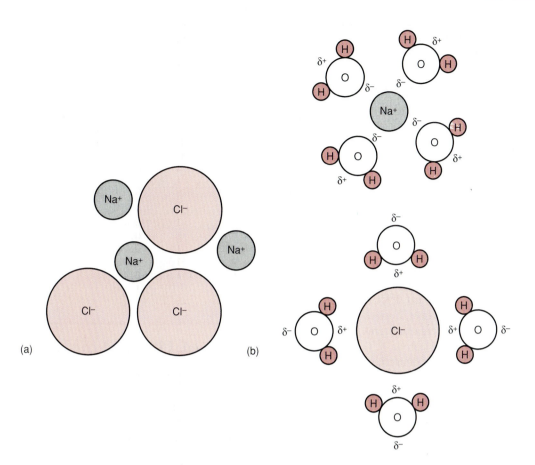

Figure 2.7 Sodium chloride (NaCl) dissolving in water. (a) A small part of a sodium chloride crystal. (b) Sodium (Na⁺) and chloride (Cl⁻) ions dissolved in water, which forms a hydration shell around the individual ions.

(a)

(b)

trical charges (figure 2.7). When the crystal comes into contact with water, however, the positively charged sodium ions attract the oxygen atoms of water molecules, which have partial negative charges (δ^-). Similarly, negatively charged chloride ions attract the partially positive (δ^+) hydrogens of water molecules. Notice that the water molecules orient around the Na⁺ and Cl⁻ in a cloud that is called a **hydration shell.** It is these hydration shells that prevent the Na⁺ and Cl⁻ ions from associating with one another and reforming a crystal.

Nonpolar molecules, such as those composed mainly of carbon and hydrogen, do not dissolve in water. Instead, they form interfaces with it, such as the interface formed in salad dressing where the polar water and nonpolar oil separate from each other. These interfaces are the sites for many important chemical reactions in animals. Thus, the inability of water to dissolve nonpolar substances is also necessary for life.

2. Water has a high boiling point or heat of vaporization. A great deal of heat energy is required to overcome the attractions between adjacent water molecules to change liquid water to a gas (vapor)—more than 580 calories are required to change 1 gram of liquid water into water vapor. When water reaches its boiling point, there is sufficient energy to break all of the hydrogen bonds between each water molecule and its neighbor so that water can escape from the surface of the liquid and enter the air. Because temperatures on earth seldom reach the boiling point of water, animals need not face the possibility of "boiling away" or being "vaporized."

3. Water has a high specific heat. Water has a high capacity to absorb heat. That is, it takes a lot of heat energy to raise the temperature of water, and conversely, much heat energy must be lost to lower its temperature. Specifically, 1 calorie of energy is required to elevate the temperature of 1 gram of water 1°C; e.g., from 36 to 37°C. Because animals are made up largely of water, they gain and lose heat relatively slowly. This retention of heat is a great protective agent for all forms of animal life because many biochemical reactions will take place only within a narrow range of temperatures.

4. Water has a high thermal conductivity. Heat energy introduced in one part of a unit of water is *rapidly* conducted throughout the rest of the unit of water. Because water is usually evenly distributed throughout an animal's body, heat produced in one part of the body spreads rapidly, preventing the development of "hot spots", which could be destructive to localized cells.

5. Water is adhesive and cohesive. *Adhesion* is the attachment of unlike substances to each other (such as water molecules to the molecules of silicon dioxide on the walls of a drinking glass). *Cohesion* is the holding together of like substances (such as water to water). On the surface of water, the water molecules show cohesion (due to their hydrogen bonds; *see figure 2.5*), making the surface appear to be covered by a "thin elastic skin." This phenomenon is known as **surface tension,** and explains why some insects can "walk on water". The adhesive and cohesive nature of water also accounts for its **capillarity,** the ability to move upward in a nar-

[9]

row space against the force of gravity. For example, water moving upward in a thin glass tube, water moving upward through a piece of porous paper, water moving to the top of a large tree, or water moving through the fine pores in the soil—all are examples of capillarity. Even though water has a high surface tension, it has a low **viscosity** (the resistance of flow by a fluid; i.e., its stickiness or thickness). This low viscosity favors ease of movement through the bodies of animals.

6. Water is a good evaporative coolant. When water changes from a liquid to a gaseous (vapor) state, its molecules escape from the body of the liquid into space. As they escape, they carry considerable heat energy with them. Many land-dwelling animals get rid of excess body heat by this type of *evaporative cooling;* e.g., panting by a dog and perspiring by a human.

7. Water has a high freezing point and a lower density as a solid than as a liquid. As water cools toward 0°C, it contracts and becomes more dense. However, unlike other liquids that become denser with decreasing temperature, water is more dense in its liquid state at 4°C than it is below 4°C, when it becomes less dense and expands to form ice crystals (figure 2.8). Ice is a regular latticework with every water molecule hydrogen-bonded to four others, unlike the small percentage of molecules that are bonded to four others in liquid water. The water molecules in ice are less densely packed than in liquid water because of the large number of hydrogen bonds. Therefore an ice crystal is larger

than the volume of water it replaces, and as a result, ice floats rather than sinks. The biological significance [10] of this property is that if solid water sank, lakes would freeze to the bottom, and many bodies of water would remain frozen nearly year round. Instead, the ice layer floats, acting as insulation to retard freezing at lower depths. This insulation effect allows many aquatic animals to survive in the chilly water beneath the ice.

Stop and Ask Yourself

13. Why is water considered the universal solvent?
14. How can a water strider walk on the surface of water?
15. Why does a dog pant?
16. Why does ice float on water?

The Molecules of Living Animals

The chemicals that enter into metabolic reactions or are produced by them can be divided into two large groups. Those that contain carbon atoms are called **organic molecules**. Those that lack carbon are called **inorganic molecules**. (A few simple molecules that contain carbon, such as CO_2, are considered inorganic.)

The most important characteristics of organic molecules depend on properties of the key element, carbon—the indispensable element for all life. The carbon atom has four electrons in its outer orbital; thus, it must share four additional electrons by covalent bonding with other atoms to fill its outer orbital with eight electrons. This unique bond-

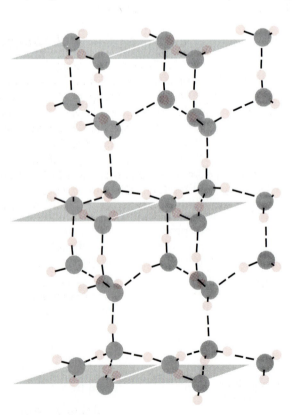

Figure 2.8 The open latticelike structure of ice. Because of the extensive hydrogen bonding, ice is less dense than liquid water and is able to float.

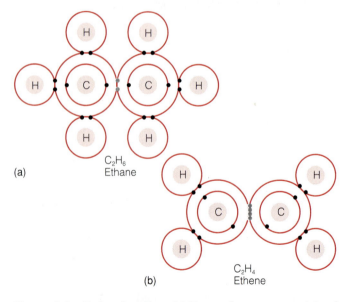

Figure 2.9 Carbon bonding. (a) Two carbon atoms are joined by a single covalent bond. The remaining three pairs of electrons are shared with hydrogen atoms in this molecule. (b) Two carbon atoms are joined by a double covalent bond. In this case, only two pairs are shared with hydrogen atoms.

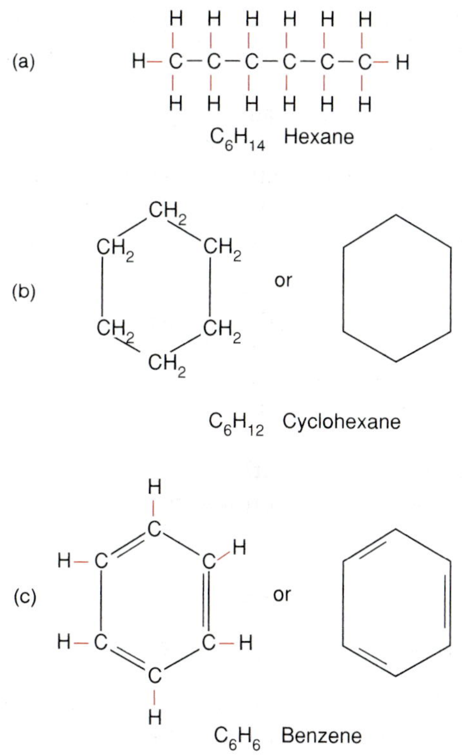

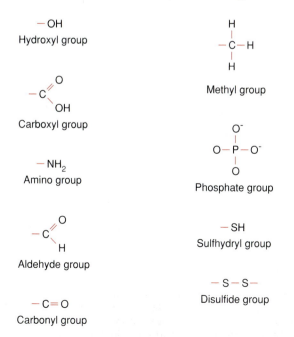

(a)

C_6H_{14} Hexane

(b)   or

C_6H_{12} Cyclohexane

(c)   or

C_6H_6 Benzene

Figure 2.10 Hydrocarbons. (a) Linear, (b) cyclic, and (c) aromatic structures.

ing requirement enables carbon to bond with other carbon atoms to form chains and rings of varying lengths and configurations, as well as to bond with hydrogen and other atoms.

With only four electrons in its outer shell, carbon atoms can form covalent bonds to fill this shell. Adjacent carbon atoms may share one or two pairs of electrons. When one pair of electrons are shared, they form a *single covalent*

— OH
Hydroxyl group

Carboxyl group

— NH₂
Amino group

Aldehyde group

— C=O
Carbonyl group

Methyl group

Phosphate group

— SH
Sulfhydryl group

— S—S—
Disulfide group

Figure 2.11 Various functional groups of organic molecules.

bond, leaving each carbon free to bond to as many as three other atoms (figure 2.9a). When two pair of electrons are shared, they form a *double covalent bond*, leaving each carbon free to bond to only two additional atoms (figure 2.9b).

Hydrocarbons are organic molecules that contain only carbon and hydrogen and most have their carbons bonded in a linear fashion, as in the molecule hexane (2.10a). The ends of some carbons, however, may be joined together to form rings, as in the cyclohexane molecule (2.10b). Some of these cyclic molecules may have a double bond between two adjacent carbon atoms, as in benzene and are known as *aromatic* compounds (figure 2.10c).

The carbon chain or ring of many organic molecules provides a relatively inactive molecular "backbone," to which *active* or *reactive* groups or atoms are attached. These are known as the *functional groups* of the molecule and are responsible for the unique chemical properties and behavior of the molecule. Some of the more important functional groups are illustrated in figure 2.11. Some classes of organic molecules can be named according to their functional groups. For example, as noted in figure 2.12, *ketones* have a carbonyl group within the carbon chain, an *organic acid* has a carboxyl group coming off of a carbon, an *aldehyde* has the aldehyde group at the end of the chain, and an *alcohol* has a hydroxyl group at one end of the chain.

The important groups of organic molecules in animals include carbohydrates, lipids, proteins, and nucleic acids. Each of these groups is now discussed.

Carbohydrates: Sources of Stored Energy

Carbohydrates are the major source of energy for animals. Most animal cells have the chemical machinery to break down the energy rich carbon-hydrogen (C—H) bonds found in these molecules. Simple carbohydrates (e.g., monosaccharides) are composed of atoms of carbon, hydrogen, and oxygen, in the ratio of 1:2:1 (e.g., CH_2O).

Carbohydrates are classified according to their molecular structure. The simplest types with short carbon chains are called **monosaccharides** (Gr. *monos*, single + *sakharon,* sugar) or *sugars.* As their name implies, monosac-

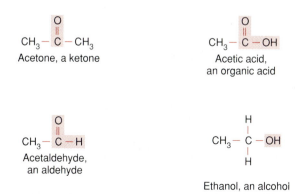

Acetone, a ketone

Acetic acid, an organic acid

Acetaldehyde, an aldehyde

Ethanol, an alcohol

Figure 2.12 Several categories of organic molecules based on functional groups (shaded).

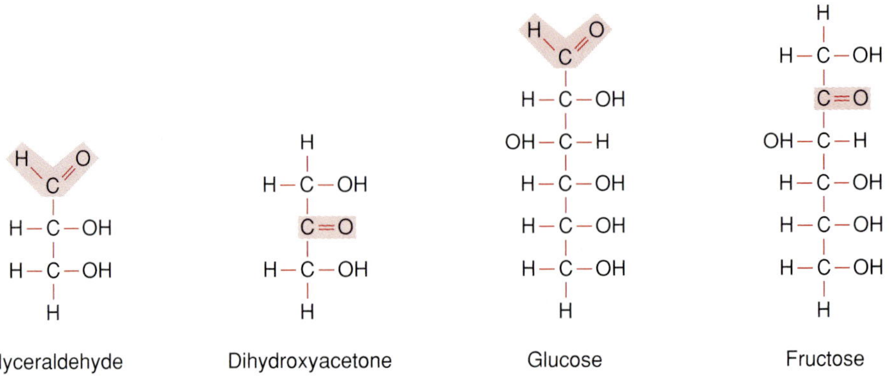

Figure 2.13 Structural formulas of four simple sugars, or monosaccharides: glyceraldehyde and dihydroxyacetone have three carbons, and glucose and fructose have six carbons. The functional groups are shaded.

Glyceraldehyde

Dihydroxyacetone

Glucose

Fructose

charides taste sweet. Monosaccharides are the building blocks of more complex carbohydrate molecules. Four common monosaccharides found in animals are glucose, fructose, glyceraldehyde, and dihydroxyacetone (figure 2.13).

Many organic compounds with the same molecular formulas have different *structures*. Such compounds are called **isomers**. Examples of isomers include glucose and fructose, both of which have the formula $C_6H_{12}O_6$, but with the atoms arranged in slightly different ways (figure 2.14).

Two monosaccharides can be combined to form a **disaccharide** (*di,* two) (figure 2.15). The covalent C-O-C link connecting the two monosaccharides is called a *glycosidic* bond. Disaccharides all have the same molecular formula, $C_{12}H_{22}O_{11}$. Examples of disaccharide isomers are sucrose, maltose, and lactose. Sucrose (table sugar) is a disaccharide formed by linking a molecule of glucose to a molecule of fructose. If a glucose molecule is bonded to its isomer, galactose, the resulting disaccharide is lactose (commonly called milk sugar). Maltose, two joined glucose subunits, gives barley seeds a sweet taste. Beer brewers ferment barley into alcohol.

Other carbohydrates are made up of many monosaccharides joined together to form **polysaccharides** (*poly,* many) (figure 2.16). Glycogen, a major storage form for glucose in animals, is an example of a polysaccharide. Be-

cause the number of glucose units within the glycogen molecule may vary, it is symbolized by the formula $(C_6H_{10}O_5)_n$, with n equal to the number of glucose units in the molecule. Other biologically important polysaccharides include *chitin* (a major component of the shells of insects and of crustaceans, such as lobsters and crabs), and plant starch.

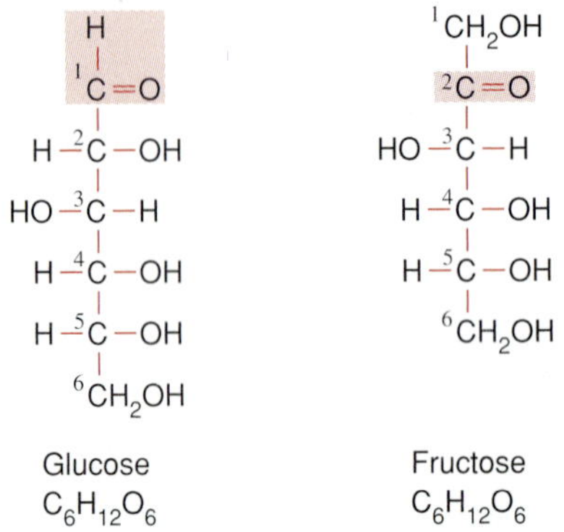

Glucose   Fructose

Sucrose

Galactose   Glucose

Lactose

Glucose   Glucose

Maltose

Figure 2.14 The structural formulas of two hexose (6 carbon) sugars. Both have the same ratio and number of atoms ($C_6H_{12}O_6$). Glucose and fructose are structural isomers of each other. Shaded areas represent functional groups.

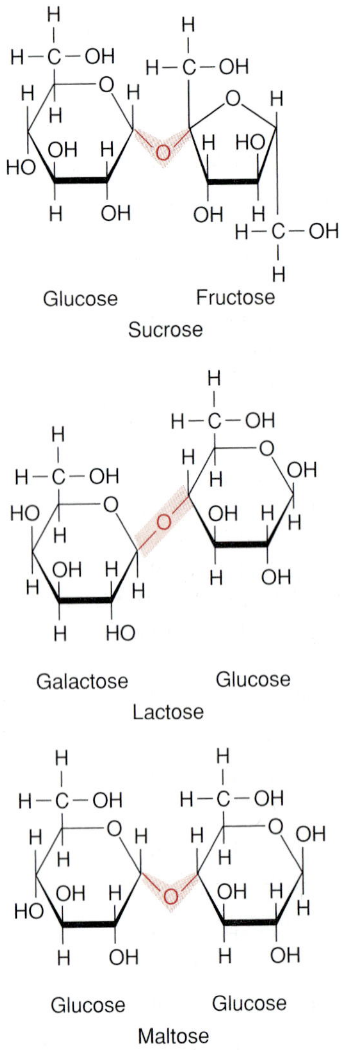

Glucose
$C_6H_{12}O_6$

Fructose
$C_6H_{12}O_6$

Figure 2.15 Three common disaccharides. Sucrose is ordinary table sugar and lactose is the sugar in milk. Maltose is found in malt. Glycosidic bonds are highlighted in color.

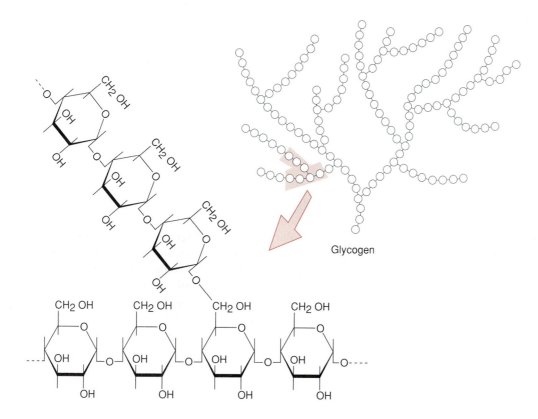

Glycogen

Lipids: Energy, Interfaces, and Signals

Lipids are nonpolar organic molecules that are insoluble in water but soluble in organic solvents, such as ether, alcohol, and chloroform. Phospholipids and cholesterol are lipids that are important constituents of cell membranes. The most common lipids in animals, however, are *fats*. Fats are used to build cell parts and to supply energy for cellular activities.

Lipid molecules are composed primarily of carbon, hydrogen, and oxygen atoms, although some may contain small amounts of phosphorous and nitrogen. They contain a much smaller proportion of oxygen than do carbohydrates, as can be illustrated by the formula for the fat, tristearin, $C_{57}H_{110}O_6$. The building blocks of fat molecules are

fatty acids and *glycerol*. Fatty acids contain long hydrocarbon chains bonded to carboxyl (—COOH) groups. Glycerol is a three-carbon alcohol with each carbon bearing a hydroxyl (—OH) group. These molecules are united so that each glycerol molecule is combined with three fatty acid molecules, each of which is joined to each of the three carbon atoms in the glycerol backbone (figure 2.17). Because there are three fatty acids, the resulting fat molecule is called a *triglyceride* neutral fat, or triacylglycerol.

Although the glycerol portion of every fat molecule is the same, there are many kinds of fatty acids, and therefore, many kinds of fats. Fatty acid molecules differ in the length of their carbon chains and in the ways the carbon atoms are combined. The most common are even-numbered chains of 14 to 20 carbons. In some cases, the carbon atoms are joined by single carbon-carbon bonds, and each carbon

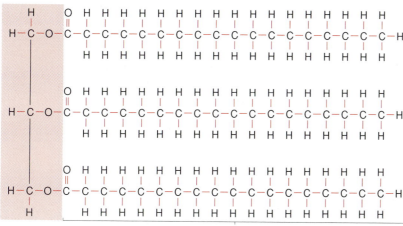

Glycerol

Fatty acids

Figure 2.17 A triglyceride is a fat molecule consisting of a glycerol molecule bonded to three fatty acid molecules.

Figure 2.18 Structural formulas for (a) saturated and (b) unsaturated fatty acids.

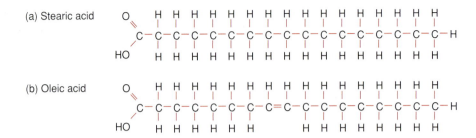

atom is bound to as many hydrogen atoms as possible. This type of fatty acid is said to be *saturated* (figure 2.18a). Other fatty acids have one or more double bonds between carbon atoms and are said to be *unsaturated* (figure 2.18b) because the double bonds replace some of the hydrogen atoms, and therefore the fatty acids contain fewer than the maximum number of hydrogen atoms. Fatty acids with numerous double bonds are said to be *polyunsaturated*.

Unsaturated fats have low melting points because their chains bend at the double bonds and the fat molecules cannot be aligned closely with one another. Consequently, [11] the fat may be fluid. A liquid fat is called an *oil*. Most plant fats are unsaturated. Animal fats, in contrast, are often saturated and occur as hard or solid fats.

A *phospholipid* molecule is similar to a fat molecule in that it contains a glycerol portion and fatty acid chains. However, the phospholipid has only two fatty acid chains. In the place of the third chain, there are phosphate (PO_4^{3-}) and nitrogen (N^+) groups. The polar phosphate and nitrogen groups are soluble in water (hydrophilic) and form the "head" of the molecule; the insoluble (nonpolar, hydrophobic) fatty acid portion forms the "tail" (figure 2.19). Phospholipids are the major structural components of membranes because of this tendency to be soluble at one end and insoluble at the other. Their importance in the structure and function of cell membranes will be discussed in chapter 3.

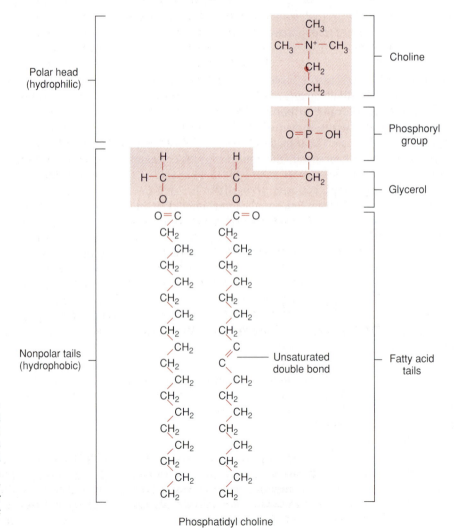

Figure 2.19 The structure of a phospholipid. Each phospholipid has a head (containing a phosphate group) and a tail (made up of two fatty acids). The head is water soluble (hydrophilic), but the tail is not (hydrophobic). In this example, choline can be replaced by other molecules, such as ethanolamine, and other fatty acids may be substituted for those shown.

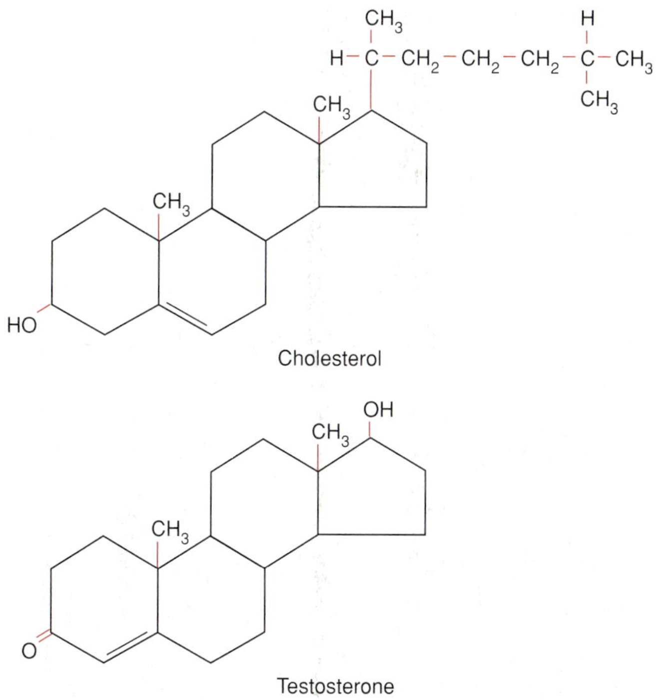

Figure 2.20 Structures of two common steroids showing the backbone of one 5-carbon ring and three 6-carbon rings. Cholesterol affects the fluidity of membranes. Testosterone is a male hormone in mammals that is responsible for the development and maintenance of secondary sex characteristics, and for the maturation and functioning of accessory sex organs.

Steroids are naturally occurring, lipid-soluble molecules composed of four carbon rings. The four rings of carbon atoms are fused together, giving a somewhat rigid structure. Three of the rings are six-sided, and the fourth is five-sided. There are a total of 17 carbons in the four rings. One important steroid is cholesterol (figure 2.20). Other biologically important steroids are vitamin D, hormones of the adrenal gland (e.g., aldosterone), the ovaries (e.g., estrogen), and the testes (e.g., testosterone) (figure 2.20).

Proteins: The Basis of Life's Diversity

In animals, **proteins** serve as major structural material, energy sources, as protection against disease in higher animals, as chemical messengers (hormones), and as cellular receptors. Some proteins play important roles in metabolic reactions by acting as biological catalysts called enzymes. They enter and speed up specific chemical reactions without being used up themselves. (Enzymes will be discussed in more detail in chapter 4.)

Proteins always contain atoms of carbon, hydrogen, nitrogen, and oxygen, and sometimes phosphorus, sulfur, or iron. The individual building blocks of proteins are called **amino acids.** Amino acids always contain an amino group ($-NH_2$), a carboxyl group ($-COOH$), a hydrogen atom,

and a functional group designated R, all bonded to a central carbon atom:

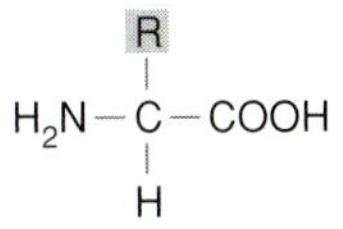

$$H_2N-\underset{\underset{H}{|}}{\overset{\overset{R}{|}}{C}}-COOH$$

The identity and unique chemical properties of each amino acid are determined by the nature of the R group linked to the central carbon atom. Twenty different amino acids occur in animals. The individual amino acids are joined together in chains by covalent bonds called **peptide bonds**. In the formation of a peptide bond, the carboxyl group of one amino acid is bonded to the amino group of another amino acid, with the elimination of water (a condensation reaction), as follows:

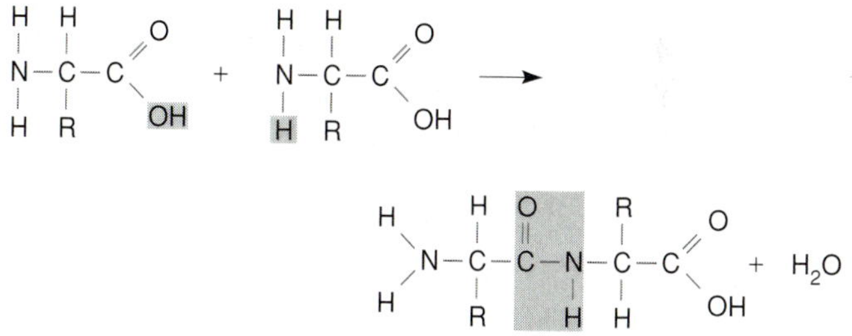

The length of the amino acid chain can vary in different proteins from less than 50 to more than 2,000 amino acids. Each kind of protein contains a specific number and kind of amino acids arranged in a particular sequence and folded up into a unique three-dimensional structure.

When two amino acids are bound together, they form a unit called a *dipeptide*; three amino acids bonded together form a *tripeptide*. When many amino acids bond together, the unit they form is a chain called a *polypeptide*. Different kinds of protein molecules have different shapes that are related to their particular functions in life processes.

Several different levels of structure can be distinguished in a protein molecule. The *primary structure* is the linear sequence of amino acids in the polypeptide chains comprising the molecule (figure 2.21a). The *secondary structure* of a protein is a repeating pattern of bonds (often hydrogen bonds) between amino acids, and it commonly takes the shape of an α-helix or pleated sheet (figure 2.21b). The *tertiary structure* results from the folding of the helix into a three-dimensional globular shape (figure 2.21c). When two protein chains associate to form a larger protein, the chains are assembled into a whole, the shape of which is called the *quaternary structure* (figure 2.21d).

Nucleic Acids: The Codes of Life

Nucleic acids are very large, complex molecules composed of sugar, phosphate, and organic bases. There are two main types of nucleic acids: **deoxyribonucleic acid (DNA)** and **ribonucleic acid (RNA)**. DNA makes up the chromosomes in the cell's nucleus and is able to replicate

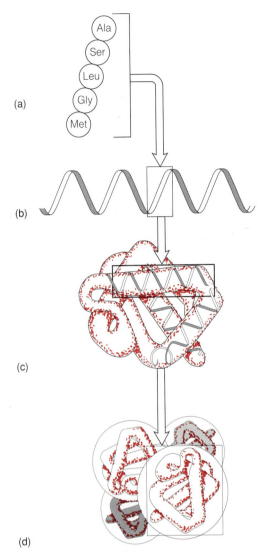

(a)

(b)

(c)

(d)

Figure 2.21 Levels of protein structure. (a) Primary structure of a protein is its linear sequence of amino acids. (b) Secondary structure is determined by the repeating configurations in the stucture of the amino acids. (c) Tertiary structure is determined by the three-dimensional folding of the secondary structure. (d) The three dimensional arrangement of polypeptides with tertiary structure gives a protein its quaternary structure.

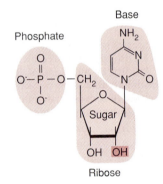

(a)   Deoxyribonucleic acid (DNA)

(b)   Ribonucleic acid (RNA)

Figure 2.22 The nucleotide units of the nucleic acids DNA (a) and RNA (b).

itself, and contains the genetic (hereditary) information of the cell. RNA may be present in either the nucleus or the cytoplasm, which is the part of the cell outside the nucleus, and acts as a carrier of information between the DNA and the cellular sites of protein synthesis. Both DNA and RNA are composed of repeating units called **nucleotides**. Each nucleotide is composed of three substances: (1) a nitrogen-containing organic base, (2) a five carbon sugar (either deoxyribose or ribose), and (3) phosphate (figure 2.22). Nucleic acids will be examined in more detail in chapters 8 and 9.

Stop and Ask Yourself

17. What is a typical carbohydrate? What is one function of a carbohydrate in an animal?
18. What is a lipid? Where would you find lipids in an animal?
19. What is the difference between a fat and an oil? How are fats used in animals?
20. What is a peptide bond?
21. What is a nucleic acid? What are several functions of nucleic acids in animals?

Summary

1. Matter is anything that has mass and occupies space. The most important particles of matter are protons, neutrons, and electrons, all of which associate to form atoms. The nucleus of an atom contains the protons and neutrons; electrons orbit around the nucleus.

2. The chemical nature of an atom is largely determined by the distribution of electrons in its outermost shell. There is a tendency for atoms to lose or gain electrons until stability in the outer shell is obtained.

3. Atoms react with other atoms to form molecules. The forces holding atoms together are called chemical bonds. A covalent bond results from the sharing of one or more pairs of electrons. A hydrogen bond is formed by the attraction of the partial positive charge of a hydrogen atom of one water molecule with the partial negative charge of the oxygen atom of another water molecule. An ionic bond results from the attraction of opposite charges.

4. The combining or breaking apart of two or more atoms that brings about a chemical change is called a chemical reaction. When two or more atoms, ions, or molecules combine to form a more complex susbtance, this process is called a combination reaction. Decomposition reactions break chemical bonds in molecules to form two or more products. When hydrolysis takes place, a molecule of water interacts with the reactant to break the reactant's bonds. Condensation reactions are hydrolytic reactions in reverse. An oxidation reaction occurs when an atom or molecule loses electrons or hydrogen atoms. When an atom or molecule gains electrons or hydrogen atoms, the process is called reduction.

5. The tendency for molecules to break up into charged ions in water is called dissociation.

6. An acid is a substance that releases hydrogen ions (protons) when dissolved in water. A base is a substance that releases hydroxyl ions (accepts protons) when dissolved in water. A salt is a compound other than water that is formed during a neutralization reaction between an acid and base.

7. The degree of alkalinity or acidity of a solution is measured by a pH scale that indicates the concentration of free hydrogen ions in water.

8. The structure of water molecules, with their polar, electrical nature, and the resulting formation of hydrogen bonds between them, gives water the following properties important in the evolution of life: (a) water is a solvent; (b) water has a high specific heat; (c) water is adhesive and cohesive; (d) water has a high boiling point; (e) water is a good evaporative coolant; and (f) water has a high freezing point and a lower density as a solid than as a liquid.

9. Simple carbohydrates are made of carbon, hydrogen, and oxygen in a 1:2:1 ratio, and they are important sources of energy for most animals. Carbohydrates are classified as monosaccharides, disaccharides, or polysaccharides, depending upon how many sugars they contain.

10. Lipids are molecules containing many more C—H bonds than carbohydrates. Fats and oils are familiar lipids. Lipids can store large amounts of energy in an animal's body.

11. Proteins are large, complex molecules composed of smaller structural units called amino acids. The amino acids are linked together via peptide bonds to form chains called polypeptides.

12. Nucleic acids are very large molecules composed of bonded units called nucleotides. The two nucleic acids are deoxyribonucleic acid (DNA) and ribonucleic acid (RNA). Nucleic acids carry the hereditary messages and regulate the synthesis of proteins.

Key Terms

acid (p. 21)
amino acids (p. 29)
anatomy (p. 16)
atomic mass (p. 17)
atomic number (p. 17)
atoms (p. 16)
base (p. 21)
biological chemistry (p. 16)
buffers (p. 22)
capillarity (p. 23)
carbohydrates (p. 25)
chemistry (p. 16)
clouds of electronegativity (p. 17)
combination reaction (p. 20)
condensation reactions (p. 21)
covalent bond (p. 18)
decomposition reactions (p. 20)
deoxyribonucleic acid (DNA) (p. 29) -
disaccharide (p. 26)
dissociation (p. 21)
electrolyte (p. 21)
electrons (p. 16)
elements (p. 16)
energy-level shell (p. 17)
half-life (p. 18)
hydration shell (p. 23)
hydrocarbons (p. 25)
hydrogen bond (p. 19)
hydrolysis reaction (p. 20)
inorganic molecules (p. 24)

ion (p. 19)
ionic bonds (p. 20)
isomers (p. 26)
isotopes (p. 17)
lipids (p. 27)
mass (p. 16)
matter (p. 16)
molecules (p. 18)
monosaccharides (p. 25)
neutrons (p. 16)
nonpolar covalent bond (p. 18)
nucleic acid (p. 29)
nucleotides (p. 30)
organic molecules (p. 24)
oxidation reaction (p. 21)
peptide bond (p. 29)
pH scale (p. 21)
physiology (p. 16)
polar covalent bond (p. 19)
polar molecule (p. 22)
polysaccharides (p. 26)
proteins (p. 29)
protons (p. 16)
radioactive (p. 17)
radioactive dating (p. 18)
radioisotopes (p. 17)
reduction (p. 21)
ribonucleic acid (RNA) (p. 29)
salt (p. 21)
steroids (p. 28)
surface tension (p. 23)
viscosity (p. 24)

■ Critical Thinking Questions

1. Considering the major elements of which all living elements are composed, is it likely that the earth's surface could have been the major chemical "breadbasket" for the origin of life? Explain.

2. What is meant by the statement, "evolution resulted in carbon-based life?"

3. How is the temperature on the earth optimal for the chemical processes of life?

4. Why should a zoologist who studies animal life also be interested in the properties of chemical bonds?

5. Water is both one of the simplest substances known and at the same time, quite complex. Discuss this statement.

6. Discuss the ways by which the dipolar nature of the water molecule results in its biological properties of capillarity, high heat of vaporization, high heat capacity, and density based on temperature.

7. The major biological molecules are polymers—long chains of subunits. What are some reasons that linear polymers are so common in living animals?

8. Which sugars are sweet and what might be the evolutionary reason for animals possessing a separate taste sense for sweetness?

9. How do the functions of proteins "emerge" as being greater than their smaller structural units (amino acids)?

■ Suggested Readings

Books

Atkins, P. W. 1987. *Molecules.* New York: W.H. Freeman.

Avers, C. J. 1986. *Molecular Cell Biology.* Reading, Mass.: Addison-Wesley.

Calvin, M., and W. A. Pryor. 1973. *Organic Chemistry of life: Readings from Scientific American.* New York: W.H. Freeman.

Henderson, L. S. 1958. *The Fitness of the Environment.* Boston: Beacon Press.

Lehninger, A. L. 1982. *Principles of Biochemistry.* New York: Worth Publishers.

Rodella, T. D. et al. 1989. *Through the Molecular Maze.* Los Altos, Calif: W. Kaufmann Press.

Stryer, L. 1988. *Biochemistry.* 3/e. New York: W.H. Freeman.

Articles

Armbruster, P., and Munzenberg, G. Creating superheavy elements. *Scientific American* May, 1989.

Doolittle, R. Proteins. *Scientific American* October, 1985.

Frieden, E. The chemical elements of life. *Scientific American* July, 1972.

Karplus, M., and A. McCammon. The dynamics of proteins. *Scientific American* April, 1986.

McPherson, A. Macromolecular crystals. *Scientific American* March, 1989.

Sharon, N. Carbohydrates. *Scientific American* November, 1980.

Sharon, N. Glycoproteins. *Scientific American* May, 1974.

Upton, A. C. The biological affects of low level ionizing radiation. *Scientific American* February, 1982.

Weinberg, R. A. The molecules of life. *Scientific American* October, 1985.

Cells, Tissues, Organs, and Systems

Concepts

1. Cells are the basic organizational units of life.
2. Eukaryotic cells exhibit a considerable degree of internal organization, with a dynamic system of membranes forming internal compartments termed organelles.
3. The structure and function of a typical cell usually applies to all animals.
4. Cells are organized into structural and functional units called tissues, organs, and systems.

Have You Ever Wondered:

[1] what the functional unit of life is?
[2] why most cells are so small?
[3] why most higher animals need "some" cholesterol in their diet?
[4] what the "fingerprints" of a cell are?
[5] why the intestinal contents of an animal (such as yourself) don't freely enter the surrounding tissue?
[6] why red blood cells can "explode" when placed in distilled water?
[7] how cells "drink" water?
[8] why the mitochondria are called the "powerhouses" of the cell?
[9] if cells have a skeleton?

These and other useful questions will be answered in this chapter.

This chapter contains underlined evolutionary concepts.

Because all animals are made of cells, the cell is as fundamental to an understanding of zoology as the atom is to an understanding of chemistry. In the hierarchy of biological organization (figure 3.1), the cell is the simplest organization of matter that exhibits the properties of life. Some organisms are single celled; others are multicellular. An animal has a body composed of many kinds of specialized cells. A division of labor among cells allows specialization into higher levels of organization (tissues, organs, and organ systems). Yet, everything that an animal does is ultimately happening at the cellular level.

This chapter will present an overview of the structure and function of a "generalized" animal cell. Because cells are open systems, they exchange and utilize matter and energy with their surroundings. This exchange and utilization will be covered in the next two chapters.

What Are Cells?

Cells are the smallest independent units of life. The German biologists Matthias Schleiden, Theodor Schwann, and Robert Virchow are credited with the early work that led to the modern cell theory. According to this theory:

1. Cells are the basic units of life on earth.
2. All organisms are constructed of cells and cell products.
3. Except for the origin of life itself, all cells arise from preexisting cells ("from life, comes life").

More recent research has added three additional tenets to the cell theory:

[1]
4. Cells are the functional units of life, in which all of the chemical reactions necessary for the reproduction and maintenance of life take place.
5. Cells of animals are interconnected so that a population of cells can function as a single unit.
6. Cells of animals reproduce, move, assume specialized shapes, and carry out the necessary functions of life.

Like the theories of evolution by natural selection, the cell theory is a cornerstone of biology.

Structurally speaking, cells are either prokaryotes or eukaryotes. The word "prokaryote" means "before nucleus," and describes cells in which DNA is localized in a region but is not bound by a membrane. All **prokaryotes** are independent, single-celled organisms, and include the bacteria and cyanobacteria (blue-green algae). Some of the more salient characteristics of a prokaryotic cell are summarized in table 3.1 and will be discussed in the rest of the chapter.

All **eukaryotes** ("true nucleus") have a membrane-bound nucleus containing DNA, which is arranged in multiple chromosomes. In addition, eukaryotic cells contain many other membrane-bound structures called organelles. An **organelle** is a structure in the cell that carries out specific functions. Eukaryotes also have a network of spe-

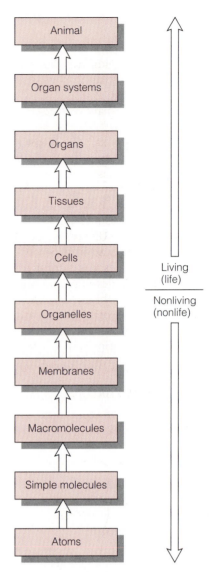

Figure 3.1 The structural hierarchy in an animal. At each level, function depends on the structural organization of that level and those below it.

cialized organelles called filaments and tubules organized into the *cytoskeleton*, which gives shape to the cell and allows intracellular movement (box 3.1).

All eukaryotic cells have three basic parts:

1. **Cytoplasm** (Gr. *kytos*, hollow vessel + *plasm*, fluid) is the portion of the cell outside the nucleus. The semifluid portion of the cytoplasm is called the *cytosol*. Suspended within the cytosol are the organelles ("little organs").
2. The **nucleus** (plural, nuclei) is the control center of the cell. It contains the chromosomes, and is separated from the cytoplasm by its own *nuclear envelope*. The **nucleoplasm** is the semifluid material in the nucleus.
3. The **plasma membrane** is the outer boundary of the cell.

Because cells vary so much in form and function, no "typical" cell exists. However, to help you learn as much

Box 3.1

The Origin of Eukaryotic Cells

The first cells were most likely very simple prokaryotic forms. The fossil record indicates that the earth is approximately 4 to 5 billion years old, and that prokaryotes may have arisen more than 3.5 billion years ago, whereas eukaryotes are thought to have first appeared about 1.5 billion years ago.

The evolution of the eukaryotic cell might have occurred when a large anaerobic (living without oxygen) amoeboid prokaryote ingested small aerobic (living with oxygen) bacteria and stabilized them instead of digesting them. This idea is known as the **endosymbiont hypothesis** and was first proposed by Lynn Margulis, a biologist at Boston University (box figure 3.1a). (**Symbiosis** is an intimate association between two organisms of different species.) According to this hypothesis, the aerobic bacteria developed into mitochondria, which are the sites of aerobic respiration and most energy conversion in eukaryotic cells. The possession of these mitochondrialike endosymbionts conferred the advantage of aerobic respiration on its host.

Flagella (whiplike structures) may have arisen through the ingestion of prokaryotes similar to spiral-shaped bacteria called spirochetes. Ingestion of prokaryotes that resembled present-day cyanobacteria (blue-green algae) could have led to the endosymbiotic development of the chloroplasts in plants.

Another hypothesis for the evolution of eukaryotic cells proposes that the prokaryotic cell membrane invaginated (folded inward) to enclose copies of its genetic material (box figure 3.1b). This invagination resulted in the formation of several double-membrane-bound entities (organelles) in a single cell. These entities could then have evolved into the eukaryotic mitochondrion, nucleus, and chloroplasts.

Although the exact mechanism for the evolution of the eukaryotic cell will never be known with certainty, the emergence of the eukaryotic cell led to a dramatic increase in the complexity and diversity of life forms on the earth. At first, these newly formed eukaryotic cells existed only by themselves. Later, however, some probably evolved into multicellular organisms in which various cells became specialized into tissues, which in turn led to the potential for many different functions. These multicellular forms would then be able to adapt to life in a great variety of environments.

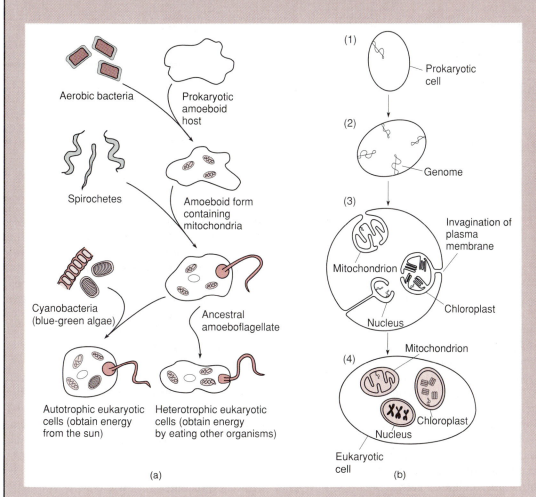

Box Figure 3.1 Two hypotheses on the evolution of the eukaryotic cell. (a) The endosymbiotic hypothesis. (b) The membrane invagination hypothesis. A prokaryotic cell (1) duplicates its genetic material (genome) (2). This is followed by invagination of the plasma membrane to form double-membrane-bound organelles and separate the individual genomes from each other (3). The nuclear genome eventually enlarges while the other organelle genomes lose many of their genes resulting in a eukaryotic cell (4).

35

Table 3.1	Comparison of Prokaryotic and Eukaryotic Cells	
Component	**Prokaryote**	**Eukaryote**
Cell wall	Present	Absent in animals (present in plants)
Centrioles	Absent	Present in animals (absent in plants)
Genetic material	Single circlular chromosome of DNA	Arranged in multiple chromosomes; DNA associated with protein
Cilia (9 + 2)	Absent	Present in some
Cytoskeleton	Absent	Present
Endoplasmic reticulum	Absent	Present
Flagellum	Often present (may be different in composition)	Present in some cells
Glycocalyx	Absent	Present
Golgi apparatus	Absent	Present
Lysosomes	Absent	Present
Mitochondria	Absent	Present
Nucleus	Absent	Present
Plasma membrane	Present	Present
Ribosomes	Present	Present
Vesicles	Present	Present

as possible about cells, figure 3.2 shows an idealized version of a eukaryotic cell and most of its component parts.

Stop and Ask Yourself

1. What are the six main points of the cell theory?
2. What are the three basic parts of the cell?
3. What is the difference between cytoplasm and cytosol?
4. How do nucleoplasm and cytoplasm differ?

Why Are Most Cells Small?

Although there are exceptions (e.g., the yolk of a chicken egg and some long nerve cells), most cells are small and can only be seen with the aid of a microscope (box 3.2). [2] One reason for this smallness is that the ratio of the volume of the cell's nucleus to the volume of its cytoplasm must not be so small that the nucleus, the major control center of the cell, cannot control the cytoplasm.

Another aspect of cell volume works to limit cell size. As the radius of a cell becomes larger, its volume increases more rapidly than its surface area (figure 3.3). The need for nutrients and the rate of waste production are proportional to the volume of the cell. The cell takes up nutrients

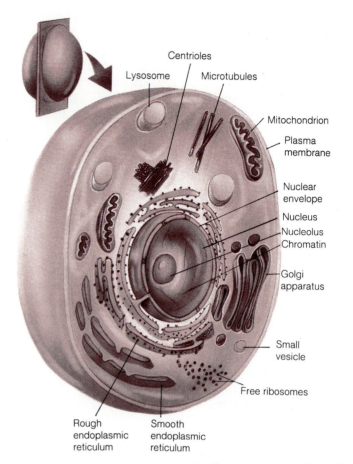

Figure 3.2 A generalized animal cell. Our understanding of the structures in this cell is based mainly on electron microscopy.

and eliminates wastes through its surface plasma membrane. If the cell volume becomes too large, the surface-to-volume ratio will be too small to carry out an adequate exchange of nutrients and wastes.

Cell Membranes

The membrane that surrounds the cell is called the *plasma membrane*. There are other membranes inside the cell that enclose some organelles (and actually make up other organelles, such as the endoplasmic reticulum) and have properties similar to the plasma membrane.

Structure of Cell Membranes

In 1972, S. Jonathan Singer and Garth Nicolson developed the *fluid-mosaic model* of membrane structure. This model is now widely accepted as an accurate representation of biological membranes. According to this model, the membrane is a double layer (bilayer) composed of proteins and phospholipids, and is fluid rather than solid. The phospholipid bilayer forms a fluid "sea" in which specific proteins float like icebergs (figure 3.4). Being fluid, the membrane is in a constant state of flux—shifting and changing,

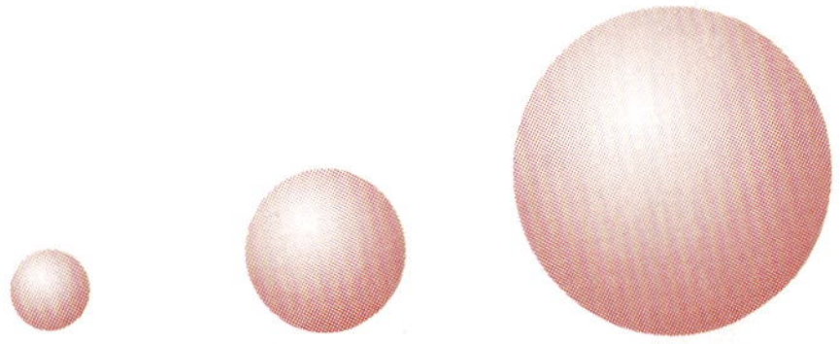

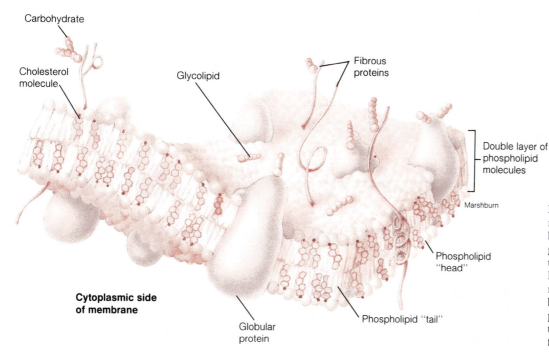

Radius (r)	1 cm	2 cm	4 cm
Surface area (A)	12.57 cm²	50.26 cm²	201.06 cm²
Volume (V)	4.19 cm³	33.51 cm³	268.08 cm³
A/V	3.0	1.50	0.75

Surface area of a sphere = $4\pi r^2$
Volume of spere = $4/3\,\pi r^3$

Figure 3.3 A portrayal of the relationship between surface area and volume. As the radius of a sphere becomes greater, its volume increases more rapidly than its surface area. A/V = surface area to volume ratio.

while retaining its uniform structure. The word *mosaic* refers to the pattern of proteins that are found dispersed on the phospholipid bilayer.

The following are important points of the fluid-mosaic model:

[3]

1. Cholesterol is present in the plasma membrane and organelle membranes of eukaryotic cells. The cholesterol molecules are embedded in the interior of the membrane and help to make the membrane less permeable to water-soluble substances. In addition, the relatively rigid structure of the cholesterol molecules (figure 3.5) help to make the membrane more stable than it would be otherwise.

2. The phospholipids have one polar end and one nonpolar end (*see figure 2.19*). The polar ends are oriented on one side toward the outside of the cell and into the fluid cytoplasm on the other side and the nonpolar ends face each other in the middle of the bilayer. The "tails" of the phospholipid molecules are attracted to each other, and are repelled by water (they are *hydrophobic*, "water-dreading"). As a result, the polar spherical "heads" (the phosphate portion) are located over the cell surfaces (outer and inner) and are "water-attracting" (they are *hydrophilic*).

3. The membrane proteins are individual molecules attached to the inner or outer membrane surface (*extrinsic proteins*) or embedded in it (*intrinsic proteins*) (*see figure 3.4*). Intrinsic proteins serve as links to sugar-protein markers on the cell surface. Some intrinsic proteins help to move ions or molecules across the membrane, and others attach the membrane to the cell's inner scaffolding (the cytoskeleton) or to various molecules outside the cell.

4. When carbohydrates unite with proteins, they form *glycoproteins*, and when they unite with lipids, they form *glycolipids* on the surface of a plasma membrane. Surface carbohydrates and portions of the proteins make up the **glycocalyx** (cell coat) (figure 3.6). The complexly arranged and distinctively shaped groups of sugar molecules (oligosaccharides) of the glycocalyx act as a molecular "fingerprint" for each cell type. The glycocalyx is necessary for cell-to-cell recognition and the behavior of certain cells and is a key component in coordinating cell behavior in animals.

[4]

In the body of animals, many cells are not in direct contact with one another because fluid-filled space (*intercellular space*) separates them. In other cases, the cells are

Carbohydrate

Cholesterol molecule

Glycolipid

Fibrous proteins

Double layer of phospholipid molecules

Marshburn

Phospholipid "head"

Cytoplasmic side of membrane

Globular protein

Phospholipid "tail"

Figure 3.4 The fluid-mosaic model of membrane structure. Intrinsic globular proteins may protrude above or below the lipid bilayer and may move about in the membrane. Extrinsic peripheral proteins are attached to the inner and outer surfaces.

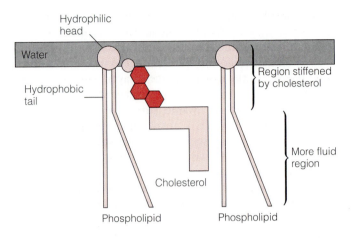

Figure 3.5 Drawing showing the arrangement of cholesterol between lipid molecules of a lipid bilayer. Cholesterol stiffens the lipid bilayer. Only half the lipid bilayer is shown; the other half is a mirror image.

tightly packed, and their plasma membranes commonly are connected by **intercellular junctions**. In one type, called a *tight junction*, the membranes of adjacent cells become fused by calcium (Ca^{2+}) and magnesium (Mg^{2+}) ions that bind proteins of the two cell surfaces (figure 3.7a). Tight junctions prevent material from moving through intercellular spaces. For example, they are functionally important where a sheet of cells forms a barrier, such as the layer of cells lining the gut (intestinal lumen) of vertebrates. These cells and their tight junctions keep the nutrients that have left the intestinal lumen and crossed the epithelial cells to blood vessels from moving back into the intestinal lumen.

[5]

Another type of intercellular junction is called a *desmosome*. A desmosome is an attachment that serves to rivet or "spot weld" adjacent cells to form a reinforced structural unit (figure 3.7b). This type of junction is commonly found in skin cells. In other cases (e.g., heart muscle cells), the membranes are connected by *gap junctions*, which are tubular channels (figure 3.7c). These channels link the cytoplasm of adjacent cells. They allow ions and various small organic molecules to be exchanged between cells.

Stop and Ask Yourself

5. What do the terms "fluid" and "mosaic" refer to in the fluid-mosaic model of an animal cell membrane?
6. What is a selectively permeable membrane?
7. What is a gap junction? What is a desmosome? What is a tight junction? What purpose does each of these serve in an animal?
8. What is the function of cholesterol in the plasma membrane? The phospholipid? The protein?

Functions of Cell Membranes

Cell membranes play important roles in: (1) the regulation of material moving into and out of the cell, and from one part of the cell to another; (2) separation of the inside of the cell from the outside; (3) separation of various organ-

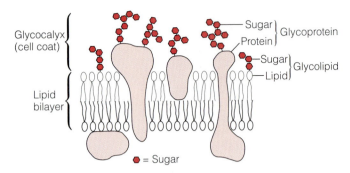

Figure 3.6 The glycocalyx. An illustration of the glycocalyx showing the glycoproteins and glycolipids. Note that all of the attached carbohydrates are on the outside of the plasma membrane.

elles within the cell; (4) providing a large surface area on which specific chemical reactions can occur; (5) separating cells from one another; and (6) serving as a site for receptors containing specific cell identification markers that differentiate one cell type from another.

The ability of the cell membrane to let some things in and keep others out is called **selective permeability** (L. *permeare* or *per*, through + *meare*, pass), and is essential for maintaining cellular homeostasis. **Homeostasis** (Gr. *homeo-*, always the same + *stasis*, standing) is the maintenance of a relatively constant internal environment despite fluctuations in the external environment. However,

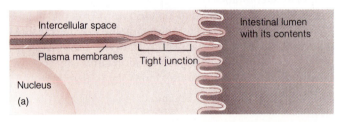

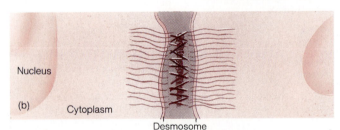

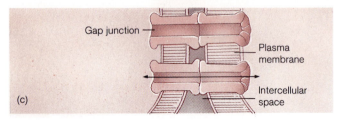

Figure 3.7 Cell junctions. (a) Diagram of epithelial cells in the intestinal lining showing how tight junctions act as barriers against movement of material through spaces between cells. (b) Desmosomes hold animal cells together; (c) gap junctions are tubular openings that allow small molecules to move from cell to cell.

before we can fully understand how substances pass into and out of cells and organelles, we must understand how the molecules of those substances are able to move from one place to another.

Movement Across Membranes

There are a number of ways that molecules can cross membranes, both using their own energy and relying on an outside source of energy. Table 3.2 summarizes the various kinds of transmembrane movement and the following sections discuss them in more detail.

Simple Diffusion

Molecules tend to move randomly (due to spontaneous molecular motion) from areas where they are highly concentrated to areas of lower concentration, until they are evenly distributed in a state of *equilibrium*. This process of mol-

ecules spreading out randomly until they are evenly distributed is called **simple diffusion** (L. *diffundre*, to spread). Simple diffusion accounts for most of the short-distance transport of substances moving into and out of cells. Figure 3.8a-d shows the diffusion of sugar particles away from a sugar cube placed in water.

Molecules dissolved in water or another solvent are called the solute. Certain small solute molecules that are nonpolar (e.g., CO_2 and O_2) can move across (through) the lipid bilayer of plasma membranes in response to a difference in the concentration of molecules on either side of the membrane (figure 3.8e). This difference in solute concentration is called the *concentration gradient*.

Facilitated Diffusion

For polar molecules (not soluble in lipids) diffusion may occur through *protein channels* (*pores*) in the lipid bilayer (figure 3.9). The mechanism for this type of passive transport is not yet fully known. However, it is generally accepted that the channel proteins offer a continuous path-

Table 3.2 Different Types of Movement Across Plasma Membranes		
Type of Movement	**Description**	**Example in the Body of an Animal**
Simple diffusion	No cell energy is needed. Molecules move "down" a concentration gradient. Molecules spread out randomly from areas of higher concentration to areas of lower concentration until they are distributed evenly—equilibrium is reached.	A frog inhales oxygen, which moves into the lungs and then diffuses into the bloodstream.
Facilitated diffusion	Carrier proteins in a plasma membrane temporarily bind with molecules, and assist their passage through the membrane. Other proteins form channels for movement of molecules through the membrane.	Specific amino acids in the gut combine with carrier proteins to pass through the gut cells into the bloodstream
Osmosis	Water molecules move through selectively permeable membranes from areas of higher concentration to areas of lower concentration	Water molecules move into a frog's red blood cell when the concentration of water molecules outside the blood cell is greater than it is inside.
Filtration	Hydrostatic pressure forces small molecules through selectively permeable membranes from areas of higher pressure to areas of lower pressure.	A frog's blood pressure forces water and dissolved wastes into the kidney tubules during the process of urine formation.
Active transport	Specific carrier proteins in the plasma membrane bind with molecules or ions to help them across the membrane "against" a concentration gradient; energy is required.	Movement of sodium ions from inside the sciatic nerve of a frog (the sodium-potassium pump).
Endocytosis	The bulk movement of material into a cell	
Pinocytosis	Plasma membrane encloses small amounts of fluid droplets and takes them into the cell.	The kidney cells of a frog take in fluid in order to maintain fluid balance.
Phagocytosis	Plasma membrane forms a pocket around a solid particle or other cell and draws it into the phagocytic cell.	The white blood cells of a frog engulf and digest harmful bacteria.
Receptor-mediated endocytosis	Extracellular molecules bind with specific receptors on a plasma membrane, causing the membrane to invaginate and draw molecules into the cell.	The intestinal cells of a frog take up large molecules from the inside of the gut.
Exocytosis	The movement of material out of a cell. Vesicle (with particles) fuses with plasma membrane and expels particles or fluids from cell through plasma membrane.	The sciatic nerve of a frog releases a chemical (neurohumor).

Box 3.2

Microscopes—Windows into the Cell

Despite the very small size of most cells, knowledge of cell structure and function has accumulated through the use of various techniques. One technique is **micros-copy**—the use of microscopes to peer into the innermost parts of a cell. Several types of microscopes are used.

The **light microscope** relies on the bending (refraction) of light rays. Light rays pass through the center of a curved lens. The farther they are from the center, the more they bend (box figure 3.2a). The *compound microscope* that is used in most zoology laboratories, contains at least two of these lenses (box figure 3.2b). All the rays eventually converge through the lens-system onto one focal point—the eye(s). To make the different parts of the cell "stand-out," specific stains are often used to highlight certain structures. Unfortunately, these staining procedures kill the cell. However, living cells can be observed through the *phase contrast micro-scope*. In this type of light microscopy, small differences in the way different parts of the cell refract light are converted to larger variations in brightness. In the *dark-field microscope,* living cells can be observed by making the field surrounding the specimen appear black while the specimen itself is brightly illuminated. The best light microscopes magnify images approximately 2,000 times.

In the **transmission electron microscope (TEM),** an electron beam is focused on a very thin section or slice of the cell by means of electromagnets (box figure 3.2c). After passing through the cell, the electron beam travels through more magnetic lenses, which magnify the image and project it onto either a fluorescent screen or photographic film. Magnifications of several hundred thousand times are possible with the TEM.

The **scanning electron microscope (SEM)** is used to study surfaces rather than thin sections of cells (box figure 3.2d). SEM photomicrographs, with their three-dimensional quality, reveal remarkable details of the surface of cells or other objects. Surfaces to be studied are first covered with a very thin layer of metal, such as gold. In the SEM, electron beams scan the surface of the spec-imen, driving off electrons from the atoms of the metal surface—called secondary electrons. The pattern of these scattered secondary electrons is then detected on a cathode ray tube like that in a television set. Maximum magnifications of the SEM are usually around twenty thousand times.

The **scanning tunneling microscope (STM)** was invented in the 1980s and can acheive magnifications of over 100 million. At this magnification, atoms on the surface of a solid can be viewed. The electrons sur-rounding the surface atoms *tunnel* or project a very short distance from the surface. The STM has a needle probe with a point so sharp that there is often only one atom at its tip. The probe is lowered towards the surface of

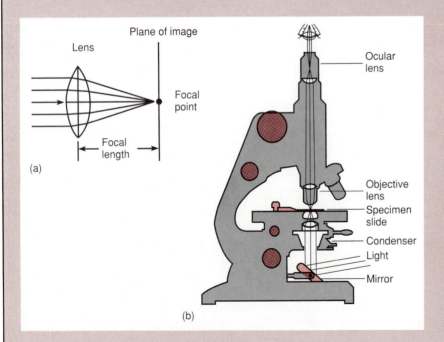

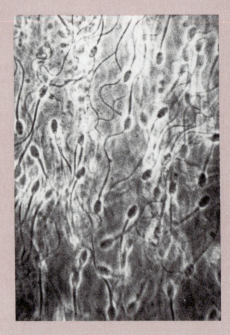

Box Figure 3.2. Types of microscopes and their images (sperm cells). (a) Focusing of light rays. (b) Compound light microscope ($\times$ 1900). (c) Transmission electron microscope ($\times$ 29,000). (d) Scanning electron microscope ($\times$ 2150).

the specimen until an electron cloud just touches the surface atoms. When a small voltage is applied between the tip and specimen, electrons flow through a narrow channel in the electron clouds. The arrangement of atoms on the surface of the specimen is determined by moving the probe tip back and forth over the surface. As the tip follows the surface contours, its motion is recorded and analyzed by a computer to create an ac-

curate, three-dimensional image of the surface atoms. The surface map can be either displayed on a computer screen or plotted on paper. The microscope's inventors, Gerd Binnig and Heinrich Rohrer, shared the 1986 Nobel Prize in Physics for their work. Interestingly, the other recipient of the prize was Ernst Ruska, the inventor of the first transmission electron microscope.

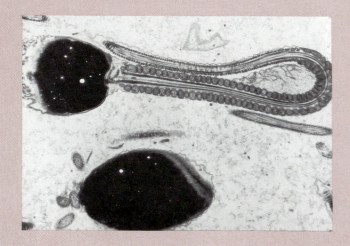

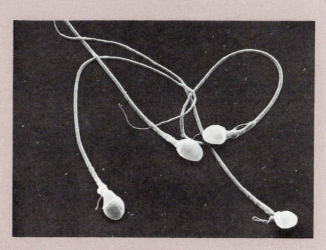

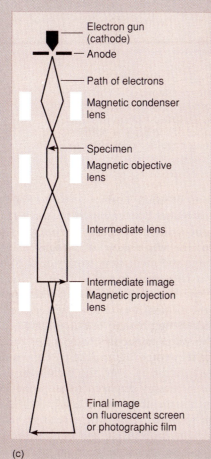

Electron gun (cathode)
Anode
Path of electrons
Magnetic condenser lens
Specimen
Magnetic objective lens
Intermediate lens
Intermediate image Magnetic projection lens
Final image on fluorescent screen or photographic film

(c)

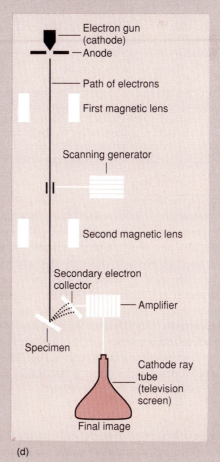

Electron gun (cathode)
Anode
Path of electrons
First magnetic lens
Scanning generator
Second magnetic lens
Secondary electron collector
Amplifier
Specimen
Cathode ray tube (television screen)
Final image

(d)

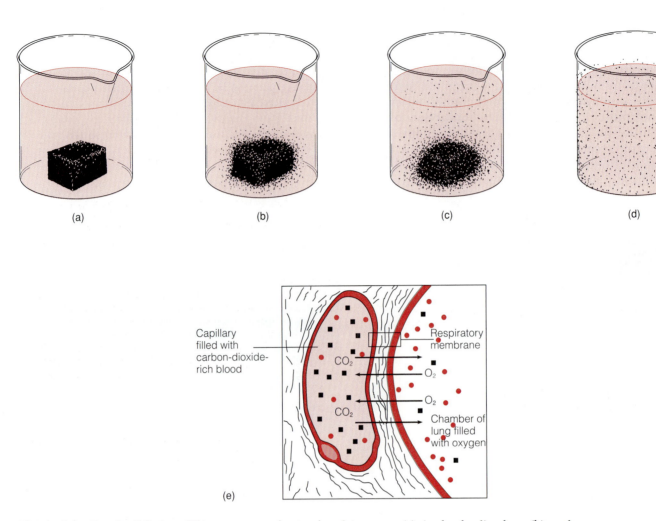

(a)　　　　　　(b)　　　　　　(c)　　　　　　(d)

Capillary
filled with
carbon-dioxide-
rich blood

Respiratory
membrane

CO_2

O_2

O_2

CO_2

Chamber of
lung filled
with oxygen

(e)

Figure 3.8 Simple diffusion. When a sugar cube is placed in water (a) it slowly dissolves (b) and disappears. As this happens, the sugar molecules diffuse from a region where they are more concentrated to a region (c) where they are less concentrated. When they are evenly distributed throughout the water, equilibrium is reached (d). (e) An example of diffusion in the lungs of a frog. Because carbon dioxide (CO_2) is in a much higher concentration in the blood capillary than in the chamber of the lung, it diffuses from the capillary to the lung chamber. Oxygen (O_2) is much higher in the lung chamber compared to the capillary; hence it diffuses from the lung chamber into the capillary.

way for specific molecules to move across the plasma membrane so that they never come into contact with the hydrophobic layer or its polar surface.

Large molecules and some of those not soluble in lipids require assistance in passing through the plasma membrane. The process used by these molecules is called **facilitated diffusion** and like simple diffusion, requires no energy input to occur. To pass through the membrane, a molecule temporarily binds with a *carrier protein* in the plasma membrane and is transported from an area of higher to one of lower concentration (figure 3.10). The rate at which faciliated diffusion can occur is limited by the number of carrier molecules in the membrane.

Osmosis

The passage of *water* through a selectively permeable membrane from an area of higher concentration to an area of lower concentration is called **osmosis** (Gr. *osmos*, pushing). The relative concentrations of water are determined

by the amount of solute in the water on either side of the membrane. A higher concentration of solute (for example sugar) on one side of the membrane means less space is available for the water molecules (figure 3.11a). Water passes through the membrane from side 2 to side 1; the membrane is impermeable to solute so it does not move across. The net effect is an increase in water concentration on side 1 (the original area of lower water concentration), until the increasing pressure of the volume of solution in 1 pushes the water molecules back across the membrane as quickly as they enter (figure 3.11b). This **osmotic pressure** of the solution in 1 on the membrane is the force (hydrostatic pressure) required to stop the *net* flow of water across the membrane into 1.

The term **tonicity** refers to the relative concentration of solutes in the fluid inside and outside the cell. Using red blood cells as an example, in an *isotonic* (Gr. *isos*, equal + *tonus*, tension) solution, the solute (salt) concentration is the same inside and outside the cell (figure

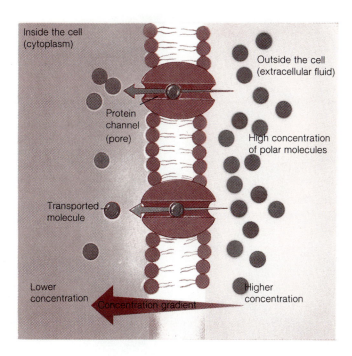

Figure 3.9 Transport proteins. Molecules can move into and out of cells through integrated channel proteins (pores) in the plasma membrane without the use of energy.

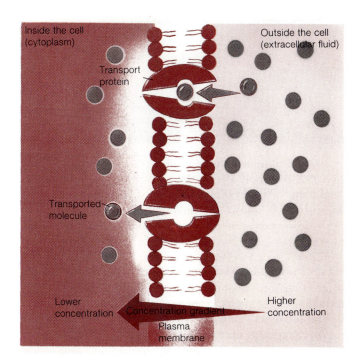

Figure 3.10 Facilitated diffusion and transport proteins. Some molecules move through the plasma membrane with the assistance of carrier proteins that transport the molecules down their concentration gradient, from a region of higher concentration to one of lower concentration. A transport protein alternates between two configurations, moving a molecule across a membrane as the shape of the protein changes.

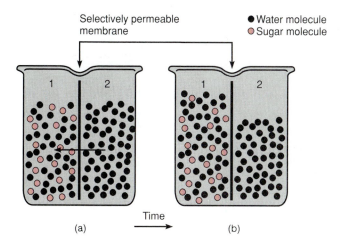

Figure 3.11 Osmosis. (a) The beaker is separated into two compartments by a selectively permeable membrane. Initially, compartment 1 contains sugar and water molecules, and compartment 2 contains only water molecules. Due to molecular motion, water will move down the concentration gradient (from 2 to 1) by osmosis. The sugar molecules remain in 1 because they are too large to pass through the membrane. (b) At equilibrium, there will be no net increase in water molecules in 1.

3.12a). The concentration of water molecules is also the same inside and outside the red blood cell; thus, water molecules move through the plasma membrane at the same rate in both directions, and there is no net movement of water in either direction.

In a *hypertonic* (Gr. *hyper*, above) solution, the solute concentration is higher outside the red blood cell than inside it. Because there is a higher concentration of water molecules inside the cell than outside, water moves *out* of the cell, which shrinks (figure 3.12b). This condition is called *crenation* in red blood cells.

In a *hypotonic* (Gr. *hypo*, under) solution, the solute concentration is lower outside the red blood cell than inside. Conversely, the concentration of water molecules is higher outside the cell than inside. As a result, water moves *into* the cell, which swells and may explode (figure 3.12c). [6]

Filtration

Filtration is a process that *forces* small molecules through selectively permeable membranes with the aid of hydrostatic (water) pressure (or some other externally applied force, such as blood pressure). For example, in the body of an animal such as a frog, filtration is evident when blood pressure forces water and dissolved molecules through the permeable walls of small blood vessels called capillaries (figure 3.13). In filtration, large molecules, such as proteins, do not pass through the smaller membrane pores. Filtration also takes place in the kidneys when water and dissolved wastes are forced out of the blood vessels into the kidney tubules by blood pressure, as the first step in the formation of urine.

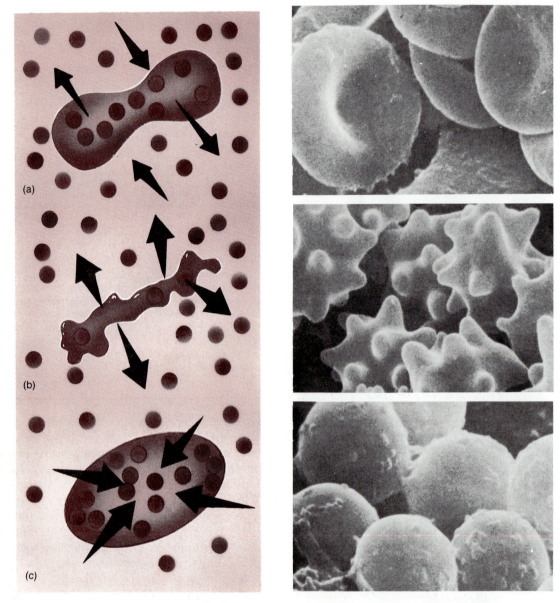

Figure 3.12 The effect of salt concentration on cell volumes. (a) An isotonic solution with the same salt concentration inside and outside the cell has no effect on the size of the red blood cell. (b) A hypertonic (high salt) solution causes water to leave the red blood cell and results in a shrunken appearance. (c) A hypotonic (low salt) solution results in an inflow of water, causing the red blood cell to swell. Arrows indicate direction of water movement.

Stop and Ask Yourself

9. How does diffusion occur in an animal cell?
10. What is osmosis? How does it differ from diffusion?
11. What is the difference between simple diffusion and facilitated diffusion?
12. What is the purpose of facilitated diffusion?

Active Transport

Active transport processes move molecules and other substances through a selectively permeable membrane against a concentration gradient, that is, from an area of higher to one of lower concentration. Because the movement is against the concentration gradient, energy is required.

The active transport process is similar to facilitated diffusion except that the transport protein in the plasma membrane must use energy to move the molecules against their concentration gradient (figure 3.14).

One active transport mechanism, the *sodium-potassium pump*, helps maintain the high concentrations of potassium and low concentrations of sodium inside nerve cells that are necessary for transmission of nervous impulses. Another, the *calcium pump*, keeps the calcium concentration many hundred times lower inside the nerve cell than outside.

Histology

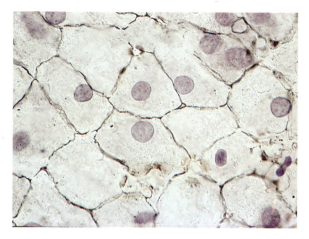

(a) Simple squamous epithelium consists of a single layer of tightly packed flattened cells (×250).

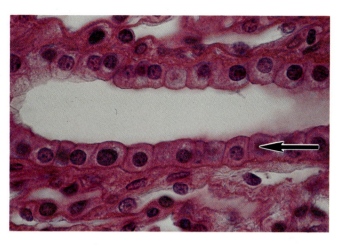

(b) Simple cuboidal epithelium consists of a single layer of tightly packed cube-shaped cells. Notice the single layer of cells indicated by the arrow (×250).

(c) Simple columnar epithelium consists of a single layer of elongated cells. The arrow is pointing to a specialized goblet cell that secretes mucus (×400).

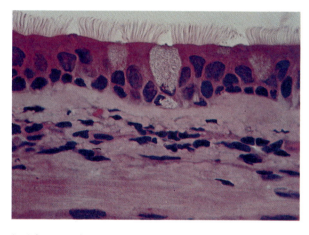

(d) Ciliated columnar epithelium. Notice the tuft of cilia at the top of each cell (×500).

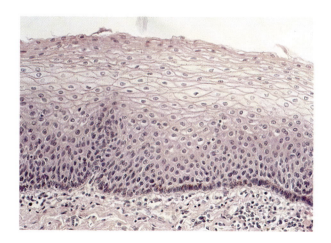

(e) Stratified squamous epithelium consists of many layers of cells (×67).

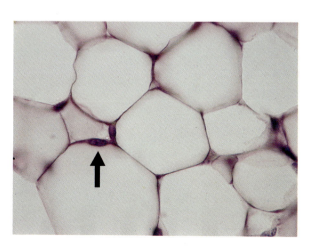

(f) Adipose tissue cells contain large fat droplets that cause the nuclei to be pushed close to the plasma membranes. The arrow is pointing to a nucleus (×250).

(Continued)

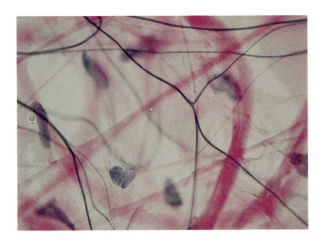

(g) Loose connective tissue contains numerous fibroblasts that produce collagenous and elastic fibers (×250).

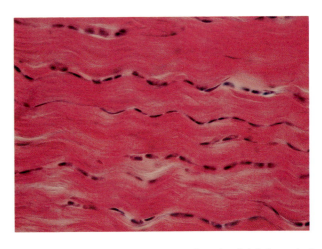

(h) Fibrous connective tissue consists largely of tightly packed collagenous fibers (×100).

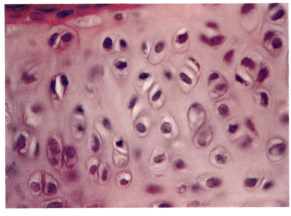

(i) Hyaline cartilage cells are located in lacunae surrounded by intercellular material containing fine collagenous fibers (×250).

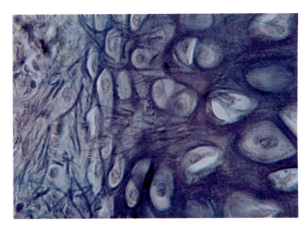

(j) Elastic cartilage contains fine collagenous fibers and many elastic fibers in its intercellular material (×100).

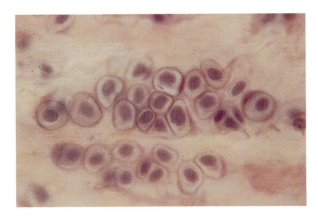

(k) Fibrocartilage contains many large collagenous fibers in its intercellular material (×195).

(l) Bone matrix is deposited in concentric layers around osteonic canals (×160).

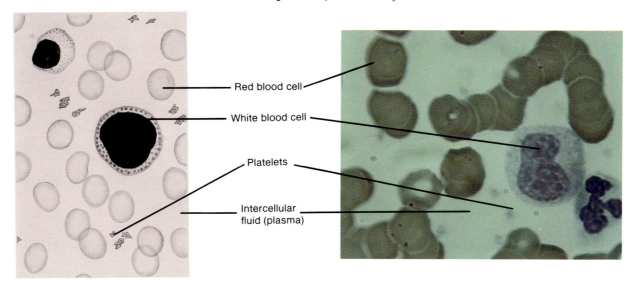

Red blood cell

White blood cell

Platelets

Intercellular
fluid (plasma)

(m) Blood is a type of connective tissue. It consists of an intercellular fluid (plasma) in which red blood cells, white blood cells, and platelets are suspended (×640).

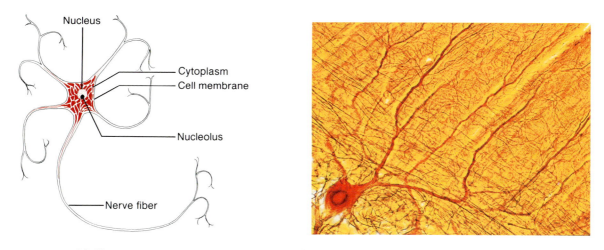

Nucleus

Cytoplasm
Cell membrane

Nucleolus

Nerve fiber

(n) Nervous tissue. Neurons in nerve tissue function to transmit impulses to other neurons or to muscles or glands (×450).

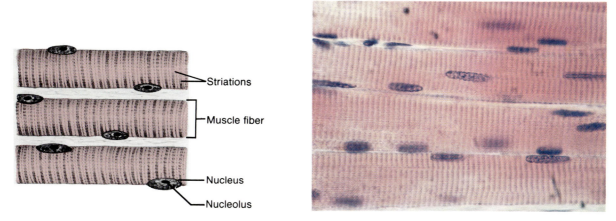

Striations

Muscle fiber

Nucleus

Nucleolus

(o) Skeletal muscle tissue is composed of striated muscle fibers (cells) that contain many nuclei (×250).

(Continued)

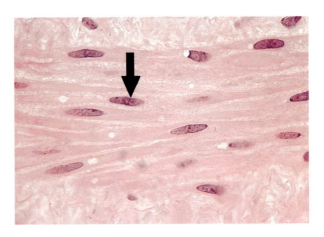

Nucleus
Cytoplasm
Cell membrane

(p) Smooth muscle is formed of spindle-shaped cells, each containing a single nucleus (arrow) (×250).

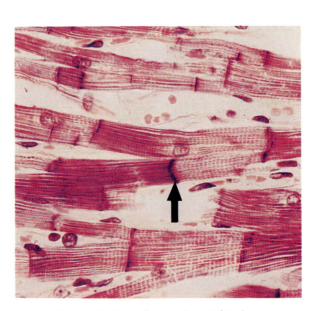

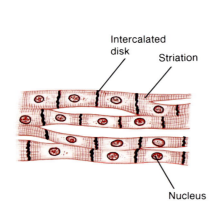

Intercalated disk
Striation
Nucleus

(q) Cardiac muscle consists of striated cells, each containing a single nucleus and specialized junctions called intercalated discs (arrow) (×400).

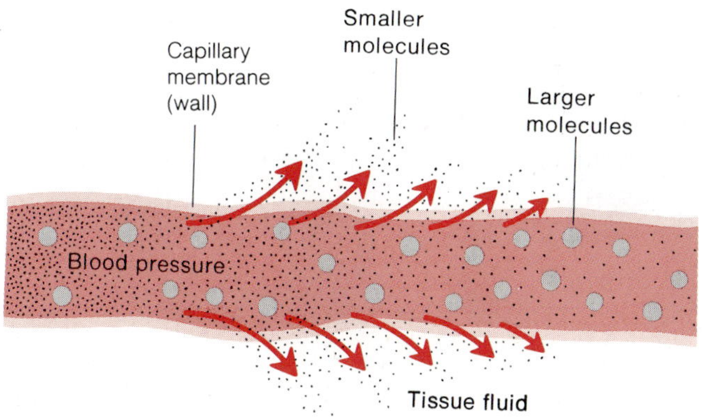

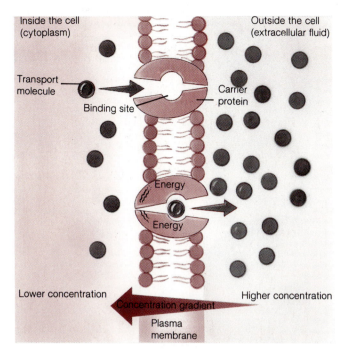

Figure 3.13 Filtration. Small molecules are forced through the wall of a capillary by the high blood pressure in the capillary. Larger molecules cannot pass through the small openings in the capillary wall and remain in the capillary. Arrows indicate the direction of small molecule movement.

Figure 3.14 Active transport. During active transport, a molecule combines with a carrier protein whose shape is altered as a result of the combination. This change in configuration, along with energy, helps move the molecule through the plasma membrane against a concentration gradient.

Endocytosis

Another process by which substances may move through the plasma membrane is called endocytosis. **Endocytosis** (Gr. *endon*, within) involves bulk movement of materials across the plasma membrane, rather than movement of individual molecules. There are three forms of endocytosis: pinocytosis, phagocytosis, and receptor-mediated endocytosis.

[7] **Pinocytosis** ("cell drinking," from Gr. *pinein*, to drink + *cyto*, cell) is nonspecific uptake of small droplets of extracellular fluid. Any small solid dissolved in the fluid is also taken into the cell. Pinocytosis occurs when a small portion of the plasma membrane becomes indented (invaginated). The open end of the invagination seals itself off, forming a small vesicle. This tiny vesicle becomes detached from the plasma membrane and moves into the cytoplasm (figure 3.15a).

Phagocytosis ("cell eating," from Gr. *phagein*, to eat + *cyto*, cell) is similar to pinocytosis except that the material taken into the cell is solid rather than liquid. Commonly, an organelle called a lysosome (see below) combines with the vesicle to form a **phagolysosome** ("digestion vacuole"), and lysosomal digestive enzymes cause the contents to be broken down (figure 3.15b). An example of phagocytosis is a white blood cell engulfing a yeast cell and digesting it (figure 3.16). The products of this digestion may then diffuse out of the phagolysosome and be used as raw materials for the nutrition of a cell. Any remaining residue may be expelled from the cell.

Receptor-mediated endocytosis involves a specific receptor on the plasma membrane that "recognizes" an extracellular molecule and binds with it (figure 3.15c). This reaction somehow stimulates the membrane to indent and create a vesicle that contains the selected molecule. A variety of important molecules are brought into cells in this manner.

Exocytosis

Proteins and other molecules produced in the cell that are destined for export (secretion) are packaged in vesicles by an organelle known as the Glogi apparatus (described in a later section). In the process of **exocytosis** (Gr. *exo*, outside), these secretory vesicles fuse with the plasma membrane and release their contents into the extracellular environment. This process adds new membrane material, which replaces that which was lost from the plasma membrane during endocytosis.

Stop and Ask Yourself

13. How do active transport and facilitated diffusion across plasma membranes differ?
14. What is the difference between endocytosis and exocytosis?
15. How does receptor-mediated endocytosis occur?
16. Where does an active-transport process occur in an animal?

Cytoplasm and Organelles

Many of the functions of a cell that are performed in the cytoplasmic compartment result from the activity of specific structures called organelles. These organelles are summarized in table 3.3 and discussed in the following sections.

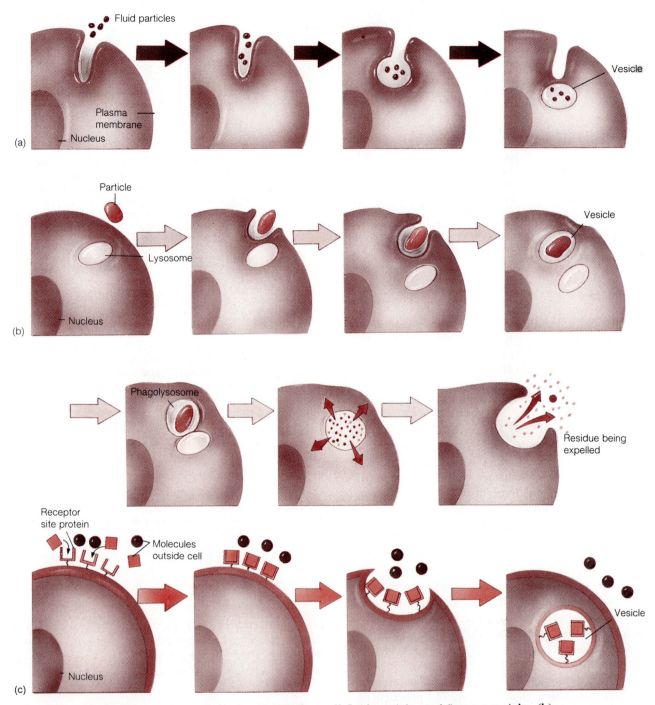

Figure 3.15 Endocytosis. (a) Pinocytosis. A cell takes in small fluid particles and forms a vesicle. (b) Phagocytosis. A cell takes in a solid particle and forms a vesicle. A lysosome combines with a vesicle, forming a phagolysosome. Lysosomal enzymes digest the particle. (c) In receptor-mediated endocytosis, a specific molecule binds to a receptor protein, inducing the formation of a vesicle.

Cytoplasm

The cytoplasm of a cell is composed of two distinct phases. The *particulate phase* consists of well-defined structures, such as membrane bound vesicles, organelles, and lipid droplets. The *aqueous phase* consists of the fluid *cytosol*, in which the above structures are suspended and in which are dissolved various molecules.

Ribosomes: Protein Workbenches

Ribosomes are the sites for protein synthesis. They contain almost equal amounts of protein and a special kind of ribonucleic acid called ribosomal RNA (rRNA). The role of ribosomes in protein synthesis is discussed in chapter 9. Some ribosomes are attached to the ER (see next section) and some float freely in the cytoplasm. Whether ribosomes

Table 3.3 Structure and Function of Cellular Components

Component	Structure/Description	Function
Centriole	Located within centrosome; contains nine triple microtubules	Forms basal body of cilia and flagella; functions in mitotic spindle formation
Cilia, flagella	Threadlike processes	Cilia move small particles past fixed cells; are a major form of locomotion in some cells; flagella propel cells
Cytoplasm	Semifluid enclosed within plasma membrane; consists of fluid cytosol and organelles	Dissolves substances; houses organelles, vesicles, inclusions
Cytoplasmic inclusions	Temporary substances in cytoplasm	Store products of cell's metabolic activities
Cytoplasmic vesicles	Membrane-bound sacs	Transport and store cellular materials
Cytoskeleton	Interconnecting microfilaments and microtubules; flexible cellular framework	Assists in cell movement; provides support; site for binding of specific enzymes
Cytosol	Fluid part of cytoplasm; enclosed within plasma membrane; surrounds nucleus	Houses organelles; serves as fluid medium for metabolic reactions
Chromosomes	Nucleic acid (DNA) makes up the different chromosomes	Controls heredity and cellular activities
Endoplasmic reticulum (ER)	An extensive membrane system extending throughout the cytoplasm from the plasma membrane to the nuclear envelope	Storage and internal transport; rough ER serves as site for attachment of ribosomes; smooth ER makes steroids
Golgi apparatus	Stacks of disklike membranes	Packaging and routing of cell's synthesized products
Lysosome	Membrane-bound sphere	Digests materials
Microfilaments	Rodlike structures containing the protein actin	Give structural support and assist in cell movement
Microtubules	Hollow cylindrical structures	Assist in movement of cilia, flagella, and chromosomes; transport system
Mitochondrion	Organelle with double, folded membranes	Converts energy into a form usable by the cell
Nucleolus	Rounded mass within nucleus; contains RNA and protein	Preassembly point for ribosomes
Nucleus	Spherical structure surrounded by a nuclear envelope; contains nucleolus and DNA	Contains DNA that controls cell's genetic program and metabolic activities
Peroxisome	Membrane bound organelle containing oxidative enzymes	Carries out metabolic reactions and destroys hydrogen peroxide, which is toxic to the cell
Plasma membrane	The outer bilayered boundary of the cell; composed of protein, cholesterol, and phospholipid	Protection; regulation of material movement; cell-to-cell recognition
Ribosomes	Contain RNA and protein; some are free and some are attached to ER	Sites of protein synthesis

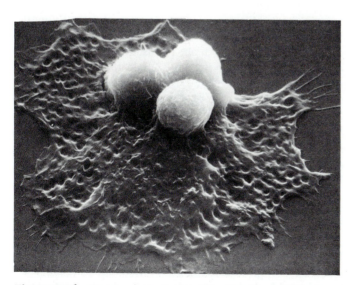

Figure 3.16 Phagocytosis. A white blood cell engulfing three cancer cells.

are free or attached, they are usually grouped in clusters that are connected by a strand of another kind of ribonucleic acid called *messenger* (mRNA). These clusters are called *polyribosomes* or *polysomes* (*see figure 3.1*).

Endoplasmic Reticulum: Production and Transport

The **endoplasmic reticulum** (**ER**) is a complex, membrane-bound labyrinth of flattened sheets, sacs, tubules, and membranes that branch and spread throughout the cytoplasm. The ER is continuous from the plasma membrane to the nuclear envelope (*see figure 3.1*) and functions as a series of channels that help various materials circulate throughout the cytoplasm. It also serves as a storage unit for enzymes and other proteins and serves as a point for attachment of ribosomes. ER that has ribosomes attached to it is called *rough ER* (figure 3.17a), and ER that does not have ribosomes attached is called *smooth ER* (figure

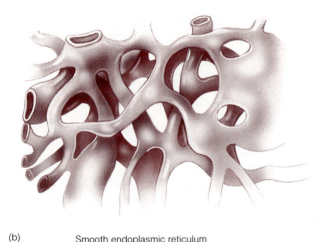

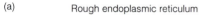

(a) Rough endoplasmic reticulum

(b) Smooth endoplasmic reticulum

Figure 3.17 Endoplasmic reticulum. (a) Rough ER is coated with ribosomes. Notice the double membrane and the lumen between each membrane. (b) Smooth ER lacks ribosomes.

3.17b). Smooth ER is the site for the production of steroids, detoxification of a wide variety of organic molecules, and storage of calcium ions in muscle cells. Most cells contain both types of ER, although the relative proportion varies among cells.

Golgi Apparatus: Packaging and Export

The **Golgi apparatus** or **complex** (named for Camillo Golgi, who discovered it in 1898), is a collection of membranes associated physically and functionally with the ER in the cytoplasm (figure 3.18a; *see figure 3.1*). It is composed of flattened stacks of membrane-bound *cisternae* (singular, *cisterna*; closed spaces serving as fluid reservoirs). The Golgi apparatus functions in the packaging and secretion of glycoproteins that are important components of the outer coating of the plasma membrane (*see figure 3.6*).

Some proteins that are synthesized by ribosomes attached to the rough ER, are sealed off in little packets called *transfer vesicles* that pass from the ER to the Golgi apparatus and fuse with it (figure 3.18b). In the Golgi appa-

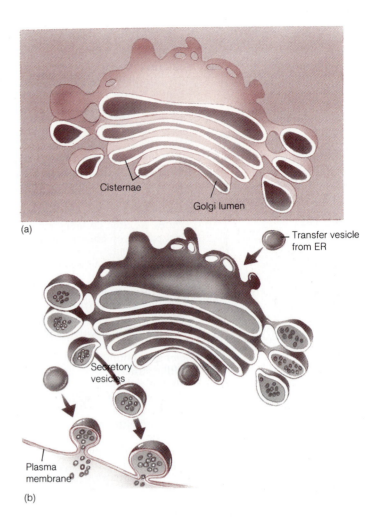

Figure 3.18 The Golgi apparatus. (a) The Golgi apparatus consists of a stack of cisternae. Notice the curved nature of the cisternae. (b) The Golgi apparatus functions in the packaging and secretion of cell products. Secretory vesicles move from the Golgi apparatus to the plasma membrane and fuse with it, releasing their contents to the outside of the cell via exocytosis.

ratus, the proteins can be concentrated, chemically modified, and compacted. Eventually, they are packaged into *secretory vesicles*, which are released into the cytoplasm close to the plasma membrane. When the vesicles reach the plasma membrane, they fuse with it and release their contents to the outside of the cell by exocytosis. Golgi apparatuses are most abundant in cells that secrete chemical substances (e.g., pancreatic cells secreting digestive enzymes and nerve cells secreting transmitter substances). As noted below, the Golgi apparatus also produces lysosomes.

Lysosomes and Peroxisomes: Digestion and Degradation

Lysosomes (Gr. *lyso*, dissolving + *soma*, body) are membrane-bound spherical organelles that contain enzymes called *acid hydrolases*, which are capable of digesting organic molecules, such as fats, proteins, and polysaccha-

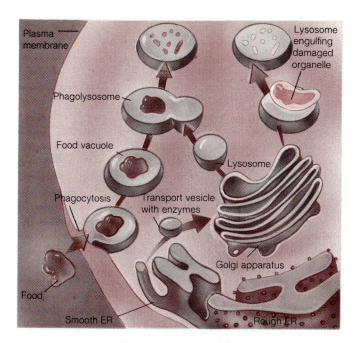

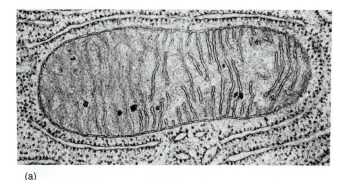

(a)

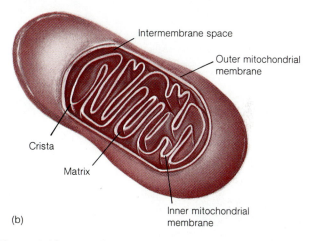

(b)

Figure 3.19 Lysosome formation and function. Lysosomes arise from the Golgi apparatus and fuse with vesicles that have engulfed foreign material to form digestive vesicles. These vesicles function in the normal recycling of cell constituents.

Figure 3.20 Mitochondrion. (a) An electron micrograph illustrating the mitochondrial membranes, cristae, and matrix. The matrix contains DNA, ribosomes, and many kinds of enzymes. (b) An artist's depiction of a "typical" mitochondrion.

rides, under acidic conditions. The enzymes are synthesized in the ER, transported to the Golgi apparatus for processing, and then secreted by the Golgi apparatus in the form of *lysosomes* or as vesicles that fuse with lysosomes (figure 3.19). Lysosomes fuse with phagocytic vesicles, thus exposing the vesicle's contents to the lysosome's enzymes.

Cells can also selectively digest portions of their own cytoplasm or organelles. When the digested materials are returned to the cytoplasm, they can be reused by the cell, recycling the cell constituents.

Peroxisomes are single, membrane-bound organelles that contain oxidative enzymes, instead of digestive enzymes and are probably formed by pinching off ER. They use these enzymes and oxygen to carry out their metabolic reactions. Peroxisomes are so named because they can form hydrogen peroxide (H_2O_2) as they oxidize various substances and they then destroy the hydrogen peroxide with their enzymes to produce water and oxygen. This destruction is crucial to the cell because peroxides are toxic. Peroxisomes are numerous in liver and kidney cells where they detoxify many compounds.

Mitochondria: Power Generators

Mitochondria (singular, mitochondrion) are double-membrane bound organelles that are spherical to elongate in shape. The outer membrane is separated from the inner membrane by a small space. The inner membrane folds and doubles in upon itself to form incomplete partitions called *cristae* (singular, crista; figure 3.20). The cristae increase the surface area available for the chemical reactions that produce usable energy for the cell. The space between the

cristae is the *matrix*. The matrix contains ribosomes, circular DNA, and other material. Because they convert energy to a usable form mitochondria are frequently called the "power generators" of the cell. Mitochondria usually multiply when a cell needs to produce more energy. The production of cellular energy will be covered in more detail in chapter 5.

[8]

Microtubules, Intermediate Filaments, and Microfilaments: The Cytoskeleton

Microtubules are hollow, slender, cylindrical structures found in animal cells (figure 3.21a). Each microtubule is made of spiraling subunits of globular proteins called *tubulins* (figure 3.21b). Microtubules function in the movement of organelles, such as secretory vesicles, and in chromosome movement during division of the cell nucleus.

They are also part of a transport system within the cell. For example, in nerve cells, they help move materials through the long nerve processes.

Microtubules are an important part of the cytoskeleton (see below) in the cytoplasm, and they are involved in the overall shape changes that cells undergo during periods of specialization.

Intermediate filaments are a chemically heterogeneous group of protein fibers, the specific proteins of which can vary with cell type (figure 3.22). These filaments help

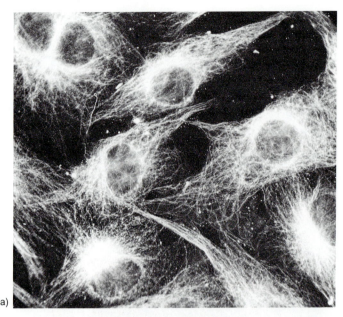

(a)

Tubulin
subunits

(b)

Figure 3.21 Microtubules and microfilaments. (a) Electron-micrograph of microtubules. (b) Each microtubule is composed of globular proteins, called tubulins. (c) Microfilaments.

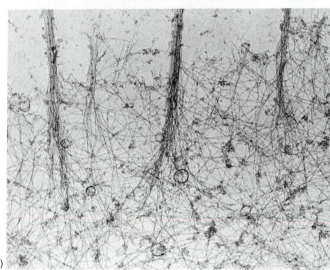

(c)

to maintain the shape of the cell, the spatial organization of organelles, and promote mechanical activities within the cytoplasm.

Microfilaments are solid strings of protein (actin) molecules (figure 3.21c). They are most highly developed in muscle cells as *myofibrils*, which help these cells shorten or contract. Actin microfilaments in nonmuscle cells provide mechanical support for various cellular structures, and help form contractile systems responsible for some cellular movements (e.g., amoeboid movement in some protozoa).

In most cells, the microtubules, intermediate filaments, and microfilaments form the flexible cellular framework [9] called the **cytoskeleton** ("cell skeleton") (*see figure 3.22*). This latticed framework extends throughout the cytoplasm, connecting the microtubules, intermediate filaments, microtubules, and various organelles in a structural and functional entity.

Endoplasmic
reticulum
Ribosome

Plasma
membrane

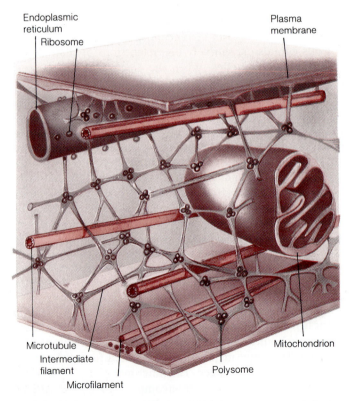

Microtubule

Intermediate
filament

Microfilament

Polysome

Mitochondrion

Figure 3.22 The cytoskeleton. A model of the cytoskeleton showing the three-dimensional arrangement of the microtubules, intermediate filaments, and microtubules.

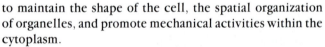

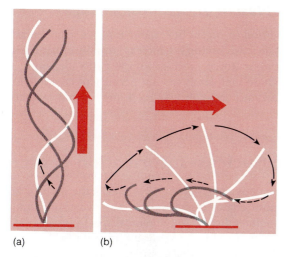

Figure 3.23 Movement of flagella and cilia. (a) Flagellar movement occurs from the base upward in the form of waves, whereas a cilium moves in two strokes (b).

Cilia and Flagella: Movement

Cilia (singular, cilium; L. "eyelashes") and **flagella** (singular, flagellum; L. "small whips") are elongated appendages on the surface of some cells. They are the means by which cells, including many unicellular organisms, propel themselves. In stationary cells, cilia or flagella move material over the cell's surface.

Although flagella are 5 to 20 times as long as cilia, and move somewhat differently, they have a similar structure (figure 3.23). Both are membrane-bound cylinders that enclose a matrix. In this matrix is an **axoneme** or **axial filament**, which consists of nine pairs of microtubules arranged in a circle around two central tubules (figure 3.24). This is called a 9 + 2 pattern of microtubules. Each microtubule pair (a doublet) also has pairs of *dynein* (protein) *arms* projecting toward a neighboring doublet and *spokes* extending toward the central pair of microtubules. Cilia and flagella move as a result of the microtubule doublets sliding along one another.

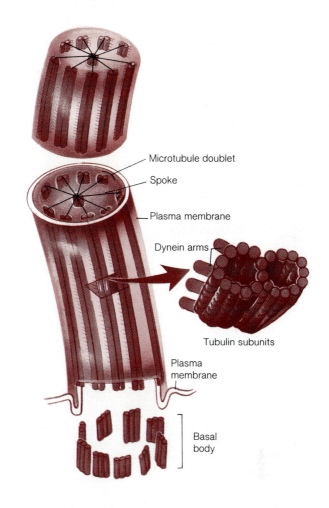

Microtubule doublet
Spoke
Plasma membrane
Dynein arms
Tubulin subunits
Plasma membrane
Basal body

Figure 3.24 Drawing of the internal structure of cilia and flagella. In cross section, the arms extend from each microtubule doublet toward a neighboring doublet, and spokes extend toward the central paired microtubules. The dynein arms push against the adjacent microtubule doublet to bring about movement.

Figure 3.24 redrawn, with permission, from Norman K. Wessels and Janet L. Hopson, *Biology.* Copyright © 1988, McGraw-Hill Publishing Company, New York, New York.

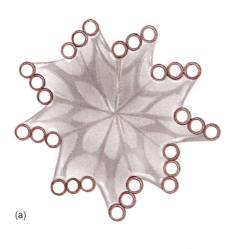

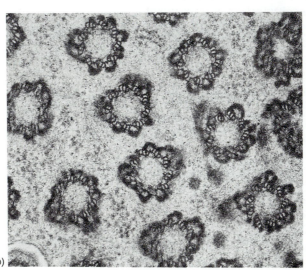

Figure 3.25 Centrioles and basal bodies. (a) An artist's sketch illustrating the triplet arrangement in the 9 + 0 pattern in a centriole. (b) An electron micrograph cross-sectional view through the basal bodies of cilia in the mammalian oviduct.

Centrioles: Specialized Microtubules

There is a specialized nonmembranous region of cytoplasm near the nucleus that is called the *centrosome* or *cell center*. The centrosome contains two organelles called **centrioles** (*see figure 3.1*) that lie at right angles to each other. Each centriole is composed of nine triplet microtubules that radiate from the center like the spokes of a wheel (figure 3.25). The centrioles are duplicated preceding cell division and are involved with the movement of the chromosomes.

In the cytoplasm at the base of each cilium or flagellum lies a short, cylindrical **basal body** (*see figure 3.24*), also made up of microtubules. However, these tubules form a 9 + 0 pattern; nine sets of three with none in the middle. Structurally, the basal body is similar to the centriole.

Cytoplasmic Inclusions: Storage

In addition to organelles, the cytoplasm also contains **cytoplasmic inclusions**, which are usually either basic food material or stored products of the cell's metabolic activities. These inclusions are not permanent components of a cell, and are constantly being destroyed and replaced.

The Nucleus: Information Center

The nucleus contains the DNA and is the control and information center of the cell. It has two major functions. The nucleus directs the chemical reactions that occur in cells by transcribing genetic information in the DNA into RNA, which then translates this specific information into proteins (e.g., enzymes) that determine the cell's specific activities. The nucleus also stores genetic information and transfers it during cell division from one cell to the next, and from one generation of organisms to the next.

Nuclear Envelope: Gateway to the Nucleus

The **nuclear envelope** is a membrane that separates the nucleus from the cytoplasm. It is continuous with the endoplasmic reticulum at a number of points. A large number (over 3000) of *nuclear pores* penetrate the surface of the nuclear envelope (figure 3.26a). These pores make it possible for materials to enter and leave the nucleus, and for the nucleus to be in direct contact with the endoplasmic reticulum (*see figure 3.1*). Nuclear pores are not simply holes in the nuclear envelope; each is composed of an ordered array of globular and filamentous granules, probably proteins (figure 3.26b). These granules form the nuclear pore complex, which governs the transport of molecules into and out of the nucleus. The size of the pores prevents DNA from leaving the nucleus.

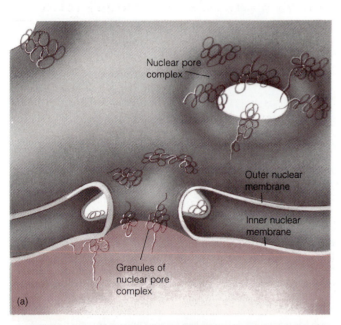

Nuclear pore complex

Outer nuclear membrane

Inner nuclear membrane

Granules of nuclear pore complex

(a)

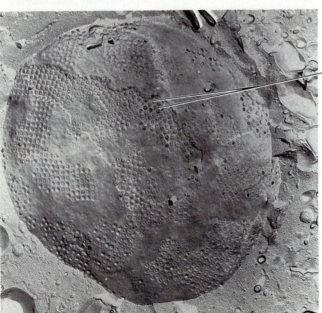

Nuclear pores

(b)

Figure 3.26 Nuclear pores. (a) An artist's interpretation of pore structure showing how the pore spans the two-layered nuclear envelope. The protein granules around the edge and in the center govern what passes through the pores. (b) An electron micrograph of the nuclear envelope showing the many pores.

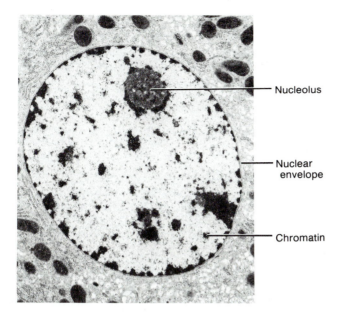

Figure 3.27 The nucleus. The nucleolus, chromatin, and nuclear envelope are visible in this nucleus.

Nucleolus

Nuclear envelope

Chromatin

Chromosomes: Genetic Containers

The nucleoplasm is the inner mass of the nucleus. In a cell that is not dividing, it contains genetic material called **chromatin.** Chromatin consists of a combination of DNA and protein and is the uncoiled, tangled mass of **chromosomes,** which contain the hereditary information in segments of DNA called *genes.* During cell division, each chromosome becomes tightly coiled making it visible when viewed through a light microscope.

Nucleolus: Preassembly Point for Ribosomes

The **nucleolus** (plural, nucleoli) is an organelle in the nucleoplasm that is present in nondividing cells (figure 3.27). Two or three nucleoli form in most cells, but there can be thousands or more in some cells (e.g., amphibian eggs). The nucleolus is not surrounded by a membrane. Nucleoli function in the synthesis of ribosomes, and there are usually proteins and RNA in many stages of synthesis and assembly. The nucleolus can be described as a preassembly point for ribosomes because their assembly is completed after leaving the nucleus through the pores of the nuclear envelope.

Stop and Ask Yourself

17. What are several functions of the nucleus?
18. Why are the pores in the nuclear envelope important?
19. What is the difference between chromatin, nucleoplasm, and chromosomes?
20. What is the function of a nucleolus?

Tissues

In an animal, individual cells differentiate during development to perform special functions. These specialized cells carry out their functions as aggregates called tissues. A **tissue** (Fr. *tissu*, woven) is a group of similar cells specialized for the performance of a common function. Animal tissues are classified as one of four types: epithelial, connective, muscular, or nervous.

Epithelial Tissue: Many Forms and Functions

Epithelial tissue exists in many structural forms. In general it either *covers* or *lines* something and is typically made up of renewable flat sheets of cells that have surface specializations adapted for their specific roles. Usually epithelial tissues are separated from underlying, adjacent tissues by a basement membrane. The typical functions of epithelial tissues are *absorption* (e.g., the lining of the small intestine), *transport* (e.g., kidney tubules), *excretion* (e.g., sweat glands), *protection* (e.g., the skin), and *sensory reception* (e.g., the taste buds in the tongue). The size, shape, and arrangement of epithelial cells are directly related to these specific functions.

Epithelial tissues are classified on the basis of shape and the number of layers present. Epithelium can be *simple*, consisting of only one layer of cells, or *stratified*, consisting of two or more layers stacked one on top of each other (colorplate 1e). The shapes of the individual epithelial cells can be flat (*squamous epithelium;* colorplate 1a), cube shaped (*cuboidal epithelium;* colorplate 1b), or columnlike (*columnar epithelium;* colorplate 1c). Columnar epithelium may also possess cilia (colorplate 1d).

Connective Tissue: Connection and Support

Connective tissues serve to support and bind together tissues. Unlike epithelial tissues, connective tissues are distributed throughout an extracellular *matrix.* This matrix contains fibers that are embedded in a ground substance, which has a consistency anywhere from liquid to solid. To a large extent, the functional properties of the various connective tissues are determined by the nature of this extracellular material.

Connective tissues are of two general types depending upon whether the fibers are loosely or densely packed. In **loose connective tissue** (colorplate 1g), the matrix contains strong, flexible fibers of the protein collagen that are interwoven with fine, elastic and reticular fibers. These fibers give loose connective tissue its elastic consistency, and make it an excellent binding tissue (e.g., binding the skin to underlying muscle tissue). **Fibrous connective tissue** (colorplate 1h) is made up of fibers that are very densely packed. The collagen fibers lie parallel to one another, creating very strong cords, such as *tendons* (which connect muscles to bones or to other muscles), and *ligaments* (which connect bones to bones).

Adipose tissue is a type of loose connective tissue that consists of large cells that store lipid (colorplate 1f). Most often the cells accumulate in large numbers to form what we commonly call fat.

Cartilage is a hard yet flexible tissue (colorplates 1i-k) that supports such structures as the outer ear, and forms the entire skeleton of such animals as sharks and rays. Cells called *chrondrocytes* lie within spaces called *lacunae* that are surrounded by a rubbery matrix. This matrix is secreted by *chondroblasts* and with the collagen fibers, gives cartilage its strength and elasticity.

Bone cells (*osteocytes, osteoblasts,* or *osteoclasts*) also lie within lacunae (colorplate 1l), but the matrix around them is heavily impregnated with calcium phosphate, making this kind of tissue very hard and ideally suited for its functions of support and protection. The structure and function of bone will be considered in more detail in chapter 32.

Blood is a connective tissue in which specialized red and white blood cells plus platelets are suspended in a fluid called *plasma* (colorplate 1m). Blood transports various substances throughout the body of animals and will be covered in more detail in chapter 36.

Muscle Tissue: Movement

Muscle tissue allows movement. The details of this contractile process will be discussed in chapter 32. Therefore, at this point we will only introduce the three kinds of muscle tissue: *smooth muscle, skeletal muscle,* and *cardiac muscle* (colorplates 1o-q).

Nervous Tissue: Communication

Nervous tissue is composed of individual cells called *neurons* (colorplate 1n). Neurons are specialized for conducting electrical impulses. Neurons receive information about changes in the external and internal environments of an animal and stimulate other tissues or organs to respond appropriately to those changes. Neurons occur in a wide variety of shapes and sizes. Interspersed among them are *glial cells* that support the neurons. Nervous tissue will be covered in more detail in chapter 37.

Organs

Organs are the functional units of an animal's body that are made up of more than one type of tissue. Examples include the heart, lungs, liver, spleen, and kidneys.

Systems

The next higher level of structional organization in animals is the system. A **system** (Gr. *systema*, being together) is an association of organs that have a common function. The systems that are found in higher vertebrate animals are the integumentary, skeletal, muscular, nervous, endocrine, circulatory, lymphatic, respiratory, digestive, urinary, and reproductive systems. These will be discussed in detail in chapters 32 through 39.

The highest level in an animal body is the organismic level. All the parts of the animal body functioning with one another contribute to the total organism—a living entity or individual.

Stop and Ask Yourself

21. What is a tissue?
22. What are some functions of epithelial tissue? Where would these tissues be found in a higher animal?
23. How are epithelial tissues classified?
24. What are the different types of connective tissue?

■ Summary

1. All animal cells have three basic parts: the nucleus, cytoplasm, and the plasma membrane.

2. Cell membranes are composed mainly of phospholipids and proteins and allow certain materials to move into and out of them. This quality is called selective permeability. Our knowledge of the plasma membrane is based on the fluid-mosaic model.

3. Some molecules use their own energy to pass through a cell membrane from areas of higher concentration to areas of lower concentration. Examples of these passive processes are: simple diffusion, facilitated diffusion, osmosis, and filtration.

4. Active transport across cell membranes require energy from the cell to move substances from areas of lower concentration to areas of higher concentration. Additional processes that move molecules across membranes are endocytosis, phagocytosis, and exocytosis.

5. The cytoplasm of a cell is composed of two phases. The particulate phase consist of membrane-bound vesicles, organelles, and various inclusions. The aqueous phase consists of the fluid cytosol.

6. Ribosomes are the sites of protein synthesis.

7. The endoplasmic reticulum (ER) is a series of channels that function in transport, enzyme and protein storage,

and provide a point of attachment for ribosomes. Two types of ER exist: smooth and rough.

8. The Golgi apparatus aids in the synthesis and secretion of glycoproteins, as well as the processing and modification of other materials (e.g., enzymes).

9. Lysosomes digest nutrients and clean away dead or damaged cell parts. Peroxisomes contain enzymes that transfer hydrogen from various substrates to oxygen, producing and then degrading hydrogen peroxide as they detoxify materials.

10. Mitochondria convert energy in food molecules to a form usable by the cell.

11. Microtubules, intermediate filaments, and microfilaments make up the cytoskeleton of the cell. The cytoskeleton functions in transport, support, and the movement of structures in the cell, such as organelles and chromosomes.

12. Centrioles assist in cell division and are involved with the movement of chromosomes during cell division.

13. Cytoplasmic inclusions are usually either food material that the cell needs or the stored products of the cell's metabolic activities.

14. The nucleus of a cell contains DNA, which controls the cell's genetic program and other metabolic activities.

15. The nuclear envelope contains many pores that allow material to enter and leave the nucleus.

16. The chromosomes in the nucleus have DNA organized into genes, which are specific DNA sequences that control and regulate the activities of the cell.

17. The nucleolus is a preassembly point for ribosomes.

18. Tissues are groups of cells with a common structure and function. Four types exist: epithelial, connective, muscular, and nervous.

19. An organ is composed of more than one type of tissue. A system consists of an association of organs.

▪ Key Terms

active transport (p. 44)
adipose tissue (p. 54)
axial filament (p. 51)
axoneme (p. 51)
basal body (p. 52)
blood (p. 54)
bone (p. 54)
cartilage (p. 54)
centrioles (p. 52)
chromatin (p. 53)
chromosomes (p. 53)
cilia (p. 51)
connective tissue (p. 53)
cytoplasm (p. 34)

cytoplasmic inclusions
   (p. 52)
cytoskeleton (p. 50)
endocytosis (p. 45)
endoplasmic reticulum (ER)
   (p. 47)
endosymbiont hypothesis
   (p. 35)
epithelial tissue (p. 53)
eukaryotes (p. 34)
exocytosis (p. 45)
facilitated diffusion (p. 42)
fibrous connective tissue
   (p. 53)

filtration (p. 43)
flagella (p. 51)
glycocalyx (p. 37)
Golgi apparatus (p. 48)
homeostasis (p. 40)
intercellular junctions
   (p. 40)
intermediate filaments
   (p. 49)
light microscope (p. 38)
loose connective tissue
   (p. 53)
lysosomes (p. 48)
microfilaments (p. 50)
microscopy (p. 38)
microtubules (p. 49)
mitochondria (p. 49)
muscle tissue (p. 54)
nervous tissue (p. 54)
nuclear envelope (p. 52)
nucleolus (p. 53)
nucleoplasm (p. 34)
nucleus (p. 34)
organelle (p. 34)
organs (p. 54)
osmosis (p. 42)

osmotic pressure (p. 42)
peroxisomes (p. 49)
phagocytosis (p. 45)
phagolysosome (p. 45)
pinocytosis (p. 45)
plasma membrane (p. 34)
prokaryotes (p. 34)
receptor-mediated
   endocytosis (p. 45)
ribosomes (p. 46)
scanning electron
   microscope (SEM)
   (p. 38)
scanning tunneling
   microscope (STM)
   (p. 38)
selective permeability
   (p. 40)
simple diffusion (p. 41)
symbiosis (p. 35)
system (p. 54)
tissue (p. 53)
tonicity (p. 42)
transmission electron
   microscope (TEM)
   (p. 38)

▪ Critical Thinking Questions

1. The cell theory states that the cell is the basic unit of life. What does this mean?

2. One of the larger facets of modern zoology can be described as "membrane biology." What are the common principles that unite the diverse functions of membranes?

3. Why is the current model of the plasma membrane called the "fluid-mosaic" model? What is the fluid and in what sense is it fluid? What makes up the mosaic?

4. If you could visualize osmosis, seeing the solute and solvent particles as individual entities, what would an osmotic gradient look like?

5. Why do some animal cells have the ability to transport materials against a concentration gradient? Could animals survive without this capability?

6. Membranes are in continuous interaction with vesicles. How does this dynamic interaction operate and what are some of the major functions that it accomplishes?

7. Why is the mitochondrion called the "power generator" of the cell?

8. What is the function of the cytoskeleton?

9. Why is the nucleus of a cell known as the "genetic message center"?

10. What physiological functions are epithelial tissues adapted to perform?

◼ Suggested Readings

Books

Alberts, B., Bray, D., Lewis, J., Raff, M., Roberts, K., and Watson, J. 1989. *Molecular Biology of the Cell*. 2nd ed. New York: Garland Publishing Company.

Darnell, J., Lodish, H., and Baltimore, D. 1986. *Molecular Cell Biology*. New York: Scientific American Books.

DeDuve, C. 1986. *A Guided Tour of the Living Cell*, vols. 1 and 2. New York: Scientific American Books.

Articles

Bretscher, M. S. How animal cells move. *Scientific American* December, 1987.

Capaldi, R. A. A dynamic model of cell membranes. *Scientific American* March, 1974.

DeDuve, C. Microbodies in the living cell. *Scientific American* May, 1983.

Luria, S. E. Colicins and the energetics of cell membranes. *Scientific American* December, 1975.

Porter, K., and Tucker, J. The ground substance of the living cell. *Scientific American* March, 1981.

Rothman, J. E. The compartmental organization of the Golgi apparatus. *Scientific American* September, 1985.

Satir, B. The final steps in secretion. *Scientific American* October, 1975.

Satir, B. How cilia move. *Scientific American* October, 1974.

Singer, S. J., and Nicolson, G. 1972. The fluid-mosaic model of the structure of cell membranes. *Science* 175: 720–31.

Weber, K., and Osborn, M. The molecules of the cell matrix. *Scientific American* October, 1985.

Staehelin, L. A., and Hull, B. E. Junctions between living cells. *Scientific American* May, 1978.

Energy and Enzymes: Life's Driving and Controlling Forces

4

Concepts

1. All life processes in a cell are driven by energy. Energy is the capacity to do work. It can exist in two forms: kinetic energy is actively involved in doing work, and potential energy is stored for future use.
2. The cell obtains energy by utilizing chemical fuel and obeying the first and second laws of thermodynamics.
3. The speed of a chemical reaction depends on the activation energy necessary to initiate it. Catalysts reduce the amount of activation energy necessary to initiate a chemical reaction and therefore speed up the reaction. Cells use specialized proteins called enzymes as biological catalysts.
4. The activity of an enzyme is affected by any factor that alters its three-dimensional shape; e.g., temperature, pH, other chemicals. An enzyme can also employ metal ions or organic molecules to facilitate its activity; these are called cofactors. Specific cofactors that are nonprotein, organic molecules are called coenzymes.
5. ATP is the universal energy currency of all cells.

Have You Ever Wondered:

[1] what a food chain is?
[2] what work is?
[3] how an animal accomplishes biological work?
[4] how the shape of an enzyme can be compared to a handshake?
[5] how quickly enzymes function?
[6] how certain drugs treat various diseases?
[7] what the energy currency of life is?

These and other useful questions will be answered in this chapter.

This chapter contains underlined evolutionary concepts.

An animal's use of the sun's radiant energy begins with the "capture" of that energy by photosynthetic plants (and certain microorganisms) that convert light energy to chemical energy in the form of carbohydrates, fats, and proteins. Because plants capture the sun's energy they are called **primary producers** (figure 4.1). A **primary consumer** is a plant-eating animal, or *herbivore* (L. *herba*, grass + *vorare*, to devour), that obtains organic molecules by eating producers or their products. They use the energy in the organic molecules to carry out all of their cellular activities, including the synthesis of complex organic molecules. Primary consumers are preyed upon by other animals, the **secondary consumers,** and

[1] so on, in what is termed a **food chain.** As each of these organisms dies, its constituents are broken down by digestion or by various **decomposers,** such as bacteria and fungi. This producer-consumer-decomposer sequence in a food chain represents a flow of both energy and matter.

Nutrients are continuously being broken down to provide the energy necessary for life. In fact, life can be viewed as a constant flow of energy, channeled by an animal, to do the work of living. This "living chemistry" is called metabolism. **Metabolism** (Gr. *metaballein,* to change) is the total of all the chemical reactions occurring in an animal's cells. It involves the acquisition and use of energy in stockpiling, breaking down, assembling, and eliminating substances to ensure the maintenance, growth, and reproduction of the animal. Metabolism consists of synthetic processes called **anabolism** (Gr. *anabole,* building up) and degrading processes called **catabolism** (Gr. *katabole,* breaking down). This chapter will focus on what energy is, how an animal uses it, and how metabolism is controlled via enzymes. The next chapter will present the network of chemical reactions that is the highway system for the flow of energy through an animal.

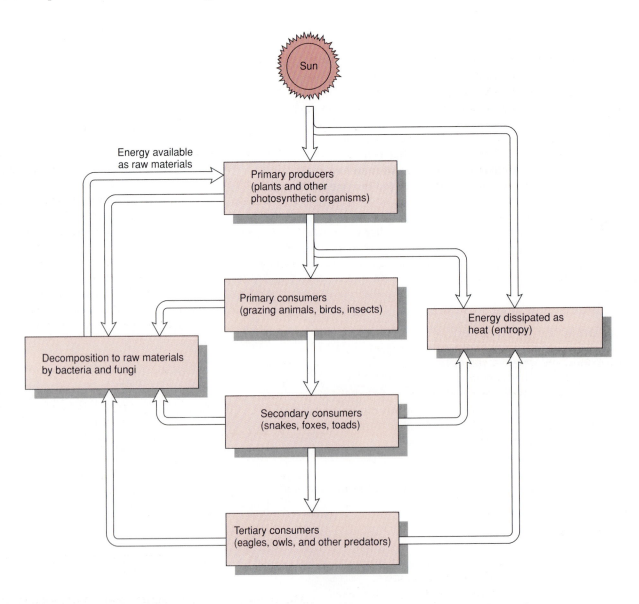

Figure 4.1 The sun's radiant energy sustains virtually all organisms. This figure represents the flow of energy and materials through living systems.

What is Energy?

[2] **Energy** is the capacity to do work. **Work** is the transfer of energy. Energy can also experience change. For example, solar (radiant) energy can be changed into the chemical energy in a plant (*see figure 4.1*). The plant can then be burned in a steam generator and the energy changed into the energy of motion (e.g., the turning of a wheel). Energy has many forms: the heat from a furnace, the sound of a jet plane, the electric current that lights a bulb, the radioactivity in a heart pacemaker, or the pull of a magnet.

Energy exists in two states: kinetic and potential. **Kinetic energy** is the energy of motion, e.g., a thundering waterfall. **Potential energy** is stored energy, e.g., a giant boulder poised on a pinnacle. In a living animal, energy in chemical bonds is a form of potential energy. It is this

[3] bond energy in organic molecules that the animal uses to accomplish biological work. Much of the work performed by an animal involves the transformation of potential energy to kinetic energy in its cells.

The most convenient way to measure energy is in terms of heat production. This is why the study of energy is called **thermodynamics** (Gr. *therme,* heat + *dynamis,* power). The most commonly employed unit of heat in an organism is the **kilocalorie** (**kcal**) or large Calorie (notice the large ''C''). A kilocalorie is the amount of heat necessary to raise 1 kilogram of water 1°C and is equal to 1000 calories. A reasonable daily intake of energy for an average person is approximately 2000 to 2500 kcal. A **calorie** (notice the small ''c'') is the amount of heat it takes to raise the temperature of 1 gram (1 cubic centimeter) of water 1°C (usually from 14.5 to 15.5°C).

Quantities of energy can also be expressed in terms of work done. In such cases, the unit of measurement is the **joule.** One calorie of heat energy is equal to 4.186 joules (abbreviated J).

Stop and Ask Yourself

1. How would you define metabolism?
2. What is energy? What is work?
3. What is the difference between kinetic and potential energy?
4. What defines a primary producer? A secondary consumer?

The Laws of Energy Transformations

Energy transformations are governed by two laws of thermodynamics. The **first law of thermodynamics,** sometimes called the law of *conservation of energy,* states that energy can neither be created nor destroyed, only transformed. What this means is that energy can change from one form to another (e.g., electrical energy passes through a hot plate to produce heat energy), or transformed from potential to kinetic energy (a squirrel eats a nut and then uses this energy to climb a tree), but it can never be lost or created. Thus, the total amount of energy in the universe must always remain constant.

The **second law of thermodynamics** states that all objects in the universe tend to become more disordered, and that the total amount of disorder in the universe is continually increasing. The measure of this degree of disorganization is called **entropy.** Consider a simple illustration. When natural gas is burned in a stove, the potential chemical energy stored in the bonds of the gas molecules is converted to light (the blue flame) and heat. Some of the heat energy can be used to boil water on the stove, and some is dissipated into the kitchen where it is no longer available to do work. This unusable energy represents increased entropy.

Activation Energy

Most chemical reactions require an input of energy to start (e.g., a match is lit, and the heat energy is used to start wood burning in a fireplace). At the level of chemical activity, it is necessary to break existing chemical bonds before it is possible to form new bonds and input energy is required to do so. In thermodynamics, this input energy is called *activation energy* (figure 4.2).

A reaction with a net release of energy is one in which the reactant contains more energy than the products. In other words, the amount of this excess energy (called ''free energy'') released into the environment is greater than the activation energy required to initiate the reaction. These reactions occur spontaneously and are called **exergonic** (L. *ex,* out + Gr. *ergon,* work) (figure 4.3a). In contrast, a chemical reaction in which the product contains more energy than the reactants, requires a greater input of energy from the environment than is released (figure 4.3b). Be-

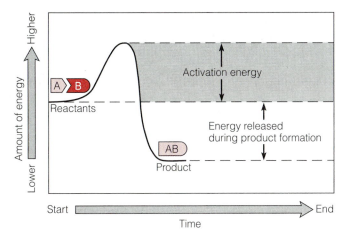

Figure 4.2 An energy diagram. Before a chemical reaction occurs, energy must be supplied to destabilize existing chemical bonds. This energy is called activation energy.

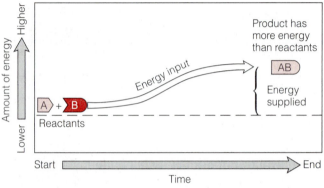

(a) Exergonic

(b) Endergonic

Figure 4.3 Exergonic and endergonic reactions. (a) In an exergonic reaction, the products have less energy than the reactant, and excess energy is released to the environment. (b) In an endergonic reaction, the product of the reaction contains more energy than the reactants, so energy input is needed for the reaction to occur.

cause these reactions do not occur spontaneously, they are called **endergonic** (Gr. *endon,* within + *ergon,* work).

The amount of reactant substance(s) converted to product substance(s) in a given period of time is called the *reaction rate.* The reaction rate of an exergonic reaction does not depend on how much energy the reaction releases but instead on the amount of activation energy that is re-

quired for the reaction to begin. The larger the activation energy of a chemical reaction, the more slowly it will occur, because fewer molecules succeed in overcoming the initial energy hurdle. However, activation energies are not fixed. For example, when stress is put on certain chemical bonds, they may break more easily. Affecting a chemical bond in a way that lowers the activation energy of a reaction is an example of **catalysis.** Any substance that peforms catalysis is called a catalyst. A **catalyst** (Gr. *kata,* down + *lysis,* a loosening) can be defined as a substance that accelerates the rate of a chemical reaction (or allows it to proceed at a lower environmental temperature) by decreasing the activation energy, without itself being used up in the reaction (figure 4.4). In cells, the catalysts are almost always enzymes.

Stop and Ask Yourself

5. What is the difference between an exergonic and endergonic chemical reaction?
6. What is meant by activation energy?
7. What are the first and second laws of thermodynamics?

Enzymes: Biological Catalysts

Enzymes are proteins having enormous catalytic power; they greatly enhance the rate at which specific chemical reactions take place. The metabolism of a living organism is organized and controlled specifically at the points where catalysis takes place. Therefore, one of the most important foundations in all of biology is that life is a series of chemical processes regulated by enzymes. An **enzyme** (Gr. *enzymos,* leavened) is a biological catalyst that is capable of accelerating a specific chemical reaction by lowering the required activation energy, but is unaltered itself in the process. In other words, the same reaction would have occured to the same degree in the absence of the catalyst, but it would have progressed at a much slower rate. Be-

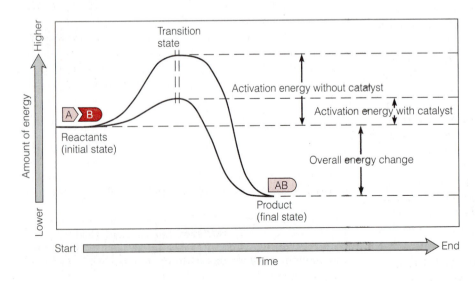

Figure 4.4 Catalysts lower the amount of activation energy required to initiate a chemical reaction. As a result, the reaction moves to completion much more quickly.

Substrates   Products

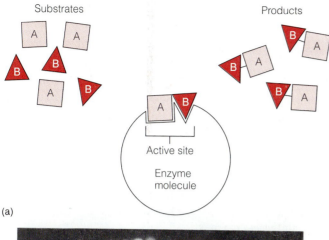

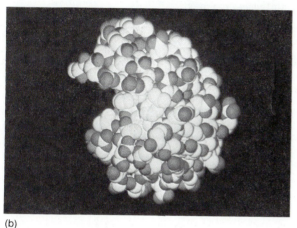

(a)

Figure 4.5 (a) Representation of an enzyme molecule with two substrate molecules fitted into its active site. The enzyme allows the substrates to react with each other more easily. (b) The active site of the enzyme can be seen as a crevice on the surface of the protein. (c) On entering the active site, the substrate induces a slight change in the shape of the protein that causes the active site to embrace the substrate. This is called induced fit.

(b)

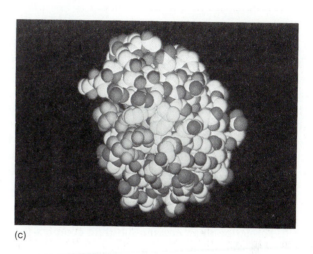

(c)

cause it is unaltered, the enzyme can be used over and over.

An enzyme is extremely selective for the reaction it will catalyze. The reactants of enzymatic reactions are called **substrates.** The preference of an enzyme for specific substrates is crucial to the metabolism of a cell. For example, cellular concentrations of many reactants must be kept at low levels in order to avoid undesirable side reactions. At the same time, concentrations must be kept high enough for the required reactions to occur at a rate that is compatible with life. As presented in the next chapter, metabolism proceeds under these seemingly conflicting conditions because enzymes channel molecules into and through specific chemical pathways (box 4.1).

With the exception of the several digestive enzymes that were the first to be discovered (e.g., pepsin and trypsin), all enzyme names end with the suffix *-ase.* Classes of enzymes are named according to their function. For example, *hydrolases* catalyze hydrolysis reactions (*see chapter 2*); *phosphatases* catalyze the removal of phosphate groups; *synthases* catalyze synthetic reactions; *dehydrogenases* remove hydrogen atoms from their substrates; and *isomerases* rearrange atoms within their substrate molecules to form structural isomers, such as glucose and fructose (*see figure 2.14*). Enzymes that catalyze the same reaction but differ from each other in amino acid composition are called *isoenzymes.*

Enzyme Structure

Enzymes are globular protein molecules that have three-dimensional shapes, with at least one surface region having an area with a crevice or pocket. This crevice occupies only a small portion of the enzyme's surface and is known as its *active site* or *catalytic site* (figure 4.5a). This site is shaped so that a substrate molecule, or several substrate molecules (depending on the reaction), fit into it in a very specific way and is held in place by weak chemical forces, such as hydrogen bonds. Binding of the substrate to the enzyme causes a change in the enzyme's shape (figure 4.5b–c). This phenomenon of change in an enzymes's shape following binding of substrate is called **induced fit.** [4] The induced fit is like a clasping handshake. The embrace of the substrate by the active site brings chemical groups of the active site into positions that enhance their ability to work on the substrate and catalyze the chemical reaction. When the reaction is completed, the product of the catalyzed reaction is released, and the enzyme resumes its initial conformation (shape), ready to catalyze another chemical reaction.

Enzyme Function

When a substrate molecule has bound to the active site of an enzyme, an **enzyme-substrate complex (ES)** is

Box 4.1

Luciferase: A Light Emitting Enzyme

Most individuals have observed the greenish-yellow glow of a firefly on a warm summer evening. This glow is produced by a reaction catalyzed by the enzyme *luciferase*. Luciferase catalyzes the breakdown of the protein luciferin. When luciferin reacts with luciferase, most of the energy is released as light rather than heat as follows:

$$\text{Luciferin} \xrightarrow[\text{Luciferase}]{\text{ATP}} \text{Light} + \text{Heat} + \text{Oxyluciferin}$$

It is this light that is observed coming from the firefly. Equally fascinating are some deep sea fish (*Photoblepharon* spp.) that have specific organs in which they maintain populations of luminescent bacteria (box figure 4.1). The bacteria use the enzyme luciferase to continuously emit light that aids the fish in movement within its dimly lit aquatic environment. In fact, the fish can even control the light from the bioluminescent bacteria by opening and closing body folds on a structure

known as the bioluminescent organ. By this control mechanism, the light can also be used in mating rituals, recognition, promoting schooling, or in the attraction of prey organisms.

Box Figure 4.1 The glowing of this fish is fueled by ATP in bioluminescent bacteria. The bacteria emit a blue-green light because of the activity of the enzyme luciferase.

---

formed. Formation of an enzyme-substrate complex is the essential first step in enzyme catalysis and can be summarized as follows:

Enzyme + Substrate ⇌ Enzyme-Substrate Complex ⇌ Products + Enzyme

Once the ES has formed, amino-acid side groups of the enzyme are placed against certain bonds of the substrate. The amino acid side groups of the enzyme react with the substrate by stressing or distorting their bonds, which lowers the activation energy needed to break the bonds. The bonds are thus broken releasing the substrates which are now free to react, producing the final products and releasing the enzyme. This entire process can take place thousands of times a second. For example, the enzyme carbonic anhydrase, found in red blood cells, catalyzes the joining of carbon dioxide and water to form carbonic acid as follows:

$$CO_2 + H_2O \underset{}{\overset{\text{carbonic}\ \text{anydrase}}{\rightleftharpoons}} H_2CO_3$$

In the presence of only one molecule of carbonic anhydrase, this reaction can occur more than 600,000 times a second. The same number of reactions would take many minutes without a catalyst. However, carbonic anhydrase is exceptionally fast; most enzyme molecules catalyze between 1 and 10,000 reactions a second. This ability of an enzyme to increase the rate of a chemical reaction can be measured by the enzyme's *turnover number*. The turnover

[5]

Table 4.1	Several Examples of Enzymes, the Reactions They Catalyze, and Their Turnover Numbers	
Enzyme	**Reaction Catalyzed**	**Turnover Number**
Amylase	Starch + $H_2O \rightarrow$ maltose	1×10^5
Catalase	$2H_2O_2 \rightarrow 2H_2O + O_2$	600×10^6
Carbonic anhydrase	$H_2O + CO_2 \rightarrow H_2CO_3$	3.6×10^6
Lactate dehydrogenase	Lactic acid $\rightarrow$ pyruvic acid + H_2	6.0×10^4
Ribonuclease	RNA + $H_2O \rightarrow$ ribonucleotides	600

number is the number of substrate molecules that one molecule of the enzyme can convert into product per minute (table 4.1).

Factors Affecting Enzyme Activity

Any condition that alters the three-dimensional shape of an enzyme will also affect its activity. Two such factors that affect enzyme activity are temperature and pH.

As presented in chapter 2, the shape of a protein is determined largely by its hydrogen bonds (*see figure 2.21*). Hydrogen bonds are easily disrupted by temperature changes. For example, most higher animals, such as birds and mammals, have enzymes that function best within a

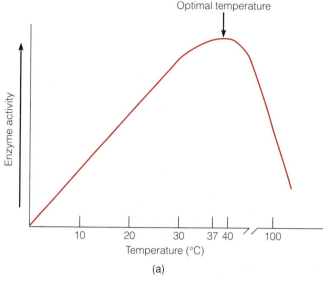

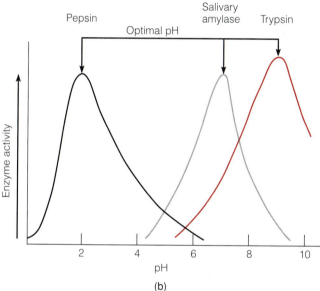

Figure 4.6 How the environment affects enzymes. (a) Effect of temperature on enzyme activity. (b) The effect of pH on three different enzymes.

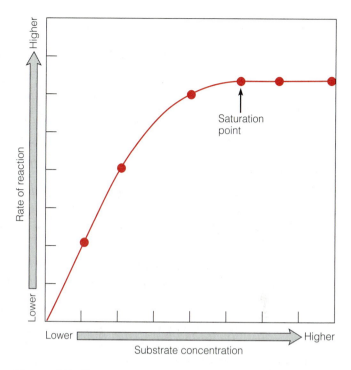

Figure 4.7 Enzyme saturation. If different concentrations of substrate are mixed with a fixed concentration of enzyme, the rate of the reaction increases with substrate concentration until a saturation point is reached. At this point, all active sites of the enzyme are continuously engaged.

relatively narrow temperature range between 35 and 40°C (figure 4.6a). Below 35°C, the bonds that determine protein shape are not flexible enough to permit the shape change that is necessary for substrate to fit into a reactive site. Above 40°C, the bonds are too weak to hold the protein in proper position and maintain its shape. When proper shape is lost, the enzyme is in essence destroyed; this loss of shape is called **denaturation.**

Most enzymes also have a pH optimum, usually between 6 and 8 (figure 4.6b). For example, when the pH is too low, the H⁺ ions combine with the R groups of the enzyme's amino acids, reducing their ability to bind with substrates. Acid environments can also denature enzymes if they are not adapted to such conditions. There are, however, some enzymes that function at a low pH. For example, pepsin is the enzyme found in the stomach of mammals and has an

optimal pH of approximately 2. The reason pepsin is able to function at such a low pH is that it has an amino acid sequence that maintains its ionic and hydrogen bonds even in the presence of large numbers of hydrogen ions (low pH). Conversely, trypsin is active in the more basic medium (pH 9) found in the small intestine of mammals. Overall, the pH optimum of an enzyme reflects the pH of the body fluid in which the enzyme is found.

Regulation of Enzyme Activity

The rate at which an enzyme converts substrates into products also depends on the enzyme and substrate concentrations. When the enzyme concentration is at a given level, the rate of product formation will increase as the substrate concentration increases. Eventually, however, a point will be reached where additional increases in substrate concentration do not result in increases in the reaction rate. When the relationship between substrate concentration and reaction rate reaches a plateau, the enzyme is said to be *saturated* (figure 4.7). For example, if one thinks of enzymes as workers and substrates as jobs, there is 100% employment when the enzyme is saturated; further availability of jobs (substrate) cannot further increase employment (conversion of substrate to product).

In addition to temperature, pH, and substrate concentration, certain chemicals can affect reaction rate by binding to an enzyme and causing it to change its three-dimensional shape. Such a change, however, is not necessarily detrimental. Using these specific chemicals, a cell can reg-

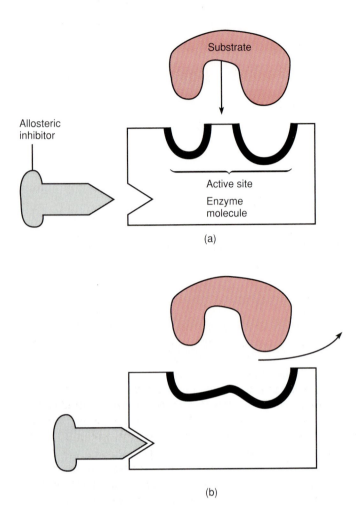

(a)

(b)

Figure 4.8 Allosteric enzyme. An enzyme has an active site(s) to which the substrate binds, and another site for the allosteric inhibitor. (a) When the inhibitor is not bound to the enzyme, the susbtrate can bind to the enzyme, and catalysis can take place. (b) When the allosteric inhibitor binds to its site, it induces a change in the three-dimensional shape of the enzyme, distorting the active site so that the substrate cannot bind.

ulate which enzymes are active and which are not at any given moment. Those chemicals that decrease an enzyme's reaction rate are called **inhibitors**; those that speed up the reaction rate are called **activators.**

When an activator or inhibitor binds to an enzyme and changes its shape, this is termed an **allosteric** (Gr. *allos,* other + *steros,* shape) **change** (figure 4.8). Enzymes usually have special binding sites (allosteric sites) for the activator and inhibitor molecules that are different from their active sites. For example, in a metabolic pathway (a series of chemical reactions), the enzyme catalyzing the third step might have an inhibitor binding site to which the molecule produced in the last step in the biochemical pathway binds. As the number of these molecules builds up, they begin to bind to the third enzyme, thus inhibiting the activity of the enzyme. After binding, the enzyme is shut down, and the biochemical pathway turned off. This **end-product** or **feedback inhibition** is just one of the many ways cells are self-regulating entities (figure 4.9) and maintain homeostasis.

In another type of enzyme control, called **competitive inhibition,** the active site itself is occupied by a molecule other than the normal substrate, preventing binding of the substrate (figure 4.10). At very high concentrations, however, the normal substrate can compete successfully for the active site(s). Cells normally do not regulate metabolism by means of competitive inhibition, although the mechanism is applied extensively in the design of drugs to treat various diseases.

[6]

Stop and Ask Yourself

8. What is an enzyme? What is a substrate?
9. How are enzymes named?
10. What is the structure of an enzyme?
11. How do enzymes function?
12. How do temperature, pH, and other factors affect enzyme function?

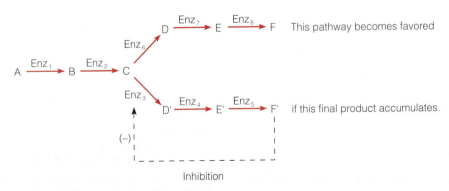

Figure 4.9 End-product inhibition in a branched metabolic pathway. Inhibition is shown by the dotted arrow and the negative sign.

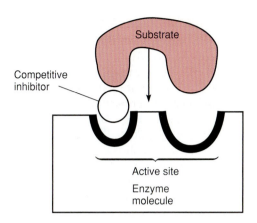

Figure 4.10 Competitive inhibition. A competitive inhibitor is shaped somewhat like the enzyme's normal substrate. It thus competes for binding in the active site.

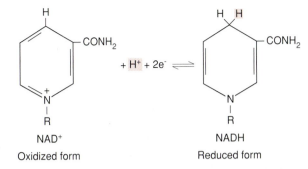

Figure 4.12 Nicotinamide adenine dinucleotide (NAD), oxidized (NAD$^+$) and reduced (NADH) forms. When the NAD is reduced, one electron is attached to the nitrogen atom at position 1. The second electron attaches to a carbon atom at position 4 and is accompanied by a hydrogen ion.

Cofactors and Coenzymes

Cofactors are metal ions such as Ca^{2+}, Mg^{2+}, Mn^{2+}, Cu^{2+}, and Zn^{2+}. Many enzymes must use these metal ions to change a nonfunctioning active site to a functioning one. In these enzymes, the attachment of a cofactor causes a shape change in the protein that allows it to combine with substrate (figure 4.11a). The cofactors of other enzymes participate in the temporary bonds between the enzyme and its substrate when the enzyme-substrate complex is formed (figure 4.11b).

Coenzymes are nonprotein, organic molecules that participate in enzyme-catalyzed reactions, often by transporting electrons, in the form of hydrogen atoms, from one enzyme to another. Many vitamins function as coenzymes or are used to make coenzymes (e.g., niacin and riboflavin). Just as a taxi transports people around a city, so coenzymes transport energy, in the form of hydrogen atoms, from one enzyme to another.

One of the most important coenzymes in the cell is the hydrogen acceptor **nicotinamide adenine dinucleotide (NAD$^+$)**. When NAD$^+$ acquires a hydrogen atom from an enzyme, it becomes reduced to NADH. (figure 4.12). The electron of the hydrogen atom contains energy that is then carried by the NADH molecule. For example, when various foods are oxidized in the cell, the cell strips electrons from the food molecules and transfers them to NAD$^+$, which is reduced to NADH. As more and more NADH molecules accumulate in the cell, the potential energy of the cell increases.

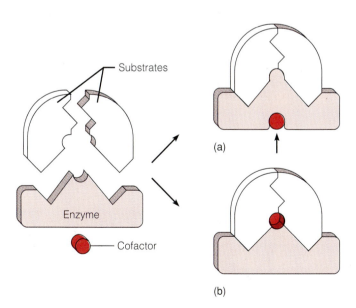

Figure 4.11 Cofactors. (a) The cofactor changes the conformation of the active site, permitting a better fit between enzyme and its substrate. (b) The cofactor participates in the temporary bonding between the active site of the enzyme and the substrate.

ATP: The Cell's Energy Currency

The major energy currency of all cells is an organic molecule called **adenosine triphosphate (ATP)**. Because ATP plays a central role as the energy currency in *all* animals, it must have evolved early in the history of life. [7]

The ability of ATP to store and release energy stems from the molecule's structure. Each ATP molecule (figure 4.13) is composed of three subunits: (1) *adenine,* an organic molecule composed of two carbon-nitrogen rings ; (2) *ribose,* a five-carbon sugar; and (3) three *phosphate* groups linked in a linear chain. The covalent bond connecting these phosphates is indicated by the "tilde" symbol (~) and represents a *high energy bond*. However, the energy is not localized in the bond itself; it is a property of the entire molecule and is simply released as the phosphate bond is broken. These bonds have a low activation energy and are broken easily. When one bond is broken, about 7.3

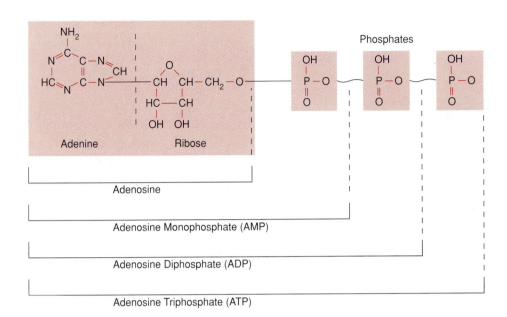

Phosphates

Adenosine

Adenosine Monophosphate (AMP)

Adenosine Diphosphate (ADP)

Adenosine Triphosphate (ATP)

Figure 4.13 The ATP molecule is the primary energy currency of the cell. It consists of an adenine portion, the sugar ribose, and three phosphates. The wavy lines connecting the last two phosphates represent high energy chemical bonds from which energy can be quickly released.

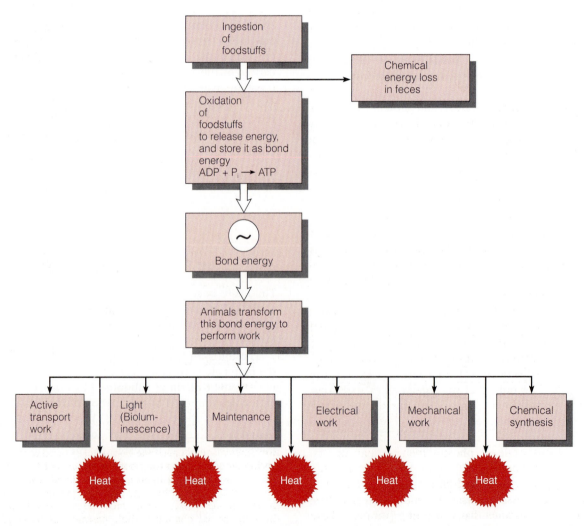

Figure 4.14 Some energy utilization pathways in animals. Notice that heat is lost in every transformation.

kilocalories (7,300 calories) are released per mole of ATP as follows:

$$ATP + H_2O \rightarrow ADP + P_i + energy \ (7.3 \ kcal/mole)$$

The energy from ATP is sufficient to drive most of the endergonic reactions of the cell. In a typical energy reaction, only the outermost of the two high-energy bonds is broken (hydrolyzed). When this happens, ATP becomes **ADP (adenosine diphosphate)**. In some cases, ADP can be hydrolyzed to **AMP (adenosine monophosphate)** as follows:

$$ADP + H_2O \rightarrow AMP + P_i + energy \ (7.3 \ kcal/mole)$$

Cells contain a reservoir of ADP and phosphate (P_i). As long as a cell is living, ATP is constantly being cleaved into ADP plus phosphate to drive the many energy-requiring processes of the cell to enable the animal to perform biological work (figure 4.14). However, ATP cannot be stored for long. Once formed, ATP lasts only a few seconds before it is used up to perform biological work. Thus, cells are constantly recycling their ADP. Using the energy derived from foodstuffs and from stored fats and starches, ADP and phosphate are recombined to form ATP, with 7.3 kilocalories of energy per mole contributed to each newly formed high energy phosphate bond.

How Cells Make Energy: An Overview

Animals make ATP in two ways: substrate-level phosphorylation and chemiosmosis. As noted above, the formation of ATP from ADP and phosphate requires the input of energy (the same 7.3 kcal that is released when ATP is hydrolyzed). Thus, to get the energy required for the synthesis of ATP from ADP and phosphate, the reaction is coupled with an exergonic reaction (figure 4.15a). A phosphate group is transferred by an enzyme to ADP from a substrate with a phosphate bond even more easily broken than that of ATP (i.e., the energy from the exergonic reaction is greater than the input of energy necessary to drive the synthesis of ATP). The generation of ATP by coupling strongly exergonic reactions with the synthesis of ATP from ADP and phosphate is called **substrate-level phosphorylation.** Substrate-level phosphorylation probably evolved very early in the history of living organisms because (1) the initial use of carbohydrates by organisms as an energy source leads to substrate-level phosphorylation; (2) the mechanism for substrate-level phosphorylation is present in most living animal cells; and (3) substrate-level phosphorylation is one of the most fundamental of all ATP-generating reactions.

Although substrate-level phosphorylation may be the oldest method of generating ATP, far more ATP is generated using another process called **chemiosmosis**. As noted in chapter 3, animals possess transmembrane channels in their mitochondrial membranes that can pump protons. These proton pumps use a flow of electrons to induce a shape

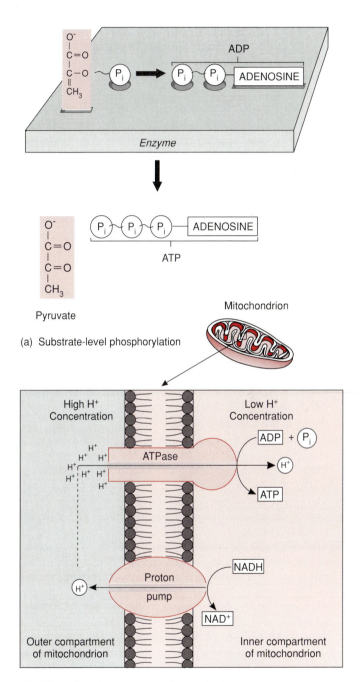

(a) Substrate-level phosphorylation

(b) Chemiosmosis

Figure 4.15 ATP generation. (a) Substrate-level phosphorylation. In this example, a phosphate group is transferred from an enzyme-substrate complex to ADP to form ATP. This reaction takes place because the high-energy bond of the substrate has higher energy than the phosphate bonds of ADP. (b) Chemiosmotic synthesis via hydrogen ions moved from one side of a membrane to the other, generating a proton gradient across the membrane. When protons move back across the membrane through special channels, their passage generates ATP from ADP and P_1.

change in the protein, which in turn causes protons to move out of the inner compartment of a mitochondrion. As the proton (H^+) concentration in the outer compartment of the mitochondrion becomes greater than the inside, the outer protons are driven through the membrane by diffusion. Within the mitochondrion, they induce the formation of ATP from ADP, phosphate, and the enzyme ATPase.

The electrons that drive the electron transport system involved in chemiosmosis are obtained from chemical bonds of food molecules in all organisms and from photosynthesis in plants. This electron-stripping process is called **cellular respiration,** or **oxidative respiration** because molecular oxygen is needed. Basically, oxidative respiration is the oxidation of food molecules to obtain energy and is the subject of the next chapter.

As a conclusion to this chapter and a bridge to the next, biologists have been questioning *what* biological molecules do for approximately 100 years. This chapter presents some facts on *how* these molecules do it via enzymes. En-zymes have been refined over 3 billion years of evolution and they have a great deal to teach us. Happily, there is some hope that the number of lessons is merely finite, for we already see examples of enzyme molecules, unrelated by evolution, but with almost identically arranged working parts. Hopefully, the next chapter will help clarify what enzymes are "trying to tell us" about the universality of energy harvesting pathways and animal life.

Stop and Ask Yourself

13. What three component parts make up the ATP molecule?
14. Why doesn't ATP accumulate in a cell?
15. What is substrate-level phosphorylation? Chemiosmosis?
16. How do cofactors function? Coenzymes?

■ **Summary**

1. Energy is the capacity to do work. Kinetic energy is the energy that is actively engaged in doing work, whereas potential energy is stored energy. Animals convert potential energy into kinetic energy to perform biological work, which is the transfer of energy.

2. The first law of thermodynamics states that the amount of energy in the universe is fixed and that energy cannot be created or destroyed; however, energy can be converted from one form to another.

3. The second law of thermodynamics states that the randomness or disorder in the universe is increasing. This process is called entropy.

4. The speed of a chemical reaction depends on the amount of activation energy required to break existing bonds. Catalysis is the chemical process of increasing the reaction rate by lowering the amount of activation energy required to initiate the reaction. Enzymes are the biological catalysts of cells.

5. Enzymes have a specific three-dimensional shape, with one or more reactive sites that bind substrates. Cells contain many different enzymes, each of which catalyzes a different chemical reaction.

6. Factors such as temperature, pH, substrate concentration, cofactors, and coenzymes can affect the reaction rate of an enzyme.

7. Cells focus their energy resources on the manufacture of ATP from ADP and phosphate, a process that requires the cell to supply 7.3 kilocalories per mole of energy, which it obtains from electrons stripped from foodstuffs. Mitochondrial membranes are the main site of ATP production. Cells use ATP to drive endergonic reactions and accomplish biological work.

8. The harvesting of chemical energy takes place via either substrate-level phosphorylation or chemiosmosis. In substrate-level phosphorylation, a phosphate group is transferred from an enzyme-substrate complex to ADP by coupling an endergonic reaction to a highly exergonic one. In chemiosmosis, there is a transport of electrons to an electron transport system in the mitochondrial membrane. In the process, a gradient of H^+ ions (protons) is established that powers the synthesis of ATP.

■ **Key Terms**

activators (p. 64)
adenosine diphosphate (ADP) (p. 67)
adenosine monophosphate (AMP) (p. 67)
adenosine triphosphate (ATP) (p. 65)
allosteric change (p. 64)
anabolism (p. 58)
calorie (p. 59)
catabolism (p. 58)
catalysis (p. 60)
catalyst (p. 60)
cellular respiration (p. 68)
chemiosmosis (p. 67)
coenzymes (p. 65)
cofactors (p. 65)

competitive inhibition (p. 64)
decomposers (p. 58)
denaturation (p. 63)
end product inhibition (p. 64)
endergonic (p. 60)
energy (p. 59)
entropy (p. 59)
enzyme (p. 60)
enzyme-substrate complex (p. 61)
exergonic (p. 59)
first law of thermodynamics (p. 59)
food chain (p. 58)
induced fit (p. 61)

■ **Critical Thinking Questions**

1. Why are the two laws of thermodynamics called laws, whereas the central organizing concept of biology, evolution, is called a theory?

2. Living organisms are constantly transforming energy via many different mechanisms. Give several examples in animals where one form of energy is transformed into another form.

3. As you read this statement, you are losing energy. At the risk of losing even more, describe what this means with respect to entropy. Based on the laws of thermodynamics, is answering this question worth it? Explain.

4. Why is ATP the major energy-carrying molecule in all living organisms?

5. Many steps in metabolism require enzymes. In turn, many enzymes require either cofactors or coenzymes to perform their function. Give several examples of cofactors and speculate why these usually are not permanently a part of the enzyme molecule.

6. Considering chemiosmosis, how does the energy-coupling function of the mitochondrial membrane compare with the transport function of the plasma membrane?

7. What is the relationship between the biological information stored in nucleic acids and proteins, and entropy? Does it require more energy to construct these information-bearing molecules than other simpler molecules? Explain.

■ **Suggested Readings**

Books

Atkins, P. 1984. *The Second Law.* New York: Freeman.

Baker, J. J. W., and Allen, G. E. 1981. *Matter, Energy, and Life.* 4th ed. Reading, Mass: Addison-Wesley.

Blum, H. F. 1962. *Time's Arrow and Evolution.* New York: Harper and Row.

Cristensen, H. N., and Cellarius, R. A. 1972. *Introduction to Bioenergetics: Thermodynamics for the Biologist.* Philadelphia: W.B. Saunders.

Fersht, A. 1985. *Enzyme Structure and Mechanism.* 2nd ed. New York: Freeman.

Articles

Hinkle, P. C., and McCarthy, R. E. How cells make ATP. *Scientific American* March, 1978.

Koshland, D. E. Protein shape and biological control. *Scientific American* October, 1973.

Kraut, J. 1988. How do enzymes work? *Science* 242: 533–539.

Leninger, A. L. How cells transform energy. *Scientific American* September, 1961.

Phillips, D. C. The three-dimensional structure of an enzyme molecule. *Scientific American* November, 1966.

5

Harvesting Energy Stored in Nutrients: Energy-Releasing Pathways

Concepts

1. All living animals harvest the energy from nutrients to fuel their metabolism with energy from ATP. These processes include glycolysis, fermentation, and oxidative respiration.
2. Glycolysis is a specific metabolic pathway that does not require oxygen and involves nine reactions that harvest chemical energy by rearranging the chemical bonds of glucose to form molecules of pyruvate and generate two molecules of ATP.
3. Animals that live in an anaerobic environment (low in oxygen) utilize fermentation to dispose of the electrons and associated hydrogen produced in glycolysis. Molecules other than oxygen (e.g., NAD^+) serve as electron acceptors.
4. In oxidative respiration, the citric acid cycle completes the breakdown of pyruvate to carbon dioxide, water, and H^+ (protons). In the mitochondrion, electrons from the H^+ are channeled to the inner mitochondrial membrane to drive proton pumps and generate ATP.
5. Lipids and proteins can also serve as nutrients for animals. These molecules are broken down to intermediates that are fed into the citric acid cycle.

Have You Ever Wondered:

[1] how the first organisms obtained energy in the absence of oxygen?
[2] why an inefficient way of harvesting energy has persisted through eons of time?
[3] how glycolysis is a metabolic memory of an animal's past?
[4] what makes bread rise?
[5] why your muscles get tired and feel as though they are "made of lead?"
[6] what the focal point of all energy generating metabolism is?
[7] why animals store excess energy in the form of fat?
[8] how cells are thrifty, expedient, and responsive in their metabolism?

These and other useful questions will be answered in this chapter.

This chapter contains underlined evolutionary concepts.

The last chapter introduced the fact that animals require a constant supply of energy to perform biological work. This energy is usually supplied by the energy rich molecule ATP (*see figure 4.13*). All animals can generate ATP by breaking down organic nutrients (carbohydrates, lipids, and proteins). The energy released is used to join ADP and phosphate (P_i) to form ATP.

In animals, the breakdown of organic nutrients, such as glucose, begins in a step-by-step series of chemical reactions called *glycolysis*. The end product of glycolysis is then further broken down either in the presence of molecular oxygen (**aerobic**)—a process called *oxidative respiration*, or in the absence of molecular oxygen (**anaerobic**)—a process called *fermentation*.

Figure 5.1 provides an overview of the entire process of catabolic metabolism. The places where ATP is generated are indicated. Note that glycolysis and fermentation (anaerobic processes) take place in the cytosol of a cell, and oxidative respiration (an aerobic process) takes place in the mitochondria. Oxidative respiration is also called *aerobic respiration* because it is an oxygen-requiring process that occurs in the cell. The terms can be used interchangeably; however, for consistency, oxidative respiration will be used in this textbook.

The reason glycolysis and fermentation occur in the cytosol is that during eukaryotic cell evolution, the enzymes that catalyze each reaction remained dissolved in the cytosol and did not localize in membranous organelles. This [1] implies an origin prior to the evolution of complex organelles. These reactions are also older in an evolutionary sense than oxidative respiration, because the reactions of the former could have occurred in the primitive environment of earth before the atmosphere contained free oxygen.

In eukaryotic cells, oxidative respiration takes place in the mitochondria, where specific enzymes are localized. In fact, the architecture of this organelle (*see figure 3.20*) is closely associated with the important mechanisms of oxidative respiration and provides an excellent example of the relationship between structure and function as will be presented later in the chapter.

This chapter presents some of the fundamental biological principles that underlie the above chemical reactions.

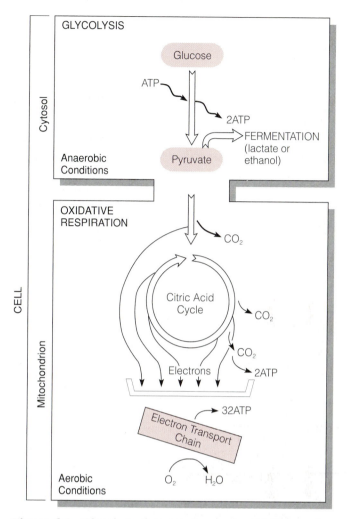

Figure 5.1 Glycolysis, fermentation, and oxidative respiration. Glycolysis begins in the cytosol and causes glucose to be broken down to pyruvate. The first step in oxidative respiration is the conversion of pyruvate to a high-energy intermediate. In the absence of oxygen, pyruvate will undergo fermentation instead of entering the citric acid cycle. The citric acid cycle removes electrons and passes them to the electron transport chain via carrier molecules. Both of these processes take place in the mitochondrion. Molecular oxygen acts as a final electron acceptor.

mentation and oxidative respiration. During glycolysis, the breakdown of the six-carbon glucose molecule to two molecules of pyruvate takes place in nine different steps (figure 5.2) as follows:

Step 1 The initiation of glycolysis requires a high-energy form of glucose. Thus, a molecule of ATP is cleaved, transferring its terminal phosphate to carbon six of the glucose molecule, forming the *activated* molecule, glucose–6–phosphate (an excellent example of an endergonic reaction). Note that a specific enzyme catalyzes this and each of the following steps.

Step 2 In step 2, the glucose–6–phosphate changes its structure and is converted to fructose–6–phosphate.

Glycolysis: The First Phase of Nutrient Metabolism

Glycolysis (Gr. *glykys*, sweet + *lyein*, to loosen) is the initial sequence of chemical reactions used by almost all cells to break the six-carbon glucose molecule (*see figure 2.14*) into two molecules of a three-carbon compound called *pyruvate* with the production of some ATP.

Splitting Glucose: The Individual Steps of Glycolysis

As noted in figure 5.1, glycolysis is a prelude for both fer-

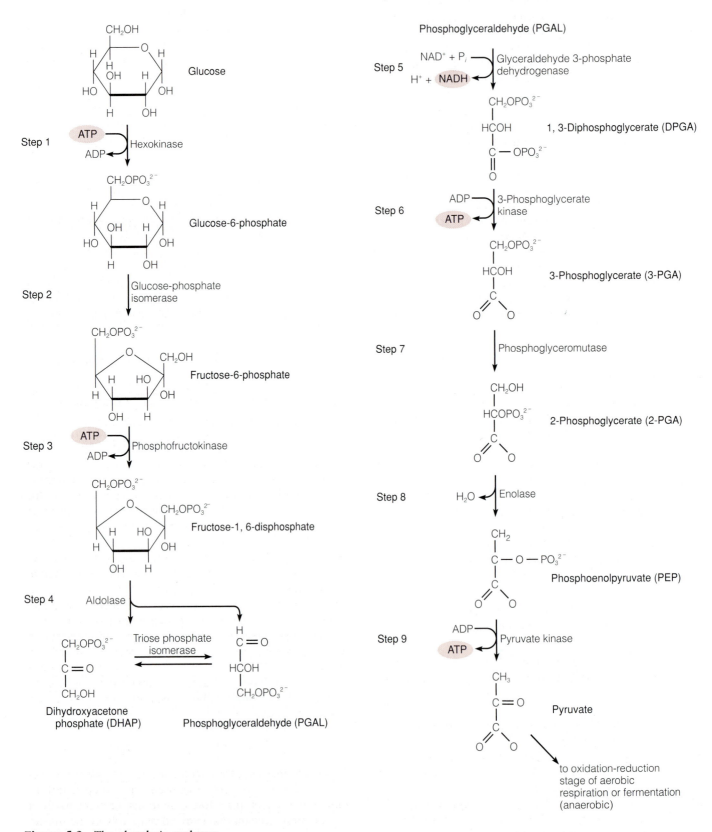

Figure 5.2 The glycolytic pathway.

Step 3   During step 3, another molecule of ATP is cleaved and its terminal phosphate is transferred to carbon number 1. This reaction produces fructose–1, 6–diphosphate. At this point, the original glucose molecule has added two phosphates, with the expenditure of 2 ATP molecules. The energy is now stored in the sugar-phosphate molecule.

Step 4   Fructose–1, 6–diphosphate is now split into two similar three-carbon molecules. One is phosphoglyceraldehyde (PGAL), and the other is its isomer, dihydroxyacetone phosphate (DHAP), which is converted immediately into another molecule of PGAL. From this point on, each step in the pathway must take place twice—once for each PGAL derived from the original glucose.

Step 5   In step 5, PGAL is both oxidized and phosphorylated. First, the aldehyde group (*see figure 2.11*) is oxidized, and NAD$^+$ accepts two electrons in the form of hydrogen atoms and in the process, becomes reduced to NADH, releasing a proton (H$^+$) (*see figure 4.12b*). At the same time, the oxidized PGAL reacts with phosphate to form the high-energy molecule 1, 3–diphosphoglycerate (DPGA). A great deal of energy is released during this exergonic reaction, which can be expressed as follows:

$$PGAL + NAD^+ + HO\text{-}P \rightarrow DPGA + NADH + energy$$
dissipated as heat

It should be noted that the phosphate needed to generate DPGA in step 5 comes from inorganic phosphate (P$_i$) dissolved in the cell's cytosol, and not from ATP.

Step 6   In step 6, both 3–phosphoglycerate (3–PGA) and a molecule of ATP are formed as one phosphate group of the DPGA phosphorylates an ADP via substrate-level phosphorylation (*see figure 4.15a*). This reaction is possible because DPGA has a higher energy content than ATP. Neither electron transport nor molecular oxygen are required. The overall balanced equation for steps 1 through 6 is:

$$glucose + 2ATP + 2HO\text{-}P + 2NAD^+ \rightarrow 2(3\text{–}PGA)$$
$$+ 2ATP + 2NADH + heat\ energy$$

(Remember that for each glucose molecule, two molecules of PGAL are formed and each must pass through steps 5 to 9; thus, the "2s" in the above equation.)

Step 7   Step 7 involves the conversion of 3–PGA into 2–phosphoglycerate (2–PGA).

Step 8   In step 8, 2–PGA is converted to phosphoenolpyruvate (PEP). PEP, like DPGA, is a high-energy phosphate compound.

Step 9   In the last step in glycolysis, the high-energy PEP is used to convert ADP to ATP via substrate-level phosphorylation with the end product, pyruvate

being formed. (Pyruvate is the ionized form of pyruvic acid.) This last reaction can be expressed as:

$$2PEP + 2ADP \rightarrow 2pyruvate + 2ATP$$

In summary, for each molecule of glucose that enters the glycolytic pathway, four molecules of ATP are formed. However, because two ATP molecules are used in steps 1 and 3, the net energy yield from glycolysis is *only* two ATP molecules. Although far from an efficient pathway, with respect to the amount of energy released, keep in mind that for hundreds of millions of years during the anaerobic first stages of life on the planet earth, glycolysis was the only way most organisms could harvest energy and generate ATP molecules.

Stop and Ask Yourself

1. What is the difference between aerobic and anaerobic respiration?
2. Where does glycolysis take place in a cell? Where does oxidative respiration take place?
3. What is the net energy yield in ATP molecules from glycolysis?
4. What is the evolutionary significance of glycolysis?

Evolutionary Perspective on Glycolysis

If glycolysis is such an inefficient way of harvesting energy, why has it persisted through so long an evolutionary period? [2] One reason might be that evolution is a slow, incremental process whereby change occurs based on past successes. When glycolysis first occurred, it gave cells possessing it a competitive advantage over those cells that did not utilize glycolysis. The biochemistry of contemporary organisms indicates that only those organisms that were capable of glycolysis survived the early competition of life on the planet earth. Obviously, later improvements in catabolism *built* on this success. During this building process, glycolysis was not discarded, but instead, used as a stepping stone for the evolution of a still more efficient process for harvesting energy. Just as successive layers of paint are added by an artist to a painting in order to obtain a "masterpiece," so evolution added another layer of chemical reactions (oxidative respiration) onto the foundation layer (glycolysis) to produce a very efficient way of harvesting energy. Looking at glycolysis in another way, most forms of animal life (including humans) carry on glycolysis [3] within their cells—a metabolic memory of an animal's evolutionary past.

Fermentation: "Life Without Air"

Fermentation (L. *fermentum*, leaven) is either an evolutionary bypass that some organisms use to keep glycolysis functioning under anaerobic conditions, or more likely, a biochemical remnant that evolved very early in the history of life when the earth's atmosphere contained little or no

---

Box 5.1

Louis Pasteur and Fermentation

Although Theordore Schwann and others had proposed in 1837 that yeast cells were responsible for the conversion of sugars to alcohol, a process they called *alcoholic fermentation,* the leading chemists of the time did not believe that microorganisms were involved. They were convinced that fermentation was due to some kind of chemical instability that degraded the sugars to alcohol. In 1856, an industralist in Lille France, where Louis Pasteur worked, requested his assistance. The industralist had a business that produced ethanol from the fermentation of beet sugars, and recent alcohol yields had declined, and the product had become sour. Pasteur discovered that the fermentation was failing because the yeast responsible for alcohol formation had been replaced by other microorganisms so that lactic acid rather

than ethanol was being produced. Between 1857 and 1860, he published several papers on fermentation. In solving this practical problem, Pasteur demonstrated that fermentations were due to the activities of specific yeasts and bacteria. His success led to a study of wine diseases and the development of pasteurization to preserve wine during storage. Pasteur's studies on fermentation continued for almost 20 years. One of his most important discoveries was that some fermentative microorganisms were anaerobic and could live in the absence of oxygen, whereas others were able to live either aerobically or anaerobically. His book, *Études sur le Vin,* published in 1866, revolutionized wine making, giving it a scientific basis for the first time.

---

oxygen. As with glycolysis, the ubiquity of fermentation is strong evidence for the common descent of organisms from primitive cells in which glycolysis and fermentation were first perfected and still persist.

In fermenation, the hydrogen atoms generated by glycolysis are donated to organic molecules as follows:

Pyruvate + NADH → reduced compound + NAD$^+$

The reduced compound can be an organic acid (e.g., acetic acid, butyric acid, propionic acid, or lactic acid), or an alcohol (ethanol).

The important point to remember about fermentation is that glucose is not completely degraded, so considerable energy still remains in the products (*see figure 5.1*). Beyond the two ATP molecules formed during glycolysis, no more ATP is produced. Fermentation serves only to regenerate NAD$^+$ (the oxidized form of NADH).

Two types of organisms can carry out fermentation: *anaerobic* and *facultatively anaerobic* ones. Anaerobic organisms include certain types of bacteria that survive only in the complete absence of molecular oxygen. Facultative organisms and tissues include certain bacteria and yeasts and cells (such as animal muscle) that can ferment nutrients when oxygen is absent to generate some ATP by providing NAD$^+$ for glycolysis. Facultatively anaerobic organisms and tissues carry out more efficient energy harvesting when oxygen is present; hence the term facultative (not obligatory). There are a number of fermentative pathways. Two of the most common produce (besides NAD$^+$) alcohol and lactic acid (box 5.1).

When yeast cells carry out fermentation, they convert pyruvate to ethanol (ethyl alcohol), with an intermediate compound called acetaldehyde (*see figure 2.12*). Acetaldehyde accepts the hydrogen from NADH, producing NAD$^+$. Yeast cells also remove a terminal CO_2 group from pyruvate

to generate CO_2. Alcoholic fermentation by yeast is very important in the brewing and wine-making industries, as well as in baking, where the carbon dioxide produced by the yeast cells is used to make bread "rise" (figure 5.3a). [4]

When the oxygen supply to skeletal muscle is insufficient (e.g., during vigorous exercise) muscle cells can use another type of fermentation and convert pyruvate to lac-

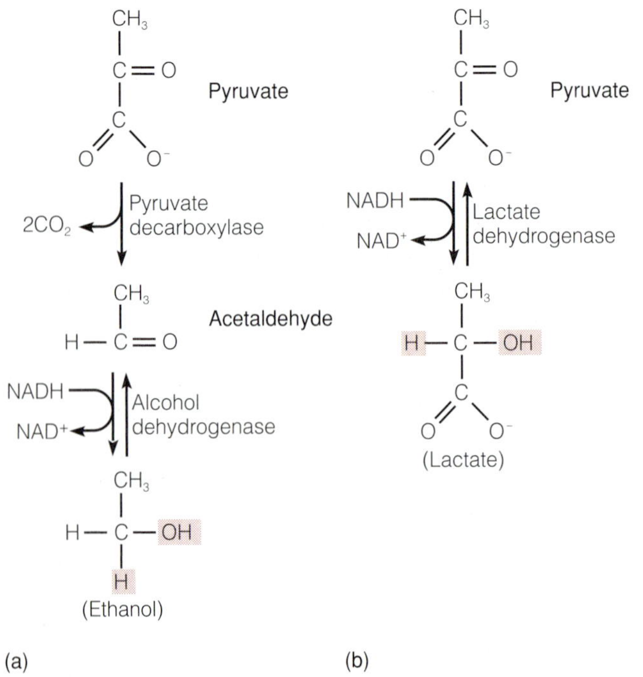

Figure 5.3 Fermentation. The purpose of fermentation is to regenerate NAD$^+$, which is needed to drive step 5 in glycolysis to ultimately obtain ATP. (a) In alcoholic fermentation, the boxes indicate places where hydrogen ions attach when electrons are donated by reduced NADH to generate NAD$^+$. (b) The pathway from pyruvate to lactate.

tate (figure 5.3b). (Lactate is the ionized form of lactic acid that occurs in solution.) The muscle cells use the enzyme lactate dehydrogenase to add the hydrogen of the NADH produced by glycolysis back to the pyruvate. This reaction produces the NAD^+ that allows glycolysis to continue as long as glucose is present. The circulatory system usually removes the lactate from the muscles as fast as it is produced to prevent toxic conditions. However, when lactate is produced faster than it can be removed, muscle contraction slows down. At this point, humans subjectively feel tired, as though their muscles are "made of lead"; muscle cramping may also occur.

[5]

Stop and Ask Yourself

5. Why has glycolysis persisted through time, even though it is an inefficient process?
6. How is fermentation a link to the past?
7. How does fermentation occur?
8. What type of organisms carry out fermentation?

Oxidative Respiration: The Big Energy Harvest

As noted above, the anaerobic generation of ATP through the reactions of glycolysis and fermentation are relatively inefficient. The end product of glycolysis (pyruvate) still contains a great deal of potential bond energy that can be harvested by further oxidation. The evolution of *oxidative respiration* in microorganisms and in the mitochondria of eukaryotic cells became possible only after molecular oxygen had accumulated in the earth's atmosphere as a result of photosynthesis. The addition of an oxygen-requiring stage to the energy-harvesting mechanisms of cells provided them with a more powerful and efficient way of extracting energy from nutrient molecules.

In oxidative respiration, the pyruvate produced by glycolysis is shunted into a metabolic pathway called the *citric acid cycle*; other products go to the electron transport chain. During this aerobic metabolism, molecular oxygen accepts electrons and is reduced to H_2O as follows:

$$\text{Pyruvate} + O_2 \rightarrow CO_2 + H_2O + 34 \text{ ATPs} + \text{heat energy}$$

During the above reaction, 34 molecules of ATP are produced for each molecule of pyruvate consumed.

Like in glycolysis, oxidative respiration is organized into a number of reactions, each catalyzed by a specific enzyme, and organized into what is called the citric acid cycle and the electron transport chain. The **citric acid cycle** is a series of ten reactions in which the pyruvate from glycolysis is oxidized to CO_2. Two electron carriers, nicotinamide adenine dinucleotide (NAD) (*see figure 4.12*) and **flavin adenine dinucleotide** (**FAD**) act as hydrogen acceptors and are reduced to NADH and $FADH_2$. During this phase of the cycle, three molecules of CO_2 are generated from each pyruvate molecule, and some energy is harvested in the form of ATP (figure 5.4). Most of the remaining energy is in the form of NADH and $FADH_2$. These two molecules are shuttled into the final phase of oxidative respiration, the *electron transport chain*. In this chain, the reduced NADH and $FADH_2$ are oxidized, and their electrons are passed along a series of oxidation-reduction steps to the final acceptor, molecular oxygen.

As previously mentioned, in eukaryotic cells, oxidative respiration takes place in the mitochondria. The citric acid cycle occurs in the mitochondrial matrix (figure 5.5). The enzymes that are used to catalyze these reactions are dissolved in the fluid matrix. (In prokaryotic cells, the enzymes occur in the cytoplasm.) The proteins (cytochromes) that bring about the reactions of the electron transport chain are bound to the inner mitochondrial membrane, which is arranged in numerous folds called *cristae* (s., crista) (in prokaryotes, they are bound to the plasma membrane). Pyruvate, oxygen, ADP, and inorganic phosphate (P_i) continuously diffuse into the mitochondrial matrix. In turn, the end products of oxidative metabolism—ATP, CO_2, and H_2O—diffuse outward into the cytosol.

Oxidation of Pyruvate

The small pyruvate molecules generated by glycolysis can pass easily through the highly permeable outer mitochondrial membrane into the outer compartment by simple diffusion (*see figure 3.8*). Pyruvate is then moved through the inner mitochondrial membrane by facilitated diffusion (which requires carrier proteins but no energy; *see figure 3.10*). Once in the mitochondrial matrix, pyruvate is oxidized (one of the three carbons of pyruvate is split off in a decarboxylation reaction) to CO_2 and an inactive form

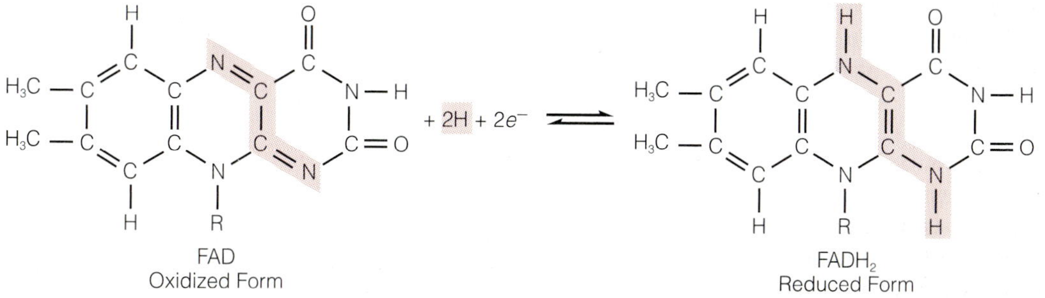

Figure 5.4 The electron carrier flavin adenine dinucleotide (FAD). The equation illustrates the oxidized form of FAD going to the reduced form ($FADH_2$). The reactive sites are shaded, and the rest of the molecule is indicated by an R.

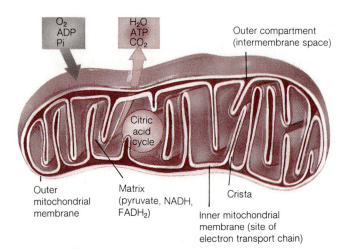

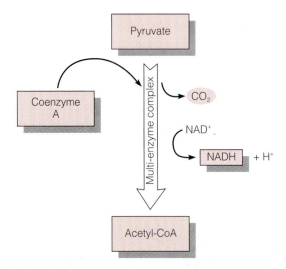

Figure 5.5 Mitochondrial architecture. Oxidative respiration occurs in the mitochondrial matrix. The reactions in the electron transport chain are carried out by molecular complexes that are an integral part of the inner membrane.

Figure 5.6 The oxidation of pyruvate. Because this reaction generates NADH, it is a significant source of energy. Its end product, acetyl-CoA, is the starting molecule for the citric acid cycle. Overall, as three-carbon pyruvate enters the mitochondrion, enzymes split away its COO^- group, which departs as CO_2. At the same time, enzymes transfer a hydrogen atom and two electrons to NAD^+, forming NADH and releasing two protons ($2H^+$). The two-carbon molecule that remains is linked to coenzyme A, forming acetyl-CoA.

of acetate. This reaction is complex and involves an assembly of enzymes called a *multi-enzyme complex*. In the process, NAD^+ is reduced to NADH, and the molecule of acetate becomes temporarily attached to a molecule of *coenzyme A (CoA)* to form *acetyl-CoA,* a high-energy compound (figure 5.6). The overall reaction is as follows:

$$\text{Pyruvate} + NAD^+ + CoA \rightarrow \text{Acetyl-CoA} + NADH + CO_2$$

The Citric Acid Cycle: A Metabolic Clearinghouse

The citric acid cycle is shown in general outline form in figure 5.7. The first reaction involves the two-carbon acetyl group of the acetyl-CoA molecule that was generated from each molecule of pyruvate. It reacts with a four-carbon compound called oxaloacetate and produces the six-carbon compound, citrate. Coenzyme A is regenerated in this reaction and can thus, participate in the oxidation of another molecule of pyruvate.

Overall, the reactions of the citric acid cycle generate two molecules of CO_2 and strip away four pairs of electrons (carried with hydrogen). The CO_2 diffuses out of the mitochondrion as a waste product of oxidative respiration (*see figure 5.5*). At the various sites indicated in figure 5.7, the stripped electrons are transferred by enzymes to NAD^+ or FAD transport molecules. During the oxidation of succinyl-CoA to succinate, one molecule of ATP is formed. Because the citric acid cycle must operate twice for each molecule of glucose catabolized, this means that two ATPs are formed directly from the cycle. As the reactions continue, succinate is converted to fumarate, malate, and eventually oxaloacetate, completing one turn of the cycle.

After two turns of the cycle, all six carbons in the orig-

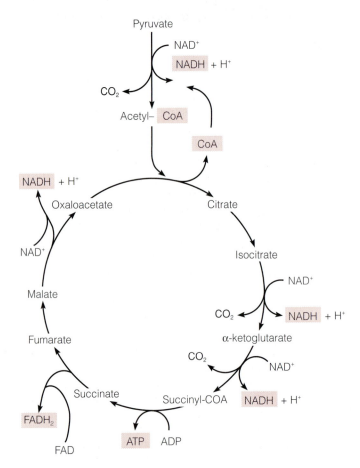

Figure 5.7 Main events of the citric acid cycle, molecular conversions, and energy carriers.

inal molecule of glucose have been oxidized, and some of the energy stored in the glucose has been transferred to four molecules of ATP (two during glycolysis and two during two turns of the citric acid cycle). The rest of the energy is in the NADH and FADH$_2$ molecules. As illustrated in figure 5.7, the oxidation of two molecules of pyruvate to acetyl-CoA produces two molecules of NADH, and the citric acid cycle produces six molecules of NADH and two molecules of FADH$_2$ from one molecule of glucose. In addition, glycolysis has generated two molecules of NADH (*see figure 5.2*). During the last phase of oxidative respiration, the electron transport chain will harvest the large amount of energy locked in these transport molecules to form many more ATP molecules.

In addition to oxidizing glucose, the citric acid cycle serves as a "clearing house" for other aspects of metabolism. For example, products from the breakdown of lipids and proteins can also enter the oxidation pathway in the citric acid cycle via acetyl-coenzyme A and can be metabolized to release energy. In addition, the various intermediates of the citric acid cycle can also diffuse out of the mitochondrion into the cytosol of the cell and serve as precursors for the synthesis of biological compounds. Thus, [6] the citric acid cycle is a focal point in the metabolism that occurs in most cells.

The Electron Transport Chain: An Energy Cascade

The last phase of oxidative respiration consists of the **electron transport chain**. This chain is a metabolic energy cascade, e.g., electrons flowing down a series of steps that begins with the oxidation of NADH and FADH$_2$ and ends with the reduction of O$_2$ to H$_2$O. Along the way, free-energy is released and used to generate ATP. For the first time in oxidative respiration, molecular oxygen plays an important role as a final electron acceptor for both NADH and FADH$_2$ as follows:

$$NADH + H^+ + \tfrac{1}{2}O_2 \rightarrow H_2O + NAD^+$$
$$FADH_2 + \tfrac{1}{2}O_2 \rightarrow H_2O + FAD$$

Once NAD$^+$ and FAD are regenerated, they can accept more electrons from glycolysis and the citric acid cycle. In addition, the proton that is released when NADH is oxidized joins another proton (that was released in an earlier reaction) to provide the two hydrogen atoms in H$_2$O.

When NADH and FADH$_2$ are oxidized, the electrons they give up are passed along the electron transport chain, by a series of electron carriers. As electrons move along the carriers, they release a small amount of free-energy because the carriers are at progressively lower energy states (figure 5.8). The electron carriers that make up the electron transport chain are flavoprotein (FP), ubiquinone (coenzyme

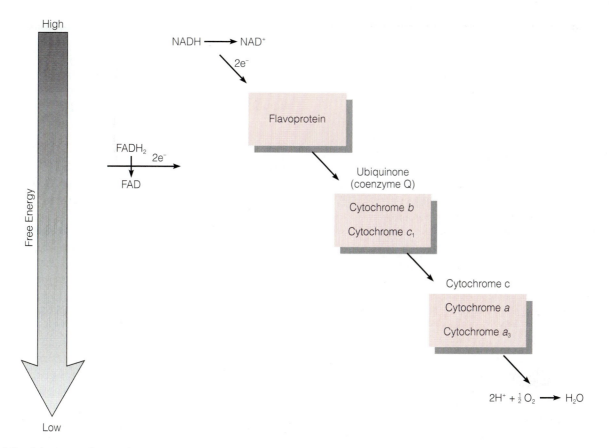

Figure 5.8 Schematic drawing of the electron transport chain. Note that the energy level drops as electrons progress down the chain, in the form of an energy cascade, reaching O$_2$ when water is formed.

Q), and the cytochromes (cytochrome *b*, *c*₁, *c*, *a*, and *a*₃). **Cytochromes** are pigmented proteins with an iron containing *heme* in the center. When electrons pass from one cytochrome to the next, the cytochromes are alternately reduced and oxidized. When the last cytochrome in the chain (cytochrome *a*₃) is reduced, it reacts with free protons (H⁺) and molecular oxygen (O₂) to form water (H₂O).

Electron Flow and the Proton Gradient

The passage of electrons down the electron transport chain provides energy for ATP synthesis in a unique way. All but two of the electron carriers are parts of large multienzyme complexes; ubiquinone (coenzyme Q) and cytochrome *c* act only as connecting links between complexes (figure 5.9).

The three multienzyme complexes must assemble *precisely* for the chain to operate. When properly assembled in the inner mitochondrial membrane, the chain accepts electrons, pumps protons, generates H₂O from protons and O₂, and stores energy for ATP synthesis. The multienzyme complexes move about within the membrane, as do the connecting carriers, ubiquinone and cytochrome *c*, though

they move about ten times as fast as the enzyme complexes. It is only when these carriers *chance* to collide that pairs of electrons are transferred from one to the other. Despite this chance meeting, electron pairs are received and donated about once every 5 to 20 milliseconds.

In 1961, the British biochemist, Peter Mitchell, proposed a mechanism whereby ATP is formed by chemiosmotic coupling. In his model, electrons move through the electron transport chain accompanied by a *proton-pumping* mechanism (*see figure 4.15b*) that sets up an energy gradient across the inner mitochondrial membrane.

The essential features of Mitchell's chemiosmotic coupling model is shown in figure 5.9. NADH and FADH₂ donate electrons to the transport chain. As the electrons move down the electron transport chain, a hydrogen ion (proton) gradient is created across the inner mitochondrial membrane as follows. Some carriers in the electron transport chain must carry H⁺ ions (protons) with the electrons. To do so, they pick up the protons (H⁺) from the mitochondrial matrix. If the next carrier in line accepts only electrons, the protons must be released. When they are, the protons are not released back into the mitochondrial matrix but are pumped across the inner mitochondrial membrane to the outer compartment (between the two mitochondrial membranes). As shown in figure 5.9, protons are pumped at three locations in the electron transport chain. As a result of this pumping of protons, the concentration of protons is low in the matrix and high in the outer compartment. A small amount of the energy released as the electrons move down the chain is used to pump the protons across the membrane. The remaining energy is stored in the proton gradient because the difference in proton (H⁺) concentrations on either side of the inner mitochondrial membrane creates a chemical and electrical imbalance across the membrane. This imbalance represents potential energy.

ATP Synthesis

Eventually, some of the protons that have become concentrated in the outer compartment begin to diffuse back across the inner mitochondrial membrane. In this way the potential energy becomes available to enzyme (ATPase) complexes in the membrane that function in ATP synthesis.

Figure 5.10 illustrates one of the large complexes located in the inner mitochondrial membrane. They form channels for the diffusion of protons back through the membrane into the mitochondrial matrix. As protons move through, the energy is used to form ATP from ADP and P₍ᵢ₎. The specific details of the way in which ATPase uses the energy from the proton gradient to synthesize ATP are still unknown. The number of ATP molecules produced by the transport of electrons by NADH and FADH₂ varies. As figure 5.8 shows, electrons from FADH₂ enter the electron transport chain at a lower energy state than do those from NADH. Therefore, for every molecule of NADH, 3 ATPs can be produced, but only 2 for each FADH₂, because less energy can be harvested from the FADH₂ electrons. The electro-

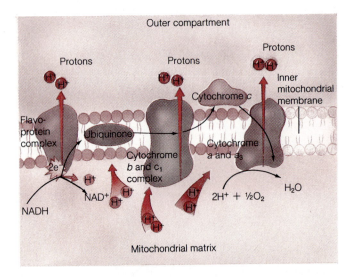

Figure 5.9 The electron transport chain. For each NADH oxidized, about six protons pass from the mitochondrial matrix to the outer compartment through all three multienzyme complexes as electrons flow from NADH to oxygen—their final acceptor. Although illustrated in a stationary position, the components of this chain move about freely in the membrane and electron transfer occurs when they collide by chance.

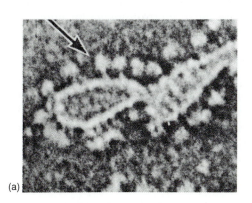

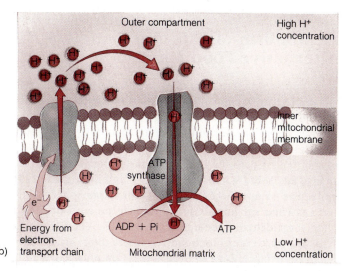

Figure 5.10 Mitchell's chemiosmotic coupling model. (a) Electron micrograph of mitochondrial cristae. (b) In this artist's model, the relationship between the proton gradient created by the electron transport chain and the synthesis of ATP is shown.

chemical gradient that is established is also responsible for several other functions. At other sites on the inner mitochondrial membrane, as H^+ (protons) pass through, they are accompanied by pyruvate that is en route to the citric acid cycle. At still other sites, P_i accompanies the H^+ en route to ATP production sites. Overall, proton pumping and reentry are central to several important mitochondrial functions, and illustrate once again, the correlation between structure and function in a living cell. Indeed, without the large-scale ATP production by mitochondria, life would have to be at a "snails pace," and most animals present on the earth today would never have evolved.

The Energy Score for Oxidative Respiration: A Balance Sheet

The cell obtains a net gain of 36 ATPs from the breakdown of each glucose molecule (table 5.1). Four ATP molecules are produced during glycolysis but two of them are used in the glycolytic reactions. The two molecules of NADH formed during glycolysis yield six ATPs, but two are used in the reactions that transport the NADH electrons across the inner mitochondrial membrane into the matrix to enter the citric acid cycle. Two more ATPs are produced in the citric acid cycle, as well as 8 NADH and 2 $FADH_2$ molecules. The eight NADH from the citric acid cycle yield 24 ATP molecules and the oxidation of the two molecules of $FADH_2$ produced during the citric acid cycle yields four more ATPs. The net gain from all these reactions is 36 molecules of ATP. As table 5.1 illustrates, the electron transport chain is more efficient at producing ATP for the cell's activities, although the importance of glycolysis for ATP production in anaerobic conditions should not be overlooked and cannot be replaced by the aerobic reactions of the citric acid cycle and electron transport.

Table 5.1	Approximate ATP Yield From the Complete Aerobic Oxidation of 1 Molecule of Glucose		
Process	**ATP Produced**	**NADH, FADH$_2$ Produced**	**ATP Used**
Glycolysis	4 ATP	2 NADH	−2 ATP
Entrance of NADH to citric acid cycle			−2 ATP
Citric Acid Cycle	2 ATP	8 NADH, 2 FADH$_2$	
Electron transport chain:			
3 ATP for each of 2 NADH generated via glycolysis	6 ATP		
3 ATP for each of 8 NADH generated via citric acid cycle	24 ATP		
2 ATP for each of 2 FADH$_2$ via citric acid cycle	4 ATP		
Total	40 ATP		−4 ATP
Net ATP	36 ATP		

Metabolism of Fats and Proteins: Alternative Food Molecules

Even though the catabolism of glucose is the most common metabolic pathway in cells, animals also consume fats and proteins, which may be used to harvest energy. Recall from chapter 2 that fats are built of long-chain fatty acids and glycerol, and are called triglycerides. The initial catabolism of a fat begins with the hydrolysis of the triglycerides (via an enzyme called a *lipase*) to glycerol and three fatty acid molecules (figure 5.11). The glycerol is phosphorylated and can enter the glycolytic pathway at the level of phosphoglyceraldehyde (*see step 4, figure 5.2*). The free fatty acids move into the mitochondrion where their carbons are removed, two at a time, to form acetyl-CoA plus NADH and $FADH_2$. The acetyl-CoA is then oxidized by the citric acid cycle, and the NADH and $FADH_2$ that are produced are oxidized via the electron transport chain. It is of interest that [7] 1 gram of fat provides about 2.5 times more ATP energy than does either 1 gram of carbohydrate or protein, because there are more hydrogen atoms per unit weight of fat than in carbohydrates or protein. This is why many animals store energy in the form of fat in adipose tissue (colorplate 1f).

Proteins are initially broken down by animals to yield individual amino acids. Some of these are distributed throughout the body and used for the synthesis of new proteins. Other amino acids are transported in the blood or extracellular fluid and comprise the *amino acid pool*. If needed for fuel, these amino acids can be further degraded by removal of the amino group to yield ammonia. This process is called a **deamination reaction** and can be illustrated as follows:

$$R-CH-COOH \ + \ H_2O \ \longrightarrow \ R-C-COOH \ + \ NH_3 \ + \ H_2$$

| Amino acid | water | keto acid | ammonia | hydrogen |

In deamination, an amino group is replaced by an oxygen atom to form a keto acid. The keto acid can then be funneled into the citric acid cycle (figure 5.11).

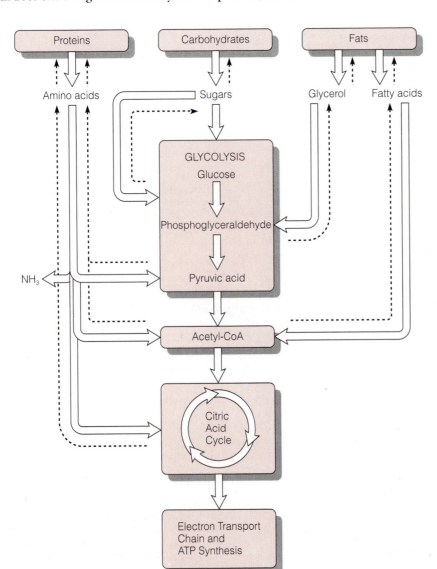

Figure 5.11 Catabolism of various food molecules. While only certain molecules can feed directly into the citric acid cycle (e.g., acetyl-CoA), proteins, carbohydrates, and fats can also be broken down into constituent parts and used as fuel for respiration. The dashed arrows indicate that the system can also work in anabolic (synthetic) reactions.

Other amino acids are changed by deamination into various metabolic intermediates, depending upon the structure of the amino acid. For example, some fragments are converted to pyruvate, others to acetate or acetyl-CoA, and still others to citric acid cycle intermediates. Eventually, the carbon skeleton of the amino acid is dismantled and oxidized to CO_2. On the average, 1 gram of protein yields about the same amount of energy (about 4 kilocalories) as does 1 gram of glucose. The ammonia that is produced from the complete catabolism of an amino acid is highly toxic and must be excreted. The various ways in which different animals rid their bodies of toxic wastes will be discussed later.

Control of Metabolism

The cell is very efficient. It does not waste energy making more of a substance than it needs. For example, if there is an overabundance of a certain amino acid in the amino acid pool, the anabolic pathway that synthesizes that amino acid from an intermediate in the citric acid cycle is turned off. The most common mechanism for this control is via **feedback inhibition.** In feedback inhibition, the end product of the anabolic pathway inhibits the enzyme that catalyzes a key step in the pathway (*see figure 4.9*).

The cell can also control its catabolism. For example, if a cell (e.g., a muscle cell) is working very hard and its ATP concentration begins to decrease, oxidative respiration increases. When there is ample ATP to meet demand, oxidative respiration slows down, sparing valuable organic molecules for other necessary functions. As is the case for anabolism, control is based on regulating the activity of enzymes at strategic points in the catabolic pathway. As a result, cells are thrifty, expedient, and responsive in their metabolism.

[8]

One of the main controlling "switches" in oxidative respiration is the enzyme phosphofructokinase, which catalyzes step 3 of glycolysis (*see figure 5.2*). By controlling the rate of this step, a cell can either speed up or slow down this entire metabolic process. Phosphofructokinase is an allosteric enzyme (*see figure 4.8*) with receptor sites for specific inhibitors and activators. It is inhibited by ATP and stimulated by either ADP or AMP. Phosphofructokinase is very sensitive to the energy needs of the cell and the ratio of ATP to ADP or AMP (figure 5.12). If ATP begins to accumulate, this enzyme shuts down glycolysis. If ADP or AMP begin to accumulate, phosphofructokinase becomes active and turns on the glycolytic pathway. Phosphofructokinase is also inhibited by citric acid in the cytosol. It is this control pathway that helps synchronize the rates of glycolysis and the citric acid cycle. For example, when citric acid begins to accumulate, glycolysis slows down, and the supply of acetyl-CoA to the citric acid cycle is reduced. Conversely, if citric acid consumption increases, glycolysis accelerates and meets the needed demand for more acetyl-CoA.

The Metabolic Pool

The degradative chemical reactions (catabolism) of glycolysis and the citric acid cycle do more than just harvest energy for ATP production. They also constitute a *metabolic pool* that supplies materials for synthesis (anabolism) of many important cellular components (*see figure 5.11*). Overall, it is the balance between catabolism and anabolism that maintains homeostasis in the living cell, and in turn, the whole animal. For example, glycolysis and the citric acid cycle are open systems. An open system is one that

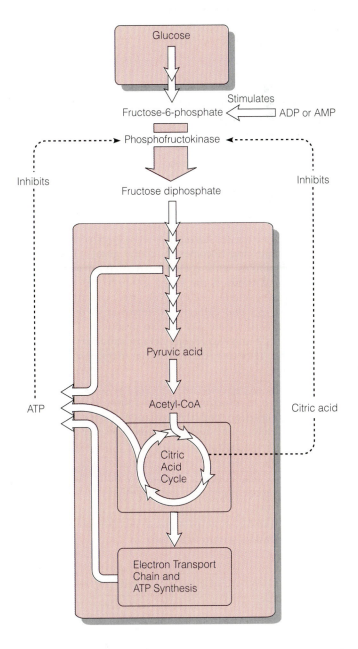

Figure 5.12 Control of cellular respiration. The allosteric enzyme, phosphofructokinase, responds to activators and inhibitors to regulate the rate of glycolysis and the citric acid cycle. It is stimulated by ADP and AMP but inhibited by ATP and citric acid.

has a two-way flow of materials into and out of it. Various compounds enter the pathways at different points so that carbohydrates, fats, and proteins can all be oxidized. At the same time, some of the intermediates of these pathways can be withdrawn from the energy-harvesting machinery and used in synthesis reactions. Thus, the products of glycolysis and the citric acid cycle are all part of a metabolic pool whereby materials are either added or withdrawn.

Stop and Ask Yourself

13. How many ATP molecules are produced during complete aerobic metabolism of one molecule of glucose?
14. How does the flow of electrons generate ATP molecules?
15. How do fats serve as an alternative food molecule?
16. What is feedback inhibition?

■ Summary

1. During glycolysis, glucose is broken down anaerobically to yield two molecules of pyruvate, which is processed further in oxidative respiration. Glycolysis harvests chemical energy by rearranging the chemical bonds of glucose to form two pyruvate molecules and two molecules of ATP.

2. Organisms living in anaerobic environments require a mechanism to dispose of the hydrogen and the associated electrons produced in glycolysis. They can be donated to a number of other compounds in a process called fermentation.

3. Each pyruvate formed during glycolysis loses a carbon dioxide molecule and becomes an acetyl group. The acetyl group combines with coenzyme A to form acetyl-CoA. Coenzyme A transfers the acetyl group to the citric acid cycle. During one turn of the cycle, the equivalents of the acetyl group's two carbons are removed in the form of carbon dioxide. Some ATP is also produced. During the process, NAD$^+$ and FAD accept electrons and become NADH and FADH$_2$. They carry these electrons to the electron transport chain.

4. Most of the energy harvested during oxidative respiration is produced by electron transport followed by chemiosmotic synthesis. In total, the oxidation of one molecule of glucose results in the net production of 36 ATP molecules.

5. Many other metabolic pathways feed into glycolysis and the citric acid cycle from the metabolic pool, enabling cells to use many organic compounds other than glucose as food sources to generate usable energy to perform biological work.

6. Feedback inhibition helps regulate metabolism and maintain homeostasis within organisms.

■ Key Terms

aerobic (p. 71)
anaerobic (p. 71)
citric acid cycle (p. 75)
cytochromes (p. 78)
deamination reaction (p. 80)
electron transport chain (p. 77)
feedback inhibition (p. 81)
fermentation (p. 73)
flavin adenine dinucleotide (FAD) (p. 75)
glycolysis (p. 71)

■ Critical Thinking Questions

1. The aerobic metabolism of glucose generates exactly the same chemical end products and the same amount of energy as when glucose is burned. However, when a cell "burns" glucose, it does so in many small steps. Why has evolution divided glucose catabolism into so many small steps?

2. Each step of glycolysis and the citric acid cycle involves a specific enzyme. How do so many enzymes work together? What is a major principle underlying the cooperation of enzymes?

3. In anaerobic catabolism of glucose, further conversions of pyruvate to lactate do not yield any more usable energy. What, then, is the advantage of these conversions?

4. Why is anaerobic respiration wasteful and potentially harmful to an animal?

5. What relationship does the chemiosmotic pump have to the concept of potential energy? To the concept of kinetic energy?

Suggested Readings

Books

Becker, J. J. W., and Allen, G. E. 1981. *Matter, Energy, and Life.* 4th ed. Reading, MA: Addison-Wesley.

Becker, W. M. 1977. *Energy and the Living Cell: An Introduction to Bioenergetics.* New York: Harper & Row, Publishers, Inc.

Articles

Dickerson, E. Cytcohrome *c* and the evolution of energy metabolism. *Scientific American* March, 1980.

Hinkle, P. C., and McCarty, R. E. How cells make ATP. *Scientific American* March, 1978.

Krebs, H. A. 1970. The history of the tricarboxyclic acid cycle. *Perspectives in Biology and Medicine* 14:154–170.

Levine, M. H., Stammers, D., and Stuart, D. 1978. Structure of pyruvate kinase and similarities with other enzymes: Possible implications for protein taxonomy and evolution. *Nature* 271:626–630.

Racker, E. 1980. From Pasteur to Mitchell, a hundred years of bioenergetics. *Federation Proceedings* 39:210–215.

Schulman, R. G. NMR spectroscopy of living cells. *Scientific American* January, 1983.

The Continuity of Life

Part II

The ability to reproduce is a fundamental property of living organisms. All organisms age and eventually die. For the species, however, reproduction provides a potential means of increasing the size of a population, and results in continual replacement of aging individuals with young, vigorous ones. In addition, sexual reproduction produces individuals with new genetic combinations, which increases genetic diversity, and makes long term survival of the species more likely. Sexual reproduction also leads to individual variation—and variation is the foundation for evolution.

Reproductive processes involve specific activities of individual cells. In unicellular organisms (protists), one individual divides to produce two individuals. Most animals produce individual reproductive cells, such as eggs and sperm. Characteristics are transmitted from one generation to the next by specific genetic mechanisms. Each offspring has a new set of genes that is somewhat different from each of the parent's.

An individual's genetic makeup is expressed through developmental processes and continues throughout life in the form of growth, maturation, and aging. Developmental processes are also involved in the continual replacement of body cells, in healing of injuries, and in specific responses to infection and disease. However, developmental processes gone awry, such as in abnormal growth, can be life threatening.

Chapters 6 through 11 present the cell division mechanisms involved in reproduction and development, and the means by which genetic information is transmitted and expressed.

6

The Principles of Cell Division and Reproduction

Concepts

1. The cell cycle consists of four phases: G_1, a period of normal metabolism; S, the phase of DNA replication; G_2, a brief period of further cell growth; mitosis (nuclear division) and cytokinesis (cytoplasmic division). G_1, S, and G_2 are referred to as interphase, and mitosis and cytokinesis are known as the mitotic phase (M).
2. The process that ensures an orderly and accurate distribution of chromosomes during the cell cycle is mitosis, which can be divided into four phases: prophase, metaphase, anaphase, and telophase. Cytokinesis often begins during late anaphase or early telophase and soon after mitosis has ended, two daughter cells are evident.
3. Meiosis is a special type of cell division that reduces the chromosome number by one-half. Meiosis allows for the random distribution of parental chromosomes to the offspring.
4. Gamete formation involves meiosis and the formation of specialized reproductive cells called sperm and eggs.
5. Asexual reproduction, in which new organisms arise from mitotic processes, is usually more rapid than sexual reproduction.

Have You Ever Wondered:

[1] how cell division is related to a motion picture?
[2] why in cell division, DNA undergoes untwisting, unzipping, rezipping, and retwisting?
[3] how long a DNA molecule in a human cell is?
[4] what is happening to a cell that looks like someone tied a cord around it and pulled the rope tight?
[5] why cells are grown in culture, outside of an animal's body?
[6] what causes genetic variation in sexually reproducing animals?
[7] how the earliest organisms reproduced?
[8] how turkeys can reproduce without sperm?
[9] what one of the greatest disadvantages is to asexual reproduction?
[10] how male mollies (fish) produce baby mollies?
[11] why sexual reproduction is a "key" to evolution?

These and other useful questions will be answered in this chapter.

This chapter contains underlined evolutionary concepts.

he cellular capacity to handle large amounts of DNA (the information-carrying molecule) emerged more than a billion years ago among the ancestors of modern-day eukaryotes. This capacity was based on two occurrences: (1) the "packaging" of DNA and its associated proteins into compact structures called chromosomes, and (2) the development of an apparatus that could move the chromosomes. Eukaryotic cell division evolved as a result of these two developments.

One of the most intriguing activities of a cell is its ability to reproduce itself by a process called *cell division*. In some cases, an entire new organism is reproduced, as when the unicellular protozoan called *Amoeba proteus* divides to form duplicate offspring. In other cases, cell division enables a multicellular organism, such as yourself, to grow and develop from a single cell. In fact, even after you are fully grown, cell division continues to function in renewal and repair to maintain homeostasis.

DNA controls cell division. What is most remarkable about this control is the faithfulness with which genetic programs are passed from one generation to the next. A cell preparing to undergo cell division first copies all of its genes and then allocates them equally to its two daughter cells. The key to biological inheritance—the bridge to the next generation—is in this division process.

This chapter focuses on how eukaryotic cells produce daughter cells. The careful regulation of daughter cell production is a theme that will be repeated throughout this chapter.

The Cell Cycle is a Dynamic Process

Cell division occurs in all animals. Cells divide in two basic stages: **mitosis** is division of the nucleus, and **cytokinesis** (Gr. *kytos,* hollow vessel + *kinesis,* motion) is division of the cytoplasm. However, cell division is just one phase in the life history of a cell. Between divisions (*interphase*), the cell must grow and carry out its various metabolic processes. The **cell cycle** is that period from the time a cell is produced until it goes through mitosis (figure 6.1).

The cell cycle is a dynamic process with four consecutive phases that can be expressed as follows:

$$G_1 \rightarrow S \rightarrow G_2 \rightarrow M$$

The G_1 (first gap) phase represents the early growth phase of the cell. In many organisms, this phase occupies the major portion of the cell's life-span. The S (DNA synthesis) phase involves DNA duplication, whereby a replica of the genome (the total genetic constitution of an organism) is synthesized. The G_2 (second gap) phase prepares the cell for genomic separation. It includes replication of the mitochondria and other organelles, synthesis of microtubules and protein that will make up the mitotic spindle fibers, and chromosome condensation. The M (mitotic) phase is represented by the assembly of the microtubules, their binding to the chromosomes, the separation of the sister

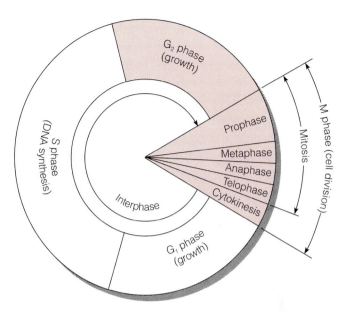

Figure 6.1 The life cycle of a eukaryotic cell. During the *S* (synthesis) phase, the chromosomes replicate, resulting in two identical copies called sister chromatids. During G_1, normal components of the cell are synthesized and metabolism occurs, often resulting in cell growth. During G_2, metabolism and growth continue until the mitotic phase is reached. This drawing is generalized, and there is a great deal of variation in the length of different stages from one cell to the next.

chromosomes, reformation of nuclear membranes in daughter cells, and cytokinesis. Cytokinesis usually begins late in mitosis when the cytoplasm of the cell divides, creating two daughter cells.

Cells continually cycle through these phases, although different cells may cycle at different rates. Table 6.1 presents the approximate cell cycles of some specialized cells.

Interphase and Mitosis: Partitioning the Hereditary Material

Often erroneously described as a resting period between cell divisions, **interphase** (L. *inter,* between) (includes G_1, S, and G_2 phases) is actually a period of great metabolic

Table 6.1	Approximate Cell Cycles of Some Specialized Human Cells
Cell Type	**Approximate Life Span**
Bone marrow (blood-forming)	10 hours
Egg cells	2 days
Stomach lining cells	2 days
Sperm	2–3 days
Cells lining large intestine	3–4 days
White blood cells	13 days
Skin cells	19–34 days
Liver cells	18 months
Nerve cells*	Lifetime of body

*Nerve cells do not reproduce after early childhood.

activity that occupies about 90% of the total duration of the cell cycle. It is the period during which the normal activities of the cell take place. Interphase also sets the stage for cell division, because DNA replication is completed during the *S* phase of interphase.

Mitosis has received more attention from biologists than any other part of the eukaryotic cell cycle. The separation of chromosomes has long fascinated biologists and will continue to do so for years to come, until this most fundamental of all biological events is fully understood. By convention, mitosis is divided into four phases: prophase, metaphase, anaphase, and telophase. In a dividing cell, however, the process is actually continuous, with each phase smoothly flowing into the next phase. There are no start and stop positions; the phases make up a continuum (figure 6.2). If you were to watch a film on mitosis, you would see that each drawing in figure 6.2 is merely a single frame of a continuous, changing drama. [1]

Interphase: Preparing the Scene

Before a cell divides, the DNA in its nucleus must produce a perfect copy of itself. This process is called *replication*, because the double-stranded DNA makes a replica, or du-

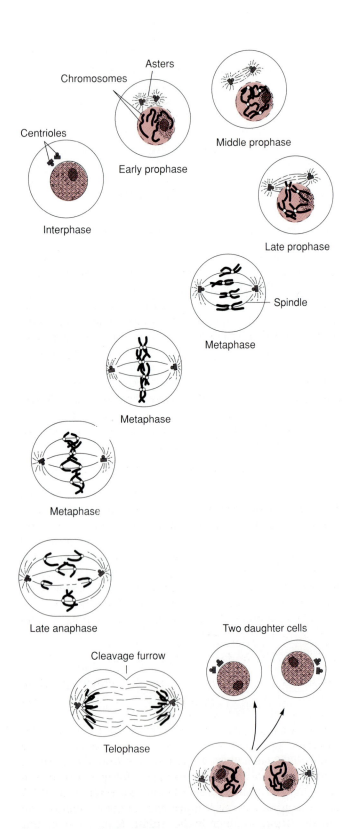

Figure 6.2 The continuum of mitosis and cytokinesis.

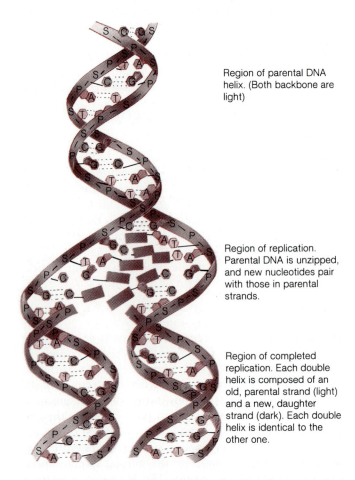

Region of parental DNA helix. (Both backbone are light)

Region of replication. Parental DNA is unzipped, and new nucleotides pair with those in parental strands.

Region of completed replication. Each double helix is composed of an old, parental strand (light) and a new, daughter strand (dark). Each double helix is identical to the other one.

Figure 6.3 The replication of DNA, showing the untwisting of the original double strand, and the formation and twisting of the new double strands.

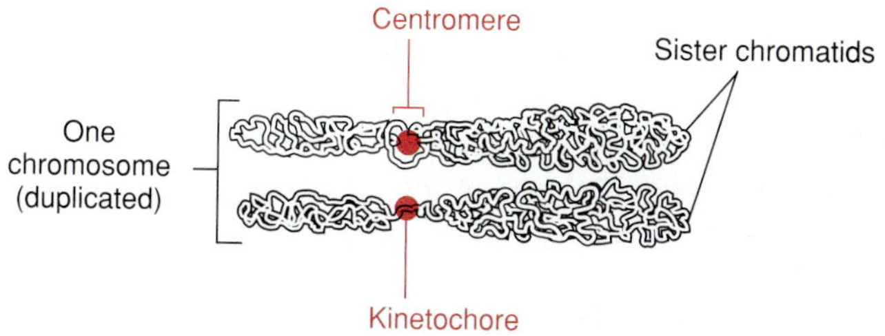

Figure 6.4 Duplicated chromosomes. Each parent chromosome replicates to make two geneti_''ly_ identical sister chromatids attached at a region called _   entromere.

plicate, of itself. Replication is essential to ensure that each daughter cell receives the *same* genetic material as is present in the parent cell.

During replication, the twisted double strands of DNA begin to replicate by untwisting (figure 6.3). The weak hydrogen bonds joining the purine and pyrimidine bases break, and the DNA molecule begins to "unzip" itself (*see figure 2.5*). As soon as new bonds are formed, the untwisted strands begin to twist again. Apparently, the activities of

[2] untwisting, unzipping, rezipping, and retwisting all take place at the same time and ensure that each newly formed double helix is exactly like the other one. The final result is a pair of *sister chromatids* (figure 6.4). A **chromatid** is a copy of a chromosome produced by replication. Each chromatid is attached to its other copy, or sister, at a point of constriction called a centromere. The **centromere** is a specific DNA sequence of about 220 nucleotides and has a specific location on any given chromosome. Bound to each centromere is a disk of protein called a **kinetochore.** The kinetochore will eventually serve as an attachment site for the microtubules of the mitotic apparatus.

After replication, the chromosomes are fully extended.

[3] For example, the total length of the DNA strands in a human nucleus is greater than the length of the entire human body, but the strands are too thin to be seen with the light microscope. As the cell cycle moves into the G_2 phase, however, the chromosomes begin the process called *condensation*. During condensation, each sister chromatid becomes folded into a very tightly compacted structure. During the G_2 phase, the cell also begins to assemble the structures that it will later use to move the chromosomes to opposite poles (ends) of the cell. For example, centrioles replicate, and there is extensive synthesis of α and β tubulins, which are the proteins that make up the microtubules (*see figure 3.21*).

Prophase: Formation of the Mitotic Apparatus

The first phase of mitosis, **prophase** (Gr. *pro,* before + phase), begins when chromosomes become visible with the light microscope as threadlike structures. (The term mitosis refers to this emergence; it comes from the Greek, *mitos,* which means thread.) By the time the cell cycle enters prophase, the chromatids are well folded, and the nucleoli and nuclear envelope begin to break up and their components are resorbed into the endoplasmic reticulum

(figure 6.5a). The newly formed spindle microtubules move between the separating centrioles, and the two centriole pairs move apart.

By the end of prophase, the centriole pairs have moved to opposite poles of the cell. It is the position of the centrioles at the poles that determines the direction in which the cell divides. When the centrioles reach the poles of the cell, they radiate an array of microtubules called an **aster** (L. *aster,* little star). The asters brace each centriole against the plasma membrane, mechanically stiffening the point of microtubular attachment. Between the centrioles, the microtubules form a spindle of fibers that extends from pole to pole. The asters, spindle, centrioles, and microtubules are collectively called the **mitotic apparatus**.

As prophase continues, a second group of microtubules grows out from the kinetochore to the poles of the cell. These kinetochore microtubules connect each sister chromatid to the poles of the spindle.

Metaphase: Separation of Sister Chromatids

As the dividing cell moves into **metaphase** (Gr. *meta,* after + phase), the chromatids begin to align at the center of the cell, along the metaphase, or equatorial plate (figure 6.5b). With the help of the microtubules attached to the individual kinetochores, the chromatids become neatly arrayed in the center of the cell, each centromere equidistant from each pole of the cell.

Toward the end of metaphase, the centromeres divide and free the two sister chromatids from their attachment to each other, although they remain aligned next to each other. After the centromeres divide, the sister chromatids are considered full-fledged chromosomes.

Anaphase: Movement of the Chromosomes

Of all the phases of mitosis, **anaphase** (Gr. *ana,* back again + phase) is the shortest time period (figure 6.5c). Each chromosome moves apart from its copy (the sister chromatid that had been attached to it) as it is pulled toward its respective pole. When the chromosomes are being moved to the poles, they can appear straight, U-shaped, or J-shaped, depending on the location of the centromere region. Anaphase ends when all the chromosomes have moved to the poles of the cell; each pole now has a complete set of chromosomes.

Telophase: Re-Formation of Nuclei

Telophase (Gr. *telos,* end + phase) begins once the chromosomes arrive at the opposite poles of the cell. During telophase, the mitotic spindle is disassembled (figure 6.5d). The α and β tubulin monomers from the microtubules are now used to construct the cytoskeleton of the new cells. A nuclear envelope re-forms around each set of chromosomes, which begin to uncoil for gene expression to begin. One of the first genes to be expressed is the rRNA gene that results in the reappearance of the nucleolus. The cell also begins to pinch in the middle. Mitosis is over, but cell division is not.

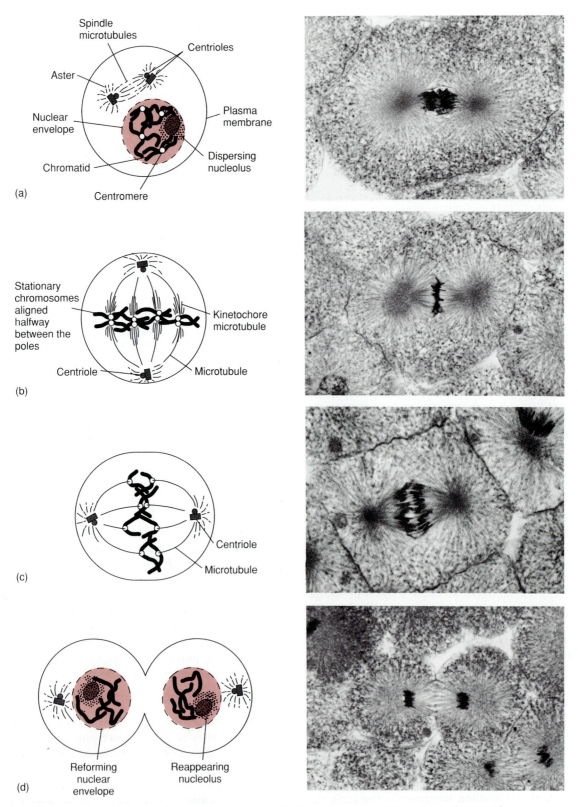

Figure 6.5 Stages of mitosis. (a) During prophase, the nucleoli disappear, and the sister chromatids become folded into discrete chromosomes. In the cytoplasm, the centrioles move apart, and the mitotic apparatus begins to form. (b) At metaphase, the centriole pairs are now at opposite poles of the cell. The chromatids line up in the center of the cell, arranged so their centromeres all lie in one plane. (c) Anaphase begins when the paired chromatids move apart. (d) At telophase, daughter nuclei begin to form at the two poles of the cell where the chromosomes have gathered.

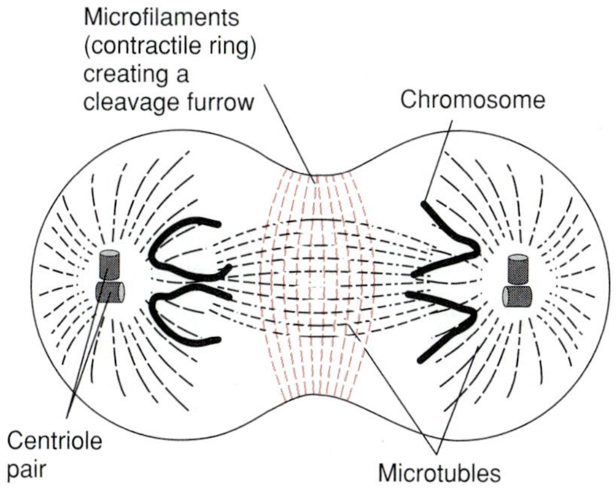

Figure 6.6 Cytokinesis. Diagram of the spatial relationship between the contractile ring of microfilaments and the microtubules of the mitotic spindle.

Cytokinesis: Partitioning the Cytoplasm

The final phase of cell division is *cytokinesis,* in which the cytoplasm divides (figure 6.6). Cytokinesis usually starts sometime during late anaphase or early telophase. Before cytokinesis, the two newly formed nuclei still share the same cytoplasmic compartment but are positioned at opposite ends of the cell. Cellular separation is accomplished by a pinching of the plasma membrane by a contracting belt of microfilaments called the *contractile ring.* The ring lies midway between the two spindle poles and asters, and the microfilaments of the ring are linked to the inner surface of the plasma membrane so that their contractile activity pulls the cell surface inward. The *cleavage furrow,* located where the pinching occurs, looks as though a cord [4] were tied around the circumference of the cell and pulled tight. Two new, genetically identical daughter cells are formed, each about one-half the size of the original parent cell.

Control of Cell Division

The timing and rate of cell division in different parts of an animal are critical to normal growth, development, and maintenance. Although many questions about the control of cell division remain unanswered, biologists have learned a great deal by studying cells grown outside an animal's body. The technique of growing cells outside the body is called either *tissue culture* or *cell culture.* Research with [5] cells grown in tissue culture has identified certain factors that can either inhibit or stimulate cell division. For example, certain hormones and other chemicals called *growth factors* are vital for stimulating normal cell division. Conversely, cell division can be inhibited by blocking protein synthesis, cutting off essential nutrients, or by allowing cells to become overcrowded. When some cells

touch other cells, *contact inhibition* occurs by an unknown mechanism, and cell division stops. Many cells produce **chalones** (Gr. *chalon,* relaxing), which are chemicals that inhibit division in other cells when a certain cell density has been reached. If cell density decreases, the chalone concentration decreases, and cell division resumes.

Stop and Ask Yourself
1. Why is cell division referred to as a continuum?
2. What is the function of the mitotic spindle?
3. What happens during each of the following phases of mitosis: prophase, metaphase, anaphase, and telophase?
4. What is cytokinesis? How does it occur?

Meiosis: The Basis of Sexual Reproduction

The vast majority of animals do not simply divide in half to reproduce, as do single cells, but instead undergo sexual reproduction. *Sexual reproduction* requires a genetic contribution from two different sex cells. Egg and sperm cells are specialized sex cells called **gametes** (Gr. *gamete,* wife; *gametes,* husband). In animals, a *male gamete* (sperm) unites with a *female gamete* (egg) during fertilization to form a single cell called a **zygote** (Gr. *zygotos,* yoked together). The fusion of gametes is called **syngamy** (Gr. *gamos,* marriage). The zygote is the first cell of the new animal. Each of the two gametes contributes one-half the genetic information to the zygote.

To maintain a constant and stable number of chromosomes in the next generation, animals that reproduce sexually must produce gametes with one-half the chromosome number of their ordinary body cells (called **somatic cells**). All of the cells in the body, except for the egg and sperm cells, have the diploid (2N) number of chromosomes. A diploid set of chromosomes contains a species-specific number of chromosome pairs. The chromosomes are in pairs because each cell contains one chromosome of each type from one parent and another chromosome of each type from the other parent. Humans have 46 total chromosomes (2N), comprising 23 pairs of chromosomes. The necessary reduction of chromosome number is accomplished by a type of cell division called **meiosis** (Gr. *meiosis,* dimunition). Meiosis occurs in the cells of the ovaries and testes and reduces the number of chromosomes to one-half the diploid (2N) number. The cells that result (eggs and sperm) are said to be haploid (N), or have the haploid number of chromosomes. Haploid cells have only one of each type of chromosome (only one member of each pair of chromosomes found in diploid cells). When the nuclei of the two gametes combine during fertilization, the diploid number is restored.

Meiosis begins after the G_2 phase in the cell cycle—after DNA replication has occurred. Two successive nuclear divisions take place, designated *meiosis I* and *meiosis II*. Each division has a prophase, metaphase, anaphase, and telophase. The microtubular spindles function in meiosis just as they do in mitosis. The result of mitosis is two identical daughter cells, each with the same number of chromosomes as the parent cell, whereas the two nuclear divisions of meiosis result in four daughter cells, each with one-half the number of chromosomes as the parent cell. Moreover, these daughter cells are not genetically identi-

cal. Like mitosis, meiosis is a continuous process, and only for convenience do biologists divide it into the following phases.

The First Meiotic Division: Prophase I to Telophase I

During interphase, each of the chromosomes has already replicated to form two sister chromatids joined at their centromeres (*see figure 6.4*). In prophase I, DNA folds more and more tightly, and becomes visible under a light microscope (figure 6.7a). Because a cell has a copy of each

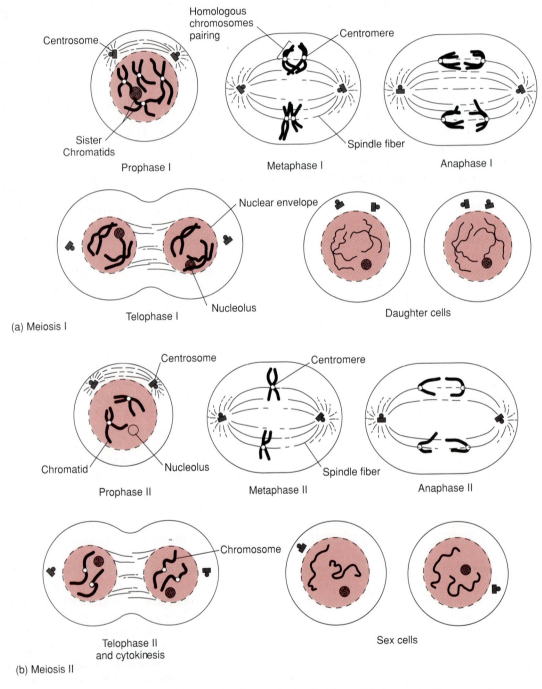

Figure 6.7 Meiosis and cytokinesis. (a) Stages in the first meiotic division. (b) Stages in the second meiotic division.

type of chromosome from each original parent cell, it contains the haploid number of chromosome pairs. The chromosomes of each pair code for similar traits, e.g., hair color, eye color. Chromosomes that carry genes for the same traits are called **homologous chromosomes**, are the same length, and have a similar staining pattern, making them identifiable as matching pairs. During prophase I, homologous chromosomes line up side-by-side in a process called **synapsis** (Gr. *synapsis,* conjunction) forming a *tetrad* of chromatids (a bivalent). The tetrad thus contains the two homologous chromosomes, each with its copy, or sister chromatid (figure 6.8). A network of protein and RNA

is laid down between the sister chromatids of the two homologous chromosomes. This network holds the sister chromatids in a precise union so that each gene is located directly across from its sister gene on the homologous chromosome. An analogy would be zipping up a zipper. The resulting complex is called a **synaptonemal complex**. Within this complex, the DNA duplexes of each chromatid unwinds (*see figure 6.3*), and single strands of DNA pair with their complementary member from the other homologue.

Synapsis also initiates a series of events called **crossing-over**, whereby segments of DNA are exchanged between the nonsister chromatids of the two homologous chromosomes in a tetrad (figure 6.8a,b). This process effectively redistributes genetic information among the paired homologous chromosomes and produces new combinations of genes on the various chromatids in homologous pairs. Thus, each chromatid ends up with new combinations of instructions for a variety of traits. Crossing-over is a form of **genetic recombination** and is a major source of genetic variation in a population of a given species. It will be discussed in more detail in chapter 8.

The point of crossing-over can be seen under the light microscope as an X-shaped structure called a **chiasma** (pleural, chiasmata; Gr. cross) (figure 6.8). The presence of a chiasma indicates that two nonsister chromatids of paired homologous chromosomes have exchanged parts. Apparently, the continued attachment of nonsister chromatids at chiasmata plays a role in aligning homologues at the equatorial plate during metaphase I.

In metaphase I, the microtubules form a spindle apparatus just as in mitosis (*see figures 6.5; 6.7*). However, unlike mitosis where homologous chromosomes do not pair, each pair of homologues lines up in the center of the cell.

When all chromosome pairs are in the center of the cell, all those inherited from one parent do not necessarily line up on the same side as those inherited from the other parent. Using only three homologues as an example, any of the alignments shown in figure 6.9 are possible. Any one of these four alignments is as likely to occur as any other. This random alignment of chromosomes at metaphase I is another major source of genetic variation in sexually reproducing animals. [6]

Anaphase begins when homologous chromosomes separate and begin to move toward each pole (*see figure 6.7a*). Because the chromosomes of each homologous pair are moving in opposite directions, each daughter cell will receive one chromosome from each homologous pair. However, sister chromatids do not part at this stage because there is no centromere division in the first meiotic division. Thus, each of these chromosomes still consists of two chromatids and is called a *dyad*. Each pole of the cell now has a complete set of chromosomes: one member of each homologous pair. Because the orientation of each pair of homologous chromosomes in the center of the cell is random (*see figure 6.9*), the specific chromosomes that each pole receives from each pair of homologues is also random.

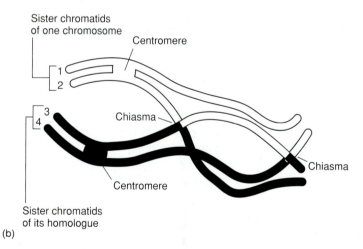

(a)

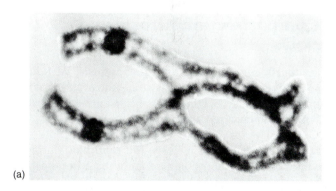

(b)

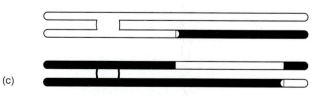

(c)

Figure 6.8 Synapsis and crossing-over. (a) A pair of homologous chromosomes in late prophase. The sister chromatids are visible, as are the centromeres. (b) A diagram illustrating how crossing-over occurs, and the location of chiasmata where nonsister chromatids remain temporarily attached. (c) The same chromosomes diagrammatically separated to show the result of crossing-over.

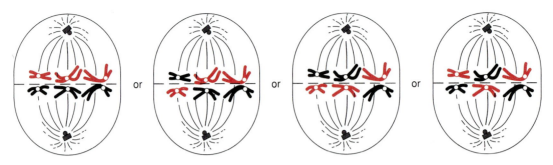

Figure 6.9 Chance alignment of paired homologous chromosomes during metaphase I of meiosis.

Meiotic telophase is similar to mitotic telophase in that the parent cell divides into two daughter cells via cytokinesis (*see figure 6.7a*). During this phase, nuclear envelopes appear around the chromosome sets, the nucleoli reappear, and the spindle fibers disappear. The transition to the second nuclear division is called *interkinesis*. Cells proceeding through interkinesis do not replicate their DNA. After a time period of variable length, meiosis II occurs.

The Second Meiotic Division: Prophase II to Telophase II

The second meiotic division (meiosis II) resembles an ordinary mitotic division (*see figure 6.7b*). The phases are referred to as prophase II, metaphase II, anaphase II, and telophase II. During prophase II, the spindle apparatus appears, and the chromosomes line up on the equatorial plate during metaphase II. During anaphase II, centromere division finally occurs, and the sister chromatids separate to become individual chromosomes. At the end of telophase II and cytokinesis, the final products of these two divisions of meiosis are four new "division products." In most animals, each of these "division products" is haploid and may function directly as a gamete (sex cell).

Meiotic Errors

Errors in meiosis may result in genetic problems in the gametes. As a result, fertilization involving a genetically abnormal gamete perpetuates the problem in all of the cells of the animal that is produced. One such meiotic error is triomy 21 (Downs syndrome). This and other meiotic errors will be discussed in more detail in chapter 8.

Stop and Ask Yourself

5. What are gametes?
6. What is synapsis and the synaptonemal complex?
7. What is the importance of crossing over?
8. In meiosis, why don't sister chromatids part during anaphase I?
9. What happens during the second meiotic division?

A Comparison of Mitosis and Meiosis

Figure 6.10 compares mitosis and meiosis. Although meiosis involves two nuclear divisions, the events that are unique to meiosis all occur during the first meiotic division and can be summarized as follows:

1. During prophase I of meiosis, the replicated chromosomes pair with their homologues in the process called synapsis, producing closely associated complexes called tetrads. Crossing-over occurs between nonsister (homologous) chromatids, and genetic material is exchanged. This event is visible under the light microscope by the appearance of chiasmata. Neither synapsis nor crossing-over occurs during mitosis.
2. During metaphase I of meiosis, tetrads consisting of homologous pairs of chromosomes (rather than individual chromosomes as in mitosis) line up in the center of the cell.
3. At anaphase I of meiosis, homologous chromosomes separate and move to opposite poles of the cell. Sister chromatids do not separate, as they do in mitosis. Instead, the sister chromatids of each chromosome move to the same pole of the cell.

The second meiotic division, meiosis II, is identical to mitosis in that sister chromatids are separated; however, the final outcome of meiosis is a reduction in the number of chromosomes per cell to one-half (haploid) of the original number (diploid).

Gamete Formation: Gametogenesis

During the life cycle of most animals, some diploid cells undergo meiosis and haploid gametes are formed in a process called **gametogenesis.** In the testes of the male, the type of gametogenesis that produces sperm is called **spermatogenesis;** in the ovaries of the female, the production of ova (eggs) is called **oogenesis.**

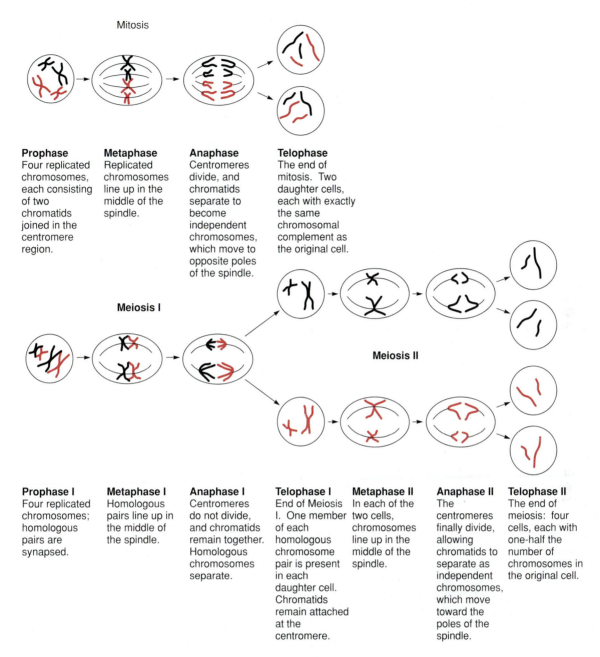

Figure 6.10 A comparison of mitosis and meiosis. Meiosis involves two nuclear divisions, resulting in four daughter cells.

Spermatogenesis

Spermatogenesis produces mature sperm as follows (figure 6.11a):

1. Spermatogenesis begins in unspecialized germ cells called *spermatogonia* (singular, *spermatogonium*). The spermatogonium increases in size and becomes a *primary spermatocyte.*
2. A primary spermatocyte undergoes meiosis (meiotic division I) and produces two smaller *secondary spermatocytes.*
3. Both secondary spermatocytes undergo a second meiotic division (meiotic division II) to form *spermatids.* (Notice that each primary spermatocyte gives

rise to four spermatids, each with the haploid number of chromosomes.)
4. The spermatids develop into mature sperm (spermatozoa) without undergoing any further cell division.

Oogenesis

Oogenesis, the maturation of ova (singular ovum) or eggs, differs from spermatogenesis in several ways and proceeds as follows (figure 6.11b):

1. Oogenesis begins in unspecialized germ cells called *oogonia,* (singular, *oogonium*), which undergo mitosis.

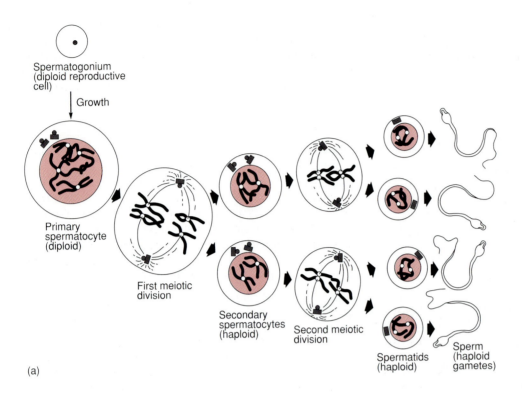

Spermatogonium
(diploid reproductive
cell)

Growth

Primary
spermatocyte
(diploid)

First meiotic
division

Secondary
spermatocytes
(haploid)

Second meiotic
division

Spermatids
(haploid)

Sperm
(haploid
gametes)

(a)

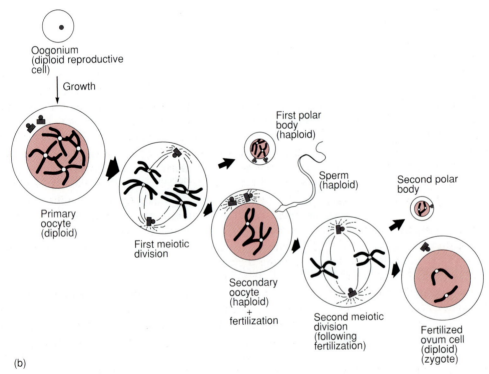

Oogonium
(diploid reproductive
cell)

Growth

Primary
oocyte
(diploid)

First meiotic
division

Secondary
oocyte
(haploid)
+
fertilization

First polar
body
(haploid)

Sperm
(haploid)

Second polar
body

Second meiotic
division
(following
fertilization)

Fertilized
ovum cell
(diploid)
(zygote)

(b)

Figure 6.11 Gametogenesis. (a) Spermatogenesis is a meiotic process that involves two successive divisions. (b) In oogenesis, meiosis is completed following fertilization of the secondary oocyte.

2. An oogonium develops and grows into a *primary oocyte*, which contains the diploid number of chromosomes. The primary oocyte undergoes a first meiotic division to produce two daughter cells of unequal size.

3. The larger of the daughter cells is the *secondary oocyte*. It contains almost all of the food-rich cytoplasm of the primary oocyte, which provides nourishment for the developing ovum.

4. The smaller of the two daughter cells is a *first polar body,* and is essentially only a nucleus. It may divide again, but eventually it degenerates.

5. In some animals, if the secondary oocyte is fertilized, it begins to go through a second meiotic division, producing a haploid ovum and another tiny polar body is formed. This *second polar body* is also destined to disintegrate. When the haploid sperm and ovum nuclei

merge (syngamy), the ovum becomes a zygote. Notice that in oogenesis, each primary oocyte gives rise to only one functional gamete instead of four as in spermatogenesis.

Asexual Reproduction

In the biological sense, *reproduction* means producing offspring that may (or may not) be exact copies of the parents. Reproduction is part of a *life cycle,* a recurring frame of events in which individuals grow, develop, and reproduce according to a program of instruction encoded in the DNA they inherit from their parents. One of the two major types of reproduction that occur in the biological world is asexual reproduction.

[7] The first organisms to evolve probably reproduced by pinching in two, much as do the simplest organisms that exist today. This is a form of *asexual reproduction,* which is reproduction without the union of gametes or sex cells. In the first 2 billion years or more of evolution, forms of asexual reproduction were probably the only means by which the primitive organisms could increase their numbers. While asexual reproduction is effective in increasing the numbers of a species, those species reproducing asexually tend to evolve very slowly because all offspring of any one individual will be alike, providing less genetic diversity for evolutionary selection.

Asexual reproduction is common among the animallike protists (one-celled eukaryotic organisms, such as protozoa), as well as in lower invertebrates (animals lacking backbones), such as sponges, jellyfishes, flatworms, and many segmented worms. The ability to reproduce asexually is often correlated with a marked capacity for regeneration.

Fission

Protists and some multicellular animals may reproduce by fission. **Fission** is the division of one cell into two (figure 6.12a). In this process, the cell pinches in two by an inward furrowing of the plasma membrane. *Binary fission* is where the division is equal; each offspring contains approximately equal amounts of protoplasm and associated structures. Binary fission is common in protozoa; for some, it is their only means of reproduction.

In fission, the plane of division may be asymmetrical, transverse, or longitudinal, depending on the species. For example, the multicellular, free-living flatworms, such as

the common planarian, reproduce by transverse fission (figure 6.12b). Some flatworms and annelids reproduce by forming numerous constrictions along the length of the body; a chain of daughter individuals results (figure 6.12c). This type of asexual reproduction is called *multiple fission.*

Budding

Another method of asexual reproduction found in lower invertebrates is **budding.** For example, in the cnidarian *Hydra* and many species of sponges, certain cells divide rapidly and develop on the body surface to form a bud (figure 6.12d). The bud cells proliferate to form a cylindrical structure, which develops into a new animal, usually breaking away from the parent. If the buds remain attached to the parent, they form a colony. A **colony** is a group of closely associated individuals of one species.

Fragmentation

Fragmentation is a type of asexual reproduction whereby a body part is lost and then regenerates into a new organisms. For example, in sea anemones, as the organism moves, small pieces break off from the adult and develop into new individuals (figure 6.12e).

Parthenogenesis

Certain insects, flatworms, roundworms, lizards, lobsters, and turkeys can reproduce without sperm and normal fertilization. These animals carry out what is called **parthenogenesis** (Gr. *parthenos,* virgin + *genesis,* production). (However, most parthenogenetic animals also can reproduce sexually at some point in their life history.) Parthenogenesis is a spontaneous activation of a mature egg, followed by normal egg divisions and subsequent embryonic development. In fact, mature eggs of species that do not undergo parthenogenesis can sometimes be activated to develop without fertilization by pricking them with a needle, by exposing them to high calcium levels, or by altering their temperature. [8]

Because parthenogenetic eggs are not fertilized, they do not receive male chromosomes. One would thus expect the offspring to have only a haploid set of chromosomes. In some animals, however, meiotic division is suppressed, so the diploid number is conserved. In other animals, meiosis occurs, but an unusual mitosis returns the haploid embryonic cells to the diploid condition.

Overall, animals that reproduce parthenogenically have substantially less genetic variability than do animals with chromosome sets from two parents. This condition may be an advantage for animals that are well adapted to a relatively stable environment. However, in meeting the challenges of a changing environment, parthenogenetic animals may have less flexibility, which may explain why this form of reproduction is relatively uncommon.

Parthenogenesis also plays an important role in the social organization in colonies of certain bees, wasps, and

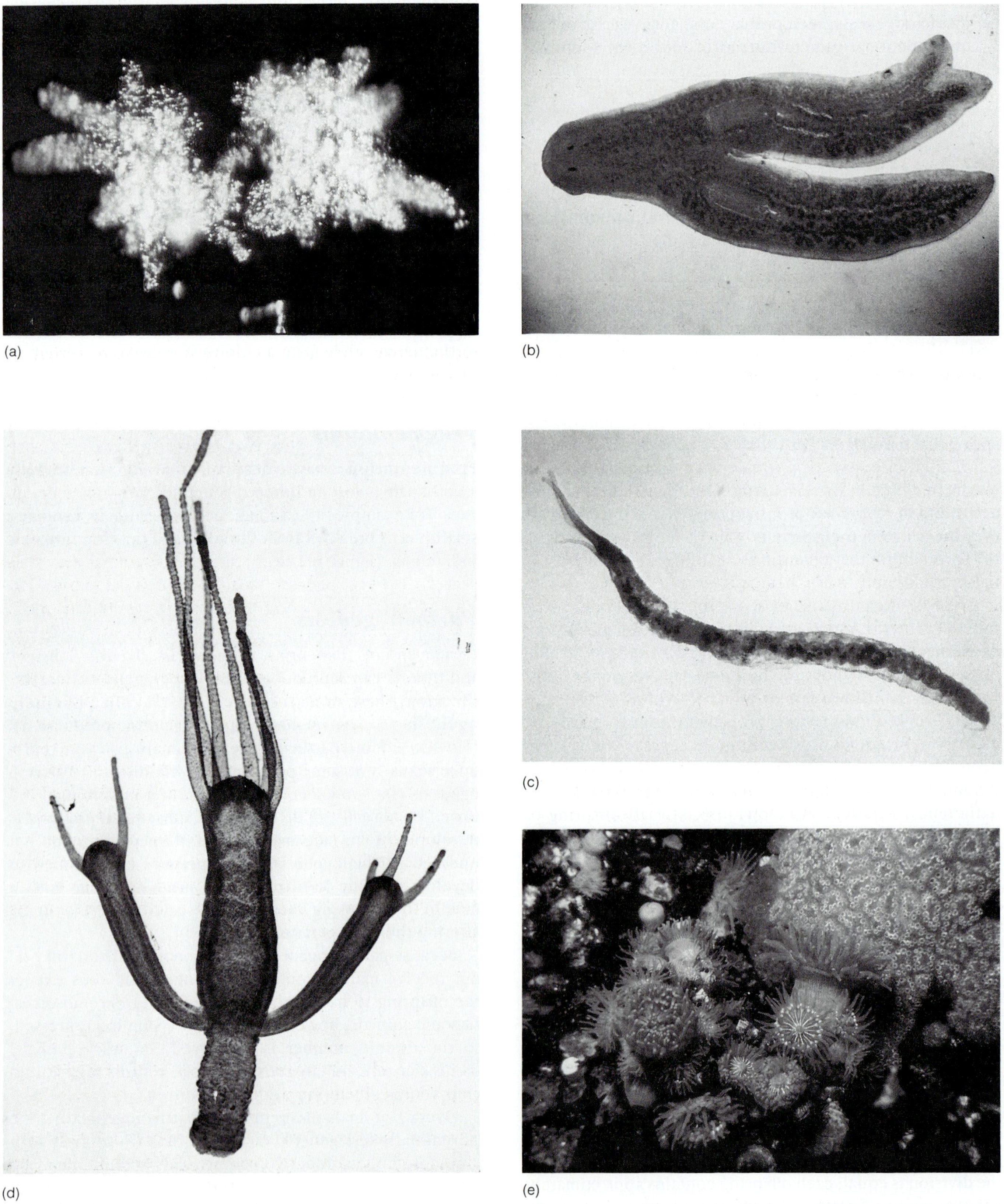

Figure 6.12 Asexual reproduction. (a) *Amoeba,* a one-celled eukaryote, undergoes fission to form two individual organisms. (b) Planarian worms undergoing transverse fission. (c) The annelid, *Autolysis,* undergoing various constrictions. (d) A hydra with a developing bud. (e) Small sea anemones produced by fragmentation.

ants. In these insects, large numbers of males (drones) are produced parthenogenetically, whereas sterile female workers and reproductive females (queens) are produced sexually.

Asexual Reproduction: Advantages and Disadvantages

The predominance of asexual reproduction in protists and some animals can be partially explained by the environment in which they live. The marine environment is usually very stable. Stable environments may favor this form of reproduction because a combination of genes that matches the relatively unchanging environment well would be an advantage over a greater number of gene combinations, many of which would not match well with the environment. In other habitats, asexual reproduction is seasonal. The season during which asexual reproduction occurs coincides with the period when the environment is predictably hospitable. Under such conditions, it is advantageous for the organism to produce asexually a large number of progeny with identical characteristics. A large number of organisms, well adapted to a given environment, can be produced even if only one individual is present.

Without the tremendous genetic variability bestowed by meiosis and sexual processes, however, a population of genetically identical organisms stands a greatly increased chance of being devastated by a single disease or environmental insult, such as a long drought. A given line of asexually reproducing organisms can cope with a changing environment only through the relatively rare spontaneous mutations (alterations in genetic material) that prove to be beneficial. Paradoxically, however, most mutations are detrimental or lethal, and herein lies one of the greatest dis-

[9]

advantages of asexual reproduction: all such mutations will be passed on to every offspring along with the normal, unmutated genes. Consequently, the typical asexual organism may have only one "good" copy of each hereditary unit (gene); the one on the homologous chromosome may be a mutated form that is nonfunctional or potentially lethal.

Sexual Reproduction

In *sexual reproduction,* the offspring have unique combinations of genes inherited from the two parents. Offspring of a sexual union are somewhat different from their parents and siblings—they have genetic diversity. Each new individual represents a combination of traits derived from two parents, because the union of two gametes in syngamy,

or fertilization, unites one gamete from each parent. Each gamete is haploid and carries only one member of each homologous pair of chromosomes characteristically found in the cells of that species. The diploid zygote that results from syngamy contains a double set of chromosomes, one set having been contributed by each parent. In due course, the zygote will divide and form an embryo that will develop into a new individual.

Variant Forms of Sexual Reproduction

Usually, sexual reproduction involves the fusion of gametes from a male and female parent. However, some sexually reproducing animals occasionally depart from this basic reproductive mode and exhibit variant forms of sexual reproduction.

Hermaphroditism (Gr. *hermaphroditos,* an organism with the attributes of both sexes) occurs when an animal has both functional male and female reproductive systems. This dual sexuality is sometimes called the **monoecious** (Gr. *monos,* single + *oikos,* house) condition. Although some hermaphrodites fertilize themselves, most also mate with another member of the same species (e.g., earthworms and sea slugs). When this occurs, each animal serves as both male and female—donating and receiving sperm. Hermaphroditism is especially beneficial to sessile (attached) animals (e.g., barnacles) that may only occasionally encounter the opposite sex.

Another variation of hermaphroditism, **sequential hermaphroditism,** occurs when an animal is one sex during one phase of its life cycle and the opposite sex during another phase. Hermaphrodites are either *protogynous* (Gr. *protous,* first + *gyne,* women) or *protandrous* (Gr. *protos,* first + *andros,* man). In protandry, an animal is a male during its early life history and becomes a female later in the life history. The reverse is true for protogynous animals. A change in the sex ratio of a population is one factor that can induce sequential hermaphroditism, which is common in oysters, and a number of fishes, including anemone fishes and the common aquarium molly.

[10]

Sexual Reproduction: Advantages and Disadvantages

New combinations of traits can arise more rapidly in sexually reproducing animals because of recombination. The resulting genetic diversity or variability increases the chances of the species surviving sudden environmental changes. Furthermore, variation is the foundation for evolution. In contrast to the way asexual reproduction can retain mutations, copies of deleterious and lethal mutations tend to be eliminated from sexually reproducing populations.

[11]

Sexual reproduction also has some disadvantages. For example, a mammal that cannot reproduce asexually can never bequeath its own exact set of genetic material to its progeny. As noted earlier in this chapter, meiosis bestows on the progeny a reassortment of maternal and paternal chromosomes. Thus, the same mixing processes that created the successful gene combinations in the adult, work to dismantle it partially in the offspring.

Stop and Ask Yourself

13. How does each of the following types of asexual reproduction occur: fission, budding, and fragmentation?
14. How can parthenogenesis occur?
15. What are some advantages and disadvantages of asexual reproduction?
16. Does hermaphroditism have beneficial aspects? Explain.
17. What are some advantages and disadvantages of sexual reproduction?

■ Summary

1. The replication of DNA and its subsequent allocation to daughter cells during mitosis involves a number of phases, collectively called the cell cycle. The cell cycle is that period from the time a cell is produced until it goes through mitosis.

2. Mitosis maintains the parental number of chromosome sets in each daughter nucleus. It separates the sister chromatids of each (duplicated) chromosome for distribution to daughter nuclei.

3. Interphase represents about 90% of the time of the eukaryotic cell cycle.

4. The first phase of mitosis is prophase, during which the mitotic apparatus forms. At the end of prophase, the nuclear envelope disintegrates, and microtubules attach each of the sister chromatids to the two poles of the cell.

5. The second phase of mitosis is metaphase. During this phase, the sister chromatids align along the metaphase plate in the center of the cell. At the end of metaphase, the centromeres joining each pair of sister chromatids split, freeing each chromatid, although the sister chromatids remain next to each other.

6. The third phase of mitosis is anaphase, during which the chromosomes and their copies (the sister chromatids that have been joined to them) are pulled to opposite poles of the cell by the microtubules.

7. The final phase of mitosis is telophase, during which the mitotic apparatus is disassembled, the nuclear envelope re-forms, and the chromosomes unfold in the nucleoplasm.

8. Most cells undergo cytokinesis by being pinched in two by a belt of microfilaments and microtubules.

9. Control of cell division is the result of a balance between factors that stimulate cell division and factors that inhibit it, including cessation of protein synthesis, depletion of nutrients, and contact inhibition.

10. Meiosis is a special form of nuclear division that occurs during gamete formation in most eukaryotes. It consists of a single replication of the chromosomes and two chromatids

four cells, each with half the original number of chromosomes.

11. In the life cycle of most animals, certain diploid cells undergo gametogenesis to form haploid gametes (sperm in males and eggs in females). Fusion of a sperm and an egg nucleus at fertilization produces a new diploid cell (zygote), which undergoes mitosis and cytokinesis to develop into a new individual.

12. Asexual reproduction occurs in some eukaryotes by fission, budding, fragmentation, or parthenogenesis and is usually advantageous in relatively unchanging environments.

13. Sexual reproduction increases the genetic diversity of a population by recombination in parental cells and is advantageous in changing environments.

■ Key Terms

anaphase (p. 89)
aster (p. 89)
budding (p. 97)
cell cycle (p. 87)
centromere (p. 89)
chalones (p. 91)
chiasma (p. 93)
chromatid (p. 89)
colony (p. 97)
crossing-over (p. 93)
cytokinesis (p. 87)
fission (p. 97)
fragmentation (p. 97)
gametes (p. 91)
gametogenesis (p. 94)
genetic recombination (p. 93)

hermaphroditism (p. 99)
homologous chromosomes (p. 93)
interphase (p. 87)
kinetochore (p. 89)
meiosis (p. 91)
metaphase (p. 89)
mitosis (p. 87)
mitotic apparatus (p. 89)
monoecious (p. 99)
oogenesis (p. 94)
parthenogenesis (p. 97)
prophase (p. 89)
sequential hermaphroditism (p. 99)
somatic cells (p. 91)
spermatogenesis (p. 94)

synapsis (p. 93)

synaptonemal complex
(p. 93)

syngamy (p. 91)

telophase (p. 89)

zygote (p. 91)

■ Critical Thinking Questions

1. Humans are diploid organisms and are biased in favor of diploidy over haploidy. However, what are the biological advantages of having diploid cells? Are there advantages for some organisms being haploid?

2. If the function of meiosis were only to make haploid cells from a diploid cell, this could be accomplished in a single cell division; i.e., cancelling chromosome replication and separating homologous chromosomes would produce two haploid gametes from each diploid cell. If this could occur, what important aspect of meiosis would be omitted from this simple procedure? Is reproduction life's death-defying escape from inevitable destruction? Explain.

3. Which do you think evolved first, meiosis or mitosis? Why? What do you think may have been some of the stages in the evolution of one from the other?

4. How can aberrant forms of reproduction be advantageous to an organism? Give several examples.

5. How do new genetic combinations arise during meiosis?

■ Suggested Readings

Books

John, P. C. L. (ed). 1985. *The Cell Cycle*. New York: Cambridge University Press.

Mitichison, J. 1972. *Biology of the Cell Cycle*. New York: Cambridge University Press.

Prescott, D. M. 1976. *Reproduction of Eukaryotic Cells*. New York: Academic Press.

Zimmerman, A. M., and Forer, A. 1981. *Mitosis/Cytokinesis*. New York: Academic Press.

Articles

Beams, S. W., and Kessel, R. G. 1976. Cytokinesis: a comparative study of cytoplasm division in animal cells. *American Scientist* 64:279–290.

Benditt, J. Genetic skeleton. *Scientific American* July, 1988.

Chandley, A. C. 1988. Meiosis in man. *Trends in Genetics* 4:79–83.

Inoue, S. 1981. Cell division and the mitotic spindle. *Journal Cell Biology* 91:132s–137s.

John, B. 1976. Myths and mechanisms of meiosis. *Chromosoma* 54: 295–325.

Mazia, D. The cell cycle. *Scientific American* January, 1974.

McIntosh, J. R., and McDonald, K. L. The mitotic spindle. *Scientific American* October, 1989.

Pardee, A. B. 1978. Animal cell cycle. *Annual Review Biochemistry* 47: 715–750.

Sloboda, R. D. 1980. The role of microtubules in cell structure and cell division. *American Scientist* 68:290–298.

Silberner, J. Off on switch for cell division found. *Science News* August 24, 1985.

Stahl, F. Genetic recombination. *Scientific American* February, 1987.

Yanieshevsky, R. M., and Stein, G. H. 1981. Regulation of the cell cycle in eukaryotic cells. *Internation Review Cytology* 69:223–259.

7

Inheritance Patterns

Concepts

1. Modern genetics began with the work of Gregor Mendel. The principles that he described explain the inheritance patterns of many animal traits.
2. Certain mathematical tools are used by biologists to aid in the analysis of genetics problems.
3. Rapid developments in 20th century genetics have expanded upon the work of Gregor Mendel. Geneticists have discovered that:
   (a) many alternative forms of a gene may exist in a population,
   (b) these alternative forms may interact in different ways,
   (c) some traits are determined by interactions between many gene pairs, and
   (d) the environment influences the expression of genes.

Have You Ever Wondered:

[1] why modern genetics is so important to biology and our modern society?
[2] how an Augustinian monk ever got started crossing garden peas?
[3] why mathematics is such a valuable tool in genetics?
[4] why there are so many expressions of some traits?
[5] why some traits, such as height, show a range of variation, whereas other traits show all-or-none expression?
[6] how the environment influences the expression of some genes?

These and other useful questions will be answered in this chapter.

This chapter contains underlined evolutionary concepts.

Genetics (Gr. *gennan*, to produce) is the study of the transmission of biological information from one generation to a following generation, and of how information is expressed in an organism. Humans have been aware for centuries that certain traits are passed from parents to offspring and have used that knowledge in selective breeding of domesticated plants and animals.

The importance of genetics to modern societies, however, surpasses anything that ancient animal breeders could have imagined. Modern genetics began with the work of Gregor Mendel (1822–1884). Since then, geneticists have been performing crosses, characterizing the genetic material (DNA), and investigating the functions of DNA.

The genetic revolution has had a tremendous effect on biology. Genetics can explain why offspring are, in some ways, similar to their parents, whereas in other ways, they are very different. Genetic mechanisms also help explain how species change over evolutionary time. Genetic and evolutionary themes are interdependent in biology, and biology without either would be unrecognizable from its present form.

[1] Emerging genetic technologies are also forcing our society to come face-to-face with some very important decisions regarding what, if any, limits should be placed on genetic manipulations. The benefits that accrue from genetic technologies are impressive. Genetic technologies permit the mass production of hormones, antibodies, or other proteins that have been available only in very small quantities and at very high costs. It may be possible to dramatically increase agricultural production by manipulating the genetic material of crops and livestock. During the lifetime of most of the readers of this text, it will probably become possible to insert properly functioning genetic material into humans with defective genes.

Even though genetic technologies hold great promise for the future, they must not be pursued without consideration of possible negative effects. The scientific community and society in general, must deal with questions relating to the effects of genetically engineered bacteria and other organisms on the environment, the use of genetically engineered organisms for waging war, and the moral and ethical dilemma faced by anyone who tries to decide what kinds of genetic manipulations may be performed on humans.

In light of these concerns we devote three chapters of this textbook to genetics. Many of the genetic principles that can be traced back to the beginnings of modern genetics are covered in this chapter. The chromosomal basis of inheritance is covered in chapter 8. Modern concepts of genes and gene functions and some of the modern technologies that allow humans to manipulate genes are covered in chapter 9.

The Birth of Modern Genetics

Gregor Mendel was born in 1822 in an area of Europe now known as Czechoslovakia (figure 7.1). He was the son of

Figure 7.1 Gregor Mendel (1822–1884): the father of genetics.

a farmer and grew up knowing about the cultivation and propagation of crop plants. He entered an Augustinian monastery in Brunn (Brno, Czechoslovakia) and became a priest at the age of 25. He was then given the opportunity to study physics, mathematics, and natural sciences at the University of Vienna in Austria. His studies in Vienna were completed in 1854, and Mendel returned to Brunn where he began teaching.

In 1857, Mendel began experiments with the garden pea, *Pisum sativum.* Mendel was not the first scientist to study inheritance; however, his approach was unique. He [2] selected an experimental organism whose pollination could be easily controlled, he selected seven clearly defined traits to study, he carried his crosses through multiple generations, and he approached his experiments as a mathematician, as well as a biologist. Mendel published his work in 1866, but the importance of his conclusions was unrecognized by scientists of his day. Mendel died in 1884. The rediscovery and confirmation of Mendel's writings in 1900 propelled genetics into the 20th century.

Stop and Ask Yourself

1. What is genetics?
2. Why is a basic understanding of genetics an important part of any biologist's training?
3. In what way was Mendel's approach to the study of genetics unique?

Mendelian Inheritance Patterns

The fruit fly, *Drosophila melanogaster,* has become a classic tool for studying inheritance patterns. Its utility stems from its ease of handling, its short life cycle, and its easily recognized characteristics.

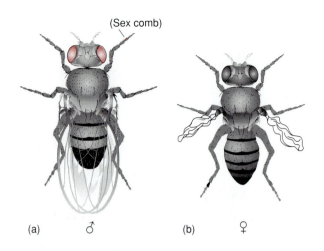

(Sex comb)

(a) ♂   (b) ♀

Figure 7.2 Distinguishing sexes and phenotypes of *Drosophila melanogaster.* (a) A male with wild-type wings and wild-type eyes. (b) A female with vestigial wings and sepia eyes. In contrast to the female, the posterior aspect of the male's abdomen has a wide dark band and a rounded tip.

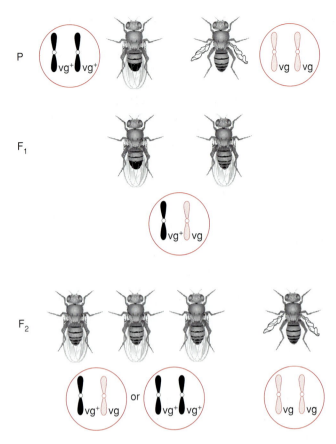

Figure 7.3 A cross between parental flies with wild-type wings and vestigial wings, carried through two generations.

When any fruit fly trait is studied, comparisons are always made to a wild-type fly. If a fly has a characteristic similar to that found in wild flies, it is said to have the wild-type expression of that trait. (In the examples used below, wild-type wings lay over the back at rest, and extend past the posterior tip of the body, and wild-type eyes are red.) Numerous mutations from the wild-type body form have been described, such as vestigial wings (reduced, shriveled wings) and sepia (brown) eyes (figure 7.2).

Segregation

Mendel's Principle of Segregation can be illustrated by crossing wild-type fruit flies with flies having vestigial wings. (These flies come from stocks of flies that have been inbred for generations to be sure they bred true for wild-type wings or vestigial wings.) These flies are members of the *parental generation* and one finds that all of their offspring (progeny) have wild-type wings. These wild-type progeny are members of the first generation of offspring, or the *first filial (F_1) generation* (figure 7.3). If these flies are allowed to mate with each other, the progeny that result are members of the *second filial (F_2) generation.* Approximately 1/4 of these flies have vestigial wings and 3/4 have wild wings (figure 7.3). Note that the vestigial characteristic, though present in the parents, disappears in the F_1 generation and reappears in the F_2 generation. In addition, the ratio of wild-type flies to vestigial winged flies in the F_1 generation is approximately 3:1. We would also find that reciprocal crosses would yield similar results. (Reciprocal crosses involve the same characteristics, but the investigator reverses the sexes of the individuals introducing a particular expression of the trait into the cross.)

From results like these, Mendel concluded that traits were determined by heritable "factors," which we now call **genes** (Gr. *genos,* race). Genes that determine the expression of a particular trait can exist in alternative forms called **alleles** (Gr. *allelos,* each other). The vestigial allele is present in the F_1 generation in our example, and, even though it is masked by the wild-type allele for wing shape, it retains its uniqueness because it is expressed again in some members of the F_2 generation. Those factors that can hide the expression of another factor Mendal called **dominant** and **recessive** factors were those whose expression could be masked. In our example, the wild-type allele is dominant because it can mask the expression of the vestigial allele, which is therefore recessive.

The occurrence of flies with masked traits brings up two more important points that Mendel noticed in his pea experiments.

The visual expression of alleles may not always indicate the underlying genetic make-up of the animal. This visual expression is called the **phenotype** and the genetic make-up is called the **genotype**. In our example, the flies of the F_1 had the same phenotype as one of the parents, but they differed genotypically because they carried both a dominant and recessive allele. They are hybrids, and because this cross concerns only one trait, it is called a **monohybrid cross** (Gr. *monos,* one + L. *hybrida,* offspring of two kinds of parents).

An organism is said to be **homozygous** (L. *homo,* same + Gr. *zygon,* paired) if it carries two identical genes for a given trait, and is **heterozygous** (Gr. *heteros*, other) if the genes are different. Thus, in our example, all members of the parental generation, are homozygous because we chose true-breeding flies to cross. All members of the F_1 generation are heterozygous.

Crosses are often diagrammed using symbols to represent different alleles. A letter or letters are chosen that are descriptive of the trait in question. Often, the first letter of the description of the dominant allele is used. In fruit flies, and other organisms where all mutants are compared to a wild-type, the symbol is taken from the allele that was derived by a mutation from the wild condition. A superscript "+" next to the symbol represents the wild-type allele. A capital letter means that the allele being represented is dominant, and a lower case letter means that the allele being represented is recessive. Thus, we can represent our orignal cross as follows:

	Wild-type		Vestigial
P generation	vg^+vg^+	X	$vgvg$
		↓	
F_1 generation		vg^+vg	

Matings of the F_1 flies resulted in a ratio of three wild-type flies to one vestigial fly in the F_2 generation. Mendel recognized that a 3:1 ratio would result when two alleles, each present in a separate gamete, combined randomly at fertilization (*see chapter 6*).

During the formation of gametes, the alleles in each parent are incorporated into separate gametes. Recall that during meiosis I, homologous chromosomes move toward opposite poles of the cell, and the resulting gametes have only one member of each chromosome pair (*see figure 6.7*). In our example, one gene for wing condition of the wild-type parent is on each member of one homologous pair, and all gametes produced by this fly end up with one wild-type gene. Similarly, the pair of vestigial genes present in the vestigial fly are segregated into separate gametes during meiosis. Mendel's **Principle of Segregation** states that pairs of hereditary factors are distributed between gametes during gamete formation. (Mendel, of course, knew nothing of the cellular basis of gamete formation). Fertilization results in the random combination of gametes and brings homologous chromosomes together again.

The analysis of crosses becomes easier when one approaches it in a systematic fashion. The *Punnett square* is a tool used by geneticists to help predict the results of crosses. A cross of two F_1 flies from our example will be used to illustrate the use of a Punnett Square (figure 7.4). The first step is to determine the kinds of gametes produced by each parent. One of the two axes of a square is designated for each parent, and the different kinds of gametes produced by each parent are placed along the appropriate axis. Combining gametes in the interior of the square tells us the results of random fertilization. Figure 7.4 shows that the F_1 flies are heterozygous, with one wild-type allele and one vestigial allele. The F_2 generation is shown on the in-

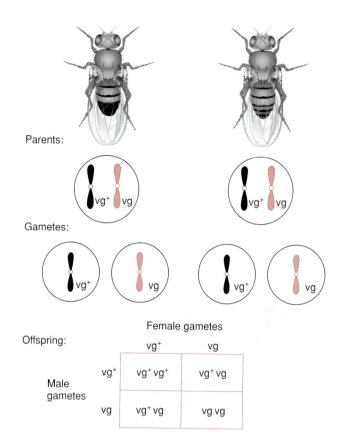

Figure 7.4 A Punnett square can be used to help predict the results of a cross. After determining the kinds of gametes produced by each member of a cross, gametes produced by each sex are placed along the axes of a square. Results of matings are analyzed by combining gametes and counting results.

side of the Punnett square. There are two phenotypes, in a ratio of 3:1.

The *phenotypic ratio* expresses the results of a cross according to the relative numbers of progeny in each visually distinct class (e.g., 3 wild-type : 1 vestigial). The Punnett square has thus explained in another way the F_2 results in figure 7.3. It also shows that there are three different genotypes in F_2 individuals. The *genotypic ratio* expresses the results of a cross according to the relative numbers of progeny in each genotypic category (e.g., 1 vg^+vg^+ : 2 vg^+vg : 1 $vgvg$). A 3:1 phenotypic ratio and a 1:2:1 genotypic ratio are typical results of monohybrid crosses. The results of crosses rarely achieve phenotypic or genotypic ratios exactly. Instead, the results approximate these expected ratios and generally are closer to the expected values as more progeny are produced. You have now seen that Mendel's Principle of Segregation is an explanation of how members of a pair of genes are distributed into separate gametes. At fertilization, the paired condition is restored and results in predictable phenotypic and genotypic ratios in the next generation.

Independent Assortment

In his crosses, Mendel also followed two pairs of characteristics. We can make similar crosses using flies that have

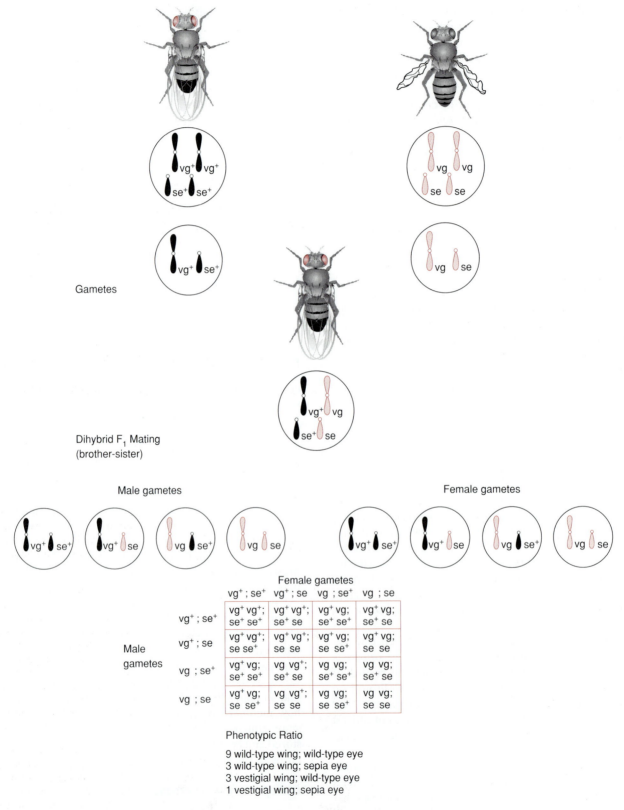

Gametes

Dihybrid F₁ Mating
(brother-sister)

Male gametes   Female gametes

	Female gametes			
	vg⁺ ; se⁺	vg⁺ ; se	vg ; se⁺	vg ; se
vg⁺ ; se⁺	vg⁺ vg⁺; se⁺ se⁺	vg⁺ vg⁺; se⁺ se	vg⁺ vg; se⁺ se⁺	vg⁺ vg; se⁺ se
vg⁺ ; se	vg⁺ vg⁺; se se⁺	vg⁺ vg⁺; se se	vg⁺ vg; se se⁺	vg⁺ vg; se se
vg ; se⁺	vg⁺ vg; se⁺ se⁺	vg vg⁺; se⁺ se	vg vg; se⁺ se⁺	vg vg; se⁺ se
vg ; se	vg⁺ vg; se se⁺	vg vg⁺; se se	vg vg; se se⁺	vg vg; se se

Male
gametes

Phenotypic Ratio

9 wild-type wing; wild-type eye
3 wild-type wing; sepia eye
3 vestigial wing; wild-type eye
1 vestigial wing; sepia eye

Figure 7.5 Constructing a Punnett square for a cross involving two characteristics. Note that every gamete has one allele for each trait and that all combinations of alleles for each trait are represented.

vestigial wings and sepia eyes and flies that are wild for these characteristics. Sepia eyes are dark brown, and wild-type eyes are red. The results of crosses carried through two generations are shown in figure 7.5.

Note that in the parental generation, the flies are homozygous for the traits in question, and each parent produces only one kind of gamete. Gametes have one allele for each trait. Because each parent produces only one kind of gamete, fertilization must result in offspring heterozygous for both traits. The F_1 flies have the wild-type phenotype, thus we know that wild-type eyes are dominant to sepia eyes. The F_1 flies are hybrids, and because the cross involves two characteristics, it is a **dihybrid cross** (Gr. *di,* two + L. *hybrida,* offspring of two kinds of parents).

Gregor Mendel recognized the 9:3:3:1 ratio as what would be expected if genes obey certain probability laws. He concluded that during gamete formation, the distribution of genes determining one trait did not influence how genes determining the other trait were distributed. In our example, this means that an F_1 gamete that ends up with a vg^+ gene for wing condition may also have either the *se* or se^+ gene for eye color, as shown in the F_1 gametes of figure 7.5. Note that all combinations of the eye color and wing condition genes are present, and all combinations are equally likely. This example illustrates Mendel's **Principle of Independent Assortment,** which states that during gamete formation pairs of factors segregate independently of one another.

Mendel's Principle of Independent Assortment is explained by events of meiosis (*see figure 6.7*). Cells pro-duced during meiosis have one member of each homologous pair of chromosomes. Independent assortment simply means that when homologous chromosomes line up and then segregate, the behavior of one pair of chromosomes does not influence the behavior of any other pair (figure 7.6). Notice that after meiosis, maternal and paternal chromosomes are distributed randomly among cells.

The accurate determination of all possible gametes produced by participants in a cross is critically important in learning to handle genetics problems. You must remember that: (1) all gametes must have one (and only one) allele for each trait, and (2) all possible combinations of alleles for the traits being considered must be represented in the gametes.

Trihybrid Crosses

It is also possible to follow three pairs of alleles in genetics crosses. As you might guess, increasing the number of traits considered makes genetic analyses more difficult. The number of possible different gametes produced by organisms heterozygous for multiple traits increases according to the following formula:

$$g = (2)^n.$$

g = the number of different gametes produced by the heterozygous organism.

n = the number of gene pairs being considered.

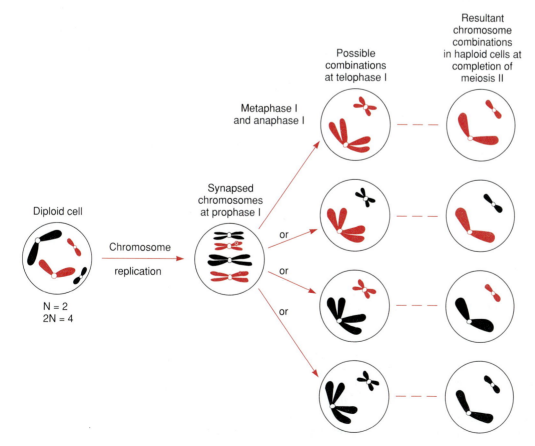

Figure 7.6 Independent assortment of chromosomes during meiosis. Maternal and paternal chromosomes are distinguished by color. Homologous chromosomes are indicated by similar size. During the first meiotic division, one homologous pair of chromosomes (and hence the genes this pair carries) is segregated without regard to the movements of any other homologous pair. Thus there are all possible combinations of large and small chromosomes in the cells at telophase I, and in the cells at the completion of telophase II. Most organisms have more than two pairs of homologous chromosomes in each cell. As the number of homologous pairs increase, the number of different kinds of gametes also increases.

Thus, a trihybrid would produce 2³ or eight different kinds of gametes. Punnett square analysis would result in an 8 by 8 square containing 64 cells and a 27:9:9:9:3:3:3:1 ratio of phenotypes in the offspring.

Testing Phenotypes

The wild-type flies in the F_2 generation of our monohybrid cross had two different genotypes. How could we determine the genotype of any one wild-type fly? One way is by carrying out a **testcross**, which involves crossing the phenotypically dominant fly to a fly that shows the recessive phenotype. (Remember that any organism that shows a recessive trait is homozygous for that trait.) The results of a testcross will be different depending on whether the phenotypically dominant fly is homozygous or heterozygous. A testcross with a fly homozygous for a dominant trait will always yield phenotypically dominant offspring and a testcross with a fly heterozygous for a dominant trait will always yield a 1:1 ratio of phenotypically dominant to phenotypically recessive offspring (table 7.1).

Stop and Ask Yourself

4. What is Mendel's Principle of Segregation? What events of meiosis does this principle reflect?
5. What is Mendel's Principle of Independent Assortment? What events of meiosis does it reflect?
6. How would you define the following terms: homozygous, heterozygous, phenotype, genotype, and allele?
7. What is a testcross?

Probability

[3] The application of some simple mathematics to genetic analysis can help make problem solving much easier. *Probability* is a measure of the certainty of an event. The probability of an event is the proportion of times an event occurs in a large number of trials. For example, tossing an ordinary coin has two possible results—the coin will land on its "heads" side or its "tails" side. Because both events are equally likely, the probability of each is one-half. Similarly, rolling a die has six equally likely results, each with a probability of one-sixth. The probability of all possible outcomes of an experiment is always one, and thus the probability of any one event is always between zero and one.

The ratio of phenotypes that one obtains from a Punnett square is simply a statement of the chances of an offspring having a particular phenotype. In a dihybrid cross, for example, the probability of an offspring being phenotypically dominant for both traits is 9/16. (The "9" is derived from the 9 in the 9:3:3:1 ratio, and the "16" is the sum of all possible outcomes. This probability could also be expressed as 0.56.)

Table 7.1 A Comparison of the Results of a Testcross of Heterozygous and Homozygous Dominant Flies

Dominant Phenotype (wild-type wings; wild-type eyes)	
If homozygous (vg^+vg^+)	**If heterozygous** (vg^+vg)
$vgvg \times vg^+vg^+$	$vgvg \times vg^+vg$
Gametes: vg vg^+	vg vg^+, vg
Offspring: $vgvg^+$ all wild-type wings.	$vgvg^+, vgvg$ 1 wild-type wings: 1 vestigial wings.

Addition Rule

The addition rule states that the probability of one of two (or more) mutually exclusive events is the sum of their individual probabilities. For example, the probability of getting a 2 OR a 6 when rolling a die is 1/6 + 1/6 or 1/3. In a cross involving flies heterozygous for the eye and wing traits described above, the probability that a single fly will have wild-type eyes and wild-type wings OR wild-type eyes and vestigial wings is 9/16 + 3/16, or 3/4. "OR" is emphasized above because when it occurs in a probability statement, it is a clear indication that the addition rule should be applied.

Multiplication Rule

The multiplication rule will be especially helpful for predicting results in crosses involving more than two traits. It states that the probability of independent events happening in a particular sequence, or simultaneously, is equal to the product of their individual probabilities. For example, the probability of rolling a die and getting a 2 on the first roll AND a 6 on the second roll is 1/6 × 1/6 or 1/36. Similarly, the probability of getting two offspring in the above dihybrid cross, the first having both wild-type traits AND the second with wild-type eyes and vestigial wings is 9/16 × 3/16 or 27/256. "AND" is emphasized above because when it occurs in a probability statement, the multiplication rule should be applied.

Simplifying Genetics

A cross involving two or more independently assorting traits can be viewed as two or more simultaneous crosses involving one trait, and the multiplication rule can be used to make analysis easier. For example, a dihybrid cross can be viewed as the product of two monohybrid crosses:

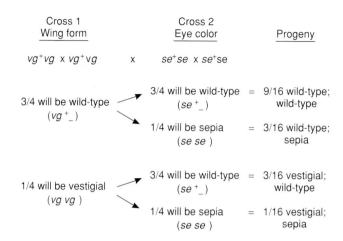

Cross 1 Wing form		Cross 2 Eye color	Progeny

$vg^+vg \times vg^+vg \quad \times \quad se^+se \times se^+se$

This system could easily be used to analyze the results of a trihybrid cross by adding a column for the third pair of heterozygous alleles to the right of the eye color trait.

Frequently, one is only interested in knowing the probability of getting one particular phenotype in a cross. For example, what is the probability of getting an offspring of the A_, B_, cc phenotype in the cross *Aa Bb Cc × Aa Bb Cc*? Again, we can view the cross as the product of three monohybrid crosses. The probability of obtaining:

$$A\_ \text{ from } Aa \times Aa = 3/4,$$
$$\text{and } B\_ \text{ from } Bb \times Bb = 3/4,$$
$$\text{and } cc \text{ from } Cc \times Cc = 1/4.$$

Therefore, the probability of A_, B_, cc is 3/4 × 3/4 × 1/4, or 9/64.

Stop and Ask Yourself

8. What mathematical procedure would you use if you were asked to find the probability of event "X" OR event "Y" occurring?
9. What procedure would you use to find the probability of event "X" AND event "Y" occurring?
10. What is the range of probabilities that is possible for any event?

Other Inheritance Patterns

Traits that we have considered so far have been determined by two genes, where one allele was dominant to a second. In this section, you will find that there are often many alleles in a population, not all traits are determined by an interaction between a single pair of dominant or recessive genes, and various internal and external environmental factors can affect the phenotype.

Multiple Alleles

Traits are determined in one individual by two genes, one carried on each chromosome of a homologous pair. Unlike

Table 7.2   Genotypes and Phenotypes in the ABO Blood Groups

Genotype(s)	Phenotype
$I^A I^A$, $I^A i^O$	A
$I^B I^B$, $I^B i^O$	B
$I^A I^B$	A and B
$i^O i^O$	O

the situation in an individual, a population may have many different alleles that have the potential to contribute to the phenotype of any member of the population. These are called **multiple alleles.** Genes for a particular trait are located at the same position on a chromosome. The gene's position on the chromosome is called its **locus** (L. *loca*, place). As will be presented in chapters 12 to 14, the diversity of alleles in a population is very important in supplying the phenotypic variation upon which evolution can act. [4]

Numerous human loci have multiple alleles. The familiar ABO blood types are determined by three alleles, symbolized I^A, I^B, and i^O. The combinations of alleles that determine a person's phenotype are shown in table 7.2. Note that i^O is recessive to I^A and to I^B. I^A and I^B, however, are neither dominant nor recessive to each other. When I^A and I^B are present together, both are expressed.

The Rh blood type is also determined by multiple alleles. It was discovered during World War II that matching ABO blood types did not always ensure a successful transfusion. Another blood cell trait, first discovered in the Rhesus monkey, was found to be responsible for transfusion rejections when the ABO blood type was matched. The Rh system is inherited as if it is determined by two alleles with Rh positive being dominant to Rh negative. In fact, the system is much more complex. It now appears that there are in excess of thirty alleles determining the Rh phenotype!

Incomplete Dominance and Codominance

Incomplete dominance is an interaction between two alleles that are expressed more-or-less equally, and the heterozygote is different from either homozygote. In cattle, the alleles for red coat color and for white coat color interact to produce an intermediate called roan. Because neither the red nor the white allele is dominant, we use upper case letters and a prime or a superscript to represent genes. Thus red cattle would be symbolized RR, white cattle would be symbolized R'R', and roan cattle would be symbolized RR'.

Codominance occurs when the heterozygote expresses the phenotypes of both homozygotes. Thus, in the ABO blood types, the $I^A I^B$ heterozygote expresses both alleles.

Another example of codominance is the inheritance of sickle cell anemia. Sickle cell anemia arose in Africa as a result of a mutation in the genes coding for one of the proteins making up hemoglobin. The normal allele is symbolized Hb^A, and the sickle allele is symbolized Hb^S. When present in the homozygous condition, the sickle cell allele results in severe sickling of red blood cells, and the anemia that results is usually fatal. Even though the Hb^s allele is very detrimental in the homozygous state, it is much less dangerous when heterozygous. The heterozygote has about equal numbers of Hb^A and Hb^s hemoglobin molecules in his/her cells, and sickling only develops under oxygen stress. The anemia is not only less severe, but the heterozygote is less susceptible to infections by the organism that causes malaria. Prior to malarial control programs in Africa, the heterozygote had a better chance of surviving than either homozygote!

Crosses involving incomplete dominance and codominance are analyzed in exactly the same manner as the previous crosses discussed. Mendelian rules are obeyed, but the results are interpreted differently. For example, a cross between a roan bull and a roan cow would be analyzed as follows:

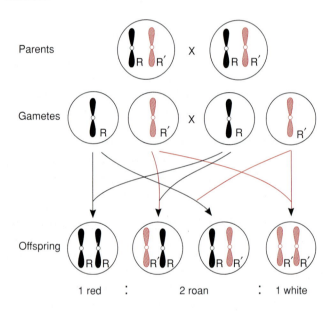

1 red : 2 roan : 1 white

The 1:2:1 genotypic ratio is now also the phenotypic ratio.

Modifier Genes, Epistasis, and Pleiotropy

Many genes, called **modifier genes,** enhance or dilute the effects of genes at other loci. Coat color in mammals, for example, is determined by genes from at least five interacting loci. Genes at A and B loci determine the basic color of the hair, and those at the D locus may modify the expression of the genes at the other two loci. If the genes at the D locus are homozygous dominant or heterozygous, the genes at the A and B loci are fully expressed. If the genes at the D locus are homozygous recessive, however, the expression of the genes at the A and B loci is diluted. For example, chestnut coat color of horses is determined by a particular combination of genes at A and B loci. When present with homozygous recessive genes at the D locus, the chestnut genotype is transformed to cremello (nearly white).

The same example can be used to describe another kind of gene interaction. Coat color in mammals is also influenced by a gene at the C locus. When the recessive allele is present in the homozygous state, the expression of other coat color loci is blocked, and the mammal is white. This interaction is an example of **epistasis** (Gr. *epi,* upon + *stasis,* to stand), which occurs when genes at one locus prevent the expression of genes at a second locus.

A gene does not always influence only one trait. **Pleiotropy** (Gr. *pleion,* greater + L. *tropica,* to turn) is a condition in which a gene has two or more phenotypic effects. *Phenylketonuria* is a genetic disease caused by a recessive gene that results in a deficiency of the enzyme phenylalanine hydroxylase. Phenylketonuriacs are unable to metabolize the amino acid phenylalanine, and the resultant accumulation of phenylalanine and phenylpyruvic acid causes a variety of phenotypic effects including: mental retardation, larger than normal heads, and lightened hair color. The pleiotropic effect of many genes reflects the fact that most genes code for enzymes. The modification of a single enzyme may interrupt many processes dependent on a particular biochemical pathway.

Quantitative Traits: Polygenes

In all traits considered thus far, genes produced discrete variations in phenotypes. It is easy to distinguish a vestigial-winged fruit fly from a wild-type winged fruit fly, or a person with type A blood from a person with type O blood. Such traits display *discontinuous variation.* Many other traits, however, vary by small increments and must be categorized by some kind of quantitative measure (e.g., weight or height must be measured). Traits such as height, skin color, weight, and intelligence display *continuous variation* over a range of phenotypes.

Traits that display continuous variation are often determined by an interaction of multiple loci, where each locus [5] adds a small increment to a phenotype. A mating between very tall and very short individuals (figure 7.7a) would produce offspring intermediate in height (figure 7.7b). A mating between two individuals that are heterozygous at each locus produces a range of phenotypes (figure 7.7c). Plotting the frequency of each phenotype in the phenotypic range produces the bell-shaped distribution that is commonly seen in polygenic inheritance in a population (figure 7.7d). Because it requires quantitative measurements to distinguish between phenotypes, traits like these are often called **quantitative traits.** Similarly, the participation of multiple loci in determining these traits prompts the use of the term **polygenes** when referring to the genes involved.

The steplike transitions between phenotypes seen in figure 7.7d are made smaller as more loci are added to a cross. As discussed in the next section, no genes are isolated from external and internal environments. Environmental conditions often influence the degree to which genes are expressed and, therefore, they make transitions between phenotypes of quantitative traits smoother.

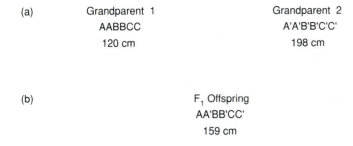

(a)

Grandparent 1
AABBCC
120 cm

Grandparent 2
A'A'B'B'C'C'
198 cm

(b)

F₁ Offspring
AA'BB'CC'
159 cm

(c) Possible F₂ Offspring:

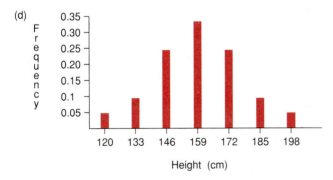

Female gametes

	A'B'C'	A'BC'	A'B'C	AB'C'	ABC'	A'BC	AB'C	ABC
A'B'C'	198	185	185	185	172	172	172	159
A'BC'	185	172	172	172	159	159	159	146
A'B'C	185	172	172	172	159	159	159	146
AB'C'	185	172	172	172	159	159	159	146
ABC'	172	159	159	159	146	146	146	133
A'BC	172	159	159	159	146	146	146	133
AB'C	172	159	159	159	146	146	146	133
ABC	159	146	146	146	133	133	133	128

Male gametes

(d)

Height (cm): 120, 133, 146, 159, 172, 185, 198 — Frequency (bar graph)

Figure 7.7 An example of how polygenic inheritance can produce continuous variation. Assume three loci are involved with the determination of height, and assume that each gene designated with a prime adds 13 cm to a base height of 120 cm. (a) Grandparents showing the two phenotypic extremes. (b) Offspring in the next generation are intermediate in height. (c) A mating between one member of that second generation and a person of similar stature could produce offspring that are very tall, very short, or anywhere between those extremes. (The numbers in the Punnett square were derived by multiplying 13 by the number of primed alleles in a cell and adding that product to 120 cm) (d) Note in the bar graph that the distribution between extremes approaches a normal (bell-shaped) curve. Slight environmental variations would smooth the steplike transitions from one phenotypic class to another.

Environmental Effects and Gene Expression

It is tempting to envision our genetic makeup as a set of blueprints that dictates our final form, just like the blueprints of a building dictate the final form of a home. That picture is simplistic. For most traits, the outcome of a set of genetic instructions is the product of a genotype interacting with the environment. One can think of the interaction as a genotype establishing a more-or-less fixed potential; and whether or not that potential will be realized depends upon important environmental influences. For example, nutritional factors may influence whether or not a person achieves his/her genetic potential as regards physical or mental stature. [6]

It is not too difficult to investigate the effects of the environment on gene expression in animals that can be manipulated experimentally. One simply maintains animals of identical genotypes in different environments and observes the effect of the environment on the phenotype. Numerous traits in the fruit fly are influenced by the environment. For example, the number of facets (lenses) in certain compound eye mutants is affected by the temperature at which cultures are maintained.

Nature/Nurture and Intelligence

It is more difficult to study the effects of the environment on gene expression in humans. Scientists are not able to experimentally manipulate human matings or human environments. One of the controversies in human genetics concerns whether one's genes or environment is primarily responsible for determining intelligence. In spite of the fact that there are problems in using IQ tests to make cross-cultural comparisons, they do provide useful data when comparisons are made within a more limited sample. Results from studies using techniques similar to those described above have given conflicting impressions. Although it is not possible to give an accurate evaluation of the percent contribution of genotype versus environment, it is clear that both factors influence intelligence. The nature of stimuli and the encouragement given to a young child, educational expectations of a particular socioeconomic group, and dietary considerations, have been implicated in influencing intelligence.

Stop and Ask Yourself

11. What are multiple alleles?
12. Define the term "locus."
13. How would you define the following terms: epistasis, modifier genes, pleiotropy, and polygenes?
14. How do quantitative inheritance patterns explain the occurrence of continuous variation in some traits?
15. What environmental factors do you think influence one's intelligence?

■ Summary

1. Genetics is the study of the transmission of information from one generation to a following generation. Genetic research of the last 20 years has influenced both science and society.

2. Gregor Mendel worked with garden peas and concluded that traits are determined by pairs of factors (genes), and that one gene may be dominant to another (the recessive gene). These genes are passed unaltered from parents to offspring.

3. Mendel's Principle of Segregation states that pairs of genes are distributed between gametes during gamete formation when homologous chromosomes are distributed to different gametes during meiosis.

4. Mendel's Principle of Independent Assortment states that during gamete formation, pairs of genes segregate independently of one another. This is a result of meiotic processes in which members of one homologous pair of chromosomes are not influenced by the movements of any other pair of chromosomes.

5. Applying the probability rules to genetic analyses helps one predict the results of genetic crosses.

6. Populations may have many alternative expressions of a gene at any locus. Human traits, like the ABO and Rh blood groups, are examples of traits determined by multiple alleles.

7. Incomplete dominance is an interaction between two alleles where the alleles contribute more-or-less equally to the phenotype. The heterozygote is intermediate between the parental extremes.

8. Codominance is an interaction between two alleles where both alleles are expressed in the heterozygote.

9. Modifier genes enhance or dilute the effects of genes at other loci.

10. Epistatic genes block the expression of genes at other loci.

11. Pleiotropy is the condition in which genes have two or more phenotypic effects.

12. Quantitative traits vary by small increments, displaying continuous variation. These traits are the result of genes at many loci (polygenes) adding small increments of expression to a phenotype.

13. Most traits are not determined only by the genotype of an organism. Rather, the environment influences the degree to which a genotype will be expressed.

■ Key Terms

alleles (p. 104)
codominance (p. 109)
dihybrid cross (p. 107)
dominant (p. 104)
epistasis (p. 110)
genes (p. 104)
genetics (p. 103)
genotype (p. 104)
heterozygous (p. 105)
homozygous (p. 104)
incomplete dominance (p. 109)
locus (p. 109)

modifier genes (p. 110)
monohybrid cross (p. 104)
multiple alleles (p. 109)
phenotype (p. 104)
pleiotropy (p. 110)
polygenes (p. 110)
Principle of Independent Assortment (p. 107)
Principle of Segregation (p. 105)
quantitative traits (p. 110)
recessive (p. 104)
testcross (p. 103)

■ Critical Thinking Questions

1. Supply the genotypes, as completely as possible, for the parents and progeny of the following crosses:

   (a) wild-type wing; sepia eye × wild-type wing; wild-type eye =
   3 wild-type wing; sepia eye : 1 vestigial wing; sepia eye : 3 wild-type wing; wild-type eye : 1 wild-type wing; sepia eye,

   (b) wild-type wing; sepia eye × wild-type wing; wild-type eye =
   1 wild-type wing; sepia eye : 1 wild-type wing; wild-type eye,

   (c) vestigial wing; wild-type eye : wild-type wing; sepia eye =
   all wild-type wing; wild-type eye,

   (d) vestigial wing; wild-type eye × wild-type wing; sepia eye =
   1 vestigial wing; wild-type eye : 1 vestigial wing; sepia eye : 1 wild-type wing; sepia eye : 1 wild-type wing; wild-type eye.

2. Assuming the parental genotypes *AaBbCc* × *AaBbcc*, what is the probability that:

   (a) the next offspring will be genotypically *AABBCc*?

   (b) the next offspring will be genotypically *aaBbcc* or *AAbbCc*?

   (c) the next offspring will be phenotypically *ABc*?

3. The following progeny are the result of a cross between two fruit flies. Unfortunately, the phenotypes of the parental flies were not recorded. Formulate an hypothesis regarding the genotypes of the parental flies. (Hint: consider the ratio between wing phenotypes separately from eye phenotypes in formulating your hypothesis.)

   Progeny
   293 wild-type wing; wild-type eye
   310 wild-type wing; sepia eye
   97 vestigial wing; wild-type eye
   100 vestigial wing; sepia eye

4. Indicate the *ABO* genotypes of the members of the following family. (You may not have enough information to fill in all genotypes completely.)

   Grandfather—Type *A* ___ Grandfather—Type *O* ___
   Grandmother—Type *AB* ___ Grandmother—Type *B* ___
         ↓   ↓
       Father—Type *A*   ×   Mother—Type *B*
       ___   ___
                       ↓
   Son₁—Type *A*   Daughter—Type *B*   Son₂—Type *AB*
   ___   ___   ___

5. Do you think that Mendel's conclusions regarding assortment of genes for two traits would have been any different if he had used traits encoded by genes carried on the same chromosome? Explain.

6. Explain why the phenotypic range of polygenic traits shows a gradual transition between phenotypes rather than the stair-step transitions shown in figure 7.7d.

7. Design an experiment that could be used to determine whether or not a particular trait in fruit flies is influenced by an environmental factor, such as temperature, light, or nutritional state. Why is the investigation of environmental effects more difficult in human genetics studies than in studies of fruit flies?

■ Suggested Readings

Books

Ehrlich, P. R., and Feldman, S. S. 1977. *The Race Bomb: Skin Color, Prejudice, and Intelligence.* New York: Quandrangle.

Cummings, M. R. 1988. *Human Heredity.* Minnesota: West Publishing Co.

Dunn, L. C. 1965. *A Short History of Genetics.* New York: McGraw-Hill.

Farnsworth, M. W. 1988. *Genetics,* 2nd ed. New York: Harper and Row Publishers.

Iltis, H. 1932. *Life of Mendel.* (Transl. by E. Paul and C. Paul). Boston: W.W. Norton.

Klug, W. S., and Cummings, M. R. 1988. *Concepts of Genetics,* 2nd ed. Columbus, OH: Charles E. Merrill Publishing Company.

Levitan, M. 1988. *Textbook of Human Genetics,* 3rd ed. New York: Oxford Univ. Press.

Stern, C., and Sherwood, E. R. (eds.). 1966. *The Origin of Genetics, A Mendel Source Book.* San Francisco: W.H. Freeman. (Contains translations of Mendel's classic papers as well as other papers by early geneticists.)

Strickberger, M. W. 1985. *Genetics,* 3rd ed. New York: Macmillan Publishing Company.

Suzuki, D. T., Griffiths, A. J. F., Miller, J. H., and Lewontin, R. C., 4th ed. 1989. *An Introduction to Genetic Analysis.* San Francisco: W.H. Freeman and Company.

Tamarin, R. H. 1986. *Principles of Genetics,* 2nd ed. Boston: Prindle, Weber and Schmidt.

Articles

Belmont, L., and Marolla, F. A. 1973. Birth order, family size, and intelligence. *Science* 182:1096–1101.

Bouchard, T. J., Jr., and McGue, M. 1981. Familial studies of intelligence: a review. *Science* 212:1055–1059.

Diamond, J. Blood, genes, and malaria. *Natural History* February, 1989.

Murray, J. How the leopard gets its spots. *Scientific American* March, 1988.

8

Chromosomes and Gene Linkage

Concepts

1. The genetic material is in the form of chromatin, and during mitosis and meiosis, it is organized into chromosomes. Chromosomes may be represented differently in males and females; however, the number of chromosomes is constant for a given species.
2. Genes that are on the same chromosome (linked) tend to be inherited together. Inheritance patterns differ depending upon whether genes are linked on sex chromosomes or on autosomes and the distance between linked genes. Linkage relationships are used to help determine the position of loci on chromosomes.
3. Human pedigree analysis allows geneticists to determine how a trait is inherited.
4. Changes in chromosome number and structure can occur. The pheontypic consequences of such changes depend upon the amount of genetic material lost or duplicated.

Have You Ever Wondered:

[1] why there are approximately equal numbers of males and females in the world?
[2] how geneticists map chromosomes?
[3] why some traits, such as color-blindness, are more common in males than in females?
[4] how "sex checks" of Olympic athletes are carried out?
[5] whether or not females can express pattern baldness?
[6] what causes genetic diseases, such as Down Syndrome?

These and other useful questions will be answered in this chapter.

This chapter contains underlined evolutionary concepts.

Neither Gregor Mendel nor any other geneticist of the time had any inkling of the cellular mechanisms by which traits were passed from generation to generation. Although the cell theory was well accepted by the mid 1800s, little was known of cellular processes. The nucleus had been described; however, it was not until the early 1900s that an understanding of the functions of the nucleus began to emerge.

A few years before Mendel's death, Friedrich Schneider (1831–1890) and Walther Flemming (1843–1905) described darkly staining nuclear threads, now called chromosomes. Around the turn of the century, Edouard van Beneden (1849–1922) announced that the number of chromosomes was usually constant in each cell of an organism. The only exception occurred during gamete formation, when the number of chromosomes was reduced by one-half. Not long after this discovery, the involvement of the nucleus in fertilization was described, and the chromosomal theory of inheritance was announced. These discoveries were quickly followed by those of Thomas Hunt Morgan (1866–1945) and his coworkers showing that genes occur in linear arrays on chromosomes. The purpose of this chapter is to examine how genes are arranged on chromosomes, and to examine how these arrangements influence patterns of inheritance.

Eukaryotic Chromosomes

For most of the life of a cell, the genetic material is in the form of **chromatin** (Gr. *chroma,* color), which is loosely organized DNA, protein, and RNA. During mitosis and meiosis, chromatin is coiled and folded into chromosomes so that it can be precisely distributed between cells.

Chromosomes are usually studied as they appear during metaphase of mitosis. Metaphase chromosomes consist of two rodlike structures, called chromatids joined at the centromere (*see figure 6.4*). Two chromatids of a chromosome are often thought of as being exact copies of one another, but there are occurrences that may result in one chromatid having different genes than the chromatid to which it is joined (its "sister chromatid"). In addition to the centromere, other constrictions, called secondary constrictions, may occur along the length of a chromosome. When a secondary constriction occurs near the tip of a chromatid, the portion of the chromosome at the tip of the chromatid is called a **satellite.** Secondary constrictions usually are regions of a chromosome responsible for reorganizing the nucleolus after cell division is completed.

Recall that chromosomes of most animals exist in homologous pairs, which carry genes that code for the same traits and are visually similar. In particular, chromosome size and centromere position help geneticists recognize homologous pairs.

Heterochromatin and Euchromatin

Geneticists have found that not all chromatin is the same because all regions of a chromosome are not equally active.

Some human genes may be active only after adolescence is achieved. In other cases, entire chromosomes may not function in particular cells. These *inactive regions* are said to be **heterochromatic,** whereas *active regions* of chromosomes are said to be **euchromatic.** Certain regions of chromosomes, especially around the centromere, are heterochromatic in all preparations from a particular species. These regions contain constitutive heterochromatin, and produce dark banding patterns with certain staining procedures. Banding patterns can also be used to identify particular chromosomes. Regions of chromosomes that alternate between heterochromatic and euchromatic states are said to contain facultative heterochromatin.

Organization of DNA and Protein

DNA of eukaryotic organisms is coiled around a core of four pairs of *histone proteins,* a combination called a **nucleosome.** Approximately 200 nucleotide pairs wrap in a coil around the histones, which are designated H2A, H2B, H3, and H4. An H1 spacer or linker histone is associated with the DNA between nucleosomes. The nucleosome chain is folded into a cylindrical coil, resembling a *solenoid.* Further folding results in formation of the chromosomes we can recognize during mitosis and meiosis (figure 8.1).

The processes involved in the condensation of chromatin into chromosomes (prophase of mitosis and prophase I of meiosis), the unfolding of chromatin (telophase of mitosis and telophase II of meiosis), and the replication of DNA (*S* stage of interphase) are complex indeed. Unfortunately, little is known of how these processes are controlled.

Sex Chromosomes and Autosomes

In the early 1900s, attention turned to the cell to find a chromosomal explanation for determination of maleness or femaleness. The first evidence for a chromosomal basis for sex determination came from work with the insect *Protenor.* One darkly staining chromosome of *Protenor,* called the X chromosome, is represented differently in males and females. All somatic (body) cells of males have one X chromosome (*XO*), and all somatic cells of females have two X chromosomes (*XX*). Similarly, one-half of all sperm contain a single X and one-half contain no X. All female gametes contain a single X. This pattern suggests that fertilization of an egg by an X-bearing sperm will result in female offspring, and fertilization of an egg by sperm with no X chromosome will result in a male offspring. As illustrated in figure 8.2, this sex determination system explains the approximately 50:50 ratio of females to males in this insect species. Chromosomes that are represented differently in females than in males and function in sex determination are called **sex chromosomes.** Chromosomes that are alike and not involved in determining sex are called **autosomes** (Gr. *autus,* self + *soma,* body).

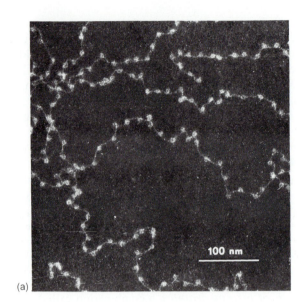

(a)

Figure 8.1 The organization of eukaryotic chromosomes. (a) An electron micrograph showing a strand of chromatin. Nucleosomes appear as beads arranged sequentially on a thread of DNA. (b) A nucleosome consists of four pairs of histone proteins wound by a strand of DNA. Linker proteins are associated with DNA between adjacent nucleosomes. (c) The nucleosome/DNA chain is wound into a solenoid, which then undergoes folding.

Figure 8.1 redrawn, with permission, from W. M. Becker, *The World of the Cell.* Copyright © 1986, Benjamin-Cummings Publishing Company, Inc., Menlo Park, California.

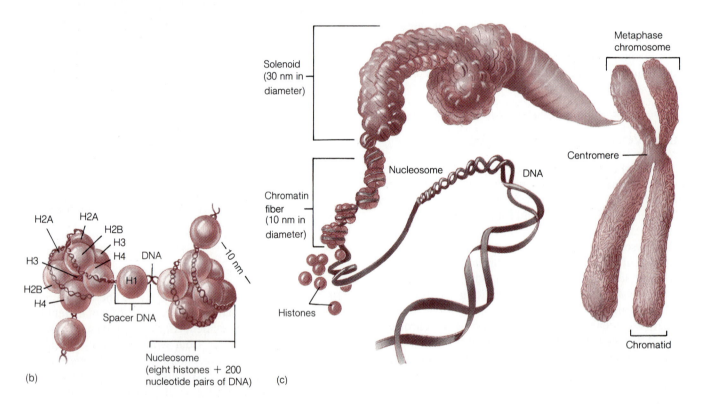

(b)

(c)

The system of sex determination described for *Protenor* is called the *X-O system.* It is the simplest system for determining sex because it involves only one kind of chromosome. Many other animals, however, have an *X-Y system* of sex determination. In the *X-Y* system, males and females have an equal number of chromosomes, but the male is *XY* and the female is *XX.* The sex chromosomes may look quite different. Figure 8.3 shows the X and Y chromosomes of male fruit flies (*Drosophila melanogaster*) and humans. This mode of sex determination also results in approximately equal numbers of male and female offspring:

[1]

	Sperm	
	X	Y
Egg X	XX	XY

1 female : 1 male

It is a mistake to think that the number of X chromosomes, or the presence of a Y chromosome are the only genetic factors that determine sex. The *XO* or *XY* status of an individual is often referred to as the "chromosomal sex" of an individual. Genes that are not a part of sex chromosomes can influence, and even reverse the chromosomal sex of some animals.

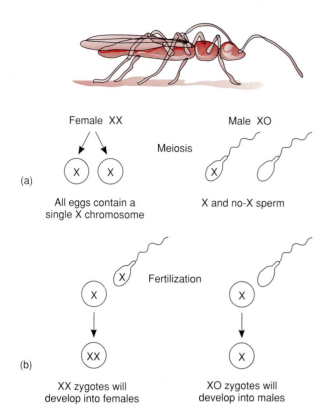

Figure 8.2 The *XO*-system of sex determination for the insect *Protenor*. (a) In females, all cells except gametes possess two X chromosomes. During meiosis, homologous X chromosomes segregate and all eggs contain one X chromosome. Males possess one X chromosome per cell. Meiosis results in one-half of all sperm cells having one X, and one-half of all sperm cells having no X. (b) Fertilization results in one-half of all offspring having one X chromosome—these offspring are males, and one-half of all offspring having two X chromosomes—these are females.

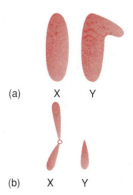

Figure 8.3 Sex chromosomes. (a) The sex chromosomes of a male fruit fly and male human (b). Even though X and Y chromosomes carry different genes, and are of different size and shape, they function during meiosis as a homologous pair. Pairing during prophase I of meiosis is less precise than for autosomal pairs.

Table 8.1	Chromosome Numbers of Selected Animals	
Common Name	**Scientific Name**	**Number of Chromosomes**
Fruit fly	*Drosophila melanogaster*	8
Frog	*Rana pipiens*	26
Honeybee	*Apis mellifera*	32
Hydra	*Hydra vulguria attenuata*	32
Cat	*Felis domesticus*	38
Rat	*Rattus norvegicus*	42
Human	*Homo sapiens*	46
Chimpanzee	*Pan troglodytes*	48
Dog	*Canis familiaris*	78
Chicken	*Gallus domesticus*	78
Carp	*Cyprinus carpio*	104

Number of Chromosomes

Even though the number of chromosomes is constant within a species, there is great variation in chromosome number among species (table 8.1). One cannot predict, based on size or complexity of the organism, how many chromosomes a species will have.

Chromosomes are present in sets. By far, the most common condition in animals is for one copy of each pair of homologous chromosomes to be present. This is called the **diploid** (Gr. *di*, two + *eoides*, doubled) condition. Some animals have one member of each pair of chromosomes in each cell and are said to be **haploid** (Gr. *hapl*, single) (e.g., male honeybees and some rotifers).

It is unusual for animals to have more than the diploid number of chromosomes. When this condition does exist, it is called **polyploidy** (Gr. *polys*, more). The upset in numbers of sex chromosomes apparently interferes with reproductive success. When polyploidy does occur (brine shrimp, snout beetles, some flatworms, and some sow bugs), it is often accompanied by asexual reproduction.

Stop and Ask Yourself

1. What is heterochromatin? What is euchromatin?
2. What is the difference between chromatin and chromosomes? What is a nucleosome?
3. Why are there approximately equal numbers of males and females in human offspring?

Linkage Relationships and Chromosome Mapping

Within a few years of the rediscovery of Mendel's work, William Bateson and R. C. Punnett discovered that certain traits of the pea violated the Principle of Independent Assortment. Rather than getting a 9:3:3:1 ratio of progeny from a dihybrid cross, they got a 7:1:1:7 ratio. Bateson and Punnett suggested that their results could be explained by a physical connection between the genes. We now know

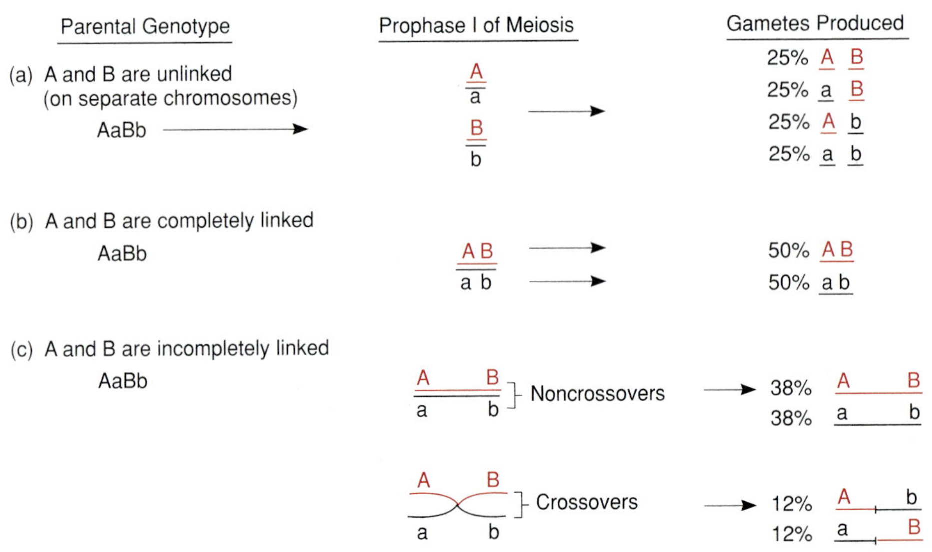

Parental Genotype	Prophase I of Meiosis	Gametes Produced

(a) A and B are unlinked
(on separate chromosomes)

AaBb ⟶

$\dfrac{A}{a}$

$\dfrac{B}{b}$ ⟶

25% A B
25% a B
25% A b
25% a b

(b) A and B are completely linked

AaBb

$\dfrac{A\ B}{a\ b}$ ⟶

50% A B
50% a b

(c) A and B are incompletely linked

AaBb

$\dfrac{A\quad B}{a\quad b}$ ⎤ Noncrossovers ⟶ 38% A   B
                                              38% a   b

A   B ⎤ Crossovers ⟶ 12% A___b
a   b   12% a___B

Figure 8.4 The effects of gene linkage on the production of different types of gametes. (a) As described in chapter 7, unlinked genes assort into gametes independently of one another if they are on separate chromosomes. (b) Complete linkage occurs when loci are so close together on a chromosome that crossing-over is very unlikely. (c) Incomplete linkage occurs when loci are further from one another, and crossing-over results in recombinant genotypes.

that many genes are carried on the same chromosome and are called **linked genes.** Because they tend to be inherited together, these genes belong to a **linkage group.** Genes that are on the same chromosome are carried together to one pole of the cell during anaphase I of meiosis, and do not assort independently, although the chromosomes do. Linked genes, however, are not linked forever, because crossing-over can occur (*see figure 6.8*). In crossing-over, parts of homologous chromosomes can be exchanged and genetic variability among offspring can be increased.

Autosomal Linkage

Genes that are linked to autosomal chromosomes tend to be inherited together, and the pattern of inheritance is similar in both males and females. Occasionally, genes are so closely linked that they are always inherited together. This is called **complete linkage.** More often, however, there is some crossing-over between linked genes, and such genes are said to show **incomplete linkage.** Figure 8.4 illustrates the kinds of gametes that would be produced by individuals in a dihybrid cross assuming: no linkage, complete linkage, and incomplete linkage. Crossing-over takes place in incomplete linkage, although the progeny showing the genotypes resulting from crossing-over (*Aabb* and *aaBb*) are less common than the genotypes where the linkage between genes has been retained (*AaBb* and *aabb*).

Crossing-over of linked genes occurs during prophase I of meiosis and is also called recombination. Exactly where recombination occurs along a chromosome is random, but it is not accidental. A complex of proteins, called the synaptonemal complex helps homologous chromosomes pair in a gene-to-gene fashion, and proteins within that complex, called **recombination nodules,** are apparently responsible for snipping and rejoining homologous chromosomes. If chromatid fragments from homologous chromosomes are rejoined after snipping, crossing-over has occurred.

Chromosome Mapping

In 1911, Thomas Hunt Morgan and his coworkers suggested a relationship between recombination of linked genes and the linear distance between genes on a chromosome. They suggested that as the distance between two genes increases, the frequency of crossing-over between those genes increases. If crossing-over is equally likely at every point along a chromosome, and the distance between loci A and C is twice that between A and B, then the chances of crossing-over someplace between A and C are twice as great as the chances of crossing-over between A and B.

This observation explains the difference between completely linked and incompletely linked genes. Completely linked genes are so close together on a chromosome that the chances of crossing-over between them is very small. Incompletely linked genes are further apart on the chromosome, and thus crossing-over occurs more frequently.

This observation also suggests that recombination frequency could be used as a relative measure of the distance between two loci on a chromosome. *Recombination frequency* is the fraction of all offspring that show crossing over between two loci. Recombination frequencies have been used by geneticists to construct **chromosome maps,** also called **linkage maps.** Rather than depicting the absolute distance between loci in discrete units of measurement, chromosome maps show relative distances using units of measurement called *map units (mu).* Map units are calculated by expressing recombination frequencies as percentages, where one map unit equals 1% recombination.

Maps of entire chromosomes are gradually established as geneticists determine distances between adjacent loci. For example, if the distances between three loci A, B, and C are known, as well as the distances between A, B, and D, then a map of all four loci can be established (figure 8.5). Figure 8.6 is a map of many of the loci found on the X chromosomes of *Drosophila melanogaster* (box 8.1).

[2]

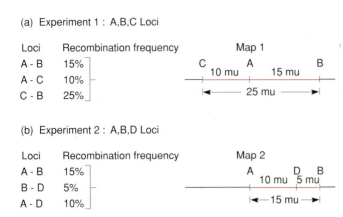

(a) Experiment 1 : A,B,C Loci

Loci	Recombination frequency
A - B	15%
A - C	10%
C - B	25%

Map 1

(b) Experiment 2 : A,B,D Loci

Loci	Recombination frequency
A - B	15%
B - D	5%
A - D	10%

Map 2

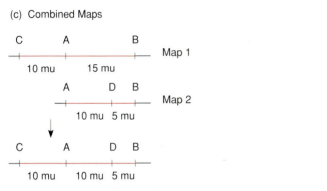

(c) Combined Maps

Figure 8.5 Chromosome maps of many loci are established by combining the results of two or more mapping experiments, gradually adding additional loci. In the experiments shown here, determining the relative distances between A, B, and C loci (a), and determining the relative distances between A, B, and D loci (b), allows one to construct a map showing the relative positions of all four loci (c).

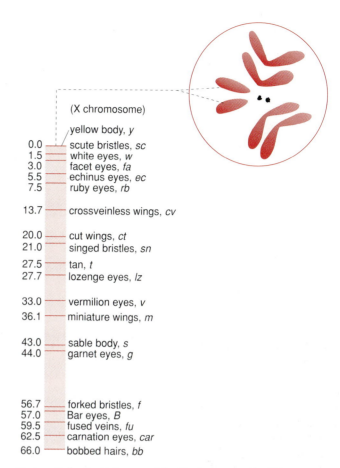

(X chromosome)

	yellow body, *y*
0.0	scute bristles, *sc*
1.5	white eyes, *w*
3.0	facet eyes, *fa*
5.5	echinus eyes, *ec*
7.5	ruby eyes, *rb*
13.7	crossveinless wings, *cv*
20.0	cut wings, *ct*
21.0	singed bristles, *sn*
27.5	tan, *t*
27.7	lozenge eyes, *lz*
33.0	vermilion eyes, *v*
36.1	miniature wings, *m*
43.0	sable body, *s*
44.0	garnet eyes, *g*
56.7	forked bristles, *f*
57.0	Bar eyes, *B*
59.5	fused veins, *fu*
62.5	carnation eyes, *car*
66.0	bobbed hairs, *bb*

Figure 8.6 Partial map of the X chromosome of *Drosophila melanogaster*. Map units are shown to the left of the chromosome, and the name of each locus is indicated on the right.

Sex Linkage

Genes linked to sex chromosomes do not necessarily influence traits having to do with sexual characteristics, and inheritance patterns for sex-linked genes are different from those for autosomally linked genes.

X Linkage The primary consideration concerning X linkage has to do with the fact that females usually have two X chromosomes, and males usually have a single X chromosome. Therefore, females have two genes for each trait. Males, however, have only one gene to consider, and are said to be **hemizygous** for X-linked traits. Genes in the hemizygous state are always expressed, regardless of whether they are dominant or recessive because there is no allele on a homologous chromosome to interact with.

In crosses involving X linkage, it is important to remember that the genes involved are carried on the X chromosome. In symbolizing these genes, it is customary to use an "*X*" to symbolize the X chromosome and superscript letters to symbolize the gene (e.g., X^a or X^A). A "*Y*" is used to designate the other sex chromosome.

An example will illustrate important considerations when working with crosses involving X linkage. The white

eye color in *Drosophila* is produced by a recessive X-linked allele. Note that the two crosses in figure 8.7 involve white-eyed flies and red-eyed (wild-type) flies in the parental generation. The results of the two crosses illustrate how X linkage can affect the F_1 flies. In cross A, it is the female that introduces the white allele, and in cross B, it is the male that introduces the white eye allele. The results of a cross differ depending on the sex of the parent that introduces the mutant gene into a cross. Therefore, it is important in crosses involving X linkage to keep track of the sex and phenotype of parents and offspring.

X-linked, recessive traits tend to be expressed in male offspring. If the mother is heterozygous (a carrier), she will not show the trait, but approximately one-half of her sons will (*see figure 8.7, Cross A, F_2 and Cross B, F_2*). Male offspring always receive the X chromosome from their mother and the Y chromosome from their father, and thus X-linked traits tend to be passed to males from their mothers.

As illustrated in Cross A, X-linked recessive genes are often expressed in males of alternate generations. A father that shows a recessive, X-linked trait will pass the gene to all of his daughters. The daughters will then be carriers and may pass the gene to one-half of their sons.

[3]

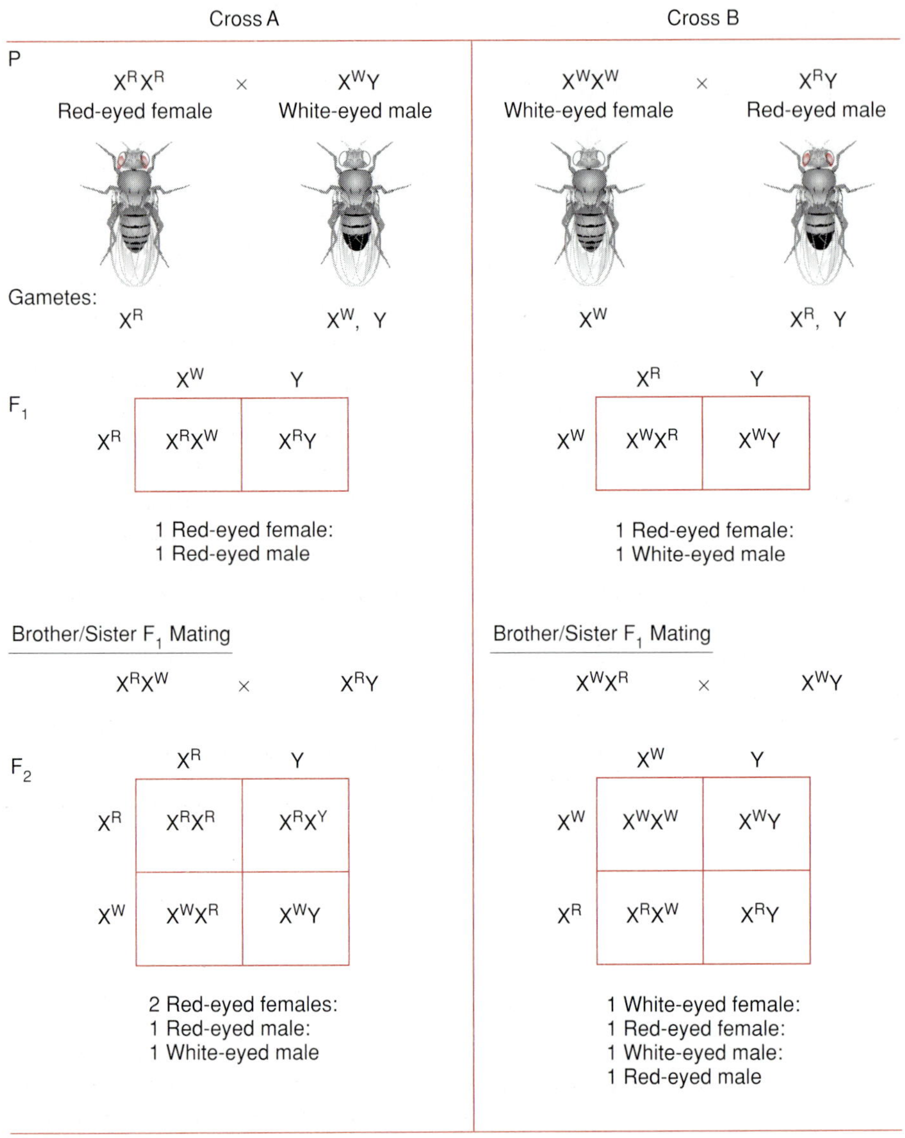

Cross A

P

$X^R X^R$ × $X^W Y$
Red-eyed female   White-eyed male

Gametes: X^R X^W, Y

F₁ → **F$_1$**

	X^W	Y
X^R	$X^R X^W$	$X^R Y$

1 Red-eyed female:
1 Red-eyed male

Brother/Sister F$_1$ Mating

$X^R X^W$ × $X^R Y$

F$_2$

	X^R	Y
X^R	$X^R X^R$	$X^R X^Y$
X^W	$X^W X^R$	$X^W Y$

2 Red-eyed females:
1 Red-eyed male:
1 White-eyed male

Cross B

$X^W X^W$ × $X^R Y$
White-eyed female   Red-eyed male

Gametes: X^W X^R, Y

	X^R	Y
X^W	$X^W X^R$	$X^W Y$

1 Red-eyed female:
1 White-eyed male

Brother/Sister F$_1$ Mating

$X^W X^R$ × $X^W Y$

	X^W	Y
X^W	$X^W X^W$	$X^W Y$
X^R	$X^R X^W$	$X^R Y$

1 White-eyed female:
1 Red-eyed female:
1 White-eyed male:
1 Red-eyed male

Figure 8.7 Patterns of inheritance of X-linked traits. Crosses A and B are called reciprocal crosses because they differ only in regard to the sex of the parent that introduces a trait (white eye) into the cross. These crosses illustrate several important considerations for working with crosses involving X linkage.

X-linked traits show up more frequently in males than in females. For a female offspring to show an X-linked trait, her father must show the trait and her mother must at least be a carrier (Cross B), or the gene in question must be dominant.

There are many human traits that are determined by genes carried on the X chromosome. One of these is red-green color blindness. Another, less common trait is hemophilia or "bleeder's disease."

The Lyon Hypothesis   The presence of unequal numbers of X chromosomes in males and females presents an interesting question. Why is it that females, with two X's, do not produce twice the amount of X-linked gene products than do males? Another interesting and related observation is that females, known to be heterozygous for an X-linked, recessive trait, occasionally express the recessive allele! Both of these observations were explained by Mary Lyon based upon her work with mice.

In 1961, she suggested that at an early stage of development, one of the two X chromosomes present in mammalian body cells, except in some ovarian tissues, is inactivated. Much earlier, Murray L. Barr (1908–) discovered a densely staining patch of chromatin present in nuclei of cells of females but not males. Lyon suggested that each **Barr body** is really an X chromosome that has become entirely inactive, or heterochromatic. This X inactivation occurs very early in development, and the X that is inactivated is determined by chance. Because inactivation is random, it should not be surprising that an occasional female, known to be heterozygous, should express a recessive X-linked trait. Barr bodies are used in diagnosing certain diseases involving sex chromosome abnormalities and they have also been used to check the sex of athletes competing in Olympic events that are open to only one sex. The procedure involves staining and examining cells scraped from the inside of the mouth.

[4]

Box 8.1

Linkage Analysis in Humans

The determination of linkage relationships in humans is more difficult than in animals such as *Drosophila melanogaster*. Some evidence for linkage relationships on the X chromosome can be derived by studying family histories. However, only a few X-linked loci have been mapped in this fashion. In recent years, however, one technique has proven especially useful in studying human linkage groups. *Somatic cell hybridization* involves stimulating cells from two tissue cultures to fuse in culture dishes, forming a new tissue culture line that is distinct from either parental line.

When mouse tumor cells and human fibroblast cells are cultured together with inactivated (made incapable of invading cells) Sendai virus, plasma membranes fuse and eventually the mouse and human chromosomes are incorporated into a single nucleus. Because cell fusion occurs in a relatively small proportion of cells, mouse/human hybrid cells must be separated from the mixture of mouse, human, and mouse/human cells. Human cells grown under these conditions eventually die. Mouse tu-

mor cells, however, will survive and reproduce. To separate mouse cells from hybrid cells special mouse tumor cell lines have been developed that are unable to synthesize some substance necessary for survival. Growth of the mixed mouse and mouse/human hybrid cells on a medium deficient in the essential substance causes the mouse cells to die. Because the human genes in the mouse/human hybrids can synthesize the essential substance, they are the only cells to survive.

To associate certain human genes with particular human chromosomes, researchers take advantage of another occurrence. When mouse/human hybrid cells are cultured, human chromosomes are randomly lost. Cytological examination of the cells can determine which chromosome is lost and various chemical studies can then determine what enzyme(s) or other protein(s) is no longer being produced. Thus particular traits can be associated with particular chromosomes. Other genetic techniques can then be used to map these traits.

Y Linkage   Y linkage is rare because the Y chromosome is small and carries few genes except those determining maleness. The pattern of transmission for Y-linked traits, when they occur, is relatively easy to diagnose. Y-linked traits are passed from a father to all of his sons because all sons must receive their father's Y chromosome. One example is hypertrichosis of the ear, which is manifested in hairy ear lobes.

Sex-Influenced Traits

Autosomally determined traits can also be expressed differently in the two sexes. These **sex-influenced traits** are mentioned briefly in this section on sex linkage because inheritance patterns can superficially resemble patterns for X-linked traits. Consider, for example, the hypothetical family history in figure 8.8, showing a distribution of pattern baldness. As with X-linked traits, a majority of those affected with baldness are males; however, unlike X linkage, the transmission is from father to son. The explanation for the preponderance of affected males lies in gene physiology rather than gene transmission. The gene for pattern baldness acts as a dominant gene in males because of high levels of testosterone, the male sex hormone. Females, on the other hand, express the trait only when the gene is

homozygous because no testosterone is present. When the genes are expressed, they result in extreme thinning of hair, not complete baldness. Because the dominance relationships of the two alleles are not fixed, upper and lower case letters cannot be used to symbolize alleles.

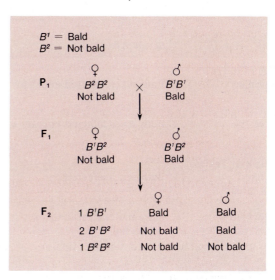

Figure 8.8   The expression of pattern baldness, a sex-influenced trait.

[5]

Stop and Ask Yourself

4. What is a linkage group?
5. What accounts for the fact that a male with a single, recessive, X-linked allele usually shows the recessive trait?
6. What accounts for the fact that females heterozygous for an X-linked trait usually show the dominant phenotype, but occasionally show the recessive phenotype?
7. What is the difference between sex-linked traits and sex-influenced traits?

Human Pedigree Analysis

Human geneticists cannot experimentally manipulate their subjects. Therefore, information about particular traits must be obtained from other sources, family histories in particular. The investigation and analysis of human pedigrees is often a first step in human genetic analysis.

Pedigree charts can be constructed from birth records and personal remembrances. In these charts, family generations are indicated by successive rows of symbols and are numbered with roman numerals. Individuals in a generation are numbered with arabic numerals, with the firstborn on the left and the lastborn on the right. Frequently roman and arabic numerals are omitted from pedigree charts. One can still use numbers to refer to specific individuals. ("III,5" would refer to the fifth individual from the left in the third generation.) Males are represented either by squares or the standard male (♂) sex symbol. Females are represented by circles or the standard female (♀) sex symbol. Matings are indicated by horizontal lines connecting male and female symbols. Offspring are shown by lines connecting matings with individuals in a subsequent generation. Individuals affected with the trait in question are indicated by shading the symbol. Carriers are indicated with a dot inside a symbol or a partially shaded symbol.

In analyzing human pedigrees, one looks at the distribution patterns of affected and unaffected individuals. Table 8.2 summarizes some of the patterns one can expect to see in pedigree charts. Sometimes pedigrees are misleading because they are incomplete, there is intermarriage in a family, or because a recessive trait is determined by a gene that is very common in a population. Whenever possible, look for individual matings that support a hypothesis (*see chapter 1*) regarding how a trait is inherited. These are called *critical matings*. For example, study the pedigree in figure 8.9 and find matings between unaffected parents that give rise to affected offspring. The only way for this to occur would be if the trait were recessive. Note also that approximately equal numbers of males and females are affected, and that traits are passed from father to son. Thus one can rule out X or Y linkage. Other inheritance patterns are shown in figures 8.10 and 8.11.

Stop and Ask Yourself

8. What is a critical mating?
9. What would constitute a critical mating in a pedigree showing an autosomal recessive trait?
10. What pattern of inheritance would you expect in a pedigree showing a Y-linked trait?

Table 8.2 Interpreting Human Pedigrees			
	Sexes Affected	**Distribution Through Generations**	**Critical Matings**
Autosomal Traits			
Dominant	Both sexes affected equally	Every generation is affected—often very common	Look for the absence of matings proving other inheritance patterns
Recessive	Both sexes affected equally	Skips generations—often fewer affected than not affected	Affected offspring can be born to unaffected parents
Sex-Linked Traits			
X-linked Dominant	Both sexes affected	Usually present in every generation	All daughters of affected males are affected. No sons are affected if mother is unaffected
Recessive	More males than females affected	Skips generations	All sons of affected females are affected. Approximately one-half of sons of carrier females are affected
Y-Linked	Only males affected	Every generation containing males shows affected males	All sons of affected fathers are affected

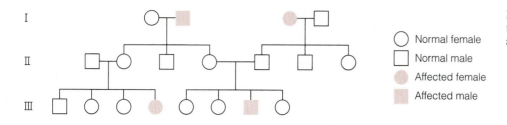

Figure 8.9 A pedigree showing a typical inheritance pattern for an autosomal recessive trait.

○ Normal female
□ Normal male
● Affected female
■ Affected male

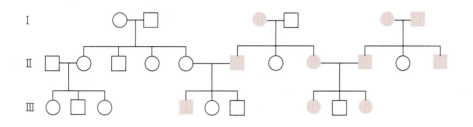

Figure 8.10 A pedigree showing a typical inheritance pattern for an autosomal dominant trait.

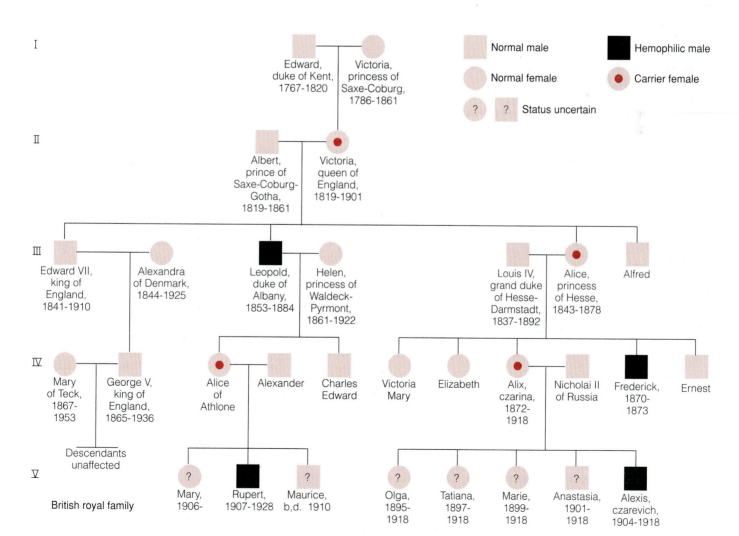

Figure 8.11 The inheritance of hemophilia in some European royal families shows a typical pattern for X-linked, recessive traits.

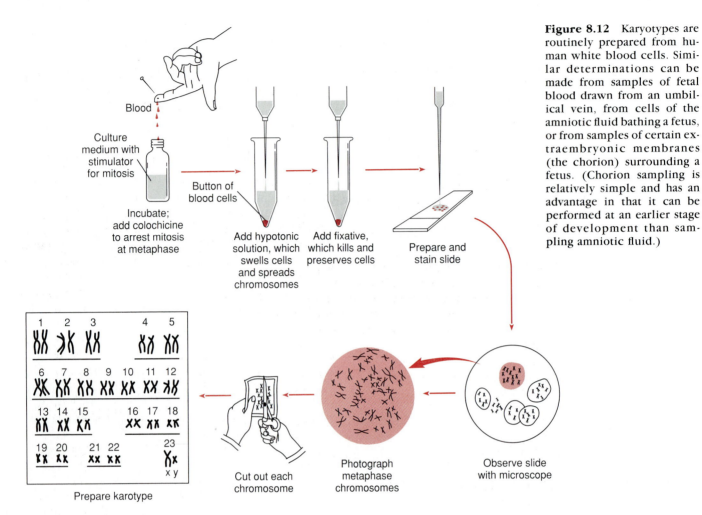

Figure 8.12 Karyotypes are routinely prepared from human white blood cells. Similar determinations can be made from samples of fetal blood drawn from an umbilical vein, from cells of the amniotic fluid bathing a fetus, or from samples of certain extraembryonic membranes (the chorion) surrounding a fetus. (Chorion sampling is relatively simple and has an advantage in that it can be performed at an earlier stage of development than sampling amniotic fluid.)

Changes in Chromosome Number and Structure

The genetic material of a cell can change, and changes are important because they increase genetic variability and help ensure survival in changing environments. One kind of genetic rearrangement, recombination, has already been presented. Crossing-over permits linked genes to recombine and experience different genetic environments. Larger, cytologically visible changes include changes in chromosome number and structure.

Detecting Number and Structure Changes

Modern tissue culture techniques permit biologists to examine chromosomes of humans and other mammals. With tissue culture techniques, it is possible to obtain large numbers of rapidly dividing cells. Preparing and examining slides and photographs of human metaphase chromosomes is called **karyotyping** (Gr. *karyo*, nucleus + *typos*, image) (figure 8.12). In karyotyping, a small quantity of blood is taken from a subject or from the umbilical tissues of a fetus. This blood is cultured in a medium that promotes the rapid division of certain white blood cells. After about

48 hours of free cell division, the drug colchicine is added to prevent cells from progressing past metaphase. Because most cells do not progress beyond metaphase, large numbers of cells with fully condensed chromosomes accumulate. Addition of a hypotonic solution causes cells to swell and chromosomes to spread within the cells, preventing excessive overlapping of chromosomes. The cells are killed, applied to a slide, and stained. Well-spread metaphase chromosomes are analyzed by computer-aided techniques or photographed. Chromosome images are arranged in order of decreasing size, by centromere position, and by banding patterns, and then numbered 1 to 22 plus X and Y (figure 8.12). Arranging the chromosomes in numbered homologous pairs, allows detection of changes in the expected structure and number of the chromosomes. These techniques are applied in diagnosing changes in chromosome numbers and chromosome structure in both fetuses and postnatal individuals.

Changes in the number of X chromosomes can be easily detected by checking the number of Barr bodies present in cheek epithelial cells. The number of Barr bodies is always one less than the number of X chromosomes. Therefore, a female with no Barr bodies would be *XO*, a female with two Barr bodies would be *XXX*, and a male with one or more Barr bodies would have two or more X's as well as a Y (box 8.2).

Genetic Engineering and Biotechnology

(a) Preparation of an electrophoretic gel for the analysis of nucleic acids.

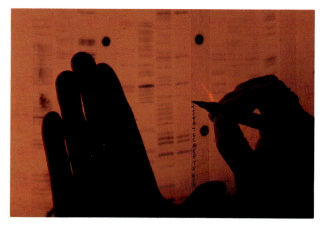

(b) Analysis of gel electrophoresis results and determination of DNA base sequences.

(c) Hybridoma cells to be used in the production of monoclonal antibodies are stored at -94°F (-34°C).

(d) Purification of a product by column chromatography.

(e) The production of interferon in a large fermenter.

Genetic Engineering and Biotechnology

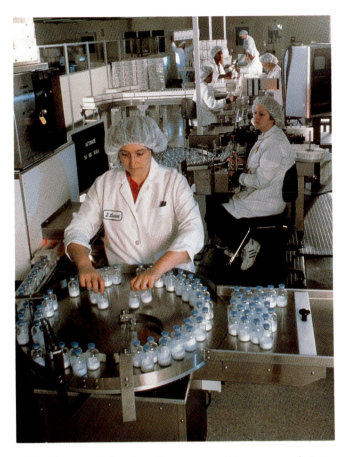

(a) At Genentech, bottles of human growth hormone are being produced by recombinant DNA techniques.

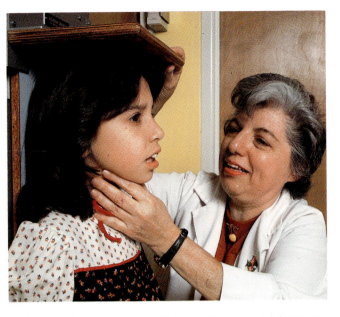

(b) A growth spurt thanks to administration of growth hormone produced by genetic engineering.

(c) Cotton plants like that on the left have been engineered to produce naturally occurring proteins that inhibit insect pests, such as the caterpillar shown damaging the plant on the right. These advances can reduce the farmer's dependence on chemical insecticides.

(d) Supermouse. Genetic engineers produced a mouse twice as big as a normal mouse by inserting a gene for human growth hormone into one of the chromosomes.

Box 8.2

Recent Advances in Genetic Technologies

Advances in genetics have come very rapidly in the last few decades (colorplates 2 and 3). Important developments have been technologies involved with biochemical detection of carrier individuals for recessive traits; technologies involved with screening the unborn for genetic diseases (*amniocentesis, chorionic villi sampling, fetoscopy,* and *ultrasound*); technologies involved with promoting fertilization (artificial insemination and in vitro fertilization); and technologies that allow gene manipulations (covered briefly in chapter 9).

The public is understandably concerned about the impact of genetic technologies on life in our world. Any major newspaper has daily accounts of the application of some genetic technology. In some articles we are reminded that our society must quickly come to terms with societal implications of genetic technologies. For example, the 1987 controversy surrounding Baby M forced the courts to decide whether or not contracts for surrogate motherhood are legally binding. Other articles indicate an expanding demand for genetic services. Since 1970, the demand for technologies involved in prenatal detection of genetic diseases has increased dramatically. In 1970, relatively few major hospitals were performing more than one amniocentesis each month. Now, amniocentesis is performed routinely and major hospitals perform thousands each year.

Rapidly expanding genetic technologies present important challenges for our society's theologians, philosophers, and ethicists. But they also present great opportunities for students interested in genetics. Not only are there opportunities in research and teaching, but public demand for genetic services has opened the rapidly growing field of genetic counseling. Most major hospitals have trained teams of genetic counselors to inform patients of the chances of an offspring having a genetic disease, to help them understand the genetic consequences of particular conditions, the alternatives available to them, and to help patients through any difficult decisions that must be made. An important job of genetic counselors is to continue to give advice and support after decisions have been made regarding particular courses of action.

Numerous colleges and universities across the country have programs that train genetic counselors. For more information on these careers you might want to consult one of the following references:

Clinical Genetic Service Centers, A National Listing. 1980. U.S. Dept. of Health and Human Services. DHHS Publication No. (HSA) 80–5135.

Lynch, H. T., Fain, P., and Marrero, K. 1980. *International Directory of Genetic Services,* 6th ed. March of Dimes. White Plains, New York.

Porter, I. H. and Hook, E. B. eds. 1979. *Service and Education in Medical Genetics.* Academic Press.

Variation in Number

Changes in chromosome number may involve entire sets of chromosomes, as in polyploidy, which was discussed earlier. **Aneuploidy** (Gr. *a,* without), on the other hand, involves the addition or deletion of one or more chromosomes, not entire sets. The addition of one chromosome to the normal 2n chromosome number (2n + 1) is called a trisomy (Gr. *tri,* three + ME. *some,* a group of), and the deletion of a chromosome from the normal 2n chromosome number (2n − 1) is called a monosomy (Gr. *monos,* single).

Aneuploidy is usually caused by errors during meiosis. **Nondisjunction** occurs when a homologous pair fails to segregate during meiosis I, or when chromatids fail to separate at meiosis II (figure 8.13). Gametes produced will either be deficient in one chromosome or have an extra chromosome. If one of these gametes is involved in fertilization with a normal gamete, the monosomic or trisomic condition will result.

[6]

Trisomy 21, or Down Syndrome, is a human trisomy involving chromosome number 21. Symptoms include short and broad hands, short stature, mental retardation, and heart problems. The average life expectancy is about 16 years. Maternal age and possibly paternal age, are important factors in the frequency of nondisjunction of these, and possibly other chromosomes. Women between the ages of 35 and 39 are seven times more likely to have a Down child than women 15 to 19 years of age. The risk increases to a 20 times higher frequency for 40 to 45-year-old women, and a 50 times higher frequency for women older than 45. This may be an important consideration for couples who marry but, for various reasons, delay having children. Other aneuploid variations usually result in severe consequences involving mental retardation and sterility.

Variation in Structure

DNA is a remarkably stable molecule. However, changes in its structure do occur. If these changes are cytologically invisible, they are called point mutations, which will be considered in the next chapter. Other changes may involve breaks in chromosomes. After breaking, pieces of chromosomes may be lost, or they may reattach, but not necessarily in their original position. The result is a chromosome that may have a different sequence of genes, multiple copies of genes, or missing genes. All of these changes can occur spontaneously. They can also be induced with various environmental agents, such as ionizing radiation and certain chemicals.

Translocations involve transfers of genetic material between nonhomologous chromosomes. Translocations may take a number of forms; the most common form involves a single break in each of two chromosomes and an exchange of broken pieces (figure 8.14a). Because translocations involve shifting, not deleting or adding genetic material, they may not be too serious.

In some instances, however, translocations have been associated with certain phenotypic abnormalities. A form of Down Syndrome, called translocation Down, results from a translocation. If the bottom third of chromosome 21 is translocated to another chromosome (most often chromosome 14), a gamete could end up with both the normal chromosome 21 and a portion of chromosome 21 being carried on chromosome 14. When fertilization occurs, the

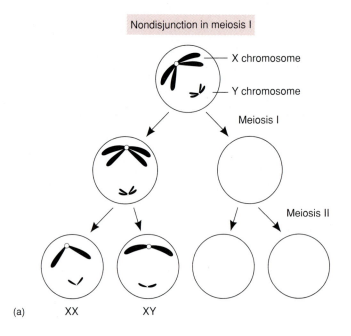

(a) Nondisjunction in meiosis I

- X chromosome
- Y chromosome

Meiosis I

Meiosis II

XX   XY

Nondisjunction of X-containing sperm in meiosis II

- X chromosome
- Y chromosome

Meiosis I

Meiosis II

(b)   Y   Y   XX

Figure 8.13 The results of primary and secondary nondisjunction in sperm formation. (a) Primary nondisjunction occurs in meiosis I and results in both the X and Y chromosomes ending up in one secondary spermatocyte. A normal second meiotic division results in one-half of all sperm cells having both X and Y chromosomes. The other one-half of all sperm cells lack any sex chromosomes. (b) Secondary nondisjunction occurs after a normal first meiotic division. Failure of the chromatids of the X chromosomes, for example, to separate in the second division means that one-fourth of the sperm cells will have no sex chromosomes, one-fourth will have two X chromosomes, and one-half will have two Y chromosomes (from the normal separation of Y chromatids).

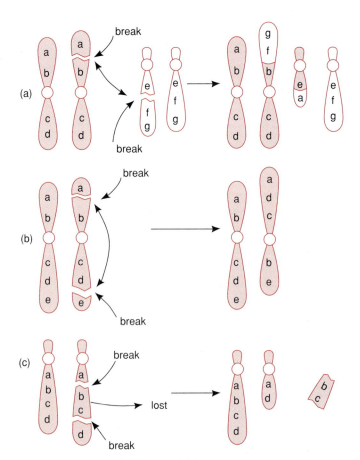

Figure 8.14 Common kinds of changes in chromosome structure. (a) Translocations involve transfers of genetic material between nonhomologous chromosomes. (b) Inversions occur when a region of a chromatid is flip-flopped, resulting in gene order changes. (c) Deletions result from the loss of a segment of a chromosome.

zygote receives three copies of some of the genes carried on chromosome 21. Another translocation that results in the movement of an arm of human chromosome 22 to chromosome 9 is the first chromosomal abnormality associated with a form of cancer. This translocated chromosome, called the Philadelphia Chromosome (Ph1), occurs in about 96% of cases of chronic myeloid leukemia. Myeloid leukemia is a malignancy of blood forming tissues (bone marrow) that causes a proliferation of certain white blood cells.

Inversions involve two breaks in a chromosome followed by the broken ends rejoining, with the central segment flip-flopped (figure 8.14b). The amount of DNA is not changed, but the order of some genes is reversed. Usually, there are no serious phenotypic effects; however, the frequency of crossing-over within an inversion is reduced.

Deletions result from breaks in chromosomes, followed by the loss of a portion of the chromosome. Deletions usually result from two breaks and the loss of an intervening segment of a chromosome (figure 8.14c). The effects of deletions increase in severity as the lost sections become larger. A human genetic disease, called Cri-du-chat (Fr.,

cry of the cat) results from a loss of part of chromosome 5. Symptoms include a very small head, wide-set eyes, mental retardation, and a catlike cry.

Duplications occur when portions of chromosomes are present in multiple copies. Duplications are difficult to detect in humans, although they probably exist. They are known in many experimental organisms. The effects of duplications are probably less severe than for deletions. In fact, duplicated regions of chromosomes are believed to have had important evolutionary consequences. Because the duplicated genes are extra copies, they are not required for expression of the phenotype. Mutations of duplicated genes may be tolerated and may even prove beneficial.

Stop and Ask Yourself

11. What procedure is used in human karyotyping?
12. What is aneuploidy and how is it likely to occur?
13. How would you describe the following kinds of chromosomal rearrangements: translocations, inversions, deletions, duplications?

Summary

1. Chromosome size, banding patterns, and centromere position can be used to identify particular chromosomes in microscopic preparations. Regions of chromosomes that contain active genes are said to be euchromatic. Regions of chromosomes, or entire sets of chromosomes, that contain inactive genes are said to be heterochromatic.

2. Eukaryotic chromosomes are complexly folded associations of DNA and histone proteins.

3. The sex of an animal is determined by the presence or absence of certain chromosomes that are represented differently in males and females. The X-O and the X-Y systems of sex determination are most common.

4. Numbers of chromosomes are constant for a given species. Although some examples of haploidy and polyploidy are cited, diploidy is most common in animals.

5. Genes carried on the same chromosome are said to be linked because they tend to be inherited together.

6. Crossing-over (recombination) between loci further apart on a chromosome will occur more frequently than crossing over between closely linked loci. Percentage of recombination is used as a relative index of the distance between linked loci.

7. X-linked traits show up more frequently in males than females because males are hemizygous for any X-linked gene and males will show the trait regardless of whether the trait is determined by a dominant or a recessive gene.

8. Y-linked traits are transmitted exclusively from father to son.

9. Geneticists often use pedigree charts to analyze inheritance patterns. Critical matings support or refute a hypothesis regarding how a trait is inherited.

10. Aneuploidy occurs when one or more, but not entire sets of, chromosomes are added to or deleted from a genome. Aneuploidy results from nondisjunction during the first or second meiotic division. Changes in chromosome number almost always result in severe phenotypic consequences.

11. Changes in chromosome structure include translocations, inversions, deletions, and duplications. The severity of the phenotypic effect depends upon the amount of genetic material lost or duplicated.

Key Terms

aneuploidy (p. 125)
autosomes (p. 115)
Barr body (p. 119)
chromatin (p. 115)
chromosome maps (p. 118)
complete linkage (p. 118)
deletions (p. 127)
diploid (p. 117)
duplications (p. 127)
euchromatic (p. 115)
haploid (p. 117)
hemizygous (p. 119)
heterochromatic (p. 115)
incomplete linkage (p. 118)
inversions (p. 127)
karyotyping (p. 124)
linkage group (p. 118)
linkage maps (p. 118)
linked genes (p. 118)
nondisjunction (p. 125)

nucleosome (p. 115)

polyploidy (p. 117)

recombination nodules
    (p. 118)

satellite (p. 115)

sex chromosomes (p. 115)

sex-influenced traits
    (p. 121)

translocation (p. 126)

■ Critical Thinking Questions

1. Coat color in cats is determined by X-linked alleles, such that black (X^B) is dominant to orange (X^b). Tortoise shell cats, however, display patches of orange and black, and they are almost always female. Explain how the Lyon Hypothesis can account for these observations.

2. Nondisjunction during the formation of a human sperm may result in *XX, YY, XY,* or no *X-* or *Y*-containing sperm depending upon when nondisjunction occurs. Diagram the meiotic events that could result in each of the above sperm. Assuming that each sperm fertilized a normal egg, how many Barr bodies would be present in the resulting zygotes?

3. How would the inheritance pattern for an X-linked, dominant trait be different from that of an X-linked, recessive trait?

4. As are the chances of getting any number from 1 to 6 when tossing a fair die, the chances of crossing-over occurring at points along a chromosome are equally likely. Use the die analogy to explain why the chances of crossing-over between two genes on a chromosome increases as the distance between the genes increases.

5. In recent years, many couples have delayed starting a family until both partners have established a career. What factors should young couples consider before making their decision about when to start a family.

6. Explain why it is important that regions of chromosomes are not continually active.

■ Suggested Readings

Books

Farnsworth, M. W. 1988. *Genetics,* 2nd ed. New York: Harper and Row Publishers.

Levitan, M. 1988. *Textbook of Human Genetics.* New York: Oxford University Press.

Rothwell, N. 1988. *Understanding Genetics.* New York: Oxford University Press.

Russell, P. J. 1986. *Genetics.* Boston: Little, Brown and Company.

Suzuki, D. T., Griffiths, A. J. F., Miller, J. H., and Lewontin, R. C. 4th ed. 1989. *An Introduction to Genetic Analysis.* San Francisco: W.H. Freeman and Company.

Articles

Beardsley, T. Sex switch: the gene determining maleness in human embryos is found. *Scientific American* March, 1988.

D'Eustachio, P., and Ruddle, R. H. 1983. Somatic cell genetics and gene families. *Science* 220:919.

Fuchs, F. Genetic amniocentesis. *Scientific American* June, 1980.

Kornberg, R. D., and Klug, A. The nucleosome. *Scientific American* February, 1981.

McKusick, V. A. The mapping of human chromosomes. *Scientific American* April, 1981.

Marx, J. L. 1985. Putting the human genome on the map. *Science* 229:150.

Murray, A. W., and Szostak, J. W. Artificial chromosomes. *Scientific American* November, 1987.

Patterson, D. The causes of Down Syndrome. *Scientific American* August, 1987.

Roberts, L. 1988. Chromosomes: the ends in view. *Science* 240:982–983.

Turnbull, A. 1988. Woman enough for the games? *New Scientist* 119:61–65.

Zeidler, J. P. 1988. Automated chromosome analysis. *Nature* 334:635.

Molecular Genetics: Ultimate Cellular Control

9

Concepts

1. Deoxyribonucleic acid (DNA) is the genetic material of the cell. Its double-helix structure suggests how it is able to replicate itself, and how it can code for the sequences of amino acids that make proteins.
2. The production of proteins involves two processes. Transcription is the production of a messenger RNA (mRNA) molecule that is complementary to a gene in DNA. Translation is the assembly of proteins at ribosomes based on the genetic information in the transcribed messenger RNA.
3. Although substantial evidence for changing gene activity exists, the mechanisms for gene regulation in eukaryotic organisms are not well understood.
4. Point mutations can drastically alter proteins. Some mutations occur spontaneously; however, various environmental agents can increase mutation rates.
5. Modern genetic technologies are being used to manipulate DNA. Recombinant DNA techniques are used to insert DNA into bacteria to produce large quantities of a desired protein. Gene insertion is the introduction of copies of desired genes into an organism lacking those genes.

Have You Ever Wondered:

[1] what the "master molecule of life," DNA, looks like?
[2] how DNA replicates itself in order to be distributed to daughter cells during cell division?
[3] how DNA codes for protein?
[4] why genes coding for some proteins (e.g., sex hormones) can be active at some times in your life but not at other times?
[5] how many genes humans possess?
[6] why genes "jump?"
[7] why most mutations are harmful?
[8] what happens when a mutation occurs?
[9] what biologists are doing in recombinant DNA experiments?
[10] if families with histories of serious genetic diseases can hope to be free of those diseases in the future?

These and other useful questions will be answered in this chapter.

This chapter contains underlined evolutionary concepts.

Everyone reading this book has a unique combination of physical and personality traits. Many of these characteristics are traceable to genetic differences, and genetic differences are manifested in the kinds of proteins present in each reader. Some of these proteins contribute to observable traits, such as eye color, hair color, and height. Many other proteins contribute to one's uniqueness in more subtle, but just as important ways. These proteins are enzymes (*see chapter 4*) that regulate rates of chemical reactions in organisms. Within certain environmental limits, we are who we are by the proteins that we synthesize! The purpose of this chapter is to explore the biochemical answers to such questions as: What is a gene? How do genes control the expression of traits? How are the activities of genes controlled over the life of an organism? Much of the thrust of modern genetics revolves around questions such as these.

DNA: The Genetic Material

Twentieth-century biologists involved in the search for the genetic material realized that a molecule that serves as the genetic material must have certain characteristics in order to explain properties of life. First, the genetic material must be able to code for the sequence of amino acids in proteins and control protein synthesis. Second, it must be able to replicate (produce an exact copy of) itself prior to cell division, so that daughter cells receive complete sets of genetic instructions. Third, the genetic material must be in the nucleus of eukaryotic cells. (The participation of the nucleus in fertilization and the chromosomal basis of heredity were described in the early 1900s.) Fourth, it must be able to change over time to account for evolutionary change. Only one molecule, DNA (deoxyribonucleic acid), fulfills all of these requirements.

The Double-Helix Model

Two kinds of molecules participate in the production of proteins. By the early 1900s, it was known that both molecules are found in the nucleus and both are based on a similar building block, the nucleotide, giving them their name—nucleic acids. One of these molecules, **deoxyribonucleic acid or DNA,** is the genetic material and the other, **ribonucleic acid or RNA,** is produced in the nucleus and moves to the cytoplasm, where it participates in protein synthesis. The study of the function of DNA and RNA is called **molecular genetics.**

Nucleic-Acid Structure   DNA and RNA are large molecules made up of subunits called nucleotides (figure 9.1; *see figure 2.22*). A nucleotide consists of a nitrogen-containing organic base, either in the form a double ring (**purine**) or a single ring (**pyrimidine**). Nucleotides also contain a pentose (five-carbon) sugar and a phosphate (PO_4). The nucleotides of DNA and RNA, however, differ

in several ways. Both DNA and RNA contain the purine bases *adenine* and *guanine,* and the pyrimidine base *cytosine.* The second pyrimidine in DNA, however, is *thymine,* whereas in RNA it is *uracil.* A second difference in the nucleotides of DNA and RNA involves the sugar. The pentose of DNA is *deoxyribose,* and in RNA it is *ribose.* A third important difference between DNA and RNA is that DNA is a double-stranded molecule and RNA is single stranded.

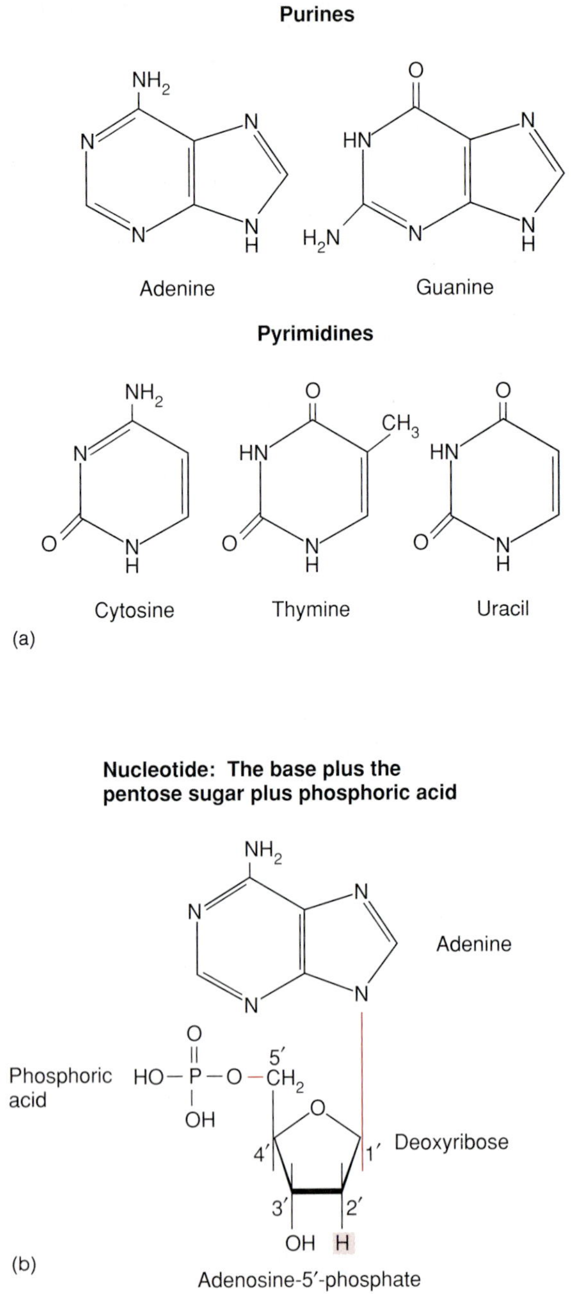

(a)

(b)

Figure 9.1   Components of nucleic acids. (a) The nitrogenous bases found in DNA and RNA. (b) Nucleotides are formed by attaching a nitrogenous base to the 1′ carbon of a pentose sugar and a phosphoric acid to the 5′ carbon of the sugar. (Carbons of the sugar are numbered with primes to distinguish them from the carbons of the nitrogenous base.) The sugar in DNA is deoxyribose, and the sugar in RNA is ribose. In ribose, an hydroxyl group (−OH) would replace the shaded hydrogen.

[1] The key to understanding the function of DNA is knowing how nucleotides are linked together into a three-dimensional structure. Research into the structure of DNA culminated in 1953 with the description of DNA by James Watson, a U.S. biologist, and Francis Crick, from Great Britain. They proposed that the DNA molecule was ladderlike, with the rails of the ladder consisting of alternating sugar-phosphate groups (figure 9.2). The phosphate of a nucleotide attaches at the fifth (5′) carbon of deoxyribose. Adjacent nucleotides attach to one another by a bond between the phosphate of one nucleotide and the third (3′) carbon of deoxyribose. The two strands are held together by the pairing of nitrogenous bases between strands. Adenine (a purine) is hydrogen bonded to its complement, thymine (a pyrimidine), and guanine (a purine) is hydrogen bonded to its complement, cytosine (a pyrimidine) (figure 9.3). Each strand of DNA is oriented such that the 3′ carbons of deoxyribose in one strand are oriented in the opposite directions from the 3′ carbons in the other strand. The strands' terminal phosphates are therefore at opposite ends, and the DNA molecule is thus said to be **antiparallel** (Gr. *anti,* against + *para,* beside + *allelon,* of one another). The entire molecule is twisted into a right-handed helix, with one complete spiral every 10 base pairs (figure 9.3). Investigations have found that this description of DNA is accurate for DNA from nearly all organisms.

DNA Replication in Eukaryotes

An important aspect of DNA replication was worked out in the late 1950s by Matthew Meselson and Franklin Stahl. They elegantly demonstrated that each DNA strand serves as a template for a new strand. One strand can serve as a template for the synthesis of a second strand, because the [2] pairing requirements between purine and pyrimidine bases dictate positioning of nucleotides in a new strand. Thus, each new molecule contains one strand from the old molecule and one newly synthesized strand. Because one-half of the old molecule is conserved in the new molecule, DNA replication is said to be *semiconservative.*

DNA replication in eukaryotic cells occurs in many chromosomes simultaneously, involves numerous enzymes, and results in the production of histone proteins that associate with the DNA in chromatin (*see figure 8.1*). Once the

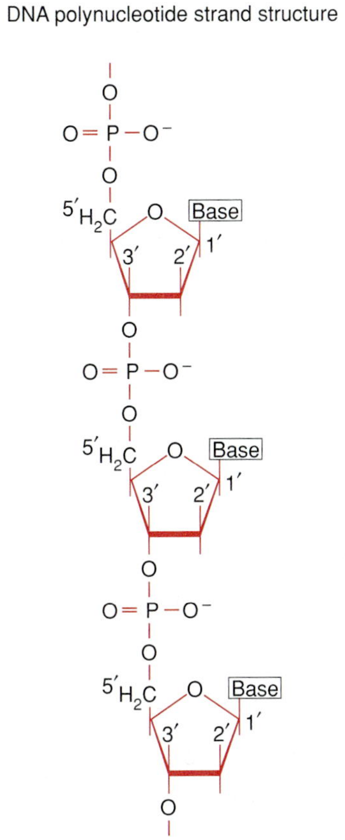

DNA polynucleotide strand structure

Figure 9.2 Nucleotides of one strand of nucleic acid are joined by linking the phosphate of one nucleotide to the 3′ carbon of an adjacent nucleotide. The 3′ and 5′ carbons of DNA are used to determine the direction a polynucleotide strand runs. The 5′ end of this strand is oriented up and the 3′ end is oriented down.

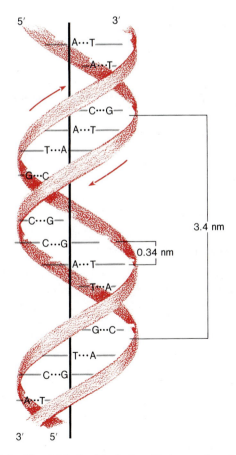

Figure 9.3 The DNA double helix. Hydrogen bonds are indicated with dotted lines between the nitrogenous bases. There are three hydrogen bonds between cytosine and guanine and two between thymine and adenine. The antiparallel orientation of the two strands is indicated by using the 3′ and 5′ carbons at the ends of each strand and the arrows indicating the 5′ to 3′ direction.

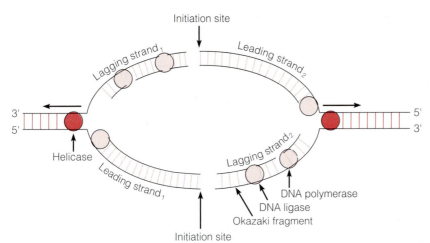

Figure 9.4 A diagrammatic representation of DNA replication. Recall that replication begins simultaneously at many sites along the length of a chromosome. Notice that synthesis of the leading strand is continuous from the initiation site, and that synthesis of the lagging strand is discontinuous from the initiation site. Replication is bidirectional from the initiation site. Notice that the two leading strands (and the lagging strands) produced as enzymes move from the initiation site are on opposite sides of the molecule.

strands have been separated by enzymes, replication of DNA is initiated simultaneously at many sites along the strands. The replication units, which are the segments of DNA copied at each initiation site, are called *replicons* (figure 9.4). Two new strands are synthesized simultaneously, going in both directions from the initiation site of a replicon. *DNA polymerases*—the enzymes that assemble nucleotides into DNA strands—copy an existing strand, beginning at the 3' end and working toward the 5' end. Because the replicated strand is complementary to the old strand, the new DNA strands are synthesized from their 5' end to their 3' end. DNA polymerase can begin only at the 3' end of a DNA segment, therefore, continuous replication can occur only in one direction from an initiation site. This segment of a strand is called the *leading strand.* The other strand, called the *lagging strand,* is copied in small, *Okazaki fragments.* DNA replication of the lagging strand is said to be discontinuous. Enzymes, called *ligases,* link the Okazaki fragments, as well as the replicons from other replication sites.

Stop and Ask Yourself

1. How are the nucleotides of DNA and RNA similar? How are they different?
2. Why is the DNA molecule said to be antiparallel?
3. Why is DNA replication said to be semiconservative?
4. How do the following concern DNA replication: Replicons? DNA polymerase? Leading strand? Lagging strand? Okazaki fragments? Ligases?

Genes In Action

The important relationship between genes and enzymes was worked out by George Beadle and Edward Tatum in the 1930s. The culmination of their research was the one gene-one enzyme hypothesis, stating that one gene codes for one specific enzyme. This hypothesis has undergone some revisions as more has become known about proteins, and today is known as the *one-gene-one polypeptide theory.* This early work indicated to later researchers that the genetic material must carry instructions for the synthesis

of proteins. A gene can be defined as a sequence of bases in DNA that codes for the synthesis of one polypeptide and genes must somehow transmit their information from the nucleus to the cytoplasm, where protein synthesis occurs. The *central dogma* of molecular genetics describes the relationship among the steps from DNA to the production of a protein. The synthesis of an RNA molecule is called **transcription** (L. *trans,* across + *scriba,* to write), and the uncoding of the message at the ribosome is called **translation** (L. *trans,* to transfer + *latere,* to remain hidden).

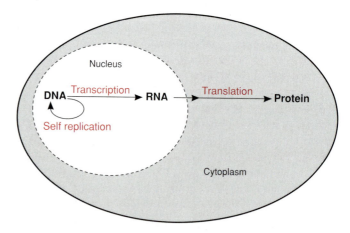

Three Kinds of RNA

There are three kinds of RNA, each with a specific role to play in protein synthesis. All three kinds of RNA are produced in the nucleus from DNA. **Messenger RNA (mRNA)** is a linear strand that is responsible for carrying a set of genetic instructions for synthesizing proteins to the cytoplasm. **Transfer RNA (tRNA)** picks up amino acids in the cytoplasm, carries them to ribosomes, and helps position them for incorporation into a polypeptide. **Ribosomal RNA (rRNA)**, along with proteins, make up ribosomes.

The Genetic Code

DNA must code for the 20 different amino acids that are found in all organisms. The information-carrying capabilities of DNA reside in the sequence of nitrogenous bases. The genetic code is represented by a sequence of three

[3]

bases—a *triplet* code (figure 9.5). Each three-base combination is called a **codon,** and close examination of figure 9.5 reveals that more than one codon can specify the same amino acid because there are 64 possible codons, but only 20 amino acids. This characteristic of the code is referred to as **degeneracy.** Usually, when the first two bases are determined, the amino acid is determined. Note that not all codons code for an amino acid. The base sequences UAA, UAG, and UGA are all stop signals. As a period indicates the end of a sentence, the stop codons indicate where polypeptide synthesis should end. The base sequence AUG codes for the amino acid methionine, which serves as a start signal.

The genetic code is often referred to as a *universal code* because of its remarkable constancy in virtually all life forms. Exceptions to the code have been found only in the DNA of mitochondria and a few single-celled organisms.

Second position

First position	U	C	A	G	Third position
U	UUU Phe	UCU Ser	UAU Tyr	UGU Cys	U
	UUC	UCC	UAC	UGC	C
	UUA Leu	UCA	UAA STOP	UGA STOP	A
	UUG	UCG	UAG	UGG Trp	G
C	CUU Leu	CCU Pro	CAU His	CGU Arg	U
	CUC	CCC	CAC	CGC	C
	CUA	CCA	CAA Gln	CGA	A
	CUG	CCG	CAG	CGG	G
A	AUU Ile	ACU Thr	AAU Asn	AGU Ser	U
	AUC	ACC	AAC	AGC	C
	AUA	ACA	AAA Lys	AGA Arg	A
	AUG Met	ACG	AAG	AGG	G
G	GUU Val	GCU Ala	GAU Asp	GGU Gly	U
	GUC	GCC	GAC	GGC	C
	GUA	GCA	GAA Glu	GGA	A
	GUG	GCG	GAG	GGG	G

Glu = Glutamic acid	Trp = Tryptophan
Phe = Phenylalanine	Pro = Proline
Gly = Glycine	Ser = Serine
Val = Valine	Cys = Cysteine
Leu = Leucine	Arg = Arginine
Ile = Isoleucine	Asn = Aparagine
His = Histidine	Gln = Glutamine
Lys = Lysine	Tyr = Tyrosine
Met = Methionine	Asp = Aspartic acid

Figure 9.5 The genetic code as reflected in 64 messenger RNA codons. The first base of the triplet is on the left side of the figure. The second base is at the top, and the third base is on the right side of the figure. The abbreviations used for amino acids are shown above. In addition to coding for the amino acid methionine, the AUG codon also serves as the initiator codon.

Transcription

As is indicated by the central dogma, the genetic information in DNA is not translated directly into proteins. It first is transcribed into messenger RNA (mRNA), one of the three types of RNA. Transcription involves numerous enzymes that unwind a region of a DNA molecule, initiate and end mRNA synthesis, and modify the mRNA after transcription has been completed. Unlike DNA replication, only one or a few genes are exposed, and only one of the two DNA strands (the *coding strand*) is transcribed (figure 9.6).

One of the important enzymes of this process is *RNA polymerase.* (In eukaryotes there are actually three RNA polymerases.) After a section of DNA is unwound, RNA polymerase recognizes a specific sequence of DNA nucleotides called the *promoter.* It attaches at the promoter and begins joining nucleotides at the 3′ end of the DNA strand, as in DNA replication. The mRNA, therefore, forms in the 5′ to 3′ direction as do the new DNA strands during replication. In RNA, the same pairing of complementary bases that is seen in DNA occurs, except that in RNA, the base uracil replaces the base thymine as a complement to adenine. Thus everywhere adenine occurs in the DNA being transcribed, uracil will be the complementary base added to the mRNA strand. *Terminator sequences,* which are DNA nucleotide sequences consisting of strings of guanine-cytosine base pairs followed by strings of adenine thymine base pairs, signal the end of transcription. Certain proteins may also be involved with chain termination.

Newly transcribed mRNA must be modified before leaving the nucleus to carry out protein synthesis. Some base sequences in newly transcribed mRNA do not code for proteins. These are called *noncoding sequences.* Some of these occur at the ends of mRNA and are thought to help in the recognition of mRNA by the ribosome and in maintaining the stability of the mRNA at the ribosome. Base sequences of mRNA that code for proteins are called *coding sequences.* In eukaryotic organisms, the coding sequences of a gene, called **exons,** are broken up by noncoding regions called **introns.** The whole mRNA strand is called the **primary transcript.**

Posttranscriptional modification involves cutting out introns so that the mRNA coding region can be read continuously at the ribosome (figure 9.7). Postranscriptional modification occurs on *spliceosomes,* which consist of proteins and small nuclear ribonucloproteins symbolized snRNP's (pronounced "snurps").

Translation

Translation is the synthesis of proteins at the ribosomes in the cytoplasm, based on the genetic information contained in the transcribed mRNA.

Amino Acids and tRNA Another type of RNA, called transfer RNA (tRNA) is important in the translation process. It brings the different amino acids coded for by the mRNA into alignment so that a polypeptide can be

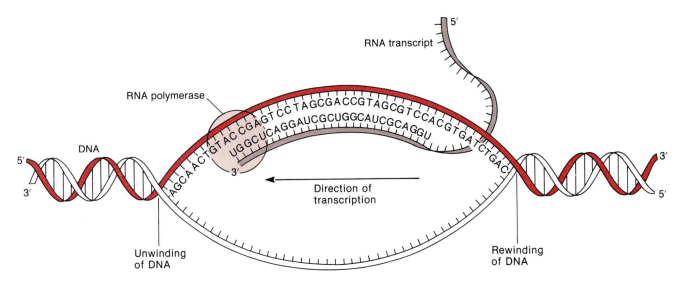

Figure 9.6 Transcription involves the production of a messenger RNA molecule that is complementary to the coding strand of the segment of DNA. Note that transcription occurs from the 3' end of a gene to the 5' end of the gene.

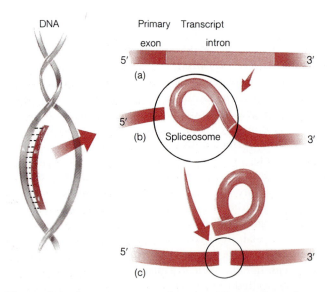

Figure 9.7 Posttranscriptional events. The product of transcription, the primary transcript, has one or more introns that must be removed before mRNA can leave the nucleus. Spliceosomes, containing snRNA (snurps) and certain polypeptides are the sites where the introns are removed and the exons are spliced together. (a) The primary transcript. (b) At the spliceosome, the intron is cleaved at the 5' end and forms a lariatlike loop. (c) The exons are joined.

made. Transfer RNA is frequently represented as a two-dimensional cloverleaf (figure 9.8a), although the molecule actually undergoes helical twisting (figure 9.8b). Complementary pairing of bases across the molecule maintains tRNA's configuration. The shape of tRNA has important functional implications. The presence of some unusual bases (unusual, that is, for nucleic acids), such as dihydrouridine, pseudouridine, and ribothymidine, disrupts the normal base pairing and forms important loops in the

molecule. The unpaired bases in the loop nearest the 3' end of the molecule (the "t loop") facilitate the positioning of tRNA at the ribosome. The center loop (the "anticodon loop") has a sequence of three unpaired bases called the **anticodon**. Pairing of the codon of the mRNA strand and its complementary anticodon of the tRNA positions the amino acid being carried by tRNA. The loop near the 5' end of the tRNA is the recognition site for the enzyme (aminoacyl synthetase) that attaches an amino acid to tRNA. The 3' end of the tRNA is the amino acid attachment site (figure 9.8).

Each different amino acid is carried by a different tRNA. Aminoacyl synthetase recognizes both the tRNA and the amino acid and catalyzes the attachment of the amino acid to the tRNA, a process that requires the expenditure of one ATP. The bond between the amino acid and the tRNA retains much of the energy that was in the ATP, storing it for later use in linking this amino acid to another.

Ribosomes, mRNA, and Peptide Bonds Ribosomes, the sites of protein synthesis, consist of large and small subunits that organize the pairing between the codon and the anticodon. There are several sites on the ribosome that serve as binding sites for mRNA and tRNA. At the initiation of translation, mRNA binds to a small subunit that is separate. Attachment of the mRNA requires that the initiation codon (AUG) of mRNA be aligned with the *P (peptidyl)* site of the ribosome (figure 9.9). A tRNA with a complementary anticodon for methionine binds to the mRNA, and a large subunit joins, forming a complete ribosome.

The formation of a polypeptide is ready to begin. There is another site, the *A (aminoacyl)* site, which is next to the P site. A second tRNA, whose anticodon is complementary to the codon in the A site is positioned. There are now two tRNA molecules with their attached amino acids side-by-side in the P and A sites. This step requires enzyme aid (elongation factors) and energy, in the form of guanine

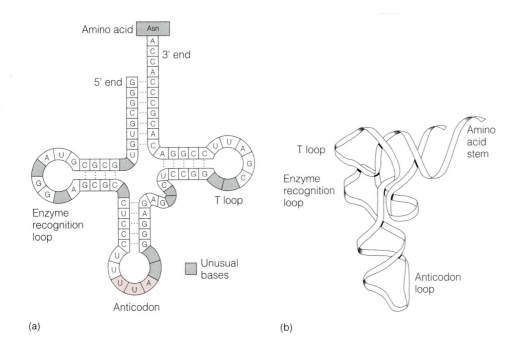

(a)

(b)

Figure 9.8 (a) A diagrammatic representation of transfer RNA. An amino acid attaches to the 3' end of the molecule. The ACC base sequence at this end of the molecule is consistent in all tRNAs. The "t loop" helps position tRNA at the ribosome. The anticodon is the sequence of three bases that pairs with the codon in mRNA, thus positioning the amino acid that tRNA carries. The enzyme recognition loop positions tRNA in aminoacyl synthetase, the enzyme that attaches the correct amino acid to the tRNA. (b) A more accurate representation of tRNA when its three dimensional folding is considered.

triphosphate (GTP). An enzyme (*peptidyl transferase*), which is actually a part of the larger ribosomal subunit, breaks the bond between the amino acid and tRNA in the P site and catalyzes the formation of a *peptide bond* between that amino acid and the amino acid in the A site. (Recall the energy for this reaction had been previously conferred to the bond between the amino acid and tRNA by ATP.)

The mRNA strand then moves along the ribosome a distance of one codon. This step requires more enzymatic action by elongation factors and more energy from GTP. The tRNA with two amino acids attached to it that was in the A site is now in the P site. A third tRNA can now enter the exposed A site. This process continues until the entire mRNA has been translated and a polypeptide chain has been synthesized. Often many ribosomes will be translating a single mRNA strand simultaneously. Translation ends when a termination codon (e.g., UAA) is encountered. Proteins called release factors, as well as the hydrolysis of GTP to GDP + P_i, are involved in release of the ribosome from the mRNA.

One would expect as many different tRNAs as there are different codons, but that is not quite the case. One tRNA may serve more than one codon, as long as these codons are specifying the same amino acid. The explanation for this phenomenon is known as the **wobble hypothesis.** The third base of the anticodon has more freedom in pairing than other bases and may pair with bases other than those predicted by the pairing rules. In other words, there is a slight "wobble" in pairing at that third position.

In spite of the complexities of translation, the process is very rapid. Approximately 20 peptide bonds are formed every second. An average-sized protein can be synthesized in about 15 seconds at a cost of four high-energy bonds per peptide bond.

Posttranslational Modifications

The proteins that are synthesized in the translation process have a variety of destinations in an animal. Some are secreted from a cell, some are incorporated into lysosomes, some make up components of membranes, and some make up the structural framework of the cell. Most proteins are synthesized as preproteins. *Preproteins* have a sequence of amino acids (called a *leader*) attached at one end that helps direct them to their destination. The leader is often a series of amino acids that promotes the movement of an otherwise lipid-insoluble protein through a cellular membrane.

Protein synthesis often occurs on the surface of endoplasmic reticulum (ER). The attachment of the ribosome to the ER is an active process that occurs as the peptide chain grows and involves special proteins called signal recognition particles. The positioning of protein synthesis on the ER allows preproteins to move into the ER as the protein is being translated. Once the preprotein or a portion of it is through the membrane, the leader is cleaved from the molecule, and, when the translation is finished, the protein can be moved to the Golgi apparatus for packaging into a secretory vesicle or a lysosome (figure 9.10).

Stop and Ask Yourself

5. The genetic code is a degenerate code. What does this mean?
6. How does a primary transcript differ from the mRNA molecule that leaves the nucleus?
7. What is the anticodon? What is the function of other "loops" of tRNA?
8. What is the role of the following in protein synthesis: A site? P site? Preprotein?

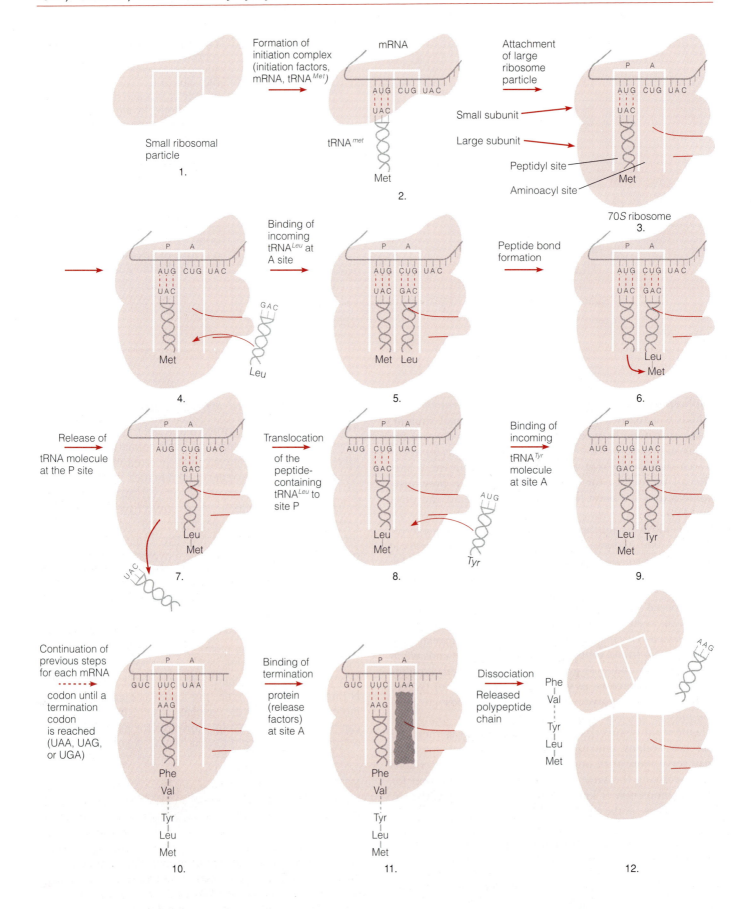

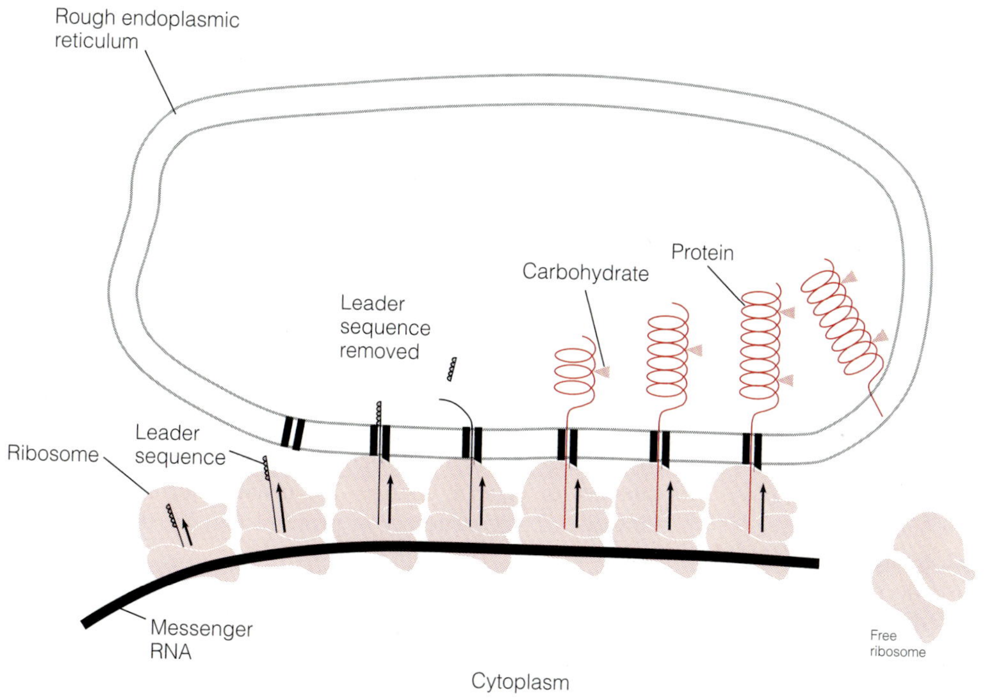

Figure 9.10 Posttranslational events. Preproteins often move into the endoplasmic reticulum during translation. The leader sequence promotes entry into the ER, and is then cleaved from the molecule. Other molecules, such as carbohydrates, that are important in the function of the protein may be attached in the ER. The protein can then be transported to the Golgi apparatus for packaging.

Control of Gene Expression in Eukaryotes

The gene control processes that function in prokaryotic organisms are well understood. Genes of prokaryotes occur on circular DNA molecules, and related genes are usually linked into functional units called operons. Operons consist of **structural genes,** which are genes that code for the desired end product; and **regulator genes,** which control transcription of the structural genes. Gene control processes of eukaryotic organisms are less well known.

Many genes are functional only during particular stages in the life history of an animal. Consider, for example, the changes that humans go through only during puberty. Other genes are turned on and off more frequently. Aldosterone, for example, is a steroid hormone produced in the cortex (outer regions) of the adrenal gland. This hormone functions at the kidney to promote reabsorption of sodium from the urine and secretion of potassium into the urine. Aldosterone enters kidney cells by diffusion, combines with a cytoplasmic receptor, and enters the nucleus to alter gene activity. This section describes what is known about the processes that turn eukaryotic genes on and off.

[4]

Figure 9.9 (left page) The events of translation. Translation begins by a methionine-tRNA associating with the P site of the smaller ribosomal subunit and the initiation codon of mRNA associated with that subunit. The larger ribosomal subunit attaches, and a tRNA carrying the next amino acid enters the A site. A peptide bond is formed between the two amino acids, freeing the tRNA in the P site. The ribosome moves down the mRNA, opening the A site, which is now associated with the next codon. This process continues until an mRNA stop signal (in this case UAA) is encountered. Protein release factors enter the A site and the ribosome dissociates, releasing the polypeptide chain.

Cytological Evidence of Changing Gene Activity

Recall that regions of chromosomes, or entire chromosomes, exist in inactive, or heterochromatic states. Some regions are always (constitutively) heterochromatic and probably contribute to the structural stability of a chromosome. Other regions of chromosomes are heterochromatic in some preparations, but at other times in the life of a cell they may be euchromatic (active). (Recall the discussion of the inactivation of the X chromosome in chapter 8.) Heterochromatin that can sometimes be active is evidence that genes are switched on and off in the life of a cell.

Other evidence of changing levels of gene activity is seen in cells (oocytes) undergoing meiosis in the ovaries of many vertebrates (*see figure 6.11*). After fertilization, cell division is very rapid, and DNA is replicating itself for upcoming divisions. Transcription must occur prior to fertilization to provide the mRNA needed during early development. Portions of chromosomes unfold to form lateral DNA loops of various sizes. These loops are regions where transcription is occurring. Chromosomes in this condition are called *lampbrush chromosomes.*

A third indication of changing gene activity is observed in cells of larval flies. DNA of fly larvae undergoes repeated replications without intervening cell divisions. This DNA is stacked into giant *polytene chromosomes.* The polytene state may promote rapid RNA synthesis required during larval development. Clusters of looped DNA, called *chromosomal puffs,* are produced in polytene chromosomes during RNA synthesis (*see figures 11.13; 11.14*). The location of these puffs is different on different chromosomes, and is frequently different on the same chromosomes from different tissues or at different stages of larval development.

Biochemical Evidence

It might be surprising to learn that there is far more DNA in the nucleus than is needed to code for all the proteins of an organism. It is estimated that humans have between 100,000 and 150,000 structural genes; however, humans have enough DNA to account for 2 million genes! What is the purpose of all of the extra DNA? Surely some of this extra DNA consists of regulatory genes.

About 25% of eukaryotic DNA consists of base sequences that are repeated hundreds or thousands of times in the genome. Some of this *repetitive DNA* is constitutive heterochromatin. Other repetitive DNA is transcribed into ribosomal RNA. The rest consists of base sequences of unknown function.

The remaining 75% of eukaryotic DNA is *single-copy DNA*. It consists of base sequences that occur only once in the genome and some is untranslated. Some untranslated DNA codes for mRNA introns. Scientists do not know why intron DNA exists. It has been suggested that intron DNA may play a role in regulating gene expression. This idea is supported by experiments in which removal of intron DNA from a gene prevented translation of the gene. Other scientists suggest that these untranslated sequences may play a role in promoting recombination between genes. Intron DNA probably does not account for all of the untranslated, single copy DNA. It is likely that much of the remaining DNA consists of regulatory genes.

Nucleoproteins and Gene Regulation Histone proteins contribute to nucleosome particles. They are relatively homogeneous and probably do not function in gene regulation. Other nuclear proteins, called *nonhistone proteins,* are more diverse and may have important roles in gene regulation. These proteins may bind to specific regulatory genes to promote or prevent structural gene transcription.

Transposable Elements: "Jumping Genes"

Gene regulation in eukaryotes is further complicated by firm evidence that the location of some genes is not fixed. This phenomenon has been documented most thoroughly in studies of the genes that code for kernel color of corn. **Transposons** are genes that move from position to position on one chromosome or that move between chromosomes. The functions of transposons are unclear, but it is very likely that they are involved with regulating or changing the function of genes close to the points of transposon insertion. Transposons have been described in fruit flies (*Drosophila*), and will undoubtedly be found in many other animals.

Stop and Ask Yourself

9. How do the following suggest that the activity of genes of eukaryotic organisms changes: Lampbrush chromosomes? Chromosomal puffs?

10. What are transposons?

Mutations

An important requirement of the genetic material is that it must account for evolutionary change. Changes in the structure of genes and chromosomes are called mutations. Changes in chromosome number and cytologically visible changes in structure were described in chapter 8. Changes in nucleotide sequences are called **point mutations,** and may result from the replacement, addition, or deletion of nucleotides. Mutations are always random events. Although certain environmental factors may change mutation rates, there is no way to predict what genes will be affected or what the nature of the change will be. Some mutations may be unnoticed or even beneficial; however, the consequences of genetic changes are usually negative because they disturb the structure of proteins that are the products of millions of years of evolution.

Estimates of mutation rates for a given gene are generally within the range of one mutation every 10^5 to 10^8 replications (one mutation of a given gene per 100,000 to 100,000,000 gametes). When one considers that humans have about 100,000 structural genes, it means that every gamete carries one new mutation, and every zygote carries two new mutations (one from the egg and one from the sperm). Fortunately, most of these mutations are recessive and must be paired with a similar mutation from a second individual to be expressed.

Kinds of Point Mutations

Animals are protected from the negative effects of base changes when the changes occur in the third position of a DNA triplet. Because many amino acids are coded for by the first two bases of a DNA triplet, the third base can vary and not change the amino acid specified. Thus, roughly 30% of all single base changes will never manifest themselves in altered protein structure and are referred to as *silent mutations.*

On the other hand, base changes in the first or second position of a DNA code, and sometimes base changes at the third position, will result in the substitution of one amino acid for another. Sickle cell anemia (*see chapter 7*) results in reduced ability of hemoglobin to carry oxygen and the tendency of affected red blood cells to get stuck in, and clog, capillary beds. These dramatic changes in circulatory physiology result from a single base substitution in one of the polypeptides of hemoglobin. Valine (GUG) is substituted for glutamic acid (GAG) at one position. Mutations that result in single amino acid changes are called *missense mutations.*

Mutations that result in the most severe changes in protein structure are the result of the addition or deletion of a base. These mutations are called *frameshift mutations,* because adding or deleting a base causes all codons after the change to be misread. This kind of change is likely to create a useless protein. Stop signals may not be read, causing translation beyond the normal termination point, or a stop signal might be created in the middle of a gene, causing premature termination (figure 9.11).

Box 9.1

In Vitro Fertilization

The failure to produce viable ova is not the only kind of problem that contributes to some women's inability to conceive and bear children. Even when viable ova are produced, problems can arise if oviducts are blocked, and ova cannot move down the oviduct to the uterus. Alternatively, low sperm count, a common male problem, is also a cause of infertility. What these two problems have in common is that couples experiencing either still have an opportunity to give birth to their own children through *in vitro fertilization*.

In vitro fertilization (IVF) is the fertilization of mature ova, previously extracted by syringe from an ovary, in petri dishes containing an appropriate medium. After fertilization, embryos begin development. At the blastocyst stage, they can be placed in the uterus, where implantation can occur. (The blastocyst, a hollow ball of cells, normally develops as the embryo passes down the oviduct.) After implantation, development and birth proceed as usual. Because the oviduct is bypassed in this

procedure, the blockage is of no consequence. Low sperm counts are also much less of a problem with IVF, because high sperm loss during the trip between the vagina and the oviduct does not occur.

In spite of its obvious benefits, IVF has its critics. Critics see some applications—like extracting ova from one woman every month, fertilizing those ova with sperm from a selected donor, and implanting embryos into surrogate mothers—as possible abuses of the technology. They also point out that current procedures involve the extraction of more ova than will be implanted as embryos in the uterus. What is to be done with the remaining embryos? Perhaps they should be frozen for implantation at a later date. Alternatively, they are extremely valuable in research into mechanisms of eukaryotic gene control. Their use in research could help us gain a more complete understanding of the origins of many cancers. Obviously, there are no easy, widely accepted answers to questions like these.

Point of nucleotide
insertion in (b) and (c)

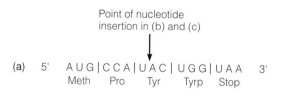

(a) 5′ A U G | C C A | U A C | U G G | U A A   3′
       Meth   Pro   Tyr   Tyrp   Stop

(b) 5′ A U G | C C A | U C A | C U G | G U A 3′
       Meth   Pro   Ser   Ser   Val

(c) 5′ A U G | C C A | U A A | C U G | G U A 3′
       Meth   Pro   Stop   Ser   Val

Figure 9.11 A segment of mRNA showing that insertion of a nucleotide can cause frameshift mutations. (a) Normal mRNA base sequence. (b) After the insertion of "C" the stop signal is deleted and the sequence of amino acids following the point of insertion is altered. (c) After the insertion of "A" a premature stop signal occurs.

Causes of Mutations

Although mutation rates can be increased by various environmental agents, mutations are not unique to our modern world. Mutations have occurred throughout the history of life on earth. Many mutations are said to be *spontaneous mutations* because they have no obvious cause. Many of these are a result of base pairing errors during replication, and result in a substitution of one base pair for another.

A variety of mutagenic agents affect mutation rates. Thousands of chemicals used in industry and in science laboratories are suspected mutagens. Some of these, such as 5-bromodeoxyuridine, are *base analogs*. They have structures similar to normal DNA bases, but may pair differently during replication and cause a base substitution. Other chemicals change the structure of certain bases. Nitrous acid (HNO_2), a powerful mutagen, converts adenine to a guaninelike compound, and after replication, an A-T base pair would be changed to a G-C base pair.

Electromagnetic radiation (ultraviolet light, X rays, and gamma rays) and nuclear radiation (alpha and beta particles, protons, and neutrons) are also mutagenic. Damage is proportional to the dose received and is cumulative. (That is, repeated small doses have the same effect as a single large dose, as long as the total dose is the same.) Ultraviolet light from the sun, for example, easily penetrates the outer epidermal cell layers and can cause base substitutions in DNA of cells in the germinal layers of the epidermis. These changes are a primary cause of skin cancer.

DNA Repair

Fortunately, enzyme systems are present that help repair damage done to DNA. When a base change occurs on one DNA strand, it can be detected by comparing the changed strand to the unaltered complementary strand. Most known repair systems consist of enzymes that cut out altered base sequences and replace them with the proper nucleotides.

Deficiencies in these enzyme repair systems can cause an increased incidence of some kinds of cancer. Xeroderma pigmentosum is an inherited deficiency in the enzymes that repair damage done by ultraviolet light, and results in an increased incidence of skin cancer.

Stop and Ask Yourself

11. Why is a base change in the third position of a DNA triplet less likely to change a protein's structure than a change in the first or second position of a DNA triplet?
12. How does the fact that DNA is a double helix protect against the negative effects of point mutations?

Applications of Genetic Technologies

Technologies that have allowed scientists to understand life are now being applied in the manipulation of life. Most scientists and many laypersons would agree that applications of these technologies have great potential for improving the quality of life on earth. Scientists can now assay individuals for some genetic abnormalities before birth, detect some abnormalities hidden in the genome of prospective parents, manipulate fertilization (box 9.1), and manipulate DNA (colorplates 2 and 3). However, there are numerous ethical considerations involved with the application of these technologies. The remainder of this chapter focuses on two applications of genetic technologies, and briefly discusses ethical dilemmas presented by each application.

Recombinant DNA

[9] **Recombinant DNA** techniques involve the incorporation of human, or other, DNA into bacteria in hopes that the bacteria will become a factory for the desired protein. For example, most of the insulin used to treat diabetes is now produced using recombinant DNA.

Recombinant DNA techniques differ depending on whether one is interested in producing a protein of prokaryotic or eukaryotic origin. In either case, however, a vector must be used to incorporate foreign DNA into the bacteria. One kind of vector is a plasmid. *Plasmids* are small, extrachromosomal circles of DNA that occur naturally in bacteria. They replicate independently from a bacterial chromosome, can be isolated by gel electrophoresis,

and can be incorporated back into bacteria using a calcium-enriched medium. Because plasmids are self-replicating, many copies of an engineered plasmid will be present in each bacterial cell. If one wishes to isolate prokaryotic DNA for transfer, the procedure involves cutting the DNA into fragments using *restriction enzymes* (figure 9.12). The same kind of enzyme is used to cleave plasmids (figure 9.12). Most restriction enzymes leave unpaired bases at the ends of DNA fragments. Because the same enzyme is used on the prokaryotic DNA and on the plasmids, the unpaired bases of the DNA fragments and the unpaired bases of the plasmids will be complementary. When plasmid DNA and DNA fragments are mixed, and an enzyme called *DNA ligase* is added, the fragments and the plasmid DNA may join. The plasmids can then be reincorporated into bacteria.

After incorporating DNA into bacteria, the investigator must find the bacterial copies that contain the plasmids with the desired gene. Not all plasmids will have incorporated the desired gene and not all bacteria will have incorporated plasmids. If the plasmid has genes that convey resistance to an antibiotic, culturing bacteria on a medium containing that antibiotic will select for bacteria containing the plasmid. Out of those bacteria that survive, only a few will have the desired gene. One technique for isolating these bacteria involves isolating mRNA from a cell culture actively synthesizing the desired protein and incorporating a radioactive label into that mRNA. The mRNA has a base sequence complementary to the desired gene, and one can induce hybridization (bonding) between mRNA and DNA. This procedure is done by transferring bacterial colonies to special filter paper, treating the filter paper with alkali to separate DNA strands, and flooding the filter paper with labeled mRNA. The filter paper is then placed on photographic film, and the hybridized DNA/mRNA will expose the film, identifying the bacterial colony containing the desired gene. Bacteria from that colony can be cultured, and their protein product recovered.

Techniques used for incorporating eukaryotic DNA into plasmids are slightly different. They begin by isolating mRNA rather than DNA, because mRNA from a cell undergoing protein synthesis will not have intron base sequences. An enzyme called *reverse transcriptase* is used to produce DNA from the mRNA. This DNA, called *complementary DNA (cDNA)*, can be incorporated into plasmids and into bacteria in the usual fashion. Culturing these bacteria can result in large quantities of protein at relatively low cost.

The Controversy   No one doubts that recombinant DNA technologies can provide supplies of valuable proteins and useful bacteria. Early concerns that envisioned monsters emerging from laboratories resulted in very strict National Institutes of Health (NIH) Guidelines regarding the kinds of research that could be done and the conditions under which it could be conducted. Monsters have not materialized, and as a result NIH guidelines for laboratory applications have been relaxed. Now, concern centers around recombinant organisms that eventually need to be tested outside the laboratory. Even though the scientists

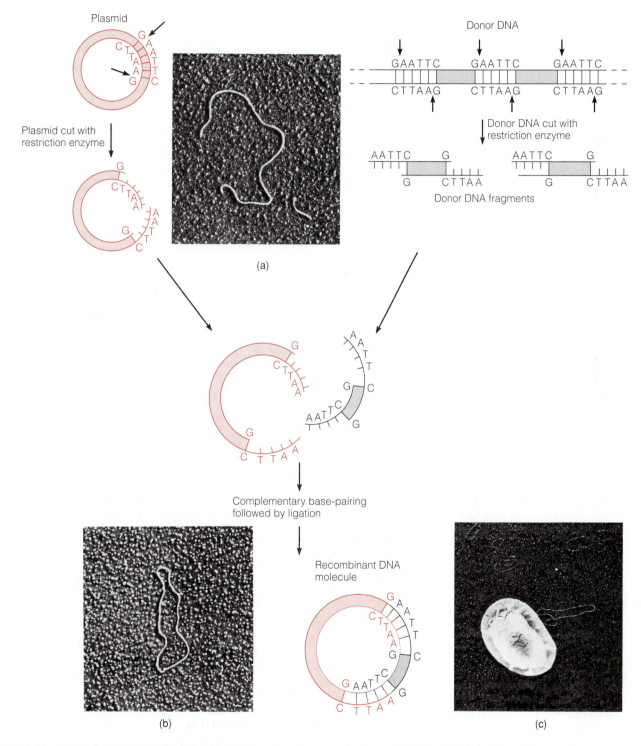

Figure 9.12 Recombinant DNA. (a) These techniques involve cutting plasmid DNA and donor DNA with restriction enzymes. This process leaves the ends of DNA fragments with complementary base sequences. (The electron micrograph shows a cleaved plasmid and a fragment of host DNA.) (b) Plasmid and host fragments are joined by ligase enzymes. (c) The plasmid, now containing host DNA is introduced into bacteria. (Note the plasmid lying next to the bacterium in the electron micrograph.)

involved may take precautions to insure that bacteria will have little chance of harming the environment, many people (including some environmental scientists) have concerns about remote possibilities of the uncontrolled proliferation of engineered organisms. Thus, even though guidelines for laboratory applications have been relaxed, guidelines for field tests are stringent.

Gene Insertion

Gene insertion involves the introduction of copies of desired genes into an organism that lacks those genes. In human applications, gene insertion, or gene therapy, could [10] be used to insert normal genes into a patient with defective genes. In other applications, it could involve the insertion of genes for disease resistance, faster growth, or increased size into plant or animal stocks. Progress is being made with these technologies in experimental organisms.

The basic strategy of gene insertion is to use recombinant DNA techniques to isolate a desirable gene and incorporate these genes into specially engineered viruses, called *retroviruses*. Retroviruses can insert DNA into a human cell, but they do not damage the cell. Treatment could involve taking a bone-marrow sample from a patient and culturing those cells with engineered viruses. After the cells have been allowed to proliferate in culture, they could be infused back into a patient intravenously, with the hope that enough transformed cells would be present to counter the effects of a genetic abnormality. Because this kind of treatment involves somatic (body) cells rather than germ (gamete producing) cells, this cure would not be passed on to future generations. Certain blood-borne diseases, and diseases like Lesch-Nyhan Syndrome, might be successfully treated in this fashion. (Lesch-Nyhan Syndrome is presently incurable. Its symptoms include self-mutilation, compulsive use of obscenities, and spitting and vomiting.)

Germ-line gene therapy is much more difficult. It involves introducing genes into early developmental stages of animals (colorplate 3). These genes could be incorporated into gametes and, thus, could eradicate genetic threats to a family's future.

The Controversy   There is no doubt that gene insertion will become a reality and that it has tremendous potential for good. Ethical concerns regard decision-making processes involved with these manipulations. Some individuals question whether humans have any business altering the human genome. Others simply wonder who will be trusted to make decisions about what genes are "good" and what genes are "bad."

Stop and Ask Yourself

13. How are the following used in recombinant DNA techniques: Restriction enzymes? Plasmids? DNA ligase? Reverse transcriptase?
14. What is gene insertion? Differentiate between somatic and germ-line gene therapy.

◼ Summary

1. Deoxyribonucleic acid (DNA) is the hereditary material of the cell. Ribonucleic acid (RNA) participates in protein synthesis.
2. Nucleic acid building blocks are called nucleotides. Nucleotides consist of a nitrogenous (purine or pyrimidine) base, a phosphate, and a pentose sugar.
3. DNA consists of a double helix, in which the sugar-phosphate groups form the rails of the helix, and nitrogenous bases pair across the molecule, being held together by hydrogen bonds.
4. The replication of DNA is semiconservative. During replication, the DNA strands separate, and each strand serves as a template for a new strand. One-half of the old molecule is conserved in each new molecule.
5. A gene can be defined as a series of nucleotides of DNA that code for a single polypeptide.
6. Proteins are produced in a cell as a result of two processes. Transcription involves the production of a messenger RNA molecule from a DNA molecule. Translation involves the movement of messenger RNA to the cytoplasm where transfer RNA and ribosomes are involved with linking amino acids in a proper sequence to produce a polypeptide.
7. Proteins are often produced as preproteins. Sequences of amino acids attached at one end of the protein help direct proteins to their destination.
8. Although there is good evidence for changing gene activity, the control processes are poorly understood in eukaryotes.
9. Transposable elements, or transposons, are genes that move from position to position on one chromosome or that move between chromosomes.
10. Point mutations are changes in nucleotide sequences in DNA. Mutations can occur spontaneously, although a variety of mutagenic agents affect mutation rates. These mutagens include various chemicals and electromagnetic and nuclear radiation.
11. Recombinant DNA techniques involve the incorporation of human, or other, DNA into bacteria in hopes that the bacteria will produce a desired protein. Gene insertion involves the introduction of desired genes into an organism that lacks those genes.

Key Terms

anticodon (p. 134)

antiparallel (p. 131)

codon (p. 133)

degeneracy (p. 133)

deoxyribonucleic acid
(DNA) (p. 130)

exons (p. 133)

gene insertion (p. 142)

introns (p. 133)

messenger RNA (mRNA)
(p. 132)

molecular genetics (p. 130)

point mutations (p. 138)

posttranscriptional
modification (p. 133)

primary transcript (p. 133)

purine (p. 130)

pyrimidine (p. 130)

recombinant DNA (p. 140)

regulator genes (p. 137)

ribonucleic acid (RNA)
(p. 130)

ribosomal RNA (rRNA)
(p. 132)

structural genes (p. 137)

transcription (p. 132)

transfer RNA (tRNA)
(p. 132)

translation (p. 132)

transposons (p. 138)

wobble hypothesis (p. 135)

Critical Thinking Questions

1. Meselson and Stahl found that DNA replication was semi-conservative. Explain the results they would have gotten in their experiments if replication were conservative. (That is, if replication resulted in one entirely new molecule and conserved the parental molecule.)

2. Explain why degeneracy in the genetic code lessens the impact of mutations that result in base substitutions. Does degeneracy also lessen the impact of frameshifts? Explain.

3. Given the fact that most mutations are recessive and usually do not show up in natural populations, explain why zoos are concerned with the loss of genetic diversity by excessive inbreeding within the confines of a zoo.

4. Imagine that you are a prospective parent, and that it is the year 2010. You are told by a genetic counselor that you and your spouse both carry a rare recessive gene that could result in a severely deformed child. The chance of this trait showing up in your child is one in four. You have four choices.

   a. You may decide not to have a child.

   b. You may adopt a child.

   c. You may take your chances with nature.

   d. You may have corrective genes inserted into your own (your spouse's) ovum, which would then undergo *in vitro* fertilization with your spouse's (your) sperm. You can then be confident of having a normal child.

Which option would you choose? Why?

Suggested Readings

Books

Emery, A. E. H. 1984. *An Introduction to Recombinant DNA.* New York: John Wiley and Sons.

Keller, E. F. 1983. *A Feeling for the Organism: The Life and Work of Barbara McClintock.* San Francisco: W.H. Freeman.

Krimsky, S. 1982. *The Social History of the Recombinant DNA Controversy.* Cambridge Massachusetts: MIT Press.

Lewin, B. 1990. *Genes,* 4th ed. New York: John Wiley & Sons.

Russell, P. J. 1986 *Genetics.* Boston: Little, Brown, and Company.

Strickberger, M. W. 1985. *Genetics,* 3rd ed. New York: Macmillan.

Articles

Cech, T. R. RNA as an enzyme. *Scientific American* November, 1986.

Cohen, S. N., and Shapiro, J. A. Transposable genetic elements. *Scientific American* February, 1980.

De Duve, C. 1988. The second genetic code. *Nature* 333:117–118.

Donelson, D. E., and Turner, M. J. How the trypanosome changes its coat. *Scientific American* February, 1985.

Doolittle, R. F. Proteins. *Scientific American* October, 1985.

Lake, J. A. The Ribosome. *Scientific American* August, 1981.

Laskey, R. A. 1988. A role for the nuclear envelope in controlling DNA replication within the cell cycle. *Nature* 332:546–548.

Ptashne, M. How gene activators work. *Scientific American* January, 1989.

Radman, M., and Wagner, R. The high fidelity of DNA duplication. *Scientific American* August, 1989.

Roberts, L. 1988. The race for the cystic fibrosis gene. *Science* 240:141–144.

Special focus: Genetic engineering. 1984. *American Biology Teacher,* 46(8).

Special focus: Genetic engineering. 1984. *American Biology Teacher,* 46(9).

Steitz, J. A. ''Snurps.'' *Scientific American* June, 1988.

Vaughan, C. 1988. Second thoughts on second genetic code. *Science News* 133:341.

Weintraub, H.M. Antisense RNA and DNA. *Scientific American* January, 1990.

10

Descriptive Embryology

Concepts

1. Embryology is the study of the development of an organism from the fertilized egg to the time that all major organ systems have formed. The focus of descriptive embryology is on describing developmental stages of an animal, and the focus of experimental embryology is on understanding the cellular controls for development.
2. Development of a new animal begins with fertilization of an egg by a sperm. Fertilization events help insure that fertilization is successful, that multiple fertilization is prevented, and that the egg is prepared for development.
3. Patterns of development of different animals have important common themes. Some of the unique developmental patterns of different animals arise from differences in how yolk is distributed in animal eggs.
4. The development of echinoderms illustrates how eggs with very little, evenly distributed yolk may develop.
5. The development of amphibians illustrates how a moderate amount of unevenly distributed yolk influences development.
6. The embryos of reptiles, birds, and mammals require adaptations for long developmental periods in terrestrial environments.

Have You Ever Wondered:

[1] why, although millions are produced, only one sperm usually penetrates an egg during fertilization?
[2] why the study of fertilization could lead to effective and safe birth control techniques?
[3] why echinoderms (sea stars and sea urchins) are often used to study animal development?
[4] why frog eggs are darkly pigmented on top and lightly pigmented on the bottom?
[5] what a chicken egg contains?
[6] what mammals lay eggs?

These and other useful questions will be answered in this chapter.

This chapter contains underlined evolutionary concepts.

How does one explain the complexity of form and function found in adult animals? Can all of this complexity arise from a single egg? Most biologists of the seventeenth and eighteenth centuries believed that gametes contained miniaturized versions of all of the elements present in the adult. This concept was known as **preformation.** Preformationists divided themselves into two groups. The *ovists* held that the future organism was preformed in the ovum, and the *animalculists* believed that the future organism (the homuncule or animalcule) was preformed in the sperm.

In the middle of the eighteenth century, the competing theory of **epigenesis** (Gr. *epi,* on + L. *genesis,* birth) gained in popularity. Biologists who favored this theory maintained that the egg contained the material from which the embryo is gradually built—much like a carpenter builds a building from lumber, nails, bricks, and mortar. The essence of this theory was described by Aristotle during the fourth century B.C., and depended upon an unexplained "creative principle" to direct the assembly.

In separate experiments, Wilhelm Roux (1888) and Hans Driesch (1892) set out to determine whether epigenesis or preformation was correct. Both allowed a fertilized egg to divide to the two-cell stage. Roux, using amphibian (frog, toads, salamanders) embryos, killed one of the two cells with a hot needle. Driesh, using echinoderm (sea stars, sea urchins, sea cucumbers) embryos, completely separated the divided cells. If an entire animal developed from a single cell, then epigenesis would be supported. If a portion of the animal developed, then preformation would be favored. What was the result? Interestingly, Roux described the formation of a half embryo (figure 10.1a), and Dreisch found that each cell retained the potential to develop into an entire organism (figure 10.1b). We now know that Driesh was the more correct of the two, and that the killed cell, still attached to Roux's developing amphibian embryo, probably altered the development of the untreated cell.

This episode in history marks an important turning point in **embryology** (Gr. *embryo,* to be full + *logos,* discourse), the study of the development of an organism from the fertilized egg to the time that all major organ systems are formed. Early studies were descriptive in nature, documenting the stages an organism goes through in embryological development. Such material is the subject of this chapter. The work of Roux, Driesch, and others was the beginning of experimental embryology (or developmental biology), which focuses on understanding the cellular basis of development and is the subject of chapter 11.

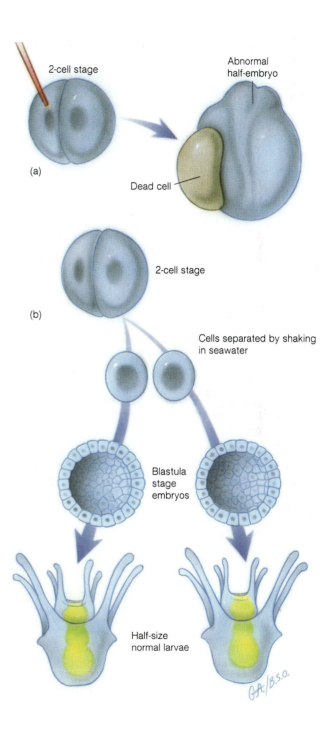

Figure 10.1 A comparison of the results of experiments by Wilhelm Roux and Hans Driesch. (a) A "hemiembryo" was produced by Wilhelm Roux by killing one blastomere of a two-celled amphibian embryo. (b) Driesh found that separating cells of a two-celled embryo resulted in the development of two small, but otherwise normal, larvae.

Fertilization

A variety of mechanisms are used by different animals to bring opposite sexes together, and to induce the release of eggs and/or sperm. Similarly, once gametes have been released, there are different mechanisms that increase the likelihood of fertilization.

For fertilization to occur, a sperm must penetrate an egg. Most eggs have a jellylike coating of protein, or protein and polysaccharide (mucopolysaccharide) around the egg. To penetrate this coating, the sperm of most animals possess enzymes called *sperm lysins.* These enzymes may be associated with a specialized organelle at the head of the sperm, called the **acrosome,** or with the plasma membrane of the anterior tip of the sperm. When sperm contacts the gel coat, the acrosome releases lysins that dissolve a pathway for the sperm (figure 10.2). In humans, even 80 million sperm per ejaculation often means reduced fertility, because sperm lysins from many sperm are required to dissolve the gel coat of an egg. The acrosome of some species reorganizes into an acrosomal process after releasing lysins (figure 10.3). Just outside the egg plasma membrane is a layer called the vitelline layer. *Egg binding proteins* on the surface of the acrosomal process bind to sperm attachment proteins on the vitelline layer of the egg plasma membrane. Acrosomal and egg plasma membranes then fuse.

Egg Activation

The fusion of acrosomal and egg membranes is the beginning of *egg activation.* Egg activation is a series of biochemical changes in the egg that ensure the completion of fertilization and initiates embryonic development. The events of egg activation have been studied extensively in echinoderms, and some of the findings of that work are discussed below.

Membrane and Cortical Events Some of the earliest changes in the zygote occur at the plasma membrane, and in the outer region of the cell cytoplasm (called [1] the *cortex*). The purpose of these early changes is to ensure fertilization by a single sperm cell.

After contact by sperm, microvilli from the plasma membrane of the ovum wrap around a single sperm. The sperm then is drawn into the egg by contraction of microfilaments in egg cytoplasm (*see figures 10.2; 10.3*).

A second series of events defends against multiple fertilization. Within milliseconds of contact by the first sperm, sodium ions (Na+) diffuse to the interior of the cell from the surrounding medium, and calcium ions (Ca2+) are released from the endoplasmic reticulum. These ionic changes make the plasma membrane unresponsive to other sperm cells, and initiate the formation of a protective envelope around the egg, called the **fertilization membrane.**

The fertilization membrane forms as granules in the cortex discharge into the region between the egg plasma membrane and the vitelline layer (figure 10.3c). Enzymes released by the cortical granules loosen the vitelline layer's contact with the plasma membrane. The granules allow water to enter the space between the vitelline layer and the egg plasma membrane, causing the vitelline layer to lift off the egg. Proteins of the cortical granules thicken and strengthen the vitelline layer. All of these reactions are completed in one or two minutes following fertilization. The above reactions are important, because multiple fertilization (**polyspermy**) will normally result in genetic imbalances, and a nonviable embryo.

Other important changes occur in the egg cortex. After sperm penetration, the cortical layer thickens, and rotational and sliding movements of the outer egg cytoplasm begin. In amphibians, these cortical changes result in the formation of a **gray crescent** on the egg, opposite the point of sperm penetration, between the animal and vegetal hemispheres. As you will see, the gray crescent has an important influence on later development.

Metabolic and Nuclear Events

Prior to nuclear fusion (called syngamy; *see chapter 6*), other nuclear events usually occur. Postfertilization changes help prepare the zygote for the mitotic divisions that ensue. Calcium ion release from the endoplasmic reticulum is accompanied by the transport of sodium and hydrogen ions out of the egg. This ion movement raises the intracellular pH and initiates the following changes in zygote physiology. DNA replication occurs, and, in most species investigated, rates of protein synthesis increase. Rapid protein synthesis following fertilization meets the needs of new cells for enzymes and structural proteins, including tubulin, a protein used in the formation of the mitotic spindle, and histones, proteins that contribute to the structure of chromosomes. There is little mRNA synthesis in the zygote. Instead, existing maternal (egg) mRNA is activated, and it directs the formation of the bulk of the proteins synthesized in early stages of embryonic development. This influence is called *maternal dominance* (box 10.1).

The vertebrate egg undergoes the first meiotic division in the ovary. The second meiotic division occurs after sperm penetration, and results in the formation of a polar body and the egg nucleus (the female pronucleus). The sperm nucleus, following dissociation of the flagellum and other accessory structures, is called the male pronucleus. Both pronuclei migrate toward the center of the egg, and meet at the mitotic spindle, which has been assembled in preparation for the first mitotic division of the zygote.

The region where meiosis has been completed is referred to as the **animal pole.** It contains less yolk, more mitochondria, more ribosomes, and is more metabolically active than the opposite, **vegetal pole** of the egg.

Medical Applications

As scientists learn more of these fertilization reactions, medical applications will follow. For example, knowledge of the processes occurring at the egg cell membrane that [2] ensure fertilization by a single sperm cell, may allow the development of techniques that prevent fertilization. Fertilization interactions may also be the key to some infertility problems.

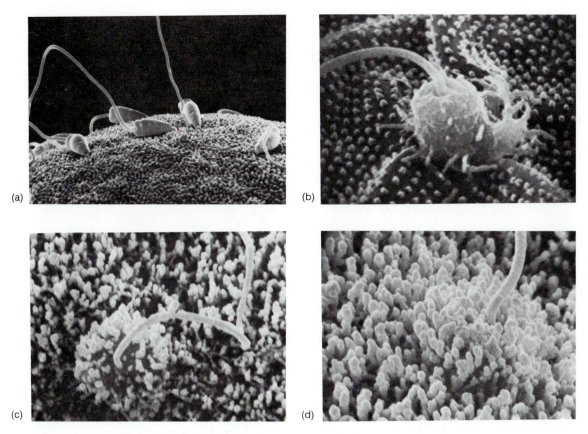

Figure 10.2 Scanning electron micrographs of the fertilization of sea urchin eggs. The fertilization membrane was removed before photomicrographs (c) and (d) were taken. (a) Sperm contacting the egg surface. (b) Microvilli of the egg cell membrane begin to enfold the head of the sperm. (c and d) Sperm penetration being completed.

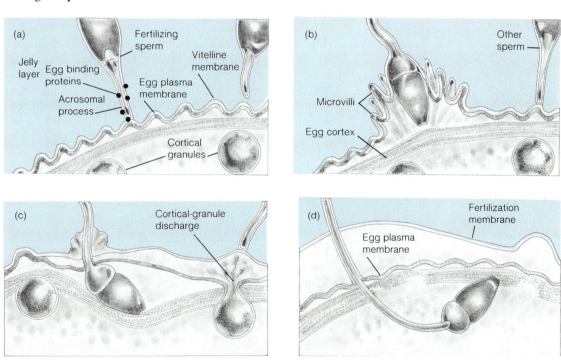

Figure 10.3 The fertilization of echinoderm eggs. (a) Acrosomal process contacts the vitelline membrane. (b) Sperm lysins have created a hole in the vitelline membrane. Changes in the permeability of the plasma membrane to sodium ions initiate changes in the membrane that permit only a single sperm to enter. (c) Cortical granules discharge, and the vitelline membrane begins to rise off the egg plasma membrane. (d) The fertilization membrane is formed.

Box 10.1

Egg and Sperm—Equal Partners?

The influence of egg cytoplasm on development was presented in chapter 8. Another indication of the importance of egg cytoplasm is the fact that development of some organisms can be induced without fertilization (Parthenogenesis is discussed in chapter 6).

Examples of natural parthenogenesis are found in the life histories of freshwater animals, such as rotifers (*see chapter 19*) and cladocerans (*see chapter 22*). Parthenogenesis allows females to produce all-female offspring that produce more all-female offspring, quickly filling a lake or pond in the spring. The development of haploid, male bees (drones) is another well-known example of natural parthenogenesis. In vertebrates, natural parthenogenesis is less common, although it does occur in some birds and lizards.

Natural parthenogenesis does not necessarily result in haploid offspring. Sometimes gametes are produced by mitosis rather than meiosis. In other cases, chroma-

tids disjoin at the second meiotic division, but get incorporated into the same gamete. In still other cases, the first duplication of DNA in a developing egg is not followed by mitosis and cytokinesis.

Artificial parthenogenesis is the experimental initiation of development of an unfertilized egg. It is done by exposing eggs to hypotonic or hypertonic salt solutions, weak acids, or temperature shocks; poking eggs with a needle; electrically shocking eggs; or by shaking eggs. Often (as with mammals), development is aborted after early developmental stages. In spite of development not being carried to completion, artificial parthenogenesis reinforces two concepts concerning the egg: (1) The egg cytoplasm contains a wealth of resources not present in the sperm and (2) development is initiated by changes in the cortex that result from physical and chemical effects of sperm penetration.

Stop and Ask Yourself

1. What is the difference between descriptive and experimental embryology?
2. What is the role of the following in fertilization: Acrosome? Egg binding proteins? Fertilization membrane?
3. What membrane and cortical events occur during egg activation? What is their purpose?
4. What changes occur in the egg nucleus and cytoplasm after sperm penetration is completed? What is their purpose?

Egg Types and Cleavage Patterns

Billions of cells that make up adult animals arise from the zygote. **Cleavage** is a term used to refer to mitotic divisions that occur during embryonic development, and the cells produced by it are called **blastomeres** (Gr *blasto*, sprout + *mero*, part). The first mitotic division of the zygote results in two daughter cells. These blastomeres divide in synchrony, producing a four-celled embryo. As more blastomeres are produced, divisions become asynchronous. Throughout early embryology, there is no overall increase in the size of the embryo. Instead, blastomeres become smaller, and the proportion of DNA to cytoplasm increases.

Quantity and Distribution of Yolk

Egg sizes and cleavage patterns differ among animal species because of differences in the quantity and distribution of yolk in an egg. Yolk is a mixture of proteins, lipids, and

glycogen and serves as the food reserve of the developing embryo. The quantity of yolk present is related to the length of independent development of the animal. Independent development is the development that occurs between the time of fertilization and the point at which the animal begins feeding on its own, or the point at which the embryo is nourished by a parent. Thus, the bird egg is relatively large because it has a large supply of yolk to sustain the embryo up to the time of hatching. The egg of a mammal, on the other hand, is very small. The mammalian embryo needs only enough food reserve to survive for the few days it takes to travel from the oviduct to uterus, where the embryo is nourished by the uterine wall and eventually the placenta.

Cleavage Patterns

The influence of yolk on cleavage patterns is evidenced by the relative size of the blastomeres that result from cleavage and the rate of cleavage. Eggs that have evenly distributed yolk usually have cleavage patterns that result in uniformly sized blastomeres. Eggs that have unevenly distributed yolk have cleavage patterns that result in unequal blastomeres.

Cleavages that completely divide an egg are said to be **holoblastic** (Gr. *holo*, whole + *blasto*, sprout). In eggs with moderate to large quantities of unevenly distributed yolk, the cleavages slow, or stop, when yolk is encountered (figure 10.4a,b). If cleavages cannot completely divide the embryo, the cleavages are said to be **meroblastic** (Gr. *mero*, part). The development of these embryos occurs around, or on top of, the yolk (figure 10.4c).

Other cleavage patterns result from differences in the orientation of the mitotic spindle during mitosis, and the degree to which the fate of early blastomeres is predetermined.

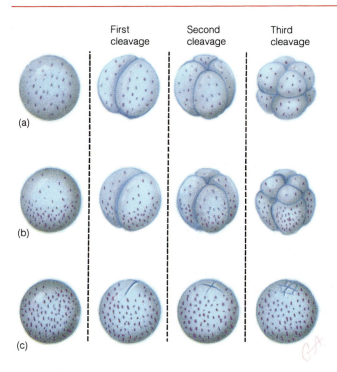

First cleavage   Second cleavage   Third cleavage

(a)

(b)

(c)

Figure 10.4 Early cleavage patterns. (a) Holoblastic cleavage of an embryo containing a small quantity of evenly distributed yolk results in uniformly sized blastomeres. (b) Holoblastic cleavage of an embryo containing moderate to large quantities of unevenly distributed yolk results in unequal blastomeres. (c) Meroblastic cleavage occurs when large quantities of yolk prevent the embryo from dividing completely.

The Primary Germ Layers and Their Derivatives

Tissues and organs of animals arise from layers, or blocks, of embryonic cells called *primary germ layers.* Their development from a nondescript form in the early embryo to their form in late embryonic through adult stages is called **differentiation.** One layer, **ectoderm** (Gr. *ektos,* outside + *derm,* skin), gives rise to the outer body wall. **Endoderm** (Gr. *endo,* within) forms the inner lining of the digestive cavity. **Mesoderm** (Gr. *meso,* in the middle) gives rise to tissues that are located between ectoderm and endoderm. Undifferentiated mesoderm (called **mesenchyme**) develops into muscles, blood and blood vessels, skeletal elements, and (other) connective tissues (table 10.1).

Stop and Ask Yourself

5. What is the influence of yolk on cleavage patterns?
6. What is the difference between holoblastic and meroblastic cleavages?
7. What tissue is derived from each of the following germ layers: Ectoderm? Endoderm? Mesoderm?
8. What is mesenchyme?

Table 10.1   The Primary Germ Layers and Their Derivatives in Vertebrates

Germ Layer	Derivative
Ectoderm	Epidermis of the skin
	Nervous tissue
	Sensory organs
Endoderm	Gut tract lining
	Digestive glands
	Respiratory tract lining
Mesoderm	Connective tissues
	Circulatory system
	Bones, tendons, ligaments
	Dermis of the skin
	Excretory structures
	Reproductive structures
	Muscle

Echinoderm Embryology

The embryology of echinoderms has been extensively studied. Echinoderms are easily maintained in laboratories, and the ease with which one can obtain mature sperm and eggs makes them convenient models to illustrate principles of early animal development.

The eggs of echinoderms have relatively little yolk, and the yolk is evenly distributed throughout the egg. Cleavages are holoblastic, and result in similarly sized blastomeres (figure 10.5). In just a few hours, a solid ball of small cells, called the **morula** (L. *morum,* mulberry) (figure 10.5e), is produced.

As cell division continues, cells pull away from the interior of the embryo. A fluid-filled cavity, the **blastocoel,** is produced, and the cells form a single layer around the cavity. The embryo is now in the form of a hollow sphere and is called a **blastula** (figure 10.5f). In sea urchins, development through the blastula stage takes place within the fertilization membrane. When the cells of the blastula develop cilia, the blastula breaks out of the fertilization membrane and begins to swim. Late in the blastula stage, groups of cells break free of the animal end of the embryo, and position themselves within the blastocoel. These cells, called *primary mesenchyme,* will form skeletal elements (called spicules) of the embryo.

The next series of events in echinoderm embryology are called **gastrulation.** The first sign of gastrulation is the invagination of cells at a point in the vegetal half of the embryo (figure 10.5g). The point of invagination is the **blastopore,** and it will eventually form the anal opening of the larva. During invagination, an embryonic gut, the **archenteron** (Gr. *archeo,* ancient + *enteron,* gut), elongates and reduces the size of the blastocoel (figure 10.5h).

During gastrulation, the embryo also begins to lengthen and assumes a pyramidal shape. Although adult echinoderms do not have head and tail ends, larvae have a preferred direction of movement. The end of an animal that meets the environment during locomotion is called the

[3]

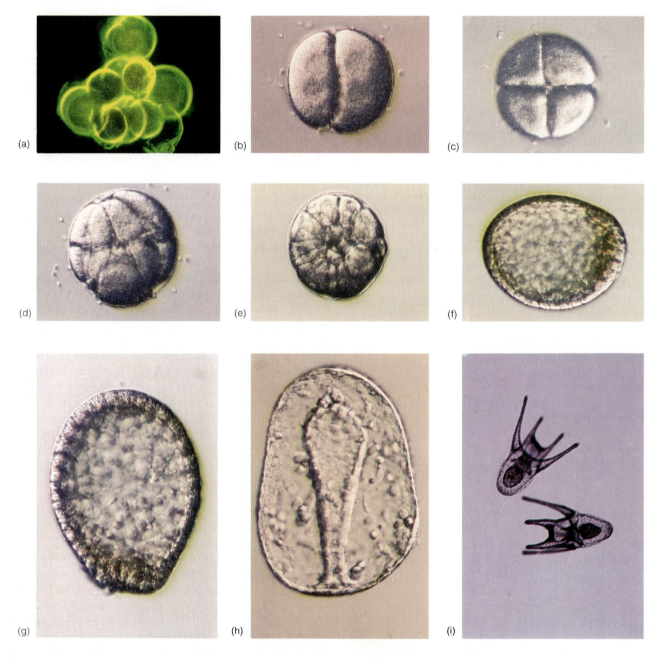

Figure 10.5 Stages in the development of an echinoderm. (a) Unfertilized eggs. (b–d) Early cleavages. The first two cleavages occur along the animal/vegetal axis. The third cleavage occurs at right angles to the previous cleavages, midway between animal and vegetal poles. An eight cell stage with approximately equal blastomeres results. (e) The morula. (f) The blastula—a hollow sphere whose walls are a single cell-layer thick. (g) Gastrulation begins with invagination at a point on the surface of the blastula. (h) As gastrulation continues, the archenteron enlarges. (i) The pluteus. Spicules have formed, and the gut track is complete.

anterior end, and is where the head of most animals is located. The opposite end is the posterior end. The shape changes that occur during gastrulation result in the establishment of the anterior/posterior axis of the embryo.

Finally, a body cavity, or coelom, forms from outpockets of the archenteron, and the gut breaks through the anterior body wall. The opening thus produced is the mouth. Both of these events are more significant than the brief mention given them here. In addition, the blastopore of some animals' embryos forms the mouth, and the opposite opening forms the anus.

The cell movements that begin in gastrulation are the result of groups of cells changing their shapes, all at the same time. These shape changes are mediated by contractile microfilaments (figure 10.6). The result of these precise, coordinated changes is the transformation of a single-layered sphere of cells into an embryo with several layers of cells. Cells of the body surface are ectodermal, those lining the coelom are mesodermal, and those lining the archenteron are endodermal. The progressive development of an animal's form that begins in gastrulation is called **morphogenesis** (Gr. *morph,* form + *genesis,* birth). In the case of the sea urchin, these changes result in the development of a *pluteus larva* that swims freely in the sea and feeds on even smaller plants and animals.

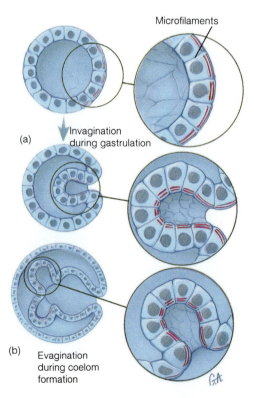

(a) Invagination during gastrulation

Microfilaments

(b) Evagination during coelom formation

Figure 10.6 Morphogenesis involves the coordinated movement and shape changes in embrological cells. (a) Coordinated microfilament contractions on the inner surface of a tissue layer can bring about evagination. (b) Similar contractions on the outer surfaces of a tissue can result in invagination. Contractile microfilaments, like the ones depicted here, are also used in cytokinesis (cytoplasmic division), the movement of some single-celled organisms, and muscle contraction.

Stop and Ask Yourself

9. What are the changes that occur in an echinoderm egg that lead to the formation of the blastula?
10. How does the archenteron of an echinoderm gastrula form? What is the fate of the blastopore?
11. What is morphogenesis?

Vertebrate Embryology

Early stages of vertebrate (animals with a vertebral column or "backbone") development are similar to those described for echinoderms. The differences that exist are the result of longer developmental periods, and for some vertebrates, adaptations for development on land.

The Chordate Body Plan

Vertebrates are members of the phylum Chordata, and all chordates can be characterized by certain structures. The endpoint of our study of vertebrate embryology will be the point at which most of these characteristic structures have been formed (figure 10.7).

The nervous system of chordates develops from ectoderm, and is dorsal and hollow. The first evidence of a developing nervous system is the formation of the *neural tube.* Nervous tissue proliferates anteriorly into a brain.

The *notochord* is the primary axial supportive structure present in all chordate embryos, as well as many adults. It is a flexible, yet supportive structure that lies just beneath the dorsal hollow nerve chord, mesodermal in origin, and consists of vacuolated cells packed into a connective tissue sheath.

In addition to the notochord and the dorsal hollow nerve chord, all chordates possess pharyngeal gill slits and a post-anal tail at some point in their life history.

Amphibian Embryology

The eggs of most amphibians are laid in watery environments and are fertilized as they are released from the female (figure 10.8). Frog eggs have a pigmented animal pole. Because the vegetal pole is heavily laden with yolk, the eggs rotate in their jelly coats, so that the less dense, darkly pigmented animal pole is oriented up. This rather simple series of events has interesting adaptive significance. Given that amphibian eggs usually develop with little care or protection from the parents, one might wonder how they escape detection by predators. Certainly many do not. The pigmentation, however, helps camouflage the developing embryos. When viewed from below, the light color of the vegetal end of floating eggs blends with the sky above. When viewed from above, the dark color of the animal end blends with the bottom of the pond, lake, or stream. The

[4]

Neural tube

Notochord

(b)

Gut

Pharyngeal
gill slits

(a)

Figure 10.7 The chordate body plan. The development of all chordates involves the formation of a neural tube, the notochord, gill slits, and a postanal tail. Derivatives of all three primary germ layers are present. (a) Side view. (b) Cross section.

dark pigment of the animal pole also absorbs heat from the sun, and the warming that results may promote development.

Initial Cleavages The first cleavage of the amphibian embryo, like that of echinoderms, is longitudinal. It begins at the animal pole, and divides the gray crescent in half. If the first cleavage is experimentally forced to pass to one side of the gray crescent, only the blastomere with the gray crescent will develop normally. Because of the large amount of yolk in the vegetal end of the egg, cleavages are slower than in the animal end. The amphibian morula, therefore, consists of many small cells at the animal end

of the embryo and fewer, larger cells at the vegetal end of the embryo (figures 10.9a-d and 10.10a-c). The amphibian blastula forms in much the same way as the echinoderm blastula, except that the yolky vegetal cells cause the blastocoel to form in the animal half of the embryo (figure 10.9e). Unlike that of echinoderms, the blastula wall of amphibians is composed of multiple cell layers.

Gastrulation The cells of the blastula that will develop into specific structures are grouped together on the surface of the blastula. Gastrulation involves the movement of some of these cells into the interior of the embryo. Embryologists have used dyes or carbon particles to mark the surface of the blastula, and then have followed the movements of these cells during gastrulation. *Fate maps* can be constructed to show what will happen to groups of cells on the surface of the blastula during gastrulation. Embryonic cells are designated according to their fate by "presumptive notochord," "presumptive endoderm," and so forth.

The first sign that gastrulation is beginning is the formation of a groove between the gray crescent and the vegetal region of the embryo. This groove is the slitlike blastopore. The animal-pole margin of the blastopore is called the *dorsal lip of the blastopore*. Cells at the bottom of the groove move to the interior of the embryo, and the groove spreads transversely (figures 10.9f and 10.10d). This groove is similar to that which occurs during echinoderm blastopore formation. In amphibians, however, superficial cells now begin to roll over the dorsal lip of the blastopore in a process called *involution*. Cells spread from the animal pole toward the blastopore, and replace those moving into the interior of the embryo. In the process, the ends of the slitlike blastopore continue to spread transversely and downward toward the vegetal pole, until one end of the slit meets and joins the opposite end of the slit. A ringlike blastopore now surrounds the protruding, yolk-filled cells near the vegetal end of the embryo (figure 10.10e). These

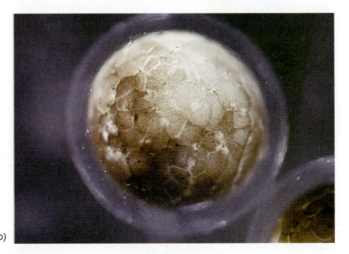

(a)

(b)

Figure 10.8 Fertilization of amphibian eggs. (a) Egg release and fertilization in frogs occurs when the male (above) mounts and grasps the female (below). This positioning is called amplexus. Eggs are fertilized by the male as they are released by the female. (b) The zygote of a frog shortly after fertilization. Each zygote has a jellylike coat. Note the pigmented animal pole.

protruding cells are called the **yolk plug.** Eventually, the lips of the blastopore contract to completely enclose the yolk. The blastopore is said to have "closed."

During the closing of the blastopore, two other movements occur. First, the spreading of cells from the animal pole toward the dorsal lip of the blastopore, and the rolling of cells into the blastopore form the archenteron. As these mesodermal and endodermal cells roll into the interior of the embryo, the archenteron becomes larger, and the blastocoel becomes smaller (figure 10.9f,g). Because the large yolk-filled cells of the vegetal end are less active in these movements, their fate (making up the floor of the gut tract) is accomplished somewhat passively. Second, not only does gastrulation result in a spreading and thinning of ectodermal cells toward the blastopore, but it also results in ectoderm spreading over the entire embryo, a process called **epiboly.**

Mesoderm Formation   Some of the last cells to roll over the dorsal lip into the blastopore are presumptive notochord and presumptive mesoderm (figure 10.9h). Initially, these cells make up the dorsal lining of the archenteron near the blastopore. Later, they detach from the endoderm, and move to a position between the endoderm and ectoderm in the region of the dorsal lip of the blastopore. This mesoderm, referred to as **chordamesoderm,** spreads anteriorly (the embryo is now beginning to elongate) and laterally (to either side) between ectoderm and endoderm. Chordamesoderm is thicker dorsally in a region that will differentiate into notochord. As mesoderm spreads, it also thickens along the sides of the embryo. These thickenings are called **somites,** and are visible externally as a row of bumps on either side of the embryo (figure 10.9k). As mesoderm continues to spread ventrally,

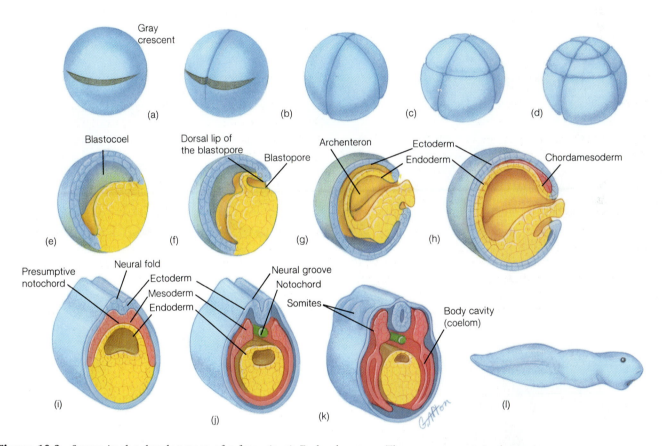

Figure 10.9   Stages in the development of a frog. (a–c) Early cleavages. The gray crescent is shown in (a). The first and second cleavages usually begin at the animal pole, and are at right angles to one another. The second cleavage begins before the first cleavage is completed. The third cleavage is horizontal. Because of the dense, yolky cytoplasm at the vegetal pole, this third cleavage occurs closer to the animal pole, and results in unequal blastomeres. (d) The morula. (e) The blastula. Note the location of the blastocoel near the animal pole, and the multiple cell layers of the blastula. (f–g) Gastrulation. Involution begins as cells move into the interior of the embryo, forming a slitlike blastopore. The blastopore is extended to the sides, and toward the vegetal pole. Ectodermal cells spread and thin over the surface of the embryo, gradually covering the vegetal end. The ends of the blastopore join, forming a yolk plug, and the blastopore eventually closes. (h) Chordamesoderm begins the formation of the notochord, and spreads laterally. (i–j) The embryo lengthens during gastrulation, and when gastrulation is completed, nerve cord formation begins. The neural plate develops, its edges upfold, and the neural tube is formed. (k) The mesoderm proliferates, forming somites. Mesoderm along the sides of the embryo splits, forming the body cavity, or coelom. (l) The larval stage, or tadpole, gradually uses up the yolk stored in endodermal cells and begins to feed.

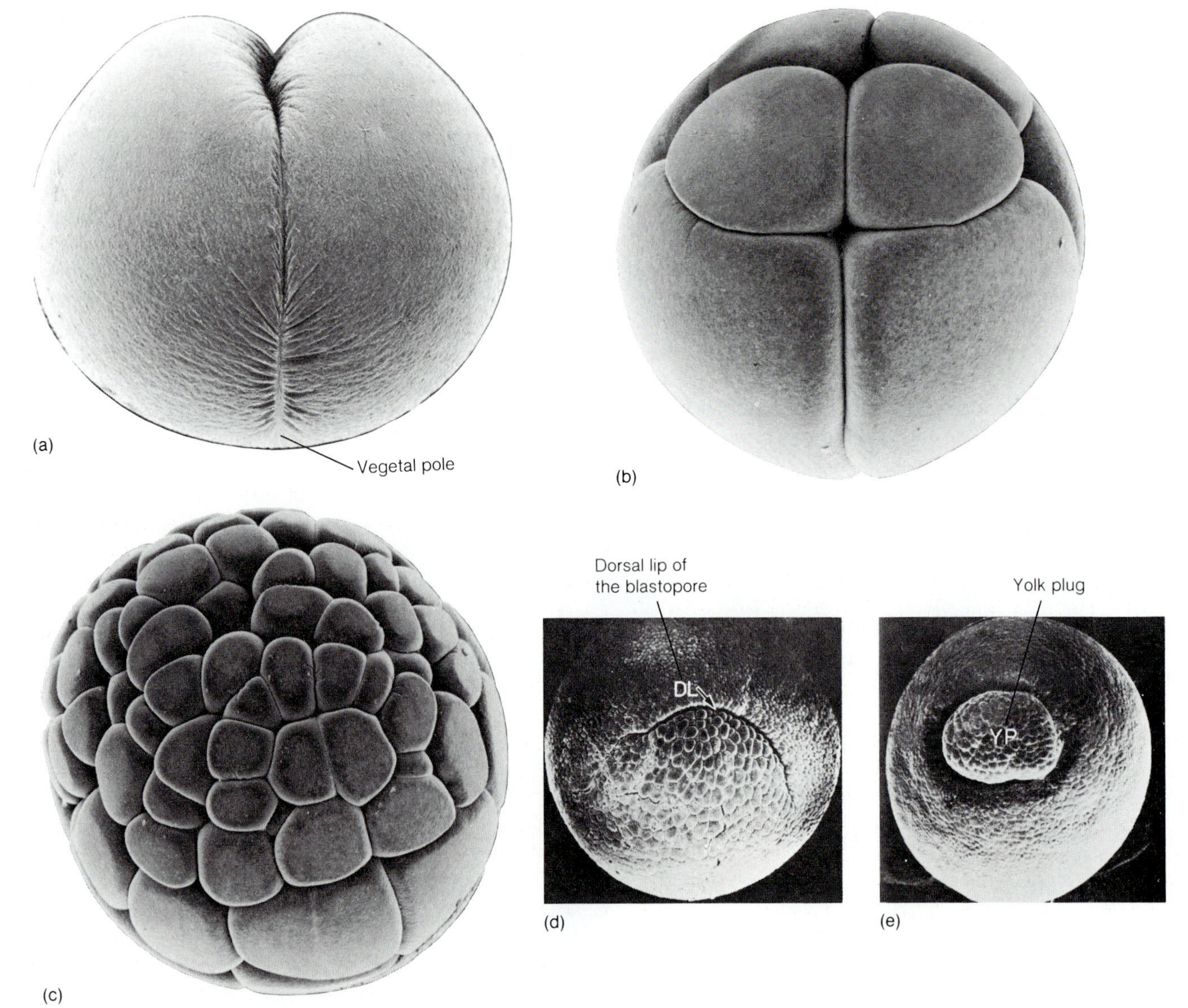

(a)

Vegetal pole

(b)

(c)

Dorsal lip of the blastopore

DL

(d)

Yolk plug

YP

(e)

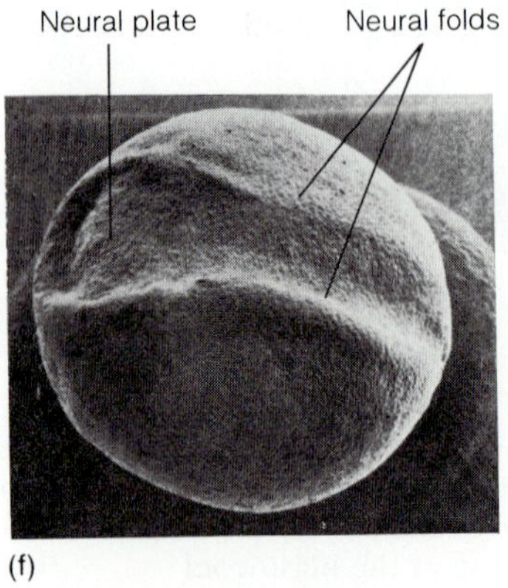

Neural plate   Neural folds

(f)

Figure 10.10 Scanning electron micrographs of frog development. (a–b) Early cleavages. (c) The morula. (d) Early gastrulation, showing a slitlike blastopore. (e) The ends of the blastopore have met, the blastopore is closing, and the yolk plug is visible. (f) The formation of the neural tube.

it splits to form the body cavity (coelom) and the mesodermal lining of the body wall and gut.

Neural Tube Formation   During late gastrulation, external changes along the upper surface of the embryo begin the formation of the neural tube—a process called **neurulation.** After gastrulation is completed, an oval-shaped area on the dorsal side (the future backside) of the embryo marks the presumptive neural tube. This region is called the *neural plate* (figure 10.9i). Microfilaments in neural plate cells cause a flattening and thickening of the neural plate. The edges of the neural plate roll up and over the midline of the neural plate. These longitudinal ridges, called *neural folds,* meet dorsally to form the neural tube (figures 10.9i,j; 10.10f). The portion of the neural tube that will become the brain is the last to close.

With further development of the mesoderm, the amphibian embryo will gradually take on the form of a tadpole larva. Yolk in cells lining the floor of the gut is gradually depleted, and the larva will begin feeding on algae and other plant material.

Development in Terrestrial Environments

Development of reptiles (class Reptilia), birds (class Aves), and mammals (class Mammalia) occurs on land rather than in the water. Development on land requires that the embryo be protected from desiccation, and in these animals, a series of extraembryonic membranes serve that purpose. The longer developmental periods of these animals reflect the fact that they lack independent larval stages. (Larval stages, such as those of amphibians, allow individuals to achieve increased complexity in spite of short embryonic periods.)

Avian Embryology

Chicken development can be used to model the development of birds and reptiles. What is commonly referred to as the "egg" of a chicken is, in reality, the true egg, plus a variety of membranes that protect the egg. The yellow portion of the chicken egg is the single cell produced in the chicken ovary. This egg is released into the oviduct, where fertilization may occur. Following fertilization, membranes and fluids are deposited around the egg (figure 10.11a). A vitelline membrane covers the surface of the true egg. The "white" consists of water, and a protein called *albumen*. This watery environment protects the egg from mechanical damage and drying. Albumen is a source of nutrients (in addition to the yolk of the egg), and is eventually consumed during development. Two denser strands of albumen (called chalazas) attach at the inside of the shell and at the egg, and keep the egg suspended in the center of the watery albumen. The shell is made of calcium carbonate impregnated with protein. Thousands

[5]

of tiny pores (40 to 50 microns in diameter) in the shell permit exchange of gases between the embryo and the outside. On the inside of the shell are two shell membranes. An air pocket is formed between these membranes, at the rounded end of the shell. The air sac enlarges during development, as air moves through pores in the shell to replace water loss. As hatching approaches, the chick penetrates the air sac with its beak, the lungs inflate, and the chick begins to breathe from the air sac, while still exchanging gases across vascular extraembryonic membranes.

Early Cleavages and Gastrulation Cleavage of the chicken egg is meroblastic (figure 10.11b). A small disk of approximately 60,000 cells at the animal end of the egg develops and is called the **blastoderm.** The blastoderm is raised off the yolk, leaving a fluid-filled space analogous to the blastocoel of the amphibian blastula. The proliferation and movement of cells of the blastoderm result in a sorting of cells into two layers. The **epiblast** (Gr. *epi*, upon + *blast*, sprout) is the outer layer of cells, and the **hypoblast** (Gr. *hypo*, below) is the inner layer of cells (figure 10.12a-c). The movements of blastoderm cells are the beginning of gastrulation. The chicken egg is released from the reproductive tract of the female at about this time.

A medial, linear invagination, called the **primitive streak,** gradually extends anteriorly (figure 10.12d). A depression, called **Henson's node,** forms at the anterior margin of the primitive streak and marks the beginning of an inward migration of epiblast cells, comparable to involution of the amphibian gastrula. The primitive streak is, therefore, analogous to the dorsal lip of the blastopore. This migration occurs during a dramatic posterior movement of Henson's node. Migrating cells form mesoderm and endoderm, and what is left of the epiblast on the surface of the embryo is the ectoderm. The three germ layers are now arranged on the surface of the yolk.

Following gastrulation, notochordal cells are separated from the overlying neural ectoderm, and the neural tube

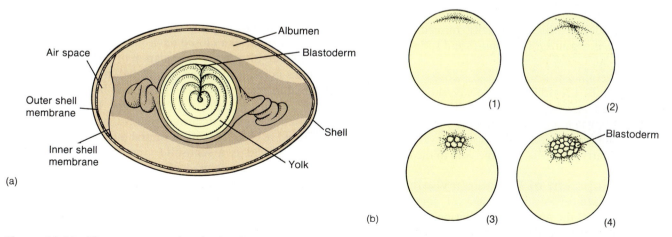

Figure 10.11 The structure and early development of the chicken egg. (a) The egg and the egg membranes. Active cytoplasm is a small area on top of the yolky cytoplasm. Albumen is an accessory food source, and, along with egg membranes, protects the embryo and prevents the embryo from drying. (b) Initial cleavages result in the formation of a patch of cells at the animal pole of the egg, called the blastoderm.

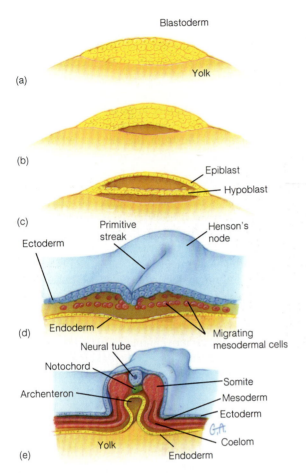

Figure 10.12 Gastrulation in the chick embryo. (a–b) Meroblastic cleavage results in the formation of the blastoderm. (c) Cells of the blastoderm rise off the yolk and rearrange into two layers—epiblast and hypoblast. (d) Gastrulation results from the migration of epiblast cells into a longitudinal groove called the primitive streak. These migrating cells form mesoderm and endoderm, and cells remaining in the epiblast form ectoderm. (e) The embryo lifts off the yolk when the margins of the embryo grow downward and meet below the embryo.

forms as described earlier for the amphibian embryo. In addition, mesoderm, which originally formed as solid blocks of cells, is organized into somites and splits to form the coelom.

Further development results in the embryo lifting off the yolk (figure 10.12e). This occurs when the margins of the embryo grow downward and meet below the embryo. A connection between the embryo and the yolk is retained and is called the *yolk stalk*. Blood vessels develop in the yolk stalk and carry nutrients from the yolk to the embryo.

The Development of Extraembryonic Membranes Extraembryonic membranes of amniotes include the yolk sac, the amnion, the chorion, and the allantois (figure 10.13a,b). Reptiles and birds have a large quantity of yolk that becomes enclosed by a **yolk sac.** The yolk sac develops from a proliferation of the endoderm and mesoderm around the yolk. The yolk sac is highly vascular and distibutes nutrients to the developing embryo.

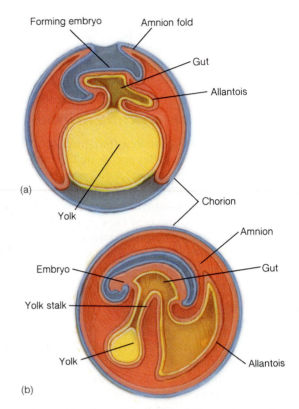

Figure 10.13 The formation of extraembryonic membranes. (a) The embryo lifts off the yolk as hypoblast proliferates around the yolk to form the yolk sac. The amnion begins forming before the allantois. (b) Amnion folds fuse above the embryo enclosing the embryo in the amnion, and forming a continuous chorion on the inside of the shell. The allantois increases in size as uric acid accumulates. The yolk is being used up.

Following the neural tube stage, the ectoderm and mesoderm on both sides of the embryo lifts off the yolk and grows dorsally over the embryo. As these membranes meet dorsally, they fuse and form an inner **amnion** and an outer **chorion.** The amnion encloses the embryo in a fluid-filled sac. This amniotic cavity provides protection against shock and drying. The chorion is nearer the shell, becomes highly vascular, and aids in gas exchange.

Development in a closed environment presents a problem of waste disposal. Accumulation of nitrogenous waste products in the embryo, or unconfined in the shell would be lethal for the embryo. The immediate breakdown product of proteins is highly toxic ammonia. This ammonia is converted to a less toxic form, uric acid, which is excreted and stored in the **allantois,** a ventral outgrowth of the gut tract. Uric acid is a semisolid, and thus little water is wasted. The allantois gradually enlarges during development to occupy the region between the amnion and the chorion. In addition, the allantois becomes highly vascular and functions with the chorion in gas exchange.

Mammalian Embryology

Most mammalian eggs have very little yolk. As a result, cleavages are holoblastic and early development of most mammals is initially different from that just described for the chicken. This difference is soon overshadowed by impressive similarities, including the development of a primitive streak, Henson's node, and extraembryonic membranes.

Similarities in the development of mammals to that of reptiles and birds are even more impressive when one considers primitive mammals. The duckbilled platypus and the spiny anteater lay eggs, and their embryos are nourished by large quantities of yolk stored within the egg. They are oviparous, which means that they are born from eggs laid outside the body of the mother. Initial cleavages of their eggs are meroblastic. Opossums and kangaroos have eggs that contain some yolk, but the yolk is not used during development. After fertilization, the yolk is ejected from the embryo. Even though the embryos are nourished within the uterus, the young are born very early, and develop within a marsupium. Marsupial mammals seem to represent a transition between oviparous and viviparous (embryos develop and are nourished in the female reproductive system) mammals.

[6]

Over the years, knowledge of mammalian developmental patterns has been relatively slow to emerge because of the difficulty in observing development when the embryo is concealed within the uterus. Recent advances with culturing mammalian embryos have resulted in progress in this area. Human development is representative of mammalian development, and is discussed in chapter 39.

The Fate of Mesoderm

Following gastrulation in birds, reptiles, and mammals, all three primary germ layers have formed. Of the three layers, the fate of the mesoderm is the most complex. Mesoderm forms all of the supportive tissues of vertebrates, including connective tissues (bone, cartilage, and blood) and muscle. These supportive tissues are frequently associated with derivatives of other primary germ layers. For example, the inner lining of the gut is endodermal, but mesodermally derived structures, such as smooth muscle, blood, and blood vessels, make up the bulk of that system.

Stop and Ask Yourself

12. How would you define involution, epiboly, yolk plug, dorsal lip of the blastopore, and chordamesoderm?
13. What are the initial phases of mesoderm formation in amphibians?
14. How does a chicken egg provide a suitable environment for terrestrial development?
15. In what way(s) is the development of primitive mammals similar to the development of birds?

Summary

1. Embryology is the study of the development of an organism from the fertilized egg to the point that all major organ systems are formed. Descriptive embryology involves the documentation of these changes. Experimental embryology is the study of the cellular controls of development.

2. One of the goals of embryology is the description of the formation and fates of the primary germ layers: ectoderm, endoderm, and mesoderm.

3. Fertilization is the process by which gametes are brought together, forming a diploid zygote. A variety of mechanisms have evolved to insure that fertilization takes place.

4. Eggs are activated by contact with sperm. Changes in the egg cytoplasm ensure that a single sperm penetrates the egg, male and female pronuclei are positioned for mitosis, and the cell is metabolically ready for mitosis.

5. Cleavage patterns are influenced by the quantity and distribution of yolk. Eggs that contain little yolk divide into blastomeres that are equal in size. Eggs that contain larger amounts of yolk either do not divide completely, or divide into blastomeres that are unequal in size.

6. Echinoderm development is representative of the development of a zygote that contains very little yolk. Cleavages are holoblastic and equal. Gastrulation occurs by invagination of the blastula. Mesoderm and the coelom form from simple outpockets of the archenteron.

7. Frog eggs have a large amount of unequally distributed yolk. As a result, their cleavage is holoblastic and unequal. The blastocoel forms near the animal pole, and gastrulation occurs by involution. Chordamesoderm forms between ectoderm and endoderm, and develops into somites, coelomic linings, and the notochord. Changes in the ectoderm include the formation of the tubular nerve cord.

8. The development of reptiles, birds, and mammals shows adaptations for terrestrialism, including the formation of extraembryonic membranes.

9. The initial cleavages of a chicken egg result in the formation of a disk of cytoplasm, the blastoderm, at the animal end of the embryo. Gastrulation involves the movement of cells toward a medial longitudinal groove, called the primitive streak.

10. Mammalian development has been modified from that of reptiles and birds by the loss of yolk.

11. The fate of mesoderm is more complex than the fate of either ectoderm or endoderm.

■ Key Terms

acrosome (p. 147)
allantois (p. 156)
amnion (p. 156)
animal pole (p. 147)
archenteron (p. 149)
blastocoel (p. 149)
blastoderm (p. 155)
blastomeres (p. 148)
blastopore (p. 149)
blastula (p. 149)
chordamesoderm (p. 153)
chorion (p. 156)
cleavage (p. 148)
differentiation (p. 149)
ectoderm (p. 149)
embryology (p. 145)
endoderm (p. 149)
epiblast (p. 155)
epiboly (p. 153)
epigenesis (p. 145)

fertilization membrane
  (p. 147)
gastrulation (p. 149)
gray crescent (p. 147)
Henson's node (p. 155)
holoblastic (p. 148)
hypoblast (p. 155)
meroblastic (p. 148)
mesenchyme (p. 149)
mesoderm (p. 149)
morphogenesis (p. 151)
morula (p. 149)
neurulation (p. 154)
polyspermy (p. 147)
preformation (p. 145)
primitive streak (p. 155)
somites (p. 153)
vegetal pole (p. 147)
yolk plug (p. 153)
yolk sac (p. 156)

■ Critical Thinking Questions

1. In what sense is fertilization a random event? Why do you think random fertilization is adaptive for a species?

2. Evaluate the following statement. "The primary differences between early cleavages of most animal embryos are the result of the quantity and distribution of yolk."

3. Many terrestrial vertebrates have eggs that are protected by a shell and have large quantities of yolk. Why do most mammalian eggs lack these features?

■ Suggested Readings

Books

Balinsky, B. 1982 (6th ed.). *An Introduction to Embryology.* Philadelphia: W.B. Saunders Company.

Cohen, J. 1977. *Reproduction.* London: Butterworths.

Giudice, G. 1986. *The Sea Urchin Embryo: A Developmental Biological System.* New York: Springer-Verlag.

Hopper, A. F. and N. H. Hart. 1985. *Foundations of Animal Development,* 2nd ed. Boston: Oxford University Press.

Karp, B., and Berrill, N. J. 1981. *Development.* 2d ed. New York: McGraw-Hill.

Saunders, J. W., Jr. 1982. *Developmental Biology.* New York: Macmillan.

Slack, J. M. W. 1983. *From Egg to Embryo: Determinative Events in Early Development.* Cambridge: Cambridge University Press.

Articles

Cooke, J. 1988. The early embryo and the formation of body pattern. *American Scientist* 76:3535–42.

Epel, D. 1980. Fertilization. *Endeavour* 4:26.

Epel, D. The program of fertilization. *Scientific American* November, 1977.

Gerhart, J., Black, S., Scharf, S., Gimlich, R., Vincent, J. P., Danilchik, M, Rowning, B., and Roberts, J. 1986. Amphibian early development. *BioScience* 36:541–549.

Gould, M., and Stephano, J. L. 1987. Electrical responses of eggs to acrosomal protein similar to those induced by sperm. *Science* 235:1654–1657.

Rahn, H., Ar, A., and Paganelli, V. How bird eggs breathe. *Scientific American* February, 1979.

Wassaman, P. M. 1987. The biology and chemistry of fertilization. *Science* 235:553–560.

Wassaman, P. M. Fertilization in mammals. *Scientific American* December, 1988.

Modern Developmental Biology

Concepts

1. The development of a complex animal from a single cell is one of the most impressive occurences in biology. Determination is the final selection of cells from among several alternatives.
2. Determined cells must undergo differentiation. Differentiation is heralded by the appearance of specific proteins and messenger RNA molecules that are found only in certain types of cells.
3. The creation of form is a process called morphogenesis. A typical vertebrate limb has an array of fingers or toes, a set of bones in the wrist or ankle regions, and includes muscles, tendons, nerves, blood vessels, and skin. The pattern in which these structures are arranged is created during development. The process by which it unfolds is called pattern formation.
4. The processes of determination, differentiation, morphogenesis, and pattern formation are the result of gene regulation at the level of DNA processing, RNA synthesis and processing, and protein synthesis. Hormones also influence the developmental process.
5. Aging and the death that follows can also be regarded as developmental processes.
6. Cancer cells usually have uncontrolled growth.

Have You Ever Wondered:

[1] how cells communicate with each other?
[2] why blood vessels are always located in the same place in different people?
[3] how the fused eyelids of kittens and puppies open after birth?
[4] what makes a leg a leg and an arm an arm?
[5] how it is possible that the five-digit limb of land vertebrates evolved into bat wings or whale flippers?
[6] what a homeobox and "homeo-madness" are?
[7] why chromosomes "puff?"
[8] why humans and animals age?

These and other useful questions will be answered in this chapter.

This chapter contains underlined evolutionary concepts.

The development of a complex animal from a single cell (the zygote) is one of the most impressive occurences in biology. The zygote is said to be **totipotent,** meaning it can give rise to all the different kinds of specialized cells found in the adult animal.

This chapter presents some of the mechanisms by which cells become organized into an animal form and develop their specific functional specializations. Development is viewed as a series of events leading from the more general to the more specific, and eventually to the death of the animal.

Determination: Commitment to a Type of Differentiation

Chapter 10 introduced the fact that embryonic cells differentiate and participate in morphogenesis. Before this is accomplished, however, most cell lines become *committed* or *determined* to a specific type of differentiation, such as a nerve cell, red blood cell, or intestinal lining cell. **Determination** is the final selection by cells of a single developmental pathway from among several alternatives.

Once a subpopulation of cells is determined, they are stable and usually do not change with cell division. Because this determination is passed from one cell to its progeny during cell division, it is an *inherited* state of gene control. However, despite the normal permanence of determination, changes in cells can occur.

Change in Determination

A good example of a change in determination occurs in fruit flies (*Drosophila* spp). Inside a fruit fly larva are small sacs of cells called *imaginal discs.* Each disc is composed

of cells determined to become some part of the fly (figure 11.1). The Swiss developmental biologist Ernst Hadorn investigated the determined state of these discs by causing them to divide repeatedly. His culture chambers were the abdominal cavities of adult flies (figure 11.2). After serial transplantations of the genital discs, he observed that once

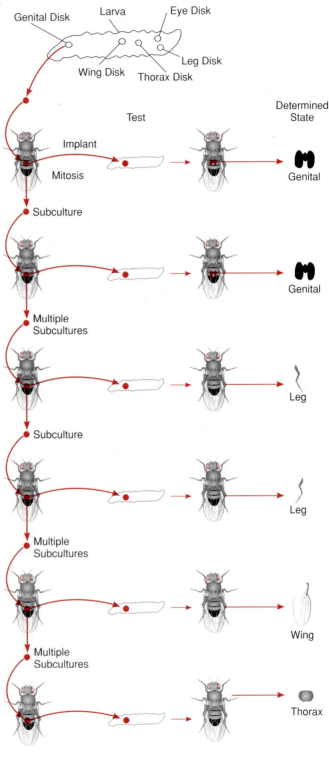

Figure 11.2 Transdetermination sequences. Fragments of imaginal discs are cultured in the abdomen of adult female flies. Some possible transdetermination pathways are shown on the right.

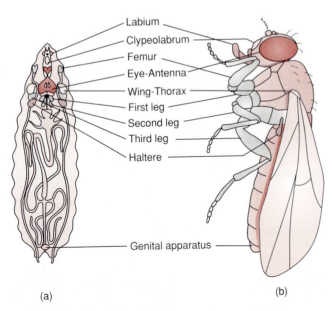

Figure 11.1 Imaginal discs. (a) Location of imaginal discs in the larva of a fruit fly, and (b) the corresponding parts of the adult that they form.

Figure 11.2 redrawn, with permission, from Norman K. Wessels and Janet L. Hopson, *Biology.* Copyright © 1988, McGraw-Hill Publishing Company, New York, New York.

a sufficient number of cell divisions had taken place, these cells gave rise to leg tissue. He called this process **transdetermination** because the cells had shifted from one determined state (genital apparatus) to another (leg). This new state remained stable for several more subcultures when another transdetermination occurred. In place of leg tissue, wing tissue formed. After further divisions, thorax tissue appeared. The reasons for transdetermination are still not understood, but related experiments on vertebrate embryos, as covered next, tell us more about determination.

The Cytoplasm and Determination Stability

During the 1950s, Robert Briggs and Thomas J. King at Indiana University used **nuclear transplantation** to study determination. They transplanted nuclei from various embryonic blastula cells of the frog *Rana pipiens,* into eggs whose nuclei had been removed (enucleated eggs). These experiments tested the capacity of cells at advanced stages of development to interact with mature egg cytoplasm in directing developmental processes (figure 11.3). Their results indicated that even though the frog blastula cells are

partially determined, they still carry a full set of genes capable of directing complete development to a tadpole.

John B. Gurdon obtained similar results in three successive nuclear transplants using nuclei of the African clawed frog, *Xenopus laevis* (figure 11.4). He found that nuclei from fully differentiated intestinal epithelial cells could direct development of a zygote to a normal tadpole. In fact, intestinal epithelial nuclei could be used as donors to establish clones of frogs, all derived from the initial nuclear transplant. His results suggested that (1) in a determined and differentiated epithelial cell, all the genes required for other determined states and modes of differentiation are present; and (2) when the nucleus of a determined cell is removed and exposed to the cytoplasm of

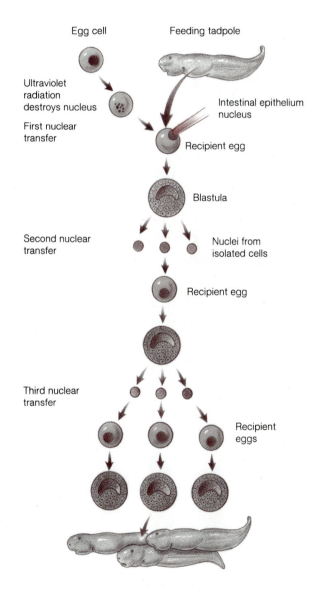

Figure 11.4 John B. Gurdon's serial transplant experiment. After a second transfer, nuclei that are mitotically descended from an original intestinal epithelium nucleus can interact with an enucleated egg's cytoplasm to produce normal development. The tadpoles that result are nuclear clones because they bear identical nuclear genetic information to each other and to the original individual.

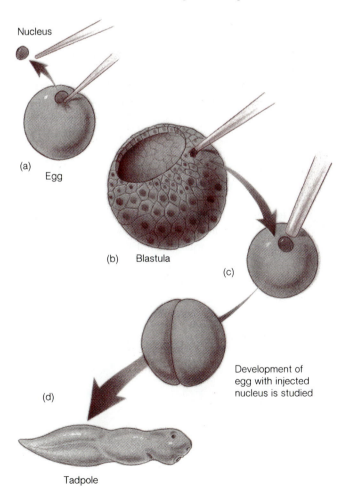

Figure 11.3 The nuclear transplantation technique. (a) The nucleus from a frog egg is removed. (b) Cells are taken from the blastula, and (c) injected into the enucleated egg. (d) The egg then develops into a tadpole.

another cell type, the original state of determination is lost. Thus, although determination is a stable condition, it is dependent on the cytoplasm of the cell. Exposure to a different cytoplasm "deprograms" the nucleus and permits it to participate in the entire developmental process.

Stop and Ask Yourself

1. Why is the zygote said to be totipotent?
2. How can developmental determination be defined?
3. How can changes in determination occur?
4. How does the cytoplasm affect the stability of determination?

Differentiation: Building the Cell Phenotype

Just because cells are determined does not mean they are fully functional. Determined cells must also undergo differentiation. **Differentiation** occurs when the cells acquire the molecular and structural features that equip them for their specialized functions. It includes the appearance of specific proteins and messenger RNA molecules that are found only in certain types of cells.

The Causes of Differentiation

One of the most exciting discoveries of the early part of the century was that adjacent cells and tissues interact with each other in ways that influence organ development. In 1921, the German biologist Hans Spemann, showed that the chordamesoderm underlying the ectoderm of the newt (a tailed amphibian) gastrula not only *induces* the ectoderm to develop into the nervous system, but sets in motion an entire series of events that produce the larva of this amphibian. Spemann was awarded the Nobel Prize in Physiology or Medicine in 1937 for his work on inductive phenomena.

Spemann and coworkers also showed that in the newt, cells of the dorsal lip of the blastopore give rise to the chordamesoderm. The chordamesoderm underlies the dorsal ectoderm that differentiates into neural plate and neural tube. When the dorsal lip of the blastopore of a heavily pigmented species of newt was transplanted into the blastocoel of a lightly pigmented species, at gastrulation it came to underlie the belly ectoderm of the host embryo (figure 11.5). The researchers found that secondary axial structures (brain and spinal cord) formed at the transplantation site (figure 11.5). Spemann concluded that the implanted dorsal-lip cells *induced* or directed the surrounding host ectoderm to form these secondary axial structures. He called the dorsal-lip tissue an *organizer,* because it appeared to organize a set of structures in a host embryo.

Biologists have also discovered molecular signals that influence differentiation and morphogenesis. These "inducer molecules" initiate a response by binding with receptor molecules on the surface of responding cells. Recent studies have suggested that differentiating cells [1] maintain physical contact with each other through gap junctions (*see figure 3.7c*) and/or specific plasma membrane glycoproteins called **cell adhesion molecules (CAMs)**. As their names suggest, these molecules mediate adhesion between particular cell types. For example, neural CAMs and neuroglia CAMs are located on the surface of neurons, and liver CAMs are found on liver cells. Together with inducers, CAMs seem to be the signals that cells use to associate and develop into tissues that characterize organs and organ systems in an animal.

Morphogenesis: How Cells Organize into Functional Units

Complex morphological events occur in specific sets of cells, such as those forming the eye, heart, or lungs. These changes give rise to an organ with a definitive form in a

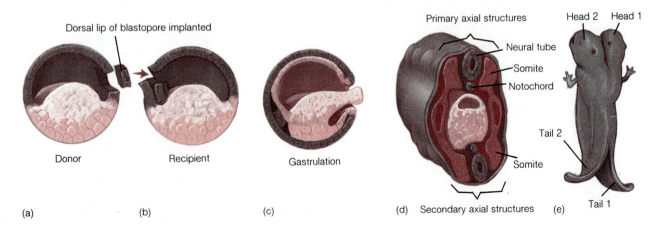

Figure 11.5 The organizer-induction experiment. (a) Chordamesodermal tissue from the dorsal lip of the blastopore (the donor) is implanted into the blastocoel of a recipient (b). (c) Gastrulation proceeds, and (d) the implant develops as a secondary axial structure. (e) The ectoderm near the transplanted chordamesoderm responds to the inducing tissue by forming a new head and tail.

process called **morphogenesis** (Gr. *morphe,* form + *gennan,* to produce). The form must be perfect if the organ's function is to be carried out. The many types of morphogenesis in an embryo are the result of at least six basic mechanisms that will now be discussed.

Single-Cell Migrations

Single cells may migrate from one site to another in embryos, and set up cell populations that will form specific tissues and organs. For example, in the vertebrate embryo, *neural-crest cells* arise at the site where the neural tube forms. Figure 11.6 illustrates the normal pathways that these cells follow. Cells that will form skin cells migrate through one pathway. Other cells stop and differentiate at various levels along another pathway to form parts of the nervous system.

Moving and Stationary Cells: Interactions

The movement of single cells is also affected by large extracellular molecules composed of sugars and proteins. Examples of such molecules include collagen, fibronectin, laminin, and hyaluronic acid. These molecules function in two ways. They provide a ground substance to which migrating cells can stick, and they mark a trail on/through which the migrating cells can move.

An example of this type of movement is the developing cornea (part of the outermost layer of the eyeball). Neural-crest cells migrate to the edge of the area where the cornea is forming but do not enter the cornea's cellular matrix, which is made up primarily of collagen fibers (figure 11.7a). Endothelial cells then secrete hyaluronic acid into the matrix (figure 11.7b). The acid acts like a "guiding light" that allows the neural-crest cells to migrate into the matrix of the cornea. The same endothelial cells then se-

crete the enzyme hyaluronidase, which digests the hyaluronic acid, leaving the nerual-crest cells immobilized and fixed in the matrix where they proceed to differentiate further (figure 11.7c).

These extracellular molecules can also apply mechanical forces in the developing embryo. When some cells move, they may adhere to these molecules, which act as a substratum. By adhering, the cells push backwards so they can move forward. An analogy would be the tires of a tractor digging into the soil (traction) so that the tractor can move forward. In this way, embryonic cells exert traction on their substratum and can actually deform the substratum much in the same way that the tractor tire would leave a rut in soft soil.

In the embryo, most mesodermal cells do not move very far. However, by exerting force on their substratum, they can deform the substratum in tracts, or pathways, contributing to the development of tendons, ligaments, muscles and even blood vessels. These simple mechanical activities [2] of embryonic cells lead to mechanical events and morphological ordering in the developing embryo.

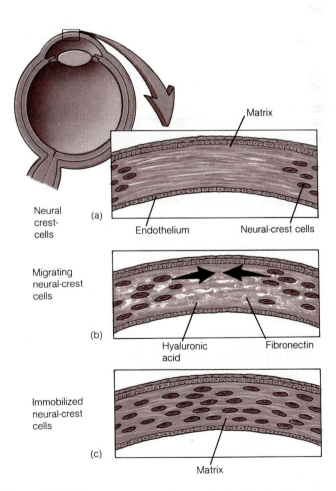

Figure 11.7 The environmental control of cell migration. (a) Neural-crest cells fail to enter the matrix in the forming cornea. (b) After hyaluronic acid is secreted by the endothelium, the cells migrate in. (c) When the enzyme hyaluronidase is secreted, hyaluronic acid is digested and the cells become immobilized in the matrix.

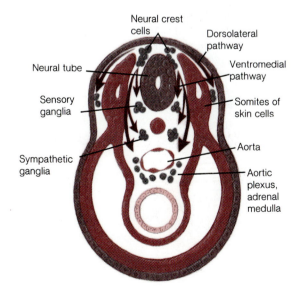

Figure 11.6 Single-cell migration. The pathways of neural-crest migration. This migration demonstrates the cell locomotion and guidance that is part of morphogenesis in the embryo.

Figure 11.7 redrawn, with permission, from Norman K. Wessels and Janet L. Hopson, *Biology.* Copyright © 1988, McGraw-Hill Publishing Company, New York, New York.

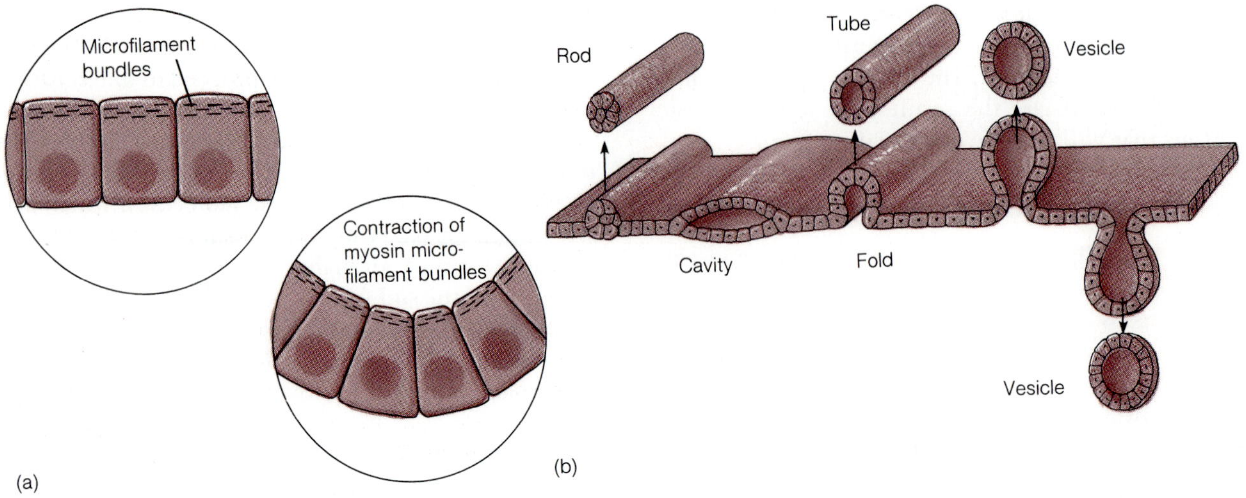

Microfilament bundles

Contraction of myosin microfilament bundles

(a)

Rod

Tube

Vesicle

Cavity

Fold

Vesicle

(b)

Figure 11.8 How cells change shape. (a) Microfilaments associated with myosin interact with actin to contract one end of the cell forcing it into a wedge shape. The change in the shape of the cells alters the shape of the epithelium as a whole. (b) Transformation of an epithelial sheet as it deforms into folds, bulges, invaginations, and projections. As projections extend further and pinch off, tubes and spherical vesicles can form.

Movement of Entire Cell Populations

Many organs of the body (e.g., kidneys, lungs, eyes, brain) are formed by inward and outward folding of large populations of cells. These foldings are caused by the activity of the cell adhesions and the cytoskeleton (*see figure 3.22*). For example, contraction of microfilaments may cause contraction of one end of a cell, forcing it into a wedge shape (figure 11.8a). If adjacent microfilaments are also contracted, there may be a change in a whole sheet of cells to form rods, tubes, and vesicles (figure 11.8b).

Localized Relative Growth

Another morphogenetic mechanism involves differences in growth rates of cells. The rate of cell division at a particular site can be increased or decreased relative to nearby cells. For example, during the early development of the vertebrate limb, cell division in the surrounding body wall slows, compared to the higher rates at the site of leg or arm formation.

Localized Cell Death

For some cells, degeneration and controlled death are part of their normal morphogenesis. For example, kittens and [3] puppies are born with their eyes shut. From the time of their formation until just after birth, the eyelids consist of an unbroken layer of epidermal cells. After birth, cell stretching occurs in a thin line across each eyelid in response to some internal signal and causes these epidermal cells to die. As the dead cells degenerate, a slit forms in the skin, and the upper and lower lids part.

Morphogenetic death is also involved in separating the digits of both hands and feet during human development (figure 11.9). Cells between the digits degenerate and die during development, resulting in separate digits.

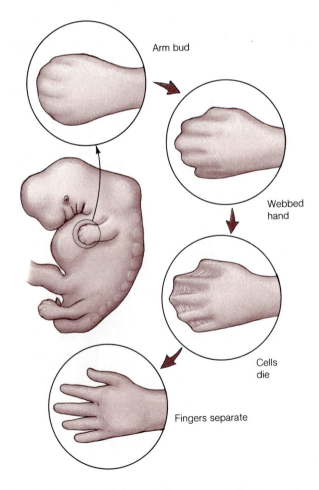

Arm bud

Webbed hand

Cells die

Fingers separate

Figure 11.9 The progressive formation of the human hand. Cell degeneration and death are involved in forming the final structure of the hand.

Extracellular Matrix: Cellular Cement

Large amounts of *extracellular matrix* contribute to the shape of bones, tendons, cartilage, corneas, and others. Filamentous molecules, such as collagen and large sugar polymers, are the major components of this matrix. The matrix surrounds many embryonic cells, helps cement the cells in place, and helps determine the size and shape of nearby tissues.

Stop and Ask Yourself

5. What is morphogenesis?
6. How can moving and stationary cells interact?
7. What may result when an entire population of cells moves?
8. How can relative growth and localized death affect morphogenesis?

Principles of Pattern Formation

A typical vertebrate limb has an array of fingers or toes, a set of bones in the wrist or ankle regions, and includes muscles, tendons, nerves, blood vessels, and skin. The pattern in which these structures are arranged is created during development. The process by which it unfolds is called **pattern formation.** To make a simple analogy, all countries of the world have flags, and in most cases, these flags have distinct patterns that allow them to be easily identified. For example, the United States has stars and stripes, whereas France has the tricolor. Both flags are the same size, are made from the same material, and have the same colors, but they are easily distinguishable because their colors are arranged in different, definite shapes. The French tricolor flag is divided into three equal stripes, one red, one white, and one blue; the United States flag has the same three colors but is arranged as stars and stripes. The flags can be distinguished because of their different patterns.

Similarly, the various designs of different vertebrate limbs develop in characteristic patterns that allow them to be distinguished (box 11.1). The importance of pattern formation can be more readily grasped by looking at the structure of our own arms and legs. Both structures consist of a series of bones, muscles, joints, and ligaments. These tissues, in turn, are made up of cells. In fact, no cell type is found in your leg that is not also found in your arm. What makes a leg a leg and an arm an arm is how these cell types and the tissues they make up are arranged within each limb—the same way the flags of France and the United States can be distinguished.

[4]

The Vertebrate Limb

The mechanisms that control the creation of spatial patterns in animals must be at work during development, *before* the final form of the particular structure is apparent. Just as a flag is a defined size and has boundries in which the pattern of colors and shapes are created, the vertebrate embryo can be thought of as a patchwork of discrete, bounded regions, each of which will develop into a particular structure. A limb can thus be thought of as a pattern within a boundry. Because morphogenesis of wings has been thoroughly studied in the chick embryo, it will be used as an example of pattern formation in the vertebrate limb.

Each chick limb begins development as a small buldge called a *limb bud*, which occurs on one side of the embryonic body (figure 11.10a). A limb bud consists of a core of mesoderm enclosed in a jacket of skin ectoderm. One strip on this ectodermal jacket is thickened to produce the *apical ectodermal ridge* (AER) (figure 11.10b). The AER

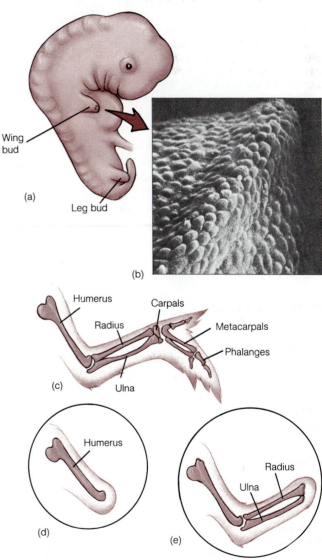

Figure 11.10 Chick embryo limb bud development. (a) Location of wing and leg buds at 3 days. (b) Scanning electron micrograph of the apical ectodermal ridge on a chick embryo's wing bud. (c) The end product of normal limb development. (d) Results of early surgical removal of the apical ectodermal ridge. (e) Results of later surgical removal of the apical ectodermal ridge.

Box 11.1

Vitamin A Causes Major Change in Pattern Formation of Limb Structure

One novel way biologists study the cellular and molecular controls of pattern formation is to use specific chemicals that cause demonstrable changes in limb patterns during development and regeneration. Using this approach, Malcolm Maden in London has recently demonstrated that vitamin A (retinoic acid) significantly changes the limb pattern in developing limbs of amphibians.

If an axolotl (the larval stage of the amphibian, *Ambystoma* spp.) is placed in fresh water after a limb has been amputated at the level of the wrist, regeneration processes produces a new limb (box figure 11.1a). However, if the axolotl is placed in water containing vitamin A (retinoic acid) after a limb has been amputated at the level of the wrist, regeneration processes cease. When the axolotl is returned to fresh water, some regeneration occurs, but the pattern of tissue in the regenerating limb is not normal. Instead of regenerating the parts initially removed, a complete limb is created distal (further from the midline of the body) to the plane of the amputation. Thus, if the initial amputation was through the wrist, instead of regenerating a hand, the vitamin A-treated limb produces a humerus, radius, ulna, wrist, and hand (box figure 11.1b). Apparently, vitamin A resets the positional values of the distal cells to produce more proximal (closer to the midline of the body) cells and structures.

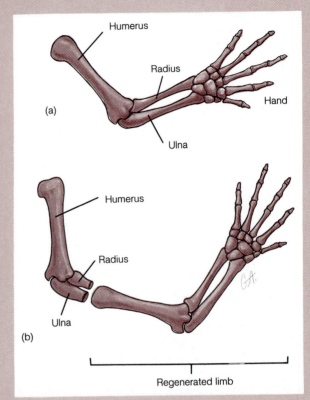

Box figure 11.1 Pattern formation of limb structure. (a) Normal regeneration and pattern formation in fresh water. (b) Abnormal pattern formation in water containing vitamin A.

---

is important for normal limb development (figure 11.10c) because surgical removal of the ridge terminates the outward growth of the bud. An incomplete limb produced by very early removal of the AER contains only proximal limb structures (figure 11.10d). If the AER is removed slightly later, a limb with more distal skeletal parts develops (figure 11.10e). Thus, there seems to be a progressive laying down of the limb's normal proximal-to-distal pattern that depends on the presence of the apical ectodermal ridge.

What the above pattern formation indicates is that developmental events act in concert to coordinate limb development. An error in one event may lead to extra digits, missing digits, or incorrect orientation of parts. However, the long term consequences of such errors may not be totally detrimental. Instead, these errors might contribute to the variations in limb structure that are at the center of evolutionary change. Without a doubt, the evolution of the [5] five-digit limb of land vertebrates into bat wings, whale flippers, or horse legs was possible because mutations affected one of the control systems of limb formation. Thus, evolutionary alterations in body structure can result from

heritable changes in the developmental process of individual animals.

Regulatory or Homeotic Genes Help Control Pattern Formation

As presented in chapters 8 and 9, structural genes code for proteins, whereas other genes code for the nonmessenger RNAs (tRNA and rRNA). Another class of genes, called **regulatory** or **homeotic genes,** helps determine body shape by controlling the developmental fate of groups of cells. Apparently these genes work by turning structural genes on and off.

Regulatory genes were first identified in fruit flies, where a mutation in such a gene might produce a fly with such bizarre characteristics as an extra set of wings (figure 11.11) or legs growing from the head in place of antennae. A specific DNA base sequence almost 200 nucleotides long was found repeated at least seven times in each of the fruit [6] fly regulatory genes. This sequence has been called the **homeobox.** Since their discovery, homeoboxes have attracted a great deal of attention, to the point of engendering what some biologists have called "homeo-madness." Homeoboxes are widely distributed, occurring in species ranging from sea urchins to mammals, including humans. In all these animals, the genes containing the homeobox seem to serve as "master" genetic elements, interacting with a cascade of other genes to send a cell down a specific developmental pathway.

The homeobox appears to be located in the coding sequences of genes, so it is presumably translated into protein. Part of the homeobox sequence resembles known genes for DNA-binding proteins, suggesting that the product of the homeobox is a protein that regulates other genes. Thus, the homeotic (regulatory) genes may switch on or off genes involved in complex morphogenetic events.

Overall, regulatory genes may prove to be keys to understanding how an animal is constructed by developmental processes, as well as keys to some changes in body structure that have been important in evolution, such as the duplication of body segments, the hallmark of evolution in animals whose bodies are built of many segments.

The result of regulatory genes acting as they do *early* in the development of an animal, can have profound effects on the organization and appearance of an animal. What this means is a minor gene change (mutation) can result in an alteration in morphogenesis that allows a radical change in environmental adaptation. A relatively small number of regulatory genes, acting early on morphogenetic processes, may well be the key that produces one animal form instead of another. In contrast to earlier beliefs that mutations in hundreds of genes, perhaps occurring over millions of years, were needed to generate a complex characteristic, such as a new pair of legs on a thoracic segment, our growing knowledge of regulatory and morphogenetic processes provides us with a greater understanding of other evolutionary scenarios.

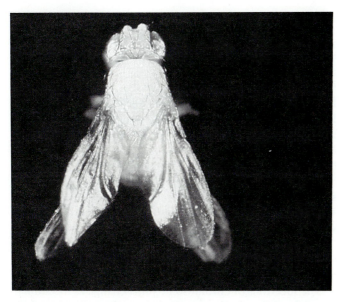

Figure 11.11 Mutation of a regulatory gene. Photograph of a mutant fly with complete duplication of the second thorax segment with wings.

Stop and Ask Yourself

9. How does the apical ectoderm ridge function in pattern formation?
10. How are developmental positions determined in the embryo?
11. How do regulatory genes help control pattern formation?
12. What is a homeobox?

Gene Regulation in Development

The processes of determination, differentiation, morphogenesis, and pattern formation all depend on specific proteins and macromolecules that may regulate these processes. The mechanisms that regulate the activity of the genes for making a cell's characteristic proteins and macromolecules will be the subject of this section. Some of the most likely control mechanisms are summarized in figure 11.12.

DNA Processing

Three gene control mechanisms occur at the level of DNA processing: gene amplification, gene rearrangement, and selective gene loss.

Gene amplification   The genes that code for ribosomal RNA are repeated many times in the genomes of higher animals. These multiple copies of rRNA are in the genome of every cell in the animal, where they form the core of the nucleolus. It is in the nucleolus that ribosomal subunits are assembled. One special type of cell, the oocyte (the devloping egg), synthesizes a million or more additional copies of the rRNA genes, which exist as extrachromosomal circles of DNA. This selective synthesis of additional DNA is called **gene amplification** and is an excellent way of enhancing expression of the rRNA genes. Gene amplification enables the oocyte to make tremendous numbers of ribosomes that will serve in the sudden burst of protein synthesis that occurs immediately after the egg is fertilized. Gene amplification occurs in the oocytes of many amphibians and insects; however, as a rule, most other cells do not have to make extra copies of specific genes when they differentiate.

Gene rearrangement   In higher animals, **gene rearrangement** has been demonstrated only in the developing immune system. It is a unique mechanism whereby pieces of genes are moved around in cells to produce a variety of combined genes. This mechanism allows a relatively small number of gene pieces to be rearranged in a great variety of ways to generate the many hundreds of thousands of antibodies needed to defend the animal (immunity and antibodies will be discussed in detail in chapter 36).

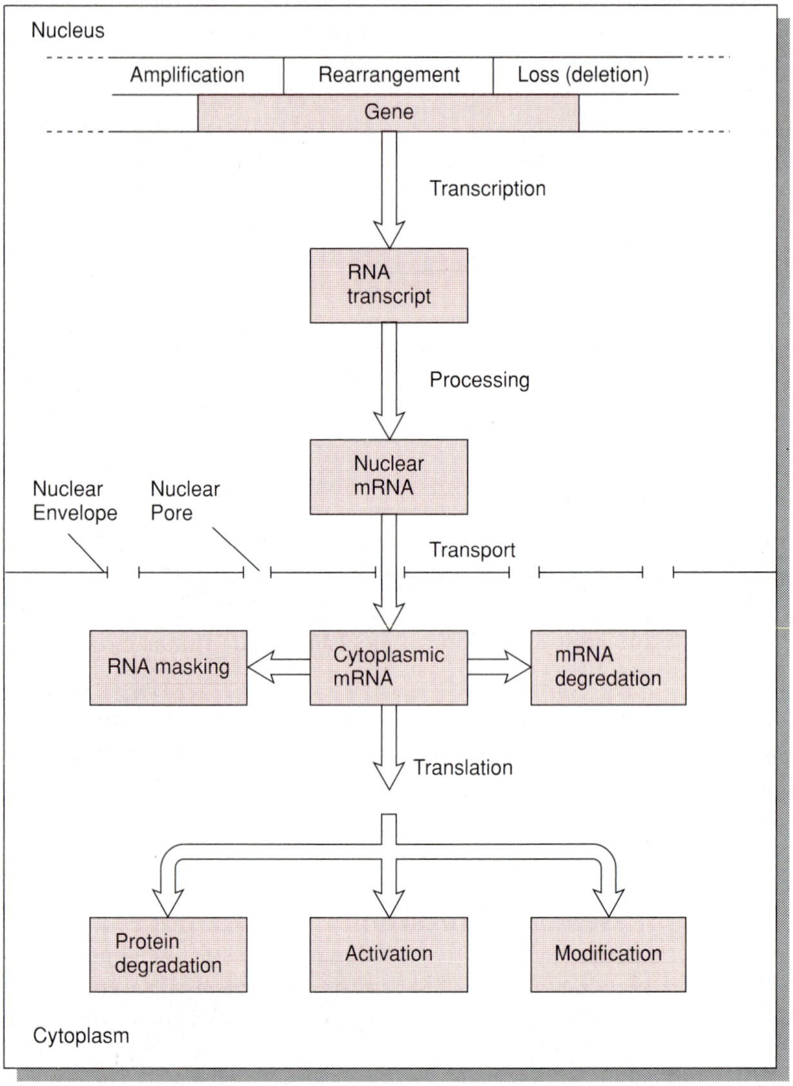

Figure 11.12 Gene regulation in development. Control points for gene expression and regulation in a eukaryotic cell.

Selective gene loss In some animals, whole chromosomes or parts of chromosmes (genes) are eliminated from certain cells early in embryonic development. This selective loss is called **chromosome diminution** (the act of reduction). For example, in the roundworm *Ascaris* (*see figure 19.8*), parts of chromosomes are discarded during embryonic development. Breakage occurs at multiple sites along the chromosomes—places where the chromatin packing is especially tight—creating many chromosome fragments.

RNA Synthesis and Processing

Although important, gene amplification, rearrangement, and deletion at the level of DNA processing do not occur in most types of differentiating cells. A more common process called **differential gene activation** occurs when some genes function and others do not during the synthesis and processing of RNA.

Production of mRNA One model system used to study differential gene activation during development involves special chromosomes found in certain insect cells.

These cells grow very large, and as they grow, their genetic material replicates repeatedly. Up to ten sequential replications produce nuclei with as many as 1024 sets of chromosomes (normal diploid body cells have two sets). This repeated replication without mitosis produces giant **polytene** (multistranded) **chromosomes**, because the replicated DNA and associated proteins of each chromosome remain together as one unit. When stained, each polytene chromosome has characteristic patterns of dark and light bands (figure 11.13).

Genetic activity in polytene chromosomes involves a specific change in chromosome organization. The many strands of a particular chromosome loosen and loop outward, forming a "puff" (figure 11.14). Chromosome puffs function as sites of messenger RNA synthesis (transcription). The structure of the puffs allows the RNA polymerase enzyme responsible for RNA synthesis to operate efficiently.

The locations of chromosome puffs change as the animal develops; some puffs disappear and others form at new sites. The shifting puffs are visual indicators of the selective switching on and off of specific genes during development.

[7]

Molecular mechanism of transcriptional control

In the DNA of higher animals, regulatory proteins interact with specific nucleotide sequences called regulatory sites to control the transcription of the DNA to mRNA. These regulatory proteins can function as either activators (promotors) or repressors (inhibitors). Because in most eukaryotic cells the vast majority of DNA is not

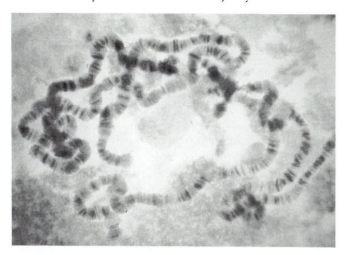

Figure 11.13 Giant polytene chromosomes from a *Drosophila* salivary gland stained to illustrate the banding pattern.

expressed, most genes are inactive. Thus, most regulatory proteins act to turn on transcription of specific genes rather than to turn them off (*see chapter 9*).

The sites on the DNA molecule where RNA polymerase binds and transcription control occurs are called promoter regions. Such sites are located "upstream" from the gene. (Because RNA polymerase reads from the 5' toward the 3' end of a gene, "upstream" means toward the 5' end of the gene.) Eukaryotic cells contain three different polymerases for transcribing DNA into RNA, and there are two different kinds of promotors for each of them. Polymerase I and II (figure 11.15a) recognize specific gene sequences located "upstream" from the point on the DNA molecule where RNA synthesis begins. Polymerase I transcribes genes for ribosomal RNA and messenger RNA, and polymerase II transcribes genes coding for proteins. Polymerase III (figure 11.15b) initially binds to a site in the middle of the gene it transcribes, "downstream" from the start of RNA synthesis. It apparently recognizes a regulatory protein already bound there. Polymerase III transcribes the genes for various small RNA molecules, such as transfer RNA.

Posttranscriptional control

Processing of transcribed RNA provides another point at which gene expression can be controlled. Gene expression can be measured

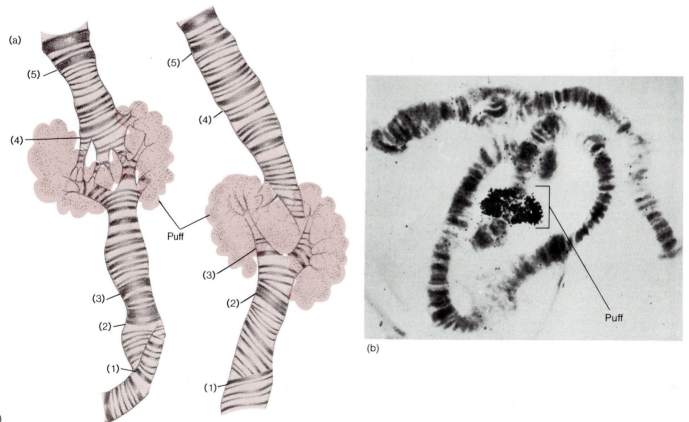

Figure 11.14 Chromosome puffs. (a) A short section of the same chromosome taken from two different cells of the fly, *Trichocladius*. Puffs are clearly seen. The bands are numbered for reference to show that different parts of the chromosome are puffed in different cells, indicating that different parts of the genome are being transcribed in each cell. (b) The stained bands are unpuffed in active genes. The puffs are sites of intensive RNA synthesis.

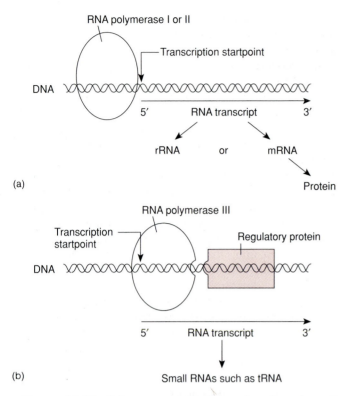

(a)

(b)

Figure 11.15 Polymerase recognition sites in eukaryotic cells. (a) RNA polymerase I or II transcription starting point, and (b) RNA polymerase III starting point.

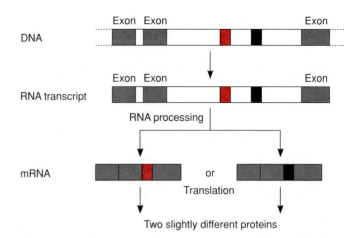

Figure 11.16 Variable RNA processing. The same gene and same RNA transcript can produce two different mRNAs by alternate joining processes. The two mRNAs are then translated into two slightly different proteins.

by the amount of protein, or transfer and ribosomal RNA, that a cell makes. However, every time a gene is transcribed into RNA, it is not necessarily expressed. Much happens between transcription and the final appearance of a protein. Because a eukaryotic cell has a nuclear envelope that separates translation from transcription, other possibilities for controlling gene expression appear in the cytoplasm (*see figure 11.12*).

As described in chapter 9, eukaryotic cells must process RNA transcripts (mRNA, tRNA, rRNA) before they can act as fully functional RNA. The RNA segments representing the introns of the genes must be deleted and the coding segments (exons) joined together. Different patterns of joining can lead to the production of different proteins from the same initial RNA transcript (figure 11.16). This mechanism provides an exception to the one gene-one polypeptide rule discussed in chapter 9. It is not known to be very common but has interesting evolutionary implications, especially if found to be more common.

Regulation of Protein Synthesis

Regulation of gene activity can also occur at the level of translation of mRNA into protein. The lifetime of mRNA in the cytoplasm is an important factor in controlling protein synthesis. Some mRNA in eukaryotes is broken down by enzymes in the cytoplasm in a matter of minutes; other mRNA may last for weeks. The different amounts of time that various mRNA molecules function will determine how much protein synthesis each directs.

In other cases, mRNA may accumulate, but translation is delayed until some control signal activates it. For example, the mRNA that is used for protein synthesis during cleavage is synthesized by the egg nucleus prior to fertilization. This mRNA is stored in the cytoplasm of an egg in the form of an inactive "masked messenger"—a messenger that is not translated until fertilization. By synthesizing large amounts of specific mRNA, and delaying translation until a signal is given, a developing cell can respond to a stimulus with an explosive burst of synthesis of specific gene products.

Even after protein synthesis has occurred, the expression of the gene can be regulated by mechanisms affecting the specific proteins. During differentiation, the newly produced proteins can be modified. Parts of the protein can be "chopped off" or other parts added. It is only when the various alterations are complete, that the protein is used by the cells of the developing embryo.

Another means by which levels of a specific protein are regulated in a cell is by a process called *degradation,* in which a protein is broken down. As soon as some proteins are manufactured, they are subject to destruction. Short-lived proteins have certain amino acids at their N-terminal (amino) end. Arginine, asparatic acid, lysine, leucine, and phenylalanine have protein half-lives (the time it takes for half the protein to disappear) of 2 to 3 minutes; glutamic acid and tyrosine have half-lives of 10 minutes; and glutamine and isoleucine have half-lives of 30 minutes. Long-lived proteins, which may exist for 20 to 30 hours, have methionine, serine, alanine, threonine, valine, or glycine at the N-terminal end. This mechanism indicates that cells have evolved a mechanism to ensure rapid turnover of some proteins and persistence in others. Thus, the amount of a specific protein present in a cell is the result of the balance between the synthesis and degradation of the protein. By speeding up or slowing down either synthesis or degradation, gene expresion can be indirectly regulated.

Hormonal Control of Development

As noted in the above sections, during morphogenesis, many cell types differentiate. These cells are organized into tissues and organs; each organ then grows to a predetermined size to form a properly proportioned adult body. Each tissue and organ is responsible for controlling its own growth, especially in the early stages of development. As the adult form becomes more complex, however, some

overall coordination of growth and tissue homeostasis is required. This control of continued growth and differentiation is provided by the chemical messengers called hormones. A **hormone** is a chemical messenger produced in one part of the body and specifically influencing certain activities of cells in another part of the body. (Hormones will be discussed in more detail in chapter 39).

One example of hormonal control of differentiation is the dramatic changes in body form that occur during **metamorphosis** (Gr. *meta,* after + *morphosis,* bringing into shape) in insects and many amphibians. Metamorphosis is the developmental transformation from the larval to the adult form. The familiar conversion of tadpole to frog is caused by the thyroid hormone thyroxine (figure 11.17). If thyroid secretion is inhibited in tadpoles, metamorphosis does not occur. Conversely, metamorphosis can be accelerated experimentally by adding thyroxine to the water around tadpoles.

Insect metamorphosis has been studied in greater detail (figure 11.18). Larvae of butterflies and moths, called caterpillars, bear no resemblance to the adults of the species (colorplate 16). The caterpillar is merely a feeding machine that voraciously engorges itself. After hatching, a cat-

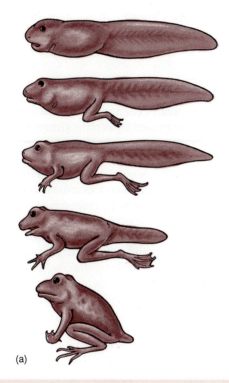

(a)

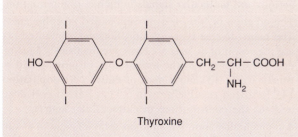

(b)

Figure 11.17 Metamorphosis in the frog, *Rana pipiens.* (a) Stages showing the gradual appearance of froglike features. (b) The molecular structure of the iodine-containing thyroid hormone, thyroxine.

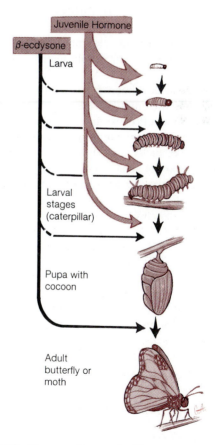

Figure 11.18 Hormonal control of insect metamorphosis. β-ecdysone is present during much of the larva's life and is responsible for the molts the larva undergoes—from larva to larva, larva to pupa, and pupa to adult. However, it is the level of juvenile hormone that determines which molting-stage the insect develops into. In the final metamorphic event, the pupa is transformed into an adult. The width of the arrows reflect the concentration of juvenile hormone present in the insect.

erpillar increases some 5,000-fold in size in a matter of days. Suddenly, it stops feeding and spins a cocoon, within which it metamorphoses into a pupa. The pupa then develops into an adult moth or butterfly. All these developmental changes are regulated by two hormones: β-ecdysone and juvenile hormone. When both hormones are present in the larva, it feeds and grows into a larger larva. At the end of larval life, the neurosecretory cells that secrete juvenile hormone are shut off. As a result, the tissues are exposed only to β-ecdysone, and the larva changes into a pupa within its cocoon. After the formation of the pupa, β-ecdysone is again produced in the absence of juvenile hormone. When ecdysone acts on pupal tissues, these tissues develop into adult tissues and an adult butterfly or moth eventually emerges from the cocoon.

Aging as a Stage in Development

Animals continue to change after completing their embryonic development. Some of the changes are growth and maturation (including puberty) in humans, metamorphosis from larval to adult forms in many insects, and wound healing in many animals. Similarly, aging and the death that follows are both forms of developmental change in an animal. Like other processes in development, aging may be programmed.

[8]   Two hypotheses have been proposed to explain aging. In one hypothesis, the cumulative effects of mutations and other "insults" to a cell cause its functional decline. The other hypothesis suggests that aging and death are innate properties of cells, programmed either by the expression of specific genes or by scheduled changes that affect the entire genome, such as the declining ability of DNA to replicate or a loss of the effectiveness of DNA repair mechanisms.

Cancer: Cells Out of Control

Under normal circumstances, when cells differentiate, control mechanisms limit their growth and division. On rare occasions, this genetic control over cell division becomes altered permanently. As a result, a cell divides again and again, until its offspring begin to crowd surrounding cells and interfere with tissue functions. This alteration has spawned a **tumor** (L., *tumere,* to swell): a clonal population of abnormally dividing cells.

Cells that divide excessively but remain at their original site in the body form **benign tumors**, which do not recur after removal. **Malignant tumors** (cancerous tumors), on the other hand, are very harmful and exhibit the following characteristics: they may have unusual numbers of chromosomes; many of their metabolic processes may have been deranged; and their cell surfaces are altered.

The altered cell surface of cancer cells allows these cells to escape the normal controls of cell position in the body. Normal cells use molecules on their surface to recognize one another and stick together. Cancer cells lose this ability and spread into tissues surrounding the original tumor. The spread of cancer cells beyond their original site is called **metastasis.**

Cancer can be caused by physical agents, such as X rays and by chemical agents called **carcinogens.** A carcinogen is an agent that causes gene mutations that lead to cancer. Cancer can also be caused by certain viruses. Whatever the initial cause of the cancer, the mechanisms involve the activation of **oncogenes** that are either native to the cell or introduced in viral genomes. An oncogene is any gene having the potential to induce cancerous transformations. The normal cellular genes corresponding to oncogenes (often called *protooncogenes* when the cell is in the normal noncancerous state) are thought to be key genes in the control of cell growth and differentiation.

■ Summary

1. The genetic program for development is written in the nucleotide sequences of DNA in the nucleus of the zygote. The zygote is said to be totipotent, meaning it can give rise to all the different kinds of specialized cells found in the adult animal.

2. Determination is the final selection of cells of a single developmental pathway from among several alternatives.

3. Despite the normal permanence of determination, changes or mistakes in a given cell type can occur. Examples include the transdetermined cells from the imaginal discs of fruit flies and cytoplasmic influences on the nuclei in nuclear transplantation.

4. Once a cell becomes determined, it then undergoes differentiation. Differentiation is heralded by the appearance of specific proteins and messenger RNA molecules that are found only in a certain type of cell. Inductive interactions and molecular signals influence differentiation.

5. The creation of form is a process called morphogenesis. The many diverse types of morphogenesis in an embryo are the result of at least six basic mechanisms: (1) single cell movements, (2) the interactions of moving and stationary cells, (3) movement of entire cell populations, (4) localized relative growth, (5) localized cell death, and (6) the contributions of extracellular matrix.

6. A typical vertebrate limb has an array of fingers or toes, a set of bones in the wrist or ankle regions, and includes muscles, tendons, nerves, blood vessels, and skin. The pattern in which these structures are arranged is created during development. The process by which it unfolds is called pattern formation.

Embryology and Metamorphosis of a Frog

(a) Eight-cell stage.

(b) Early blastula stage.

(c) Late gastrula (yolk plug) stage. Note the circular blastopore with yolk protruding from within the embryo.

(d) Neural groove stage.

(e) A tadpole of the American bullfrog, *Rana catesbeiana*.

(f) A tadpole of the American bullfrog undergoing metamorphosis. Note the presence of both legs and tail.

Colorplate 5
Two Interpretations of Animal Phylogeny

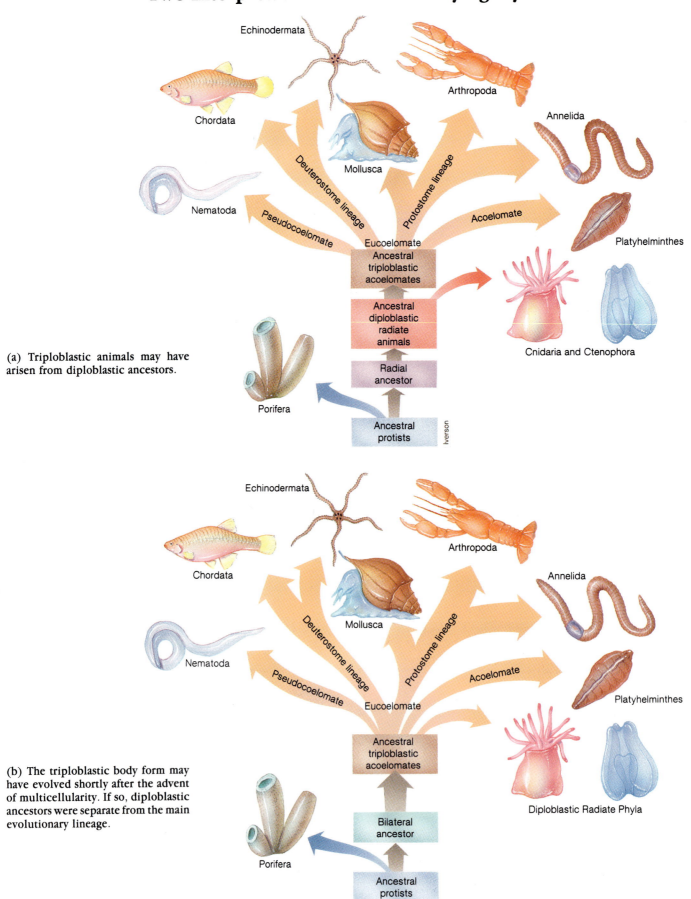

(a) Triploblastic animals may have arisen from diploblastic ancestors.

(b) The triploblastic body form may have evolved shortly after the advent of multicellularity. If so, diploblastic ancestors were separate from the main evolutionary lineage.

7. Regulatory or homeotic genes help determine body shape by controlling the developmental fate of groups of cells. Apparently these genes work by turning structural genes on or off.

8. The processes of determination, differentiation, morphogenesis, and pattern formation all depend on specific proteins and macromolecules. Some of the control mechanisms that help to regulate these molecules operate at the level of DNA processing, the level of RNA synthesis and processing, and the level of protein synthesis.

9. The control of continued growth and differentiation in an organism is provided by chemical messengers called hormones.

10. Organisms continue to change after completing their embryonic development. Aging and the death that follows are both forms of developmental change in an organism.

11. A tumor is a clonal population of abnormally dividing cells. Two types exist: benign and malignant.

■ Key terms

benign tumor (p. 172)
carcinogens (p. 172)
cell adhesion molecules (CAMs) (p. 162)
chromosome diminution (p. 168)
determination (p. 160)
differential gene activation (p. 168)
differentiation (p. 162)
gene rearrangement (p. 167)
gene amplification (p. 167)
homeobox (p. 166)
hormone (p. 171)

malignant tumor (p. 172)
metamorphosis (p. 171)
metastasis (p. 172)
morphogenesis (p. 163)
nuclear transplantation (p. 161)
oncogene (p. 172)
pattern formation (p. 165)
polytene chromosomes (p. 168)
regulatory or homeotic genes (p. 166)
totipotent (p. 160)
transdetermination (p. 161)
tumor (p. 172)

■ Critical Thinking Questions

1. In early embryonic development, gene transcription does not begin until the late gastrula stage. Protein synthesis during cleavage is exclusively directed by mRNA templates already present in the zygote cytoplasm. Thus, how is this early information directing cleavage inherited?

2. Speculate on how the linear, essentially one-dimensional information of a gene is transformed into the three-dimensional shape of an animal.

3. Explain how mistakes in development (either induced in the laboratory or in nature) can provide insight into normal development.

4. How can mutations in regulatory genes generate a human with six fingers on each hand?

5. Compare determination and differentiation. In what sense is determination inherited?

■ Suggested Readings

Books

Gilbert, S. F. 1988. *Developmental Biology.* 2/e. Sunderland, MA: Sinauer Associates, Inc.

Gurdon, J. 1974. *The Control of Gene Expression in Animal Development.* Cambridge, MA: Harvard University Press.

Hopper, A., and Hart, N. 1985. *Foundations of Animal Development.* Oxford: Oxford University Press.

Trinkaus, J. P. 1984. *Cells Into Organs.* Englewood Cliffs, NJ: Prentice Hall.

Walbot, V., and Holder, N. 1987. *Developmental Biology.* New York: Random House.

Wessels, N. K. 1977. *Tissue Interactions in Development.* Menlo Park, CA: Benjamin Cummings.

Articles

Bishop, J. M. 1987. The molecular genetics of cancer. *Science* 235:305–311.

Croce, C., and Klein, G. Chromosome translocations and cancer. *Scientific American* March, 1985.

Edelman, G. M. Topobiology. *Scientific American* May, 1989.

Edelman, G. M. Cell adhesion molecules: A molecular basis for animal form. *Scientific American* January, 1984.

Gehart, S. 1986. Amphibian early development. *BioScience* 36:541–549.

Gehring, W. J. The molecular basis of development. *Scientific American* October, 1985.

Gehring, W. J., and Hirome, Y. 1986. Homeotic genes and the homeobox. *Annual Review Genetics* 20:147–173.

Hynes, R. Fibronectins. *Scientific American* June, 1986.

Kimmel, C., and Warga, R. 1988. Cell lineage and developmental potential of cells in the zebrafish embryo. *Trends in Genetics* 4:68–74.

Marx, J. L. 1988. Evolution's link to development explored. *Science* 240(3):880–884.

Sachs, L. Growth, differentiation, and the reversal of malignancy. *Scientific American* January, 1986.

Schibler, U., and Sierra, F. 1987. Alternative promoters in developmenal gene expression. *Annual Review Genetics* 21:237–257.

Wolpert, L. Pattern formation in biological development. *Scientific American* April, 1978.

Evolution

Part III

All organisms have an evolutionary history and a future that will be influenced by evolutionary forces. Evolution is a major unifying theme in biology because it helps us understand the life-shaping processes that all species experience. Both humans and jellyfish are the products of these shared forces.

Ground-breaking studies of evolution were carried out by Charles Darwin and Alfred Wallace in the nineteenth century. Their historical work serves as an important introduction to evolutionary theory. The work of these and other pioneer evolutionary biologists is presented in chapter 12. Since these early studies, our knowledge of genetics, in particular the principles of population genetics, has greatly influenced evolutionary theory. These genetic ideas, along with the questions of direction and rate of change, are presented in chapter 13. Finally, paleontologists and other researchers are finding more pieces to the human origin puzzle. Some of this fascinating information is presented in chapter 14.

Even though part III ends with chapter 14, our coverage of evolution does not end there. In spite of organisms experiencing common evolutionary processes, evolution has resulted in a great diversity of animal life. An important goal of part IV is to present the diversity of animal life from an evolutionary perspective.

Evolution: An Historical Perspective

Concepts

1. Organic evolution is the change of a species over time.
2. Although the concept of evolution is very old, the modern explanation of how change occurs was formulated by Charles Darwin. Darwin began gathering his evidence of evolution during a worldwide mapping expedition on the H.M.S. *Beagle* and spent the rest of his life formulating and defending his ideas.
3. Darwin's theory of evolution by natural selection, although modified from its original form, is still a highly regarded account of how evolution occurs.
4. Modern evolutionary theorists apply principles of genetics, ecological theory, and geographic and morphological studies when investigating evolutionary mechanisms.
5. "Adaptation" may refer to either a process of evolutionary change or to the result of a change. In the latter sense, an adaptation is a structure or a process that increases an animal's potential to survive and reproduce in specific environmental conditions.

Have You Ever Wondered:

[1] whether the idea of evolutionary change originated with Charles Darwin?
[2] what circumstances led Charles Darwin to become a naturalist on the H.M.S. *Beagle*?
[3] how Charles Darwin's experiences in South America convinced him that evolution occurs?
[4] what prompted Charles Darwin to publish his work in 1859, 23 years after returning from his voyage?
[5] why geography is important for evolutionary biologists?

These and other useful questions will be answered in this chapter.

Questions of Earth's origin and life's origin have been on the minds of humans since prehistoric times, when accounts of creation were passed orally from generation to generation. For many people, these questions centered around concepts of purpose. Religious and philosophical writings help provide answers to questions such as: Why are we here? What is human nature really like? How do we deal with our mortality?

Many of us are also concerned with other, very different, questions of origin. How old is the planet Earth? How long has life been on Earth? How did life arise on Earth? How did the human species come into existence? Answers for these questions come from a different authority—that of scientific inquiry.

The purpose of this chapter is to present the history of the study of organic evolution, and to introduce the **theory of evolution by natural selection. Organic evolution** (L. *evolutus,* unroll), according to Charles Darwin, is "descent with modification." This statement simply means that species change over time. Evolution by itself does not imply any particular lineage or any particular mechanism, and virtually all scientists agree that the evidence for change in organisms over long time periods is overwhelming (*see chapter 1*). Further, most scientists agree that natural selection, the mechanism for evolution outlined by Charles Darwin, is one explanation of how evolution occurs. In spite of the scientific certainty of evolution and an acceptance of a general mechanism, there is still much to be learned about the details of evolutionary processes. Scientists will be debating these details for years to come (box 12.1).

Pre-Darwinian Theories of Change

[1] The idea of evolution did not originate with Charles Darwin. Some of the earliest references to evolutionary change are from the ancient Greeks. The philosophers Empedocles (495–435 B.C.) and Aristotle (384–322 B.C.) believed that humans evolved over long time periods. Empedocles described all life ultimately coming from four classical elements: fire, water, earth, and air. The mechanics for change suggested by Empedocles were very different from modern evolutionary theory. His ideas of change, however, laid the groundwork for the mechanisms proposed centuries later.

During the eighteenth century, a number of scientists made significant contributions to the ideas of evolution. Georges-Louis Buffon (1707–1788) spent many years in the study of comparative anatomy. His observations of structural variations in particular organs of related animals, and especially his observations of vestigial structures, convinced him that change must have occurred during the history of life on Earth. He believed that vestigial structures, such as the human appendix, are the remnants of once useful organs that were gradually modified by the environment and the modifications were passed on to later

generations. Erasmus Darwin (1731–1802), a physician and the grandfather of Charles Darwin, was intensely interested in questions of origin and change. He believed in the common ancestry of all organisms. Like Buffon, he suggested that organisms adapt to conditions of life, and that these adaptations are passed on to subsequent generations.

Lamarck: An Early Theory of Evolution

Jean Baptiste Lamarck (1744–1829) was a distinguished French zoologist. His contributions to zoology include important studies of the classification of animals. Lamarck, however, is remembered more for his theory of how change occurs. He believed that species are not constant, and that existing species were derived from preexisting species. Although his theory of change was not new, he published it more widely than had previously been done.

Lamarck's rather elaborate explanation of how evolutionary change occurs is called the **theory of inheritance of acquired characteristics.** Lamarck believed that organisms develop new organs or modify existing organs as environmental problems present themselves. In other words, organs change as the need arises. Lamarck illustrated this point with the often-quoted example of the giraffe (figure 12.1). He contended that ancestral giraffes had short necks, much like those of any other mammal. Straining to reach higher branches during browsing resulted in their acquiring higher shoulders and longer necks. These modifications, produced in one generation, were passed on to the next generation. Lamarck went on to state that the use of any organ resulted in that organ becoming highly developed, and that disuse resulted in degeneration. Thus, the evolution of highly specialized structures, such as vertebrate eyes, and vestigial structures, such as the human appendix, could be explained.

Lamarck published his theory in 1802 and defended it in the face of social and scientific criticism. Society in general was unaccepting of the ideas of evolutionary change, and evidence for evolution had not been developed thoroughly enough to convince most scientists that evolutionary change occurs. Thus, Lamarck was criticized in his day more for advocating ideas of evolutionary change, than for the mechanism he proposed for that change. Today he is criticized for proposing and defending a mechanism of inheritance and evolutionary change that we now know lacks reasonable supporting evidence. For a change to be passed on to the next generation, it must be based on genetic change in the germ cells. The giraffes' longer necks could not be passed on because they did not originate as changes in the genetic material. Actually, Lamarck's ideas of inheritance were not new. Others before him had advocated similar ideas, and these ideas remained prevalent for some time after Lamarck's death. Darwin even incorporated them into *On the Origin of Species.* Perhaps we too easily forget Lamarck's other accomplishments in zoology and his steadfastness in promoting the idea of evolutionary change.

Generation 1

Generation 2

Generation 3

Stop Ask Yourself

1. What is meant by the term "organic evolution?"
2. What contributions to our concept of evolution were made by: Empedocles? Buffon? Erasmus Darwin?
3. What is the theory of inheritance of acquired characteristics, and how did Lamarck use it to explain how evolution occurred?

Darwin's Early Years and His Journey

Charles Robert Darwin (1809–1882) was born on February 12, 1809. His father, like his grandfather, was a physician. During Charles Darwin's youth in Shrewsbury, England, his interests centered around dogs, collecting, and hunting birds—all popular pastimes in wealthy families of nineteenth century England. These activities captivated him far more than the traditional education he received at boarding school. In 1825, he entered medical school in Edinburgh, Scotland. For two years, he enjoyed the company of the school's well-established scientists. Darwin, however, was not interested in a career in medicine because he could not bear the sight of pain. This problem prompted his father to suggest that he train as a clergyman for the Church of England. With this in mind, Charles enrolled at Cambridge University in 1828 and graduated with honors in 1831. This training, like the medical training he received, was disappointing for Darwin. Again, his most memorable experiences were those with Cambridge scientists. During his stay at Cambridge, Darwin developed a keen interest in collecting beetles and made valuable contributions to beetle taxonomy.

Voyage of the Beagle

One of his Cambridge mentors, a botanist by the name of John S. Henslow, nominated Darwin to serve as a naturalist on a mapping expedition that was to travel around the world. He was commissioned as a naturalist on the H.M.S. *Beagle,* which set sail on December 27, 1831 on a voyage that lasted 5 years (figure 12.2). The position of naturalist was an unpaid position, and his quarters were extremely cramped. In spite of constant battles with seasickness, Darwin helped with routine seafaring tasks and made numerous collections, which he sent to Cambridge. The voyage gave Darwin ample opportunity to explore tropical rain forests, fossil beds, the volcanic peaks of South America, [2]

Figure 12.1 Jean Baptiste Lamarck believed that characteristics were acquired through need in one generation and then passed to subsequent generations. Therefore, the necks and shoulders of giraffes that got longer and higher by stretching to reach higher and higher into trees were passed on to the next generation of giraffes.

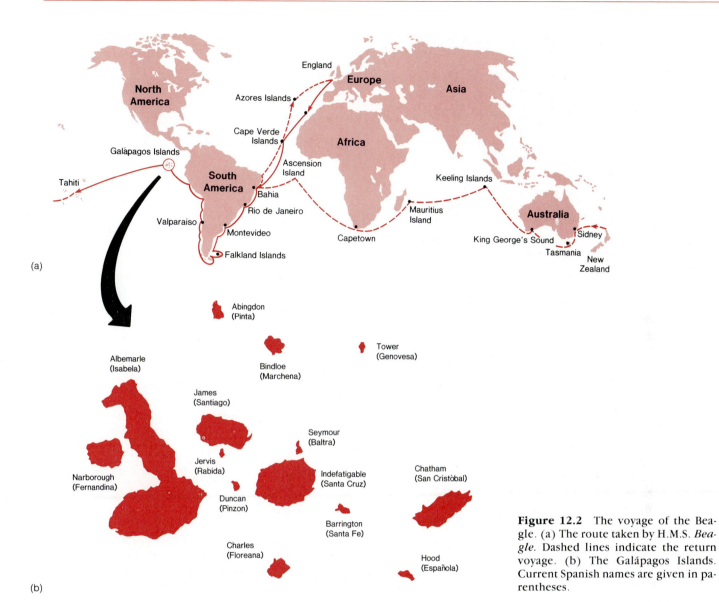

(a)

(b)

Figure 12.2 The voyage of the Beagle. (a) The route taken by H.M.S. *Beagle.* Dashed lines indicate the return voyage. (b) The Galápagos Islands. Current Spanish names are given in parentheses.

Figure 12.3 Charles Darwin in 1840, four years after returning from his voyage around the world.

Box 12.1

The Origin of Life on Earth—Life from Nonlife

Geologists estimate that Earth is between 4.5 and 5 billion years old. The oldest fossils are of bacteria that come from rocks about 3.5 billion years old. Thus, it took no more than 1 to 1.5 billion years for life to originate.

Biochemists have been investigating how life could have arisen, given the conditions that were probably on the primordial earth. The goal of this research is not necessarily to recreate life, or even to know exactly how life originated, but rather to demonstrate that life *could* have arisen from nonliving elements.

In trying to explain how life may have arisen, scientists first needed to know the conditions that existed on Earth after its formation. In 1929, J. B. S Haldane described the atmosphere of the primordial earth as a reducing atmosphere (containing little free oxygen). Initially, scientists believed that the atmosphere contained primarily hydrogen, water, ammonia, and methane. Recent studies suggest, however, that water vapor, nitrogen, carbon dioxide, hydrogen, and carbon monoxide may have been the primary gases of primitive Earth.

In 1953, Stanley Miller and Harold Urey constructed a reaction vessel, in which they duplicated the atmosphere described by Haldane (box figure 12.1). They heated the mixture to 80°C, and provided the atmosphere with an electrical spark to simulate lightning. Over the course of a week, they removed samples from their system and found that they contained a variety of common amino acids and other organic acids. Since that original experiment, many variations have been tried using different atmospheric components, and different energy sources have been successfully tested. (Although electrical discharges seem most likely, ultraviolet light,

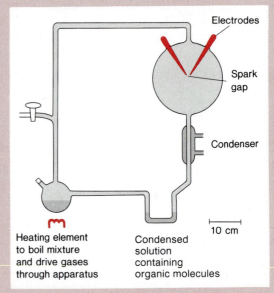

Box Figure 12.1 The Miller/Urey apparatus. When the conditions that were likely present in the atmosphere of primitive Earth are duplicated in this apparatus, many of the molecules that are the building blocks of life can be produced.

radioactivity, volcanic activity, and shock waves from tremors and meteors could have provided energy for these conversions.) Many of these variations have produced not only amino acids but also formaldehyde, hydrogen cyanide, purines and pyrimidines, organic acids, and sugars. Clearly, all monomers making up the important biomolecules can be synthesized in this fashion. Scientists believe that bodies of water on primordial Earth were filled with these elemental compounds. In a

---

and the coral atolls of the South Pacific. Most importantly, Darwin spent 5 weeks on the **Galápagos Islands,** a group of volcanic islands that lie 600 miles off the coast of Ecuador. Some of his most revolutionary ideas came from his observations on these islands. At the end of the voyage, Darwin was just 27 years old (figure 12.3), and he would spend the rest of his life examining specimens, rereading notes, making new observations, and preparing numerous publications. His most important publication, *On the Origin of Species,* revolutionized biology.

Early Development of Darwin's Ideas of Evolution

The development of Darwin's theory of evolution by natural selection was a long, painstaking process. Darwin had to become convinced that change occurs over time. Before leaving on his voyage, Darwin accepted the prevailing opinion that the earth and its inhabitants had been created 6,000 years ago. Because his observations suggested that change does occur, he realized that 6,000 years could not account for the diversity of modern species if they arose through gradual change. Once ideas of change were established in Darwin's thinking, it took about 20 years of study to conceive, and thoroughly document, the mechanism by which change occurs. Darwin died without knowing the genetic principles that support his theory.

Geology

Darwin began his voyage by reading Charles Lyell's (1779–1875) *Principles of Geology.* In this book, Lyell developed the ideas of another geologist, James Hutton into the theory of **uniformitarianism.** His theory was based on the idea that the earth is shaped today by the forces of wind, rain, rivers, volcanoes, and geological uplift—just as it has been

reducing atmosphere (no gaseous oxygen), these compounds would not be oxidized as they would be today.

The next step in the origin of life must have been the hooking together of these components to eventually make the polymers of living organisms, such as polypeptides, polynucleotides, and polysaccharides. To understand how this step may have occurred, two problems needed answers. First, polymerization involves condensation (water removal) reactions. How could water be removed if the starting compounds were in a "prebiotic watery soup?" The presence of water would favor the retention of monomers, not polymerization. One idea is that some of this "soup" became isolated in tidepools or freshwater ponds, and water was lost through evaporation until condensation reactions became possible. Another popular idea is that reacting molecules were adsorbed (concentrated) on the surfaces of clay particles where polymerization could occur. A second problem concerns the source of energy for condensation reactions because ATP was not available. Polyphosphates and carbodimides were found in the laboratory "prebiotic soup" and could serve as coupling agents. They have the ability to react with organic building blocks, and in the process, release enough energy to promote condensation into polymers.

For life to exist, polymers must be collected together and not diluted in a vast watery medium. A. I. Oparin found that polymers can coalesce (come together) in an aqueous medium to form small droplets that he called *coacervates*. These colloidal particles are often surrounded by a membranelike structure and are between 1 and 500 micrometers in diameter. Furthermore, catalytic activity can be introduced into these droplets by supplying enzymes and raw materials. S. W. Fox produced coacervatelike particles, called *microspheres*, in which he found inherent catalytic activity. It may have been at this stage that metabolic activity began.

The final steps in the origin of life are major steps, for which there are endless speculations, but little concrete evidence. In some way, nucleic acids must have been incorporated into the earliest cells, so that the processes of replication, transcription and translation could evolve. These processes were essential if living forms were to perpetuate themselves.

Life in a "prebiotic soup" was limited by the nutrients produced in the primordial environment. If life were to continue, another source of nutrients would be needed. Photosynthesis, which is the production of organic molecules using solar energy and inorganic compounds, freed living organisms from a dwindling supply of nutrients. The first photosynthetic organisms probably used hydrogen sulfide as a source of hydrogen for reducing carbon dioxide to sugars. Later, water would serve this same purpose.

When water began to supply the hydrogen for photosynthesis, a photosynthetic waste product, oxygen, began to accumulate in the atmosphere. Earth and its atmosphere slowly began to change. Ozone in the upper atmosphere began to filter ultraviolet radiation from the sun, and gaseous oxygen began to accumulate. The reducing atmosphere slowly became an oxidizing atmosphere, and at least some living forms began to utilize that oxygen. It was only about 420 million years ago, however, that enough ozone had built up to make life on land possible. Ironically, the change from a reducing atmosphere to an oxidizing atmosphere also meant that life would no longer arise nonbiologically.

in the past. Lyell and Hutton contended that it was these forces, not catastrophic events, that shaped the face of the earth over hundreds of millions of years. This book planted two important ideas in Darwin's mind. It suggested to Darwin that the earth could be much older than 6,000 years. In addition, if the face of the earth changed gradually over long periods of time, could not living forms also change during that time?

Fossil Evidence

Once the H.M.S. *Beagle* reached South America, Darwin spent time digging in the dry riverbeds of the pampas (grassy plains) of Argentina. He found the fossil remains of an extinct hippopotamuslike animal, now called *Toxodon,* and fossils of a horselike animal, *Thoantherium.* Both of these fossils were from animals that were clearly different from any other animal living in the region. Modern horses were in South America, of course, but they had been brought to the Americas by Spanish explorers in the 1500s. The fossils suggested that horses had been present and had become extinct long before the 1500s. Darwin also found fossils of giant armadillos and giant sloths. Except for their large size, these fossils were very similar to forms Darwin found living in the region.

Fossils were not new to Darwin. They were popularly believed to be the remains of animals that perished in catastrophic events, such as Noah's flood. In South America, however, Darwin understood them to be evidence that the species composition of the earth has changed. Some species became extinct without leaving any descendants. Others became extinct, but not before giving rise to new species.

Galápagos Islands

On its trip up the western shore of South America, the H.M.S. *Beagle* stopped at the Galápagos Islands. These islands are named after the very large tortoises that inhabit

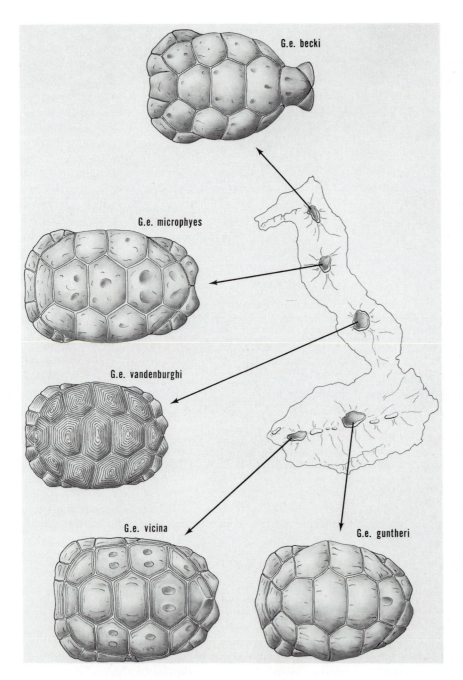

G.e. becki

G.e. microphyes

G.e. vandenburghi

G.e. vicina

G.e. guntheri

Figure 12.4 Distribution of the five subspecies of giant land tortoises (*Geochelone elephantopous*) on Albermarle Island. The five subspecies of tortoises on Albermarle Island occur at the tops of the five principal volcanoes that form the island. They evolved independently when the island was submerged so that only the volcanic tips were above water. Therefore, the tortoises were isolated from each other as if on five separate islands. They remain isolated today by deep valleys between volcanoes.

Abingdon   Albemarle   Duncan

Figure 12.5 Galápagos tortoises. Longer-necked subspecies live in drier regions and feed on high-growing vegetation. Shorter-necked subspecies live in moister regions and feed on low-growing vegetation.

them (Sp. *galápago,* tortoise). The tortoises weigh up to 250 kilograms, have shells up to 1.8 meters in diameter, and live for 200 to 250 years. It was pointed out to Darwin by the islands' governor that the shape of the tortoise shells from different parts of Albemarle Island differed (figure 12.4). Darwin noticed other differences as well. Tortoises from the drier regions had longer necks than tortoises from wetter habitats (figure 12.5). In spite of their differences, the tortoises were quite similar to each other and to the tortoises on the mainland of South America.

How could these overall similarities be explained? Darwin reasoned that the island forms were derived from a few ancestral animals that managed to travel from the mainland, across 600 miles of ocean. (Because the Galápagos Islands are volcanic islands and arose out of the sea bed, there was never any land connection with the mainland. One modern hypothesis is that tortoises floated from the mainland on mats of vegetation that regularly break free from coastal riverbanks.) Without predators on the islands, tortoises gradually increased in number.

Darwin also explained some of the differences that he saw. In dryer regions, where vegetation was sparse, tortoises with longer necks would be favored because they could reach higher to get food. In moister regions, tortoises with longer necks would not necessarily be favored and the shorter-necked tortoises could survive.

Darwin made similar observations of a group of dark, sparrowlike birds. Although he never studied them in detail, Darwin noticed that the Galápagos finches bore similarities suggestive of common ancestry. Scientists now know that Galápagos finches also descended from an ancestral species that originally inhabited the mainland of South America. The chance arrival of a few finches probably set up the first bird populations on the islands. Early finches encountered many different habitats, all empty of other birds and predators. Ancestral finches, probably seed eaters, multiplied rapidly and filled the seed-bearing habitats most attractive to them. Fourteen species of finches arose from this ancestral group, including one species found on small Cocos Island northeast of the Galápagos Islands. Each species is adapted to a specific habitat on the islands. The most obvious difference between these finches relates to dietary adaptations and is reflected in the size and shape of their beaks (figure 12.6).

Darwin's experiences in South America and the Galá- [3] pagos Islands convinced him that animals change over time and that, in the course of the history of life on earth, many kinds of animals have become extinct. Although he would not document his hypothesis for many years, these experiences helped him formulate ideas of how evolutionary change occurs. Darwin noted that females produce far more offspring than survive. Some of these offspring may possess variations that increase their chances of surviving and reproducing. Darwin thought of nature as a force that eliminates less favorable variations from a group of animals. Eventually such a process could lead to the formation of new, identifiable groups or, in cases of extreme environmental change, to extinction.

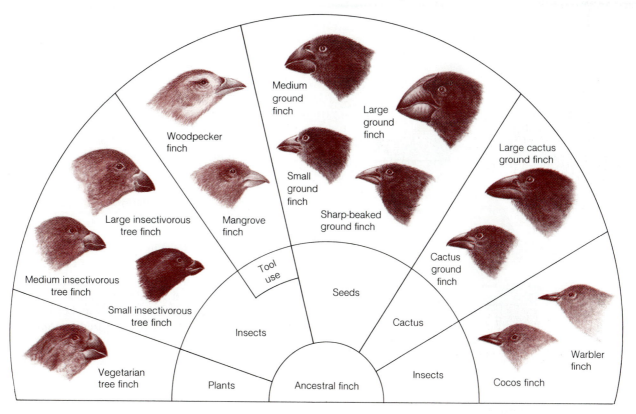

Figure 12.6 Adaptive radiation of the Galápagos finches. Ancestral finches from the South American mainland colonized the Galápagos Islands. Open habitats and few predators promoted the radiation of finches into most of the roles normally filled by birds.

The Theory of Evolution by Natural Selection

On his return to England in 1836, Darwin married his cousin, Emma Wedgwood. They lived first in London and then moved to a country residence in 1842. It was to be 17 years before Darwin published his ideas, but they were by no means idle years.

Continuing Studies

Darwin worked diligently on the notes and specimens he had collected and made new observations. He was particularly interested in the obvious success of plant and animal breeders in promoting the development of desired variations in plant and animal stocks (figure 12.7). He realized that these breeders were promoting an artificial evolution and wondered if this artificial selection of traits could have a parallel in the natural world.

Initially, Darwin was unable to find a natural process similar to artificial selection. However, in 1838, Darwin read an essay by Thomas Malthus (1766-1834) entitled *Essay on the Principle of Population*. Malthus believed that the human population has the potential to increase geometrically, doubling with each generation. However, because resources cannot keep pace with the increased demands of a burgeoning population, the influences of population-checking factors, such as poverty, wars, plagues, and famine, begin to be felt. It occurred to Darwin

Japanese Bantam

Birchen Game Bantam

Bearded Araucana

Figure 12.7 The artificial selection of domestic fowl has resulted in the diverse varieties that we see today. Breeders perpetuated the variations they desired by allowing only certain offspring to breed.

that a similar struggle to survive occurs in nature. This struggle, when viewed over generations, could serve as a means of **natural selection.** Traits that enhanced the chances of an animal's survival would be perpetuated, and traits that were detrimental would be eliminated.

Natural Selection

By 1844 Darwin had formulated, but not yet published, his ideas on natural selection. The essence of his theory is as follows:

1. All organisms have a far greater reproductive potential than is ever realized. For example, a female oyster releases about 100,000 eggs with each spawning, a female sea star releases about one million eggs each season, and a female robin may lay four fertile eggs each season. What if all of these eggs were fertilized and developed to reproductive adults by the following year? A million sea stars, each producing another million eggs, repeated over just a few generations would soon fill up the oceans! Even the four young adult robins, each producing four more robins, would result in unimaginable resource problems in just a few years.
2. Inherited variations exist. Seldom are any two individuals exactly alike. Some of these genetic variations may confer an advantage to the individual possessing them. In other instances, variations may neither harm nor help an individual. In still other instances, particular variations may be harmful. These variations can be passed on to offspring.
3. There is a constant struggle for existence. Because many more offspring are produced than can survive, many individuals will die. Which ones will survive? Darwin reasoned that the individuals that survive will be those with the traits (variations) that permit successful reproduction in a particular environment. These traits are said to be adaptive, and the struggle for existence is often called *survival of the fittest.*
4. Adaptive traits will be perpetuated in subsequent generations. Because organisms without adaptive traits are less likely to reproduce, the adaptive traits will become more frequent in a population, and nonadaptive traits will tend to be eliminated.

With these ideas, Darwin formulated a theory that explained how the tortoises and finches of the Galápagos Islands changed over time (figure 12.8). In addition, Darwin's theory explained how some animals, such as the ancient South American horses, could become extinct. What if a new environment is presented to a group of animals that are ill-adapted to that environment? Climatic changes, food shortages, and other environmental stressors could lead to extinction.

Adaptation Adaptation occurs when a change increases an animal's chance of successful reproduction. It is likely to occur when an organism encounters a new environment, and may result in the evolution of multiple new groups if an environment can be exploited in different ways. No terms in evolution have been laden with more confusion than "adaptation" and "fitness or adaptedness." **Adaptation** is sometimes used to refer to a process of change in evolution. That use of the term is probably less confusing than when "adaptation" is used to describe the result of the process of change. The confusion results from trying to define exactly what adaptation is. For our purposes, adaptations are defined as structures or processes that increase an organism's, or a species', potential to successfully reproduce in a specified environment. In a similar fashion, *adaptedness* or *fitness* is a measure of the capacity for successful reproduction in a given environment.

There has been a tendency to view every evolutionary change as an adaptation to some kind of environmental situation. The view has been that if a structure is now performing a specific function, it must have arisen for that purpose, and is, therefore, an adaptation. An extreme extension of this incorrect view is that evolutionary adaptations lead to perfection.

Alfred Russel Wallace

Alfred Russel Wallace (1823–1913) (figure 12.9) was an explorer of the Amazon Valley, and led a zoological expedition to the Malay Archipelago, which is an area of great biogeographical importance. Wallace, like Darwin, was impressed with evolutionary change and had read the writings of Malthus on human populations. In the midst of a bout with malarial fever, he synthesized a theory of evolution similar to Darwin's theory of evolution by natural selection. After writing the details of his theory, Wallace sent his paper to Darwin for criticism. Darwin recognized the similarity of Wallace's ideas, and prepared a short summary of his own theory. Both Wallace's and Darwin's papers were published in the *Journal of the Proceedings of the Linnean Society* in 1859. Darwin's insistence on having Wallace's ideas presented along with his own shows Darwin's integrity. Darwin then shortened a manuscript he had been working on since 1856 and published it as *On the Origin of Species* in November, 1859. The 1,250 copies prepared in the first printing sold out the day the book was released. [4]

In spite of the similarities in the theories of Wallace and Darwin, there were also important differences. Wallace, for example, believed that every evolutionary modification was a product of selection and, therefore, had to be adaptive for the organism. Darwin, on the other hand, was willing to admit that natural selection may not explain all evolutionary changes. He did not insist on finding adaptive significance for every modification. Further, unlike Darwin, Wallace stopped short of attributing human intellectual

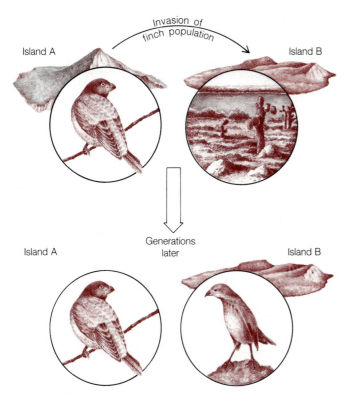

Figure 12.8 Natural selection of finches on the Galápagos Islands resulted in changes in beak shape. In this illustration, tree-feeding birds on island A invaded island B. The relatively treeless habitats of island B selected against birds most adapted to living in trees, and individuals that could exploit seeds on the ground and in low-growing vegetation were more likely to survive and reproduce. Subsequent generations of finches on island B should show ground-feeding beak characteristics.

Figure 12.9 Alfred Russell Wallace.

functions and the ability to make moral judgments to evolution. On both of these matters, Darwin's ideas are closer to the views of most modern scientists.

Wallace's work provided an important spark that motivated Darwin to publish his own ideas. The theory of natural selection, however, is usually credited to Charles Darwin. The years of work given to the theory by Darwin, and the accumulation of massive evidence for the theory, led even Wallace to attribute the theory to Darwin. Wallace wrote to Darwin in 1864:

> . . . I shall always maintain . . . (the theory of evolution by natural selection) . . . to be actually yours and yours only. You had worked it out in details I had never thought of years before I had a ray of light on the subject, . . .

Stop and Ask Yourself

4. How did the following contribute to Charles Darwin's formulation of the theory of natural selection: Uniformitarianism? South American fossils? Galápagos tortoises?
5. What are four elements of the theory of evolution by natural selection?

Evolutionary Thought After Darwin

The most significant changes that occurred in evolutionary thought began in the 1930s and have continued to the present. This period of evolutionary study is called the *period of modern synthesis*, or *neodarwinism*. As will be presented in chapter 13, the incorporation of the principles of population genetics into evolutionary theory has helped modern scientists explain the origin of the variations that are so important to evolutionary theories. These principles also help explain how allelic frequencies change as variations are selected for or against during evolution.

Biogeography

[5] In the tradition of Darwin and Wallace, biologists of the period of modern synthesis recognized the importance of geography as an explanation of the evolution and the distribution of plants and animals (*see chapter 1; figure 1.3*). One of the distribution patterns that biogeographers try to explain is how similar groups of organisms can live in places separated by seemingly uncrossable barriers. Recall

that Darwin was struck by the presence of fossil horses in South America. This distribution was puzzling because modern horses in America were introduced to America from Europe. The fossil horses must have arisen in America or arrived by some unknown means. They also became extinct and left no descendants. Biogeographers also try to explain why plants and animals, separated by geographical barriers, are often very different in spite of similar environments. Evolution may take different directions in different parts of the world, and therefore, major predators in Africa and America might be expected to be different if they had no common ancestry. Finally, biogeographers try to explain why oceanic islands, such as the Galápagos, often have relatively few, but unique, resident species. They try to document island colonization and subsequent evolutionary events, which may be very different from evolutionary events in ancestral, mainland groups.

Modern evolutionary biologists recognize the importance of geological events, such as volcanic activity, the movement of great land masses, climatic changes, and geological uplift, in creating or removing barriers to the movements of plants and animals. As scientists learned more about the geologic history of the earth, they understood more about animal distribution patterns and what factors may have played important roles in their evolution. Only in understanding how the surface of the earth came to its present form can we understand its inhabitants (box 12.2).

Evolution is one of the major unifying themes in biology because it helps explain both the unity and diversity of life. In chapter 1 you learned of the sources of evidence for evolution—*biogeography* gives evidence of prehistoric climates, habitats, and animal distribution patterns; *paleontology* provides evidence of animals that existed in the past; *comparative anatomy* leads to the description of homologous structures; and *molecular biology* provides evidence of relatedness of animals based upon their biochemical similarities. This evidence leaves little doubt that evolution has occurred in the past, and this chapter described the historical development of a theory that accounts for how evolution occurs. In chapter 13 you will learn of how the incorporation of the principles of population genetics has affected scientific concepts of the mechanism of evolution.

Stop and Ask Yourself

6. What is the period of modern synthesis?
7. What are some of the distribution patterns that biogeographers attempt to explain?
8. How has the discovery of continental drift influenced our understanding of evolution?

Box 12.2

Continental Drift

It seems remarkable, but the Earth's largest land masses are moving—about one centimeter a year! The 1960s revolutionized geology as **continental drift** became an accepted theory.

During the Permian period (about 250 million years ago), all of Earth's land masses were united into a single continent, called *Pangaea*. Soon after, Pangaea began to break up, as the huge crustal plates that carried its parts began to move apart. Approximately 200 million years ago, there were two great continents. The northern continent was called *Laurasia*, and the southern continent was called *Gondwana*. Seventy million years ago, Gondwana broke apart, followed later by the breakup of Laurasia (box figure 12.2).

Movement of these crustal plates continues today. Their study is known as *plate tectonics*. During these movements, new crustal material is thrust up from the ocean floor along the mid-Atlantic ridge, and flows in both directions from that ridge. The mid-Atlantic ridge runs from the Falkland Islands, at its southern end, to Iceland, at its northern end. As the huge crusts move away from each other in the Atlantic, old crustal material sinks back into the earth in deep oceanic trenches.

The drifting of the continents has important implications for biogeographers and paleontologists. Because oceanic barriers were created as a result of continental drift, plants and animals were separated from one another when continents separated. Evolution then proceeded independently in each biogeographical region and resulted in species characteristic of that region.

Continental drift also explains why some fossils have a worldwide distribution. Land organisms in existence during the time that all continents were united as Pangaea had access to most of the land masses of the world. It should not be surprising, therefore, to find some fossils that are similar in all parts of the world, such as the fossils that Charles Darwin found in South America that were very similar to animals living in Africa. Were it not for continental drift, this pattern would be very difficult to explain.

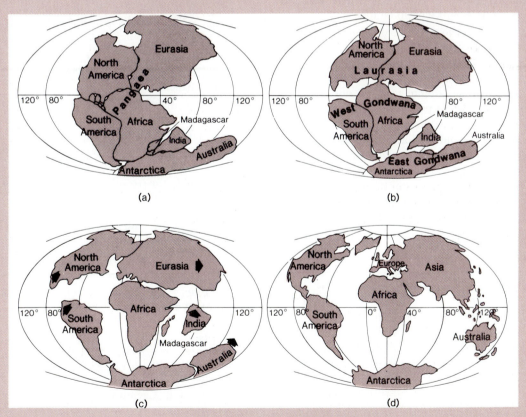

Box Figure 12.2 Continental Drift. (a) About 250 million years ago, the continents of the earth were joined into a single land mass that geologists call Pangaea. (b) About 150 million years ago, Pangaea broke up into the northern (Laurasia) and southern (Gondwana) continents. (c) This breakup was followed, about 70 million years ago, by the separation of continents in the southern hemisphere, and later by the separation of continents in the northern hemisphere. (d) The present position of continents.

Summary

1. Organic evolution is the change of a species over time.

2. Ideas of evolutionary change can be traced back to the ancient Greeks.

3. Jean Baptiste Lamark was an eighteenth century zoologist who advocated evolution, and proposed a mechanism—inheritance of acquired characteristics—to explain it.

4. Charles Darwin saw impressive evidence for evolutionary change while on a mapping expedition on the H.M.S. *Beagle*. The theory of uniformitarianism, South American fossils, and observations of tortoises and finches on the Galápagos Islands convinced Darwin that evolution occurs.

5. After returning from his voyage, Darwin began formulating his theory of evolution by natural selection. In addition to his experiences on his voyage, later observations of artificial selection and Malthus' theory of human population growth helped shape his theory.

6. Darwin's theory of natural selection includes the following elements: a. All organisms have a greater reproductive potential than is ever realized. b. Inherited variations exist. c. There is a constant struggle for existence in which those organisms that are best suited to their environment will survive. d. The adaptive traits present in the survivors will tend to be passed on to subsequent generations, and the nonadaptive traits will tend to be lost.

7. "Adaptation" may refer to a process of change or a result of change. In the latter sense, an adaptation is a structure or a process that increases an organism's potential to survive and reproduce in a given environment.

8. All evolutionary changes are not adaptive, nor do all evolutionary changes lead to perfect solutions to environmental problems.

9. Alfred Russel Wallace outlined a theory similar to Darwin's, but never accumulated significant evidence documenting his theory.

10. Modern evolutionary theorists apply principles of genetics, ecological theory, and geographic and morphological studies to solving evolutionary problems.

Key Terms

adaptation (p. 185)
continental drift (p. 187)
Galápagos Islands (p. 180)
natural selection (p. 184)
organic evolution (p. 177)

theory of evolution by
  natural selection (p. 177)
theory of inheritance of
  acquired characteristics
  (p. 177)
uniformitarianism (p. 180)

Critical Thinking Questions

1. In chapter 1 there is a discussion of the scientific method. Review that section and do the following: a. Outline a hypothesis and design a test of "inheritance of acquired characteristics." b. Define what is meant by the word "theory" in the theory of evolution by natural selection.

2. Assuming that you have already studied chapter 9, describe the implications of inheritance of acquired characteristics for the central dogma of molecular genetics.

3. Review the definition of "adaptation" in the sense of a result of evolutionary change. Imagine that two deer, A and B, are identical twins. Deer A is shot by a hunter before it had a chance to reproduce. Deer B is not shot and goes on to reproduce. According to our definition of adaptation is deer B more fit for its environment than deer A? Why or why not?

4. Why is the stipulation of "a specific environment" included in the definition of "adaptation?"

Suggested Readings

Books

Bowler, P. 1984. *Evolution: The History of an Idea.* Berkeley: University of California Press.

Darwin, C. 1894. *On the Origin of Species.* Reprint. 1975. Cambridge: Cambridge University Press.

Dodson, E. O. and Dodson, P. 1985. *Evolution: Process and Product.* Belmont: Wadsworth Publishing Company.

Eldredge, N. and Cracraft, J. 1980. *Phylogenetic Patterns and the Evolutionary Process: Method and Theory in Comparative Biology.* New York: Columbia University Press.

Futuyma, D. J. 1986. *Evolutionary Biology,* 2nd ed. Sutherland: Sinauer Associates, Inc.

Godfrey L. R. 1985. *What Darwin Began.* Old Tappan: Allyn and Bacon, Inc.

Greene, J. C. 1959. *The Death of Adam: Evolution and its Impact on Western Thought.* Ames: Iowa State University Press.

Mayr, E. 1982. *The Growth of Biological Thought: Diversity, Evolution, and Inheritance.* Cambridge: Harvard University Press.

Smith, J. M. (ed.). 1982. *Evolution Now: A Century After Darwin.* New York: W.H. Freeman and Company.

Stebbins, G. L. 1982. *Darwin to DNA, Molecules to Humanity.* New York: W.H. Freeman and Company.

Volpe, E. P. 1985. *Understanding Evolution,* 5th ed. Dubuque: Wm. C. Brown Publishers.

Articles

Ayala, F. J. The mechanisms of evolution. *Scientific American* September, 1978.

Dickerson, R. E. Chemical evolution and the origin of life. *Scientific American* September, 1978.

Herbert, S. Darwin as a geologist. *Scientific American* May, 1986.

Lack, D. Darwin's finches. *Scientific American* April, 1953.

Lewontin, R. C. Adaptation. *Scientific American* September, 1978.

Mayr, E. Evolution. *Scientific American* September, 1978.

Evolution and Gene Frequencies

13

Concepts

1. In modern genetic theory, organic evolution is defined as a change in the frequency of genes in a population.
2. The principles of modern genetics help biologists understand how variation arises. This variation increases the chances of a population's survival in changing environments.
3. Population genetics is the study of events occurring in gene pools. The Hardy-Weinberg theorem helps scientists understand the circumstances under which evolution occurs. Evolution is likely when: (a) genetic drift or neutral selection occurs, (b) gene flow occurs, (c) mutations introduce new genes into populations, or (d) natural selection occurs.
4. Balanced polymorphism occurs when two or more body forms are maintained in a population without a range of phenotypes between them.
5. The fundamental unit of classification is the species, and the process by which new species are formed is called speciation.
6. For speciation to happen, isolation must occur.
7. Different organisms, as well as structures within organisms, evolve at different rates. Evolution may also proceed in jumps rather than at a constant pace.
8. Molecular evolutionists study DNA and proteins to uncover evolutionary relationships.

Have You Ever Wondered:

[1] what population genetics is?
[2] whether or not evolution must occur?
[3] what role chance plays in evolution?
[4] how natural selection can produce distinct subpopulations in a population?
[5] how new species arise?
[6] what molecular evolution is?

These and other useful questions will be answered in this chapter.

In the process of natural selection, individuals of a population survive or die because of the fitness conferred on them by the genes that they possess. The result of natural selection (and of evolution in general) is reflected in how common, or how rare, specific genes are in subsequent generations of a population. Selection for or against a gene is indirect, because it is the protein encoded by a particular gene, not the gene itself, that confers advantage or disadvantage to an organism in a particular environment. In other words, *natural selection acts on phenotypes, not directly on individual genes.*

A more precise definition of evolution now emerges. *Organic evolution is a change in the frequency of alleles in a population.* The frequency of an allele in a population is the abundance of that particular allele in relation to the sum of all alleles at that locus. Frequency is ordinarily expressed as a decimal fraction. Another way to express the same idea is that *organic evolution is a change in the total genetic makeup of a population (the* **gene pool***).* This chapter examines evolution from this perspective and covers some of the mechanisms by which evolution occurs, as well as processes by which new species arise.

Modern Synthesis: A Closer Look

Much of your background for understanding the modern synthesis comes from studying the genetics in chapters 7 to 9 because they explain why variations between individuals exist, and how variations are passed to future generations. Genetic variation is important in evolution because some variations may confer an advantage to individuals, leading to natural selection. The potential for genetic variation in individuals of a population is unlimited. Even the relatively simple principles of inheritance described by Gregor Mendel provide for remarkable variation. In addition, crossing-over, multiple alleles, mutations, and transposable elements add to this variation. The result is that no two individuals, except identical twins, are genetically identical. Chance combinations of genes are likely to result in some individuals being better able to survive and reproduce in a given environment than other individuals.

Must Evolution Happen?

Evolution is central to all of biology, but is evolution always occurring in a particular population? As we will see, there are certainly times when the *rate* of evolution is very slow, and there are times when the *rate* of evolution is very rapid. But, are there times when evolution does not occur at all? The answer to this question lies in the theories of **population genetics,** the study of the genetic events that occur in gene pools.

[1]

The Hardy-Weinberg Theorem

In 1908, an English mathematician, Godfrey H. Hardy, and a German physician, Wilhelm Weinberg, independently derived a mathematical model describing what happens to the frequency of genes in a population over time. Their combined ideas became known as the Hardy-Weinberg theorem. It states that the mixing of genes at meiosis and their subsequent recombination will not alter the frequencies of the genes in future generations, as long as certain assumptions are met. If they are met by a population, then evolution will not occur, because the gene frequencies will not change from generation to generation, even though the specific mixes of genes in individuals may vary.

Assumptions of The Hardy-Weinberg Theorem The Hardy-Weinberg theorem has certain assumptions that must be met for it to accurately describe a population:

1. The size of a population must be very large. Large size ensures that the frequency of a gene will not change by chance alone.
2. Mating within the population must be random. Every individual must have an equal chance of mating with any other individual in the population. Expressed in a slightly different way, the choice of a mate must not be based upon similarity or dissimilarity in a given trait. If this condition is not fulfilled, then some individuals are more likely to reproduce than others, and natural selection may occur.
3. There must not be any migration of individuals into, or out of, the population. Migration may introduce new genes into the gene pool, or add or delete copies of existing genes.
4. Mutations must not occur. If they do, mutational equilibrium must exist. *Mutational equilibrium* occurs when mutation from the wild-type gene to a mutant form is balanced by mutation from the mutant form back to the wild type. In either case, no new genes are introduced into the population from this source.

These assumptions are clearly very restrictive, and few, if any, real populations meet them. The theorem, however, does provide a useful theoretical framework for examining changes in gene frequencies in populations.

Genetic Equilibrium To illustrate the Hardy-Weinberg theorem, consider a hypothetical population that meets all of the assumptions listed above. In this population, a particular autosomal locus has two alleles; allele A has a frequency of 0.8 [f(A) = 0.8], and allele a has a frequency of 0.2 [f(a) = 0.2]. Another way to say the same thing is that, in this population, 80% of the genes for this locus are allele A, and 20% are allele a. For the purpose of our example, assume dominance between the two alleles; however, it is irrelevant to the Hardy-Weinberg theorem.

Because random mating is assumed, one can easily determine the genotypic and phenotypic frequencies of the next generation using a Punnett-square type analysis (figure 13.1), or simply by applying the multiplication and addition rules of probability theory (*see chapter 7*). Recall that the probability of an allele being in any gamete is simply that allele's frequency in the population; in this case 0.8 for *A* and 0.2 for *a*. The probability of two alleles occurring together in the next generation is the product of their individual frequencies. Also recall that heterozygotes can be produced in two ways (*Aa* or *aA*), so one must also apply both the multiplication and addition rules to determine the frequency of heterozygotes in the next generation. Using these rules, genotypic frequencies for the F_1 generation are determined as follows:

$$P(AA) = P(A) \times P(A)$$
$$= 0.8 \times 0.8$$
$$= 0.64$$

$$P(Aa \text{ or } aA) = [P(A) \times P(a)] + [P(a) \times P(A)]$$
$$= [0.8 \times 0.2] + [0.2 \times 0.8]$$
$$= 0.16 + 0.16$$
$$= 0.32$$

$$P(aa) = P(a) \times P(a)$$
$$= 0.2 \times 0.2$$
$$= 0.04$$

Because we are assuming dominance, the phenotypic frequencies of the F_1 generation are:

$$f(\text{phenotype } A) = 0.64 + 0.32 = 0.96$$
$$f(\text{phenotype } a) = 0.04$$

Gene frequencies of the F_1 generation can be determined by adding the frequency of each homozygote to one-half the frequency of the heterozygote. (The frequency of an allele in the heterozygote is reduced by one half because heterozygotes contain only a single copy of each allele.)

	f(allele *A*)	f(allele *a*)
AA individuals	0.64	0
Aa individuals	1/2(0.32)=0.16	1/2(0.32)=0.16
aa individuals	0	0.04
Overall frequency	0.80	0.20

Note that the overall frequency of each allele in the F_1 generation has not changed from that assumed for the parental generation. This example illustrates that if the assumptions of the Hardy-Weinberg theorem are met, the frequency of genes does not change from generation to generation. In other words, **Hardy-Weinberg equilibrium** is achieved, and evolution does *not* occur!

The Hardy-Weinberg theorem can be stated in general mathematical terms. Considering a locus with two interacting alleles, if p = the frequency of one allele, and q = the frequency of a second allele, then ($p + q = 1$). The frequency of genotypes in the next generation can be found by algebraically expanding the binomial equation, $(p + q)^2 = p^2 + 2pq + q^2 = 1$ (figure 13.2).

The Hardy-Weinberg theorem can be extended to cases of multiple alleles, polyploidy, sex linkage, and two or more loci. All of these applications involve somewhat more sophisticated mathematics, but the conclusions are the same. Gene frequencies will remain constant if all of the assumptions of the theorem are met.

Must evolution occur? Must gene frequencies change? Theoretically no. However, the assumptions of the theorem [2] are restrictive, and natural populations probably never

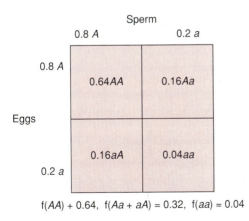

f(AA) + 0.64, f(Aa + aA) = 0.32, f(aa) = 0.04

Figure 13.1 Determining the genotypic frequencies of offspring from allelic frequencies in a parental generation.
This analysis assumes random mating, and that each allele is equally likely to be incorporated into a viable gamete. The frequency of each allele in a sperm or egg is simply the frequency of that allele in the parental population.

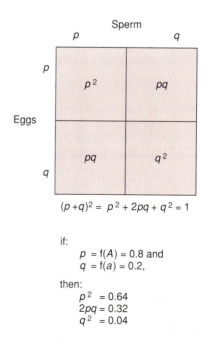

$(p + q)^2 = p^2 + 2pq + q^2 = 1$

if:

p = f(A) = 0.8 and
q = f(a) = 0.2,

then:

p^2 = 0.64
$2pq$ = 0.32
q^2 = 0.04

Figure 13.2 Expansion of the binomial $(p + q)^2 = 1$. Substitution of allelic frequencies of a parental generation for p and q will derive genotypic frequencies for the next generation.

meet them all. In fact, in the next section, we will see how, when the assumptions are not met, evolutionary change occurs.

Evolutionary Mechanisms

Evolution is neither a creative force working for progress, nor a dark force working to sacrifice individuals for the sake of the group. It is neither moral nor immoral. It has neither a goal, nor a mind to conceive a goal. Such goal-oriented thinking is said to be *teleological*. Evolution is simply a result of some individuals in a population surviving and being more effective at reproducing than others in the population, leading to changes in gene frequencies. In this section, we will examine some of the situations when the Hardy-Weinberg assumptions are not met—situations in which gene frequencies do change from one generation to the next and evolution occurs.

Population Size, Genetic Drift, and Neutral Evolution

[3] Chance often plays an important role in the perpetuation of genes in a population, and the smaller the population size, the more significant the effect of chance events may be. Fortuitous circumstances, such as a chance encounter between reproductive individuals, may promote reproduction. Some traits of a population survive, not because they convey increased fitness, but because they happened to be in gametes that were involved in fertilization. When chance events influence evolution, **genetic drift** is said to occur. Because gene frequencies are changing independently of natural selection, genetic drift is often referred to as **neutral selection.** *Genetic drift is most likely to occur in small, isolated populations.* In these populations, a chance change in gene frequency is perpetuated because the gene pool is small and over time, inbreeding may become prevalent. Genetic drift and inbreeding are likely to reduce genetic variation within a population (figure 13.3). In large populations, changes in gene frequencies in one part of the population are likely to be countered by opposing changes elsewhere in the population. Therefore, the likelihood of genetic drift occurring in small populations suggests that a Hardy-Weinberg equilibrium will not occur in such populations.

Consider the example of some rather unusual multi-legged bullfrogs discovered on a cotton farm in Mississippi. The extra legs that these frogs possess are of no selective advantage. More than likely, a mutation created a new gene that had remained hidden in the heterozygous state. Perhaps an unusually harsh winter killed a large portion of the

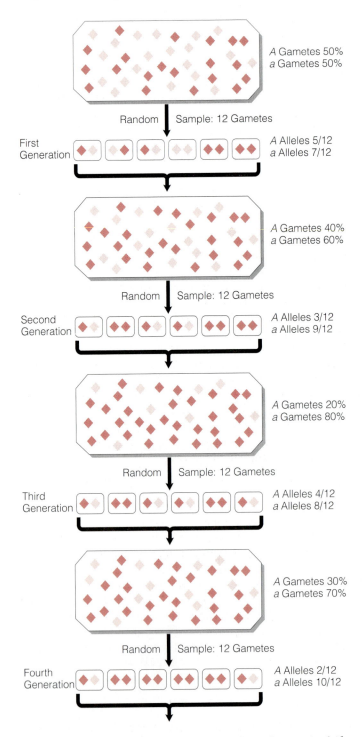

Figure 13.3 Loss of genetic diversity through genetic drift. Assume that the first generation has equal numbers of *A* and *a* alleles. Random distribution of *A* and *a* could result in gametes with a higher frequency of *a*. In each subsequent generation, *A* alleles are, by chance, slowly lost. By the fifth generation, the *A* allele has been eliminated from the population.

original population and, by chance, a disproportionate number of the survivors carried the gene for multiple legs. The following spring, multilegged bullfrogs probably began to show up. Whatever the exact sequence of events, they occurred in a very restricted population. These bullfrogs were not sharing genes with thousands of other bullfrogs.

Two special cases of genetic drift have influenced the genetic makeup of some populations. When a few individuals from a parental population colonize new habitats, they never carry a representative sample of the gene pool from which they came. The new colony that emerges from the founding individuals is likely to have a distinctive genetic makeup with far less variation than the larger population. This form of genetic drift is called the **founder effect.**

An often-cited example of the founder effect concerns the genetic makeup of the Dunkers of eastern Pennsylvania. They emigrated from Germany to the United States early in the eighteenth century and, for religious reasons, have not married outside their sect. Examination of certain traits (ABO blood type, for example) in their population reveals very different gene frequencies from the Dunker populations of West Germany. These differences are attributed to the chance absence of certain genes in the individuals that founded the original Pennsylvania Dunker population.

A similar effect can occur when the number of individuals in a population is drastically reduced. An extreme example occurs in the lemmings of Scandinavian countries. These rodents undergo extreme population cycles. Every few years, the population builds to huge numbers and then it crashes. During crash years, relatively few individuals overwinter in mountainous regions. Those that remain have only a portion of the original gene pool, and the new population that emerges in the next cycle may be genetically different from the old population. This form of genetic drift is called the **bottleneck effect.**

Gene Flow

The Hardy-Weinberg theorem assumes that no individuals enter a population from the outside (immigrate), and that no individuals leave a population (emigrate). If immigration or emigration occurs, the Hardy-Weinberg equilibrium is upset, and changes in gene frequency (evolution) occur. These changes in gene frequency from migration of individuals are called **gene flow.** Although there are certainly some natural populations for which gene flow may not be significant, most populations experience genetic changes from this source.

Mutation

Changes in the structure of genes and chromosomes are called mutations. Changes in chromosome number and structure are discussed in chapter 8, and point mutations are discussed in chapter 9.

The Hardy-Weinberg theorem assumes that no mutations occur, or that mutational equilibrium exists. Mutations, however, are a fact of life. Most importantly, *mutations are the origin of all new genes and a source of variation that may prove adaptive for an animal.* Mutation counters the loss of genetic material that results from natural selection and genetic drift, and increases the likelihood that genes will be present that allow a group to survive future environmental shocks.

Mutations are random events, and the likelihood of a mutation is not affected by the usefulness of the mutation. Organisms have no special devise to filter deleterious genetic changes from advantageous changes before they occur. Mutations occur, and animals must take the bad along with the good. The effects of mutations vary enormously. Most are either deleterious or neutral. Some mutations that are neutral in one environment may help an organism survive in another environment.

Mutational equilibrium occurs when mutation from the wild-type allele to a mutant form is balanced by mutation from the mutant back to the wild type, and has the same effect on gene frequency changes as if no mutation occurred. Mutational equilibrium rarely exists, however, and the extent to which it is absent is called mutation pressure. In other words, **mutation pressure** is a measure of the tendency for gene frequencies to change through mutation.

Natural Selection Reexamined

The theory of natural selection remains preeminent in modern biology. Natural selection occurs whenever some phenotypes are more successful at leaving offspring than other phenotypes, and the tendency for it to occur—and upset Hardy-Weinberg equilibrium—is called **selection pressure.** Although natural selection is simple in principle, it is quite diverse in actual operation.

Modes of Selection Most populations have a range of phenotypes for a given trait. They may be characterized using a bell-shaped curve, where phenotypic extremes are less common than the intermediate phenotypes. Natural selection may affect a range of phenotypes in three different ways.

Directional selection occurs when individuals at one phenotypic extreme are at a disadvantage compared to all other individuals in the population (figure 13.4a). In response to this selection, the deleterious gene(s) decreases in frequency and all other genes increase in frequency. Directional selection may occur when a mutation gives rise to a new gene, or when the environment changes to select against an existing phenotype.

A classic example of directional selection, *industrial melanism,* occurred in England during the industrial revolution. Museum records and experiments document how environmental changes affected selection against two forms of the peppered moth, *Biston betularia.*

In the early 1800s, a gray form made up about 99% of peppered moth populations. That form still predominates in nonindustrial northern England and Scotland. In industrial areas of England the gray form was replaced by a black form over a period of about 50 years. In these areas, the gray form made up only about 5% of the population, and 95% of the population was black. The gray phenotype, previously advantageous, had become deleterious.

The nature of the selection pressure was understood when it was discovered that birds prey more effectively on moths resting on a contrasting background. Prior to the industrial revolution, the gray form was favored because gray moths blended with the bark of trees on which the

moths rested. The black form contrasted with the lighter bark, and was easily spotted by birds (figure 13.5a). Early in the industrial revolution, however, factories used soft coal, and spewed soot and other pollutants into the air. Soot covered the tree trunks where the moths rested, and bird predators could easily pick out gray moths against the black background. The black form was now effectively camouflaged (figure 13.5b).

In the 1950s, the British Parliament enacted air pollution standards that have helped reduce soot in the atmosphere. As would be expected, the gray form of the moth has had a small, but significant, increase in frequency.

Another form of natural selection may occur when circumstances select against individuals of an intermediate phenotype (figure 13.4b). **Disruptive selection** produces [4] distinct subpopulations. Consider, for example, what could happen in a population of snails having a range of shell colors between white and dark brown. Their marine tidepool habitat provides two background colors. The sand, made up of pulverized mollusc shells, is white, and rock outcroppings are brown. In the face of predation by shorebirds, what phenotypes are going to be most common? Although white snails may not actively select a white background, those present on the sand are less likely to be preyed upon than intermediate phenotypes on either sand or rocks. Similarly, brown snails are less likely to be preyed upon than intermediate phenotypes on either substrate. Thus, two distinct subpopulations, one white and one brown, could be produced through disruptive selection.

A third form of natural selection occurs when both phenotypic extremes are deleterious. This process is called **stabilizing selection,** and results in a narrowing of the phenotypic range (figure 13.4c). During long periods of environmental constancy, new variations that arise, or new combinations of genes that occur, are unlikely to result in more fit phenotypes than the genes that have allowed a population to survive for thousands of years, especially when the new variations are at the extremes of the phenotypic range.

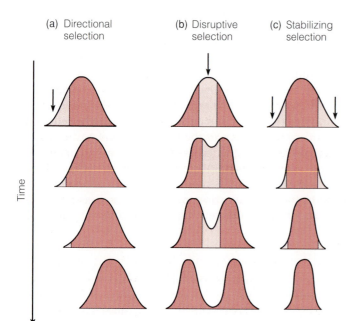

(a) Directional selection (b) Disruptive selection (c) Stabilizing selection

Time

Figure 13.4 Modes of selection. (a) Directional selection occurs when individuals at one phenotypic extreme are selected against. It results in a shift in phenotypic distribution toward the advantageous phenotype. (b) Disruptive selection occurs when an intermediate phenotype is selected against and results in the production of distinct subpopulations. (c) Stabilizing selection occurs when individuals at both phenotypic extremes are selected against, and results in a narrowing at both ends of the range.

(a)

(b)

Figure 13.5 Two body forms of the peppered moth, *Biston betularia.* There is a gray and black form in each picture. (a) Prior to the industrial revolution, the black form (at bird's eye level) was easily spotted by bird predators, and the gray form (lower right of picture) was camouflaged. (b) In industrial regions, after the industrial revolution, selection was reversed because of soot covering the bark of trees where moths rested. Note how clearly the gray form can be seen, whereas the black form (near oak leaf) is almost invisible.

A good example of stabilizing selection is the horseshoe crab (*Limulus*), which occurs along the Atlantic coast of the United States (figure 13.6). Comparison of the fossil record with living forms indicates that this species has changed very little over 200 million years. Apparently, the combination of characteristics present in this species represents a very successful combination of characteristics for the horseshoe crab's environment.

Neutralist-Selectionist Controversy Few biologists doubt that both natural selection and neutral selection occur. An interesting debate among evolutionists, however, concerns the extent to which each is operating in natural populations. Neutralists attribute far greater significance to genetic drift in promoting the spread of new genes through populations than selectionists do.

Although no consensus has emerged, it seems clear that both processes operate, but may not be equally important in all circumstances. For example, during long periods when environments are relatively constant, and stabilizing selection is acting on phenotypes, neutral selection may operate at a molecular level. Certain genes could be randomly established in a population. Occasionally, however, the environment shifts, and directional or disruptive selection begins to operate, resulting in gene frequency changes (often fairly rapid).

The neutralist-selectionist controversy illustrates the kind of debate occurring among evolutionists. These debates concern the mechanics of evolution and are the foundations of science. They lead to experiments that will ultimately present a clearer picture of evolution.

Balanced Polymorphism and Heterozygote Superiority

Polymorphism occurs in a population when two or more distinct forms exist without a range of phenotypes between them. **Balanced polymorphism** (Gr. *poly,* many +

morphe, form) occurs when different phenotypic expressions are maintained at relatively stable frequencies in the population and may resemble a population in which disruptive selection operates.

Sickle-cell anemia (*see chapters 7 and 9*) results from a change in the structure of hemoglobin. The red blood cells of persons with the disease are misshapen, and their ability to carry oxygen is reduced. In the heterozygous state, there are roughly equal quantities of normal and sickled cells. Sickle-cell heterozygotes occur in some African populations with a frequency as high as 0.4 (figure 13.7). The maintenance of the sickle-cell heterozygotes and both homozygous genotypes at relatively unchanging frequencies makes this trait an example of a balanced polymorphism.

Why hasn't such a seemingly deleterious gene been eliminated by natural selection? The sickle-cell gene is most common in regions of Africa that are heavily infested with the malarial parasite, *Plasmodium falciparum.* Heterozygotes are less susceptible to malarial infections; if infected, they experience less severe symptoms than without sickled cells. Individuals homozygous for the normal allele are at a disadvantage because they experience more severe malarial infections, and individuals homozygous for the sickle-cell allele are at a disadvantage because they suffer from severe anemia caused by the sickled cells. The heterozygotes, who experience symptoms of anemia only when stressed, are more likely to survive than either homozygote. This system is also an example of *heterozygote superiority,* which occurs when the heterozygote is more fit than either homozygote. Heterozygote superiority can lead to balanced polymorphism because perpetuation of the alleles in the heterozygote condition maintains both alleles at a higher frequency than would be expected if natural selection acted only on the homozygous phenotypes.

Figure 13.6 The horseshoe crab, *Limulus,* is found along the eastern coast of the United States. Its present body form is nearly identical to fossils that are 200 million years old. It is an excellent example of stabilizing selection.

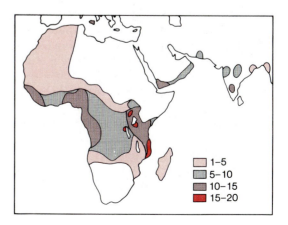

Figure 13.7 The incidence of sickle-cell anemia in Africa. Shading on the map indicates the percent of the population with the disease. The regions where incidence of the disease is highest also have the highest rates of malarial infestations.

Figure 13.8 Sandhill cranes have elaborate courtship rituals that prepare two individuals for mating. A change in species-specific behavior patterns in one deme could result in ethological isolation, and lead to speciation.

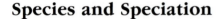

Species and Speciation

Taxonomists classify organisms into groups based upon similarities and differences among them. This classification system is discussed in chapters 1 and 15. The fundamental unit of classification is the species. Unfortunately, it is difficult to formulate a universally applicable definition of species. According to a biological definition, a **species** is a group of populations in which genes are actually, or potentially, exchanged through multiple generations.

Although this definition is concise, it has problems associated with it. Taxonomists often work with morphological characteristics, and the reproductive criterion must be assumed based upon morphological and ecological information. Also, some organisms do not reproduce sexually. Obviously, other criteria need to be applied in these cases. Another problem concerns fossil material. Paleontologists describe species of extinct organisms, but how are they to test the reproductive criterion? Finally, populations of similar organisms may be so isolated from each other that the exchange of genes is geographically impossible. To test a reproductive criterion, biologists can transplant individuals to see if mating can occur. (This procedure, however, is often difficult and sometimes illegal.) Under such circumstances, however, one can never be certain that mating of transplanted individuals would really occur if animals were together in a natural setting.

Rather than trying to establish a definition of a species that solves all these problems, it is probably better to simply understand the problems associated with the biological definition. In describing species, taxonomists use morphological, physiological, and ecological criteria, realizing that all of these have a genetic basis.

There may be a great deal of variation in a species. Genetically distinct populations within a species are called *subspecies* or *races*. The subspecies concept has been hotly debated, because there is disagreement as to how much genetic variation warrants the division of a species into subspecies. Many biologists believe that the criteria are so arbitrary that the subspecies concept should be abandoned.

Sometimes species variations are correlated with geographical differences. If so, the species is said to form a *cline*. For example, eleven subspecies of grass frogs (*Rana pipiens*) occur from Vermont to Louisiana. (Members of a cline do not have to be classified as separate subspecies.) Although some interbreeding may occur in transition areas where ranges of adjacent subspecies overlap, more distant populations do not interbreed. Clines present another problem for the biological definition of a species. In some cases, animals from the extremes of the geographic range are incapable of successful mating when brought together, but animals in each transition area along the cline can exchange genes. Theoretically, genes can be passed throughout the geographic range and the assemblage must be considered a single species.

The processes by which two or more species are formed [5] from a single stock are called **speciation.** *A requirement of speciation is that subpopulations become isolated from one another.* When this occurs, natural selection and genetic drift can result in evolution taking a different course in each subpopulation. At some point, the two groups become reproductively isolated from each other and cannot successfully breed. **Reproductive isolation** can occur in different ways.

Premating isolation mechanisms prevent mating from taking place. For example, if courtship behavior patterns of two animals are not mutually appropriate, mating will not occur (figure 13.8). Other isolating mechanisms prevent successful fertilization and development, even though mating may have occurred. These are called **postmating isolation** mechanisms. For example, conditions in the reproductive tract of a female may not support the sperm of another individual, and these individuals will be reproductively isolated from each other. Other premating and postmating isolation mechanisms are described in table 13.1.

Table 13.1 Mechanisms Promoting Reproductive Isolation

Premating Isolation

Ecological isolation—Individuals occupy different niches within a particular range. Individuals may be ecologically isolated by habitat, food preferences, daily activity patterns, and the like.

Seasonal isolation—Individuals may have different breeding periods.

Ethological isolation—Specific behavior patterns of individuals may prevent mating.

Mechanical isolation—The structure of genitalia may prevent mating.

Character isolation —Young imprint on parental color patterns and behaviors. If any individual does not match another's imprinted image of what a mate should look like, mating may not occur.

Postmating Isolation Mechanisms

Gametic isolation—The failure of fertilization, even though mating has occurred. If the pH or other conditions in a female's reproductive tract does not support survival of sperm, fertilization is unlikely.

Developmental isolation—Even if fertilization occurs, a zygote may be inviable. Relatively small differences in chromosome numbers or chromosome structure between two species will often cause the death of the zygote.

Hybrid inviability—Hybrids are often sterile because mismatched chromosomes cannot synapse properly during meiosis. Thus, isolation is delayed one generation. (A well-known example of this is the mule, which is the progeny of a male donkey and a mare. The few instances of female mules giving birth are believed to be the result of the chance assortment of grandparental chromosomes into gametes.)

Allopatric Speciation

One way for isolation to occur is for subpopulations to become geographically isolated from one another. Geographic isolation can lead to speciation and, when it does, it is called **allopatric** (Gr. *allos,* other + *patria,* fatherland) **speciation.** For example, geologic and climatic changes may permanently separate members of a population. Evolution in these separate populations may result in members not being able to mate successfully with each other, even if experimentally reunited. Allopatric speciation is believed by many biologists to be the most common kind of speciation (figure 13.9).

The finches that Darwin saw on the Galápagos islands are a classic example of allopatric speciation, as well as an important evolutionary process called adaptive radiation. **Adaptive radiation** occurs when a number of new forms diverge from an ancestral form, usually in response to the development of a highly adaptive trait or the opening of major new habitats.

Fourteen species of finches evolved from the original finches that colonized the Galápagos Islands. It is likely that ancestral finches were distributed between a few of

the islands of the Galápagos, having emigrated from the mainland. Populations became isolated on various islands over a period of time. Even though the original population probably displayed some genetic variation, even greater variation arose as groups became isolated. The original finches were seed eaters and, after their arrival, they probably filled their preferred habitats very rapidly. Variations within the original finch population may have allowed

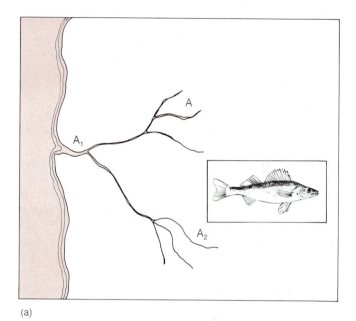

(a)

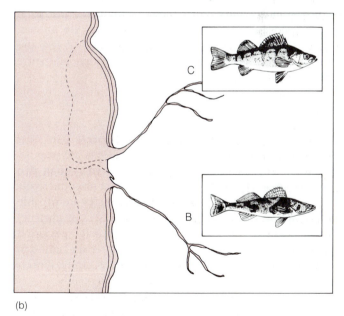

(b)

Figure 13.9 Allopatric speciation can occur when a geographic barrier divides a population. (a) In this hypothetical example, a population of freshwater fish in a river drainage system is divided into three subpopulations: A, A_1, and A_2. Genetic exchanges may occur between A and A_1, and between A_1 and A_2. Exchanges are less common between A and A_2. (b) A rise in the level of the ocean forces the breakup of A_1, and separates A and A_2 into separate populations. Genetic drift and different selection pressures in the two populations may eventually result in the formation of species B and C.

some birds to exploit new islands and habitats where no finches had been. Mutations changed the genetic composition of the isolated finch populations, introducing further variations. Natural selection favored the retention of the variations that happened to promote successful reproduction.

The combined forces of isolation, mutation, and natural selection, allowed the finches to diverge into a number of species with specialized feeding habits (*see figure 12.6*). Of the 14 species of finches, six have beaks specialized for crushing seeds of different sizes. Others feed on flowers of the prickly pear cactus or in the forests on insects and fruit.

Parapatric Speciation

Another form of speciation is called **parapatric** (Gr. *para*, beside) **speciation,** which occurs in small, local populations, called **demes.** For example, all of the frogs in a particular pond, or all of the sea urchins in a particular tidepool, make up a deme. Individuals of a deme are more likely to breed with one another than with other individuals in the larger population; and, because they experience the same environment, they are subject to similar selection pressures. Demes are not completely isolated from each other, because individuals, developmental stages, or gametes can move between demes of a population. On the other hand, the relative isolation of a deme may mean that it experiences different selection pressures than other members of the population. If so, speciation can occur. Although most evolutionists theoretically agree that parapatric speciation can occur, no certain cases are known. Parapatric speciation is therefore considered of less importance in the evolution of animal groups than allopatric speciation.

Sympatric Speciation

A third kind of speciation is called **sympatric** (Gr. *sym*, together) **speciation.** Sympatric populations are populations that occupy overlapping ranges. Even though two populations are sympatric, they still may be reproductively isolated from one another. Many plant species are capable of producing viable forms with multiple sets of chromosomes. Such events could lead to sympatric speciation among groups in the same habitat. Sympatric speciation in animals, however, is generally considered to rarely, if ever, occur.

Stop and Ask Yourself

8. Why is isolation necessary for speciation to occur?
9. What form of premating isolation is likely to promote allopatric speciation?
10. What is postmating isolation? What are three forms it may take?
11. What are sympatric, parapatric, and allopatric speciation?
12. What is adaptive radiation?

Rates Of Evolution

Charles Darwin perceived evolutionary change as occurring gradually over millions of years. This concept, called **phyletic gradualism,** has been a traditional interpretation of the tempo, or rate, of evolution.

Some contemporary evolutionists, led by Stephen J. Gould of Harvard and Niles Eldredge of the American Museum of Natural History, have suggested that evolutionary

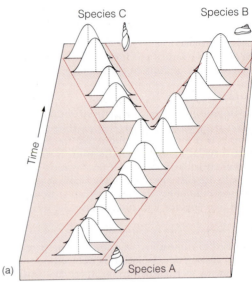

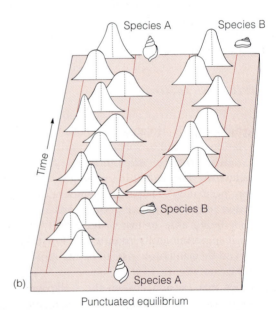

Figure 13.10 A comparison of phyletic gradualism and punctuated equilibrium. (a) Phyletic gradualism is a model of evolution in which gradual changes occur over very long time periods. Notice that a continuous series of intermediate phenotypes are present as species B and species C gradually diverge from species A. (b) Punctuated equilibrium is a model of evolution in which long periods of stasis are interrupted by rapid periods of change. Notice the very rapid divergence of species B from species A, after which species B enters a period of relative stasis.

change may occur very rapidly. Studies of the fossil record show that many species do not change significantly over millions of years. These periods of *stasis* (Gr. *stasis,* standing still), or equilibrium, are interrupted when a group encounters an ecological crisis. Over the next 10,000 to 100,000 years, a variation might be selected for that previously would have been selectively neutral or disadvantageous or a whole set of previously occupied habitats might become available. (Events, such as geological uplift or climatic changes, that occur in 10,000 to 100,000 years are almost instantaneous in an evolutionary time frame.) This geologically brief period of change "punctuates" the previous million or so years of equilibrium, and eventually settles into the next period of stasis. In this model, the periods of stasis are characterized by stabilizing selection, and the periods of change are characterized by directional or disruptive selection (figure 13.10).

One advantage of this **punctuated equilibrium model** is its explanation for why the fossil record does not always show transitional stages between related organisms. Gradualists attribute the absence of transitional forms to the fact that fossilization is an unlikely event, and therefore, many transitional forms disappeared without leaving a fossil record. If punctuated equilibrium occurred, however, intermediate forms probably did not exist. The rapid pace (geologically speaking) of evolution resulted in "jumps" from one form to another.

The debate between phyletic gradualists and those favoring punctuated equilibrium is, like the neutralist/selectionist debate, an example of the debates occurring among evolutionary biologists and geologists. Few evolutionists try to argue gradualism or rapid change to the exclusion of the other. Most recognize that both processes occur in evolution. The debate has become one of which tempo characterizes *most* evolutionary change.

Molecular Evolution

Many evolutionists study changes in animal structure and function that are observable on a large scale—for example, changes in the shape of a bird's beak or in the length of an animal's neck. All evolutionary change, however, results from changes in the sequence of bases in DNA and amino acids in proteins. Molecular evolutionists investigate evolutionary relationships between organisms by studying DNA and proteins.

[6]

Cytochrome *c* is a protein present in the pathways of cellular respiration in all eukaryotic organisms (table 13.2). Organisms shown to be closely related from using other investigative tools, have similar cytochrome *c* molecules. About 20 of the 100 amino acids are the same in every organism. One region, consisting of a series of 11 amino acids, is the same in all species investigated. Where there are differences, similar amino acids are substituted for one another. (For example, polar amino acids are substituted for polar amino acids, and nonpolar amino acids are substituted for nonpolar amino acids.)

Table 13.2	Amino Acid Differences in Cytochrome *c* from Different Organisms
Organisms	**Number of Variant Amino Acid Residues**
Cow and sheep	0
Cow and whale	2
Horse and cow	3
Rabbit and pig	4
Horse and rabbit	5
Whale and kangaroo	6
Rabbit and pigeon	7
Shark and tuna	19
Tuna and fruit fly	21
Tuna and moth	28
Yeast and mold	38
Wheat germ and yeast	40
Moth and yeast	44

The fact that cytochrome *c* has changed so little during hundreds of millions of years does not mean that mutations occur more slowly in genes coding for cytochrome *c* than in other genes. Rather, it suggests that mutations of the cytochrome *c* gene are nearly always detrimental, and will be selected against. Selectively neutral changes, and selectively advantageous changes have been calculated to have occurred at a rate of 1% change every 20 million years! Because it has changed so little, cytochrome *c* is said to have been conserved evolutionarily.

Not all proteins are conserved as rigorously as cytochrome *c*. Although variations in highly conserved proteins can be used to help establish evolutionary relationships between distantly related organisms, less-conserved proteins are useful for looking at relationships between more closely related animals. Because some proteins are conserved and others are not, the best information regarding evolutionary relationships is obtained by comparing as many proteins as possible in any two species and combining this information with evidence from paleontology, comparative anatomy, and biogeography (*see chapter 1*).

Gene Duplication

Recall that most mutations are selected against. However, if an extra copy of a gene is present, one copy may be modified and, as long as the second copy is furnishing the essential protein, the organism is likely to survive. *Gene duplication,* the accidental duplication of a gene on a chromosome, is one way that extra genetic material can arise (*see chapter 8*).

Vertebrate hemoglobin and myoglobin are believed to have arisen from a common ancestral molecule. Hemoglobin carries oxygen in the blood stream, and myoglobin is an oxygen storage molecule in muscle. The ancestral molecule probably carried out both functions. However, about one billion years ago, gene duplication followed by modification of one gene resulted in the formation of two polypeptides: myoglobin and hemoglobin. Further gene duplications over the last 500 million years probably explain the fact that most vertebrates, other than primitive fish,

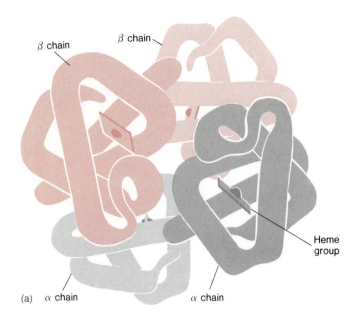

(a) α chain

β chain

β chain

Heme group

α chain

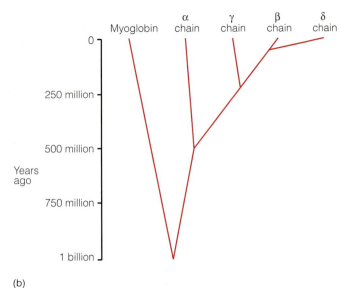

(b)

Figure 13.11 The evolution of vertebrate hemoglobin. (a) Hemoglobin consists of a polypeptide of about 150 amino acids and a heme group. In most vertebrates, the molecule has a quaternary structure with four polypeptide/heme subunits. Oxygen binds reversibly with an iron atom held by each heme group. (b) An ancestral molecule may have served oxygen storage functions in muscle and oxygen transport functions in blood. About 1 billion years ago, gene duplication probably gave rise to two genes coding for proteins of hemoglobin and myoglobin. Today, lampreys and hagfish still have hemoglobin containing a single polypeptide. Subsequent gene duplication about 500 million years ago gave rise to genes coding for two hemoglobin polypeptides, called the α and β *chains*. Two of each of these chains associated during the evolution of fish. Further gene duplication probably accounts for the additional two polypeptides found in some mammalian hemoglobin.

have hemoglobin molecules consisting of four polypeptides (figure 13.11a,b).

Mosaic Evolution

A previous section described how rates of evolution can vary both in populations and in molecules and structures. A species might be thought of as a mosaic of different molecules and structures that have evolved at different rates. Some molecules or structures are conserved in evolution, others change more rapidly. The basic design of a bird provides a simple example. All birds are easily recognizable as birds because of highly conserved structures, such as feathers, beaks, and a certain body form. Particular parts of birds, however, are less conservative and have a higher rate of change. Wings have been modified for hovering, soaring, and swimming. Similarly, legs have been modified for wading, swimming, and perching. These are examples of **mosaic evolution.**

Stop and Ask Yourself

13. How do phyletic gradualism and punctuated equilibrium models differ?
14. How can amino acid sequencing and DNA base sequencing provide evidence of evolutionary relationships?
15. What does it mean to say that cytochrome *c* is an evolutionarily conserved protein?

■ Summary

1. Organic evolution is a change in the frequency of genes in a population.
2. Unlimited genetic variation, in the form of new genes and new combinations of genes, increases the chances that a population will survive future environmental changes.
3. Population genetics is the study of events occurring in gene pools. The Hardy-Weinberg theorem describes the fact that, given certain assumptions, gene frequencies of a population remain constant from generation to generation.

4. The assumptions of the Hardy-Weinberg theorem, when unmet, define circumstances under which evolution will occur. a. Fortuitous circumstances may allow only certain genes to be carried into the next generation. Such chance variations in gene frequencies are called genetic drift or neutral selection. b. Gene frequencies may change as a result of individuals immigrating into, or emigrating from, a population. c. Mutations are the source of new genetic material for populations. Mutational equilibrium rarely exists, and thus, mutations usually result in changing gene

frequencies. d. The tendency for gene frequencies to change, due to differing fitness, is called selection pressure. Selection may be directional, disruptive, or stabilizing.

5. Balanced polymorphism occurs when two or more body forms are maintained in a population. Heterozygote superiority can lead to balanced polymorphism.

6. According to a biological definition, a species is a group of populations within which there is potential for exchange of genes. There are significant problems associated with the application of this definition.

7. In order for speciation to happen, reproductive isolation must occur. Speciation may occur sympatrically, parapatrically, or allopatrically, although most speciation events are believed to be allopatric.

8. In allopatric speciation, premating isolation may prevent mating from occurring and/or postmating isolation may prevent the development of fertile offspring, if mating has occurred.

9. Phyletic gradualism is a model of evolution that depicts change as occurring gradually, over millions of years. Punctuated equilibrium is a model of evolution that depicts long periods of stasis interrupted by brief periods of relatively rapid change.

10. The study of molecular evolution, using rates of molecular change, helps establish evolutionary interrelationships between organisms.

11. A duplicated gene may be modified by mutation, and, by chance, come to serve a function other than its original role.

12. Any species can be thought of as a mosaic of different molecules and structures that have evolved at differing rates.

■ Key Terms

adaptive radiation (p. 197)
allopatric speciation (p. 197)
balanced polymorphism (p. 195)
bottleneck effect (p. 193)
demes (p. 198)
directional selection (p. 193)
disruptive selection (p. 194)
founder effect (p. 193)
gene flow (p. 193)
gene pool (p. 190)
genetic drift (p. 192)
Hardy-Weinberg equilibrium (p. 191)
mosaic evolution (p. 200)
mutation pressure (p. 193)
neutral selection (p. 192)

parapatric speciation (p. 198)
phyletic gradualism (p. 198)
population genetics (p. 190)
postmating isolation (p. 196)
premating isolation (p. 196)
punctuated equilibrium model (p. 199)
reproductive isolation (p. 196)
selection pressure (p. 193)
speciation (p. 196)
species (p. 196)
stabilizing selection (p. 194)
sympatric speciation (p. 198)

■ Critical Thinking Questions

1. Can natural selection act upon variations that are not inherited? (Consider for example, deformities that arise from contracting a disease.) If so, what is the effect of that selection upon subsequent generations?

2. In what way does overuse of antibiotics and pesticides increase the likelihood that these chemicals will eventually become ineffective? This is an example of which one of the three modes of natural selection?

3. What are the implications of the "bottleneck effect" for wildlife managers, who try to help endangered species, like the whooping crane, recover from near extinction?

4. What kinds of factors influence rates of evolution in animals? Would you expect rates of speciation to differ for taxa that have been in existence for hundreds of millions of years, as compared to taxa that have been in existence for only tens of millions of years?

■ Suggested Readings

Books

Bowler, P. 1984. *Evolution: The History of an Idea.* Berkeley: University of California Press.

Dodson, E. O. and Dodson, P. 1985. *Evolution: Process and Product.* Belmont: Wadsworth Publishing Company.

Futuyma, D. J. 1986. *Evolutionary Biology,* 2nd ed. Sunderland: Sinauer Associates, Inc.

Godfrey L. R. 1985. *What Darwin Began.* Old Tappan: Allyn and Bacon, Inc.

Hecht, M. K., Wallace, B., and Prance, G. T. (eds). 1967–1988. *Evolutionary Biology,* Vols. 1–22. New York: Plenum Press.

Mayr, E. 1982. *The Growth of Biological Thought: Diversity, Evolution, and Inheritance.* Cambridge: Harvard University Press.

Stebbins, G. L. 1982. *Darwin to DNA, Molecules to Humanity.* New York: W.H. Freeman and Company.

Volpe, E. P. 1985. *Understanding Evolution,* 5th ed. Dubuque: Wm C. Brown Publishers.

Articles

Amato, I. 1987. Tics in the tocks of molecular clocks; comparing the DNA, RNA and proteins of different species may reveal the entire tree of life, but obstacles are emerging. *Science News* 131:74–75.

Callagan, C. A. 1987. Instances of observed speciation. *The American Biology Teacher* 49(1):34–36.

Carson, H. L. November 1987. The process whereby species originate. *BioScience* 37:715–720.

Coyne, J. A. and Barton, N. H. 1988. What do we know about speciation? *Nature* 331:485–486.

May, R. R. The evolution of ecological systems. *Scientific American* September, 1987.

Myers, N. 1985. The ends of the lines. *Natural History* 94:2–6.

Sheldon, P. 1988. Making the most of evolution diaries. *New Scientist* 117:52–54.

14

Human Evolution

Concepts

1. Primates arose about 65 million years ago as insectivores that foraged in trees and on the ground. Today, two primate suborders are recognized: Prosimii and Anthropoidea.
2. Hominids (members of the family Hominidae) are characterized by an upright posture and bipedalism (walking on two feet). Bipedalism conferred advantages to hominids, and was accompanied by numerous modifications of the skeleton and other body parts. These modifications are used by paleontologists in identifying ancient hominids from skeletal fragments.
3. The evolution of hominids was influenced strongly by climatic changes.
4. Divergence between Old World monkeys and apes occurred in excess of 30 million years ago. Divergence between great apes and hominids occurred 5 to 10 million years ago.
5. Members of the hominid lineage include four species of *Australopithecus*, as well as *Homo habilis*, *Homo erectus*, and *Homo sapiens*. Although much is known about the life-styles and phylogeny of prehistoric humans, numerous controversies result from varying interpretations of scientific evidence.
6. Since the appearance of *Homo sapiens*, cultural evolution has been a major factor in shaping human life on earth.

Have You Ever Wondered:

[1] what the primates are?
[2] how closely related humans are to gorillas and chimpanzees?
[3] what the advantages of bipedal locomotion are?
[4] who "Lucy" was?
[5] whether or not tool use is unique to humans?
[6] how old the species *Homo sapiens* is?
[7] what cultural evolution is?
[8] how paleontologists find evidence of spoken language?

These and other useful questions will be answered in this chapter.

efore modern theories of evolution, questions about our origins absorbed human thought, and fueled fires of debate. Even the originators of theories of natural selection, Charles Darwin and Alfred Russel Wallace, differed on whether or not natural selection could account for the human form. The suggestion that chimpanzees, gorillas, and orangutans might share ancestry with humans horrified Victorian Europe and brought ridicule to nineteenth century evolutionists.

Prior to the twentieth century, debates regarding human origins had little direct evidence supporting their various claims. Early in this century, however, paleontologists began an active search for fossil evidence of prehistoric humans. Today, many resources have been added to the investigations. Paleontologists continue to hunt fossils, but in addition, paleoecologists study the environmental constraints placed on early humans, molecular biologists look for biochemical evidence of human ancestry, *taphonomists* study the way that bones and artifacts become buried, physiologists study the energetics of the primate design, and ethologists study the behavior of social primates. Although many details are still missing, these scientists are continually supplying more information about our past.

Primate Characteristics

Biological classification was introduced briefly in chapter 1 and it is covered in more detail in chapter 15. It is a system of naming in which animals are grouped into hierarchical rankings that reflect degrees of relatedness. These ranks are, from general to specific, kingdom, phylum, class, order, family, genus, species.

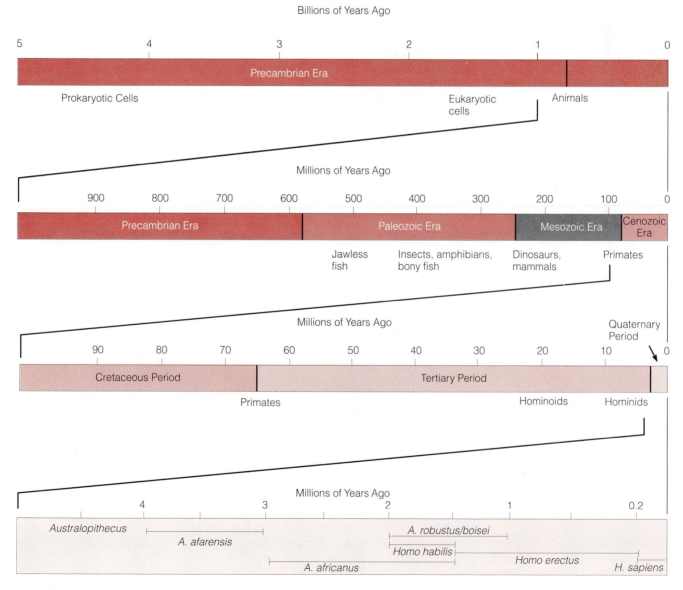

Figure 14.1 Geological time, with special reference to important events of hominoid evolution. The first prokaryotic cells arose about 5 billion years ago, and the first animals arose about 0.8 billion years ago. Primates arose at the boundary of the Cenozoic and Mesozoic eras about 65 million years ago. Hominoids arose 20 million years ago and hominids between 5 and 10 million years ago. Most hominid evolution occured in the last 5 million years.

Primates is one of numerous orders in the vertebrate [1] class Mammalia. In addition to humans, it includes the monkeys, apes (orangutans, gorillas, and chimpanzees), lemurs, and tarsiers. Their mammalian characteristics include hair, a muscular diaphragm that separates the thoracic and abdominal cavities, regulation of body temperature by metabolic processes, and nourishment of young during fetal development via embryonic and uterine membranes and after birth via mammary glands.

The fossil record documenting early primate evolution is incomplete. Therefore, the following scenario is an untested, but highly regarded, hypothesis for the origin of major primate characteristics.

Primates arose late in the Cretaceous period, about 65 million years ago (figure 14.1). They were probably insectivores (insect eaters) that ran along tree branches, tree trunks, and across the ground—similar to modern tree shrews (figure 14.2). Arboreal (tree dwelling) habitats and a shift to diurnal (daytime) activity favored the visual sense as the primary means for locating food, exploring the environment, and negotiating uncertain footing. *Binocular vision* was an important primate development. The eyes of most mammals are on the sides of the head, but the eyes of primates are on the front of the head. This location provides an overlapping field of view and improved depth perception. The eyes of primates also are large, are protected by bony orbits of the skull, and (with the exception of nocturnal forms) have receptors for color vision. As the visual sense was emphasized, the sense of smell became less important, as evidenced by the snout of primates, which is reduced in comparison to that of other mammals.

Numerous adaptations of primate limbs can also be traced to arboreal, insectivorous origins. The center of gravity of most primates has shifted from the front legs to the rear legs. Hindlimbs are muscular, often longer than forelimbs, and dominant in locomotion. This *hindlimb dominance* results in primates resting in a sitting-up position, often propped up or balanced by their tail. Both forelimbs and hindlimbs are capable of extensive rotation, which aids in clinging to branches and moving through trees. Although a few primates have claws, most have *nails* that protect the ends of long digits. Sensitive *foot* and *hand pads* are useful in exploring the arboreal environment. *Friction ridges,* which are folds in the outer layers of the skin of hands and feet, aid in clinging to tree branches. In many primates, the medial digits of the feet (the thumb and great toe) oppose the remaining digits. *Opposable thumbs* allow the hand (and in most primates the feet as well) to close around a branch or other object.

During primate evolution, the brain increased in size and complexity. Increased reliance on the visual sense was accompanied by an increase in the part of the brain devoted to vision and a reduction in the part devoted to the sense of smell (olfaction). Muscle coordination involved with climbing, jumping, and swinging through trees required extensive development of motor regions of the brain. Complex social structures characteristic of primates require a degree of nervous development unparalleled by other an-

Figure 14.2 Modern tree shrews, *Tupaia glis*. These insectivores are believed to be similar to ancestral, shrewlike animals that gave rise to the primate lineage.

imals. Much of the long prenatal (before birth) development of primates is accounted for by the length of time it takes for the brain to develop. Even so, the size of the head, compared to the size of the birth canal, places a limit on the length of intrauterine development. Thus, a long period of postnatal development is required for full brain development.

Classification

Classification of the primates is not entirely settled. A traditional interpretation (table 14.1) divides the order into two suborders, **Prosimii** (L. *prae,* before + *simia,* ape) and **Anthropoidea** (Gr. *anthropo,* human).

Suborder Prosimii

Of the prosimians, lemurs are confined to the island of Madagascar, off the eastern coast of Africa. They can be as large as a cat; but their long snout, large mobile ears, and long bushy tails give them the general appearance of a squirrel (figure 14.3a). Lorises are found in continental Africa and Asia. They are small, have short faces, and lack tails. Tarsiers are found in Indonesia and the Philippines (figure 14.3b). The single species (*Tarsius syrichta*) is kitten sized, has well-developed hind legs that it uses for jumping from branch to branch, and has huge eyes in relation to the size of its head. When it is at rest, it clings vertically to a branch or trunk of a tree.

Suborder Anthropoidea

Members of the superfamily Ceboidea are called the **New World monkeys** and are found in Central and South America (figure 14.4). It is not uncommon to see a new-world monkey hanging by its distinctive *prehensile* (L. *prehensus,* to be seized), or grasping, *tail.* Members of the superfamily Cercopithecoidea are the **Old World monkeys** and are found in Africa, Asia, Japan, and the Philippines (figure 14.5a-b). Old World monkeys may use their tails as a balancing aid, but they are not prehensile. All monkeys are diurnal, and they move by running on all four legs across the tops of branches. New and Old World monkeys are believed to have arisen from a common ancestor and diverged when populations were separated by the south Atlantic Ocean, during the separation of South America and Africa 35 to 50 million years ago (*see box 12.2*).

The remaining primates are the apes (superfamily Hominoidea), which are also diurnal. The family Hylobatidae contains the lesser apes, including the gibbons (*Hylobates*). They live in Southeast Asia and are the most arboreal of all apes (figure 14.6a). They can run swiftly across the tops of branches, or swing from branch to branch, while suspended by hands and long arms. (This swinging locomotion is called *brachiation.*) Members of the the family Pongidae include the African and Asian apes and are often called "great apes." The Orangutan (*Pongo*) is a large, arboreal Asian ape, and although it has the same general design of the gibbon, its large size prevents much brachiation (figure 14.6b). The gorilla (*Gorilla*) and chimpanzee (*Pan*) are African apes (figure 14.6c). Gorillas and chimpanzees are basically terrestrial, and use a form of locomotion called "knuckle walking", in which the weight supported by the front appendages is borne on the anterior surfaces of flexed digits. The last hominoid family, Hominidae, contains the human species, *Homo sapiens.* The term "hominid" refers to members of the family Hominidae, and the term "hominoid" refers to members of the superfamily Hominoidea.

We now know that humans are more closely related to great apes than previously thought. Karyotype analysis (*see figure 8.12*) of humans and great apes reveals that 18 of

Table 14.1	The Classification of Primates*	
Taxon	**Common Name**	**Distribution**
Suborder Prosimii		
Superfamily Lemuroidea	Lemurs, aye-ayes	Madagascar
Superfamily Lorisoidea	Lorises	Africa and Asia
Superfamily Tarsoidea	Tarsiers	Southeast Asia
Suborder Anthropoidea		
Superfamily Ceboidea	New World monkeys	Central and South America
Superfamily Cercopithecoidea	Old World monkeys	Africa and Asia
Superfamily Hominoidea		
Family Hylobatidae	Gibbons, siamangs	Southeast Asia
Family Pongidae	Orangutan, ramamorphs, chimpanzee, gorilla,	Southeast Asia
	Proconsul, Dryopithecus	Africa
Family Hominidae	*Australopithecus,*	Africa
	Homo	Worldwide

*The classification of the primates is controversial. Some zoologists think that similarities in the reproductive tract and brain of tarsiers and monkeys warrants placing tarsiers in a new superfamily (Tarsioidea) in the Anthropoidea. Growing sentiment supports separating Asian apes (family Pongidae) and African apes (family Panidae or Gorillidae) because, based on genetic analysis, African apes are more closely related to hominids than they are to Asian apes. Other zoologists favor including African apes and hominids in one subfamily and Asian apes is a second subfamily of the family Hominidae.

(a)

(b)

the 23 chromosomal pairs are identical in all species. (Great apes actually have 24 pairs of chromosomes. The extra pair resulted from two arms of chromosome number two separating to form two pairs.) The other five pairs differ only by inversions and translocations that have occurred (*see figure 8.14*). DNA analysis also links humans and great apes. There is a 1.2% difference in DNA base sequences of chimpanzees and humans, and a 1.4% difference between gorillas and humans. Percent difference in the base composition of DNA in different species is called the *genetic distance* between the species. Chimpanzees and humans are as genetically similar as chimpanzees and gorillas! The genetic distance between humans and orangutans, on the other hand, is 2.2%, suggesting that orangutans are further removed from an ape/human common ancestry. A 1% genetic difference is typically found between members of the same genus in other animal groups. The divergence of the great ape and human lineages is believed to have occurred between 5 and 10 million years ago. [2]

This information has led some scientists to suggest that all great apes belong in one family, Hominidae. In such a family, Ponginae would be a subfamily containing the orangutans, and Homininae would be a subfamily containing gorillas, chimpanzees, and humans.

Stop and Ask Yourself

1. How did the following characteristics adapt early primates for an arboreal, insect-eating life style: Binocular vision? Friction ridges? Long tails? Opposable thumbs?
2. What are the two suborders of Primates? What are examples of each suborder? Why do some researchers believe that all apes should be included in a single family?

Figure 14.3 Suborder Prosimii. (a) A ring-tailed lemur, *Lemur catta*. Notice the eyes partially directed towards the front, allowing some stereoscopic vision; longer hind limbs than forelimbs; and somewhat elongated digits. (b) The tarsier, *Tarsius syrichta*, tends to sit upright in trees, with its head bent at a right angle to the vertebral column. The large eyes and flattened snout, allow good binocular vision, and the elongated fingers and toes are used to grasp branches.

Figure 14.4 A familiar New World monkey, the spider monkey (*Ateles*), in the superfamily Ceboidea.

(a)

(b)

(c)

(a)

(b)

Figure 14.5 Old World monkeys, superfamily Cercopithecoidea. (a) The Sykes monkey (*Cercopithecus*). (b) Olive baboons (*Papio*).

Figure 14.6 Some familiar apes in the superfamily Hominoidea. (a) The gibbon (*Hylobates*). (b) The orangutan (*Pongo*). (c) The chimpanzee (*Pan*).

Some Hominid Characteristics

It is a luxury for paleontologists to have more than a few fragments of fossilized bones with which to work. Wind, water, and scavenging animals rarely leave complete skeletal parts. Paleontologists must identify fragments of fossilized human bones, distinguish human bones from bones of closely related species, and distinguish human bones from the myriad bones that may have been dragged to a campsite by human hunters or other animals.

Many of the differences between hominids and other hominoids are the result of humans having *upright posture* and being **bipedal** (walking on two, rather than four, appendages). Apes may rise on their hind legs to get a better view of the landscape, and they may walk a short distance on two legs, but their most efficient form of locomotion is quadrupedal walking or brachiation. Bipedal locomotion is advantageous for three reasons. (1) Bipedalism frees hands to carry tools, food, firewood, and other items. (2) It provides a higher line of sight, which allows an animal to survey the landscape for food or predators. (3) Bipedalism, although less efficient for fast locomotion than quadrupedalism, is as energetically efficient as quadrupedalism for slower walking.

[3]

Bipedal locomotion requires adaptations for balancing, and adaptations that permit the weight of the body to be supported by two, rather than four, appendages. Axial support for vertebrates is provided by the vertebral column. In nonhuman primates, the vertebral column is arched slightly, and the bulk of the body is slung below the column. Walking upright with such a vertebral column would result in most of the body mass being anterior to the spine. The human vertebral column is curved, which brings the center of gravity more in line with the axis of support (figure 14.7a). In addition, the series of bones that make up the vertebral column become larger from the neck to the pelvis, as the force of compression increases.

A number of modifications of the head and neck are also adaptations for bipedal locomotion. In quadrupedal animals, the head is held more-or-less horizontally on the end of neck vertebrae. To support the skull, powerful neck muscles run between the vertebral column and the back of the skull. Elongate processes, called spinous processes, on vertebrae in the neck and chest regions and large, ridged surfaces on the back of the skull provide points of attachment for these muscles (figure 14.7b). The head of bipedal humans, however, is balanced on top of the vertebral column, and its weight is borne by the vertebrae. The spinous processes of neck vertebrae and the ridges of bone on the back of the skull are reduced in humans. The balancing of the skull on top of the vertebral column also results in the spinal cord exiting the skull from below, rather than from

behind. The opening in the skull for the spinal cord, the *foramen magnum,* is thus shifted ventrally.

The face of humans is less protruding than that of other apes. This design helps to balance the skull on the vertebral column, and results from a reduction in the size of teeth, and a reduction in the size of the bones that contribute to the large ridges above the eyes (the *supraorbital ridges*). Enlargement of the brain is reflected in the expansion of the anterior portion of the skull.

The bones of the appendages are also modified for upright, bipedal locomotion. The forelimbs of hominids are not used for knuckle walking or brachiation, and therefore, are shorter than in other apes. The pelvis is short and wide,

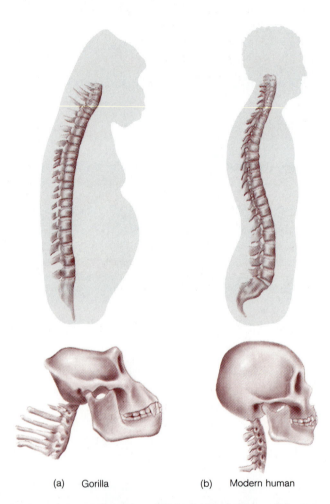

(a)   Gorilla   (b)   Modern human

Figure 14.7 A comparison of the axial skeleton of humans and the gorilla. (a) The large spinous processes of the cervical (neck) vertebrae of the gorilla serve as points of attachment for muscles that support the head. (b) The greater curvature of the human vertebral column places the vertebrae in line with the body's center of gravity. In humans, the foramen magnum (arrows) is located directly below the skull, so that the skull is balanced on the vertebral column. In gorillas (and other nonhuman apes), the foramen magnum is located nearer the back of the skull. Notice the heavy ridges above the eyes of the gorilla, and the ridges of bone at the back of the skull that serve as points of attachment for the muscles that support the head.

which transmits weight directly to the legs, maintains the size of the birth canal, and provides surfaces for the attachment of leg muscles (figure 14.8b,c). The upper bone in the hominid leg (the femur) is angled at the knee toward axis of the body (figure 14.8b). Angling of the femur places the feet under the center of gravity while walking, resulting in the smooth stride of hominids, compared to the bipedal "waddle" of other apes. Another obvious difference in appendages between hominids and other hominoids is that the foot of hominids is not used for grasping; the great toe is not opposable (figure 14.8a). Instead, the foot is narrowed and bears weight at the heel, the great toe, and the small toe—forming a stable tripod.

The inside of a skull contains information about the brain that it formerly housed. Plastic casts or natural earthen casts of the inside of a skull (endocasts) can be made to use for study of evolutionary changes in the brain. During evolution, the regions of the brain associated with memory, emotions, muscular control, and sensory pro-

cessing were the focus of change. In addition, complex social interactions in members of the primate lineage selected for rapid brain evolution. Brain development, in turn, promoted tool construction and use, language, complex subsistence patterns, and even more social complexity. Cranial capacity (brain size) has been an important criterion in the classification of hominoids.

Stop and Ask Yourself

3. What are the advantages of bipedalism for hominids?
4. In the evolution of bipedalism, how were the following modified: The vertebral column? The head and neck? The appendages and pelvis?
5. What aspects of primate life probably selected for increased intelligence?

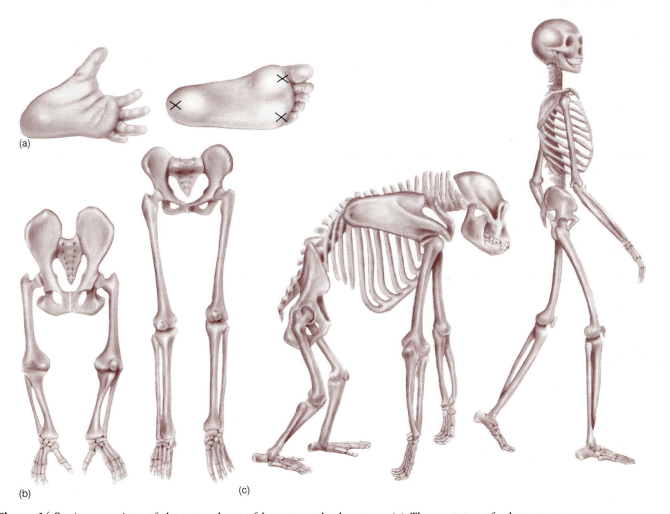

(a)

(b)   (c)

Figure 14.8 A comparison of the appendages of humans and other apes. (a) The great toe of a human is not opposable; rather, it is parallel to the other toes. Its weight-bearing surfaces (marked by ''x'') form a stable tripod. (b) The femur of humans is angled toward the body's center of gravity. The angle formed between the femur, and a line across the knee, is called the valgus angle, and is smaller in humans than in other apes. (c) The pelvic girdle of humans is relatively short, and transfers the weight of the upper body directly to the legs. In other apes, more weight is borne by the arms, and the pelvis is more elongate.

Hominid Evolution

The evolution of hominids was strongly influenced by geography and climate. By the time the first apes appeared, continental drift (*see box 12.2*) had separated the major continents of the earth. Apes were restricted to Asia and Africa, which were largely tropical at the time. Climate and geography changed, however, and these changes probably supplied the selection pressures that fostered evolutionary change. About 20 million years ago, global temperatures turned sharply cooler. Temperate regions expanded, and seasons became more pronounced. Geological uplift created highlands and dry belts across eastern Africa. Between 5 and 7 million years ago, global temperatures fell further. The formation of the west Antarctic ice sheet dropped sea levels and virtually dried up the Mediterranean Sea. As a result, the continuous tropical forests of Africa were broken into a mosaic of forest and vast grassland (savannah). Many primates, accustomed to feeding on fruit, insects, and leaves, faced extinction. During this time, the divergence between hominids and the other African great apes occurred, probably in the transition areas between deep tropical forests and savannah. Eventually, the primates that were present on the grasslands began to exploit grains, tubers, and grazing animals. Later stages of hominid evolution were affected by climatic changes associated with cycles of glaciation, which have occurred at approximately 100,000 year intervals over the last 2 million years.

Prehominid Ancestors

Fossils of 30 to 35 million-year-old, cat-sized primates probably represent the point of divergence between the monkey and ape lineages (figure 14.9). *Aegyptopithecus*

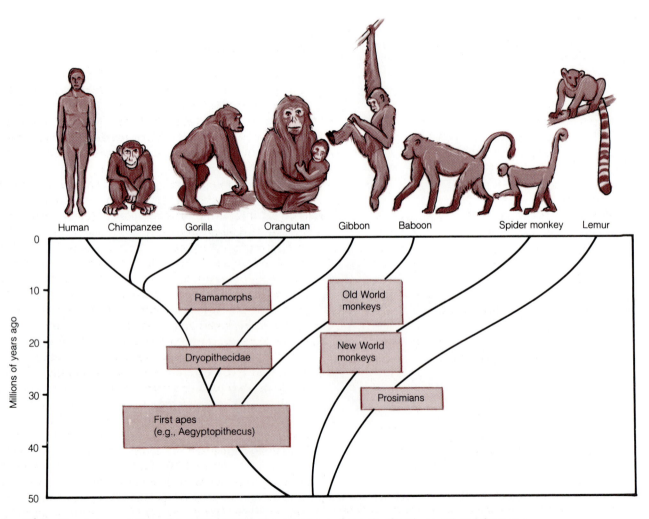

Figure 14.9 One interpretation of hominoid evolution. Evidence from the fossil record and from molecular biology suggests that the divergence of hominids and the African apes was more recent (5 to 10 million years ago) than previously thought. A very large gap in the fossil record exists between about 4 and 20 million years ago. Fossils from this time period, if they exist, could reveal much about early evolution of apes.

zeuxis (found in Egypt) shows characteristics intermediate between monkeys and apes. Later fossils (20 to 25 million years old) with definite apelike characteristics have been placed in the family Dryopithecidae (figure 14.10).

One group of Miocene apes, the "ramamorphs," includes the genera *Ramapithecus, Sivapithecus,* and the very large *Gigantopithecus.* They are of special interest because they show characteristics once presumed to be hominid. Fossils of *Ramapithecus* existed 14 million years ago, so this time was suggested as the time of the split between hominids and chimps and gorillas. In the early 1980s, however, a number of researchers have pointed out characteristics of ramamorphs that strongly suggest affinities to orangutans. Others suggested that the ramamorphs might be ancestral to both hominids and orangutans. Regardless of this debate, there is general agreement that the ramamorphs are not hominid. This distinction fits well with biochemical evidence of an ape/hominid divergence between 5 and 10 million years ago, rather than 14 million years ago.

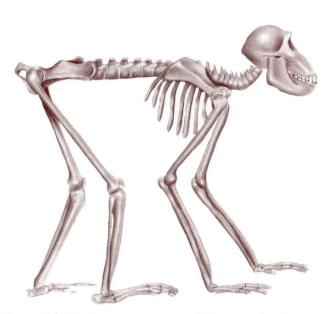

Figure 14.10 A reconstruction of *Proconsul africanus* (Family Dryopithecidae) from bones found by Mary Leakey, Alan Walker, and Martin Pickford. This individual was a young female that lived 18 million years ago. Its mixture of monkey and ape characteristics suggests that it may be similar to the common ancestor of Old World monkeys and humans.

Australopithecus

The oldest hominid fossils are placed into the genus ***Australopithecus*** (ME. *austr,* dawn + Gr. *pithek,* ape). They came from the Hadar region of Ethiopia (3.6 million years old), and Laetoli in Tanzania (3.7 million years old) (figure 14.11). Slightly older records were left when three bipedal hominids walked across 20 m of volcanic ash at Laetoli 3.75 million years ago (figure 14.12).

There are four australopithecine species recognized, however, some scientists feel that these assignments need revision. The group of oldest fossils (mentioned above) were named *Australopithecus afarensis.* Some of these were part of a 40% complete skeleton dubbed "Lucy" that was found at Hadar, Ethiopia. A second species, *Australopithecus africanus,* has been described from South Africa and East Africa. These fossils are approximately 2 million years old. The two australopithecine species are *Australopithecus robustus* and *Australopithecus boisei.* Fossils of the former have been found in South Africa and East Africa, and those of the latter have been found in East Africa. These two species were more robust than either *Australopithecus afarensis* or *Australopithecus africanus* (1.75 m tall and 60 kg compared to 1.3 m tall and 40 kg).

Australopithecines had cranial capacities between 400 and 600 cm³. (The smaller value is about the brain size of a modern chimpanzee and compares to a value for modern humans of 1,400 to 1,700 cm³.) There is no doubt that australopithecines were hominids. Their faces were less protruding than those of apes, and comparison of the teeth and jaws (the *dental arcade*) of chimpanzees, australopithecines, and modern humans shows australopithecines with characteristics intermediate between apes and humans (figure 14.13). The structure of the spine, pelvis, and leg

[4]

Casablanca sequence (Morocco)
Ain Hanech (Algeria)
Hadar (Ethiopia)
Melka Kunture (Ethiopia)
Gona (Ethiopia)
Omo (Ethiopia)
Gadeb (Ethiopia)
Koobi Fora (Kenya)
Chesowanja (Kenya)
Peninj (Tanzania)
Olduvai Gorge (Tanzania)
Laetoli (Tanzania)
Sterkfontein and Swartkrans (South Africa)

Figure 14.11 Southern and eastern Africa is often described as the cradle of humanity. Numerous sites (cited in the text) where important hominid fossils have been discovered are shown here.

Figure 14.11 redrawn, with permission, from N. Toth, "The First Technology." Copyright © 1967 by Scientific American, Inc. All rights reserved.

bones leaves little doubt that australopithecines were bipedal.

Australopithecines were probably never very far from trees. Their long forelimbs and curved finger and toe bones are apelike features that suggest significant time spent above the ground. The flat, grinding teeth of australopithecine fossils show wear patterns that suggest a diet consisting primarily of fruit (figure 14.14).

Australopithecine Phylogeny   Currently, there are three interpretations of early hominid phylogeny. Many believe that *Australopithecus afarensis* is ancestral to other australopithecines and to the lineage (*Homo*) that leads to modern humans (figure 14.15a). If so, the *Homo* lineage would have appeared between 2 and 3 million years ago. Others, notably the contemporary paleontologists

Richard and Mary Leakey, believe that the *Homo* lineage is much older. They believe that some early fossils, now assigned to *Australopithecus afarensis,* are really a very early species of *Homo.* If this is the case, there must be an undiscovered common ancestor for the *Australopithecus* and *Homo* lineages (figure 14.15b). A third interpretation derives the *Homo* lineage through *Australopithecus africanus* (figure 14.15c). Extinction of the australopithecine lineage occurred slightly less than 1 million years ago.

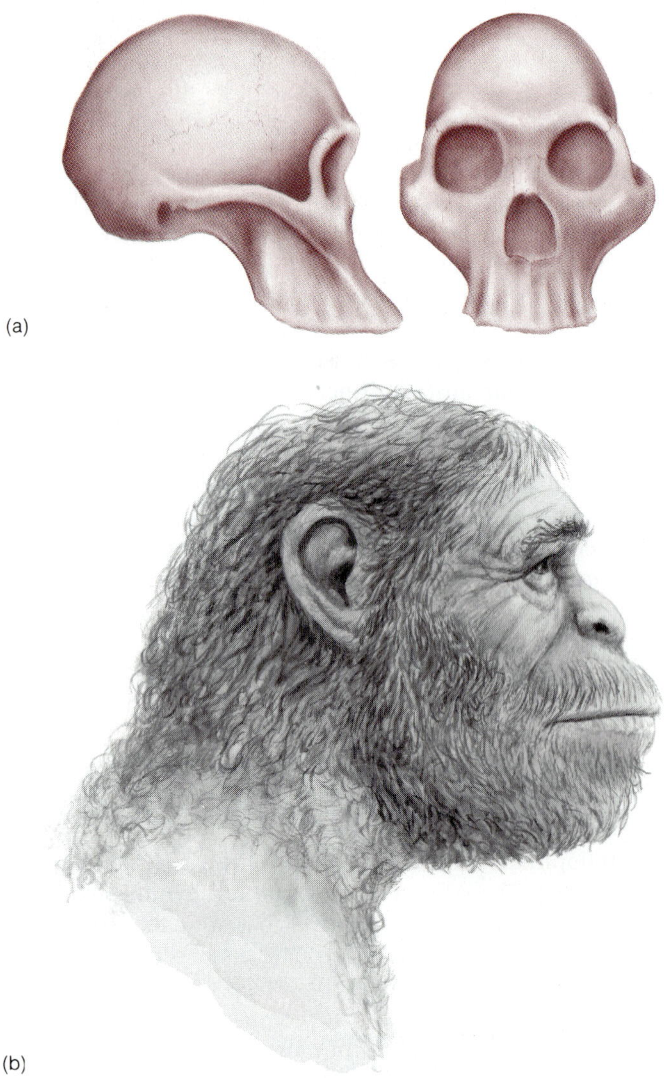

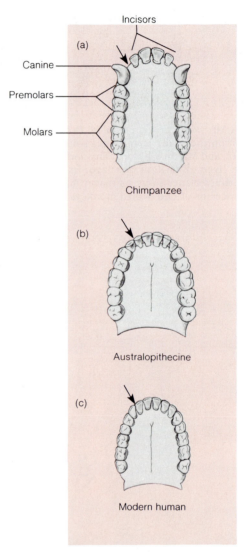

Figure 14.13   A comparison of the dental arcades of (a) a chimpanzee, (b) an australopithecine, and (c) a modern human. A chimpanzee's jaw is rectangular in outline, has large canine teeth, and relatively smaller molars. A gap between incisors and canines, called the diastema (arrows), is relatively large in chimpanzees. In humans, the jaw is rounded, molars are larger relative to incisors and canines, and the diastema is gone. Australopithecines are intermediate in all of these characteristics. (Approximately 45% of australopithecine skulls show a diastema.)

(a)

(b)

Figure 14.12   The skull (a) and an artist's reconstruction (b) of *Australopithecus africanus.*

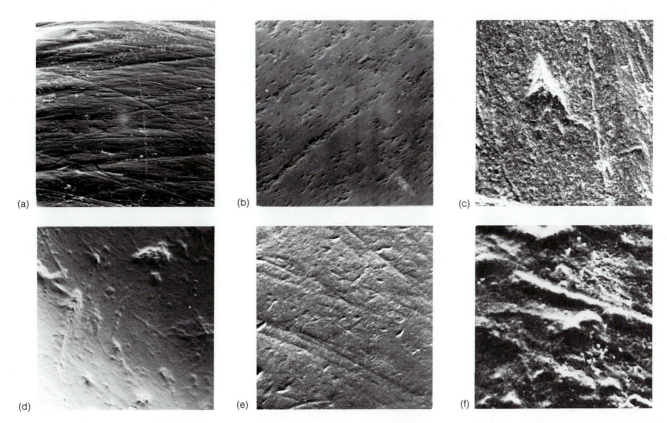

Figure 14.14 Electron micrographs of microwear patterns of the teeth of (a) a grazer (white rhinoceros), (b) a carnivore (cheetah), (c) a bone-crunching scavenger (hyena), and (d) a fruit eater (orangutan). The microwear patterns of (e) *Australopithecus* teeth are most similar to those of fruit-eating animals. (f) *Homo erectus* teeth show deep scratches and pits, consistent with wear patterns of a meat eater. Some of the pitting, however, was probably due to soil particles attached to underground tubers.

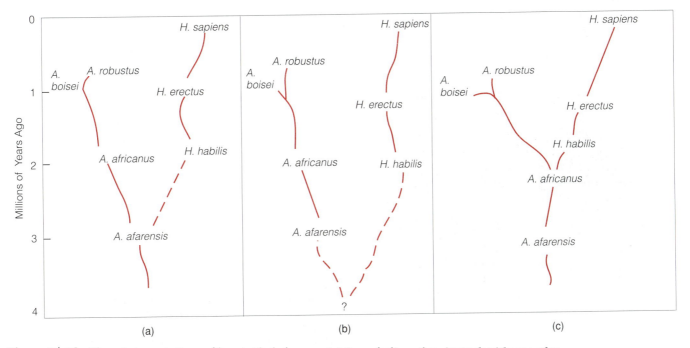

Figure 14.15 Three interpretations of hominid phylogeny. (a) Some believe that *Australopithecus afarensis* is ancestral to other australopithecines, and to the *Homo* lineage. (b) According to the ''Leakey Hypothesis,'' the *Homo* lineage may be much older than was previously thought and may have an unknown ancestor in common with the australopithecines. (c) Others believe that *Australopithecus africanus* was ancestral to later australopithecines and to the *Homo* lineage.

Homo habilis

In the early 1960s, and again in the early 1970s, some important fossil discoveries were made in Kenya and Tanzania by Louis and Mary Leakey and their son, Richard. One of these finds was a fragmented, but nearly complete skull—generally referred to as skull 1470 (figure 14.16). These fossils range in age from 1.75 to 1.9 million years old. The Leakeys' named these hominids **Homo babilis** (L. *homo,* human + *habilis,* skillful) and used the antiquity of these fossils to support their contention that the *Homo* lineage is much older than previously assumed.

Homo habilis was similar in overall size to *Australopithecus,* though the higher, rounder cranium gives it a more modern appearance. The cranial capacity ranged in size from 650 to 800 cm³, significantly larger than that of australopithecines. Molars and premolars were reduced, compared to australopithecines, and wear patterns on the teeth suggest that *Homo habilis* was primarily a fruit eater. Curved fingers and powerful, grasping hands indicate that this hominid was a climber. *Homo habilis* walked bipedally and probably made frequent trips into trees for food and to escape predators. It most likely took refuge in trees at nightfall, much like modern baboons and chimpanzees.

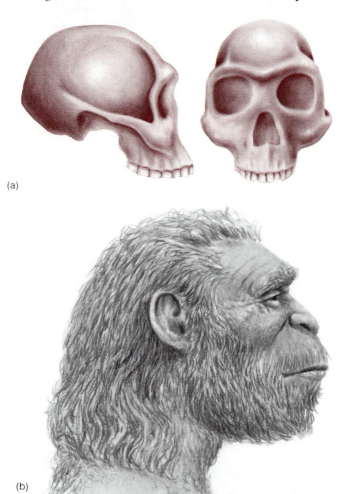

(a)

(b)

Figure 14.16 *Homo babilis.* (a) Fragments of skull 1470 were discovered near Lake Turkana, Kenya in 1972. (b) An artist's reconstruction of *Homo babilis.*

Figure 14.17 Oldowan tools include hammerstones, choppers, polyhedrons, scrapers, and discoids.

The use of tools is not unique to humans. Chimpanzees [5] modify and use twigs to help extract termites from nests. They wield sticks and rocks in threatening displays. Australopithecines probably did the same. Records of primitive tool making are very incomplete, because wood is less likely to be preserved than rock. However, when tools began to be made from rocks, important records were left. Primitive stone "tool kits" date to 2.5 million years ago (figure 14.17a). These primitive tools were probably made by *Homo babilis* with relatively few blows on a rock by a second rock, called a hammerstone, and belong to a tool technology called the **Oldowan industry.** That *Homo babilis* made, and used primitive tools is also suggested by their relatively large brain and manipulative hands—thus the name "skillful human." Although *Homo babilis* apparently did not make meat a substantial part of its diet, it may have scavenged dead animals for meat, hide, and tendons.

The time of the extinction of *Homo habilis,* about 1.6 million years ago, corresponds to the time when another species, *Homo erectus,* first appeared. *Homo habilis* was probably the ancestor of *Homo erectus.*

Homo erectus

The oldest **Homo erectus** (L. *homo,* human + *erectus,* upright) fossils of are from 1.6 million-year-old deposits in Koobi Fora, which possibly was the birthplace of this hominid (*see figure 14.11*). *Homo erectus* spread rapidly, leaving 750,000-year-old remains in China ("Peking Man") and Indonesia. Recent fossils from England and Germany are dated about 200,000 years old (figure 14.18).

The earliest cranial fossils of *Homo erectus* have a capacity of 800 to 900 cm³. By 200,000 years ago, the brain had increased to 1,100 cm³. The dentition of *Homo erectus* was more modern than that of *Homo babilis.* Except for its prominent supraorbital ridges, sloping forehead, shorter stature, and somewhat heavier bones, *Homo erectus* would have appeared very modern.

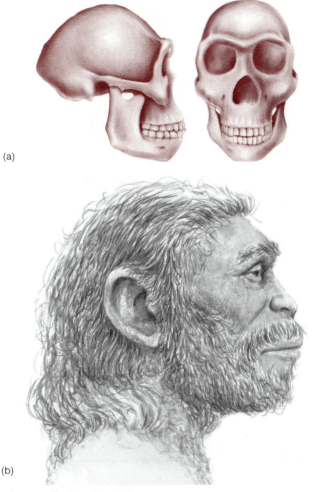

(a)

(b)

Figure 14.18 The skull (a) and an artist's reconstruction (b) of *Homo erectus.*

Figure 14.19 The Acheulean-style tools include hand axes, cleavers, and flakes.

One site in China had ash accumulations six meters thick! Campsites become more organized between 200,000 and 1.5 million years ago. The earliest campsites were marked only by scattered stones and fractured bones. Later campsites contained organized bone accumulations, and (in France) a 3 by 9 m hut. Campsites such as these are the products of intelligent, socially interdependent hominids.

Homo erectus existed for more than 1 million years, far longer than our species has been in existence. The changes in hominid life-styles that occurred during that time were as significant as the advent of bipedalism and language (discussed later in this chapter).

Homo sapiens

The origin of our species, *Homo sapiens* (L. *homo,* human + *sapiens,* wise), is shrouded in mystery. Few human fossils have been found in the critical period between 200,000 and 400,000 years ago. The earliest fossils that are undisputedly from members of our species are about 100,000 years old. A gradual transition from *Homo erectus* is a widely accepted hypothesis for the origin of *Homo sapiens.* The traditional classification of *H. sapiens* includes the two subspecies described next. This classification is not universally agreed upon. Recent research has led some authorities to conclude that these "subspecies" are actually separate species that were probably capable of hybridization. [6]

Neanderthals In 1856, fossils of ***Homo sapiens neanderthalis,*** were found in Neander Valley near Dusseldorf, Germany (figure 14.20). Neanderthals were robust, as evidenced by the large cross-sectional area of bones and stout points of muscle attachment. Skull bones were thick, with prominent supraorbital ridges. The face projected from a sloping forehead to a broad nose and massive jaws. A brain of about 1450 cm³ was equal in size to the brain of contemporary humans. Below the neck, Neanderthals and contemporary humans would have been difficult to distinguish.

Dispersal required more than movement to new areas because, *Homo erectus* was pioneering the habitation of temperate and even arctic climates. This first, great human dispersal could not have happened without numerous other pioneering achievements.

The remains of *Homo erectus* are found at former campsites along with scattered animal bones and artifacts. (*Artifacts* include any objects modified by humans.) That *Homo erectus* hunted is evidenced by the bones of large cattle, elephants, and deer found at their campsites; by wear patterns on their teeth (*see figure 14.14*); and by cutting tools found at campsites. Tools of *Homo erectus* belong to a technology called the **Acheulean industry** (figure 14.19; box 14.1). Smaller stone flakes served as knives; larger, pointed tools were used as hand axes (hallmarks of Acheulean industry), and truncated tools were used as cleavers. The use of each of these tools is documented by characteristic markings left on animal bones. *Homo erectus* campsites also document the first use of fire by hominids. The oldest evidence of fire use is from 1.4 million-year-old sites in Kenya. (This claim, however, is disputed.) In China, France, and Hungary, hearths and ash are commonly associated with *Homo erectus* campsites.

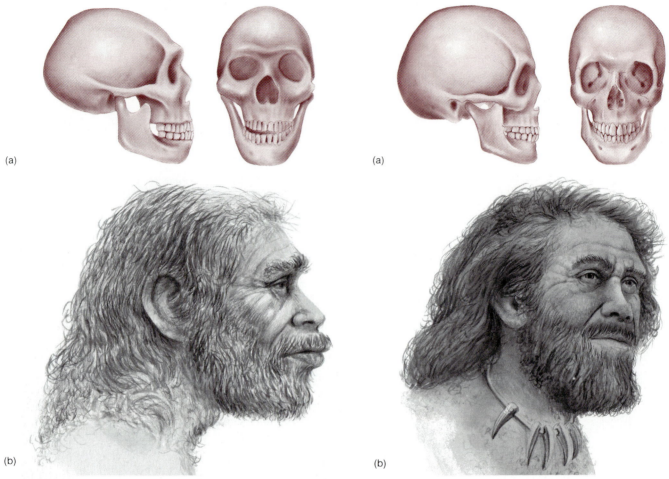

Figure 14.20 The skull (a) and an artist's reconstruction (b) of *Homo sapiens neanderthalensis.*

Figure 14.21 The skull (a) and an artist's reconstruction (b) of *Homo sapiens sapiens.*

The place of origin of Neanderthals is disputed. Some researchers believe Neanderthals arose in South Africa, others believe they arose in the Mideast. By 70,000 years ago, Neanderthals were living in Europe and England, in spite of arctic conditions brought on by glaciation. They dwelt in caves, in shelters made of wood, or, when no wood was present, in shelters made from the bones and skin of the mammoths and wooly rhinoceroses that they hunted. A diversity of refined tools, made of stone, bone, antler, and ivory, show that Neanderthals had a strong sense of style and order. Evidence of clothing also has been found at these campsites.

Neanderthals were intelligent, they had an awareness of self, and a sense of spiritual existence. At Le Moustier, France, a teenage boy was buried on his right side, as if in a sleeping position—head lying on forearm. A pile of flints served as a pillow, a finely crafted stone ax was at his side, and a supply of meat was nearby, evidenced by the wild cattle bones present. All of these were apparently for use in a coming journey.

Cro-Magnons Neanderthals disappeared about 40,000 years ago in eastern Europe and the Mideast, and about 35,000 years ago in western Europe. The race of humans to which we belong, ***Homo sapiens sapiens,*** took Neanderthal's place (figure 14.21). The origins of earliest members of the Cro-Magnon race is puzzling. Recent evidence from studies of human mitochondrial DNA suggests that our race is at least 200,000 years old and that we trace our ancestry to African populations of *Homo erectus.* Though this conculsion is not undisputed, it implies that Cro-Magnons spread throughout Africa, Europe and Asia, and that Cro-Magnons would have been contemporaries of Neanderthals. Eventually, Neanderthals disappeared, possibly because of direct conflict with Cro-Magnons, or as a result of competition for food or other resources. A second scenario for the rise of the Cro-Magnon race now seems less likely. In this "local continuity hypothesis" the Neanderthals are envisioned as a transitional race, which eventually gave rise to modern humans.

Cro-Magnons had reduced supraorbital ridges, steep foreheads, high and round crania, and prominent chins. More refined stone working appeared with Cro-Magnons. Thin blades, awls, and points are among about 100 identifiable implements found at their campsites.

The prehistoric art that has been found is the work of

Cro-Magnons. Some of their stone working lost utilitarian value and took on symbolic meanings. They carved antlers and ivory and were responsible for European cave paintings. The images in these paintings are of animals hunted by Cro-Magnons (figure 14.22). Cave paintings were probably associated with rituals designed to bring good fortune to the hunters, because caves show little other evidence of human habitation.

Homo sapiens sapiens was responsible for a second major dispersal by humans. This dispersal went as far as the New World and Australia. During the last ice age, sea levels dropped about 100 m and exposed a land bridge, called Beringia, between Siberia and Alaska (figure 14.23). Beringia was a treeless tundra, but it was rich in grasses. Grazing animals, such as mammoths, bison, elk, and antelope crossed to North America. Predators, including humans, followed. An artifact, called a *clovis point,* is common at 11,000-year-old North American sites (figure 14.24). Other artifacts from the Yukon are tentatively dated at 27,000 years ago. Migration to Australia occurred 32,000 years ago, and would have meant sea trips of up to 90 km.

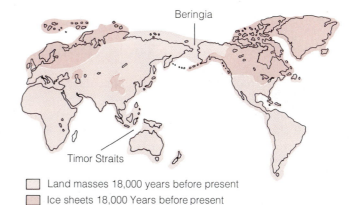

Land masses 18,000 years before present

Ice sheets 18,000 Years before present

Figure 14.23 At the climax of the last glaciation, approximately 18,000 years ago, the land mass of Beringia was exposed and allowed migration of early, modern humans to North America. Migration to Australia would have involved a sea journey.

Stop and Ask Yourself

6. What climatic changes took place about 20 million years ago that fostered hominid evolution?
7. What modern primates are the likely descendants of the ramamorphs?
8. What are the three interpretations of australopithicine phylogeny?
9. What hominid(s): First used fire? Were the first to be associated with Oldowan industry? Were the first to be associated with Acheulean industry? Were the first artists? Were responsible for major dispersals? Lived in caves?

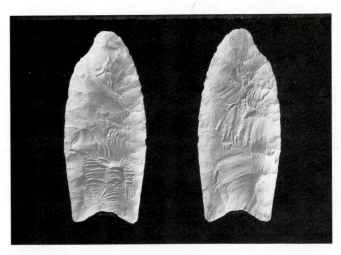

Figure 14.24 Clovis points, such as these, have been found throughout North America. They are clear evidence that humans were in North America 11,000 years ago. Still older artifacts have been reported from parts of Canada.

Figure 14.22 A Cro-Magnon cave painting from a cave at Lascaux, France. Paintings like these may have been associated with ceremonies to promote hunting success.

Cultural Evolution

At what point in hominoid evolution can we begin using the term "human?" In previous discussions, the term was used sparingly in reference to members of the family Hominidae. An important criterion in defining "human" is the development of culture. **Culture** is a system of learned behaviors, symbols, beliefs, institutions, and technology characteristic of a group that is transmitted through generations. Although a rudimentary capacity for culture exists in other animals, culture is generally limited to humans.

Cultural evolution is said to have occurred when cultures change over time. It is evolution in the sense of change, but it may not be organic evolution because it can be based entirely upon learning rather than genetic changes. Cultural evolution may be "Lamarckian" because

[7]

Box 14.1

Tools, Handedness, and Intelligence

The earliest tools may have been unmodified stones, used for pounding plant material during food processing. Accidental fracturing of these rocks would have produced sharp flakes of stone. At some point, the flakes may have been found useful for cutting plant material, shaping wooden implements, and butchering scavenged animals. Later, flakes may have been intentionally produced. When one stone, called a hammerstone, is used to strike another stone from which a tool is made, sharp flakes are produced (box figure 14.1). The stone that is struck is called the core. The flakes that chip off in the process were once thought to have been waste. It is now believed that the flakes were just as important in early tool technologies as the finished core. Flakes and finished cores are shown in figure 14.19.

Early hominids must have developed a certain mental capacity and manual dexterity prior to the advent of tool technologies. These earliest tool makers had to conceive of flake production and begin to plan ahead so that lava rock could be transported from stream beds to campsites and butchering sites. Later, however, tool technologies must have promoted the selection of increasing mental abilities. Increased standardization of tools required that the tool maker "see" the product in an unmodified piece of lava and systematically extract that product.

Nicholas Toth, a contemporary anthropologist, found that the flakes produced during the making of single-faced scrapers differed, depending upon whether the worker was right-handed or left-handed. After examining ancient stone flakes, he concluded that a tendency for handedness in humans was developing between 1.4 and 1.9 million years ago. (Modern humans are 90% right-handed.) This finding is significant because handedness is associated with differentiation of the two halves of the human brain. In most modern humans, the left cerebral hemisphere controls muscles on the right side of the body and dominates language and time sequencing processes. The making of advanced stone tools requires sequencing of types of blows with a hammerstone. A possible correlation between advanced stone technologies and language development is discussed later in this chapter. The right side of the brain controls muscles on the left side of the body, and is specialized for spatial patterning.

These findings suggest that natural selection may have favored the evolution of a brain with increased capacity for conceiving of, and planning for the future. The evolution of these kinds of intellectual functions must have preceded the dispersal of *Homo erectus* to colder climates. The fashioning of transport vessels, and the development of techniques for winter food storage would have been essential for successful dispersal.

Box figure 14.1 (facing page) Bifacial flaking was a technique used in Oldowan toolmaking. (a) A hammerstone was held in one hand and used to strike a rock core held in the other hand. (b) After a flake was struck off, the core was turned so that the flake scar (1) could serve as a platform for striking off the next flake (2). (c,d) Further strikes yielded a core with a bifacial edge (3). Flakes also had sharp edges and were used as smaller cutting tools.

practices are acquired in one generation and passed to subsequent generations. On the other hand, genetic changes can certainly influence learning. It is very difficult to determine the extent to which cultural changes are genetically based.

Cultural evolution often occurs very rapidly because ideas, technologies, and art are carried with individuals as they migrate. The time frame for cultural evolution is, therefore, measured in hundreds of years rather than millions of years. The changes that have occurred in hominids in the last few hundred thousand years have been largely a result of cultural evolution. When faced with environmental problems, humans have sought shelter, learned to use new food sources, and often moved to new territories.

Human history is marked by three periods during which specific life-styles were prominent. Rapid cultural changes, cultural revolutions, mark the transitions between these periods.

Hunter Gatherers

Prior to hunting and gathering, the first of the three distinctive life-styles, primates were primarily fruit eaters. Although they often may have existed in social groups, each individual probably foraged for its own food. Food sharing may have begun as early as 2 million years ago, and may have marked the beginning of the hunting and gathering life-styles for hominids. Hunting and gathering was certainly common by the time of *Homo erectus,* and may have

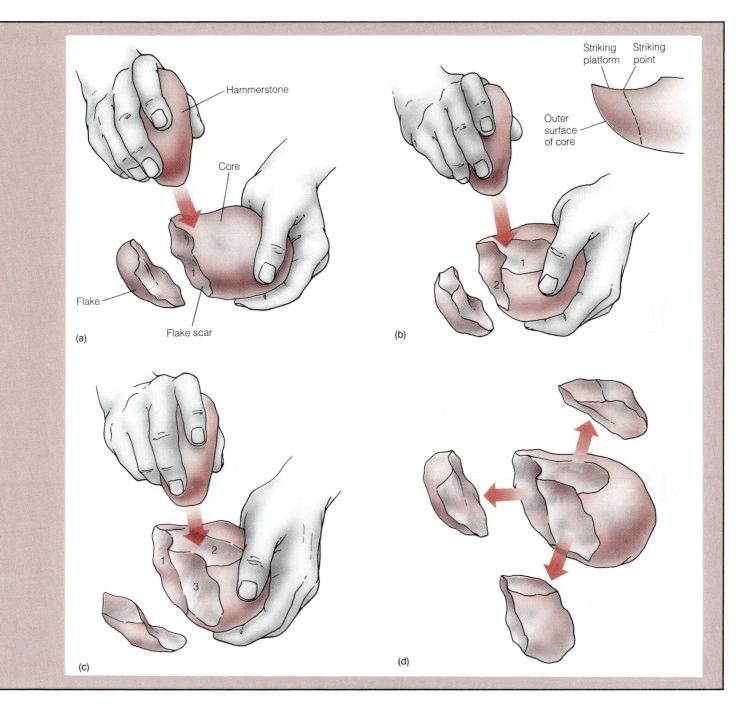

(a) Hammerstone · Core · Flake · Flake scar

(b) Striking platform · Striking point · Outer surface of core

(c)

(d)

been a way of life for *Homo habilis*. We can look at ancient campsites and the few contemporary cultures that retain this life-style for clues to its nature.

Primitive **hunters and gatherers** were probably grouped into communities of 25 to 30 individuals. In a hunting and gathering existence, approximately 10 km² are required to support each person in a group. As group size increases beyond 25, it becomes more difficult to travel the distances required to supply food for the group. Food consisted primarily of plant material, supplemented by meat. Temporary campsites served as a home base. Women and children probably gathered roots, berries, and the like around the vicinity of the camp. Men ranged further when hunting for meat. The result was that food could be gath-

ered from an area larger than a single person could cover. Technologies were expanded to include containers of bark or skin, digging sticks to help extract tubers from the ground, and stone implements for killing and processing animals.

Language Even though spoken language does not leave a fossil record, there are clear indications that language evolved in hunters and gatherers. Evidence for language development comes from two sources. First, cranial casts of primitive skulls can be examined for evidence of vocal control centers. The development of the regions of the brain devoted to speech in modern humans leave the skull slightly vaulted. *Homo habilis* (skull 1470) has these [8] characteristics. The development of some potential for speech, however, does not necessarily mean that spoken language existed. Human artifacts also give indications of spoken language. Tools show increasing sophistication and standardization through time. Rule systems for tool construction seem to have developed. It is unlikely that complex rule systems could be perpetuated only by modeling what one sees. Similarly, the development of art is often cited as evidence of spoken language. Art, defined in an archeological context as abstract artifacts, conveys meaning only in a cultural context, and that meaning is most effectively conveyed through language.

By 40,000 years ago, all of these activities were present and rapidly expanding. Language faculties were probably well developed by this time. Simple language was likely present earlier, and communication in the form of gestures and facial expressions must have occurred in very early primates.

As suggested above, language allowed early humans to communicate about objects in the environment and to store and classify information in memory. It facilitates instruction and allows humans to bargain, discuss, and argue. The development of language was probably the most significant event in our cultural evolution.

Agriculture

The next major life-style change and step in cultural evolution was a shift toward an agricultural economy, the **agricultural revolution.** Scarce resources (brought on by increasing population or over-hunting) influenced the transition to farming. Whereas hunters and gatherers require about 10 km² of land per person, agriculturally based economies require only about 0.5 km² of land per person. (Irrigation reduces this requirement to 0.1 km².) Alternatively, the agricultural revolution may have been a response to expanding commerce between neighboring groups. Increased pressure to produce food for trading may have launched humans into intensive farming.

Earliest agriculture may have involved opportunistic, small-scale cultivation of grain crops to supplement hunting and gathering economies. Even small-scale agriculture would require that nomadic life-styles be modified. As campsites became permanent communities, the transition to large-scale agriculture probably occurred rapidly. The agricultural revolution began 10,000 years ago in the Mideast, in the region known as the fertile crescent—Israel, Jordan, Syria, Turkey, and Iraq. It began in China 7,000 years ago, and in Central America about 5,000 years ago. This revolution transformed small communities into large, urban areas. The domestication of some animals and the advent of metallurgy (about 8,000 years ago), made agricultural practices (and also warfare) more efficient. As people migrated, they took agricultural practices with them. All too often, they clear-cut land to make new land available for their crops, leaving depleted soils behind them.

Industrialization

The **industrial revolution** began in the eighteenth century and ushered in a third life-style change and stage in cultural evolution. One cannot help but gaze with astonishment at the results of this third era in human history. The earth is marked with both the rewards and the scars of the industrial revolution. Many changes have promoted the health and welfare of at least some humans. On the other hand, no animal has so drastically marred the face of the earth, or threatened so many species, including our own, with extinction. This era is a reminder that biological evolution and cultural evolution have another common feature—progress is not guaranteed by either.

Before leaving the topic of cultural evolution, brief mention should be made of a revolution in progress. Chapters 7 to 9 describe the nature of the genetic revolution, and many believe that it will impact our culture just as strongly as the agricultural or industrial revolutions. Only the advantage of a historical perspective will reveal its full impact.

Stop and Ask Yourself

10. What is culture?
11. In what way is cultural evolution different from biological evolution?
12. What was the life style of hunter-gatherers? Which hominids lived that life-style?
13. Where, when, and why did the agricultural revolution begin?

■ Summary

1. Humans are members of the vertebrate order, Primates. Ancestral primates arose about 65 million years ago as arboreal insectivores.
2. Primates are classified into two suborders. Prosimii includes the lemurs, lorises, and tarsiers. Anthropoidea includes the New World monkeys, Old World monkeys, and apes.
3. Hominids are characterized by an upright posture and bipedalism.
4. Numerous skeletal modifications go hand in hand with bipedalism.
5. Hominid evolution was fostered by a cooling and drying climate that transformed tropical forests into a mosaic of forest and savannah.

6. The divergence between apes and Old World monkeys occurred 20 to 30 million years ago. The gorilla/chimpanzee and hominid lineages diverged between 5 and 10 million years ago.

7. Members of the genus *Australopithecus* are the earliest known hominids. They originated in Africa 3.6 million years ago, and became extinct 1 million years ago. They, or an unknown common ancestor of the australopithicine and *Homo* lineages, may be ancestral to all later hominids.

8. *Homo habilis* existed between 1.6 and 2 million years ago. (Some believe this genus is much older.) It had a large brain and used primitive tools.

9. *Homo erectus* originated in Africa about 1.6 million years ago and spread into Europe and Asia. Many accomplishments are associated with *Homo erectus,* including the extensive use of fire, fashioning of specialized tools, and highly organized campsites. Its extinction occurred about 500,000 years ago.

10. Fossil records show that our species is at least 100,000 years old. Neanderthals were replaced by Cro-Magnons about 35,000 years ago. Further refinements in stone working and art and dispersals to North America and Australia are a part of Cro-Magnon heritage.

11. Humans are characterized by culture, which is a system of learned behavior, beliefs, institutions and technology that is transmitted through generations.

12. The change of cultures through time is called cultural evolution.

13. Human prehistory and history is marked by hunter-gatherer, agricultural, and industrial economies.

Key Terms

Acheulean industry (p. 215)	*Homo sapiens sapiens*
agricultural revolution	(p. 216)
(p. 220)	hunters and gatherers
Anthropoidea (p. 205)	(p. 219)
Australopithecus (p. 211)	industrial revolution
bipedal (p. 208)	(p. 220)
cultural evolution (p. 219)	New World monkeys
culture (p. 219)	(p. 205)
Homo erectus (p. 214)	Old World monkeys
Homo habilis (p. 214)	(p. 205)
Homo sapiens	Oldowan industry (p. 214)
neanderthalis (p. 215)	primates (p. 204)
	Prosimii (p. 205)

Critical Thinking Questions

1. Why were bipedalism and increased brain size important in human evolution? Is it important to try to determine which (if either) came first? Why or why not?

2. Do you think tool use selected for increased intelligence, or increased intelligence (perhaps selected for by social behaviors) promoted tool use? Explain.

3. Suppose that some newly discovered tools were dated at 3.5 million years before present. What implications would

that discovery have for interpretations presented in figure 14.15?

4. Discussions regarding the degree to which our genetic make up affects our behavior are a source of heated debate among sociobiologists. Why is this controversy significant for those who study human evolution. (See the books by Wilson and Kitcher, cited below.)

Suggested Readings

Books

Campbell, B. 1988. *Humankind Emerging*, 5th ed. Glenview: Scott, Foresman & Co.

Campbell, B. 1985. *Human Evolution*. New York: Aldine Publishing Company.

Delson, E., ed. 1985. *Ancestors: The Hard Evidence.* New York: Alan R. Liss.

Gould, S. J. 1981. *The Mismeasure Of Man*. New York: W.W. Norton.

Kitcher, P. 1985. *Vaulting Ambition: The Quest for Human Nature*. Cambridge: MIT Press.

Leakey, M. D. 1984. *Disclosing The Past: An Autobiography.* New York: Doubleday.

Lewin, R. 1984. *Human Evolution: An Illustrated Introduction*. New York: W.H. Freeman and Co, Inc.

NSF Mosaic Reader. 1983. *Human Evolution*. Garden City Park: Avery Publishing Group Inc.

Wilson, E. O. 1978. *On Human Nature*. Cambridge: Harvard University Press.

Articles

Cartmill, M., Pilbeam, D., and Isaac, G.. 1986. One hundred years of paleoanthropology. *Amercian Scientist* 74:410–419.

Gould, S. J. 1987. Bushes all the way down, we are all products of a recent African twig. *Natural History* 96:12–16.

Lewin, R. 1987. Four legs bad, two legs good. *Science* 235:969–971.

Lewin, R. 1987. The origin of the modern human mind. *Science* 236:668–670.

Lewin, R. 1987. Africa: cradle of modern humans. *Science* 237:1292–1295.

Lewin, R. 1988. Modern human origins under close scrutiny. *Science* 239:1240–1241.

Lovejoy, C. O. Evolution of human walking. *Scientific American* February, 1989.

Mereson, A. 1988. Monkeying around with the relatives. *Discover* 9:26–26.

Pilbeam, D. The descent of hominoids and hominids. *Scientific American* March, 1984.

Simons, E. L. 1989. Human origins. *Science* 245:1343–1350.

Toth, N. The first technology. *Scientific American* April, 1987.

Turner, C. G., II. Teeth and prehistory in Asia. *Scientific American* February, 1989.

Walker, A. and Teaford, M. The hunt for *Proconsul. Scientific American* January, 1989.

Animallike Protists and Animalia

Part IV

One person can never fully appreciate the impressive diversity in the animal kingdom. Zoologists, therefore, must specialize—devoting their lives to the study of particular animal groups. Knowledge of all aspects of the biology of animals is invaluable because it reveals the delicate balances in nature and gives clues to how we can preserve those balances. Furthermore, as zoologists learn more about animal groups, information that directly affects human welfare emerges. For example, from research on the nervous systems of squid and cockroaches comes much of what we know about nerve cells and many nervous disorders. In spite of the work of generations of zoologists, there remains a wealth of unanswered questions about animals. Many species, especially in the tropics, have not been described. Some species may hold keys to unlocking the secrets of cancer, AIDS, and other diseases. Other species may provide insight into managing world resources.

Chapters 15 to 31 present an overview of the known animallike protists and animal phyla. Specifically, chapter 15 is an introduction to animal taxonomy, which is the study of the naming of organisms, and to the basic organization of animal bodies. Chapter 16 then covers the animallike protists and chapters 17 to 31 survey the animal phyla. These chapters are the beginning of an exciting journey into the diversity of the animal kingdom. Perhaps one of these chapters will captivate your attention and you will join the thousands of zoologists who have spent their lives studying a portion of the animal kingdom.

15

Animal Classification, Organization, and Phylogeny

Concepts

1. Order in nature allows systematists to name animals and establish evolutionary relationships among them.
2. All organisms can be placed in one of five kingdoms based on whether their cells are prokaryotic or eukaryotic; whether they are truly multicellular or not, and whether they get their food through absorption, ingestion, or autotrophy.
3. Animal body plans can be categorized based upon how cells are organized into tissues, and how body parts are distributed within and around an animal.
4. Evolutionary relationships are represented by evolutionary tree diagrams.

Have You Ever Wondered:

[1] why zoologists use scientific names, rather than common names, for the animals they study?
[2] why zoology courses often cover organisms such as *Amoeba* and *Paramecium* when these organisms are really not animals?
[3] why sedentary animals, such as sea anemones, do not have heads and tails?
[4] why some animals (e.g., jellyfish) never have highly muscular bodies?
[5] why evolutionary tree diagrams have become standard features of most zoology textbooks?

These and other useful questions will be answered in this chapter.

This chapter contains underlined evolutionary concepts.

 One of the cornerstones of science that is virtually unchallenged is that there is order in nature. The order found in living systems is a natural consequence of the shared evolutionary processes that influence life. The purpose of this chapter is to describe how the classification of animals and the basic organization of their bodies reflect that order.

Classification of Organisms

One of the characteristics of modern humans is our ability to communicate with a spoken language. Language not only allows us to communicate, but it also helps us encode and classify concepts, objects, and organisms that we encounter. To make sense out of life's diversity, we need more than just names for organisms. A potpourri of over a million animal names is of little use to anyone. To be useful, a naming system must reflect the order present in nature. **Systematics** (Gr. *systema*, system + *ikos*, body of facts) or **taxonomy** (Gr. *taxis*, arrangement + L. *nominalis*, belonging to a name), is the study of the classification (naming and grouping) of organisms. These studies result in the description of new species, and in the organization of animals into groups (taxa) based upon degree of evolutionary relatedness.

A Taxonomic Hierarchy

Our modern classification system is rooted in the work of Carolus Linnaeus (1707–1778). His binomial system (*see chapter 1*) is still used today. Linnaeus also recognized that different species could be grouped into broader categories based on shared characteristics. A group of animals that shares a particular set of characteristics forms an assemblage called a **taxon.** For example, a housefly (*Musca domestica*), though obviously unique, shares certain characteristics with other flies. (The most important of these

Table 15.1	Taxonomic Categories of a Human and a Dog	
Taxon	**Human**	**Domestic Dog**
Kingdom	Animalia	Animalia
Phylum	Chordata	Chordata
Class	Mammalia	Mammalia
Order	Primates	Carnivora
Family	Hominidae	Canidae
Genus	*Homo*	*Canis*
Species	*sapiens*	*familiaris*

being a single pair of wings.) Based on these similarities, all true flies form a logical, higher group. Further, all true flies share certain characteristics with bees, butterflies, and beetles. Thus, these animals form an even higher taxonomic group. They are all insects.

Linnaeus recognized five taxa. Modern taxonomists use those five, and have added two other major taxa. They are arranged hierarchically (from broader to more specific): **kingdom, phylum, class, order, family, genus,** and **species** (figure 15.1; table 15.1). Even though Linnaeus's work was before the time of Charles Darwin (1809–1882), many of his groupings reflect evolutionary relationships. Morphological similarities between two animals are often the result of a common evolutionary history. Thus, as Linnaeus grouped animals according to shared characteristics, he also was describing evolutionary relationships. Ideally, members of the same taxonomic group are more closely related to each other than to members of different taxa. The hierarchical nature of these groupings is shown in figure 15.1.

Above the species level, there are no precise definitions of what constitutes a particular taxon. (The species concept was discussed in chapter 12.) Disagreements as to whether two species should be grouped into the same taxon, or

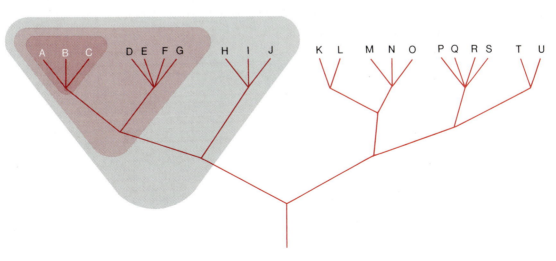

Figure 15.1 An example of how a group of species are classified. Hypothetical species A to U are members of a single class. Species A to J are members of the same order. Species A to G are members of the same family. Species A to C are members of the same genus.

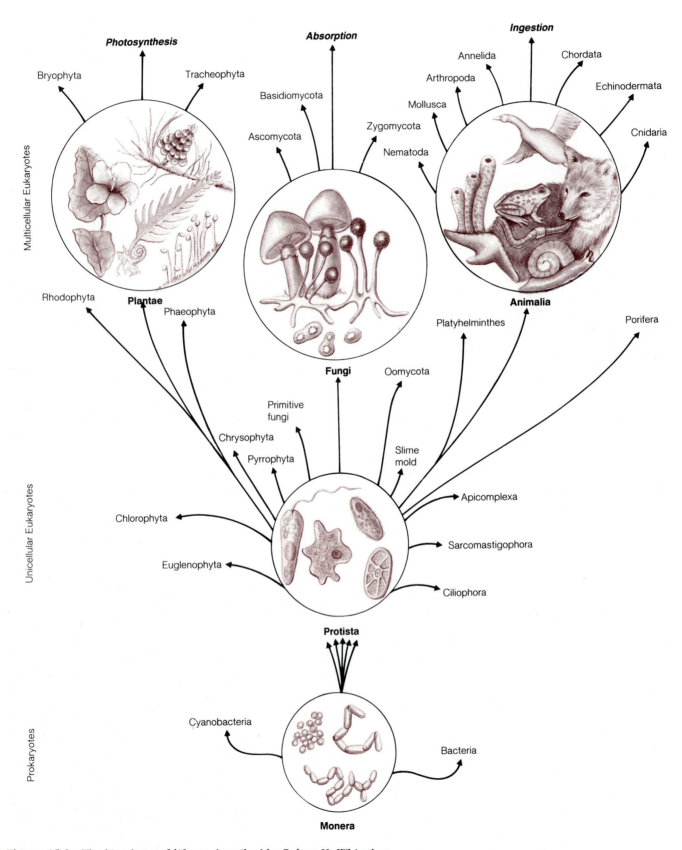

Figure 15.2 The kingdoms of life, as described by Robert H. Whittaker.

different taxa are common. Some biologists (splitters) prefer to have taxonomic categories with a few, closely related species included. Others (lumpers) prefer categories with more species.

Nomenclature

[1] Do you call certain freshwater crustaceans crawdads, crayfish, or crawfish? Do you call a common sparrow an English sparrow, a barn sparrow, or a house sparrow? The binomial system of nomenclature brings order to a chaotic world of common names. There are two problems with common names. First, common names vary from country to country, and from region to region within a country. Some species have literally hundreds of different common names. Biology transcends regional and nationalistic boundaries, and so must the names of what biologists study. Second, many common names refer to taxonomic categories higher than the species level. A superficial examination will simply not distinguish most different kinds of pillbugs (class Crustacea, order Isopoda) or most different kinds of crayfish (class Crustacea, order Decapoda). A common name, even if one recognizes it, often does not specify a particular species.

The binomial system of nomenclature is universal, and one always knows what level of classification is involved in any description. No two kinds of animals are given the same species name, as required by the *International Code of Zoological Nomenclature;* therefore the confusion caused by common names is avoided. When writing the scientific name of an animal, the genus begins with a capital letter, the species designation begins with a lowercase letter, and the entire scientific name is italicized or underlined. Thus, the scientific name of humans is written *Homo sapiens,* and when the species is understood, it can be abbreviated *H. sapiens.*

Kingdoms of Life

The earliest system of classification included two kingdoms of organisms, plants and animals. This system must have seemed very natural, because most organisms encountered by early taxonomists were easily assigned to one of those two categories. Certain technological advances, especially the refinement of microscopes, have shown that many organisms do not fit neatly into either the plant or the animal kingdom. *Euglena,* for example, is a single celled-organism that moves through the water and lacks rigid cell boundaries (animallike characteristics), yet it has chloroplasts and can carry out photosynthesis (a plantlike characteristic). Similarly, what should be done with bacteria and fungi? Early taxonomists classified them as plants.

The advent of the electron microscope allowed scientists to discover two fundamentally different kinds of cells. Even though bacteria and cyanobacteria (blue-green algae) had been considered plants, their cells are prokaryotic, that is they lack a membrane-bound nucleus and other membranous organelles (*see table 3.1*). All organisms except the bacteria and cyanobacteria, are eukaryotic. This fundamental difference is recognized in the classification of living organisms.

Five Kingdoms In recent years, a system of classification that uses five kingdoms has received widespread acceptance (figure 15.2). It distinguishes between kingdoms based on cellular organization and mode of nutrition.

Kingdom Monera. Members of the kingdom **Monera** are the bacteria and the cyanobacteria (figure 15.3). They are distinguished from all other organisms by one characteristic—they are prokaryotic.

Kingdom Protista. Members of the kingdom **Protista** are eukaryotic and consist of single cells, or colonies of cells (figure 15.4). As discussed later, colonies are loose associations of cells that show very little interdependence. Included in this kingdom are the protozoa (animallike protists), including *Amoeba, Paramecium,* and many others.

Kingdom Plantae. Members of the kingdom **Plantae** are eukaryotic, multicellular, and autotrophic (figure 15.5). Autotrophic organisms produce their own food, usually by photosynthesis. They are also characterized by walled cells and are usually nonmotile.

Kingdom Fungi. Members of the kingdom **Fungi** are also eukaryotic and multicellular (figure 15.6). They are distinguished from plants by being absorptive heterotrophs. They feed by digesting dead organisms extracellularly and absorbing the breakdown products. Like plants, they have walled cells and are usually nonmotile.

Kingdom Animalia. Members of the kingdom **Animalia** are eukaryotic, multicellular, and ingestive heterotrophs. They feed by ingesting other organisms or parts of other organisms. Their cells lack walls and they are usually motile (figure 15.7). This text is primarily devoted to

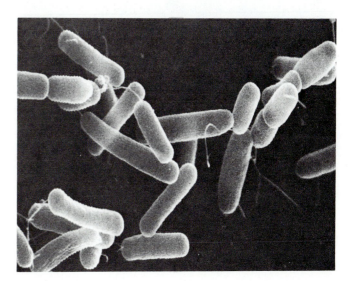

Figure 15.3 Bacilli (note the flagella). Scanning electron micrograph of a common bacterial shape.

Figure 15.4 *Stentor*, a common animallike protist.

Figure 15.6 A fungus. The mushroom, *Lepiota naucina*.

Figure 15.5 This ladyslipper (*Cypripedium reginae*) is a member of the kingdom Plantae.

Figure 15.7 Members of the kingdom Animalia. A female white-tailed deer and her fawn.

the kingdom Animalia. The next chapter, however, covers the animallike protists. Its inclusion is a part of a tradition that originated with the two-kingdom classification system. Protists were once considered a phylum (Protozoa) in the animal kingdom, and general zoology courses usually include them.

Animal Systematics

As in any human endeavor, disagreements have arisen in animal systematics. These disagreements revolve around methods of investigation and whether or not data may be used in describing distant phylogenetic relationships. Three contemporary schools of systematics exist.

Evolutionary systematics is the oldest of the three approaches to systematics. It is sometimes called the "traditional approach," although it has certainly changed since the beginnings of animal systematics. Evolutionary systematists believe that the order of the animal kingdom is a result of evolution, and that their goal as systematists is to reconstruct evolutionary pathways. Results of the work of these systematists are frequently portrayed on phylogenetic trees, where organisms are ranked in a hierarchical fashion. A basic assumption is that organisms closely related to an ancestor will resemble that ancestor more closely than they resemble distantly related organisms. Two kinds of similarities between organisms are recognized. (See also the discussion of homology and analogy in chapter 1.) *True similarities,* or homologies, are resemblances that result from common ancestry. An example is the similarity in the arrangement of bones in the wing of a bird and the arm of a human (*see figure 1.7*). *False similarities,* or analogies, are resemblances that result from organisms adapting under similar evolutionary pressures. The latter process is sometimes called *convergent evolution.* The similarity between the wings of birds and insects is a false similarity.

Numerical taxonomy emerged during the 1950s and 1960s and represents the opposite end of the spectrum from evolutionary systematics. The founders of numerical taxonomy believed that the criteria for grouping taxa had become too arbitrary and vague. They tried to make taxonomy more objective. Numerical taxonomists use mathematical models and computer aided techniques to group samples according to overall similarity. There is no attempt to distinguish between true and false similarities. Numerical taxonomists admit that false similarities exist. They contend, however, that false similarities will be overshadowed by the numerous true similarities used in data analysis. Numerical taxonomists do not try to weigh the importance of individual traits. Instead, numerical taxonomists work to analyze as many traits as possible. A second major difference between evolutionary systematics and numerical taxonomy is that numerical taxonomists limit discussion of evolutionary relationships to closely related taxa.

Phylogenetic systematics (**cladistics**) represents a third approach to animal systematics. As is true for evolutionary systematists, cladists believe that the goal of systematics is the reconstruction of animal phylogenetic relationships. Cladists differentiate between true and false similarities and attempt to distinguish between them in their work. In addition, cladists distinguish between true similarities that are derived from a remote common ancestor, and those of more recent origin. Cladists contend that the latter are more useful in phylogenetic studies because they are more likely to be retained in all members of a given group. For example, mammals characteristically have hair, three middle ear bones, and mammary glands. These characteristics are relatively recently derived from a common ancestor, and are fairly constant within the class. Other characteristics, such as pharyngeal gill pouches and a postanal tail, are shared with distant ancestors and are more sporadically observed. Because it is very difficult to identify unique characteristics in distantly related organisms, cladists concentrate on describing evolutionary relationships among closely related groups. **Cladograms** are diagrams depicting the history of taxa (figure 15.8).

Stop and Ask Yourself

1. What is animal systematics?
2. What are the two parts of a scientific name? What format should be used when writing species names in a term paper?
3. What does it mean when one says that our classification system is heirarchical?
4. Why are common names inadequate for scientific purposes?

Patterns of Organization

One of the most strikingly ordered series of changes in evolution is reflected in body plans in the animal kingdom and the protists. We can look at evolutionary changes in animal body plans and see what might be likened to a road map through a mountain range. What is most easily depicted are the starting and ending points and a few of the "attractions" along the route. What one cannot see from this perspective are the torturous curves and grades that must be navigated, and the extra miles that must be traveled, as one tries uncharted back roads. Unlike what we depict on a grand scale, evolutionary changes do not always mean "progress" and increased complexity. Evolution frequently results in backtracking, in experiments that fail, and in inefficient or useless structures. Evolution results in frequent dead ends, even though the route to that dead end may be filled with grandeur. The account that follows is a look at patterns of animal organization. As far as evolutionary pathways are concerned, view this account as an inexplicit road map through the animal kingdom. On a grand scale, it can be viewed as portraying evolutionary trends, but it should never be thought of as depicting evolutionary pathways.

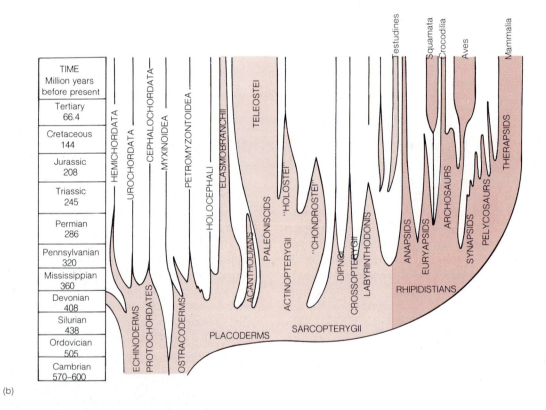

Figure 15.8 The phylogenetic relationships of reptiles, birds, and mammals as derived from phylogenetic systematics (cladistics) and evolutionary systematics are shown in the shaded areas. (a) A cladogram is constructed by identifying points at which two groups diverged. Animals that share a branching point are included in the same taxon. Some shared characteristics are indicated here. (b) A phylogenetic tree derived from evolutionary systematics depicts the degree of divergence since branching from a common ancestor, which is indicated by the time periods on the vertical axis. The width of the branches indicates the number of recognized genera for a given time period.

Symmetry

The bodies of animals and protists are organized into almost infinitely diverse forms. Within this diversity, however, certain patterns of organization can be described. The concept of symmetry is fundamental to understanding animal organization. **Symmetry** describes how the parts of an animal are arranged around a point or an axis (table 15.2).

Asymmetry, which is the absence of a central point or axis around which body parts are equally distributed, is characteristic of most protists and many sponges (figure 15.9). Asymmetry cannot be said to be an adaptation to anything or advantageous to an organism. Asymmetrical organisms do not develop complex communication, sensory, or locomotor functions. It is clear, however, that protists and animals whose bodies consist of aggregates of cells have flourished.

[3] A sea anemone can move along a substrate, but only very slowly (colorplate 14a). How is it to gather food? How does it detect and protect itself from predators? For this animal, a blind-side would leave it vulnerable to attack and cause it to miss many meals. The sea anemone, as is the case for most sedentary animals, has sensory and feeding structures uniformly distributed around its body. Sea anemones do not have distinct head and tail ends. Instead, one point of reference is the end of the animal that possesses the mouth (the oral end), and a second point of reference is the end opposite the mouth (the aboral end). Animals such as the sea anemone are said to be radially symmetrical. **Radial symmetry** is the arrangement of body parts such that *any plane* passing through the central oral-aboral axis divides the animal into mirror images (figure 15.10). Radial symmetry is often modified by the arrangement of some structures in pairs, or in other combinations, around the central oral-aboral axis. Paired arrangements of some structures in radially symmetrical animals is referred to as biradial symmetry. The arrangement of structures in five's around a radial animal is referred to as pentaradial symmetry.

Although the sensory, feeding, and locomotor structures found in radially symmetrical animals could never be called "simple," one never sees structures comparable to the complex sensory, locomotor, and feeding structures found in many other animals. The evolution of such structures in radially symmetrical animals would require repeated distribution of very specialized structures around the animal. The inability to evolve this degree of specialization may be one disadvantage of the radial design. On the other hand, the lack of highly specialized structures may be because they confer little advantage to sedentary animals. **Bilateral symmetry** is the arrangement of body parts such that there is only *a single plane,* passing dorsoventrally through the longitudinal axis of an animal, that divides the animal into right and left mirror images (figure 15.11). Bilateral symmetry is characteristic of active, crawling, or swimming animals. Because bilateral animals move primarily in one direction, one end of the animal is continually encountering the environment. The end that meets the environment is usually where complex sensory,

nervous, and feeding structures evolve and develop. These developments result in the formation of a distinct head, and are called **cephalization** (Gr. *kephale*, head). The *anterior* end of an animal is where cephalization occurs. *Posterior* is opposite anterior; it is the animal's tail end. Other important terms of direction and terms describing body planes and sections are applied to bilateral animals. These terms are used for locating body parts relative to a point of reference or an imaginary plane passing through the body (tables 15.2, 15.3; *see figure 15.11*).

Table 15.2	Animal Symmetry
Term	**Meaning**
Asymmetry	The arrangement of body parts without a central axis or point (e.g., the sponges).
Bilateral symmetry	The arrangement of body parts such that *a single plane* passing dorso-ventrally through the longitudinal axis divides the animal into right and left mirror images (e.g., the vertebrates).
Radial symmetry	The arrangement of body parts such that *any plane* passing through the oral/aboral axis divides the animal into mirror images (e.g., the cnidarians). Radial symmetry can be modified by the arrangement of some structures in pairs, or other combinations, around the central axis (e.g., biradial symmetry in the ctenophorans and pentaradial symmetry in the echinoderms).

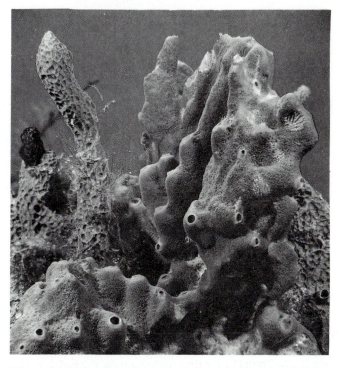

Figure 15.9 Sponges display a cell-aggregate organization, and as seen in this brown volcano sponge (*Hemectyon ferox*), many are symmetrical.

Figure 15.10 Radially symmetrical animals, such as this white anemone (*Sagartia modesta*), can be divided into equal halves by any plane that passes through the oral/aboral axis.

Table 15.3	Terms of Direction
Term	**Description**
Aboral	The end opposite the mouth
Oral	The end containing the mouth
Anterior	The head end; usually the end of a bilateral animal that meets its environment
Posterior	The tail end
Caudal	Toward the tail
Cephalic	Toward the head
Distal	Away from the point of attachment of a structure on the body (e.g., the toes are distal to the knee)
Proximal	Toward the point of attachment of a structure on the body (e.g., the hip is proximal to the knee)
Dorsal	The back of an animal; usually the upper surface; synonymous with posterior for animals that walk upright
Ventral	The belly of an animal; usually the lower surface; synonymous with anterior for animals that walk upright
Inferior	Below a point of reference (e.g., the mouth is inferior to the nose in humans)
Superior	Above a point of reference; (e.g., the neck is superior to the chest)
Lateral	Away from the plane that divides a bilateral animal into mirror images
Medial (median)	On or near the plane that divides a bilateral animal into mirror images

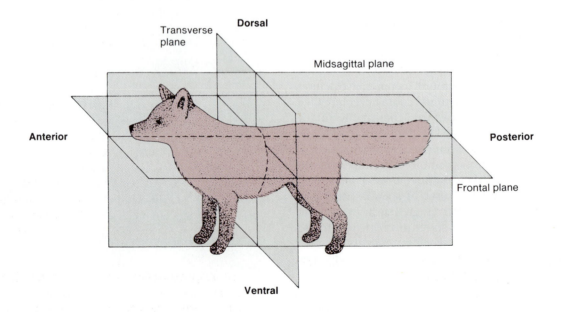

Figure 15.11 Planes and terms of direction that are useful in locating parts of a bilateral animal are indicated. A bilaterally symmetrical animal, such as this fox, has only one plane of symmetry. An imaginary midsagittal plane is the only plane through which the animal could be cut to yield mirror-image halves.

Other Patterns of Organization

In addition to body symmetry, there are other recognizable patterns of animal organization. In a broad context, these patterns of organization may reflect evolutionary trends. As explained earlier, however, one should not view these trends as exact sequences in animal evolution.

The Unicellular (Cytoplasmic) Level of Organization

Organisms whose bodies consist of single cells or cellular aggregates display the unicellular level of organization. Unicellular body plans are characteristic of the Protista. Some zoologists prefer to use the designation "cytoplasmic" to emphasize the fact that all living functions are carried out within the confines of a single plasma membrane. It is a mistake to consider unicellular organization "simple." All unicellular organisms must provide for the functions of locomotion, food acquisition, digestion, water regulation, sensory perception, and reproduction in a single cell.

Cellular aggregates (colonies) consist of loose associations of cells in which there is little interdependence, cooperation, or coordination of function—therefore, cellular aggregates cannot be considered tissues (*see chapter 3*). In spite of the absence of interdependence, there is some division of labor in these organisms. Some cells may be specialized for reproductive, nutritive, or structural functions.

Diploblastic Organization

Cells are organized into tissues in most animal phyla. **Diploblastic** (Gr. *dis,* twice + *blaste,* to sprout) organization is the simplest, tissue-level organization (figure 15.12). Body parts are organized into layers that are derived embryologically from [4] two tissue layers. Ectoderm gives rise to the *epidermis,* the outer layer of the body wall. Endoderm gives rise to the *gastrodermis,* the tissue that lines the gut cavity. Between the epidermis and the gastrodermis is a noncellular layer called *mesoglea.* In some diploblastic organisms, cells occur in the mesoglea but they are always derived from ectoderm or endoderm.

The cells in each tissue layer are functionally interdependent. The gastrodermis consists of nutritive (digestive) and muscular cells, and the epidermis contains epithelial and muscular cells. You may already be familiar with the feeding movements of *Hydra,* or the swimming movements of a jellyfish. These kinds of functions are only possible when groups of cells cooperate, showing tissue-level organization.

Triploblastic Organization

Animals described in chapters 18 to 31 are **triploblastic** (Gr. *treis,* three + *blaste,* to sprout). That is, their tissues are derived from three embryological layers. As with diploblastic animals, ectoderm forms the outer layer of the body wall, and endoderm lines the gut. In addition to these two layers, a third embryological layer is sandwiched between the ectoderm and endoderm. This layer is mesoderm, which gives

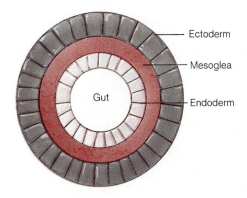

Figure 15.12 The diploblastic body plan. Diploblastic animals have tissues derived from ectoderm and endoderm. Between these two layers is a noncellular mesoglea.

rise to supportive, contractile, and circulatory cells. Triploblastic animals have an organ system level of complexity. Tissues are organized together to form excretory, nervous, digestive, reproductive, and circulatory systems. Triploblastic animals are usually bilaterally symmetrical (or have evolved from bilateral ancestors) and are relatively active animals.

Triploblastic animals are organized into several subgroups based upon the presence or absence of a body cavity and, for those that possess one, the kind of body cavity present. A body cavity, or **coelom** (Gr. *kilos,* hollow) is a fluid-filled cavity in which the internal organs can be suspended and separated from the body wall. Body cavities are advantageous because they:

1. Provide more room for organ development.
2. Provide more surface area for diffusion of gases, nutrients, and wastes into and out of organs.
3. Provide an area for storage.
4. Often act as hydrostatic skeletons.
5. Provide a vehicle for eliminating wastes and reproductive products from the body.
6. Facilitate increased body size.

Of these, the hydrostatic skeleton deserves further comment. Body-cavity fluids give support while allowng the body to remain flexible. Hydrostatic skeletons can be illustrated with a water-filled balloon. It is rigid yet flexible. Because the water in the balloon is incompressible, squeezing one end causes the balloon to lengthen. Compressing both ends causes the middle of the balloon to become fatter. In a similar fashion, body-wall muscles, acting on coelomic fluid, are responsible for movement and shape changes in many animals.

The Triploblastic Acoelomate Design. Triploblastic animals whose mesodermally derived tissues form a relatively solid mass of cells between ectodermally and endodermally derived tissues are referred to as being **acoelomate** (Gr. *a,* without + *kilos,* hollow) (figure 15.13a). Some cells between the ectoderm and endoderm of acoelomate animals are loosely organized, undifferentiated cells

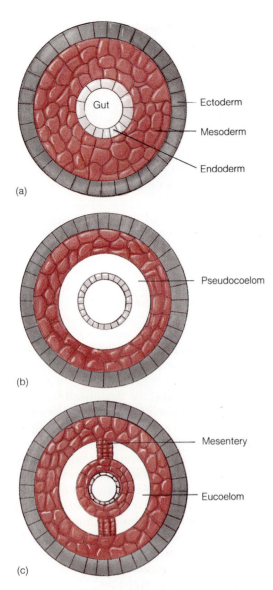

Figure 15.13 Triploblastic body plans. Triploblastic animals have tissues derived from ectoderm, mesoderm, and endoderm. (a) The triploblastic acoelomate design. (b) The triploblastic pseudocoelomate design. Note that there is no mesodermal lining on the gut track. (c) The triploblastic eucoelomate design. The eucoelom is completely surrounded by mesodermally derived tissues.

called *parenchyma*. Parenchymal cells, however, are not specialized for a particular function.

The Triploblastic Pseudocoelomate Design. A **pseudocoelom** (Gr. *pseudes,* false) is a body cavity not entirely lined by mesoderm (figure 15.13b). No muscular or connective tissues are associated with the gut tract, no mesodermal sheet covers the inner surface of the body wall, and no membranes suspend organs in the body cavity. Embryologically, the pseudocoelom is derived from the blastocoel of the embryo (*see chapter 10*). Lumping of pseudocoelomates together is somewhat artificial because the pseudocoelom may be their only common feature.

The Triploblastic Eucoelomate Design. A **eucoelom** (Gr. *eus,* true) is a body cavity that is completely surrounded by mesoderm (figure 15.13c). The inner body wall is lined by a thin mesodermal sheet, the *peritoneum,* and visceral organs are lined on the outside by a mesodermal sheet, the *serosa.* The peritoneum and the serosa are continuous, and suspend visceral structures in the body cavity. These suspending sheets are called *mesenteries.* Having mesodermally derived tissues, such as muscle and connective tissue, associated with internal organs enhances the function of virtually all internal body systems. In the chapters that follow, you will find many variations on the triploblastic, eucoelomate design.

Stop and Ask Yourself

5. How would you distinguish between radial symmetry and bilateral symmetry? Why is radial symmetry advantageous for sedentary animals?
6. Why does cephalization usually accompany bilateral symmetry?
7. What kinds of tissues form from the mesoderm?
8. What advantages do body cavities confer to coelomate animals?

Evolutionary Relationships and Tree Diagrams

Phylogenetic relationships are frequently represented by evolutionary tree diagrams (colorplate 5). There are problems associated with the use of tree diagrams, however. One of these is that such diagrams usually depict relationships between levels of classification above species. (Most used in this text depict relationships between phyla and classes.) It is correctly argued that it is misleading to depict larger taxonomic groups as ancestral because evolution occurs in species groups, not phyla and classes. A second problem associated with evolutionary tree diagrams is that when a taxon is represented as being ancestral, one has a tendency to visualize modern representatives of that taxon, rather than (often hypothetical) ancestral species. Modern representatives of any group of animals should be visualized as being at the tips of a "tree branch," and they may be very different from ancestral species. We use modern representatives to help visualize general characteristics of an ancestral species, but never to depict details of structure, function, or ecology. In spite of these problems, tree diagrams persist in scientific literature because they can help one appreciate evolutionary relationships and time scales.

[5]

Stop and Ask Yourself

9. Why is it misleading to depict taxonomic groups higher than species on evolutionary tree diagrams?
10. Are the animals depicted along the trunk of an evolutionary tree diagram those that we see today? Why or why not?

Summary

1. Systematics is the study of the classification of organisms. The binomial system of classification originated with Carolus Linnaeus and is used throughout the world in classifying organisms.

2. The classification system used in this book is convenient because it addresses problems of how to classify "borderline" groups—those with some animal and some plant features. It uses cellular structure, general level of organizational complexity, and modes of nutrition to group organisms into one of five kingdoms: Monera, Protista, Fungi, Animalia, and Plantae.

3. There are three modern approaches to systematics. They are evolutionary systematics, numerical taxonomy, and phylogenetic systematics or cladistics. Evolutionary systematists and cladists believe that the ultimate goal of systematics is to establish broad evolutionary relationships. Numerical taxonomists believe that knowing details of distant evolutionary relationships is largely impossible and, therefore, prefer to spend their time describing relationships between closely related taxa.

4. The bodies of animals are organized into almost infinitely diverse forms. Within this diversity, however, certain patterns of organization can be described. Symmetry describes how the parts of an animal are arranged around a point or an axis.

5. Other patterns of organization reflect how cells are associated together into tissues, and how tissues are organized into organs and organ systems.

6. Evolutionary tree diagrams are useful for depicting evolutionary relationships, but their limitations must be understood.

Key Terms

acoelomate (p. 233)
Animalia (p. 227)
asymmetry (p. 231)
bilateral symmetry (p. 231)
cephalization (p. 231)
cladograms (p. 229)
class (p. 225)
coelom (p. 233)
diploblastic (p. 233)
eucoelom (p. 234)
evolutionary systematics (p. 229)
family (p. 225)
Fungi (p. 227)
genus (p. 225)
kingdom (p. 225)
Monera (p. 227)

numerical taxonomy (p. 229)
order (p. 225)
phylogenetic systematics (cladistics) (p. 229)
phylum (p. 225)
Plantae (p. 227)
Protista (p. 227)
pseudocoelom (p. 234)
radial symmetry (p. 231)
species (p. 225)
symmetry (p. 231)
systematics (p. 225)
taxon (p. 225)
taxonomy (p. 225)
triploblastic (p. 233)

Critical Thinking Questions

1. In one sense, our classification system above the species level is artificial. In another sense, however, it is real. Explain this paradox.

2. Explain why protists, such as *Euglena,* presented problems for a two-kingdom classification system.

3. Why are virtually all radially symmetrical animals found in marine environments?

4. Describe the usefulness of evolutionary tree diagrams in zoological studies. Describe two problems associated with their use.

Suggested Readings

Books

Barnes, R. D. 1987. *Invertebrate Zoology,* 5th ed. Philadelphia: W.B. Saunders Company.

Eldredge, N. and Cracraft, J. 1980. *Phylogenetic Patterns and the Evolutionary Process, Method and Theory in Comparative Biology.* New York: Columbia University Press.

Hyman, L. H. 1940. *The Invertebrates.* New York: McGraw-Hill Book Company, Inc.

Margulis, L. and Schwartz, K. V. 1987. *Five Kingdoms: An Illustrated Guide to the Phyla of Life on Earth,* 2nd ed. San Francisco: W.H. Freeman and Co. Publishers.

Romer, A. S. and Parsons, T. S. 1986. *The Vertebrate Body,* 6th ed. Philadelphia: W.B. Saunders Company.

Wiley, E. O. 1981. *Phylogenetics: The Theory and Practice of Phylogenetic Systematics.* New York: John Wiley and Sons.

Articles

Duellman, W. E. 1985. Systematic Zoology: slicing the Gordian knot with Ockham's razor. *American Zoologist* 25:751–762.

Gould, S. J. 1976. The five kingdoms. *Natural History* 85(6):30.

Mayr, E. 1981. Biological classification: toward a synthesis of opposing methodologies. *Science* 241:510–516.

Whittaker, R. H. 1969. New concept of kingdoms of organisms. *Science* 163:150–160.

Animallike Protists

Concepts

1. The kingdom Protista is a polyphyletic group with origins in ancestral members of the kingdom Monera.
2. Protists display unicellular organization. All functions must be carried out within the confines of a single plasma membrane.
3. The animallike protists include members of three phyla: Sarcomastigophora, Apicomplexa, and Ciliophora.
4. Certain members of the above phyla have had and continue to have important influences on human health and welfare.

Have You Ever Wondered:

[1] whether or not all protist phyla can be traced back to a single moneran ancestor?
[2] whether protists should be considered single cells or entire organisms?
[3] what causes "red tides?"
[4] what protist causes dysentery in humans?
[5] what protist has caused more deaths in armies throughout history than actual combat?
[6] why sandboxes should always be kept covered when not in use?
[7] what group of protists has members that are considered the most complex of all protists?

These and other useful questions will be answered in this chapter.

This chapter contains underlined evolutionary concepts.

Evolutionary Perspective

Where are your "roots?" Although most people are content to go back into their family tree a few hundred years, biologists look back millions of years to the origin of all life forms. The fossil record indicates that virtually all protist and animal phyla living today were present during the Cambrian period, about 550 million years ago (*see table 1.1*). Unfortunately, there is little fossil evidence of the evolutionary pathways that gave rise to these phyla. Instead, evidence is gathered from examining the structure and function of living species. The "evolutionary perspective," in chapters 16 to 31 presents hypotheses regarding the origin of protist and animal phyla. These hypotheses seem reasonable to most zoologists, however, they are untestable, and alternative interpretations can be found in the scientific literature.

Ancient members of the kingdom Monera were the first living organisms on this planet. The Monera gave rise to the Protista late in the Precambrian period, about 1.5 billion years ago. The Endosymbiotic Hypothesis is one of a number of explanations of how this could have occurred (*see box 3.1*). Most biologists agree that the protists probably arose from more than one ancestral moneran group. [1] Depending on the classification system used, there are between 7 and 45 phyla of protists recognized today. These phyla represent numerous evolutionary lineages. When groups of organisms are believed to have had separate origins, they are said to be **polyphyletic** (Gr. *polys,* many + *phylon,* race). Some protists are plantlike because they are primarily autotrophic (they produce their own food). Others are animallike because they are primarily heterotrophic (they feed upon other organisms). This chapter covers three phyla of animallike protists.

Life Within a Single Plasma Membrane

Protists display **unicellular** (**cytoplasmic**) **organization.** [2] Their single-celled organization does not necessarily imply that they are simple organisms. Often, they are more complex than any particular cell in higher organisms. In some protistan phyla, individuals may group together to form colonies, which are associations of individuals that are not dependent upon one another for most functions. Protistan colonies, however, can become very complex, with some individuals becoming so specialized that it becomes difficult to tell whether one is observing a colony or a multicellular organism.

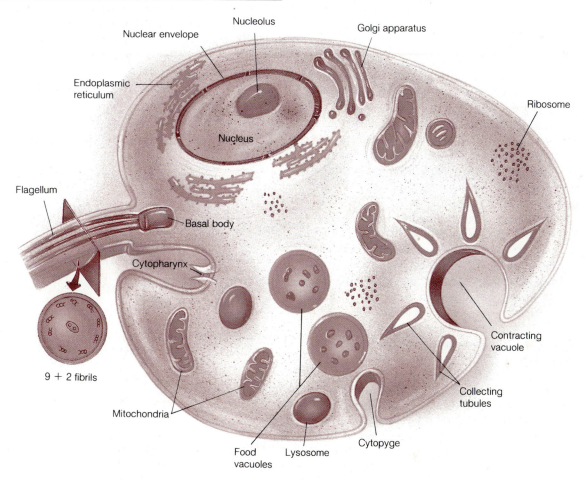

Figure 16.1 A generalized protist.

Maintaining Homeostasis

Specific functions are carried out in protists by organelles that are similar to the organelles of other eukaryotic cells (figure 16.1; *see figure 3.2*). Some protistan organelles, however, reflect specializations for unicellular lifestyles.

The plasma membrane of many protists is underlaid by a regular arrangement of microtubules. Together, they are called the **pellicle.** The pellicle is rigid enough to maintain the shape of the protist, but it is also flexible.

The cytoplasm of protists is differentiated into two regions. The portion of the cytoplasm just beneath the pellicle is called **ectoplasm** (Gr. *ectos,* outside + *plasma,* to form). It is relatively clear and viscous (firm). The inner cytoplasm is called **endoplasm** (Gr. *endon,* within). It is usually granular and more fluid. The conversion of cytoplasm between these two states is important in one kind of locomotion and is discussed later.

Most marine protists have salt concentrations similar to their environments. Freshwater protists, however, must regulate the water and salt concentrations of their cytoplasm. Water enters freshwater protists by osmosis because of higher salt concentrations in the protist than in the environment. This excess water is removed by **contractile vacuoles.** In some protists, contractile vacuoles are formed by coalescence of smaller vacuoles. In others, the vacuoles are permanent organelles that are filled by collecting tubules that radiate into the cytoplasm (figure 16.1). The contraction of microfilaments has been implicated in the emptying of contractile vacuoles.

Protists use a variety of feeding strategies. *Photoautotrophic* protists contain chlorophyll and use the energy in sunlight to produce food from carbon dioxide and water. *Chemolithoautotrophic* protists obtain energy from reactions involving inorganic compounds. *Heterotrophic* protists ingest other organisms or products of other organisms—either by absorbing dissolved nutrients by active transport or by ingesting whole or particulate food through endocytosis (*see figure 3.15*). In some protists, food may be ingested in a specialized region analogous to a mouth, called the **cytopharynx.** Digestion and transport of food occurs in **endocytic (food) vacuoles,** or vesicles that form during endocytosis. Digestion is mediated by enzymes and acidity changes. Endocytic vacuoles fuse with enzyme-containing lysosomes and circulate through the cytoplasm, distributing the products of digestion. After digestion is completed, the vacuoles are called **egestion vacuoles.** They release their contents by exocytosis, sometimes at a specialized region of the plasma membrane or pellicle called the **cytopyge.**

Because protists are small, they have a large surface area in proportion to their volume. This high surface-to-volume ratio facilitates two other maintenance functions. Gas exchange involves acquiring quantities of oxygen needed for cellular respiration and eliminating the carbon dioxide that is produced as a by-product. Excretion is the elimination of the nitrogenous by-products of protein metabolism. The principal nitrogenous waste in protists is ammonia. Both gas exchange and excretion occur by diffusion across the plasma membrane or pellicle.

Reproduction

Protists reproduce asexually and sexually. One of the simplest and most common forms of asexual reproduction in protists is **binary fission.** In binary fission, mitosis produces two nuclei that are distributed into two similar-sized individuals when the cytoplasm divides. During cytoplasmic division, some organelles are also divided to ensure that each new protist will possess all organelles. Depending on the group of protists, division of the cytoplasm may be longitudinal or transverse (figure 16.2 and 16.3).

Other forms of asexual reproduction are common. During *budding,* mitosis is followed by the incorporation of one nucleus into a cytoplasmic mass that is much smaller than the parent cell. **Multiple fission** or **schizogony** (Gr. *schizein,* to split) occurs when a large number of daughter cells are formed from the division of a single protist. Schizogony begins with multiple mitotic divisions in a mature individual. When a certain number of nuclei have been produced, cytoplasmic division results in the separation of each nucleus into a new cell.

Sexual reproduction involves the formation of gametes and the subsequent fusion of gametes to form a zygote. In most protists, the sexually mature individual is haploid.

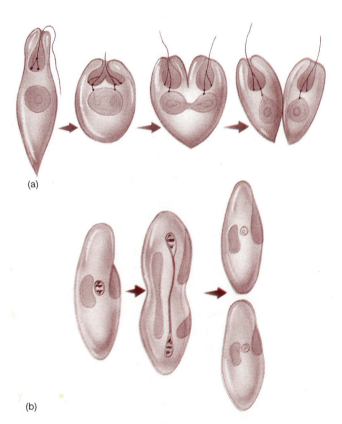

(a)

(b)

Figure 16.2 Binary fission begins with mitosis. Cytoplasmic division divides some surface features between the two cells and results in two similarly sized protists. Binary fission is longitudinal in some protists (a) and transverse in others (b).

Figure 16.2 redrawn, with permission, from *Living Invertebrates* by Pearse/Buchsbaum, copyright by The Boxwood Press.

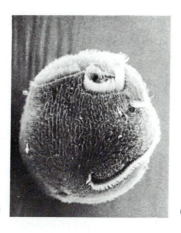

Figure 16.3 Binary fission of the ciliate, *Stentor coeruleus*. Fission includes the division of some surface features (a,b), in this case cilia modified into a bandlike structure called a membranelle (M). F designates the frontal field, and the beginning of a fission furrow is shown by the arrow in (b). Fission is completed by division of the cytoplasm (c,d).

Gametes are produced by mitosis, and meiosis occurs following union of gametes. Ciliated protists are an exception to this pattern. Specialized forms of sexual reproduction will be discussed as protist groups are discussed.

Symbiotic Lifestyles

Symbiotic lifestyles are important for many protists and animals. *Symbiosis* occurs when one organism lives in an intimate association with another. For many protists, these interactions involve a form of symbiosis called *parasitism*, in which one organism lives in or on a second organism, called a *host*. The host is harmed but usually survives, at least long enough for the parasite to complete one or more life cycles.

The relationships between a parasite and its host(s) are often complex. Some parasites have life histories involving multiple hosts. The *definitive host* is the host that harbors the sexual stages of the parasite. The sexual stages may produce offspring that enter another host. This second host is called an *intermediate host*, and asexual reproduction occurs here. More than one intermediate host and more than one immature stage may be involved in some life cycles. For the life cycle to be completed, the final, asexual stage must have access to a definitive host.

Other kinds of symbiosis involve relationships that do not harm the host. *Commensalism* is a symbiotic relationship in which one member of the relationship benefits, and the second member is neither benefited nor harmed. *Mutualism* is a symbiotic relationship in which both members of the relationship benefit.

Stop and Ask Yourself

1. Ancestral members of what kingdom of organisms gave rise to the Protista?
2. What is the function of the following structures: Pellicle? Contractile vacuole? Cytopharynx? Egestion vacuole?
3. What are three forms of asexual reproduction in protists?
4. Why are gametes produced by mitosis rather than by meiosis in most protists?

Phylum Sarcomastigophora

Members of the phylum Sarcomastigophora (sar'ko-mas-ti-gof'o-rah) (Gk. *sarko*, fleshy + *mastigo*, whip + *phoros*, to bear) possess pseudopodia (see the discussion of amebae below) and/or one or more flagella. The phylum is divided into three subphyla. The subphyla Mastigophora and Sarcodina are discussed in this section. With over 18,000 described species, Sarcomastigophora is the largest phylum of protists (table 16.1).

Flagellar Locomotion

Members of the subphylum Mastigophora (mas-ti-gof'o-rah) use flagella in locomotion. Movements of flagella may be two-dimensional, whiplike movements or helical movements that result in the protist being pushed or pulled through its medium (figure 16.4).

Flagella originate from basal bodies (kinetosomes) and are organized in a 9 + 2 arrangement of microtubules (figure 16.5; *see figure 3.24*). Nine pairs of microtubules are arranged around a core of two microtubules. Each pair, called a doublet, consists of A and B microtubules. The A microtubule bears a pair of dynein arms that extends toward the B microtubule of an adjacent doublet. A radial spoke runs between each doublet and the sheath that surrounds the two central tubules. The membrane of the flagellum is an extension of the plasma membrane.

Table 16.1	Classification of the Phyla Sarcomastigophora, Apicomplexa, and Ciliophora*

Phylum Sarcomastigophora (sar′ko-mas-ti-gof′o-rah)
Protists that possess flagella, pseudopodia, or both that are used for locomotion and feeding; single type of nucleus.

Subphylum Mastigophora (mas-ti-gof′o-rah)
One or more flagella used for locomotion; autotrophic or heterotrophic.

Class Phytomastigophora (fi′to-mas-ti-go-for′ah)
Chloroplasts usually present; mainly autotrophic, though some heterotrophic. *Euglena, Volvox.*

Class Zoomastigophora (zo′o-mas-ti-go-for′ah)
Lack chloroplasts; heterotrophic or saprozoic. *Trypanosoma, Trichonympha, Trichomonas.*

Subphylum Sarcodina (sar-ko-din′ah)
Pseudopodia for movement and food gathering; naked or with shell or test.

Superclass Rhizopoda (ri-zop′o-dah)
Lobopodia, filopodia, reticulopodia, or no distinct pseudopodia. *Amoeba, Entamoeba, Arcella, Difflugia.*

Class Lobosa (lo-bo′sah)
Pseudopodia mostly lobopodia.

Superclass Actinopoda (ak′ti-nop″o-dah)
Spherical, planktonic; axopodia supported by microtubules; includes three classes of radiolarians and one class of heliozoans. *Actinophrys, Actinosphaerium.*

Phylum Apicomplexa (a″pi-kom-plex′ah)
Parasitic protists with an apical complex used for penetrating host cells; cilia and flagella lacking, except in certain reproductive stages.

Class Sporozoa (spor′o-zo″ah)
Spores or oocysts present; flagella or cilia present in microgametes. *Eimeria, Plasmodium, Toxoplasma.*

Phylum Ciliophora (sil-i-of′or-ah)
Protists with simple or compound cilia at some stage in the life history; heterotrophs with a well-developed cytopharynx and feeding organelles; at least one macronucleus and one micronucleus present; includes three classes that are distinguished from one another by patterns of cilia. *Paramecium, Colpidium, Stentor, Euplotes, Vorticella.*

*A partial listing. The taxonomy of the protists is currently unsettled. Students should not be surprised to see other classification schemes presented in other textbooks.

The movement of flagella results from a sliding of A and B microtubules relative to each other. The sliding of tubules on one side of a flagellum or cilium causes that side to lengthen and the cilium or flagellum to bend away from the sliding microtubules. The forces responsible for microtubule sliding seem to arise from the dynein arms of the A microtubule interacting with the B microtubule of an adjacent doublet (figure 16.5b).

Class Phytomastigophora

The subphylum Mastigophora is divided into two classes. Members of the class Phytomastigophora (fi′to-mas-ti-go-for′ah) (Gk. *phytos,* plant) possess chlorophyll and one or two flagella. Phytomastigophorans are responsible for producing a large portion of the food in marine food webs. Much of the oxygen in our atmosphere also comes from photosynthesis by these marine protists.

Marine phytomastigophorans include the dinoflagellates (figure 16.6). Dinoflagellates have one flagellum that wraps around the organism in a transverse groove. The primary action of this flagellum causes the protist to spin on its axis. A second flagellum is a trailing flagellum that pushes the organism forward. In addition to chlorophyll, many dinoflagellates contain xanthophyll pigments, which give these protists a golden-brown color. At times, dinoflagellates become so numerous that they color the water. One genus, *Ptychodiscus,* has representatives that produce tox- [3]

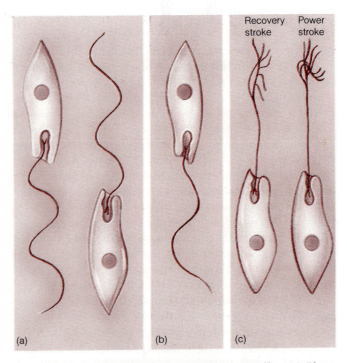

Figure 16.4 Patterns of movement of flagella. (a) Planar waves. The protist moves in a direction opposite the direction of wave propagation. (b) Helical waves add spin to the body. (c) Some protists restrict flagellar movements to the tip of the flagellum.

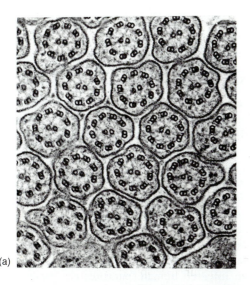

(a)

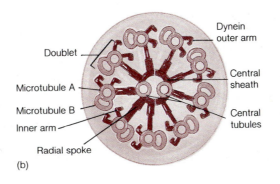

Doublet

Dynein outer arm

Microtubule A

Central sheath

Microtubule B

Inner arm

Central tubules

Radial spoke

(b)

Figure 16.5 The 9 + 2 organization of microtubules in cilia and flagella. (a) Cross section through cilia. (b) Diagrammatic representation of the ultrastructure of cilia and flagella. Dynein outer arms on the A microtubules interact with adjacent B microtubules to cause a sliding between A and B microtubules of one doublet. Sliding of microtubules on one side of the cilium or flagellum causes that side to lengthen and the cilium or flagellum bends toward the opposite side.

Figure 16.6 Scanning electron micrograph of two species of dinoflagellates, *Peridinium* in the upper left and *Ceratium* in the lower half.

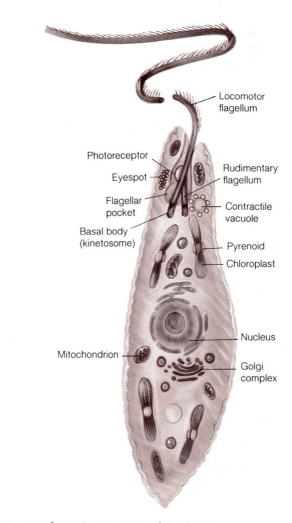

Locomotor flagellum

Photoreceptor

Eyespot

Rudimentary flagellum

Flagellar pocket

Contractile vacuole

Basal body (kinetosome)

Pyrenoid

Chloroplast

Nucleus

Mitochondrion

Golgi complex

Figure 16.7 The structure of *Euglena*.

ins. Periodic "blooms" of these organisms are called "red tides" and result in fish kills along the continental shelves. Human deaths may result from consuming tainted molluscs or fish. Entire companies of Japanese soldiers were reportedly killed during World War II from eating contaminated fish.

Euglena is a phytomastigophoran found in fresh water (figure 16.7). Each chloroplast has a *pyrenoid,* which synthesizes and stores polysaccharides. If cultured in the dark, euglenoids feed by absorption. Some euglenoids (e.g., *Peranema*) lack chloroplasts and are always heterotrophic.

Euglena orients toward light of certain intensities. An *eyespot* is a pigment cup covering a photoreceptor at the base of the flagellum. The eyespot permits light to strike the photoreceptor from only one direction, allowing *Euglena* to orient in relation to a light source.

Euglenoid flagellates are haploid and reproduce by longitudinal binary fission. Sexual reproduction in these species is unknown.

Volvox is a colonial flagellate consisting of up to 50,000 cells embedded in a spherical, gelatinous matrix (figure 16.8a). Cells possess two flagella, which cause the colony to roll and turn gracefully through the water (figure 16.8b).

Although most cells of *Volvox* are relatively unspecialized, reproduction is dependent upon certain specialized cells. Asexual reproduction occurs in the spring and summer when certain cells withdraw to the watery interior of the parental colony and form daughter colonies. When the parental colony dies and ruptures, daughter colonies are released.

Sexual reproduction in *Volvox* occurs during autumn. Some species are dioecious (having separate sexes), other species are monoecious (having both sexes in the same colony). In autumn, specialized cells differentiate into macrogametes or microgametes. Macrogametes are large, filled with nutrient reserves, and nonmotile. Microgametes form as a packet of flagellated cells that leaves the parental colony and swims to a colony containing macrogametes. The packet then breaks apart and syngamy occurs between macro- and microgametes. The zygote, an overwintering stage, secretes a resistant wall around itself and is released when the parental colony dies. Because the parental colony consists of haploid cells, the zygote must undergo meiosis to reduce the chromosome number from the diploid zygotic condition. One of the products of meiosis then undergoes repeated mitotic divisions to form a colony consisting of just a few cells. The other products of meiosis degenerate.

This colony is released from the protective zygotic capsule in the spring.

Class Zoomastigophora

Members of the class Zoomastigophora (zo′o-mas-ti-go-for′ah) (Gk. *zoion,* animal) lack chloroplasts and, therefore, are heterotrophic. Some members of this class are important parasites of humans.

One of the most important species of zoomastigophorans is *Trypanosoma brucei.* This species is divided into three subspecies, *T. b. brucei, T. b. gambiense,* and *T. b. rhodesiense.* The first of these three subspecies is a parasite of nonhuman mammals of Africa. The latter two cause sleeping sickness in humans. All subspecies use tsetse flies (*Glossina*) as intermediate hosts and as infective agents. When a tsetse fly bites an infected human, parasites are picked up with the meal of blood. Trypanosomes multiply asexually in the gut of the fly for about 10 days, then migrate to the salivary glands. While in the fly, the trypanosomes are transformed, in 15 to 35 days, through a number of body forms. When the infected tsetse fly bites another vertebrate host, the parasites travel with salivary secretions into the blood of a new definitive host. The parasites multiply asexually in the new host and are again transformed through a number of body forms. Parasites may live in the blood, lymph, spleen, central nervous system, and cerebrospinal fluid (figure 16.9).

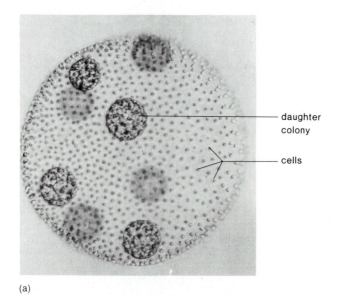

daughter colony

cells

(a)

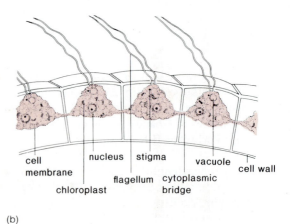

cell membrane

nucleus stigma

vacuole cell wall

chloroplast

flagellum cytoplasmic bridge

(b)

Figure 16.8 *Volvox,* a colonial flagellate. (a) A *Volvox* colony showing asexually produced daughter colonies. (b) An enlargement of a portion of the body wall.

Symptoms of infection in humans include the development of a small sore at the site of infection. This sore heals in a few days, but lymph nodes become swollen as they become congested with parasites. Swelling is especially prominent in the neck. During slave-trading years, this sign (now called Winterbottom sign) was recognized by traders. Realizing that it meant certain death for an infected person, the slave had a good chance of being thrown overboard. Later symptoms include fever, increased swelling of lymph nodes, pain, headache, and weakness.

Infections of *Trypanosoma brucei rhodesiense* are more protracted. Trypanosomes enter the central nervous system and cause general apathy, mental dullness, and lack of coordination. "Sleepiness" develops and the infected individual may fall asleep during normal daytime activities. Death results from any combination of the above symptoms or from heart failure, malnutrition, or other weakened conditions. If detected early, sleeping sickness is curable. However, if an infection has advanced to the central nervous system, recovery is unlikely.

Pseudopodia and Ameboid Locomotion

Members of the subphylum Sarcodina (sar'ko-din"ah) are the amebae (singular ameba). When feeding and moving, they form temporary cell extensions called **pseudopodia** (singular pseudopodium) (Gr. *pseudes,* false + *podion,* little foot). Pseudopodia exist in a variety of forms. *Lobopodia* (Gr. *lobos,* lobe) are broad cell processes containing ectoplasm and endoplasm and are used for loco-motion and engulfing food (figure 16.10a). *Filopodia* (L. *filum,* thread) contain ectoplasm only and provide a constant two-way streaming that delivers food in a conveyor-belt fashion (figure 16.10b). *Reticulopodia* (L. *reticulatus,* netlike) are similar to filopodia, except that they branch and rejoin to form a netlike series of cell extensions (figure 16.10c). *Axopodia* (L. *axis,* axle) are thin and filamentous and are supported by a central axis of microtubules. The cytoplasm covering the central axis is adhesive and movable. Food caught on axopodia can be delivered to the central cytoplasm of the ameba (figure 16.10d).

The plasma membrane of an ameba apparently has adhesive properties, because new pseudopodia attach to the substrate as they are formed. Older pseudopodia are less firmly attached than newer pseudopodia and lose contact with the substrate at the end opposite the direction of movement. The plasma membrane also seems to slide over the underlying layer of cytoplasm when an ameba moves. The plasma membrane may be "rolling" in a way that is (very roughly) analogous to a bulldozer track rolling over its wheels. A thin fluid layer between the plasma membrane and the ectoplasm may facilitate this rolling.

As an ameba moves, the fluid endoplasm flows forward into an advancing pseudopodium. As it reaches the tip of a pseudopodium, endoplasm is converted into ectoplasm.

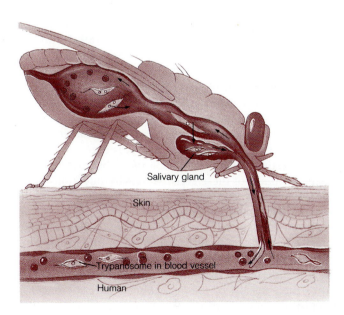

Figure 16.9 The life cycle of *Trypanosoma brucei.* When a tsetse fly feeds on a vertebrate host, trypanosomes enter the vertebrate's circulatory system with the fly's saliva. Trypanosomes multiply in the circulatory and lymphatic systems by binary fission. When a tsetse fly bites this vertebrate host again, trypanosomes move into the gut of the fly and undergo binary fission. Trypanosomes then migrate to the fly's salivary glands, where they are available to infect a new host.

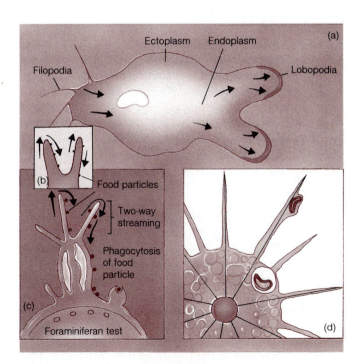

Figure 16.10 Variations in pseudopodia. (a) Lobopodia contain both ectoplasm and endoplasm and are used for locomotion and engulfing food. (b) Filopodia contain ectoplasm only and provide constant two-way streaming that delivers food particles to a protist in a conveyor-belt fashion. (c) Reticulopodia are similar to filopodia except that they branch and rejoin to form a netlike series of cell extensions. (d) Axopodia deliver food to the central cytoplasm.

Figure 16.9 redrawn, with permission, from *Living Invertebrates* by Pearse/Buchsbaum, copyright by The Boxwood Press.

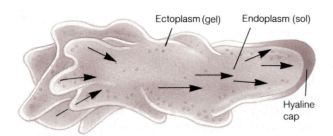

Figure 16.11 The mechanism of ameboid locomotion. Endoplasm (sol) flows into an advancing pseudopodium. At the tip of the pseudopodium endoplasm changes into ectoplasm (gel). At the opposite end of the ameba, ectoplasm changes into endoplasm and begins flowing in the direction of movement.

At the same time, ectoplasm near the opposite end is converted into endoplasm and begins flowing forward (figure 16.11).

The mechanism for the conversion between fluid endoplasm and viscous ectoplasm involves the interactions between the contractile proteins, actin and myosin. Interactions between these proteins are also responsible for muscle contraction in animals, and the presence of actin and myosin in protists and animals is evidence of evolutionary ties between the two groups.

Class Lobosa

The most familiar amebae belong to the genus *Amoeba* (figure 16.12). These amebae are naked and are normally found on substrates in shallow water of freshwater ponds, lakes, and slow-moving streams, where they feed on other protists. Food is engulfed by *phagocytosis,* a process that involves the cytoplasmic changes described earlier for ameboid locomotion. In the process food is incorporated into food vacuoles (*see figure 3.15b*). Binary fission occurs when an ameba reaches a certain size limit. As with other amebae, no sexual reproduction is known to occur.

Other members of the class Lobosa (lo-bo′sah) possess a test or shell. *Tests* are protective structures secreted by the cytoplasm. They may be calcareous (made of calcium carbonate), proteinaceous (made of protein), siliceous (made of silica [SiO$_2$]), or chitinous (made of chitin—a polysaccharide). Other tests may be composed of sand or other debris cemented into a secreted matrix. Usually one or more openings in the test allow pseudopodia to be extruded. *Arcella* is a common freshwater, shelled ameba. It has a brown, proteinaceous test that is flattened on one side and domed on the other. Pseudopodia project from an opening on the flattened side. *Difflugia* is another common freshwater, shelled ameba (figure 16.13). Its shell is vase shaped and is composed of mineral particles embedded in a secreted matrix.

[4] Many amebae are symbiotic. *Entamoeba histolytica,* the only severely pathogenic ameba of humans, causes one form of dysentery. Dysentery is marked by inflammation and ulceration of the lower intestinal tract, accompanied by a debilitating diarrhea that includes blood and mucus.

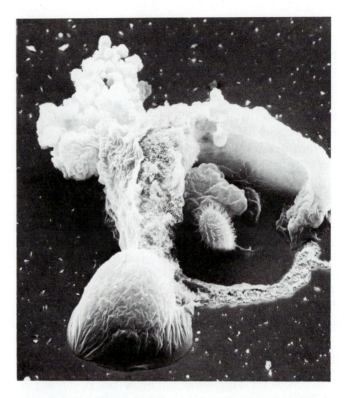

Figure 16.12 Scanning electron micrograph of an ameba encircling its prey, a ciliate.

Amebic dysentery is a worldwide problem that plagues humans in crowded, unsanitary conditions.

A significant problem in the control of *Entamoeba histolytica* is the fact that an individual can be infected and contagious without experiencing symptoms of the disease. Amebae live in the folds of the intestinal wall, feeding on starch and mucoid secretions. Amebae are passed from one host to another in the form of cysts that are transmitted by fecal contamination of food or water. Amebae leave the cysts after a host ingests contaminated food or water and take up existence in the intestinal wall.

In its invasive (pathogenic) form, *Entamoeba histolytica* releases lysosomes, whose enzymes digest or "hydrolyze" intestinal cells. The ameba then absorbs the predigested products of these cells. Secondary infections of the liver, lungs, and skin may occur, and bacterial infections of ulcerated areas complicate medical treatments. This invasive form of ameba never forms cysts and hence, symptomatic hosts never pass the infection to other people.

Whether or not a particular infection will become virulent is determined by complex interactions between the host's diet, the pH of the gut contents, and the bacterial flora of the gut. Any change in these factors may result in a change from nonvirulency to virulency. For example, the change in diet that results from travel to foreign countries may mark the onset of virulency.

Other symbiotic amebae of humans are nonpathogenic. *Entamoeba coli* is a harmless intestinal commensal that feeds on bacteria, protozoa, and yeast. Because the life cycle and the mode of transmission of *Entamoeba coli* is

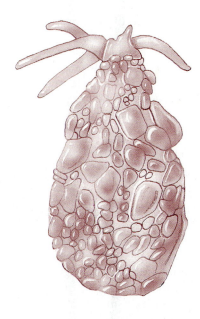

Figure 16.13 *Difflugia,* a common shelled ameba.

nearly identical to *Entamoeba histolytica,* infection with the former indicates that conditions are favorable for contracting the latter. A third symbiotic ameba, *Entamoeba gingivalis,* is a common, harmless symbiont of the human mouth. It can usually be detected by microscopic examination of scrapings from the cheek or gum surfaces inside the mouth.

Foraminiferans, Heliozoans, and Radiolarians

Foraminiferans (commonly called forams) are a group of amebae whose members are primarily marine (figure 16.14). Foraminiferans possess reticulopodia and secrete a shell that is mostly calcium carbonate. As foraminiferans grow, they secrete new, larger chambers that remain attached to the older chambers. Test enlargement follows a symmetrical pattern that may result in a straight chain of chambers or spiral arrangement that resembles a snail shell. Many of these shells may reach relatively large sizes. The so-called Mermaid's pennies, found in Australia, may be several cm in diameter.

Foraminiferan tests have left an abundant fossil record that began in the Cambrian period. Foram tests make up a large component of marine sediments, and the accumulation of foram skeletons on the floor of primeval oceans has resulted in some of our limestone and chalk deposits. The white cliffs of Dover are one example of a foraminiferan-chalk deposit. Oil geologists use fossilized forams to identify geologic strata when taking exploratory cores.

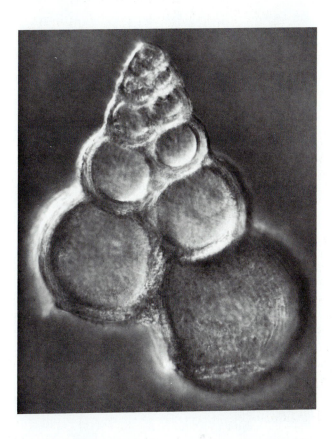

Figure 16.14 Foraminiferan test (*Gümbelina*).

Heliozoans are freshwater amebae that are either planktonic or live attached by a stalk to some substrate. (The plankton of a body of water consists of those organisms that float freely in the water.) Heliozoans are either naked or enclosed within a shell that contains openings for axopodia (figure 16.15a).

Radiolarians are planktonic, marine amebae. They are relatively large; some colonial forms may reach several cm in length. They possess a skeleton (usually siliceous) of either long, movable spines and needles or a highly sculptured and ornamented lattice (figure 16.15b). When radiolarians die, their skeletons drift to the ocean floor. Some of the oldest known fossils of eukaryotic organisms are radiolarians.

Stop and Ask Yourself

5. How and when does sexual reproduction occur in *Volvox*?
6. How do trypanosomes gain entry to their human hosts?
7. What mechanism explains ameboid locomotion?
8. Why will a person experiencing symptoms of amebic dysentery never pass the infection to a second individual?

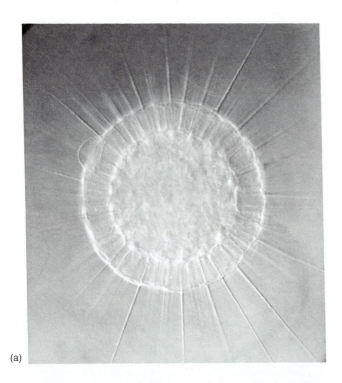

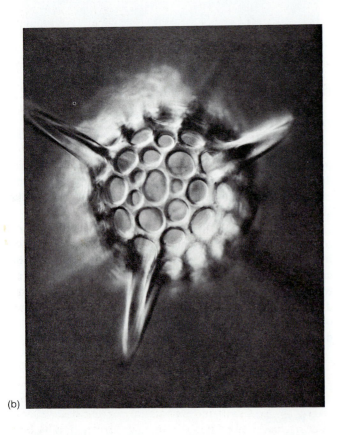

Figure 16.15 Heliozoan and radiolarian tests. (a) *Actinosphaerium sol.* This genus contains a number of common species of heliozoans. (b) The radiolarian *Hexaconthium.*

Phylum Apicomplexa

Members of the phylum Apicomplexa (a″pi-kom-plex′ah) are all parasites. They possess a characteristic complex of structures, called the *apical complex,* which aids in penetration of host cells. Apicomplexans possess a single type of nucleus and, except for certain reproductive stages, lack cilia and flagella (*see table 16.1*).

Class Sporozoa

The most important species in this phylum are members of the class Sporozoa (spor′o-zo″ah). The class name derives from the fact that most sporozoans produce a resistant spore or oocyst following sexual reproduction. Some members of this class, called coccidians, cause a variety of diseases in domestic animals and humans.

Although there is considerable variability in life cycles of sporozoans, certain generalizations are possible. Many are intracellular parasites, and their life cycles may be divided into three phases. *Schizogony* is multiple fission of an asexual stage in host cells, resulting in the formation of many more (usually asexual) individuals, called merozoites, that leave the host cell and infect many other cells.

Some of the merozoites produced as described above undergo *gametogony,* which begins the sexual phase of the life cycle. In so doing, the parasite forms either microgametocytes or macrogametocytes. Microgametocytes undergo multiple fission to produce biflagellate microgametes that emerge from the infected host cell. The macrogametocyte develops directly into a single macrogamete. Fertilization of the macrogamete by a microgamete produces a zygote, which is enclosed by a cyst (frequently resistant), the oocyst.

The zygote undergoes meiosis, and the resulting cells divide repeatedly by mitosis. This process, called *sporogony,* occurs in the oocyst and produces many rodlike sporozoites. Sporozoites infect the cells of a new host when the oocyst is ingested by the new host, or sporozoites are otherwise introduced (e.g., by a mosquito bite).

One sporozoan genus, *Plasmodium,* causes malaria and has been responsible for more human suffering than most other diseases. Malaria has been known since antiquity. Accounts of the disease go back as far as 1550 B.C. Malaria was a significant contributor to the failure of the Crusades [5] during the medieval era and it has contributed more to the devastation of armies than has actual combat.

The life cycle of *Plasmodium* involves vertebrate and mosquito hosts (figure 16.16). Schizogony occurs first in liver cells and later in red blood cells, and gametogony also occurs in red blood cells. Gametocytes are taken into a mosquito during a meal of blood and subsequently fuse. The zygote penetrates the gut of the mosquito and is transformed into an oocyst. Sporogony forms haploid sporozoites that may enter a new host when the mosquito bites the host.

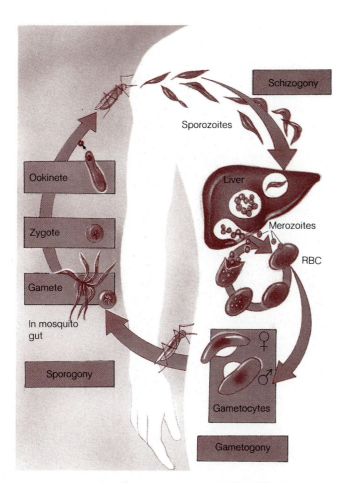

Figure 16.16 The life cycle of *Plasmodium.* Schizogony occurs in liver cells and later, in the red blood cells of humans. Gametogony occurs in red blood cells. Microgametes and macrogametes are taken into a mosquito during a blood meal and fuse to form zygotes. Zygotes penetrate the gut of the mosquito and form the oocysts. Meiosis and sporogony form many haploid sporozoites that may enter a new host when the mosquito bites the host.

The symptoms of malaria recur periodically and are called *paroxysms* (box 16.1). Chills and fever are correlated with the maturation of parasites, the rupture of red blood cells, and the release of sequestered metabolites.

Four species of *Plasmodium* are the most important human malarial species. *P. vivax* causes benign malaria in which the paroxysms recur every 48 hours. This species is temperate and has been nearly eradicated in many parts of the world. *P. falciparum* causes malignant malaria, which is the most virulent form of malaria in humans. Paroxysms recur about every 48 hours. It was once worldwide but is now mainly tropical and subtropical. It remains one of the greatest killers of humanity, especially in Africa. *P. malariae* is worldwide in distribution and causes a type of malaria, with paroxysms that recur every 72 hours. *P. ovale* is the rarest of the four human malarial species, and is primarily tropical in distribution.

Other members of the class Sporozoa also cause important diseases. Coccidiosis is a disease of poultry, sheep, cattle, and rabbits. Two genera, *Isospora* and *Eimeria,* are

particularly important parasites of poultry. Yearly losses to the poultry industry in the United States from coccidiosis have approached 35 million dollars. Toxoplasmosis is a disease of mammals, including humans, and birds. Sexual reproduction of *Toxoplasma* occurs primarily in cats. In all other hosts, asexual reproduction is the rule. Infections occur when oocysts are ingested with food contaminated by cat feces, or when meat containing encysted sporozoites is eaten raw or poorly cooked. Most infections in humans are asymptomatic and once infected an effective immunity develops. However, if a woman is infected near the time of pregnancy, or during pregnancy, *congenital toxoplasmosis* may develop in a fetus. Congenital toxoplasmosis is a major cause of stillbirths and spontaneous abortions. Fetuses that survive frequently show signs of mental retardation and epileptic seizures. There is no cure for congenital toxoplasmosis. Steps can be taken to avoid infections by *Toxoplasma.* Precautions include: keeping stray and pet cats away from children's sandboxes; using sandbox covers; and awareness, on the part of couples considering having children, of the potential dangers of eating raw or very rare pork, lamb, and beef.

[6]

Stop and Ask Yourself

9. Members of which protist genus have caused the most human suffering since antiquity?
10. What are the three stages in the life cycle of *Plasmodium*?
11. How are infections of *Toxoplasma* acquired?
12. What are the dangers of toxoplasmosis for a fetus?

Phylum Ciliophora

The phylum Ciliophora (sil-i-of'or-ah) includes some of the most complex protists (*see table 16.1*). Members of the Ciliophora all have cilia at some time in their life history. All ciliates have a large polyploid nucleus that controls daily functions and one or more small nuclei that are the genetic reserve of the cell. Unlike most other protists, ciliates are diploid, and meiosis precedes syngamy. Ciliates are widely distributed in freshwater and marine environments. A few ciliates are symbiotic.

[7]

Cilia and Other Pellicular Structures

Cilia are generally similar to flagella except that they are much shorter and are widely distributed over the surface of the protist. Ciliary movements are coordinated. In some protists, pairs of cilia occur in rows. Rows of cilia beat slightly out of phase with one another so that ciliary waves called *metachronal waves,* periodically pass over the surface of the ciliate (figure 16.17). Many ciliates can rapidly reverse the direction of ciliary beating, which changes the direction of the ciliary waves and the direction of movement. Ciliary reversal is dependent on ionic changes across

Box 16.1

Malaria Control—A Glimmer of Hope

The following quotation is from *The Lake Regions of Central Africa* by Sir Richard Burton (1821–1890). Sir Richard Burton was an adventurer whose visits to the far east and Africa brought him into contact with the greatest killer of humanity, *Plasmodium falciparum*.

> The approach of malignant fever is very insidious. An attack begins mostly with an ordinary chill, attended by no unusual or marked symptoms. Sometimes the patient has had a light chill a day or two before this, which he has neglected. Sometimes he has felt slightly unwell for ten or fourteen days; has complained of loss of appetite and general weakness; but as these symptoms are not very marked, they are very apt to be overlooked, especially with newcomers.
>
> The real attack may begin with a chill or with a fever, but its effects are, in either case, at once evident in a peculiarly yellow skin and haggard countenance. In fever there is profuse perspiration, a rush of blood to the head, high and irregular pulse, and general prostration. Sometimes the body is hot, but dry. Thirst is urgent, but the stomach rejects whatever is drunk.
>
> If the paroxysm of fever returns, it is with renewed force, and the third attack is commonly fatal. Before death the patient becomes insensible; there is violent vomiting, which is, in fact, regurgitation of ingesta, mixed with green and yellow fluid. Immediately after the chill, and even before this has passed off, the urine becomes dark red or black. The pulse is very irregular, the breathing slow and finally the patient sinks away into a state of coma, and dies without a struggle.

Malaria is still a fact of life—especially in Africa. Nearly 300 million people are afflicted annually, and one-third of them die. The fight against this disease has centered on two fronts: elimination of the mosquito species known to carry *Plasmodium* parasites and treatment of infected persons with antimalarial drugs. In the fight against mosquitoes, DDT (dichloro-diphenyl-trichloro-ethane) was employed successfully in the 1950s and 1960s. Its use greatly reduced the incidence of malaria by the middle 1960s. Mosquito-control programs, however, began to lose their effectiveness, largely because mosquitoes developed resistance to DDT. DDT has also been found to be a highly persistent pesticide. It retains its toxicity for many years and can build up to lethal levels in aquatic and terrestrial environments (see *chapter 42*). Its use became less effective and more expensive—both in economic and environmental terms. Other pesticides are now being used, but mosquito resistance is becoming a problem with these pesticides too.

Even more serious is the development of resistance by *Plasmodium* parasites to antimalarial drugs, such as chloroquine. Chloroquine is now ineffective against Asian populations of *Plasmodium* and may eventually be ineffective in Africa.

As bleak as this picture sounds, scientists and health officials are optimistic that malaria will eventually be conquered. Recent advancements in molecular biology and immunology are being employed in the development of antimalarial vaccines. Some vaccines are now being tested. Researchers have found that the complexity of the disease presents special problems in the development of vaccines. Not only do each of the *Plasmodium* species require a separate vaccine, but each stage of the life cycle of a single species requires a separate immunological component in a vaccine.

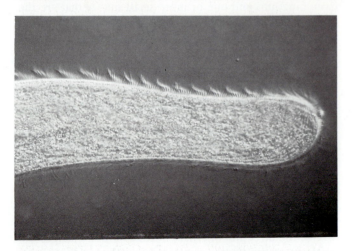

Figure 16.17 Metachronal waves in the ciliate *Spirostomum*.

the plasma membrane—analogous to the mechanism by which impulses are conducted along animal nerve and muscle cells. The coordination of cilia in metachronal waves results from coupling of adjacent cilia. When one cilium beats, neighboring cilia are induced to beat because of extracellular, mechanical interactions between cilia.

Basal bodies (kinetosomes) of adjacent cilia are interconnected with an elaborate network of fibers that are believed to anchor the cilia and give shape to the protist.

The evolution of some ciliates has resulted in the specialization of cilia (figure 16.18). Cilia may cover the outer surface of the protist (figure 16.18a) They may be joined together to form *cirri,* which are used in movement (figure 16.18b). Alternatively, cilia may be lost from large regions of a ciliate (figure 16.18c).

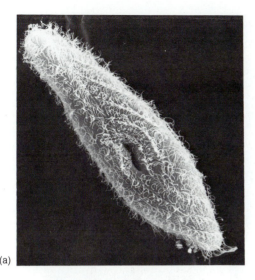

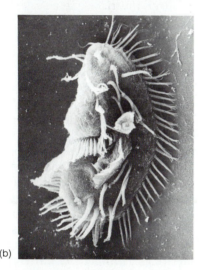

Figure 16.18 Photomicrographs showing variations in ciliature. (a) The cilia of *Paramecium* are relatively unspecialized. (b) The cilia of some ciliates are modified into organelles. The large ciliary organelles of *Oxytricha* are called cirri. Cirri are used in moving across a substrate. (c) In attached species, such as this vorticellid, *Campanella,* cilia may be lost over large regions of the ciliate. Here, cilia are used in feeding and are restricted to the oral region.

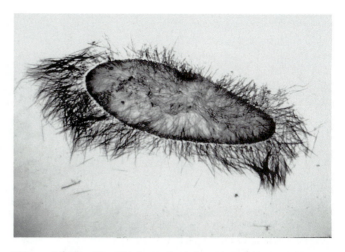

Figure 16.19 The discharged trichocysts of *Paramecium.*

Trichocysts are pellicular structures primarily used for attachment. They are rodlike or oval organelles oriented perpendicular to the plasma membrane. In *Paramecium,* they have a "golf tee" appearance (figure 16.19). Trichocysts can be discharged from the pellicle and, after discharge, they remain connected to the body by a sticky thread.

Nutrition

Some ciliates, such as *Paramecium,* have a ciliated *oral groove* along one side of the body. Cilia of the oral groove sweep small, organic particles toward the cytopharynx where a food vacuole is formed. When a food vacuole reaches an upper size limit, it breaks free and circulates through the endoplasm.

Some free-living ciliates prey upon other protists or small animals. Prey capture is usually a case of fortuitous contact. *Didinium* is a ciliate that feeds principally on *Paramecium,* a prey item that is bigger than itself. The *Didinium* forms a temporary opening that can greatly enlarge to consume its prey (figure 16.20).

Suctorians are ciliates that live attached to their substrate by a stalk. They possess tentacles to which prey adhere. Their prey, often ciliates or amebae, are paralyzed by secretions of the tentacles. The tentacles digest an opening in the pellicle of the prey, and prey cytoplasm is drawn into the suctorian through tiny channels in the tentacle. The mechanism for this probably involves tentacular microtubules (figure 16.21).

Genetic Control and Reproduction

Ciliates have two kinds of nuclei. A large, polyploid **macronucleus** regulates the daily metabolic activities. One or more smaller **micronuclei** serve as the genetic reserve of the cell.

Asexual reproduction of ciliates occurs by transverse binary fission and occasionally by budding. Budding occurs

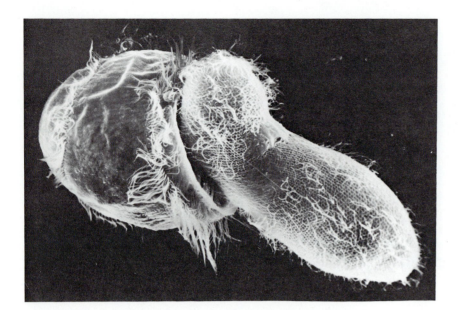

Figure 16.20 *Didinium* feeding on *Paramecium*.

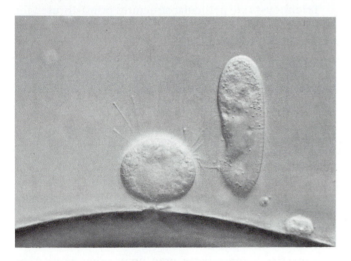

Figure 16.21 Suctorian feeding. A ciliate is held by the knobbed tip of a tentacle. Tentacles discharge enzymes that immobilize the prey and dissolve the pellicle. Pellicles of tentacle and prey fuse, and the tentacle enlarges and invaginates to form a feeding channel. Prey cytoplasm is moved down the feeding channel and incorporated into endocytic vacuoles at the bottom of the tentacle.

in suctorians and results in the formation of ciliated, free-swimming "larvae" that attach to the substrate and take the form of the adult.

Sexual reproduction of ciliates occurs by **conjugation** (figure 16.22). The partners involved are called conjugants. Many species of ciliates have numerous mating types, not all of which are compatible with one another. Compatible mating types are called *syngens*. The initial contact between individuals apparently occurs randomly, and adhesion is facilitated by sticky secretions of the pellicle. Fusion of ciliate plasma membranes occurs and lasts for several hours.

The macronucleus does not participate in the genetic exchange that follows. Instead, the macronucleus breaks up during or after micronuclear events and is reformed

from micronuclei of the daughter ciliates. Assuming that a ciliate enters into this process with a single diploid micronucleus, meiosis occurs, and four haploid pronuclei are produced. Three of these pronuclei degenerate. The remaining pronucleus undergoes mitosis to produce two genetically identical pronuclei. One of these is stationary and the other migrates to the opposite conjugant in a mutual exchange of pronuclei. Fusion of pronuclei in each of the conjugants results in a single, diploid micronucleus, called a *zygotic nucleus,* in each of the conjugants. The conjugants then separate. If conjugation begins in a species with two micronuclei (e.g., *Paramecium aurelia*), the initial meiotic divisions result in eight haploid pronuclei and seven degenerate. The single remaining pronucleus undergoes mitosis, and mutual exchange occurs as previously described.

After separation, the *exconjugants* undergo a series of nuclear divisions to restore the nuclear characteristics of the particular species, including the formation of a macronucleus from one or more micronuclei. These events are accompanied by cytoplasmic divisions.

Symbiotic Ciliates

Most ciliates are free-living; however, some are commensalistic or mutualistic and a few are parasitic. *Balantidium coli* is an important parasitic ciliate that lives in the large intestines of humans, pigs, and other mammals. At times, it is a ciliary feeder; at other times, it produces proteolytic enzymes that digest host epithelium, causing a flask-shaped ulcer. (Its pathology resembles that of *Entamoeba histolytica.*) *B. coli* is passed from one host to another in cysts that are formed as feces begin to dehydrate in the large intestine. Fecal contamination of food or water is the most common form of transmission. It is potentially worldwide in distribution, but is most common in the Philippines.

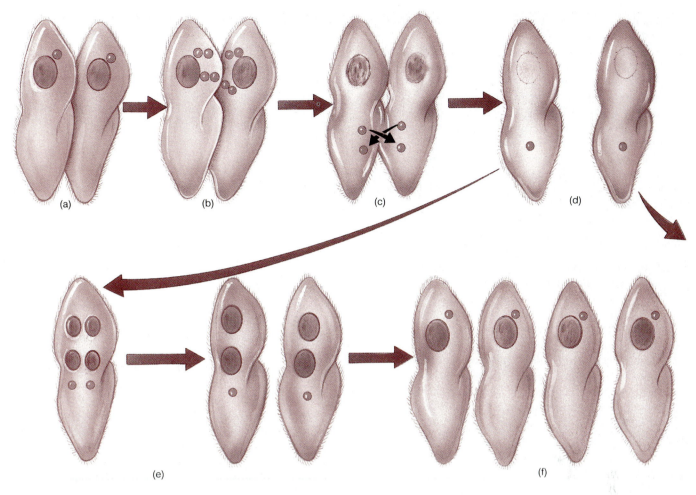

Figure 16.22 Conjugation in *Paramecium*. (a) Random contact brings individuals of opposite mating types together. (b) Meiosis results in four haploid pronuclei. (c) Three pronuclei and the macronucleus degenerate. Mitosis and mutual exchange of pronuclei is followed by fusion of pronuclei. (d–f) Separation of conjugants is followed by nuclear divisions that restore nuclear characteristics of the species. These events may be accompanied by cytoplasmic divisions.

Stop and Ask Yourself

13. How do cilia differ from flagella? How are cilia controlled?
14. What are trichocysts used for in ciliates?
15. What occurs during conjugation?
16. When does meiosis occur in ciliates—before or after syngamy?

Summary

1. Protists are a polyphyletic group that arose about 1.5 billion years ago from the Monera. The evolutionary pathways leading to modern protists are uncertain.

2. Protists are both single cells and entire organisms. Many of their functions are carried out by organelles specialized for the unicellular lifestyle.

3. Many protists live in symbiotic relationships with other organisms, often in a host-parasite relationship.

4. Members of the phylum Sarcomastigophora possess pseudopodia and/or one or more flagella.

5. Flagella have a 9 + 2 arrangement of microtubules. Sliding microtubules causes a flagellum to bend. A variety of

flagellar movements push or pull a protist through its medium.

6. Members of the class Phytomastigophora are plantlike and include the genera *Euglena* and *Volvox*. Members of the class Zoomastigophora are animallike and include *Trypanosoma*, which causes sleeping sickness.

7. Amebae use pseudopodia for feeding and locomotion. Ameboid locomotion involves interactions between the proteins actin and myosin.

8. Members of the class Lobosa include the freshwater genera *Amoeba*, *Arcella*, and *Difflugia*, and the symbiotic genus *Entamoeba*. Foraminiferans and radiolarians are common marine amebae.

9. Members of the phylum Apicomplexa are all parasites. The phylum includes *Plasmodium* and *Toxoplasma*, which cause malaria and toxoplasmosis, respectively.

10. Many apicomplexans have a three-part life cycle involving schizogony, gametogony, and sporogony.

11. The phylum Ciliophora contains some of the most complex of all protists. Its members possess cilia, a macronucleus, and one or more micronuclei.

12. Cilia are coordinated by mechanical coupling of cilia and can be specialized for different kinds of locomotion.

13. Sexual reproduction occurs in ciliates by conjugation. Diploid ciliates undergo meiosis of the micronuclei to produce haploid pronuclei that are exchanged between two conjugants.

■ Key Terms

binary fission (p. 238)
conjugation (p. 250)
contractile vacuoles
   (p. 238)
cytopharynx (p. 238)
cytopyge (p. 238)
ectoplasm (p. 238)
egestion vacuoles (p. 238)
endocytic (food) vacuoles
   (p. 238)
endoplasm (p. 238)

macronucleus (p. 249)
micronuclei (p. 249)
multiple fission
   (schizogony) (p. 238)
pellicle (p. 238)
polyphyletic (p. 237)
pseudopodia (p. 243)
trichocysts (p. 249)
unicellular (cytoplasmic)
   organization (p. 237)

■ Critical Thinking Questions

1. If it is impossible to know for certain the evolutionary pathways that gave rise to protist and animal phyla, do you think it is worth constructing hypotheses about those relationships? Why or why not?

2. In what ways are protists similar to animal cells? In what ways are they different?

3. If sexual reproduction is unknown in *Euglena*, how do you think this lineage of organisms has survived through evolutionary time? (Recall that sexual reproduction provides the genetic variability that allows species to adapt to environmental changes.)

4. What is a logical strategy for controlling a disease such as malaria? How do malarial control programs come into conflict with responsible ecological practices?

5. If you were travelling out of this country and you were concerned about contracting amebic dysentery, what steps could you take to prevent acquiring the disease? How would the precautions differ if you were going to a country where malaria was a problem?

■ Suggested Readings

Books

Barnes, R. D. 1987. *Invertebrate Zoology.* 5th ed. Philadelphia: Saunders College Publishing.

Corliss, J. D. 1979. *The Ciliated Protozoa: Characterization, Classification, and Guide to the Literature.* 2nd ed. Elmford, New York: Pergamon Press, Inc.

Farmer, J. N. 1980. *The Protozoa: Introduction to Protozoology.* St. Louis: C.V. Mosby Co.

Grell, K. G. 1973. *Protozoology.* Berlin: Springer-Verlag.

Levandowsky, M., and Hunter, S. H. 1979. *Biochemistry and Physiology of Protozoa.* New York: Academic Press. Vol I-III.

Laybourn-Parry, J. 1984. *A Functional Biology of Free-Living Protozoa.* Berkeley: University of California Press.

Nisbet, B. 1984. *Nutrition and Feeding Strategies in Protozoa.* London: Croom Helm Ltd.

Pennak. R. W. 1989. *Freshwater Invertebrates of the United States.* 3rd ed. New York: John Wiley and Sons, Inc.

Schmidt, G. D. and Roberts, L. S. 1989. *Foundations of Parasitology.* 4th ed. St. Louis: C.V. Mosby Co.

Articles

Corliss, J. 1984. The kingdom Protista and its 45 phyla. *Biosystems* 17:87–126.

Cox, F. E. G. 1988. Which way for malaria? *Nature* 331:486–487.

Hawking, F. The clock of the malaria parasite. *Scientific American* June, 1970.

Kerr, R. A. 1986. Shaping new tools for paleoceanographers. *Science* 234:427–428.

Lazarides, E., and Revel, J. P. The molecular basis of cell movement. *Scientific American* May, 1979.

Levine, N. D. 1980. A newly revised classification of the protozoa. *Journal Protozoology* 27:37–58.

Miller, L. H., Howard, R. J., Carter, R., Good, M. F., Nussenzwieg, V., and Nussenzwieg, R. S. 1986. Research toward malaria vaccines. *Science* 234:1349–1355.

Satir, P. How cilia move. *Scientific American* October, 1974.

Tangley, L. 1987. Malaria: fighting the African scourge. *BioScience* 37(2):94–98.

Phyla Porifera, Cnidaria, and Ctenophora

(a) Phylum Porifera, class Demospongia. Two orange puffball sponges (*Tethya aurantia*) surrounded by anemones (phylum Cnidaria, class Anthozoa).

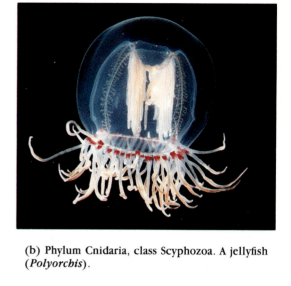

(b) Phylum Cnidaria, class Scyphozoa. A jellyfish (*Polyorchis*).

(c) Phylum Cnidaria, class Anthozoa. A California hydrocoral (*Allopora californica*).

(d) Phylum Ctenophora. Comb jellyfish (*Pleurobranchia*).

Phylum Arthropoda

(b) A brown recluse spider (*Loxosceles reclusa*) is recognized by the dark brown, violin-shaped mark on the dorsal aspect of its prosoma.

(a) Class Arachnida. A black widow spider (*Lactrodectus*) is recognized by its shiny black body with a red hourglass pattern on its abdomen.

(d) Class Malacostraca. An amphipod beachhopper (*Orchestoidea californiana*).

(c) Class Branchiopoda. A cladoceran water flea (*Daphnia*). Note the eggs carried under the carapace.

(f) Class Cirripedia. Stalked barnacles (*Lepas*).

(e) A lobster (*Homarus americanus*).

Multicellular and Tissue Levels of Organization

17

Concepts

1. How multicellularity originated in animals, and whether animals are monophyletic, diphyletic, or polyphyletic, are largely unknown.
2. Animals whose bodies consist of aggregations of cells, but whose cells do not form tissues, are found in the phyla Mesozoa, Placozoa, and Porifera.
3. Animals that show diploblastic, tissue-level organization are found in the phyla Cnidaria and Ctenophora.
4. Members of the phylum Cnidaria are important in zoological research because of their relatively simple organization and their contribution to coral reefs.

Have You Ever Wondered:

[1] how multicellularity could have arisen in the animal kingdom?
[2] how natural sponges are prepared for use in cleaning applications?
[3] what value the intricate branching canal systems are to a sponge?
[4] whether cells of a sponge body can communicate with one another?
[5] how soft-bodied cnidarians support themselves?
[6] why one should avoid touching blue, gas-filled floats washed up on beaches of temperate and tropical waters?
[7] which jellyfish should be avoided when swimming in coastal waters?
[8] what organisms are responsible for the formation of coral reefs?

These and other useful questions will be answered in this chapter.

This chapter contains underlined evolutionary concepts.

Evolutionary Perspective

Animals with multicellular and tissue levels of organization have captured the interest of scientists and laypersons alike. A description of some members of the phylum Cnidaria, for example, could fuel a science fiction writer's imagination.

From a distance I was never threatened, in fact I was infatuated with its beauty. A large, inviting, bright blue float lured me closer. As I swam nearer I could see that hidden from my previous view was an infrastructure of tentacles, some of which dangled nearly nine meters below the water's surface! The creature seemed to consist of many individuals and I wondered whether or not each individual was the same kind of being because, when I looked closely, I counted eight different body forms!

I was drawn closer and the true nature of this creature was painfully revealed. The beauty of the gas-filled float hid some of the most hideous weaponry imaginable. When I brushed against those silky tentacles I experienced the most excruciating pain. Had it not been for my life vest, I would have drowned. Indeed, for some time, I wished that had been my fate.

This fictitious account is not far from reality for swimmers of tropical waters who have come into contact with *Physalia physalis,* the Portuguese man-of-war (*see figure 17.15*). In organisms such as *Physalia physalis,* cells are grouped together, specialized for various functions, and are interdependent. This chapter covers five animal phyla with multicellular organization that varies from a loose association of cells to cells organized into two distinct tissue layers. These phyla include the Mesozoa, Placozoa, Porifera, Cnidaria, and Ctenophora.

Origins of Multicellularity

The origins of multicellularity are shrouded in mystery. Many zoologists believe that multicellularity could have [1] arisen as dividing cells remained together, in the fashion of many colonial protists. Although there are a number of variations of this hypothesis, they are all treated here as the **colonial hypothesis.** Ancient, flagellated protists are often described as possible ancestral forms because flagellated protists may be colonial. (The formation of colonies, however, is not limited to flagellated protists.) One variation of the colonial hypothesis explains the derivation of two tissue layers by invagination of a spherical, colonial flagellate similar to *Volvox* (figure 17.1a; *see figure 16.8*). The mechanism depicted in this hypothesis is reminiscent of an embryonic process called gastrulation.

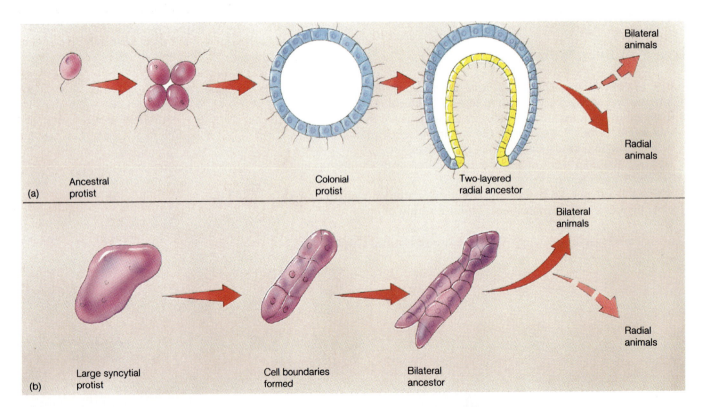

Figure 17.1 Two hypotheses regarding the origin of multicellularity. (a) The colonial hypothesis. Multicellularity may have arisen when cells produced by a dividing protist remained together. Invagination of cells could have formed a second cell layer. (b) The syncytial hypothesis. Multicellularity could have arisen when plasma membranes were formed within the cytoplasm of a large, multinucleate protist.

A second proposed mechanism is called the **syncytial hypothesis** (figure 17.1b). **A syncytial cell is a large, multinucleate cell. The formation of plasma membranes in the cytoplasm of a syncytial protist could have formed a small, multicellular organism.** This hypothesis is supported by the fact that syncytial organization occurs in some protist phyla.

Animal Origins

A fundamental question concerning animal origins is whether animals are *monophyletic* (derived from a single ancestor), *diphyletic* (derived from two ancestors), or *polyphyletic* (derived from many ancestors). The view that animals are polyphyletic is attractive to a growing number of zoologists. The nearly simultaneous appearance of all animal phyla in fossils from the Precambrian and Cambrian periods, about 550 million years ago, is difficult to explain if animals are monophyletic. If this view is correct, more than one explanation of the origin of multicellularity could be possible and more than one body form could be ancestral. Conversely, the impressive similarities in cellular organization in all animals support the view that all or most animals are derived from a single ancestor. For example, asters are formed during mitosis in most animals, certain cell junctions are similar in all animal cells, flagellated sperm are produced by most animals, and proteins that accomplish movement are similar in most animal cells. These common features are difficult to explain, assuming

polyphyletic origins. If one assumes one or two ancestral lineages, then only one or two hypotheses regarding the origin of multicellularity can be correct (colorplate 5).

Phylum Mesozoa

There are approximately 50 species in the phylum Mesozoa (mez'o-zo"ah) (Gr. *mesos,* middle + *zoin,* animal). As larvae and/or adults, mesozoans are parasites of other marine invertebrates. They have two layers: an inner layer of reproductive cells and an outer layer of ciliated cells. Mesozoans are dioecious (sexes are separate) with complex life histories involving both sexual and asexual cycles. Some zoologists now believe that the mesozoans actually represent two phyla, Orthonectida and Dicyemida (figure 17.2).

Phylum Placozoa

Members of the phylum Placozoa (plak'o-zo"ah) (Gr. *plak,* flattened + *zoin,* animal) are small (2 to 5mm) marine organisms. The phylum's one species, *Tricoplax adherans,* was first described in the 1800s, largely forgotten, and then rediscovered in the 1960s on algae from the Red Sea. *Tricoplax* consists of two epithelial layers with fiber cells (possibly locomotor in function) between them. The upper epithelium consists of flagellated cells and transparent spheres of fatty material. The lower epithelium consists of flagellated cylinder cells and gland cells of nutritive function (figure 17.3). When *Tricoplax* feeds, it forms a temporary, ventral digestive cavity by raising its body off the substrate and secreting enzymes into the cavity. The similarity of this animal to a hypothetical ancestor proposed in the syncytial hypothesis of animal origins causes some

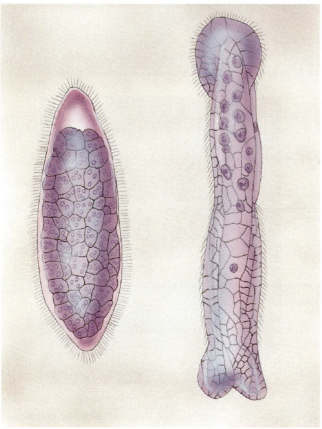

(a)   (b)

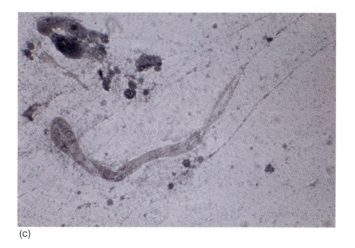

(c)

Figure 17.2 Phylum Mesozoa. (a) *Rhopalura,* an orthnectid parasite of clams. (b) *Pseudicyema,* a dicyemid parasite of cuttlefish. (c) A mesozoan from the tissues of an octopus.

Figure 17.3 Phylum Placozoa. *Tricoplax adherans* consists of two epithelial layers with fiber cells sandwiched in between those layers.

zoologists to wonder if this group could be closely related to the first animals.

Phylum Porifera

The Porifera (po-rif'er-ah) (L. *porus,* pore + *fera,* to bear), or sponges, are animals consisting of loosely organized cells, but with no well-defined tissues. They are asymmetrical (or sometimes radially symmetrical), sessile, and mostly marine (figure 17.4; table 17.1). Sponges vary in size from less than a cm to a mass that would fill one's arms. Certain sponges were used by ancient Greeks for painting, washing, and padding armor. Although synthetic sponges are common in modern homes, natural sponges are preferred for most professional applications.

Cell Types, Body Wall, and Skeletons

In spite of their relative simplicity, sponges are more than colonies of independent cells. As in all animals, sponge cells are specialized for particular functions. This organization is often referred to as *division of labor.*

Thin, flat cells, called **pinacocytes,** line the outer surface of a sponge. Pinacocytes may be mildly contractile, and their contraction may change the shape of some sponges. In a number of sponges, some pinacocytes are specialized into tubelike, contractile **porocytes,** which can regulate water circulation (figure 17.5a). Openings through porocytes are pathways for water moving through the body wall.

Just below the pinacocyte layer of a sponge is a jellylike layer referred to as the *mesohyl.* Certain cells, called *mesenchyme cells,* are found moving about in the mesohyl.

(a)

(b)

(c)

Figure 17.4 Many sponges are brightly colored, commonly with hues of red, orange, green, or yellow. (a) *Verongia.* (b) *Axiomella.* (c) *Spinosella.*

Table 17.1 Classification of the Porifera

Phylum Porifera (po-rif'er-ah)
The animal phylum whose members are sessile and either asymmetrical or radially symmetrical; body organized around a system of water canals and chambers; cells not organized into tissues or organs.

Class Calcispongiae (kal-si-spun'je-e)
Usually small sponges with monaxon, triaxon, or tetraaxon calcium carbonate spicules; three body forms. *Leucoselenia* and *Scypha.*

Class Hyalospongiae (hi'ah-lo-spun'je-e)
Triaxon siliceous spicules that are sometimes fused into an intricate lattice; cup or vase shaped; scyconoid body form; found at 450 to 900 m depths in tropical West Indies and eastern Pacific. Glass sponges.

Class Demospongiae (de-mo-spun'je-e)
Brilliantly colored sponges with monaxon or tetraaxon siliceous spicules or spongin; leuconoid body form; up to 1 m in height and diameter. Includes one family of freshwater sponges, Spongillidae, and the bath sponges, family Spongidae.

Class Sclerospongiae (skler'o-spun'je-e)
Leuconoid sponges with siliceous spicules and spongin, as well as an encasement of calcium carbonate; restricted in distribution to coral reefs in Caribbean waters. Coralline sponges.

Mesenchyme cells are ameboid and may be specialized for reproduction, secreting skeletal elements, transporting food, storing food, and forming contractile rings around openings in the sponge wall.

Below the mesohyl and lining an inner chamber(s) are choanocytes, or collar cells. **Choanocytes** (Gr. *choane,* funnel + *cyte,* cell) are flagellated cells that have a collarlike ring of microvilli surrounding a flagellum. Microfilaments connect the microvilli, forming a netlike mesh within the collar. The flagellum creates water currents through the sponge, and the collar filters microscopic food particles from the water (figure 17.5b). The presence of choanocytes in sponges suggests an evolutionary link between the sponges and a group of protists called choanoflagellates. This link is discussed further at the end of this chapter.

Sponges are supported by a skeleton that may consist of microscopic needlelike spikes called **spicules.** Spicules are formed by mesenchyme cells, are made of calcium carbonate or silica, and may take on a variety of shapes (figure 17.6). Alternatively, the skeleton may be made of **spongin,** a fibrous protein made of collagen. (Collagen is an important element in supportive tissues of animals.) A commercial sponge is the spongin skeleton of a sponge that has been dried, beaten, and washed until all cells are removed. [2]

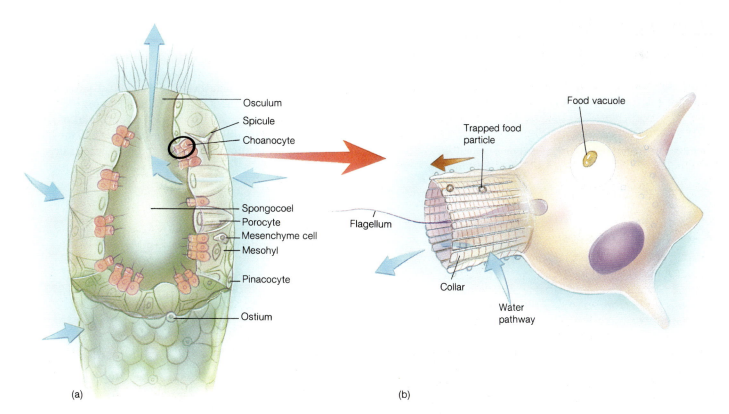

(a) (b)

Figure 17.5 Morphology of a simple sponge. (a) In this example, pinacocytes form the outer body wall, and mesenchyme cells and spicules are found in the mesohyl. Ostia are formed by porocytes that extend through the body wall. (b) Choanocytes are flagellated cells with a flagellum that is surrounded by a collar of microvilli that traps food particles. Food is moved toward the base of the cell, where it is incorporated into a food vacuole.

Water Currents and Body Forms

The life of a sponge depends upon the water currents created by choanocytes. Water currents bring food and oxygen to a sponge and carry away metabolic and digestive wastes. The way in which food filtration and circulation are accomplished is reflected in the body forms present in the phylum. Zoologists have described three sponge body forms.

The simplest and least common sponge body form is the **ascon** (figure 17.7a). Ascon sponges are vaselike. *Ostia* are the outer openings of porocytes and lead directly to a chamber called the *spongocoel*. Choanocytes line the spongocoel, and their flagellar movements draw water into the spongocoel through the ostia. Water exits the sponge through the *osculum,* which is a single, large opening at the top of the spongocoel.

In the **sycon** body form, the sponge wall appears folded (figure 17.7b). Water enters a sycon sponge through ostia, however, the ostia are not the openings of porocytes. Rather, they are the openings of invaginations of the body wall, called *incurrent canals.* Pores in the body wall connect incurrent canals to *radial canals,* and the radial canals lead to the spongocoel. Choanocytes line radial canals (rather than the spongocoel), and the beating of choanocyte flagella moves water from ostia, through incurrent and radial canals, to the spongocoel, and out the osculum.

Leucon sponges have an extensively branched canal system (figure 17.7c). Water enters the sponge through ostia and moves through branched incurrent canals, which lead to chambers that are lined by choanocytes. Canals leading away from the chambers are called *excurrent canals.* Proliferation of chambers and canals has resulted in a small spongocoel, and often multiple exit points (oscula) for water leaving the sponge.

In complex sponges, increased surface area for choanocytes results in large volumes of water being moved through the sponge and greater filtering capabilities. Although the evolutionary pathways in the phylum are complex and incompletely described, most pathways have resulted in the leuconoid body form. [3]

Maintenance Functions

Sponges feed on particles that are in the 0.1 to 50 μm size range. Their food consists of bacteria, microscopic algae, protists, and other suspended organic matter. Large populations of sponges play important roles in reducing turbidity of coastal waters. A single leuconoid sponge, 1 cm in diameter and 10 cm high, can filter in excess of 20 l of water every day!

Choanocytes filter small, suspended food particles. Water passes through their collar near the base of the cell then moves into a sponge chamber at the open end of the collar. Suspended food is trapped on the collar and moved along microvilli to the base of the collar, where it is in-

Figure 17.6 Photomicrograph of sponge spicules.

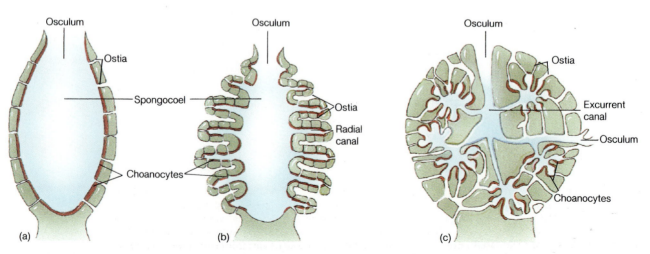

Figure 17.7 Sponge body forms. (a) An asconoid sponge. Choanocytes line the spongocoel in ascon sponges. (b) A syconoid sponge. The body wall of sycon sponges appears folded. Choanocytes line radial canals that open into the spongocoel. (c) A leuconoid sponge. The proliferation of canals and chambers has resulted in the loss of the spongocoel as a distinct chamber. Multiple oscula are frequently present.

corporated into a food vacuole (*see figure 17.5b*). Digestion occurs in the food vacuole by lysosomal enzymes and pH changes. Some food is passed undigested to ameboid mesenchyme cells, which distribute it to other cells.

Filtration is not the only way that sponges feed. Larger food particles (up to 50 μm in size) may be phagocytized by pinacocytes lining incurrent canals. Nutrients dissolved in seawater may also be absorbed by active transport.

Because of extensive canal systems and the circulation of large volumes of water through sponges, all sponge cells are in close contact with water. Thus, the loss of nitrogenous wastes (principally ammonia) and gas exchange occur by diffusion.

Sponges do not have nerve cells to coordinate body functions. Most reactions are the result of individual cells responding to a stimulus. For example, water circulation through some sponges is at a minimum at sunrise and at a maximum just before sunset because light inhibits the constriction of porocytes and other cells surrounding ostia, keeping incurrent canals open. Other reactions, however, [4] suggest some communication between cells. For example, the rate of water circulation through a sponge can drop suddenly without any apparent external cause. This reaction can be due only to choanocytes ceasing activities more-or-less simultaneously and implies some form of internal communication. The nature of this communication is unknown. Chemical messages transmitted by ameboid mesenchyme cells and electrical conduction over cell surfaces are possible control mechanisms.

Reproduction

Most sponges are monoecious (both sexes occur in the same individual) but do not usually undergo self-fertilization because they produce eggs and sperm at different times. Certain choanocytes lose their collars and flagella and undergo meiosis to form flagellated sperm. Other choanocytes (and mesenchyme cells in some sponges) probably undergo meiosis to form eggs. Eggs are retained in the mesohyl of the parent. Sperm cells exit one sponge through the osculum(a) and enter another sponge with the incurrent water. Sperm are trapped, incorporated into a vacuole and transported to an egg by choanocytes.

In most sponges, early development occurs in the mesohyl. Cleavage of a zygote results in the formation of a flagellated larval stage. (A **larva** is an immature stage that may undergo a dramatic change in structure before attaining the adult body form.) The larva breaks free and is carried out of the parent sponge by water currents. After no more than 2 days of a free swimming existence, the larva settles to the substrate and begins development of the adult body form (figure 17.8a,b).

Asexual reproduction of freshwater and some marine sponges involves the formation of resistant capsules containing masses of mesenchyme cells. These capsules, called **gemmules,** are released when the parent sponge dies in the winter and can survive both freezing and drying (figure 17.8c). When favorable conditions return in the spring, ameboid cells stream out of a tiny opening, called the *micropyle,* and organize themselves into a sponge.

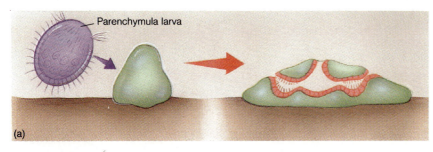

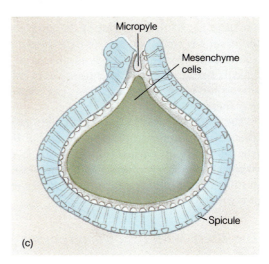

Figure 17.8 Development of sponge larval stages. (a) Most sponges have a parenchymula larva. Flagellated cells cover most of the outer surface of the larva. After the larva settles and attaches, the outer cells lose their flagella, move to the interior, and form choanocytes. Interior cells move to the periphery and form pinacocytes. (b) Some sponges have an amphiblastula larva, which is hollow and has one half of the larva composed of flagellated cells. On settling, the flagellated cells invaginate into the interior of the embryo and will form choanocytes. Nonflagellated cells overgrow the choanocytes and form the pinacoderm. (c) Gemmules are resistant capsules containing masses of mesenchyme cells. Gemmules are released when a parent sponge dies (e.g., in the winter) and ameboid mesenchyme cells form a new sponge when favorable conditions return.

Some sponges possess remarkable powers of regeneration. If a sponge is squeezed through silk cloth, the sponge is broken into clumps of cells. Random reaggregation of ameboid cells forms new sponges.

Stop and Ask Yourself

1. How do the "colonial" and "syncytial" hypotheses account for the origin of multicellularity?
2. What phylum contains animals that consist of two epithelial layers and feed by the formation of a temporary digestive cavity?
3. What are three kinds of cells found in the Porifera and what are their functions?
4. What is the path of water circulating through an ascon sponge? Through a sycon sponge?

Phylum Cnidaria or Coelenterata

Members of the phylum Cnidaria (ni·dar′e-ah) (Gr. *knide,* nettle) or Coelenterata (s-len′te-rat′ah) (Gr. *koilos,* hollow + *enteron,* gut) possess radial or biradial symmetry. Biradial symmetry is a modification of radial symmetry in which a single plane, passing through a central axis, divides the animal into mirror images. It results from the presence of a single or paired, structure in a basically radial animal and differs from bilateral symmetry in that there is no distinction between dorsal and ventral surfaces. Radially symmetrical animals have no anterior or posterior ends. Thus, terms of direction are based on the position of the mouth opening. Recall that the end of the animal that contains the mouth is the *oral end,* and the opposite end is the *aboral end.* Radial symmetry is advantageous for sedentary animals because sensory receptors are evenly distributed around the body. These organisms can respond to stimuli that come from all directions.

The Cnidaria are mostly marine and occasionally represent a significant part of the total mass of living material (the biomass) in marine (especially reef) environments (table 17.2). Many zoologists find cnidarians ideal subjects to study because of their functional simplicity and powers of regeneration.

The Body Wall and Nematocysts

Cnidarians possess diploblastic, tissue-level organization (*see figure 15.12*). Cells are organized into tissues which carry out specific functions and all cells are derived from two embryological layers. The ectoderm of the embryo gives rise to an outer layer of the body wall, called the **epidermis,** and the inner layer of the body wall, called the **gastrodermis,** is derived from endoderm (figure 17.9). Cells of the epidermis and gastrodermis are differentiated into a number of cell types that function in protection, food gathering, coordination, movement, diges-

---

Table 17.2   Classification of the Cnidaria

Phylum Cnidaria (ni·dar′e-ah)
The animal phylum whose members are characterized by radial or biradial symmetry, diploblastic organization, a gastrovascular cavity, and nematocysts.

Class Hydrozoa (hi′dro-zo″ah)
Nematocysts present on the epidermis; gametes produced epidermally and always released to the outside of the body; no wandering mesenchyme cells in mesoglea; medusae usually with a velum; some freshwater species. *Hydra, Obelia, Gonionemus, Physalia.*

Class Scyphozoa (si′fo-zo″ah)
Medusa prominent in the life history; gametes gastrodermal in origin and released into the gastrovascular cavity; nematocysts present in the gastrodermis as well as epidermis; polyp small; medusa lacks a velum; mesoglea with wandering mesenchyme cells of epidermal origin. *Aurelia.*

Class Cubozoa (ku′bo-zo″ah)
Medusa prominent in life history; polyp small; gametes gastrodermal in origin; medusa cuboidal in shape with tentacles that hang from each corner of the bell. *Chironex.*

Class Anthozoa (an′tho-zo″ah)
Colonial or solitary polyps; medusae absent; nematocysts present in the gastrodermis; gametes gastrodermal in origin; gastrovascular cavity divided by mesenteries that bear nematocysts; internal biradial or bilateral symmetry present. Anemones and corals. *Metridium.*

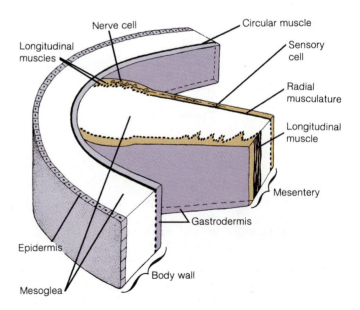

Figure 17.9 The body wall of a cnidarian (class Anthozoa). Cnidarians are diploblastic (two tissue layers). The epidermis is derived embryologically from ectoderm and the gastrodermis is derived embryologically from endoderm. Between these layers is mesoglea.

tion, and absorption. Between the epidermis and gastrodermis is a jellylike layer called **mesoglea.** Cells in the mesoglea of some cnidarians have their origin in either the epidermis or the gastrodermis. The mesoglea is 90% water; the remainder is inorganic salts and organic materials. (In comparison, your tissues are about 65% water.)

One kind of cell, though not uniquely cnidarian, is characteristic of the phylum. Epidermal and/or gastrodermal cells called **cnidocytes** produce structures called nematocysts, which are used for attachment, defense, and feeding. A **nematocyst** is a fluid-filled capsule enclosing a coiled, hollow tube (figure 17.10). The capsule is capped at one end by a lidlike *operculum.* The cnidocyte has a modified cilium, called a *cnidocil.* When the cnidocil is stimulated, the operculum is forced open and the coiled tube is discharged—as one would evert a sweater sleeve that had been turned inside out. A combination of mechanical and chemical stimuli is most effective in initiating nematocyst discharge. For example, a prey organism brushing against a cnidocil may cause a nematocyst to discharge; however, a clean glass rod touching a cnidocil usually does not initiate discharge. The strength of the stimulus required to cause nematocyst discharge varies and is partially controlled by nerve endings associated with the bases of cnidocytes.

Nearly 30 kinds of nematocysts have been described.

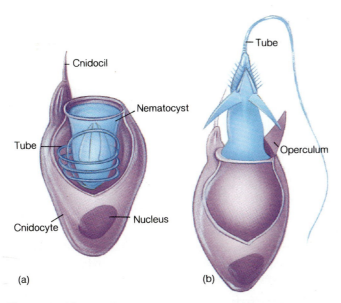

Figure 17.10 Cnidocyte structure and nematocyst discharge. (a) A nematocyst develops in a capsule in the cnidocyte. The capsule is capped at its outer margin by an operculum (lid) that is displaced upon discharge of the nematocyst. The triggerlike cnidocil is responsible for nematocyst discharge. (b) A discharged nematocyst. When the cnidocil is stimulated, there is a rapid (osmotic) influx of water, causing the nematocyst to evert, first near its base, and then progressively along the tube from base to tip. The tube revolves at enormous speeds as the nematocyst is discharged. In nematocysts that are armed with barbs, the advancing tip of the tube is aided in its penetration of the prey as barbs spring forward from the interior of the tube and then flick backward along the outside of the tube.

Nematocysts used in food gathering and defense may discharge a long tube armed with spines that penetrates the prey. The spines have hollow tips that discharge paralyzing toxins. Other nematocysts contain unarmed tubes that wrap around prey or a substrate. Still other nematocysts have sticky secretions that help the animal to anchor itself. Six or more kinds of nematocysts may be present in one individual.

Polymorphism

Most cnidarians are polymorphic, meaning that they possess two body forms in their life histories (figure 17.11). The **polyp** is usually asexual and sessile. It is attached to a substrate at the aboral end, has a cylindrical body, called the *column,* and a mouth surrounded by food-gathering tentacles. The **medusa** (plural medusae) is dioecious and free swimming. It is shaped like an inverted bowl and has tentacles dangling from its margins. The mouth opening is centrally located facing downward, and the medusa swims by gentle pulsations of the body wall. The mesoglea is more abundant in a medusa than in a polyp, giving the former a jellylike consistency.

Maintenance Functions

The gastrodermis of all cnidarians lines a blind-ending cavity, called the **gastrovascular cavity.** This cavity serves as a digestive chamber, as a cavity for the exchange of respiratory gases and metabolic wastes, and as a cavity into which reproductive products may be discharged. Both food and digestive wastes enter and leave the gastrovascular cavity through the mouth.

The food of most cnidarians consists of very small crustaceans. Nematocysts entangle and paralyze prey, contractile cells in the tentacles cause the tentacles to shorten, and food is drawn toward the mouth. As food enters the gastrovascular cavity, gastrodermal *gland cells (see figure 17.9)* secrete lubricating mucus and enzymes, which reduce food to a soupy broth. Certain gastrodermal cells, called *nutritive-muscular* cells phagocytize partially digested food and incorporate it into vacuoles, where digestion is completed. Nutritive-muscular cells also have circularly oriented contractile fibers that help move materials into or out of the gastrovascular cavity by peristaltic contractions. During peristalsis, ringlike contractions move along the body wall, pushing contents of the gastrovascular cavity ahead of them, expelling undigested material through the mouth.

Cnidarians derive most of their support from the buoyancy of water around them. In addition, they have a hydrostatic skeleton to aid in support and movement. A **hydrostatic skeleton** is water or body fluids confined in a cavity of the body and against which contractile elements of the body wall act. In the Cnidaria, the water-filled gastrovascular cavity acts as a hydrostatic skeleton. Certain cells of the body wall, called *epithelio-muscular cells,* are contractile and aid in movement. When a polyp closes its

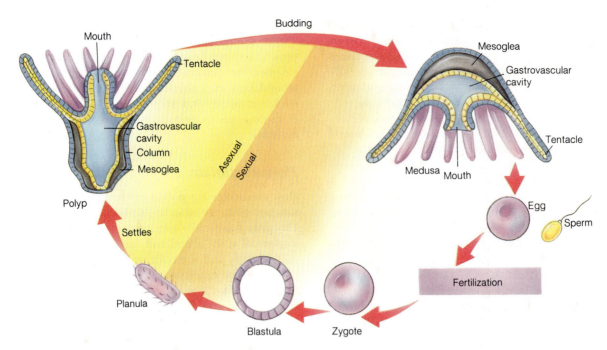

Figure 17.11 A generalized cnidarian life cycle showing alternation between medusa and polyp body forms. Dioecious medusae produce gametes that may be shed to the outside of the organism, where fertilization takes place. Early development forms a ciliated planula larva. After a brief free swimming existence, the planula settles to the substrate and forms a polyp. Budding of the polyp produces additional polyps and medusa buds. Medusae break free of the polyp and swim away. The polyp or medusa of many species is either lost or reduced, and the sexual and asexual stages have been incorporated into one body form.

mouth (to prevent water from escaping) and contracts longitudinal epithelio-muscular cells on one side of the body, the polyp bends toward that side. If these cells contract while the mouth is open, water escapes from the gastrovascular cavity, and the polyp collapses. Contraction of circular epithelio-muscular cells causes constriction of a part of the body and, if the mouth is closed, water in the gastrovascular cavity is compressed, and the polyp elongates.

Polyps use a variety of forms of locomotion. They may move by somersaulting from base to tentacles and from tentacles to base again, or move in an inchworm fashion, using their base and tentacles as points of attachment. Polyps may also glide very slowly along a substrate while attached at their base or walk on their tentacles.

Medusae move by swimming and floating. Most horizontal movements are from being passively carried by water currents and wind. Vertical movements are the result of swimming. Contractions of circular and radial epithelio-muscular cells cause rhythmic pulsations of the bell and drive water from beneath the bell, propelling the medusa through the water.

Cnidarian nerve cells have been of interest to zoologists for many years because they may be the most primitive nervous elements in the animal kingdom. By studying these cells, zoologists may gain insight into the evolution of animal nervous systems. Nerve cells are located below the epidermis, near the mesoglea, and interconnect to form a two-dimensional *nerve net*. This net conducts impulses around the body in response to a localized stimulus. The extent to which an impulse spreads over the body depends upon the strength of a stimulus. For example, a weak stimulus applied to a polyp's tentacle may cause the tentacle to be retracted. A strong stimulus at the same point may cause the entire polyp to withdraw from the stimulus.

Sensory structures of cnidarians include receptors for perceiving touch and certain chemicals and are distributed throughout the body. More specialized receptors are located at specific sites on a polyp or medusa. Receptors for equilibrium and balance and photoreception are described later in this chapter.

Because cnidarians have large surface area to volume ratios, all cells are a short distance from the body surface, and oxygen, carbon dioxide, and nitrogenous wastes can be exchanged by diffusion.

Reproduction

Most cnidarians are dioecious. Sperm and eggs may be released into the gastrovascular cavity or to the outside of the body. In some instances, eggs are retained in the parent until after fertilization.

A blastula forms early in development, and migration of surface cells to the interior fills the embryo with cells that will eventually form the gastrodermis. The embryo elongates to form a ciliated, free-swimming larva, called a *planula*. The planula attaches to a substrate, interior cells split to form the gastrovascular cavity, and a young polyp develops (figure 17.11).

Medusae are nearly always formed by budding from the body wall of a polyp, and polyps may form other polyps by budding. Buds may detach from the polyp or they may remain attached to the parent to contribute to a colony of individuals. Variations on this general pattern will be discussed in the survey of cnidarian classes that follows.

Stop and Ask Yourself

5. What are the three layers of the cnidarian body wall? Which layer may contain cells that do not originate in that layer?
6. What is a nematocyst? What are several functions of nematocysts?
7. How would you characterize the nervous organization of cnidarians?
8. What is a hydrostatic skeleton? What cavity of a cnidarian serves as the hydrostatic compartment?

Class Hydrozoa

Hydrozoans (hi'dro-zo"ah) are small, relatively common cnidarians. The vast majority are marine, but this is the one cnidarian class with freshwater representatives. Most hydrozoans have life cycles that display distinct polymorphism; however, some are strictly polypoid, and others are predominately medusoid.

Hydrozoans can be distinguished from other cnidarians by three features (*see table 17.2*). Nematocysts are only in the epidermis; gametes are epidermal and released to the outside of the body rather than into the gastrovascular cav-

ity; and the mesoglea never contains ameboid mesenchyme cells.

Most hydrozoans have colonial polyp forms, some of which may be specialized for feeding, producing medusae by budding, or defending the colony. In *Obelia,* a common marine cnidarian, the planula develops into a feeding polyp, called a **gastrozooid** (gas'tra-zo'oid) (figure 17.12). The gastrozooid has tentacles, feeds on microscopic organisms in the water, and secretes a skeleton, called the *perisarc,* around itself. The perisarc is made of protein and chitin and covers all of the gastrozooid, except for the tentacles.

Growth of an *Obelia* colony results from budding of the original gastrozooid. Rootlike processes grow into and horizontally along the substrate. They anchor the colony and give rise to branch colonies. The entire colony has a continuous gastrovascular cavity, a continuous body wall, a continuous perisarc, and is a few cm high. Gastrozooids are the most common type of polyp in the colony; however, some polyps produced during the growth of an *Obelia* colony are called gonozooids. A **gonozooid** (gon'o-zo'oid) is a reproductive polyp that produces medusae by budding. *Obelia*'s small medusae are formed on a stalklike structure of the gonozooid. When medusae mature, they break free of the stalk and swim out an opening at the end of the gonozooid. Medusae reproduce sexually to give rise to more colonies of polyps.

Gonionemus (figure 17.13a) is a hydrozoan in which the medusa stage predominates. It lives in shallow marine waters, where it is often found clinging to seaweeds by adhesive pads on its tentacles. The biology of *Gonionemus* is typical of most hydrozoan medusae. The margin of the *Gonionemus* medusa projects inward to form a shelflike

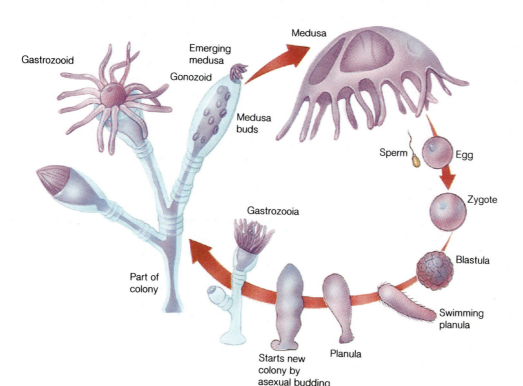

Figure 17.12 *Obelia* structure and life cycle. This hydrozoan displays alternation between polyp and medusa stages. Unlike *Obelia,* the majority of colonial hydrozoans have medusae that remain attached to the parental colony and gametes or larval stages are released from the medusa through the gonozooid. The medusa is often degenerate and may be little more than gonadal specializations in the gonozooid.

lip, called the *velum*. A velum is found on most hydrozoan medusae but is absent in all other cnidarian classes. The velum concentrates water expelled from beneath the medusa to a smaller outlet, creating an efficient jet-propulsion system. The mouth is at the end of a tubelike *manubrium* that hangs from the medusa's oral surface. The gastrovascular cavity leads from the inside of the manubrium into four radial canals that extend to the margin of the medusa. Radial canals are connected at the margin of the medusa by an encircling ring canal.

In addition to a nerve net, *Gonionemus* has a concentration of nerve cells, called a nerve ring, that encircles the margin of the medusa. The *nerve ring* coordinates swimming movements and is often likened to a primitive central nervous system. (Central nervous systems are concentrations of nervous tissues involved with receiving information from sensory receptors and initiating responses to that information.) Embedded in the mesoglea around the margin of the medusa are sensory structures called statocysts (figure 17.13b). **Statocysts** consist of a small sac surrounding a calcium carbonate concretion called a *statolith*. When *Gonionemus* tilts, the statolith moves in response to the pull of gravity, and nerve impulses are initiated, which may change position or swimming behavior.

Gonads of *Gonionemus* medusae hang from the oral surface, below the radial canals. *Gonionemus* is dioecious, and the gametes are shed directly into sea water. A planula larva develops and attaches to the substrate, eventually forming a polyp (about 5 mm tall). The polyp reproduces by budding to make more polyps and medusae.

Hydra is a common freshwater hydrozoan that is found hanging from the underside of floating plants in clean streams and ponds (figure 17.14). *Hydra* lacks a medusa stage and reproduces asexually by budding from the side of the polyp. Hydras are somewhat unusual hydrozoans, because sexual reproduction occurs in the polyp stage. Testes are conical elevations of the body surface that form from mitosis of certain epidermal cells called *interstitial cells*. Sperm form by meiosis in cysts in the testes. Mature sperm exit the testes through temporary openings in the end of a cyst. Ovaries also form from interstitial cells. One large egg is formed per ovary. During egg formation, yolk is incorporated into the egg cell from gastrodermal cells. As ovarian cells disintegrate, the egg is left attached to the body wall by a thin stalk of tissue. After fertilization and early development, epithelial cells lay down a resistant chitinous shell. The embryo drops from the parent, overwinters, hatches in the spring, and develops into an adult.

Large oceanic hydrozoans belong to the order Siphonophora. These colonies are associations of numerous polypoid and medusoid individuals. Some polyps, called dactylozooids, possess a single, long (up to 9 m) tentacle armed with nematocysts that are used in capturing prey. Other polyps are specialized for digesting prey. Various medusoid individuals form swimming bells, sac floats, oil floats, leaflike defensive structures, and gonads.

Physalia, commonly called the Portuguese man-of-war, is a very large, colonial siphonophore (figure 17.15). It

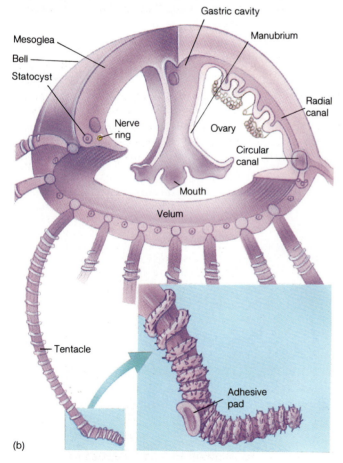

(a)

(b)

Figure 17.13 (a) A *Gonionemus* medusa. (b) The structure of *Gonionemus*.

Figure 17.13 redrawn, with permission, from *Living Invertebrates* by Pearse/Buchsbaum, copyright by The Boxwood Press.

Figure 17.14 A *Hydra* that has just fed on a small crustacean and captured a second prey. The swollen body wall indicates the presence of food in the gastrovascular cavity.

Figure 17.15 *Physalia physalis,* the Portuguese man-of-war. The bluish float is about 12 cm long, and the nematocyst-laden tentacles can be up to 9 m long. Note the fish that has been captured by the tentacles.

[6] lacks swimming capabilities and moves at the mercy of wind and waves. Its nematocyst-laden dactylozooids are lethal to small vertebrates and dangerous to humans. During feeding, fish are drawn by the dactylozooids to within ten cm of the float, where feeding polyps begin digestion. Digestion leaves only the skeletal remains of the fish.

Class Scyphozoa

Members of the class Scyphozoa (si′fo-zo″ah) are all marine and are called "true jellyfish" because the dominant stage in their life history is the medusa (figure 17.16). Unlike hydrozoan medusae, scyphozoan medusae lack a velum, the mesoglea contains ameboid mesenchyme cells, nematocysts occur in the gastrodermis as well as the epidermis, and gametes are gastrodermal in origin (*see table 17.2*).

Many scyphozoans are harmless to humans; others can deliver unpleasant and even dangerous stings. *Mastigias quinquecirrha,* the so-called stinging nettle, is a common Atlantic scyphozoan whose populations increase in late summer and become hazardous to swimmers (figure 17.16a). A rule of thumb for swimmers is to avoid helmet-shaped jellyfish with long tentacles and fleshy lobes hanging from the oral surface. [7]

(a)   (b)

Figure 17.16 Representative scyphozoans. (a) *Mastigias quinquecirrha.* (b) *Aurelia.*

Aurelia is a common scyphozoan in both Pacific and Atlantic coastal waters of North America (figure 17.16b). The margin of its medusa has a fringe of tentacles and is divided by notches. The mouth of *Aurelia* leads to a stomach with four *gastric pouches,* which contain nematocyst-laden *gastric filaments.* Radial canals lead from gastric pouches to the margin of the bell. In *Aurelia,* but not all scyphozoans, the canal system is extensively branched and leads to a ring canal around the margin of the medusa. Gastrodermal cells of all scyphozoans possess cilia for the continuous circulation of seawater and partially digested food.

Aurelia is a plankton feeder. At rest, it sinks slowly in the water, and microscopic animals are trapped in mucus on the epidermal surfaces. This food is carried by cilia to the margin of the medusa. Four fleshy lobes, called *oral lobes,* hang from the manubrium and are used to scrape food from the margin of the medusa (figure 17.17a). Cilia on the oral lobes carry food to the mouth.

In addition to general sensory receptors located on the epidermis, *Aurelia* has eight specialized structures, called rhopalia, located in the notches at the margin of the medusa. Each *rhopalium* (figure 17.17b) consists of two sensory pits (presumed to be olfactory) and a statocyst. Photoreceptors, called ocelli, are also associated with rhopalia. *Aurelia* displays a distinct phototaxis, coming to the surface at twilight and descending to greater depths during bright daylight.

Scyphozoans are dioecious. *Aurelia's* eight gonads are located in the gastric pouches, two per pouch. Gametes are released into the gastric pouches. Sperm swim through the mouth to the outside of the medusa. In some scyphozoans, eggs are fertilized in the female's gastric pouches, and early development occurs there. In *Aurelia,* eggs lodge in the oral lobes, where fertilization and development to the planula stage occurs.

The planula develops into a polyp called a *scyphistoma* (figure 17.18). The scyphistoma lives a year or more, during which time budding produces miniature medusae, called *ephyrae.* Repeated budding of the scyphistoma results in ephyrae being stacked on the polyp like saucers piled on top of one another. After ephyrae are released, they gradually attain the adult form.

Class Cubozoa

The class Cubozoa (ku'bo-zo"ah) was formerly classified as an order in the Scyphozoa. The medusa is cuboidal, and tentacles hang from each of its corners. Polyps are very small and, for some species, polyps are unknown. Cubozoans are very active swimmers and feeders in warm tropical waters. Some possess dangerous nematocysts (figure 17.19).

Class Anthozoa

Members of the class Anthozoa (an'tho-zo"ah) are colonial or solitary and lack medusae. They include anemones, corals, sea fans, and sea pansies. Anthozoans are all marine and are found at all depths.

Anthozoan polyps differ from hydrozoan polyps in three respects. (1) The mouth of an anthozoan leads to a *pharynx,* which is an invagination of the body wall that leads into the gastrovascular cavity. (2) The gastrovascular cavity is divided into sections by *mesenteries* that bear nematocysts and gonads on their free edges. (3) The mesoglea contains ameboid mesenchyme cells (*see table 17.2*).

Externally, anthozoans appear to show perfect radial symmetry. Internally, the mesenteries and other structures convey biradial, or even bilateral, symmetry to members of this class.

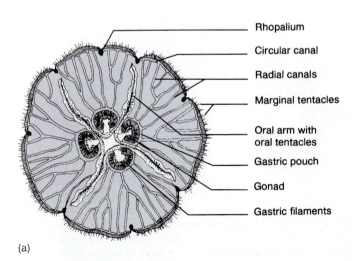

(a)

Rhopalium
Circular canal
Radial canals
Marginal tentacles
Oral arm with oral tentacles
Gastric pouch
Gonad
Gastric filaments

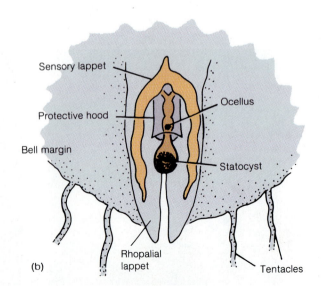

(b)

Sensory lappet
Protective hood
Bell margin
Ocellus
Statocyst
Rhopalial lappet
Tentacles

Figure 17.17 *Aurelia.* (a) Internal structure. (b) A section through a rhopalium of *Aurelia.* Each rhopalium consists of two sensory (olfactory) lappets, a statocyst, and a photoreceptor called an ocellus.

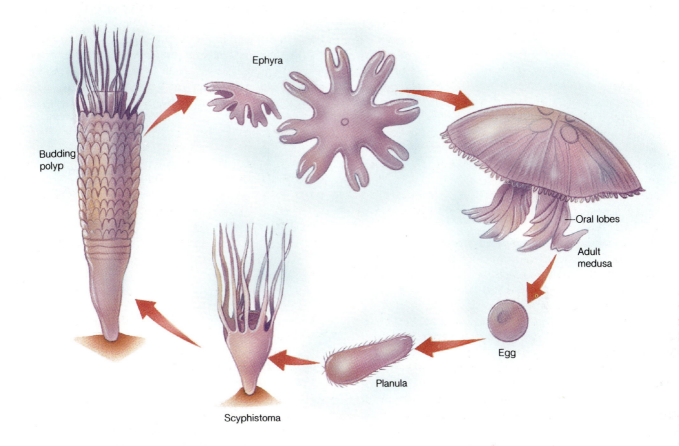

Figure 17.18 *Aurelia* life history. *Aurelia* is dioecious and, like all scyphozoans, the medusa predominates in the life history of the organism. The planula develops into a polyp called a scyphistoma, which produces young medusae, or ephyrae, by budding.

Figure 17.19 Class Cubozoa. The sea wasp, *Chironex fleckeri*. Note the cuboidal medusa and the tentacles that hang from the corners of the bell. *Chironex fleckeri* has caused more human suffering and death off Australian coasts than the Portuguese man-of-war has in any of its home waters. Death from heart failure and shock is not likely unless one is repeatedly stung.

(a)

(b)

Figure 17.20 Representative sea anemones. (a) Giant sea anemone (*Anthopleura xanthogrammica*). (b) This sea anemone lives in a mutualistic relationship with a hermit crab (*Petrochirus diobenes*). Hermit crabs lack a heavily armored exoskeleton over much of their bodies and seek refuge in empty snail shells. When this crab outgrows its present home it will take its anemone with it to a new snail shell. This anemone, riding on the shell of the hermit crab, has a degree of mobility that is unusual for other anemones. The crab, in turn, is protected from predators by the anemone's nematocysts.

Sea anemones are solitary, frequently large, and colorful (figure 17.20a). Some attach to solid substrates, some burrow in soft substrates, and some live in symbiotic relationships (figure 17.20b). The polyp attaches to its substrate by a *pedal disc* (figure 17.21). An *oral disc* contains the mouth and hollow, oral tentacles. At one or both ends of the slitlike mouth is a *siphonoglyph,* which is a ciliated tract that circulates water through the gastrovascular cavity to maintain the hydrostatic skeleton.

Mesenteries are arranged in pairs. Some attach at the body wall at their outer margin and to the pharynx along their inner margin. Other mesenteries attach to the body wall, but are free along their entire inner margin. Openings in mesenteries near the oral disc permit circulation of water between compartments set off by the mesenteries. The free lower edges of the mesenteries form a trilobed *mesenterial filament.* Mesenterial filaments bear nematocysts, cilia that

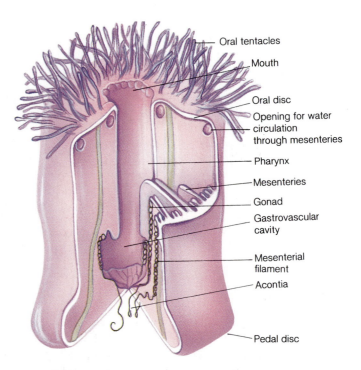

Figure 17.21 The structure of the anemone, *Metridium.*

aid in water circulation, gland cells that secrete digestive enzymes, and absorptive cells that absorb products of digestion. Many anemones have thread-like *acontia* that bear nematocysts in the gastrovascular cavity. Acontia are defensive and can be extruded through small openings in the body wall or through the mouth when an anemone is threatened.

Muscle fibers are largely gastrodermal. Longitudinal muscle bands are restricted to the mesenteries. Circular muscles are in the gastrodermis of the column. When threatened, anemones contract their longitudinal fibers, allowing water to escape from the gastrovascular cavity. This action causes the oral end of the column to fold over the oral disc, and the anemone appears to collapse. The reestablishment of the hydrostatic skeleton depends upon gradual uptake of water into the gastrovascular cavity via the siphonoglyphs.

Anemones are capable of limited locomotion. Movement is accomplished by gliding on their pedal disc, crawling on their side, and walking on their tentacles. When disturbed, some "swim" by thrashing their body or tentacles. Some anemones float using a gas bubble held within folds of the pedal disc.

Anemones feed on invertebrates and fish. Tentacles capture prey and draw it toward the mouth. Radial muscle fibers in the mesenteries open the mouth to receive the food.

Both sexual and asexual reproduction are shown by anemones. In asexual reproduction, a piece of pedal disc may break away from the polyp and grow into a new individual in a process called *pedal laceration.* Alternatively, longitudinal or transverse fission may divide one individual into two, with missing parts being regenerated. Unlike other

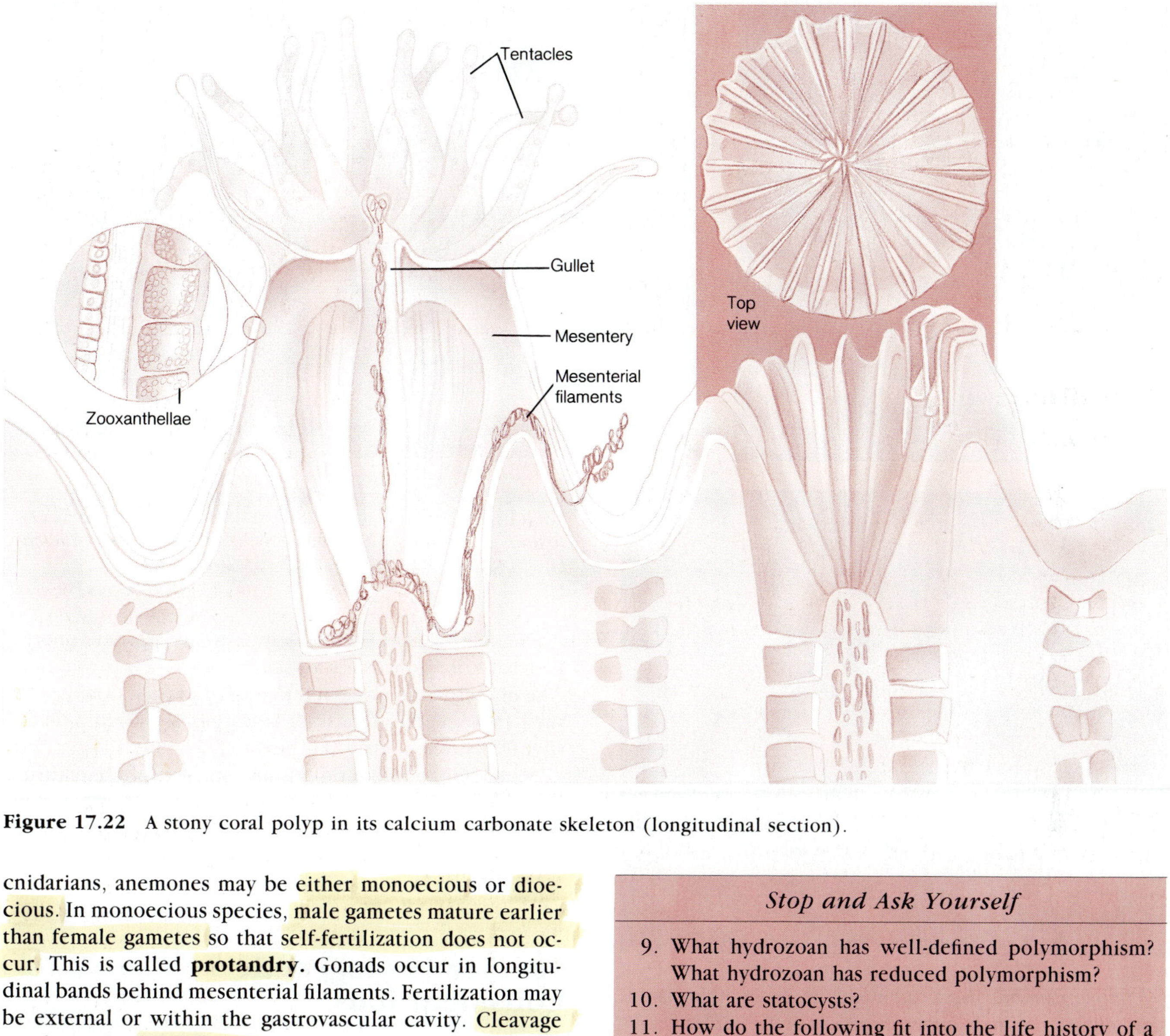

Figure 17.22 A stony coral polyp in its calcium carbonate skeleton (longitudinal section).

cnidarians, anemones may be either monoecious or dioecious. In monoecious species, male gametes mature earlier than female gametes so that self-fertilization does not occur. This is called **protandry.** Gonads occur in longitudinal bands behind mesenterial filaments. Fertilization may be external or within the gastrovascular cavity. Cleavage results in the formation of a planula, which develops into a ciliated larva that settles to the substrate, attaches, and eventually forms the adult.

[8] Other anthozoans are coral-forming animals. Stony corals are responsible for the formation of coral reefs (box 17.1) and, except for lacking siphonoglyphs, are similar to the anemones. Their common name derives from a cuplike calcium carbonate exoskeleton secreted around their base and the lower portion of their column, by epithelial cells (figure 17.22). When threatened, polyps retract into their protective skeletons. Sexual reproduction is similar to that of anemones, and asexual budding produces other members of the colony.

The colorful octacorallian corals are common in warm waters. They have eight pinnate (featherlike) tentacles, eight mesenteries, and one siphonoglyph. The body walls of members of a colony are connected, and mesenchyme cells secrete an internal skeleton of protein or calcium carbonate. Sea fans, sea whips, red corals, and organ-pipe corals are members of this group (figure 17.23).

Stop and Ask Yourself

9. What hydrozoan has well-defined polymorphism? What hydrozoan has reduced polymorphism?
10. What are statocysts?
11. How do the following fit into the life history of a scyphozoan: Scyphistoma? Ephyra? Planula?
12. What is protandry? How does it apply to the Anthozoa?

Phylum Ctenophora

Animals in the phylum Ctenophora (ti-nof'er-ah) (Gr. *kteno,* comb + *phoros,* to bear) are called sea walnuts or comb jellies (table 17.3). There are approximately 90 described species, all of which are marine. As do the Cnidaria, ctenophores possess a form of radial (i.e., biradial) symmetry, are diploblastic with a gastrovascular cavity, and have an epidermal nerve net. These similarities suggest close evolutionary relationships between cnidarians and ctenophores. Most have a spherical form, although several groups are flattened and/or elongate.

Figure 17.23 Representative octacorallian corals. (a) Red gorgonian (*Lophogorgia chilensis*). (b) Fleshy sea pen (*Ptilosaurus gurneyi*). (c) Organ pipe coral (*Tubipora musica*). (d) Purple sea fan (*Gorgonia ventalina*).

Pleurobranchia is monoecious, as are all ctenophores. Two bandlike gonads are associated with the gastrodermis. One of these is an ovary and the other a testis. Gametes are shed through the mouth, fertilization is external, and a slightly flattened larva develops.

Many ctenophores are luminescent, with light-emitting cells associated with the gastrodermis. Luminescent ctenophores may occur in large swarms, creating spectacular sights on late summer evenings.

Pleurobranchia has a spherical or ovoid, transparent body about 2 cm in diameter. It occurs in the colder waters of the Atlantic and Pacific oceans (figure 17.24). *Pleurobranchia*, like most ctenophorans, has eight meridional bands of cilia, called **comb rows,** that run between the oral and aboral poles. Comb rows are locomotor structures that are coordinated through a statocyst at the aboral pole. *Pleurobranchia* normally swims with its aboral pole oriented downward. Tilting is detected by the statocyst, and the comb rows adjust the animal's orientation. Two long, branched tentacles arise from pouches near the aboral pole. Tentacles possess contractile fibers that retract the tentacles and adhesive cells, called **colloblasts,** which are used for prey capture (figure 17.24c).

Ingestion occurs as the tentacles wipe the prey across the mouth. The mouth leads to a branched gastrovascular canal system. Some canals are blind; however, two small, anal canals open to the outside near the apical sense organ. Thus, unlike the cnidarians, ctenophores have an anal opening. Some undigested wastes are eliminated through these canals, and some are probably also eliminated through the mouth.

Further Phylogenetic Considerations

The evolutionary position of the phyla covered in this chapter is subject to debate. If the animal kingdom is polyphyletic, then all phyla could have had separate origins, although scientists who believe in multiple origins agree that

Table 17.3 Classification of the Ctenophora

Phylum Ctenophora (ti-nof′er-ah)
The animal phylum whose members are biradially symmetrical, diploblastic, usually ellipsoid or spherical in shape, possess colloblasts, and have meridionally arranged comb rows.

  Class Tentaculata (ten-tak′u-lata)
  With tentacles that may or may not be associated with sheaths, into which the tentacles can be retracted. *Pleurobranchia.*

  Class Nuda (nuda)
  Without tentacles; flattened; a highly branched gastrovascular cavity. *Beroe.*

the number of independent origins is probably small. Many zoologists believe it to be at least diphyletic, with the Porifera being derived separately from all other phyla. The similarity of poriferan choanocytes and choanoflagellate protists suggests evolutionary ties between these groups. Although some zoologists believe that choanoflagellate like protists could have given rise to phyla other than the Porifera, most do not. One thing that nearly everyone agrees upon, however, is that the Porifera are evolutionary "dead ends." They gave rise to no other animal phyla.

If two origins are assumed, the origin of the nonporiferan lineage is also debated. One interpretation is that the ancestral animal was derived from a radially symmetrical ancestor, which in turn may have been derived from a colonial flagellate similar in form to *Volvox* (*see figure 16.8*). If this is true, then the *radiate phyla* (Cnidaria and Ctenophora) could be closely related to that ancestral group (colorplate 5). Other zoologists contend that bilateral symmetry is the ancestral body form, and a bilateral ancestor gave rise to both the radiate phyla and bilateral phyla (colorplate 5). In this interpretation, the radiate phyla are further removed from the base of the evolutionary tree.

The evolutionary relationships of the cnidarian classes to each other and to the ctenophorans is also debated. The classical interpretation is that primitive Hydrozoa were the ancestral radial animals. Zoologists that favor this hypothesis believe that the polyp stage originated as a larval stage and that the medusa, because it is sexual, represents the primitive adult stage. The Anthozoa may then have evolved by incorporating sexual reproduction into the polyp and eliminating the medusa stage. Likewise, the Scyphozoa could have evolved from a primitive hydrozoan stock through reduction of the polyp stage. This hypothesis is not universally agreed upon. Other interpretations depict the Scyphozoa or the Anthozoa as the primitive cnidarian stock.

There is very little disagreement that the Ctenophora and the Cnidaria are closely related. Some zoologists think of these phyla as being derived from a common ancestor. Others believe that the Ctenophora were derived from an ancient cnidarian medusa.

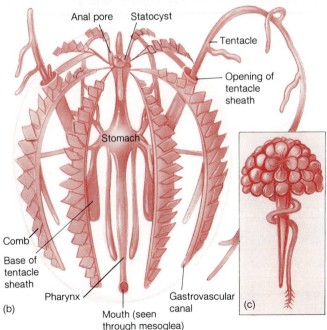

Figure 17.24 (a) The ctenophore *Mnemiopsis*. (b) The structure of *Pleurobranchia*. (c) Colloblasts consist of a hemispherical sticky head that is connected to the core of the tentacle by a straight filament. A contractile spiral filament coils around the straight filament. Straight and spiral filaments prevent struggling prey from escaping. The animal usually swims with the oral end forward or upward.

Figure 17.24 redrawn, with permission, from *Living Invertebrates* by Pearse/Buchsbaum, copyright by The Boxwood Press.

Stop and Ask Yourself

13. What characteristics are shared by the Ctenophora and the Cnidaria?
14. What are colloblasts?
15. How are a statocyst and comb rows used by ctenophores to maintain an upright position in the water?
16. The sponges are considered evolutionary "dead ends." Why?

Box 17.1

Coral Reefs

Not all corals build coral reefs. Those that do not are called *ahermatypic corals* and often live at great depths in cold seawater. *Hermatypic corals* are reef-building species. Coral reefs are built as calcium carbonate skeletons of one generation of stony corals are built upon the skeletons of preceding generations. It requires millions of years for massive reefs, such as those found in warm, shallow waters of the Indian Ocean, the south Pacific Ocean, and the Caribbean Sea to develop. Reef formation requires constantly warm (20° C), shallow water (less than 90 m), and constant salinity near 3.5%.

Most reef building activities are the result of stony corals living in a mutualistic relationship with a group of dinoflagellates called **zooxanthellae** (*see figure 17.22*). Stony corals depend on photosynthetic activities of zooxanthellae as a principal source of carbohydrates. Predatory activities serve mainly as a source of protein for polyps. Zooxanthellae also promote exceptionally high rates of calcium deposition. Calcium carbonate is laid down as a result of the conversion of carbon dioxide (CO_2) and water (H_2O) to carbonic acid (H_2CO_3). This reaction is facilitated by the enzyme carbonic anhydrase, which is present in coral polyps. Carbonic acid is then quickly converted to carbonate (CO_3^{2-}) without an enzyme catalyst. Calcium from seawater combines with carbonate to form calcium carbonate ($CaCO_3$). As zooxanthellae carry on photosynthesis, they remove CO_2 from the environment of the polyp. Associated pH changes induce the precipitation of dissolved $CaCO_3$ as aragonite (coral limestone). It is thought that the 90 m depth limit for reef building corresponds to the limits to which sufficient light penetrates to support dinoflagellate photosynthesis.

Certain algae, called **coralline algae,** live outside the coral organisms and create their own calcium carbonate masses. These algae contribute to the reef by cementing together larger coral formations.

Reef formations can extend hundreds of meters below the ocean's surface, however, only the upper and outer layer includes living material. Most of the reef formation consists of skeletons of previous generations of stony corals. (The depth of the reef mass is evidence of changing oceanic levels during glacial periods and of the sub-sidence of the ocean floor.) In addition to the outer layer of photosynthetic and cnidarian life forms, the reef supports a host of other organisms, including fishes, molluscs, arthropods, echinoderms, and sponges. The exceptionally high productivity of reef communities depends upon the ability of reef organisms to recycle nutrients rather than to lose them to the ocean floor.

There are three types of coral reefs. *Fringing reefs* are built up from the sea bottom so close to a shoreline that no navigable channel exists between the shoreline and the reef. This reef formation frequently creates a narrow, shallow lagoon between the reef and the shore. Surging water creates frequent breaks and irregular channels through these reefs. *Barrier reefs* are separated from shore by wide, deep channels. The Great Barrier Reef of Australia is 1700 km long with a channel 20 to 50 m deep and up to 48 km wide. (The Great Barrier Reef actually consists of a number of different reef forms, including barrier reefs.) *Atolls* are circular reefs that enclose a lagoon in the open ocean. One hypothesis regarding their origin, first described by Charles Darwin, is that atolls were built up around an island that later sank.

Box Figure 17.1 A fringing reef surrounding an island off the eastern coast of Australia. Note the human-made shipping channel cut through the reef.

Summary

1. Although the origin of multicellularity in animals is unknown, the colonial hypothesis and the syncytial hypothesis are explanations of how animals could have arisen. Whether the animal kingdom had origins in one, two, or many ancestors is debated.

2. Members of the phylum Mesozoa are parasites of marine invertebrates and have a two cell-layer design.

3. Placozoa are marine animals with two epithelial layers. They feed by forming a temporary digestive cavity.

4. Animals in the phylum Porifera are the sponges. Cells of sponges are specialized to perform functions, such as creating water currents, filtering food, producing gametes, forming skeletal elements, and lining the sponge body wall.

5. Sponges circulate water through their body to bring in food and oxygen and to carry away wastes and reproductive products. Evolution has resulted in most sponges having complex canal systems and large water-circulating capabilities.

6. Members of the phylum Cnidaria are radially or biradially symmetrical and possess diploblastic, tissue-level organization. Cells are specialized for food gathering, defense, contraction, coordination, digestion, and absorption.

7. The Hydrozoa differ from members of other cnidarian classes in that hydrozoans have no nematocysts in their gastrodermis, have ectodermal gametes, and mesoglea without mesenchyme cells. Most hydrozoans have well-developed polyp and medusa stages.

8. The class Scyphozoa contains the jellyfish. The polyp stage of scyphozoans is usually very small.

9. Members of the class Cubozoa are found in warm, tropical waters. Some possess dangerous nematocysts.

10. The Anthozoa lack the medusa stage. They include sea anemones and corals.

11. Members of the phylum Ctenophora are biradially symmetrical and diploblastic. They are characterized by bands of cilia, called comb rows.

12. Most authorities agree that the Porifera originated separately from other animals. The Cnidaria and Ctenophora are closely related phyla and may be closely related to ancestral animals.

Key Terms

ascon (p. 258)
choanocytes (p. 257)
cnidocytes (p. 261)
colloblasts (p. 270)
colonial hypothesis (p. 254)
comb rows (p. 270)
coralline algae (p. 272)
epidermis (p. 260)
gastrodermis (p. 260)
gastrovascular cavity (p. 261)
gastrozooid (p. 263)
gemmules (p. 259)
gonozooid (p. 263)
hydrostatic skeleton (p. 261)
larva (p. 259)
leucon (p. 258)
medusa (p. 261)
mesoglea (p. 261)
nematocyst (p. 261)
pinacocytes (p. 256)
polyp (p. 261)
porocytes (p. 256)
protandry (p. 269)
spicules (p. 257)
spongin (p. 257)
statocysts (p. 264)
sycon (p. 258)
syncytial hypothesis (p. 255)
zooxanthellae (p. 272)

Critical Thinking Questions

1. If most animals are derived from a single ancestral stock, and if that ancestral stock was radially symmetrical, would the "colonial hypothesis" or the "syncytial hypothesis" of animal origins be more attractive to you? Explain.

2. Why do most zoologists believe that animals have at least two origins?

3. Colonies were defined in chapter 16 as "loose associations of independent cells." Why are sponges considered to have surpassed that level of organization? In your answer, compare a sponge with a colonial protist like *Volvox*.

4. Compare and contrast the mechanisms used by *Volvox*, freshwater sponges, and freshwater hydras for surviving harsh winter conditions of temperate lakes.

5. Most sponges and sea anemones are monoecious, yet sexual reproduction usually occurs between separate individuals. Why is this advantageous for these animals? What ensures that self fertilization does not occur in sea anemones?

6. Do you think that polymorphism is advantageous for cnidarians? Explain.

7. Why do most zoologists believe that ctenophores and cnidarians are closely related?

Suggested Readings

Books

Barnes, R. D. 1987. *Invertebrate Zoology.* 5th ed. Philadelphia: Saunders College Publishing.

Dunn, D. F. 1982. Cnidaria. In S. P. Parker (ed.). *Synopsis and Classification of Living Organisms,* Vol 1. New York: McGraw-Hill Book Co.

Harbison, G. R., and Madin, L. P. 1982. Ctenophora. Parker, S. P. (ed.). *Synopsis and Classification of Living Organisms,* Vol. 1. New York: McGraw-Hill Book Co.

Articles

Adey, W. H. 1987. Food production in low-nutrient seas; bringing tropical ocean deserts to life. *BioScience* 37:340–348.

Bilbaut, A., Hernandes-Nicaise, M. L., Leech, C. A., and Meech, R. W. 1988. Membrane currents that govern smooth muscle contraction in a ctenophore. *Nature* 331:533–535.

Brownlee, S. 1987. Jellyfish aren't out to get us. *Discover* 8:42–52.

Goreau, T. F., Goreau, N. I., and Goreau, T. J. Corals and coral reefs. *Scientific American* August, 1979.

Herbert, S. Darwin as a geologist. *Scientific American* May, 1986.

Lenhoff, H. M., and Lenhoff, S. G. Trembly's polyps. *Scientific American* April, 1988.

Ricciuti, E. R. 1986. A genuine monster. *Audubon* 88:22–24.

Roberts, L. 1988. Corals remain baffling. *Science* 239:256.

Stuller, J. 1988. With the gales in their sails. *Audubon* 90:84–85.

Wilkinson, C. R. 1987. Interocean differences in size and nutrition of coral reef sponge popualtions. *Science* 236:1654–1657.

18

The Triploblastic, Acoelomate Body Plan

Concepts

1. The acoelomates are represented by the phyla Platyhelminthes, Nemertea, and Gnathostomulida. These phyla are phylogenetically important because they are transitional between radial animals and the more complex bilateral animals. Three important characteristics appeared initially in this group: bilateral symmetry, a true mesoderm that gives rise to muscles and other organs, and a nervous system with a primitive brain and nerve cords.
2. Because the mesodermal mass completely fills the area between the outer epidermis and digestive tract, these animals lack a body cavity; hence their name, acoelomates.
3. The phylum Platyhelminthes is a large group of dorsoventrally flattened animals commonly called the flatworms.
4. Members of the class Turbellaria are mostly free living. Representatives of the class Trematoda include the subclasses Aspidogastrea, Monogenea, and Digenea. The Cestoidea (the tapeworms) are exclusively vertebrate parasites.
5. The phylum Nemertea (proboscis or ribbon worms) contains predominantly marine, elongate, burrowing worms that possess digestive and vascular systems.
6. The phylum Gnathostomulida (the gnathostomulids) is a small, recently discovered group of minute marine worms living anaerobically.
7. Acoelomates may have evolved from a primitive organism resembling a modern turbellarian.

Have You Ever Wondered:

[1] what you call a larva that is sexually mature?
[2] why monoecious animals exchange sperm?
[3] how parasites called "flukes" got their name?
[4] how monogenetic flukes got their name?
[5] why *Fasciola hepatica* is called the sheep liver fluke?
[6] how 10-meter-long tapeworms feed, even though they lack a mouth and digestive tract?

These and other useful questions will be answered in this chapter.

This chapter contains underlined evolutionary concepts.

Evolutionary Perspective

Members of the phyla Platyhelminthes, Nemertea, and Gnathostomulida are the first animals to exhibit bilateral symmetry (*see figure 15.11*) and a body organization more complex than that of the cnidarians. All the animals covered in this chapter are triploblastic (have three primary germ layers), acoelomate (without a coelom), and classified into three phyla: (1) the phylum Platyhelminthes includes the flatworms that are either free living (e.g., turbellarians) or parasitic (e.g., flukes and tapeworms); (2) the phylum Nemertea includes a small group of elongate, unsegmented, soft-bodied worms that are mostly marine; and (3) the phylum Gnathostomulida is a recently discovered group of about 100 species of minute, anaerobic worms that are found in mud and sand.

The evolutionary relationship of the major phylum in this chapter (Platyhelminthes) to other phyla is controversial. One interpretation is that the triploblastic acoelomate body plan is an important intermediate between the radial, diploblastic plan and the triploblastic coelomate plan (colorplate 5a). The flatworms would thus represent an evolutionary side branch from a hypothetical triploblastic acoelomate ancestor. Evolution from radial ancestors could have involved a larval stage that became sexually mature in its larval body form. Larval sexual maturation is

[1] called **neoteny** (Gr. *neos,* new + *teinein,* to extend) and has occurred many times in animal evolution.

Other zoologists envision the evolution of the triploblastic, acoelomate plan from a bilateral ancestor (colorplate 5b). Primitive acoelomates, similar to flatworms, would have preceded the radiate phyla, and the radial, diploblastic plan would be secondarily derived.

The recent discovery of a small group of worms (Lobatocercebridae, Annelida) that shows both flatworm and annelid characteristics (annelids are a group of coelomate animals, such as the earthworm) suggests that the acoelomate body plan is a secondary characteristic. In this case, the flatworms would represent a side branch that resulted from the loss of a body cavity.

Phylum Platyhelminthes: The Flatworms

Animals in the phylum Platyhelminthes (Gr. *platys,* flat + *hemins,* worm) are dorsoventrally flattened, exhibit both cephalization and bilateral symmetry, are triploblastic and acoelomate. Their mesodermally derived tissues include a loose tissue called **parenchyma** (Gr. *parenck,* anything poured in beside) that fills spaces between other more specialized tissues, organs, and the body wall. This is the first

phylum covered that has organ-level organization—a significant evolutionary advancement over the tissue level of organization.

When a digestive tract is present in flatworms, it has a single opening, a mouth, and ends blindly. Thus, the mouth must serve for both ingestion of food and egestion of digestive wastes. Flatworms have an excretory system that helps maintain water and ionic balance. Most flatworms are monoecious (hermaphroditic) and have well-developed reproductive systems. These animals range in adult size from 1 mm or less to many meters in length.

The almost 13,000 species of Platyhelminthes are divided into three classes (table 18.1): (1) the Turbellaria consist of mostly free-living flatworms, whereas the (2) Trematoda and (3) Cestoidea contain solely parasitic species.

Table 18.1 Classification of the Platyhelminthes into Three Classes

Phylum Platyhelminthes (plat″e-hel-min′thez) Flatworms; bilateral acoelomates.

Class Turbellaria (tur′bel-lar′e-a) Mostly free living and aquatic; external surface usually ciliated; predaceous; possess rhabdites, protrusible proboscis, frontal glands, and many mucous glands; mostly hermaphroditic. *Convoluta, Notoplana, Dugesia.*

Class Trematoda (trem′a-to′da) Trematodes (digenetic flukes); all are parasitic; several holdfast devices present; have complicated life cycles involving both sexual and asexual reproduction.

Subclass Aspidogastrea Mostly endoparasites of molluscs; possess large opisthaptor; most lack an oral sucker. *Aspidogaster, Cotylaspis.*

Subclass Monogenea Mostly ectoparasites on vertebrates; one life-cycle form in only one host; bear opisthaptor. *Disocotyle, Polystoma.*

Subclass Digenea Adults endoparasites in vertebrates; at least two different life-cycle forms in two or more hosts; have oral sucker and acetabulum. *Schistosoma, Fasciola, Clonorchis.*

Class Cestoidea (ses-toid′da) All parasitic with no digestive tract; have great reproductive potential; tapeworms

Subclass Cestodaria Body not subdivided into proglottids; larva in crustaceans, adult in fish. *Amphilina.*

Subclass Eucestoda True tapeworms; body divided into scolex, neck, and strobila; strobila composed of many proglottids; both male and female reproductive systems in each proglottid; adults in the digestive tract of vertebrates. *Protocephalus, Taenia, Echinococcus, Taeniarhynchus, Diphyllobothrium.*

Class Turbellaria: The Free-Living Flatworms

Members of the class Turbellaria (L. *turbellae*, a commotion + *aria*, like) are mostly free-living bottom dwellers in freshwater and marine environments where they crawl on stones, sand, or vegetation. Over 300 species have been described. Turbellarians feed as predators and scavengers. The few terrestrial turbellarians known live in the humid tropics and subtropics. Although most turbellarians are less than 10 mm long, a few are very large (e.g., the terrestrial, tropical ones may reach 60 cm in length). Coloration is mostly in shades of black, brown, and gray, although some groups display brightly colored patterns (figure 18.1).

Body Wall   The triploblastic acoelomate characterization of the flatworms describes the general structure of the turbellarian body wall. As in the Cnidaria, the ectodermal derivatives include an epidermis that is in direct contact with the environment (figure 18.2). Some epidermal cells are ciliated and others contain microvilli. A basement membrane of connective tissue separates the epidermis from mesodermally derived tissues. An outer layer of circular muscle and an inner layer of longitudinal muscle lie beneath the basement membrane. Other muscles are located dorsoventrally and obliquely between the dorsal and ventral surfaces. Between the longitudinal muscles and the gastrodermis are the loosely organized parenchymal cells.

The innermost tissue layer is the endodermally derived gastrodermis. It consists of a single layer of cells that comprise the digestive cavity. The gastrodermis secretes enzymes that aid in digestion and absorb the end products of digestion.

On the ventral surface of the body wall there are several types of glandular cells that are of epidermal origin. **Rhabdites** are rodlike cells that swell and form a protective

defensive mucous sheath around the body. **Adhesive glands** open to the epithelial surface and produce a chemical that attaches part of the turbellarian to a substrate. **Releaser glands** secrete a chemical that dissolves the attachment as needed.

Locomotion   Turbellarians are primarily bottom dwellers that glide over the substrate. They move using cilia and muscular undulations. As they move, turbellarians lay down a sheet of mucus that aids in adhesion and helps the cilia gain traction. The densely ciliated ventral surface and the flattened body of turbellarians enhance the effectiveness of this locomotion.

Digestion and Nutrition   One order of free-living marine turbellarians, the Acoela, lacks a digestive cavity (figure 18.3a). The mouth opens through a pharynx into an area of loosely packed parenchyma cells rather than into a blind digestive tract that is characteristic of other turbellarians. This blind cavity varies from a simple, unbranched chamber in the order Macrostomida, to a highly branched system of digestive tubes in the orders Tricladida and Polycladida (figure 18.3b,d,e). Others (the order Lecithoepitheliata) have digestive tracts that are lobed or diverticulated (figure 18.3c). From an evolutionary perspective, highly branched digestive systems are an advancement that results in more gastrodermis being close to the sites of digestion and absorption and reduces the distance the nutrients must diffuse. This design is especially important in some of the larger turbellarians and partially compensates for the absence of a circulatory system.

The turbellarian pharynx functions as an ingestive organ. It varies in design from a simple ciliated tube to a complex organ developed from the folding of muscle layers. In the latter, the free end of the tube lies in a pharyngeal sheath

Figure 18.1   This marine flatworm, (*Prostheceraeus bellustriatus*) shows brilliant markings and bilateral symmetry. This tiger flatworm inhabits the warm shallow waters around the Hawaiian Islands.

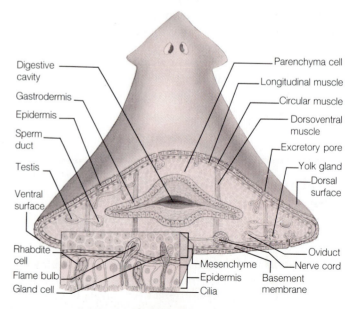

Digestive cavity
Gastrodermis
Epidermis
Sperm duct
Testis
Ventral surface
Rhabdite cell
Flame bulb
Gland cell

Parenchyma cell
Longitudinal muscle
Circular muscle
Dorsoventral muscle
Excretory pore
Yolk gland
Dorsal surface
Oviduct
Nerve cord
Mesenchyme
Epidermis
Cilia
Basement membrane

Figure 18.2   Platyhelminthes. Cross section through the body wall of a sexually mature turbellarian showing the relationships of the various body structures.

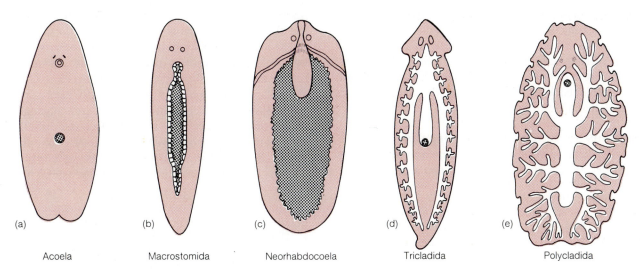

(a)	(b)	(c)	(d)	(e)
Acoela	Macrostomida	Neorhabdocoela	Tricladida	Polycladida

Figure 18.3 Digestive systems in some of the various orders of turbellarians (a-e).

and can project out of the mouth when feeding (figure 18.4).

Most turbellarians, such as the common planarian, are carnivores and feed on small, live invertebrates or scavenge on larger, dead animals; some are herbivores and feed on algae that they scrape from rocks. They can detect the presence of food from a considerable distance by means of sensory cells (*chemoreceptors*) located on their heads.

Digestion of the food is partially extracellular. Enzymes are secreted from pharyngeal glands that help break down food into smaller units that can be taken into the pharynx. Once inside the digestive cavity, small units of food are engulfed by phagocytic cells that line it, and digestion is completed in intracellular vesicles.

Exchanges With the Environment   There are no respiratory organs in the turbellarians; thus, respiratory gases (CO_2 and O_2) are exchanged by diffusion through the body wall. Most metabolic wastes (e.g., ammonia) are also removed by diffusion through the body wall. Osmoregulation, the regulation of water and ions in the body, is carried out by organs called **protonephridia** (Gr. *protos,* first + *nephros,* kidney; singular, protonephridium). Protonephridia are a network of fine tubules that run the length of the turbellarian, along each of its sides (figure 18.5a). Numerous, fine side branches of the tubules originate in the mesenchyme as tiny enlargements called **flame cells** (figure 18.5b). Flame cells have numerous cilia that project into the lumen of the tubule. The tubule wall surrounding

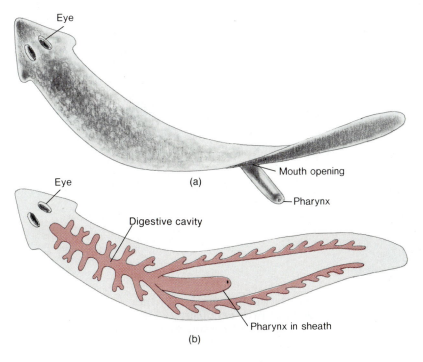

Figure 18.4   The turbellarian pharynx. (a) A planarian turbellarian with its pharynx extended in the feeding position, and (b) retracted within the pharyngeal sheath.

the flame cell is perforated by slitlike *fenestrations* (openings). The beating of the cilia drives fluid down the tubule, creating a negative pressure in the tubule. As a result, fluid from the surrounding tissue is sucked through the fenestrations into the tubule. The tubules eventually merge and open to the outside of the body wall through a minute opening called a **nephridiopore.**

Nervous System and Sense Organs

The most primitive type of flatworm nervous system, found in worms in the order Acoela (e.g., *Convoluta* spp.), is a subepi-

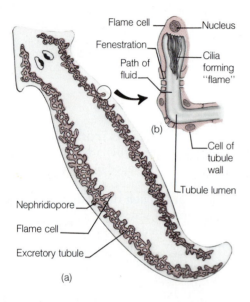

Figure 18.5 Protonephridial (excretory) system in a turbellarian. (a) The system lies in the mesenchyme and consists of a network of fine tubules that run the length of the animal on each side and open to the surface by minute excretory pores called nephridiopores. (b) Numerous fine, side branches from the tubules originate in the mesenchyme in enlargements called flame cells.

dermal nerve plexus (figure 18.6a). This plexus resembles the nerve net of cnidarians. A statocyst in the anterior end functions as a *mechanoreceptor* (a receptor that is excited by mechanical pressure) that detects the position of the turbellarian's body in reference to the direction of gravity. Turbellarians in the order Polycladida also have a nerve net but it shows more centralization with cerebral ganglia (figure 18.6b). The nervous system of most other turbellarians, such as the planarian *Dugesia,* consists of a subepidermal nerve net and several pairs of long nerve cords (figure 18.6c). The nerve cords are connected by lateral branches called *commissures* (points of union). Nerve cords and their commissures give a ladderlike appearance to the turbellarian nervous system. These neurons are organized into sensory (going to the primitive brain), motor (going away from the primitive brain), and association (connecting) types—an important evolutionary advance with respect to the nervous system. Anteriorly, the nervous tissue is concentrated into a pair of *cerebral glanglia* (singular, glanglion) that is called a primitive brain.

Turbellarians are capable of responding to a variety of stimuli in their external environment. Many tactile and sensory cells are distributed over the body. These cells detect touch, water currents, and chemicals. **Auricles** (or sensory lobes) may project from the side of the head (figure 18.6c). Chemoreceptors that aid in the location of food are especially dense in these auricles.

Most turbellarians have two simple eyespots called **ocelli** (*see figure 18.4*). These eyespots are used in orienting the animal with respect to the direction of light. (Most turbellarians are negatively phototactic and move away from light.) Each eyespot (ocellus) consists of a cuplike depression lined with black pigment (figure 18.6d). Photoreceptor nerve endings in the cup are part of the neurons that leave the eye and connect with a cerebral ganglion.

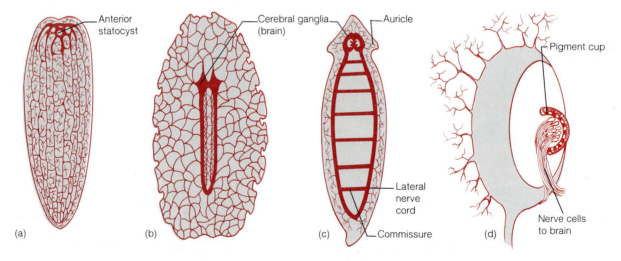

Figure 18.6 Nervous systems in Turbellaria. (a) *Convoluta* has a nerve net with a statocyst. (b) The nerve net in a turbellarian in the order Polycladida has cerebral glanglia and two lateral nerve cords. (c) The cerebral ganglia and nerve cords in the planarian, *Dugesia.* (d) Illustration of a section through the eye (ocellus) of a turbellarian.

Figures 18.5 and 18.6 redrawn, with permission, from *Living Invertebrates* by Pearse/Buchsbaum, copyright by The Boxwood Press.

Reproduction and Development   Many turbellarians reproduce asexually by transverse fission. Fission usually begins as a constriction behind the pharynx (figure 18.7). The two (or more) animals that result from fission are called **zooids** and they will regenerate missing parts after separating from each other. Sometimes, the zooids remain attached until they have attained a fairly complete degree of development, at which time, they detach as independent individuals.

Turbellarians, such as the common planarian, are monoecious, and reproductive systems arise from the mesodermal tissues in the parenchyma. Numerous paired testes lie along each side of the worm and are the site of sperm production. Sperm ducts (*vas deferens*) lead to a *seminal vesicle* (a sperm storage organ) and a protrusible *penis* (figure 18.8). The penis projects into a *genital chamber*. The female system has one to many pairs of *ovaries. Oviducts* lead from the ovaries to the genital chamber, which opens to the outside through the genital pore.

[2] Even though turbellarians are monoecious, reciprocal sperm exchange between two animals is usually the rule. This cross-fertilization ensures greater genetic diversity than does self-fertilization. During cross-fertilization, the penis of each individual is inserted into the copulatory sac of the partner. After copulation, sperm move from the copulatory sac to the genital chamber and then through the oviducts to the ovaries, where fertilization occurs. Yolk may either be directly incorporated into the egg during egg formation, or yolk cells may be laid around the zygote as it passes down the female reproductive tract past the *vitellaria* (yolk glands).

Zygotes are laid with or without a gellike mass. In many turbellarians, zygotes are enclosed in a hard capsule called a **cocoon** (Latin, *coccum,* egg shell). These cocoons are attached to the substrate by a stalk and contain several embyros per capsule. Two kinds of capsules are laid. Summer capsules hatch in 2 to 3 weeks, and immature animals emerge. Autumn capsules have thick walls that can resist freezing and drying, and hatch after overwintering.

Development of most turbellarians is direct—a gradual series of changes transform the embryos into adults. A few turbellarians have a free-swimming stage called a **Muller's larva.** It has ciliated arms that it uses in feeding and locomotion. The larva eventually settles to the substrate and develops into a young turbellarian.

Stop and Ask Yourself

1. What are several general characteristics of the phylum Plathyhelminthes?
2. What is a cocoon? Muller's larva? Zooid?
3. How does a turbellarian move? Feed? Reproduce?

Class Trematoda: The Flukes

The approximately 8,000 species of parasitic flatworms in the class Trematoda (Gr. *trematodes,* perforated form) are collectively called **flukes,** a word that describes their wide, flat shape. Almost all adult flukes are parasites of vertebrates, whereas immature stages may be found in vertebrates, invertebrates, or encysted on plants. Many species are of great economic and medical importance. [3]

Most flukes are flat and oval to elongate and range from less than 1 mm to 6 cm in length (figure 18.9). They feed on host cells and cell fragments. The digestive tract includes a mouth and muscular, pumping pharynx. Posterior to the pharynx, the digestive tract divides into two blind-ending, variously branched cecae (singular, cecum). Some flukes may supplement their feeding by absorbing nutrients across their body wall.

The structure of the body wall is similar for all flukes and represents an evolutionary adaptation to the parasitic

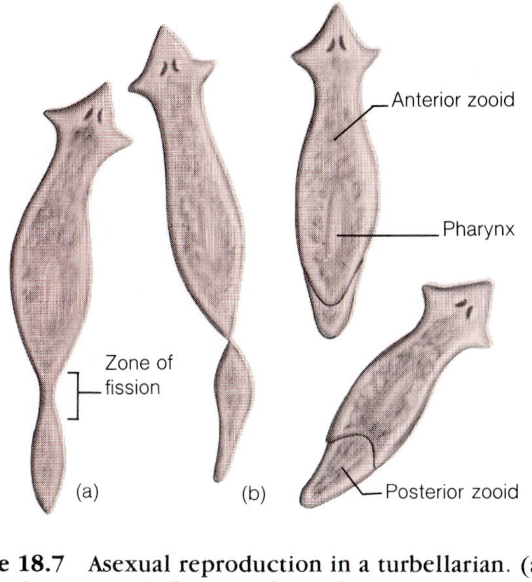

Figure 18.7   Asexual reproduction in a turbellarian. (a) Just before division and (b) just after. The posterior zooid will soon develop a head, pharynx, and other structures.

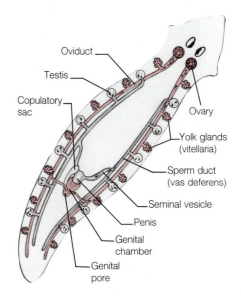

Figure 18.8   The reproductive system of a turbellarian includes both male and female structures.

Figure 18.8 redrawn, with permission, from *Living Invertebrates* by Pearse/Buchsbaum, copyright by The Boxwood Press.

way of life. The epidermis consists of an outer layer called the **tegument** (figure 18.10), which forms a syncytium (a continuous layer of fused cells). The outer zone of the tegument consists of an organic layer of proteins and carbohydrates called the *glycocalyx*. The glycocalyx aids in the transport of nutrients, wastes, and gases across the body wall, and protects the fluke against enzymes and the host's immune system. Also found in this zone are microvilli that facilitate nutrient exchange. Cytoplasmic bodies that contain the nuclei and most of the organelles lie below the basement membrane. Slender cell processes called *cytoplasmic bridges* connect the cytoplasmic bodies with the outer zone of the tegument.

Subclass Aspidogastrea   This subclass consists of a small group of flukes that are primarily internal parasites (*endoparasites*) of molluscs. All aspidogastreans are characterized by the presence of a large, oval holdfast organ called the **opisthaptor** that covers the entire ventral surface of the animal (figure 18.11). The opisthaptor functions as an extremely strong attachment organ and is usually subdivided by ridges or septae. The oral sucker, characteristic of most other trematode mouths, is absent. The life cycle of aspidogastreans may involve only one host (a mollusc), or two hosts. In the latter case, the final host is usually a vertebrate that becomes infected by ingesting a mollusc that contains immature aspidogastreans.

Subclass Monogenea   Monogenetic flukes are so named because there is but one generation in their life cycle; i.e., one adult develops from one egg. Monogeneans are mostly external parasites (*ectoparasites*) of freshwater and marine fishes, where they attach to the gill filaments and feed on epithelial cells, mucus, or blood. Attachment is faciliated by a large, posterior opisthaptor (figure 18.12). When eggs are produced, they leave the adult and contain one or more threads. These threads are sticky and attach the eggs to the fish gill. Eventually, a ciliated larva called [4]

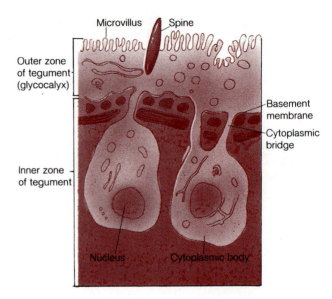

Figure 18.10   Drawing showing the fine structure of the tegument of a fluke. This figure represents an evolutionary design that is highly efficient at absorbing nutrients.

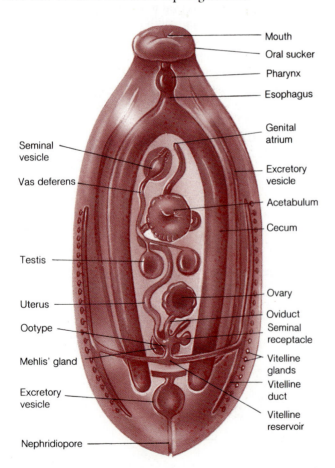

Figure 18.9   A stylized and generalized fluke (digenetic trematode).

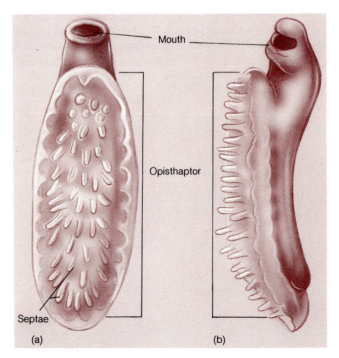

Figure 18.11   A representative aspidogastrean fluke. (a) Ventral and (b) lateral views showing the large opisthaptor and numerous septae.

Figures 18.9 and 18.11 redrawn, with permission, from P. E. Lutz, *Invertebrate Zoology.* Copyright © 1986, Benjamin/Cummings Publishing Company, Menlo Park, California.

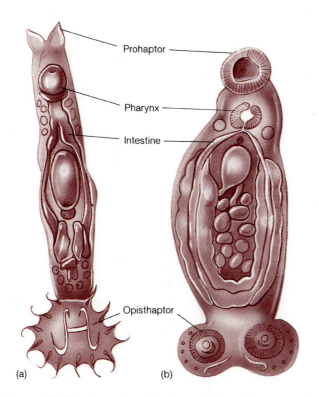

Figure 18.12 Drawing of two monogeneid trematodes.
(a) *Gyrodactylus* and (b) *Sphyranura*.

an **oncomiracidium** hatches from the egg and swims to another host fish, where it attaches by its opisthaptor and develops into an adult.

Subclass Digenea The vast majority of all flukes (approximately 6,000 species) belong to the subclass Digenea. In this subclass, at least two different forms, an adult and one or more larval stages develop—a characteristic from which the name of the subclass was derived. Because digenetic flukes require at least two different hosts to complete their life cycles, these animals possess the most complex life cycles of any platyhelminth. As adults, they are all endoparasites in the bloodstream, digestive tract, ducts of the digestive organs, or other visceral organs in a wide variety of vertebrates that serve as definitive, or final, hosts. The one or more intermediate hosts (the hosts that harbor immature flukes) may harbor several larval stages. The adhesive organs are two large suckers. The anterior sucker is the **oral sucker** and surrounds the mouth. The other sucker, the **acetabulum,** is located ventrally on the middle portion of the body (*see figure 18.9*).

Some Important Trematode Parasites of Humans Several digenetic flukes are important parasites of humans and have certain basic stages in their life cycles. The eggs of trematodes are oval and usually possess a lidlike hatch called an **operculum** (figure 18.13a). When

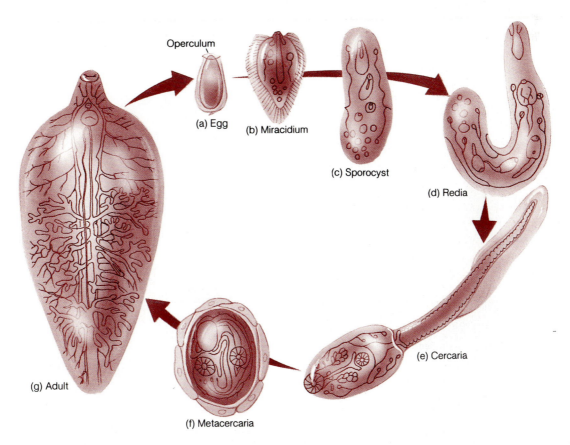

Figure 18.13 Parts a–g present a generalized life cycle of the digenetic trematode, *Fasciola hepatica* (the common liver fluke).

an egg reaches fresh water, the operculum opens and a cilated larva called a **miracidium** (plural, miracidia) swims out (figure 18.13b). The miracidium swims until it finds a suitable first intermediate host (a snail). The miracidium penetrates the snail, loses its cilia, and develops into a **sporocyst** (figure 18.13c). (Alternately, the miracidium may remain in the egg and hatch after being eaten by a snail.) Sporocysts are baglike and contain embryonic cells that develop into either **daughter sporocysts**, or **rediae** (singular, redia) (figure 18.13d). At this point in the life cycle, asexual reproduction first occurs. From a single miracidium, hundreds of daughter sporocysts, and in turn, hundreds of rediae can form. Embryonic cells in each daughter sporocyst or redia produce hundreds of the next larval stage, called **cercariae** (singular, cercaria) (figure 18.13e). A cercaria has a digestive tract, suckers, and a tail. Cercariae leave the snail and swim freely until they encounter a second intermediate or final host, which may be a vertebrate, invertebrate, or a plant. The cercaria penetrates this host and encysts as a **metacercaria** (plural, metacercariae) (figure 18.13f). When the second intermediate host is eaten by the definitive host, the metacercaria excysts and develops into an adult (figure 18.13g). Three representative human digenetic flukes will now be described.

The Chinese liver fluke, *Clonorchis sinensis,* is a common parasite of humans in the Orient, where over 50 million people are infected. The adult lives in the bile passageways of the liver, where it feeds on epithelial tissue and blood (figure 18.14a). Embryonated eggs are released by the adults into the common bile duct, make their way

to the intestines, and are eliminated with feces (figure 18.14b). The miracidia are released when a snail ingests the eggs. Following sporocyst and redial stages, cercariae emerge. If the cercaria contacts a fish (the second intermediate host), it penetrates the epidermis of the fish, loses its tail, and encysts. The metacercaria develops into an adult when a human eats raw or poorly cooked fish.

Fasciola hepatica is called the sheep liver fluke (*see figure 18.13a*) because it is common in sheep-raising areas and uses sheep or humans as its definitive host. Instead of living in bile passageways, this fluke lives in the liver. Eggs work their way through the liver and pass via bile passageways to the intestines, from which they are eliminated. When eggs are deposited in fresh water, they hatch, and the miracidia must locate the proper species of snail. If a snail is found, micacidia penetrate the snail's soft tissue and develop into sporocysts that develop into rediae and give rise to cercariae. After the cercariae emerge from the snail, they encyst on aquatic vegetation. Sheep or other animals become infected when they graze on the aquatic vegetation. Humans may become infected with *Fasciola hepatica* by eating a freshwater plant called watercress that contains the encysted metacercaria.

Schistosomes are blood flukes that are of vast medical significance. The impact these flukes have had on history is second only to *Plasmodium* (malaria). They infect over 200 million people throughout the world. Infections are most common in Africa (*Schistosoma mansoni*), South and Central America (*S. haematobium*), and Southeast Asia (*S. japonicum*). The adult dioecious worms live in the bloodstream of humans (figure 18.15a). The male fluke is shorter

[5]

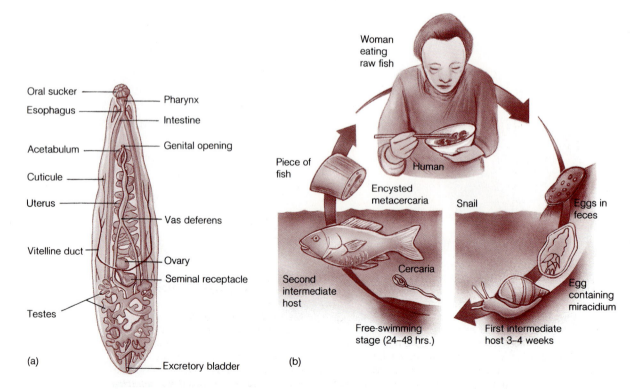

(a)

Oral sucker — Pharynx
Esophagus — Intestine
Acetabulum — Genital opening
Cuticule —
Uterus —
— Vas deferens
Vitelline duct —
— Ovary
— Seminal receptacle
Testes —
— Excretory bladder

(b)

Woman eating raw fish
Human
Piece of fish
Encysted metacercaria
Snail
Eggs in feces
Second intermediate host
Cercaria
Egg containing miracidium
Free-swimming stage (24–48 hrs.)
First intermediate host 3–4 weeks

Figure 18.14 The Chinese liver fluke, *Clonorchis sinensis.* (a) Dorsal view and (b) life cycle.

and thicker than the female, and the sides of the male's body are curved under to form a canal along the ventral surface ("schistosoma" means "split body"). The female fluke is long and slender, and is carried in the canal of the male (figure 18.15b). Copulation is continuous, and the female produces thousands of eggs. Each egg contains a spine that aids it in moving through host tissue until it is eliminated in either the feces or urine (figure 18.15c). Unlike other flukes, schistosome eggs lack an operculum. The miracidium escapes through a slit that develops in the egg when the egg reaches fresh water (figure 18.15d). The miracidium seeks a snail (figure 18.15e), penetrates it, develops into a sporocyst (figure 18.15f), redia (18.15g), and forked-tailed cercariae (figure 18.15h). The cercariae leave the snail and penetrate the skin of a human (figure 18.15i).

Penetration is aided by anterior digestive glands that secrete digestive enzymes. Once in a human, the cercariae lose their tails and develop into adults in the intestinal veins, skipping the metacercaria stage (box 18.1).

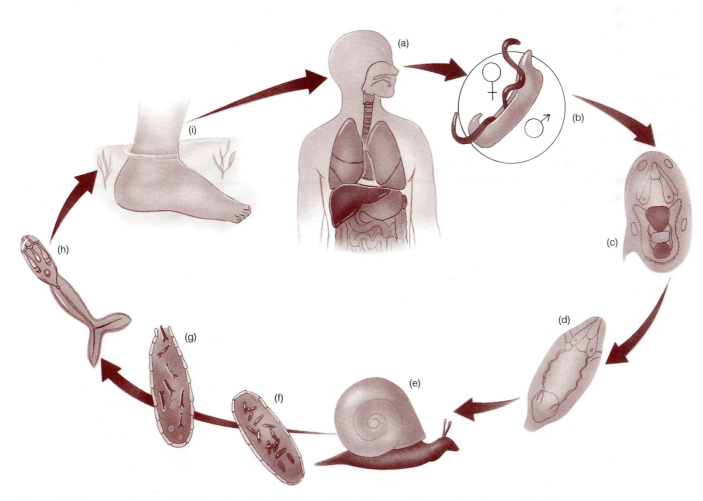

Figure 18.15 A representative life cycle of a schistosome fluke. The cycle begins in a human (a) when the female fluke lays eggs (b–c) in the thin walled, small vessels of the small intestine. Secretions from the eggs weaken the walls, and the blood vessels rupture, releasing eggs into the intestinal lumen. From there the eggs leave the body. If they reach fresh water, the eggs hatch into ciliated, free-swimming larvae called miracidia (d). A miracidium burrows into the tissues of an aquatic snail (e), losing its cilia in the process and develops into a sporocyst, then redia that multiplies asexually (f,g). Eventually forked-tailed larvae (cercariae) are produced (h). After the cercariae leave the snail, they actively swim about. If they encounter human skin (i), they attach to it and release tissue-degrading enzymes. The larvae enter the body and migrate to the circulatory system, where they mature. They end up at the intestinal vessels, where sexual reproduction takes place and the cycle begins anew.

Class Cestoidea: The Tapeworms

The most highly specialized class of flatworms are members of the class Cestoidea (Gr. *kestos*, girdle + *eidos*, form), commonly called either tapeworms or cestodes. All of the approximately 3,500 adult species are endoparasites that reside in the digestive system of vertebrates. Their color is often white with shades of yellow or grey. Adult tapeworms range in size from 1 mm to 15 m in length.

Adult tapeworms are characterized by two unique adaptations to their parasitic way of life. (1) Tapeworms lack a mouth and digestive tract in all of their life-cycle stages; they absorb nutrients directly across their body wall (tegument). (2) Most adult tapeworms consist of a long series of repeating segments called **proglottids.** Each proglottid contains a complete set of reproductive structures.

As with most endoparasites, adult tapeworms live in a very stable environment. The intestinal tract of a vertebrate has very few environmental variations that would require the development of great anatomical or physiological complexity in any single tapeworm body system. Homeostasis (internal constancy) of a tapeworm is maintained by the physiology of the tapeworm's host. In adapting to such a specialized environment, tapeworms have lost some of the structures believed to have been present in ancestral turbellarians. Tapeworms are, therefore, a good example of the fact that evolution does not always result in greater complexity.

Subclass Cestodaria Representatives of the subclass Cestodaria are all endoparasites in the intestine and coelom of primitive fish. About 15 species have been identified. They possess some digenetic trematode features in that only one set of both reproductive systems is present in each animal, some bear suckers, and their bodies are not divided into units or proglottids as in other cestodes. Yet the complete absence of a digestive system, the presence of larval stages similar to those of cestodes, and the presence of mesenchymal muscle cells all suggest strong phylogenetic affinities with other cestodes.

Subclass Eucestoda Almost all of the more than 3,500 species of cestodes belong to the subclass Eucestoda and are called true tapeworms. They represent the ultimate degree of specialization of any parasitic animal. The body is divided into three regions (figure 18.16a). At one end is a holdfast structure called the **scolex** that contains circular or leaflike suckers and sometimes a rostellum of hooks (figure 18.16b). It is via the scolex that the tapeworm firmly anchors itself into the intestinal wall of its definitive vertebrate host.

Posteriorly, the scolex narrows to form the *neck*. Transverse constrictions in the neck give rise to the third body region, the **strobila** (Gr. *strobilus,* anything twisted up). The strobila consists of a series of linearly arranged segments. These segments (proglottids) function primarily as reproductive units. As a tapeworm grows, new proglottids are added in the neck region, and older proglottids are gradually pushed posteriorly. As they move posteriorly, proglottids mature and begin producing eggs. Thus, anterior proglottids are said to be *immature,* those in the midregion of the strobila are *mature,* and those at the posterior end that have accumulated eggs are *gravid.*

The outer body wall of tapeworms consists of a tegument similar in structure to that of trematodes (*see figure 18.10*). It plays a vital role in the absorption of food because tapeworms have no digestive system. The tegument even absorbs some of the host's own enzymes to facilitate digestion.

With the exception of the reproductive systems, the body systems of tapeworms are reduced in structural complexity. The nervous system consists of only a pair of lateral nerve cords that arise from a nerve mass in the scolex and extend the length of the strobila. A protonephridial system also runs the length of the tapeworm (*see figure 18.5*).

Tapeworms are monoecious and most of their physiology is devoted to producing large numbers of eggs. Each proglottid contains a complete set of reproductive organs (figure 18.16a). Numerous testes are scattered throughout the proglottid and deliver sperm via a duct system to a copulatory organ called a *cirrus.* The cirrus opens through a *genital pore,* which is an opening shared with the female system. The male system of a proglottid usually matures before the female system, so that copulation usually occurs with another mature proglottid of the same tapeworm or with another tapeworm in the same host. As previously mentioned, the avoidance of self-fertilization leads to hybrid vigor.

[6]

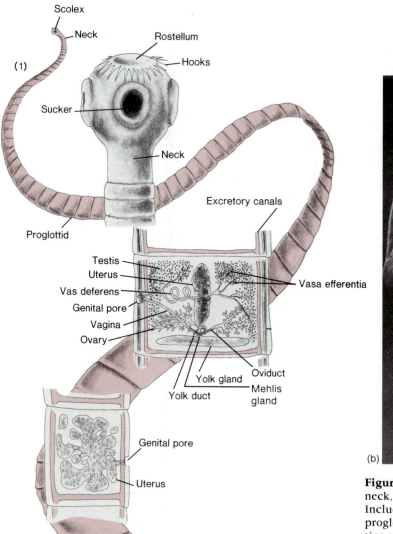

(a)

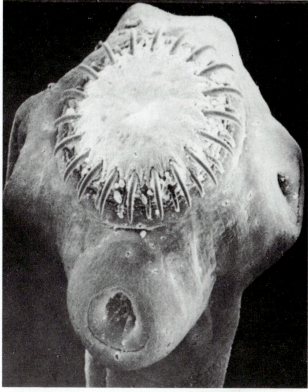

(b)

Figure 18.16 A tapeworm. (a) Diagram showing the scolex, neck, and proglottids of the pork tapeworm, *Taenia solium.* Included are detailed views of the scolex and neck, a mature proglottid with a complete set of male and female reproductive structures, and a gravid proglottid that is nearly filled by an expanded uterus containing eggs. (b) A scanning electron micrograph of the scolex of a cestode.

Eggs are produced in a single pair of ovaries. Sperm stored in a *seminal receptacle* fertilize eggs as they move through the oviduct. After passing the vitelline (yolk) gland to pick up yolk cells, the eggs move into the *ootype* which is an expanded region of the oviduct that shapes the capsules around the eggs. The ootype is surrounded by Mehlis's gland, which aids in the formation of the egg capsule. Most tapeworms have a blind-ending *uterus* where eggs accumulate (figure 18.16a). As eggs accumulate, the reproductive organs degenerate; thus, gravid proglottids can be thought of as "bags of eggs." Eggs are released when gravid proglottids break free from the end of the tapeworm and pass from the host with the feces. In a few tapeworms, the uterus opens to the outside of the worm, and eggs are released into the intestine of the host. Because the proglottids are not continuously lost in these worms, the adult tapeworms usually become very long.

Some Important Tapeworm Parasites of Humans   The most common tapeworm of humans is the beef tapeworm, *Taeniarhynchus saginatus* (figure 18.17). Adults live in the small intestine and may reach lengths of 3 m. About 80,000 eggs per proglottid are released as proglottids break free of the adult worm. As an egg develops, it forms a six-hooked (hexacanth) larva called the **onchosphere.** As cattle graze on pastures contaminated with human feces, the oncospheres (or proglottids) are ingested by the intermediate host. Digestive enzymes of the cattle free the oncospheres, and the larvae use their hooks to bore through the intestinal wall into the bloodstream. The bloodstream carries the larvae to skeletal muscles, where they encyst and form a fluid-filled bladder called a **cysticercus** (plural, cysticerci) or **bladder worm.** When a human eats infected meat (termed "measly beef") that is raw or improperly cooked, the cysticercus is released from the meat, the scolex attaches to the human intestinal wall, and the tapeworm matures.

A closely related tapeworm, *Taenia solium* (the pork tapeworm), has a life cycle similar to *Taeniarhynchus saginatus,* except that the intermediate host is the pig. The

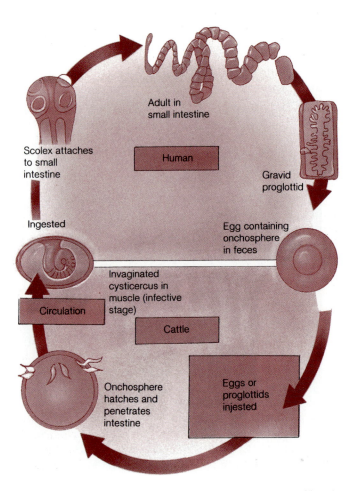

Figure 18.17 The life cycle of the beef tapeworm, *Taeniarhynchus saginatus.*

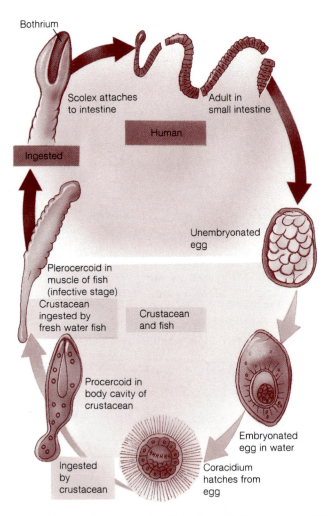

Figure 18.18 The life cycle of the broad fish tapeworm, *Diphyllobothrium latum.*

strobila has been reported as being 10 m long, but 2 to 3 m is more common. The pathology is more serious in the human than in the pig. Oncospheres are frequently released from gravid proglottids before the proglottids have had a chance to leave the small intestine of the human host. When these larvae hatch, they move through the intestinal wall, enter the bloodstream and are distributed throughout the body where they eventually encyst in human tissue. The disease that results is called **cysticercosis.**

The broad fish tapeworm, *Diphyllobothrium latum,* is relatively common in the northwestern parts of North America. This tapeworm has a scolex with two longitudinal groves (*bothria*; singular, bothrium, figure 18.18) that act as holdfast structures. The adult worm may attain a length of 10 m and shed up to a million eggs a day. Many proglottids release eggs through uterine pores. When eggs are deposited in fresh water, they hatch, and ciliated larvae called **coracidia** (singular, coracidium) emerge. These coracidia swim about until they are ingested by small crustaceans called copepods. The larvae shed their ciliated coats in the copepods and develop into **procercoid larvae.** When copepods are eaten by fish, the procercoids burrow into the muscle of the fish and become **pleurocercoid larvae.** Larger fish that eat smaller fish become similarly infected with pleurocercoids. When infected, raw or poorly cooked fish are eaten by a human (or other carnivore), the pleurocercoids attach to the small inestine and grow into adult worms.

Stop and Ask Yourself

8. What is a proglottid? A scolex? A strobila?
9. How do tapeworms obtain nutrients?
10. What is the life cycle of a typical tapeworm?
11. What is each of the following: Coracidium? Cysticercus? Procercoid larva? Pleurocercoid larva?

Phylum Nemertea: The Probosis Worms

Nemerteans (Gr. *nemertes,* one of the Nereids, unerring one) are long, flattened worms. Most of the approximately 900 species are found in marine mud and sand. Like the turbellarians, nemerteans are triploblastic, acoelomate, and

Table 18.2 Classification of the Nemertea and Gnathostomulida

Phylum Nemertea (nem·er′te·a)

  Class Anopla (an′o·pla)
  Proboscis without stylet; mouth posterior to the brain.
  Carinina, Tubulanus, Lineus.

  Class Enopla (en′o·pla)
  Proboscis almost always with stylet; mouth anterior to
  brain and opens into a ryhnchocoel. *Prostoma,*
  Tetrastemma.

Phylum Gnathostomulida (nath′o·sto·myu′lid-a)
Common in beach sands; gnathostomulids.
Austrognathia, Gnathostomula.

possess a ciliated epidermis. Mucous glands are abundant throughout the epidermis. Body musculature is organized into two or three layers. Nemerteans possess a nervous system and sensory structures that are similar to those of turbellarians (figure 18.19). Adult worms range in size from a few millimeters to several centimeters in length. Most nemerteans are pale yellow, orange, green, or red. The classification of the nemerteans is given in table 18.2.

When they move, nemerteans glide on a trail of mucus. Cilia and peristaltic contractions of body muscles provide the propulsive forces.

The most distinctive feature of nemerteans is a long proboscis that is held in a sheath called a **rhynchocoel.** The proboscis may be tipped with a barb called a *stylet.* Carnivorous species use the proboscis to capture annelid (segmented worms) and crustacean prey.

Unlike the other flatworms, nemerteans have a complete one-way digestive tract. They have a mouth for ingesting food and an anus for eliminating digestive wastes. This characteristic enables mechanical breakdown of food, digestion, absorption, and feces formation to proceed sequentially from an anterior to posterior direction—a major evolutionary innovation found in all later bilateral animals.

Another major innovation found in all later animals evolved first in the nemerteans—a *circulatory system* consisting of two lateral blood vessels and often, tributary vessels that branch from lateral vessels. However, no heart is present, and contractions of the walls of the large vessels help to propel blood along. Blood does not circulate but simply moves forward and backward through the longitudinal vessels. Blood cells are present in some species and presumably help distribute oxygen, metabolic wastes, and nutrients throughout the worm. It is this combination of blood vessels with their capacity to serve local tissues, and a one-way digestive system with its greater efficiency at processing nutrients, that allows nemerteans to reach lengths up to several meters, much larger than most other flatworms.

Nemerteans are dioecious. Male and female reproductive structures develop from parenchymal cells along each side of the body. External fertilization results in the formation of a helmet-shaped, ciliated **pilidium larva.** After

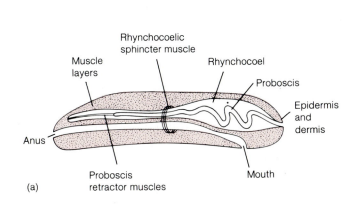

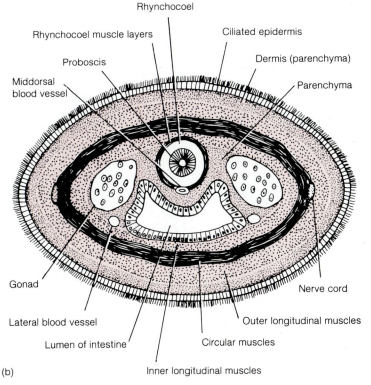

Figure 18.19 Diagram of nemerteans. (a) Longitudinal section, showing the tubular gut. (b) Cross section through the body.

a brief free-swimming existence, the larva develops into a young worm that settles to the substrate and begins feeding.

Phylum Gnathostomulida: The Gnathostomulids

This phylum was recently discovered (1956), and its approximately 100 species of minute, anaerobic worms are especially abundant on the Atlantic coast of North America. They are characteristically found in mud and sand, where they live in the spaces between sand grains.

Gnathostomulids (Gr. *gnathos*, jaw + *stoma*, mouth) are generally less than 1 mm in length, elongate, cylindrical, and mostly transparent (figure 18.20a). There is an anterior *head,* a constricted *neck,* and a *trunk* that terminates in a tapered *tail.* The body wall is ciliated, each epidermal cell having a single cilium whose beat is reversible. Scattered mesenchymal cells and delicate longtitudinal and circular muscle strands are located between the epidermis and gut. Locomotion is by ciliary action, and body muscles provide serpentine movements.

A ventral *mouth* at the base of the neck is equipped with a pair of lateral *jaws* (figure 18.20b), which give the phylum its name (Gr. *gnathos*, jaw). Food consists of micro-organisms that are scraped off sand grains and passed into the blind-end digestive tract. There is an anterior ganglion, several lateral nerve cords, but no circulatory or protonephridial systems.

Most gnathostomulids are monoecious, although only one reproductive system may be functional in a single animal. The female system contains a single *ovary* and an associated sperm storage sac, or *bursa.* The male system contains a pair of *testes* and a *copulatory organ* that may bear a stylet. Fertilization is followed by the release of a single large egg, usually by the rupture of the body wall.

Further Phylogenetic Considerations

Controversies surrounding the relationships of the Platyhelminthes to other animal phyla were discussed at the beginning of this chapter. Zoologists who believe that the flatworm design is central to animal evolution envision an ancestral flatworm that was probably similar to the plan described for turbellarians. From that ancestor, Gnathostomulida, Nemertea, Turbellaria, Trematoda, and Cestoidea evolved. Conversely, if the triploblastic acoelomate design is not central to the evolutionary hypothesis, the various phyla covered in this chapter represent evolutionary regression.

More conclusive evidence exists linking the parasitic flatworms to ancient free-living ancestors. The divergence between the free-living and parasitic way of life probably occurred in the Cambrian period, 600 million years ago. The first flatworm parasites were probably associated with primitive molluscs, arthropods, and echinoderms. It must have been much later that they acquired the vertebrate hosts and complex life cycles that have been described in this chapter.

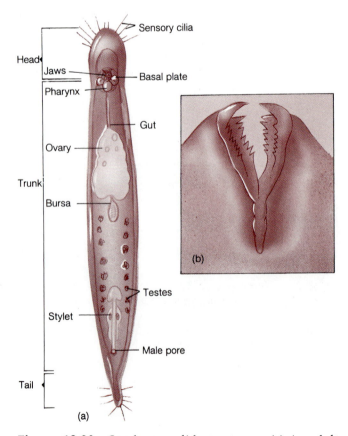

Figure 18.20 Gnathostomulid structures. (a) An adult worm. (b) The paired lateral jaws.

Stop and Ask Yourself

12. Why are the nemerteans called probosis worms?
13. What is a rhynchocoel?
14. What major body system developed for the first time in the nemerteans?
15. How can a typical gnathostomulid be described?

Summary

1. The free-living Platyhelminthes, members of the class Turbellaria, are small, bilaterally symmetrical animals with some cephalization and an acoelomate type of body.

2. Most turbellarians move entirely by cilia and are predators and scavengers. Digestion is initially extracellular and then intracellular.

3. Protonephridia are present in many flatworms and are involved in osmoregulation. A primitive brain and nerve cords are present.

4. Turbellarians are monoecious with the reproductive systems adapted for internal fertilization and egg deposition.

5. There are three subclasses of flukes (Aspidogastrea, Monogenea, and Digenea) and most are external or internal parasites of vertebrates. A gut is present and most flukes are monoecious. The mongenetic flukes are mostly ectoparasites of fish. The digenetic flukes constitute the largest group of parasitic flatworms.

6. Cestodes, or tapeworms, are gut parasites of vertebrates. They are structurally more specialized than flukes, having a scolex with attachment organs, a neck region, and a strobila, which consists of a chain of segments (proglottids) budded off from the neck region. A gut is absent, and the reproductive system is repeated in each proglottid.

7. Nemerteans are similar to flatworms but can be much larger. They are predatory on other invertebrates, which they capture with a unique probosis. They have a one-way digestive tract, a blood-vascular system, and the sexes are separate.

8. The Gnathostomulida are a small phylum of minute, acoelomate worms that live in anaerobic marine sediments.

Key Terms

acetabulum (p. 281)
adhesive glands (p. 276)
auricles (p. 278)
bladder worm (p. 285)
cercariae (p. 282)
cocoon (p. 279)
coracidia (p. 286)
cysticercosis (p. 286)
cysticercus (p. 285)
daughter sporocysts, or
  rediae (p. 282)
flame cells (p. 277)
flukes (p. 279)
metacercariae (p. 282)
miracidium (p. 282)
Muller's larva (p. 279)
neoteny (p. 275)
nephridiopore (p. 278)
ocelli (p. 278)

onchosphere (p. 285)
oncomiracidium (p. 281)
operculum (p. 281)
opisthaptor (p. 280)
oral sucker (p. 281)
parenchyma (p. 275)
pilidium larva (p. 287)
pleurocercoid larvae
  (p. 286)
procercoid larvae (p. 286)
proglottids (p. 284)
protonephridia (p. 277)
releaser glands (p. 276)
rhabdites (p. 276)
rhynchocoel (p. 287)
scolex (p. 284)
sporocyst (p. 282)
strobila (p. 284)
tegument (p. 280)
zooids (p. 279)

Critical Thinking Questions

1. Describe the morphological and developmental similarities and differences between gnathostomulids and turbellarians.

2. How do parasitic flatworms evade their host's immune system?

3. How would a zoologist go about documenting the complex life cycle of a digenetic trematode?

4. Describe some of the key features of acoelomate animals.

5. Why has it been so difficult to eradicate schistosome parasites from areas where they are so prevalent?

6. Describe the evolutionary relationship between free-living flatworms and parasitic flatworms.

7. Why is cephalization an evolutionary advancement? Give several examples to prove your point.

8. Why must some parasites produce such large numbers of eggs?

9. Explain how a radial ancestor could have given rise to both the Cnidaria and Platyhelminthes.

Suggested Readings

Books

Barnes, R. D. 1987. 5/e. *Invertebrate Zoology*. Philadelphia: Saunders.

Pearse, V., and Buchsbaum, M. 1987. *Living Invertebrates*. Palo Alto: Blackwell/Boxwood.

Gibson, R. 1972. *Nemerteans*. London: Hutchinson University Library.

Gibson, R. 1982. *British Nemerteans: Keys and Notes for Identification of the Species*. New York: Cambridge University Press.

Hyman. L. H. 1951. *The Invertebrates*. Vol. II. *Platyhelminthes and Rhynchocela*. New York: McGraw-Hill.

Riser, N. W., and Morse, M. P. (eds). 1974. *Biology of the Turbellaria*. New York: McGraw-Hill.

Articles

Riedl, R. J. 1979. Gnathostomulida from America. *Science* 163: 445–446.

Roe, P., and Norenburg, J. L. (eds). 1985. Symposium on the comparative biology of nermetines. *American Zoologist* 25 (1):1–151.

Sterrer, W. 1972. Systematics and evolution within the Gnathostomulida. *Systematic Zoology* 21:151–161.

19

The Pseudocoelomate Body Plan: Aschelminths

Concepts

1. There are nine phyla that are grouped together under the aschelminths: Gastrotricha, Rotifera, Kinorhyncha, Nematoda, Nematomorpha, Acanthocephala, Loricifera, Priapulida, and Entoprocta. Because most of these phyla have had a separate evolutionary history, this grouping is mostly one of convenience.
2. The major unifying aschelminth feature is a pseudocoelom. The pseudocoelom is a type of body cavity (space) that develops from the blastocoel (or the primitive cavity in the embryo) and is not fully lined by mesoderm, as in the true coelomates. In the pseudocoelomates, the muscles and other structures of the body wall and internal organs are in direct contact with fluid in the pseudocoelom.
3. Other generally occurring aschelminth features include a complete digestive tract, a muscular pharynx, constant cell numbers, protonephridia, cuticle, and adhesive glands.

Have You Ever Wondered:

[1] why worms molt?
[2] how rotifers got their name?
[3] in what animal the reproductive organ called a penis first developed?
[4] what the most abundant animals on Earth are?
[5] why worms move in an undulating, wavelike fashion?
[6] why all nematodes are round?
[7] what the most common roundworm parasite in the United States is?
[8] why you should not eat improperly cooked pork products?
[9] what causes the disease elephantiasis?
[10] what causes heartworm disease in dogs?
[11] why horsehair worms were thought to arise from horses tails?
[12] how spiny-headed worms got their name?
[13] what the most recently described animal phylum is?

These and other useful questions will be answered in this chapter.

This chapter contains underlined evolutionary concepts.

Evolutionary Perspective

The nine different phyla that are grouped for convenience as the aschelminths (Gr. *askos,* bladder + *helmins,* worm) are very diverse animals, they have obscure phylogenetic affinities, and their fossil record is meager. Two hypotheses have been proposed for their phylogeny. The first hypothesis contends that the phyla are related based on the presence of the following structures: a pseudocoelom, a cuticle, a muscular pharynx, and adhesive glands. The second hypothesis contends that the various aschelminth phyla are not related to each other. The absence of any single unique feature found in all groups strongly suggests an independent evolution for each phylum. The similarities that are present among living aschelminths may simply be the result of convergent evolution as these various animals adapted to similar environments.

The correct phylogeny may actually be something between the two hypotheses. Accordingly, all phyla are probably distantly related to each other based on the few anatomical and physiological features they share. True convergent evolution may have also produced some visible nonphylogenetic similarities, but each phylum probably arose from a common acoelomate ancestor, and diverged very early in evolutionary history (figure 19.1). Such an ancestor might have been a primative, ciliated, acoel turbellarian (*see figure 18.3a*), which would indicate that the first aschelminths were ciliated, acoelomate, marine, lacked a cuticle, and probably were monoecious.

General Characteristics

The aschelminths are the first assemblage of animals to possess a distinct body cavity, but one that lacks the peritoneal (mesodermal) linings and membranes called *mesenteries* that are found in more advanced animals. As a result, the various internal (visceral) organs lie free in the cavity. Such a cavity is called a *pseudocoelom (see figure 15.13b),* and the animals are called *pseudocoelomates.* The pseudocoelom is often fluid-filled, serves as a cavity for circulation,

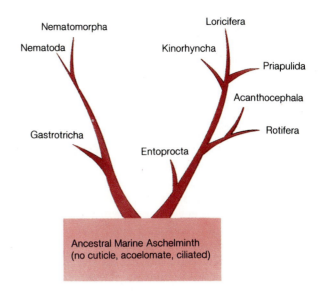

Figure 19.1 The possible phylogenetic relationships among the nine aschelminth phyla.

aids in digestion, and acts as an internal hydrostatic skeleton that functions in locomotion.

Most aschelminths (the acanthocephalans and nematomorphs are exceptions) have a complete tubular digestive tract that extends from an anterior *mouth* to a posterior *anus.* This tube-within-a-tube body plan was first encountered in the nemerteans (*see figure 18.19*) and is characteristic of almost all other higher animals. It permits, for the first time, the mechanical breakdown of food, digestion, absorption, and feces formation to proceed sequentially and continually from an anterior to posterior direction—an evolutionary advancement over the blind-ending digestive system. Most aschelminths also have a specialized muscular *pharynx* that is adapted for feeding.

Many aschelminths show **eutely,** a condition in which the number of cells or nuclei in syncytia are constant both for the entire animal and for a given organ, for all the animals of that species (box 19.1). For example, the number of body (somatic) cells in all adult nematodes, *Caenorhabditis elegans,* is 959, and the number of cells in the pharynx of every worm in the species is precisely 80.

Box 19.1

An Application of Eutely

The science of aging is called **gerontology** (Gr. *gerontos,* old man). The eutelic characteristic of aschelminths makes them excellent research animals for studies on aging because: (1) none of the cellular systems of aschelminths are being continually renewed; (2) aschelminths are devoid of the capacity to repair cells; and (3) because there is an exact number of cells present, their exact lineage is known. In a eutelic animal, cellular longevity appears to be a simple, measurable parameter, and the onset of aging can be easily studied. Some of the characteristics of aging that have been studied in these animals include the progressive disorganization of muscle and nerve cells, mitochondrial degeneration, decrease in cellular motility, accumulation of age pigments, and increase in specific gravity.

Most aschelminths are microscopic, although some may grow to a length of over a meter. They are bilaterally symmetrical, unsegmented, triploblastic, and cylindrical in cross section. An osmoregulatory system of protonephridia is found in most aschelminths and functions in water and ion homeostasis. This system is best developed in freshwater forms in which osmotic problems are the greatest. No separate blood or gas exchange systems are present. There is some cephalization, with the anterior end containing a primitive brain, sensory organs, and a mouth. The vast majority of aschelminths are dioecious. Both reproductive systems are relatively uncomplicated; life cycles are usually simple, except in the case of the parasitic animals; and cleavage is determinate (*see figure 10.4*). Cilia are generally absent from the external surface, but a thin, tough external cuticle is present for the first time in any animal group. The **cuticle** consists of scleroprotein and may bear spines, scales, or other forms of ornamentation that protect the animal and are useful to taxonomists. Some aschelminths shed this cuticle in a process called *molting* [1] or *ecdysis* in order to lengthen. Beneath the cuticle is a syncytial *epidermis* that actively secretes the cuticle. Several longitudinal muscle layers lie beneath the epidermis.

Most aschelminths are freshwater animals; only a few are found in marine environments. The nematomorphs, acanthocephalans, and many nematodes are parasitic. The remainder of this assemblage are mostly free-living; some rotifers are colonial.

3 mm in length. Gastrotrichs move over the substrate using cilia on their ventral surface. The phylum contains a single class divided into two orders.

The dorsal cuticle often contains scales, bristles, or spines. A syncytial epidermis is found beneath the cuticle, and a forked tail is often present (figure 19.2). Sensory structures include tufts of long cilia and bristles on the rounded head. The nervous system includes a brain and a pair of lateral nerve trunks. The digestive system is a straight tube with a mouth, a muscular pharynx, a stomach-intestine, and an anus. Microorganisms and organic detritus are ingested by the action of the pumping pharynx. Digestion is mostly extracellular. Adhesive glands in the forked tail secrete materials that anchor the animal to solid objects. Paired protonephridia occur in freshwater species, rarely in marine ones. Gastrotrich protonephridia, however, are morphologically different from those found in the acoelomates. The terminal cell is composed of cytoplasmic rods and possesses a single flagellum instead of the cilia found in flame cells.

Most of the marine species reproduce sexually and are hermaphroditic. Most of the freshwater species reproduce asexually by parthenogenesis; the females can lay two kinds of unfertilized eggs. Thin-shelled eggs hatch into females during favorable environmental conditions, whereas thick-shelled, resting eggs can withstand unfavorable conditions for long periods. There is no larval stage; development is direct, and the juveniles have the same form as the adults.

Phylum Gastrotricha: The Gastrotrichs

The gastrotrichs (gas·tro·tri′ks) (Gr. *gastros,* stomach + *trichos,* hair) are a small phylum of about 500 free-living marine and freshwater species that inhabit the interstitial spaces of bottom sediments. They range from 0.01 mm to

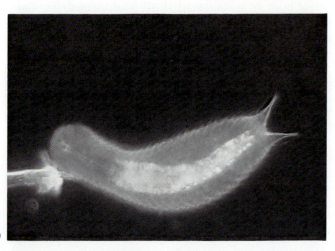

(a)

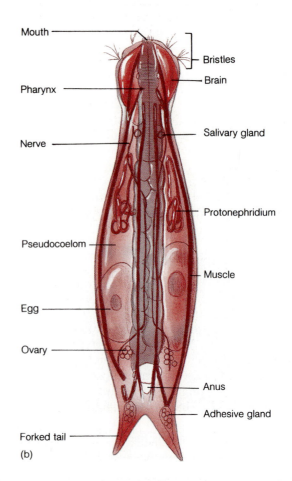

Mouth — Bristles — Brain

Pharynx

Nerve — Salivary gland

Pseudocoelom — Protonephridium

Egg — Muscle

Ovary

Anus

Adhesive gland

Forked tail

(b)

Figure 19.2 Phylum Gastrotricha. (a) Photomicrograph of a gastrotrich, *Chaetonothus.* (b) Illustration of the internal anatomy (ventral view) of a freshwater gastrotrich.

Phylum Rotifera: The Rotifers

[2] The rotifers (ro·tif'ers) (L. *rota,* wheel + *fera,* to bear) derive their name from the characteristic ciliated organ, the **corona** (crown), located around lobes on the head of these animals (figure 19.3). The cilia of the corona do not beat in synchrony; instead, each cilium is at a slightly earlier stage in the beat cycle than the preceding cilium in the sequence, creating a *metachronal* beat. A wave of beating cilia thus appears to pass around the periphery of the ciliated lobes and gives the impression of a pair of spinning wheels.

Rotifers are small animals (0.1 to 1 mm in length) that are abundant in most freshwater habitats; a few (less than 10%) are marine. There are about 2,000 species that comprise three classes (table 19.1). The body is composed of approximately 1000 cells, and the organs are eutelic. Rotifers are usually solitary free-swimming animals although some colonial forms are known, and others may occur interstitially in sediments.

External Features

The external surface of a rotifer is covered by an epidermally secreted, nonchitinous cuticle that is never molted. In many species, the cuticle is thickened to form a *lorica* or case. The cuticle and lorica provide protection and are the main support elements, although fluid in the pseudocoelom also provides hydrostatic support. The epidermis is syncytial; i.e., cell membranes between nuclei are incomplete.

The *head* contains the corona, mouth, sensory organs, digestive glands, and brain (figure 19.3b). The corona consists of a large ciliated area called the *buccal field.* The *trunk* is the largest part of a rotifer and is elongate and saclike. The *anus* occurs dorsally on the posterior trunk.

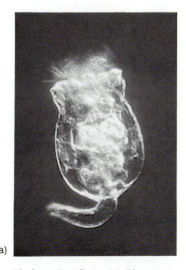

(a)

Figure 19.3 Phylum Rotifera. (a) Photomicrograph of a rotifer, *Brachionus.* (b) Illustration of the internal anatomy (dorsal view) of a typical rotifer, such as *Philodina.*

Table 19.1 Classification of the Rotifera

Phylum Rotifera (ro·tif'e·ra)
A ciliated corona surrounding a mouth; muscular pharynx (mastax) present with jawlike features; intracellular cuticle; parthenogenesis is common; both freshwater and marine species.

Class Seisonidae (sy'son·id'ea)
A single genus of marine rotifers that are commensals of crustaceans; large and elongate body with reduced corona. *Seison.*

Class Bdelloidea (Digonota) (del·oid'e·a)
Anterior end retractile and bearing two trochal disks; mastax adapted for grinding; paired ovaries; cylindrical body; males absent. *Adineta, Philodina, Rotaria.*

Class Monogononta (mon'o·go·no'ta)
Rotifers with one ovary; mastax not designed for grinding; produce mictic and amictic eggs. *Conochilus, Collotheca, Notommata.*

The posterior, narrow portion called the *foot* is set off sharply from the trunk. The terminal portion of the foot usually bears one to several toes. On the foot, at the base of the toes are many *pedal glands* whose ducts open on the toes. Secretions from these glands aid in temporary attachment of the foot to a substratum.

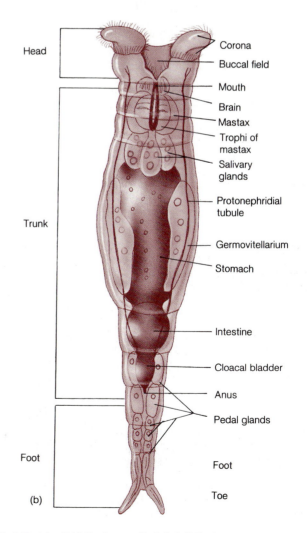

(b)

Figure 19.3b redrawn, with permission, from P. E. Lutz, *Invertebrate Zoology.* Copyright © 1986, Benjamin/Cummings Publishing Company, Menlo Park, California.

Feeding and the Digestive System

Most rotifers feed on small microorganisms and suspended organic material. The coronal cilia create a current of water that brings food particles to the mouth. The mouth contains a unique structure called the **mastax.** The mastax is a muscular organ in which food is ground. The inner walls of the mastax contain several large structures called *trophi* (figure 19.3b). The trophi vary in morphological detail and are used by taxonomists to distinguish species.

From the mastax, food passes through a short, ciliated *esophagus* to the ciliated *stomach. Salivary* and *digestive glands* secrete digestive enzymes into the pharynx and stomach. The complete digestion (extracellular) of food and its absorption occur in the stomach. In some species, a short ciliated *intestine* extends posteriorly and becomes a *cloacal bladder,* which receives water from the protonephridia and eggs from the ovaries, as well as digestive waste. The cloacal bladder opens to the outside via an anus at the base of the foot.

Other Visceral Systems

All visceral organs lie in a pseudocoelom that is filled with fluid and interconnecting ameboid cells. Osmoregulation is accomplished by protonephridia that empty into the cloacal bladder. Rotifers, like other pseudocoelomates, exchange gases and dispose of nitrogenous wastes across body surfaces. The nervous system is composed of two lateral nerves and a bilobed, ganglionic *brain* that is located on the dorsal surface of the mastax. Sensory structures include numerous ciliary clusters and sensory bristles concentrated on either one or more short antennae or the corona. One to five photosensitive eyespots may be found on the anterior end.

Reproduction and Development

Rotifers are dioecious and reproduce sexually, although several types of parthenogenesis occur in most species. Smaller males appear only sporadically in the Monogononta, and no males are known in the Bdelloidea. Most rotifers have a single ovary and an attached syncytial *vitellarium,* which produces yolk. The ovary and vitellarium often fuse to form a single *germovitellarium (see figure 19.3b).* The vitellarium produces yolk that is incorporated into the eggs. After fertilization, each egg travels through a short oviduct to the cloacal bladder and out its opening.

In males, the mouth, cloacal bladder, and other digestive organs are either degenerate or absent. A single testis produces sperm that travel through a ciliated vas deferens to the gonopore. Male rotifers typically have an eversible *penis* that ejects sperm, like a hypodermic needle, into the pseudocoelom of the female. [3]

In the class Seisonidea, the females produce haploid eggs that must be fertilized to develop into either males or females. In the Bdelloidea, all females are parthenogenetic and produce diploid eggs that hatch into diploid females. In the class Monogononta, two different types of eggs are produced (figure 19.4). **Amictic eggs** (thin-shelled

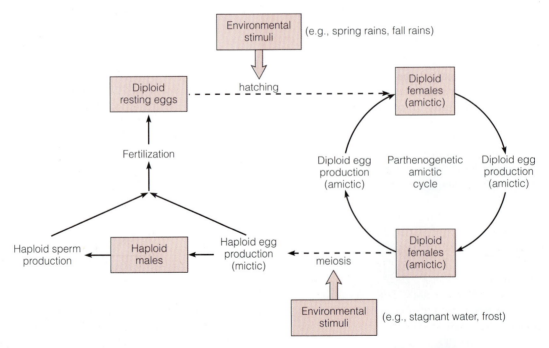

Figure 19.4 The life cycle of a monogonont rotifer. Dormant, diploid, resting eggs hatch in response to environmental stimuli (e.g., melting snows and spring rains) to begin a first amictic cycle. Other environmental stimuli (e.g., population density, stagnating water) later stimulate the production of haploid mictic eggs that lead to the production of dormant eggs that carry the species through the summer (e.g., when the pond dries up). With autumn rains, there is a second amictic cycle. Frost stimulates the production of mictic eggs again, and the eventual dormant resting eggs that are produced carry the rotifer over the winter.

summer eggs) are produced by mitosis, are diploid (2n), cannot be fertilized, and directly develop into amictic females. Thin-shelled, **mictic eggs** are haploid. If the mictic egg is not fertilized, it develops parthenogenetically into a male; if fertilized, mictic eggs secrete a thick, heavy shell and become *dormant* or *resting winter eggs.* Dormant eggs always hatch with melting snows and spring rains into amictic females which begin a first amictic cycle, building up large populations quickly. By early summer, some females have begun to produce mictic eggs, males appear, and dormant eggs are produced. Another amictic cycle, as well as the production of more dormant eggs, occurs before the yearly cycle is over. Dormant eggs are often dispersed by winds or birds, accounting for the unique distribution patterns of many rotifers. Most females lay either amictic or mictic eggs, but not both. Apparently, during oocyte development, the physiological condition of the female determines whether her eggs will be amictic or mictic.

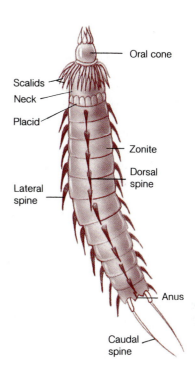

Figure 19.5 Phylum Kinorhyncha. An illustration of the external anatomy of an adult in longitudinal view.

Stop and Ask Yourself

1. How do the cilia of a rotifer beat?
2. How can the anatomy of a rotifer be described?
3. How do rotifers feed? Reproduce?
4. What is the difference between a mictic and an amictic rotifer egg?

Phylum Kinorhyncha:
The Kinorhynchs

Kinorhynchs (kin'o-rink's) are small (less than 1 mm long), elongate worms found exclusively in marine environments, where they live in mud and sand. Because they have no external cilia or locomotor appendages, they simply burrow through the mud and sand with their snout. In fact, the phylum takes its name (Kinorhyncha, Gr. *kinein,* motion + *rhynchos,* snout) from this method of locomotion. The phylum Kinorhyncha contains about 130 known species that are divided into two orders.

The body surface of a kinorhynch is devoid of cilia and is composed of 13 definite segments called **zonites** (figure 19.5). The *head,* represented by zonite 1, bears the mouth, an oral cone, and spines. The *neck,* represented by zonite 2, contains spines called **scalids** and plates called **placids.** The head can be retracted into the neck. The *trunk* consists of the remaining 11 zonites and terminates with the anus. Each trunk zonite bears a pair of lateral spines and one dorsal spine.

The body wall consists of a cuticle, epidermis, and two pairs of muscles: a dorsolateral and ventrolateral pair. The pseudocoelom is large and contains ameboid cells.

A complete digestive system is present, consisting of a mouth, buccal cavity, muscular pharynx, esophagus, stomach-intestine (where digestion and absorption take place),

and anus. Most kinorhynchs feed on diatoms, algae, and organic matter.

A pair of protonephridia is located in zonite 10. The nervous system consists of a brain and single ventral nerve cord with a ganglion (a mass of nerve cells) in each zonite. Sensory organs are represented by eyespots and sensory bristles in some species. Kinorhynchs are dioecious, and have paired gonads. The male gonopore is surrounded by several spines that may be used in copulation. The young hatch into larvae that do not show all of the zonites. As the larvae grow and molt, the adult morphology appears. Once adulthood is attained, molting no longer occurs.

Phylum Nematoda: The Nematodes
or Roundworms

Nematodes (nem-a-to'des) (Gr. *nematos,* thread) are the most abundant animals on Earth—there may be some 5 billion in every acre of fertile garden soil. Zoologists estimate that there are anywhere from 10,000 to 500,000 roundworm species, feeding on every conceivable source of organic matter—from rotting substances to the living tissues of other invertebrates, vertebrates, and plants (box 19.2). They range in length from microscopic to several meters. Many nematodes are parasites of plants or animals; others are free-living in marine, freshwater, or soil habitats.

Except in their sensory structures, nematodes lack cilia; a characteristic they share with arthropods. Also in common with some arthropods, the sperm of nematodes is ameboid. Two classes of nematodes are recognized: Adenophorea and

[4]

Figure 19.5 redrawn, with permission, from P. E. Lutz, *Invertebrate Zoology.* Copyright © 1986, Benjamin/Cummings Publishing Company, Menlo Park, California.

Box 19.2

The Ecology of Soil Nematodes

Nematodes are incredibly abundant and diverse in soils. Some are parasitic on the roots of various plants, where they do considerable damage. Their large reproductive potential is generated at the expense of the plants on whose tissues they feed. The result is millions of dollars worth of damage to garden, truck farm, woody, and ornamental plants.

The vast majority of soil nematodes, however, are free living. They move between soil particles and are very important in the ecology of the soil. For example, some nematodes are voracious feeders on soil bacteria and fungi; they help control the populations of these microorganisms. Other nematodes feed, in turn, on the microbial feeders and also play an important role in biological control. Still other nematodes are intimately important in the entire process of decomposition. Many of these species are omnivorous or saprophytic (decompose organic matter). The abundant soil nematodes are fed upon by soil arthropods and earthworms. Overall, nematodes are very essential to the energy flow and nutrient cycling in soil ecosystems.

Secernentea. The Adenophoreans have a poorly developed excretory system and lack external, glandular sensory structures called phasmids. Secerenteans possess both an excretory system and phasmids (table 19.2).

External Features

A typical nematode body is slender, elongate, cylindrical, and tapered at both ends (figure 19.6a,b). Much of the success of nematodes is due to their outer, noncellular, collagenous *cuticle* (figure 19.6c) that is continuous with the foregut, hindgut, sense organs, and parts of the female reproductive system. The cuticle may be either smooth or contain spines, bristles, papillae (small, nipplelike projections), warts, or ridges, all of which are of taxonomic significance. Three primary layers make up the cuticle: cortical, matrix, and basal layers. The cuticle functions to maintain internal hydrostatic pressure, provide mechanical protection, and aid in resisting digestion by the host in parasitic species. During maturation, the cuticle is usually molted four times.

Beneath the cuticle is the *epidermis*, or hypodermis, which projects into the pseudocoelom (figure 19.6d). The epidermis may be syncytial, and its nuclei are usually located in the four *epidermal cords* (one dorsal, one ventral, and two lateral) that project inward. The longtitudinal muscles are the principal means of locomation in nematodes. Contraction of these muscles results in undulatory waves that pass from the anterior to posterior end of the animal. [5]

Some nematodes have *lips* surrounding the mouth, and some species bear *spines* or *teeth* on or near the lips. In others, the lips have disappeared. Some roundworms have *head shields* that afford protection. Sensory organs include amphids, phasmids, or ocelli. **Amphids** are anterior depressions in the cuticle that contain modified cilia and function in chemoreception. **Phasmids** are located near the anus and also function in chemoreception. Paired *ocelli* (eyes) are present in aquatic nematodes.

Internal Features

The nematode pseudocoelom is a spacious, fluid-filled cavity that contains the visceral organs and forms a hydrostatic skeleton. All nematodes are round because of the equal outward force generated in all directions by the body muscles contracting against the pseudocoelomic fluid (figure 19.6d). [6]

Feeding and the Digestive System

Depending on the environment, nematodes are capable of feeding on a wide variety of foods—they may be carnivores, herbivores, omnivores, saprobes that consume decomposing organisms, or parasitic species that feed on blood and other tissue fluids of their hosts.

Nematodes have a complete digestive system consisting of a mouth, which may have teeth, jaws, or stylets; buccal cavity; muscular pharynx; long tubular intestine, where digestion and absorption occur; a short rectum; and an anus. Hydrostatic pressure in the pseudocoelom is responsible for the passage of food through the alimentary canal.

Table 19.2   Classification of the Nematodes
Phylum Nematoda (nem-a-to′da) Nematodes, or roundworms.

Class Secernentea (Phasmidea) (ses-er-nen′te-a)
    Paired glandular or sensory structures called phasmids
    in the tail region; similar pair of structures (amphids)
    poorly developed in anterior end; excretory system
    present; both free-living and parasitic species. *Ascaris,*
    Enterobius, Rhabditis, Turbatrix, Necator,
    Wuchereria.

Class Adenophorea (Aphasmidia) (a-den″o-for′e-a)
    Phasmids absent; most free living, but some parasitic
    species occur. *Dioctophyme, Trichinella, Trichuris.*

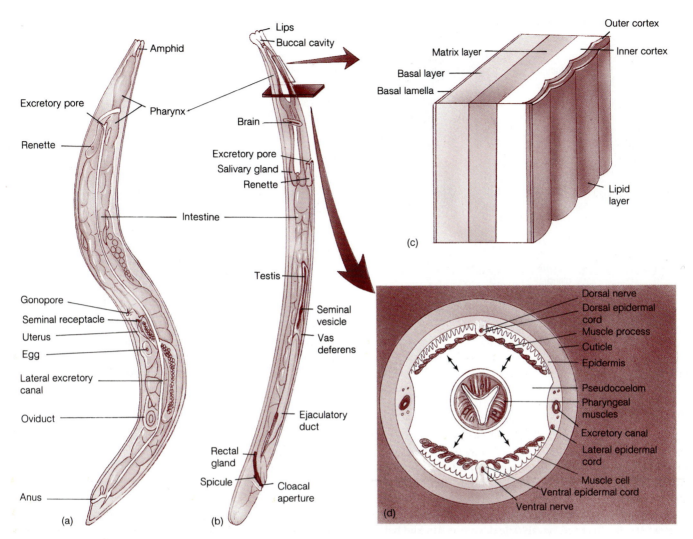

Figure 19.6 Phylum Nematoda. (a) Internal anatomical features of a female and (b) male *Rhabditis*. (c) A section through a nematode cuticle showing the various layers. (d) A cross-section through the region of the muscular pharynx of a nematode. The hydrostatic pressure (arrows) in the pseudocoelom acts to maintain the rounded body shape of a nematode and also to collapse the intestine, which aids in food and waste material moving from the mouth to the anus.

Other Visceral Systems

Nematodes accomplish osmoregulation and excretion of nitrogenous waste products (ammonia, urea, uric acid) with two unique systems. The *glandular system* (figure 19.7a) is found in aquatic species, and consists of ventral gland cells, called **renettes,** that are located posterior to the pharynx. Each gland absorbs waste material from the pseudocoelom and empties it to the outside through an excretory pore. Parasitic nematodes have a more advanced system, called the *tubular system* (figure 19.7b), that develops from the renette system. In this system, the renettes unite to form a large canal, which opens to the outside via an excretory pore.

The nervous system consists of a circumpharyngeal brain (*see figure 19.6b*). Nerves extend anteriorly and posteriorly; many connect to each other via commissures. Certain neuroendocrine secretions are involved in growth, molting, cuticle formation, and metamorphosis.

Reproduction and Development

Most nematodes are dioecious, with the males being smaller than the females. The long, coiled gonads lie free in the pseudocoelom.

The female system consists of a pair of convoluted *ovaries* (figure 19.8a). Each ovary is continuous with an oviduct whose proximal end is swollen to form a *seminal receptacle*. Each oviduct becomes a tubular *uterus*; each uterus unites to form a *vagina* that opens to the outside through a *gonopore*.

The male system consists of a single *testis*, which is continuous with a *vas deferens* that eventually expands into a *seminal vesicle* (figure 19.8b). The seminal vesicle connects to the cloaca. Males are commonly armed with a posterior flap of tissue called a *bursa,* which contains copulatory *spicules*. These curved spicules aid the male in the transfer of sperm to the female gonopore during copulation.

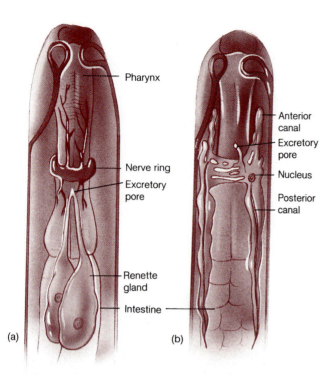

(a)

(b)

Figure 19.7 Two types of nematode excretory systems. (a) Glandular, as found in *Rhabditis*. (b) Tubular, as found in *Ascaris*.

After copulation, each fertilized egg is moved to the gonopore by hydrostatic forces in the pseudocoelom (*see figure 19.6d*). The number of eggs produced varies with the species; some nematodes produce only several hundred, whereas others may produce 200,000 daily. Some nematodes give birth to larvae. The development and hatching of the eggs are influenced by external factors, such as temperature and moisture. Hatching produces a larva that has most adult structures. The larva undergoes four molts (the shedding of the cuticle), although in some species, the first one or two molts may take place in the egg before hatching.

Some Important Nematode Parasites of Humans

Parasitic nematodes possess a number of evolutionary adaptations to their way of life. These include a high fecundity rate, life cycles that ensure transmission from one host to another, an enzyme-resistant cuticle, resistant eggs, and encysted larvae. The life cycles of nematodes are not as complicated as of cestodes or trematodes because only one (or at the most two) host is involved. Discussions of the life cycles of five important human parasites follow.

Ascaris lumbricoides: The Intestinal Roundworm
It has been estimated that as many as 800 million people throughout the world may be infected with this nematode. Adult *Ascaris* (Gr. *askaris,* intestinal worm) live in the small intestine of humans. Large numbers of eggs are produced and exit with the feces (figure 19.9). A

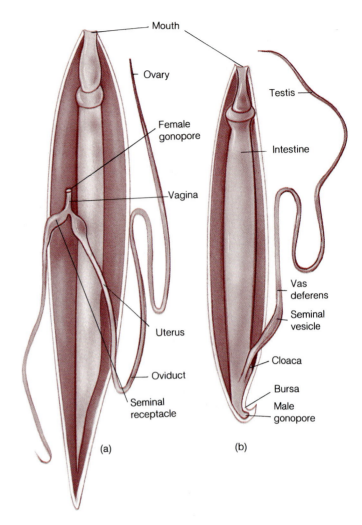

(a)

(b)

Figure 19.8 Illustrations showing the reproductive systems of a (a) female and (b) male nematode such as *Ascaris.* The sizes of the reproductive systems are exaggerated to show details.

first-stage larva develops rapidly in the egg, molts and matures into a second-stage larva—the infective stage. When a human ingests these eggs, they hatch in the intestine, the larvae penetrate the intestinal wall and are carried, via the circulation, to the lungs. They molt twice in the lungs, migrate up the trachea and are swallowed. The worms then attain sexual maturity in the intestine, mate, and begin egg production.

Enterobius vermicularis: The Human Pinworm
Pinworms (*Enterobius*; Gr. *enteron*, intestine + *bios,* life) are the most common roundworm parasite in the United States. The adults are located in the lower (cecal) region of the large intestine. At night, gravid females migrate out of the cecum to the perianal area, where they deposit eggs containing a first-stage larva (figure 19.10). When humans ingest these eggs, the eggs hatch, the larvae molt four times in the small intestine and migrate to the large intestine. Adults mate, and females begin egg production in a short period of time.

[7]

Figures 19.7 and 19.8 redrawn, with permission, from P. E. Lutz, *Invertebrate Zoology.* Copyright © 1986, Benjamin/Cummings Publishing Company, Menlo Park, California.

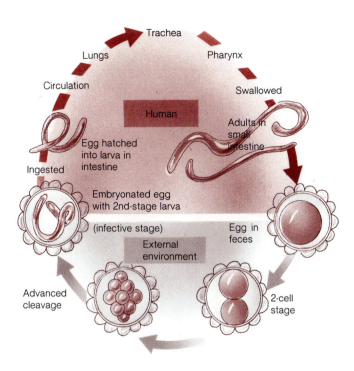

Figure 19.9 The life cycle of *Ascaris lumbricoides.* Eggs pass out of the human host in the feces. When a human consumes the embryonated eggs, they hatch in the intestine, reach the circulation, are carried to the lungs where they molt twice, migrate up the trachea, are swallowed, and mature in the small intestine. Mating occurs, completing the life cycle.

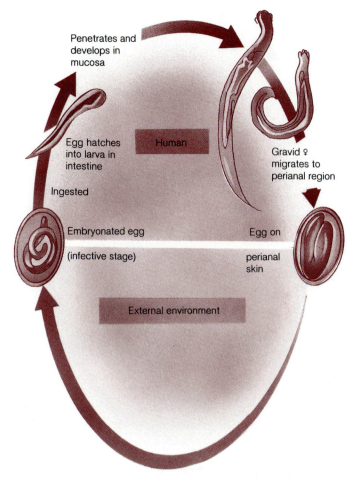

Figure 19.10 The life cycle of *Enterobius vermicularis.* At night, gravid females migrate out of the anus of a human to the perianal skin and deposit their eggs in the external environment. When consumed by a human, the eggs hatch in the intestine, the larvae grow into adults, migrate to the large intestine, mate, and complete the life cycle.

Necator americanus: **The New World Hookworm**   The New World or American hookworm, *Necator americanus* (L. *necator,* killer) is widely distributed in the southern United States. This nematode is so named because its anterior end forms a "hook." The adults live in the small intestine, where they hold onto the intestinal wall with teeth and feed on blood and tissue fluids (figure 19.11). Individual females may produce as many as 10,000 eggs daily, which pass out of the body in the feces. The eggs hatch on warm, moist soil and release a small *rhabditiform* larva. It molts and becomes the infective *filariform* larva. Humans become infected when the filariform larva penetrates the skin, usually on the foot. (Outside defecation and subsequent walking barefoot through the immediate area maintains the life cycle in humans.) After the larva burrows through the skin, it reaches the circulatory system. The rest of its life cycle is similar to that of *Ascaris* (*see figure 19.9*).

Trichinella spiralis: **The Porkworm**   Adult *Trichinella* (Gr. *trichinos,* hair) live in the mucosa of the small intestine of humans and other carnivores (e.g., the pig). In the intestine, adult females give birth to young larvae that then enter the circulatory system and are carried to skeletal (striated) muscles of the same host (figure 19.12). The young larvae encyst in the skeletal muscles and remain infective for many years. Infective meat (muscle) must be ingested by another host to continue the life cycle. Humans most often become infected by eating im-

properly cooked pork products. Once ingested, the larvae excyst in the stomach and make their way to the small intestine, where they molt four times and develop into adults. The disease casued by this nematode is called **trichinosis.**

Wuchereria **spp.: The Filarial Worms**   In tropical countries, over 250 million humans are infected with filarial (L. *filium,* thread) worms. Two representative examples are *Wuchereria bancrofti* and *W. malayi.* These elongate, threadlike nematodes live in the lymphatic system, where they block the vessels. Because lymphatic vessels return tissue fluids to the circulatory system, when the filiarial nematodes block these vessels, fluids tend to accumulate in peripheral tissues. It is this fluid accumulation that causes the enlargement of various appendages, a condition called **elephantiasis** (figure 19.13).

Adult worms live in the lymphatic vessels, where they copulate and produce larvae called *microfilariae* (figure 19.14). The microfilariae are released into the bloodstream of the human host and migrate to the skin at night. When

[8]

[9]

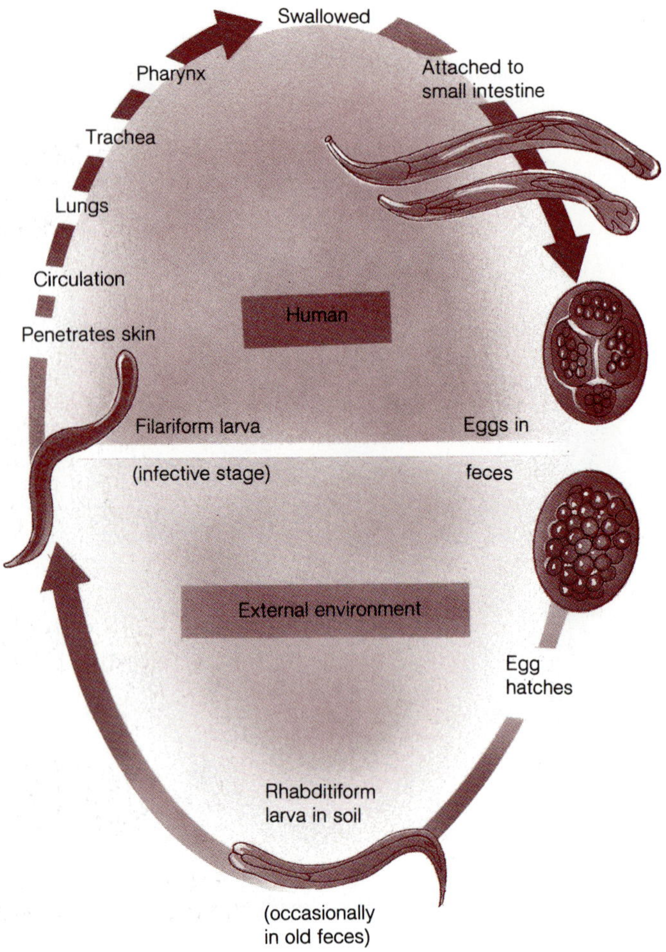

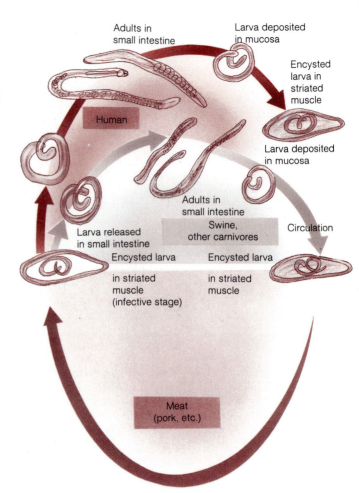

Figure 19.11 The life cycle of *Necator americanus.* Eggs pass out of the human in the feces. If they reach warm, fertile soil, they hatch into rhabditiform larvae, which molt into infective filiform larvae. When these larvae come into contact with humans, they penetrate the skin and are carried to the lungs by the circulation. They molt, migrate up the trachea, and are swallowed. Upon reaching the small intestine, they molt two more times and grow into adults. The adults mate and complete the life cycle.

Figure 19.12 The life cycle of *Trichinella spiralis.* When a human consumes raw or improperly cooked pork, the encysted larvae excyst in the stomach and reach the small intestine. They grow into adults, mate, and produce larvae. The larvae migrate to skeletal muscle, where they encyst and complete the life cycle.

a mosquito feeds on a human, it ingests the microfilariae. The microfilariae migrate to the mosquito's thoracic muscles, where they molt twice and become infective. When the mosquito takes another meal of blood, the infective third-stage larvae are injected into the blood of the human host through the mosquito's proboscis. The final two molts take place as the larvae enter the lymphatic vessels.

A filarial worm prevalent in the United States is *Dirofilaria immitis,* a parasite of dogs. The adult worms live in the heart and large arteries of the lungs; the infection is
[10] called **heartworm disease.** Once established, these filarial worms are difficult to eliminate; prevention with heartworm medicine is advocated for all dogs.

Stop and Ask Yourself

5. What is the function of the nematode cuticle?
6. What are some internal structures of nematodes? Some external structures?
7. What are the two types of excretory systems found in nematodes?
8. What are the life cycles of the following human parasites: *Ascaris*? *Enterobius*? *Necator*? *Trichinella*? and *Wuchereria*?

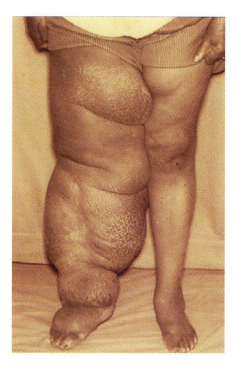

Figure 19.13 An example of elephantiasis caused by the filarial worm, *Wuchereria bancrofti.*

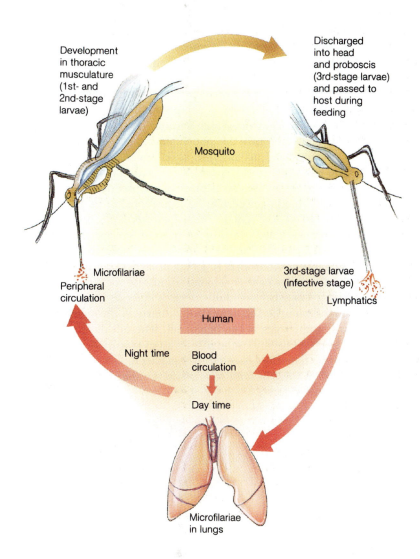

Development
in thoracic
musculature
(1st- and
2nd-stage
larvae)

Discharged
into head
and proboscis
(3rd-stage larvae)
and passed to
host during
feeding

Mosquito

Microfilariae
Peripheral
circulation

3rd-stage larvae
(infective stage)

Lymphatics

Human

Night time

Blood
circulation

Day time

Microfilariae
in lungs

Figure 19.14 The life cycle of *Wuchereria* spp. When an infected mosquito takes a meal of blood, it injects the infective third-stage larvae into the blood and lymphytatic systems. The larvae reach the lymphatic vessels where they grow into adults, mate, and produce microfilariae which accumulate in the lungs. At night, the microfilariae leave the lungs and enter the peripheral circulation. When another mosquito feeds on the host's blood, it ingests the microfilariae, which migrate to its thoracic musculature and develop into first- and second-stage larvae to complete the life cycle.

Phylum Nematomorpha: The Hairworms

[11] Nematomorphs (nem'a-to-mor'phs) (Gr. *nema*, thread + *morphe*, form) are a small group of long worms commonly called either **hairworms** or **horsehair worms.** The hairlike nature of these worms is so striking that they were formerly thought to arise spontaneously from the hairs of a horse's tail in drinking troughs or other stock-watering places. The adults are free living, but the juveniles are all parasitic in arthropods. They have a worldwide distribution and can be found in both running and standing water. About 250 species are distributed in two orders.

The body of a nematomorph is extremely long and threadlike and has no distinct head (figure 19.15). The body wall is composed of a thick cuticle; a cellular epidermis is present with longitudinal cords and a muscle layer of longitudinal fibers. The pseudocoelom is reduced, and because the adults do not feed, the digestive tract is reduced. The nervous system contains an anterior nerve ring and a vental cord.

Nematomorphs have separate sexes; two long gonads extend the length of the body. After copulation, the eggs are deposited in water. When an egg hatches, a small larva that has a protrusible proboscis armed with spines emerges. Terminal stylets are also present on the proboscis. The larva must quickly enter an arthropod (e.g., beetles, cockroaches) host, either by penetrating the host or being eaten. Lacking a digestive system, the larva feeds by absorbing material directly across its body wall. Once mature, the worm leaves its host only when the arthropod is near water. Sexual maturity is attained during the free-living adult phase of the life cycle.

Phylum Acanthocephala: The Spiny-Headed Worms

Adult acanthocephalans (a-kan'tho-sef'a-lans) (Gr. *akantha*, spine or thorn + *kephale*, head) are all endoparasites in the intestinal tract of vertebrates. Two hosts are required to complete the life cycle. The juveniles are parasites of crustaceans and insects, and the adults live in vertebrates, especially fish. Acanthocephalans are generally small (less than 40 mm long), although one important species that occurs in pigs, *Macracanthorhynchus hirudinaceus,* can be up to 80 cm long. The body of the adult is elongate and composed of a short anterior *proboscis,* a *neck* region, and a *trunk* (figure 19.16a). The proboscis is covered with [12] recurved spines (figure 19.16b); hence the name spiny-headed worms. The retractible proboscis provides the means of attachment in the host's intestine. Females are always larger than males and there are about 1,000 species.

The body wall of acanthocephalans is covered by a living syncytial tegument that has been adapted to the parasitic way of life. A glycocalyx consisting of mucopolysaccha-

Figure 19.15 Phylum Nematomorpha. Photomicrograph of two adult worms.

rides and glycoproteins covers the tegument and protects against host enzymes and immune defenses. No digestive system is present; food is absorbed directly through the tegument from the host by specific membrane transport mechanisms and pinocytosis. Protonephridia may be present. The nervous system is composed of a ventral, anterior ganglionic mass from which anterior and posterior nerves arise. Sensory organs are poorly developed.

In acanthocephalans, the sexes are separate, and the male has a protrusible penis. Fertilization is internal, and development of eggs takes place in the pseudocoelom. The biotic potential of certain acanthocephalans is great; for example, gravid female *Macroacanthocephalus hirudinaceus* may conatin up to 10 million embryonated eggs. The eggs pass out of the host with the feces and must be eaten by certain insects (e.g., cockroaches or grubs), or by aquatic crustaceans (e.g., amphipods, isopods, ostracods). Once in the invertebrate, the larva emerges from the egg and is now called an **acanthor.** It burrows through the gut wall, and lodges in the hemocoel, where it develops into an **acanthella** and eventually into a **cystacanth.** When the intermediate host is eaten by a mammal, fish, or bird, the cystacanth excysts and attaches to the intestinal wall with its spiny proboscis.

Stop and Ask Yourself

  9. Why are nematomorphs called horsehair worms?
10. What is the life cycle of a nematomorph?
11. What is unique about the tegument of an acanthocephalan?
12. What is the life cycle of an acanthocephalan?

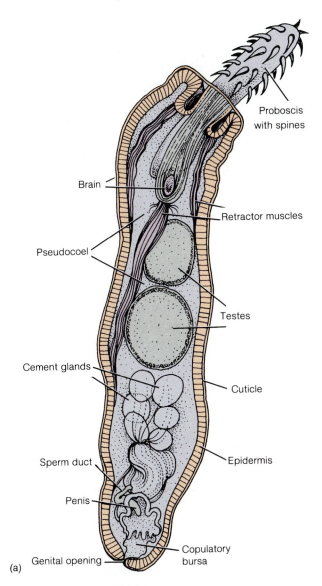

(a)

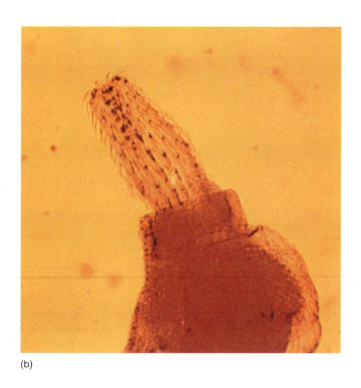

(b)

Figure 19.16 Phylum Acanthocephala. (a) Drawing of an adult male, dorsal view. (b) Proboscis of a spiny-headed worm.

Phylum Loricifera: The Loriciferans

[13] The phylum Loricifera (lor'a-sif-er-a) (L. *lorica,* clothed in armor + *fero,* to bear) is the most recent phylum to be described. Its first members were identified and named in 1983. Lorciferans occur in interstitial spaces in marine gravel. The characteristic species is *Nanaloricus mysticus.* It is a small, bilaterally symmetrical worm that has a spiny head called an **introvert,** a thorax, and an abdomen surrounded by a *lorica* (a loose fitting protective case) (figure 19.17). Both the introvert and thorax can be retracted into the anterior end of the lorica. The introvert bears eight *oral stylets* that surround the oral mouth. The lorical cuticle is periodically molted. A pseudocoelom is present and contains a short digestive system, brain and several ganglia.

Lociferans are dioecious with paired gonads. A distinctively shaped **Higgins larva** is produced that contains most of the adult organs and systems. The larva molts several times before reaching adulthood.

Phylum Priapulida: The Priapulids

The priapulids (pri'a-pyu-li-ds) (Gr. *priapos,* phallus + *ida,* pleural suffix) are a small group (only 13 species) of marine worms found in cold waters. They live buried in the mud and sand of the sea floor, where they feed on small annelids and other invertebrates.

The elongate body (figure 19.18) is cylindrical in cross section, and ranges in length from 2 mm to about 8 cm.

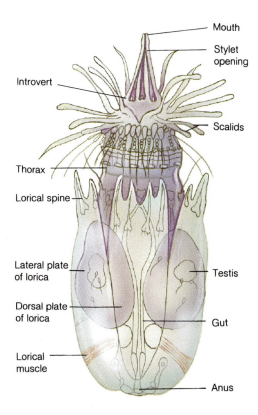

Figure 19.17 Phylum Loricifera. The anatomy of an adult male (dorsal view).

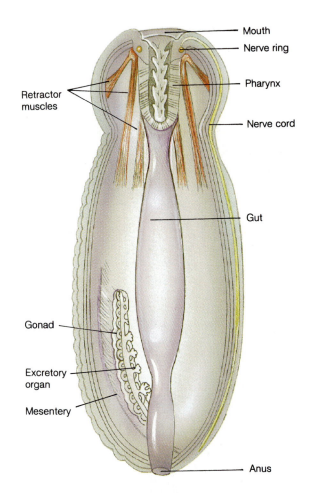

Figure 19.18 Phylum Priapulida. The internal anatomy of a priapulid.

The anterior part of the body is an introvert, which can be drawn into the longer, posterior trunk. The introvert functions in burrowing and is surrounded by spines. The muscular body is covered with a thin cuticle that bears spines. The trunk bears superficial annuli, but there is no external or internal evidence of segmentation. A straight digestive system is suspended in a large pseudocoelom that acts as a hydrostatic skeleton. In some species, the pseudocoelom contains ameboid cells that probably function in gas transport. The nervous system consists of a nerve ring around the pharynx and a single midventral nerve cord. The sexes are separate but not distinguishable. A pair of gonads is suspended in the pseudocoelom and shares a common duct with the protonephridia. The duct opens near the anus, and gametes are shed into the sea. Fertilization is external, and the eggs eventually sink to the bottom, where the larvae develop into adults. The cuticle is repeatedly molted throughout life. The most commonly encountered species is *Priapulus caudatus.*

Phylum Entoprocta: The Entoprocts

The entoprocts (en′to-prok′s) (Gr. *entos,* within + *proktos,* anus) comprise a small phylum of about 100 marine species of sedentary filter feeders. They are either solitary or colonial and are found in coastal waters. One group is commensal on the body surface of various invertebrates. Most entoprocts are microscopic in size. Entoprocts may form large, matlike colonies on rocks. An individual entoproct consists of a muscular stalk bearing a cup-shaped **calyx** with a crown of ciliated tentacles (figure 19.19). The stalk is surrounded by a chitinous cuticle and may bear an attachment disk with adhesive glands. The pseudocoel is filled with loose connective tissue. Entoprocts are filter feeders, and when they feed, the cilia on the tentacles drive food into the mouth. The digestive tract forms a U-shaped gut located in the calyx. Also found in the calyx are a pair of protonephridial tubules that open through a single pore into the mouth. The nervous system consists of a small, central ganglion and radiating nerves. Exchange of gases occurs across the body surface.

Entoprocts reproduce by asexual budding and also show sexual reproduction. Most entoprocts are hermaphroditic, but usually eggs and sperm are produced at different times in one animal. Sperm are released freely into the water, and fertilization occurs internally. Embryos develop in a brood chamber, from which free-swimming larvae are released. Eventually, the larvae settle to the substratum and develop into adults.

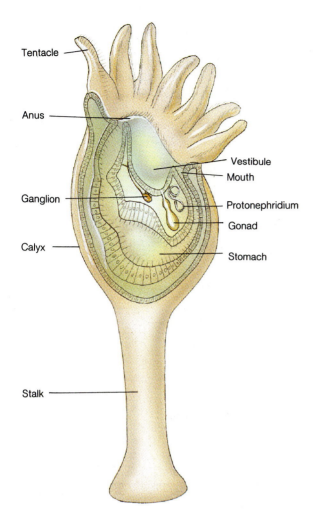

Tentacle

Anus

Vestibule

Mouth

Ganglion

Protonephridium

Gonad

Calyx

Stomach

Stalk

Figure 19.19 Phylum Entoprocta. Some morphological features of a typical entoproct.

Figure 19.19 redrawn, with permission, from *Living Invertebrates* by Pearse/Buchsbaum, copyright by The Boxwood Press.

Further Phylogenetic Considerations

The aschelminths are clearly a diverse assemblage of animals. Despite the common occurence of a cuticle, pseudocoelom, a muscular pharynx, and adhesive glands there are no distinctive features that unite the various phyla.

The gastrotrichs show some distant relationships to the acoelomates. For example, many gastrotrichs lack a body cavity, are monoecious, are small in small size, and their

ventral cilia may have been derived from the same ancestral sources as those of the turbellarian flatworms.

The rotifers have certain features in common with the acoelomates. For example, the rotifer mastax resembles the jaws of gnathostomulids. The protonephridia of rotifers closely resemble those of some freshwater turbellarians, and it is generally believed that rotifers originated in freshwater habitats. Both flatworms and rotifers have separate ovaries and vitellaria. Rotifers probably had their origins from the earliest acoelomates and may have had a common bilateral, metazoan ancestor.

The kinorhynchs, acanthocephalans, loriciferans, and priapulids all have a spiny anterior end that can be retracted; thus, they are probably related. Loriciferans and kinorhynchs appear to be most closely related.

The affinities of the nematodes to other phyla are vague. No other living group is believed to be closely related to these worms. Nematodes probably evolved in freshwater habitats and then colonized the oceans and soils. There is speculation that the ancestral nematodes were sessile, attached at the posterior end, with the anterior end protruding upward into the water. The nematode cuticle, feeding structures, and food habits probably preadapted these worms for parasitism. In fact, free-living species could become parasitic without substantial anatomical or physiological changes.

Nematomorphs may be more closely related to nematodes than to any other group by virtue of both groups being cylindrical in shape, having a cuticle, dioecious, and sexually dimorphic. However, because the larval form of some nematomorphs has a resemblance to the Priapulida, the exact affinity to the nematodes is questionable.

The phylogenetic position of the entoprocts is still controversial. Some zoologists consider entoprocts more closely related to a phylum of coelomates called the Ectoprocta (*see chapter 24*). However, because entoprocts have a pseudocoelom and protonephridia similar to flatworms and rotifers, they are discussed with the aschelminths.

Stop and Ask Yourself
13. What are the characteristics of a typical loriciferan?
14. How can a typical priapulan be described?
15. How can a typical entoproct be described?
16. What evidence is there that the individual aschelminth phyla exhibit diverse phylogenetic relationships?

Summary

1. Nine phyla are grouped together as the aschelminths. Most have a well-defined pseudocoelom, a constant number of body cells or nuclei (eutely), protonephridia, and a complete digestive system with a well-developed pharynx. No

organs are developed for gas exchange or circulation. The body is covered with a cuticle that may be molted. Longitudinal muscles are often present in the body wall.

2. The phylogenetic affinities among the nine phyla and with other phyla are uncertain.

3. Gastrotrichs are microscopic, aquatic animals with a head, neck, and trunk. Numerous adhesive glands are present. The group is generally hermaphroditic, although males are rare and female parthenogenesis is common in freshwater species.

4. The majority of rotifers inhabit fresh water. The head of these animals bears a unique ciliated corona used for locomotion and food capture. Males are smaller than females and unknown in some species. Females may develop parthenogentically.

5. Kinorhynchs are minute worms living in marine habitats. Their bodies are comprised of 13 zonites, which have cuticular scales, plates, and spines.

6. Nematodes live in aquatic and terrestrial environments; many are parasitic and of medical and agricultural importance. They are all elongate, slender, and circular in cross section. Two sexes are present.

7. The Nematomorpha are threadlike and free-living in fresh water. They lack a digestive canal.

8. Acanthocephalans are also known as the spiny-headed worms because of their spiny proboscis. All are endoparasites in vertebrates.

9. The phylum Loricifera was described in 1983. These microscopic animals have a spiny head and thorax and are found between gravel grains in marine environments.

10. The phylum Priapulida contains only 13 known species of cucumber-shaped, wormlike animals that live buried in the bottom sand and mud in marine habitats.

11. The phylum Entoprocta contains about 100 species of sessile or sedentary filter feeders; most are hermaphroditic.

■ Key Terms

acanthella (p. 302)	heartworm disease (p. 300)
acanthor (p. 302)	Higgins larva (p. 303)
amictic eggs (p. 294)	horsehair worms (p. 302)
amphids (p. 296)	introvert (p. 303)
aschelminths (p. 291)	mastax (p. 294)
calyx (p. 304)	mictic eggs (p. 295)
corona (p. 293)	phasmids (p. 296)
cuticle (p. 292)	placids (p. 295)
cystacanth (p. 302)	renettes (p. 297)
elephantiasis (p. 299)	scalids (p. 295)
eutely (p. 291)	trichinosis (p. 299)
gerontology (p. 291)	zonites (p. 295)
hairworms (p. 302)	

■ Critical Thinking Questions

1. Discuss the limitations placed on shape changes in nematodes by the structure of the body wall.

2. What characteristics set the Nematomorpha apart from the Nematoda? What characteristics do the Nematomorpha share with the Nematoda?

3. In what respects are the kinorhynchs like nematodes? Like rotifers?

4. How are gastrotrichs related to the rotifers?

5. Compare and contrast a loriciferan with a priapulid.

6. How can a pseudocoelom be described?

7. Of the nine phyla covered in this chapter, which one is most successful? Explain your answer.

8. Describe five characteristics of pseudocoelomate animals.

9. Discuss the phylogeny of the Aschelminths.

■ Suggested Readings

Books

Bogitsh, B., and Cheng, T. 1990. *Human Parasitology.* Philadelphia: Saunders.

Poinar, G. O. 1983. *The Natural History of Nematodes.* Englewood Cliffs, N.J.: Prentice-Hall.

Schmidt, G. D., and Roberts, L. S. 1989. *Foundations of Parasitology.* 4/e. St. Louis: Times Mirror Mosby.

Zuckerman, B. M. (ed.) 1980. *Nematodes as Biological Models.* vol II: *Aging and Other Model Systems.* New York: Academic Press.

Articles

D'Hondt, J. L. 1971. Gastroticha. *Annual Review Oceanography and Marine Biology* 9:141–150.

Gilbert, J. J. 1984. To build a worm. *Science* 84(5):62–70.

Jennings, J. B., and Gibson, R. 1969. Observations on the nutrition of seven species of rhynchocoelan worms. *Biological Bulletin* 136:405–410.

Kristensen, R. M. 1983. Loricifera, a new phylum with Aschelminthes characters from the meiobenthos. *Zeitschrift Zoologie Systumatiks Evolution-Fforschung* 21:163–180.

Moore, J. Parasites that change the behaviour of their host. *Scientific American* January, 1984.

Walsh, J. 1979. Rotifers, nature's water purifiers. *National Geographic* 155:286–292.

Molluscan Success

20

Concepts

1. Triploblastic animals are often assembled into two groups, the protostomes and deuterostomes, based on certain developmental features.
2. Molluscs have a true coelom (eucoelom), as well as a head-foot, visceral mass, mantle, and a mantle cavity. Most also have a radula.
3. Members of the class Gastropoda are the snails and slugs. Their bodies are modified by torsion and shell coiling.
4. Clams, oysters, mussels, and scallops are members of the class Bivalvia. They are modified for filter feeding and are often found burrowing in aquatic substrates.
5. The class Cephalopoda includes the octopods, squids, cuttlefish, and nautili. They are the most complex of all invertebrates and are adapted for predatory life-styles.
6. Other molluscs include members of the classes Scaphopoda (tooth shells), Monoplacophora, Aplacophora (solenogasters), and Polyplacophora (chitons).
7. The exact relationship of molluscs to other animal phyla is debated. Specializations of molluscs have obscured evolutionary relationships among molluscan classes.

Have You Ever Wondered:

[1] what characteristics can be used to identify molluscs?
[2] what torsion is and why it is adaptive for a snail?
[3] how a snail extends its tentacles?
[4] how a pearl is formed?
[5] how a clam feeds?
[6] how some bivalves burrow through limestone and coral?
[7] what invertebrate preys on whales?
[8] what molluscs have the most advanced nervous and sensory functions among invertebrates?

These and other useful questions will be answered in this chapter.

This chapter contains underlined evolutionary concepts.

Evolutionary Perspective

Octopods, squids, and cuttlefish (the cephalopods) can be considered the invertebrate world's most adept predators. Predatory lifestyles have resulted in the evolution of large brains (by invertebrate standards), complex sensory structures (by any standards), rapid locomotion, grasping tentacles, and tearing mouthparts. In spite of these adaptations, cephalopods rarely make up a major component of any community. Once numbering about 9,000 species, the class Cephalopoda now includes only about 550 species.

Zoologists do not know why the cephalopods have declined so dramatically. Cephalopods may have been outcompeted by vertebrates, because the vertebrates were also making their appearance in prehistoric seas, and some vertebrates acquired active, predatory life-styles. Alternatively, the cephalopods may have declined simply because of random evolutionary events.

The same has not been the case for all molluscs. This group has, as a whole, been very successful. If success is measured by numbers of species, the molluscs can be considered twice as successful as vertebrates! The vast majority of its nearly 100,000 living species belong to two classes: Gastropoda, the snails and slugs; and Bivalvia, the clams and their close relatives.

Molluscs are triploblastic, as are all the remaining animals covered in this textbook. In addition, they are the first animals described in this textbook that possess a *eucoelom*, although the eucoelom of molluscs is only a small cavity (the *pericardial cavity*) surrounding the heart and gonads. A eucoelom is a body cavity that arises in mesoderm and is lined by a sheet of mesoderm called the peritoneum (*see figure 15.13c*).

Relationships To Other Animals

Comparative embryology is the study of similarities and differences in early development of animals. Events in embryology of animals may be similar because of shared ancestry; however, similarities in development can also reflect adaptations of distantly related, or unrelated, species to similar environments. Comparative embryologists, therefore, have a difficult task of sorting *homologous* developmental sequences from *analogous* developmental sequences (*see chapter 1*). This difficulty is well illustrated in the attempt to determine the phylogenetic relationships of the molluscs and other phyla.

The phyla that are described in the following chapters are divided into two large groups, sometimes called "superphyla," and many of the reasons for this separation stem from comparative embryology. Unfortunately, there are numerous exceptions to the following generalizations. Thus, some zoologists question whether or not these two groups represent true evolutionary assemblages.

Protostomes (spiralians) include animals in the phyla Mollusca, Annelida, Arthropoda, and others. The developmental characteristics that unite these phyla are shown in figure 20.1a. One characteristic is the pattern of the early cleavages of the zygote. In *spiral cleavage* (thus the term "spiralians"), the spindle is oriented oblique to the animal/vegetal axis of a cell (*see chapters 6 and 10*). For example, division of the four-celled embryo produces an eight-celled embryo in which the upper tier of cells is twisted out of line with the lower cells. A second characteristic that is common to many protostomes is that early cleavage is *determinate*, meaning that the fate of cells is established very early in development. If blastomeres of a two- or four-celled embryo are separated, none will develop into a complete organism. A third characteristic is reflected in the term "protostome" (Gr. *protos*, first + *stoma*, mouth). The blastopore, which forms during an embryonic process called gastrulation (*see chapter 10*), usually remains open and forms the mouth. Other attributes of many protostomes include a top-shaped larva called a **trochophore larva** and a pattern of coelom and mesoderm formation, called schizocoelous, in which the mesoderm splits to form the coelom.

The other group, the **deuterostomes (radialians)**, include animals in the phyla Echinodermata, Hemichordata, Chordata, and others. The developmental characteristics that unite these phyla are shown in figure 20.1b. *Radial cleavage* occurs when the mitotic spindle is oriented perpendicular to the animal/vegetal axis of a cell and results in blastomeres oriented directly over one another. Cleavage is *indeterminate*, meaning that the fate of blastomeres is determined late in embryology and, if blastomeres are separated, they can develop into entire individuals. In deuterostomes (Gr. *deutero*, second + *stoma*, mouth), the blastopore either forms the anus, or the blastopore closes, and the anus forms in the region of the blastopore. Mesoderm and coelom formation also differ from that of protostomes, with the mesoderm pinching off as pouches of the gut, and the eucoelom forming from the mesodermal pouches. A kidney-bean-shaped larva, called a *diplerula* larva, is present in a few deuterostomes.

Origin of the Eucoelom

There are a number of hypotheses regarding the origin of the eucoelom. These hypotheses are important because they influence how one pictures the evolutionary relationships among triploblastic phyla. These two hypotheses, and their evolutionary implications, are discussed next.

One hypothesis for the origin of the eucoelom is the schizocoel hypothesis (Gr. *schizen*, to split + *koilos*, hollow). It is patterned after the method of mesoderm development and eucoelom formation in many protostomes (*see figure 20.1a*), in which all mesoderm is derived from a particular ectodermal cell of the blastula. Mesoderm derived from this cell fills the area between ectoderm and endoderm. The eucoelom arises from a splitting of this mesoderm. If the eucoelom formed in this way during evolution, mesodermally derived tissues would have preceded the eucoelom, implying that a triploblastic, acoelomate (flatworm) design could be the forerunner of the eucoelomate body plan (colorplate 5).

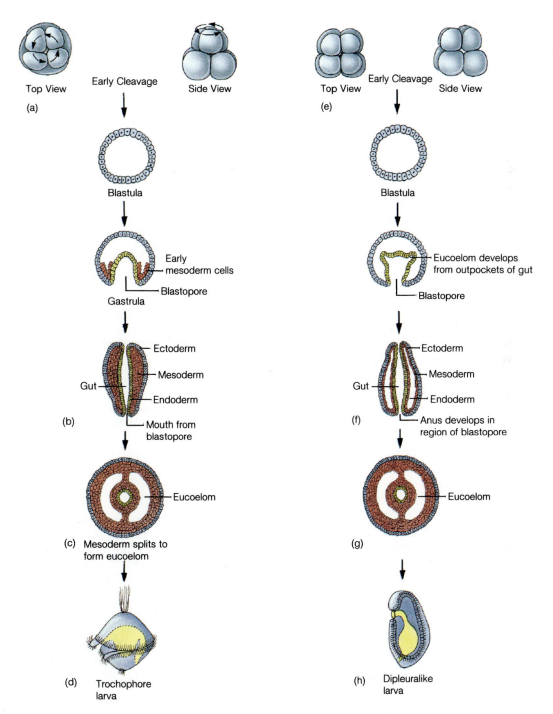

Figure 20.1 Triploblastic eucoelomate animals are grouped according to common developmental characteristics. Protostomes (spiralians) are characterized by spiral and determinate cleavage (a), a mouth that forms from an embryonic blastopore (b), schizocoelous eucoelom formation (c), and a trochophore larva (d). Deuterostomes (radialians) are characterized by radial and indeterminate cleavage (e), an anus that forms in the region of the embryonic blastopore (f), enterocoelous eucoelom formation (g), and a dipleurula (or similar) larva (h).

The enterocoel hypothesis (Gr. *enteron*, gut + *koilos*, hollow) suggests that the eucoelom may have arisen as outpocketings of a primitive gut tract. This hypothesis is patterned after the method of eucoelom formation in deuterostomes (other than vertebrates; *see figure 20.1b*). The implication of this hypothesis is that mesoderm and the eucoelom formed from the gut of a diploblastic animal. If this is true, the triploblastic, acoelomate design would be secondarily derived by mesoderm filling the body cavity of a eucoelomate animal.

Unfortunately zoologists may never know which, if either, of these hypotheses is accurate. Some zoologists believe that the eucoelom may have arisen more than once in different evolutionary lineages, in which case, more than one explanation could be correct.

Molluscan Characteristics

Molluscs range in size and body form from the giant squid, measuring 18 m in length, to the smallest garden slug, less than 1 cm long. In spite of this diversity, the phylum Mollusca (mol-lus'kah) (L. *molluscus*, soft) is not difficult to characterize (table 20.1).

[1] The body of a mollusc is divided into two main regions—the head-foot and the visceral mass (figure 20.2). The **head-foot** is elongate with an anterior head, containing the mouth and certain nervous and sensory structures, and an elongate foot, used for attachment and locomotion. A **visceral mass** contains the organs of digestion, circulation, reproduction, and excretion and is attached at the dorsal aspect of the head-foot.

The **mantle** of a mollusc usually attaches to the visceral mass, enfolds most of the body, and may secrete a shell that overlies the mantle. (Modifications of the mantle are described in discussions that follow.) The *shell* of a mollusc is secreted in three layers (figure 20.3). The outer layer of the shell is called the *periostracum*. This protein layer is secreted by mantle cells at the outer margin of the mantle. The middle layer of the shell, called the *prismatic layer,* is the thickest of the three layers and consists of calcium carbonate mixed with organic materials. It is also secreted by cells at the outer margin of the mantle. The inner layer of the shell, the *nacreous layer,* forms from

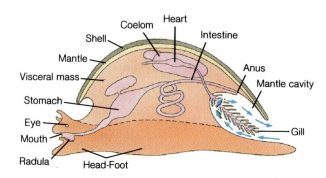

Figure 20.2 All molluscs possess three features unique to the phylum. The head-foot is a muscular structure usually used for locomotion and sensory perception. The visceral mass contains organs of digestion, circulation, reproduction, and excretion. The mantle is a sheet of tissue that enfolds the rest of the body and secretes the shell.

Table 20.1 Classification of the Mollusca
Phylum Mollusca (mol-lus'kah) The eucoelomate animal phylum whose members possess a head-foot, visceral mass, mantle, and mantle cavity. Most molluscs also possess a radula and a shell.
Class Gastropoda (gas-trop'o-dah) Shell, when present, usually coiled; body symmetry distorted by torsion; some monoecious species. *Nerita, Orthaliculus, Helix.*
Class Bivalvia (bi"val've-ah) Body enclosed in a shell consisting of two valves, hinged dorsally; no head or radula; wedge-shaped foot. *Anodonta, Mytilus, Venus.*
Class Cephalopoda (sef'ah-lop'o-dah) Foot modified into a circle of tentacles and a siphon; shell reduced or absent; head in line with the elongate visceral mass. *Octopus, Loligo, Sepia, Nautilus.*
Class Scaphopoda (ska-fop'o-dah) Body enclosed in a tubular shell that is open at both ends; tentacles; no head. *Dentalium.*
Class Monoplacophora (mon'o-pla-kof"o-rah) Molluscs with a single arched shell; foot broad and flat; certain structures serially repeated. *Neopilina.*
Class Aplacophora (a'pla-kof"o-rah) Shell, mantle, and foot lacking; wormlike; head poorly developed; burrowing molluscs. *Neomenia.*
Class Polyplacophora (pol'e-pla-kof'o-rah) Elongate, dorsoventrally flattened; head reduced in size; shell consisting of eight dorsal plates. *Chiton.*

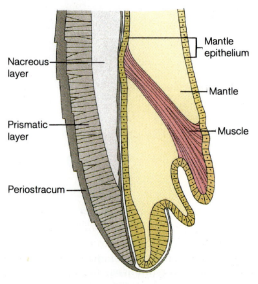

Figure 20.3 A transverse section of a bivalve shell and mantle shows the three layers of the shell and the portions of the mantle responsible for secretion of the shell.

Figure 20.3 redrawn, with permission, from *Living Invertebrates* by Pearse/Buchsbaum, copyright by The Boxwood Press.

thin sheets of calcium carbonate alternating with organic matter. The nacreous layer is secreted by cells along the entire epithelial border of the mantle. Secretion of nacre causes the shell to grow in thickness.

Between the mantle and the foot is a space called the **mantle cavity.** The mantle cavity opens to the outside and functions in gas exchange, excretion, elimination of digestive wastes, and release of reproductive products.

The mouth of most molluscs possesses a rasping structure, called a **radula,** which consists of a chitinous belt and rows of posteriorly curved teeth (figure 20.4). The radula overlies a fleshy, tonguelike structure supported by a cartilaginous **odontophore.** Muscles associated with the odontophore permit the radula to be protruded from the mouth. Muscles associated with the radula move the radula back and forth over the odontophore. Food is scraped from a substrate and passed posteriorly to the digestive tract.

Stop and Ask Yourself

1. What characteristics unite the protostomes? The deuterostomes?
2. Why does the enterocoel hypothesis of the origin of the eucoelom suggest that eucoelomate animals evolved from diploblastic animals?
3. How would you characterize the body of a generalized mollusc?
4. What is the mantle cavity of a mollusc used for?

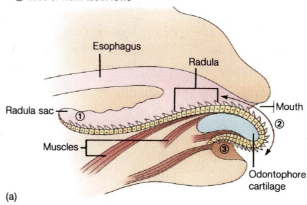

① Secretion of tooth rows
② Tooth rows in active use
③ Loss of worn tooth rows

(a)

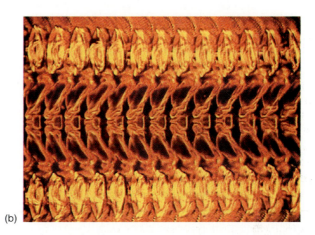

(b)

Figure 20.4 Radular structure. (a) The radular apparatus lies over the cartilaginous odontophore. Muscles attached to the radula move the radula back and forth over the odontophore. (b) Scanning electron micrographs of radular teeth arrangement of the marine snail, *Nerita.* Tooth structure is an important taxonomic characteristic for zoologists who study molluscs.

Class Gastropoda

The class Gastropoda (gas-trop'o-dah) (Gr. *gaster,* gut + *podos,* foot) includes the snails, limpets, and slugs. With over 35,000 living species (*see table 20.1*), it is the largest and most varied molluscan class. Its members occupy a wide variety of marine, freshwater, and terrestrial habitats. Most people give gastropods little thought unless they encounter *Helix pomatia* (escargot) in a French restaurant or are pestered by garden slugs and snails. One important impact of gastropods on the lives of many humans is that gastropods are intermediate hosts for some medically important trematode parasites of humans (*see chapter 18*).

Torsion

One of the most significant modifications of molluscan design in the gastropods occurs early in gastropod development. **Torsion** is a 180°, counterclockwise twisting of the visceral mass, the mantle, and the mantle cavity. After torsion, the gills, anus, and openings from the excretory and reproductive systems are positioned just behind the head and nerve cords, and the digestive tract is twisted into a U shape (figure 20.5).

The adaptive significance of torsion is speculative, however, three advantages are plausible. First, without torsion, withdrawal into the shell would proceed with the foot entering first, and the more vulnerable head entering last. With torsion, the head enters the shell first and is followed by the foot, leaving the head less exposed to potential predators. In some snails, protection is enhanced by a proteinaceous covering, called an **operculum,** on the dorsal, posterior margin of the foot. When the foot is drawn into the mantle cavity, the operculum closes the opening of the shell. Another advantage of torsion concerns an anterior opening of the mantle cavity that allows clean water from in front of the snail to enter the mantle cavity, rather than water contaminated with silt stirred up by the snail's crawl-

[2]

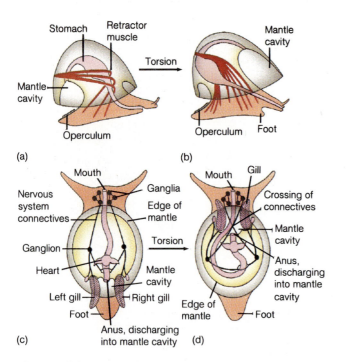

(a)

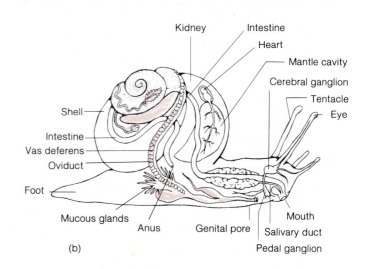

Figure 20.5 Torsion in gastropods. (a) A pretorsion gastropod larva. Note the posterior opening of the mantle cavity and the untwisted digestive tract. (b) After torsion, the digestive tract is looped, and the mantle cavity opens near the head. The foot is drawn into the shell last, and the operculum closes the shell opening. (c) A hypothetical (adult) ancestor shows the arrangement of internal organs prior to torsion. (d) Modern adult gastropods have an anterior opening of the mantle cavity and the looped digestive tract.

Figure 20.6 Gastropod structure. (a) A land (pulmonate) gastropod (*Orthaliculus*). (b) Internal structure of a generalized gastropod.

ing. An additional advantage to torsion derives from the twist in the mantle's sensory organs around to the head region. This positioning allows a snail greater sensitivity to stimuli coming from the direction in which it moves.

The major problem associated with torsion is that the anus and nephridia empty dorsal to the head and create potential fouling problems. However, a number of evolutionary adaptations seem to circumvent this problem. In more primitive gastropods, including many fossil species, there is a deep notch in the shell and mantle above the head. This notch allows water to exit the mantle cavity nearer the posterior margin of the mantle cavity, carrying waste products out of the mantle cavity further from the head. Some gastropods accomplish a similar feat with a series of excurrent openings through the mantle and shell posterior to the head. In other gastropods, an increasingly asymmetrical design shifts the inhalent portion of the mantle cavity to the left and the exhalent portion of the mantle cavity to the right, so that exhalent water flushes wastes anteriorly and laterally. This latter change is associated with evolutionary changes in shell coiling described below. Still other gastropods undergo *detorsion,* in which the embryo undergoes a full 180° torsion and then untwists approximately 90°. The mantle thus opens on the right side of the body, behind the head.

Shell Coiling

The earliest fossil gastropods possessed a shell that was coiled in one plane. This design is not common in later fossils, probably because growth resulted in an increasingly cumbersome shell. (Some modern snails, however, have secondarily returned to this shell form.)

Most modern snail shells are asymmetrically coiled into a more compact form, with successive coils or *whorls* slightly larger than, and ventral to, the preceding whorl (figure 20.6a). This pattern leaves less room on one side of the visceral mass for certain organs, and it is thought that organs that are now single were probably paired ancestrally. This design will be described further when particular body systems are described.

Locomotion

Nearly all gastropods have a flattened, ciliated foot that is covered with gland cells and used to creep across the substrate (figure 20.6b). The smallest gastropods use cilia to

(a) (b)

Figure 20.7 Carnivorous snails. Most carnivorous snails have an extensible proboscis, which is used to penetrate the soft tissues of prey. The radula at the tip of the proboscis is used to tear apart prey tissues. (a) This Florida horse conch, *Pleuroplaca gigantia*, is ingesting another gastropod. (b) *Conus* is a predatory snail whose single radular tooth is modified into a hollow, barbed spear used to inject a neurotoxin into prey. If carelessly handled, some cone shells can give humans painful stings.

propel themselves over a mucous trail. Larger gastropods use waves of muscular contraction that move over the foot. The foot of some gastropods is modified for clinging, as in abalones and limpets, or for swimming, as in sea butterflies and sea hares.

Feeding and Digestion

Most gastropods feed by scraping algae or other small, attached organisms from their substrate. Others are herbivores that feed on larger plants, scavengers, parasites, or predators.

The anterior portion of the digestive tract may be modified into an extensible *proboscis*, which contains the radula. This structure is important for some predatory snails that must extract animal flesh from hard-to-reach areas (figures 20.7 and 20.8a,b). The digestive tract of gastropods, like that of most molluscs, is ciliated. Food is trapped in mucous strings and incorporated into a rotating mucoid mass called the **protostyle.** The protostyle extends to the stomach and is rotated by cilia. Enzymes and acid are released into the stomach from a *digestive gland* located in the visceral mass, and food trapped on the protostyle is freed and digested. Wastes are formed into fecal pellets in the intestine.

Other Maintenance Functions

Gas exchange always involves the mantle cavity. Primitive gastropods had two gills; modern gastropods have lost one gill because of coiling. Some gastropods have a rolled extension of the mantle, called a **siphon,** that serves as an inhalent tube. Species that burrow extend the siphon to the surface of the substrate to bring in water. Gills are lost or reduced in land snails (pulmonates), but these snails have a richly vascular mantle that is used for gas exchange

between blood and air. Contractions of the mantle help circulate air and water through the mantle cavity.

Gastropods, as do most molluscs, have an **open circulatory system.** During part of its circuit around the body, blood leaves the vessels and directly bathes cells in tissue spaces called *sinuses*. Molluscs typically have a heart consisting of a single, muscular ventricle and two auricles. Most gastropods have lost one member of the pair of auricles because of coiling.

In addition to transporting nutrients, wastes, and gases, the blood of molluscs acts as a hydraulic skeleton. A **hydraulic skeleton** uses blood confined to tissue spaces for support. A mollusc uses its hydraulic skeleton to extend body structures by contracting muscles distant from the extending structure. For example, snails have sensory tentacles on their heads, and if the tentacle is touched, it can be rapidly withdrawn by retractor muscles. However, no antagonistic muscles exist to extend the tentacle. Extension is accomplished more slowly by contracting distant muscles to squeeze blood into the tentacle from adjacent blood sinuses.

[3]

The nervous system of primitive gastropods is characterized by six ganglia located in the head-foot and visceral mass (*see figure 20.6b*). The nerves that link these ganglia are twisted by torsion. The evolution of the gastropod nervous system has resulted in the untwisting of nerves and the concentration of nervous tissues into fewer, larger ganglia, especially in the head.

Gastropods have well-developed sensory structures. Eyes may be at the base or at the end of tentacles, they may be simple pits of photoreceptor cells, or consist of a lens and cornea. Statocysts are in the foot. *Osphradia* are chemoreceptors in the anterior wall of the mantle cavity that detect sediment and chemicals in inhalent water or air. The osphradia of predatory gastropods help detect prey.

Figure 20.8 Variations in the gastropod body form. (a) This heteropod (*Carinaria*) is a predator that swims upside down in the open ocean. Its body is nearly transparent. The head is at the left, and the shell is below and to the right. (b) Colorful nudibranchs have no shell or mantle cavity. The projections on the dorsal surface are used in gas exchange. In some nudibranchs, the dorsal projections are armed with nematocysts acquired from cnidarian prey that the nudibranchs use for protection. Nudibranchs prey on sessile animals, such as soft corals and sponges. (c) A tiger cowrie (*Cyraaca tigris*). Cowries have been collected for many years because of their lustrous shell. The largest whorl of a cowrie shell curls into the shell, leaving a slitlike opening and a flattened underside. When this snail is undisturbed, the mantle is extruded and covers the outer surface of the shell. Shiny nacre is thus added to the outside of the shell as well as to the inside.

Primitive gastropods possessed two *nephridia*. In modern species, the right nephridium has disappeared, probably because of shell coiling. The nephridium consists of a sac with highly folded walls and is connected to the reduced eucoelom, the pericardial cavity. Excretory wastes are derived largely from fluids filtered and secreted into the eucoelom from the blood. The nephridium modifies this waste by selectively reabsorbing certain ions and organic molecules. The nephridium opens to the mantle cavity or, in land snails (pulmonates), on the right side of the body adjacent to the mantle cavity and anal opening. Ammonia is the primary excretory product for aquatic species, because they have access to water in which to dilute the toxic ammonia. Without the ability to dilute ammonia, it must be converted to a less toxic form—uric acid. Because uric acid is relatively insoluble in water and less toxic, it can be excreted in a semisolid form, which helps conserve water.

Reproduction and Development

Many marine snails are dioecious. Gonads lie in spirals of the visceral mass (*see figure 20.6*b). Ducts discharge gametes into the sea, where external fertilization occurs.

Many other snails are monoecious, and internal, cross-fertilization is the rule. Copulation may result in mutual sperm transfer, or one snail may act as the male and the other as the female. A penis has evolved from a fold of the body wall, and portions of the female reproductive tract have become glandular and secrete a protective jelly or capsule around the fertilized egg. Some monoecious snails are protandric. In these species, testes develop first, and after they degenerate, ovaries mature.

Eggs are shed singly or in masses for external fertilization. Internally fertilized eggs are deposited in gelatinous strings or masses. Eggs of terrestrial snails are large and yolky, deposited in moist environments, such as forest-floor leaf litter, and may be encapsulated by a calcareous shell. Spiral cleavage results in a free-swimming trochophore larva that develops into another free-swimming larva with foot, eyes, tentacles, and shell, called a **veliger larva.** Sometimes, the trochophore is suppressed, and the veliger is the primary larva. Torsion occurs during the veliger stage, followed by settling and metamorphosis to the adult.

Class Bivalvia

With close to 30,000 species, the class Bivalvia (bi″val′ve-ah) (L. *bis*, twice + *valva*, leaf) is the second largest molluscan class. This class includes the clams, oysters, mussels, and scallops (*see table 20.1*). These animals are covered by a sheetlike mantle and a shell consisting of two valves (hence the class name), and they are compressed laterally. Many bivalves are edible, and some form pearls. Because most bivalves are filter feeders, their greater value is in the removal of bacteria from polluted water.

Shell and Associated Structures

The two convex halves of the shell are called **valves.** Along the dorsal margin of the shell is a proteinaceous hinge and a series of tongue-and-groove modifications of the shell, called *teeth,* that prevent the valves from twisting (figure 20.9). The oldest part of the shell is the **umbo,** a swollen area near the anterior margin of the shell. Although bivalves appear to have two shells, embryologically the shell forms as a single structure. The shell is continuous along its dorsal margin, but the mantle, in the region of the hinge, secretes relatively greater quantities of protein and relatively little calcium carbonate. The result is an elastic *hinge ligament.* The elasticity of the hinge ligament opens the valves when certain muscles are relaxed.

Adductor muscles are located at either end of the dorsal half of the shell and are responsible for closing the shell. Anyone who has tried to force apart the valves of a bivalve shell knows the effectiveness of these muscles. Adductor muscles consist of two kinds of muscle fibers that are histologically similar to smooth muscle tissue (colorplate 1p) of vertebrates. *Fast fibers* close the valves quickly and *catch fibers,* although they contract slowly, maintain a contracted state with little expenditure of energy. Metabolites that fatigue other kinds of muscles accumulate very slowly in catch muscles. This mechanism is important for bivalves, because the primary defense of most bivalves against predatory sea stars is to tenaciously refuse to open their shell. In chapter 25, you will see how sea stars are adapted to meet this defense strategy.

The mantle of bivalves is attached to the shell around the adductor muscles and near the margin of the shell. If a sand grain or a parasite becomes lodged between the shell and the mantle, the mantle secretes nacre around the irritant, gradually forming a pearl. Highest quality pearls are formed by the Pacific pearl oysters *Pinctada margaritifera* and *Pinctada mertensi.* [4]

Gills, Filter Feeding, and Digestion

The adaptation of bivalves to sedentary, filter-feeding lifestyles includes the loss of the head and radula and, except for a few bivalves, the gills have become greatly expanded and covered with cilia. Gills form folded sheets (lamellae) with one end attached to the foot and the other end attached to the mantle. The mantle cavity ventral to the gills is the inhalent region, and the cavity dorsal to the gills is the exhalent region (figure 20.10a). Cilia move water into the mantle cavity through an *incurrent opening* of the mantle. Sometimes this opening is at the end of a *siphon,* which is an extension of the mantle. A bivalve buried in the substrate can extend its siphon to the surface and still feed and exchange gases. Water moves from the mantle cavity into small pores in the surface of the gills, and from there, into vertical channels in the gills, called *water tubes.* In moving through water tubes, blood and water are in close proximity, and gases are exchanged by diffusion (figure 20.10b). Water exits the bivalve through a part of the mantle cavity at the dorsal aspect of the gills, called the *su-*

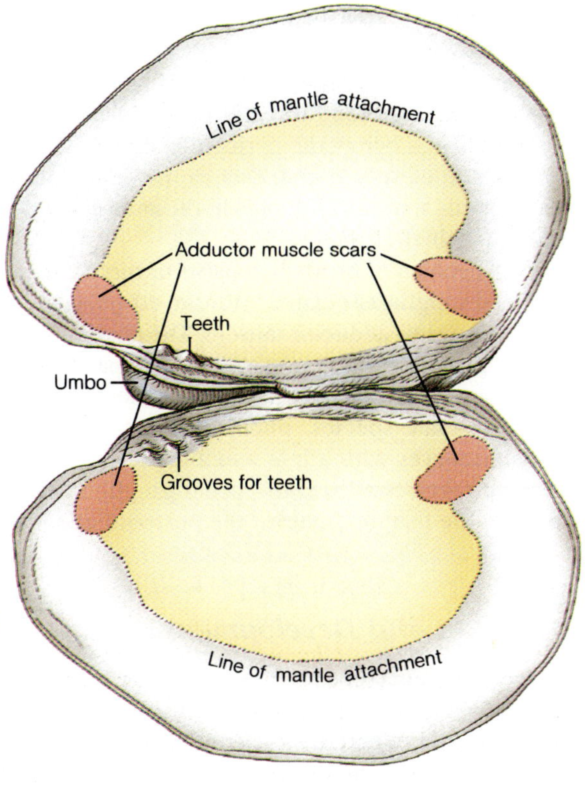

Figure 20.9 Inside view of a bivalve shell.

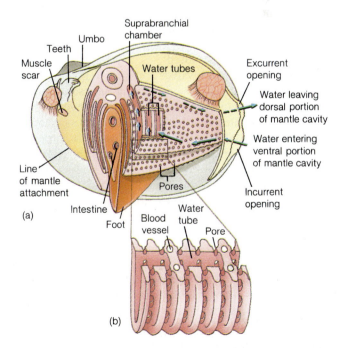

Figure 20.10 The lamellibranch gill of a bivalve. (a) Solid arrows indicate incurrent water currents. Dashed lines indicate excurrent water currents. Food is filtered as water enters water tubes through pores in the gills. (b) Horizontal section through a portion of a gill. Water passing through a water tube passes in close proximity to blood. Gas exchange occurs between water and blood in the water tubes.

Figure 20.10 redrawn, with permission, from *Living Invertebrates* by Pearse/Buchsbaum, copyright by The Boxwood Press.

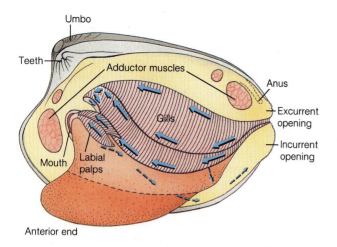

Figure 20.11 Bivalve feeding. Arrows show the path of food particles after being filtered by the gills. Dashed arrows show the path of particles rejected by the gills and the labial palps.

prabranchial chamber and an *excurrent opening* in the mantle (*see figure 20.10a*).

[5] Cilia covering the gills not only create water currents, but they also filter suspended food from the water, transport food toward the mouth, and sort filtered particles (figure 20.11). Cilia covering leaflike **labial palps,** located on either side of the mouth, also sort filtered particles. Small particles are carried by cilia into the mouth, and larger particles are moved to the edges of the palps and gills. This rejected material, called *pseudofeces,* falls, or is thrown, onto the mantle and is transported toward the inhalent opening by a ciliary tract on the mantle. Pseudofeces are washed from the mantle cavity by water rushing out when the valves are forcefully closed.

The digestive tract of bivalves is similar to that of other molluscs (figure 20.12a). Food entering the esophagus is entangled in a mucoid food string, which extends to the stomach and is rotated by cilia lining the digestive tract. A consolidated mucoid mass, the **crystalline style**, projects into the stomach from a diverticulum, called the style sac (figure 20.12b). Enzymes for carbohydrate and fat digestion are incorporated into the crystalline style. Cilia of the style sac rotate the style against a chitinized **gastric shield.** Abrasion of the style against the gastric shield and acidic conditions in the stomach dislodge enzymes. As the crystalline style rotates, the mucoid food string winds around the crystalline style, and is pulled further into the stomach from the esophagus. Food particles in the food string are dislodged by this action and the lower pH in the stomach. Further sorting separates fine particles from undigestible coarse materials. The latter are sent on to the intestine. Partially digested food from the stomach enters a *digestive gland*, where extracellular digestion, absorption, and intracellular digestion occur. Undigested wastes in the digestive gland are carried back to the stomach and then to the intestine by cilia. The intestine empties through the anus near the excurrent opening, and feces are carried away by excurrent water.

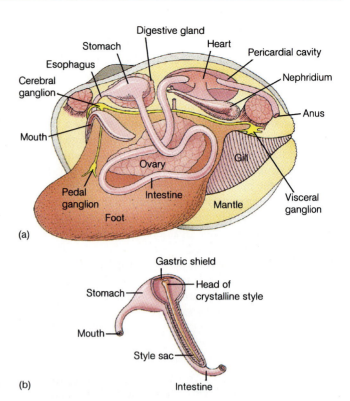

Figure 20.12 Bivalve structure. (a) Internal structure of a bivalve. (b) The bivalve stomach showing the crystalline style and associated structures.

Other Maintenance Functions

The circulatory system of bivalves is an open system. Blood flows from the heart to tissue sinuses, nephridia, gills, and back to the heart (figure 20.13). In all bivalves, the mantle is an additional site for oxygenation. In some bivalves, a separate aorta delivers blood directly to the mantle. Two nephridia are located below the pericardial cavity (the eucoelom). Their duct system connects to the eucoelom at one end and opens at nephridiopores in the anterior region of the suprabranchial chamber.

The nervous system of bivalves consists of three pairs of interconnected ganglia associated with the esophagus, the foot, and the posterior adductor muscle. The margin of the mantle is the principle sense organ. It always has general sensory cells and it may have sensory tentacles and photoreceptors. In some species, photoreceptors are in the form of complex eyes with a lens and a cornea. Other receptors include statocysts near the pedal ganglion and an osphradium in the mantle, beneath the posterior adductor muscle.

Reproduction and Development

Most bivalves are dioecious. A few are monoecious, and some of these species are protandric. Gonads are located in the visceral mass, where they surround the looped intestine. Ducts of these gonads open directly to the mantle cavity or by the nephridiopore to the mantle cavity.

Figure 20.11 redrawn, with permission, from *Living Invertebrates* by Pearse/Buchsbaum, copyright by The Boxwood Press.

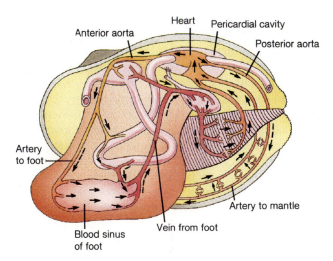

Figure 20.13 labels: Anterior aorta, Heart, Pericardial cavity, Posterior aorta, Artery to foot, Artery to mantle, Vein from foot, Blood sinus of foot

Figure 20.13 Bivalve circulation. Blood flows from the single ventricle to tissue sinuses through anterior and posterior aortae. Blood from tissue sinuses flows to the nephridia, to the gills, and then enters the auricles of the heart. In all bivalves, the mantle is an additional site for oxygenation. In some bivalves, a separate aorta delivers blood to the mantle. This blood returns directly to the heart. The ventricle of bivalves is always folded around the intestine. Thus, the pericardial cavity (the eucoelom) encloses the heart and a portion of the digestive tract.

Figure 20.13 redrawn, with permission, from *Living Invertebrates* by Pearse/Buchsbaum, copyright by The Boxwood Press.

External fertilization occurs in most bivalves. Gametes exit through the suprabranchial chamber of the mantle cavity and the exhalent opening. Development proceeds through trochophore and veliger stages (figure 20.14a,b). When the veliger settles, the adult form is assumed.

The largest families of freshwater bivalves brood their young. Fertilization occurs in the mantle cavity by sperm brought in with inhalent water. Members of the family Sphaeridae brood their young in maternal gills through reduced trochophore and veliger stages. Young clams are shed from the gills. Members of the families Unionidae and Mutelidae brood their young through a modified veliger stage called a **glochidium**, which is parasitic on fishes (figure 20.14c). These larvae possess two tiny valves, and some species have toothlike hooks. Larvae exit through the exhalent aperture and sink to the substrate. Most of these will die. If a fish contacts a glochidium, however, the larva attaches to the gills, fins, or another body part and begins to feed on host tissue. The fish forms a cyst around the larva. After a period of larval development, during which it begins acquiring its adult structures, the miniature clam falls from its host and takes up its filter-feeding life-style. The glochidium acts as a dispersal stage for an otherwise sedentary animal.

Bivalve Diversity

Bivalves are found in nearly all aquatic habitats. They may live completely or partially buried in sand or mud, attached to solid substrates, or burrowed into submerged wood, coral, or limestone.

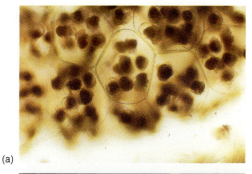
(a)

(b)

(c)

Figure 20.14 Larval stages of bivalves. (a) A trochophore larva. (b) A veliger. (c) A glochidium.

The mantle margins of burrowing bivalves are frequently fused to form distinct openings in the mantle cavity. This fusion helps to direct the water washed from the mantle cavity during burrowing and helps keep sediment from accumulating in the mantle cavity (figure 20.15).

Some surface-dwelling bivalves are attached to the substrate either by proteinaceous strands called *byssal threads,* which are secreted by a gland in the foot, or by cementation to the substrate. The former method is used by the common mussel *Mytilus,* and the latter by oysters (figure 20.16).

Boring bivalves live beneath the surface of limestone, clay, coral, wood, and other substrates. Boring begins after the larvae settle to the substrate and it occurs by mechanical abrasion of the substrate by the anterior margin of the valves. Physical abrasion is sometimes accompanied by acidic secretions from the mantle margin that dissolve limestone. As the bivalve grows, portions of the burrow recently bored are larger in diameter than other, usually external, portions of the burrow. Thus, the bivalve is often imprisoned in its rocky burrow.

[6]

Stop and Ask Yourself

5. What are three advantages conferred to gastropods by torsion?
6. What is a hydraulic skeleton?
7. How are the adductor muscles of a bivalve specialized to discourage predatory sea stars?
8. How do bivalves feed?

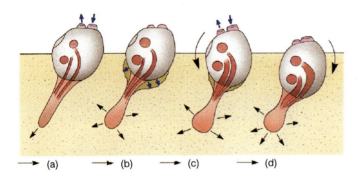

Figure 20.15 Bivalve burrowing. (a) Adductor muscles relax, and the valves push against the substrate to anchor the bivalve. The foot is extended by hydraulic pressure and pedal protractor muscles, which extend from each side of the foot to each valve. Circular and transverse muscles contract to make the foot into a narrow, probing structure. Blue arrows show movement of water into and out of the mantle cavity. (b) The tip of the foot is dilated with blood to form a second anchor, and the valves are closed. Water from the mantle cavity washes substrate away from the bivalve when the valves close (lower blue arrows). (c–d) Pedal retractor muscles contract and pull the bivalve downward. The process is repeated as necessary.

Figure 20.16 Bivalvia. Some bivalves, such as these oysters, attach to their substrate by cementation. Usually the lower (attached) valve is larger than the upper valve.

Class Cephalopoda

The class Cephalopoda (sef'ah-lop'o-dah) (L. *cephalic*, head + Gr. *podos*, foot) includes the octopods, squids, cuttlefish, and nautili (figure 20.17; *see table 20.1*). They are the most complex molluscs and, in many ways, the most complex invertebrates. The anterior portion of their foot has been modified into a circle of *tentacles* or *arms* that are used for prey capture, attachment, locomotion, and copulation (figure 20.18). The foot is also incorporated into a *funnel* that is associated with the mantle cavity and used for jetlike locomotion. The molluscan body plan is further modified in that the cephalopod head is in line with the visceral mass. Cephalopods have a highly muscular mantle that encloses all of the body except the head and tentacles. The mantle acts as a pump for bringing large quantities of water into the mantle cavity.

Shell

Ancestral cephalopods probably had a conical shell. The only living cephalopods that possess a complete shell are the nautili (*see figure 20.17a,b*). They have a coiled shell that is subdivided by septa. As the nautilus grows, it moves forward, secreting new shell around itself and leaving an empty septum behind. Only the last chamber is occupied. These chambers are fluid filled when they are formed. Septa are perforated by a cord of tissue called a *siphuncle,* which absorbs fluids by osmosis and replaces them with metabolic gases. The amount of gas in the chambers is regulated to alter the buoyancy of the animal.

In all other cephalopods, the shell is reduced or absent. In cuttlefish, the shell is internal and laid down in thin layers, leaving small, gas-filled spaces that increase buoyancy. Cuttlefish shell, called cuttlebone, has been used to make powder for polishing and has been fed to pet birds to supplement their diet with calcium. The shell of a squid is reduced to an internal, chitinous structure called the *pen*. In addition to this reduced shell, squid also have cartilaginous plates in the mantle wall, neck, and head that support the mantle and protect the brain. The shell is absent in octopods.

Locomotion

As predators, cephalopods depend upon their ability to move quickly using a jet-propulsion system. The mantle of cephalopods contains radial and circular muscles. When circular muscles contract, they decrease the volume of the mantle cavity and close collarlike valves to prevent water from moving out of the mantle cavity between the head and the mantle wall. Water is thus forced out of a narrow funnel. Muscles attached to the funnel control the direction of the animal's movement. Radial mantle muscles bring water into the mantle cavity by increasing the volume of the mantle cavity. Posterior fins act as stabilizers in squid and also aid in propulsion and steering in cuttlefish. "Flying squid" (family Onycoteuthidae) have been clocked at speeds of 20 knots (about 20 mph). Octopods are more sedentary animals. They may use jet-propulsion in an escape response, but normally they crawl over the substrate using their tentacles.

(a)

(b)

(c)

(d)

Figure 20.17 Cephalopoda. (a) A chambered nautilus. (b) A chambered nautilus shell cut to show internal chambers. (c) The cuttlefish *Sepia*. Cuttlefish are capable of rapid color changes and are often iridescent and luminescent. The ink sac of *Sepia* contains the pigment sepia, used for many years by artists. (d) *Octopus*.

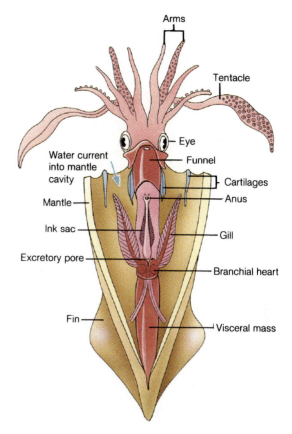

Figure 20.18 Internal structure of the squid, *Loligo*.

Feeding and Digestion

Cephalopod prey are located by sight and are captured with tentacles, which bear adhesive cups. In squid, these cups are reinforced with tough protein around their margins and sometimes possess small hooks (figure 20.19).

Jaws and a radula are present in all cephalopods. The jaws are powerful, beaklike structures used for tearing food, and the radula rasps food, forcing it into the mouth cavity.

Cuttlefish and nautili feed on small invertebrates on the ocean floor. Octopods are nocturnal hunters and feed on snails, fish, and crustaceans. Octopods have salivary structures that are used to inject poison into their prey. Squid feed on fish and shrimp, which they kill by biting across the back of the head. Giant squid even prey upon the young of sperm whales—just as adult sperm whales prey upon the young of giant squid.

The digestive tract of cephalopods is muscular, and peristalsis (coordinated muscular waves) replaces ciliary action in moving food. Most digestion occurs in a stomach and a large cecum. Digestion is primarily extracellular with large digestive glands supplying enzymes. An intestine ends at the anus, and wastes are carried out of the mantle cavity with exhalent water (*see figure 20.18a*).

Other Maintenance Functions

Cephalopods, unlike other molluscs, have a closed circulatory system. Blood is confined to vessels throughout

[7]

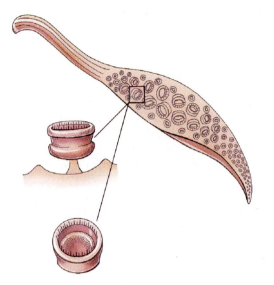

Figure 20.19 A cephalopod tentacle with suction cups.

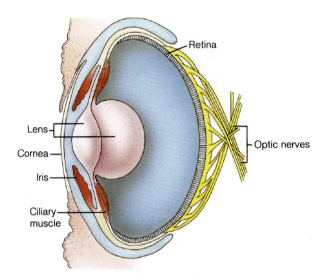

Figure 20.20 The cephalopod eye is immovable in a supportive and protective socket of cartilages. It contains a rigid, spherical lens. An iris in front of the lens forms a slitlike pupil that can open and close in response to varying light conditions. Note that the optic nerves come off the back of the retina.

its circuit around the body. Capillary beds connect arteries and veins, and exchanges of gases, nutrients, and metabolic wastes occur across capillary walls. In addition to having a heart consisting of two auricles and one ventricle, cephalopods have contractile arteries and structures called *branchial hearts*. The latter are located at the base of each gill and help move blood through the gill. These modifications increase blood pressure and the rate of blood flow—necessary for active animals with relatively high metabolic rates. Large quantities of water circulate over the gills at all times.

Greater excretory efficiency is achieved in the cephalopods because of the closed circulatory system. A close association of blood vessels with nephridia allows both filtration and secretion of wastes directly from the blood into the excretory system.

[8] The cephalopod nervous system is unparalleled by any other invertebrate. Cephalopod brains are large, and their evolution is directly related to cephalopod predatory habits and their dexterity. The brain is formed by a fusion of ganglia. It has large areas devoted to controlling muscle contraction (swimming movements and sucker closing), sensory perception, and functions such as memory and decision making. Research on cephalopod brains has provided insight into human brain functions. The ability to learn and remember, to make decisions, and to reason and make moral choices is the result of circuits involving thousands of nerve cells. Scientists use the simpler cephalopod system to help sort out this complex circuitry (box 20.1).

Among the nerve cells emerging from the brain are *giant nerve cells* that are used in cephalopod escape responses. These giant nerve cells eventually synapse with other giant nerve cells that innervate mantle musculature. Rapid conduction along these pathways stimulates contraction of mantle muscles used in escape responses.

The eyes of octopods, cuttlefish, and squids are similar in design to vertebrate eyes (figure 20.20). (This similarity is an excellent example of convergent evolution.) In contrast to the vertebrate eye, nerve cells leave the eye from the outside of the eyeball, so no blindspot exists. Like many aquatic vertebrates, cephalopods focus by moving the lens back and forth. Cephalopods can form images, distinguish shapes, and discriminate some colors. The nautiloid eye is less complex. It lacks a lens, and the interior is open to seawater, thus it acts as a pinhole camera.

Cephalopod statocysts respond to gravity and acceleration and are located in cartilages next to the brain. Osphradia, which are chemoreceptors in the mantle cavity, are present only in *Nautilus*. Tactile receptors and chemoreceptors are widely distributed over the body.

Cephalopods have pigment cells called **chromatophores.** Tiny muscles attach to these pigment cells and, when these muscles contract, the chromatophores expand and quickly change the color of the animal. Color changes, in combination with ink discharge, function in alarm responses. In defensive displays, color changes may spread in waves over the body to form large, flickering patterns. Color changes may also help cephalopods to blend with their background. The cuttlefish, *Sepia,* can even make a remarkably good impression of a checkerboard background. Color changes are also involved with courtship displays. Some species combine chromatophore displays with bioluminescence.

All cephalopods possess an ink gland that opens just behind the anus. Ink is a brown or black fluid containing melanin. Discharged ink confuses a predator, allowing the cephalopod to escape. *Sepiola* reacts to danger by darkening itself with chromatophore expansion prior to releasing ink. After ink discharge, *Sepiola* changes to a lighter color again to assist its escape.

Figure 20.19 redrawn, with permission, from *Living Invertebrates* by Pearse/Buchsbaum, copyright by The Boxwood Press.

Box 20.1

Octopus Learning

The nervous systems of cephalopods have proven invaluable to understanding the functions of complex nervous systems. Although nearly all cephalopods have been used in brain research, octopods are especially useful animals because they are relatively easy to maintain in laboratory tanks, are easily trained, and can withstand and recover from various surgical procedures. Certain discoveries from research with octopods have led to advancements in human neurosurgery.

Biologists at the Stazione Zoologica, in Naples Italy, work on octopus learning using the common European octopus, *Octopus vulgaris.* Since the late 1940s, John Z. Young, Brian B. Boycott, Martin J. Wells, and Norman S. Sutherland have led this research.

One of the series of experiments performed at Stazione Zoologica involved training an octopus not to feed on a crab when the crab was presented to the octopus with certain combinations of Plexiglas figures. When a crab was placed into a laboratory tank with an octopus, the normal response was to attack and grab the crab with one or more arms, return with the crab to the octopus' den, and then kill and feed on the crab. Training involved presenting a Plexiglas square, or other figure, to the octopus along with a crab. When the octopus attacked the crab in the presence of the Plexiglas figure, the octopus was given an electric shock. After only six training sessions over 2 days, the octopus learned the combinations of crab and figures that would result in not being shocked.

To determine which parts of the brain were responsible for learning, portions of the brain were surgically removed from untrained octopods, and after their recovery from surgery, training sessions were conducted. Regions of the brain called vertical lobes and the superior frontal lobes proved to be the learning and memory centers for optic stimuli. The greater the portion of

these regions that was removed, the more difficult learning and remembering became. It seems that the total number of available cells is the most important determinant of memory functions.

The establishment of memory in the octopus apparently involves two phases. A short-term memory is established by reinforcement training over relatively short time periods. Periodic reinforcement of what is in short-term memory results in changes in the brain and the incorporation of the information into long-term memory. The nerve pathways that are activated in incorporating information into these memory systems are being studied; however, physiologists are far from being able to accurately describe physiological processes occurring during learning and remembering.

If we are ever to understand higher nervous functions in humans, we must initially rely on systems simpler than our own. The octopus has played, and will continue to play, an important part in deciphering this complex system of nerve cells.

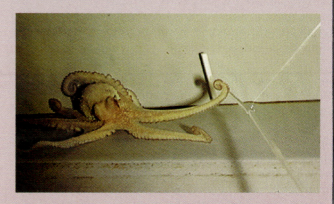

Box figure 20.1 An *Octopus bimaculoides* with a tentacle around a Plexiglas plate that it is being taught to recognize.

Reproduction and Development

Cephalopods are dioecious. Gonads are located in the dorsal portion of the visceral mass. The male reproductive tract consists of testes and structures for encasing sperm in packets called **spermatophores.** The female reproductive tract produces large, yolky eggs and is modified with glands that secrete gellike cases around eggs. These cases frequently harden on exposure to seawater.

One tentacle of male cephalopods, called the **hectocotylus,** is modified for spermatophore transfer. In *Loligo* and *Sepia,* the hectocotylus has several rows of smaller suckers capable of picking up spermatophores. Prior to copulation, a male removes spermatophores through his

funnel. Male and female tentacles intertwine, and the male inserts his hectocotylus through the funnel of the female and deposits a spermatophore near the opening to the oviduct. Spermatophores have an ejaculatory mechanism that frees sperm from the baseball-bat-shaped capsule. Eggs are fertilized as they leave the oviduct and deposited singly or in stringlike masses. They are usually attached to some substrate, such as the ceiling of an octopod's den. Octopods tend eggs during development by cleaning them of debris with their arms and squirts of water.

Development of cephalopods occurs in the confines of the egg membranes, and the hatchlings are miniatures of adults. Young are never cared for after hatching.

Stop and Ask Yourself

9. How does the jet-propulsion system of a squid work?
10. In what way is the mechanism of movement of food in the cephalopod digestive tract different from that in other molluscs?
11. What is a closed circulatory system?
12. What is a chromatophore? A spermatophore?

Class Monoplacophora

Members of the class Monoplacophora (mon′o-pla-kof″o-rah) (Gr. *monos*, one + *plak*, plate + *phoros* to bear) possess an undivided, arched shell; a broad, flat foot; and serially repeated pairs of gills and foot retractor muscles. They are dioecious; however, nothing is known of their embryology. This group of molluscs was known only from fossils until 1952, when a limpetlike monoplacophoran, named *Neopilina*, was dredged up from a depth of 3520 m off the Pacific coast of Costa Rica (figure 20.22). The significance of *Neopilina* in theories of molluscan phylogeny is covered at the end of this chapter.

Class Scaphopoda

Members of the class Scaphopoda (ska-fop′o-dah) (Gr. *skaphe*, boat + *podos*, foot) are called tooth shells or tusk shells. There are over 300 species, and all are burrowing marine animals that inhabit moderate depths. Their most distinctive characteristic is a conical shell that is open at both ends. The head and foot project from the wider end of the shell, and the rest of the body, including the mantle, is greatly elongate and extends the length of the shell (figure 20.21). Scaphopods live mostly buried in the substrate with head and foot oriented down, with the apex of the shell projecting into the water above. Incurrent and excurrent water enters and leaves the mantle cavity through the opening at the apex of the shell. Functional gills are absent, and gas exchange occurs across mantle folds. Scaphopods have a radula and tentacles, which they use in feeding on foraminiferans. Sexes are separate, and trochophore and veliger larvae are produced.

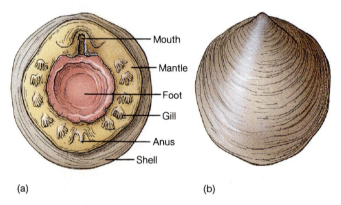

Figure 20.22 *Neopilina*, a monoplacophoran. (a) Ventral view. (b) Dorsal view.

Figure 20.23 Aplacophora. A scanning electron micrograph of the soelenogaster *Meiomenia*. The body is covered by flattened, spinelike calcareous spicules. The ventral groove shown here may be formed from a rolling of the mantle margins. *Meiomenia* is approximately 2 mm long.

Figure 20.21 Scaphopoda.

(a)

Figure 20.24 Polyplacophora. (a) Dorsal view of a polyplacophoran. Note the shell consisting of eight valves and the mantle extending beyond the margins of the shell. (b) A ventral view of a polyplacophoran. The mantle cavity is the region between the mantle and the foot. (c) Internal structure.

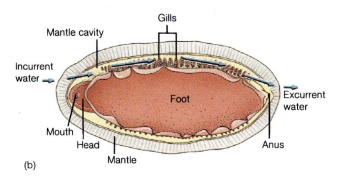

(b)

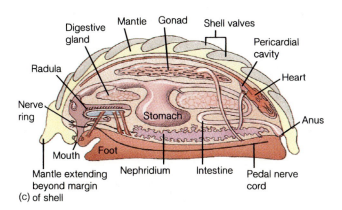

(c)

Class Aplacophora

Members of the class Aplacophora (a′pla-kof″o-rah) (Gr. *a,* without + *plak,* plate + *phoros,* to bear) are called solenogasters (figure 20.23). There are about 250 species of these relatively long, cylindrical molluscs that lack a shell and crawl on their ventral surface. Their nervous system is ladderlike and reminiscent of the flatworm design, causing some to suggest that this group may be closely related to the ancestral molluscan stock. One small group of aplacophorans contains burrowing species that feed on microorganisms and detritus and possess a radula and nephridia.

The majority of these molluscs, however, lack nephridia and a radula, are mostly surface dwellers on corals and other substrates, and are carnivores, frequently feeding on cnidarian polyps.

Class Polyplacophora

The class Polyplacophora (pol′e-pla-kof″o-rah) (Gr. *polys,* many + *plak,* plate + *phoros,* to bear) contains the chitons. Chitons are common inhabitants of shallow marine waters, wherever hard substrates occur. Chitons were used for food by early Native Americans. They have a fishy flavor, however, are tough to chew, and difficult to collect.

Chitons are easy to recognize. They have a reduced head, a flattened foot, and a shell that is divided into eight articulating dorsal valves (figure 20.24a). The broad foot is covered by a muscular mantle that extends beyond the margins of the shell and foot (figure 20.24b). The mantle cavity is restricted to the space between the margin of the mantle and the foot. Chitons crawl over their substrate in a manner similar to gastropods. The muscular foot promotes secure attachment to a substrate, which allows chitons to withstand strong waves and tidal currents. When chitons are disturbed, the edges of the mantle are applied tightly to the substrate, and contraction of foot muscles raises the middle of the foot, creating a powerful vacuum that holds the chiton to its substrate. Articulations in the shell allow chitons to roll into a ball when dislodged from the substrate.

A linear series of gills are located in the mantle cavity on each side of the foot. Currents of water, created by cilia on the gills enter below the anterior mantle margins and exit posteriorly. Openings of the digestive, excretory, and reproductive tracts are located near the exhalent area of the mantle cavity, and products of these systems are carried away with exhalent water.

Most chitons feed on attached algae. A chemoreceptor, the subradular organ, is extended from the mouth to detect food, which the radula rasps from the substrate. Food is trapped in mucus and enters the esophagus by ciliary action. Extracellular digestion occurs in the stomach, and wastes are sent on to the intestine.

The nervous system is ladderlike, with four anterior/posterior nerve cords and numerous transverse nerves. A nerve ring encircles the esophagus. Sensory structures include osphradia, tactile receptors on the mantle margin, chemoreceptors near the mouth, and statocysts in the foot. In some chitons, photoreceptors dot the surface of the shell.

Sexes are separate in chitons. A swimming trochophore settles and metamorphoses into an adult without passing through a veliger stage.

Further Phylogenetic Considerations

Fossil records of molluscan classes indicate that the phylum is over 500 million years old. Although molluscs have protostome affinities, zoologists do not know the exact relationship of this phylum to other animal phyla. The discovery of *Neopilina* (class Monoplacophora) in 1952 seemed to revolutionize ideas regarding the position of the molluscs in the animal phyla. The most striking feature of *Neopilina* was a segmental arrangement of gills, excretory structures, and nervous system. Because annelids and arthropods (*see chapters 22 and 23*) also have a segmental arrangement of body parts, monoplacophorans were considered a "missing link" between other molluscs and the annelid-arthropod evolutionary line. This link was further supported by the fact that molluscs, annelids, and arthropods share certain protostomate characteristics (*see figure 20.1a*), and chitons also show a repetition of some body parts.

Most zoologists now agree that the segmentation seen in some molluscs is very different from that of annelids and arthropods. Although information on the development of the serially repeating structures in *Neopilina* is not available, no serially repeating structure in any other mollusc develops in an annelid/arthropod fashion. Segmentation is likely not an ancestral molluscan characteristic. Many zoologists now believe that molluscs diverged from ancient triploblastic stocks independently of any other phylum (colorplate 5). Other zoologists maintain that, in spite of the absence of annelidlike segmentation in molluscs, protostomate affinities still tie the molluscs to the annelid/arthropod line. Whichever is the case, the relationship of the molluscs to other animal phyla is distant.

The diversity of body forms and life-styles in the phylum Mollusca is an excellent example of adaptive radiation. Molluscs probably began as slow-moving, marine, bottom dwellers. The development of unique molluscan features allowed them to diversify relatively quickly. By the end of the Cambrian period, some were filter feeding, some were burrowing, and others were swimming and chasing prey. Later, some molluscs became terrestrial and invaded many habitats, including tropical rain forests and arid deserts. During their evolution, molluscs have acquired some structures and lost other structures in a way that reflects specializations for particular life-styles. In the molluscs, as in many other animal groups, adaptive radiation has obscured evolutionary relationships among members.

Stop and Ask Yourself

13. How would you characterize members of the class Scaphopoda?
14. What mollusc has serially repeated pairs of gills and an undivided, arched shell?
15. What molluscs have eight dorsal, articulating plates?
16. What is the significance of *Neopilina* in hypotheses concerning molluscan phylogeny?

■ Summary

1. Triploblastic animals are divided into two groups, which may represent evolutionary assemblages. Protostomes, or spiralians, include members of the phyla Mollusca, Annelida, Arthropoda, and others. Deuterostomes, or radialians, include members of the phyla Echinodermata, Hemichordata, Chordata, and others.

2. Theories regarding the origin of the eucoelom influence how zoologists interpret evolutionary relationships among triploblastic animals.

3. Molluscs are characterized by a head-foot, a visceral mass, a mantle, and a mantle cavity. Most molluscs also have a radula.

4. Members of the class Gastropoda are the snails and slugs. They are characterized by torsion and often have a coiled shell. Like most molluscs, they use cilia for feeding, have an open circulatory system, well-developed sensory structures, and nephridia. Gastropods may be either monoecious or dioecious.

5. The class Bivalvia includes the clams, oysters, mussels, and scallops. They lack a head and are covered by a sheetlike mantle and a shell consisting of two valves. Most bivalves use expanded gills for filter feeding and most are dioecious.

6. Members of the class Cephalopoda are the octopods, squids, cuttlefish, and nautili. Except for the nautili, they have a reduced shell. The anterior portion of their foot has been modified into a circle of tentacles. Cephalopods have a closed circulatory system, highly developed nervous and sensory systems, and they are efficient predators.

7. Other molluscs include tooth shells (class Scaphopoda), *Neopilina* (class Monoplacophora), solenogasters (class Aplacophora), and chitons (class Polyplacophora).

8. Some zoologists believe that the molluscs are derived from the annelid/arthropod lineage. Others believe that they arose from triploblastic stocks independently of any other phylum. Adaptive radiation in the molluscs has obscured some evolutionary relationships among members.

■ Key Terms

chromatophores (p. 320)	odontophore (p. 311)
closed circulatory system (p. 319)	open circulatory system (p. 313)
comparative embryology (p. 308)	operculum (p. 311)
crystaline style (p. 316)	protostomes (spiralians) (p. 308)
deuterostomes (radalians) (p. 308)	protostyle (p. 313)
gastric shield (p. 316)	radula (p. 311)
glochidium (p. 317)	siphon (p. 313)
head-foot (p. 310)	spermatophores (p. 321)
hectocotylus (p. 321)	torsion (p. 311)
hydraulic skeleton (p. 313)	trochophore larva (p. 308)
labial palps (p. 316)	umbo (p. 315)
mantle (p. 310)	valves (p. 315)
mantle cavity (p. 311)	veliger larva (p. 314)
	visceral mass (p. 310)

■ Critical Thinking Questions

1. What major problem is associated with the use of comparative embryology in phylogenetic studies?

2. Compare and contrast hydraulic skeletons of molluscs with hydrostatic skeletons of cnidarians and pseudocoelomates.

3. Review the functions of body cavities presented in chapter 15. Which of those functions would apply to the eucoelom of molluscs? What additional function(s), if any, are carried out by the eucoelom of molluscs?

4. Make a table that lists the variety of modifications of the molluscan mantle, the function of each modification, and the class in which each modification occurs.

5. Torsion and shell coiling are often confused by students. Describe each and their effects on gastropod structure and function.

6. In what ways is the closed circulatory system advantageous for cephalopods?

■ Suggested Reading

Books

Barnes, R. D. 1987. *Invertebrate Zoology.* 5th ed. Philadelphia: Saunders College Publishing.

Boss, K. J. 1982. Mollusca. In S. P. Parker (ed.). *Synopsis and Classification of Living Organisms,* vol 1. New York: McGraw-Hill Book Co., Inc.

Cousteau, J., and Diol'e, P. 1973. *Octopus and Squid, the Soft Intelligence.* Garden City: Doubleday & Co.

Hyman, L. H. 1940. *The Invertebrates: Mollusca.* New York: McGraw-Hill Book Co.

Moynihan, M. H., and Rodaniche, A. F. 1977. Communication, crypsis, and mimicry among cephalopods. In T. A. Sebeok (ed.). *How Animals Communicate.* Indiana University Press.

Pennak, R. W. 1989. *Freshwater Invertebrates of the United States.* 2nd ed. New York: John Wiley & Sons, Inc.

Russell-Hunter, W. D. 1979. *A Life of Invertebrates.* New York: Macmillan Publishing Co., Inc.

Pearse, V., Pearse, J., Buchsbaum, M., and Buchsbaum, R. 1987. *Living Invertebrates.* New York: Blackwell Scientific Publications.

Solem, A. 1974. *The Shell Makers: Introducing Mollusks.* New York: John Wiley & Sons, Inc.

Articles

Boycott, B. B. Learning in the octopus. *Scientific American* March, 1965.

Feder, H. M. Escape responses in marine invertebrates. *Scientific American* July, 1972.

Heslinga, G. A. and Fitt, W. K. 1987. The domestication of reef-dwelling clams. *BioScience* 37:332–339.

Linsle, R. M. 1978. Shell form and evolution of gastropods. *American Scientist* 66:432–441.

Lemche, H. 1957. A new living deep-sea mollusk of the Cambro-Devonian class Monoplacophora. *Nature* 179:413.

Raeburn, P. 1988. Ancient Survivor. *National Wildlife* 26(5):36–38

Roper, C. R. E., and Boss, K. J. The giant squid. *Scientific American* April, 1982.

Vogel, S. 1988. How organisms use flow-induced pressures. *American Scientist,* 76:92–94.

Ward, P., Greenwald, L., and Greenwald, O. E. The buoyancy of the chambered nautilus. *Scientific American* October, 1980.

Willows, H. O. D. Giant brain cells in mollusks. *Scientific American* February, 1971.

Yonge, C. M. Giant clams. *Scientific American* April, 1975.

21

Annelida: The Metameric Design

Concepts

1. Members of the phylum Annelida are the segmented worms. The relationships of annelids to lower triploblastic animals is debated.
2. Metamerism has important influences on virtually every aspect of annelid structure and function.
3. Members of the class Polychaeta are annelids that have become adapted to a variety of marine habitats. Some live in or on marine substrates; others live in burrows or are free swimming. They are characterized by the presence of parapodia and numerous, long setae.
4. Members of the class Oligochaeta are found in freshwater and terrestrial habitats. They lack parapodia and have a few, short setae.
5. The class Hirudinea contains the leeches. They are predators in freshwater, marine, and terrestrial environments. Body-wall musculature and the eucoelom are modified from the pattern found in the other annelid classes. These differences influence locomotor and other functions of leeches.
6. The ancient polychaetes are probably the ancestral stock from which modern polychaetes, oligochaetes, and leeches evolved.

Have You Ever Wondered:

[1] what worm is the basis of a great communal feast?
[2] what metamerism is?
[3] what the usefulness of the fan of a fanworm is?
[4] why the blood of some marine worms is green?
[5] why epitoky is advantageous for some marine worms?
[6] why an earthworm is so difficult to extract from its burrow?
[7] why leeches should be referred to as predators rather than parasites?
[8] why earthworms are found in soil around deciduous vegetation?

These and other useful questions will be answered in this chapter.

This chapter contains underlined evolutionary concepts.

Evolutionary Perspective

[1] At the time of the November full moon, on islands near Samoa in the South Pacific, natives rush about preparing for one of their biggest yearly feasts. In just one week, the sea will yield a harvest that can be scooped up in nets and buckets. Worms by the millions will transform the ocean into what one writer has called a "vermicelli soup!" Celebrants will gorge themselves on worms that have been cooked or wrapped in breadfruit leaves. The Samoan palolo worm (*Eunice viridis*) spends its entire adult life in coral burrows at the sea bottom. Each November, one week after the full moon, this worm emerges from its burrow, and specialized body segments devoted to reproduction break free and float to the surface, while the rest of the worm is safe on the ocean floor. The surface water is discolored as gonads release their countless eggs and sperm. The natives' feast is shortlived, however; these reproductive swarms last only 2 days and do not recur for another year.

The Samoan palolo is a member of the phylum Annelida (ah-nel'i-dah) (L. *annellus,* ring). Other members of this phylum include the soil-building earthworms, predatory leeches, and countless marine worms (table 21.1).

Relationships to Other Animals

Annelids are protostomes (colorplate 5). Protostome characteristics, such as spiral cleavage, a mouth derived from an embryonic blastopore, schizocoelous coelom formation, and trochophore larvae are present in most members of the phylum. (Certain exceptions will be discussed later.) Annelids also have a complete digestive tract and display segmental organization of most body parts. The origin of this diverse phylum, like that of most other phyla, occurred at least as early as the Precambrian period, more than 600 million years ago. Unfortunately, there is little evidence documenting the evolutionary pathways that resulted in the first annelids.

There are a number of hypotheses that account for annelid origins. These hypotheses are tied into hypotheses regarding the origin of the eucoelom (*see chapter 20*). If one assumes a schizocoelous origin of the eucoelom, as many zoologists believe, then the annelids evolved from ancient flatworm stock. On the other hand, if an enterocoelous coelom origin is correct, then annelids evolved from ancient diploblastic animals, and the triploblastic, acoelomate body may have arisen from a triploblastic, eucoelomate ancestor. The recent discovery of a worm, *Lobatocerebrum,* which shares annelid and flatworm characteristics, has lent support to the enterocoelous origin hypothesis. *Lobatocerebrum* is classified as an annelid based on the presence of certain segmentally arranged excretory organs, an annelidlike body covering, a complete digestive tract, and an annelidlike nervous system. As do flatworms, however, it has a ciliated epidermis and is acoelomate. Some zoologists believe that *Lobatocerebrum* illustrates how the triploblastic, acoelomate design could have been derived from the annelid lineage.

Metamerism and Tagmatization

When one looks at an earthworm, one of the first characteristics noticed is the organization of the body into a series of ringlike segments. What is not obvious, however, is that the body is divided internally as well. Each segment has its own excretory, circulatory, and nervous elements. Segmental arrangement of body parts in an animal is called [2] metamerism (Gr. *meta,* after + *mere,* part).

Metamerism has profound influences on virtually every aspect of annelid structure and function, such as the anatomical arrangements of organs that are coincidentally associated with metamerism. For example, the compartmentalization of the body has resulted in each segment having its own excretory, nervous, and circulatory structures. Two related functions, however, are probably the primary adaptive features of metamerism: flexible support and efficient locomotion. These functions depend on the metameric arrangement of the eucoelom and can be understood by examining the development of the eucoelom and the arrangement of body-wall muscles.

During embryonic development, the body cavity of annelids arises by a segmental splitting of a solid mass of mesoderm that occupies the region between ectoderm and endoderm on either side of the embryonic gut tract. Enlargement of each cavity forms a double-membraned *septum* on the anterior and posterior margin of each eucoelomic space and dorsal and ventral mesenteries associated with the digestive tract (figure 21.1).

Muscles also develop from the mesodermal layers associated with each segment. A layer of circular muscles lies below the epidermis, and a layer of longitudinal muscles, just below the circular muscles, runs between the septa

Table 21.1	Classification of the Phylum Annelida

Phylum Annelida (ah-nel'i-dah)
The phylum of triploblastic, eucoelomate animals whose members are metameric (segmented), long, and cylindrical or oval in cross section. Annelids have a complete digestive tract and a ventral nerve cord. The phylum is divided into three classes.

Class Polychaeta (pol″e-ket'ah)
The largest annelid class; mostly marine; head with eyes and tentacles; parapodia bear numerous setae; monoecious or dioecious; development frequently involves a trochophore larval stage. *Nereis, Arenicola, Sabella.*

Class Oligochaeta (ol″i-go-ket'ah)
Few setae and no parapodia; no distinct head; monoecious with direct development; primarily in fresh water or terrestrial. *Lumbricus, Tubifex.*

Class Hirudinea (hi′roo-din″eah)
Leeches; bodies with 34 segments; each segment subdivided into annuli; anterior and posterior suckers present; parapodia and setae absent. *Hirudo.*

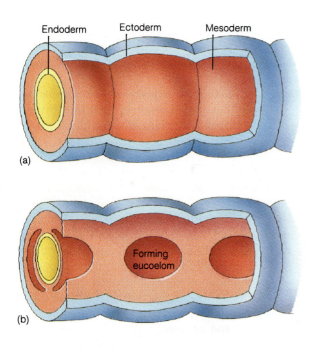

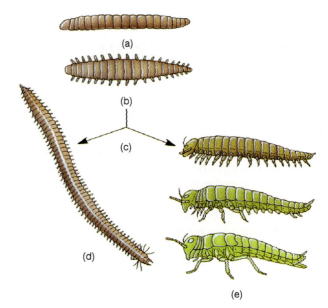

Figure 21.2 A possible sequence in the evolution of the annelid arthropod line from a hypothetical wormlike ancestor. (a) A wormlike prototype. (b) Paired, metameric appendages develop. (c) Divergence of the annelid and arthropod lines. (d) Paired appendages develop into parapodia of ancestral polychaetes. (e) Extensive tagmatization results in specializations characteristic of the arthropods. A head is a sensory and feeding tagma, a thorax is a locomotor tagma, and an abdomen contains visceral organs.

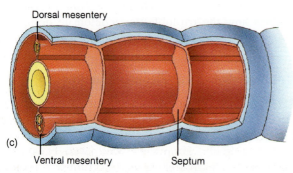

Figure 21.1 Development of metameric, eucoelomic spaces in annelids. (a) A solid mesodermal mass separates ectoderm and endoderm in early embryological stages. (b) Two cavities in each segment form from a splitting of the mesoderm (schizocoelous coelom formation). (c) These cavities spread in all directions. Enlargement of the eucoelomic sacs leaves a thin layer of mesoderm applied against the outer body wall (the parietal peritoneum) and the gut tract (the visceral peritoneum) and forms dorsal and ventral mesenteries. Anterior and posterior expansion of the coelom in adjacent segments results in the formation of the double-membraned septum that separates annelid metameres.

that separate each segment. In addition, some annelids (some polychaetes) have oblique muscles and others (the leeches) have dorsoventral muscles.

The segmental arrangement of eucoelomic spaces and muscles creates *hydrostatic compartments* and makes possible a variety of locomotor and supportive functions not possible in nonmetameric animals that utilize a hydrostatic skeleton. Each segment can be controlled independently of distant segments, and muscles can act as antagonistic pairs within a segment. The constant volume of coelomic fluid provides a skeleton against which muscles operate. Resultant localized changes in the shape of groups of segments provides the basis for swimming, crawling, and bur-

rowing. These functions are described further in the coverage of each class that follows.

A second advantage of metamerism is that it lessens the impact of injury. If one or a few segments are injured, adjacent segments, set off from injured segments by septa, may be able to maintain nearly normal functions, which increases the likelihood that the worm, or at least a part of it, will survive the trauma.

A third advantage of metamerism is that it permits the modification of certain regions of the body for specialized functions, such as feeding, locomotion, and reproduction. The specialization of body regions in a metameric animal is called **tagmatization** (Gr. *tagma,* arrangement). Although it is best developed in the arthropods, some annelids also display tagmatization. (The arthropods include animals such as insects, spiders, mites, ticks, and crayfish).

The origin of metamerism has been the subject of speculation for many years. Hypotheses are based on even more circumstantial evidence than those regarding the origin of the eucoelom, and zoologists may never be able to resolve which, if any, of the hypotheses is correct.

Virtually all zoologists agree that, because of similarities in the development of metamerism in the two groups, annelids and arthropods are closely related. Other common features include triploblastic eucoelomate organization, bilateral symmetry, a complete digestive tract, and a ventral nerve cord. As usual, there is little fossil evidence documenting ancestral pathways that led from a common ancestor to the earliest representatives of these two phyla. Zoologists are confident that the annelids and arthropods evolved from a marine, wormlike, bilateral ancestor that possessed metameric design. Figure 21.2 depicts a sequence of evolutionary changes that may have given rise to these two phyla.

Stop and Ask Yourself:

1. What are two hypotheses regarding the origin of the phylum Annelida?
2. What is metamerism? What is tagmatization?
3. What are three advantages of metameric organization?
4. What other phylum is closely related to the annelids?

Class Polychaeta

Members of the class Polychaeta (pol″e-ket′ah) (Gr. *polys*, many + *chaite*, hair) are mostly marine, and are usually between 5 and 10 cm long (*see table 21.1*). With more than 5,300 species, Polychaeta is the largest of the annelid classes. Polychaetes have become adapted to a variety of habitats. Many live on the ocean floor, under rocks and shells, and within the crevices of coral reefs. Other polychaetes are burrowers and move through their substrate by peristaltic contractions of the body wall. A bucket of intertidal sand normally yields vast numbers and an amazing variety of these burrowing annelids. Other polychaetes construct tubes of cemented sand grains or secreted organic materials. Mucus-lined tubes serve as protective retreats and feeding stations.

External Structure and Locomotion

In addition to metamerism, the most distinctive feature of polychaetes is the presence of lateral extensions called **parapodia** (Gr. *para*, beside + *podion*, little foot). Parapodia are usually laterally compressed, with an upper lobe, called a *notopodium,* and a lower lobe, called a *neuropodium.* Parapodia are supported by chitinous rods, and numerous setae project from the parapodia, giving them their class name. *Setae* are bristles that are secreted from invaginations of the distal ends of parapodia. They aid locomotion by digging into the substrate and are also used to hold a worm in its burrow or tube.

The **prostomium** (Gr. *pro,* before + *stoma,* mouth) of a polychaete is a lobe that projects dorsally and anteriorly to the mouth and contains numerous sensory structures, including eyes, antennae, palps, and ciliated pits or grooves, called nuchal organs. The first body segment, the **peristomium** (Gr. *peri,* around), surrounds the mouth cavity and bears sensory tentacles or cirri (figure 21.3).

The epidermis of polychaetes consists of a single layer of columnar cells that secrete a protective, nonliving *cuticle.* Some polychaetes have epidermal glands that secrete luminescent compounds.

Figure 21.3 *Nereis virens.* External structure.

Various species of polychaetes are capable of walking, fast crawling, or swimming. To do so, the longitudinal muscles on one side of the body act antagonistically to the longitudinal muscles on the other side of the body so that undulatory waves move along the length of the body from the posterior end toward the head. The propulsive force is the result of parapodia and setae acting against the substrate or water. Parapodia on opposite sides of the body are out of phase with one another. When longitudinal muscles on one side of a segment contract, the parapodial muscles on that side also contract, stiffening the parapodium and protruding the setae for the power stroke (figure 21.4a). As a polychaete changes from a slow crawl to swimming, the period and amplitude of undulatory waves increase (figure 21.4b).

Burrowing polychaetes push their way through sand and mud by peristaltic contractions of the body wall or by eating their way through the substrate. In the latter, organic matter in the substrate is digested and absorbed and undigestible materials are eliminated via the anus.

Feeding and the Digestive System

The digestive tract of polychaetes is a straight tube and is suspended in the body cavity by mesenteries and septa. The anterior region of the digestive tract is modified into a *proboscis,* which can be everted through the mouth by special protractor muscles and coelomic pressure. Retractor muscles bring the proboscis back into the peristomium. When the proboscis is everted, paired jaws are opened and may be used for seizing prey. Predatory polychaetes usually do not leave their burrow or coral crevice. When prey approaches a burrow entrance, the anterior portion of the worm is quickly extended, the proboscis is everted, and the prey is pulled back into the burrow. Some polychaetes

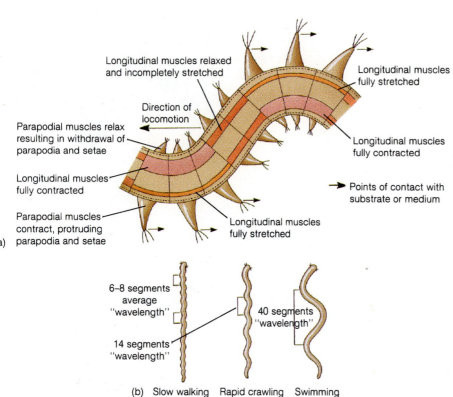

Longitudinal muscles relaxed and incompletely stretched

Longitudinal muscles fully stretched

Direction of locomotion

Parapodial muscles relax resulting in withdrawal of parapodia and setae

Longitudinal muscles fully contracted

Longitudinal muscles fully contracted

Points of contact with substrate or medium

Parapodial muscles contract, protruding parapodia and setae

Longitudinal muscles fully stretched

(a)

6–8 segments average "wavelength"

14 segments "wavelength"

40 segments "wavelength"

(b) Slow walking   Rapid crawling   Swimming

Figure 21.4 Polychaete locomotion. (a) Dorsal view of a primitive polychaete showing the antagonism of longitudinal muscles on opposite sides of the body and the resultant protrusion and movement of parapodia. (b) Both the period and amplitude of locomotor waves increase as a polychaete changes from a "slow walk" to a swimming mode.

have poison glands at the base of the jaw. Other polychaetes are herbivores and scavengers and use jaws for tearing food. Deposit-feeding polychaetes, e.g., *Arenicola,* extract organic matter from the marine sediments they ingest. The digestive tract consists of a *pharynx* that when everted, forms the proboscis; a storage sac, called a *crop*; a grinding *gizzard*; and a long, straight *intestine.* These are similar to digestive organs of earthworms (*see figure 21.13*). Organic matter is digested extracellularly, and the inorganic particles are passed through the intestine and released as "castings."

Many sedentary and tube-dwelling polychaetes are filter feeders. They usually lack a proboscis but possess other specialized feeding structures. Some tube dwellers, called fanworms, possess tentacles that form a funnel-shaped fan (figure 21.5). Distal, pinnate branches of the tentacles are called *pinnules.* Cilia on the tentacles circulate water

[3]

through the fan, trapping food particles in eddies formed between, and in front of, the pinnules. Trapped particles are carried along a food groove at the axis of the tentacle. During transport, a sorting mechanism rejects the largest particles and transports the finest particles to the mouth. Another filter feeder, *Chaetopterus,* lives in a U-shaped tube and secretes a mucous bag that collects food particles, which may be as small as 1 micron (μ). The parapodia of segments 14 through 16 are modified into fans that create filtration currents. When full, the entire mucous bag is ingested.

Elimination of digestive waste products can be a problem for tube-dwelling polychaetes. Those that live in tubes

Figure 21.5 Photograph of a fanworm (*Spirobranchus giganteus*).

that open at both ends simply have wastes carried away by water circulating through the tube. Those that live in tubes that open at one end must turn around in the tube to defecate, or they may use ciliary tracts along the body wall to carry feces to the tube opening.

Polychaetes that inhabit substrates rich in dissolved organic matter can absorb as much as 20 to 40% of their energy requirements across their body wall. This method of feeding occurs in other animal phyla too, but rarely accounts for more than 1% of energy needs.

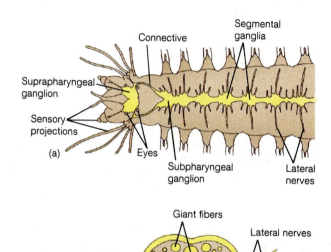

Figure 21.6   The circulatory system of a polychaete.

Gas Exchange and Circulation

Respiratory gases of most annelids simply diffuse across the body wall, and parapodia increase the surface area for these exchanges. In many polychaetes, the surface area for gas exchange is further increased by parapodial gills.

The circulatory system of polychaetes is a closed system. Oxygen is usually carried in combination with molecules called *respiratory pigments*. Respiratory pigments are usually dissolved in the plasma rather than contained in blood cells. Blood may be colorless, green, or red, depending on the presence and/or type of respiratory pigment.

[4]

Contractile elements of polychaete circulatory systems consist of a *dorsal aorta* that lies just above the digestive tract and propels blood from rear to front, and a *ventral aorta* that lies ventral to the digestive tract and propels blood from front to rear. Running between these two vessels are two or three sets of *segmental vessels* that receive blood from the ventral aorta and break into capillary beds in the gut and body wall. Capillaries coalesce again into segmental vessels that deliver blood to the dorsal aorta (figure 21.6).

Nervous and Sensory Functions

Nervous systems are similar in all three classes of annelids. The annelid nervous system includes a pair of *suprapharyngeal ganglia*, which are connected to a pair of *subpharyngeal ganglia* by *commissures* that run dorsoventrally along either side of the pharynx. A double *ventral nerve cord* runs the length of the worm along the ventral margin of each coelomic space, and there is a paired *segmental ganglion* in each segment. The double ventral nerve cord and paired segmental ganglia may be fused to varying extents in different taxonomic groups. Lateral nerves emerge from each segmental ganglion, supplying the body-wall musculature and other structures of that segment (figure 21.7a).

Segmental ganglia are responsible for coordinating swimming and crawling movements in isolated segments. (Anyone who has used portions of worms as live fish bait can confirm that the head end—the pharyngeal ganglia—is

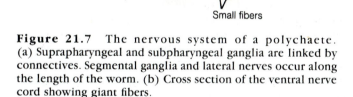

Figure 21.7   The nervous system of a polychaete. (a) Suprapharyngeal and subpharyngeal ganglia are linked by connectives. Segmental ganglia and lateral nerves occur along the length of the worm. (b) Cross section of the ventral nerve cord showing giant fibers.

Figure 21.7 redrawn, with permission, from *Living Invertebrates* by Pearse/Buchsbaum, copyright by The Boxwood Press.

not necessary for coordinated movements.) Each segment acts separate from, but is closely coordinated with, neighboring segments. The subpharyngeal ganglia help mediate locomotor functions requiring coordination of distant segments. The suprapharyngeal ganglia probably controls motor and sensory functions involved with feeding and sensory functions associated with forward locomotion.

In addition to small-diameter fibers that help coordinate locomotion, the ventral nerve cord also contains *giant fibers* (figure 21.7b). Annelid giant fibers are involved with protective or escape reactions. For example, a harsh stimulus at one end of a worm, such as a fish hook, causes a very rapid withdrawl from the stimulus. Giant fibers are approximately 50 μ in diameter and conduct impulses at 30 m/sec (as opposed to 0.5 m/sec in the smaller, 4-μ-diameter annelid fibers).

Sensory structures of polychaetes include eyes, nuchal organs, ciliated sense organs, statocysts, and tactile receptors. Two to four pairs of eyes occur on the surface of the prostomium. They vary in complexity from a simple cup of receptor cells to structures made up of a cornea, lens, and vitreous body. Most polychaetes react negatively to increased light intensities. Fanworms, however, react negatively to decreasing light intensities. If shadows cross them, fanworms retreat to their tubes. This response is believed to help protect fanworms from passing predators. *Nuchal organs* are pairs of ciliated sensory pits or slits in the head region. They are innervated by nerves from the suprapharyngeal ganglia and are thought to be chemoreceptors for food detection. Statocysts are found in the head region of polylchaetes, and the body wall is covered by ciliated tubercles, ridges, and bands, all of which contain receptors for tactile senses.

Excretion

Annelids excrete ammonia, and because ammonia diffuses readily into the water, most nitrogen excretion probably occurs across the body wall. Excretory organs of annelids are more active in regulating water and salt balances, although these abilities are limited. Most marine polychaetes, if presented with extremely diluted seawater, cannot survive the osmotic influx of water and the loss of salts that results. The evolution of efficient osmoregulatory abilities has allowed only a few polychaetes to invade freshwater.

The excretory organs of annelids, like those of many invertebrates, are called nephridia. Two types of nephridia are found in annelids. A *protonephridium* consists of a tubule with a closed bulb at one end and a connection to the outside of the body at the other end. Protonephridia have a tuft of flagella in their bulbular end that drives fluids through the tubule (figure 21.8a and *see figure 18.5*). Some primitive polychaetes possess paired, segmentally arranged protonephridia that have their bulbular end projecting through the anterior septum into an adjacent segment and the opposite end opening through the body wall at a *nephridiopore*.

Most polychaetes possess a second kind of nephridium, called a metanephridium. A **metanephridium** consists of an open, ciliated funnel, called a *nephrostome,* that projects through an anterior septum into the eucoelom of an adjacent segment. At the opposite end, a tubule opens through the body wall at a nephridiopore or occasionally through the intestine (figure 21.8b,c). There is usually one pair of metanephridia per segment, and tubules may be extensively coiled, with one portion dilated into a bladder. A capillary bed is usually associated with the tubule of a metanephridium for active transport of salts between the blood and the nephridium (figure 21.8d).

Some polychaetes also have chloragogen tissue associated with the digestive tract. This tissue functions in amino acid metabolism in all annelids and will be described further in a later section.

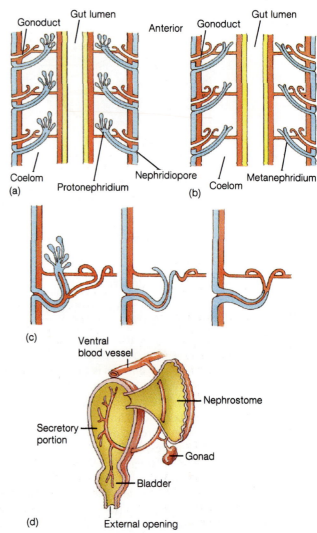

Figure 21.8 Annelid nephridia. (a) The protonephridium. The bulbular ends of this nephridium contain a tuft of flagella that drives wastes to the outside of the body. In primitive polychaetes a gonoduct (coelomoduct) carries reproductive products to the outside of the body. (b) The metanephridium. An open ciliated funnel (the nephrostome) drives wastes to the outside of the body. (c) In modern annelids, the gonoduct and the nephridial tubules undergo varying degrees of fusion. (d) Nephridia of modern annelids are closely associated with capillary beds for secretion, and nephridial tubules may have an enlarged bladder.

Regeneration, Reproduction, and Development

All polychaetes have remarkable powers of regeneration. They can replace lost parts, and some species have break points that allow worms to sever themselves when grabbed by a predator. Lost segments are later regenerated.

Some polychaetes reproduce asexually by budding or by transverse fission; however, sexual reproduction is much more common. Most polychaetes are dioecious. Gonads develop as masses of gametes and project from the coelomic peritoneum. Primitively, gonads occur in every body segment, but most polychaetes have gonads limited to spe-

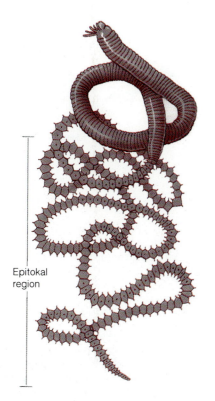

Figure 21.9 *Palola viridis*, showing the epitokal region.

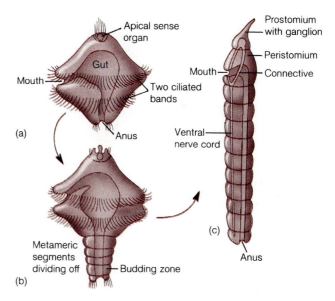

Figure 21.10 Polychaete development. (a) A trochophore. (b) A later planktonic larva showing the development of body segments. As more segments develop the larva will settle to the substrate. (c) A juvenile worm.

cific segments. Gametes are shed into the coelom where they mature. Mature female worms are often packed with eggs. Gametes may leave worms by entering nephrostomes of metanephridia and exiting through the nephridiopore, or they may be released, in some polychaetes, after the worm ruptures. In these cases, the adult soon dies. Only a few polychaetes have separate gonoducts, a condition that is believed to be primitive (*see figure 21.8a–c*).

Fertilization is external in most polychaetes, although copulation occurs in a few species. One of the most unique copulatory habits has been reported in *Platynereis megalops* from Woods Hole, Massachusetts. Toward the end of their lives, male and female worms cease feeding, and their gut tracts begin to degenerate. At this time, gametes have accumulated in the body cavity. During sperm transfer, male and female worms coil together, and the male inserts his anus into the mouth of the female. Because the digestive tracts of earth worms have degenerated, sperm are transferred directly from the male's coelom to the egg-filled coelom of the female. This method ensures fertilization of most eggs, and after fertilization is accomplished, the female sheds eggs from her anus. Both worms die after copulation.

Epitoky is the formation of a reproductive individual (an *epitoke*) that differs from the nonreproductive form of the species (an *atoke*). Frequently, an epitoke has a body that is modified into two body regions. Anterior segments carry on normal maintenance functions, and posterior segments are enlarged and filled with gametes. The epitoke may have modified parapodia for more efficient swimming (figure 21.9).

At the beginning of this chapter, there was an account of the reproductive swarming habits of *Eunice viridis* (the Samoan palolo), and one culture's response to those swarms. Similar swarming occurs in other species, usually in response to changing light intensities and lunar periods. The Atlantic palolo, for example, swarms at dawn during the first and third quarters of the July lunar cycle.

Swarming of epitokes is believed to accomplish at least three things. First, because nonreproductive individuals remain safe below the surface waters, predators cannot devastate an entire population. Second, external fertilization requires that individuals become reproductively active at the same time and in close proximity to one another. Swarming ensures that large numbers of individuals will be in the right place at the proper time. Finally, swarming of vast numbers of individuals for brief periods provides a banquet for predators. However, because vast numbers of prey are available for only short periods during the year, predator populations cannot increase beyond the limits of their normal diets. Therefore, predators can dine gluttonously and still leave epitokes that will yield the next generation of animals. [5]

Spiral cleavage of fertilized eggs may result in planktonic trochophore larvae that bud segments near the anus. Larvae eventually settle to the substrate (figure 21.10). As growth proceeds, newer segments continue to be added posteriorly. Thus the anterior end of a polychaete is the oldest end. Many other polychaetes lack a trochophore and display direct development or metamorphosis from another larval stage.

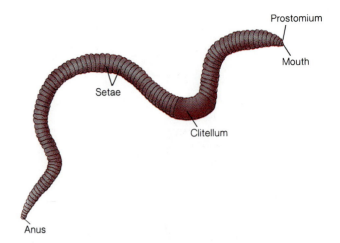

Figure 21.11 External structures of the earthworm.

Class Oligochaeta

The class Oligochaeta (ol"i-go-ket'ah) (Gr. *oligos*, few + *chaite*, hair) has over 3,000 species that are found throughout the world in freshwater and terrestrial habitats (*see table 21.1*). Aquatic species live in shallow water, where they burrow in mud and debris. Terrestrial species are found in soils with high organic content, rarely leaving their burrows. In hot, dry weather they may retreat to depths of 3 m below the surface. Soil-conditioning habits of earthworms are well known (box 21.1). *Lumbricus terrestris* is commonly used in zoology laboratories because of its large size. It was introduced to the United States from northern Europe and has flourished. Common native species like *Eisenia foetida* and various species of *Allolobophora* are smaller. A few oligochaetes are estuarine, and some are marine.

External Structure and Locomotion

Oligochaetes have setae, but fewer than are found in polychaetes (thus the derivation of the class name). Oligochaetes lack parapodia, because parapodia and long setae would interfere with their burrowing life styles, although they do have short setae on their integument. The prostomium consists of a small lobe or cone in front of the mouth, which lacks sensory appendages. A series of segments in the anterior half of an oligochaete is usually swollen into a girdlelike structure called the **clitellum** that is used for mucous secretion during copulation and cocoon formation (figure 21.11). As in the polychaetes, the body is covered by a nonliving, secreted cuticle.

Oligochaete locomotion involves the antagonism of circular and longitudinal muscles in groups of segments. Neurally controlled waves of contraction move from rear to front.

Segments bulge and setae protrude when longitudinal muscles are contracted, providing points of contact with the burrow wall. In front of each region of longitudinal muscle contraction, circular muscles contract, causing the setae to retract, and the segments to elongate and push forward. Contraction of longitudinal muscles in segments behind a bulging region causes those segments to be pulled forward. Thus segments move forward relative to the burrow as waves of muscle contraction move anteriorly on the worm (figure 21.12).

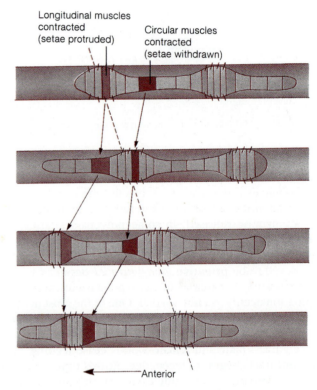

Figure 21.12 Earthworm locomotion. Arrows designate activity in specific segments of the body, and the broken lines indicate regions of contact with the substrate.

Burrowing is the result of coelomic hydrostatic pressure being transmitted to the prostomium. As an earthworm pushes its way through the soil, it uses expanded posterior segments and protracted setae to anchor itself to its burrow wall. (Any child pursuing fishing worms experiences the effectiveness of this anchor when trying to extract a worm from its burrow.) Contraction of circular muscles transforms the prostomium into a conical wedge, 1 mm in diameter at its tip. Contraction of body-wall muscles generates coelomic pressure that forces the prostomium through the soil. During burrowing, earthworms swallow considerable quantities of soil.

[6]

Box 21.1

Soil Conditioning by Earthworms

Earthworms have had an inestimable impact on the development of our planet's soil. For the past 100 million years earthworms have evolved with deciduous vegetation. Leaf fall and the death of land plants have provided a massive food source for earthworms and other soil-inhabiting organisms. The excrement, death, and decay of these organisms builds the organic constituents of our soils, and burrowing by earthworms aerates the soil and improves drainage.

In typical grassland and woodland soils, earthworms reach populations of hundreds of animals per m². They often dominate the invertebrate biomass (the total mass of invertebrate animals) in a region. Earthworms ingest soil as they feed on organic matter and as they burrow through the soil. Charles Darwin estimated that 15 tons of soil/acre/year passed through earthworm bodies. Recent estimates, based on more accurate estimates of earthworm populations, give even more impressive tillage figures of nearly 40 tons/acre/year!

Earthworms are surface feeders that emerge at night to feed on leaf fragments and other plant debris. Some of these fragments are ingested immediately, others are carried into burrows. Of the plant matter ingested by earthworms, less than 10% is incorporated into worm tissues. The rest passes through the digestive tract, is incorporated into castings (fecal material and soil), and released deeper in the soil. Earthworm castings are also rich in ammonia.

Earthworms function as vegetation shredders in our soils. Shredding vegetation and incorporating it into fecal material increases the surface area of plant matter by several orders of magnitude and hastens its eventual decomposition by bacteria and fungi. One study demonstrated the role of shredders in the breakdown of plant litter by soil animals. Leaf-filled nylon bags were buried in the soil. Some bags had a 0.5 mm mesh size, which excluded all earthworms and other large invertebrates. Other bags had a mesh size of 7 mm, which allowed all invertebrates to enter. The rate of breakdown of leaf litter was reduced by approximately two thirds in the small mesh bags.

Feeding and the Digestive System

Oligochaetes are scavengers and feed primarily on fallen and decaying vegetation, which they drag into their burrows at night. The digestive tract of oligochaetes is tubular and straight (figure 21.13). The mouth leads to a muscular *pharynx.* In the earthworm, *pharyngeal muscles* attach to the body wall. The pharynx acts as a pump for ingesting food. The mouth pushes against food and the pharynx pumps the food into the esophagus. The *esophagus* is narrow and tubular and is frequently expanded to form a stomach, crop, or gizzard; the latter two are common in terrestrial species. A *crop* is a thin-walled storage structure, and a *gizzard* is a muscular, cuticle-lined grinding structure. *Calciferous glands* are evaginations of the esophageal wall that rid the body of excess calcium absorbed from food. Calciferous glands also have an important function in regulating the pH of body fluids. The *intestine* is a straight tube and is the principle site of digestion and absorption. The surface area of the intestine is increased substantially by a dorsal fold of the lumenal epithelium called the *typhlosole* (figure 21.14). The intestine ends at the anus.

Gas Exchange and Circulation

Both respiratory and circulatory functions are as described for polychaetes. Some segmental vessels are expanded and may be contractile. In the earthworm, for example, expanded segmental vessels surrounding the esophagus propel blood between dorsal and ventral blood vessels and anteriorly in the ventral vessel toward the mouth. Even though these are sometimes called "hearts," the main propulsive structures are the dorsal and the ventral vessels (*see figure 21.13*). Branches from the ventral vessel supply the gut and body wall.

No gills are present in oligochaetes. Freshwater oligochaetes of the family Tubificidae are "blood red" because of high concentrations of hemoglobin in their blood. Hemoglobin allows them to live in environments that contain very low oxygen concentrations, such as at the bottoms of deep lakes and in polluted lakes and streams, where oxygen levels are very low.

Nervous and Sensory Functions

The ventral nerve cords and all ganglia of oligochaetes have undergone a high degree of fusion. Other aspects of nervous structure and function are essentially the same as those described earlier for polychaetes. As with polychaetes, giant fibers mediate escape responses. An escape response results from the stimulation of either the anterior or the posterior end of a worm. An impulse conducted to the opposite end of the worm initiates the formation of an anchor, and longitudinal muscles contract to quickly pull the worm away from the stimulus.

Oligochaetes lack well-developed eyes, which should not be surprising given their subterranean life-style. It is not unusual for animals living in perpetual darkness to be without well-developed eyes. Other oligochaetes have simple pigment-cup ocelli, and all have a "dermal light sense"

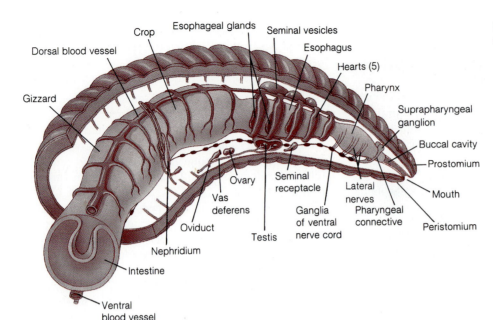

Figure 21.13 Internal structures of the earthworm.

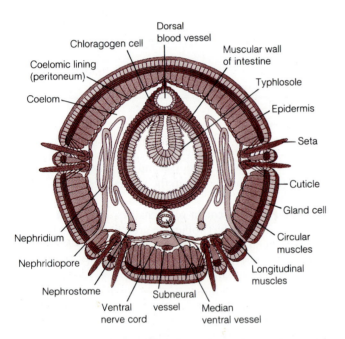

Figure 21.14 Earthworm cross section. The nephrostomes shown here would actually be associated with the next anterior segment. The thickness of the cuticle is exaggerated.

Figure 21.14 redrawn, with permission, from *Living Invertebrates* by Pearse/Buchsbaum, copyright by The Boxwood Press.

that arises from photoreceptor cells scattered over the dorsal and lateral surfaces of the body. Scattered photoreceptor cells mediate a negative phototaxis in strong light (evidenced by movement away from the light source) and a positive phototaxis in weak light (evidenced by movement toward the light source).

Oligochaetes are sensitive to a wide variety of chemical and mechanical stimuli. Receptors for these stimuli are free nerve endings and other general receptors scattered over the surface of the body, especially around the prostomium.

Excretion

Oligochaetes use metanephridia for excretion and for salt and water regulation. As with polychaetes, funnels of metanephridia are associated with the segment just anterior to the segment containing the tubule and the nephridiopore. Nitrogenous wastes include ammonia and urea. Oligochaetes excrete copious amounts of very dilute urine, although they retain vital salts, which is very important for organisms living in environments where water is plentiful but essential ions are limited.

Oligochaetes (as well as other annelids) possess chloragogen tissue that surrounds the dorsal blood vessel and lies over the dorsal surface of the intestine (*see figure 21.14*). **Chloragogen tissue** acts similarly to the vertebrate liver. It is a site of amino acid metabolism. Deamination of amino acids, and the conversion of ammonia to urea occurs there. Chloragogen tissues also convert excess carbohydrates into energy-storage molecules of glycogen and fat.

Reproduction and Development

All oligochaetes are monoecious, and mutual sperm exchange occurs during copulation. One or two pairs of testes and one pair of ovaries are located on the anterior septum of certain anterior segments. Both the sperm ducts and the oviducts have ciliated funnels at their proximal ends to draw gametes into their respective tubes.

Accessory saclike structures (seminal vesicles, seminal receptacles, and ovisacs) function in gamete storage prior to or after copulation (figure 21.15). Testes are closely associated with three pairs of *seminal vesicles*, which are sites for maturation and storage of sperm prior to their release. *Seminal receptacles* receive sperm during copulation. A pair of very small *ovisacs*, associated with oviducts, are sites for the maturation and storage of eggs prior to egg release.

(a)

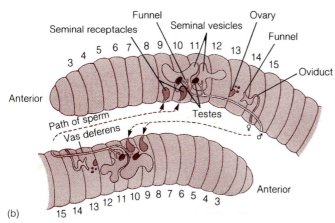

(b)

Figure 21.15 Earthworm reproduction. (a) Earthworms mating. (b) Mating earthworms showing arrangements of reproductive structures and the path taken by sperm during sperm exchange (shown by arrows).

During copulation of *Lumbricus,* two worms line up facing in opposite directions, with the ventral surfaces of their anterior ends in contact with each other. This orientation lines up the clitellum of one worm with the genital segments of the other worm. Worms are held in place by a mucous sheath, secreted by the **clitellum**, that envelopes the anterior halves of both worms. Some species also have penile structures and genital setae that help maintain contact between worms. In *Lumbricus,* sperm are released from the sperm duct and travel along the external, ventral body wall in sperm grooves formed by the contraction of special muscles. Muscular contractions along this groove help propel sperm toward the openings of the seminal receptacles. In other oligochaetes, copulation results in the alignment of sperm duct and seminal receptacle openings and transfer of sperm is direct. Copulation lasts 2 to 3 hours, after which the worms separate.

Following copulation, the clitellum forms a cocoon for the deposition of eggs and sperm. The cocoon consists of mucoid and chitinous materials that encircle the clitellum. A food reserve, albumin, is secreted into the cocoon by the clitellum, and the worm begins to back out of the cocoon. Eggs are deposited in the cocoon as the cocoon passes the

openings to the oviducts, and sperm is released as the cocoon passes the openings to the seminal receptacles. Fertilization occurs in the cocoon and, as the worm continues backing out, the ends of the cocoon are sealed, and the cocoon is deposited in moist soil.

Spiral cleavage is modified, and no larva is formed. Hatching occurs in 1 to a few weeks, depending on the species, when young worms emerge from one end of the cocoon.

Asexual reproduction also occurs in freshwater oligochaetes. It consists of transverse division of a worm, followed by regeneration of missing segments.

Stop and Ask Yourself

8. How does an earthworm move across a substrate? How does an earthworm burrow?
9. What is the function of calciferous glands? The typhlosole? Chloragogen tissue?
10. How would you describe the method of sperm transfer and egg deposition in earthworms?
11. In what way is development of an oligochaete different from that of a polychaete?

Class Hirudinea

The class Hirudinea (hi′roo-din″eah) (L. *hirudin,* leech) contains approximately 500 species of leeches (*see table 21.1*). Most leeches are freshwater; others are marine, or completely terrestrial. Leeches prey on other invertebrates or feed on body fluids of vertebrates (box 21.2).

External Structure and Locomotion

Leeches lack parapodia and head appendages. They are dorsoventrally flattened and taper anteriorly. Leeches have 34 segments, but the segments are difficult to distinguish externally because they have become secondarily divided. Several secondary divisions, called *annuli,* are in each true segment. Anterior and posterior segments are usually modified into suckers (figure 21.16).

Modifications of body-wall musculature and the coelom influence patterns of locomotion in the leeches. The musculature of leeches is more complex than that of other annelids. A layer of oblique muscles is present between the circular and longitudinal muscle layers. In addition, dorsoventral muscles are responsible for the typical leech flattening. The coelom of leeches has lost its metameric partitioning. Septa are lost, and the coelom has been invaded by connective tissue. The result is a series of interconnecting sinuses.

These modifications have resulted in altered patterns of locomotion. Rather than being able to utilize independent coelomic compartments, the leech has a single hydrostatic cavity and uses it in a looping type of locomotion. The mechanics of this locomotion is described in figure 21.17. Leeches also swim using undulations of the body.

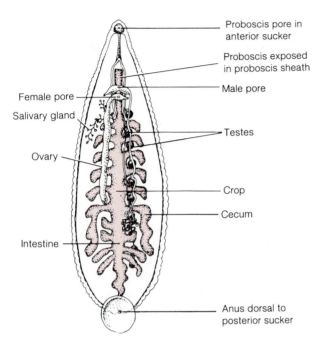

Figure 21.16 Internal structure of a leech. Each true segment is subdivided by annuli and the eucoelom is not subdivided by septae.

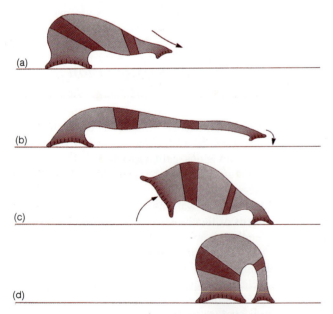

Figure 21.17 Leech locomotion. (a,b) Attachment of the posterior sucker causes reflexive release of the anterior sucker, contraction of circular muscles, and relaxation of longitudinal muscles. This muscular activity compresses fluids in the single hydrostatic compartment, and the leech extends. (c,d) Attachment of the anterior sucker causes reflexive release of the posterior sucker, the relaxation of circular muscles, and the contraction of longitudinal muscles, causing body fluids to expand the diameter of the leech. The leech shortens, and the posterior sucker again attaches.

Feeding and the Digestive System

Many leeches feed on body fluids or the entire bodies of other invertebrates. Some feed on blood of vertebrates, including human blood. Leeches are sometimes called parasites; however, the association between a leech and its host is relatively brief. Therefore, describing leeches as predatory is probably more accurate. Leeches are also not species specific, as are most parasites. (Leeches are, however, class specific. That is, a leech that preys upon a turtle may also prey on an alligator, but probably would not prey on a fish or a frog.) [7]

The mouth of a leech opens in the middle of the anterior sucker. In some leeches, the anterior digestive tract is modified into a protrusible proboscis, lined inside and outside by cuticle. In others, the mouth is armed with three chitinous jaws. While feeding, a leech attaches to its prey by the anterior sucker and either extends its proboscis into the prey or uses its jaws to slice through host tissues. Salivary glands secrete an anticoagulant called "hirudin" that prevents blood from clotting.

Behind the mouth is a muscular pharynx that pumps body fluids of the prey into the leech. The esophagus follows the pharynx and leads to a large stomach with lateral cecae. Most leeches ingest large quantities of blood or other body fluids and gorge their stomachs and lateral cecae, increasing their body mass two to ten times. After engorgement, a leech can tolerate periods of fasting that may last for months. The digestive tract ends in a short intestine and the anus (*see figure 21.16*).

Gas Exchange and Circulation

Gas exchange occurs across the body wall. The basic annelid circulatory design is retained in some leeches, but in most it is highly modified, and vessels are replaced by coelomic sinuses. Coelomic fluid has taken over the function of blood and, with the exception of two orders, respiratory pigments are lacking.

Nervous and Sensory Functions

The nervous system of leeches is similar to that of other annelids. Ventral nerve cords are unfused, except at the ganglia. The suprapharyngeal and subpharyngeal ganglia and the pharyngeal connectives are all fused into a nerve ring that surrounds the pharynx. There is also a similar fusion of ganglia at the posterior end of the animal.

A variety of epidermal sense organs are widely scattered over the body. Most leeches have photoreceptor cells located in pigment cups (two to ten) along the dorsal surface of the anterior segments. Normally, leeches are negatively phototactic, but when searching for food, the behavior of some leeches changes, and they become positively phototactic, which increases the likelihood of contacting prey that happens to pass by.

Hirudo, the medicinal leech, has a well-developed temperature sense. This sense helps the leech detect the higher body temperature of its mammalian prey. Other leeches are attracted to extracts of prey tissues.

All leeches have sensory cells with terminal bristles in a row along the middle annulus of each segment. These sensory cells, called *sensory papillae,* are of uncertain function but are taxonomically important.

Excretion

Leeches have 10 to 17 pairs of metanephridia, one per segment in the middle segments of the body. Metanephridia are highly modified and possess, in addition to the nephrostome and tubule, a capsule that is believed to be involved with the production of coelomic fluid. Chloragogen tissue is proliferated through the body cavity of most leeches.

Reproduction and Development

All leeches reproduce sexually and are monoecious. None are capable of asexual reproduction or regeneration. They have a single pair of ovaries and from four to many testes. Leeches have a clitellum that includes three body segments, and it can be seen in the spring when most leeches breed.

Sperm transfer and egg deposition usually occur in the same manner as described for oligochaetes. A penis may assist the transfer of sperm between individuals. Leeches that lack a penis may transfer sperm by expelling a packet of sperm (called a *spermatophore*) from one leech through the integument and into the body cavity of another. Cocoons are deposited in the soil or are attached to underwater objects. There are no larval stages, and the offspring are mature by the following spring.

Further Phylogenetic Considerations

Although the origins of the phylum as a whole are speculative and somewhat controversial, the evolutionary relationships among members of the three major annelid classes are as clear as those in any other phylum of animals. Polychaetes are the most primitive of the three annelid classes, as evidenced by basic metamerism, spiral cleavage, and trochophore larval stages in some species. (Some zoologists contend that the occurrence of the latter is too variable in the class to be considered a part of the evidence of the ancestral status of the class.) Adaptive radiation of polychaetes has resulted in the great diversity of modern polychaetes.

Some members of the ancient polychaete stock invaded fresh waters, which required the ability to regulate the salt and water content of body fluids. It was from this group that the oligochaetes probably evolved. Initially, oligochaetes were strictly freshwater, and many remain in that habitat; however, during the Cretaceous period, approximately 100 million years ago, oligochaetes invaded moist, terrestrial environments. This period saw the climax of the giant land reptiles but, more importantly, it was a time of [8] proliferation of flowering land plants. The reliance of modern earthworms on deciduous vegetation can be traced back to the exploitation of this food source by their ancestors. As described earlier, terrestrial oligochaetes deserve a large share of the credit for developing the soils of this planet. A few oligochaetes have secondarily invaded marine environments.

Some of those early freshwater oligochaetes also gave rise to the Hirudinea. As with the oligochaetes, some leeches colonized marine habitats from fresh water.

Stop and Ask Yourself

12. How is basic annelid metamerism modified in leeches?
13. How is the body-wall musculature and the body cavity of a leech used in locomotion?
14. Why is it more accurate to call leeches predators than parasites?
15. Which annelid class is thought to be ancestral to the other two classes?

■ Summary

1. The origin of the Annelida is largely unknown. A primary diagnostic characteristic of the annelids is metamerism.
2. Metamerism allows efficient utilization of separate coelomic compartments as a hydrostatic skeleton for support and movement. Metamerism also lessens the impact of injury and makes tagmatization possible.
3. Members of the class Polychaeta are mostly marine and possess parapodia with numerous setae. Locomotion of polychaetes involves the antagonism of longitudinal muscles on opposite sides of the body, which creates undulatory waves along the body wall and causes parapodia to act against the substrate.
4. Polychaetes may be predators, herbivores, scavengers, or filter feeders.
5. The nervous system of polychaetes usually consists of a pair of suprapharyngeal ganglia, subpharyngeal ganglia, and double ventral nerve cords that run the length of the worm.
6. Polychaetes have a closed circulatory system. Oxygen is carried by respiratory pigments dissolved in blood plasma.
7. Either protonephridia or metanephridia are used in excretion in polychaetes.
8. Most polychaetes are dioecious, and gonads develop from coelomic epithelium. Fertilization is usually external. Epitoky occurs in some polychaetes.
9. Development of polychaetes usually results in a planktonic trochophore larva that buds off segments near the anus.
10. The class Oligochaeta includes primarily freshwater and terrestrial annelids. Oligochaetes possess few setae and they lack a head and parapodia.
11. Oligochaetes are scavengers that feed on dead and decaying vegetation. Their digestive tract is tubular, straight, and frequently has modifications for storage, grinding, and increasing the surface area for secretion and absorption.
12. Oligochaetes possess metanephridia. Chloragogen tissue is a site for the formation of urea from protein metabolism and synthesis and storage of glycogen and fat.
13. Oligochaetes are monoecious and exchange sperm during copulation.
14. Members of the class Hirudinea are the leeches. Complex arrangements of body-wall muscles and the loss of septa influences patterns of locomotion.
15. Leeches are predatory and feed on body fluids or entire bodies of other invertebrates, or the blood of vertebrates.
16. Leeches are monoecious, and reproduction and development occur as in oligochaetes.
17. Ancestral polychaetes are the annelids from which the other two major classes evolved. Some members of this ancient polychaete stock invaded fresh water and gave rise to early freshwater oligochaetes. These oligochaetes gave rise to modern freshwater oligochaetes, terrestrial oligochaetes, and leeches.

■ Key Terms

chloragogen tissue (p. 336)   parapodia (p. 329)
clitellum (p. 337)   peristomium (p. 329)
epitoky (p. 333)   prostomium (p. 329)
metamerism (p. 327)   tagmatization (p. 328)
metanephridium (p. 332)

■ Critical Thinking Questions

1. What evidence is there that would link the annelids and arthropods in the same evolutionary line?

2. Distinguish between a protonephridium and a metanephridium. Name a class of annelids whose members may have protonephridia. What other phylum have we studied whose members also had protonephridia? Do you think that metanephridia would be more useful for a coelomate or an acoelomate animal? Explain.

3. Contrast and compare specific patterns of locomotion in the three classes of annelids.

4. Compare and contrast reproductive and developmental strategies of polychaetes, oligochaetes, and leeches.

5. What adaptations do earthworms have for living in perpetual darkness?

■ Suggested Readings

Books

Barnes, R. D. 1987. *Invertebrate Zoology.* 5th ed. Philadelphia: Saunders College Publishing.

Brinkhurst, R. O., and Jamieson, B. B., 1972. *Aquatic Oligochaeta of the World.* Toronto: Toronto University Press.

Pearse, V., Pearse, J., Buschsbaum, M., and Buchsbaum, R. 1987. *Living Invertebrates.* New York: Blackwell.

Pettibone, M. H. 1982. Annelida. In S. P. Parker (ed.) *Synopsis and Classification of Living Organisms.* Vol 2. New York: McGraw-Hill Book Co.

Russell-Hunter, W. D. 1979. *A Life of Invertebrates.* New York: Macmillan Publishing Co., Inc.

Articles

Amato, I. 1986. Leech swimming: the neural story. *Science News* 130: 342–343.

Brown, S. C. 1975. Biomechanics of water pumping by *Chaetopterus variopedatus* Renier: skeletomusculature and mechanics. *Biological Review* 149: 136–156.

Conniff. R. 1987. The little suckers have made a comeback. *Discover* 8:84–93.

Lent, C. M., and Dickinson, M. H. The neurobiology of feeding in leeches. *Scientific American*, June, 1988.

Mangum, C. 1970. Respiratory physiology in annelids. *American Scientist* 58(6): 641–647.

Nicholls, J. G., and Van Essen, D. The nervous system of the leech. *Scientific American* January, 1974.

Nordell, D. 1988. Milking leeches for drug research. *New Scientist* 117: 43–44.

Seymour, M. K. 1969. Locomotion and coelomic pressure in *Lumbricus. Journal of Experimental Biology,* 51:47.

Wald, G., and S. Rayport. 1977. Vision in annelid worms. *Science,* 196: 1434–1439.

22

The Arthropods: Blueprint for Success

Concepts

1. Arthropods have been successful in almost all habitats on the earth. Some ancient arthropods were the first animals to live most of their lives in terrestrial environments.
2. Metamerism with tagmatization, an exoskeleton, and metamorphosis have contributed to the success of arthropods.
3. Members of the subphylum Trilobita are extinct arthropods that were a dominant form of life in the oceans between 345 and 600 million years ago.
4. Members of the subphylum Chelicerata have a body divided into two regions and have chelicerae. The class Merostomata contains the horseshoe crabs. The class Arachnida contains the spiders, mites, ticks, and scorpions. Some ancient arachnids were the first terrestrial arthropods, and modern arachnids have numerous adaptations for terrestrial life. The class Pycnogonida contains the sea spiders.
5. Animals in the subphylum Crustacea have biramous appendages and two pairs of antennae. The class Branchiopoda includes the fairy shrimp, brine shrimp, and water fleas. The class Malacostraca include the crabs, lobsters, crayfish, and shrimp, and the classes Copepoda and Cirrepedia include the copepods and barnacles, respectively.

Have You Ever Wondered:

[1] what the most abundant animal is?
[2] how an arthropod grows within the confines of a rigid exoskeleton?
[3] how arthropods were preadapted for terrestrialism?
[4] why some spiders go ballooning?
[5] what two spiders found in the United States are dangerous to humans?
[6] what causes the bite of a chigger to itch so badly?
[7] what mite lives in the hair follicles of most readers of this textbook?
[8] what crustaceans colonize the hulls of ships?

These and other useful questions will be answered in this chapter.

This chapter contains underlined evolutionary concepts.

Evolutionary Perspective

[1] What animal species has the greatest number of individuals? The answer can only be an educated guess; however, many zoologists would argue that one of the many species of small (1 to 2 mm) crustaceans, called copepods, that drift in the open oceans must have this honor (figure 22.1). Copepods have been very successful, feeding on the vast photosynthetic production of the open oceans. After only 20 minutes of towing a plankton net behind a slowly moving boat (at the right location and during the right time of year), one can collect over 3 million copepods—enough to solidly pack a 2 gallon pail! Copepods are food for fish, such as herring, sardines, mackerel, as well as for whale sharks and the largest mammals, the blue whale and its relatives. Humans benefit from copepod production by eating fish that feed on copepods. (Unfortunately, we use a small fraction of the total food energy in these animals. In spite of one-half of the earth's inhabitants lacking protein in their diet, humans process into fishmeal most of the herring and sardines caught. The fishmeal is then fed to poultry and hogs. In eating the poultry and hogs, we lose over 99 percent of the original energy present in the copepods!)

Copepods are one of many groups of animals belonging to the phylum Arthropoda (ar'thra-po'dah) (Gr. *arthro*, joint + *podos*, foot). Crayfish, lobsters, spiders, mites, scorpions, and insects are also arthropods. About 750,000 species of arthropods have been described, which means that three out of every four kinds of animals are arthropods! In this chapter and chapter 23, you will discover the many ways in which some arthropods are considered among the most successful of all animals.

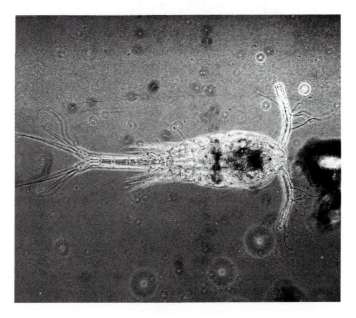

Figure 22.1 Copepods, such as *Calanus*, may be the most abundant animals in the world.

Classification and Relationships to Other Animals

As discussed in chapter 21, arthropods and annelids are closely related to each other. Shared protosome characteristics, such as the development of the mouth from the blastopore and schizocoelous coelom formation, as well as other common characteristics, such as the presence of a

---

Table 22.1   Classification of the Phylum Arthropoda*

Phylum Arthropoda (ar'thra-po'dah)
Animals that show metamerism with tagmatization, a jointed exoskeleton, and a ventral nervous system.

Subphylum Trilobita (tri"lo-bit'ah)
All extinct; lived from Cambrian to Carboniferous periods; bodies are divided into three longitudinal lobes; head, thorax, and abdomen present; one pair of antennae and biramous appendages.

Subphylum Chelicerata (ke-lis"e-ra'tah)
Body usually divided into prosoma and opisthoma; first pair of appendages (chelicerae) for feeding.

Class Merostomata (mer'o-sto'mah-tah)
Aquatic, with book gills on opisthoma. Two subclasses: Eurypterida, a group of extinct arthropods called giant water scorpions, and Xiphosura, the horseshoe crabs. *Limulus.*

Class Arachnida (ah-rak'ni-dah)
Mostly terrestrial, with book lungs or tracheae; usually four pairs of walking legs in adults. Spiders, scorpions, ticks, mites, harvestmen.

Class Pycnogonida (pik'no-gon;"i-dah)
Reduced abdomen; no special respiratory or excretory structures; four to six pairs of walking legs; common in all oceans. Sea spiders.

Subphylum Crustacea (krus-tas'eah)
Head with two pairs of antennae, one pair of mandibles, and two pairs of maxillae; biramous appendages.

Class Remipedia (ri-mi-pe'de-ah)
A single species of cave-dwelling crustaceans from the Bahamas; body with approximately 30 segments that bear uniform, biramous appendages.

Class Cephalocarida (sef'ah-lo-kar'i-dah)
Small (3 mm) marine crustaceans with uniform, leaflike, triramous appendages.

Class Branchiopoda (brang'ke-o-pod'ah)
Flattened, leaflike appendages used in respiration, filter feeding, and locomotion; found mostly in fresh water. Fairy shrimp, brine shrimp, clam shrimp, water fleas.

Class Malacostraca (mal-ah-kos'trah-kah)
Appendages may be modified for crawling, feeding, swimming. Lobsters, crayfish, crabs.

Class Copepoda (ko'pepod'ah)
Maxillipeds modified for feeding. Copepods.

Class Cirripedia (sir'i-ped'eah)
As adults, these animals are sessile, marine, and enclosed by calcium carbonate valves. Barnacles.

*Subphylum Uniramia not included.

ventral nerve cord and metamerism, give evidence of common ancestry.

Zoologists, however, disagree about the evolutionary relationships among the arthropods. Many zoologists believe that it is not one phylum, but three. These ideas are discussed at the end of chapter 23. The arthropods are treated in this textbook as members of a single phylum. Living arthropods are divided into three subphyla: Chelicerata, Crustacea, and Uniramia. All members of a fourth subphylum, Trilobita, are extinct (table 22.1). Trilobita, Chelicerata, and Crustacea are discussed in this chapter and the Uniramia are discussed in chapter 23.

Metamerism and Tagmatization

Three aspects of arthropod biology have been important in contributing to their success. One of these is *metamerism* (*see chapter 20*). Metamerism of arthropods is most evident externally, because the arthropod body is often composed of a series of similar segments, each bearing a pair of appendages (*see figure 21.2e*). Internally, however, the body cavity of arthropods is not divided by septa, and most organ systems are not metamerically arranged. The reason for the loss of internal metamerism is speculative; however, the presence of metamerically arranged hydrostatic compartments would be of little value in the support or locomotion of animals enclosed by an external skeleton (discussed below).

As discussed in chapter 21, metamerism permits the specialization of regions of the body for specific functions. This specialization is called *tagmatization*. Tagmatization

is well developed in arthropods. Body regions, called *tagmata* (singular, *tagma*) are specialized for feeding and sensory perception, locomotion, and visceral functions.

The Exoskeleton

Arthropods are enclosed by an external, jointed skeleton, called an **exoskeleton** or **cuticle.** The exoskeleton is often cited as the major reason for arthropod success. It provides

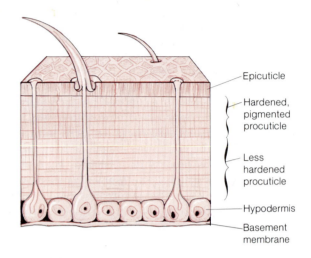

Figure 22.2 The arthropod exoskeleton. The epicuticle is made of waxy lipoprotein and is impermeable to water. The inner layer of the procuticle is hardened by calcium carbonate deposition and/or sclerotization. Chitin, a tough, leathery polysaccharide and several kinds of proteins make up the bulk of the procuticle. The entire exoskeleton is secreted by the hypodermis.

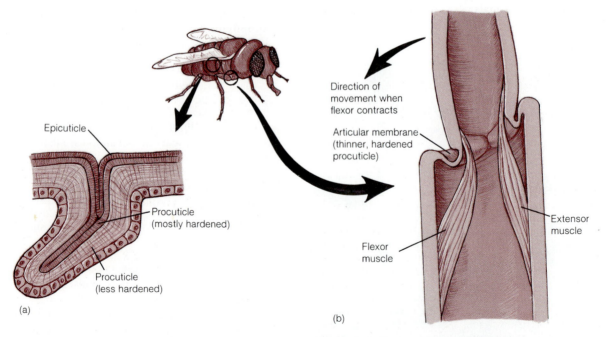

Figure 22.3 Modifications of the exoskeleton. (a) Invaginations of the exoskeleton result in firm ridges and bars when the procuticle in the region of the invagination remains thick and hard. These are used as muscle attachment sites. (b) Regions where the procuticle is thinned are flexible and form membranes and joints.

structural support, protection, impermeable surfaces for prevention of water loss, and a system of levers for muscle attachment and movement.

The exoskeleton covers all body surfaces and invaginations of the body wall, such as the anterior and posterior portions of the gut tract. It is nonliving and is secreted by a single layer of epidermal cells (figure 22.2). The epidermal layer is sometimes called the *hypodermis* because, unlike other epidermal tissues, it is covered on the outside by exoskeleton rather than being directly exposed to air or water.

The exoskeleton is composed of two layers. The *epicuticle* is the outermost layer. Because it is made of a waxy lipoprotein, it is impermeable to water and serves as a barrier to microorganisms and pesticides. The bulk of the exoskeleton is below the epicuticle and is called the *procuticle*. (In crustaceans, the procuticle is sometimes called the *endocuticle*.) The procuticle is composed of **chitin**, a tough, leathery polysaccharide and several kinds of proteins (*see chapter 2*). Hardening of the procuticle is accomplished through a process called sclerotization and

sometimes by impregnation with calcium carbonate. *Sclerotization* involves the formation of chemical bonds between protein chains. In insects and most other arthropods, this bonding occurs in the outer portion of the procuticle. Hardening of the exoskeleton of crustaceans is accomplished by sclerotization and by the deposition of calcium carbonate in the middle regions of the procuticle. Some proteins give the exoskeleton resiliency. When it is distorted, energy is stored. Stored energy can be used in activities such as flapping wings and jumping. The inner portion of the procuticle is not hardened.

Hardening in the procuticle provides armorlike protection for arthropods, but it also necessitates a variety of adaptations to allow arthropods to live and grow within its confines. Invaginations of the exoskeleton form firm ridges and bars for muscle attachment. Another modification of the exoskeleton is the formation of joints. A flexible membrane, called an *articular membrane* is present in regions where the procuticle is thinner and less hardened (figure 22.3). Other modifications of the exoskeleton include sensory receptors, called *sensilla*, that are in the form of pegs, bristles, and lenses, and modifications of the exoskeleton that permit gas exchange.

Growth of an arthropod would be virtually impossible unless the exoskeleton were periodically shed; such as in the molting process called **ecdysis.** Ecdysis is divided into three stages that involve (1) the detachment of the epidermal layer from the exoskeleton (figure 22.4a); (2) the digestion of old procuticle, and the secretion of new procuticle and epicuticle, (figure 22.4b); and (3) stretching by air or water intake and the secretion of additional epicuticle through pores in the procuticle (figure 22.4c). The arthropod is vulnerable to predators for a few hours or days after ecdysis, because the new exoskeleton is relatively soft. The animal usually remains hidden until the exoskeleton is hardened and pigments are deposited in the procuticle. All of these changes are controlled by the nervous and endocrine systems; the controls will be discussed in more detail later.

[2]

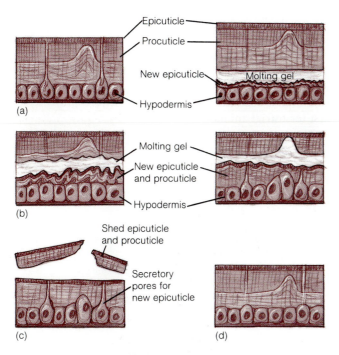

Figure 22.4 Events of ecdysis. (a) During preecdysis, the hypodermis detaches from the exoskeleton, and the space between the old exoskeleton and the hypodermis is filled with a fluid called molting gel. (b) The hypodermis begins secreting a new epicuticle, and a new procuticle is formed as the old procuticle is digested. The products of digestion are incorporated into the new procuticle. Note that the new epicuticle and procuticle are wrinkled beneath the old exoskeleton to allow for increased body size after ecdysis. (c) Ecdysis occurs when the animal swallows air or water, and the exoskeleton splits along predetermined ecdysal lines. The animal pulls out of the old exoskeleton. (d) After ecdysis, the new exoskeleton hardens by calcium carbonate deposition, and/or sclerotization, and pigments are deposited in the outer layers of the procuticle. Additional material is added to the epicuticle.

Metamorphosis

A final characteristic that has contributed to arthropod success is a reduction of competition between adults and immature stages because of metamorphosis. *Metamorphosis is a radical change in body form and physiology that occurs as an immature stage,* usually called a *larva,* becomes an adult. The evolution of arthropods has resulted in an increasing divergence of body forms, behaviors, and habitats between immature and adult stages. Adult crabs, for example, are usually found prowling the sandy bottoms of their marine habitats for live prey or decaying organic matter, whereas larval crabs live and feed in the plankton. Similarly, the caterpillar that feeds on leafy vegetation eventually develops into a nectar-feeding adult butterfly or moth. Having different adult and immature stages means that they will not compete with each other for food or living space. In some arthropod and other animal groups, larvae also serve as the dispersal stage.

Subphylum Trilobita

Members of the subphylum Trilobita (tri″lo-bit′ah) (Gr. *tri*, three + *lobos*, lobes) were a dominant form of life in the oceans from the Cambrian period (600 million years ago) to the Carboniferous period (345 million years ago). They crawled along the substrate feeding on annelids, molluscs, and dead and decaying organic matter. The body of trilobites was oval, flattened, and divided into three longitudinal regions (figure 22.5). All body segments articulated so the trilobite could roll into a ball to protect its soft ventral surface. Most fossilized trilobites are found in this position. Trilobite appendages consist of two lobes. The inner lobe served as a walking leg, and the outer lobe bore spikes or teeth that may have been used in digging or swimming. Because they possessed two lobes or *rami*, these appendages are called **biramous appendages.**

Stop and Ask Yourself

1. What are the four subphyla of arthropods?
2. What three aspects of structure and function have been important in arthropod success?
3. How does metamorphosis reduce competition between adult and immature forms?
4. In what way is the name "Trilobita" descriptive of that group of animals?

Subphylum Chelicerata

One arthropod lineage, the subphylum Chelicerata (ke-lis″e-ra′tah) (Gr. *chele*, claw + *ata*, plural suffix) includes familiar animals, such as spiders, mites, and ticks, and less familiar animals, such as horseshoe crabs and sea spiders. These animals have two tagmata. The *prosoma* or *cephalothorax* is a sensory, feeding, and locomotor tagma. It usually bears eyes, but unlike other arthropods, never has antennae. Paired appendages attach to the prosoma. The first pair, called **chelicerae** are usually pincerlike, or *chelate*, and are most often used in feeding. The second pair, called *pedipalps*, are usually sensory, but may also be used in feeding, locomotion, or reproduction. Pedipalps are followed by paired walking legs. Posterior to the prosoma is the *opisthosoma* or *abdomen,* which contains digestive, reproductive, excretory, and respiratory organs.

Class Merostomata

Members of the class Merostomata (mer′o-sto′mah-tah) are divided into two subclasses. The Xiphosura are the horseshoe crabs, and the Eurypterida are the giant water scorpions (figure 22.6). The latter are extinct, having lived from the Cambrian period (600 million years ago) to the Permian period (280 million years ago). Eurypterids are an important group in an evolutionary context. As discussed in the next section, their descendants were probably the first terrestrial animals.

There are only four species of horseshoe crabs living today; one species, *Limulus polyphemus*, is widely distributed in the Atlantic Ocean and the Gulf of Mexico (fig-

Figure 22.5 Trilobite structure. The body of a trilobite was divided into three longitudinal sections (thus the subphylum name). It was also divided into three tagmata. A head, or cephalon, bore a pair of antennae and eyes. The trunk, or thorax, bore appendages used in swimming or walking. A series of posterior segments formed the pygidium, or tail.

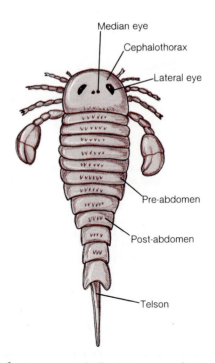

Figure 22.6 A eurypterid, *Euripterus remipes*.

ure 22.7a). Horseshoe crabs scavenge sandy and muddy substrates for annelids, small molluscs, and other invertebrates. They have remained virtually unchanged for over 250 million years and were cited in chapter 13 as an example of stabilizing selection.

The cephalothorax of horseshoe crabs is covered with a hard, horseshoe-shaped *carapace*. The chelicerae, pedipalps and first three pairs of walking legs are chelate and are used for walking and food handling. The last pair of appendages has leaflike plates at their tips and are used for locomotion and for digging (figure 22.7b).

The abdomen of a horseshoe crab bears a long, unsegmented telson. If a horseshoe crab is flipped over by wave action, it arches its telson and opisthoma dorsally, causing the animal to roll to its side and flip right side up again. The first pair of abdominal appendages cover genital pores and are called *genital opercula*. The remaining five pairs of appendages are *book gills*. The name is derived from the resemblance of these plate-like gills to the pages of a closed book. Gas exchange between the blood and water occurs as blood circulates through the book gills. Horseshoe crabs have an open circulatory system, as do all arthropods. Blood circulation in horseshoe crabs is similar to that described later in this chapter for arachnids and crustaceans.

Horseshoe crabs are dioecious. During reproductive periods, males and females congregate in intertidal areas. The male mounts the female and grasps her with his pedipalps. Eggs are fertilized by the male as they are shed from the female into depressions in the sand that have been excavated by her. Fertilized eggs are covered with sand and develop unattended.

Class Arachnida

Members of the class Arachnida (ah-rak′ni-dah) (Gr. *arachne*, spider) are some of the most misrepresented members of the animal kingdom. Their reputations as fearsome and grotesque creatures vastly exaggerate reality. The majority of spiders, mites, ticks, scorpions, and related forms are either harmless or very beneficial to humans.

Most zoologists believe that arachnids arose from the eurypterids and were the first arthropods to invade terrestrial environments. The earliest fossils of aquatic scorpions date back to the Silurian period (405 to 425 million years ago), fossils of terrestrial scorpions date from the Devonian period (350 to 400 million years ago), and fossils of all other arachnid groups are present by the Carboniferous period (280 to 345 million years ago).

Water conservation is a major concern for any terrestrial [3] organism, and ancestral arachnids were preadapted for terrestrialism by their relatively impermeable exoskeleton. *Preadaptation* occurs when a structure present in members of a species proves useful in promoting reproductive success when an individual encounters new environmental situations. Later adaptations included evolution of efficient excretory structures, internal surfaces for gas exchange, appendages modified for locomotion on land, and greater deposition of wax in the epicuticle.

Form and Function

Most arachnids are carnivores. Small arthropods are usually held by the chelicerae while enzymes from the gut tract pour over the prey. Partially digested food is then taken into the mouth. The gut tract of arachnids is divided into three regions. The anterior portion is called the *foregut*, and the posterior portion is called the *hindgut*. Both develop as infoldings of the body wall and are lined with cuticle. A portion of the foregut is frequently modified into a pumping *pharynx*, and the hindgut is frequently a site

(a)

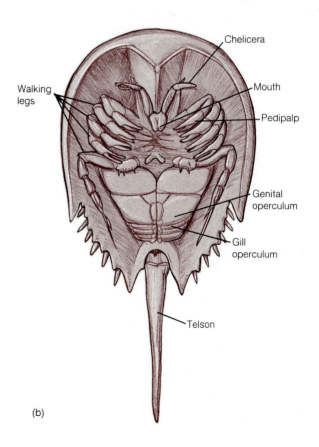

Chelicera

Mouth

Walking legs

Pedipalp

Genital operculum

Gill operculum

Telson

(b)

Figure 22.7 *Limulus polyphemus.* (a) Horseshoe crab dorsal and ventral surfaces. (b) Ventral view.

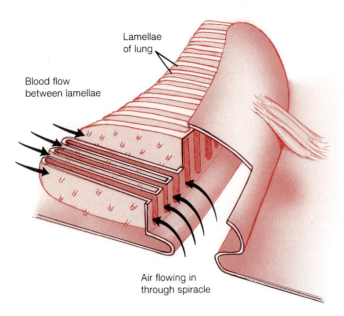

Lamellae
of lung

Blood flow
between lamellae

Air flowing in
through spiracle

Figure 22.8 An arachnid book lung. Air and blood moving on opposite sides of a lamella of the lung exchange respiratory gases by diffusion.

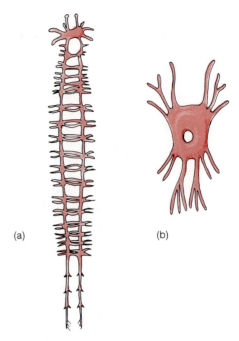

(a)

(b)

Figure 22.9 Arthropod nervous systems. (a) Nervous systems of some arthropods, such as primitive crustaceans, have paired segmental ganglia and paired ventral nerve cords. (b) In many arthropod groups, specialization has resulted in the concentration of nervous tissue into a few ganglia, or a single ganglion. The nervous system of an arachnid (mite) is shown here.

of water reabsorption. The *midgut* is between the foregut and hindgut. It is noncuticular and lined with secretory and absorptive cells. Lateral diverticula increase the area available for absorption and storage.

Arachnids use coxal glands and/or malpighian tubules for excreting nitrogenous wastes. **Coxal glands** are paired, thin-walled, spherical sacs bathed in the blood of body sinuses. Nitrogenous wastes are absorbed across the wall of the sacs, transported into a long, convoluted tubule, and excreted through excretory pores at the base of the posterior appendages. Arachnids that are adapted to dry environments possess blind-ending diverticula of the gut tract that arise at the juncture of the midgut and hindgut. These tubules are called **malpighian tubules**. They absorb waste materials from the blood, and empty them into the gut tract. Excretory wastes are then eliminated with digestive wastes. The major excretory product of arachnids is uric acid. As discussed in chapter 38, uric acid excretion is advantageous for terrestrial animals because it is excreted as a semisolid with little water loss.

Gas exchange also occurs with minimal water loss because arachnids have few exposed respiratory surfaces. Some arachnids possess structures, called **book lungs**, that are assumed to be modifications of the book gills found in the Merostomata. Book lungs are paired invaginations of the ventral body wall that are folded into a series of leaflike lamellae (figure 22.8). Air enters the book lung through a slitlike opening called a **spiracle** and circulates between lamellae. Diffusion of respiratory gases occurs between the blood moving among the lamellae and the air in the lung chamber. Other arachnids possess a series of branched, chitin-lined tubules that deliver air directly to body tissues. These tubule systems, called **tracheae** (s., trachea), open to the outside through spiracles that are located along the

ventral or lateral aspects of the abdomen. (Tracheae are also present in insects but had a separate evolutionary origin. Aspects of their physiology will be described in chapter 23.)

The circulatory system of arachnids, like that of most arthropods, is an open system in which blood is pumped by a dorsal contractile vessel (usually called the *dorsal aorta* or "*heart*") and is released into tissue spaces. In arthropods, the eucoelom is reduced to cavities surrounding the gonads and sometimes the coxal glands. Large tissue spaces, or sinuses, are derived from the blastocoel and are called the **hemocoel.** Blood bathes the tissues and then returns to the dorsal aorta through openings in the aorta called *ostia*. Arachnid blood contains the dissolved respiratory pigment hemocyanin and has ameboid cells that aid in clotting and body defenses.

The nervous system of all arthropods is ventral and, in ancestral arthropods, must have been laid out in a pattern similar to that of the annelids (figure 22.9a; *see figure 21.7a*). With the exception of scorpions, the nervous system of arachnids is greatly concentrated by fusion of ganglia (figure 22.9b).

The body of an arachnid is supplied with a variety of sensory structures. Mechanoreceptors are sensilla (modifications of the cuticle that serve as sensory receptors) that detect displacement of body parts. *Setae* are hairlike, cuticular modifications that may be set into a membranous socket. Displacement of a seta initiates an impulse in an associated nerve cell (figure 22.10a). Vibration receptors are very important to some arachnids. Spiders that use webs

Figure 22.9 redrawn, with permission, from *Living Invertebrates* by Pearse/Buchsbaum, copyright by The Boxwood Press.

Class Insecta

(a) Order Hemiptera. A water strider (*Gerris*).

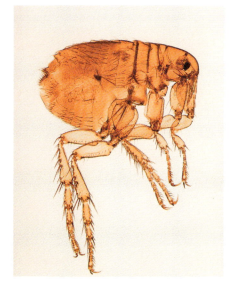

(b) Order Siphonaptera. A flea (*Pulex irritans*).

(c) Order Trichoptera. A caddis fly larva (*Neotrichia*) in its stone case.

(d) An adult caddis fly.

(e) Order Coleoptera. A dung beetle cuts a ball out of a mass of dung and rolls it to an underground burrow.

(f) Order Homoptera. A treehopper feeds on the sap of plants.

Phylum Echinodermata

(a) Class Asteroidea. A sea star with one arm lifted exposing tube feet.

(b) A sea star feeding on a mussel.

(c) Class Ophiuroidea. A brittle star (*Ophiopholis aculeata*).

(d) Class Ophiuroidea. A basket star.

(e) Class Echinoidea. A sea urchin (*Strongylocentrotus*).

(f) Class Echinoidea. A sand dollar burying itself in the sand.

(g) Class Holothuroidea. A sea cucumber (*Parastichopus californicus*).

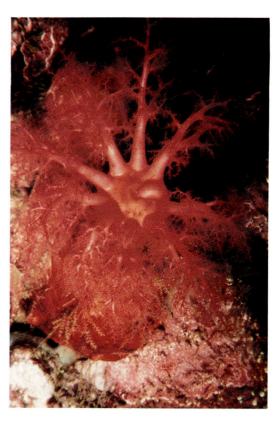

(h) Elaborate tentacles on the sea cucumber, *Cucumaria frondosa.*

(i) Class Crinoidea. A sea lily (*Cenocrinus*) photographed in deep water off the coast of Jamaica.

(j) Class Crinoidea. A feather star (*Comanthina*).

Phylum Chordata; Subphyla Urochordata and Vertebrata

(a) The urochordate *Ciona intestinalis*, an attached tunicate.

(b) Class Agnatha. A lamprey (*Petromyzon marinus*). Note the sucking mouth and teeth used to feed on other fish.

(c) Class Chondrichthyes. A gray reef shark (*Carcharhinus*).

(d) Class Chondrichthyes. A manta ray (*Manta hamiltoni*) with two remoras (*Remora remora*) attached to its ventral surface.

(e) A lesser electric ray (*Narcine brasiliensis*).

(f) A bullseye stingray (*Urolophus concentricus*).

(g) The egg case of a skate opened to show the embryo and yolk sac.

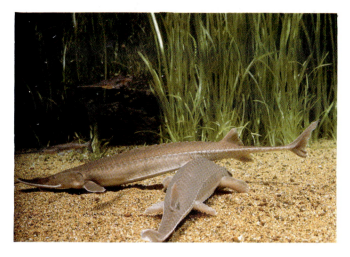

(h) Class Osteichthyes. A shovelnose sturgeon (*Scaphirhyn-chus platorynchus*). Sturgeons are covered anteriorly by heavy bony plates and posteriorly by scales.

(i) The distinctive rostrum of a paddle fish (*Polydon spathula*) is densely innervated with sensory structures that are probably used to detect minute electric fields.

(j) A marine teleost, the lionfish (*Pterois*), has extremely venomous spines.

(k) Bottom fish, such as this winter flounder (*Pseudopleuronectes americanus*), have both eyes on one side of the head, and they often rest on their side fully or partially buried on the substrate.

(l) Freshwater teleosts, such as this speckled darter (*Etheostoma stigmaeum*), are common in temperate streams.

(m) Cichlid fish, including this harlequin cichlid (*Cichlasoma festae*), are common in tropical fresh waters.

Class Amphibia

(a)

(b)

(a) Eggs, (b) larva, and (c, below) adult of the spotted salamander, *Ambystoma maculatum*. Larvae are herbivores, and
adults feed on worms and small arthropods.

(c)

(d) Order Gymnophiona. A caecilian.

(e) Order Caudata. A salamander, the redspotted newt (*Notophthalamus viridescens*).

(f) Order Caudata. Blue Ridge Spring salamander (*Gyrinophilus danielsi*).

(g) Order Anura. Tree frog (*Hyla andersoni*).

(h) Order Anura. Leopard frog (*Rana pipiens*).

(i) Order Anura. American toad (*Bufo americanus*).

Class Reptilia

(a) Order Testudines. A Galápagos tortoise, *Geochelone*.

(b) Green sea turtles (*Chelonia mydas*) nest every 2 to 4 years and migrate many miles to nesting beaches in the Caribbean and South Atlantic oceans.

(c) Order Squamata. A chameleon (*Chameleo chameleon*) using its tongue to capture prey. Note the prehensile tail.

(d) The gila monster (*Heloderma suspectum*) is a poisonous lizard of southwestern North America.

(e) A diamondback rattlesnake (*Crotalus atrox*) in its striking posture. Note the opening to a pit organ just below and slightly anterior to the eye.

(f) An amphisbaenian "worm lizard" (*Amphisbaena alba*), sometimes called a two-headed snake.

(g) Order Crocodilia. Alligators and caimans are broad snouted, and the fourth tooth on either side of their lower jaw remains inside the mouth when the jaw is closed. A female American alligator (*Alligator mississippiensis*) tending to her nest.

(h) True crocodiles have narrow to broad snouts. When their jaws are closed, the fourth tooth on either side of the lower jaw fits into a groove on the outside of the lower jaw. A Nile crocodile (*Crocodylus niloticus*) is shown here.

Class Mammalia

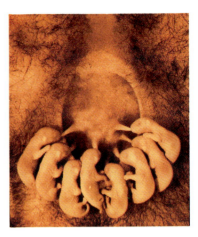

(b) Opossum young nursing in a marsupial pouch.

(a) Order Marsupialia; Infraclass Metheria. An opossum (*Didelphis marsupialis*) with young.

(c) Order Insectivora; Infraclass Eutheria. A common mole (*Talpa europeae*).

(d) Order Edentata. A sloth (*Bradypus*) in a Brazilian rain forest.

(e) Order Cetacea. A humpback whale (*Megaptera nodosa*).

(f) Order Sirenia. A manatee.

(g) Order Carnivora. An Arctic fox (*Alopex lagopus*).

(h) Order Artiodactyla. An endangered species, the key deer (*Odocoileus virginianus clavium*).

(j) Order Rodentia. A flying squirrel (*Glaucomys*).

(i) Order Perissodactyla. The black rhino (*Diceros bicornis*), another endangered species. There are fewer than 4,000 individuals living in Africa today.

(k) Order Primates. A lowland gorilla (*Gorilla gorilla graueri*).

(l) Order Primates. A mandrill (*Mandrillus sphinx*).

Support and Movement

(a) The hydroskeleton of sea anemones (*Corynactis californica*) allows them to shorten or close when longitudinal muscles contract, or to lengthen or open when circular muscles contract.

(b) A cicada nymph (*Platypedia*) leaves its old exoskeleton as it molts. This exoskeleton provides external support for the body and attachment sites for muscles.

(c) Exercise is a beneficial form of stress that causes this person's muscles to grow larger and stronger.

(d) A green frog (*Rana clamitans*) catching a grasshopper with its long, sticky tongue.

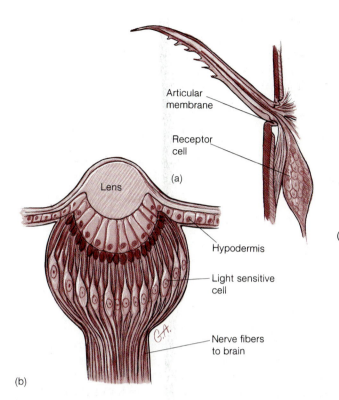

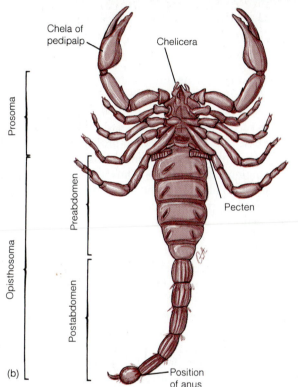

Figure 22.10 Arthropod seta and eye (ocellus). (a) A seta is a hairlike modification of the cuticle set in a membranous socket. Displacement of the seta initiates a nerve impulse in a receptor cell (sensillum) associated with the base of the seta. (b) The lens of this spider eye is a thickened, transparent modification of the cuticle. Below the lens and hypodermis are light-sensitive sensillae that contain pigments that convert light energy into nerve impulses.

Figure 22.11 Order Scorpionida. (a) A scorpion captures its prey using chelate pedipalps. Venom from its sting paralyzes the prey prior to feeding. (b) Ventral view of a scorpion.
Figure 22.11b redrawn, with permission, from *Living Invertebrates* by Pearse/Buchsbaum, copyright by The Boxwood Press.

to capture prey, for example, determine both the size of the insect and its position on the web by the vibrations the insect makes while struggling to free itself. The chemical sense of arachnids is comparable to taste and smell in vertebrates. Small pores in the exoskeleton, are frequently associated with peglike, or other, modifications of the exoskeleton, allow chemicals to stimulate nerve cells. Arachnids possess one or more pairs of eyes (figure 22.10b). These eyes are used primarily for detecting movement and changes in light intensity. The eyes of some hunting spiders probably form images.

Arachnids are dioecious. Paired genital openings are on the ventral side of the second abdominal segment. Sperm transfer is usually indirect. The male packages sperm in a spermatophore, after which it is transferred to the female. Courtship rituals confirm that individuals are of the same species, attract a female to the spermatophore, and position the female to receive the spermatophore. In some taxa, copulation occurs, and sperm transfer is accomplished via a modified pedipalp of the male. Development is direct, and the young hatch from eggs as miniature adults. Many arachnids tend their developing eggs and young during and after development.

Order Scorpionida   Members of the order Scorpionida (skor″pe-ah-ni′dah) are the scorpions. Ancestral scorpions are the oldest terrestrial arthropods. Modern species are common from tropical and subtropical areas to the temperate regions of middle North America. Scorpions are secretive and nocturnal, spending most of the daylight hours hidden under logs and stones.

Scorpions have small chelicerae that project anteriorly from the front of the carapace. A pair of enlarged, chelate pedipalps are posterior to the chelicerae. The opisthoma is divided. An anterior *preabdomen* contains the slitlike openings to book lungs, comblike tactile receptors called

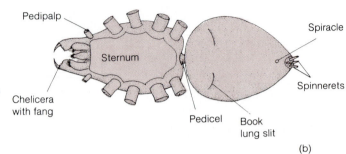

Figure 22.12 External structure of a spider.

pectines, and genital openings. The *postabdomen* (commonly called the tail) is narrower than the preabdomen and is curved dorsally and anteriorly over the body at rest. At the tip of the postabdomen is a sting. The sting has a bulbular base that contains venom-producing glands and a hollow, sharp, barbed point. Smooth muscles eject venom during stinging (figure 22.11). Only a few scorpions have venom that is highly toxic to humans. Species in the genera *Androctonus* (northern Africa) and *Centuroides* (Mexico, Arizona, and New Mexico) have been responsible for human deaths. Scorpions from the midwestern areas of North America give stings comparable to wasp stings.

Prior to reproduction, male and female scorpions have a period of courtship that may last for hours or days. Male and female scorpions face each other, extend their abdomens high into the air, and circle each other. The male seizes the female with his pedipalps, and they repeatedly walk backward and then forward. The lower portion of the male reproductive tract forms a spermatophore that is deposited on the ground. During courtship, the male positions the female so that the genital opening on her abdomen is positioned over the spermatophore. Downward pressure of the female's abdomen on a triggerlike structure of the spermatophore causes sperm to be released into the female's genital chamber.

Most arthropods are **oviparous**; females lay eggs that develop outside the body. Female scorpions, however, brood eggs in their reproductive tract. Some scorpion species are **ovoviviparous**; internal development occurs, although large, yolky eggs provide all the nourishment for development. Other species are **viviparous** meaning the embryos are nourished by nutrients provided by the mother. Eggs develop in diverticula of the ovary that are closely associated with diverticula of the digestive tract. Nutrients pass from the digestive tract diverticula to the developing embryos. Development requires several months, and more than 60 young are brooded. After birth, the young crawl onto the mother's back, where they remain for about 1 week.

Order Araneae With about 34,000 species, the order Araneae (ah-ran'a-e) is the largest group of arachnids (figure 22.12). The prosoma of spiders bears chelicerae with poison glands and fangs. Pedipalps are leglike, and in males, are modified for sperm transfer. There are eight eyes on the dorsal, anterior margin of the carapace.

The prosoma is attached to the opisthoma by a slender, waistlike *pedicel*. The abdomen is swollen or elongate and contains openings to the reproductive tract, book lungs, and trachea. It also has six to eight conical projections, called spinnerets, that are associated with silk glands. The protein that forms silk is emitted as a liquid but hardens as it is drawn out. Silk is both strong and resilient. In addition to being formed into webs for capturing prey (box 22.1), silk is used to line retreats, to lay a safety line that is fastened to the substrate to interrupt a fall, and to wrap eggs into a case for development. Silk lines produced by young spiders are caught by air currents and serve as a dispersal mechanism. They have been known to carry spiders at great altitudes for hundreds of miles. This behavior is called *ballooning*. [4]

Most spiders feed on insects and other arthropods. A few feed on small vertebrates. Once captured in webs or by hunting, prey are paralyzed and sometimes wrapped in silk. Enzymes are introduced through a puncture in the body wall and predigested products are sucked into the spider's digestive tract by a pumping esophagus. The venom of most spiders is harmless to humans. Black widow spiders (*Lactrodectus*) and brown recluse spiders (*Loxosceles*) are exceptions (colorplate 7a,b). [5]

Mating of spiders involves complex behaviors that include chemical and tactile signals. Chemicals called pheromones are deposited by a female on her web or on her body to attract a male. (*Pheromones* are chemicals released into the environment by one individual to create a behavioral change in another member of the same species.) A male may attract a female by plucking the strands of a female's web. The pattern of plucking is species specific and helps identify and locate a potential mate. The tips of a

Box 22.1

The Orb Weavers

Some of the most beautiful and intricate spider webs are produced by members of the family Araneidae, the orb weavers. Many species are relatively large, and many readers will be familiar with the black and yellow *Argiope*, or writing spider.

Orb weavers construct intricate circular webs that would be the envy of many human weavers. Web construction begins with the spider playing out a thread that is carried by the wind to a nearby surface. Alternatively, the spider may fix one end of the web, drop to the ground, and climb to a second point of attachment. The first thread is pulled tight, attached, and reinforced with additional lines (box figure 22.1a,b). The spider drops a perpendicular line from the center of the first line to form a "Y" or "T." Additional lines help to anchor the web. Radii are then placed from the center

outward, like the spokes of a wheel (box figure 22.1c–i).

To complete the web, a spiraling thread of silk is constructed. First, a temporary spiral is constructed that begins at the center of the web and is worked toward the periphery (box figure 22.1j). The permanent spiral is constructed from outside to inside. As the permanent spiral is laid down, the temporary spiral is removed (box figure 22.1k,l). The spacing of lines in the spiral is determined by the leg span of the species involved and influences the size of the prey captured.

A web is not a permanent construction. When webs become wet with rain or dew, or when they age, they lose their stickiness. The entire web, or at least the adhesive spiral, is eaten and replaced.

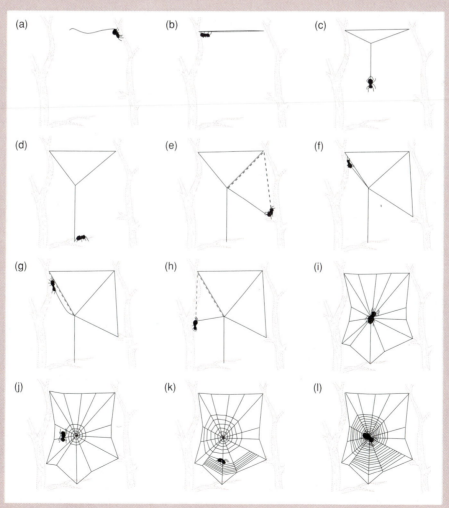

Box figure 22.1 Various stages (a-l) in the construction of an orb-web.

male's pedipalps possess a bulblike reservoir with an ejaculatory duct and a penislike structure called an *embolus*. Prior to mating, the male fills the reservoir of his pedipalps by depositing sperm on a small web and then collecting sperm with his pedipalps. During mating, a pedipalp is engorged with blood, the embolus is inserted into the female's reproductive opening, and sperm are discharged. The female deposits up to 3,000 eggs in a silken egg case, which is sealed and attached to webbing, placed in a retreat, or carried about by the female.

Order Opiliones

Members of the order Opiliones (o′pi-le″on-es) are the harvestmen or daddy longlegs. The prosoma is broadly joined to the abdomen, and thus the body appears ovoid. Legs are very long and slender. Many harvestmen are omnivores (they feed on a variety of plant and animal material) and others are strictly predators. Prey are seized by pedipalps and ingested as described for other arachnids. Digestion is both external and internal. Sperm transfer is direct, as males have a penislike structure. Females have a tubular ovipositor that is projected from a sheath at the time of egg laying. Hundreds of eggs are deposited in damp locations on the ground.

Order Acarina

Members of the order Acarina (ak′ar-i″nah) are the mites and ticks. Many are ectoparasites (parasites on the outside of the body) on humans and domestic animals. Others are free living. Of all arachnids, acarines have had the greatest impact on human health and welfare.

Mites are 1 mm or less in length. The prosoma and abdomen are fused and covered by a single carapace. Mouthparts are carried on an anterior projection called the *capitulum*. Chelicerae and pedipalps are variously modified for piercing, biting, anchoring, and sucking, and adults have four pairs of walking legs.

Free-living mites may be herbivores or scavengers. Herbivorous mites, such as spider mites, cause damage to ornamental and agricultural plants. Scavenging mites inhabit soil, leaf litter, and include some pest species that feed on flour, dried fruit, hay, cheese, and animal fur (figure 22.13a).

Parasitic mites usually do not remain permanently attached to their hosts, but feed for a few hours or days and then drop to the ground. One mite, the notorious chigger or red bug (*Trombicula*), is a parasite during one of its larval stages on all groups of terrestrial vertebrates (figure 22.13b). Host skin is enzymatically broken down and sucked up by a larva, causing local inflammation and intense itching at the site of the bite. Unless the chigger is scratched off earlier, the larva will drop from the host after about 10 days and then molt to the next immature stage, called a nymph. Nymphs eventually molt to adults, and both nymphs and adults feed on insect eggs. [6]

A few mites are permanent ectoparasites. The follicle mite, *Demodex folliculorum*, is very common (but harmless) in hair follicles of most of the readers of this text. Itch mites cause scabies in humans and other animals. *Sarcoptes scabei* is the human itch mite. It tunnels in the epidermis of human skin where females lay about 20 eggs each day. Secretions of the mites cause skin irritation, and infections are acquired by contact with an infected individual. [7]

Ticks are ectoparasites during their entire life history. They may be up to 3 cm in length but are otherwise similar to mites (figure 22.14). Hooked mouthparts are used to attach to their hosts and to feed on blood. The female ticks, whose bodies are less sclerotized than those of males, expand when engorged with blood. Copulation occurs on the host, and after feeding, females drop to the ground to lay eggs. Eggs hatch into six-legged immatures called seed ticks. Immatures feed on host blood and drop to the ground for each molt. Some ticks transmit diseases to humans and domestic animals. For example, *Dermacentor andersoni* transmits the bacteria that cause Rocky Mountain spotted fever and tularemia, and *Ixodes dammini* transmits the bacterium that causes Lyme disease.

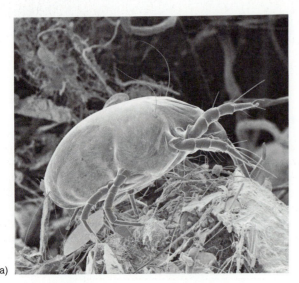

(a)

(b)

Figure 22.13 Order Acarina. Some common mites. (a) *Dermatophagoides farinae* is common in homes and grain storage areas. It is believed to be a major cause of dust allergies. (b) The chigger (*Trombicula*). This larval mite is notorious for its irritating bite.

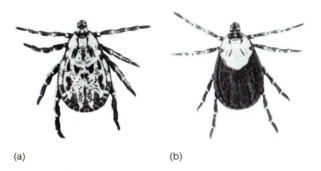

(a)   (b)

Figure 22.14 Order Acarina. A male (a) and a female (b) tick, *Dermacentor andersoni*. The female's body is less sclerotized and when she feeds, her body becomes engorged with blood. People walking in tick-infested regions should examine themselves regularly and remove any ticks found on their skin, because ticks can transmit diseases, such as Rocky Mountain spotted fever, tularemia and Lyme disease. (c) *Ixodes dammini*, the tick that transmits the bacterium that causes Lyme disease.

(c)

Class Pycnogonida

Members of the class Pycnogonida (pik′no-gon″i-dah) are the sea spiders. All are marine and are most common in cold waters (figure 22.15). Pycnogonids live on the ocean floor and are frequently found feeding on cnidarian polyps and ectoprocts. Some sea spiders feed by sucking up prey tissues through a proboscis. Others tear at prey with their chelicerae.

Pycnogonids are dioecious. Gonads are U-shaped, and branches of the gonads extend into each leg. Gonopores are located on one of the pairs of legs. Eggs are released by the female, and as the male fertilizes the eggs, they are cemented into spherical masses and attached to a pair of elongate appendages of the male, called *ovigers*, where they are brooded until hatching.

Figure 22.15 Class Pycnogonida. Sea spiders, such as *Pycnogonum,* are often found in intertidal regions feeding on cnidarian polyps.

Stop and Ask Yourself

5. How are the members of the subphylum Chelicerata characterized?
6. What are book lungs? What animals have them?
7. What characteristic of arachnids preadapted them for terrestrial environments?
8. What are three ways that spiders use silk?

Subphylum Crustacea

Some of the members of the subphylum Crustacea (krus-tas′eah) (L. *crustaceus*, hard shelled,) such as crayfish, shrimp, lobsters, and crabs, are familiar to nearly everyone. In addition, there are many lesser-known, but very common, taxa. These include copepods, cladocerans, fairy shrimp, isopods, amphipods, and barnacles. Except for some members of one order, crustaceans are all aquatic.

Crustaceans differ from other living arthropods in two ways. They have two pairs of antennae, whereas all other arthropods have one pair or none. In addition, crustaceans possess biramous appendages, each of which consists of a basal segment, called the *protopod*, with two rami attached. The medial ramus is the *endopod* and the lateral ramus is the *exopod* (figure 22.16). A similar condition was described for the trilobites and is evidence that the trilobites were ancestral to the crustaceans.

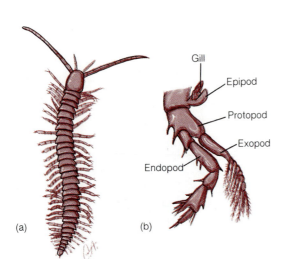

Figure 22.16 The crustacean body form. (a) Primitive crustaceans are believed to have resembled this small, marine, cave-dwelling crustacean (class Remipedia). Note the series of similar appendages along the length of the body. (b) An appendage from a larval lobster shows the generalized biramous structure. A protopod attaches to the body wall. An exopod (a lateral ramus) and an endopod (a medial ramus) attach at the end of the protopod. In modern crustaceans, both the distribution of appendages along the length of the body and the structure of appendages are modified for specialized functions.

Class Malacostraca

Malacostraca (mal-ah-kos'trah-kah) (Gr. *malakos*, soft + *ostreion*, shell) is the largest class of crustaceans. It includes crabs, lobsters, crayfish, shrimp, mysids, shrimplike krill, isopods and amphipods.

The Order Decapoda (dek-i-pod'ah) is the largest order of crustaceans and includes shrimp, crayfish, lobsters, and crabs. Shrimp have a laterally compressed, muscular abdomen and pleopods that are used for swimming. Lobsters, crabs, and crayfish are adapted to crawling on the surface of the substrate (colorplate 7e). The abdomen of crabs is greatly reduced and is held flexed beneath the cephalothorax.

Crayfish are often used to illustrate general crustacean structure and function. They are convenient to study because of their relative abundance and large size. The body of a crayfish is divided into two regions. A *cephalothorax* is derived from developmental fusion of a sensory and feeding tagma (the head) with locomotor tagma (the thorax). The exoskeleton of the cephalothorax extends laterally and ventrally to form a shieldlike *carapace*. The *abdomen* is posterior to the cephalothorax, has locomotor and visceral functions, and in crayfish, takes the form of a muscular "tail."

Paired appendages are present in both body regions (figure 22.17). The first two pairs of cephalothoracic appendages are the *first* and *second antennae*. The third through the fifth pairs of appendages are associated with the mouth. During crustacean evolution, the third pair of appendages became modified into chewing or grinding structures

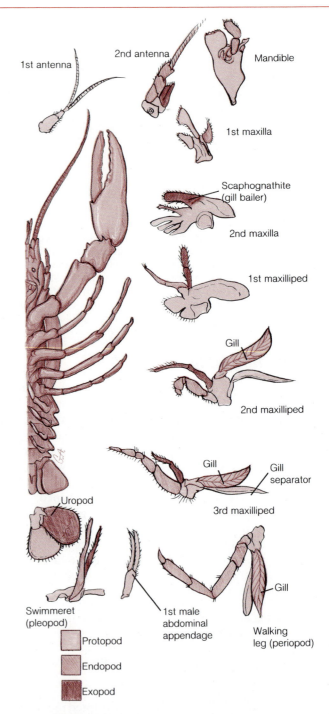

Figure 22.17 Crayfish appendages. Ventral view of a crayfish. Appendages removed and arranged in sequence.

called **mandibles.** The fourth and fifth pairs of appendages, called *maxillae*, are used for food handling. The second maxilla bears a gill and a thin, bladelike structure, called a *scaphognathite*, used to circulate water over the gills. The sixth through the eighth cephalothoracic appendages are called *maxillipeds* and are derived from the thoracic tagma. They are accessory sensory and food handling appendages. Each also bears a gill. Appendages 9 to 13 are thoracic appendages called *pereiopods*. The first pereiopod, known as the *cheliped*, is enlarged and chelate (pincherlike) and used in defense and capturing food. All but the

Figure 22.16 and 22.17 redrawn, with permission, from *Living Invertebrates* by Pearse/Buchsbaum, copyright by The Boxwood Press.

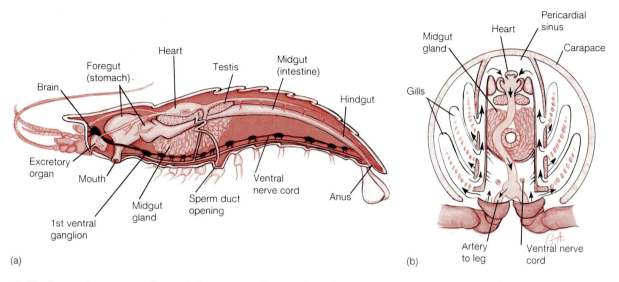

Figure 22.18 Internal structure of a crayfish. (a) Lateral view of a male. In the female, the ovary is located in the same place as the testis of the male, but the gonoducts open at the base of the third periopods. (b) Cross section of the thorax in the region of the heart. In this diagram, gills are shown attached higher on the body wall than they actually occur in order to show the path of blood flow through them.

Figure 22.18 redrawn, with permission, from *Living Invertebrates* by Pearse/Buchsbaum, copyright by The Boxwood Press.

last pair of appendages of the abdomen are called *pleopods* (*swimmerets*) and are used for swimming. In females, developing eggs are attached to pleopods, and the embryos are brooded until after hatching. In males, the first two pairs of pleopods are modified into *gonopods* (*claspers*) that are used for sperm transfer during copulation. The abdomen ends in an extension called the *telson*. The telson bears the anus and is flanked on either side by flattened appendages of the last segment, called *uropods*. The telson and uropods make an effective flipperlike structure used in swimming and in escape responses.

All crustacean appendages, except the first antennae, have presumably evolved from an ancestral biramous form, as evidenced by their embryological development, in which they arise as simple two-branched structures. (First antennae develop as uniramous appendages and later acquire the branched form. The crayfish and their close relatives are unique in having branched first antennae.) Structures, such as the biramous appendages of a crayfish, whose form is based upon a common ancestral pattern and that have similar development in the segments of an animal, are said to be **serially homologous.**

Crayfish prey upon other invertebrates, eat plant matter, and scavenge dead and dying animals. The *foregut* includes an enlarged stomach, part of which is specialized for grinding. A *digestive gland*, empties digestive enzymes into the stomach. Small particles are sent to the *midgut* (intestine) and into the digestive gland, where further digestion and absorption occur. Digestion is mostly extracellular. A short *hindgut* ends in the anus and is important in water and salt regulation (figure 22.18a).

As described above, the gills of a crayfish are attached to the bases of some cephalothoracic appendages. Gills are located in a *branchial (gill) chamber,* the space between the carapace and the lateral body wall (figure 22.18b). Water is driven anteriorly through the branchial chamber by the beating of the scaphognathite of the second maxilla. Oxygen and carbon dioxide are exchanged between blood and water across the gill surfaces, and oxygen is carried in blood plasma by a respiratory pigment, *hemocyanin.*

Circulation in crayfish is similar to that of most arthropods. Dorsal, anterior, and posterior arteries lead away from a muscular heart. Branches of these vessels empty into sinuses of the hemocoel. Blood returning to the heart collects in a ventral sinus and enters the gills before returning to the *pericardial sinus*, which surrounds the heart (figure 22.18b).

Crustacean nervous systems show trends similar to those in annelids and arachnids. Primitively, the ventral nervous system is ladderlike. Higher crustaceans show a tendency toward centralization and cephalization. In crayfish, there are *supraesophageal* and *subesophageal ganglia* that receive sensory input from receptors in the head and control head appendages. There is a fusion of the ventral nerves and segmental ganglia, and giant neurons in the ventral nerve cord function in escape responses (figure 22.18a). When impulses are conducted posteriorly along giant fibers of a crayfish, powerful abdominal flexor muscles of the abdomen contract alternately with weaker extensor muscles, causing the abdomen to flex (the propulsive stroke) and then extend (the recovery stroke). The telson and uropods form a paddlelike "tail" that propels the crayfish posteriorly.

In addition to antennae, the sensory structures of crayfish include compound eyes, simple eyes, statocysts, chemoreceptors, proprioceptors, and tactile setae. Chemical receptors are widely distributed over the appendages and the head. Many of the setae covering the mouthparts and antennae are chemoreceptors that are used in sampling food and detecting pheromones. A single pair of *statocysts* is located at the bases of the first antennae. A statocyst is a pitlike invagination of the exoskeleton that contains setae and a group of cemented sand grains called a *statolith*. Movements of the crayfish cause the statolith to move and displace setae. Statocysts provide information regarding movement, orientation with respect to the pull of gravity, and vibrations of the substrate. Because the statocyst is cuticular, it is replaced with each molt. Sand is incorporated into the statocyst when the crustacean is buried in sand. Other receptors involved with equilibrium, balance, and position senses are tactile receptors on the appendages and at joints. When a crustacean is crawling or resting, stretch receptors at the joints are stimulated. Tilting is detected by changing patterns of stimulation. These widely distributed receptors are very important to most crustaceans, because many lack statocysts.

Crayfish have *compound eyes* that are mounted on movable eyestalks. The lens system consists of 25 to 14,000 individual receptors called *ommatidia*. Compound eyes also occur in insects, and their physiology is discussed in chapter 23. Larval crustaceans have a single, median photoreceptor consisting of a few sensilla. These simple eyes, called *ocelli*, allow larval crustaceans to orient toward or away from the light. Many larvae are planktonic and use their ocelli to orient toward surface waters.

The endocrine system of a crayfish controls functions such as ecdysis, sex determination, and color change. Endocrine glands release chemicals called *hormones* into the blood, where they circulate and cause responses at certain target tissues. In crustaceans, endocrine functions are closely tied to nervous functions. Nervous tissues that produce and release hormones are called *neurosecretory tissues*. *X-organs* are neurosecretory tissues located in eyestalks of crayfish. Associated with each X-organ is a *sinus gland* that accumulates and releases the secretions of the X-organ. Other glands, called *Y-organs*, are not directly associated with nervous tissues. They are located near the bases of the maxillae. Both the X-organ and the Y-organ control ecdysis. The X-organ produces *molt-inhibiting hormone*, and the sinus gland releases it. The target of this hormone is the Y-organ. As long as molt-inhibiting hormone is present, the Y-organ is inactive. Under certain conditions, molt-inhibiting hormone release is prevented and the Y-organ releases *ecdysone* hormone, leading to molting. (These "certain conditions" are often complex and species specific. They include factors such as nutritional state, temperature, and photoperiod.) Other hormones that facilitate molting have also been described. These include, among others, a molt-accelerating factor.

Another endocrine function is mediated by *androgenic glands*, located in the cephalothorax of males. (Females possess rudiments of these glands during development, but the glands never mature.) Normally, androgenic hormone(s) promote the development of testes and male characteristics, such as gonopods. Removal of androgenic glands from males results in the development of female sex characteristics, and if androgenic glands are experimentally implanted into a female, she will develop testes and gonopods.

Many other crustacean functions are probably regulated by hormones. Some that have been investigated include: the development of brooding structures of females in response to ovarian hormones, the seasonal regulation of ovarian functions, the regulation of heart rate by eyestalk hormones, and the regulation of body color changes by eyestalk hormones.

The excretory organs of crayfish are called *antennal glands* (green glands) because they are located at the bases of the second antennae. In other crustaceans, they are called maxillary glands because they are located at the base of the second maxillae. In spite of their name, they are not glands. They are structurally similar to the coxal glands of arachnids and they presumably had a common evolutionary origin. Excretory products are formed by filtration of blood. Ions, sugars, and amino acids are reabsorbed in the tubule before the dilute urine is excreted. As with most aquatic animals, ammonia is the primary excretory product. However, crayfish do not rely solely on the antennal glands to excrete nitrogenous waste. Diffusion of ammonia across thin parts of the exoskeleton is very important. Even though it is toxic, ammonia is water soluble and rapidly diluted by water. All freshwater crustaceans face a continual influx of fresh water and loss of ions. Thus, the elimination of excess water and the reabsorption of ions become extremely important functions. Gill surfaces are important in ammonia excretion and water and ion regulation (osmoregulation).

Crayfish, and all other crustaceans except the barnacles, are dioecious. Gonads are located in the dorsal portion of the thorax, and gonoducts open at the base of the third (females) or fifth (males) periopods. Mating occurs just after a female has molted. The male turns the female onto her back and deposits nonflagellated sperm near the openings of the female's gonoducts. Fertilization occurs after copulation, as the eggs are shed. The eggs are sticky and are fastened securely to the female's pleopods. Movement of the pleopods over the eggs keeps them aerated. The development of crayfish embryos is direct, with young hatching as miniature adults. In many other crustaceans, a planktonic, free-swimming larva called a *nauplius* is present (figure 22.19a). In some, the nauplius develops into a miniature adult. In crabs and their relatives, a second larval stage called a *zoea* is present (figure 22.19b). When all adult features are present, except sexual maturity, the immature is called the *postlarva*.

Members of the order Isopoda (i'so'pod'ah) are the "pillbugs." Isopods are dorsoventrally flattened. Aquatic species may live in freshwater or marine substrates where they scavenge for food. A few isopods have become modified as ectoparasites of fish and use modified appendages for clinging to and feeding on their host. Terrestrial isopods are the only land-dwelling crustaceans. They live under rocks and logs and in leaf litter, where they are protected

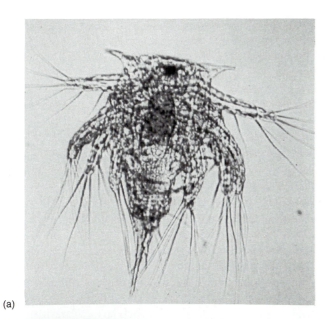

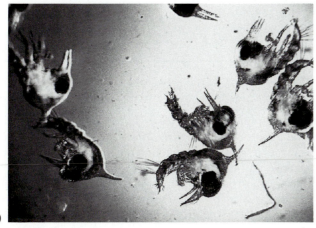

Figure 22.19 Crustacean larvae. (a) Nauplius larva of a barnacle. (b) Zoea larva of a crab.

from desiccation. As do their aquatic relatives, terrestrial isopods excrete ammonia as a waste product of metabolism. This characteristic is unusual because ammonia is toxic, and terrestrial organisms ordinarily convert it to less toxic uric acid or urea. Terrestrial isopods have less wax incorporated into their epicuticle than other terrestrial arthropods. When disturbed or threatened with drying, isopods roll into a ball to protect themselves or to reduce water loss from their delicate ventral surface—thus the name "pillbug."

Members of the order Amphipoda (am-fi-pod′ah) are easily distinguished from isopods by a laterally compressed body that gives them a shrimplike appearance (colorplate 7d). Amphipods move by crawling or swimming on their side along the substrate. (Hence the common names sideswimmers and scuds.) Some species are modified for burrowing, climbing, or jumping. (The latter include the semiterrestrial beach fleas.) This order contains freshwater and marine species that feed as scavengers and a few species that are parasites.

Class Branchiopoda

Members of the class Branchiopoda (bran′ke-o-pod′ah) (Gr. *branchio*, gill + *podos*, foot) are primarily found in fresh water. All branchiopods possess flattened, leaflike appendages that are used in respiration, filterfeeding, and locomotion.

Fairy shrimp and brine shrimp comprise the order Anostraca (an-ost′ra-kah). Fairy shrimp are usually found in temporary ponds formed by spring thaws and rains. Eggs are brooded, and when the female dies, and the temporary pond begins to dry, the embryos become dormant in a resistant capsule. Embryos lay on the forest floor until the pond fills again the following spring, at which time they hatch into nauplius larvae. Dispersal may occur if embryos are carried to other locations by animals, wind, or water currents. Their short and uncertain life cycle is an adaptation to living in ponds that dry up. The vulnerability of these slowly swimming and defenseless crustaceans probably explains why they live primarily in temporary ponds, a habitat that contains few larger predators. Brine shrimp also form resistant embryos. They live in salt lakes and ponds (e.g., Great Salt Lake in Utah).

Members of the order Cladocera (kla·dos′er-ah) are called water fleas (colorplate 7c). Their bodies are covered by a large carapace, and they swim using their second antennae, which they repeatedly thrust downward to create a jerky, upward locomotion. Females reproduce parthenogenetically (without fertilization) in spring and summer and can rapidly populate a pond or lake. Eggs are brooded in an egg case beneath the carapace. At the next molt, the egg case is released and either floats or sinks to the bottom of the pond or lake. In response to decreasing temperature, changing photoperiod, or decreasing food supply, females produce eggs that develop parthenogenetically into males. Sexual reproduction then occurs and produces resistant "winter eggs" that overwinter and hatch in the spring.

Class Copepoda

Members of the class Copepoda (ko′pe-pod′ah) (Gr. *kope*, oar + *podos*, foot) include some of the most abundant crustaceans (recall the introduction to copepods at the beginning of this chapter). There are both marine and freshwater species. Copepods possess a cylindrical body and a median ocellus that develops in the nauplius stage and persists into the adult stage. The first antennae (and the thoracic appendages in some) are modified for swimming, and the abdomen is free of appendages (*see figure 22.1*). Most copepods are planktonic and use their second maxillae for filter feeding. Their importance in marine food webs was noted in the introduction to this chapter. A few copepods live on the substrate, a few are predatory, and others are fish parasites.

Class Cirripedia

Members of the class Cirripedia (sir′i-ped′eah), the barnacles, are sessile and highly modified as adults (figure 22.20; colorplate 7f). They are exclusively marine and are

Box 22.2

Sacculina, A Highly Modified Parasite

Barnacles of the order Rhizocephala are parasites. Many are very similar to free-living barnacles but others, such as *Sacculina,* are some of the most highly modified of all animal parasites. Not only do adults not look like barnacles, they are difficult to recognize as animals and more closely resemble a fungus. Larval stages, however, disclose the true identity of this parasite.

The life cycle of *Sacculina* begins when a cypris larva attaches by its first antennae to a seta on a limb of a crab. The larva moves to a membranous area and bores through the crab exoskeleton. Once inside the body, the larva loses its exoskeleton, and dedifferentiated cells move through the blood to the midgut. The parasite then grows throughout the hemocoel and branches into a myceliumlike mass of parasite tissue. When the crab molts, a brood sac containing the parasite's eggs is formed in the flexed abdomen of the crab—in the same position

that the female crabs normally brood their own young.

Early research indicated that all crabs parasitized by *Sacculina* were apparently females. It was later discovered that in fact males were parasitized and transformed into females by the parasite! It was once thought that the destruction of the testes caused the sex change, but it is now known that the parasite destroys the androgenic gland. Just as experimental removal of the androgenic glands transforms males to females, so does parasitization.

The crab cares for the parasite's brood sac as if it were its own. Fertilization occurs when a male cypris larva introduces sperm-forming tissue into the parasite's brood sac. Nauplius larvae are released from the brood sac and they metamorphose into cypris larvae. Parasitism prevents further molting by the crab, results in sterility, and usually causes the crab's death.

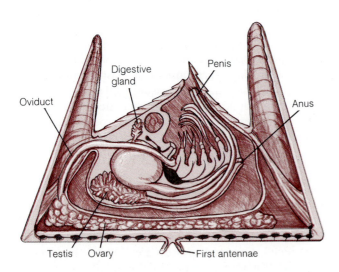

Figure 22.20 Class Cirripedia. Internal structure of an acorn (stalkless) barnacle.

very successful. Most barnacles are monoecious. The planktonic nauplius of barnacles is followed by a planktonic larval stage, called a *cypris larva,* which has a bivalved carapace. Cypris larvae attach to the substrate by their first antennae and metamorphose to adults. In the process of metamorphosis, the abdomen is reduced, and the gut tract becomes U shaped. Thoracic appendages are modified for filtering and moving food into the mouth. The larval carapace is covered by calcareous plates in the adult stage.

Barnacles attach to a variety of substrates, including rock [8] outcroppings, the bottom of ships, whales, and other animals. Some barnacles attach to their substrate by a long stalk and are called gooseneck barnacles. Others are nonstalked and are called acorn barnacles. Barnacles that colonize the bottom of ships reduce both ship speed and fuel efficiency. Much time, effort, and money has been devoted to research on how to keep ships free of barnacles.

Some barnacles have become highly modified parasites (box 22.2). The evolution of parasitism in barnacles is probably a logical consequence of living attached to other animals.

Further Phylogenetic Considerations

After studying this chapter, it should be clear that the arthopods have been very successful. The diverse body forms and life styles of copepods, crabs, lobsters, crayfish, and barnacles are an example of adaptive radiation (*see figure 12.6*). Few aquatic environments are without some crustaceans.

The subphylum Chelicerata is a very important group of animals from an evolutionary standpoint even though they are less numerous in terms of numbers of species and individuals than are many of the crustacean groups. Their arthropod exoskeleton and the evolution of excretory and respiratory systems that minimize water loss, resulted in ancestral members of this subphylum becoming the first terrestrial animals. Chelicerates, however, are not the only terrestrial arthropods. In terms of numbers of species and numbers of individuals, chelicerates are dwarfed in terrestrial environments by the fourth arthropod lineage—the insects and their relatives. This lineage and the evolutionary relationships within the entire phylum are the subject of the next chapter.

Stop and Ask Yourself

9. What evidence supports the hypothesis that trilobites were ancestral to the crustaceans?
10. What structures of a crayfish are serially homologous? What does this mean?
11. What are neurosecretory tissues? How are they involved in regulating crustacean metamorphosis?
12. What crustaceans belong to the class Branchiopoda and reproduce by parthenogenesis?

■ Summary

1. Arthropods and annelids are closely related animals. Living arthropods are divided into three subphyla: Chelicerata, Crustacea, and Uniramia. All members of a fourth subphylum, Trilobita, are extinct.

2. Arthropods have three distinctive characteristics: they are metameric and display tagmatization, they possess an exoskeleton, and many undergo metamorphosis during development.

3. Members of the extinct subphylum Trilobita had bodies that were oval and flattened and consisted of three tagmata and three longitudinal lobes. Appendages were biramous.

4. The subphylum Chelicerata has members whose bodies are divided into a prosoma and an opisthosoma. They also possess a pair of mouth appendages called chelicerae.

5. The horseshoe crabs and the giant water scorpions belong to the class Merostomata.

6. The class Arachnida includes spiders, mites, ticks, and scorpions. Ancestral arachnids were probably the first terrestrial arthropods. Their exoskeleton partially preadapted this group for its terrestrial habitat.

7. The sea spiders are the only members of the class Pycnogonida.

8. The subphylum Crustacea contains animals characterized by two pairs of antennae and biramous appendages. All crustaceans, except for some isopods, are aquatic.

9. Members of the class Branchiopoda have flattened, leaflike appendages. Examples are fairy shrimp, brine shrimp, and water fleas.

10. The class Malacostraca includes the crabs, lobsters, crayfish, shrimp, isopods and amphipods. This is the largest crustacean class in terms of numbers of species.

11. Members of the class Copepoda include the copepods.

12. The class Cirrepedia contains the barnacles.

■ Key Terms

biramous appendages (p. 346)
book lungs (p. 348)
chelicerae (p. 346)
chitin (p. 345)
coxal glands (p. 348)
ecdysis (p. 345)
exoskeleton or cuticle (p. 344)
hemocoel (p. 348)

malpighian tubules (p. 348)
mandibles (p. 354)
oviparous (p. 350)
ovoviviparous (p. 350)
serially homologous (p. 355)
spiracle (p. 348)
tracheae (p. 348)
viviparous (p. 350)

Critical Thinking Questions

1. What is tagmatization, and why is it advantageous for metameric animals?
2. In spite of being an armorlike covering, the exoskeleton permits movement and growth. Explain how this is accomplished.
3. Why is the arthropod exoskeleton often cited as the major reason for arthropod success?
4. Explain why excretory and respiratory systems of ancestral arachnids probably preadapted these organisms for terrestrial habitats.
5. What group of arthropods do you think contained the ancestors of crustaceans? What evidence supports your hypothesis?
6. Barnacles are obviously very successful arthropods. What do you think could have selected for their highly modified body form?

Suggested Readings

Books

Bliss, D. E. (ed.). 1982. *The Biology of Crustacea,* vols. 1–5. San Diego: Academic Press, Inc.

Bristow, W. S. 1971. *The World of Spiders.* Minneapolis: William Collins Sons & Co., Ltd.

Foelix, R. F. 1982. *Biology of Spiders.* Cambridge: Harvard University Press.

Kaston, B. J. 1972. *How to Know the Spiders.* 2nd ed. Dubuque: Wm. C. Brown Co. Publishers.

King, P. E. 1973. *Pycnogonids.* New York: St. Martins Press, Inc.

Levi, H. W. 1982. Crustacea. In S. P. Parker (ed.). *Synopsis and Classification of Living Organisms,* vol. 2. New York: McGraw-Hill Book Co.

McDaniel, B. 1979. *How to Know the Ticks and Mites.* Dubuque: Wm. C. Brown Co. Publishers.

Pennak, R. W. 1989. *Freshwater Invertebrates of the United States,* 3rd ed. New York: John Wiley and Sons, Inc.

Savory, T. H. 1977. *Arachnida.* 2nd ed. San Diego: Academic Press, Inc.

Articles

Burgess, J. W. Social spiders. *Scientific American* March, 1976.

Kaston, B. J. 1964. The evolution of spider webs. *American Zoologist* 4:191–207.

Lizotte, R. S., and Rovner, J. S. 1988. Nocturnal capture of fireflies by lycosid spiders: visual versus vibratory stimuli. *Animal Behaviour* 36:1809–1815.

Miller, J. A. 1984. Spider silk, stretch, and strength. *Science News* 125: 391.

Quicke, D. 1988. Spiders bite their way towards safer insecticides. *New Scientist* 120:38–41.

Robinson, M. H. 1987. In a world of silken lines, touch must be exquisitely fine. *Smithsonian* 18:94–102.

Salmon, M. 1971. Signal characteristics and acoustic detection by fiddler crabs, *Uca rapax* and *Uca pugilator. Physiological Zoology* 44:210.

Savory, T. H. Hidden lives. *Scientific American* February, 1968.

Stowe, M. K., Tumilinson, J. H., and Heath, R. R. 1987. Chemical mimicry: bolas spiders emit components of moth prey species sex pheromones. *Science* 236:964–968.

Wells, M. S. 1988. Effects of body size and resource value on fighting behaviour in a jumping spider. *Animal Behaviour* 36:321–326.

Wicksten, M. D. Decorator crabs. *Scientific American* February, 1980.

Yager, J. 1981. Remipedia, a new class of Crustacea from a marine cave in the Bahamas. *Journal of Crustacean Biology* 1:328–333.

The Insects and Myriapods: Terrestrial Triumphs

Concepts

1. Flight, along with other arthropod characteristics, has resulted in insects becoming one of the most successful groups of terrestrial animals.
2. Members of the classes Diplopoda, Chilopoda, Pauropoda, and Symphyla are the myriapods.
3. Members of the class Insecta or Hexapoda are characterized by three pairs of legs, and they often have wings.
4. Adaptations for living on land are reflected in many aspects of insect structure and function.
5. Insects have important effects on human health and welfare.
6. Some zoologists believe that the arthropods actually should be divided into three phyla.

Have You Ever Wondered:

[1] why insects have been so successful on land?
[2] how millipedes can use hydrogen cyanide as a repellant without destroying their own tissues?
[3] how fast the wings of a midge beat?
[4] how far a human could leap if one could jump, relative to body size, the same distance a flea can jump?
[5] what an insect sees?
[6] how insects communicate over long distances?
[7] how the social organization in a honeybee hive is regulated?

These and other useful questions will be answered in this chapter.

This chapter contains underlined evolutionary concepts.

Evolutionary Perspective

By almost any criterion, the insects have been enormously successful. Although there are numerous freshwater and parasitic species, the success of insects has largely been due to their ability to exploit terrestrial habitats.

During the late-Silurian and early-Devonian periods (about 400 million years ago) terrestrial environments were largely uninhabited by animals. Low-growing herbaceous plants and the first forests were beginning to flourish, and enough ozone had accumulated in the upper atmosphere to filter ultraviolet radiation from the sun. Animals with adaptations that permitted life on land had a wealth of photosynthetic production available, and unlike in marine habitats, had little competition with other animals for resources. However, the problems associated with terrestrial life were substantial. Support and movement outside of a watery environment were difficult on land, as were water, ion, and temperature regulation.

In chapter 22, the arachnids were presented as the first terrestrial animals. The exoskeleton preadaped them for life on land because of its supporting and water-conserving properties. The insects, however, followed and soon became the dominant form of animal life on land. Why did insects predominate on land relative to other arthropods? No one knows for certain, but insect dominance is probably due to the evolution of flight. Insects are the only animals other than birds and bats that fly. This ability allowed them to use widely scattered food resources, to invade new habitats, and to escape unfavorable environments. Flight—along with other generalized arthropod characteristics, [1]

Table 23.1 Classification of Uniramous Arthropods

Phylum Arthropoda (ar'thra-po'dah)
Animals with metamerism and tagmatization, a jointed exoskeleton, and a ventral nervous system.

   Subphylum Uniramia (yoo'ne-ram'eah) (L. *unis*, one + *ramis*, branch)
Head with one pair of antennae and one pair of mandibles; all appendages uniramous.

      Class Diplopoda (dip'le-pod'ah)
Two pairs of legs per apparent segment; body round in cross section. Millipedes.

      Class Chilopoda (ki'le-pod'ah)
One pair of legs per segment; body oval in cross section. Centipedes.

      Class Pauropoda (por'e-pod'ah)
Small (0.5 to 2mm), soft-bodied animals; 11 segments; nine pairs of legs; live in leaf mold. Pauropods.

      Class Symphyla (sim-fi'lah)
Small (2 to 10 mm); long antennae; centipedelike; 10 to 12 pairs of legs; live in soil and leaf mold. Symphylans.

      Class Insecta (in-sekt'ah) or **Hexapoda** (hex'sah-pod'ah)
Three pairs of legs; often winged (usually two pairs); body with head, thorax, and abdomen; mandibulate mouthparts variously adapted. Insects.

         Subclass Apterygota (ap-ter-i-go'tah)
Primitively wingless insects; pregenital abdominal appendages; ametabolous metamorphosis; indirect sperm transfer.

            Order Collembola (col-lem'bo-lah)
Antennae with 4 to 6 segments; compound eyes absent; abdomen with 6 segments, with springing appendage on fourth segment. Springtails.

            Order Protura (pro-tu'rah)
Minute, with cone-shaped head; antennae, compound eyes, and ocelli absent; abdominal appendages on first three segments. Proturans.

            Order Diplura (dip-lu'rah)
Head with many segmented antennae; compound eyes and ocelli absent; cerci multisegmented or forcepslike. Diplurans.

            Order Microcoryphia (mik-ro-kor'if-eah)
Tapering, cylindrical abdomen with long ceri; active, jump when disturbed. Jumping bristletails.

            Order Thysanura (thi-sa-nu'rah)
Tapering abdomen; scales on body; terminal cerci; long antennae. Silverfish.

         Subclass Pterygota (ter-i-go'tah)
Insects descendant from winged ancestors. No pregenital abdominal appendages.

            Superorder Exopterygota (ek-op-ter-i-go'tah)
Paurometabolous (or hemimetabolous) metamorphosis; wings develop as external wing pads; direct sperm transfer.

               Order Ephemeroptera (e-fem-er-op'ter-ah)
Elongate, abdomen with 2 or 3 tail filaments; wings held above the body at rest; short antennae. Mayflies.

               Order Odonata (o-do-nat'ah)
Elongate, membranous wings; abdomen long and slender; compound eyes occupy most of head. Dragonflies, damselflies.

               Order Grylloblattaria (gril-lo-blat-tar'eah)
Slender, elongate and wingless; antennae long and slender; found in glacier edges and ice caves, often high altitudes. Rock crawlers.

               Order Phasmida (fas'mi-dah)
Body elongate and sticklike; wings reduced or absent; some tropical forms are flattened and laterally expanded. Walking sticks, leaf insects.

               Order Orthoptera (or-thop'ter-ah)
Forewing long, narrow, and leathery; hindwing broad and membranous; chewing mouthparts. Grasshoppers, crickets, katydids.

               Order Mantodea (man-to'deah)
Prothorax long; prothoracic legs long and armed with strong spines for grasping prey; voracious predators. Mantids.

               Order Blattaria (blat-tar'eah)
Body oval and flattened; head concealed from above by a shieldlike extension of the prothorax. Cockroaches.

such as the exoskeleton, metamorphosis, and high reproductive potential—permitted insects to become one of the dominant classes of organisms on earth.

The insects make up one of five classes in the subphylum Uniramia (table 23.1). The four noninsect classes (discussed next) are grouped into a convenient, nontaxonomic grouping called the myriapods (Gr. *myriad,* ten thousand + *podus,* foot).

Class Diplopoda

The class Diplopoda (dip'lah-pod'ah) (Gr. *diploos,* twofold + *podus,* foot) contains the millipedes. Millipedes have 11 to 100 trunk segments that have been derived from an embryological and evolutionary fusion of primitive me-

tameres. An obvious result of this fusion is the occurrence of two pairs of appendages on each apparent trunk segment. Each segment is actually the fusion of two segments. Fusion is also reflected internally by two ganglia, two pairs of ostia, and two pairs of tracheal trunks per apparent segment. Most millipedes are round in cross section, although some are more flattened (figure 23.1a).

Millipedes are worldwide in distribution and are nearly always found in or under leaf litter, humus, or decaying logs. Their epicuticle does not contain much wax, and therefore, their choice of habitat is important to prevent drying. Their many legs, simultaneously pushing against the substrate, help millipedes bulldoze through the habitat. Millipedes feed on decaying plant matter using their mandibles in a chewing or scraping fashion. A few millipedes have mouthparts modified for sucking plant juices.

Order Plecoptera (ple-kop'ter-ah)
Long antennae; weak mouthparts; two pairs of membranous wings; multisegmented cerci; nymphs aquatic. Stoneflies.

Order Embiidina (em-bi-i-din'ah)
Elongate; live in silk tunnels; head with long antennae, compound eyes, and chewing mouthparts; males winged, females lack wings. Webspinners.

Order Dermaptera (der-map'ter-ah)
Elongate; biting mouthparts; multisegmented antennae; abdomen with unsegmented forcepslike cerci. Earwigs.

Order Zoraptera (zo-rap'ter-ah)
Antennae with 9 segments; biting mouthparts; some lack wings. Zorapterans.

Order Isoptera (i-sop'ter-ah)
White and wingless (except in dispersal stages); antlike except abdomen broadly joins thorax; social. Termites.

Order Psocoptera (so-cop'ter-ah)
Head with long antennae and specialized chewing mouthparts; prothorax reduced; wings may be absent. Booklice.

Order Phthiraptera (fthi-rap'ter-ah)
Small, wingless ectoparasites of birds and mammals; often divided into two orders, treated here as suborders. Lice.

Suborder Anoplura (an-o-plu'rah)
Wingless; body dorsoventrally flattened; narrow head; broad thorax and abdomen; piercing-sucking mouthparts; ectoparasites of mammals. Sucking lice.

Suborder Mallophaga (mal-of'ah-gah)
Wingless; body dorsoventrally flattened; broad head; chewing mouthparts; most are bird ectoparasites. Chewing lice.

Order Hemiptera (hem-ip'ter-ah)
Proximal portion of forewing sclerotized, distal portion membranous; sucking mouthparts hang ventrally off anterior margin of head (prognathous). Bugs.

Order Homoptera (ho-mop'ter-ah)
Wings entirely membranous; mouthparts appear to hang ventrally off posterior margin of head (hypognathous). Cicadas, leaf hoppers, aphids, whiteflies.

Order Thysanoptera (thi-sa-nop'ter-ah)
Small bodied; sucking mouthparts; wings narrow and fringed with long setae; plant pests. Thrips.

Superorder Endopterygota (en-dop-ter-i-go'tah)
Holometabolous metamorphosis; wings develop internally; direct sperm transfer.

Order Neuroptera (neu-rop'ter-ah)
Wings membranous; hind wings held rooflike over body at rest; antennae long and threadlike. Lacewings, snakeflies, antlions.

Order Coleoptera (ko-le-op'ter-ah)
Forewings sclerotized, forming a cover (elytra) over the abdomen; hindwings membranous; chewing mouthparts; the largest insect (animal) order. Beetles.

Order Mecoptera (me-kop'ter-ah)
Head with broad rostrum, long antennae, biting mouthparts; wings membranous. Scorpion flies.

Order Strepsiptera (strep-sip'ter-ah)
Males free living with well-developed wings and reduced mouthparts; females lack wings; females and larvae are parasites of other insects. Stylopoids.

Order Trichoptera (tri-kop'ter-ah)
Mothlike with setae-covered antennae; chewing mouthparts; wings covered with setae and held rooflike over abdomen; larvae aquatic and often dwell in cases that they construct. Caddis flies.

Order Lepidoptera (lep-i-dop'ter-ah)
Wings broad and covered with scales; mouthparts formed into a sucking tube. Moths, butterflies.

Order Diptera (dip'ter-ah)
Mesothoracic wings well developed; metathoracic wings reduced to knoblike halteres; sucking or lapping mouthparts. Flies.

Order Siphonaptera (si-fon-ap'ter-ah)
Laterally flattened, sucking mouthparts; jumping legs; parasites of birds and mammals. Fleas.

Order Hymenoptera (hi-men-op'ter-ah)
Wings (when present) membranous with few veins; well-developed ovipositor, sometimes modified into a sting; mouthparts modified for biting and lapping; social and solitary species. Ants, bees, wasps.

(a)

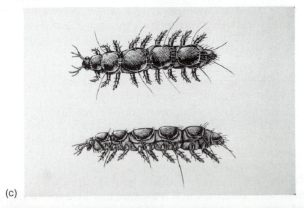

(b)

(c)

(d)

Figure 23.1 The myriapods. (a) A millipede. (b) A centipede that has captured a lizard. (c) A pauropod. (d) A symphylan.

Millipedes roll into a ball when faced with desiccation or when they are disturbed. Many also possess *repugnatorial glands* that produce hydrogen cyanide, which is repellant to other animals. Hydrogen cyanide is not synthesized and stored as hydrogen cyanide because it is very caustic and would destroy millipede tissues. Instead, a precursor compound and an enzyme are mixed as they are released from separate glandular compartments. Repellants increase the likelihood that the millipede will be dropped unharmed and decrease the chances that the same predator will try to feed on another millipede.

[2]

Sperm are transferred to the female millipede with modified trunk appendages of the male, called gonopods, or in spermatophores. Eggs are fertilized as they are laid and hatch in several weeks. Immatures acquire more legs and segments with each molt until they reach adulthood.

Class Chilopoda

Members of the class Chilopoda (ki′lah-pod′ah) (Gr. *cheilos,* lip + *podus,* foot) are the centipedes. They spend most of their time scurrying about the surfaces of logs, rocks, or other forest-floor debris. As do millipedes, most centipedes lack a waxy epicuticle, and therefore, are found in moist habitats. Their bodies are oval in cross section, and they have a single pair of long legs on each of their 15 or more trunk segments. The last pair of legs is usually modified into long sensory appendages, or in a few, into pinching structures.

Centipedes are fast-moving predators. Food usually consists of small arthropods, earthworms, and snails; however, some feed on frogs and rodents (figure 23.1b). Poison claws (modified first trunk appendages called *maxillipeds*) are used to kill or immobilize prey. Maxillipeds, along with mouth appendages, hold the prey as mandibles chew and ingest the food. The venom of most centipedes is harmless to humans, although some larger centipedes have bites that are comparable to wasp stings; a few deaths have been reported from large, tropical species.

Reproduction of centipedes may involve courtship displays in which the male lays down a silk web using glands at the posterior tip of the body. A spermatophore is placed in the web and picked up by the female who introduces the spermatophore into her genital opening. Eggs are fertilized as they are laid. A female may brood and guard eggs by wrapping her body around the eggs, or they may be deposited in the soil. Young are similar to adults except that they have fewer legs and segments. Legs and segments are added with each molt.

Class Pauropoda

Members of the class Pauropoda (por′o-pod′ah) (Gr. *pauros,* small + *podus,* foot) are soft-bodied animals with 11 segments that live in forest-floor litter (figure 23.1c). They feed on fungi, humus, and other decaying organic matter. Their very small size and thin, moist exoskeleton allow gas exchange across the body surface and diffusion of nutrients and wastes in the body cavity. Sperm transfer occurs with a spermatophore. Eggs are laid singly or in clusters in humus. The young hatch with three pairs of legs, and legs are added with subsequent molts.

Class Symphyla

Members of the class Symphyla (sim-fil'ah) (Gr. *sym*, same + *phyllos*, leaf) are small arthropods (2 to 10 mm in length) that occupy soil and leaf mold and superficially resemble centipedes (figure 23.1d). They lack eyes and have 12 leg-bearing trunk segments. The posterior segment may have one pair of spinnerets or long sensory bristles. Symphylans normally feed on decaying vegetation; however, some species are pests of vegetables and flowers.

The reproductive behavior of one genus, *Scutigerella*, is well known. The male deposits a spermatophore on the tip of a stalk. A female picks it up and stores the sperm in special pouches in her mouth. The female removes eggs from her gonopore with her mouth and fertilizes them as she attaches them to moss or lichens.

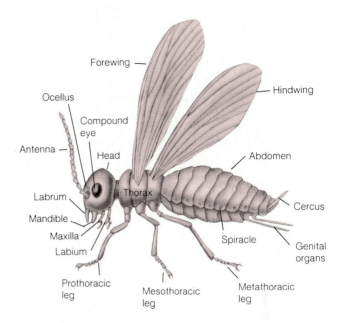

Figure 23.2 The external structure of a generalized insect. **Figure 23.2** redrawn, with permission, from *Living Invertebrates* by Pearse/Buchsbaum, copyright by The Boxwood Press.

Stop and Ask Yourself

1. What probably accounts for the dominance of insects on land over other arthropods?
2. What classes of arthropods are collectively called the myriapods?
3. How would you characterize members of the following classes: Diplopoda? Chilopoda? Pauropoda? Symphyla?
4. How do millipedes discourage predators?

Class Insecta or Hexapoda

Members of the class Insecta (in-sekt'ah) (L. *insectum*, cut into) or Hexapoda (Gr. *hexa*, six + *podus*, feet) are, in many ways, the most successful land animals. Order-level classification varies depending on the authority consulted. One system is shown in table 23.1. Representatives of selected orders are shown in colorplate 8. In spite of obvious diversity, there are common features that make insects easy to recognize. Many insects have wings, and virtually all adults have three pairs of legs.

External Structure and Locomotion

The body of an insect is divided into three tagmata: head, thorax, and abdomen (figure 23.2). The head bears a single pair of antennae, mouthparts, compound eyes, and 0, 2, or 3 ocelli. The thorax consists of three segments. They are,

from anterior to posterior, the **prothorax**, the **mesothorax**, and the **metathorax.** One pair of legs attaches along the ventral margin of each thoracic segment, and a pair of wings, when present, attaches at the dorsal margin of the mesothorax and metathorax. Wings have thickened, hollow veins for increased strength. The thorax also contains two pairs of spiracles, which are openings to the tracheal system. Most insects have 10 or 11 abdominal segments, each of which has a lateral fold in the exoskeleton that allows the abdomen to expand when the insect has gorged itself or when it is full of mature eggs. Each abdominal segment has a pair of spiracles. Also present are genital structures used during copulation and egg deposition, and sensory structures called cerci. Gills are present on abdominal segments of certain aquatic insects.

Insect Flight   The great diversity of insects and insect habitats is accompanied by diversity in how insects move. From an evolutionary perspective, however, flight is the most important form of insect locomotion. Insects were the first animals to fly. One of the most popular hypotheses on the origin of flight states that wings may have evolved from rigid, lateral outgrowths of the thorax that probably served as protection for the legs or spiracles. Later, these fixed lobes could have been used in gliding from the top of tall plants to the forest floor. The ability of the wing to flap, tilt, and fold back over the body probably came later.

Another requirement for flight was the evolution of limited thermoregulatory abilities. *Thermoregulation* is the ability to maintain body temperatures at a level different from environmental temperatures. Achieving relatively high body temperatures, perhaps 25°C or greater, is needed for flight muscles to contract rapidly enough for flight. (Thermoregulatory mechanisms are discussed later in this chapter.)

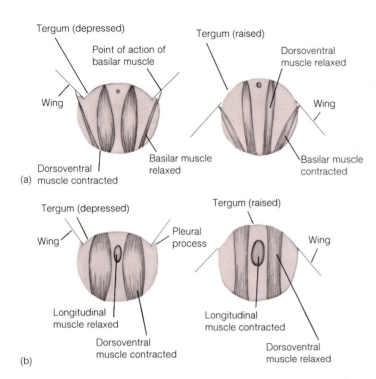

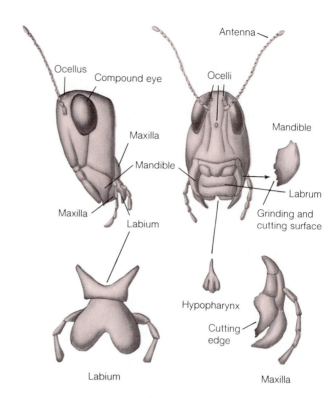

Figure 23.3 Insect flight. (a) Muscle arrangements for the direct or synchronous flight mechanism. Note that muscles responsible for the downstroke attach at the base of the wings. (b) Muscle arrangements for an indirect or asynchronous flight mechanism. Wings move up and down as a result of muscles changing the shape of the thorax.

Figure 23.4 The head and mouthparts of a grasshopper. All mouthparts except the labrum are derived from segmental appendages. The labrum is a sensory upper lip. The mandibles are heavily sclerotized and used for tearing and chewing. The maxillae have cutting edges and a sensory palp. The labium forms a sensory lower lip. The hypopharynx is a sensory, tonguelike structure.

Some insects use a **direct** or **synchronous flight** mechanism, in which a downward thrust of the wings results from muscles acting on the bases of the wings. The upward thrust of the wings is accomplished by muscles attaching dorsally and ventrally on the exoskeleton (figure 23.3a). The synchrony of direct flight mechanisms comes from the fact that each wingbeat must be preceded by a nerve impulse to the flight muscles.

Other insects use an **indirect** or **asynchronous flight** mechanism. Muscles act to change the shape of the exoskeleton for both upward and downward strokes of the wings. The upward thrust of the wing is produced by dorsoventral muscles pulling the dorsal exoskeleton downward. The downward thrust occurs when longitudinal muscles contract and cause the exoskeleton to arch upward (figure 23.3b). The power and velocity of these strokes are enhanced by the resilient properties of the exoskeleton. During a wingbeat, the thorax is deformed, and in the process, energy is stored in the exoskeleton. At a critical point midway into the downstroke, stored energy reaches a maximum, and at the same time, resistance to wing movement suddenly decreases. The wing then "clicks" through the rest of the cycle using energy stored in the exoskeleton. Asynchrony of this flight mechanism arises from the fact that there is no one-to-one correspondence between nerve impulses and wingbeats. A single nerve impulse can result [3] in approximately 50 cycles of the wing, and frequencies of 1,000 cycles per second (cps) have been recorded in

some midges! The asynchrony between wingbeat and nerve impulses is dependent on flight muscles being stretched during the "click" of the thorax. The stretching of longitudinal flight muscles during the upward beat of the wing initiates the subsequent contraction of these muscles. Similarly, stretching during the downward beat of the wing initiates subsequent contraction of dorsoventral flight muscles. Indirect flight muscles are frequently called *fibrillar flight muscles*.

Simple flapping of wings is not enough for flight. The tilt of the wing must be controlled to provide lift and horizontal propulsion. In most insects, muscles that control wing tilting attach to sclerotized plates at the base of the wing.

Other Forms of Locomotion Locomotion across the ground or other substrate is accomplished by walking, running, jumping, or swimming. When walking, insects have three or more legs on the ground at all times, creating a very stable stance. When running, fewer than three legs may be in contact with the ground. A fleeing cockroach (order Blattaria) reaches speeds of about 5 km/hr, although it seems much faster when trying to outrun one. The apparent speed is the result of their small size and ability to quickly change directions. Jumping insects, such as grasshoppers (order Orthoptera), usually have long

Supraesophageal ganglion
Foregut
Dorsal aorta
Ovary
Oviduct
Rectum
Anus
Subesophageal ganglion
Mouth
Salivary gland
Gastric cecae
Nerve cord
Midgut
Malpighian tubule
Hindgut
Genital chamber
Sperm receptacle
Ovipositor

Figure 23.5 Digestive system of a generalized insect. Salivary glands produce enzymes, but may be modified for the secretion of silk, anticoagulants, or pheromones. The crop is an enlargement of the foregut that is used for storing food. The proventriculus is a grinding structure. Gastric cecae secrete digestive enzymes. The intestine and the rectum are modifications of the hindgut that function in absorbing water and the products of digestion.

metathoracic legs that generate large, propulsive forces. Energy for a flea's (order Siphonaptera) jump is stored as elastic energy of the exoskeleton. Muscles that flex the legs distort the exoskeleton. A catch mechanism holds the legs in this "cocked" position until special muscles release the catches and allow the stored energy to quickly extend the legs. This action hurls the flea for distances that exceed 100 times its body length. A comparable distance for a human broad jumper would be the length of two football fields!

[4]

Nutrition and the Digestive System

The diversity of insect feeding habits parallels the diversity of insects themselves. Figure 23.4 shows the head and mouthparts of an insect such as a grasshopper or cockroach. An upper, liplike structure is called the *labrum.* It is sensory, and unlike the remaining mouthparts, is not derived from segmental, paired appendages. *Mandibles* are sclerotized, chewing mouthparts. The *maxillae* often have cutting surfaces and bear a sensory palp. The *labium* is a sensory lower lip. The digestive tract, as in all arthropods, consists of a *foregut,* a *midgut,* and a *hindgut* (figure 23.5). Enlargements for storage and diverticula that secrete digestive enzymes are common. Variations on this plan are specializations for sucking or siphoning plant or animal fluids (figure 23.6).

Gas Exchange

Gas exchange with air requires a large surface area for the diffusion of gases. In terrestrial environments, these surfaces are also avenues for water loss. Respiratory water loss in insects, as in some arachnids, is reduced through the invagination of respiratory surfaces to form highly branched systems of chitin-lined tubes, called *tracheae.* Tracheal systems apparently arose separately in the two groups.

Tracheae open to the outside of the body through *spiracles,* which are usually provided with some kind of closure device to prevent excessive water loss. Spiracles lead to tracheal trunks that branch, eventually giving rise to smaller branches, the tracheoles. *Tracheoles* end intracellularly and are especially abundant in metabolically active

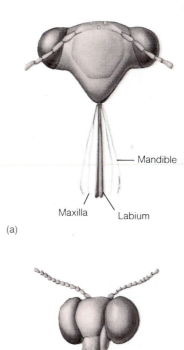

Mandible
Maxilla
Labium
(a)

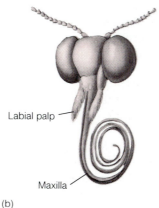

Labial palp
Maxilla
(b)

Figure 23.6 Specializations of insect mouthparts. (a) Mouthparts of a cicada (order Homoptera) are specialized for feeding on plant juices. The labium serves as a sheath for the mandibles and maxillae, which are inserted into plant tissues. The labium is reduced. (b) The sucking mouthparts of a butterfly consist of modified maxillae that coil when not in use. Mandibles, labia, and the labrum are reduced in size. In both (a) and (b), a portion of the anterior digestive tract is modified as a muscular pump for drawing liquids through the mouthparts.

Figure 23.5 and 23.6 redrawn, with permission, from *Living Invertebrates* by Pearse/Buchsbaum, copyright by The Boxwood Press.

tissues, such as flight muscles. No cells are more than 2 or 3 μ from a tracheole (figure 23.7).

Most insects have ventilating mechanisms that move air into and out of the tracheal system. For example, alternate compression and expansion of the larger tracheal trunks by contracting flight muscles ventilates the tracheae. In some insects, carbon dioxide produced by metabolically active cells is sequestered in the hemocoel as bicarbonate ions (HCO_3^-). As oxygen diffuses from the tracheae to the body tissues, and is not replaced by carbon dioxide, a vacuum is created that draws more air into the spiracles. This process is called *passive suction*. Periodically, the sequestered bicarbonate ions are converted back into carbon dioxide, which escapes through the tracheal system.

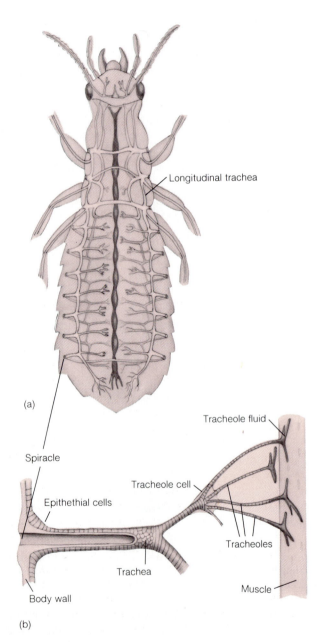

(a)

Longitudinal trachea

Spiracle

Epithethial cells

Tracheole fluid

Tracheole cell

Tracheoles

Trachea

Muscle

Body wall

(b)

Figure 23.7 The tracheal system of an insect. (a) Major tracheal trunks. (b) Tracheoles end in cells, and the terminal portions of tracheoles are fluid filled.

The rate of diffusion of oxygen across tracheole walls increases as a tissue's metabolic rate increases. In a resting tissue, much of a tracheole is fluid filled. As the tissue becomes more active, metabolites accumulate in them. Fluids are drawn from the tracheoles into the surrounding tissues because accumulating metabolites alter the osmotic characteristics of tissue fluids. Air then replaces the tracheolar fluids. Oxygen delivery to tissues is increased because oxygen diffuses more quickly through air than water.

Circulation and Temperature Regulation

The circulatory system of insects is similar to that described for other arthropods, although the blood vessels are less well developed. Blood distributes nutrients, hormones, and wastes, and ameboid blood cells participate in body defense and repair mechanisms. Blood is not important in gas transport, but as noted above, some insects temporarily store bicarbonate ions in the hemocoel.

As described earlier, temperature regulation is a requirement for flying insects. Virtually all insects warm themselves by basking in the sun or resting on warm surfaces. Because they use external heat sources in temperature regulation, insects are generally considered to be *ectotherms*. Other insects, especially some moths, can generate heat by rapid contraction of flight muscles, a process called *shivering*. Metabolic heat generated in this way can raise the temperature of thoracic muscles from near 0° to 30°C. Because some insects rely to a limited extent on metabolic heat sources, they have a variable body temperature and are sometimes called *heterotherms*. Insects are also able to cool themselves by seeking cool, moist habitats. Honeybees can cool a hive by beating their wings at the entrance of the hive, thus circulating cooler outside air through the hive.

Stop and Ask Yourself

5. How can you recognize an insect as being an anthropod?
6. Why was the evolution of thermoregulatory abilities an important step in the evolution of flight?
7. What is indirect or asynchronous flight?
8. How do gases move across tracheole walls, and how does the accumulation of metabolites in tissues affect the rate of gas exchange?

Nervous and Sensory Functions

The nervous system of insects is similar to the pattern described for annelids and other arthropods. The supraesophageal ganglion is associated with sensory structures of the head. It is joined by connectives to the subesophageal ganglion, which innervates the mouthparts and salivary glands and has a general excitatory influence on other body parts. Segmental ganglia of the thorax and abdomen undergo various degrees of fusion in different taxa. Insects also possess a well-developed visceral nervous system that innervates the gut tract, sex organs, and heart.

Box 23.1

A Moth/Bat Story

Bats, some of the most successful nighttime hunters, feed on flying insects and rely on **echolocation** for finding their prey. During echolocation, bats emit ultrasonic sounds (sound at frequencies too high to be heard by humans) that reflect from flying insects back to the bats' unusually large external ears. Using this information, bats can determine the exact location of an insect and can even distinguish the kind of insect.

Noctuid moths, a prey item of these bats, possess an effective escape behavior. The tympanal organs of noctuid moths are sensitive to sounds in the 3,000 to 150,000 cps frequency range, which encompasses frequencies characteristic of a bat's cries. A weak stimulus from a bat a long distance away results in a moth flying away from the source of the sound. A stronger stimulus results in the moth's flight becoming very erratic. Often, strong stimuli result in the moth diving straight for the ground, a reflexive behavior advantageous for the moth, because sound echoing off the insect becomes indistinguishable from sound echoing off objects on the ground.

The stereotyped nature of a moth's response is explained by the structure and placement of its tympanal organs. They are located on either side of the metathoracic segment, and their bilateral placement helps a moth determine the direction of incoming sound. Sound arriving from a moth's right side strikes the right tympanal organ more strongly, because the moth's body shades the left tympanal organ from sound. Thus, the moth can determine the approximate location of the predator. In addition, each receptor consists of a cuticular membrane overlying an air-filled cavity. Nerve impulses are initiated when sound waves displace the cuticular membrane and one of two sensory cells associated with the inside of the membrane. One cell, called the A_1 cell, is stimulated by relatively low-energy sound waves. These sounds are made by a bat so far away that sounds echoing off the moth would be undetectable by the bat. In this situation, a moth turns away from the source of the sound. A second cell, called the A_2 cell, is stimulated by high-energy sound waves. These sounds are made by a bat near enough to detect the moth. The response of the moth to the activity of the A_2 cell is erratic flight and/or a dive toward the ground.

The apparent complexity of some insect behavior is deceptive. It may seem as if insects make conscious decisions in their actions; however, this is seldom the case. As a noctuid moth's evasive responses to a bat's cries, most insect behavior patterns are reflexes programmed by specific interconnections of nerve cells.

Recent research has demonstrated that insects are capable of some learning and have a memory. For example, bees (order Hymenoptera) instinctively recognize flowerlike objects by their shape and ability to absorb ultraviolet light, which makes the center of the flower appear dark. If a bee is rewarded with nectar and pollen, it learns other characteristics of the flower. Most important among these characteristics is odor, followed by color and shape. Bees that feed once at artificially scented feeders choose that odor in 90% of subsequent feeding trials. Odor is a very reliable cue for bees because it is more constant than color and shape. The latter may be damaged by wind, rain, and herbivores.

Sense organs of insects are similar to those found in other arthropods, although they are usually specialized for functioning on land. Some receptors, called *sensilla*, consist of sensory cells and modifications of the cuticle. Other receptors are located internally.

Mechanoreceptors perceive physical displacement of the body or one of its parts. *Setae* are distributed over the mouthparts, antennae, and legs (*see figure 22.10a*). Displacement of setae may occur as a result of touch, air movements, and vibrations of the substrate. Stretch receptors at the joints, on other parts of the cuticle, and on muscles monitor posture and position.

Hearing is a mechanoreceptive sense in which airborne pressure waves displace certain receptors. All insects can respond to pressure waves with generally distributed setae; others have specialized receptors. For example, **Johnston's organs** are found in the base of the antennae of most insects, including mosquitoes and midges (order Diptera). Antennae of these insects are covered with long setae that vibrate when struck by certain frequencies of sound. The vibrating setae cause the antenna to move in its socket, stimulating sensory cells. Sound waves in the frequency range of 500 to 550 cps attract and elicit mating behavior of male mosquitoes (*Aedes aegypti*). These waves are in the range of the sounds produced by the wings of females. **Tympanal organs** are found in the legs of crickets and katydids (order Orthoptera), in the abdomen of grasshoppers (order Orthoptera) and some moths (order Lepidoptera), and in the thorax of other moths. Tympanal organs consist of a thin, cuticular membrane covering a large air sac. The air sac acts as a resonating chamber. Just under the membrane are sensory cells that detect pressure waves. Grasshopper tympanal organs can detect sounds in the range of 1,000 to 50,000 cps. (The human ear can detect sounds between 20 and 20,000 cps.) Bilateral placement of tympanal organs allows insects to discriminate the direction and origin of a sound (box 23.1).

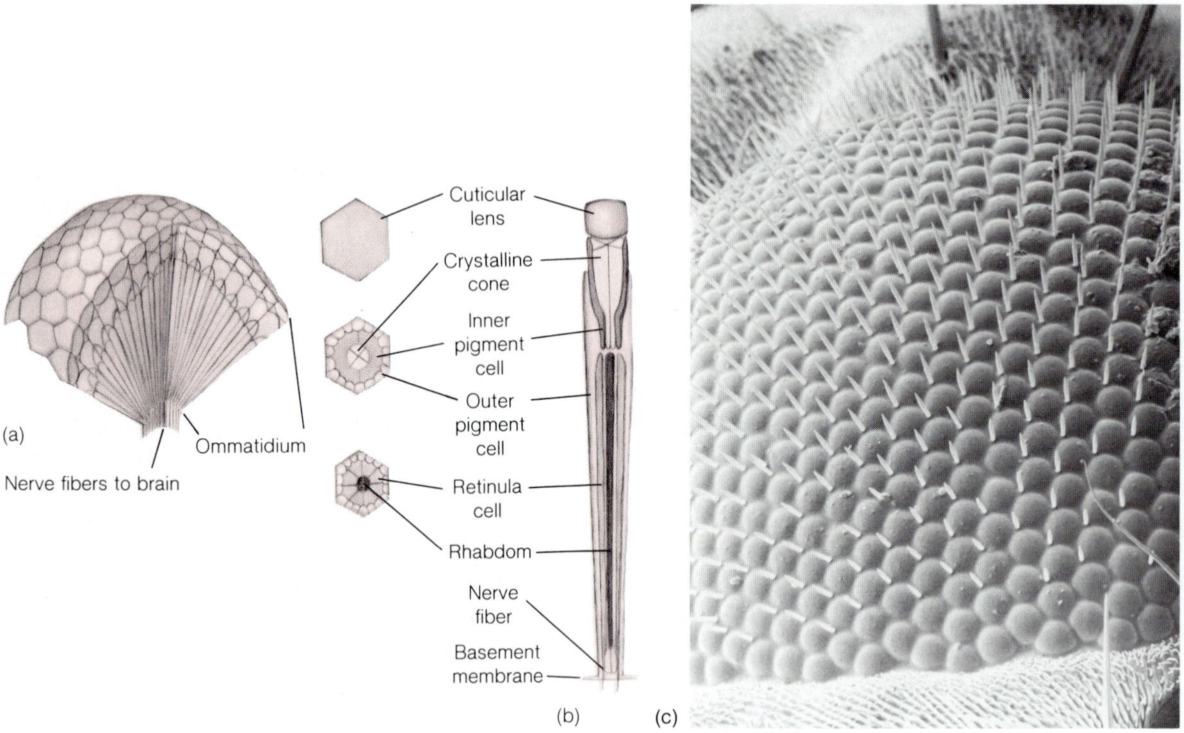

Figure 23.8 The compound eye of an insect. (a) Each facet of the eye is the lens of a single sensory unit called an ommatidium. (b) The structure of an ommatidium. The lens and the crystalline cone serve as light-gathering structures. Retinula cells have light-gathering areas, called the rhabdom. Pigment cells prevent light in one ommatidium from reflecting into adjacent ommatidia. In insects that are active at night, the pigment cells are often migratory, and pigment can be concentrated around the crystalline cone. In these insects, low levels of light from widely scattered points can excite an ommatidium. (c) Scanning electron micrograph of the compound eye of a fruitfly (*Drosophila*).

Figure 23.8 redrawn, with permission, from *Living Invertebrates* by Pearse/Buchsbaum, copyright by The Boxwood Press.

Chemoreception is used in feeding, selection of egg-laying sites, mate location, and in some insects, social organization. Chemoreceptors are usually abundant on the mouthparts, antennae, legs, and ovipositors and take the form of hairs, pegs, pits, and plates that have one or more pores leading to internal nerve endings. Chemicals diffuse through these pores and bind to and excite nerve endings. The role of intraspecific chemicals, called pheromones, is discussed later in this chapter.

All insects are capable of detecting light and may use light in orientation, navigation, feeding, or other functions. *Compound eyes* are well developed in most adult insects. They are similar in structure and function to those of other arthropods, although they probably are not homologous (of common ancestry) with those of crustaceans and chelicerates. Compound eyes consist of a few to 28,000 receptors, called **ommatidia**, that are fused into a multifaceted eye. The outer surface of each ommatidium is a *lens* and is one facet of the eye. Below the lens is a *crystalline cone*. The lens and the crystalline cone serve as light-gathering structures. Certain cells of an ommatidium, called *retinula cells*, have special light-collecting areas, called the *rhabdom*. The cells of the rhabdom convert light energy into

nerve impulses. *Pigment cells* surround the crystalline cone, and sometimes the rhabdom, and prevent the light that strikes one rhabdom from reflecting into an adjacent ommatidium (figure 23.8).

Although many insects form an image of sorts, the concept of an image has no real significance for most species. The compound eye is better suited for detecting movement. [5] Movement of a point of light less than 0.1° can be detected as light successively strikes adjacent ommatidia. For this reason, bees prefer flowers blowing in the wind, and predatory insects prefer moving prey. Compound eyes detect wavelengths of light that the human eye cannot detect, especially in the ultraviolet end of the spectrum. In some insects, compound eyes also detect polarized light, which may be used for navigation and orientation (box 23.2).

Ocelli consist of 500 to 1,000 receptor cells beneath a single cuticular lens (*see figure 22.10b*). The function of ocelli is not well understood. Their functions seem to complement the functions of compound eyes. For example, experimental covering of ocelli decreases the responsiveness of some insects to being shadowed, as might occur by a predatory bird.

Box 23.2

Communication in Honeybees

The exploitation of food sources by honeybees has been studied for decades, but its study still offers important challenges for zoologists. One of these areas of research concerns the extent to which honeybees communicate the location of food to other bees. A worker bee that returns to a hive laden with nectar and pollen stimulates other experienced workers to leave the hive and visit pollen and nectar sources that are productive. Inexperienced workers are also recruited to leave the hive and search for nectar and pollen, but stronger stimuli are needed to elicit their searching behavior. In the darkness of the hive, the incoming bee performs what researchers have described as a round dance and a waggle dance. Throughout the dancing, other workers contact the dancing bee with their antennae and mouthparts, picking up the odors associated with pollen, nectar, and other objects in the vicinity of the incoming bee's food source. During the dance, the incoming bee moves first in a semicircle to the left, then in a straight course to the starting point. Next, she follows a semicircle to the right, and then another straight course to the starting point. During the linear parts of the dance, the abdomen of the bee moves in a waggling fashion (box figure 23.2). These stimuli apparently encourage inexperienced workers to leave the hive and begin searching for food. As described in the text, their search relies heavily upon olfaction, and workers tend to be attracted to pollen and nectar similar to that brought back to the hive by the dancing bee.

The round and waggle dances may also convey information on location of a food source. Biologists have found that information regarding the direction and distance of a food source from the hive are contained in the dance. The angle that the waggle dance makes with the vertical of the comb approximates the angle between the sun and the food source (box figure 23.2). Similarly, the number of straight line runs per unit time, the duration of sounds made during the dance, and the number of waggles during the dance vary with the distance of the food source from the hive.

These observations indicate that bees communicate information regarding distance, direction, and kind of food to other bees when returning from a foraging trip. Thus, the exploitation of pollen and nectar is a very efficient process and is one source of evidence of the highly evolved nature of the honeybee colony.

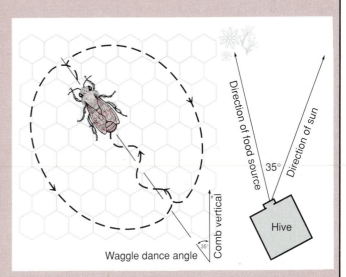

Box Figure 23.2   The waggle dance of the honeybee.

Excretion

Malpighian tubules and the rectum are the primary insect excretory structures. Malpighian tubules end blindly in the hemocoel and open to the gut tract at the junction of the midgut and the hindgut. The inner surface of their cells is covered with microvilli. Various ions are actively transported into the tubules, and water passively follows. Uric acid is secreted into the tubules and then into the gut tract, as are amino acids and salts, (figure 23.9). In the rectum, water, certain ions, and other materials are reabsorbed, and the uric acid is eliminated. As described in chapter 22, excretion of uric acid is advantageous for terrestrial animals because it is accompanied by little water loss. There is, however, an evolutionary trade-off to consider. The conversion of primary nitrogenous wastes (ammonia) to uric acid is energetically costly. It has been estimated that nearly one-half of the food energy consumed by a terrestrial insect is used to process metabolic wastes! In aquatic insects, ammonia simply diffuses out of the body into the surrounding water.

Chemical Regulation

Many physiological functions of insects, such as cuticular sclerotization, osmoregulation, egg maturation, cellular metabolism, gut peristalsis, and heart rate, are controlled by the endocrine system. As in all arthropods, ecdysis is under neuroendocrine control. In insects, the subesophageal ganglion and two endocrine glands, the *corpora allata* and the *prothoracic glands* control these activities.

Neurosecretory cells of the subesophageal ganglion manufacture *brain hormone* or *prothoracicotropic hormone*. Brain hormone travels in neurosecretory cells to a structure called the *corpora cardiaca*. It is released into the hemocoel by the corpora cardiaca and stimulates the prothoracic gland to secrete *ecdysone*. Ecdysone initiates the reabsorption of the inner portions of the procuticle and the formation of the new exoskeleton. Other hormones are also involved in ecdysis. The recycling of materials absorbed from the procuticle, changes in metabolic rates, and pigment deposition are a few of probably many functions controlled by hormones. Ecdysone of all arthropods is similar, although not identical. Experimental application of insect ecdysone will initiate ecdysis in other arthropods and visa versa.

Little is known of the external triggers that stimulate the release of brain hormone. A meal of blood and the resulting distension of the abdomen is the trigger that initiates molting in the bedbug, *Cimex* (order Hemiptera). In other insects, molting is regulated by changes in photoperiod or temperature.

During immature stages, the corpora allata produces and releases small amounts of *juvenile hormone*. The amount of juvenile hormone circulating in the hemocoel determines the nature of the next molt. Large concentrations of juvenile hormone result in a molt to a second immature stage. Intermediate concentrations of juvenile hormone result in a molt to a third immature stage. Low concentrations of juvenile hormone result in a molt to the adult stage. Decreases in the level of circulating juvenile hormone also lead to the degeneration of the prothoracic gland so that in most insects, no further molts occur once adulthood is reached. Interestingly, once the final molt has been accomplished, the level of juvenile hormone increases again, but now it promotes the development of accessory sexual organs, the synthesis of yolk, and the maturation of eggs.

Pheromones are chemicals released by an animal that cause behavioral or physiological changes in another member of the same species. Many different uses of pheromones by insects have been described (table 23.2). A pheromone is a molecule, or a specific mixture of molecules, that is produced by an insect or found in the insect's environment and modified. Pheromones are often so specific that the stereoisomer (chemical mirror image) of the pheromone may be ineffective in initiating a response. They may be carried several km by wind or water, and a few molecules falling on a chemoreceptor of another individual may be enough to elicit a response. [6]

All these characteristics of pheromone responses are illustrated by certain beetles (order Coleoptera) that live and breed in the bark of pine trees. The attack on a tree is a joint attack by large numbers of bark beetles and is advantageous because it weakens the tree making it unable to produce large quantities of a potentially toxic resin. The attack may be initiated by members of one sex, in which case, the beetles begin boring into the tree and releasing a pheromone that attracts other beetles. The first pheromone of pine bark beetles to be identified was a 1:1:1 mixture of three complex alcohols modified from tree products. The ratio of components in the mixture is critical to pheromone activity and is believed to help reduce potential confusion of an insect on encountering similar chemicals of tree, rather than beetle, origin. A responding beetle may follow an airborne pheromone trail for many km to the entrance of a burrow. Combinations of sex pheromones and auditory stimuli are emitted by a responding beetle at the entrance of a burrow occupied by a beetle

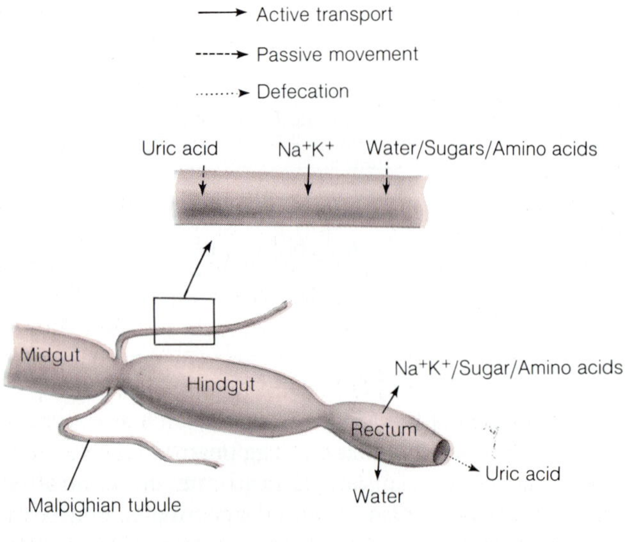

Figure 23.9 Malpighian tubules remove nitrogenous wastes from the hemocoel. Various ions are actively transported across the outer membrane of the tubule. Water follows these ions into the tubule and carries amino acids, sugars, and some nitrogenous wastes along passively. Some water, ions, and organic compounds are reabsorbed in the basal portion of the Malpighian tubules and the hindgut; the rest are reabsorbed in the rectum. Uric acid moves into the hindgut and is excreted.

Table 23.2   Functions of Insect Pheromones
Sex pheromones—Excite or attract members of the opposite sex; accelerate or retard sexual maturation. Female moths produce and release pheromones that attract males.
Caste regulating pheromones—Used by social insects to control the development of individuals in a colony. The amount of "royal jelly" fed a female bee larva will determine whether the larva will become a worker or a queen.
Aggregation pheromones—Produced to attract individuals to feeding or mating sites. Certain bark beetles aggregate on pine trees during an attack on a tree.
Alarm pheromones—Warn other individuals of danger; may cause orientation toward the pheromone source and elicit a subsequent attack or flight from the source. When one is stung by one bee, other bees in the area are alarmed and are likely to attack.
Trailing pheromones—Laid down by foraging insects to help other members of a colony identify the location and quantity of food found by one member of the colony. Ants can often be observed trailing to and from a food source. The pheromone trail is reinforced each time an ant travels over it.

that originally released the pheromone. These signals identify the species and sex of the responding beetle. If the responding beetle is the correct species and sex, and has arrived before another beetle, chemical and auditory stimuli from the beetle in the burrow welcome the responding beetle. Thus, the pheromones emitted by these beetles function as sex pheromones and aggregation pheromones. They are very specific mixtures of compounds, and can function over long distances.

Stop and Ask Yourself

9. What evidence indicates that bees are capable of learning?
10. Compound eyes are particularly well suited for what visual function?
11. What are the endocrine functions that regulate ecdysis in insects?
12. What are pheromones? What are three kinds of functions they serve in insects?

Reproduction and Development

One of the reasons for the success of insects is their high reproductive potential. Reproduction in terrestrial environments, however, has its risks. Temperature, moisture, and food supplies vary with the season. Fertilization requires highly evolved copulatory structures because gametes will dry quickly on exposure to air. In addition, mechanisms are required to bring males and females together at appropriate times.

Sexual maturity is regulated by complex interactions between internal and external environmental factors. Internal regulations include interactions between endocrine glands (primarily the corpora allata) and reproductive organs. External regulating factors may include the quantity and quality of food. For example, the eggs of mosquitoes (order Diptera) do not mature until after the female takes a meal of blood, and the number of eggs produced is proportional to the quantity of blood ingested. Photoperiod (the relative length of daylight and darkness in a 24-hour period) is used by many insects for timing reproductive activities, because it can be used to anticipate seasonal changes. Population density, temperature, and humidity also influence reproductive activities.

A few insects, including silverfish (order Thysanura) and spingtails (order Collembola) have indirect fertilization. The male deposits a spermatophore that is picked up later by the female. Most insects have complex mating behavior that is used to locate and recognize a potential mate, to position a mate for copulation, or to pacify an aggressive mate. Mating behavior may involve the use of pheromones, visual signals (fireflies, order Coleoptera) and auditory signals (cicadas, order Homoptera; and grasshoppers, crickets, and katydids, order Orthoptera). Once other stimuli have brought the male and female together, tactile stimuli from the antennae and other appendages help position the insects for mating.

Sperm transfer is usually accomplished by abdominal copulatory appendages of the male, and sperm are stored in an outpocketing of the female reproductive tract. Eggs are fertilized as they leave the female and are usually laid near the larval food supply. Females may use an ovipositor to deposit eggs in or on some substrate.

Insect Development and Metamorphosis

Insect evolution has resulted in the divergence of immature and adult body forms and habits. For insects in the superorder Endopterygota (*see table 23.1*), immature stages, called **instars,** have become a time of growth and accumulation of reserves for the transition to adulthood. The adult stage is associated with reproduction and dispersal. In these orders, there is a tendency for insects to spend a greater part of their lives in juvenile stages. The developmental patterns of insects reflect degrees of divergence between immatures and adults and are classified into three (or sometimes four) categories.

In insects that display **ametabolous** (Gr. *a,* without + *metabolos,* change) **metamorphosis,** the primary differences between adults and larvae are body size and degree of sexual development. Both adults and larvae are wingless. The number of molts in the ametabolous development of a species is variable and, unlike most other insects, molting continues after sexual maturity is reached. Silverfish (order Thysanura) have ametabolous metamorphosis.

Paurometabolous (Gr. *pauros,* small) **metamorphosis** involves a species-specific number of molts between egg and adult stages, during which immatures gradually take on the adult form. Occasionally, this development is spread over many years (box 23.3). The external development of wings (except in those insects, such as lice, that have secondarily lost wings), the attainment of adult body size and proportions, and the development of genitalia occurs during this time. Immatures are called **nymphs.** Grasshoppers (order Orthoptera) and milkweed bugs (order Hemiptera) show paurometabolous metamorphosis (figure 23.10).

Some authors use an additional classification for insects that have a series of gradual changes in their development, but whose immature form is much different from the adult form (e.g., mayflies, order Ephemeroptera; dragonflies, order Odonata). This kind of development is called **hemimetabolous** (Gr. *hemi,* half) **metamorphosis,** and the immatures are called **naiads** (L. *naiad,* water nymph).

In **holometabolous** (Gr. *holos,* whole) **metamorphosis,** immatures are called *larvae* because they are very different from the adult in body form, behavior, and habitat (figure 23.11). There is a species-specific number of larval instars, and the last larval molt forms the **pupal stage.** The pupal stage is a time of apparent inactivity but is actually a time of radical cellular change, during which all characteristics of the adult insect take form. The pupal stage may be enclosed in a protective case. A **cocoon** is constructed partially or entirely from silk by the last larval instar (e.g., moths, order Lepidoptera). The **chrysalis** (e.g., butterflies, order Lepidoptera) and **puparium** (e.g.,

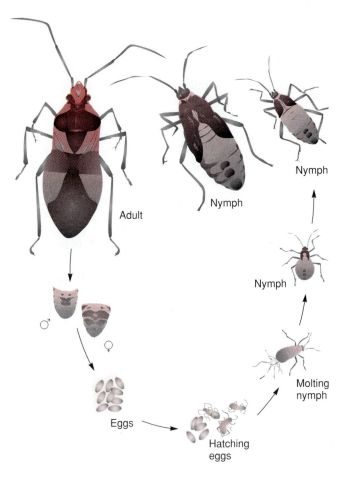

Figure 23.10 Paurometabolous development of the milkweed bug, *Oncopeltus fasciatus* (order Hemiptera). Eggs hatch into nymphs. Note the gradual increase in size of the nymphs and the development of external wing pads. In the adult stage, the wings are fully developed, and the insect is sexually mature.

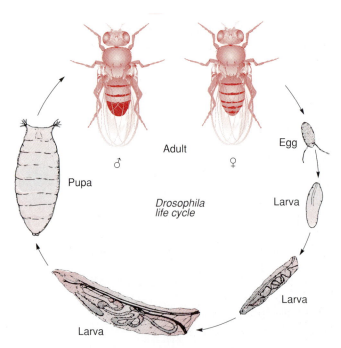

Figure 23.11 Holometabolous development of the fruit fly, *Drosophila melanogaster* (order Diptera). The egg hatches into a larva that is very different in form and habitat from the adult. After a certain number of larval instars, the insect pupates. During the pupal stage, all characteristics of the adult are formed.

flies, order Diptera) are the last larval exoskeletons and are retained through the pupal stage. Other insects (e.g., mosquitoes, order Diptera) have pupae that are unenclosed by a larval exoskeleton, and the pupa may be active. The final molt to the adult stage usually occurs within the cocoon, chrysalis, or puparium, and the adult then exits, frequently using its mandibles to open the cocoon or other enclosure. This final process is called *emergence* or *eclosion*.

Insect Behavior

Insects have many complex behavior patterns. Most of these are innate (genetically programmed). For example, a newly emerged queen in a honeybee hive will search out and try to destroy other queen larvae and pupae in the hive. This behavior is innate because no experiences taught the potential queen that her survival in the hive required the death or dispersal of all other potential queens. Similarly, no experience taught her how queen-rearing cells differ from the cells containing worker larvae and pupae. Some insects are capable of learning and remembering and these abilities play important roles in insect behavior. (Recall the description of how bees learn to recognize certain flowers.)

Social Insects The evolution of social behavior has occurred in many insects, and is particularly evident in those insects that live in colonies. Usually, different members of the colony are specialized, often structurally as well as behaviorally, for performing different tasks. Social behavior is most highly evolved in the bees, wasps, and ants (order Hymenoptera) and termites (order Isoptera). Each kind of individual in an insect colony is called a **caste.** Often, three castes are present in a colony. Reproductive females are referred to as *queens. Workers* may be male or female and are involved with support, protection, and maintenance of the colony. Their reproductive organs are often degenerate. Reproductive males inseminate the queen(s) and are called *kings* or *drones.*

Honeybees (order Hymenoptera) have all three of the above castes in their colonies (figure 23.12). A single queen lays all the eggs. Workers are female and they construct the comb out of wax that they produce. They also gather nectar and pollen, feed the queen and drones, care for the larvae, and guard and clean the hive. These tasks are divided among workers according to age. Younger workers take care of jobs around the hive, and older workers forage for nectar and pollen. Except for those that overwinter, workers live for about 1 month. Drones develop from unfertilized eggs, do not work, and are fed by workers until they leave the hive to attempt mating with a queen.

(a)

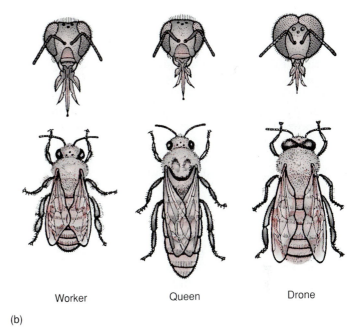

Worker   Queen   Drone

(b)

The honeybee caste system is controlled by pheromones released by the queen. Workers lick and groom the queen and other workers. In so doing, they pick up and pass to other workers a caste-regulating pheromone. This pheromone inhibits the rearing of new queens by workers. As the queen ages, or if she dies, the amount of caste-regulating pheromone in the hive decreases. As the pheromone decreases, workers begin to feed the food for queens (''royal jelly''), to several female larvae that are developing in the hive. This food contains chemicals that promote development of queen characteristics. The larvae that receive royal jelly develop into queens, and as they emerge, the new queens begin to eliminate each other until only one remains. The queen that remains goes on a mating flight and returns to the colony, where she will live for several years.

The evolution of social behavior with many individuals leaving no offspring and where individuals are sacrificed for the perpetuation of the colony has puzzled evolutionists for many years. It may be explained by the concepts of *kin selection* and *altruism,* which will be discussed in chapter 40.

Insects and Humans

Only about 0.5% of insect species affect human health and welfare. Many have provided valuable services throughout human history. Commercially valuable insect products, such as wax, honey, and silk have been utilized by humans for thousands of years (box 23.4). Insects are responsible for the pollination of approximately 65% of all plant species. Insects and flowering plants have coevolutionary relationships that directly benefit humans. The annual value of insect-pollinated crops is estimated in the billions of dollars.

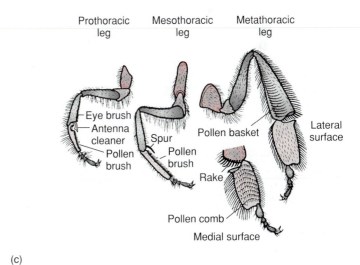

(c)

Figure 23.12 Honeybees (order Hymenoptera). (a) A honeybee (*Apis mellifera*). (b) Honeybees have a social organization consisting of three castes. The castes are distinguished by overall body size, as well as the size of their eyes. (c) The prothoracic legs of a worker bee are used to clean pollen from the antennae and body. The mesothoracic legs are used for gathering pollen from body parts. The inner surface of metathoracic legs have setae, called the pollen comb, that remove pollen from the mesothoracic legs and the abdomen. Pollen is then compressed into a solid mass by being squeezed in a pollen press and moved to a pollen basket on the outer surface of the leg, where the pollen is carried.

Figure 23.12b and 23.12c redrawn, with permission, from *Living Invertebrates* by Pearse/Buchsbaum, copyright by The Boxwood Press.

Box 23.3

The Longest-Lived Insects

Many of the readers of this book are 20 years of age or younger. For humans, that is a life just beginning, but for most insects 20 years spans 20, 40, or more lifetimes. For the 17-year "locust," however, 20 years represents approximately one lifetime. These longest-lived of all insects are not really locusts, but are cicadas and belong to the order Homoptera (box figure 23.3). (True locusts are short-horned grasshoppers in the order Orthoptera.) Cicadas were first referred to as "locusts" in 1634 by Pilgrims who thought they were the same kind of insects that were responsible for plagues described in the Bible.

The reason for the mistaken identity is that both kinds of insects can appear very quickly in outbreak proportions and they are frequently gone as suddenly as they appear. True locust plagues come and go quickly because the grasshoppers are migratory. While they are in an area, food crops can be devastated. Cicadas, on the other hand, emerge from the ground for a brief adult existence. After mating, female cicadas lay eggs in the twigs of trees, which they slit with their ovipositors. Adults die after mating and do not feed. After 1 or 2 months of development in the twig, the nymphs drop to the ground, burrow into the soil, and begin to feed on juices that they extract from the roots of the tree. After passing through five nymphal stages, immature cicadas emerge many years later through emergence tunnels in the soil and molt to the adult form on the trunk of the tree.

The longest-lived cicadas are the periodic cicadas. The emergence of northern broods of *Magicicada* occurs 17 years after the eggs are laid, and the emergence of southern broods occurs 13 years after the eggs are laid. In any locality, the emergence is impressively coordinated, and in the span of a few days, the air is filled with the call of cicadas looking for mates. Emergences are not coordinated over large geographical ranges, however. Throughout North America there are cicada broods emerging nearly every year. The method by which emergences are coordinated is poorly understood. Soil temperature seems to be a crucial factor in influencing emergence in the final year.

A number of hypotheses have been proposed to explain why periodicity evolved in cicadas. A popular hypothesis is that during mass emergences, the risk for any one cicada of capture by a predator decreases as cicada density increases. Predators, chiefly birds, may dine gluttonously for a few weeks, but because prey are around for only a short time every 13th or 17th year, predators cannot depend upon this food source. The likelihood of an individual cicada avoiding attack by a predator may be greater in a situation of high prey density.

Not all cicadas are periodic. Some species have members that emerge each year in a locality. These species have life cycles that generally last fewer than 10 years.

Box Figure 23.3 The periodic cicada, *Magicicada septendecim.*

Insects also serve as agents of biological control. The classic example of the regulation of a potentially harmful insect by another insect is the control of cottony-cushion scale by vedalia (lady bird) beetles. The scale insect, *Icerya purchasi,* was introduced into California in the 1860s. Within 20 years, the citrus industry in California was virtually destroyed. The vedalia beetle (*Vedalia cardinalis*) was brought to the United States in 1888 and 1889 and cultured on trees infested with scale. In just a few years, the scale was under control, and the citrus industry began to recover.

Biological controls are often not noticed until a nonselective pesticide is applied over an area, and the natural balances between predator and prey are disturbed. This situation occurred in California when DDT was used to kill codling moths in walnut orchards. In addition to killing codling moths, the insect predator of frosted scale was also killed. Frosted scale populations increased to outbreak proportions.

There are many other beneficial insects. Soil-dwelling insects play important roles in aeration, drainage, and turnover of soil, and they promote decay processes. Other insects serve important roles in food webs. Insects are used in teaching and research and have contributed to advances in genetics, population ecology, and physiology. Insects have also given endless hours of pleasure to those who collect them and enjoy their beauty.

Box 23.4

"Killer Bees?"

The "African bee," *Apis mellifera andersoni,* is common in most parts of Africa. It is a small bee and is adapted to warm climates with extended dry seasons. In its home range, it has many enemies, including humans and birds. Aggressive behavior and frequent swarming have allowed colonies to escape predation and survive drought.

In 1956, a few *Apis mellifera andersoni* queens were imported to Brazil in hopes of breeding these queens with local stocks to create bees better adapted to tropical climates than were the local bees. (Bees of the Americas were imported from Europe in the 1600s.) A few of these queens escaped captivity and hybridized with local bees. (It is really not accurate to refer to the hybridized bees as "African bees," "Africanized bees," or "killer bees." Their reputation as "killers" has been exaggerated. Most authorities now refer to hybridized bees as Brazilian bees.) Many of the qualities that allowed the African bee to succeed in Africa allowed these Brazilian bees to spread 100 to 200 mi. per year. Their frequent swarming and their ability to nest in relatively open shelters gives a distinct selective advantage over the bees imported

from Europe many years earlier. By 1969, Brazilian bees had spread to Argentina; by 1973, to Venezuela; and by early 1980, they had crossed the Panama Canal. Mexico presented few barriers to their spread toward the southern United States, and they are now approaching southern California and southern Texas. How far will Brazilian bees ultimately spread? An educated guess may be made by looking at the distribution of the African bee in Africa. African bees cannot overwinter outside of tropical regions. Similarly, the Brazilian bee will likely be limited in America to the warmer latitudes.

The primary implications of the spread of Brazilian bees has less to do with threats to human health than to the health of the beekeeping industry. The unpleasant temperament of these bees can be dealt with by wearing protective clothing. More formidable problems associated with the culture of these bees include their tendency to swarm and their lower productivity. Swarming for a beekeeper means the loss of bees and a lowered honey production. Frequent swarming makes profitable beekeeping almost impossible.

Some insects, however, are parasites and vectors of disease. Parasitic insects include body and pubic lice (order Anoplura). Other insects transmit disease-causing microorganisms, nematodes, and flatworms. The impact of insect-transmitted diseases, such as malaria, yellow fever, bubonic plague, encephalitis, leishmaniasis, and typhus, has changed the course of history (*see box 16.1*).

Other insects are pests of domestic animals and plants. Some reduce the health of domestic animals and the quality of animal products. Insects feed on crops and transmit diseases of plants, such as Dutch elm disease, potato virus, and asters yellow. Worldwide, annual lost revenue from insect damage to crops or insect-transmitted diseases, is approximately 3 billion dollars.

Further Phylogenetic Considerations

Phylogenetic relationships within the arthropods are not uniformly agreed upon. The traditional approach has been to view the arthropods as a single phylum, which implies a common (monophyletic) origin. For many years, the phylum Arthropoda was divided into three subphyla: Trilobita, Chelicerata, and Mandibulata. The latter two designations were descriptive of the mouthparts found in those groups.

The crustaceans, insects, and myriapods possess mandibles, whereas the arachnids and the merostomates possess chelicerae. The use of these subphylum designations implies that all arthropod mandibles had a common evolutionary origin, and that the biramous appendages of crustaceans had a common origin with the uniramous appendages of insects. It is now clear, based upon both embryological and paleontological studies, that at least the former assumption, and probably the latter as well, is wrong. Although the mandibles of insects and myriapods are homologous, they are not homologous to crustacean mandibles. The presence of mandibles in crustaceans and in insects and myriapods is an example of convergent evolution. Thus, the mandibulate arthropods are more accurately divided into two subphyla: Crustacea and Uniramia.

A related, and perhaps more fundamental, question that is debated concerns whether or not the arthropod taxa represent fundamentally different evolutionary lineages. Many zoologists believe that the living arthropods should be divided into three separate phyla: Chelicerata, Crustacea, and Uniramia. A polyphyletic origin of these groups implies convergent evolution of remarkably similar arthropodan features in all three (or at least two of three) phyla. Although there is evidence for dual origins of tracheae, mandibles, and compound eyes, convergence in all other arthropod traits seems unlikely to many zoologists.

Most zoologists view the trilobites as an important ancestral group. It is possible to envision crustaceans and arachnids arising from the trilobites, but it is more difficult to envision a similar origin for myriapods and insects. Questions regarding the evolutionary relationships within the arthropods are difficult to answer and will probably remain unanswered until new fossils are discovered or data from molecular studies provide more information on ancestral arthropods.

Stop and Ask Yourself

13. What is the sequence of changes that occurs in the life history of a holometabolous insect?
14. What are three castes in a honeybee hive and how is the caste system regulated?
15. What is the nature of the debate regarding arthropod phylogeny?

Summary

1. During the Devonian period, insects began to exploit terrestrial environments. Flight, the exoskeleton and metamorphosis, are probably keys to insect success.
2. Myriapods include four classes of arthropods. Members of the class Diplopoda are the millipedes and are characterized by double trunk segments. The centipedes are in the class Chilopoda. They are characterized by a single pair of legs on each of their 15 segments and a body that is oval in cross section. The class Pauropoda contains soft-bodied animals that feed on fungi and decaying organic matter in forest-floor litter. Members of the class Symphyla are centipedelike arthropods that live in soil and leaf mold, where they feed on decaying vegetation.
3. Animals in the class Insecta are characterized by a head with one pair of antennae, compound eyes, and ocelli; a thorax with three pairs of legs and usually two pairs of wings; and an abdomen that is mostly free of appendages.
4. Insect flight involves a direct (synchronous) flight mechanism or an indirect (asynchronous) flight mechanism.
5. Mouthparts of insects are adapted for chewing or sucking, and the gut tract may be modified for pumping, storage, digestion, and water conservation.
6. Gas exchange occurs through a tracheal system.
7. The insect nervous system is similar to that of other arthropods. Sensory structures include tympanal organs, compound eyes, and ocelli.
8. Malpighian tubules transport uric acid to the digestive tract. Conversion of nitrogenous wastes to uric acid conserves water but is energetically expensive.
9. Hormones regulate many insect functions, including ecdysis and metamorphosis. Pheromones are chemicals emitted by one individual that alter the behavior of another member of the same species.
10. Insect adaptations for reproduction on land include resistant eggs, external genitalia, and mechanisms that bring males and females together at appropriate times.
11. Metamorphosis of an insect may be ametabolous, paurometabolous, hemimetabolous, or holometabolous. Metamorphosis is controlled by neuroendocrine and endocrine secretions.
12. Insects show both innate and learned behavior.
13. Relatively few insect species affect humans. Some are beneficial to humans, and a few are parasites and/or transmit diseases of humans or agricultural products.
14. Whether the arthropods represent a monophyletic group or a polyphyletic group is a question that is still being debated.

Key Terms

ametabolous metamorphosis (p. 373)
caste (p. 374)
chrysalis (p. 373)
cocoon (p. 373)
direct or synchronous flight (p. 366)
hemimetabolous metamorphosis (p. 373)
holometabolous metamorphosis (p. 373)
indirect or asynchronous flight (p. 366)
instars (p. 373)
Johnston's organs (p. 369)
mesothorax (p. 365)
metathorax (p. 365)
naiads (p. 373)
nymphs (p. 373)
ommatidia (p. 370)
paurometabolous metamorphosis (p. 373)
prothorax (p. 365)
pupal stage (p. 373)
puparium (p. 373)
tympanal organs (p. 369)

Critical Thinking Questions

1. What are the problems associated with living and reproducing in terrestrial environments? Explain how insects overcome these problems.
2. List as many examples as you can think of where insects communicate with each other. In each case, what is the form and purpose of the communication?
3. In what way does holometabolous metamorphosis reduce competition between immature and adult stages?
4. What roles do larvae play in the life history of holometabolous insects?
5. What aspect of the evolution of insect social behavior has been difficult to explain?
6. Some biologists think that the arthropods are a polyphyletic group. What does that mean? What would polyphyletic origins require in terms of the origin of the exoskeleton and its derivatives?

■ **Suggested Readings**

Books

Borror, D. J., Tripplehorn, C. A., and Johnson, N. F. 1989. *An Introduction to the Study of Insects*. 6th ed. Philadelphia: Saunders College Publishing.

DeBach, P. 1974. *Biological Control by Natural Enemies*. New York: Cambridge University Press.

Dethier, V. G. 1976. *The Hungry Fly*. Cambridge: Harvard University Press.

Gillot, Cedric. 1980. *Entomology*. New York: Plenum Press.

Horridge, G. A. (ed.) 1975. *The Compound Eye and Vision of Insects*. Atlanta: Clarendon Group, Inc.

Jaques, H. E. 1974. *How to Know the Insects*. 2nd ed. Dubuque: Wm. C. Brown Company Publishers.

Matthews, T. W. and Matthews, J. R. 1978. *Insect Behavior*. New York: John Wiley & Sons, Inc.

McCafferty, W. P. 1981. *Aquatic Entomology: A Fisherman's and Ecologist's Illustrated Guide to Insects and Their Relatives*. Providence: Science Books International.

Wigglesworth, V. B. 1982. *Principles of Insect Physiology*. 7th ed. New York: John Wiley & Sons, Inc.

Wilson, E. O. 1971. *The Insect Societies*. Cambridge: Harvard University Press.

Articles

Buck, J., and Buck, E. Synchronous fireflies. *Scientific American* May, 1976.

Camhi, J. M. The escape system of the cockroach. *Scientific American* December, 1980.

Eberhard, W. G. Horned beetles. *Scientific American* March, 1980.

Franks, N. 1989. Army ants: a collective intelligence. *American Scientist* 77(2):138–145.

Funk, D. H. The mating of tree crickets. *Scientific American* August, 1989.

Heinrich, B., and Bartholomew, G. A. The ecology of the African dung beetle. *Scientific American* November, 1979.

Holldobler, B. 1976. Tournaments and slavery in a desert ant. *Science* 192:912–914.

Holldobler, B., and Haskins, C. P. 1977. Sexual calling behavior in primitive ants. *Science* 195:793–794.

Holdobler, B. K., and Wilson, E. O. Weaver ants. *Scientific American* December, 1977.

Horridge, G. A. The compound eye of insects. *Scientific American* July, 1977.

Jones, J. C. The feeding behavior of mosquitoes. *Scientific American* June, 1978.

Manzel, R., and Erber, J. Learning and memory in bees. *Scientific American* July, 1978.

McMasters, J. H. 1989. The flight of the bumblebee and related myths of entomological engineering. *American Scientist* 72(2):164–169.

Milne, L. J., and Milne, M. The social behavior of burying beetles. *Scientific American* August, 1978.

Milne, L. J., and Milne, M. Insects of the water surface. *Scientific American* April, 1978.

Schneider, D. The sex-attractant receptors of moths. *Scientific American*, July, 1974.

Seeley, T. D. How honeybees find a home. *Scientific American* October, 1982.

Simmons, L. W. 1988. The calling song of the field cricket, *Gryllus bimaculatus* (De Geer): constraints on transmission and its role in intermale competition and female choice. *Animal Behaviour* 36:380–394.

Von Frisch, K. 1974. Decoding the language of the bee. *Science* 185:663–668.

Waterhouse, D. F. The biological control of dung. *Scientific American* April, 1974.

Wright, R. H. Why mosquito repellents repel. *Scientific American* July, 1975.

24

Other Eucoelomate Phyla

Concepts

1. The three lophophorate phyla are eucoelomate (possess a true coelom), bilateral, protostome animals. These phyla (Brachiopoda, Ectoprocta, Phoronida) are grouped together because they possess a crown of tentacles called a lophophore. The lophophore surrounds only the mouth and is specialized for sedentary filter feeding. Members of the Brachiopoda and Phoronida possess a vascular system that helps distribute nutrients and gases. Phoronids also have blood cells that contain hemoglobin for transporting oxygen.
2. The lesser protostome phyla (Echiura, Pentastomida, Onychophora, Pogonophora, Sipuncula, Tardigrada) are all eucoelomate protostomes; however, some have deuterostome characteristics in their embryonic stages. They are grouped together because they all probably evolved from the same annelid/arthropod line even though their relationships to each other are obscure.

Have You Ever Wondered:

[1] why brachiopods are commonly called lampshells?
[2] why bryozoans are called moss animals?
[3] why echinurans are commonly called spoon worms?
[4] what is unique about pogonophores called beard worms?
[5] why sipunculans are called peanut worms?
[6] what animals may be an evolutionary transition between annelids and arthropods?
[7] how humans become infected with tongue worms?
[8] what water bears are?
[9] how some animals can enter a period of suspended animation?

These and other useful questions will be answered in this chapter.

This chapter contains underlined evolutionary concepts.

Evolutionary Perspective

The common food-catching organ (lophophore) that the Brachiopoda, Ectoprocta, and Phoronida possess offers some evidence for their evolutionary similarity; however, each phylum probably evolved independently as evidenced by their unique life-styles. As do deuterostomes, these phyla, the lophophorates, have a coelom that is divided into compartments. The lack of definitive head structures is probably the result of their sessile filter feeding. Their embryology shows the deuterostome characteristic of radial clevage, and the protostome characteristic of a blastopore that becomes the mouth. Some zoologists thus consider the lophophorates phylogenetically intermediate between deuterostomes and protostomes.

All three phyla are sessile (attached). They have a reduced head, possess a U-shaped digestive tract, secrete a protective covering, and have members with a trochophore larval stage. Because the above characteristics are correlated with a sessile existence, the fact that these phyla share them may represent evolutionary convergence rather than a close evolutionary relationship.

The six phyla grouped together as the lesser protostomes (Echiura, Pogonophora, Sipuncula, Onychophora, Pentastomida, and Tardigrada) are all eucoelomate protostomes; however, deuterostome characteristics exist in their embryonic stages. The nephridia, thin flexible cuticle, nonjointed appendages, and the structure of their body wall suggest a relationship to the annelids. Conversely, the presence of a reduced coelom, a chitinous cuticle that is molted, modified feeding appendages, an open circulatory system with a dorsal heart, and tracheae for gas exchange suggest a relationship to the arthropods. Because their relationship to each other and to the major protostome phyla is problematic, they probably evolved at different times from annelid/arthropod ancestors.

The Lophophorates

As noted above, the three phyla (Brachiopoda, Ectoprocta [Bryozoa], Phoronida) discussed in this section have one major homologous anatomical feature—the **lophophore** (lo′fe-for) (Gr. *lophos*, crest or tuft + *phorein*, to bear). The lophophore is a circumoral (around the mouth) body region characterized by a circular or U-shaped ridge with either one or two rows of ciliated, hollow tentacles (figure 24.1). The anus, when present, lies outside the circle of tentacles. The coelom is divided into two compartments: the metocoel and mesocoel. The *metocoel* is subdivided by septa and extends into the mantle (the soft extension of the body wall) to form a series of *mantle canals* (ciliated tubules) that end near the outer edge of the mantle. The *mesocoel* extends into the hollow tentacles of the lophophore. The lophophore functions as a food-collecting organ, a surface for gas exchange, and it has sensory cells

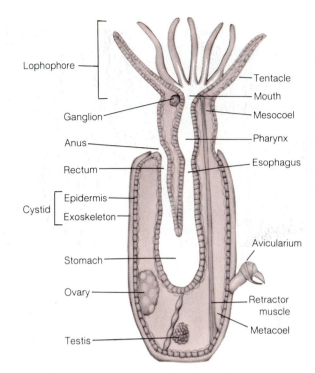

Figure 24.1 An ectoproct zooid with the lophophore extended.

receptive to chemicals and touch that are concentrated on its tentacles. The lophophore can usually be extended for feeding or withdrawn for protection. All lophophorates are sessile or sedentary filter feeders and live in a secreted chitinous or calcareous tube. Nearly all of the approximately 4,500 species are marine; a few ectoprocts are commonly found in fresh water.

Phylum Brachiopoda: The Brachiopods, or Lampshells

The brachiopods (brak-i-op′ods) (Gr. *brachion*, arm + *podos*, foot) bear a superficial resemblance to the bivalve molluscs because they have a *bivalved, calcareous* and/or *chitinous shell* that is secreted by a mantle that encloses nearly all of the body. However, unlike the left and right valves in molluscs, the brachiopods have dorsal and ventral valves. In addition, molluscs filter with their gills, whereas brachiopods use a lophophore. Brachiopods are commonly called **lampshells** because they resemble ancient Roman oil lamps. [1]

Brachiopods are exclusively marine; most species live from the tidal zone to the edge of the continental shelves (about 200 m deep). There are about 300 living species in the phylum, which contains two classes (table 24.1). In the Articulata, the valves are composed primarily of calcium carbonate and have a hinge with interlocking teeth (figure 24.2a). The Inarticulata have unhinged valves that are composed primarily of calcium phosphate (24.2b). The Inarticulata valves are held together only by muscles. Most members of both classes have a stalked *pedicel* that is usually attached to a hard surface. Some, such as *Lingula,* have

(a)

(b)

Figure 24.2 Phylum Brachiopoda. (a) Articulata brachiopod opened to show the two attached valves. (b) Inarticulata brachiopods (*Lingula cuncata*) in feeding position in their burrows.

Table 24.1   Classification of the Brachiopoda
Phylum Brachiopoda (brak-i-op'o-da). Attached, often stalked marine lophophorates with dorsal and ventral valves. Lamp shells.
Class Articulata (ar-tic-yu'la-ta) Valves locking together with a hinge; lophophore usually with internal skeleton; no anus. *Lacazella, Hemithyris, Magellania.*
Class Inarticulata (in-ar-tic'yula-ta) Lack a hinge between the valves; no internal skeleton in the lophophore; anus present. *Lingula, Glottidia.*

a muscular pedicel used for burrowing and anchoring in mud or sand.

The large horseshoe-shaped lophophore in the anterior mantle cavity bears long, ciliated tentacles used in respiration and feeding (figure 24.3). The cilia set up water currents that carry food particles (mainly organic detritus and algae) between the valves and over the lophophore into the mouth.

There are two types of digestive systems in the brachiopods. The Articulata have a blind intestine and no anus. Undigested materials are periodically expelled from the mouth into the mantle cavity. The Inarticulata have an intestine that opens posteriorly to the outside via an anus.

The coelom in brachiopods contains one or two pairs of ciliated nephridia that open into the coelom and act to excrete metabolic wastes and expel gametes.

Most brachiopods have an open circulatory system that consists of a small contractile heart suspended in the coelom above the stomach. Some of the cells (*coelomocytes*) in the coelomic fluid contain hemerythrin, a respiratory pigment that transports oxygen.

The nervous system includes a small ganglion around the esophagus and several nerves extending from the ganglion. There are no well-developed sense organs, although the edge of the mantle has sensory lobes that detect touch and chemicals in the water.

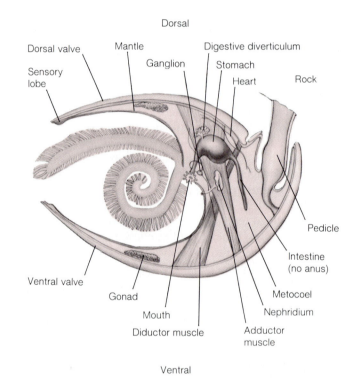

Figure 24.3 The internal anatomy of an articulate brachiopod. Contraction of the adductor muscles closes the valves; contraction of the diductor muscles raises the dorsal valve.

Figure 24.3 redrawn, with permission, from *Living Invertebrates* by Pearse/Buchsbaum, copyright by The Boxwood Press.

A solitary, dioecious brachiopod reproduces sexually by releasing gametes from multiple gonads into the metacoel and discharging them into the water by the nephridia. Fertilization is usually external, and the nonfeeding, ciliated, free-swimming larva (which resembles the trochophore larva) is planktonic before settling and developing into an adult. Development is similar to deuterostomes, with radial, mostly equal, holoblastic cleavage, and enterocoelous coelom formation. In the Inarticulata, the juvenile resembles a small brachiopod with a coiled pedicel in the mantle

Table 24.2 Classification of the Ectoprocta

Phylum Ectoprocta (ek-to-prok′ta); (Bryozoa) (bry-o-zo′a).
Freshwater and marine lophophorates that form encrusting, branching colonies with an exoskeleton.

Class Stenolaemata (sten-o-la′mat-a)
Exclusively marine with narrow zooids and a circular lophophore; all with calcified body walls; one living order. *Crisia, Lichenopora, Tubulipora.*

Class Gymnolaemata (gym-no-la′mat-a)
Mainly marine with cylindrical zooids and a circular lophophore. Two orders. *Victorella, Bugula, Electra.*

Class Phylactolaemata (phy-lak-to′la-mat-a)
Exclusively in fresh water with cylindrical zooids and a horseshoe-shaped lophophore. *Cristatella, Fredericella, Plumatella.*

(a)

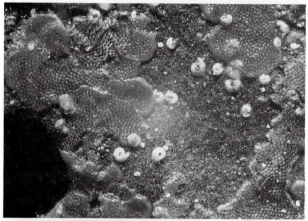
(b)

cavity. There is no metamorphosis; when the juvenile settles to the bottom, the pedicel attaches to a solid object, and adult existence begins.

Stop and Ask Yourself

1. How do brachiopods differ from molluscs?
2. What is the difference between the Articulata and Inarticulata?
3. What is the function of a lophophore?
4. How do brachiopods reproduce?

Phylum Ectoprocta (Bryozoa): The Ectoprocts, or Moss Animals

The ectoprocts (ek-to-proks) (Gr. *ektos,* outside + *proktos,* anus) or bryozoans superficially resemble hydroids or corals. Bryozoa (Gr. *bryon,* moss + *zoon,* animal) means
[2] **moss animals** and refers to the mosslike appearance of the colonies. The name ectoprocta is used to distinguish this group of eucoelomate animals, with the anus located outside the ring of tentacles, from the entoprocts in which the anus is within the ring of tentacles (*see figure 19.20*). There are about 4,000 living species of ectoprocts belonging to three classes (table 24.2). All species are aquatic (both freshwater and marine) and less than 1.5 mm in length.

Each body, or **zooid,** has a circular or horsehoe-shaped lophophore and is covered with a thin cuticle that encloses a calcified exoskeleton (*see figure 24.1*). The feeding body (lophophore, digestive tract, muscles, nerves) is called the *polypide,* the exoskeleton plus body wall (epidermis) is the *cystid,* and the secreted, nonliving part (exoskeleton) is the *zooecium* (Gr. *zoo,* animal + *oceus,* house). Ectoprocts have an eversible lophophore that can be withdrawn into the body. Contraction of the retractor muscle rapidly withdraws the lophophore, whereas contraction of the muscles encircling the body wall exert pressure on the coelomic fluid, everting the lophophore.

(c)

Figure 24.4 Phylum Ectoprocta. Ectoprocts have a wide variety of shapes. (a) Upright branching form attached at a base (*Crisia*). (b) Thin encrusting sheets (*Eurystomella*). (c) Lacy, delicate folds (*Philolopora*).

Ectoprocts grow by budding. Thin portions of the body wall grow out as small vesicles or tubes and contain a complete zooid. The different budding patterns reflect the genetics of the individual animal and factors such as current flow and substrate. These factors determine the colony shape (e.g., thin sheets, convoluted folds, massive corallike heads, upright tangles) (figure 24.4). Each colony can contain about two million zooids.

When feeding, ectoprocts extend the lophophore and spread the tentacles out in the shape of a funnel. Cilia on the tentacles collect food particles from the overlying water and direct them into the mouth. The ingested material passes through a U-shaped gut. Digestion occurs in the relatively large stomach, and has both extra- and intracellular components. However, unlike the brachiopods and phoronids (discussed below), ectoprocts lack circulatory and excretory systems. Gaseous exchange is across the body surface. The nervous system consists of a dorsal ganglion, a circumpharyngeal nerve ring, and a subepidermal nerve net. No sense organs are present. In some species, small nerves may extend through exoskeletal pores, allowing some coordination between adjacent zooids. In addition, these pores allow the exchange of materials between adjacent zooids via the coelomic fluid.

As do hydroids and some other colonial organisms (*see chapter 17*), many ectoprocts display polymorphism. Most of the zooids of a colony (termed *autozooids*) feed and have a well-developed lophophore; other zooids (termed *heterozooids*) do not feed but offer protection to the colony from other organisms, structural support, nutrient stores, or protection for gametes and/or embryos. For example, one type of heterozooid may contain an *avicularium* (L. *aves,* bird) that resembles a bird beak and can snap at small, invading microorganisms (*see figure 24.1*).

Most ectoprocts are monoecious. In some species, heterozooids produce either eggs or sperm in different colonies. In others, both sperm and eggs are produced in the same autozooid in simple gonads. Sperm are released into the coelomic cavity, exit through pores in the tips of the tentacles, and are caught by the tentacles of other colonies. Eggs are fertilized as they are released and brooded in the coelom; some species have a modified *ovicell* in which the embryo develops. Marine species have radial cleavage and a free-swimming, ciliated larva that bears some resemblance to the trochophore of molluscs and annelids. This larva swims for a variable period of time, depending on the species, and then sinks and attaches to a rock or other substrate and grows into a zooid. A colony is formed by budding.

Some freshwater ectoprocts produce a dormant stage called a statoblast (figure 24.5). A **statoblast** is a hard, resistant capsule containing a mass of germinative cells. Statoblasts are asexually produced and accumulate in the metacoel. They can survive long periods of freezing and drying, enabling a colony to survive many years in seasonally variable lakes and ponds. Some float and are carried downstream, or are blown or carried from pond to pond, spreading ectoprocts over a large area. When environmental conditions become favorable, the statoblasts hatch and give rise to new polypides and eventually new colonies.

One interesting characteristic of ectoprocts is the life-span of the polypide of a given zooid, which normally does not persist for more than several weeks. Degeneration reduces the polypide to one or two dark masses called **brown bodies** (box 24.1). Following regression, a small group of undifferentiated cells completely regenerates a new polypide.

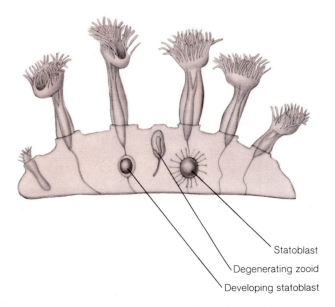

Statoblast
Degenerating zooid
Developing statoblast

Figure 24.5 An ectoproct colony forming statoblasts.

(a)

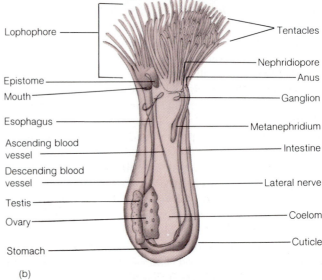

Lophophore — Tentacles
— Nephridiopore
Epistome — — Anus
Mouth — — Ganglion
Esophagus — — Metanephridium
Ascending blood vessel — — Intestine
Descending blood vessel — — Lateral nerve
Testis — — Coelom
Ovary — — Cuticle
Stomach —

(b)

Figure 24.6 Phylum Phoronida. (a) A sand-encrusted tube with the upper end of the phoronid protruded, showing the lateral whorls of the lophophore (*Phoronopsis viridis*). (b) A longitudinal view of the internal anatomy of a phoronid.

Figure 24.6 redrawn, with permission, from *Living Invertebrates* by Pearse/Buchsbaum, copyright by The Boxwood Press.

Box 24.1

The Unusual Ectoproct Brown Bodies

Most marine ectoprocts undergo a cyclic phenomenon of polypide regression to form brown bodies and then subsequent renewal (box figure 24.1). Brown body formation is thought to be the final product of aging in these animals.

Most marine ectoprocts live for 6 to 72 days. One of the first signs of regression is the appearance of orange-brown inclusions in the stomach wall. The tentacles then degenerate, followed by other parts of the body until all that remains is one or two brown bodies in a cyst. However, a few viable cells remain. As they aggregate and begin to grow, a completely restored polypide is produced in the cyst. As a result, the new polypide is completely rejuvenated following the almost total regression of the old polypide, and the brown bodies are either defecated or left in the coelom.

Many of these changes are similar to those that occur in the cells of aging vertebrates. Histochemical and electron microscopy studies have documented a number of similar characteristic alterations in most cellular organelles in both ectoprocts and vertebrates. What happens as these brown bodies form is a generalized tissue response that rejuvenates the ectoproct zooid. This response thus opposes cellular senescence and is a form of "immortality."

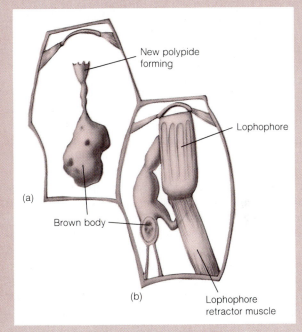

Box Figure 24.1 Ectoproct brown bodies.

Box figure 24.1 redrawn, with permission, from P. E. Lutz, *Invertebrate Zoology.* Copyright © 1986, Benjamin/Cummings Publishing Company, Menlo Park, California.

Phylum Phoronida: The Phoronids

The phoronids (fo-ron-ids) consist of about one dozen marine species divided between two genera: *Phoronis* and *Phoronopsis*. These animals live in permanent, chitinous tubes either buried in muddy or sandy sediments, or attached to solid surfaces (figure 24.6a). A few species bore into mollusc shells or calcareous rock. Generally, only the tentacles extend into the overlying water. Most phoronids are small—less than 20 cm long.

The adult phoronid body consists of an anterior lophophore with two parallel rings of long tentacles (figure 24.6b). The tentacles of the lophophore are filled with coelomic fluid that serves as a hydrostatic skeleton to hold them upright. The cilia on the tentacles drive water into the ring of tentacles from the top of the lophophore and out through the narrow spaces between the tentacles. Suspended food particles are directed toward the mouth. A flap of tissue called the *epistome* (Gr. *epi,* around + *stome,* mouth) covers the mouth.

Internally, phoronids possess a pair of metanephridia that empty near the anus (via the nephridiopore) and function to remove metabolic wastes as well as sperm and eggs from the body. The digestive tract consists of a simple tube that leads from the mouth to the esophagus, makes a U-turn, and then continues through the intestine to empty to the outside via the anus, which is located near the mouth.

Although there is no distinct heart, the longitudinal blood vessels are contractile and pump blood containing a type of hemoglobin through individual vessels to each tentacle. Respiratory exchange occurs through the surface of the tentacles.

The nervous system consists of a nerve ring around the esophagus and a small dorsal ganglion near the anus. A single lateral nerve extends down the body and branches to the trunk muscles, coordinating the rapid retraction of the animal into its tube (e.g., if disturbed).

Some phoronids reproduce asexually by budding and transverse fission; however, the majority are hermaphroditic. The gonads are located in the coelom. Gametes pass from the coelom through the nephridiopore to the tentacles. Cross-fertilization is the rule, and the zygotes are either protected among the coils of the lophophore or released into the sea. Cleavage has both spiral and radial characteristics. In most species, a free-swimming larva called the **actinotroch** develops and feeds on plankton while drifting in the sea. It eventually settles to the bottom, metamorphoses, and begins to grow ventrally to form the body of the sedentary adult. As the animal grows, it burrows into the substrate. The body wall contains gland cells that eventually secrete the chintinous tube.

Stop and Ask Yourself

5. How do ectoprocts differ from endoprocts?
6. What is the difference between an autozooid and a heterozooid?
7. What is the anatomy of a typical phoronid?
8. How do phoronids reproduce?

The Lesser Protostomes

The remainder of this chapter covers six small phyla of eucoelomates that probably stemmed from various points along the annelid/arthropod line—hence the term lesser protostomes. Three of the phyla (Echiura, Pogonophora, Sipuncula) are all bottom-dwelling marine worms that have some affinities to the annelids. For example, the Echiura and Pogonophora have a structure (proboscis) that is used for burrowing and feeding. The Pogonophora live in tubes, lack a digestive tract, and have long food-gathering tentacles. The Onychophora, Pentastomida, and Tardigrada can be grouped together because they have unjointed limbs with claws and a cuticle that is molted, as in the arthropods.

The lesser protostomes are a good example of the fact that many animals are invariably difficult to place neatly into the categories of human-made classification schemes. Nevertheless, they draw our attention to what must be an intriguing phylogenetic linkage between the Annelida and the Arthropoda.

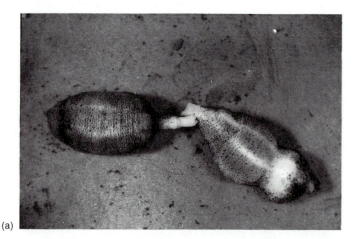

(a)

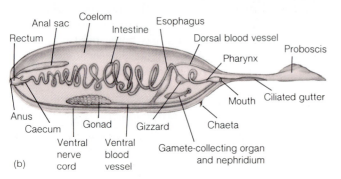

(b)

Phylum Echiura: The Spoon Worms

The echiurans (ek-ee-yur′ans) (Gr. *echis*, serpent + *oura*, tail) consist of about 130 species of marine animals that have a worldwide distribution. The phylum Echiura is divided into three orders. One echiuran, *Urechis caupo*, can be easily cultured in the laboratory and has been studied extensively by developmental biologists.

Echiurans usually live in shallow waters where they either burrow in mud or sand, or live protected in rock crevices. The soft body is covered only by a thin cuticle. As a result, the animals keep to the safety of their burrows or crevices even when feeding. An echiuran feeds by sweeping organic material into its spatula-shaped proboscis that contains a ciliated gutter. The proboscis can be extended for a considerable distance, but it can never be retracted into the body. Echinurans are sometimes called **spoon worms** because of the spatulate nature of the proboscis. In some species, the proboscis is several times the length of the body. The mouth is situated near the base of the proboscis, and just behind the mouth is located a pair of chitinous *chaetae* that are used for burrowing. Individual echiurans are from 15 to 50 cm in length, but the extensible proboscis may increase their length up to 2 m. [3]

The body of an echiuran consists of a large fluid-filled coelom and is divided into a trunk and proboscis (figure 24.7). Coelomic fluid contains phagocytic amebocytes and blood cells with hemoglobin. The muscular body wall surrounds the large coelom, in which a long, coiled intestine is located. A simple closed circulatory system is present, as well as a ventral nerve cord that extends into the proboscis, several pairs of metanephridia, and a pair of anal sacs that empty into the anus at the end of the worm. The exchange of gases takes place through the body wall and proboscis.

All echiurans are dioecious, and sexual dimorphism is extreme in some species. The eggs or sperm do not complete their development in the single ventral gonad but are released into the coelom. After they mature, they are collected by special *collecting organs* and then released into the sea water, where fertilization occurs, giving rise to free-swimming trochophore larvae. The early development of echiurans is similar to annelids with spiral cleavage. However, later development diverges from the annelid pattern in that no segmentation occurs.

Figure 24.7 Phylum Echiura. (a) Photograph of *Listriolabus pelodes* showing the long proboscis. (b) Internal structures of an echiuran.

Figure 24.7 redrawn, with permission, from *Living Invertebrates* by Pearse/Buchsbaum, copyright by The Boxwood Press.

Phylum Pogonophora: The Pogonophores, or Beard Worms

The pogonophores (po'go-nof'e-rs) (Gr. *pogon,* beard + *phora,* bearing) or **beard worms** are tube-dwelling marine worms distributed throughout the world's oceans, especially along the continental slopes. They are named for the thick tuft of white or reddish tentacles that distinguish this group of about 120 species divided among three orders.

[4]

The slender, delicate body is protected in a secreted chitinous tube consisting of a series of rings, which the worm adds to as it grows (figure 24.8a). The tubes are embedded in soft marine sediments in cold, deep (over 100 m), nutrient-poor waters (box 24.2). They range in length from about 10 cm to over 2 m.

Pogonophores have no mouth or digestive tract. Nutrient uptake is via the outer cuticle and from the endosymbiotic bacteria that they harbor in the posterior part (*trophosome*) of the body. These bacteria are able to fix carbon dioxide into organic compounds that both the host and symbiont can use.

The pogonophore body consists of three main regions (figure 24.8b): a short forepart (*prosoma*) bearing many ciliated tentacles, a long trunk, and a short, segmented posterior end. Glandular regions in the forepart and trunk secrete the chitinous tube material, adhesives, and mucus. The trunk is unsegmented and marked with *ciliary bands* and *papillae* (small bumps) that aid in anchoring the worm in its tube. A segmented posterior end (*opisthosoma*) functions primarily in burrowing and anchoring the body to the tube. The tentacles are serviced by blood vessels and function as the primary gas-exchange surface. The closed circulatory system is well developed, with a distinct heart and blood vessels. Both the blood and coelomic fluid contain large amounts of hemoglobin that makes the blood and worms bright red. The anterior section of the body contains a coelom that extends into the tentacles. A brain is located anteriorly with a giant axon that is probably involved in the rapid-withdrawl response.

Very little is known about pogonophoran reproduction and development. In general, there are separate sexes, and sperm are packaged into spermatophores before being released by a male. The spermatophores perhaps reach the tubes of neighboring females by floating. After fertilization, a solid blastula develops following radial cleavage.

Phylum Sipuncula: The Sipunculans, or Peanut Worms

The sipunculans (sigh-pun'kyu-lans) (Gr. L. *sipunculus,* little siphon) or **peanut worms** (because of their peanut shape when disturbed) consist of about 350 species of unsegmented, coelomate, burrowing worms found in oceans throughout the world. These worms live in mud, sand, or any protected retreat, and are whitish yellow or tan. Their burrows may be mucus lined, but sipunculans do not construct true tubes as do pogonophorans. They range in length from about 2 mm to 75 cm.

[5]

The body of a sipunculan is composed of an anterior *introvert* and a posterior trunk (figure 24.9a). When the introvert is extended, the anterior portion with its ciliated tentacles surrounds the mouth. The tentacles contain grooves whose cilia sweep food particles toward the mouth. Food consists of small, bottom-dwelling invertebrates, detritus, or particulate organic material. It passes from the mouth to the esophagus, which is connected to a long, U-shaped intestine arranged in a spiral coil (figure 24.9b). Anteriorly, the intestine ends at an anus that opens to the outside near the introvert.

The body wall is annelidlike, and a large, fluid-filled coelom runs the entire length of the trunk. Contraction of body-wall muscles against the coelomic fluid causes the hydraulic extension of the introvert. Special dorsal and ventral retractor muscles pull the introvert back. Sipunculans

(a)

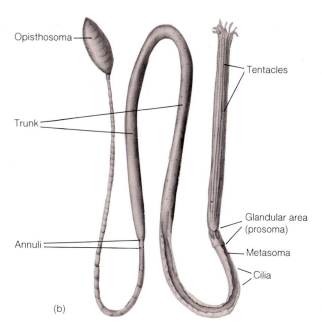

Opisthosoma

Trunk

Annuli

Tentacles

Glandular area (prosoma)

Metasoma

Cilia

(b)

Figure 24.8 Phylum Pogonophora. (a) Giant red pogonophorans inside their tubes. (b) External features of *Lambellisabella pachyptila.*

Figure 24.8 redrawn, with permission, from P. E. Lutz, *Invertebrate Zoology.* Copyright © 1986, Benjamin/Cummings Publishing Company, Menlo Park, California.

Box 24.2
Hydrothermal Vent Communities

What some have called the oceanographic discovery of the century was made in 1977 by a Woods Hole Oceanographic Institute expedition to the Galápagos Rift, in the Pacific ocean (box figure 24.2a). The rift (an opening made by splitting) is over 2,700 m (over 1.5 miles) below the surface and part of an extensive midoceanic ridge system that has developed where the tectonic plates of the earth's crust are moving apart. In such places, a flow of lava (magma) occasionally emerges and **hydrothermal vents** spew out hot water rich in hydrogen sulfide and other minerals.

One unusual finding of the expedition was that the life of a vent community is based not on the "rain" of material generated by the producers in the surface zones but on a rich community of chemolithotrophic bacteria that derive all of the energy they need from the oxidation of inorganic compounds, such as the hydrogen sulfide.

They live in total darkness here because the vents are far below the level of light penetration.

The expedition also noted that ground around each vent was covered with many clams, crabs, polychaete annelids, and one species (*Riftia pachyptila*) of pogonophore (box figure 24.2b). Like other pogonophores, *Riftia* is nourished in part by the endosymbiotic bacteria found in its trophostome. These bacteria can oxidize hydrogen sulfide to sulfate and reduce carbon dioxide to organic compounds, which nourish both the symbiont and host. The worms' hemoglobin carries oxygen and sulfide, tightly bound to another protein. This chemical binding keeps these two molecules from reacting in an unproductive fashion before they are delivered to the bacteria. They also protect the host's tissue from the toxic hydrogen sulfide. These vent communities are among the few on earth that do not depend on input solar energy for life.

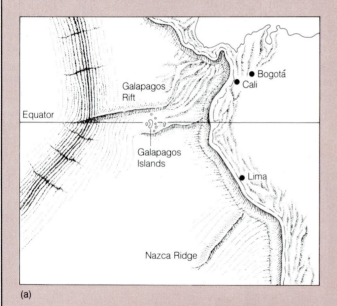

(a)

(b)

Box Figure 24.2 Hydrothermal vent communities. (a) The Galápagos Rift is the site of an extensive ocean-bottom community living on inorganic nutrients issuing from hydrothermal vents. (b) Life around these vents in the Galápagos Rift. Tube worms and crabs can be seen close to the vent.

(a)

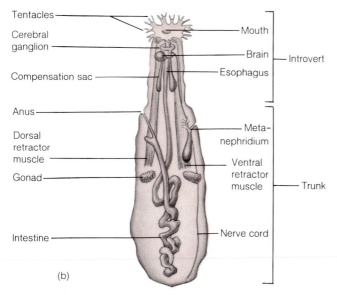

(b)

Figure 24.9 Phylum Sipuncula. (a) Sipunculid worms (*Phascolosoma agassizi*). (b) A typical sipunculan with the introvert extended.

lack gas exchange and circulatory organs. The coelomic fluid, however, contains many cells called *coelomocytes* that contain the respiratory pigment hemerythrin that transports oxygen.

The tentacles are hollow and fluid filled but do not connect to the coelom. Instead, their channels join a circumpharyngeal ring canal. Muscular *compensation sacs* extend from the ring canal; their muscular contractions force fluid into the tentacles and extend them hydrostatically.

A pair of large metanephridia are located in the anterior trunk. They regulate coelomic fluid volume, remove nitrogenous wastes, and serve as an outlet for gametes. An unusual type of cell, called an **urn** is present in large numbers in the coelomic fluid. Some urns are fixed to the peritoneum, whereas others are free to move about, gathering waste material. They eventually carry the waste material to the nephridia where they exit through the nephridiopore, carrying the waste material with them. The anterior nervous system is annelidlike, with a supraesophageal brain and a ventral nerve cord that runs the length of the trunk, but it

lacks ganglia. Sensory cells and chemoreceptors are located on the introvert and tentacles.

Sipunculids are dioecious. Gonads are attached to the coelomic wall and liberate their gametes into the coelom. After maturity, the gametes escape into the sea water via the metanephridia. Fertilization is external, cleavage is spiral, and development is either direct (no larva) or it may produce a free-swimming trochophore larva. The larva eventually settles to the bottom and grows into an adult. In a few species, asexual reproduction can also occur by transverse fission—the posterior part of the parent constricts to give rise to a new individual.

Stop and Ask Yourself

9. What is unique about the proboscis of a spoon worm?
10. Why are the pogonophores called beard worms?
11. Why are sipunculans also called peanut worms?

Phylum Onychophora: The Onychophorans, Velvet Worms or Walking Worms

The onychophorans (on-y-kof'o-rans) (Gr. *onyx,* claw + *pherein,* to bear), also known as **velvet** or **walking worms,** are free-living terrestrial coelomates that live in certain humid, tropical regions. Their ancestors have been considered an evolutionary transition between annelids and arthropods because of their many similarities to both phyla. [6] These interesting worms may live up to 6 years. More than 100 species have been described, with *Peripatus* being the best known genus.

The body of an onychophoran is more or less cylindrical and ranges from 1 to 15 cm in length (figure 24.10). The anterior end consists of two large antennae and a ventral mouth. The mouth is surrounded by oral papillae and claw-like mandibles similar to but not homologous with those of arthropods. The legs vary in number from 14 to 43 pairs; each leg has a pair of terminal claws. The entire surface of the body is covered by large and small tubercles that are covered by small scales and arranged in rings or bands. Onychophorans are blue, green, orange, or black.

Onychophorans usually come out at night and move by using their unjointed legs to crawl. Most species are predaceous and feed on small invertebrates. In order to capture fast-moving prey, onchophorans secrete an adhesive slime (produced in their adhesive gland) from their oral papillae. Some species can eject a stream of slime with enough force to strike a prey animal 50 cm away. The slime hardens immediately, entangling the prey which is then masticated with the mandibles and sucked into the mouth. The food passes down a chitin-lined pharynx and esophagus and into the large, straight intestine, where digestion and absorption take place. The hindgut empties to the outside through the chitin-lined anus.

(a)

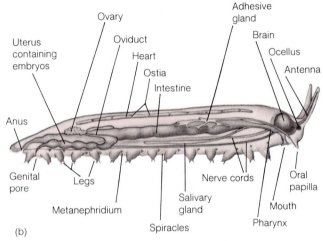

(b)

Figure 24.10 Phylum Onychophora. (a) *Peripatus.* (b) Lateral view of the internal anatomy of a female.

The sexes are separate. The ovaries consist of a pair of fused, elongate organs connected to an oviduct, uterus, and a genital pore. In males, there are two testes and two genital tracts. Near their exit from the body, these two tracts fuse to form a single tube in which sperm are formed into spermatophores. The male and female genital pores are located on the ventral part of the worms. Fertilization is internal. Onychophorans are either oviparous or viviparous. The oviparous species lay large, yolky eggs, each enclosed in a shell, in moist places; cleavage is spiral. The viviparous species retain the embryos in the uterus.

Phylum Pentastomida: The Pentastomids, or Tongue Worms

The pentastomids (pent-ta-stom′ids) (Gr. *pente,* five + *stoma,* mouth) or **tongue worms** are all endoparasitic in the lungs or nasal passageways of carnivorous vertebrates (figure 24.11). Ninety percent of pentastomids hosts are reptiles (e.g., snakes, lizards, crocodiles). There are about 90 species; two of the more well-known genera are *Linguatula* and *Raillietiella.*

The pentastomid body is elongate and wormlike, 2 to 15 cm in length, and is not differentiated into regions. The surface of the worm is covered by a thick, chitinous cuticle that gives the worm an annulated appearance. The cuticle is molted periodically during larval development. Beneath the cuticle are the circular and longitudinal striated muscles. The anterior end bears five protuberances; four are legs that contain a hook for attachment, and one is a *snout* that bears a mouth. When feeding, a pentastomid grasps the host tissue with its clawed legs, inflicting a wound. The mouth is applied to the wound, and blood and tissue fluid are sucked into the worm.

Pentastomids contain a hemocoel, as do the arthropods, but the coelom surrounds only the gonads. The straight digestive tract consists of a mouth, esophagus, intestine, rectum, and anus. Because pentastomids are endoparasites, they lack organs for respiration, excretion, and circulation. The nervous system consists of a brain and a ventral nerve cord with ganglia—just as in the arthropods.

Pentastomids are dioecious, and females are larger than males. Gonads are unpaired, and gonopores open to the outside. Male and female pentastomids mate in the final host. Following internal fertilization, females lay millions of shelled eggs, which pass out in the host's nasal secretions, saliva, or feces. If the eggs are eaten by one of a variety of vertebrates, intermediate hosts, the larvae develop, after molting several times. The larva is characterized by having four to six arthropodlike jointed appendages. When the intermediate host is eaten by a final host, the larva is freed by the digestive enzymes and it migrates up the esophagus to the lungs, trachea, or nasal sinuses.

Humans sometimes become infected with pentastomids [7] by ingesting food or water contaminated with eggs. These parasitic infections are common in parts of South America, Africa, and the Middle and Far East. Because humans are not eaten by a final host, the larvae eventually become encysted and die.

Phylum Tardigrada: The Tardigrades, or Water Bears

The tardigrades (tar-di-gray′ds) (L. *tardus,* slow + *gradus,* step) are commonly called **water bears** because of their [8] body shape and legs and the way they lumber over aquatic vegetation (figure 24.12a). These small animals (less than 1 mm in length) live in marine interstitial areas, in freshwater detritus, and in the water film on terrestrial lichens, liverworts, and mosses. They are yellowish, red-brown, or grayish blue. There are about 500 species; the most common genera are *Echiniscus, Echiniscoides,* and *Macrobiotus.*

The body of a tardigrade is elongate, cylindrical, unsegmented, and has four pairs of legs that have claws (figure 24.12). The head is continuous with the trunk. The entire body is covered with a proteinaceous cuticle that is periodically molted. There are separate muscle strands located beneath the cuticle that facilitate movement of the legs and body. A poorly defined hemocoel contains the viscera. The coelom is confined to a space around the gonads.

(a)

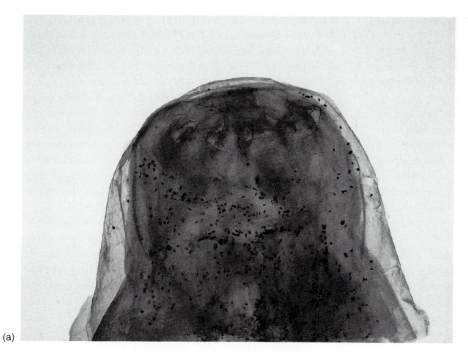

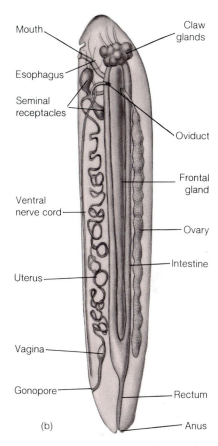

Mouth

Esophagus

Seminal
receptacles

Ventral
nerve cord

Uterus

Vagina

Gonopore

Claw
glands

Oviduct

Frontal
gland

Ovary

Intestine

Rectum

Anus

(b)

Figure 24.11 Phylum Pentastomida. (a) Close-up view of the hooks used by pentastomids to attach to their hosts. (b) Internal anatomy of a female.

(a)

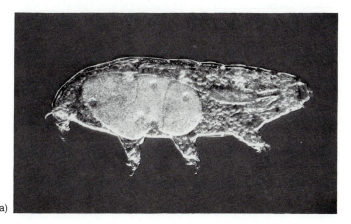

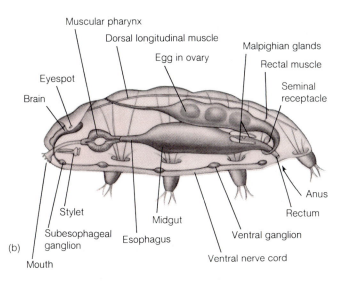

Muscular pharynx

Dorsal longitudinal muscle

Egg in ovary

Malpighian glands

Rectal muscle

Seminal
receptacle

Eyespot

Brain

Anus

Rectum

Stylet

Midgut

Ventral ganglion

Subesophageal
ganglion

Esophagus

Ventral nerve cord

Mouth

(b)

Figure 24.12 Phylum Tardigrada. (a) Photograph of a tardigrade. (b) Internal anatomy of a typical tardigrade.

Tardigrades are mostly herbivorous. The anterior end of the body contains a mouth with a pair of *stylets* that pierce plant tissue. The plant juices are sucked out by the muscular pumping pharynx. The pharynx leads to an esophagus, which continues as the midgut where absorption takes place. The remainder of the digestive tract is a short rectum and anus.

At the junction of the midgut and rectum are organs called **malpighian glands** that are excretory in function. Tardigrades lack organs for gas exchange, circulation, and osmoregulation. Respiratory exchange takes place across the general body surface. Their nervous system consists of a multilobed brain joined to a subesophageal ganglion that branches into a double ventral nerve cord with four ganglia. Sensory organs include a pair of eyespots, sensory bristles, and spines, especially on the head.

Tardigrades are dioecious with a single gonad dorsal to the midgut. A single oviduct or sperm duct empties into a gonopore. Fertilization is internal. Several dozen ornate eggs are laid by the female. After about 2 weeks, a juvenile hatches from the egg, molts, and develops into an adult. In some moss-dwelling species, males are rare or have never been observed, and parthenogenic reproduction presumably occurs.

[9] Tardigrades (as well as nematodes and many rotifers) are able to enter a period of suspended animation termed **cryptobiosis** (Gr. *kryptos,* hidden + *bios,* life). This ability offers great survival benefit to these animals that live in habitats where conditions can suddenly become adverse. If a tardigrade begins to dry out (desiccate), it contracts into a shape that produces an ordered packing of organs and tissues to minimize mechanical damage caused by the desiccation. Overall metabolism slows. When rehydration occurs, the above events are reversed. Interestingly, repeated periods of cryptobiosis can extend a life-span of approximately 1 year to 60 to 70 years.

Further Phylogenetic Considerations

The phoronids are often considered to be the ancestors of the brachiopods because the embryology of these two lophophorate phyla is somewhat similar. Furthermore, both have ciliated tentacles with an upstream ciliary system. However, some zoologists believe that brachipods are deuterostomes because they have radial cleavage and enterocoelous coelom formation. If this distinction is correct, the brachiopod and bryozoan lophophores would represent convergent structures.

Both sipunculans and echiurans bear a number of resemblances to annelids and to each other; however, the lack of metameric segmentation in adults clearly distinguishes the Sipuncula and Echiura from the Annelida. The brief metameric condition that exists during the embryonic development of echiurans suggests that they diverged from the presumed ancestral annelid line later than did the sipunculans. Based on their metameric opisthosoma, pogonophorans seem to also be closely related to the annelids.

Onychophorans are probably descendants of a metameric group of animals that gave rise to the arthropods. They may be one of the earliest phyla with arthropod characteristics that diverged from the segmented protostome lineage that gave rise to the annelids and arthropods.

Most zoologists agree that pentastomids are protostomes closely related to the arthropods, because both have a cuticle, segmented body wall, hemocoel, jointed paired legs in the larvae, lack of cilia, similar sperm, and undergo molting.

The relationship of the tardigrades to other groups is uncertain. They possess several aschelminth features, such as eutely, parthenogenesis, a triradiate pharynx, and variable egg types. However, they also have several arthropod and annelid features (e.g., similar nervous systems and cuticles).

Stop and Ask Yourself

12. Why are the tardigrades called water bears?
13. Why are the phoronids considered ancestors of the brachiopods?
14. How can the Sipuncula be distinguished from the Annelida?
15. How are pentastomids related to arthropods?

Summary

1. The Brachiopoda, Ectoprocta, and Phoronida are lophophorates because they all have a lophophore—a crown of ciliated tentacles surrounding the mouth. They have a U-shaped digestive tract, a free-swimming larva, and are sessile as adults.

2. Brachiopods possess a mantle that covers the body, and secretes a dorsal and ventral shell.

3. Ectoprocts are small, colonial animals. Each individual lives in a zooecium. A zooecium is a secreted exoskeleton.

4. Phoronids live in tubes in shallow water where they feed by extending their lophophores into the water.

5. Echiurans are small, burrowing marine worms that feed on organic material with a long proboscis.

6. Pogonophores live deep in the ocean along rifts where they build tubes. They do not have a digestive system and instead use endosymbiotic bacteria to generate carbon compounds.

7. Sipunculans are small, burrowing marine worms that possess an eversible introvert that is covered with tentacles and used for feeding.

8. Onychophorans are small, motile worms found in humid, tropical environments. They are metameric and crawl with unsegmented, clawed appendages. They may be similar to arthropod ancestors.

9. Pentastomids are endoparasites in the lungs and nasal passages of various vertebrates, where they feed on blood and tissue fluids.

10. Tardigrades are very small animals that live in moist environments, where they feed primarily on plant juices. They possess eight unjointed legs and a nonchitinous cuticle that is molted. They are capable of undergoing cryptobiosis when environmental conditions become unfavorable.

Key Terms

actinotroch (p. 385)
beard worms (p. 387)
brown bodies (p. 384)
cryptobiosis (p. 392)
hydrothermal vents (p. 388)
lampshells (p. 381)
lophophore (p. 381)
malpighian glands (p. 392)
moss animals (p. 383)

peanut worms (p. 387)
spoon worms (p. 386)
statoblast (p. 384)
tongue worms (p. 390)
urn (p. 389)
velvet or walking worms (p. 389)
water bears (p. 390)
zooid (p. 383)

Critical Thinking Questions

1. How are the pentastomids unique when compared to the other phyla covered in this chapter?

2. How does cryptobiosis offer a survival value?

3. Give three distinguishing characteristics for each of the following phyla: Echiura, Ectoprocta, and Brachiopoda.

4. What are some characteristics the three lophophorate phyla have in common?

5. Describe the anatomical similarities and differences between ectoprocts and endoprocts.

6. In what respects are the sipunculans similar to the pseudocoelomates? How do they differ?

7. How do sulfur-reducing bacteria play a role in the life of *Riftia*?

8. How are many of the pecularities of the ectoprocts related to miniaturization?

9. What evidence exists that the onychophorans are an ancient phylum?

10. How are sipunculans related to the annelids? How do they differ?

Suggested Readings

Books

Rice, M. E., and Todorovic, M. (eds). 1975. *Proceedings of the International Symposium on the Biology of the Sipuncula and Echiura.* Washington, D.C.: American Museum of Natural History.

Woollacott, R. M., and Zimmer, R. C. (eds). 1977. *Biology of the Bryozoans.* New York: Academic Press.

Articles

American Society of Zoologists. 1977. Biology of the lophophorates. *American Zoologist* 17(1):3–150.

Ballard, R. D., and Grassle, J. F. 1979. Incredible world of the deep-sea rifts. *National Geographic* 156:680–705.

Crowe, J. H., and Cooper, A. F. Cryptobiosis. *Scientific American* December, 1971.

Edmond, J. M., and Von Damm, K. Hot springs on the Ocean Floor. *Scientific American* January, 1983.

Felbeck, H. 1981. Chemoautotrophic potential of the hydrothermal vent tube worm, *Riftia pachyptilia* Jones (Vestimentifera). *Science* 213:336–338.

Gould, S. J., and Calloway, C. B. 1980. Clams and brachiopods—ships that pass in the night. *Paleobiology* 6:383–341.

Hackman, R. H., and Goldberg, M. 1975. *Peripatus:* its affinities and its cuticle. *Science* 190:528–583.

Nielson, C. 1977. The relationships of the Entoprocta, Ectoprocta, and Phoronida. *American Zoologist* 17:149–150.

Richardson, J. R. Brachiopods. *Scientific American* September, 1986.

25

The Echinoderms

Concepts

1. Echinoderms are a part of the deuterostome evolutionary lineage. They are characterized by pentaradial symmetry, a calcium carbonate internal skeleton, and a water-vascular system.
2. Although there are many classes of extinct echinoderms, living echinoderms are divided into six classes. These are Asteroidea—the sea stars, Ophiuroidea—brittle stars and basket stars, Echinoidea—sea urchins and sand dollars, Holothuroidea—sea cucumbers, Crinoidea—sea lilies and feather stars, and Concentricycloidea—sea daisies.
3. Pentaradial symmetry of echinoderms probably developed during the evolution of sedentary life-styles, in which the water-vascular system was used in suspension feeding. Later, evolution resulted in some echinoderms becoming more mobile, and the water-vascular system came to be used primarily in locomotion.

Have You Ever Wondered:

[1] what evidence links echinoderms and chordates to a common evolutionary pathway?
[2] why most sea stars have five arms, rather than four or six?
[3] how a sea star can open the shell of a clam?
[4] how brittle stars use their snakelike arms?
[5] what sea urchins use to chew through rock and coral?
[6] how sea cucumbers "spill their insides" for a predator or a collector?
[7] what the most recently discovered class of living echinoderms is?
[8] what the original function of the water-vascular system was?

These and other useful questions will be answered in this chapter.

This chapter contains underlined evolutionary concepts.

Evolutionary Perspective

If one could visit 400 million-year-old Paleozoic seas, one would see representatives of nearly every phylum studied in the previous nine chapters of this textbook. In addition, one would observe many representatives of the phylum Echinodermata (i-ki′na-dur″ma-tah) (Gr. *echinos*, spiny + *derma*, skin + *ata*, to bear). Many ancient echinoderms were attached to their substrate and probably lived as filter feeders—a feature found in only one class of modern echinoderms. Today, we know this phylum by the relatively common sea stars, sea urchins, sand dollars, and sea cucumbers. In terms of numbers of species, echinoderms may seem to be a dying phylum. Studies of fossil records indicate that about 12 of 18 classes of echinoderms have become extinct. That does not mean, however, that living echinoderms are of minor importance. Members of three classes of echinoderms have flourished and often make up a major component of the biota of marine ecosystems (table 25.1). Occasionally some of these taxa proliferate to nuisance proportions (box 25.1).

Table 25.1 Classification of the Phylum Echinodermata

Phylum Echinodermata (i-ki′na-dur″ma-tah).
The phylum of triploblastic, eucoelomate animals whose members are pentaradially symmetrical as adults, possess an endoskeleton covered by epithelium, and possess a water-vascular system. Pedicellarias are often present.

Class Crinoidea (kri-noi′de-ah),
Free-living or attached by an aboral stalk of ossicles; flourished in the Paleozoic era; approximately 230 living species. Sea lilies, feather stars.

Class Asteroidea (as′te-roi″de-ah)
Rays not sharply set off from central disc; ambulacral grooves with tube feet; suction discs on tube feet; pedicellarias present. Sea stars.

Class Ophiuroidea (o′fe-u-roi″de-ah)
Arms sharply marked off from the central disc; tube feet without suction discs. Brittle stars.

Class Holothuroidea (hol′o-thu-roi″de-ah)
No rays; elongate along the oral/aboral axis; microscopic ossicles embedded in a muscular body wall; circumoral tentacles. Sea cucumbers.

Class Echinoidea (ek′i-noi″de-ah)
Globular or disc shaped; no rays; movable spines; skeleton (test) of closely fitting plates. Sea urchins, sand dollars.

Class Concentricycloidea (kon-sen′ti-si-kloi″de-ah)
Two concentric water-vascular rings encircle a disclike body; no digestive system; digest and absorb nutrients across their lower surface; internal brood pouches; no free-swimming larval stage. Sea daisies.

Relationships to Other Animals

Most zoologists believe that echinoderms share a common ancestry with hemichordates and chordates. Evidence of these evolutionary ties is seen in the deuterostome characteristics that they share (*see figure 20.1*): an anus that develops in the region of the blastopore, a coelom that forms from outpockets of the embryonic gut tract (vertebrate chordates are an exception), and radial, indeterminate cleavage. Unfortunately, no fossils have been discovered that document a common ancestor for these phyla or that demonstrate how the deuterostome lineage was derived from ancestral diploblastic or triploblastic stocks. Perhaps evidence from comparative biochemistry will provide further insight into these evolutionary relationships. [1]

Although adults are radially symmetrical, it is generally accepted that echinoderms evolved from bilaterally symmetrical ancestors. Evidence for this relationship includes bilaterally symmetrical echinoderm larval stages. Certain extinct echinoderms, called carpoids, were bilaterally symmetrical or asymmetrical. Carpoids also bore a crown of tentacles similar to those of some bilateral lophophorate animals. These observations also suggest an evolutionary relationship between some lophophorate animals and the echinoderms. This claim, however, is speculative and controversial.

Echinoderm Characteristics

There are approximately 5,300 species of echinoderms. They are exclusively marine and occur at all depths in all oceans. Modern echinoderms have a form of radial symmetry, called **pentaradial symmetry**, in which body parts are arranged in fives, or a multiple of five, around an oral/aboral axis (figure 25.1a). Radial symmetry is adaptive for sedentary or slowly moving animals because it allows a uniform distribution of sensory, feeding, and other structures around the animal. Some modern echinoderms, however, have secondarily returned to a basically bilateral form.

The skeleton of echinoderms consists of a series of calcium carbonate plates and ossicles. These plates are derived from mesoderm, held in place by connective tissues, and covered by an epidermal layer. If the epidermal layer is abraded away, the skeleton may be exposed in some body regions. The skeleton is frequently modified into fixed or articulated spines that project from the body surface.

The evolution of the skeleton may be responsible for the pentaradial body form of echinoderms. The joints between two skeletal plates represent a weak point in the skeleton (figure 25.1b). By not having weak joints directly opposite one another, the skeleton is made stronger than if the joints were arranged opposite each other. [2]

The **water-vascular system** of echinoderms is a series of water-filled canals and their extensions are called *tube feet*. It originates embryologically as a modification of the coelom and is ciliated internally. The water-vascular system

(a)

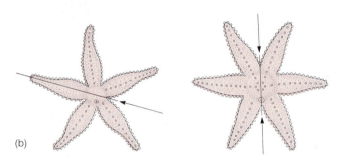

(b)

Figure 25.1 Pentaradial symmetry. (a) Echinoderms possess a form of radial symmetry, called pentaradial symmetry, in which body parts are arranged in fives around an oral/aboral axis. Note the madreporite between the bases of the two arms in the background and the tube feet on the tips of the upturned arms. (b) Comparison of hypothetical penta- and hexaradial echinoderms. The five-part organization may be advantageous because joints between skeletal ossicles are never directly opposite one another, as they would be if an even number of parts were present. Having joints on opposite sides of the body in line with each other (arrows) could make the skeleton weaker.

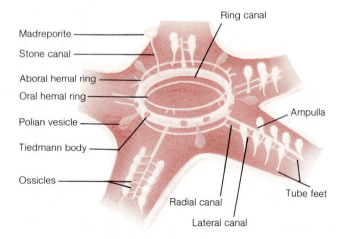

Figure 25.2 The water-vascular system of a sea star. The ring canal gives rise to radial canals to each arm, a stone canal that ends at a madreporite on the aboral surface, and often has Polian vesicles and Tiedemann bodies associated with it.

includes a *ring canal* that surrounds the mouth (figure 25.2). The ring canal usually opens to the outside or to the body cavity through a *stone canal* and a sievelike plate, called the *madreporite*. The madreporite may serve as an inlet to replace water lost from the water-vascular system and may help equalize pressure differences between the water-vascular system and the outside. *Tiedemann bodies* are swellings that are often associated with the ring canal. They are believed to be the site for production of phagocytic cells, called coelomocytes, whose functions will be described later in this chapter. *Polian vesicles* are sacs that are also associated with the ring canal and function in fluid storage for the water-vascular system.

Five (or a multiple of five) *radial canals* branch from the ring canal. Radial canals are associated with arms of star-shaped echinoderms. In other echinoderms, they may be associated with the body wall and arch toward the aboral pole. Many *lateral canals* branch off each radial canal and end at the tube feet.

Tube feet are extensions of the canal system and usually emerge through openings in skeletal ossicles (colorplate 9a). Internally, tube feet usually terminate in a bulblike, muscular *ampulla*. When an ampulla contracts, it forces water into a tube foot, which then extends. Valves prevent the backflow of water from the tube foot into the lateral canal. A tube foot often has a suction cup at its distal end. When the foot is extended and comes into contact with solid substrate, muscles of the suction cup contract and create a vacuum. In some taxa, the tube feet have a pointed or blunt distal end. These echinoderms may extend their tube feet into a soft substrate to secure contact during locomotion or to sift sediment during feeding.

The water-vascular system has other functions in addition to locomotion. As will be discussed at the end of this chapter, the original function of water-vascular systems was probably feeding, not locomotion. In addition, the soft membranes of the water-vascular system permit diffusion of respiratory gases and nitrogenous wastes across the body wall.

Class Asteroidea

The sea stars make up the class Asteroidea (as'te-roi"de-ah) (Gr. *aster*, star + *oeides*, in the form of). They are often found on hard substrata in marine environments, although some species are also found in sandy or muddy substrates. Sea stars may be brightly colored with red, orange, blue, or grey. *Asterias* is an orange sea star common along the Atlantic coast of North America and is frequently studied in introductory zoology laboratories.

Sea stars have five arms that radiate from a *central disc*. The oral opening, or mouth, is in the middle of one side of the central disc. It is normally oriented downward and is surrounded by movable *oral spines*. The aboral surface is roughened by movable and fixed spines that project from the skeleton. Thin folds of the body wall, called **dermal branchiae**, extend between ossicles and function in gas

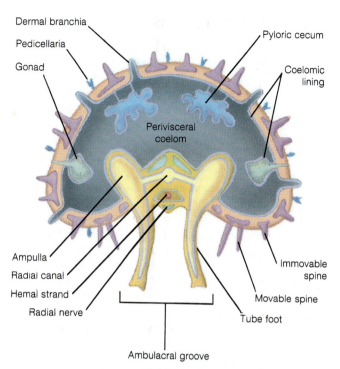

Figure 25.3 Body wall and internal anatomy of a sea star. A cross section through one arm of a sea star shows the structures of the water-vascular system and the tube feet extending through the ambulacral groove.

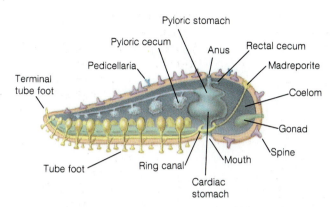

Figure 25.4 Digestive structures in a sea star.

exchange (figure 25.3). In some sea stars, the aboral surface is covered with pincherlike structures called **pedicellariae,** which are used for cleaning the body surface of debris and for protection. Pedicellariae may be attached on a movable spine or they may be immovably fused to skeletal ossicles.

An **ambulacral groove** runs the length of the oral surface of each arm and is formed by a series of ossicles in the arm. It houses the radial canal, and paired rows of tube feet protrude through the body wall on either side of the ambulacral groove. Tube feet of sea stars move in a stepping motion. Alternate extension, attachment, and contraction of tube feet move sea stars across their substrate. Tube feet are coordinated by the nervous system so that all feet move the sea star in the same direction; however, the tube feet do not move in unison. The suction discs of tube feet are effective attachment structures, allowing sea stars to maintain their position, or move from place to place, in spite of strong wave action.

Maintenance Functions

Sea stars feed on snails, bivalves, crustaceans, polychaetes, corals, detritus, and a variety of other food items. The mouth opens to a short *esophagus* and then to a large stomach that fills most of the coelom of the central disc. The stomach is divided into two regions. The larger, oral stomach, sometimes called the *cardiac stomach* receives ingested food (figure 25.4). It joins the smaller, aboral stomach, sometimes called the *pyloric stomach.* The pyloric stomach gives rise to ducts that connect to secretory and absorptive structures called *pyloric ceca.* Two pyloric ceca extend into each arm. A short intestine leads to *rectal ceca* (uncertain functions) and to a nearly nonfunctional anus, which opens on the aboral surface of the central disc.

Some sea stars ingest whole prey, which are digested extracellularly within the stomach. Undigested material is expelled through the mouth. Many sea stars feed on bivalves by forcing the valves apart. (Anyone who has tried to pull apart the valves of a bivalve shell can appreciate that this is a remarkable accomplishment.) When a sea star feeds on a bivalve, it wraps itself around the ventral margin of a bivalve (colorplate 9b). Tube feet attach to the outside of the shell, and the body-wall musculature forces the valves apart. When the valves are opened about 0.1 mm, the oral (cardiac) portion of the stomach is everted through the mouth and into the bivalve shell by increased coelomic pressure. Digestive enzymes are released, and partial digestion occurs in the bivalve shell. This digestion further weakens the bivalve's adductor muscles, and the shell eventually opens completely. Partially digested tissues are taken into the aboral (pyloric) portion of the stomach, and into the pyloric ceca for further digestion and absorption. After feeding and initial digestion, the stomach is retracted using *stomach retractor muscles.*

Transport of gases, nutrients, and metabolic wastes in the coelom occurs by diffusion and by the action of ciliated cells lining the body cavity. Gas exchange and excretion of metabolic wastes (principally ammonia) occur by diffusion across dermal branchiae, tube feet, and other membranous structures.

Sea stars possess strands of tissue that encircle the mouth near the ring canal, extend aborally near the stone canal, and run into the arms near radial canals (*see figure 25.2*). These tissues make up the *hemal system,* and although they have been likened to a vestigial circulatory system, their functions are largely unknown. They may aid in the transport of large molecules, hormones, or coelomocytes, which engulf and transport waste particles within the body.

The nervous system of sea stars consists of a *nerve ring* that encircles the mouth and *radial nerves* that extend into each arm. Radial nerves lie within the ambulacral groove,

[3]

Box 25.1

A Thorny Problem for Australia's Barrier Reef

The crown-of-thorns sea star (*Acanthaster planci*) is a common inhabitant of the South Pacific. Adults have a diameter of about 0.5 m and 13 to 16 arms (box figure 25.1). Their common name is derived from their venomous spines that can cause swelling, pain, and nausea in humans.

The crown-of-thorns sea star has become a problem in the waters off Australia and other South Pacific coasts. It feeds on coral polyps, and in one day, a single individual can extract polyps from 0.1 m² of reef. Over the last 20 years, the crown-of-thorns sea star has experienced a dramatic population increase. Thousands of sea stars have been observed moving slowly along a reef, leaving a white, almost sterile, limestone coral skeleton in their trail. They have seriously damaged over 500 km² of Australia's Great Barrier Reef.

One or more hypotheses may explain why there is a problem now, when these sea stars and coral polyps have coexisted for millions of years. Some believe that a part of the increase in sea stars may be due to the destruction of sea star predators, in particular, the giant triton gastropods. Tritons are valued for their beautiful shells, and their habitat has been disrupted by blasting to create shipping channels. Pesticides and other pollutants are also believed to be responsible for the destruction of predators of crown-of-thorns larvae. Another hypothesis suggests that the apparent increase in the sea star population may be a natural fluctuation in population size.

The reefs of the South Pacific are a source of economic wealth in the form of fisheries and tourism; therefore, the proliferation of crown-of-thorns sea stars has been the subject of intense study and control efforts. The Australian government has spent in excess of $3,000,000 and has not yet achieved satisfactory control. Control measures have included the injection of formaldehyde into adults by SCUBA divers, the erection of wire fences to divert populations away from reefs, and the use of computers to predict movements of colonies and local population explosions. One of the difficulties in these control efforts stems from the fact that sea star larvae are planktonic and are widely dispersed by oceanic currents.

The original relationship between the crown-of-thorns sea star, coral polyps, and possibly sea star predators is an interesting example of a balanced predator/prey relationship. What is unfortunate is that humans often do not appreciate such balances until they have been altered.

Box figure 25.1 The crown-of-thorns sea star (*Acanthaster*) on coral (*Pocillopora*).

just oral to the radial canal of the water-vascular system and the radial strands of the hemal system (*see figure 25.3*). Radial nerves are essential for coordinating the functions of tube feet. Other nervous elements are in the form of a *nerve net* that is associated with the body wall.

Most sensory receptors are distributed over the surface of the body and tube feet. Sea stars respond to light, chemicals, and various mechanical stimuli. They often have specialized photoreceptors at the tips of their arms. These are actually tube feet that lack suction cups but have a pigment spot surrounding a group of ocelli. Responses to light may be positive or negative depending on the species being considered.

Regeneration, Reproduction, and Development

Sea stars are well known for their powers of regeneration. Any part of a broken arm can be regenerated. In a few species, an entire starfish can be regenerated from a broken arm if the arm contains a portion of the central disc. Regeneration is a slow process, taking up to 1 year for complete regeneration. Asexual reproduction involves dividing the central disc, followed by regeneration of each half.

Sea stars are dioecious, but sexes are indistinguishable externally. Two gonads are present in each arm and increase in size to nearly fill an arm during the reproductive periods. Gonopores open between the bases of each arm.

The embryology of echinoderms has been studied extensively because of the relative ease of inducing spawning and maintaining embryos in the laboratory. External fertilization is the rule. Because gametes cannot survive long in the ocean, maturation of gametes and spawning must be coordinated if fertilization is to take place. Photoperiod (the relative length of light and dark in a 24-hour period) and temperature are environmental factors that are used to coordinate sexual activity. In addition, the release of gametes by one individual is accompanied by the release of spawning pheromones, which induce other sea stars in the area to spawn, increasing the likelihood of fertilization.

The early stages of echinoderm embryology are covered in detail in chapter 10. Embryos are planktonic, and cilia are used in swimming (figure 25.5). After gastrulation, bands of cilia differentiate, and a bilaterally symmetrical larva, called a *dipleurula larva,* is formed. This larva usually feeds on planktonic protists and is typical of all echinoderms except the crinoids.

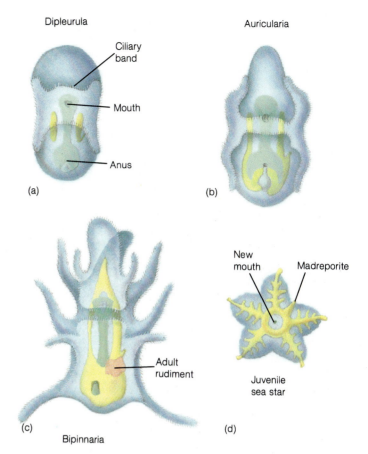

Figure 25.5 Development of a sea star. Later embryonic stages are ciliated and swim and feed in the plankton. In a few species, embryos develop from yolk stored in the egg during gamete formation. Following blastula and gastrula stages, larvae develop. The sequence of sea star larval development includes (a) dipleurula, (b) auricularia, and (c) bipinnaria larvae. (d) Settling to the substrate is followed by attachment and metamorphosis to a juvenile sea star.

Stop and Ask Yourself

1. Why do many zoologists believe that ancestral echinoderms were bilaterally symmetrical?
2. What are the functions of the water-vascular system?
3. How does a sea star feed on a clam?
4. How is the nervous tissue arranged in a sea star?

Class Ophiuroidea

The class Ophiuroidea (o'fe-u-roi'de-ah) (Gr. *ophis,* snake + *oura,* tail + *oeides,* in the form of) includes the basket stars and the brittle stars. With over 2,000 species, this is the most diverse group of echinoderms. Ophiuroids, however, are often overlooked because of their small size and their tendency to occupy crevices in rocks and coral or to cling to algae.

The arms of ophiuroids are long and, unlike those of asteroids, are sharply set off from the central disc, giving the central disc a pentagonal shape. Brittle stars have unbranched arms and most range in size from 1 to 3 cm (colorplate 9c). Basket stars have arms that branch repeatedly (colorplate 9d). Neither dermal branchiae nor pedicellariae are present in ophiuroids. The tube feet of ophiuroids lack suction discs and ampullae, and contraction of muscles associated with the base of a tube foot causes the tube foot to be extended. Unlike the sea stars, the madreporite of ophiuroids is located on the oral surface.

The water-vascular system of ophiuroids is not used for locomotion. Instead, the skeleton is modified to permit a unique form of grasping and movement. Superficial ossicles, which originate on the aboral surface, cover the lateral and oral surfaces of each arm. The ambulacral groove—containing the radial nerve, hemal strand, and radial canal—is thus said to be "closed." *Ambulacral ossicles* are

in the arm, forming a central supportive axis (figure 25.6). Successive ambulacral ossicles articulate with one another and are acted upon by relatively large muscles to produce snakelike movements, allowing the arms to be curled around a stalk of algae or hooked into a coral crevice. During locomotion, two arms are used to pull the animal along a substrate, while the other three arms trail behind. [4]

Maintenance Functions

Ophiuroids are predators and scavengers. They use their arms and tube feet in sweeping motions to collect prey and particulate matter, which are then transferred to the mouth. Some ophiuroids are filter feeders that wave their arms and trap plankton on mucus-covered tube feet. Trapped plankton is passed from tube foot to tube foot along the length of an arm until it reaches the mouth.

The mouth of ophiuroids is in the center of the central disc, and five triangular jaws form a chewing apparatus. The mouth leads to a saclike stomach. There is no intestine, and no part of the digestive tract extends into the arms (figure 25.6).

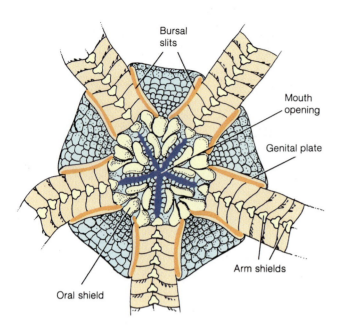

Figure 25.6 Oral view of the disc of the brittle star, *Ophiomusium*.

The coelom of ophiuroids is reduced and is mainly confined to the central disc, but it still serves as the primary means for the distribution of nutrients, wastes, and gases. Coelomocytes aid in the distribution of nutrients and the expulsion of particulate wastes. Ammonia is the primary nitrogenous waste product and it is lost by diffusion. Diffusion occurs across tube feet and membranous sacs, called *bursae,* that invaginate from the oral surface of the central disc. Slits in the oral disc, near the base of each arm, allow cilia to move water into and out of the bursae.

Regeneration, Reproduction, and Development

Ophiuroids, as sea stars, are able to regenerate lost arms. If a brittle star is grasped by an arm, the contraction of certain muscles may sever and cast off the arm—hence the common name "brittle star." This process is called *autotomy* (Gr. *autos,* self + *tomos,* to cut) and is used in escape reactions. The arm is later regenerated. Some species also have a fission line across their central disc. When split into halves along this line, two ophiuroids will be regenerated.

Ophiuroids are dioecious. Males are usually smaller than females and often are carried about by females. The gonads are associated with each bursa, and gametes are released into the bursa. Eggs may be shed to the outside, or retained in the bursa, where they are fertilized and held through early development. Embryos are protected in the bursa and are sometimes nourished by the parent. A larval stage, called an *ophiopluteus* is planktonic. Its long arms bear ciliary bands that are used to feed on plankton, and it undergoes metamorphosis before sinking to the substrate.

Class Echinoidea

The sea urchins, sand dollars, and heart urchins make up the class Echinoidea (ek'i-noi-de-ah) (Gr. *echinos,* spiny + *oeides,* in the form of). There are about 800 species widely distributed in nearly all marine environments. Sea urchins (colorplate 9e) are sometimes referred to as *regular echinoids* because they are spherical in shape and display typical pentaradial symmetry. Sea urchins are specialized for living on hard substrata, often wedging themselves into crevices and holes in rock or coral. Sand dollars and heart urchins are sometimes referred to as *irregular echinoids* because they are somewhat flattened, and their symmetry is modified to a nearly bilateral or asymmetrical form. Sand dollars and heart urchins usually live in sand or mud and burrow just below the surface (figure 25.7). They use tube feet to catch organic matter settling on them or passing over them. Sand dollars often occur in very dense beds, which favors efficient reproduction and feeding.

Sea urchins are rounded and have their oral end oriented toward the substrate. Their skeleton, called a *test,* consists of ten closely fitting plates that arch between oral and aboral ends. Five *ambulacral plates* have openings for tube feet, and alternate with five *interambulacral plates,* which have tubercles for the articulation of spines. The proximal end of each spine is a concave socket, and muscles at its base move the spine. Spines are often sharp and sometimes hollow and contain venom that is dangerous to swimmers. The pedicellariae of sea urchins have either two or three jaws and are connected to the body wall by a relatively long stalk supported by a skeletal rod (figure 25.8). They are used for cleaning the body of debris and capturing planktonic larvae, which provide an extra source of food. Pedicellariae of some sea urchins contain venom sacs and are grooved or hollow to inject poison into a predator, such as a sea star.

The water-vascular system is similar to that of other echinoderms. Radial canals run along the inner body wall between the oral and the aboral poles. Tube feet possess ampullae and suction cups, and the water-vascular system opens to the outside through many pores in one aboral ossicle that serves as a madreporite.

Echinoids move by using spines for pushing against the substrate and tube feet for pulling. Sand dollars and heart urchins use spines to help burrow in soft substrates (colorplate 9f). Some sea urchins burrow into rock and coral to escape the action of waves and strong currents. They form cup-shaped depressions and deeper burrows using the action of their chewing Aristotle's lantern, which is described next.

[5]

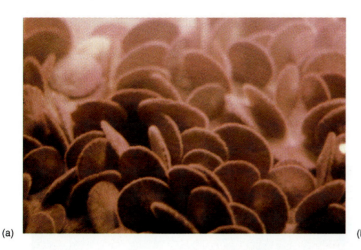
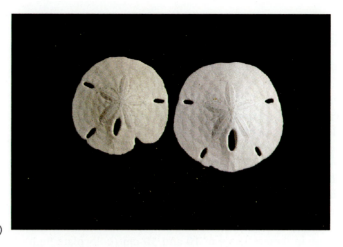

(a)

(b)

Figure 25.7 Sand dollars. (a) Sand dollars are specialized for living in soft substrates, where they often occur partially buried in large beds. (b) Sand dollar skeletons or tests. The slots in the tests of some sand dollars may aid in feeding.

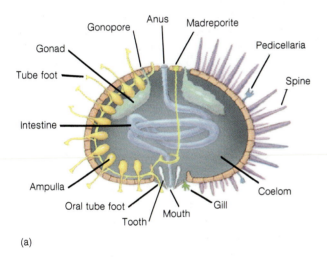

(a)

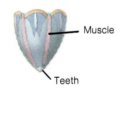

(b)

Figure 25.8 Internal anatomy of a sea urchin. (a) Sectional view. (b) Aristotle's lantern is a chewing structure consisting of about 35 ossicles and associated muscles.

Maintenance Functions

Echinoids feed on algae, bryozoans, coral polyps, and dead animal remains. Food is manipulated by oral tube feet surrounding the mouth. A chewing apparatus, called **Aristotle's lantern,** projects from the mouth (figure 25.8). It consists of about 35 ossicles and attached muscles and cuts food into small pieces for ingestion. The mouth cavity leads to a pharynx, an esophagus, and a long, coiled intestine that ends aborally at the anus (figure 25.8).

Echinoids have a large coelom, and coelomic fluids are the primary circulatory medium. Small gills, found in a thin membrane surrounding the mouth, are outpockets of the body wall and are lined by ciliated epithelium. Gas exchange occurs by diffusion across this epithelium and across the tube feet. Ciliary currents, changes in coelomic pressure, and the contraction of muscles associated with Aristotle's lantern move coelomic fluids into and out of gills. Excretory and nervous functions are similar to those described for asteroids.

Reproduction and Development

Echinoids are dioecious. Gonads are located on the internal body wall of the interambulacral plates. During breeding season, they nearly fill the spacious coelom. One gonopore is located in each of five ossicles, called genital plates, at the aboral end of the echinoid, although the irregularly shaped echinoids usually have only four gonads and gonopores. Gametes are shed into the water, and fertilization is external. Development eventually results in a *pluteus larva* that spends several months in the plankton and eventually undergoes metamorphosis to the adult.

Stop and Ask Yourself

5. How is a sea star distinguished from a brittle star?
6. What is autotomy?
7. What is Aristotle's lantern?
8. What structures of echinoids are used for gas exchange?

Class Holothuroidea

There are approximately 500 species in the class Holothuroidea (hol'o·thuroi'de·ah) (Gr. *holothourion,* sea cucumber + *oeides,* in the form of) and they are commonly called sea cucumbers. Sea cucumbers are found at all depths in all oceans, where they crawl over hard substrates or burrow through soft substrates.

Sea cucumbers have no arms, and they are elongate along the oral/aboral axis. They lie on one side, which is usually flattened as a permanent ventral side, giving them a secondary bilateral symmetry. Tube feet surrounding the mouth are elongate and referred to as *tentacles.* Most adults range in length between 10 and 30 cm (colorplate 9g,h). Their body wall is thick and muscular and lacks protruding spines or pedicellariae. Beneath the epidermis is the dermis, a thick layer of connective tissue in which ossicles are embedded. Ossicles of sea cucumbers are microscopic in size and do not function in determining body shape (figure 25.9). A circle of larger ossicles forms a *calcareous ring* and encircles the oral end of the digestive tract, serving as a point of attachment for body wall muscles (figure 25.10). Beneath the dermis is a layer of circular muscles overlying longitudinal muscles. The body wall of sea cucumbers, when boiled and dried, is known as trepang in the orient. It may be eaten as a main course item or be added to soups as flavoring and a source of protein.

The madreporite of sea cucumbers is internal, and the water-vascular system is filled with coelomic fluid. The ring canal encircles the oral end of the digestive tract and gives rise to one to ten Polian vesicles. Five radial canals and the canals to the tentacles, branch from the ring canal. Radial canals and tube feet, with suction cups and ampullae, run between the oral and aboral poles. The side of a sea cucumber resting on the substrate contains three of the five rows of tube feet, which are primarily used for attachment. The two rows of tube feet on the upper surface may be reduced in size, or they may be absent.

Sea cucumbers are mostly sluggish burrowers and creepers, although some swim by undulating their body from side to side. Locomotion using tube feet is inefficient because the tube feet are not anchored by body wall ossicles. Locomotion more commonly results from contractions of body wall muscles, producing wormlike, locomotor waves that pass along the length of the body.

Maintenance Functions

Most sea cucumbers ingest particulate organic matter using their *tentacles.* Food is trapped in mucus covering the tentacles, either as the tentacles sweep across the substrate or while tentacles are held out in sea water. The digestive tract consists of a stomach; a long, looped intestine; a rectum; and an anus (figure 25.10). Tentacles are thrust into the mouth to wipe off trapped food. During digestion, coelomocytes move across the intestinal wall, secrete enzymes to aid in digestion, and engulf and distribute the products of digestion.

The coelom of sea cucumbers is large, and the cilia of the coelomic lining circulate fluids throughout the body cavity, distributing respiratory gases, wastes, and nutrients. The hemal system of sea cucumbers is well developed, with relatively large sinuses and a network of channels containing coelomic fluids. Its primary role is food distribution.

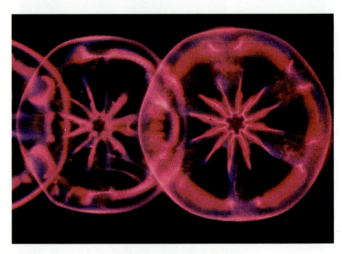

Figure 25.9 Sea cucumber ossicles. The structure of the ossicles varies among species and is a taxonomically important characteristic.

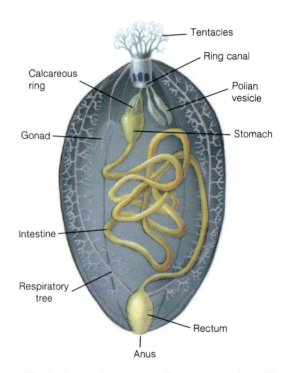

Figure 25.10 Internal structure of a sea cucumber, *Thyone.* The mouth leads to a stomach that is supported by a calcareous ring. The calcareous ring is also the attachment site for longitudinal retractor muscles of the body. Contractions of these muscles pull the tentacles into the anterior end of the body. The stomach leads to a looped intestine. The intestine continues to the rectum and anus.

Respiratory trees are a pair of tubes that attach at the rectum and branch throughout the body cavity of sea cucumbers. Water circulates into these tubes by the pumping action of the rectum. When the rectum dilates, water moves through the anus into the rectum. Contraction of the rectum, along with contraction of an anal sphincter, forces water into the respiratory tree. Water exits the respiratory tree when tubules of the tree contract. Respiratory gases and nitrogenous wastes move between the coelom and sea water across these tubules.

The nervous system of sea cucumbers is similar to that of other echinoderms, but has additional nerves supplying the tentacles and pharynx. Some sea cucumbers have statocysts, and others have relatively complex photoreceptors.

Casual examination would lead one to believe that sea cucumbers are defenseless against predators. Many sea cucumbers, however, produce toxins in their body walls that act to discourage predators. (Some of these toxins have been found to inhibit growth of cancer cells.) In other sea cucumbers, tubules of the respiratory tree, called *Cuverian tubules,* can be everted through the anus. They contain sticky secretions and toxins capable of entangling and immobilizing predators. In addition, contractions of the body wall may result in expulsion of one or both respiratory trees, the digestive tract, and the gonads through the anus. This process is called *evisceration* and is a defensive adaptation that may discourage predators. It is followed by regeneration of lost parts.

[6]

Reproduction and Development

Sea cucumbers are dioecious. They possess a single gonad, located anteriorly in the coelom, and a single gonopore near the base of the tentacles. Fertilization is usually external, and embryos develop into planktonic larvae. Metamorphosis precedes settling to the substrate. In some species, eggs are trapped by a female's tentacles as they are released. After fertilization, eggs are transferred to the body surface, where they are brooded. Although it is rare, coelomic brooding also occurs. Eggs are released to the body cavity where fertilization (by an unknown mechanism) and early development occur. The young leave by a rupture in the body wall. Sea cucumbers can also reproduce by transverse fission, followed by regeneration of lost parts.

Stop and Ask Yourself

9. What structural features of sea cucumbers result in their having a secondary bilateral symmetry?
10. How is the water-vascular system of sea cucumbers modified for feeding?
11. What is a respiratory tree?
12. What are Cuverian tubules?

Class Crinoidea

Members of the class Crinoidea (krin-oi′de-ah) (Gr. *krinon,* lily + *oeides,* in the form of) include the sea lilies and the feather stars. They are the most primitive of all living echinoderms and are very different from any covered thus far. There are approximately 630 species living today; however, an extensive fossil record indicates that many more were present during the Paleozoic era, 200 to 600 million years ago.

Sea lilies are attached permanently to their substrate by a *stalk* (figure 25.11). The attached end of the stalk bears a flattened disc or rootlike extensions that are fixed to the substrate. Disclike ossicles of the stalk appear to be stacked on top of one another and are held together by connective tissues, giving a jointed appearance. The stalk usually bears projections, or *cirri* that are arranged in whorls around the stalk (colorplate 9i). The unattached end of a sea lily is called the crown. The aboral end of the crown is attached to the stalk and is supported by a set of ossicles, called the *calyx.* Five arms also attach at the calyx. They are branched, supported by ossicles, and bear smaller branches—giving them a featherlike appearance. Tube feet are located in a double row along each arm. Ambulacral grooves on the arms lead toward the mouth. The mouth and anus open onto the upper (oral) surface.

Feather stars are similar to sea lilies except they lack a stalk and are swimming and crawling animals (figure 25.12; colorplate 9j). The aboral end of the crown bears a ring of rootlike cirri, which are used for clinging when the animal

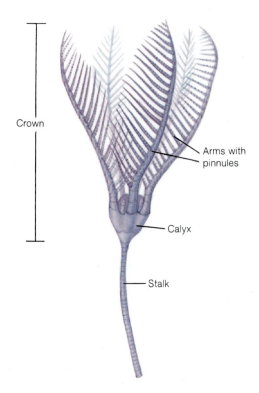

Figure 25.11 Crinoidea. A sea lily (*Ptilocrinus*).

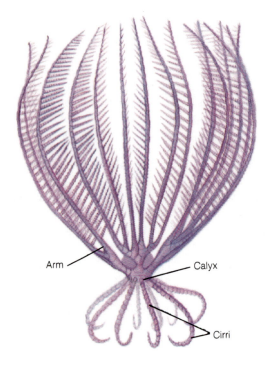

Figure 25.12 Crinoidea. A feather star (*Neometra*).

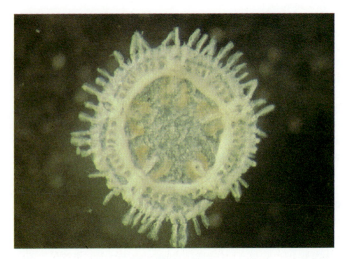

Figure 25.13 Photograph of a preserved sea daisy (*Xyloplax medusiformis*). This specimen is 3 mm in diameter.

is resting on a substrate. Swimming is accomplished by raising and lowering the arms, and crawling results from using the tips of the arms to pull the animal over the substrate.

Maintenance Functions

Circulation, gas exchange, and excretion in crinoids are similar to other echinoderms. In feeding, however, crinoids differ from other living echinoderms. They use outstretched arms for suspension feeding. When a planktonic organism contacts a tube foot, it is trapped and carried to the mouth by cilia in ambulacral grooves. Although this method of feeding is different from the way other modern echinoderms feed, it probably reflects the original function of the water-vascular system.

Crinoids lack the nerve ring found in most echinoderms. Instead, a cup-shaped nerve mass below the calyx gives rise to radial nerves that extend through each arm and control the tube feet and arm musculature.

Reproduction and Development

Crinoids, as other echinoderms, are dioecious. Gametes form from germinal epithelium in the coelom and are released by rupturing the walls of the arms. Some species spawn into sea water, where fertilization and development occur. Other species brood embryos on the outer surface of the arms. Metamorphosis occurs after larvae have attached to the substrate. As with other echinoderms, crinoids can regenerate lost parts.

Class Concentricycloidea

The class Concentricycloidea (kon-sen′tri-si-kloi″de-ah) (ME *consentrik*, having a common center + Gr. *kykloeides*, like a circle) contains a single described species, known as the sea daisy. Sea daisies have been recently discovered in deep oceans, on wood and other debris (figure 25.13). They lack arms and are less than one cm in diameter. The most distinctive feature of this species is two circular water-vascular rings that encircle the disclike body. The inner of the two rings probably corresponds to the ring canal of members of other classes because it has Polian vesicles attached. The outer ring contains tube feet and ampullae and probably corresponds to the radial canals of members of other classes. In addition, this animal lacks an internal digestive system. Instead, the surface of the animal that is applied to the substrate (decomposing wood) is covered by a thin membrane, called a *velum*, that digests and absorbs nutrients. Internally, there are five pairs of brood pouches where embryos are held during development. There are apparently no free-swimming larval stages. The mechanism for fertilization is unknown.

Further Phylogenetic Considerations

As described earlier, most zoologists believe that echinoderms evolved from bilaterally symmetrical ancestors. Radial symmetry probably evolved during the transition from active to more sedentary life styles; however, the oldest echinoderm fossils, about 600 million years old, give little direct evidence of how this transition occurred.

Ancient fossils do give clues regarding the origin of the water-vascular system and the calcareous endoskeleton. Of all living echinoderms, the crinoids most closely resemble

[8] the oldest fossils. Because crinoids use their water-vascular system for suspension feeding, it is believed that filter feeding, not locomotion, was probably the original function of the water-vascular system. As do crinoids, early echinoderms probably assumed a mouth-up position and were attached aborally. Arms and tube feet could have been used to capture food and move it to the mouth. The calcium carbonate endoskeleton may have evolved for support of extended filtering arms and for protection of these sessile animals.

Many modern echinoderms are more mobile. This free-living life-style is probably secondarily derived, as is the mouth-down orientation of most echinoderms. The mouth-down position would be advantageous for predatory and scavenging life-styles. Similarly, changes in the water-vascular system, such as the evolution of ampullae, suction discs, and feeding tentacles, can be interpreted as adaptations for locomotion and feeding in a more mobile life-style. The idea that the free-living life-style is secondary, is reinforced by the observation that some echinoderms, such as the irregular echinoids and the holothuroids, have

bilateral symmetry imposed upon a pentaradial body form.

Evolutionary relationships among echinoderm classes are obscure. The large number of extinct classes that can be traced back in excess of 500 million years, as well as the remarkable changes that have occurred in the evolution of this phylum, make statements regarding phylogenetic relationships highly speculative.

Stop and Ask Yourself

13. What is the function of the stalk, calyx, and cirri in crinoids?
14. How is the function of the crinoid water-vascular system different from that of other echinoderms?
15. How do the water-vascular systems and feeding mechanisms of sea daisies differ from those of other echinoderms?
16. What could account for the evolution of the calcium carbonate endoskeleton of echinoderms?

■ Summary

1. Echinoderms, chordates, and other deuterostomes share a common, but remote, ancestry. Modern echinoderms were probably derived from bilaterally symmetrical ancestors.
2. Echinoderms are pentaradially symmetrical, have an internal skeleton of interlocking calcium carbonate ossicles, and have a water-vascular system that is used for locomotion, food gathering, attachment, and exchanges with the environment.
3. Members of the class Asteroidea are the sea stars. They are predators and scavengers, and their arms are broadly joined to the central disc. Sea stars are dioecious, and external fertilization results in the formation of planktonic bipinnaria and brachiolaria larvae. Sea stars also have remarkable powers of regeneration.
4. The brittle stars and basket stars make up the class Ophiuroidea. Arms are sharply set off from the central disc. Ophiuroids are dioecious, and externally fertilized eggs may develop in the plankton or they may be brooded.
5. The class Echinoidea includes the sea urchins, heart urchins, and sand dollars. They have a specialized chewing structure, called Aristotle's lantern. External fertilization results in a planktonic pluteus larva.
6. Members of the class Holothuroidea include the sea cucumbers. They rest on one side, are elongate along their oral/aboral axis, and their body wall contains microscopic ossicles. Many sea cucumbers eviscerate themselves when

disturbed. Sea cucumbers are dioecious, and fertilization and development are external.
7. The class Crinoidea contains the sea lilies and feather stars. They are oriented oral side up and use arms and tube feet in suspension feeding. Crinoids are dioecious, and fertilization and development are external.
8. The class Concentricycloidea contains one recently discovered species that lives on wood and other debris in deep water.
9. Radial symmetry of echinoderms probably evolved during a transition to a sedentary, filter-feeding life-style. The water-vascular system and the calcareous endoskeleton are probably adaptations for that life-style. The evolution of a more mobile life-style has resulted in the use of the water-vascular system for locomotion and the assumption of a mouth-down position.

■ Key Terms

ambulacral groove (p. 397)
Aristotle's lantern (p. 401)
dermal branchiae (p. 396)
pedicellariae (p. 397)
pentaradial symmetry (p. 395)
respiratory trees (p. 403)
tube feet (p. 396)
water-vascular system (p. 395)

■ Critical Thinking Questions

1. What is pentaradial symmetry and why is it adaptive for echinoderms?

2. Why do zoologists think that pentaradial symmetry was not present in the ancestors of echinoderms?

3. What are the functions of the coelom in virtually all echinoderms?

4. Compare and contrast the structure and function of the water-vascular systems of asteroids, ophiuroids, echinoids, holothuroids, and crinoids.

5. In which of the above groups is the water-vascular system probably most similar in form and function to an ancestral condition? Explain your answer.

6. What physical process is responsible for gas exchange and excretion in all echinoderms? What structures facilitate these exchanges in each echinoderm class?

■ Suggested Readings

Books

Barnes. R. D. 1987. *Invertebrate Zoology,* 5th ed. Philadelphia: Saunders College Publishing.

Binyon, J. 1972. *Physiology of Echinoderms.* New York: Pergamon Press, Inc.

Fell, H. B. 1982. Echinodermata. In S. P. Parker (ed.). *Synopsis and Classification of Living Animals,* vol, 2. New York: McGraw-Hill Book Co.

Pearse, V., Pearse, J., Buchsbaum, M., and Buchsbaum, R. 1987. *Living Invertebrates.* New York: Blackwell Scientific Publications.

Articles

Birkeland, C. 1989. The faustian traits of the crown-of-thorns starfish. *American Scientist* 72(2):154–163.

Bosch, I, Rivkin R. B., and Alexander, S. P. 1989. Asexual reproduction by oceanic plantotropic echinoderm larvae. *Nature* 337:169–170.

Burnett, A. L. 1960. The mechanism employed by the starfish *Asterias forbesi* to gain access to the interior of the bivalve *Venus mercenaria. Ecology* 4:583–584.

Ewing, T. 1988. Thorny problem as Australia's coastline faces invasion. *Nature* 333:387.

Feder, H. M. Escape responses in marine invertebrates. *Scientific American* July, 1972.

Grober, M. S. 1988. Brittle-star bioluminescence functions as an aposematic signal to deter crustacean predators. *Animal Behaviour* 36:493–501.

Messing, C. G. 1988. Sea lilies and feather stars. *Sea Frontiers* 34:236–241.

Mitchell, T. 1988. Coral-killing starfish meet their mesh. *New Scientist* 120:28.

Shimek, R. B. 1987. Sex among the sessile: with the onset of spring in cool northern Pacific waters, even sea cucumbers bestir themselves. *Natural History* 96:60–63.

Hemichordata and Invertebrate Chordates

Concepts

1. Members of the phyla Echinodermata, Hemichordata, and Chordata are probably derived from a common diploblastic or triploblastic ancestor.
2. The phylum Hemichordata includes the acorn worms (class Enteropneusta) and the pterobranchs (class Pterobranchia). Hemichordates live in or on marine substrates and feed on sediment or suspended organic matter.
3. Animals in the phylum Chordata are characterized by a notochord, pharyngeal gill slits or pouches, a tubular nerve cord, and a postanal tail.
4. The urochordates are marine, and are called tunicates. They are attached or planktonic, and solitary or colonial as adults. All are filter feeders.
5. Members of the subphylum Cephalochordata are called lancelets. They are filter feeders that spend most of their time partly buried in marine substrates.
6. Motile, fishlike chordates may have evolved from sedentary, filter-feeding ancestors as a result of neoteny in a motile larval stage.

Have You Ever Wondered:

[1] what acorn worms are?
[2] what characteristics are shared by all chordates at some time in their life history?
[3] what animals deposit cellulose in their body walls?
[4] why sea squirts are classified in the same phylum (Chordata) as humans?
[5] why cephalochordates, such as amphioxus, are studied in introductory zoology laboratories?
[6] how fishlike chordates could have evolved from filter-feeding ancestors?

These and other useful questions will be answered in this chapter.

This chapter contains underlined evolutionary concepts.

Evolutionary Perspective

Some members of one of the phyla discussed in this chapter are more familiar to beginning students of zoology than members of any other group of animals. This familiarity is not without good reason, for zoologists themselves are members of one of these phyla—Chordata. Other members of these phyla, however, are much less familiar. Observations during a walk along a seashore at low tide may reveal coiled castings (sand, mud, and excrement) at the openings of U-shaped burrows, and in excavating these burrows may reveal a wormlike animal that is one of the members of a small phylum—Hemichordata. Other members of this phylum include equally unfamiliar filter feeders called pterobranchs.

While at the seashore, one could also see animals clinging to rocks exposed by low tide. At first glance, they might be described as jellylike masses with two openings at their unattached end (colorplate 10a). Some are found as solitary individuals, others live in colonies. Handling these animals may be rewarded with a stream of water squirted from their openings. Casual observations provide little evidence that these small filter feeders, called sea squirts or tunicates, are chordates. However, detailed studies have made that conclusion certain. Tunicates and a small group of fishlike cephalochordates are often called the invertebrate chordates because they lack a vertebral column.

Phylogenetic Relationships

Animals in the phyla Hemichordata and Chordata share deuterostome characteristics with echinoderms. Most zoologists, therefore, believe that ancestral representatives of these phyla were derived from a common, as yet undiscovered, diploblastic or triploblastic ancestor (colorplate 5). The chordates are characterized by a dorsal, tubular nerve cord, a notochord, pharyngeal gill slits, and a postanal tail. The only characteristics that they share with the hemichordates are gill slits and, in some species, a dorsal, tubular nerve cord. Therefore, most zoologists agree that the evolutionary ties between the chordates and hemichordates are closer than those between echinoderms and either phylum. However, chordates and hemichordates probably diverged from widely separated points along the deuterostome lineage. This generalization is supported by the diverse body forms and life-styles present in these phyla.

Phylum Hemichordata

The phylum Hemichordata (hem'i-kor-da'tah) (Gr. *hemi*, half + L. *chorda*, cord) includes the acorn worms (class Enteropneusta) and the pterobranchs (class Pterobranchia) (table 26.1). Members of both classes live in or on marine sediments.

Table 26.1 Classification of the Hemichordata and Chordata

Phylum Hemichordata (hem'i-kor-da'tah)
Widely distributed in shallow, marine, tropical waters and deep, cold waters; soft bodied and wormlike; epidermal nervous system; most with pharyngeal gill slits.

Class Enteropneusta (ent'er-op-nus''tah)
Shallow water, wormlike animals; inhabit burrows on sandy shorelines; body divided into three regions: proboscis, collar, and trunk. Acorn worms (*Balanoglossus, Saccoglossus*).

Class Pterobranchia (ter'o-brang''ke-ah)
With or without gill slits; two or more arms; often colonial, living in an externally secreted encasement. *Rhabdopleura.*

Phylum Chordata (kor-dat'ah) (L. *chorda,* cord)
Occupy a wide variety of marine, freshwater and terrestrial habitats. A notochord, pharyngeal gill slits, a dorsal tubular nerve cord, and a postanal tail all present at some time in chordate life histories.

Subphylum Urochordata (u'ro-kor-dat'ah)
Notochord, nerve cord, and postanal tail present only in free-swimming larvae; adults sessile, or occasionally planktonic, and enclosed in a tunic that contains some cellulose; marine. Sea squirts or tunicates.

Class Ascidiacea (as-id'e-as''e-ah)
All sessile as adults; solitary or colonial; colony members interconnected by stolons.

Class Larvacea (lar-vas'e-ah)
Planktonic; adults retain tail and notochord; lack a cellulose tunic; epithelium secretes a gelatinous covering of the body.

Class Thaliacea (tal'e-as''e-ah)
Planktonic; adults are tailless and barrel shaped; oral and atrial openings are at opposite ends of the tunicate; water currents produced by muscular contractions of the body wall.

Subphylum Cephalochordata (sef'a-lo-kor-dat'ah)
Body laterally compressed and transparent; fishlike; all four chordate characteristics persist throughout life. Amphioxus (*Branchiostoma*).

Class Enteropneusta

Members of the class Enteropneusta (ent'er-op-nus''tah) (Gr. *entero,* intestine + *pneustikos,* for breathing) are marine worms that usually range in size between 10 and 40 cm, although some can be as long as 2 m. There are about 70 described species, and most occupy U-shaped burrows in sandy and muddy substrates between the limits of high and low tides. The common name of the enteropneusts—acorn worms—is derived from the appearance of the *proboscis,* which is a short, conical projection at the anterior end of the worm. A ringlike *collar* is posterior to the proboscis, and an elongate *trunk* is the third division of the body (figure 26.1). Acorn worms are covered by a ciliated epidermis and gland cells. The mouth is located ventrally [1]

Table 26.1 (continued)

Subphylum Vertebrata (ver'te-bra'tah)
Notochord, nerve cord, postanal tail, and gill slits present at least in embryonic stages; vertebrae surround nerve cord and serve as primary axial support; skeleton modified anteriorly into a skull for protection of the brain.

Class Cephalaspidomorphi (sef-ah-las'pe-do-morf'e)
Fishlike; jawless; no paired appendages; cartilaginous skeleton; sucking mouth with teeth and rasping tongue. Lampreys.

Class Myxini (mik-sy-ny)
Fishlike; jawless; no paired appendages; mouth with four pairs of tentacles; olfactory sacs open to mouth cavity; 5 to 15 pairs of gill slits. Hagfishes.

Class Chondrichthyes (kon-drik'thi-es)
Fishlike; jawed; paired appendages and cartilaginous skeleton; no swim bladder. Skates, rays, sharks.

Class Osteichthyes (os'te-ik'thee-ez)
Bony skeleton; swimbladder and operculum present. Bony fish.

Class Amphibia (am-fib'e-ah)
Skin with mucoid secretions; possess lungs and/or gills; moist skin serves as respiratory organ, aquatic developmental stages usually followed by metamorphosis to an adult. Frogs, toads, salamanders.

Class Reptilia (rep-til'e-ah)
Dry skin with epidermal scales; amniotic eggs; terrestrial embryonic development. Snakes, lizards, alligators.

Class Aves (a'vez)
Scales modified into feathers for flight; efficiently regulate body temperature (endothermic); amniotic eggs. Birds.

Class Mammalia (ma-may'le-ah)
Bodies at least partially covered by hair; endothermic; young nursed from mammary glands; amniotic eggs. Mammals.

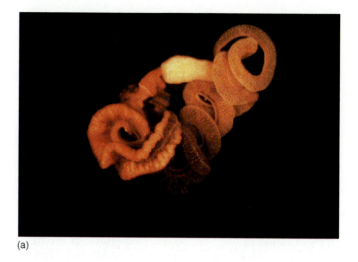

(a)

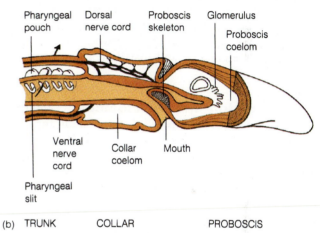

(b) TRUNK   COLLAR   PROBOSCIS

Figure 26.1 Class Enteropneusta. (a) *Saccoglossus kowalevskii*. (b) Longitudinal section showing the proboscis, collar, pharyngeal regions, and the internal structures.

between the proboscis and the collar. A variable number of gill slits, from a few to several hundred, are positioned laterally on the trunk. Gill slits are openings between the anterior region of the digestive tract, called the pharynx, and the outside of the body. Their functions in hemichordates and chordates are discussed later in this chapter.

Most of the activities of enteropneusts are restricted to the confines of their burrows. Movement and burrowing is accomplished by contractions of muscles in the proboscis.

Maintenance Functions Feeding of acorn worms is assisted by ciliary mucus. Detritus and other particles adhere to the mucus-covered proboscis. Tracts of cilia transport food and mucus posteriorly and ventrally. Ciliary tracts converge near the mouth and form a mucoid string that enters the mouth. Acorn worms may reject some substances trapped in the mucoid string by pulling the proboscis against the collar. Material to be rejected is transported along ciliary tracts of the collar and trunk and discarded posteriorly.

The digestive tract of enteropneusts is a simple tube (figure 26.1b). Digestion of food occurs as enzymes are released from diverticula of the gut, called hepatic sacs. The posterior end of the worm is extended out of the burrow during defecation. At low tide, one may see coils of fecal material, called castings, lying on the substrate at one of the openings of a burrow.

The nervous system of enteropneusts is ectodermal in origin and lies at the base of the ciliated epidermis. It consists of dorsal and ventral nerve tracts and a network of epidermal nerve cells, called a *nerve plexus*. In some species, the dorsal nerve is tubular and usually contains giant nerve fibers that rapidly transmit impulses. There are no major integrative areas. Sensory receptors are unspecialized and widely distributed over the body.

Because acorn worms are small, exchange of respiratory gases and metabolic waste products (principally ammonia) probably occur by diffusion across the body wall. In addition, respiratory gases are exchanged at the pharyngeal gill slits. Cilia associated with pharyngeal gill slits circulate water into the mouth and out of the body through gill slits. As water passes through gill slits, gases are exchanged by diffusion between water and blood sinuses surrounding the pharynx.

The circulatory system of acorn worms consists of one dorsal and one ventral contractile vessel (figure 26.1b). Blood moves anteriorly in the dorsal vessel and posteriorly in the ventral vessel. Branches from these vessels lead to open sinuses. All blood flowing anteriorly passes into a series of blood sinuses, called the glomerulus, at the base of the proboscis. Excretory wastes may be filtered through the glomerulus, into the coelom of the proboscis, and released to the outside through one or two pores in the wall of the proboscis. The blood of acorn worms is colorless, lacks cellular elements, and distributes nutrients and wastes.

Reproduction and Development Enteropneusts are dioecious. Two rows of gonads lie in the body wall in the anterior region of the trunk, and each gonad opens separately to the outside. Fertilization is external. Spawning by one worm induces others in the area to spawn—behavior that suggests the presence of spawning pheromones. Ciliated larvae, called *tornaria,* swim in the plankton for several days to a few weeks. The larvae settle to the substrate and are gradually transformed into the adult form (figure 26.2).

Class Pterobranchia

Pterobranchia (ter'o-brang"ke-ah) (Gk. *pteron,* wing or feather + *branchia,* gills) is a small class of hemichordates found mostly in deep, oceanic waters of the southern hemisphere. A few are found in European coastal waters and in shallow waters near Bermuda. There are about 20 described species of pterobranchs.

Pterobranchs are small, ranging in size from 0.1 to 5 mm. Most live in secreted tubes in asexually produced colonies. As in enteropneusts, the pterobranch body is divided into three regions. The proboscis is expanded and shield-like (figure 26.3). It secretes the tube and aids in movement in the tube. The collar possesses two to nine arms with numerous ciliated tentacles. The trunk is U shaped.

Maintenance Functions Pterobranchs use water currents generated by cilia on their arms and tentacles to filter feed. Food particles are trapped and transported by cilia toward the mouth. Although there is a single pair of pharyngeal gill slits in one genus, there is little need for either respiratory or excretory structures in animals as small as pterobranchs, because exchanges of gases and wastes occur by diffusion.

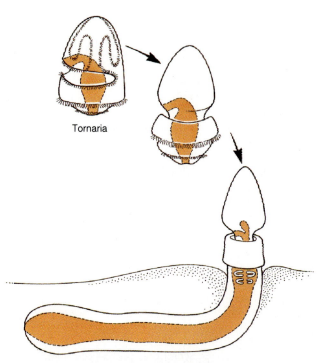

Tornaria

Recently settled enteropneust

Figure 26.2 The tornaria larva and metamorphosis of *Glossobalanus*. When larval development is complete, a tornaria locates a suitable substrate, settles, and begins to burrow and elongate.

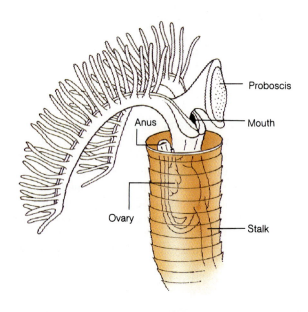

Figure 26.3 External structure of *Rhabdopleura*. Ciliated tracts on tentacles and arms direct food particles toward the mouth.

Figure 26.2 and 26.3 redrawn, with permission, from *Living Invertebrates* by Pearse/Buchsbaum, copyright by The Boxwood Press.

Box 26.1

Planktonic Tunicates

Although the description of urochordates in this textbook is based primarily upon attached ascidian tunicates, planktonic species are numerous and important in marine food webs. Dense swarms of tunicates, hundreds of kilometers wide and many meters deep, are common in the open ocean. Swarms of larvaceans (*see figure 26.4a*) have been estimated to contain up to 25,000 animals per m³! Larvaceans filter organisms as small as bacteria (0.1μm in diameter), and in turn, are fed on by other plankton feeders, such as sardines and herring.

Thaliacean tunicates also occur in large, dense swarms in the open ocean (*see figure 26.4b*). Most are aggregations of solitary individuals, however, some form spectacular luminescent colonies. Pyrosome colonies, such as the one shown in box figure 26.1, are found in many oceans. Colonies 10 m long and 1 m in diameter are common. Individuals are oriented with oral siphons pointed outward and atrial siphons directed toward the center of the colony. Ciliary currents and contractions of body-wall muscles direct water toward a central cavity of the colony and slowly move the entire colony through the water. When a part of the colony is stimulated by chemical or mechanical stimuli, it luminesces and

ceases ciliary beating. The luminescence spreads over the entire colony, and the colony stops moving. This behavior may help the colony avoid unfavorable environments or confuse or frighten predators.

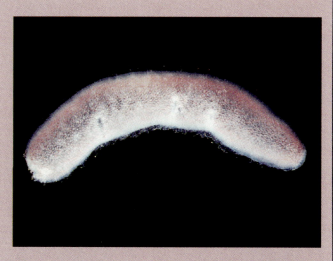

Box Figure 26.1 A pyrosome colony (*Pyrosoma spinosum*).

Reproduction and Development Asexual budding is common in pterobranchs and is responsible for colony formation. Pterobranchs also possess one or two gonads in the anterior trunk. Most species are dioecious, and external fertilization results in the development of a planulalike larva that lives for a time in the tube of the female. This nonfeeding larva eventually leaves the female's tube, settles to the substrate, forms a cocoon, and metamorphoses into an adult.

Stop and Ask Yourself

1. What are the three body regions of a hemichordate? What is the function of these regions in acorn worms? In pterobranchs?
2. How are the feeding mechanisms of acorn worms and pterobranchs similar?
3. How do respiratory and excretory functions occur in hemichordates?

Phylum Chordata

Although the phylum Chordata (kor-dat'ah) (L. *chorda*, cord) does not have an inordinately large number of species (about 45,000) its members have been very successful at

adapting to aquatic and terrestrial environments throughout the world. Sea squirts, members of the subphylum Urochordata, were briefly described in the introduction to this chapter. Other chordates include lancelets (subphylum Cephalochordata) and the vertebrates (subphylum Vertebrata) (*see table 26.1*).

It is relatively easy to characterize members of the phylum Chordata. At some time in their life history, all chordates have four easily recognized characteristics: a notochord, pharyngeal gill slits, a tubular nerve cord, and a postanal tail (*see figure 10.7a,b*). [2]

The phylum is named after the **notochord** (Gr. *noton*, the back + L. *chorda*, cord), a supportive rod that extends most of the length of the animal dorsal to the body cavity and into the tail. It consists of a connective-tissue sheath that encloses cells, each of which contains a large, fluid-filled vacuole. This arrangement gives the notochord some turgidity, which prevents compression along the anterior/posterior axis. At the same time, it is flexible enough to allow some freedom for lateral bending, as in the lateral undulations of a fish during swimming. In most adult vertebrates, the notochord is partly or entirely replaced by cartilage or bone.

Pharyngeal gill slits are a series of openings in the pharyngeal region between the digestive tract and the outside of the body. In some chordates, diverticula from the gut in the pharyngeal region never break through to form an open passageway to the outside. These diverticula are then called *pharyngeal gill pouches.* The earliest chordates used the gill slits for filter feeding; some living chordates still use them for feeding. Some chordates have developed gills in the pharyngeal pouches for gas exchange. As is the notochord, the pharyngeal gill slits of higher vertebrates are mainly embryonic features.

The **tubular nerve cord** and its associated structures are largely responsible for the success of the chordates. The nerve cord runs along the longitudinal axis of the body, just dorsal to the notochord, and is usually expanded anteriorly as a brain. This central nervous system is associated with the development of complex systems for sensory perception, integration, and motor responses.

The fourth chordate characteristic is a **postanal tail.** (a tail that extends posteriorly beyond the anal opening.) The tail is generally supported by the notochord or vertebral column.

Subphylum Urochordata

Members of the subphylum Urochordata (u'ro-kor-dat'ah) (Gr. *uro,* tail + L. *chorda,* cord) are the tunicates or sea squirts. The subphylum is divided into three classes (*see table 26.1*). The ascidians comprise the largest class. They are sessile as adults and are either solitary or colonial. The larvaceans and thaliaceans are planktonic as adults. (figure 26.4; box 26.1). In some localities, tunicates occur in large enough numbers to be considered a dominant life form.

Sessile urochordates attach their saclike bodies to rocks, pilings, hulls of ships, and other solid substrates. The unattached end of urochordates contains two *siphons* that permit sea water to circulate through the body. One siphon is the *oral siphon,* which is the inlet for water circulating through the body and is usually directly opposite the attached end of the ascidian (figure 26.5). It also serves as the mouth opening. The second siphon, the *atrial siphon* is the opening for excurrent water.

The body wall of most tunicates (L. *tunicatus,* to wear [3] a tunic or gown) is a connective-tissuelike covering, called

Figure 26.4 Subphylum Urochordata. (a) Members of the class Larvacea are planktonic and have a tail and notochord that persist into the adult stage. (b)The thaliaceans are barrel-shaped, planktonic urochordates. Oral and atrial siphons are at opposite ends of the body, and muscles of the body wall contract to create a form of weak jet propulsion.

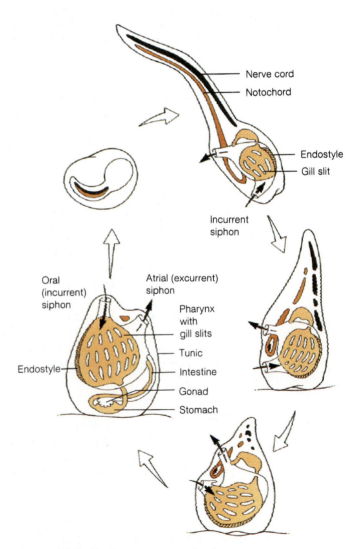

Figure 26.5 Tunicate metamorphosis.

the *tunic,* that appears gellike, but is often quite tough. It is secreted by the epidermis and is composed of proteins, various salts, and cellulose. Some mesodermally derived tissues are incorporated into the tunic, including blood vessels and blood cells. Rootlike extensions of the tunic, called *stolons,* help anchor a tunicate to the substrate and may connect individuals of a colony.

Maintenance Functions   Longitudinal and circular muscles are present below the body wall epithelium and help to change the shape of the adult tunicate. They act against the elasticity of the tunic and the hydrostatic skeleton created by sea water confined to internal chambers.

The nervous system of tunicates is largely confined to the body wall. It forms a nerve plexus with a single ganglion located on the wall of the pharynx between the oral and atrial openings (figure 26.6a). This ganglion is not vital for coordinating bodily functions. Tunicates are sensitive to many kinds of mechanical and chemical stimuli, and receptors for these senses are distributed over the body

wall, especially around the siphons. There are no complex sensory organs.

The most obvious internal structures of the urochordates are a very large *pharynx* and a cavity, called the *atrium,* that surrounds the pharynx laterally and dorsally (figure 26.6b). The pharynx of tunicates originates at the oral siphon and is continuous with the remainder of the digestive tract. The oral margin of the pharynx has tentacles that prevent large objects from entering the pharynx. The pharynx is perforated by numerous gill slits called *stigmas.* Cilia associated with the stigmas cause water to circulate into the pharynx, through the stigmas, and into the surrounding atrium. Water leaves the tunicate through the atrial siphon.

The digestive tract of adult tunicates continues from the pharynx and ends at the anus near the atrial siphon. During feeding, a mucous sheet is formed by cells of a ventral, ciliated groove, called the **endostyle** (figure 26.6b; box 26.2). Cilia move the mucous sheet dorsally across the pharynx. Food particles, brought into the oral siphon with incurrent water, are trapped in the mucous sheet and passed dorsally. Food is incorporated into a string of mucus that is moved by ciliary action into the next region of the gut tract. Digestive enzymes are secreted in the stomach, and most absorption occurs across the walls of the intestine. Digestive wastes are carried from the anus out of the atrial siphon with excurrent water.

In addition to its role in feeding, the pharynx also functions in gas exchange. Gases are exchanged between water circulating through the tunicate.

The tunicate heart lies at the base of the pharynx. One vessel from the heart runs anteriorly under the endostyle and another runs posteriorly to the digestive organs and gonads. Blood flow through the heart is not unidirectional. Peristaltic contractions of the heart may propel blood in one direction for a few beats, then the direction is reversed. The significance of this reversal is not understood. Tunicate blood plasma is colorless and contains various kinds of ameboid cells.

Excretion is accomplished by the diffusion of ammonia into water that passes through the pharynx. In addition, ameboid cells of the circulatory system accumulate uric acid and sequester it in the intestinal loop. *Pyloric glands* on the outside of the intestine are also thought to have excretory functions.

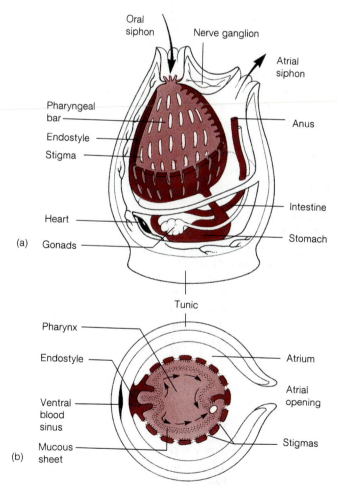

Figure 26.6 Internal structure of a tunicate. (a) Longitudinal section. Arrows show the path of water. (b) Cross section at the level of the atrial siphon. Arrows show movement of food trapped in mucus, which is produced by the endostyle.

Reproduction and Development   Urochordates are monoecious. Gonads are located near the loop of the intestine, and genital ducts open near the atrial siphon. Gametes may be shed through the atrial siphon for external fertilization, or eggs may be retained in the atrium for fertilization and early development. Although self-fertilization occurs in some species, cross-fertilization is the rule.

Development results in the formation of a tadpolelike [4] larva that possesses all four chordate characteristics. Metamorphosis begins after a brief free-swimming larval existence, during which the larva does not feed. The larva settles to a firm substrate and attaches by adhesive papillae located below the mouth. During metamorphosis, the tail is reduced by a shrinking of the outer epidermis that pulls the notochord and other tail structures internally for reorganization into adult tissues. The internal structures rotate 180°, resulting in the positioning of the oral siphon opposite the adhesive papillae and the bending of the digestive tract into a U shape (*see figure 26.5*).

Subphylum Cephalochordata

[5] Members of the subphylum Cephalochordata (sef'a-lo-kor-dat"ah) (Gr. *kephalo,* head + L. *chorda.* cord) are called lancelets. They are almost universally studied in introductory zoology courses during introductions to chordate structure and function because they so clearly demonstrate the four chordate characteristics.

There are two genera, *Branchiostoma* (amphioxus) and *Asymmetron,* and about 45 species of cephalochordates. They are distributed throughout the world's oceans in shallow waters that have clean sand substrates.

Cephalochordates are small (up to 5 cm long), tadpolelike animals. They are elongate, laterally flattened, and nearly transparent. In spite of their streamlined shape, cephalochordates are relatively weak swimmers and spend most of their time in a filter-feeding position—partly to mostly buried with their anterior end sticking out of the sand (figure 26.7a).

The notochord of cephalochordates extends from the tail to the head, giving them their name (figure 26.7b). Unlike the notochord of other chordates, most of the cells are muscle cells, making the notochord somewhat contractile. Both of these characteristics are probably adaptations to burrowing. Contraction of the muscle cells increases the rigidity of the notochord by compressing the fluids within, giving additional support when pushing into sandy substrates. Relaxation of these muscles allows increased flexibility for swimming.

Muscles are arranged on either side of the notochord in V-shaped units called *myotomes.* Myotomes on opposite sides of the body alternate with one another, and oppose each other when they contract. Waves of contraction move anteriorly to posteriorly along the body wall, causing undulations that propel the cephalochordate through the water. *Metapleural folds* are longitudinal, ventrolateral folds of the body wall that help stabilize cephalochordates during swimming. A median *dorsal fin* and a *caudal fin* also aid in swimming but they are not homologous to similar structures in some vertebrates.

An *oral hood* projects from the anterior end of cephalochordates. Ciliated, fingerlike projections, called *cirri,* hang from the ventral aspect of the oral hood and are used in feeding. The posterior wall of the oral hood bears the mouth opening that leads to a large pharynx. Numerous pairs of gill slits perforate the pharynx and are supported by cartilaginous *gill bars.* Large folds of the body wall extend ventrally around the pharynx and fuse at the ventral midline of the body, creating the *atrium,* which is a chamber that surrounds the pharyngeal region of the body. It may protect the delicate, filtering surfaces of the pharynx from bottom sediments. The opening from the atrium to the outside is called the *atriopore* (figures 26.7b).

Maintenance Functions Cephalochordates are filter feeders. During feeding, they are partially or mostly buried in sandy substrates with their mouths pointed upwards. Water is brought into the mouth by the action of cilia on the lateral surfaces of gill bars. Water passes from the pharynx, through gill slits to the atrium, and out of the body through the atriopore. Initial sorting of food occurs at the cirri. Larger materials are caught on cilia of the cirri. As these larger particles accumulate, they are thrown off by contractions of the cirri. Smaller, edible particles are pulled into the mouth with water and are collected by cilia on the gill bars and in mucus secreted by the *endostyle.*

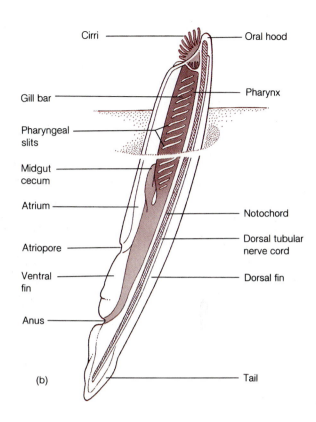

Figure 26.7 Subphylum Cephalochordata. (a) Amphioxus in its partially buried, feeding position. (b) Internal structure of amphioxus.

As in tunicates, the endostyle is a ciliated groove that extends longitudinally along the midventral aspect of the pharynx. Cilia move food and mucus dorsally, forming a food cord that is moved by cilia to the gut. A ring of cilia rotates the food cord, and in the process, food is dislodged. Digestion is both extracellular and intracellular. A diverticulum off the gut, called the *midgut cecum,* extends anteriorly. It ends blindly along the right side of the pharynx and secretes digestive enzymes. An anus is on the left side of the ventral fin.

Cephalochordates do not possess a true heart. Blood is propelled by contractile waves in the walls of major vessels. There is a single, ventral aorta, paired dorsal aortae that unite posterior to the pharynx, and branchial arteries that run between dorsal and ventral vessels in the gill bars of the pharynx. Blood contains ameboid cells and bathes tissues in open spaces.

Excretory tubules are modified coelomic cells that are closely associated with blood vessels. This arrangement suggests active transport of materials between the blood and excretory tubules.

The coelom of cephalochordates is reduced as compared to most other chordates. It is restricted to canals near the gill bars, the endostyle, and the gonads.

Reproduction and Development Cephalochordates are dioecious. Gonads bulge into the atrium from the lateral body wall. Gametes are shed into the atrium and leave the body through the atriopore. External fertilization leads to a bilaterally symmetrical larva. Larvae are free swimming, but they eventually settle to the substrate before metamorphosing to adults.

Further Phylogenetic Considerations

The evolutionary relationships between the hemichordates and chordates are difficult to document with certainty. The dorsal, tubular nerve cord and pharyngeal gill slits of hemichordates are evidence of evolutionary ties between these phyla. The short, proboscis skeleton, just dorsal to the mouth of acorn worms, has been likened to the notochord of chordates (*see figure 26.1*). Studies have suggested, however, that neither the structure nor the development of this supportive element represents homology with the chordate notochord. These similarities, therefore, are not sufficient to justify grouping hemichordates and chordates in the same phylum.

Evolutionary relationships between members of the three chordate subphyla are also speculative. As discussed in chapter 25, the earliest echinoderms were probably sessile filter feeders. The life-style of adult urochordates suggests a similar ancestry (perhaps from a common ancestor with echinoderms) for chordates (figure 26.8). The evolution of motile chordates from attached ancestors may have involved the development of a tadpolelike larva. Increased larval mobility is often adaptive for species with sedentary adults because it promotes dispersal. The evolution of motile adults could have resulted from neoteny, which is the development of sexual maturity in the larval body form. (The occurrence of neoteny is well documented in the animal kingdom, especially among amphibians.) Neoteny could have led to a small, sexually reproducing, [6]

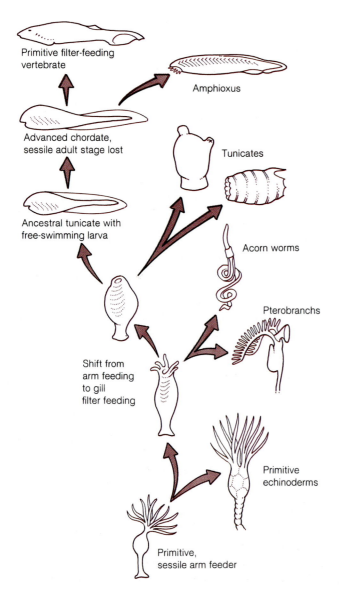

Primitive filter-feeding
vertebrate

Amphioxus

Advanced chordate,
sessile adult stage lost

Tunicates

Ancestral tunicate with
free-swimming larva

Acorn worms

Pterobranchs

Shift from
arm feeding
to gill
filter feeding

Primitive
echinoderms

Primitive,
sessile arm feeder

Figure 26.8 One interpretation of the early evolution of the chordates. The hypothetical ancestor of echinoderms, hemichordates, and chordates may have been an arm-feeding animal. A shift from arm feeding to filter feeding may have resulted in a divergence between the echinoderm and the hemichordate/chordate lineages. The neotenous development of a free-swimming chordate larva could then have resulted in the divergence between urochordates and other chordates.

fishlike chordate that could have been the ancestor of higher chordates.

The largest and most successful chordates belong to the subphylum Vertebrata. They are characterized by bony or cartilaginous *vertebrae* that completely or partially replace the notochord. A high degree of cephalization is evidenced by the development of the anterior end of the nerve cord into a brain and the development of specialized sense organs on the head. The skeleton is modified anteriorly into a *skull* or *cranium*. There are seven classes of vertebrates (*see table 26.1*). Because of their cartilaginous and bony endoskeletons, vertebrates have left an abundant fossil record. Ancient jawless fish were common in the Ordovician period, approximately 500 million years ago. Over a period of approximately 100 million years, fish became the dominant vertebrates. Near the end of the Devonian period, approximately 400 million years ago, terrestrial vertebrates made their appearance. Since that time, vertebrates have radiated into most of the earth's habitats. Chapters 27 to 31 give an account of these events.

Stop and Ask Yourself

4. What are four characteristics shared by all chordates at some time in their life history?
5. What group of chordates deposit cellulose in their body wall?
6. What is an endostyle? What functions does it have in urochordates? In cephalochordates?
7. What is the function of the midgut cecum of cephalochordates?

Summary

1. Echinoderms, hemichordates, and chordates share deuterostome characteristics and are believed to have evolved from a common diploblastic or triploblastic ancestor.

2. Members of the phylum Hemichordata include the acorn worms and the pterobranchs. Acorn worms are burrowing, marine worms, and pterobranchs are marine hemichordates whose collar possesses arms with numerous ciliated tentacles.

3. Chordates have four unique characteristics. A notochord is a supportive rod that extends most of the length of the animal. Pharyngeal gill slits are a series of openings between the digestive tract and the outside of the body. The tubular nerve cord lies just above the notochord and is expanded anteriorly into a brain. A postanal tail extends posteriorly to the anus and is supported by the notochord or the vertebral column.

4. Members of the subphylum Urochordata are the tunicates or sea squirts. Urochordates are sessile or planktonic filter feeders. Their development involves a tadpolelike larva.

5. The subphylum Cephalochordata includes small tadpolelike filter feeders that live in shallow, marine waters with clean sandy substrates. Their notochord extends from the tail into the head and is somewhat contractile.

6. The presence of gill slits and a tubular nerve cord link hemichordates and chordates to the same evolutionary lineage.

7. Chordates probably evolved from a sessile, filter-feeding ancestor. A larval stage of this sedentary ancestor may have undergone neoteny to produce a small, sexually reproducing fishlike chordate.

Key Terms

endostyle (p. 413)
notochord (p. 411)
pharyngeal gill slits
   (p. 411)
postanal tail (p. 412)
tubular nerve cord (p. 412)

Critical Thinking Questions

1. What evidence links hemichordates and chordates to the same evolutionary lineage?

2. What evidence of chordate affinities is present in adult tunicates? In larval tunicates?

3. What is neoteny? What is a possible role for neoteny in chordate evolution?

4. Discuss the possible influence of filter-feeding life-styles on early chordate evolution.

5. What selection pressures could have favored a foraging or predatory life-style for later chordates?

6. Examine the structure of the tornaria larva in figure 26.2. What deuterostome phylum, covered in an earlier chapter, has a similar larval stage? What is the name of that similar larva? In what ways is it similar? What does this suggest regarding the evolutionary relationships of acorn worms and members of this other phylum?

Suggested Readings

Books

Barnes, R. D. 1987. *Invertebrate Zoology* 5th ed. Philadelphia: Saunders College Publishing.

Barrington, E. J. W. 1965. *The Biology of Hemichordata and Protochordata.* San Francisco: W.H. Freeman and Co.

Halstead, L. B. 1968. *The Pattern of Vertebrate Evolution.* San Francisco: W.H. Freeman and Co.

Lovtrup, S. 1977. *The Phylogeny of Vertebrata.* New York: John Wiley and Sons, Inc.

Romer, A. S., and Parsons, T. S. 1986. *The Vertebrate Story,* 6th ed. Philadelphia: Saunders College Publishing.

Stahl, B. J. 1974. *Vertebrate History: Problems in Evolution.* New York: McGraw-Hill Book Co.

Russell-Hunter, W. D. 1979. *A Life of Invertebrates.* New York: Macmillan Inc.

Articles

Alldredge, A. L., and Madin, L. P. 1982. Pelagic tunicates: unique herbivores in the marine plankton. *BioScience,* 32:655–663.

Cloney, R. A. 1982. Ascidian larvae and the events of metamorphosis. *American Zoologist* 22:817–826.

Millar, R. H. 1971. The biology of ascidians. *Advances Marine Biology* 9:1–100.

27

The Fishes: Vertebrate Success in Water

Concepts

1. The earliest fossil vertebrates are 510 million-year-old ostracoderms.
2. Members of the superclass Agnatha include extinct ostracoderms, the lampreys, and the hagfishes. Agnathans lack jaws and paired appendages.
3. The superclass Gnathostomata includes the cartilaginous (class Chondrichthyes) and bony (class Osteichthyes) fishes.
4. Aquatic environments have selected for certain adaptations in fishes. These include the ability to move in a relatively dense medium, exchange gases with water or air, regulate buoyancy, detect environmental changes, regulate salt and water in their tissues, and help ensure reproductive success.
5. Adaptive radiation resulted in the large variety of fishes present today. Evolution of some fishes led to the terrestrial vertebrates.

Have You Ever Wondered:

[1] what group of fishes probably gave rise to all other fishes?
[2] how a shark's teeth are replaced?
[3] what fishes have lungs and breathe air?
[4] what fish living today is the closest relative of terrestrial vertebrates?
[5] why some sharks never stop moving?
[6] how a shark can find a flounder completely covered by sand?
[7] what environmental conditions selected for adaptations in fishes that eventually led to terrestrial vertebrates?

These and other useful questions will be answered in this chapter.

This chapter contains underlined evolutionary concepts.

Evolutionary Perspective

Over 70% of the earth's surface is covered by water, a medium that is buoyant and resistant to rapid fluctuations in temperature. Because life began in water, and living tissues are made mostly of water, it might seem that nowhere else would life be easier to sustain. This chapter describes why that is not entirely true.

One does not need to wear SCUBA gear to appreciate the fact that fishes are adapted to aquatic environments in a fashion that is unsurpassed by any other group of animals. If you spend recreational hours with hook and line, visit a marine theme park, or simply glance into a pet store when walking through a shopping mall, you can attest to the variety and beauty of fishes. This variety is evidence of adaptive radiation that began 500 million years ago and shows no sign of ceasing. Fishes not only dominate many watery environments, they are also the ancestors of all other members of the subphylum Vertebrata.

Phylogenetic Relationships

Fishes are members of the chordate subphylum Vertebrata; thus, they have vertebrae that surround their nerve cord and provide the primary axial support. They also have a skull that protects the brain (*see table 26.1*).

Even though zoologists do not know what animals were the first vertebrates, fossilized bony fragments indicate that vertebrates were present at least 510 million years ago. These fossils are from bony armor that covered animals called ostracoderms. Ostracoderms were relatively inactive filter feeders that lived on the bottom of prehistoric lakes and seas. They possessed neither jaws nor paired appendages; however, the evolution of fishes resulted in both jaws and paired appendages as well as many other structures. The results of this adaptive radiation are described in this chapter.

Did ancestral fishes live in fresh water or in the sea? The answer to this question is not simple. The first vertebrates were probably marine because ancient stocks of other deuterostom phyla were all marine. Vertebrates, however, adapted to fresh water very early, and much of the evolution of fishes occurred there. Apparently, early vertebrate evolution involved the movement of fishes back and forth between marine and freshwater environments. The majority of the evolutionary history of some fishes took place in ancient seas, and most of the evolutionary history of others occurred in fresh water. The importance of fresh water in the evolution of fishes is evidenced by the fact that over 41% of all fish species are found in fresh water, even though freshwater habitats represent only a small percentage (0.0093% by volume) of the earth's water resources.

Survey of Fishes

The taxonomy of fishes has been the subject of debate for many years. The system used in this textbook divides fishes into two superclasses based upon whether they lack jaws and paired appendages (superclass Agnatha) or possess these structures (superclass Gnathostomata) (table 27.1).

Table 27.1   Classification of Living Fishes

Subphylum Vertebrata

Superclass Agnatha (ag-nath′ah)
Lack jaws and paired appendages; cartilaginous skeleton; persistent notochord; two semicircular canals.

    Class Cephalaspidomorphi (sef-ah-las′pe-do-morf′e)
    Sucking mouth with teeth and rasping tongue; seven pairs of gill slits; blind olfactory sacs. Lampreys.

    Myxini (mik-sy-ny)
    Mouth with four pairs of tentacles; olfactory sacs open to mouth cavity; five to 15 pairs of gill slits. Hagfishes.

Superclass Gnathostomata (na′tho-sto′ma-tah)
Hinged jaws and paired appendages present; notochord may be replaced by vertebral column; three semicircular canals.

    Class Chondrichthyes (kon-drik′thi-es)
    Tail fin with large upper lobe (*heterocercal tail*); cartilaginous skeleton; lack opercula and a swim bladder or lungs. Sharks, skates, rays, ratfishes.

        Subclass Elasmobranchii (e-laz-mo′bran′ke-i)
        Cartilaginous skeleton may be partially ossified; placoid scales or no scales. Sharks, skates, rays.

        Subclass Holocephali (hol′o-sef′a-li)
        Operculum covers gill slits; lack scales; teeth modified into crushing plates; lateral-line receptors in an open groove. Ratfish.

    Class Osteichthyes (os′te-ik″the-es)
    Most with bony skeleton; single gill opening covered by operculum; pneumatic sac(s) function as lungs or swim bladders. Bony fishes.

        Subclass Dipneusti (dip-nu′ste)
        Tail fin rounded (*diphycercal tail*); fins lobed or elongate and narrow; pneumatic sacs function as lungs; teeth modified for grinding. Lungfishes.

        Subclass Crossopterygii (cros-op-te-rij′e-i)
        Paired fins with muscular lobes, and skeletal support of fins similar to that found in tetrapods; diphycercal tail; the one living species, *Latimeria,* with vestigial swim bladder; ancestral forms used pneumatic sacs as lungs; conical teeth. Lobe-finned fishes.

        Subclass Actinopterygii (ak′tin-op″te-rig-e-i)
        Paired fins supported by dermal rays; basal portions of paired fins not especially muscular; tail fin with approximately equal upper and lower lobes (*homocercal tail*); blind olfactory sacs. Ray-finned fishes.

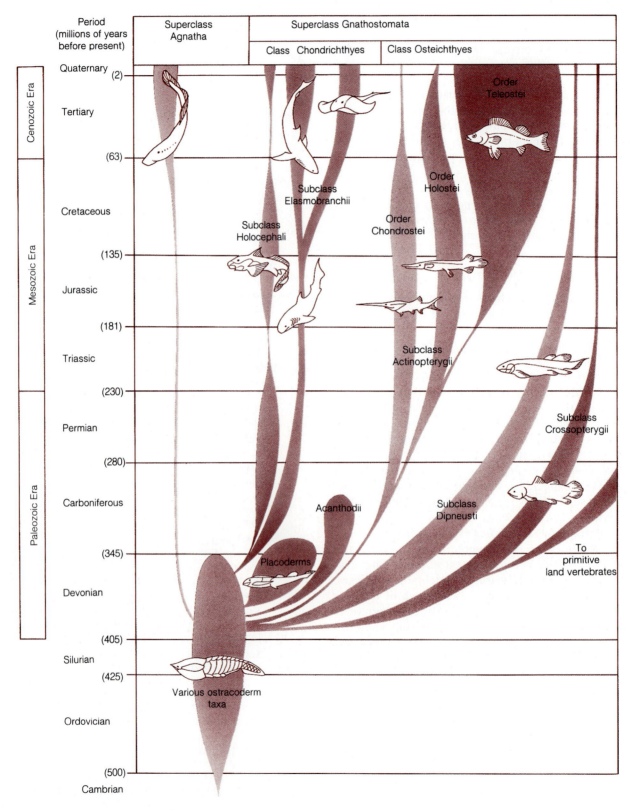

Figure 27.1 Evolution of the fishes. The relative number of species in each taxonomic group is indicated by the width of a group's branch.

Superclass Agnatha

Members of the superclass Agnatha (ag-nath′ah) (Gr. *a*, without + *gnathos*, jaw), in addition to lacking jaws and paired appendages, possess a cartilaginous skeleton and a notochord that persists into the adult stage. Ancient agnathans are believed to be ancestral to all other fishes (figure 27.1).

[1]

Ostracoderms, described briefly in the previous section, are extinct agnathans that belonged to several classes. The fossils of predatory water scorpions (phylum Arthropoda [*see figure 22.6*]) are often found with fossil ostracoderms. As sluggish as ostracoderms apparently were, bony armor was probably their only defense. Ostracoderms were bottom dwellers, often about 15 cm long (figure 27.2). Most are believed to have been filter feeders, either filtering suspended organic matter from the water or extracting annelids and other animals from muddy sediments. Bony plates around the mouths of some ostracoderms may have been used in a jawlike fashion to crack gastropod shells or the exoskeletons of arthropods.

Class Cephalaspidomorphi   Lampreys are agnathans in the class Cephalaspidomorphi (sef-ah-las′pe-do-morf′e) (Gr. *kephale*, head + *aspidos*, shield + *morphe*, form). They are common inhabitants of marine and freshwater environments in temperate regions (figure 27.3). Most adult lampreys prey on other fishes, and the larvae are filter feeders. The mouth of an adult is suckerlike and surrounded by lips that have sensory and attachment functions. Numerous epidermal teeth line the mouth and cover a movable tonguelike structure (colorplate 10b). Adults attach to prey with their lips and teeth and use their tongue to rasp away scales. Lampreys have salivary glands with anticoagulant secretions and feed mainly on the blood of their prey. Some lampreys, however, are not predatory. Members of the genus *Lampetra* are called brook lampreys. The larval stages of brook lampreys last for about 3 years, and the adults neither feed nor leave their stream. They reproduce soon after metamorphosis and then die.

Adult sea lampreys live in the ocean or the Great Lakes. Near the end of their lives, they undertake a migration that may take them hundreds of miles to a spawning bed in a freshwater stream. Once lampreys reach their spawning site, usually in relatively shallow water with swift currents, nest building begins. Lampreys make small depressions in

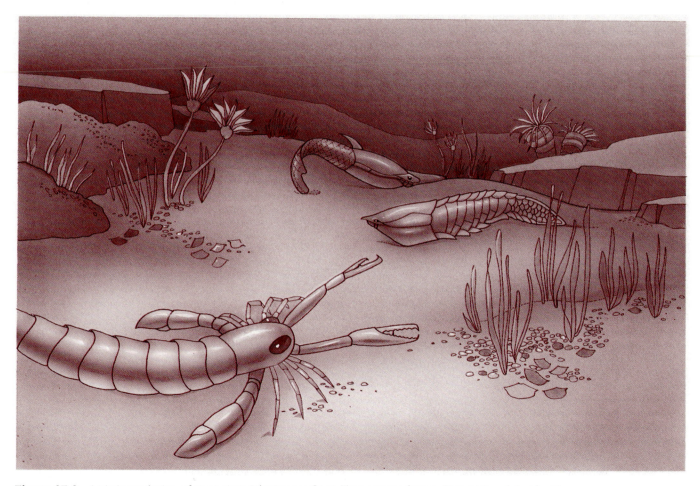

Figure 27.2   Artist's rendering of an ancient Silurian sea floor. Two ostracoderms, *Pteraspis* and *Anglaspis*, are shown with a predatory water scorpion.

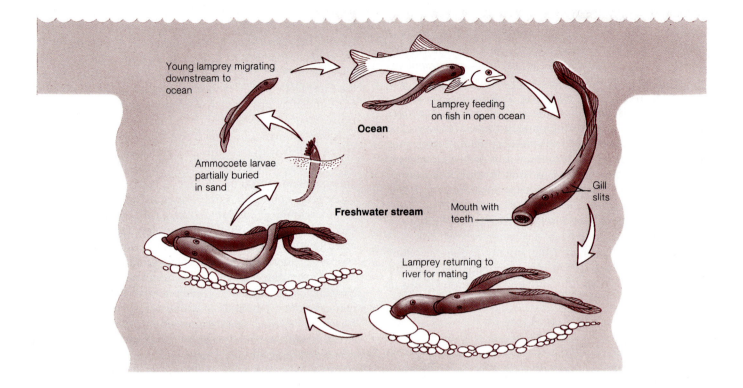

Figure 27.3 External structure and life history of a sea lamprey. Sea lampreys feed in the open ocean and near the end of their lives, migrate into freshwater streams where mating occurs. Eggs are deposited in nests on the stream bottom, and young ammocoete larvae hatch in about 3 weeks. Ammocoete larvae live as filter feeders until they attain sexual maturity.

the substrate. When the nest is prepared, a female usually attaches to a stone with her mouth. A male attaches to the female's head using his mouth, and wraps his body around the female (figure 27.3). Eggs are shed in small batches over a period of several hours, and fertilization is external. The relatively sticky eggs are then covered with sand.

Eggs hatch in approximately 3 weeks into *ammocoete larvae.* The larvae drift downstream to softer substrates, where they bury themselves in sand and mud and filter feed in a fashion similar to amphioxus (*see figure 26.7*).

Ammocoete larvae grow from 7 mm to about 17 cm over 3 years. During later developmental stages, the larvae metamorphose to the adult over a period of several months. The mouth becomes suckerlike, and the teeth, tongue, and feeding musculature develop. Lampreys eventually leave the mud permanently and begin a journey to the sea to begin life as predators. Adults will return only once to the headwaters of their stream to spawn and die (box 27.1).

Class Myxini Hagfish are members of the class Myxini (mik-sy'ny) (Gr. *myxa,* slime). Hagfish live buried in the sand and mud of marine environments, where they feed on soft-bodied invertebrates and scavenge dead and dying fish (figure 27.4). When hagfish find a suitable fish, they enter the fish through the mouth and eat the contents of the body, leaving only a sack of skin and bones. Anglers

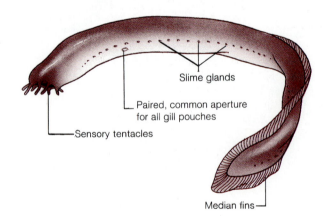

Figure 27.4 Hagfish external structure.

must contend with hagfish because hagfish will bite at a baited hook. Hagfish have the annoying habit of swallowing a hook so deeply that the hook is frequently lodged near the anus. The excessively slimy bodies of hagfish make all but the grittiest fishermen cut their lines and tie on a new hook.

Box 27.1

Lampreys and Great Lakes Fisheries

Along with many other European immigrants to the United States in the nineteenth century came many Scandinavians. Many of these Scandinavian immigrants fished for a living, and they were understandably attracted to the Great Lakes Region. Through the early part of the twentieth century, fishing was very good. Catches included a variety of smaller fishes that inhabited the shoals and bays of the Great Lakes, but the prize catches were two deep water predators—lake trout (*Salvelinus namaycush*) and whitefish (*Corygonus clupeaformis*). The yearly catch of lake trout in each of the Great Lakes in the 1930s exceeded 2,000 metric tons.

These commercial fisheries, however, were doomed. Their fate was sealed years before many immigrants even left their homes in Scandinavia. In 1829, the Welland Canal was completed. It provided a shipping route around Niagara Falls between Lakes Ontario and Erie. Niagara Falls, however, had not only been a barrier to shipping, it had also been a barrier to the sea lamprey. After the Welland Canal was completed, sea lampreys slowly worked their way from Lake Ontario to the other Great Lakes. By 1937, spawning lampreys were recorded in Lake Michigan, and by the early 1940s Great Lakes fishermen had to work very hard to bring home a single lake trout with a day's catch of predominately smaller fishes. The lamprey, like humans, had a decided preference for the larger, cold-water fish species. In 1944, the annual catch of lake trout from Lake Michigan had been reduced to less than 100 metric tons. In 1953, the annual catch of lake trout in Lake Michigan was reduced to a few hundred kilograms!

Although the invasion of lampreys into the Great Lakes brought severe economic hardship to many fishermen, it also resulted in an important success story in fishery management. In the 1950s, an intensive lamprey control program was instituted by the states bordering the Great Lakes and by Canada. Control measures involved the use of mechanical weirs that prevented spawning migrations of lampreys into the tributaries of the Great Lakes. Electrical shocking devices were employed in an attempt to kill lampreys in spawning streams. Finally, chemical control measures were employed. Lamprey populations began to decline, and by the mid 1960s lamprey control measures were considered a success.

The void left by the decline of lake trout has been filled by a sport fishery. In the late 1960s, Coho (*Concorhynchus kisutch*) and Chinook (*Oncorhynchus tshawytscha*) salmon were stocked in the Great Lakes to create a sport fishery. Survival and growth of these salmon have been remarkable, and fewer than 5% of the salmon caught are marked by lamprey wounds. Whitefish and lake trout are again being caught. To preserve this fishery, lamprey control measures will be maintained in the future to prevent large-scale growth of lamprey populations.

Stop and Ask Yourself

1. What superclass of vertebrates contains the probable ancestors of all other vertebrates?
2. How are members of the superclass Agnatha characterized?
3. How are the life-styles of adult and larval lampreys different?
4. What animals make up the class Myxini?

Superclass Gnathostomata

Two major developments in vertebrate evolution were the appearance of jaws and paired appendages. These structures are first seen in members of the superclass Gnathostomata (na'tho-sto'ma-tah) (Gr. *gnathos*, jaw + *stoma*, mouth). Jaws are used in feeding and are partly responsible for a transition to more active, predatory life-styles. *Pectoral fins* of fish are appendages that are usually located just behind the head, and *pelvic fins* are usually located ventrally and posteriorly (figure 27.5). Both sets of paired fins increase the agility of fish by giving them a more precise steering mechanism.

Two classes of gnathostomes still have living members: the cartilaginous fishes (class Chondrichthyes) and the bony fishes (class Osteichthyes). Another class, the armored fishes, or placoderms, contained the earliest jawed fishes (figure 27.6a). They are now extinct and apparently left no descendants. A fourth group of ancient, extinct fishes, the acanthodians, may be more closely related to the bony fishes (figure 27.6b).

Class Chondrichthyes Members of the class Chondrichthyes (kon-drik'thi-es) (Gr. *chondros*, cartilage + *ichthyos*, fish) include the sharks, skates, rays, and ratfish (*see table 27.1*). Most chondrichthians are carnivores or scavengers, and most are marine species. In addition to their biting mouthparts and paired appendages, chondrichthians possess epidermal placoid scales and a cartilaginous endoskeleton.

There are about 700 species in the subclass Elasmobranchii (e-laz'mo-bran'ke-i) (Gr. *elasmos*, plate metal + *branchia*, gills), which includes the sharks, skates, and rays (colorplate 10 c–f). Sharks arose from early jawed fishes midway through the Devonian period, about 375 million years ago. The absence of certain features characteristic of bony fishes (e.g., a swim bladder to regulate buoyancy, a

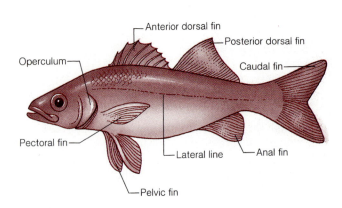

Figure 27.5   Appendages of a member of the superclass Gnathostomata.

(a)

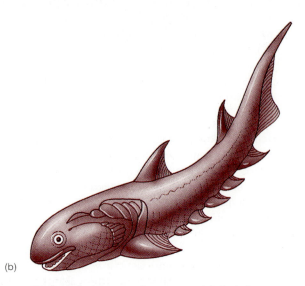

(b)

Figure 27.6   Primitive gnathostomes. (a) Artist's representation of an extinct placoderm. Note the large, hinged jaws and paired appendages. The bony armor of placoderms was usually less extensive than that of ostracoderms. (b) An acanthodian.

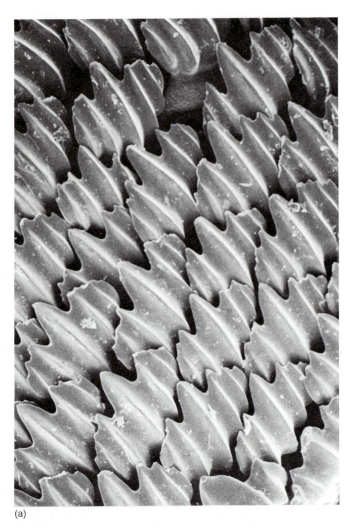

(a)

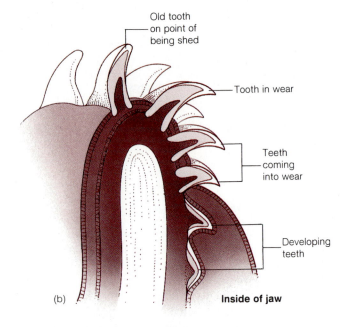

(b)

Figure 27.7   Scales and teeth of sharks. (a) A section of shark skin magnified to show posteriorly pointing placoid scales. (b) The teeth of sharks develop as modified placoid scales. Older teeth are continuously replaced by newer teeth that move from the inside to the outside of the jaw.

Box 27.2

Jaws from the Past

Paleontological records can tell scientists much about life-forms from the past. We know most about ancient animals whose bodies contained hard parts that were resistant to decay and that were more likely to fossilize. The fossil record for fishes that contained substantial quantities of bone is fairly complete. Records of fishes that lack bone are usually harder to find.

In spite of the cartilaginous skeletons of sharks, our knowledge of ancient sharks, although not perfect, is more complete than might be expected. Much of what we know of ancient sharks comes from discoveries made in Ohio. Sharks in Ohio may seem surprising at first, but during the upper Devonian (about 350 million years ago), the sea extended southwest from the St. Lawrence River region, across The Great Lakes, and down to Arkansas. The floor of this ocean, in the region now occupied by Ohio, was made of soft, deep sediments—ideal for fossilization. Some elasmobranch specimens are so well preserved that not only are cartilaginous skeletons preserved, but details of gill structures, details of muscle organization, and even remains of a last meal are sometimes found intact.

Discoveries such as these provide a wealth of information about ancient sharks. The body form of ancient sharks allowed them to become efficient predators, and that basic form is retained in most modern sharks. The shape of fossilized teeth can sometimes be used to identify the kind of shark they came from. Mineral deposits build up on teeth as they rest on the ocean floor. Assuming mineral deposits accumulate at a constant rate, it is possible to determine how long teeth have been resting on the ocean floor. Assuming that a ratio between tooth size and body size is constant for a given species, estimates of the length of a shark can be obtained from a fossilized tooth.

Reconstruction of ancient sharks from all available evidence is the job of some museum scientists. The teeth being assembled in box figure 27.2 were found in a North Carolina quarry. (Because too few teeth were found to fill out the entire jaw, false teeth were constructed from hard rubber.) They are the teeth of a 30-million-year-old *Carcharodon megalodon*, that was the ancestor of the great white shark. The largest tooth of this specimen was about 15 cm (6 in.) long. Because 2.5 cm (1 in.) of tooth equals about 3 m (10 ft.) of shark, it is estimated that this specimen was 18.5 m (60 ft.) long! The model being constructed in this photograph now hangs in the Smithsonian's National Museum of Natural History.

Box figure 27.2 These reconstructed jaws of *Carcharodon megalodon* measure 1 m by 2 m.

gill cover, and a bony skeleton) is sometimes interpreted as evidence of the primitiveness of elasmobranchs. This interpretation is mistaken, as these characteristics simply resulted from different adaptations in the two groups to similar selection pressures. Some of these adaptations are described later in this chapter.

Sharks are covered by tough skin with dermal, placoid scales (figure 27.7a). These scales project posteriorly and give the skin a tough, sandpaper texture. (In fact, dried sharkskin has been used for sandpaper.) Posteriorly pointed scales also reduce friction with the water when a shark is swimming.

The teeth of sharks are actually modified placoid scales. The row of teeth on the outer edge of the jaw is backed up by rows of teeth attached to a ligamentous band that covers the jaw cartilage inside the mouth. As the outer teeth wear and become useless, they are replaced by newer teeth moving into position from inside the jaw. In young sharks, this replacement is rapid, with a new row of teeth developing every 7 or 8 days (figure 27.7b). Crowns of teeth may be adapted for shearing prey or for crushing the shells of molluscs. [2]

Sharks range in size from less than 1 m (e.g., *Squalus,* the laboratory dissection specimen) to greater than 10 m (e.g., basking sharks and whale sharks). The largest sharks are not predatory but are filter feeders. They have gill-arch modifications that strain plankton. The fiercest and most feared sharks are the great white shark (*Carcharodon*) and the mako (*Isurus*). Extinct specimens may have reached lengths of 25 m or more (box 27.2)!

Skates and rays are specialized for life on the ocean floor. They usually inhabit shallow water, where they use their blunt teeth to feed on invertebrates. Their most obvious modification for life on the ocean floor is a lateral expansion of the pectoral fins into winglike appendages. Locomotion results from dorsoventral muscular waves that pass posteriorly along the fins. Frequently, elaborate color patterns on the dorsal surface of these animals provide effective camouflage. The sting ray (*Raja*) has a tail modified into a defensive lash—the dorsal fin persists as a venomous spine. Also included in this group are the electric rays (*Narcine*) and manta rays (*Aetobatus*) (colorplate 10d–f).

A second major group of chondrichthians, in the subclass Holocephali (hol'o-sef'a-li) (Gr. *holos*, whole + *kephalidos*, head), contains about 30 species. *Chimaera* has a large head with a small mouth that is surrounded by large lips. A narrow tapering tail has resulted in the common name "ratfish." Holocephalans diverged from other chondrichthians nearly 300 million years ago. During this time, specializations not found in other elasmobranchs have evolved. These include a gill cover, called an **operculum**, and teeth modified into large plates that are used for crushing the shells of molluscs. Holocephalans lack scales.

Class Osteichthyes Members of the class Osteichthyes (os'te-ik"the-es) (Gr. *osteon*, bone + *ichthyos*, fish) are characterized by having at least some bone in their skeleton and/or scales, an **operculum** covering the gill openings, and lungs or a swim bladder. The bony fishes have succeeded in aquatic environments as have no other group of animals. Any group that has 20,000 species and is a major life-form in most of the earth's vast aquatic habitats must be judged very successful from an evolutionary perspective.

Zoologists have not yet documented the origins of bony fishes, though most believe the bony fishes evolved from ancient agnathans. The first fossils are from late Silurian deposits (approximately 405 million years old). By the Devonian period (350 million years ago), each of the four major subclasses were in the midst of their adaptive radiations (*see figure 27.1; table 27.1*).

Members of the subclass Dipneusti (dip-nu'ste) (Gr. *di* two + *pneustikos*, of breathing) are the lungfishes. There are only three surviving genera, and all are found in regions where seasonal droughts are common. When freshwater lakes and rivers begin to stagnate and dry, these fishes use [3] lungs to breathe air (figure 27.8). Some (*Neoceratodus*) inhabit the fresh waters of Queensland, Australia. They survive stagnation by breathing air, but normally use gills and cannot withstand total drying. Others are found in freshwater rivers and lakes in tropical Africa (*Protopterus*) and tropical South America (*Lepidosiren*). They have completely lost the use of gills for gas exchange and can survive when rivers or lakes completely dry. When a lake or river has nearly dried, these lungfishes burrow into the mud. They keep a narrow air pathway open by bubbling air to the surface. After the substrate dries, the only evidence of a lungfish burrow is a small opening in the earth. Lungfishes may remain in a dormant state for 6 months or more. (*Aestivation* is a dormant state that helps an animal withstand hot, dry periods.) When rain again fills the lake or riverbed, lungfishes emerge from their burrows to feed and reproduce.

Members of the subclass Crossopterygii (cros-op-te-rij'e-i) (Gr. *krossoi*, fringe + *pteryx*, fin) are ancient fishes that possess muscular fins. The most recent fossils of crossopterygians are over 70 million years old. In 1938, however, people fishing in deep water off the coast of South Africa brought up fish that were identified as being from a group of supposedly extinct crossopterygians, called coelocanths (figure 27.9). Since then, numerous other specimens have been caught in deep water around the Comoro Islands off Madagascar.

Figure 27.8 The lungfish, *Lepidosiren paradoxa*, is a member of the subclass Dipneusti. Lungs allow the lungfish to withstand stagnation and drying of its habitat.

(a)

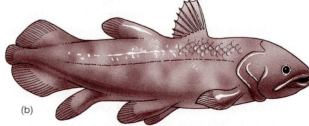

(b)

Figure 27.9 The coelacanth. *Latimeria* is the only surviving member of the subclass Crossopterygii. (a) Photograph, (b) Drawing.

[4] The discovery of this living fossil, *Latimeria chalumnae,* was a milestone event because *Latimeria* is probably the closest living fish relative of terrestrial vertebrates. It is large—up to 80 kg—and has heavy scales. Ancient coelocanths lived in freshwater lakes and rivers; thus, the ancestors of *Latimeria* must have moved from freshwater habitats to the deep sea. Other crossopterygians, called rhipidistians, became extinct before the close of the Paleozic period and are believed to have been the ancestors of ancient amphibians.

The subclass Actinopterygii (ak'tin-op"te-rig-e-i) (Gr. *aktis,* ray + *pteryx,* fin) contains fish that are sometimes called the ray-finned fishes because their fins lack muscular lobes. They usually possess **swim bladders**, which are gas-filled sacs located along the dorsal wall of the body cavity and used to regulate buoyancy. There are three infraclasses of actinopterygians that are distinguished from one another by the form of the tail, the structure of the swim bladder, the position of the fins, and the form of body scales. Zoologists now realize that there have been many points of divergence in the evolution of the Actinopterygii and that these infraclasses are not phylogenetically meaningful. The characteristics used to distinguish these groups may have evolved more than once in any of the three infraclasses. At this time, however, a more meaningful classification of actinopterygians has not been widely accepted.

The infraclass Chondrostei contains many species that lived during the Permian, Triassic, and Jurassic periods (215 to 120 million years ago), but only 25 species remain today. Ancestral chondrosteans had a bony skeleton, but living members, the sturgeons and paddlefish, have cartilaginous skeletons. Chondrosteans also have a tail with a large upper lobe.

Most sturgeons live in the sea and migrate into rivers to breed (colorplate 10g). (Some sturgeons live in fresh water but maintain the migratory habits of their marine relatives.) They are very large (up to 1,000 kg) and have bony plates covering the anterior portion of the body. Heavy scales cover the tail. The mouth of a sturgeon is small, and its jaws are weak. Sturgeons feed on invertebrates that they stir up from the sea or riverbed using their snout. Because sturgeons are valued for their caviar (eggs), they have been severely overfished.

Paddlefish are large, freshwater chondrosteans. They have a large, paddlelike rostrum that is innervated with sensory organs believed to detect weak electric fields (colorplate 10h). They swim through the water with their large mouths open, filtering crustaceans and small fishes. They are found mainly in lakes and large rivers of the Mississippi River basin and are also known from western North America.

Another actinopterygian infraclass is Holostei. Holosteans flourished in the Jurassic period and succeeded most chondrosteans. Two remaining genera occur in temperate to warm fresh waters of North America. *Lepisosteus,* the garpike, has thick scales and long jaws that it uses to catch fish. *Amia* is commonly referred to as the dogfish or bowfin.

The third infraclass of Actinopterygii, the Teleostei, or modern bony fishes, contains about 20,000 species. They have a symmetrical caudal fin and a swim bladder that has lost its connection to the digestive tract. After their divergence from marine holosteans in the late Triassic period, a remarkable evolutionary diversification occurred. Teleosts adapted to nearly every available aquatic habitat (colorplate 10i-l). The following section of this chapter—Evolutionary Pressures—describes many of these adaptations

---

Stop and Ask Yourself

5. What are characteristics of the superclass Gnathostomata? In what ways are members of this superclass considered more advanced than agnathans?
6. What class of fishes is characterized by jaws, paired appendages, a cartilaginous skeleton, and placoid scales?
7. What class of fishes is characterized by some bone in their skeleton, an operculum, and usually lungs or a swim bladder?
8. What is aestivation? What is the role of aestivation in the life of some lungfishes?

Evolutionary Pressures

Why is a fish fishlike? This apparently redundant question is unanswerable in some respects because some traits of animals are selectively neutral and, thus, neither improve nor detract from overall fitness. On the other hand, aquatic environments have physical characteristics that are important selective forces for aquatic animals. Although animals have adapted to aquatic environments in different ways, one can understand many aspects of the structure and function of a fish by studying the fish's habitat. The material presented in this section will help you appreciate the many ways that a fish is adapted for life in water.

Locomotion

Picture a young girl running full speed down the beach and into the ocean. She hits the water and begins to splash. At first, she lifts her feet high in the air between steps, but as she goes deeper, her legs encounter more and more resistance. The momentum of her upper body causes her to fall forward and she resorts to labored and awkward swimming strokes. The density of the water makes movement through it difficult and costly. For a fish, however, swimming is less energetically costly than running is for a terrestrial organism. Friction between a fish and the water is reduced by the streamlined shape of a fish and the mucoid secretions that lubricate the body surface. The buoyant properties of water also contribute to the efficiency of a fish's movement through the water. A fish needs to expend little energy in support against the pull of gravity.

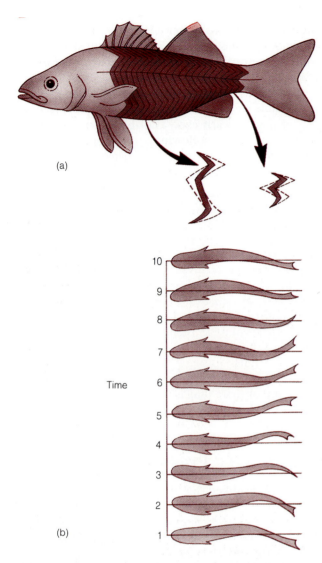

(a)

Time

10
9
8
7
6
5
4
3
2
1

(b)

Figure 27.10 Body-wall muscles and locomotion. (a) The structure of myotomes in the Actinopterygii allows each myotome to affect large regions of the body wall. Note the zigzag appearance of each myotome. (b) In forward locomotion, muscular waves pass along the body from anterior to posterior and push against the water at an angle that propels the fish forward. Wave amplitude is increased as a wave moves posteriorly and results in side-to-side sweeps of the tail. The greater the wave speed and amplitude, the greater the velocity of the fish.

Fishes move through the water using their fins and body wall to push against the incompressible surrounding water. Body-wall muscles are arranged in bundles called *myotomes* (figure 27.10a). In agnathans, myotomes are relatively unspecialized and > shaped. Anyone who has eaten a fish filet probably realizes that muscles of most fishes are arranged in a ⋝ pattern. Because these muscles extend posteriorly and anteriorly in a zig-zag fashion, contraction of each myotome can affect a relatively large portion of the body wall.

As groups of myotomes on opposite sides of the body contract alternately, the vertebral column prevents the body from collapsing along its longitudinal axis and the body flexes laterally. Waves of contraction move posteriorly with increasing amplitude and speed and push against the water at an angle that propels the fish forward (figure 27.10b).

This basic pattern of body movement is used to varying extents in fishes. Lampreys, hagfish, and eels swim inefficiently at best. Their use of fins in swimming is minimal. Very efficient, fast-swimming fishes, such as tuna and mackerel, supplement body movements with a vertical caudal (tail) fin that is tall and forked. The forked shape of the tail reduces surface area that could cause turbulence and interfere with forward movement. Many fishes can also swim backwards, but only poorly. This motion requires use of the caudal, pectoral, and pelvic fins and the reversal of the muscular waves of the body wall.

Nutrition and the Digestive System

The earliest fishes were probably filter feeders and scavengers that sifted through the mud of ancient sea floors for decaying organic matter, annelids, molluscs, or other bottom-dwelling invertebrates. Dramatic changes in the nutrition of fishes came about when the evolution of jaws transformed early fishes into efficient predators.

Most modern fishes are predators and spend much of their life searching for food. The prey that different fish eat vary tremendously. Some fishes feed on invertebrate animals floating or swimming in the plankton or living in or on the substrate. Many feed on other vertebrates. Similarly, the kinds of food that one fish eats at different times in its life varies. For example, a fish may feed on plankton as a larva but switch to larger prey, such as annelids or smaller fish, as an adult. Prey are usually swallowed whole. Teeth are often used to capture and hold prey, and some fish have teeth that are modified for crushing the shells of molluscs or the exoskeletons of arthropods. Prey capture often utilizes the suction created by closing the opercula and rapidly opening the mouth, which develops a negative pressure that sweeps water and prey inside the mouth.

Other feeding strategies have also evolved in fishes. Herring, paddlefish, and whale sharks are filter feeders. Long gill processes, called *gill rakers*, trap plankton while the fish is swimming through the water with its mouth open. Other fish, such as carp, feed on a variety of plants and small animals. A few, such as the lamprey, are parasites for at least a portion of their life. A few are primarily herbivores, feeding on plants.

The digestive tract of a fish is similar to that of other vertebrates. An enlargement, called the *stomach,* is primarily used for storing large, often infrequent, meals. In some predators, the stomach contains acidic secretions that help digest food. The *small intestine,* however, is the primary site for enzyme secretion and food digestion. Sharks and other elasmobranchs have a *spiral valve* in their intestine, and bony fish possess outpockets of the intestine, called *pyloric cecae,* which increase absorptive and secretory surfaces.

Circulation and Gas Exchange

All vertebrates have a closed circulatory system in which blood, with red blood cells containing hemoglobin, is pumped by a heart through a series of arteries, capillaries, and veins. The evolution of lungs in fishes was paralleled by changes in vertebrate circulatory systems. These changes are associated with the loss of gills, delivery of blood to the lungs, and separation of oxygenated and unoxygenated blood in the heart.

The vertebrate heart develops from four embryological enlargements of a ventral aorta. In fishes, blood flows from the venous system through the sinus venosus, the atrium, the ventricle, the conus arteriosus, and into the ventral aorta (figure 27.11a). Five afferent vessels carry blood to the gills, where the vessels branch into capillaries. Blood is collected by efferent vessels, delivered to the dorsal aorta, and distributed to the body.

Even though lungfishes are not a transitional group, they provide a good example of how the transition from gill-based circulation to lung-based circulation could have occurred. There is still circulation to gills, but a vessel to the lungs has developed as a branch off aortic arch VI (figure 27.11b). This vessel is now called the *pulmonary artery.* Blood returns to the heart through *pulmonary veins* and enters the left side of the heart. The atrium and ventricle of the lungfish heart are partially divided. These partial divisions help keep unoxygenated blood from the body separate from the oxygenated blood from the lungs. A *spiral valve* in the conus arteriosus helps direct blood from the

right side of the heart to the pulmonary artery and blood from the left side of the heart to the remaining aortic arches. Thus, in the lungfishes, we see a distinction between a *pulmonary circuit* and a *systemic circuit.*

Gas Exchange Fishes live in an environment that contains less than 2.5% of the oxygen present in air. To maintain adequate levels of oxygen in their bloodstream, fishes must pass large quantities of water across gill surfaces and extract the small amount of oxygen present in the water.

Most fishes use a muscular pumping mechanism to move the water into the mouth and pharynx, over the gills, and out of the fish through gill openings (figure 27.12). This pump is powered by muscles surrounding the pharynx and the *opercular cavity,* which is between the gills and the operculum. With the operculum closed and the mouth open, lowering the floor of the mouth draws water into the pharynx, and the size of the opercular cavity increases. Hydrostatic pressure in the opercular cavity decreases, and the water in the mouth is drawn across the gills and into the opercular cavity. Closing the mouth and raising the floor of the mouth increases hydrostatic pressure in the pharynx, and more water is forced across the gills and into the opercular cavity. The operculum opens, and the water passes out of the fish.

Some elasmobranchs and open-ocean bony fishes, such as the tuna, maintain water flow by holding their mouth open while swimming. This method is called **ram ventilation.** Elasmobranchs do not have opercula to help pump

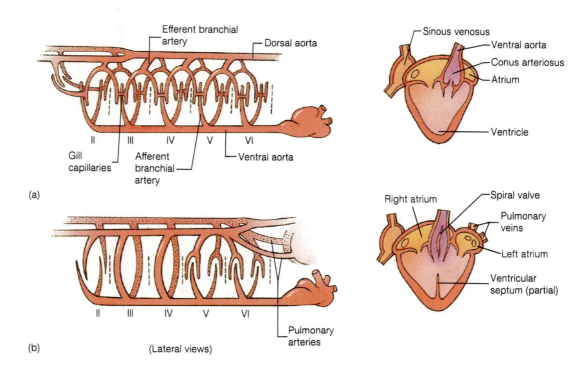

Figure 27.11 Diagrammatic representation of the circulatory systems of (a) bony fishes and (b) lungfishes. Hearts are drawn from a ventral view.

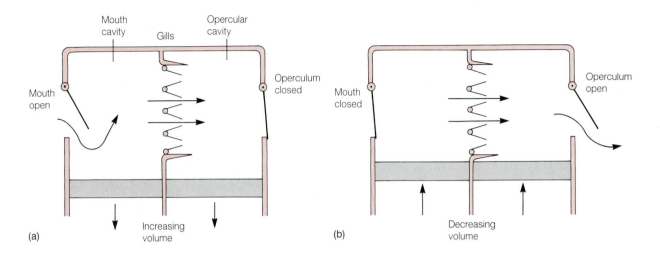

Figure 27.12 Oral and opercular pumps move water into the mouth, across gill filaments, and out the opercular opening. (a) With the operculum closed and the mouth open, the floor of the mouth is lowered. This mechanism creates negative pressure in the mouth cavity, and water is drawn into the mouth. Negative pressure in the opercular cavity draws water across the gills. (b) The closing of the mouth and the raising of the floor of the mouth create positive pressure that forces more water across the gills, in the opercular cavity, and out the opercular opening.

[5] water, and therefore, some sharks must keep moving to survive. Others can move water over their gills by a pumping mechanism similar to that described above. Rather than using an operculum in the pumping process, however, their gill bars have external flaps that close and form a cavity functionally similar to the opercular cavity of other fishes. *Spiracles* are modified gill slits that open just behind the eyes of elasmobranchs and are used as an alternate route for water entering the pharynx. Some sharks rest on the substrate and allow natural water currents to move water across their gills. The tuna has nearly immobile gill covers, and like some sharks, will die if prevented from swimming. Some fishes that use muscular pumps when stationary or when moving slowly, use ram ventilation when swimming at higher velocities.

Gas exchange across gill surfaces is very efficient. Gills are supported by **gill (visceral) arches. Gill filaments** extend from each gill arch and include vascular folds of epithelium, called **gill lamellae** (figure 27.13a,b). Blood is carried to the gills and into gill filaments in branchial arteries. The arteries break into capillary beds in gill lamellae. Gas exchange occurs as blood and water move in opposite directions on either side of the lamellar epithelium. This **countercurrent mechanism** provides very efficient gas exchange by maintaining a concentration gradient between the blood and the water over the length of the capillary bed (figure 27.13c,d).

Swim Bladders and Lungs The Indian climbing perch spends its life almost entirely on land. These fishes, as most bony fish, have gas chambers called **pneumatic sacs.** In non-teleost fishes, the pneumatic sacs connect to the esophagus or another part of the digestive tract by a *pneumatic duct*. Swallowed air enters these sacs, and gas exchange occurs across vascular surfaces. Thus, in the Indian climbing perch, lungfishes, and ancient crossopterygians, pneumatic sacs function(ed) as lungs. In other bony fishes, pneumatic sacs act as swim bladders.

Most zoologists believe that lungs are more primitive than swim bladders. Much of the early evolution of bony fishes occurred in warm, freshwater lakes and streams during the Devonian period. These bodies of water frequently became stagnant and periodically dried. Having lungs in these habitats could have meant the difference between life and death. On the other hand, later evolution of modern bony fishes occurred in marine and freshwater environments where stagnation was not a problem. In these environments, the use of pneumatic sacs in buoyancy regulation would have been adaptive (figure 27.14).

Buoyancy Regulation

Did you ever consider why it is possible for you to float in water? Water is a supportive medium, but that is not sufficient to prevent you from sinking. Even though you are made mostly of water, other constituents of tissues are more dense than water. Bone, for example, has a specific gravity twice that of water. Why is it then that you can float? Largely because of two large, air-filled sacs called lungs that allow you to float to the surface.

Fish maintain their vertical position in a column of water in four ways. One way is to incorporate low-density compounds into their tissues. Fishes (especially their livers) are saturated with buoyant oils. A second way fishes maintain vertical position is to use fins to provide lift. The pectoral fins of a shark serve as planing devices that help to create lift as the shark moves through the water. Also, the large upper lobe of a shark's caudal fin provides upward

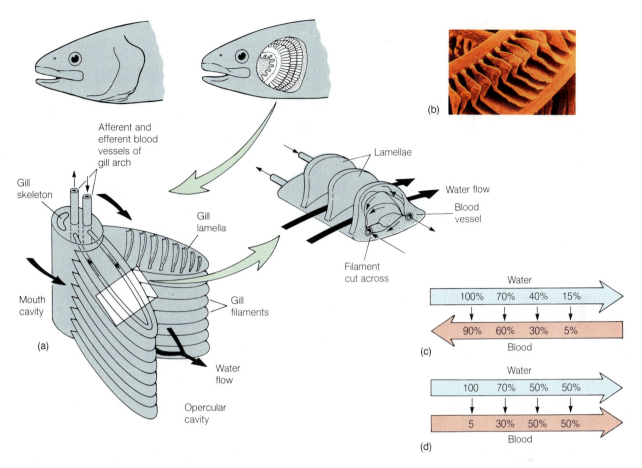

Figure 27.13 Gas exchange at the gill lamellae. (a) The gill arches under the operculum support two rows of gill filaments. Blood flows into gill filaments through afferent branchial arteries, and these arteries break into capillary beds in the gill lamellae. Water and blood flow in opposite directions on either side of the lamellae. (b) Electron micrograph of the tip of a trout gill filament showing numerous lamellae. A comparison of countercurrent (c) and parallel (d) exchanges. Water entering the spaces between gill lamellae is saturated with oxygen in both cases. In countercurrent exchange, this water encounters blood that is almost completely oxygenated, but a diffusion gradient still favors the movement of more oxygen from the water to the blood. As water continues to move between lamellae it loses oxygen to the blood, because it is continually encountering blood that has a lower oxygen concentration than is present in the water. Thus a diffusion gradient is maintained along the length of the lamella. If blood and water were to move in parallel fashion, diffusion of oxygen from water to blood would occur only until the concentration of oxygen in blood equalled the concentration of oxygen in water, and the exchange would be much less efficient.

thrust for the posterior end of the body (colorplate 10c). A third adaptation is the reduction of heavy tissues in fishes. The bones of fishes are generally less dense than those of terrestrial vertebrates. One of the adaptive features of the elasmobranch cartilaginous skeleton probably results from cartilage being only slightly heavier than water. The fourth adaptation is the *swim bladder*. Using a swim bladder, buoyancy can be regulated to meet the day-to-day needs of a fish by precisely regulating the volume of gas in it. (You can mimic this adaptation while floating in water. How well do you float after forcefully exhaling as much air as possible?)

The swim bladders of garpike, sturgeons, and other primitive bony fishes connect to the esophagus or another part of the digestive tract by the pneumatic duct. These fish gulp air at the surface to force air into their swim bladders.

Most teleosts have swim bladders that have lost a functional connection to the digestive tract. Gases (various mixtures of nitrogen and oxygen) are secreted into the swim bladder from the blood using a countercurrent exchange mechanism in a vascular network called the *rete mirabile* ("miraculous net"). Gases may be reabsorbed into the blood at the posterior end of the bladder.

Nervous and Sensory Functions

The central nervous system of fishes, as in other vertebrates, consists of a brain and a spinal cord. Sensory receptors are widely distributed over the body. In addition to generally

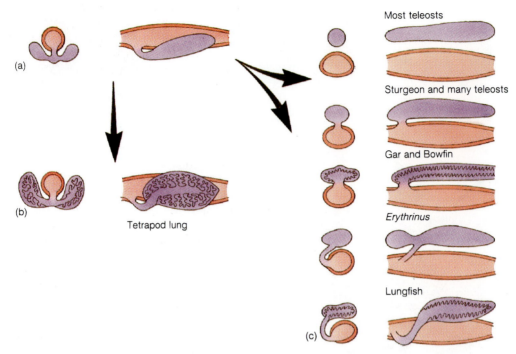

Figure 27.14 A possible sequence in the evolution of pneumatic sacs. (a) Pneumatic sacs may have originally developed from ventral outgrowths of the esophagus. Many ancient fishes probably used pneumatic sacs as lungs. (b) Primitive lungs developed further during the evolution of land vertebrates. Internal compartmentalization increases surface area for gas exchange in land vertebrates. (c) In most bony fishes, pneumatic sacs are called swim bladders, and they are modified for buoyancy regulation. Swim bladders are in a dorsal position to prevent a tendency for the fish to "belly up" in the water. Pneumatic duct connections to the esophagus are frequently lost, and gases are transferred from the blood to the swim bladder.

distributed receptors for touch and temperature, fishes possess specialized receptors for olfaction, vision, equilibrium and balance, and for detecting water movements.

Openings in the snout of fishes, called *external nares,* lead to olfactory receptors. In most fishes, receptors are located in blind-ending olfactory sacs. In a few fishes, the external nares open to nasal passages that lead to the mouth cavity. Recent research has revealed that some fishes rely heavily on their sense of smell. For example, salmon and lampreys return to spawn in the streams in which they hatched years earlier. Their migrations to these streams often involve distances of hundreds of miles and are guided by the fishes' perception of the characteristic odors of their spawning stream.

The eyes of fishes are similar in most aspects of structure to those found in other vertebrates. They are lidless, however, and the lenses are round. Focusing is accomplished by moving the lens forward or backward in the eye. (Most other vertebrates focus by changing the shape of the lens.)

Receptors for equilibrium, balance, and hearing are located in the inner ears of fishes and their functions are similar to those of other vertebrates. *Semicircular canals* detect rotational movements, and other sensory patches help with equilibrium and balance by detecting the direction of the gravitational pull. Fishes lack the outer and/or middle ear, which conducts sound waves to the inner ear in other vertebrates. As anyone who enjoys fishing knows, however, most fishes can hear. Vibrations may be passed from the water, through the bones of the skull to the middle ear, and a few fishes have chains of bony ossicles (modifications of vertebrae) that pass between the swim bladder and the back of the skull. Vibrations strike the fish, are amplified by the swim bladder, and sent through the ossicles to the skull.

Running along each side and branching over the head of most fishes is a lateral line system. The **lateral line system** consists of sensory pits in the epidermis of the skin that connect to canals that run just below the epidermis. In these pits are receptors that are stimulated by water moving against them. Lateral lines are used to detect water currents or for detecting a predator or a prey that may be causing water movements in the vicinity of the fish. Low-frequency sound may also be detected.

Electric Fishes A U.S. Navy pilot has just ejected from his troubled aircraft over shark-infested water! What

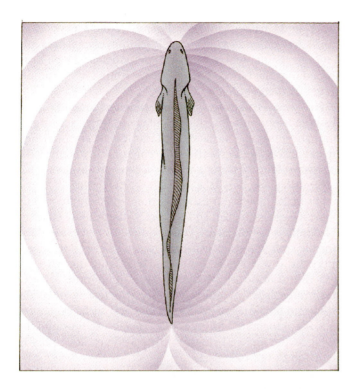

Figure 27.15 The electric field of *Gymnarchus niloticus* is used to detect the presence of prey and other objects in the fish's murky environment. Currents circulate from electric organs in its tail to electroreceptors near its head. An object in this electrical field changes the pattern of stimulation of electroreceptors.

measures can the pilot take to ensure survival under these hostile conditions? The Navy has considered this scenario. One of the solutions to the problem is a polyvinyl bag suspended from an inflatable collar. The polyvinyl bag helps conceal the downed flyer from a shark's vision and keen sense of smell. But is that all that is required to ensure protection?

All living organisms produce weak electrical fields from activities of nerves and muscles. **Electroreception** is the detection of electrical fields generated by the fish or another organism in the environment. Electroreception and/or electrogeneration has been demonstrated in over 500 species of fishes in seven families of Chondrichthyes and Osteichthyes. These fishes use their electroceptive sense for detecting prey and for orienting toward or away from objects in the environment. Although the electric receptors and the electric organs found in different groups of fishes are similar in structure, they are apparently not homologous, that is, they had separate evolutionary origins.

Nowhere is prey detection with this sense better developed than in the rays and sharks. Spiny dogfish sharks, the common laboratory specimens, locate prey by elecroreception. A shark can find and eat a flounder that is buried in sand and it will try to find and eat electrodes that are creating electrical signals similar to those emitted by the flounder. On the other hand, a shark cannot find a dead flounder buried in the sand or a live flounder covered by an insulating polyvinyl sheet. [6]

Some fishes are not only capable of electroreception but are also capable of generating electrical currents. An electric fish (*Gymnarchus niloticus*) lives in freshwater systems of Africa. Muscles near its caudal fin are modified into organs that produce a continuous electrical discharge. This current spreads between the tail and the head. Porelike perforations near the head contain electroreceptors. The electrical waves circulating between the tail and the head are distorted by objects in their field. This distortion is detected in changing patterns of stimulation of receptors (figure 27.15). The electrical sense of *Gymnarchus* is an adaptation to living in murky freshwater habitats where eyes are of limited value.

The fishes best known for producing strong electric currents are the electric eel (a bony fish) and the electric ray (an elasmobranch). The electric eel (*Electrophorus*) occurs in rivers of the Amazon basin in South America. The organs used for producing electric currents are located in the trunk of the electric eel and can deliver shocks in excess of 500 volts. The electric ray (*Narcine*) (colorplate 10e), has electric organs in its fins that are capable of producing pulses of 50 amperes at about 50 volts. Shocks produced by these fishes are sufficiently strong to stun or kill prey, discourage large predators, and teach unwary humans a lesson that will never need to be repeated.

Stop and Ask Yourself

9. What are two methods of gill ventilation used by fishes?
10. What is a countercurrent exchange mechanism, and where does it occur in fish gills?
11. Why do some fishes, such as lungfishes, need lungs?
12. What are four mechanisms used by fishes to improve their buoyancy?

Excretion and Osmoregulation

Fishes, as all animals, must maintain a proper balance of salts and water in their tissues. The regulation of these balances is called *osmoregulation* and is a major function of the kidneys and gills of fishes. Kidneys are located near the midline of the body, just dorsal to the peritoneal membrane that lines the body cavity. As with all vertebrates, the excretory structures in the kidneys are called *nephrons*. Nephrons filter bloodborne nitrogenous wastes, salts,

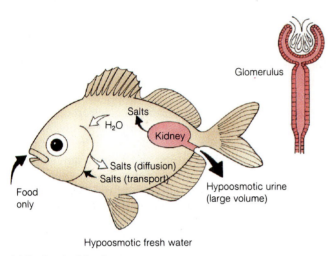

(a) Freshwater teleosts
(hypertonic blood)

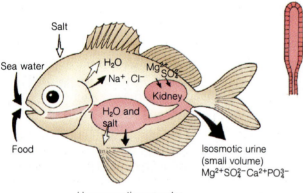

(b) Marine teleosts
(hypotonic blood)

Figure 27.16 Osmoregulation by (a) freshwater and (b) marine fishes. Large arrows indicate passive uptake or loss of water or ions. Small arrows indicate active transport processes occurring at gill membranes and kidney tubules. Insets of kidney nephrons depict adaptations within the kidney. Water, ions, and small organic molecules are filtered from the blood at the glomerulus of the nephron. Essential components of the filtrate can be reabsorbed within the tubule system of the nephron. Marine fishes conserve water by reducing the size of the glomerulus of the nephron, and thus reducing the quantity of water and ions filtered from the blood. Salts can be secreted from the blood into the kidney tubules. Marine fishes can produce urine that is isosmotic with the blood. Freshwater fishes have enlarged glomeruli and short tubule systems. They filter large quantities of water from the blood, and tubules reabsorb some ions from the filtrate. Freshwater fishes produce a hypotonic urine.

water, and small organic compounds across a network of capillaries called a *glomerulus*. The filtrate then passes into a tubule system, where essential components may be reabsorbed into the blood. The filtrate remaining in the tubule system is then excreted.

Freshwater fishes live in an environment containing few dissolved substances. Osmotic uptake of water across gill, oral, and intestinal surfaces, and the loss of essential ions by excretion and defecation are constant. To control the excess buildup of water and loss of salts, freshwater fishes never drink, and only take in water when feeding. Also, the nephrons of freshwater fishes are numerous and frequently possess large glomeruli and relatively short tubule systems. Filtration is followed by reabsorption of some salts and organic compounds. Because the tubule system is relatively short, little water is reabsorbed. Thus, large quantities of very dilute urine are produced. Even though the urine of freshwater fishes is dilute, salts are still lost through the urine and by diffusion across gill and oral surfaces. Loss of salts is compensated for by active transport of ions into the blood at the gills. Freshwater fishes also get some salts in their food (figure 27.16a).

Marine fishes face the opposite problems. Their environment contains 3.5% salts, and their tissues contain approximately 0.65% salts. Marine fishes, therefore, must combat water loss and accumulation of excess salts. They drink water and eliminate excess salts by excretion, defecation, and active transport across gill surfaces. The nephrons of marine fishes frequently possess small glomeruli and long tubule systems. Much less blood is filtered than in freshwater fishes, and water is efficiently, although not entirely, reabsorbed from the nephron (figure 27.16b).

Elasmobranchs have a unique osmoregulatory mechanism. They convert some of their nitrogenous wastes into urea in the liver. This in itself is somewhat unusual, because most fishes excrete ammonia rather than urea. Even more unusual, however, is that urea is sequestered in tissues all over the body. Enough urea is stored to make body tissues isosmotic with sea water. (That is, the concentration of solutes in a shark's tissues is essentially the same as the concentration of salts in sea water.) Therefore, the problem most marine fishes have of losing water to their environment is much less severe for elasmobranchs. Energy that does not have to be devoted to water conservation can now be used in other ways. This adaptation required the development of tolerance to high levels of urea because urea disrupts important enzyme systems in the tissues of most other animals.

In spite of this unique adaptation, elasmobranchs must still regulate the concentrations of ions in their tissues. In addition to having salt-absorbing and secreting tissues in their gills and kidneys, elasmobranchs possess a *rectal gland* that removes excess sodium chloride from the blood and excretes it into the cloaca. (A **cloaca** is a common opening for excretory, digestive, and reproductive products.)

Diadromous fishes migrate between freshwater and marine environments. Salmon and marine lampreys migrate from the sea to fresh water to spawn, and the freshwater eel (*Anguilla*) migrates from freshwater to marine environments to spawn. Diadromous migrations require that gills are capable of coping with both uptake and secretion of ions. Osmoregulatory powers needed for migration between marine and freshwater environments may not be developed in all life-history stages. Young salmon, for example, cannot enter the sea until certain cells on the gills develop salt-secreting powers.

Fishes have few problems getting rid of the nitrogenous by-products of protein metabolism. Up to 90% of nitrogenous wastes are eliminated as ammonia by diffusion across gill surfaces. Even though ammonia is very toxic, its use as an excretory product is possible in aquatic organisms because ammonia can diffuse into the surrounding water. The remaining 10% of nitrogenous wastes are excreted as urea, creatine, or creatinine. These wastes are produced in the liver and are excreted via the kidneys.

Reproduction and Development

Imagine, 45 kg of caviar from a single, 450 kg sturgeon! Admittedly, a 450 kg sturgeon is a very large fish (even for a sturgeon), but it is not unusual for a fish to produce millions of eggs in a single season. These numbers simply reflect the hazards of developing in aquatic habitats unattended by a parent. The vast majority of these millions of potential adults will never survive to reproduce. Many eggs will never be fertilized, many fertilized eggs may wash ashore and dry, many eggs and embryos may be smashed by currents and tides, and others will fall victim to predation. In spite of all of these hazards, if only four of the millions embryos of each breeding pair survive and reproduce, the population will double.

Producing overwhelming numbers of eggs, however, is not the only way that fishes increase the chances that a few of their offspring will survive. Some fishes show mating behavior that helps ensure fertilization, or nesting behavior that protects eggs from predation, sedimentation, and fouling.

Mating may occur in large schools, and the release of eggs or sperm by one individual often releases spawning pheromones that induce many other adults to spawn. Huge masses of eggs and sperm released into the open ocean help to ensure fertilization of as many eggs as possible.

The vast majority of fishes are *oviparous,* meaning that eggs develop outside the female from stored yolk. Some elasmobranchs are *ovoviviparous,* and their embryos develop in a modified oviduct of the female. Nutrients are supplied from yolk stored in the egg. Other elasmobranchs, including grey sharks and hammerheads, are *viviparous.* A placentalike outgrowth of a modified oviduct diverts nutrients from the female to the yolksacs of developing embryos. Internal development of viviparous bony fishes usually occurs in ovarian follicles, rather than in the oviduct. In guppies (*Lebistes*), eggs are retained in the ovary, and fertilization and early development occur there. Embryos are then released to a cavity of the ovary and development

Figure 27.17 A male garibaldi (*Hypsypops rubicundus*) guarding eggs. The male cultivates a nest of filamentous red algae and then entices a female to lay eggs in the nest. This male is carrying off a bat star that came too close to the nest.

continues, with nourishment coming partly from yolk and partly from secretions of the ovary.

Some fishes have specialized structures that aid in sperm transfer. Male elasmobranchs, for example, have modified pelvic fins called *claspers.* During copulation, a clasper is inserted into the cloaca of a female. Sperm travel along grooves of the clasper. Fertilization occurs in the reproductive tract of the female and usually results in a higher proportion of eggs being fertilized than in external fertilization. Thus, internal fertilization is usually accompanied by the production of fewer eggs.

In many fishes, care of the embryos is limited or nonexistent. Some fish, however, construct and tend nests (figure 27.17), and some carry embryos during development. Clusters of embryos may be brooded by being attached to some part of the body in special pouches, or they may be brooded in the mouth. Some of the best-known brooders include the seahorses and pipefish. Males of these closely related fishes carry embryos throughout development in ventral pouches. Male Brazilian catfish (*Loricaria typhys*) brood embryos in an enlarged lower lip.

Most fishes do little, if any, caring for young after they have hatched. There are exceptions, however. Short-term care of posthatching young occurs in sunfishes and sticklebacks. Male sticklebacks collect fresh plant material and accumulate it into a mass. Young take refuge in this mass. If one wanders too far from the nest, the male will snap it up in its mouth and spit it back into the nest. Sunfish males do the same for young that wander from schools of recently hatched fish. Longer-term care occurs in the Cichlidae, both those that are mouth brooded and those tended in a nest. After hatching, the young venture from the parent's mouth or nest, but the young return quickly when the parent signals danger with a flicking of the pelvic fins.

(a)

(b)

Figure 27.18 Early vertebrates at the water's edge. An artist's comparison of ancient crossopterygians (a) and the primitive amphibians (b) that may have evolved from them.

Further Phylogenetic Considerations

Two important series of evolutionary events occurred during the evolution of the Osteichthyes. One of these was an evolutionary explosion that resulted in the vast diversity of teleosts that we see today. The last one-half of this chapter should have helped you appreciate some of these events.

To appreciate the second series of events, let us again consider the lungfishes. The lungfishes' life-style represents a survival strategy that must have been important for some early Devonian fishes. The current seasonal droughts of tropical South America and Africa must be similar to, but much less geographically extensive than those in the Devonian period. Survival during these dry periods requires that an organism be able to withstand drying, either by aestivation or by moving to a larger body of water. Lungfishes withstand drying by breathing air and aestivating. That must have also been true of many freshwater Devonian fishes. However, distinctive characteristics of the lungfish skeleton lead most zoologists to conclude that the dipneustian evolutionary line gave rise to no other vertebrate taxa. Instead, the crossopterygians are thought to be a part of the evolutionary lineage that led to terrestrial vertebrates. Ancestral crossopterygians, like lungfishes, were freshwater fishes and were subject to the same evolutionary pressures present during the Devonian and Carboniferous periods. Crossopterygians had lungs and could breathe air. One group of crossopterygians, the rhipidistians may have used their muscular fins to crawl across land looking for water when their own stream dried.

Adaptations that favored survival in Devonian streams and lakes may have preadapted some rhipidistians for life that would become increasingly terrestrial. While millions of years would elapse before any vertebrate could be considered terrestrial, brief excursions onto land allowed some vertebrates to exploit resources that for the previous 50 million years, were available only to terrestrial arthropods (figure 27.18). [7]

Stop and Ask Yourself

13. Why is osmoregulation, not excretion, the major function of the kidneys of fishes?
14. What osmoregulatory problems are faced by a diadromous fish?
15. What is viviparity? How does viviparity of some elasmobranchs differ from that of some teleosts?
16. What adaptations to climatic conditions may have preadapted rhipidistian fishes for a partially terrestrial existence?

Summary

1. Members of the vertebrate superclass Agnatha were probably the ancestors of all other vertebrates.

2. Agnathans lack jaws and paired appendages and include the extinct ostracoderms, lampreys, and hagfish. Lampreys have a life history involving migrations from the open ocean, or large body of fresh water, to freshwater spawning streams. Hagfish are scavengers in marine environments.

3. The superclass Gnathostomata includes fishes with jaws and paired appendages. The class Chondrichthyes includes the sharks, skates, rays, and ratfish. The class Osteichthyes includes the bony fish.

4. There are three subclasses of Osteichthyes. Members of the subclass Crossopterygii are mostly extinct. The subclass Dipneusti includes the lungfishes, and the subclass Actinopterygii includes the ray-finned fishes. In the Actinopterygii, the superorder Teleostei contains the modern bony fishes. Members of this very large superorder have adapted to virtually every available aquatic habitat.

5. Fishes show numerous adaptations to living in aquatic environments. These adaptations include an arrangement of body-wall muscles that creates locomotor waves in the body wall, mechanisms that provide constant movement of water across gill surfaces, a countercurrent exchange mechanism to promote efficient gas exchange, buoyancy regulation, well-developed sensory receptors, including eyes, inner ears, and lateral line receptors, mechanisms of osmoregulation, and mechanisms that help ensure successful reproduction.

6. Two evolutionary lineages in the Actinopterygii are very important. One of these resulted in the adaptive radiation of modern bony fishes, the teleosts. The second evolutionary line probably diverged from the Crossopterygii. Adaptations that favored the survival of crossopterygians in early Devonian streams preadapted some crossopterygians for terrestrial habitats.

Key Terms

cloaca (p. 434)
countercurrent mechanism (p. 430)
electroreception (p. 433)
gill (visceral) arches (p. 430)
gill filaments (p. 430)
gill lamellae (p. 430)
lateral line system (p. 432)
operculum (p. 426)
pneumatic sacs (p. 430)
ram ventilation (p. 429)
swim bladder (p. 427)

Critical Thinking Questions

1. Describe the evolutionary relationships between the Agnatha, the Chondrichthyes, and the Osteichthyes.

2. What characteristic of water makes it difficult to move through, but also makes support against gravity a minor consideration? How is a fish adapted for moving through water?

3. Would it be possible for a fish to drown? Explain. Would it make a difference if the fish was an open-ocean fish, such as a tuna or a fish such as a freshwater perch?

4. Why is it a mistake to consider the cartilaginous skeleton of chondrichthians a primitive characteristic?

5. Would swim bladders with functional pneumatic ducts work well for a fish that lives at great depths? Why or why not?

6. What would happen to a deep-sea fish that was rapidly brought to the surface? Explain your answer in light of the fact that gas pressure in the swim bladders of some deep sea fishes are multiplied up to about 300 atmospheres.

Suggested Readings

Books

Foreman, R. E., Gorbman, A., Dodd, J. M., and Olsson, R. (eds.). 1985. *Evolutionary Biology of Fishes*. NATO ASI Series, Vol 103. New York: Plenum Press.

Hardisty, M. W. 1979. *Biology of the Cyclostomes*. New York: Chapman and Hall Ltd.

Hasler, A. D., and Scholz, A. T. 1983. *Olfactory Imprinting and Homing in Salmon*. New York: Springer-Verlag, Zoophysiology Series.

Moyle, P. B., and Cech, J.J. 1982. *Fishes: An Introduction to Ichthyology*. Englewood Cliffs: Prentice-Hall, Inc.

Nelson, J. S. 1984. *Fishes of the World,* 2nd ed. New York: John Wiley & Sons, Inc.

Smith, R. J. F. 1985. *The Control of Fish Migration*. New York: Springer-Verlag, Zoophysiology Series.

Steele, R. S. 1985. *Sharks of the World*. New York: Facts on File Publications.

Travola, W. N., Popper, A. N., and Fay, R.R. 1981. *Hearing and Sound Communication in Fishes*. New York: Springer-Verlag.

Tytler, P., and Calow, P. 1985. *Fish Energetics: New Perspectives*. Baltimore: John Hopkins University Press.

Articles

Fischer, E. A., and Peterson, C. W. 1987. The evolution of sexual patterns in the seabasses. *BioScience* 37:482–489.

Forey, P. L. 1988. Golden Jubilee for the coelacanth *Latimeria chalumnae*. *Nature* 336:727–732.

Fricke, Hans. 1988. Coelacanth: the fish that time forgot. *National Geographic* 173(6):824–838.

Horn, M. H., and Gibson, R. N. Intertidal Fishes. *Scientific American* January, 1988.

Levine, J. S., and MacNichol, E. F., Jr. Color vision in fishes. *Scientific American* February, 1982.

Partridge, B. L. The structure and function of fish schools. *Scientific American* June, 1982.

Shapiro, D. Y. 1987. Differentiation and evolution of sex change in fishes; a coral reef fish's social environment can control its sex. *BioScience,* 490–497.

28

Amphibians: The First Terrestrial Vertebrates

Concepts

1. Adaptations that favored survival of fishes during periodic droughts preadapted vertebrates to life on land. There were two lineages of ancient amphibians. One gave rise to modern amphibians, and the other lineage resulted in amniote vertebrates.
2. Modern amphibians belong to three orders. Caudata contains the salamanders, Gymnophiona contains the caecilians, and Anura contains the frogs and toads.
3. Although amphibians are restricted to moist habitats, most spend much of their adult life on land. Virtually all amphibian body systems show adaptations for living on land.
4. Evolution of eggs and developmental stages that were resistant to drying probably occurred in some ancient amphibians. This development was a major step in vertebrate evolution as it weakened vertebrate ties to moist environments.

Have You Ever Wondered:

[1] how the skeleton of an amphibian is adapted for life on land?
[2] how frogs, toads, and some salamanders catch prey with their tongue?
[3] why an amphibian's skin is wet?
[4] what terrestrial animals lack lungs?
[5] how a frog's ear can filter out certain frequencies of sound?
[6] what are the functions of a frog's calls?
[7] what adaptations permitted life on land?

These and other useful questions will be answered in this chapter.

This chapter contains underlined evolutionary concepts.

Evolutionary Perspective

Who, while walking along the edge of a pond or stream, has not been startled by the "plop" of an equally startled frog jumping to the safety of its watery retreat? Or, who has not marveled at the sounds of a chorus of frogs breaking through an otherwise silent spring evening? These experiences and others like them have led some to spend their lives studying members of the class Amphibia (am-fib′e-ah) (L. *amphibia,* living a double life): frogs, toads, salamanders, and caecilians. The class name implies that amphibians either move back and forth between water and land, or live one stage of their life in water and another on land. One or both of these descriptions is accurate for most amphibians.

Amphibians are the first vertebrates we have encountered that are called **tetrapods** (Gr. *tetra,* four + *podos,* foot). It is a nontaxonomic designation that applies to all vertebrates other than fishes, and adaptations for life on land are found in most tetrapods.

Phylogenetic Relationships

During the first 250 million years of vertebrate history, adaptive radiation resulted in vertebrates filling most aquatic habitats. There were many active, powerful predators in the prehistoric waters. Land, however, was free of vertebrates and except for some arthropods, was free of predators. Animals that moved around the water's edge were not likely to be prey for other animals. With lungs for breathing air and muscular fins to scurry across mud, these animals probably found ample food in the arthropods that lived there. It is no surprise that the major component of the diet of most modern amphibians is arthropods.

The origin of amphibians from ancient crossopterygians was described in chapter 27. No one knows what animal was the first amphibian, but the structure of limbs, skulls, and teeth suggests that *Ichthyostega* is probably similar to the earliest amphibians (figure 28.1). During the late Devonian and early Carboniferous periods, two lineages of early amphibians can be distinguished by details of the way the roof and the posterior portion of the skull are attached to each other. One group, called the temnospondyls, flourished into the Jurassic period. Most of this lineage became extinct, but not before giving rise to the three orders of living amphibians. This lineage is called the **nonamniote lineage.** A second group of amphibians became extinct late in the Carboniferous period. The development of an egg that was resistant to drying, an amniotic egg (*see figure 10.11*), occurred in this group. This lineage, called the **amniote lineage,** left as its descendants the reptiles, birds, and mammals (figure 28.2).

Survey of Amphibians

Amphibians occur on all continents except Antarctica but are absent from many oceanic islands. The 3,000 modern species are a mere remnant of this once-diverse group. Modern amphibians are divided into three orders: Caudata or Urodela, the salamanders; Anura, the frogs and toads; and Gymnophiona, the caecilians (table 28.1).

Order Caudata or Urodela

Members of the order Caudata (kaw′dat-ah) (L. *cauda,* tail + Gr. *ata,* to bear) or Urodela (yur′o-del″ah) (Gr. *oura,* tail + L. *dela,* carry) are the salamanders. They possess a tail throughout life, and both pairs of legs, when present, are relatively unspecialized (colorplate 11c,e,f).

Approximately 115 of the 350 described species of salamanders occur in North America. Most terrestrial salamanders live in moist forest-floor litter and have aquatic larvae. A number of families are found in caves, where constant temperature and moisture conditions create a

Figure 28.1 *Ichthyostega.* Fossils of this early amphibian were discovered in eastern Greenland in late Devonian deposits. The total length of the restored specimen is about 65 cm. Terrestrial adaptations are heavy pectoral and pelvic girdles and sturdy limbs that probably aided in lifting the body off the ground. Strong jaws suggest that it was a predator in shallow water, perhaps venturing onto shore. Other features include a skull that is similar in structure to ancient crossopterygian fishes and a finlike tail. Note that the tail fin is supported by bony rays dorsal to the spines of the vertebrae. This pattern is similar to the structure of the dorsal fins of fishes and is unknown in any other tetrapod.

Figure 28.1 redrawn, with permission, from Duellman and Treub, *Biology of Amphibians.* Copyright © 1986, McGraw-Hill Publishing Company, New York, New York.

Period (millions of years before present)

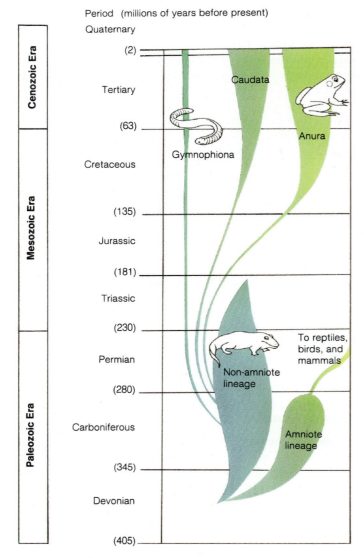

Figure 28.2 One interpretation of amphibian phylogeny. The relative number of species in each taxonomic group is indicated by the width of a group's branch.

Table 28.1 Classification of Living Amphibians
Class Amphibia (am-fib′e-ah) Skin with mucoid secretions and lacks epidermal scales, feathers, or hair; larvae usually aquatic and undergo metamorphosis to the adult; atria of the heart divided into two chambers. One cervical and one sacral vertebra.
Order Caudata (kaw′dat-ah) or **Urodela** (yur′o-del″-ah) Long tail, two pairs of limbs; lack middle ear. Salamanders, newts.
Order Gymnophiona (jim′no-fy″o-nah) Elongate, limbless; segmented by annular grooves; specialized for burrowing; tail short and pointed; rudimentary left lung. Caecilians.
Order Anura (ah-noor′ah) Tailless; elongate hind limbs modified for jumping and swimming; five to nine presacral vertebrae with transverse processes (except the first); postsacral vertebrae fused into rodlike urostyle; tympanum and larynx well developed. Frogs, toads.

but smaller. They often possess external gills, a tail fin, larval dentition, and a rudimentary tongue. Aquatic larval stages usually undergo metamorphosis into a terrestrial adult. Many other salamanders undergo incomplete metamorphosis and are *neotenic* (e.g., *Necturus*); that is, they become sexually mature while still showing larval characteristics.

Order Gymnophiona

Members of the order Gymnophiona (jim′no-fy″o-nah) (Gr. *gymnos*, naked + *ophineos*, like a snake) are the caecilians (colorplate 11d). There are about 160 described species, which are confined to tropical regions. Caecilians are wormlike burrowers that feed on worms and other invertebrates in the soil. Caecilians appear segmented because of folds in the skin that overlie separations between myotomes. They have a retractile tentacle between their eyes and nostrils. The tentacle may transport chemicals from the environment to olfactory cells in the roof of the mouth. The eyes are covered by skin; thus, caecilians are probably nearly blind.

Fertilization is internal in caecilians. Larval stages are often passed within the oviducts, where they feed on the inner lining of the oviducts by scraping it with fetal teeth. The young emerge from the female as miniatures of the adults. Other caecilians lay eggs that develop into either aquatic larvae or embryos that undergo direct development on land.

Order Anura

The order Anura (ah-noor″ah) (Gr. *a*, without + *oura*, tail) includes about 3,500 species of frogs and toads. Anurans are found in most moist environments, except in high latitudes and on some oceanic islands. A few even occur in

nearly ideal environment. Salamanders in the family Plethodontidae are the most fully terrestrial salamanders in that their eggs are laid on land, and the young hatch as miniatures of the adult. Members of the family Salimandridae are commonly called *newts*. They spend most of their lives in water and frequently retain caudal fins. Salamanders range in length from only a few centimeters to 1.5 m (the Japanese giant salamander, *Andrias japonicus*). The largest North American salamander is the hellbender (*Cryptobranchus alleganiensis*), which reaches lengths of about 65 cm.

Most salamanders have internal fertilization. Males produce a pyramidal, gelatinous spermatophore that is capped with sperm and is deposited on the substrate. Females pick up the sperm cap with the cloaca, and the sperm are stored in a special pouch, the *spermatheca*. Eggs are fertilized as they pass through the cloaca and are usually deposited singly, in clumps, or in strings. Larvae are similar to adults

Box 28.1

Poison Frogs of South America

A South American native stalks quietly through the jungle, peering into the tree branches overhead. A monkey's slight movements divulge its presence, and the hunter takes careful aim with what appears to be an almost toylike bow and arrow. The arrow sails true, and the monkey is hit. The arrow seems ineffectual at first, however, after a few moments, the monkey tumbles from the tree. Thousands of years of cultural evolution have taught these natives a deadly secret that makes effective hunting tools out of seemingly innocuous instruments.

All amphibians possess glandular secretions that are noxious or toxic to varying degrees. These glands are distributed throughout the skin and exude milky toxins designed to ward off potential predators. Toxic secretions are frequently accompanied by warning (aposematic) coloration that signals to predators the presence of noxious secretions.

Four genera of frogs (*Atopophryhnus, Colostethus, Dendrobates,* and *Phyllobates*) in the family Dendrobatidae live in tropical forests from Costa Rica to southern Brazil. The black and gold skin of *Dendrobates* (box figure 28.1) and *Phyllobates* signal their highly toxic secretions. South American natives use this toxin to tip their arrows. Frogs are killed with a stick and held over a fire. *Granular glands* in the skin release their venom, which is collected and allowed to ferment. Poisons collected in this manner are neurotoxins that prevent the transmission of nerve impulses between nerves and between nerves and muscles. Arrow tips dipped in this poison and allowed to dry contain sufficient toxin to paralyze a bird or small mammal.

Members of this family of frogs, in addition to being exploited by South American natives, have interesting reproductive habits. A female lays one to six large eggs in moist, terrestrial habitats. The female promptly abandons the eggs, but the male visits the clutch regularly and guards the eggs. The eggs hatch after approximately 2 weeks, and the tadpoles wiggle onto the male's back. The male then transports the tadpoles from the egg-laying site to water, where they are left to develop. The tadpoles metamorphose to the adult body form after approximately 6 weeks.

Box figure 28.1 A poison arrow frog (*Dendrobates pumilo*).

very dry deserts. Adults lack tails, and caudal (tail) vertebrae are fused into a rodlike structure called the *urostyle*. Hind limbs are very long and muscular and end in webbed feet.

Anurans have diverse life histories. Fertilization is almost always external, and eggs and larvae are typically aquatic. Larval stages, called *tadpoles,* have well-developed tails. Their plump bodies lack limbs until near the end of their larval existence. Unlike adults, the larvae are herbivores and possess a proteinaceous, beaklike structure used in feeding. Anuran larvae undergo a drastic and rapid metamorphosis from the larval to the adult body form.

The distinction between "frog" and "toad" is more vernacular than scientific. "Toad" usually refers to Anurans with relatively dry and warty skin, and they are more terrestrial than other members of the order. These characteristics are found in a number of distantly related taxa. True

toads belong to the family Bufonidae. Other familiar anurans include the leopard frog (*Rana pipiens*), the tree frog (*Hyla andersoni*), and the American toad (*Bufo americanus*) (colorplate 11g–i).

Stop and Ask Yourself:

1. What were two ancient lineages of amphibians? What groups of animals are the modern descendants of each lineage?
2. What animals are members of the order Caudata?
3. What order of amphibians is characterized by wormlike burrowing?
4. What order of amphibians is characterized by tail vertebrae fused into a urostyle?

Evolutionary Pressures

Most amphibians divide their lives between fresh water and land. This divided life is shown by adaptations to both environments that can be observed in virtually every body system. In the water, amphibians are supported by water's buoyant properties, they exchange gases with the water, and face the same osmoregulatory problems as freshwater fishes. On land, amphibians support themselves against gravity, exchange gases with the air, and tend to lose water to the air.

External Structure and Locomotion

Vertebrate skin protects against infective microorganisms, ultraviolet light, desiccation, and mechanical injury. As discussed later in this chapter, the skin of amphibians also functions in gas exchange, temperature regulation, and absorption and storage of water.

The skin of amphibians lacks a covering of scales, feathers, or hair. It is, however, highly glandular and its secretions aid in protection. These glands keep the skin moist to prevent drying. They also produce sticky secretions that help a male cling to a female during mating and produce toxic chemicals that discourage potential predators (box 28.1). The skin of many amphibians is smooth, although epidermal thickenings may produce warts, claws, or sandpapery textures, which are usually the result of keratin deposits or the formation of hard, bony areas.

Chromatophores are specialized cells in the epidermis and dermis of the skin and are responsible for skin color and color changes. Cryptic coloration, aposematic coloration (box 28.1), and mimicry are all common in amphibians. As with other vertebrates, color changes are the result of fairly rapid changes in the distribution of pigment in the chromatophores and slower developmental changes in the quantity of pigments in chromatophores. These changes are controlled by neurons and the endocrine system.

Support and Movement   Water buoys and supports aquatic animals. The skeletons of fishes function primarily in protecting internal organs, providing points of attachment for muscles, and keeping the body from collapsing during movement. In terrestrial vertebrates, however, the skeleton is modified to provide support against gravity and must be strong enough to support the relatively powerful muscles that propel terrestrial vertebrates across land.

The skull of amphibians is flattened, is relatively smaller, and has fewer bony elements than the skull of fishes. These changes lighten the skull so it can be supported out of the water. Changes in jaw structure and musculature allow a crushing force to be applied to prey held in the mouth.

The vertebral column of amphibians is modified to provide support and flexibility on land (figure 28.3). It acts somewhat like the arch of a suspension bridge by supporting the weight of the body between anterior and posterior paired appendages. Supportive processes called zyg-

[1]

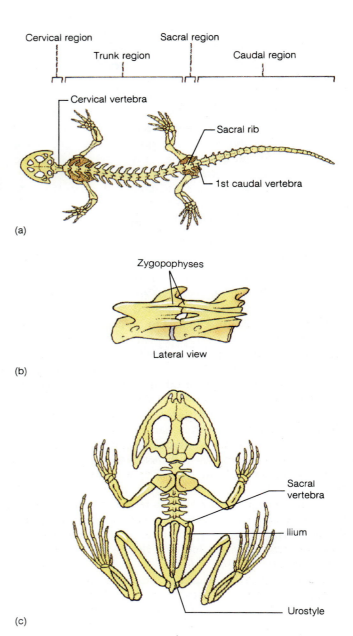

Figure 28.3   The skeletons of amphibians. (a) The salamander skeleton is divided into four regions: cervical, trunk, sacral, and caudal. (b) Interlocking processes, called zygapophyses prevent twisting between vertebrae. (c) The skeleton of a frog shows adaptations for jumping. Note the long back legs and the firm attachment of the back legs to the vertebral column through the ilium and urostyle.

apophyses on each vertebra prevent twisting. Unlike fishes, amphibians have a neck. The first vertebra is a *cervical vertebra,* which moves against (articulates with) the back of the skull and allows the head to nod vertically. The last trunk vertebra is a *sacral vertebra.* The ribs are used to anchor the pelvic girdle to the vertebral column to provide increased support. A ventral plate of bone, called the *sternum,* is present in the anterior, ventral trunk region and provides support for the forelimbs and protection for internal organs. It is reduced or absent in the Anura.

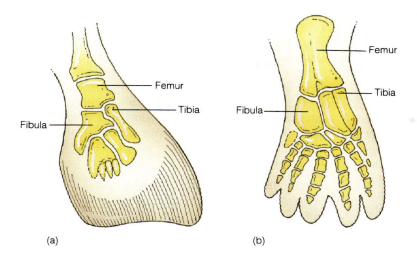

(a)　　　　　(b)

Figure 28.4 A comparison between the fin bones of a crossopterygian (a) and the limb bones of a tetrapod (b). This comparison suggests that the basic arrangement of bones seen in tetrapod limbs was already present in primitive fishes.

The origin of the bones of vertebrate appendages is not precisely known; however, similarities in the structures of the bones of the amphibian appendages and the bones of the fins of crossopterygians suggest possible homologies (figure 28.4a,b). The presence of joints at the shoulder, hip, elbow, knee, wrist, and ankle allows freedom of movement and better contact with the substrate. The pelvic girdle of amphibians consists of three bones (the *ilium, ischium,* and *pubis*) that attach pelvic appendages firmly to the vertebral column. These bones, which are present in all tetrapods but not fishes, are important for support on land.

Tetrapods depend more on appendages than the body wall for locomotion. Thus, body-wall musculature is reduced, and appendicular musculature predominates. (Contrast, for example, what one eats in a fish dinner as compared to a plate of frog legs.)

Salamanders employ a relatively unspecialized form of locomotion that is reminiscent of the undulatory waves that pass along the body of a fish. Aquatic salamanders also use this form of locomotion, and terrestrial salamanders employ it to escape from predators. At other times, terrestrial salamanders move by a pattern of limb and body movements in which the alternate movement of appendages results from muscle contractions that throw the body into a curve to advance the stride of a limb (figure 28.5). Caecilians move in an accordionlike movement in which adjacent parts of the body are pushed or pulled forward at the same time. The long hindlimbs and the pelvic girdle of anurans are modified for jumping. The dorsal bone of the pelvis (the ilium) extends anteriorly and is securely attached to the vertebral column, and the urostyle extends posteriorly and attaches to the pelvis. These skeletal modifications stiffen the posterior half of the anuran. Long hind limbs and powerful muscles form an efficient lever system for jumping. Elastic connective tissues and muscles attach the pectoral girdle to the skull and vertebral column and function as shock absorbers for landing on forelimbs (*see figure 28.3c*).

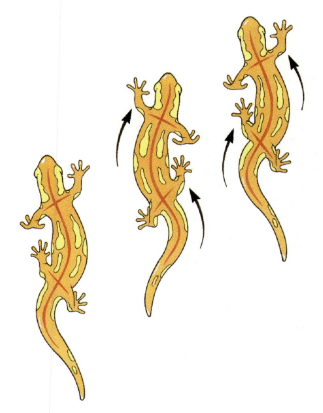

Figure 28.5 The pattern of leg movement in salamander locomotion.

Nutrition and the Digestive System

Most adult amphibians are carnivores that feed on a wide variety of invertebrates. The diets of some anurans, however, are more diverse. For example, a bullfrog will prey on small mammals, birds, and other anurans. The main factors that determine what amphibians will eat are prey size and availability. Larvae are herbivorous and feed on algae and other plant matter. Most amphibians locate their prey by sight and simply wait for prey to pass by. Olfaction plays an important role in prey detection by aquatic salamanders and caecilians.

Many salamanders are relatively unspecialized in their feeding methods, using only their jaws to capture prey. Anurans and plethodontid salamanders, however, use their tongue and jaws in a flip-and-grab feeding mechanism (figure 28.6). A true tongue is first seen in amphibians. (The "tongue" of fishes is simply a fleshy fold on the floor of the mouth. Fish food is swallowed whole and not manipulated by the "tongue.") The tongue of amphibians is attached at the anterior margin of the jaw and lies folded back over the floor of the mouth. Mucous and buccal glands on the tip of the tongue exude sticky secretions. When a [2] prey comes within range, an amphibian lunges forward and flicks out its tongue. The tongue turns over, and the lower

jaw is depressed. The fact that the head can tilt on its single cervical vertebra aids in aiming the strike. The tip of the tongue entraps the prey, and the tongue and prey are flicked back inside the mouth. All of this may happen in 0.05 to 0.15 second! The prey is held by pressing it against teeth on the roof of the mouth, and the tongue and other muscles of the mouth push food toward the esophagus. The eyes sink downward during swallowing, and help force food toward the esophagus. Digestive processes are similar to those of other vertebrates (*see chapter 37*).

Circulation, Gas Exchange, and Temperature Regulation

The circulatory system of amphibians shows remarkable adaptations for a life that is divided between aquatic and terrestrial habitats. The separation of pulmonary and systemic circuits is less efficient in amphibians than in lungfishes (figure 28.7; *see figure 27.11b*). The atrium is partially divided in urodeles and completely divided in anurans. The ventricle has no septum. A *spiral valve* is present in the conus arteriosus or ventral aorta and helps direct blood into pulmonary and systemic circuits. As discussed later, gas exchange occurs across the skin of amphibians, as well as at the lungs. Therefore, blood entering the right side of the heart is nearly as well oxygenated as blood entering the heart from the lungs! When an amphibian is completely submerged, all gas exchange occurs across the skin and other moist surfaces; therefore, blood coming into the right atrium has a higher oxygen concentration than blood returning to the left atrium from the lungs. Under these circumstances, blood vessels leading to the lungs constrict, reducing blood flow to the lungs and conserving energy. This adaptation is especially valuable for those frogs and salamanders that overwinter in the mud at the bottom of a pond.

Fewer aortic arches are present in adult amphibians than in fishes. After leaving the conus arteriosus, blood may enter the *carotid artery* (aortic arch III), which takes blood to the head; the *systemic artery* (aortic arch IV), which takes blood to the body; or the *pulmonary artery* (aortic arch VI).

Figure 28.6 Flip-and-grab feeding of the toad, *Bufo americanus.*

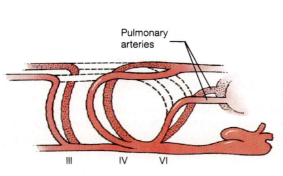

Pulmonary arteries

III   IV   VI

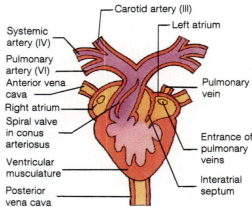

Carotid artery (III)
Left atrium
Systemic artery (IV)
Pulmonary artery (VI)
Anterior vena cava
Right atrium
Spiral valve in conus arteriosus
Pulmonary vein
Ventricular musculature
Entrance of pulmonary veins
Interatrial septum
Posterior vena cava

Figure 28.7 Diagrammatic representation of an anuran circulatory system. The heart is drawn in a ventral view. The blood flow is shown to the left. The roman numerals indicate the various aortic arches.

In addition to a vascular system that circulates blood, amphibians have a well-developed lymphatic system of blind-ending vessels that return fluids, proteins, and ions filtered from capillary beds in tissue spaces to the circulatory system. Water absorbed across the skin is also transported in the lymphatic system. Unlike other vertebrates, amphibians have contractile vessels, called *lymphatic hearts,* that pump fluid through the lymphatic system. Lymphatic spaces are present between body-wall muscles and the skin. These spaces transport and store water absorbed across the skin.

Gas Exchange   Terrestrial animals need to expend much less energy moving air across gas-exchange surfaces than do aquatic organisms because air contains more than 20 times more oxygen per unit volume than does water. On the other hand, exchanges of oxygen and carbon dioxide require moist surfaces, and exposure of respiratory surfaces to air may result in rapid water loss.

Anyone who has searched pond and stream banks for frogs knows that the skin of amphibians is moist. Amphibian skin is also richly supplied with capillary beds. These two [3] factors permit the skin to function as a respiratory organ. Gas exchange across the skin is called **cutaneous respiration** and can occur either in water or on land. Even gill-less amphibians need to surface for air infrequently, if at all. This ability allows a frog to spend the winter in the mud at the bottom of a pond. In salamanders, 30 to 90% of gas exchange occurs across the skin. Gas exchange also occurs across the moist surfaces of the mouth and pharynx. This is called **buccopharyngeal respiration** and accounts for 1 to 7% of total gas exchange.

[4]   Most amphibians, except for plethodontid salamanders, possess lungs (figure 28.8a). The lungs of salamanders are relatively simple sacs. The lungs of anurans are subdivided, increasing surface area for gas exchange. Pulmonary (lung) ventilation occurs by a **buccal pump** mechanism. Muscles of the mouth and pharynx create a positive pressure to force air into the lungs (figure 28.8b–e).

Cutaneous and buccopharyngeal respiration have a disadvantage in that the absolute contribution of these exchange mechanisms to total gas exchange is relatively constant. There is no way to increase the quantity of gas exchanged across these surfaces when the metabolic rate increases. Lungs, however, compensate for this shortcoming. As environmental temperature and activity increase, lungs contribute more to total gas exchange. At 5°C, approximately 70% of gas exchange occurs across the skin and mouth lining of a frog. At 25°C, the absolute quantity of oxygen exchanged across external body surfaces does not change significantly, but because pulmonary respiration is increased, exchange across skin and mouth surfaces accounts for only about 30% of total oxygen exchange.

Amphibian larvae and some adults respire using external gills. Three pairs of gills are supported by cartilaginous rods that are formed between embryonic gill slits. At metamorphosis, the gills are usually reabsorbed, gill slits close, and lungs become functional.

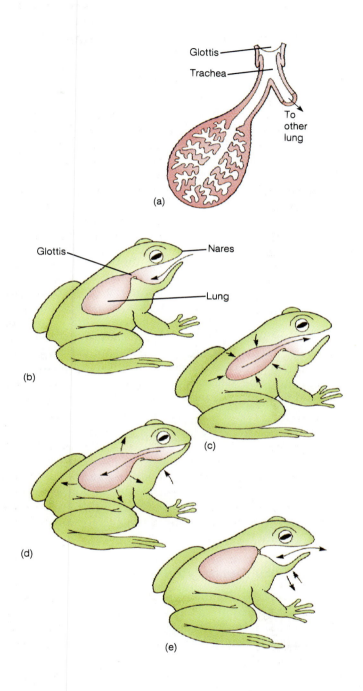

Figure 28.8   Amphibian lung structure, buccal pump, and buccopharygeal ventilation. (a) Lung structure of a frog. (b) With the opening of the respiratory tract (the glottis) closed, the floor of the mouth is lowered, and air enters the mouth cavity. (c) The glottis opens, and the elasticity of the lungs and contraction of the body wall forces air out of the lungs, over the top of air just brought into the mouth. (d) The mouth and nares are closed, and the floor of the mouth is raised, forcing air into the lungs. (e) With the glottis closed, oscillations of the floor of the mouth exchange air in the mouth cavity to facilitate buccopharyngeal respiration.

Temperature Regulation Amphibians are ectothermic. (They depend upon external sources of heat to maintain body temperature [*see chapter 38*].) Any poorly insulated aquatic animal, regardless of how much metabolic heat is produced, will lose heat as quickly as it is produced because of the powerful heat-absorbing properties of water. Therefore, when amphibians are in water, they take on the temperature of their environment. On land, however, their body temperatures can differ from that of the environment.

Temperature regulation is mainly behavioral. Some cooling results from evaporative heat loss. In addition, many amphibians are nocturnal and remain in cooler burrows or under moist leaf litter during the hottest part of the day. Amphibians may warm themselves by basking in the sun or on warm surfaces. Body temperatures may be raised 10°C above the air temperature. Basking after a meal is common, because increased body temperature increases the rate of all metabolic reactions—including digestive functions, growth, and the deposition of fats necessary to survive periods of dormancy. As discussed later in this chapter, basking and evaporative cooling also increase rates of water loss.

Amphibians often experience wide daily and seasonal temperature fluctuations, and therefore, have correspondingly wide temperature tolerances. Critical temperature extremes for some salamanders lie between −2°C and 27°C, and for some anurans between 3°C and 41°C.

Stop and Ask Yourself

5. What are the functions of amphibian skin?
6. What chambers are present in a frog's heart?
7. Why is the incomplete separation of atria and ventricles believed to be adaptive for an amphibian?
8. What are four forms of gas exchange in amphibians?

Nervous and Sensory Functions

The nervous system of amphibians is similar to that of other vertebrates. The brain of adult vertebrates develops from three embryological subdivisions. In amphibians, the *forebrain* contains olfactory centers and regions that regulate color change and visceral functions. The *midbrain* contains a region called the *optic tectum,* in which sensory information is assimilated and motor responses are initiated. Visual sensory information is also processed in the midbrain. The *hindbrain* functions in motor coordination and in regulating heart rate and respiratory movements.

Many amphibian sensory receptors are widely distributed over the skin. Some of these are simply bare nerve endings that respond to heat, cold, and pain. The lateral line system is similar in structure to that found in fishes, and it is present in all aquatic larvae, aquatic adult salamanders, and some adult anurans. Lateral line organs are distributed singly or in small groups along the lateral and dorsolateral surfaces of the body, especially the head.

These receptors respond to low frequency vibrations in the water and movements of the water relative to the animal. On land, however, lateral line receptors are less important.

Chemoreception is an important sense for many amphibians. Chemoreceptors are located in the nasal epithelium, the lining of the mouth, on the tongue, and over the skin. Olfaction is used in mate recognition, as well as detecting noxious chemicals and in locating food.

Vision is one of the most important senses in amphibians because they are primarily sight feeders. (Caecilians are an obvious exception.) A number of adaptations allow the eyes of amphibians to function in terrestrial environments (figure 28.9). The fact that the eyes of some amphibians (i.e., anurans and some salamanders) are located on the front of the head provides the binocular vision and well-developed depth perception necessary for capturing prey. Other amphibians with smaller lateral eyes (some salamanders) lack binocular vision. The lower eyelid is movable and functions to clean and protect the eye. Much of it is transparent and is called the **nictitating membrane.** When the eyeball is retracted into the orbit of the skull, the nictitating membrane is drawn up over the cornea. In addition, orbital glands lubricate and wash the eye. Together, eyelids and glands keep the eye free of dust and other debris. The lens is large and nearly round. It is set back from the cornea and is surrounded by a fold of epithelium called the iris. The iris can dilate or constrict to control the size of the pupil.

Focusing, or *accommodation,* involves bending (refracting) light rays to a focal point on the retina. Light waves moving from air across the cornea are refracted because of the change in density between the two media. Further refraction is accomplished by the lens. As are the eyes of most tetrapods, the amphibian eye is focused on distant objects when the eye is at rest. To focus on near objects, the lens must be moved forward by a *protractor lentis muscle* (figure 28.9). Receptors called *rods* and *cones* are found in the *retina.* Because cones are associated with color vision in some other vertebrates, their occurrence suggests that amphibians are capable of distinguishing between some wavelengths of light. The extent to

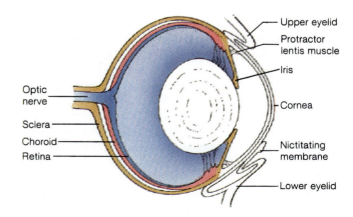

Upper eyelid
Protractor lentis muscle
Iris
Optic nerve
Cornea
Sclera
Choroid
Nictitating membrane
Retina
Lower eyelid

Figure 28.9 Longitudinal section of the eye of the leopard frog, *Rana pipiens.*

which color vision is developed is unknown. The neuronal interconnections in the retina are very complex and allow an amphibian to distinguish between flying insect prey, shadows that may warn of an approaching predator, and background movements, such as blades of grass moving with the wind.

The auditory system of amphibians is clearly an evolutionary adaptation to life on land. It transmits both substrateborne vibration and, in anurans, airborne vibrations. The ears of anurans consist of a *tympanic membrane,* a *middle ear* and an *inner ear.* The tympanic membrane is a piece of integument stretched over a cartilaginous ring that receives airborne vibrations and transmits these vibrations to the middle ear, which is a chamber beneath the tympanic membrane. Abutting the tympanic membrane is a middle ear ossicle (bone) called the *stapes* (columella). The opposite end of the stapes abuts the membrane of the *oval window,* which stretches between the middle and inner ears. High frequency (1,000 to 5,000 Hz) airborne vibrations striking the tympanic membrane are transmitted through the middle ear via the stapes and cause pressure waves in the fluids of the inner ear, which stimulate receptor cells (figure 28.10). A second small ossicle, the *operculum,* also abuts the oval window. Substrateborne vibrations transmitted through the front appendages and the pectoral girdle cause this ossicle to vibrate. The resulting pressure waves in the inner ear stimulate a second patch of sensory cells that is sensitive to low-frequency (100 to 1,000 Hz) sounds. Muscles attached to the operculum and stapes can lock either or both of these ossicles, allowing an anuran to screen out either high or low frequency sounds. This mechanism is adaptive because low

[5]

and high frequency sounds are used in different situations by anurans. Mating calls are high frequency sounds that are of primary importance for only a part of the year (breeding season). At other times, low frequency sounds may warn of approaching predators.

Salamanders lack a typanum and middle ear. They live in streams, ponds, caves, and beneath leaf litter. They have no mating calls, and the only sounds they hear are probably low-frequency vibrations transmitted through the substrate and skull to the stapes and inner ear.

The sense of equilibrium and balance is similar to that described for fishes in the previous chapter. The inner ear of amphibians has semicircular canals that help detect rotational movements and other sensory patches that respond to gravity. The latter detect linear acceleration and deceleration.

Excretion and Osmoregulation

The kidneys of amphibians lie on either side of the dorsal aorta on the dorsal wall of the body cavity. A duct leads to the *cloaca,* and a storage structure, the urinary bladder, is a ventral outgrowth of the cloaca.

The nitrogenous waste product excreted by amphibians is either ammonia or urea. Amphibians that live in fresh water excrete ammonia. It is the immediate endproduct of protein metabolism, therefore, no energy is expended converting it into other products. The toxic effects of ammonia are avoided by its rapid diffusion into the surrounding water. Amphibians that spend more time on land excrete urea that is produced from ammonia in the liver. Although urea is less toxic than ammonia, it still requires relatively large quantities of water for its excretion. Unlike ammonia, urea can be stored in the urinary bladder. Some amphibians excrete ammonia when in water and urea when on land.

One of the biggest problems faced by amphibians is osmoregulation. In water, amphibians face the same osmoregulatory problems as freshwater fishes. They must rid the body of excess water and conserve essential ions. Amphibian kidneys produce large quantities of hypotonic urine, and the skin and walls of the urinary bladder transport Na$^+$, Cl$^-$, and other ions into the blood.

On land, amphibians must conserve water. Adult amphibians do not replace water by intentional drinking, nor do they have the impermeable integument characteristic of other tetrapods or kidneys capable of producing a hypertonic urine. Instead, amphibians limit water loss by behavior that reduces exposure to desiccating conditions. Many terrestrial amphibians are nocturnal. During daylight hours, they retreat to areas of high humidity, such as under stones, in logs, in leaf mulch, or in burrows. Water loss on nighttime foraging trips must be compensated for by water uptake across the skin while in the retreat. Diurnal amphibians usually live in areas of high humidity and rehydrate themselves by entering the water. Many amphibians reduce evaporative water loss by reducing the amount of body surface exposed to air. They may curl their bodies and tails into tight coils and tuck their limbs close to their

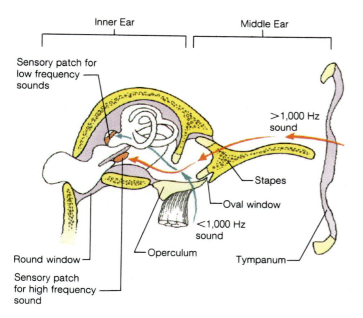

Figure 28.10 The ear of an anuran, posterior view. Solid arrows show the pathway of low frequency sounds via the operculum. Open arrows show the pathway of high frequency sound via the stapes.

Figure 28.11 The daytime sleeping posture of the green tree frog, *Hyla cinerea*. Exposed surface area is reduced by the closely tucked appendages.

(a)

(b)

Figure 28.12 The Australian burrowing frog, *Cyclorana alboguttatus,* (a) in its burrow and water-retaining skin and (b) emerging from its protective skin.

bodies (figure 28.11). Individuals may form closely packed aggregations to reduce overall surface area.

Some amphibians have protective coverings that reduce water loss. Hardened regions of skin are resistant to water loss and may be used to plug entrances to burrows or other retreat openings to maintain high humidity in the retreat. Other amphibians prevent water loss by forming cocoons that encase the body during long periods of dormancy. Cocoons are made from outer layers of the skin that detach and become parchmentlike. These cocoons open only at the nares or the mouth and have been found to reduce water loss 20 to 50% over noncocooned individuals (figure 28.12).

Paradoxically, the skin—the most important source of water loss—is also the most important structure for rehydration. When an amphibian flattens its body on moist surfaces, the skin, especially in the ventral pelvic region, absorbs water. The permeability and vascularization of the skin and its epidermal sculpturing are all factors that promote water reabsorption. Minute channels increase surface area and spread water over surfaces not necessarily in direct contact with water.

Amphibians can also temporarily store water. Water accumulated in the urinary bladder and lymph sacs can be selectively reabsorbed to replace evaporative water loss. Amphibians living in very dry environments can store volumes of water equivalent to 35% of their total body weight.

Reproduction, Development, and Metamorphosis

Amphibians are dioecious, and ovaries and testes are located near the dorsal body wall. Fertilization is usually external, and because the developing eggs lack any resistant coverings, development is tied to moist habitats, usually water. A few anurans have terrestrial nests that are kept moist by being enveloped in foam or by being located near the water and subjected to flooding. In a few species, larval stages are passed in the egg membranes, and the immatures hatch into an adultlike body. The main exception to external fertilization in amphibians is the salamanders. Only about 10% of all salamanders have external fertilization. All others use spermatophores, and fertilization is internal. Eggs may be deposited in soil or water or retained in the oviduct during development. All caecilians have internal fertilization and 75% have internal development. Amphibian development has been studied extensively and usually includes the formation of larval stages called tadpoles (colorplate 11a–c). Amphibian tadpoles often differ from the adults in mode of respiration, form of locomotion, and diet. These differences reduce competition between adults and larvae.

The timing of reproductive activities is determined by interactions between internal (largely hormonal) controls and extrinsic factors. In temperate regions, breeding periods are seasonal and occur during spring and summer. In temperate areas, temperature seems to be the most important environmental factor that induces physiological changes associated with breeding. In tropical regions, breeding of amphibians is correlated with rainy seasons.

Courtship behavior helps individuals locate breeding sites, identify potential mates, prepare individuals for reproduction, and ensure that eggs are fertilized and deposited in locations that promote successful development.

Salamanders rely primarily on olfactory and visual clues in courtship and mating, whereas in anurans, vocalizations by the male and tactile cues are important. Many species congregate in one location during times of intense breeding activity. Calls by males are usually species specific and they function in the initial attraction between mates. Once initial contact has been made, tactile cues become more important. The male grasps the female—his forelimbs around her waist—so that they are oriented in the same direction, and the male is dorsal to the female (*see figure 10.8*). This positioning is called **amplexus** and may last from 1 to 24 hours. During amplexus, the male releases sperm as the female releases eggs.

Little is known of caecilian breeding behavior. Males possess an intromittent organ that is a modification of the cloacal wall, and fertilization is internal.

Vocalization   Sound production is primarily a reproductive function of male anurans. *Advertisement calls* attract females to breeding areas, and announce to other males that a given territory is occupied. Advertisement calls are species specific, and any one species has a very limited repertoire of calls. They may also help induce psychological and physiological readiness to breed. *Reciprocation calls* are given by females in response to male calls to indicate receptiveness of a female. *Release calls* inform a partner that a frog is incapable of reproducing. They are given by unresponsive females during attempts at amplexus by a male, or by males that have been mistakenly identified as female by another male. *Distress calls* are not associated with reproduction, but are given by either sex in response to pain or being seized by a predator. These calls may be loud enough to cause a predator to release the frog. The distress call of the South American jungle frog, *Leptodactylus pentadactylus,* is a loud scream similar to the call of a cat in distress.

[6]

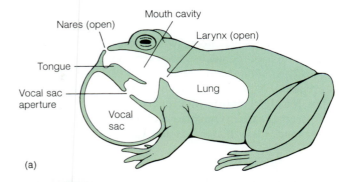

(b)

Figure 28.13   Anuran vocalization. (a) Generalized vocal apparatus of anurans. (b) Inflated vocal sac of the Great Plains toad, *Bufo cognatus.*

Figure 28.13a redrawn, with permission, from Duellman and Treub, *Biology of Amphibians.* Copyright © 1986, McGraw-Hill Publishing Company, New York, New York.

The sound-production apparatus of frogs consists of the larynx and its vocal cords. This laryngeal apparatus is well developed in males, who also possess a vocal sac. In the majority of frogs, vocal sacs develop as a diverticulum from the lining of the buccal cavity (figure 28.13). Air from the lungs is forced over the vocal cords and cartilages of the larynx, causing them to vibrate. Muscles control the tension of the vocal cords and are responsible for regulating the frequency of the sound. Vocal sacs act as resonating structures and increase the volume of the sound.

The use of sound to attract mates is especially useful in organisms that occupy widely dispersed habitats and must come together for breeding. Because many species of frogs often converge at the same pond for breeding, finding a mate of the proper species could be chaotic. Vocalizations help to reduce the chaos.

Parental Care   Parental care increases the chances of any one egg developing, but it requires large energy expenditures on the part of the parent. The most common form of parental care in amphibians is attendance of the egg clutch by either parent. Maternal care occurs in species with internal fertilization (predominantly salamanders and caecilians), and paternal care may occur in species with external fertilization (predominantly anurans). It may involve aeration of aquatic eggs, cleaning and/or moistening of terrestrial eggs, protection of eggs from predators, or removal of dead and infected eggs.

Transport of eggs may occur when development occurs on land. Females of the genus *Pipa* carry eggs on their back (figure 28.14a). *Rheobatrachus* females carry developing eggs and tadpoles in their stomach, and the young emerge from the female's mouth (figure 28.14b)! The ultimate form of parental care is viviparity and ovoviviparity, which occurs primarily in salamanders and caecilians.

Metamorphosis   Metamorphosis is a series of abrupt structural, physiological, and behavioral changes that transform a larva into an adult. The time required for metamorphosis is influenced by a variety of environmental conditions, including crowding and food availability. Most directly, however, metamorphosis is under the control of neurosecretions of the hypothalamus, hormones of the anterior lobe of the pituitary gland (the adenohypophysis), and the thyroid gland. A hormone of the adenohypophysis, *prolactin,* is secreted during the larval stages and promotes the growth of larval structures. A portion of the hypothalamus called the *median eminence* gradually develops during the larval stage, and its secretions cause the adenohypophysis to secrete increasing quantities of thyroid stimulating hormone (TSH). TSH inhibits the release of prolactin and stimulates the release of thyroxine (T_4) and triiodothyronine (T_3) from the thyroid gland. The release of these thyroid hormones initiates the changes associated with metamorphosis.

Morphological changes associated with metamorphosis of caecilians and salamanders are relatively minor. Reproductive structures develop, gills are lost, and a caudal fin (when present) is lost. In the Anura, however, changes from the tadpole to the small frog are more dramatic (figure 28.15). Limbs and lungs develop, the tail is reabsorbed, the skin thickens, and marked changes in the head and digestive tract (associated with a new mode of nutrition) occur.

Neoteny in amphibians can be explained based upon mechanisms of metamorphosis. Some salamanders are neotenic because of a failure of cells to respond to thyroid hormones, whereas others are neotenic because of a failure to produce the hormones associated with metamorphosis. Should environmental conditions change, they are able to metamorphose to the adult form.

Further Phylogenetic Considerations

One unresolved controversy concerning amphibian phylogeny is the relationship among the three orders of modern amphibians. Some zoologists place anurans, urodeles, and caecilians into a single subclass, Lissamphibia. This placement implies a common ancestry for modern amphibians and suggests that they are more closely related to each other than to any other group. Supporters of this classification point to common characteristics, such as the stapes/operculum complex, the importance of the skin in gas exchange, and aspects of the structure of the skull and teeth, as evidence of this close relationship. Other zoologists think that modern amphibians were derived from at least two nonamniotic lineages. They note that fine details of other structures, such as the vertebral column, are different enough in the three orders to suggest separate origins. This controversy is not likely to be settled soon.

In the next three chapters, our attention will turn to descendants of the amniote lineage. A group called anthracosaurs are often cited as amphibian ancestors of these animals, but support for this conclusion is weak. Three sets of evolutionary changes occurred in amphibian lineages that allowed movement onto land. Two of these occurred early enough that they are found in all amphibians. One was the set of changes in the skeleton and muscles that allowed free movement on land. A second change involved a jaw mechanism and moveable head that permitted effective exploitation of insect resources on land. The third set of changes occurred in the amniote lineage—the development of an egg that was resistant to drying. Although the *amniotic egg* is not completely independent of water, a series of extraembryonic membranes form during development that protect the embryo from desiccation, store wastes, and promote gas exchange. In addition, this egg has a leathery or calcified shell that is protective, yet porous enough to allow exchanges with the environment. These evolutionary events eventually resulted in the remaining three vertebrate groups: reptiles, birds, and mammals. [7]

Stop and Ask Yourself:

9. What region of the amphibian brain integrates sensory information and initiates motor responses?
10. What respiratory and excretory adaptations do amphibians possess that promote life in terrestrial environments?
11. What are four functions of anuran vocalizations?
12. In what ways are amniotic eggs adaptive for life on land?

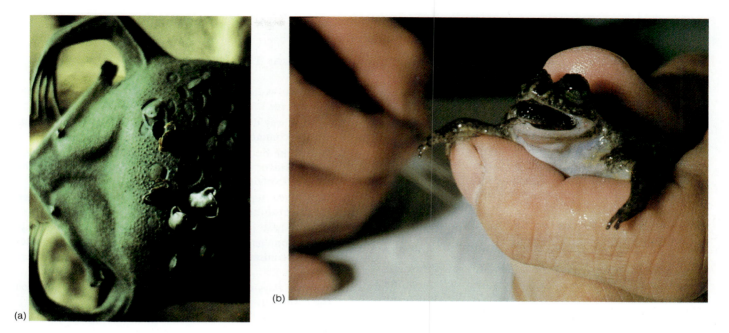

Figure 28.14   Parental care of young. (a) Female *Pipa* with young emerging from her back. (b) Female *Rheobatrachus* with young emerging from her mouth.

Figure 28.15   Events of metamorphosis in the frog. (a) Before metamorphosis. Prolactin secretion, controlled by the hypothalamus and the adenohypophysis, promotes the growth of larval structures. (b–d) Metamorphosis. The median eminence of the hypothalamus develops and initiates the secretion of thyroid stimulating hormone (TSH). TSH begins to inhibit prolactin release. TSH causes the release of large quantities of T_4 and T_3, which promote the growth of limbs, reabsorption of the tail, and other changes of metamorphosis, resulting eventually in a young, adult frog.

Summary

1. Terrestrial vertebrates are called tetrapods and probably arose from crossopterygians. Two lineages of ancient amphibians diverged. The nonamniote lineage gave rise to the three orders of modern amphibians. The amniote lineage gave rise to reptiles, birds, and mammals.

2. Members of the order Caudata (Urodela) are the salamanders. They are widely distributed, usually have internal fertilization, and may have aquatic larvae or direct development.

3. The order Gymnophiona contains the caecilians. They are tropical, wormlike burrowers. They have internal fertilization and many are viviparous.

4. Frogs and toads comprise the order Anura. They lack tails and possess adaptations for jumping and swimming. External fertilization results in tadpole larvae, which metamorphose to adults.

5. The skin of amphibians is moist and functions in gas exchange, water regulation, and protection.

6. Skeletal and muscular systems of amphibians are adapted for movement on land.

7. Amphibians are carnivores that capture prey in their jaws or by using their tongue.

8. The circulatory system of amphibians is modified to accommodate the presence of lungs, gas exchange at the skin, and loss of gills in most adults.

9. Gas exchange is cutaneous, buccopharyngeal, and pulmonary. Pulmonary ventilation is accomplished by a buccal pump.

10. Sensory receptors of amphibians, especially the eye and ear, are adapted for functioning on land.

11. Amphibians excrete ammonia or urea. Ridding the body of excess water when in water and conserving water when on land are functions of the kidneys, the skin, and behavior.

12. Reproductive habits of amphibians are diverse. Many have external fertilization and development. Others have internal fertilization and development. Courtship, vocalizations, and parental care are common in some amphibians. Metamorphosis is under the control of the nervous and endocrine systems.

13. The evolution of an egg that is resistant to drying occurred in the amniote lineage, which is represented today by reptiles, birds, and mammals.

Key Terms

amniote lineage (p. 439)
amplexus (p. 449)
buccal pump (p. 445)
buccopharyngeal respiration (p. 445)
cutaneous respiration (p. 445)
nictitating membrane (p. 446)
nonamniote lineage (p. 439)
tetrapods (p. 439)

Critical Thinking Questions

1. How are skeletal and muscular systems of amphibians adapted for life on land?

2. In what way do insects, amphibians, and flowering plants share a common evolutionary history?

3. Would the buccal pump be more important for an active amphibian or for one that is becoming inactive for the winter? Explain your answer.

4. Why is the separation of oxygenated and nonoxygenated blood at the heart not very important for amphibians?

5. Explain how the skin of amphibians is used in cooling, protection, gas exchange, and water regulation. Under what circumstances might cooling interfere with water regulation?

6. In what ways could anuran vocalizations have influenced the evolution of that order?

Suggested Readings

Books

Duellman, W. E., and Trueb, L. 1986. *Biology of Amphibians.* New York: McGraw-Hill Book Co.

Goin, C. J., and Goin, O. B. 1978. *Introduction to Herpetology,* 3rd ed. San Francisco: W.H. Freeman & Sons.

King, F. W., and Behler, J. 1979. *The Audubon Society Field Guide to North American Reptiles and Amphibians.* New York: Alfred A. Knopf, Inc.

Pough, F. H., Heiser, J. B., and McFarland, W. N. 1989. *Vertebrate Life,* 3rd ed. New York: Macmillan Publishing Company.

Walker, W. F. 1987. *Functional Anatomy of the Vertebrates: An Evolutionary Perspective.* Philadelphia: W.B. Saunders.

Articles

Gorniak, G. C., and Gans, C. 1982. How does the toad flip its tongue? Test of two hypotheses. *Science* 216: 1335–1337.

Hanken, J. 1989. Development and evolution in amphibians. *American Scientist* 77(4):336–343.

Jaeger, R. G. 1988. A comparison of territorial and non-territorial behaviour in two species of salamanders. *Animal Behavior* 36:307–400.

Milner, A. 1989. Late Extinctions of amphibians. *Nature* 338:117.

Ryan, M. J. 1990. Signals, species, and sexual selection. *American Scientist* 78(1):46–52.

Verrell, P. 1988. The chemistry of sexual persuasion. *New Scientist,* 118: 40–43.

Wilczynski, W., and Brenowitz, E. A. 1988. Acoustic cues mediate inter-male spacing in a neotropical frog. *Animal Behaviour* 36: 1054–1063.

Reptiles: The First Amniotes

29

Concepts

1. Adaptive radiation of primitive amniotes resulted in the anapsid, diapsid, and synapsid lineages of reptiles.
2. The class Reptilia is divided into four orders: Testudines includes the turtles; Squamata includes the lizards, the snakes, and the worm lizards; Rhynchocephalia includes a single species, *Sphenodon punctatus*; and Crocodilia includes the alligators and crocodiles.
3. Reptiles possess adaptations that allow many members of the class to spend most of their lives apart from standing or flowing water. These include adaptations for support and movement, feeding, gas exchange, temperature regulation, excretion, osmoregulation, and reproduction.
4. Two reptilian evolutionary lineages gave rise to other vertebrate classes: Aves and Mammalia.

Have You Ever Wondered:

[1] what living animals are most closely related to dinosaurs?
[2] why turtles are vulnerable to extinction?
[3] how a chameleon captures prey?
[4] what animal "walks at both ends?"
[5] why a lizard's tail breaks easily?
[6] why reptiles divert blood away from their lungs?
[7] what a median eye is?

These and other interesting questions will be answered in this chapter.

This chapter contains underlined evolutionary concepts.

Evolutionary Perspective

The earliest members of the class Reptilia (rep-til'e-ah) (L. *reptus,* to creep) were the first vertebrates to possess amniotic eggs. Thus, their development could be separate from standing or flowing water. Numerous other adaptations have allowed members of this class to flourish on land. Living representatives include turtles, lizards, snakes, worm lizards, crocodilians, and the tuatara (table 29.1).

Nowhere in biology is the ancient history of a group of animals more impressively documented by the fossil record than for the reptiles. Adaptive radiation of early amniotes began in the late Carboniferous and early Permian periods. This time coincided with adaptive radiation of terrestrial insects, the major prey of early amniotes. One of the ways that the early lineages of amniotes are distinguished is by the structure of the skull, particularly the modifications in jaw muscle attachment. Paleontologists have described three major lineages of early reptiles.

Reptiles in the **anapsid lineage** (Gr. *an,* without + *hapsis,* arch) lack openings, or fenestra, in the temporal (posterio-lateral) region of the skull (figure 29.1). The earliest reptiles were anapsids, and this lineage is represented today by the turtles (figure 29.2). Changes in turtles have occurred in their long evolutionary history, but the fundamental form of their skull and shell is recognizable in 200 million-year-old fossils.

Reptiles in the **diapsid lineage** (Gr. *di,* two) had upper and lower openings in the temporal region of the skull. Diapsids underwent extensive adaptive radiation in the Mesozoic era. A group called the archosaurs include the di-

nosaurs and numerous other extinct reptiles (box 29.1). Crocodilians are the only living reptiles that are a part of this group. This diapsid group also includes the reptilian ancestors of birds. The other major diapsid group contained the lepidosaurs and numerous extinct reptiles, including the aquatic ichthyosaurs and plesiosaurs. Living reptiles that are a part of this lineage include snakes, lizards, and the tuatara. [1]

Reptiles in the **synapsid lineage** (Gr. *syn,* with) had one opening in the temporal region of the skull. Although there are no living reptilian descendants of this group, they are important because a group of synapsids, called therapsids, or mammallike reptiles, gave rise to the mammals.

Table 29.1 Classification of Living Reptiles

Class Reptilia (rep-til'e-ah)
Skin dry, with epidermal scales; skull with one point of articulation with the vertebral column (occipital condyle); respiration via lungs; metanephric kidneys; internal fertilization; amniotic eggs.

Order Testudines (tes-tu'din-ez) or Chelonia (ki-lo'ne-ah)
Teeth absent in adults and replaced by a horny beak; body short and broad; shell consisting of a dorsal carapace and ventral plastron. Turtles.

Order Squamata (skwa-ma'tah)
Recognized by specific characteristics of the skull and jaws (temporal arch reduced or absent and quadrate movable or secondarily fixed); the most successful and diverse group of living reptiles. Snakes, lizards, worm lizards.

Order Rhynchocephalia (rin'ko-se-fay'le-ah)
Contains very primitive, lizardlike reptiles; well-developed parietal eye. A single species, *Sphenodon punctatus,* survives in New Zealand. Tuataras.

Order Crocodilia (krok'o-dil'e-ah)
Elongate, muscular, and laterally compressed; tongue not protrusible; complete ventricular septum. Crocodiles, alligators, caimans, gavials.

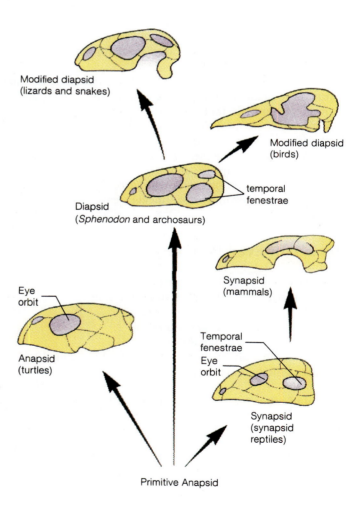

Figure 29.1 The evolution of the skulls of amniotes. Early amniote lineages can be distinguished by the presence or absence, and number, of openings in the temporal region of the skull.

Survey of Reptiles

Reptiles are characterized by a skull with one surface (condyle) for articulation with the first neck vertebra, respiration by lungs, metanephric kidneys, internal fertilization, and amniotic eggs. Reptiles also have dry skin with keratinized epidermal scales. *Keratin* is a resistant protein found in epidermally derived structures of amniotes. It is protective and when it is chemically bonded to phospholipids, it prevents water loss across body surfaces. Members of three of the four orders described below are found on all continents except Antarctica. It is only in tropical and subtropical environments, however, that reptiles are a dominant part of any major ecosystem. There are 17 orders of reptiles, but members of most orders are extinct. The four orders containing living representatives are described next (*see table 29.1*).

Order Testudines or Chelonia

Members of the order Testudines (tes-tu′din-ez) (L. *testudo,* tortise), or Chelonia (ki-lo′ne-ah) (Gr. *chelone,* tortise) are the turtles (colorplate 12a,b). There are about 225 species of turtles in the world, and they are characterized by a bony shell with limbs articulating internally to the ribs and a keratinized beak rather than teeth. The dorsal portion of the shell is the *carapace,* which is formed from a fusion of vertebrae, expanded ribs, and bones formed in the dermis of the skin (box 29.2). The bone of the carapace is covered by keratin. The ventral portion of the shell is the *plastron.* It is formed from bones of the pectoral girdle and dermal bone, and is also covered by keratin (figure 29.3). In some turtles, such as the North American box turtles (*Terrapene*), the shell has flexible areas, or hinges, that allow the anterior and posterior edges of the plastron to be raised. This design closes the shell

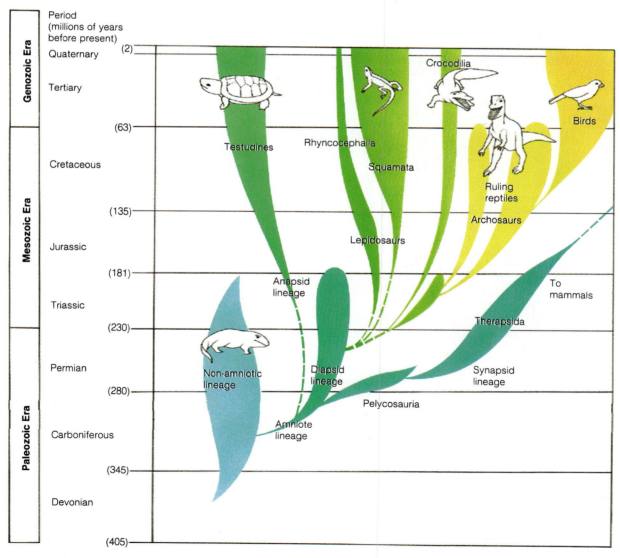

Figure 29.2 Evolutionary relationships among the amniotes. The relative number of species in each taxonomic group is indicated by the width of a group's branch.

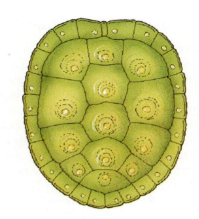

(a)

Figure 29.3 The skeleton of a turtle. (a) Dorsal view of the carapace. (b) Ventral view of the carapace and appendicular skeleton. The carapace is composed of fused vertebrae, expanded ribs, and dermal bone and is covered by keratin. (c) Dorsal view of the plastron. (d) Ventral view of the plastron. The plastron is formed from dermal bone and bone of the pectoral girdle. It is also covered by keratin.

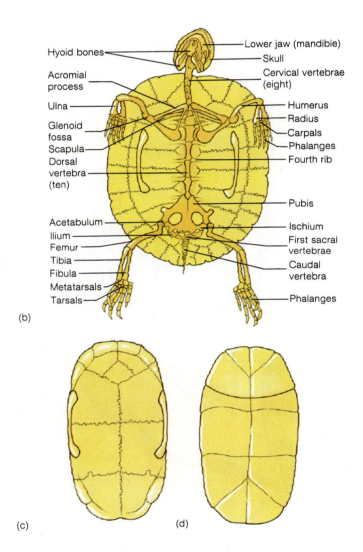

Hyoid bones
Lower jaw (mandible)
Skull
Acromial process
Cervical vertebrae (eight)
Ulna
Humerus
Radius
Glenoid fossa
Carpals
Scapula
Phalanges
Dorsal vertebra (ten)
Fourth rib
Pubis
Acetabulum
Ilium
Ischium
Femur
First sacral vertebrae
Tibia
Fibula
Caudal vertebra
Metatarsals
Tarsals
Phalanges

(b)

(c)   (d)

openings when the turtle is withdrawn into the shell. Turtles have eight cervical vertebrae that can be articulated into an "S-shaped" configuration, which allows the head to be drawn into the shell.

Turtles have long life spans. Most reach sexual maturity after 7 or 8 years and live 14 or more years. Large tortoises of the Galápagos Islands may live in excess of 100 years. All turtles are oviparous. Females use their hind limbs to excavate nests in the soil. Clutches of 5 to 100 eggs are laid and covered with soil. Development takes from 4 weeks to 1 year, and eggs are not attended by the parent during development. The young are independent of the parent at hatching.

In recent years, turtle conservation programs have been enacted. Slow rates of growth and long juvenile periods make turtles vulnerable to extinction in the face of high [2]

Box 29.1

Collision or Coincidence?

Extinction is the eventual fate of all species. Indeed, 99% of all species ever on the earth are now extinct. Often, the rate of extinction is low; however, on several occasions in the history of the earth, rates of extinction have increased dramatically, resulting in the total extinction of many successful taxa. For example, at the Permian/Triassic boundary, a mass extinction resulted in the loss of 80 to 90% of all animal species. About 65 million years ago, the dinosaurs, along with 50% of all other animal species became extinct. This Tertiary/Carboniferous extinction occurred over a period of tens of thousands to several million years. From a geological perspective, these extinctions are "sudden." How the age of the reptiles ended has been the subject of speculation for many years. Various hypotheses involving catastrophic or gradual change have been proposed.

The impact of asteroids or periods of intense volcanic activity are catastrophic events that have been suggested as causes of mass extinctions. Both events may have injected large quantities of dust into the air that shaded the earth's surface, reducing photosynthetic production, and thus, food for animals.

The presence of the element iridium in rock strata from the Cretaceous/Tertiary boundary and other periods of mass extinction is the primary evidence supporting the catastrophic hypotheses. Iridium is a primarily extraterrestrial element that is deposited during asteroid impacts with the earth. Some deposits may also result from volcanic activity. Opponents of the catastrophic hypotheses do not deny that asteroid impacts or periods of volcanic activity resulted in iridium deposits on the earth. They question, however, whether the catastrophes are responsible for mass extinctions. They point out that the paleontological record indicates that extinctions are not as abrupt as implied in catastrophic hypotheses. Extinctions apparently occurred over tens of thousands of years, not tens of years. In addition, catastrophic events would be expected to affect all animal groups more-or-less equally, which was not the case with the Cretaceous/Tertiary extinction. For example, dinosaurs became extinct, but crocodiles, turtles, birds, and early mammals did not.

There are numerous hypotheses that propose gradual, selective changes as explanations for mass extinction. Some of these involve climatic changes. Such changes could have been induced by continental drift. In the Cretaceous period, 70% of the present land area was covered by warm, shallow seas. By the end of the Cretaceous period, these seas were reduced to 15% of the present land area, resulting in the reduction of habitat for shallow-water marine organisms, a decrease in atmospheric temperatures, and dissection of land areas by newly formed rivers. Climatic and habitat changes such as these could have resulted in extinctions over periods of tens of thousands of years.

Regardless of what hypothesis (or hypotheses) of mass extinction is correct, this question is a good example of how interest fueled by controversy stimulates scientific inquiry. The gradualism/catastrophism debate has led to new, innovative ideas on the origin, evolution, and extinction of taxa.

---

mortality rates. Turtle hunting and predation on young turtles and turtle nests by dogs and other animals has severely threatened some species. Predation on nests and young is made more serious by the fact that certain beaches are used year after year by nesting sea turtles. Conservation of sea turtles is complicated by the fact that they have ranges of thousands of square kilometers of ocean, so that protective areas must include waters under the jurisdiction of many different nations.

Order Squamata

The order Squamata (skwa-ma'tah) (L. *squama,* scale + *ata,* to bear) is divided into three suborders. Ancestral members of these suborders originated in the diapsid lineage about 150 million years ago and diverged into numerous modern forms.

Suborder Sauria—The Lizards   There are about 3,300 species in the suborder Sauria (sawr'e-ah) (Gr. *sauro,* lizard). In contrast to snakes, lizards usually have two pairs of legs. The few that are legless retain remnants of a pectoral girdle and sternum. Lizards vary in length from only a few cm to as large as 3 m. Many lizards live on surface substrates and retreat under rocks or logs when necessary. Others are burrowers or tree dwellers. Most lizards are oviparous; some are ovoviviparous or viviparous. Eggs are usually deposited under rocks, debris, or in burrows.

Geckos, commonly found on the walls of human dwellings, are short and stout. They are nocturnal, and unlike most lizards, are capable of clicking vocalizations. Their large eyes, with pupils that contract to a narrow slit during the day and dilate widely at night, are adapted for night vision. Adhesive disks on their digits aid in clinging to trees and walls.

Iguanas have robust bodies, short necks, and distinct heads. This group includes the marine iguanas of the Galápagos Islands and the flying dragons (*Draco*) of Southeast Asia. The latter have lateral folds of skin that are supported

Box 29.2

Bone and Scales

Bone is the primary skeletal tissue of vertebrates. In addition to making up the skeleton, it also is present in the scales of some vertebrates.

Developmentally, bone is derived from two sources. *Dermal or membrane bone* forms many of the superficial, flat bones of the skull and some bones in the pectoral girdle. These bones were especially numerous in the roof of the skull of early vertebrates. Dermal bone is formed in the connective tissues of the dermis of the skin. During its formation, bone-forming cells called *osteoblasts* line up along connective-tissue fibers and begin depositing bone. Bony fibers coalesce into the latticework that makes up a flat bone.

Endochondral bone forms many of the long bones and the ventral and posterior bones of the skull. It develops by replacing the cartilage that formed early in development (box figure 29.2a). The cartilage-based bone grows during development. As it does, osteoblasts lay down a collar of bone around the middle region, the *diaphysis.* Cartilage cells in the diaphysis begin to break down, beginning the formation of a cavity called the *marrow cavity.* Eventually, this cavity will be filled with bone marrow, in which blood cells are formed. Bone formation proceeds toward the ends. The cartilage near the end of the bone, however, continues to grow and results in further bone elongation. Each end of the bone is an *epiphysis.* In mammals, a secondary ossification center occurs in each epiphysis and forms bony caps on the ends of the bone. However, a cartilaginous plate, the *epiphyseal plate,* remains between the epiphysis and the diaphysis and is the site of cartilage growth and bone elongation. Bone growth continues until maturity. Except for thin cartilages at the ends of bones that provide gliding surfaces for joints, all cartilage is replaced by bone at maturity, and growth stops.

The scales of fishes are composed, in part, of dermal bone. Osteoblasts in the dermis of the skin lay down a core of bone. Other dermal cells lay down a layer of *dentine,* which is similar to bone, around the bony core. Then, epidermal cells lay down a covering of *enamel* (box figure 29.2b). Enamel is one of the hardest tissues in the vertebrate body and also occurs in teeth.

The scales of the skin of reptiles and the legs of birds are formed entirely in the epidermis of the skin and do not contain bone. These scales are composed of many layers of epidermal cells (box figure 29.2c). Keratin and phospholips are incorporated into the outer, horny layers of a scale to reduce water loss across the skin. As you will see in chapter 30, the feathers of birds are modified epidermal scales.

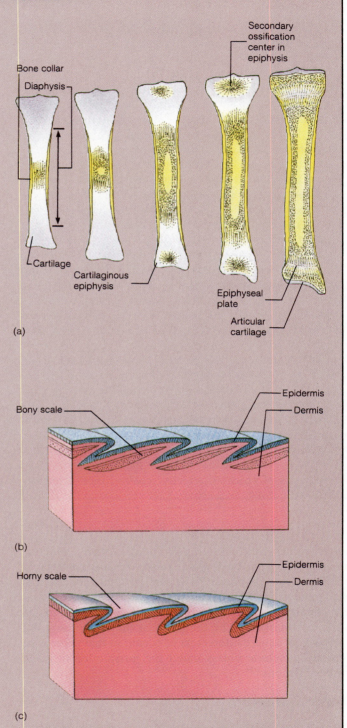

Box Figure 29.2 (a) The formation of endochondral bone. (b) Dermal scales of a fish. (c) Epidermal scales of a reptile.

by ribs. Like the ribs of an umbrella, the ribs of *Draco* can be expanded to form a gliding surface. When this lizard launches itself from a tree it can glide 30 m or more!

[3] Another group of iguanas, the chameleons, is found mainly in Africa and India. They are adapted to arboreal life-styles and use a long, sticky tongue to capture insects (colorplate 12c). *Anolis,* or the "pet-store chameleon," is also an iguanid, but is not a true chameleon. Chameleons and *Anolis* are well known for their ability to change color in response to illumination, temperature, or their behavioral state.

The only venomous lizards are the gila monster (*Heloderma suspectum*) and the Mexican beaded lizard (*Heloderma horridum*) (colorplate 12d). These heavy-bodied lizards are native to southwestern North America. Venom is released into grooves on the surface of teeth and introduced into prey as the lizard chews. Lizard bites are seldom fatal to humans.

Suborder Serpentes—The Snakes   There are about 2,300 species in the suborder Serpentes (ser-pen'tez) (L. *serpere,* to crawl). Although the vast majority of snakes are not dangerous to humans, about 300 species are venomous. Worldwide, about 30,000 to 40,000 people die from snake bites each year. Most of these deaths are in Southeast Asia. In the United States, fewer than 100 people die each year from snake bites.

Snakes are elongate and lack limbs, although vestigial pelvic girdles and appendages are sometimes present (figure 29.4). Snakes possess skull adaptations that facilitate swallowing large prey. Other differences between lizards and snakes include the mechanism for focusing the eyes and the morphology of the retina. Elongation and narrowing of the body has resulted in the reduction or loss of the left lung and displacement of the gallbladder, the right kidney, and often the gonads. Most snakes are oviparous, although a few, such as the New-World boas, give birth to live young.

The evolutionary origin of the snakes is debated. The earliest fossils are from 135 million-year-old Cretaceous deposits. Some zoologists believe that the earliest snakes were burrowers. Loss of appendages and changes in eye structure could be adaptations similar to those seen in caecilians (colorplate 11d). The loss of legs could also be adaptive if early snakes were aquatic or lived where densely tangled vegetation was common.

Suborder Amphisbaenia—Worm Lizards
There are about 135 species in the suborder Amphisbaenia (am'fis-be'ne-ah) (Gr. *amphi,* double + *baen,* to walk). They are specialized burrowers that live in soils of Africa, South America, the Caribbean, and the Mideast (colorplate 12f). Most are legless, and their skulls are wedge or shovel shaped. They are distinguished from all other vertebrates by the presence of a single median tooth in the upper jaw. The skin of amphisbaenians has ringlike folds called *annuli* and is loosely attached to the body wall. Muscles of the skin cause it to telescope and bulge outward, forming an anchor against a burrow wall. They move easily forward or backward, thus the suborder name. Amphisbaenians feed on worms and small insects and are oviparous. [4]

Order Rhynchocephalia

The one surviving species of the order Rhynchocephalia (rin'ko-se-fay'le-ah) (Gr. *rhynchos,* snout + *kephale,* head) is the tuatara (*Sphenodon punctatus*) (figure 29.5). This superficially lizardlike reptile is frequently referred to as a "living fossil," as it is virtually unchanged from extinct relatives that were present at the beginning of the Mesozoic era, nearly 200 million years ago. It is distinguished from other reptiles by tooth attachment and structure. Two rows of teeth on the upper jaw and a single row of teeth in the lower jaw produce a shearing bite that can decapitate a small bird. Formerly more widely distributed in New Zealand, the tuatara fell prey to human influences and domestic animals. It is now present only on remote

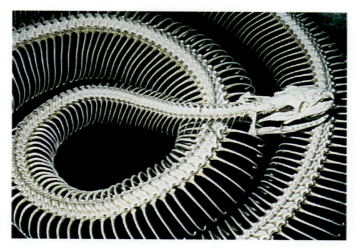

Figure 29.4   The skeleton of a snake may contain 200 or more vertebrae. Additional joints between vertebrae make the body very flexible.

Figure 29.5   The tuatara (*Sphenodon punctatus*).

offshore islands and is protected by New Zealand law. It is oviparous and shares underground burrows with ground-nesting seabirds. Tuataras venture out of their burrows at dusk and dawn to feed on insects or occasionally small vertebrates.

Order Crocodilia

There are 21 species in the order Crocodilia (krok′o-dil′e-ah) (Gr. *krokodeilos,* lizard). Along with dinosaurs, crocodilians are derived from the archosaur group of diapsids and distinguished from other reptiles by certain skull characteristics: openings in the skull in front of the eye, triangular rather than circular eye orbits, and laterally compressed teeth. Living crocodilians include the alligators, crocodiles, gavials, and caimans (colorplate 12g,h).

Crocodilians have not changed much over their 170-million-year history. The snout is elongate and often used to capture food by a sideways sweep of the head. The nostrils are at the tips of the snout, so the animal can breathe while mostly submerged. Air passageways of the head lead to the rear of the mouth and throat, and a flap of tissue near the back of the tongue forms a water-tight seal that allows breathing without inhaling the water in the mouth. A plate of bone, called the *secondary palate,* evolved in the archosaurs and separates the nasal and mouth passageways. The tail is muscular, elongate, and laterally compressed. It is used to swim, in offensive and defensive maneuvers, and to attack prey. Teeth are used only for seizing prey. Food is swallowed whole, but if a prey item is too large, crocodilians tear apart prey by holding onto a limb and rotating their bodies wildly until the prey is dismembered. The stomach is gizzardlike, and crocodilians swallow rocks and other objects as abrasives for breaking apart ingested food. Crocodilians are oviparous and display parental care of hatchlings that parallels that of birds. Nesting behavior and parental care may trace back to the common ancestor of both groups.

Stop and Ask Yourself

1. What are the three evolutionary lineages that diverged from the earliest amniotes?
2. What order of reptiles lacks teeth, has ribs and vertebrae incorporated into a bony shell, and is oviparous?
3. What adaptations for burrowing are shown by worm lizards?
4. What is a secondary palate? Why is it adaptive for crocodilians?

Evolutionary Pressures

The contrast between the life-styles of most amphibians and reptiles is striking! To appreciate this one might consider a lizard common to deserts of southwestern United States, the chuckwalla (*Sauromalus obesus*) (figure 29.6).

Chuckwallas survive during late summer when temperatures exceed 40°C (104°F) and when arid conditions result in the withering of plants and blossoms upon which chuckwallas browse. To withstand these hot and dry conditions, chuckwallas disappear below ground and aestivate. Temperatures moderate during the winter, but little rain falls, so life on the desert surface is still not possible for the chuckwalla. The summer's sleep, therefore, merges into a winter's sleep. The chuckwalla will not emerge until March when rain falls, and the desert explodes with greenery and flowers. The chuckwalla browses and drinks, storing water in large reservoirs under its skin. Chuckwallas are not easy prey. If threatened, the nearest rock crevice becomes a chuckwalla's refuge. The chuckwalla inflates its lungs with air, increasing its girth and wedging itself against the rock walls of its refuge. Friction of its body scales against the rocks make it nearly impossible to dislodge.

Adaptations displayed by chuckwallas are not exceptional for reptiles. This section discusses some adaptations that make life apart from water possible.

External Structure and Locomotion

Unlike that of amphibians, the skin of reptiles has no respiratory functions. Their skin is thick, dry, and keratinized. Scales may be modified for various functions. The large belly scales of snakes, called *gastrosteges* (Gr. *gaster,* stomach + *stegos,* cover), provide contact with the substrate during locomotion. Although the skin of reptiles is much less glandular than that of amphibians, secretions include pheromones that function in sex recognition and defense.

All reptiles periodically shed the outer, epidermal layers of the skin in a process called *ecdysis.* (The term ''ecdysis'' is also used for a similar, though unrelated, process in arthropods [*see figure 22.4*].) Because the blood supply to the skin does not extend into the epidermis, the outer epidermal cells lose contact with the blood supply and die. Movement of lymph between the inner and outer epidermal

Figure 29.6 The chuckwalla (*Sauromalus obesus*).

layers loosens the outer epidermis. Ecdysis is generally initiated in the head region, and in snakes and many lizards, the epidermal layers come off in one piece. In other lizards, smaller pieces of integument flake off. The frequency of ecdysis varies from one species to another, and it is greater in juveniles than adults.

The chromatophores of reptiles are primarily dermal in origin and function much like those of amphibians. Cryptic coloration, mimicry, and aposematic coloration occur in reptiles. Color and color change also function in sex recognition and thermoregulation.

Support and Movement   There are modifications in the skeletons of snakes, amphisbaenians, and turtles; however, in its general form, the skeleton of reptiles is based on one inherited from ancient amphibians. The skeleton is highly ossified to provide greater support. The skull is longer than that of amphibians, and a plate of bone, the secondary palate, partially separates the nasal passages from the mouth cavity (figure 29.7). As described earlier, the secondary palate evolved in archosaurs, where it was an adaptation for breathing when the mouth was full of water or food. It is also present in other reptiles, although developed to a lesser extent. Longer snouts also permit greater development of olfactory epithelium and increased reliance on the sense of smell.

Reptiles have more cervical vertebrae than do amphibians. The first two cervical vertebrae provide greater freedom of movement for the head. An *atlas* articulates with a single condyle on the skull and facilitates nodding. An *axis* is modified for rotational movements. The atlas and axis are followed by a variable number of cervical vertebrae that provide additional flexibility for the neck.

The ribs of reptiles may be highly modified. Those of turtles and the flying dragon were described previously. The ribs of snakes have muscular connections to large belly scales to aid locomotion. The cervical vertebrae of cobras have ribs that may be flared in aggressive displays.

Two or more sacral vertebrae attach the pelvic girdle to the vertebral column. The caudal vertebrae of many lizards possess a vertical plane of fracture. When a lizard is grasped by the tail, caudal vertebrae can be broken, and a portion of the tail is lost. Tail loss, or *autotomy,* is an adaptation that allows a lizard to escape from a predator's grasp, or the disconnected, wiggling piece of tail may distract a predator from the lizard. The lost portion of the tail is later regenerated.

[5]

Locomotion in primitive reptiles is similar to that of salamanders. The body is slung low between paired, stocky appendages, which extend laterally and move in the horizontal plane. The limbs of other reptiles are more elongate and slender. Limbs are held closer to the body, the knee joint is rotated anteriorly, and the elbow joint is rotated posteriorly. The body is thus higher off the ground and weight is supported vertically. Many prehistoric reptiles were *bipedal,* meaning that they walked on hind limbs. They had a narrow pelvis and a heavy outstretched tail for balance. Bipedal locomotion freed the front appendages, which became adapted for prey capture or flight in some animals.

The loss of limbs in snakes was accompanied by greater use of the body wall during locomotion. In *curvilinear locomotion,* the body is thrown into a series of curves that push backward against the substrate (figure 29.8a). *Rectilinear locomotion* is a slow, gliding movement used by large, heavy snakes or by some snakes when stalking prey. Belly scales are acted upon by muscles attached to scales and ribs. The skin and belly scales in one region of the body are pulled forward, contact the substrate, and provide temporary points of attachment. Then the body behind these points of attachment is pulled forward. Waves of contraction pass anteriorly to posteriorly along the snake and result in a straight-line movement (figure 29.8b). In a burrow, snakes may use *concertina locomotion.* The anterior part of the body is bent in a series of S-shaped waves, which contact the side of the burrow. As these waves move posteriorly, the anterior region of the body straightens. The anterior part of the body then bends again to make contact with the side of the burrow, and the posterior portion of the body is pulled forward (figure 29.8c). *Sidewinding locomotion* allows rapid travel over smooth or loose substrates. The head and anterior body are lifted off the substrate and swung through an angle of 90°. The head returns to the ground anterior and lateral to its previous contact point. Behind the head, the body is progressively lifted off the ground, posteriorly to anteriorly, and swung forward to positions nearer the head (figure 29.8d).

Nutrition and the Digestive System

Most reptiles are carnivores, although turtles will eat almost anything organic. The tongues of turtles and crocodilians are nonprotrusible and aid in swallowing. The sticky tongues of some lizards and the tuatara are used to capture prey as do some anurans. The extension of the tongue of chameleons exceeds their body length.

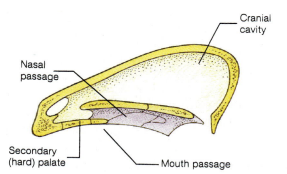

Figure 29.7   A sagittal section of the skull of a synapsid reptile showing the secondary palate that separates the nasal and mouth cavities. Extension of the bones of the anterior skull forms the anterior portion of the secondary palate (the hard palate) and skin and soft connective tissues form the posterior portion of the secondary palate (the soft palate).

Cranial cavity

Nasal passage

Secondary (hard) palate

Mouth passage

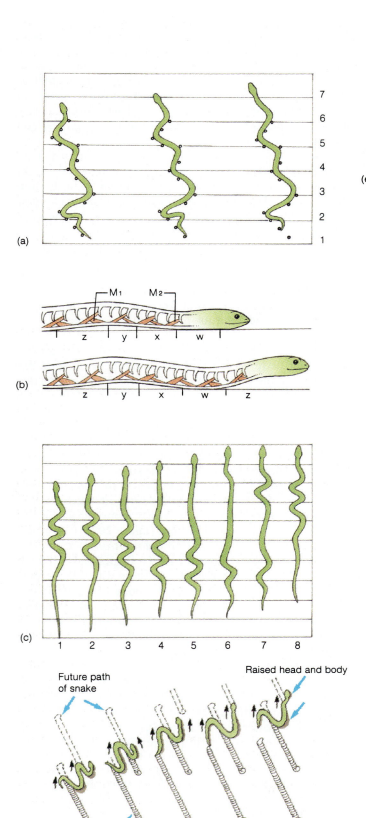

(a)

(b)

(c)

(d)

Future path of snake

Raised head and body

Imprints of belly scales

(e)

Figure 29.8 Snake locomotion. (a) Curvilinear locomotion. The body is thrown into a series of curves. Curves press backward on the substrate to push the snake forward. The points at which the snake is exerting force are shown in color. (b) Rectilinear locomotion. Ventral scales make contact with the substrate, and body-wall muscles anchored posteriorly to the point of contact (m_1) pull the ribs and the rest of the body anteriorly. Anteriorly-directed muscles (m_2) then pull the ventral skin forward to a second contact point. (c) Concertina locomotion. Lateral flexures make contact with the side of a burrow. Anteriorly, flexures are straightened and the head is pushed forward. When the snake is extended, new flexures form anteriorly, and the posterior end of the body is pulled forward. (d) Sidewinding locomotion. Loops of the body are raised off the ground and placed down again ahead of the snake. Only two or three points of contact are made with the substrate and are noted here by the shaded portion of the snake and the imprints of belly scales. Points of contact move smoothly from anterior to posterior along the snake. (e) A sidewinder (*Crotalus cerastes*) moving across a sand dune.

Probably the most remarkable adaptations of snakes involve modifications of the skull for feeding. The bones of the skull and jaws are loosely joined to each other and may spread apart to ingest prey much larger than a snake's normal head size (figure 29.9a). The bones of the upper jaw are movable on the skull, and the halves of both of the upper and lower jaws are loosely joined anteriorly by ligaments. Therefore, each half of the upper and lower jaws can be moved independently of one another. After a prey is captured, opposite sides of the upper and lower jaws are alternately thrust forward and retracted. Posteriorly pointing teeth prevent the escape of the prey and help force the food into the esophagus. The *glottis*, the respiratory opening, is far forward so that the snake can breathe while slowly swallowing its prey.

Vipers (family Viperidae) possess hollow fangs on the maxillary bone at the anterior margin of the upper jaw (figure 29.9b). These fangs connect to a venom gland that inject venom when the viper bites. The maxillary bone (upper jaw bone) of vipers is hinged so that when the snake's mouth is closed, the fangs fold back and lie along the upper jaw. When the mouth opens, the maxillary bone rotates and causes the fangs to swing down (figure 29.9c). Because the fangs project outward from the mouth, vipers may strike at objects of any size. Rear-fanged snakes (family Colubridae) possess grooved rear teeth. Venom is channeled along these grooves and worked into prey to quiet them during swallowing. These snakes usually do not strike and most are harmless; however, the African boomslang (*Dispholidus typus*) has caused human fatalities. Coral snakes, sea snakes, and cobras have fangs that are rigidly attached to the upper jaw in an erect position. When the mouth is closed, the fangs fit into a pocket in the outer gum of the lower jaw. Fangs are grooved or hollow, and venom is injected by contraction of muscles associated with venom glands. Some cobras can "spit" venom at their prey, and, if not washed from the eyes, it may cause blindness.

Venom glands are modified salivary glands. Most snake venoms are mixtures of neurotoxins and hemotoxins. The venoms of coral snakes, cobras, and sea snakes are primarily neurotoxins that attack nerve centers and cause respiratory paralysis. The venoms of vipers are primarily hemotoxins. They break up blood cells and attack blood-vessel linings.

Circulation, Gas Exchange, and Temperature Regulation

The circulatory system of reptiles is based on that of amphibians. Because reptiles are on average, larger than amphibians, their blood must travel under higher pressures to reach distant body parts. To take an extreme example, the blood of *Brachiosaurus* had to be pumped a distance of about 6 m from the heart to the head—mostly uphill (box 29.3)! (The blood pressure of a giraffe is about double that of a human to move blood the 2 m from heart to head.)

Like amphibians, reptiles possess two atria that are completely separated in the adult and have veins from the body and lungs emptying into them. Except for turtles, the sinus venosus is no longer a chamber but has become a patch of cells that act as a pacemaker. The ventricle of most reptiles is incompletely divided (figure 29.10). (Only in crocodilians is the ventricular septum complete.) The ventral aorta and the conus arteriosus divide during development and become three major arteries that leave the heart. A pulmonary artery leaves the ventral side of the ventricle and takes blood to the lungs. Two systemic arteries, one from the ventral side of the heart and the other from the dorsal side of the heart, take blood to the lower body and the head.

Blood low in oxygen enters the ventricle from the right atrium and leaves the heart through the pulmonary artery and moves to the lungs. Blood high in oxygen enters the ventricle from the lungs via pulmonary veins and the left atrium, and leaves the heart through left and right systemic arteries. The incomplete separation of the ventricle permits shunting of some blood away from the pulmonary circuit to the systemic circuit by constriction of muscles associated with the pulmonary artery. This design is advantageous because virtually all reptiles breathe intermittently. When turtles are withdrawn into their shell, their method of lung ventilation cannot function. They may also stop breathing during diving. During periods of apnea ("no breathing"), the flow of blood to the lungs is limited, conserving energy and permitting more efficient use of the pulmonary oxygen supply.

(a)

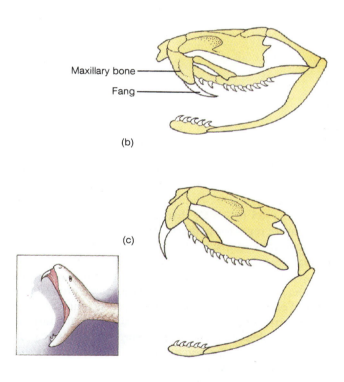

Maxillary bone

Fang

(b)

(c)

Figure 29.9 Feeding adaptations of snakes. (a) A copperhead (*Ankistrodon*) ingesting a prey. The bones of the skull are joined by flexible joints that allow them to separate during feeding. (b) The skull of a viper. The hinge mechanism of the jaw allows upper and lower bones on one side of the jaw to slide forward and backward alternately with bones of the other side. Posteriorly curved teeth hold prey as it is worked toward the esophagus. (c) Note that the maxillary bone, into which the fang is embedded, swings forward when the mouth is opened.

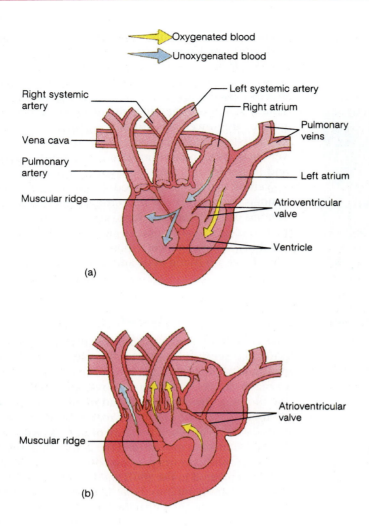

Oxygenated blood
Unoxygenated blood

Right systemic artery
Left systemic artery
Right atrium
Vena cava
Pulmonary veins
Pulmonary artery
Muscular ridge
Left atrium
Atrioventricular valve
Ventricle

(a)

Muscular ridge
Atrioventricular valve

(b)

Figure 29.10 The heart and major arteries of a lizard. (a) When the atria contract, blood enters the ventricle. An atrioventricular valve prevents the mixing of oxygenated and unoxygenated blood across the incompletely separated ventricle. (b) When the ventricle contracts, a muscular ridge closes to direct oxygenated blood to the systemic arteries and unoxygenated blood to the pulmonary artery.

Gas Exchange Reptiles exchange respiratory gases across internal respiratory surfaces to prevent the loss of large quantities of water. A larynx is present; however, vocal cords are usually absent. The respiratory passages of reptiles are supported by cartilages, and lungs are partitioned into spongelike, interconnected chambers. Lung chambers provide a large surface area for gas exchange.

Lung ventilation occurs in most reptiles by a negative pressure mechanism. Expanding the body cavity by a posterior movement of the ribs and the body wall decreases pressure in the lungs and draws air into the lungs. Air is expelled by elastic recoil of the lungs and forward movements of the ribs and body wall, which compress the lungs. The ribs of turtles are a part of their shell; thus, movements of the body wall to which they are attached are impossible. Turtles exhale by contracting muscles that force the viscera upward, compressing the lungs. They inhale by contracting muscles that increase the volume of the visceral cavity, creating negative pressure to draw air into the lungs.

Temperature Regulation Unlike aquatic animals, terrestrial animals may be faced with temperature

extremes (−65 to 70°C) that are incompatible with life. Temperature regulation, therefore, is important for animals that spend their entire lives out of water. Most reptiles use external heat sources for thermoregulation and are, therefore, *ectotherms*. Exceptions include monitor lizards and brooding Indian pythons. Female pythons coil around their eggs and elevate their body temperature as much as 7.3°C above the air temperature using metabolic heat sources.

Some reptiles can survive wide temperature fluctuations (e.g., −2 to 41°C for some turtles). To sustain activity, however, body temperatures are regulated within a narrower range, between 25 and 37°C. If that is not possible, the reptile usually seeks a retreat where body temperatures are likely to remain within the range compatible with life.

Most thermoregulatory activities of reptiles are behavioral, and they are best known in the lizards. To warm itself, a lizard may orient itself at right angles to the sun's rays, often on a surface inclined toward the sun and may press its body tightly to a warm surface to absorb heat by conduction. To cool itself, a lizard will orient its body parallel to the sun's rays, seek shade or burrows, or assume an erect posture (legs extended and tail arched) to reduce conduction from warm surfaces. In hot climates, many reptiles are nocturnal.

Various physiological mechanisms are also used to regulate body temperature. As temperatures rise, some reptiles begin panting, which releases heat through evaporative cooling. (Because their skin is dry, little evaporative cooling occurs across the skin of reptiles.) Marine iguanas divert blood to the skin while basking in the sun and warm up quickly. On diving into the cool ocean, however, heart rate and blood flow to the skin are reduced, which slow heat loss. Chromatophores also aid in temperature regulation. Dispersed chromatophores (thus a darker body) increase the rate of heat absorption.

In temperate regions, many reptiles withstand cold temperatures of winter by entering torpor when body temperatures and metabolic rates decrease. Individuals that are usually solitary may migrate to a common site to spend the winter. Heat loss from these groups, called *hibernacula,* is reduced because of a reduction in total surface area of many individuals clumped together compared to widely separated animals. Unlike true hibernators, body temperatures of reptiles in torpor are not regulated, and if the winter is too cold or the retreat is too exposed, the animals can freeze and die. Cold death is an important source of mortality for temperate reptiles.

Nervous and Sensory Functions

The brain of reptiles is similar to the brains of other vertebrates. The cerebral hemispheres are somewhat enlarged compared to those of amphibians. This increased size is associated with an improved sense of smell. The optic lobes and the cerebellum are also enlarged, which reflects increased reliance on vision and more refined coordination of muscle functions.

The complexity of reptilian sensory systems is evidenced by a chameleon's method of feeding. Its protruding eyes that swivel independently each have a different field of view. Initially, the brain keeps both images separate, but when an insect is spotted, both eyes converge on the prey. Binocular vision then provides the depth perception used to determine whether or not the insect is within range of the chameleon's tongue (colorplate 12c).

Vision is the dominant sense in most reptiles and their eyes are similar to those of amphibians (*see figure 28.9*). Snakes focus on nearby objects by moving the lens forward. Contraction of the iris places pressure on the gellike vitreous body in the posterior region of the eye, and displacement of this gel pushes the lens forward. In all other reptiles, focusing on nearby objects occurs when the normally elliptical lens is made more spherical, as a result of ciliary muscles pressing the ciliary body against the lens. Reptiles have a greater number of cones than do amphibians and probably have well-developed color vision.

Upper and lower eyelids, a nictitating membrane, and a blood sinus protect and cleanse the surface of the eye. In snakes and some lizards, the upper and lower eyelids become fused in the embryo to form a protective window of clear skin, called the *spectacle.* (During ecdysis, the outer layers of the spectacle become clouded and impair the vision of snakes.) The blood sinus, which is at the base of the nictitating membrane, swells with blood to help force debris to the corner of the eye, where it may be rubbed out. Horned lizards squirt blood from their eyes by rupturing this sinus in a defensive maneuver to startle predators.

Some reptiles possess a **median (parietal) eye** that develops from outgrowths of the roof of the forebrain. In the tuatara, it is an eye with a lens, a nerve, and a retina. In other reptiles, the parietal eye is less developed. Parietal eyes are covered by skin and probably cannot form images. They can, however, differentiate light and dark periods and are used in orientation to the sun.

There is variation in the structure of reptilian ears. The ears of snakes detect substrate vibrations. They lack a middle-ear cavity, a eustachian tube, and a tympanic membrane. A bone of the jaw articulates with the stapes and receives substrate vibrations. Snakes can also detect airborne vibrations. In other reptiles, a tympanic membrane may be on the surface or in a small depression in the head. The inner ear of reptiles is similar to that of amphibians.

Olfactory senses are better developed in reptiles than in amphibians. In addition to the partial secondary palate providing more surface for olfactory epithelium, most reptiles (all but adult crocodilians) possess blind-ending pouches that open through the secondary palate into the mouth cavity. These pouches, called **Jacobson's organs,** are best developed in snakes and lizards. The protrusible, forked tongues of snakes and lizards are accessory olfactory organs that are used to sample airborne chemicals. A snake's tongue is flicked out and then moved to the Jacobson's organs where odor molecules are perceived. In turtles and the tuatara, Jacobson's organs are used to taste objects held in the mouth.

Rattlesnakes and other pit vipers have heat-sensitive **pit organs** on each side of the face between the eye and nostril (colorplate 12e). These depressions are lined with sensory epithelium and are used to detect objects with temperatures different from the snake's surroundings. Pit vipers are usually nocturnal, and their pits are used to locate small, warm-blooded prey.

Stop and Ask Yourself

5. How is the skin of a reptile adapted for preventing water loss?
6. How might a snake move across very loose or smooth substrates? How would a snake move in an underground burrow?
7. How are the tongues of reptiles used in feeding? In sensory perception?
8. How is the heart of a reptile adapted for shunting blood away from the pulmonary circulation? Under what circumstances does this shunting occur?

Box 29.3
Changing Perceptions of Ancient Life-Styles

In the years prior to 1960, paleontologists worked on assembling ancient lineages from fossil remains based upon gross anatomical features. Now, it is possible to derive information about the life-styles of extinct animals from markings of blood vessels, muscles, and tendons present on fossils. In the 1970s, Robert T. Bakker began challenging previous perceptions of dinosaurs as lumbering giants and overgrown lizards. Bakker contended that that view of dinosaurs is difficult to reconcile with fossils, indicating that many dinosaurs were bipedal or quadrupedal or at least held their bodies well off the ground. Reptiles achieved many locomotor advancements long before similar advancements ever appeared in mammals. Unlike modern reptiles, dinosaurs had highly vascular bones, a condition typical of endothermic animals. Finally, many believe it unlikely that very large, strictly ectothermic animals could warm up rapidly enough to maintain high activity levels. Bakker proposed that many, if not all dinosaurs were endothermic.

Other scientists discount ideas that dinosaurs were endothermic. They maintain that dinosaurs may have been able to maintain active life-styles, not in spite of their large size, but because of it. As ectotherms increase in size, they resemble endotherms. Large body mass results in slower warming, but it also results in slower heat loss. Further, the climate during the Mesozoic Era was considerably warmer than it is now. Even though dinosaurs may have been ectotherms, they could have had stable body temperatures and an active life-style. Opponents of the endothermy hypothesis point out that some ectotherms have highly vascular bones. They also note that large endotherms usually have obvious cooling devices (e.g., the ears of an elephant), and these were not found on most large dinosaurs.

There are no clear winners in this debate. There may not have been a single thermoregulatory strategy for the dinosaurs. The largest dinosaurs could have been ectothermic and, although not swift and agile, they would still have been able to find food and avoid predators. On the other hand, maintenance of speed and agility in smaller dinosaurs that lived in cool climates may have required endothermy.

Excretion and Osmoregulation

The kidneys of embryonic reptiles are similar to those of fishes and amphibians. Life on land, increased body size, and higher metabolic rates require kidneys capable of processing wastes with little water loss. The embryonic kidney is replaced during development by a kidney with many more blood-filtering units, called nephrons. The functional kidneys of adult reptiles are called *metanephric kidneys.* Their function depends on a circulatory system that delivers more blood at greater pressures to filter larger quantities of blood.

Uric acid is the principal excretory product of most reptiles. It is nontoxic and being relatively insoluble in water, it will precipitate in the excretory system. Water is reabsorbed by the bladder or the cloacal walls, and the uric acid can be stored in a pastelike form. Utilization of uric acid as an excretory product also made possible the development of embryos in terrestrial environments because nontoxic uric acid can be concentrated in egg membranes.

In addition to water reabsorption in the excretory system, internal respiratory surfaces and relatively impermeable exposed surfaces reduce evaporative water loss. The behaviors that help regulate temperature also help conserve water. Nocturnal habits and avoiding hot surface temperatures during the day by burrowing reduce water loss. When water is available, many reptiles (e.g., chuckwallas) store large quantities of water in lymphatic spaces under the skin or in the urinary bladder. Many lizards possess *salt glands* below the eyes that are used to rid the body of excess salt.

Reproduction and Development

Vertebrates could never be truly terrestrial until their reproduction and embryonic development became separate from standing or running water. For vertebrates, internal fertilization and the amniotic egg made complete movement to land possible (*see figure 10.13*). The amniotic egg, however, is not completely independent of water. Pores in the eggshell that permit gas exchange also allow water to evaporate. Amniotic eggs require significant energy expenditures by parents. Parental care occurs in some reptiles and may involve maintaining relatively high humidity around the eggs. These eggs are often supplied with large quantities of yolk for long developmental periods, and parental energy and time is sometimes invested in post-hatching care of dependent young.

Accompanying the development of amniotic eggs is the necessity for internal fertilization. Fertilization must occur in the reproductive tract of the female before protective egg membranes are laid down around an egg. All male reptiles, except tuataras, possess an intromittent organ for introducing sperm into the reproductive tract of a female. Lizards and snakes possess paired *hemipenes* that are located at the base of the tail and are erected by being turned inside out, like the finger of a glove.

Gonads lie in the abdominal cavity. In males, a pair of ducts deliver sperm to the cloaca. After copulation, sperm may be stored in a *seminal receptacle* in the reproductive tract of the female. Secretions of the seminal receptacle nourish and arrest the activity of the sperm. Sperm may be stored for up to 4 years in some turtles, and up to 6 years in some snakes! In temperate latitudes, sperm can be stored over winter. Copulation may take place in the fall when individuals congregate in hibernacula, and fertilization and development occur in the spring when temperatures favor successful development. Fertilization occurs in the upper regions of the oviduct, which leads from the ovary to the cloaca. Glandular regions of the oviduct are responsible for secreting albumen and the eggshell. The shell is usually tough yet flexible. In some crocodilians, the eggshell is calcareous and rigid, like eggshells of birds.

Parthenogenesis has been described in six families of lizards and one species of snakes. In these species, no males have been found. Parthenogenesis in reptiles may arise as a result of hybridization between two ancestral species. It is probably maintained, in spite of the genetic uniformity that results, because parthenogenetic populations have higher reproductive potential than bisexual populations. A population that suffers high mortality over a cold winter can repopulate its habitat rapidly because all surviving individuals can produce offspring.

Reptiles often have complex reproductive behaviors that may involve males actively seeking out females. As in other animals, courtship functions in sexual recognition and behavioral and physiological preparation for reproduction. Head-bobbing displays by some male lizards reveal bright patches of color on the throat and enlarged folds of skin. Courtship in snakes is based primarily on tactile stimulation. It involves tail-waving displays that are followed by the male running his chin along the female, entwining his body around her, and creating wavelike contractions that pass posteriorly to anteriorly along his body. Recent research indicates that sex pheromones are also used by lizards and snakes. Vocalizations are important only in crocodilians. During the breeding season, males are hostile and may bark or cough as territorial warnings to other males. Roaring vocalizations also attract females, and mating occurs in the water.

After they are laid, the eggs are usually abandoned (figure 29.11). Virtually all turtles bury their eggs in the ground or in plant debris. Other reptiles lay their eggs under rocks, in debris, or in burrows. About 100 species of reptiles have some degree of parental care of eggs. One example is the American alligator, *Alligator mississippiensis* (colorplate 12g). The female builds a mound of mud and vegetation about 1 m high and 2 m in diameter. The center of the mound is hollowed out and filled with mud and debris. Eggs are deposited in a cavity in the center of the mound and covered. The female remains in the vicinity of the nest throughout development to protect the eggs from predation. She frees hatchlings from the nest in response to their high-pitched calls and picks them up in her mouth to transport them to water. She may scoop shallow pools for the young and remain with them for up to 2 years. Young feed on scraps of food dropped by the female when she feeds and on small vertebrates and invertebrates that they catch on their own.

Further Phylogenetic Considerations

The diapsid and synapsid lineages of ancient reptiles split from ancient amniotes about 280 million years ago and are ancestral to animals described in the next two chapters (*see figure 29.1*). Archosaurs of the diapsid lineage not only included the dinosaurs and gave rise to crocodilians, but also gave rise to two groups of fliers. The pterosaurs (Gr. *pteros,* wing + *sauros,* lizard) ranged from sparrow size to animals with wing spans of 13 m. Their membranous wings were supported by an elongation of the fourth finger, their sternum was adapted for the attachment of flight muscles, and their bones were hollow to lighten the skeleton for flight. As presented in chapter 30, these adaptations are paralleled by, though not identical to, adaptations in the birds—the descendants of the second lineage of flying archosaurs.

The synapsid lineage eventually gave rise to the mammals. The legs of synapsids were relatively long and held their body off the ground. Teeth and jaws were adapted for effective chewing and tearing. Additional bones were incorporated into the middle ear. These and other mammal-like characteristics developed between the Carboniferous and Triassic periods. The "Evolutionary Perspective" of chapter 31 describes more about the nature of this transition.

Figure 29.11 These fence lizards (*Sceloporus undulatus*) are hatching from their leathery eggs.

Stop and Ask Yourself

9. Why is uric acid an adaptive excretory product for reptiles?
10. What groups of reptiles contain parthenogenetic species?
11. In what group of reptiles are vocalizations an important part of reproductive activities?
12. What reptilian lineage gave rise to the birds? To the mammals?

■ Summary

1. The earliest amniotes are classified as reptiles. Anapsid, diapsid, and synapsid evolutionary lineages diverged from ancient amniotes.

2. The order Testudines contains the turtles. Turtles have a bony shell and lack teeth. All are oviparous.

3. The order Squamata contains the lizards, snakes, and worm lizards. Lizards usually have two pairs of legs and most are oviparous. Snakes lack developed limbs and have skull adaptations for swallowing large prey. Worm lizards are specialized burrowers. They have a single median tooth in the upper jaw and most are oviparous.

4. The order Rhynchocephalia contains one species, the tuatara. It is found only on remote islands of New Zealand.

5. The order Crocodilia contains alligators, crocodiles, caimans, and gavials. They have a well-developed secondary palate and display nesting behaviors and parental care.

6. The skin of reptiles is dry and keratinized, and provides a barrier to water loss. It also has epidermal scales and chromatophores.

7. The reptilian skeleton is modified for support and movement on land. Loss of appendages in snakes is accompanied by greater use of the body wall in locomotion.

8. Reptiles have a tongue that may be used in feeding. Bones of the skull of snakes are loosely joined and spread apart during feeding.

9. The circulatory system of reptiles is divided into pulmonary and systemic circuits and functions under relatively high blood pressures. Blood may be shunted away from the pulmonary circuit during periods of apnea.

10. Gas exchange occurs across convoluted lung surfaces. Ventilation of lungs occurs by a negative-pressure mechanism.

11. Reptiles are ectotherms and mainly use behavioral mechanisms to thermoregulate.

12. Vision is the dominant sense in most reptiles. Parietal eyes, ears, Jacobson's organs, and pit organs are important receptors in some reptiles.

13. Because uric acid is nontoxic and relatively insoluble in water, it can be stored and excreted as a semisolid. Internal respiratory surfaces and dry skin also promote conservation of water.

14. The amniotic egg and internal fertilization permit development on land. They are accompanied by significant energy expenditure on the part of the parent.

15. Some reptiles use visual, olfactory, and auditory cues for reproduction. Parental care is important in crocodilians.

16. Descendants of the diapsid evolutionary lineage include the birds. Descendants of the synapsid lineage are the mammals.

■ Key Terms

anapsid lineage (p. 454)
diapsid lineage (p. 454)
Jacobson's organs (p. 465)
median (parietal) eye (p. 465)
pit organs (p. 465)
synapsid lineage (p. 454)

■ Critical Thinking Questions

1. What kind of evidence has been used to establish evolutionary relationships of early reptiles?

2. What characteristics of the life history of turtles makes them vulnerable to extinction? What steps do you think should be taken to protect endangered turtle species?

3. Certain lizards, snakes, and worm lizards are superficially similar. Why are they grouped into different suborders?

4. What might explain the fact that parental care is common in crocodilians and birds?

5. Make a list of the adaptations that make life on land possible for a reptile. Explain why each is adaptive.

6. The incompletely divided ventricle of reptiles is sometimes portrayed as an evolutionary transition between the heart of primitive amphibians and the completely divided ventricles of birds and mammals. Do you agree with this portrayal? Why or why not?

■ Suggested Readings

Books

Bakker, R. T. 1986. *The Dinosaur Heresies.* New York: William Morrow & Co, Inc.

Goin, C. J., Goin, O. B., and Zug, G. R. 1978. *Introduction to Herpetology,* 3rd ed. San Francisco: W.H. Freeman & Co.

Schmidt, K. P., and Inger, R. F. 1975. *Living Reptiles of the World.* New York: Doubleday & Co., Inc.

Thomas, R. D. K., and Olson, E. C. (eds). 1980. *A Cold Look at the Warm-Blooded Dinosaurs.* AAAS Selected Symposium Series. Boulder: Westview Press, Inc.

Wilford, J. N. 1985. *The Riddle of the Dinosaur.* New York: Alfred A. Knopf.

Articles

Bakker, R. T. Dinosaur renaissance. *Scientific American* April, 1975.

Cole, J. Unisexual lizards. *Scientific American* January, 1984.

Crews, D. Courtship in unisexual lizards: a model for brain evolution. *Scientific American* December, 1987.

Deaming, C., and Ferguson, F. 1989. In the heat of the nest. *New Scientist* 121:33–38.

Gaffney, E. S., Hutchison, J. H., Jenkings, F. A., Jr., and Meeker, L. J. 1987. Modern turtle origins: the oldest known cryptodire. *Science* 237:289–291.

Gibbons, J. W. 1987. Why do turtles live so long? *BioScience* 37(4):262–268.

Gutzke, W. H. N., and Crews, D. 1988. Embryonic temperature determines adult sexuality in a reptile. *Nature* 332:832–834.

Langston, W., Jr. Pterosaurs. *Scientific American* February, 1981.

Newman, E. A., and Hartline, P. H. 1982. The infrared "vision" of snakes. *Scientific American* March 246:116–127.

Russell, D. A. 1982. The mass extinctions of the late Mesozoic. *Scientific American* January, 1982.

Birds: Feathers, Flight, and Endothermy

30

Concepts

1. Fossils of the earliest birds clearly show reptilian features. Fossils of *Archaeopteryx* give clues to the origin of flight in birds.
2. Integumentary, skeletal, muscular, and gas exchange systems of birds are adapted for flight and endothermic temperature regulation.
3. Large regions of the brain of birds are devoted to integrating sensory information.
4. Complex mating systems and behavior patterns increase the chances of offspring survival.
5. Migration and navigation allow birds to live, feed, and reproduce in environments that are favorable to survival of adults and young.

Have You Ever Wondered:

[1] what ancient animal is the earliest known bird?
[2] how flight evolved in birds?
[3] how a bird can sleep without falling from its perch?
[4] how the wings create enough lift to get the bird off the ground?
[5] how an osprey diving for a fish keeps the fish in focus throughout the dive?
[6] how birds navigate during migration?

These and other useful questions will be answered in this chapter.

This chapter contains underlined evolutionary concepts.

Evolutionary Perspective

Drawings of birds on the walls of caves in southern France and Spain, bird images of ancient Egyptian and ancient American cultures, and the bird images in Biblical writings are evidence that humans have marveled at birds and bird flight for thousands of years. From the early drawings of flying machines by Leonardo da Vinci (1490) to the first successful powered flight by Orville Wright on December 17, 1903, humans have tried to take to the sky and experience what it would be like to soar like a bird.

Just as impressive as flight is the ability of birds to navigate long distances between breeding and wintering grounds. For example, Arctic terns have a migratory route that takes them from the Arctic to the Antarctic and back again each year, a distance of approximately 35,000 km (22,000 mi.) (figure 30.1). Their rather circuitous route takes them across the northern Atlantic Ocean, to the coast of Europe and Africa, and then across vast stretches of the southern Atlantic Ocean before reaching their wintering grounds.

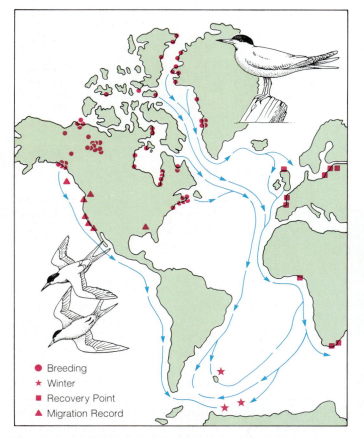

● Breeding
★ Winter
■ Recovery Point
▲ Migration Record

Phylogenetic Relationships

Birds are members of the class Aves (a'ves) (L. *avis,* bird). The major characteristics of this class concern adaptations for flight, including appendages modified as wings, feathers, endothermy, a high metabolic rate, a vertebral column modified for flight, and bones that are lightened by numerous air spaces. In addition, modern birds possess a horny bill and lack teeth.

The similarities between birds and reptiles are so striking that birds are often referred to as "glorified reptiles." Like the crocodilians, birds have descended from ancient archosaurs (*see figure 29.2*). Other flying reptiles in this evolutionary lineage (e.g., pterosaurs and pterodactyls) are ruled out of bird ancestry because these reptiles lost an important avian feature, the clavicle, long before birds appeared. (The clavicles, or "wishbone" serves as one of the attachment points for flight muscles.) Rather than having feathered wings, the flight surfaces of the wings of these primitive reptiles were membranous folds of skin.

Archaeopteryx and the Evolution of Flight

In 1861, one of the most important vertebrate fossils was found in a slate quarry in Bavaria, Germany (figure 30.2). It was a fossil of a pigeon-sized animal that lived during the Jurassic period, about 150 million years ago. It had a long, reptilian tail and clawed fingers. The complete head of this specimen was not preserved, but imprints of feathers on the tail and on short rounded wings were the main evidence that this was the fossil of an ancient bird. It was named *Archaeopteryx* (Gr. *archaios,* ancient + *pteron,* wing). Sixteen years later, a more complete fossil was discovered, revealing teeth in beaklike jaws. Four later discoveries of *Archaeopteryx* fossils have reinforced ideas of reptilian ancestry for birds. [1]

Figure 30.1 The Arctic Tern and its migration route. Arctic Terns breed in northern North America, Greenland, and the Arctic. Migrating birds cross the Atlantic Ocean on their trip to wintering grounds in Antarctica. In the process, they fly about 22,000 miles each year.

Interpretations of the life-style of *Archaeopteryx* have been important in the development of hypotheses on the origin of flight. The clavicles (wishbone) of *Archaeopteryx* are well developed and probably provided points of attachment for wing muscles. The sternum, another site for attachment of flight muscles in modern birds, is less developed. These observations indicate that *Archaeopteryx* may have been a glider and that flapping flight was very weakly developed.

[2] Some zoologists think that the clawed digits of the wings may have been used to climb trees and cling to branches. A sequence in the evolution of flight may have involved jumping from branch-to-branch or branch-to-ground. At some later point, gliding evolved. Still later, weak flapping, supplemented gliding and finally wing-powered flight evolved.

Other zoologists note that the structure of the hindlimbs suggests that *Archaeopteryx* may have been bipedal, running and hopping along the ground. Its wings may have functioned in batting flying insects out of the air or in trapping insects and other prey against the ground. The teeth and claws, which resemble talons of modern predatory birds, may have been used to grasp prey. Wings would have been useful in providing stability during horizontal jumps when pursuing prey and they would also have allowed flight over short distances. The benefits of such flight may have led eventually to wing-powered flight.

Diversity of Modern Birds

Archaeopteryx provides the only evidence of the transition between reptiles and birds. We do not know, however, whether or not *Archaeopteryx* is the direct ancestor of modern birds. There are a variety of fossil birds found for the period between 150 million and 70 million years ago. Some of these birds were large, flightless birds; others were adapted for swimming and diving, and some were fliers. Most, like *Archaeopteryx,* had reptilelike teeth. Most of the lineages represented by these fossils became extinct, along with the dinosaurs, at the end of the Mesozoic era.

Some of the few birds that survived into the Cenozoic era were the ancestors of modern, toothless birds. Adaptive radiation has resulted in about 9,100 species of living birds, which are divided into about 27 orders (table 30.1). (The number of orders varies depending on the classification system used.) The orders are distinguished from one another by characteristic behaviors, songs, and anatomical differences (figure 30.3).

Stop and Ask Yourself

1. What are characteristics of the class Aves?
2. What characteristics of *Archaeopteryx* are reptilelike?
3. What are two hypotheses for the origin of flight in birds? What aspects of the structure of *Archaeopteryx* are used to support each hypothesis?

(a)

(b)

Figure 30.2 *Archaeopteryx,* an ancient bird. (a) *Archaeopteryx* fossil. (b) Artist's representation. Some zoologists think that *Archaeopteryx* was a ground dweller rather than the tree dweller depicted here.

Evolutionary Pressures

Virtually every body system of a bird shows some adaptation for flight. Endothermy, feathers, acute senses, long flexible necks, and lightweight bones are a few of the many adaptations described in this section.

External Structure and Locomotion

The covering of feathers on a bird is called the *plumage.* Feathers have two primary functions essential for flight. They form the flight surfaces that provide lift and aid steering, and they prevent excessive heat loss, permitting the endothermic maintenance of high metabolic rates. Feathers also have roles in courtship, incubation, and waterproofing.

Feathers develop in a fashion similar to epidermal scales of reptiles, and this similarity is one source of evidence that demonstrates the evolutionary ties between reptiles and birds (figure 30.4; *see box 29.2*). Only the inner pulp of feathers contains dermal elements, such as blood vessels, which supply nutrients and pigments for the growing

Table 30.1 Classification of the Birds

Class Aves (a′ves) (L. *avis,* bird)
Adaptations for flight include: foreappendages modified as feathered wings, endothermic, high metabolic rate, neck flexible, posterior vertebrae fused, and bones lightened by numerous air spaces. The skull is lightened by a reduction in bone and the presence of a horny beak that lacks teeth. The birds.

Order Sphenisciformes (sfe-nis′i-for′mez)
Heavy bodied; flightless, flipperlike wings for swimming; well insulated with fat. Penguins.

Order Struthioniformes (stroo′the-oni-for′mez)
Large, flightless birds; wings with numerous fluffy plumes. Ostriches.

Order Rheiformes (re′i-for′mez)
Large, flightless birds; degenerate wings with soft, loose plumes. Rheas.

Order Casuariiformes (kaz′u-ar′e-i-for′mez)
Wings reduced; plumage coarse and hairlike. Cassowaries, emus.

Order Apterygiformes (ap′te-rij′i-for′mez)
Long, decurved beak with nostril near tip; wings absent; plumage hairlike. Kiwis.

Order Tinamiformes (tin-am′i-for′mez)
Moderate size; bill depressed and arched. Tinamous.

Order Gaviiformes (ga′ve-i-for′mez)
Strong, straight beak; diving adaptations include legs far back on body, bladelike tarsus, webbed feet, and heavy bones. Loons.

Order Podicipediformes (pod′i-si-ped′i-for′mez)
Wings short; plumage soft and dense; feet webbed with flattened nails. Grebes.

Order Procellariiformes (pro-sel-lar-e-i-for′mez)
Tubular nostrils, large nasal glands; wings long and narrow. Albatrosses, shearwaters, petrels.

Order Pelecaniformes (pel′e-can-i-for′mez)
Four toes joined in common web; nostrils rudimentary or absent; large gular sac. Pelicans, boobies, cormorants, anhingas, frigatebirds.

Order Ciconiiformes (si-ko′ne-i-for′mez)
Neck long, often folded in flight; long-legged waders. Herons, egrets, storks, wood ibises, flamingos.

Order Anseriformes (an′ser-ifor′mez)
South American screamers, ducks, geese and swans. The latter possess a wide flat bill and an undercoat of dense down. Webbed feet. Worldwide.

Order Falconiformes (fal′ko-ni-fir′mez)
Strong, hooked beak; wings large; raptorial feet. Vultures, secretarybird, hawks, eagles, ospreys, falcons.

Order Galliformes (gal′li-for′mez)
Short beak; short, concave wings; feet and claws strong. Curassows, grouse, quail, pheasants, turkeys.

Order Gruiformes (gru′i-for′mez)
Order characteristics variable and not diagnostic. Marsh birds including cranes, limpkins, rails, coots.

Order Charadriiformes (ka-rad′re-i-for′mez)
Order characteristics variable. Shorebirds, gulls, terns, auks.

Order Columbiformes (co-lum′bi-for′mez)
Dense feathers loosely set in skin; well-developed crop. Pigeons, doves, sandgrouse.

Order Psittaciformes (sit′ta-si-for′mez)
Maxilla hinged to skull; tongue thick; fourth toe reversible; usually brightly colored. Parrots, lories, macaws.

Order Cuculiformes (ku-koo′li-for′mez)
Fouth toe reversible; skin soft and tender; Plantain-eaters, roadrunners, cuckoos.

Order Strigiformes (strij′i-for′mez)
Large head with fixed eyes directed forward; raptorial foot. Owls.

Order Caprimulgiformes (kap′ri-mul′ji-for′mez)
Owllike head and plumage, but weak bill and feet; beak with wide gape; insectivorous. Whip-poor-will, other goatsuckers.

Order Apodiformes (a-pod′i-for′mez)
Long wings; weak feet. Swifts, hummingbirds.

Order Coliiformes (ka-li′i-for′mez)
Long tail; weak rounded wings; dense plumage. Mousebirds.

Order Trogoniformes (tro-gon′i-for′mez)
Beak short, flat, decurved, and serrate; tail long with squarish feathers; brilliantly colored. Trogons.

Order Coraciiformes (kor′ah-si′ah-for′mez)
Large head; large beak; metallic plumage. Kingfishers, todies, bee-eaters, rollers.

Order Piciformes (pis′i-for′mez)
Beak usually long and strong; legs and feet strong with fourth toe permanently reversed in woodpeckers. Woodpeckers, toucans, honeyguides, barbets.

Order Passeriformes (pas′er-i-for′mez)
Largest avian order;. 69 families of perching birds; perching foot; variable external features. Swallows, larks, crows, titmice, nuthatches, and many others.

---

feather. As feathers mature, their blood supply is cut off, and the feathers become dead, keratinized, epidermal structures seated in epidermal invaginations of the skin called *feather follicles.*

The most obvious feathers are *contour feathers,* which cover the body, wings, and tail (figure 30.5). Contour feathers consist of a *vane* with its inner and outer webs, and a supportive shaft. Feather *barbs* branch off the shaft, and *barbules* branch off the barbs. Barbules of adjacent barbs overlap one another. The ends of barbules are locked together with hooklike *hamuli.* Interlocking barbs keep contour feathers firm and smooth.

Birds maintain a clean plumage to rid the feathers and skin of parasites. *Preening,* which is done by rubbing the beak over the feathers, keeps the feathers smooth, clean, and in place. Hamuli that become dislodged can be re-hooked by running a feather through the beak. Secretions from an oil gland at the base of the tail of many birds are spread over the feathers during preening to keep the plumage water repellant and supple. The secretions also lu-

Period (millions of years before present)

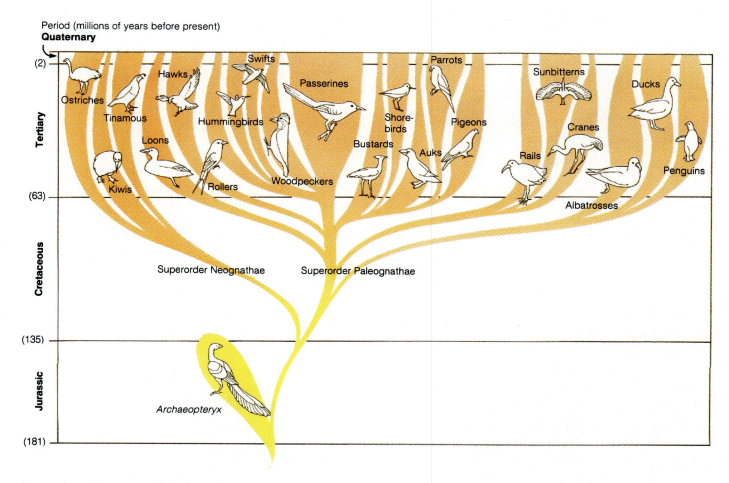

Figure 30.3 Phylogeny of the major bird orders. The relative number of species in each taxonomic group is indicated by the width of a group's branch.

bricate the beak and legs to prevent chafing. *Anting* is a maintenance behavior common to many songbirds and involves picking up ants in the beak and rubbing them over the feathers. The formic acid secreted by ants is apparently toxic to feather mites.

Most colors in a bird's plumage are produced by feather pigments deposited during feather formation. Other colors, termed *structural colors,* arise from irregularities on the surface of the feather that diffract white light. For example, blue feathers are never blue because of the presence of blue pigment. A porous, nonpigmented outer layer on a barb reflects blue wavelengths of light. The other wavelengths pass into the barb and are absorbed by the dark pigment melanin. Iridescence results from the interference of light waves caused by a flattening and twisting of barbules. An example of iridescence is the perception of interchanging colors on the neck and back of hummingbirds and grackles. Color patterns are involved in cryptic coloration, species and sex recognition, and sexual attraction.

Mature feathers receive constant wear; thus, all birds undergo a periodic renewal of their feathers by shedding and replacing them in a process called **molting.** The timing of molt periods varies in different taxa. The following is a typical molting pattern for songbirds. After hatching, a chick is covered with *down.* Down is replaced with juvenile feathers at the *postnatal molt.* A *postjuvenile molt* usually occurs in the fall and results in plumage similar to that of the adult. Once sexual maturity is attained, a *prenuptial molt* occurs in late winter or early spring, prior to the breeding season. A *postnuptial molt* usually occurs between July and October. Flight feathers are frequently lost in a particular sequence so that birds are not wholly deprived of flight during molt periods. However, many ducks, coots, and rails cannot fly during molt periods and hide in thick marsh grasses.

The Skeleton The bones of most birds are lightweight yet strong. Some bones, such as the humerus (fore-

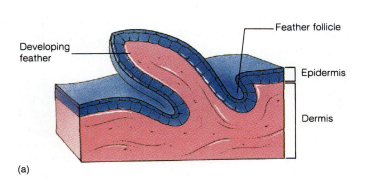

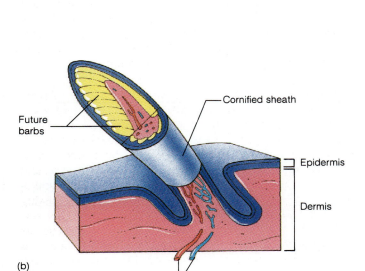

Figure 30.4 The formation of bird feathers during embryonic development. (a) Feathers form from the epidermal evaginations. (b) Later in development, the blood supply to the feather is cut off and the feather becomes a dead, keratinized epidermal structure seated in a feather follicle.

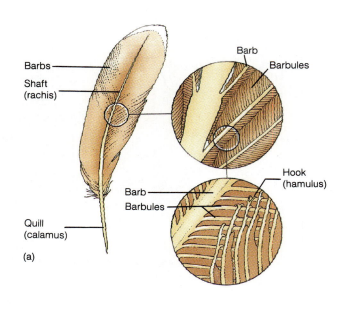

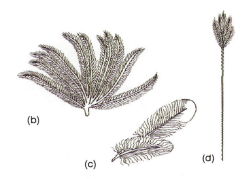

Figure 30.5 Anatomy of selected feather types. (a) Anatomy of a contour feather showing enlargements of barbs and barbules. (b) A down feather. (c) A contour feather with an aftershaft. (d) A filoplume.

arm bone), have large air spaces and internal strutting (reinforcing bony bars), which increase strength (figure 30.6c). (Engineers take advantage of this same principle. They have discovered that a hollow girder is stronger than a solid girder of the same weight.) Birds also have a reduced number of skull bones, and teeth are replaced with a lighter, keratinized sheath covering a beak. The demand for lightweight bones for flight is countered in some birds with other requirements. Some aquatic birds (e.g., loons) have dense bones, which help reduce buoyancy during diving.

The appendages involved in flight cannot manipulate nesting materials or feed young. These activities are possible because of the beak and very flexible neck. The cervical vertebrae have saddle-shaped articular surfaces that permit great freedom of movement. In addition, the first cervical vertebra (the atlas) has a single point of articulation with the skull (the occipital condyle), which permits a high degree of rotational movement between the skull and the neck. (The single occipital condyle is another characteristic shared with reptiles.) This flexibility allows the beak and neck to function as a fifth appendage.

The pelvic girdle, vertebral column, and ribs are strengthened for flight. The *thoracic region* is indicated by the presence of ribs, which attach to thoracic vertebrae. The ribs have posteriorly directed *uncinate processes* that overlap the next rib to strengthen the rib cage (figure 30.6a). (Uncinate processes are also present on the ribs of most reptiles and are additional evidence of their common ancestry.) Posterior to the thoracic region is the *lumbar*

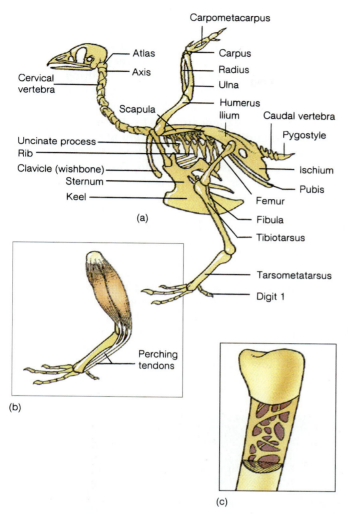

Figure 30.6 The bird skeleton. (a) The skeleton of a pigeon. (b) Perching tendons run from the toes across the back of the ankle joint, which cause the foot to grip a perch. (c) The internal structure of the humerus.

region. The *synsacrum* is formed by the fusion of the posterior thoracic vertebrae, all the lumbar and sacral vertebrae, and the anterior caudal vertebrae. Fusion of these bones helps maintain the proper flight posture and supports the hind appendages during landing, hopping, and walking. The posterior caudal vertebrae are fused into a *pygostyle,* which helps support the tail feathers that are important in steering.

The *sternum* of most birds bears a large, median keel for the attachment of flight muscles. (Exceptions to this include some flightless birds, such as ostriches.) It attaches firmly to the rest of the axial skeleton by the ribs. Paired *clavicles* are fused medially and ventrally into a furcula (wishbone).

The appendages of birds have also been modified. Some bones of the front appendages of a bird have been lost or fused and serve as points of attachment of flight feathers. The rear appendages are used for hopping, walking, running, and perching. *Perching tendons* run from the toes

across the back of the ankle joint to muscles of the lower leg. When the ankle joint is flexed, as in landing on a perch, tension on the perching tendon is increased, and the foot grips the perch (figure 30.6b). This automatic grasp helps a bird perch even while sleeping. The muscles of the lower leg can increase the tension on this tendon, for example, when an eagle grasps a fish in its talons. [3]

Muscles The largest, strongest muscles of most birds are the flight muscles. They attach to the sternum and clavicle and run to the humerus. The muscles of most birds are adapted physiologically for flight. Flight muscles must contract quickly and fatigue very slowly. These muscles have large numbers of mitochondria and produce large quantities of ATP to provide the energy required for flight, especially long-distance migrations. Domestic fowl have been selectively bred for massive amounts of muscle that is well liked by humans as food, but is poorly adapted for flight because it contains rapidly contracting fibers with few mitochondria and poor vascularization.

Flight The wings of birds are adapted for different kinds of flight. However, regardless of whether a bird soars, glides, or has a rapid flapping flight, the mechanics of staying aloft are similar. Bird wings form an **airfoil.** The anterior margin of the wing is thicker than the posterior margin. The upper surface of the wing is slightly convex, and the lower surface is flat or slightly concave. Air passing over the wing travels further and faster than air passing under the wing, decreasing air pressure on the upper surface of the wing and creating lift (figure 30.7a). The lift created by the wings must overcome the bird's weight, and the forces that propel the bird forward must overcome the drag created by the friction of the bird moving through the air. Lift can be increased by increasing the angle the leading edge of the wing makes with the oncoming air (the *angle of attack.*) As the angle of attack increases, however, the flow of air over the upper surface becomes turbulent, reducing lift (figure 30.7b). Turbulence can be reduced by forming slots at the leading edge of the wing through which air can flow rapidly, thus smoothing air flow once again. *Slotting* the feathers at the wing tips and the presence of an alula on the anterior margin of the wing reduce turbulence. The *alula* is a group of small feathers supported by bones of the second digit. During takeoff, landing and hovering flight, the angle of attack is increased, and the alula is elevated. During soaring and fast flight, the angle of attack is decreased, and slotting is reduced (figure 30.7c,d). [4]

Most of the propulsive force is generated by the distal part of the wing. Because it is further from the shoulder joint, the distal part of the wing moves further and faster than the proximal parts of the wing. During the downstroke (the powerstroke), the leading edge of the distal part of the wing is oriented slightly downward and creates a thrust somewhat analogous to the thrust created by a propeller on an airplane (figure 30.7e). During the upstroke (the recovery stroke), the distal part of the wing is oriented upward to decrease resistance. Feathers on a wing overlap so that on the downstroke, air presses the feathers at the wing margins together, allowing little air to pass between them, enhancing both lift and propulsive forces. Feathers part slightly on the upstroke, allowing air to pass between them, which reduces resistance during the recovery stroke.

The tail of a bird serves a variety of balancing, steering, and braking functions during flight. During horizontal flight, spreading the tail feathers increases lift at the rear of the bird and causes the head to dip for descent. Closing the tail feathers has the opposite effect. Tilting the tail sideways causes the bird to turn. When a bird lands, its tail is deflected upward, serving as an air brake.

Different kinds of flight are used by different birds or by the same bird at different times. During *gliding flight,* the wing is stationary, and a bird loses altitude. Waterfowl coming in for a landing use gliding flight. *Flapping flight* generates the power for flight and is the most common type of flying. Many variations in wing shape and flapping patterns result in species-specific speed and maneuverability. *Soaring flight* allows some birds to remain airborne with little energy expenditure. During soaring, wings are essentially stationary, and the bird utilizes updrafts and air currents to gain altitude. Hawks, vultures, and other soaring birds are frequently observed circling along mountain valleys, soaring downwind to pick up speed and then turning upwind to gain altitude. As the bird slows and begins to lose altitude, it turns downwind again. The wings of soarers are wide and slotted to provide maximum maneuverability at relatively low speeds (figure 30.8a). Oceanic soarers,

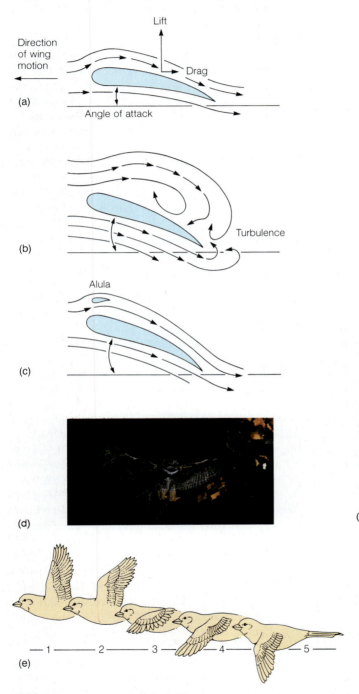

Figure 30.7 Mechanics of bird flight. (a) A bird's wing acts as an airfoil. Air passing over the top of the wing travels farther and faster than air passing under the wing, creating lift. (b) Increasing the angle of attack increases lift but also increases turbulence. (c) Turbulence is reduced by the alula. (d) Note the alula on the wings of the great horned owl. (e) The orientation of the wing during a downstroke.

Figure 30.8 Bird flight. (a) The soaring flight of a Galápagos hawk. (b) A hovering ruby-throated hummingbird.

such as albatrosses and frigate birds, have long, narrow wings that provide maximum lift at high speeds, but they compromise maneuverability and ease of takeoff and landing (box 30.1.)

Hummingbirds and gulls perform *hovering flight.* Gulls hover by taking advantage of winds deflected upward by waves or cliffs. They can maintain a stationary position in the air with few wing movements. Hummingbirds hover in still air (figure 30.8b). They fan their wings back and forth (50 to 80 beats/second) to remain suspended in front of a flower or feeding station.

Stop and Ask Yourself:

4. Describe feather maintenance behavior.
5. What adaptations of the bird skeleton promote flight?
6. How do bird wings provide lift and forward propulsion during flight?
7. What function is served by the alula?

Nutrition and the Digestive System

Most birds have ravenous appetites! This appetite supports a high metabolic rate that makes endothermy and flight possible. For example, hummingbirds feed almost constantly during the day. In spite of high rates of food consumption, their rapid metabolism often cannot be sustained overnight and they may become torpid, with reduced body temperature and respiratory rate, until they can feed again in the morning.

Bird beaks and tongues are modified for a variety of feeding habits and food sources (figure 30.9). The tongue of a woodpecker is barbed for extracting grubs from the bark of trees (*see figure 37.10d*). Sapsuckers excavate holes in trees and use a brushlike tongue for licking the sap that accumulates in these holes. The tongues of hummingbirds and other nectar feeders are rolled into a tube and used for extracting nectar from flowers.

In many birds, a diverticulum of the esophagus, called the *crop,* is a storage structure that allows birds to quickly ingest large quantities of locally abundant food and then seek safety while digesting their meal. The crop of pigeons produces "pigeon's milk," a cheesy secretion formed by the proliferation and sloughing of cells lining the crop. Young pigeons (squabs) are fed pigeon's milk until they are able to eat grain. Cedar waxwings, vultures, and birds of prey use their esophagus for similar storage functions. Crops are less well developed in insect-eating birds because insectivorous birds feed throughout the day on sparsely distributed food.

The stomach of birds is modified into two regions. The *proventriculus* secretes gastric juices that initiate digestion (figure 30.10). The *ventriculus* (*gizzard*) has muscular walls to abrade and crush seeds or other hard materials. Sand and other abrasives may be swallowed to aid digestion. The bulk of enzymatic digestion and absorption occurs in the small intestine, aided by secretions from the pancreas

and liver. Paired *ceca* may be located at the union of the large and small intestine. These blind-ending sacs contain bacteria that aid in the digestion of cellulose. Undigested food is usually eliminated through the cloaca; however, owls form pellets of bone, fur, and feathers that are ejected from the ventriculus through the mouth. Owl pellets accumulate in and around owl nests and are useful in studying their food habits.

It is common practice to group birds by their feeding habits. It is somewhat artificial, however, because birds may eat different kinds of food at different stages in their life history, or they may change diets simply because of changes in food availability. Robins, for example, feed largely on worms and other invertebrates when these foods are available. In the winter, however, robins may feed on berries.

In some of their feeding habits, birds have come into direct conflict with human interests. Bird damage to orchard and grain crops is tallied in the millions of dollars each year. Flocking and roosting habits of some birds, such

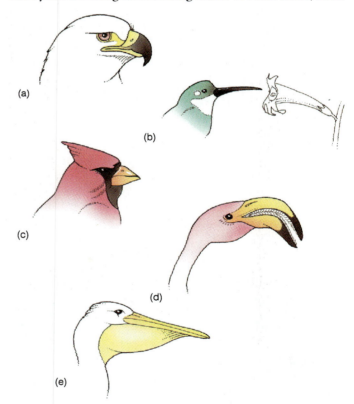

Figure 30.9 Some specializations of bird beaks. (a) The beak of an eagle is specialized for tearing prey. (b) The beak of this hummingbird matches the length and curvature of the flower from which it extracts nectar. (c) The thick, powerful beak of this cardinal is used to crack tough seeds. (d) The beak of a flamingo is used for straining food from the water in a head-down feeding posture. The upper and lower mandibles are fringed with large bristles. As water is sucked into the beak, larger particles are filtered and left outside. Inside the beak, smaller algae and animals are filtered on tiny inner bristles. The tongue is used to remove food from the bristles. (e) The lower mandible of this pelican is covered by a distensible gular membrane for storing fish scooped from the water.

Box 30.1

Sailors' Curse—Gliders' Envy

They are known by many, not very complimentary names. Dutch sailors called them mollymawks ("stupid gull"), the English called them goonies (another reference to stupidity), and the Japanese called them bakadori ("foul birds"). The names of albatrosses are probably the result of inexpressive facial features and awkward movements on land. The albatross's reputation for stupidity is accompanied by another reputation. Their appearance alongside a ship was believed to be a sure sign of changing winds. As the old sailor with "a long grey beard and glittering eye" discovered, to kill an albatross brought extremely bad luck. (*The Rime of the Ancient Mariner* by Samuel Coleridge, 1798.) His deadly aim with a crossbow caused the winds to die, and all sailors on the becalmed ship, except the mariner, died of thirst. The mariner was forced to sail on alone with the albatross hung around his neck.

> And I had done an hellish thing
> And it would work 'em woe:
> For all averr'd, I had kill'd the Bird
> That made the Breeze to blow.
> Ah wretch! said they, the bird to slay,
> That made the breeze to blow!

The origin of the superstition that associated the albatross with breezes is not difficult to understand. The albatross has relatively poorly developed flight muscles and relies primarily on soaring flight and wind to keep aloft (box figure 30.1a). Most species of albatross, are found around the Antarctic where breezes are almost constant, and they can launch themselves from cliffs into the air with minimal flapping flight. They soar swiftly down-wind, picking up speed and losing altitude. Just above the water's surface, they turn sharply into the wind and use the oncoming wind to soar higher. When air speed drops, they turn again to move downwind. Under favorable winds, an albatross can follow a ship for many miles, zig-zagging upwind and downwind, without flapping their long, narrow wings. The sight of a soaring albatross usually does mean a favorable sailing breeze!

When grounded in calm winds, an albatross experiences great difficulty becoming airborne again. They must run along the ground, flapping their wings, until air speed adequate for takeoff has been achieved (box figure 30.1b). Laysan albatrosses on Midway Island use U.S. Navy runways for taxiing before takeoff, and they soar on thermals above the island, creating serious problems for aircraft.

Albatrosses feed on fish and invertebrates near the ocean's surface and on refuse tossed from ships. After a courtship dance and mating, a single egg is laid in a mud nest. Incubation may last as long as 85 days, and parental care another 3 to 9 months. After leaving the nest, young albatrosses depart the nesting grounds and circle their newly discovered world many times. (For most albatrosses "their world" is the entire southern hemisphere!) Sexual maturity is reached after about 7 years, and some may live for up to 30 years.

(a)

(b)

Box figure 30.1   (a) An albatross in flight. (b) This albatross has just become airborne.

as European starlings and redwing blackbirds, concentrate millions of birds in local habitats, and fields of grain can be devastated. Recent monocultural practices tend to aggravate problems with grain-feeding birds by encouraging the formation of very large flocks.

In spite of commonly held beliefs, the impact of birds of prey on poultry, game birds, and commercial fisheries is minimal. Unfortunately, birds of prey have been killed with guns and poisons because of the mistaken impression that they are responsible for significant losses.

Circulation, Gas Exchange, and Temperature Regulation

The circulatory system of birds is similar to that of reptiles, except that the heart has completely separated atria and ventricles, resulting in separate pulmonary and systemic circuits. In vertebrate evolution, the sinus venosus has undergone gradual reduction in size. It is a separate chamber in fishes, amphibians, and turtles, and receives blood from the venous system. In other reptiles, it is a group of cells in the right atrium that serves as the pacemaker for the heart. In birds, the sinus venosus also persists only as a patch of pacemaker tissue in the right atrium. The heart is relatively large (up to 2.4% of total body weight) and its rate of beating is rapid. Rates in excess of 1,000 beats/minute have been recorded for hummingbirds under

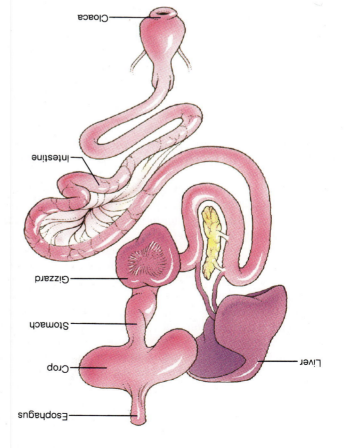

Figure 30.10 The digestive system of a pigeon.

stress. Larger birds have relatively smaller hearts and slower heart rates. The heart rate of an ostrich, for example, varies between 38 and 176 beats per minute. A large heart, rapid heart rate, and complete separation of oxygenated from unoxygenated blood are important adaptations for delivering the large quantities of blood required for endothermy and flight.

Gas Exchange Because of high metabolic rates associated with flight, birds have a greater rate of oxygen consumption than any other vertebrate. When vertebrates inspire and expire, air passes into and out of respiratory passageways. Ventilation is interrupted during exhalation, and there is a considerable quantity of "dead air" in the lungs because not all air is forced out during expiration. Because of their unique structure, ventilation of bird lungs is nearly continuous during the inspiratory and expiratory cycles and provides a nearly continuous movement of fresh air over respiratory surfaces. The quantity of "dead air" in the lungs is sharply reduced compared to other vertebrates.

External nares lead to nasal passageways and the pharynx. A slitlike *glottis* is the opening to the respiratory tract. The *trachea* is supported by bone and cartilage. A larynx is undifferentiated, but a special voicebox, called the *syrinx*, is located where the trachea divides into *bronchi*. The muscles of the syrinx and bronchi, as well as characteristics of the trachea are responsible for bird vocalizations.

The lungs of birds are made of small air tubes called *parabronchi* (figure 30.11). *Air capillaries* about 10 μm in diameter branch from the parabronchi and provide gas-exchange surfaces. Inspiration and expiration is accomplished by the expansion and compression of thin-walled *air sacs* that ramify throughout the body cavity and even penetrate some bones, such as the humerus of the wing.

Ventilation occurs by alternate compression and expansion of the air sacs during flight and other activities. When breathing, the movement of the sternum and the posterior ribs compresses the thoracic air sacs. X-ray movies of European starlings in a wind tunnel show that the wishbone is distorted when flight muscles contract. Alternate distortion and recoiling helps compress and expand the air sacs located between the bone's two shafts. During inspiration, air moves into the abdominal air sacs. At the same time, air already in the lungs moves through parabronchi to the thoracic air sacs. During expiration, the air in the thoracic air sacs moves out of the respiratory system, and the air in the abdominal air sacs moves into parabronchi. At the next inspiration, the air moves into the thoracic air sacs, and is expelled at the next expiration. It takes two ventilatory cycles to move a particular volume of gas through the respiratory system of a bird.

Thermoregulation Birds regulate their body temperatures between 38 and 45°C. Lethal extremes are lower than 32 and higher than 47°C. On a cold day, resting birds fluff their feathers to increase their insulating properties, as well as the dead air space within them. They also tuck their beaks into their feathers to reduce heat loss from the respiratory tract. The most exposed parts of a bird are the feet and tarsi which have neither fleshy muscles nor a rich blood supply. Temperatures in these extremities are allowed to drop near freezing to prevent heat loss. Countercurrent heat exchange between the warm blood flowing to the legs and feet and the cooler blood flowing to the body core from the legs and feet prevent excessive heat loss at the feet by returning heat to the body core before it goes to the extremities and is lost to the environment (see *figure 38.4*). Shivering is also used to generate heat in extreme cold. Increases in metabolism during winter months require additional food.

Some birds become torpid and allow their body temperatures to drop on cool nights. For example, whip-poorwills allow their body temperatures to drop near 16°C, and respiratory rates become very slow.

Muscular activity during flight produces large quantities of heat. Excess heat can be dissipated by panting. Evaporative heat loss from the floor of the mouth, is enhanced by fluttering vascular membranes of this region.

Nervous and Sensory Systems

A mouse skitters across the floor of a barn enveloped in the darkness of night. An owl in the loft overhead turns in the direction of the faint sounds made by tiny feet. As the mouse reaches a sack of grain, and the sounds made by hurrying feet change to a scratchy gnawing of teeth on the sack of feed, the barn owl dives for its prey (figure 30.12). Fluted tips of flight feathers make the owl's approach imperceptible to the mouse, and the owl's ears, not its eyes, provide information to guide the owl to its prey. Under similar circumstances, barn owls will successfully locate and capture prey in over 75% of attempts! This ability is just one example of the many sensory adaptations found in birds.

The forebrain of birds is much larger than that of reptiles due to the enlargement of the cerebral hemispheres, including a region of gray matter, the *corpus striatum*. The *corpus striatum* functions in visual learning, feeding, courtship, and nesting. A *pineal body* is located on the roof

Figure 30.12 A barn owl (*Tyto alba*).

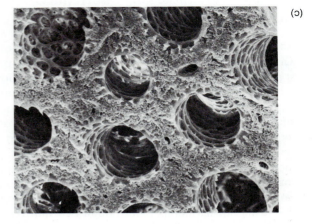

(a)

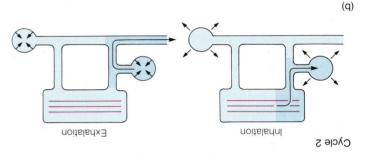

Cycle 1

Bronchus

Lung

Parabronchi

Inhalation

Thoracic sacs

Abdominal sacs

Exhalation

(b)

Cycle 2

Inhalation

Exhalation

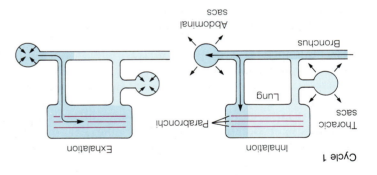

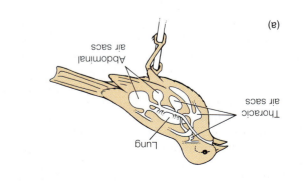

(c)

Figure 30.11 The respiratory system of a bird. (a) Air sacs branch from the respiratory tree. (b) Air flow during inspiration and expiration. Air flows through the parabronchi during both inspiration and expiration. The shading represents the movement through the lungs of one inspiration. (c) Scanning electron micrograph showing parabronchi.

Abdominal air sacs

Thoracic air sacs

Lung

of the forebrain. It appears to play a role in stimulating ovarian development and in regulating other functions influenced by light and dark periods. The *optic tectum* (the roof of the midbrain), along with the corpus striatum, plays an important role in integrating sensory functions. (The optic tectum is the primary integrative center of all vertebrates except birds and mammals.) The midbrain also receives sensory input from the eyes. As in reptiles, the hindbrain includes the cerebellum and the medulla oblongata, which coordinate motor activities and regulate heart and respiratory rates, respectively.

Vision is an important sense for most birds. The structure of the bird eye is similar to that of other vertebrates (see *figure 28.9*). The eye is usually somewhat flattened in an anterior/posterior direction; however, the eyes of birds of prey protrude anteriorly because of a bulging cornea. Birds have a unique, double-focusing mechanism. Padlike structures (similar to those of reptiles) control the curvature of the lens, and ciliary muscles change the curvature of the cornea. Double, nearly instantaneous focusing allows an [5] osprey, or other bird of prey, to remain focused on a fish throughout a brief, but breathtakingly fast, descent.

The retina of a bird's eye is thick and contains both rods and cones. Rods are active under low light intensities and cones under high light intensities. Cones are especially concentrated ($1,000,000/mm^2$) at a focal point called the *fovea*. Unlike other vertebrates, some birds have two foveae per eye. The one at the center of the retina is sometimes called the "search fovea" because it gives the bird a wide angle of monocular vision. The other fovea is at the posterior margin of the retina. It functions with the posterior fovea of the other eye to allow binocular vision. The posterior fovea is called the "pursuit fovea", because binocular vision is necessary for depth perception, which is necessary to capture prey. "Search" and "pursuit" are not meant to imply that the two foveae are found only in predatory birds. Other birds use the "search fovea" to observe the landscape below them during flight and the "pursuit" fovea when depth perception is needed, as in landing on a branch of a tree.

The position of the eyes on the head also influences the degree of binocular vision (figure 30.13). Pigeons have eyes located well back on the sides of their head, giving them a nearly 360° monocular field, but a narrow binocular field. They do not have to pursue their food (grain), and a wide monocular field of view helps them stay alert to predators while feeding on the ground. Hawks and owls have eyes further forward on the head. Their binocular field of view is increased, and their monocular field of view is correspondingly decreased.

Like reptiles, birds have a *nictitating membrane* that is drawn over the surface of the eye to cleanse and protect the eye.

Olfaction apparently plays a minor role in the lives of most birds. External nares open near the base of the beak, but the olfactory epithelium is poorly developed. Exceptions include turkey vultures, which locate their dead and dying prey largely by smell.

In contrast, hearing is well developed in most birds. The external ear opening is covered by loose, delicate feathers called *auriculars*. Middle- and inner-ear structures are similar to those of reptiles. The sensitivity of the avian ear (100 to 15,000 Hz) is similar to that of the human ear (16 to 20,000 Hz).

Excretion and Osmoregulation

Birds and reptiles face essentially identical excretory and osmoregulatory demands. Like reptiles, birds excrete uric acid, which is temporarily stored in the cloaca. Water reabsorption also occurs in the cloaca. As with reptiles, the

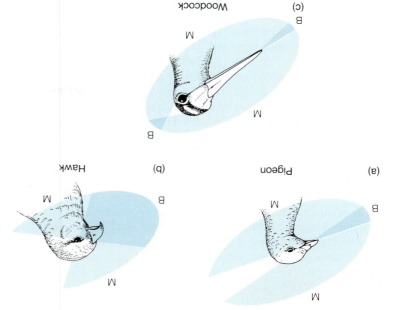

Figure 30.13 The fields of view of a pigeon (a), a hawk (b), and a woodcock (c). Woodcocks have eyes located far posteriorly and have a narrow field of binocular vision in front and behind. They can focus on predators that might be circling above them while probing mud with their long beaks. M = monocular field of view, B = binocular field of view.

Mating may follow the attraction of a mate to a territory. Female woodpeckers are attracted to the drumming by males on trees. Male ruffed grouse fan their wings on logs and create sounds that can be heard for many miles. Cranes have a courtship dance that includes stepping, bowing, stretching, and jumping displays. Mating occurs when a mate's call or posture signals readiness. It is accomplished quickly, but occurs repeatedly to assure fertilization of all the eggs that will be laid.

Over 90% of birds are **monogamous:** a single male pairs with a single female during the breeding season. Some birds (swans, geese, eagles) pair for life. Frequent mating apparently strengthens the pair bonds that develop. Monogamy is common when resources are widely and evenly distributed, and one bird cannot control access to resources. Monogamy is also advantageous because both parents usually participate in nest building and care of the young. One parent can incubate and protect the eggs or chicks while the other searches for food.

Some birds are **polygynous.** Males mate with more than one female, and the females care for the eggs and chicks. Polygyny tends to occur in species whose young are less dependent at birth and in situations where patchy resource distribution may attract many females to a relatively small breeding area. Prairie chickens are polygynous, and males display in groups called *leks.* In prairie chicken leks, the males in the center positions are preferred and attract the majority of females (figure 30.15).

A few bird species are **polyandrous** and the females mate with more than one male. Polyandry occurs in spotted sandpipers. Females are larger than males and establish and defend territories from other females. They lay a clutch of

---

excretion of uric acid conserves water and promotes development of embryos in terrestrial environments. In addition, birds have *supraorbital salt glands* that drain excess sodium chloride through the nasal openings to the outside of the body (*see box 38.2*). These are especially important in marine birds that drink seawater and feed on invertebrates containing large quantities of salt in their tissues. Salt glands can secrete salt in a solution that is about two to three times more concentrated than other body fluids. Salt glands, therefore, compensate for the kidney's inability to concentrate salts in the urine.

Stop and Ask Yourself

8. What evolutionary remnant of the vertebrate sinus venosus is found in the heart of a bird? What is the function of this remnant?
9. How are the lungs of birds adapted to provide continuous, one-way movement of air across gas exchange surfaces?
10. How does a bird cool itself?
11. What is the value of separate "search" and "pursuit" foveae for a bird of prey?

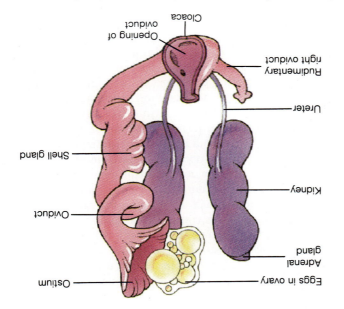

Figure 30.14 The urogenital system of a female pigeon.

Cloaca · Opening of oviduct · Rudimentary right oviduct · Ureter · Shell gland · Oviduct · Kidney · Adrenal gland · Eggs in ovary · Ostium

Reproduction and Development

Sexual activities of birds have been observed more closely than those of any other group of animals. These activities include establishing territories, finding mates, constructing nests, incubating eggs, and feeding young.

All birds are oviparous. Gonads are located in the dorsal abdominal region, next to the kidneys. Testes are paired, and coiled tubules (*vasa deferentia*) conduct sperm to the cloaca. An enlargement of the vasa deferentia, the *seminal vesicle*, is a site for temporary storage and maturation of sperm prior to mating. Testes enlarge during the breeding season. Except for certain waterfowl and ostriches, birds have no intromittent organ, and sperm transfer occurs by cloacal contact during brief mounts by the male on the female.

In females, two ovaries form during development, but usually only the left ovary fully develops (figure 30.14). A large, funnel-shaped opening of the oviduct envelopes the ovary and receives eggs after ovulation. Fertilization of the egg occurs in the upper portions of the oviduct, and the zygote is gradually surrounded by albumen secreted from glandular regions of the oviduct wall as the egg completes its passage. A shell is added by a *shell gland* in the lower region of the oviduct. The oviduct opens into the cloaca.

Territories are established by many birds prior to mating. Although size and function vary greatly among species, territories generally allow birds to mate without interference. They provide nest locations and sometimes food resources for adults and offspring. A territory that provides for a bird's needs may be more attractive to a potential mate than one

because of predation or other threats, the chances of successfully rearing young is low.

Nest construction usually begins after pair formation. It is an instinctive behavior and is usually initiated by the female. A few birds do not make nests. Emperor penguins, for example, breed on the snow and ice of Antarctica where no nest materials are available. Their single egg is incubated on the web of the foot (mostly the male's foot) tucked within a fold of abdominal skin.

Nesting Activities The nesting behavior of birds is often species specific. Some birds choose nest sites away from other members of their species, and other birds nest in very large flocks. Unfortunately, predictable nesting behavior has led to the extinction of some species of birds (box 30.2).

The number of eggs laid by a female is usually variable. Most birds incubate their eggs, and some birds have a featherless, vascularized incubation patch that helps keep the eggs at temperatures between 33 and 37°C. Eggs are turned to prevent the adherence of egg membranes in the egg and deformation of the embryo. Adults of some species sprinkle the eggs with water to cool and humidify them. The Egyptian plover carries water from distant sites in the breast

Figure 30.15 A male greater prairie chicken displaying in a lek.

eggs for each male that is attracted to and builds a nest in her territory. If a male loses his eggs to a predator, the female replaces those eggs. Polyandry results in the production of more eggs than monogamous matings. It is thought to be advantageous when food is plentiful but,

---

Box 30.2

Bright Skies and Silent Thunder

There is no sadder tale in the history of human interactions with wildlife than that of *Ectopistes migratorius*, the passenger pigeon. Although the full story of the extinction of this bird may never be known, humans will never be able to escape the responsibility for the excessive slaughter that led to the decline of a vulnerable species.

In the 19th century, flocks of passenger pigeons were so vast in eastern North America that their numbers were estimated at up to 2 billion birds. In 1813, the famed naturalist John Audubon watched a flock of passenger pigeons pass overhead for three days. The flock was said to completely darken the sky by day, and their wings sounded like thunder.

How could a species so abundant become extinct in just a few decades? What made these birds so vulnerable was their social behavior. Passenger pigeons nested, roosted in the evening, and foraged for food in huge aggregations that covered thousands of acres. It is said that trees became so laden with birds that their branches often broke under the weight. One nest site in Wisconsin was said to be 100 miles long! On feeding forays, flocks would travel hundreds of miles and strip trees of acorns, beechnuts, and other nuts. These colonies were apparently not just convenient associations, but were essential for survival. Highly organized flocks were also an im-

portant social stimulus for breeding.

Unfortunately, this behavior made the birds vulnerable. Nesting sites were predictable and hunters came to them year after year. Professional hunters are said to have traveled from nest site to nest site during the nesting season. Birds were blinded at night with lights and knocked out of trees. They were captured in huge nets or choked by burning sulfur. Their trees were felled and birds harvested. They were shot by the thousands with guns. In 1861, 14,850,000 passenger pigeons were shipped to big city markets from a single nesting site near Petoskey, Michigan.

How could any bird be expected to successfully produce and rear their young under this onslaught? The single egg tended by a pair of passenger pigeons had little chance for survival during these unfortunate years. Flocks gradually dwindled. By the late 1890s people began to realize that passenger pigeons were in trouble. In addition to rampant killing and the resultant breakup of the social organization needed for successful mating, deforestation was contributing to the passenger pigeon's decline. Laws that prohibited further killing were too late. After 1900, no passenger pigeons were seen in the wild. "Martha," the last passenger pigeon in captivity (Cincinnati Zoological Gardens), died on September 1, 1914, after living for 29 years.

feathers. The incubation period lasts between 10 and 80 days and is correlated with egg size and degree of development at hatching. One or 2 days before hatching, the young bird penetrates an air sac at the blunt end of its egg, inflates it lungs, and begins breathing. Hatching occurs as the young bird pecks the shell with a keratinized *egg tooth* on the tip of the upper jaw and struggles to free itself.

Some birds are helpless at hatching, others are more independent. Those that are entirely dependent upon their parents are said to be **altricial** (L. *altricialis*, to nourish), and they are often naked at hatching (figure 30.16a). Altricial young must be brooded constantly at first because endothermy is not developed. They grow rapidly, and when they leave the nest they are nearly as large as their parents. (For example, American robins weigh 4 to 6 grams at hatching and leave the nest 13 days later weighing 57 grams.) **Precocial** (L. *praecoct*, early ripe) young are alert and lively at hatching (figure 30.16b). They are usually covered by down, and can walk, run, swim, and feed themselves—although one parent is usually present to lead the young to food and shelter.

Young altricial birds have huge appetites and keep one or both parents continually searching for food. They may consume a mass of food that equals their own weight each day. Adults bring food to the nest or regurgitate stored food in the crop or esophagus. Vocal signals or color patterns on the beaks or throats of adults initiate feeding responses in the young. Parents instinctively feed gaping mouths, and many hatchlings have brightly colored mouth linings that attract a parent's attention. The first-hatched young is fed first and most often because it is usually largest and can stretch its neck higher than its nestmates.

Life is usually brief for birds. About 50% of eggs laid yield birds that leave the nest. Most birds, if kept in captivity, have a potential life span of 10 to 20 years. Natural longevity is much shorter. The average American robin lives 1.3 years, and the average black-capped chickadee lives less than 1 year. Mortality is high in the first year from predators and inclement weather.

Migration and Navigation

Over twenty centuries ago, Aristotle told of birds migrating to escape the winter cold and summer heat. He had the mistaken impression that some birds disappear during winter because they hibernate, and that others transmutate to another species. It is now known that birds migrate long distances. Modern zoologists study the timing of migration, the stimuli for migration, and the physiological changes that occur during migration, as well as migration routes and how birds navigate over huge expanses of land or water.

Migration (as used here) refers to periodic round trips between breeding and nonbreeding areas. Most migrations are annual, with nesting areas in northern regions and wintering grounds in the south. (Migration is more pronounced for species found in the northern hemisphere because about 70% of the earth's land is in the northern hemisphere.) Migrations occasionally involve east/west movements or altitude changes. Migration allows birds to avoid climatic extremes, and to secure adequate food, shelter, and space throughout the year.

Birds migrate in response to species-specific physiological conditions (*Zugdisposition*). Innate (genetic) clocks and environmental factors influence preparation for migration. *Photoperiod* is often cited as an important migratory cue for many birds, particularly for birds in temperate zones. Changing photoperiod initiates seasonal changes in gonadal development that often serve as migratory stimuli. Increasing day length in the spring promotes gonadal development, and decreasing day length in the fall promotes regression of gonads. In many birds, changing photoperiod also appears to promote fat deposition, which acts as an

(a)

(b)

Figure 30.16 Altricial and precocial chicks. (a) Robin feeding nestlings. Robins have altricial chicks that are helpless at hatching. (b) Killdeer have precocial chicks that are down covered and can move about.

energy reserve for migration. Many migratory birds show "migratory restlessness" (*Zugunruhe*), or an impulse for migration if caged under conditions of increasing day length. The anterior lobe of the pituitary gland and the pineal body have been implicated in mediating photoperiod responses.

The mechanics of migration are species specific. Some long-distance migrants may store fat equal to 50% of their body weight and make nonstop journeys. Other species that take a more leisurely approach to migration begin their journeys early and stop frequently to feed and rest. In clear weather, many birds fly at altitudes greater than 1,000 m, which reduces the likelihood of hitting tall obstacles. Many birds have very specific migration routes (*see figure 30.1*).

Navigation Homing pigeons have served for many years as a pigeon postal service. As long ago as the ancient Egyptian times, and as recently as World War II, pigeons were used to return messages from military ventures.

[6] Two forms of navigation are used by birds. *Route-based navigation* involves keeping track of landmarks (visual or auditory) on an outward journey so that those landmarks can be used in a reverse sequence on the return trip. *Location-based navigation* involves establishing the direction of the destination from information available at the journey's site of origin. It involves the use of sun compasses, other celestial cues, and/or the earth's magnetic field.

Birds' lenses are transparent to ultraviolet light, and their photoreceptors respond to it, allowing them to orient using the sun, even on cloudy days. This orientation cue is referred to as a *sun compass*. Because the sun moves through the sky between sunrise and sunset, birds use internal clocks to perceive that the sun rises in the east, is approximately overhead at noon, and sets in the west. The biological clocks of migratory birds can be altered. For example, birds ready for northward migration can be held in a laboratory in which the "laboratory sunrise" occurs later than the natural sunrise. When released to natural light conditions, they fly in a direction they perceive to be north, but which is really northwest. Night migrators can also orient using the sun by flying in the proper direction from the sunset.

Celestial cues other than the sun can be used to navigate. Humans recognize that in the northern hemisphere, the north star lines up with the axis of rotation of the earth. The angle between the north star and the horizon decreases as one moves toward the equator. Birds may use a similar information to determine latitude. Experimental rotations of the night sky in a planetarium have altered the orientation of birds in test cages.

There has long been speculation that birds employ *magnetic compasses* to detect the earth's magnetic field, and thus determine direction. Typically, they have been met with skepticism, but direct evidence of their existence has been uncovered. Magnets strapped to the heads of pigeons severely disorient them. European robins and a night migrator, the garden warbler, orient using the earth's magnetic field. However, no discrete magnetic receptors in birds and other animals have been found. Early reports of finding a magnetic iron, magnetite, in the head and necks of pigeons did not lead to a greater understanding of magnetic compasses. Further experiments failed to demonstrate magnetic properties in these regions. Magnetic iron has been found in a variety of tissues in a variety of animals. None is clearly associated with a magnetic sense, although the pineal body of pigeons has been implicated in the use of a sun compass and in responses to magnetic fields.

There is redundancy in bird navigational mechanisms, which suggests that under different circumstances, different sources of information are probably used.

Stop and Ask Yourself

12. Why is monogamy advantageous for most birds?
13. Why are birds with precocial young more likely to be polygynous than birds with altrical young?
14. What environmental cue is important in preparing birds for migration?
15. What is route-based migration?

Summary

1. Birds are members of the archosaur lineage. Fossils of an ancient bird, *Archeopteryx,* show reptilian affinities and give clues into the origin of flight.
2. Feathers evolved from reptilian scales and function in flight, insulation, sex recognition, and waterproofing. Feathers are maintained and periodically molted.
3. The bird skeleton is light and made more rigid by fusion of bones. The neck and beak are used as a fifth appendage.
4. Bird wings form airfoils that provide lift. Propulsive force is generated by tilting the wing during flapping. Gliding, flapping, soaring, and hovering flight are used by different birds or by the same bird at different times.
5. Birds feed on a variety of foods as reflected in the structure of the beak and other parts of the digestive tract.
6. The heart of birds consists of two atria and two ventricles. A very rapid heart rate, and rapid blood flow, support the high metabolic rate of birds.
7. The respiratory system of birds provides one-way, nearly constant, air movement across respiratory surfaces.

8. Birds are able to maintain high body temperatures endothermically because of insulating fat deposits and feathers.

9. Cerebral hemispheres of the bird are enlarged by the development of the corpus striatum. Vision is an important avian sense.

10. Birds are oviparous. Reproductive activities include the establishment and defense of territories, courtship, and nest building.

11. Eggs are usually incubated by either or both parents, and one or both parents feed the young. Altricial chicks are helpless at hatching, and precocial chicks are alert and lively shortly after hatching.

12. Migration allows some birds to avoid climatic extremes, and to secure adequate food, shelter, and space throughout the year. Photoperiod is the most important migratory cue for birds.

13. Birds use both route-based navigation and location-based navigation.

■ Key Terms

airfoil (p. 475)
altricial (p. 484)
molting (p. 473)
monogamous (p. 482)

polyandrous (p. 482)
polygynous (p. 482)
precocial (p. 484)

■ Critical Thinking Questions

1. Birds are sometimes called "glorified reptiles." Discuss why this description is appropriate?

2. What adaptations of birds promote endothermy, and flight? Why is endothermy important for birds?

3. Birds are, without exception, oviparous. Why do you think that is true?

4. What are the advantages that offset the great energy expenditure required by migration?

■ Suggested Readings

Books

Aidley, D. J. (ed.). 1980. *Animal Migration*. Cambridge: Cambridge University Press.

Baker, R. (ed.). 1981. *The Mystery of Migration*. Minneapolis: The Viking Press.

Baker, R. 1984. *Bird Navigation: The Solution of a Mystery?* New York: Holmes & Meier Publishers, Inc.

Faaborg, J. 1988. *Ornithology: An Ecological Approach*. Englewood Cliffs: Prentice-Hall.

Feduccia, A. 1980. *The Age of Birds*. Cambridge: Harvard University Press.

Gill, F. B. 1990. *Ornithology*. New York: W.H. Freeman and Company.

Peterson, R. T. 1980. *Field Guide to the Birds,* 2nd ed. Boston: Houghton Mifflin Co.

Ruppell, G. 1975. *Bird Flight*. New York: Van Nostrand Reinhold Co.

Terres, J. K. 1980. *The Audobon Society Encyclopedia of North American Birds*. New York: Alfred A. Knopf, Inc.

Welty, J. C. 1988. *The Life of Birds,* 4th ed. Philadelphia: W.B. Saunders Co.

Articles

Calder, W. A., III. The Kiwi. *Scientific American* July, 1978.

Cracraft, J. 1988. Early evolution of birds. *Nature* 335: 630–632.

Gould, J. L. 1980. The case for magnetic sensitivity in birds and bees (such as it is). *American Scientist* 68: 256–267.

Gould, J. L., and Marler, P. Learning by instinct. *Scientific American* January, 1987.

Harvey, P. H., and Partridge, L. 1988. Of cuckoo clocks and cowbirds. *Nature* 335: 630–632.

Keeton, W. T. The mystery of pigeon homing. *Scientific American* December, 1974.

Ostrom, J. H. 1979. Bird flight: how did it happen? *American Scientist* 67: 46–56.

Steinhart, P. 1989. Standing room only. *National Wildlife* 27:46–51.

Sutherland, W. J. 1988. The heritability of migration. *Nature* 334:471–472.

Wellnhofer, P. *Archaeopteryx*. *Scientific American* May, 1990.

Williams, T. C., and Williams, J. M. An oceanic mass migration of land birds. *Scientific American* October, 1978.

Mammals: Endothermy, Hair, and Viviparity

31

Concepts

1. Mammalian characteristics evolved gradually over a 200-million-year period in the synapsid lineage.
2. Two subclasses of mammals evolved during the Mezozoic era. Modern mammals include monotremes, marsupial mammals, and placental mammals.
3. The skin of mammals is thick and protective and has an insulating covering of hair.
4. Adaptations of teeth and the digestive tract allow mammals to exploit a variety of food resources.
5. Efficient systems for circulation and gas exchange support the high rates of metabolism associated with endothermy.
6. The brain of mammals has an expanded cerebral cortex that processes information from various sensory structures.
7. Metanephric kidneys permit mammals to excrete urea without excessive water loss.
8. Complex behavior patterns enhance survival.
9. Most mammals are viviparous and have reproductive cycles that help ensure internal fertilization and successful development.

Have You Ever Wondered:

[1] what group of ancient reptiles is most closely related to mammals?
[2] what group of mammals is oviparous?
[3] why mammal hair stands on end when a mammal is frightened?
[4] how teeth of mammals are specialized for different feeding habits?
[5] why the four-chambered hearts of mammals and birds are an example of convergent evolution?
[6] why salt glands are not found in mammals?
[7] why a domestic cat rubs its face on furniture around the house?

These and other useful questions will be answered in this chapter.

This chapter contains underlined evolutionary concepts.

Evolutionary Perspectives

The beginning of the Tertiary period, about 70 million years ago, was the beginning of the "age of mammals." It coincided with the extinction of many reptilian lineages, which led to the adaptive radiation of mammalian lineages. To trace the roots of the mammals, however, we must go back to the Carboniferous period, when the diapsid and synapsid lineages diverged (*see figure 29.2*).

Mammalian characteristics evolved gradually over a period of 200 million years. The early synapsids were the *pelycosaurs*. Some were herbivores; others showed skeletal adaptations that reflect increased effectiveness as predators (figure 31.1a). The anterior teeth of their upper jaw were large and were separated from the posterior teeth by a gap that accommodated the enlarged anterior teeth of the lower

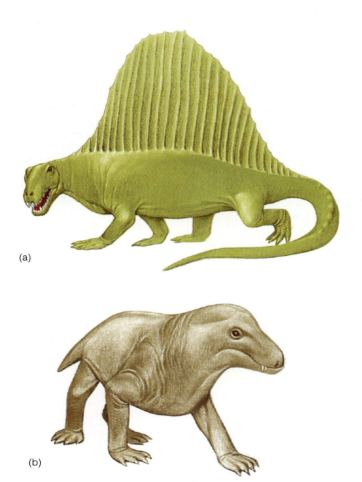

(a)

(b)

Figure 31.1 Members of the subclass Synapsida. (a) *Dimetrodon* was a 3-meter-long pelycosaur. It probably fed on other reptiles and amphibians. The large sail may have served as a recognition signal and a thermoregulatory device. (b) *Cynognathus* was a mammallike reptile that probably foraged for small animals, much like a badger does today. This badger-sized animal was a member of the order Therapsida, the stock from which mammals arose during mid-Triassic times.

jaw when the jaw was closed. The palate was arched, which gave additional strength to the upper jaw and allowed air to pass over prey held in the mouth. Their legs were longer and slimmer than those of earlier amniotes.

By the middle of the Permiman period, other successful mammallike reptiles had arisen from the pelycosaurs. They [1] were a diverse group known as the *therapsids*. Some were predators and others were herbivores. In the predatory therapsids, teeth were concentrated at the front of the mouth and enlarged for holding and tearing prey. The posterior teeth were reduced in size and number. The jaws of some therapsids were elongate and generated a large biting force when they snapped closed. The teeth of the herbivorous therapsids were also mammallike. Some had a large space, called the diastema, separating the anterior and posterior teeth. The posterior teeth had ridges (cusps) and cutting edges that were probably used to shred plant material. Unlike other reptiles, the hind limbs of therapsids were held directly beneath the body and moved parallel to the long axis of the body. Changes in the size and shape of the ribs suggest the separation of the trunk into thoracic and abdominal regions and a breathing mechanism similar to that of mammals. The last therapsids were a group called the *cycodonts* (figure 31.1b). Some of these were as large as a big dog, but most were small and little different from the earliest mammals.

The first mammals were small (less than 10 cm long) with delicate skeletons. Most of our knowledge of early mammalian phylogeny comes from the study of their fossilized teeth and skull fragments. These studies suggest that the mammals of the Jurassic and Cretaceous periods were mostly predators that fed on other vertebrates and arthropods. A few were herbivores, and others combined predatory and herbivorous feeding habits. Changes in the structure of the middle ear and the regions of the brain devoted to hearing and olfaction indicate that these senses were important during the early evolution of mammals.

Although it is somewhat speculative, some zoologists think that the small size and well-developed olfactory and auditory abilities suggest that early mammals were nocturnal. This habit may have allowed them to avoid competion with the much larger dinosaurs living at the same time and the smaller diurnal (day active [L. *diurnalis,* daily]) diapsid and synapsid reptiles. Again it is speculative, but nocturnal habits could have led to endothermy. Endothermy would have allowed small mammals to maintain body temperatures above that of their surroundings after the sun had set and the air temperature began to fall.

Diversity of Mammals

Modern members of the class Mammalia (ma-ma′le-ah) (L. *mamma,* breast) are characterized by hair, mammary glands, three middle-ear ossicles, and other characteristics listed in table 31.1. It is currently not possible to determine the extent to which all of these characteristics were developed in the earliest mammals. Although there is some disagreement among zoologists regarding subclass-level classification, most zoologists consider mid-Cretaceous (about 130 million years ago) mammals to have diverged into two subclasses (figure 31.2). Certain modern mammals, the duckbilled platypus and the echidna, are traditionally classified in the subclass Prototheria (Gr. *protos,* first + *therion,* wild beast). They are also known as monotremes (Gr. *monos,* one + *trema,* opening). This name refers to the fact that, unlike other mammals, they possess a cloaca. Monotremes are also distinguished from all other [2] mammals by the fact that they are oviparous (figure 31.3a,b). There are six species of monotremes found in Australia and New Guinea.

Members of the subclass Theria had diverged into two groups by the late Cretaceous period. The infraclass Metatheria (Gr. *meta,* after) contains the marsupial mammals. They are viviparous, but have very short gestation periods. A protective pouch, called the *marsupium,* covers the mammary glands of the female. The young crawl into the marsupium after birth, where they feed and complete development. There are about 250 species of marsupials that

Table 31.1 Classification of Mammals
Class Mammalia (ma-ma′le-ah) Mammary glands; hair; diaphragm; three middle-ear ossicles; heterodont dentition; sweat, sebaceous, and scent glands; four-chambered heart; large cerebral cortex.
Subclass Prototheria (pro′to-ther′e-ah) Members of this subclass are distinguished by very technical characteristics of the skull. The subclass contains only one order with extant species. Monotremes.
Subclass Theria (ther′e-ah) Members of this subclass are distinguished by very technical characteristics of the skull.
Infaclass Metatheria (met′ah-ther′e-ah) Viviparous, primitive placenta, young are born very early and often are carried in a marsupial pouch on the belly of the female. Marsupials.
Infaclass Eutheria (u-ther′e-ah) Complex placenta. Young develop to advanced stage prior to birth. Placentals.

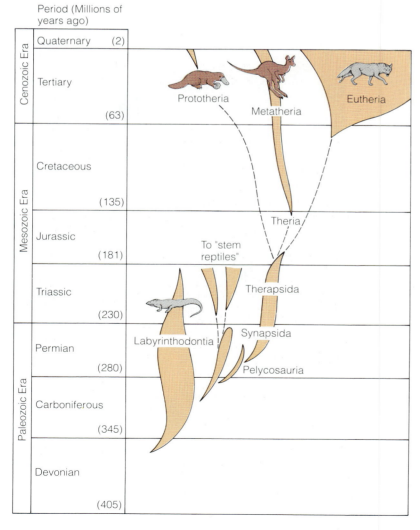

Period (Millions of years ago)

Figure 31.2 Mammalian phylogeny. Mammals evolved from ancient therapsid reptiles. The monotremes (Prototheria) diverged from early therian stocks. Marsupials (Metatheria) and placentals (Eutheria) have a more recent common ancestry. All groups of eutherian mammals are believed to have evolved from primitive insectivore stocks. The relative number of species in each taxonomic group is indicated by the width of a group's branch.

live in the Australian region and the Americas (figure 31.3c,d).

The other therian infraclass, Eutheria, (Gr. *eu,* true) contains the placental mammals. They are usually born at an advanced stage of development, having been nourished within the uterus. Exchanges between maternal and fetal circulatory systems occur by diffusion across an organ called the *placenta*, which is composed of both maternal and fetal tissue. There are about 3,800 species of eutherians that are classified into 17 orders (figure 31.4; colorplate 13; *see figures 31.13–31.15*).

Figure 31.3 Representatives of the mammalian subclasses. The subclass Prototheria. (a) A duckbilled platypus. (b) An echidna or spiny anteater. The subclass Metatheria. (c) The koala feeds on *Eucalyptus* leaves in Australia. (d) A gray kangaroo with its joey in its pouch.

Figure 31.4 Order Edentata. (a) A giant anteater (*Myrmecophaga tridactyla*). (b) An armadillo (*Dasypus novemcinctus*).

Evolutionary Pressures

Mammals are naturally distributed on all continents except Antarctica and in all oceans. The many adaptations that have accompanied their adaptive radiation are discussed in this section.

External Structure and Locomotion

The skin of a mammal, like that of other vertebrates, consists of epidermal and dermal layers. It protects from mechanical injury, invasion by microorganisms, and the sun's ultraviolet light. Skin is also important in temperature regulation, sensory perception, excretion, and water regulation (box 31.1; *see figure 32.1*).

Hair is a keratinized derivative of the epidermis of the skin and is uniquely mammalian. It is seated in an invagination of the epidermis, called a *hair follicle*. A coat of hair, called *pelage,* usually consists of two kinds of hair. Long *guard hairs* protect a dense coat of smaller, insulating *underhairs*.

Because hair is composed largely of dead cells, it must be periodically molted. In some mammals (e.g., humans), molting occurs gradually and may not be noticed. In others, hair loss occurs rapidly and may result in altered pelage characteristics. In the fall, many mammals acquire a thick coat of insulating underhair, and the pelage color may change. For example, the Arctic fox takes on a white or cream color with its autumn molt, which helps conceal the fox in a snowy environment. With its spring molt, the Arctic fox acquires a gray and yellow pelage (colorplate 13g).

Hair is also important for the sense of touch. Mechanical displacement of hair stimulates nerve cells associated with the hair root. Guard hairs may sometimes be modified into thick-shafted hairs called *vibrissae*. Vibrissae occur around the legs, nose, mouth, and eyes of many mammals. Their roots are richly innervated and very sensitive to displacement.

Air spaces in the hair shaft and air trapped between hair and the skin provide an effective insulating layer. A band of smooth muscle, called the *arrector pili muscle*, runs between the hair follicle and the lower epidermis. When the muscle contracts, the hairs stand upright, increasing the amount of air trapped in the pelage and improving its insulating properties. Arrector-pili muscles are under the control of the autonomic nervous system, which also controls a mammal's "fight-or-flight" response. In threatening [3] situations, the hair (especially on the neck and tail) stands on end and may give the perception of increased size and strength.

Hair color depends on the amount of pigment (melanin) deposited in it and the quantity of air in the hair shaft. The pelage of most mammals is dark above and lighter underneath. This pattern makes them less conspicuous under most conditions. Some mammals advertise their defenses using aposematic (warning) coloration. The contrasting markings of a skunk are a familiar example.

Pelage is reduced in large mammals from hot climates (e.g., elephants and hippopotamuses) and in some aquatic mammals (e.g., whales) that often have fatty insulation.

Claws are present in all amniote classes. They are used for locomotion and offensive and defensive behavior. Claws are formed from accumulations of keratin that cover the terminal phalanx (bone) of the digits. In some mammals, they are specialized to form nails or hooves (figure 31.5).

Glands develop from the epidermis of the skin. *Sebaceous glands* are associated with hair follicles, and their oily secretion lubricates and waterproofs the skin and hair. Most mammals also possess *sudoriferous (sweat) glands*. Small sudoriferous glands (*eccrine glands*) release watery secretions used in evaporative cooling. Larger sudoriferous glands (*apocrine glands*) secrete a mixture of salt, urea, and water, which are converted to odorous products by microorganisms on the skin.

Scent or musk glands are located around the face, feet, or anus of many mammals. These glands secrete pheromones, which may be involved with defense, species and sex recognition, and territorial behavior.

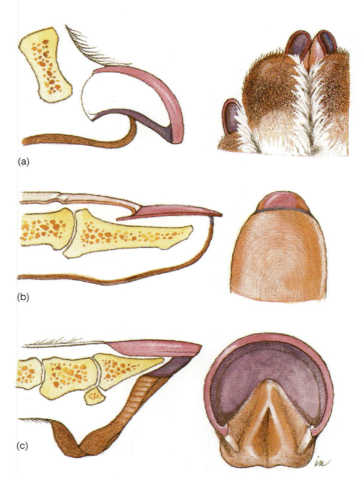

(a)

(b)

(c)

Figure 31.5 The structure of claws, nails, and hooves. (a) Claws. (b) Nails are flat, broad claws found on the hands and feet of primates and are an adaptation for arboreal habits, where grasping is essential. (c) Hooves are characteristic of ungulate mammals. The number of toes is reduced, and the animals walk or run on the tips of the remaining digits.

Box 31.1

Horns and Antlers

Horns were surely a familiar sight in prehistoric landscapes, 100 million years before they became common in mammals. *Triceratops* had three horns, one nasal horn and two above the eyes. It also had a horny shield along the posterior margin of the head. *Styracosaurus* had a 0.7-meter nasal spike. These early reptilian horns probably provided a very effective defense against fierce prehistoric carnivores.

Mammalian horns are a carryover from their reptilian heritage. They are most common in a group of hoofed mammals, the artiodactyls (e.g., cattle, sheep, and goats). A horn is a spike of bone that arises from the frontal bone of the skull and is covered with the protein keratin (box figure 31.1a,b). This bony spike, the "os cornu," grows slowly from youth to adulthood, and its marrow core is highly vascularized. Filaments of keratin arise from folliclelike structures in the skin. Keratin filaments are cemented together and completely cover the os cornu. There is no blood supply to the outer horn layers.

Horns are defensive structures. They exist in symmetrical pairs and are present in both sexes. As any farmer or rancher knows, horns are not regenerated if they are cut off. Removal of cattle horns (*polling*) is a common practice because it makes cattle much safer to handle and prevents them from harming each other. Horns are usually not shed. One exception is the pronghorn antelope (*Antilocapra americana*). Every year, a new horn grows on the os cornu beneath the old horn, and the latter is eventually pushed off.

Another kind of head ornamentation, the antler, is common in deer, elk, moose, and caribou. Antlers are highly branched structures made of bone, but are not covered by keratin. Unlike horns, they are present only in males and are shed and reformed every year (box figure 31.1c). Caribou are an exception because antlers are present in both sexes. Antlers are more recent than horns; the earliest records of antlered animals are from the Miocene epoch, and by the Pleistocene epoch, they had become common.

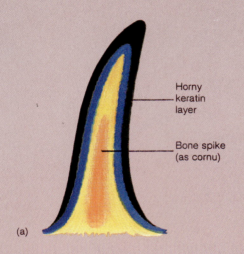

(a)
Horny keratin layer

Bone spike (as cornu)

(b)

(c)

Box Figure 31.1 Horns and antlers. (a) The structure of a horn. (b) The horn of a male bighorn sheep. (c) Development of deer antlers.

Antler development is regulated by seasonal changes in the level of the male hormone testosterone. Antlers of male elk begin to form in April as skin-covered buds from the frontal bone. The primordial cells that initiate antler growth are left behind from the previous year when the antlers were lost. Antlers begin to branch after only 2 weeks. By May, they are well formed, and by August, they are mature. Each year, antlers become more complexly branched. Throughout the spring and summer, they are covered with delicate, vascular tissue called velvet. In August, the bone at the base of the antler becomes progressively more dense and cuts off blood flow to the center of the antler. Later, blood flow to the velvet is cut off, and the velvet begins to dry. It is shed in strips as the antlers are rubbed against the ground or tree branches. Breeding activities commence after the velvet is shed, and the antlers are used in jousting matches as rival males compete for groups of females. (Rarely do these jousting matches lead to severe injury.) Selection by females of males with large antlers may explain why they can get so large in some species. (Although now extinct, the Giant Stag, *Cervis megaceros*, had antlers with a 3-meter spread and a mass of 70 kg.) Later in the fall, or in early winter, the base of the antler is weakened as bone is reabsorbed at the pedicels of the frontal bone. Antlers are painlessly cast off when an antler strikes a tree branch or other object.

Other hornlike structures are present in some mammals. Rhinos are the only perissodactyls (e.g., horses, rhinos, and tapirs) to have hornlike structures. Their "horns" consist of filamentous secretions of keratin cemented together and mounted to the skin of the head. There is no bony core, and thus they are not true horns. Rhino "horns" are prized in the Orient for their presumed aphrodisiac and medicinal properties and as dagger handles in certain mideastern cultures. These demands have led to very serious overhunting of rhinos; in many regions they are almost extinct.

The horns of giraffes are skin-covered bony knobs. Zoologists do not understand their function.

Mammary glands are functional in female mammals and are present, but nonfunctional, in males. The milk that they secrete contains water, carbohydrates (especially the sugar lactose), fat, protein, minerals, and antibodies. Mammary glands are probably derived evolutionarily from apocrine glands and usually contain substantial fatty deposits.

Monotremes have mammary glands that lack nipples. The glands discharge milk into depressions on the belly, where it is lapped up by the young. In therian mammals, mammary glands open via nipples or teats, and the young suckle for their nourishment (figure 31.6).

The Skull and Teeth The skulls of mammals show important modifications of the reptilian design. One feature used by zoologists to distinguish reptilian from mammalian skulls is the method of jaw articulation. In reptiles, the jaw articulates at two small bones at the rear of the jaw. In mammals, these bones have moved into the middle ear, and along with the stapes, form the middle-ear ossicles. Jaw articulation in mammals is by a single bone of the lower jaw.

A *secondary palate* evolved twice in vertebrates—in the diapsid lineage and in the synapsid lineage. In some therapsids, small, shelflike extensions of bone (the hard palate) partially separated the nasal and oral passageways. In mammals, the secondary palate is extended posteriorly by a fold of skin, called the soft palate, which almost completely separates the nasal passages from the mouth cavity. Unlike other vertebrates that swallow food whole or in small pieces, some mammals chew their food. The more extensive secondary palate allows mammals to breathe while chewing. Breathing needs to stop only briefly during swallowing (figure 31.7).

The structure and arrangement of teeth are important indicators of mammalian life-styles. In reptiles, the teeth

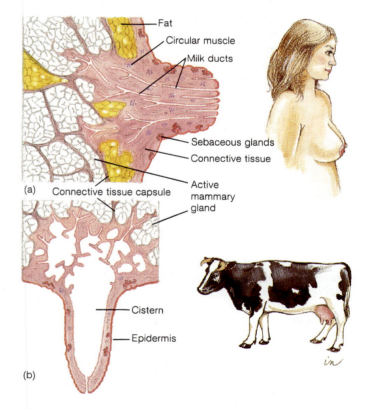

Figure 31.6 Mammary glands are specialized to secrete milk following the birth of young. (a) Many ducts lead from the glands to a nipple. Parts of the duct system are enlarged to store milk. Suckling by an infant initiates a hormonal response that causes the mammary glands to release milk. (b) Some mammals (e.g., cattle) have teats that are formed by the extension of a collar of skin around the opening of mammary ducts. Milk collects in a large cistern prior to its release. The number of nipples or teats varies with the number of young produced.

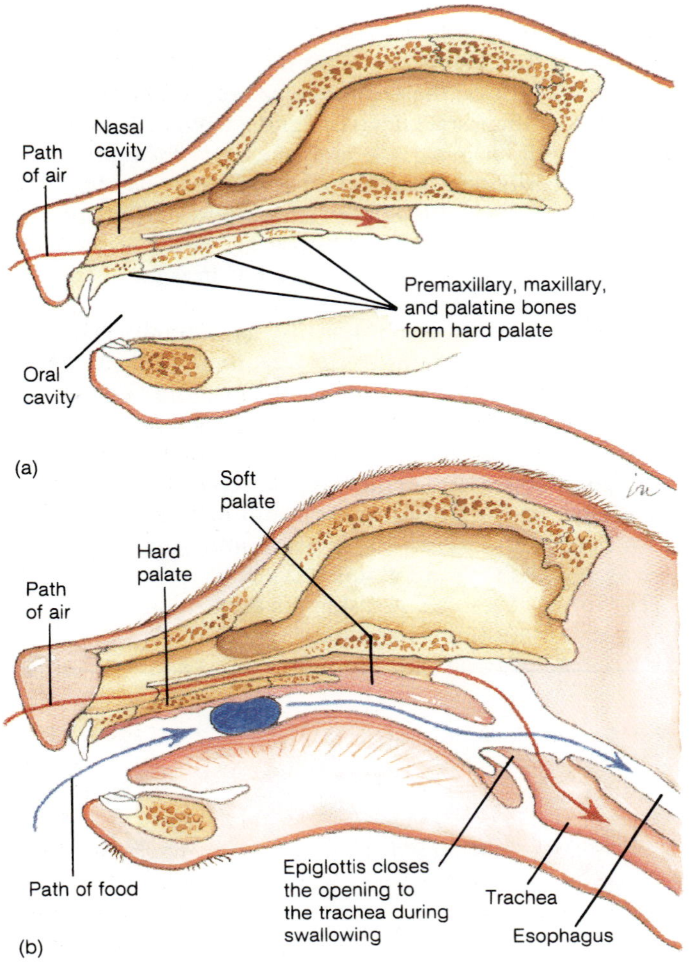

Nasal cavity

Path of air

Premaxillary, maxillary, and palatine bones form hard palate

Oral cavity

(a)

Soft palate

Hard palate

Path of air

Path of food

Epiglottis closes the opening to the trachea during swallowing

Trachea

Esophagus

(b)

Figure 31.7 The secondary palate. (a) The secondary palate of a mammal provides a nearly complete separation between nasal and oral cavities. (b) Breathing stops only momentarily during swallowing.

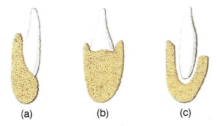

(a)   (b)   (c)

Figure 31.8 Attachment of teeth. (a,b) Reptilian teeth attach along the top or the inside of the jaw. (c) Mammalian teeth attach in sockets.

used by zoologists to characterize taxa. It is an expression of the number of teeth of each kind in the upper and lower jaws. The teeth of the upper jaw are listed above those of the lower jaw and they are indicated in following order: incisors, canine, premolars, and molars. The human dental formula is:

$$\frac{2 \cdot 1 \cdot 2 \cdot 3}{2 \cdot 1 \cdot 2 \cdot 3}$$

Mammalian teeth (dentition) may be specialized for particular diets. In some mammals, the dentition is reduced, sometimes to the point of having no teeth. Armadillos and the giant anteater (order Edentata) feed on termites and ants, and their teeth are reduced.

Some mammals (e.g., humans, order Primates; and hogs, order Artiodactyla) are omnivorous; they feed on a variety of plant and animal materials. They have anterior teeth with sharp ripping and piercing surfaces, and posterior teeth with flattened grinding surfaces for rupturing plant cell walls (figure 31.9a). [4]

Mammals that eat plant material often have flat, grinding posterior teeth and incisors, and sometimes canines, that are modified for nipping plant matter (e.g., horses, order Perissodactyla; deer, order Artiodactyla) or gnawing (e.g., rabbits, order Lagamorpha; beavers, order Rodentia) (figure 31.9b,c). In rodents, the incisors grow throughout life. Although most mammals have enamel covering the entire tooth, rodents have enamel only on the front surfaces of their incisors. They are kept sharp by slower wear in front than in back. The anterior food-procuring teeth are separated from the posterior grinding teeth by a gap, called the *diastema*. The diastema results from elongation of the snout that allows the anterior teeth to reach close to the ground or into narrow openings to procure food. The posterior teeth have a high, exposed surface (crown) and continuous growth, which allows these teeth to withstand years of grinding tough vegetation.

Canines and incisors of predatory mammals are used for catching, killing, and tearing prey. In members of the order Carnivora (e.g., coyotes, dogs, and cats), the fourth upper premolars and first lower molars form a scissorlike shearing surface, called the *carnassial apparatus,* that is used for cutting flesh from prey (figure 31.9d).

are uniformly conical. This condition is referred to as **homodont**. In mammals, the teeth are often specialized for different functions, a condition called **heterodont**. Reptilian teeth are attached along the top or inside of the jaw, whereas in mammals, the teeth are set into sockets of the jaw (figure 31.8). Most mammals have two sets of teeth during their life. The first teeth emerge before or shortly after birth and are called *deciduous* or *milk teeth*. These teeth are lost and replaced by *permanent teeth*.

There are up to four kinds of teeth in adult mammals. *Incisors* are the most anterior teeth in the jaw. They are usually chisellike and used for gnawing or nipping. *Canines* are often long, stout, and conical, and are usually used for catching, killing, and tearing prey. Canines and incisors have single roots. *Premolars* are positioned next to canines, have one or two roots, and truncated surfaces for chewing. *Molars* have broad chewing surfaces and two (upper molars) or three (lower molars) roots.

Mammalian species have characteristic numbers of each kind of adult tooth. A **dental formula** is an important tool

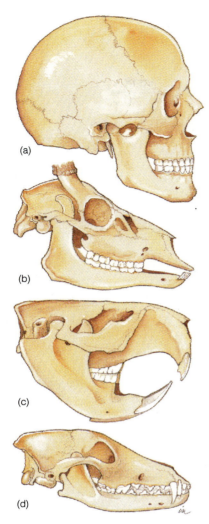

Figure 31.9 Feeding specializations of teeth. (a) An omnivore (human). (b) An herbivore, the male fallow deer (*Dama dama*). (c) A rodent, the beaver (*Castor canadensis*). (d) A carnivore, the coyote (*Canis latrans*).

The Vertebral Column and Appendicular Skeleton The vertebral column of mammals is divided into five regions. As with reptiles and birds, the first two cervical vertebrae are the atlas and axis. These are usually followed by five other cervical vertebrae. Even the giraffe and the whale have seven neck vertebrae, which are greatly elongate or compressed, respectively.

The trunk is divided into thoracic and lumbar regions, as is the case for birds. In mammals, the division is correlated with their method of breathing. The thoracic region contains the ribs, which connect to the sternum via costal cartilage and protect the heart and lungs. The articulation between the thoracic vertebrae provides the flexibility needed in turning, climbing, and lying on the side to suckle young. Lumbar vertebrae have interlocking processes that give support, but little freedom of movement.

The appendicular skeleton of mammals is rotated under the body so that the appendages are usually directly be-

neath the body. Joints usually limit the movement of appendages to a single anterior/posterior plane, causing the tips of the appendages to move in long arcs. The bones of the pelvic girdle are highly fused in the adult, a condition that is advantageous for locomotion, but presents problems during the birth of offspring. In a pregnant female, the ventral joint between the halves of the pelvis—the pubic symphysis—loosens before birth, allowing the pelvis to spread apart during birth.

Muscles Because the appendages are directly beneath the body of most mammals, the weight of the body is borne by the skeleton. Muscle mass is concentrated in the upper appendages and girdles. Many running mammals (e.g., deer, order Artiodactyla) have little muscle in their lower leg that would slow leg movement. Instead, tendons run from muscles high in the leg to cause movement at the lower joints.

Nutrition and the Digestive System

The digestive tract of mammals is similar to that of other vertebrates. There are, however, many specializations for different feeding habits. Some specializations of teeth have already been described.

Mammals are often divided into four groups based on feeding habits. These categories do not necessarily reflect evolutionary relationships. They are reflections of ecological specializations that have evolved. For example, most members of the orders Perissodactyla, Artiodactyla, and Rodentia feed primarily on plant matter. Within an order there also may be a variety of feeding habits. Most members of the order Carnivora feed on animal flesh. Other members of the order, such as bears, feed on a variety of plant and animal products.

Carnivores feed on other animals. Carnivorous habits were probably common in early mammals and remain very common. Most carnivorous mammals do not chew their food, but ingest prey whole or cut or tear chunks of flesh from prey.

Insectivores are specialized for feeding on insects, other arthropods, or soft-bodied invertebrates, and are found in the orders Insectivora (e.g, moles and shrews), Chiroptera (bats), and Tubulidentata (anteaters). Anteaters lack teeth. They use powerful forelimbs to tear into an insect nest and a long tongue covered with sticky saliva to capture prey.

Herbivores feed mostly on vegetation, but their diet also includes invertebrates inadvertently ingested while feeding. *Grazers* (e.g., horses) feed on grasses; *browsers* (e.g., deer) feed on leaves and the tips of branches of shrubs and trees; and *gnawers* (e.g., beavers) feed on bark and other plant tissues.

Specializations in the digestive tract of most herbivores reflect the difficulty of digesting food rich in cellulose. Horses, rabbits, and many rodents have an enlarged **cecum** at the junction of large and small intestines. A cecum serves as a fermentation pouch where microorganisms aid in the digestion of cellulose. Sheep, cattle, and deer are called *ruminants* (L. *ruminare,* to chew the cud). Their stomachs are modified into four chambers. The first three chambers are storage and fermentation chambers and contain microorganisms that synthesize a cellulose-digesting enzyme (cellulase). Gases produced by fermentation are periodically belched, and some plant matter (cud) is regurgitated and rechewed. Other microorganisms convert nitrogenous compounds in the food into new proteins.

Omnivores consume a wide variety of plant and animal foods. Common omnivores include bears, foxes, skunks, pigs, raccoons, and primates.

Stop and Ask Yourself:

5. What are the functions of hair in mammals?
6. What are the four kinds of glands found in mammals?
7. What is the usefulness of a dental formula?
8. What specializations of teeth and the digestive system are present in some herbivorous mammals?

Circulation, Gas Exchange, and Temperature Regulation

The hearts of birds and mammals are superficially similar. Both are four-chambered pumps that keep blood in the systemic and pulmonary circuits separate and both evolved from the hearts of ancient reptiles. Their similarities, however, are a result of adaptations to active lifestyles. The [5] evolution of similar structures in different lineages is called convergent evolution. Evolution of the mammalian heart occurred in the synapsid reptile lineage, whereas the avian heart evolved in the archosaur portion of the diapsid lineage (figure 31.10).

One of the most important adaptations in the circulatory system of eutherian mammals concerns the distribution of respiratory gases and nutrients in the fetus (figure 31.11). Exchanges between maternal and fetal blood occur across the placenta. Although there is intimate association between maternal and fetal vessels, no actual mixing of blood occurs. Nutrients, gases, and wastes simply diffuse between fetal and maternal blood supplies.

Blood entering the right atrium of the fetus is returning from the placenta and is highly oxygenated. Because fetal lungs are not inflated, resistance to blood flow through the pulmonary arteries is high. Therefore, most of the blood entering the right atrium bypasses the right ventricle and passes instead into the left atrium through a valved opening between the atria (the *foramen ovale*). Some blood from the right atrium, however, does enter the right ventricle and the pulmonary artery. Because of the resistance at the uninflated lungs, most of this blood is shunted to the aorta through a vessel connecting the aorta and pulmonary artery (the *ductus arteriosus*). At birth, the placenta is lost, and the lungs are inflated. Resistance to blood flow through the lungs is reduced, and blood flow to them increases. Flow through the ductus arteriosus decreases, and the vessel is gradually reduced to a ligament. Blood flow back to the

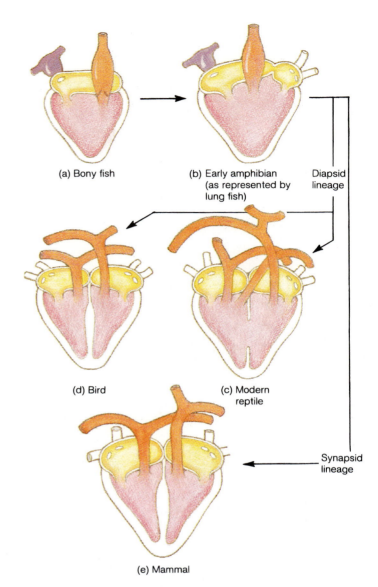

Figure 31.10 A possible sequence in the evolution of the vertebrate heart. (a) Diagrammatic representation of a bony fish heart. (b) In lungfish, partially divided atria and ventricles separate pulmonary and systemic circuits. This heart was probably similar to that in primitive amphibians and early amniotes. Two evolutionary lineages ultimately resulted in complete separation of pulmonary and systemic circuits. (c,d) One of these lineages led through diapsid reptiles to modern reptiles and birds. (e) The other led through the synapsid reptiles to the mammals.

left atrium from the lungs correspondingly increases, and the valve of foramen ovale closes and gradually fuses with the tissue separating the right and left atria.

Gas Exchange High metabolic rates are accompanied by adaptations for efficient gas exchange. The separate nasal and oral cavities and lengthening of the snout of most mammals provides an increased surface area for warming and moistening inspired air. Respiratory passageways are highly branched and provide large surface areas for gas exchange. Mammalian lungs resemble a highly vascular sponge, rather than the saclike structures of amphibians and a few reptiles.

The lungs, like those of reptiles, are inflated using a negative pressure mechanism. Unlike reptiles, however, mammals possess a muscular **diaphragm** that separates thoracic and abdominal cavities. Inspiration results from contraction of the diaphragm and expansion of the rib cage, both of which decrease the intrathoracic pressure and allow air to enter the lungs. Expiration is by elastic recoil of the lungs. Forceful exhalation can be accomplished by contraction of chest and abdominal muscles.

Temperature Regulation   Mammals are widely distributed over the earth, and some face harsh environmental temperatures. Nearly all face temperatures that require them to dissipate excess heat at some times and conserve and generate heat at other times.

Heat-producing mechanisms of mammals are divided into two categories. *Shivering thermogenesis* is muscular activity that results in the generation of large amounts of heat, but little movement. *Nonshivering thermogenesis* involves heat production by general cellular metabolism, and the metabolism of special fat deposits called *brown fat*. These heat-generating processes are discussed in more detail in chapter 38.

Heat production is effective in thermoregulation because mammals are insulated by their pelage and/or fat deposits. Fat deposits are also sources of energy to sustain high metabolic rates.

Mammals without a pelage can conserve heat by allowing the temperature of surface tissues to drop. A walrus in cold, arctic waters has a surface temperature near 0°C; however, a few centimeters below the skin surface, body temperatures are about 35°C. Upon emerging from the icy water, the skin warms quickly by increasing peripheral blood flow. Most tissues cannot tolerate such rapid and extreme temperature fluctuations. Further investigations are likely to reveal some very unique biochemical characteristics of these skin tissues.

Even though most of the body of arctic mammals is unusually well insulated, appendages often have thin coverings of fur as an adaptation to changing thermoregulatory needs. Even in winter, an active mammal sometimes produces more heat than is required to maintain body temperature. Patches of poorly insulated skin allow excess heat to be dissipated. During periods of inactivity or extreme cold, however, heat loss from these exposed areas must be reduced, often by assuming heat-conserving postures. Mammals sleeping in cold environments conserve heat by tucking poorly insulated appendages and their faces under well-insulated body parts.

Countercurrent heat-exchange systems may help regulate heat loss from exposed areas (figure 31.12). Arteries passing peripherally through the core of an appendage are surrounded by veins that carry blood back toward the body. When blood returns to the body through these veins, heat

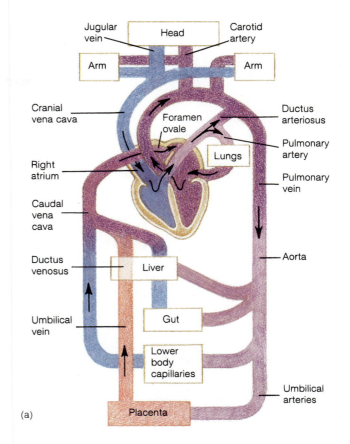

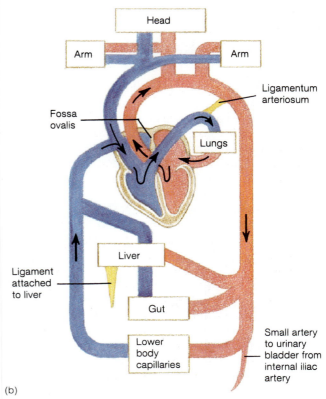

Figure 31.11   The circulatory patterns of (a) fetal, and (b) adult mammals. Highly oxygenated blood is shown in red and poorly oxygenated blood is shown in blue.

is transferred from arterial blood to venous blood and returned to the body rather than lost to the environment. When excess heat is produced, blood is shunted away from the countercurrent veins toward peripheral vessels, and excess heat is radiated to the environment.

Mammals have few problems getting rid of excess heat in cool, moist environments. Heat can be radiated into the air from vessels near the surface of the skin or lost by evaporative cooling from either sweat glands or respiratory surfaces during panting.

Hot, dry environments present far greater problems because evaporative cooling may upset water balances. The large ears of jackrabbits and elephants are used to radiate heat. Small mammals often avoid the heat by remaining in burrows during the day and foraging for food at night. Other mammals seek shade or watering holes for cooling.

Winter Sleep and Hibernation   Mammals react in various ways to environmental extremes. Caribou migrate to avoid extremes of temperature, and wildebeest migrate to avoid seasonal droughts. Other mammals retreat to burrows under the snow where they become less active, but are still relatively alert and easily aroused—a condition called *winter sleep*. For example, bears and raccoons retreat to dens in winter. Their body temperatures and metabolic rates decrease somewhat, but they do not necessarily remain inactive all winter.

Hibernation is a period of winter inactivity in which the hypothalamus of the brain slows the metabolic, heart, and respiratory rates. True hibernators include the monotremes (echidna and duckbill platypus) and many members of the Insectivora (e.g., moles and shrews), Rodentia (e.g., chipmunks and woodchucks), and Chiroptera (bats). In preparation for hibernation, mammals usually accumulate large quantities of body fat. After retreating to a burrow or a nest, the hypothalamus sets the body's thermostat to about 2°C. The respiratory rate of a hibernating ground squirrel falls from 100 to 200 breaths/minute to about 4 breaths/minute. The heart rate falls from 200 to 300 beats/minute to about 20 beats/minute. During hibernation, a mammal may lose one-third to one-half of its body weight. Arousal from hibernation occurs by metabolic heating, frequently using brown fat deposits, and it takes several hours to raise body temperature to near 37°C.

Nervous and Sensory Functions

The basic structure of the vertebrate nervous system is retained in mammals. The development of complex nervous and sensory functions goes hand-in-hand with active lifestyles and is most evident in the enlargement of the cerebral hemispheres and the cerebellum of mammals. The enlargement of the *cerebral cortex* (*neocortex*) is accompanied by most integrative functions being shifted to this region.

In mammals, the sense of touch is well developed. Receptors are associated with the bases of hair follicles and are stimulated when a hair is displaced.

Olfaction was apparently an important sense in early mammals, because fossil skull fragments show elongate snouts, which would have contained olfactory epithelium. Cranial casts of fossil skulls show enlarged olfactory regions. Olfaction is still an important sense for many mammals. Olfactory stimuli can be perceived over long distances during either the day or night and are used to locate food, recognize members of the same species, and avoid predators.

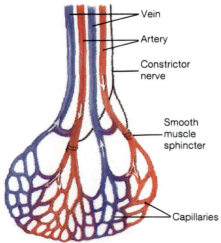

Vein
Artery
Constrictor nerve
Smooth muscle sphincter
Capillaries

Countercurrent vessels

Figure 31.12 Countercurrent heat exchangers conserve body heat in animals adapted to cold environments. Systems similar to the one depicted here are found in the legs of reindeer and in the flippers of dolphins. Venous blood retuning from an extremity is warmed by heat transferred from blood moving peripherally in arteries. During winter, the lower part of a reindeer's leg may be at 10°C while body temperature is about 40°C. Arrows indicate direction of blood flow.

Auditory senses were similarly important to early mammals. More recent adaptations include an ear flap (the *pinna*) and the auditory tube leading to the tympanum that directs sound to the middle ear. The sensory patch of the inner ear that contains receptors for sound is long and coiled and is called the *cochlea*. This design provides more surface area for receptor cells and allows mammals greater sensitivity to pitch and volume than is present in reptiles. Cranial casts of early mammals show well-developed auditory regions.

Vision is an important sense in many mammals, and the structure of the eye is similar to that described for other vertebrates. Accommodation occurs by changing the shape of the lens (*see figure 34.9*). Color vision is less well developed in mammals than in reptiles and birds. The fact that the retinas of most mammals are dominated by rods supports the hypothesis that early mammals were nocturnal. Primates, squirrels, and a few other mammals have well-developed color vision (box 31.2.).

> ### *Stop and Ask Yourself:*
>
> 9. In what way are the hearts of birds and mammals an example of convergent evolution?
> 10. What are two ways that mammals can generate metabolic heat?
> 11. How is excessive heat loss from poorly insulated legs of a reindeer prevented?
> 12. What is the difference between hibernation of a ground squirrel and the winter sleep of a bear?

Excretion and Osmoregulation

Mammals, like all amniotes, have a metanephric kidney. Unlike reptiles and birds, which excrete mainly uric acid, mammals excrete urea. Urea is less toxic than ammonia and does not require large quantities of water in its excretion. Unlike uric acid, however, urea is highly water soluble and cannot be excreted in a semisolid form; thus, some water is lost. Excretion in mammals is always a major route for water loss.

In the nephron of the kidney, fluids and small solutes are filtered from the blood through the walls of a group of capillarylike vessels, called the glomerulus. The remainder of the nephron consists of tubules that reabsorb water and essential solutes and secrete particular ions into the filtrate.

The primary adaptation of the mammalian nephron is a portion of the tubule system called the *loop of the nephron* (*see figure 38.10*). The transport processes in this loop and the remainder of the tubule system allow mammals to produce urine that is 2 to 22 times more concentrated than blood (e.g., beavers and Australian hopping mice, respectively). Nasal and orbital salt glands of birds and reptiles [6] are not needed in mammals because of the salt-concentrating abilities of the kidney.

Water loss varies greatly depending on activity, physiological state, and environmental temperature. Water is lost in urine, feces, evaporation from sweat glands and respiratory surfaces, and during nursing. Mammals in very dry environments have numerous behavioral and physiological mechanisms to reduce water loss. The kangaroo rat , named for its habit of hopping on large hind legs, is capable of extreme water conservation (figure 31.13). It is native to the southwestern deserts of the United States and survives without drinking water. Its feces are almost dry, and evaporative water loss is reduced by its nocturnal habits. Respiratory water loss is minimized by condensation as warm air in the respiratory passages encounters the cooler nasal

(a)

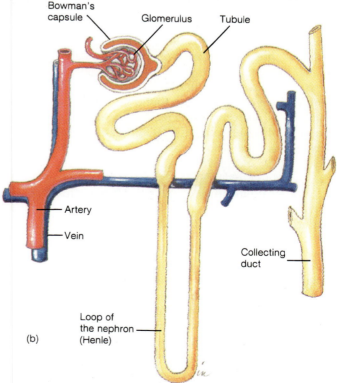

(b)

Figure 31.13 Order Rodentia. (a) The kangaroo rat (*Dipodomys ordii*). (b) The long loop of the nephron of this desert animal acts to conserve water, preventing dehydration.

Box 31.2

Mammalian Echolocation

Imagine a pool's water made so murky that a human is only able to see a few centimeters below the surface. Also imagine a clear Plexiglas sheet with a dolphin-sized opening in the middle, dividing the pool in half. At one end of the pool is an eager dolphin; at the other end is a trainer. The trainer throws a dead fish into the water, and on signal, the dolphin unhesitatingly finds its way through the murky water to the opening in the Plexiglas and then finds the fish at the other end of the pool.

Although the dolphin in the above account was trained to find the fish, it relied on a sense that it shares with a few other mammals. Toothed whales, bats, and some shrews use the return echoes of high-frequency sound pulses to locate objects in their environment. This mechanism is called *echolocation*.

Echolocation has been studied in bats more completely than in any other group of mammals. The Italian scientist Lassaro Spallanzani discovered in the late 1700s that blinded bats could navigate successfully at night, whereas bats with plugged ears could not. Spallanzani believed that echoes of the sounds made by beating wings were used in echolocation. In 1938, however, ultrasonic bat cries (inaudible to humans) were electronically recorded, and their function in echolocation was described.

Insect-eating bats navigate through their caves and the night sky and locate food by echolocation. During normal cruising flight, ultrasonic (100 KHz to 20KHz) "clicks" are emitted approximately every 50 milliseconds. As insect prey is detected, the number of clicks/second increases, the duration between clicks decreases, and the wavelength of the sound decreases, increasing directional precision and making small flying insects more easily detected. On final approach, the sound becomes buzzlike, and the bat scoops up the insect with its wings or in the webbing of its hind legs (box figure 31.2).

Modifications of the bat ear and brain allow bats to perceive faint echoes of their vocalizations and to pre-

Box Figure 31.2 Bat echolocation. A bat capturing a mealworm tossed into the air, just before the mealworm hits the water.

cisely determine direction and distance. Enlarged ear flaps funnel sounds toward particularly thin eardrums and very sensitive ear ossicles. The auditory regions of the bat brain are very large, and special neural pathways enhance a bat's ability to determine the direction of echoes.

Bats must distinguish echoes of their own cries from the cries themselves and from other noises. Leaflike folds of the nostrils direct sound emitted from the nostrils forward, rather than in all directions from the head, much like the megaphone of a cheerleader. The ears, therefore, receive little stimulation from direct vocalizations. Fat and blood sinuses surrounding the middle and inner ears reduce transmission of sound from the mouth and pharynx. Some bats temporarily turn off their hearing during sound emission by making the ear ossicles insensitive to sound waves and then turn on their hearing an instant later when the reflected sound is returning to the bat.

passages. Excretory water loss is minimized by a diet low in protein, which reduces the production of urea. The nearly dry seeds that the kangaroo rat eats are rich sources of carbohydrates and fats. Metabolic oxidation of carbohydrates produces water as a by-product.

Behavior

Mammals have complex behaviors that enhance survival. Visual cues are often used in communication. The bristled fur, arched back, and open mouth of a cat communicates

a clear message to curious dogs or other potential threats. A tail-wagging display of a dog has a similarly clear message. A wolf defeated in a fight with other wolves lies on its back and exposes its vulnerable throat. Similar displays may allow a male already recognized as being subordinate to another male to avoid conflict within a social group.

Pheromones are used to to recognize members of the same species, members of the opposite sex, and the reproductive state of a member of the opposite sex. Pheromones may also induce sexual behavior, help establish and recognize territories, and ward off predators. The young of

Figure 31.14 Order Carnivora. California sea lions on a beach during the breeding season. The adult males in the foreground are vocalizing and posturing.

Figure 31.15 Order Artiodactyla. Male elk (*Cerrus canadensis*) in the breeding season. These two large bull elk are involved in a shoving match that will help determine which male will establish its breeding territory. The dominant bull will attract, defend, and herd a group of females, called a harem.

many mammalian species recognize their parents, and parents recognize their young, by smell. Bull elk smell the rumps of females during the breeding season to recognize those in their brief receptive period. They also urinate on their own belly and underhair to advertise their reproductive status to females and other males. Male mammals urinate on objects in the environment to establish territories and to allow females to become accustomed to their odors. Rabbits and rodents spray urine on a member of the opposite sex to inform the second individual of the first's readiness to mate. Skunks use chemicals to ward off predators.

Auditory and tactile communication are also important in the lives of mammals. Herd animals are kept together and remain calm as long as the array of familiar sounds (e.g., bellowing, hooves walking over dry grasses and twigs, and rumblings from ruminating stomachs) are uninterrupted. Unfamiliar sounds may cause alarm and flight.

Vocalizations and tactile communication are important in primate social interactions. Tactile communication ranges from precopulatory "nosing" that occurs in many mammals to grooming. Grooming does much more than help maintain a healthy skin and pelage. It reinforces important social relationships within primate groups.

Territoriality Many mammals mark and defend certain areas from intrusion by other members of the same [7] species. (When cats rub their face and neck on us or on furniture in our homes, we like to think the cat is being affectionate. Cats, however, are really staking claim to their territory, using odors from facial scent glands.) Some territorial behavior attracts females to, and excludes other males from favorable sites for mating and rearing young.

Male California sea lions (*Zalophus californianus*) establish territories on shorelines where females come to give birth to young. For about 2 weeks, males engage in vocalizations, displays, and sometimes serious fighting to stake claim to favorable territories. Older, dominant bulls are usually most successful in establishing territories, and young bulls generally swim and feed just offshore. When they arrive at the beaches, females select a site for giving birth. Selection of the birth site also selects the bull that will father next year's offspring. Mating occurs approximately 2 weeks after the birth of the previous year's offspring. Development is arrested for the 3 months during which the recently born young do most of their nursing. This mechanism is called *embryonic diapause*. Thus, even though actual development takes about 9 months, the female carries the embryo and fetus for a period of 1 year (figures 31.14; 31.15).

Reproduction and Development

In no other group of animals has viviparity developed to the extent it has in mammals. It requires a large expenditure of energy on the part of the female during development and on the part of one or both parents caring for young after they are born. Viviparity is advantageous because females are not necessarily tied to a single nest site, but can roam or migrate to find food or a proper climate. Viviparity is accompanied by the evolution of a portion of the reproductive tract where the young are nourished and develop. In viviparous mammals, the oviducts are modified into one or two *uteri* (singular, uterus).

Reproductive Cycles Most mammals have a definite time or times during the year in which eggs mature and are capable of being fertilized. Reproduction usually occurs when climatic conditions and resource characteristics favor successful development. Mammals living in environments with few seasonal changes and those that exert considerable control of immediate environmental conditions (e.g., humans) may reproduce at any time of the year.

However, they are still tied to physiological cycles of the female that determine when eggs can be fertilized.

Most female mammals undergo an **estrous** (Gr. *oistros,* a vehement desire) **cycle,** which includes a time during which the female is behaviorally and physiologically receptive to the male. During the estrous cycle, hormonal changes stimulate the maturation of eggs in the ovary and induce ovulation (release of one or more mature ova from an ovarian follicle). A few mammals (e.g., rabbits, ferrets, and mink) are *induced ovulators*; ovulation is induced by coitus.

Hormones also mediate changes in the uterus and vagina. As the ova are maturing, the inner lining of the uterus proliferates and becomes more vascular in preparation for receiving developing embryos. Proliferation of vaginal mucosa is accompanied by external swelling in the vaginal area and increased glandular discharge. During this time, males show heightened interest in females, and females are receptive to males. If fertilization does not occur, the above changes in the uterus and vagina are reversed until the next cycle begins. No bleeding or sloughing of uterine lining usually occurs.

Many mammals are *monestrous* and go through only a single yearly estrous cycle that is sharply seasonal. Wild dogs, bears, and sea lions are monestrous; domestic dogs are *diestrous*. Other mammals are *polyestrous*. Rats and mice have estrous cycles that are repeated every 4 to 6 days.

The *menstrual cycle* of female humans, apes, and monkeys is similar to the estrous cycle in that it results in a periodic proliferation of the inner lining of the uterus and is correlated with the maturation of an ovum. If fertilization does not occur before the end of the cycle, *mensus*—the sloughing of the uterine lining—occurs. Human menstrual and ovarian cycles are described in chapter 39.

Fertilization usually occurs in the upper one-third of the oviduct within hours of copulation. In a few mammals, fertilization may be delayed. In some bats, for example, coitus occurs in autumn, but fertilization is delayed until spring. Females store sperm in the uterus for periods in excess of 2 months. This example of **delayed fertilization** is apparently an adaptation to winter dormancy. Fertilization can occur immediately after females emerge from dormancy rather than having to wait until males attain their breeding state.

In many other mammals, fertilization occurs right after coitus, but development is arrested after the first week or two. This **embryonic diapause** was described previously for sea lions, and also occurs in some bats, bears, martens, and marsupials. The adaptive significance of embryonic diapause varies with species. In the sea lion, embryonic diapause allows the mother to give birth and mate within a short interval, but not have her resources drained by both nursing and pregnancy. It also allows young to be born at a time when resources favor their survival. In some bats, it allows fertilization to occur in the fall before hibernation, but birth is delayed until resources become abundant in the spring.

Modes of Development   Monotremes are oviparous. Eggs are released from the ovaries with large quantities of yolk. After fertilization, shell glands in the oviduct deposit a shell around the egg. Female echidnas incubate eggs in a ventral pouch. Platypus eggs are laid in their burrows.

All other mammals nourish young by a placenta through at least a portion of their development. Nutrients are supplied from the maternal bloodstream, not yolk.

In marsupials, most nourishment for the fetus comes from "uterine milk" secreted by uterine cells. Some nutrients diffuse from maternal blood into a highly vascular yolk sac that makes contact with the uterus. This connection in marsupials is a primitive placenta. The **gestation period** (the length of time young develop within the female reproductive tract) varies between 8 and 40 days in different species. The short gestation period is a result of the inability to sustain the production of hormones that maintain the uterine lining. After birth, tiny young crawl into the marsupium, and attach to a nipple, where they suckle for an additional 60 to 270 days.

In eutherian mammals, the embryo implants deeply into the uterine wall. Embryonic and uterine tissues grow rapidly and become highly folded and vascular, forming the placenta. Although maternal and fetal blood do not mix, nutrients, gases, and wastes diffuse between the two bloodstreams. Gestation periods of eutherian mammals vary widely between 20 days (some rodents) and 19 months (the African elephant). Following birth, the placenta and other tissues that surrounded the fetus in the uterus are expelled as "afterbirth." The newborns of many species are helpless at birth (e.g., humans); others can walk and run shortly after birth (e.g., deer and horses).

Stop and Ask Yourself:

13. What adaptation of the kidney allows mammals to excrete urine that is hypertonic to the blood?
14. What are the advantages of viviparity for mammals? In what way is viviparity costly?
15. What is embryonic diapause?
16. Why is the gestation period of a marsupial so short?

Summary

1. Mammalian characteristics evolved in the synapsid lineage over a period of about 200 million years. Mammals evolved from a group of synapsids called therapsids.

2. Modern mammals include the monotremes, marsupial mammals, and placental mammals.

3. Hair functions in sensory perception, temperature regulation, and communication.

4. Sebaceous, sudoriferous, scent, and mammary glands are present in mammals.

5. The teeth and digestive tracts of mammals are adapted for different feeding habits.

6. The heart of mammals has four chambers, and circulatory patterns are adapted for viviparous development.

7. Mammals possess a diaphragm that ventilates the lungs.

8. Metabolic heat production, insulating pelage, and behavior are used in mammalian thermoregulation.

9. Mammals react to unfavorable environments by migration, winter sleep, and hibernation.

10. The nervous system of mammals is similar to that of other vertebrates. Olfaction and hearing were important for early mammals. Vision and hearing, and smell are the dominant senses in many modern mammals.

11. The nitrogenous waste of mammals is urea, and the kidney is adapted for excreting a concentrated urine.

12. Mammals have complex behavior to enhance survival.

13. Most mammals have specific times during the year when reproduction occurs. Estrous or menstrual cycles are present in female mammals.

Key Terms

cecum (p. 495)
delayed fertilization (p. 502)
dental formula (p. 494)
diaphragm (p. 497)
embryonic diapause (p. 502)
estrous cycle (p. 502)
gestation period (p. 502)
heterodont (p. 494)
homodont (p. 494)

Critical Thinking Questions

1. Why is tooth structure important in the study of mammals?

2. What does the evolution of secondary palates have in common with the evolution of completely separated, four-chambered hearts?

3. Why is the classification of mammals by feeding habits not particularly useful to phylogenetic studies?

4. Under what circumstances is endothermy disadvantageous for a mammal?

5. Discuss the possible advantages of embryonic diapause for marsupials that live in climatically unpredictable regions of Australia.

Suggested Reading

Books

Boitani, L., and Bartoli, S. 1983. *Simon and Schuster's Guide to Mammals.* New York: Simon and Schuster.

Chapman, J. A., and Feldhamer, G. A. (eds). 1982. *Wild Mammals of North America: Biology, Management, and Economics.* Baltimore: The John Hopkins University Press.

Eisenberg, J. F. 1981. *The Mammalian Radiations: An Analysis of Trends in Evolution, Adaptation, and Behavior.* Chicago: University of Chicago Press.

Griffiths, M. 1978. *The Biology of Monotremes.* San Diego: Academic Press, Inc.

Pough, F. H., Heiser, J. B., and McFarland, W. N. 1989. *Vertebrate Life,* 3rd ed. New York: Macmillan Publishing Co.

Savage, R. J. G., and Long, M. T. 1986. *Mammal Evolution.* New York: Facts On File Publications.

Stonehouse, B., and Gilmore, D. (eds). 1977. *The Biology of Marsupials.* New York: Macmillan Publishing Co.

Vandenbergh, J. G. 1983. *Pheromones and Reproduction in Mammals.* San Diego: Academic Press.

Vaughan, T. A. 1978. *Mammalogy.* Philadelphia: W.B. Saunders Company.

Walker, W. F. 1987. *Functional Anatomy of the Vertebrates.* Philadelphia: CBS College Publishing, Saunders College Publishing.

Young, J. Z., and Hobbs, M. J. 1975. *The Life of Mammals: Their Anatomy and Physiology,* 2nd ed. New York: Oxford University Press.

Articles

Griffiths, M. The Platypus. *Scientific American* May, 1988.

Hinds, D. S., and MacMillen, R. E. 1984. Energy scaling in marsupials and eutherians. *Science* 225: 73–74.

Jenkins, F. A., Crompton, A. W., and Downs, W. R. 1983. Mesozoic mammals from Arizona: new evidence on mammalian evolution. *Science* 222:1233–35.

Kanwisher, J. W., and Ridgeway, S. H. The physiological ecology of whales and porpoises. *Scientific American* June, 1983.

Marshall, L. G., Webb, S. D., Sepkoski, J. J., and Raup, D. M. 1982. Mammalian evolution and the great American interchange. *Science* 215:1351–1357.

Sloan, R. E. 1986. Gradual dinosaur extinction and simultaneous ungulate radiation in the Hell Creek formation. *Science* 232:629–633.

Suga, N. Biosonar and neural computations in bats. *Scientific American* June, 1990.

Wheatley, D. 1988. Whale size quandary. *Nature* 336:626.

Zapol, W. M. Diving adaptations of the Weddell Seal. *Scientific American* June, 1987.

Form and Function

Part V

Is the whole of an animal, such as a human, equal to the sum of its parts? A superficial answer would be "yes." However, the structure and function of an animal is never as simple as this answer implies. A body is composed of many parts (e.g., cells, tissues, organs, and systems), yet rarely are any of these parts independent of one another. Simple additive relationships fail to describe adequately the interactions between the body's parts. Instead, an animal is the product of many complex interactions. In understanding the structure and function of any system, one only begins to understand the whole animal. Cells, tissues, organs, and systems all interact to maintain a steady homeostatic state compatible with life. Ultimately, one needs to look inward to see the genetic potential of the animal, outward to see how environmental constraints limit the fulfillment of that potential, and backward in time to see the evolutionary pressures that shaped the particular animal.

Parts I through IV of this textbook examine animal life at molecular, cellular, genetic, developmental, and taxonomic levels. Throughout these various parts, the evolutionary forces and pressures that shaped the vast array of animal life forms were presented, concluding with five chapters on the vertebrates. Part V (chapters 32 through 39) continues this coverage of animal life by presenting an overview of the various organ systems: integumentary, skeletal, muscular, nervous and sensory, endocrine, circulatory, lymphatic, respiratory, digestive, urinary, and reproductive. For continuity and interest, discussion centers on the human body. However, each chapter contains information about important differences in the system or systems described in other animal groups that you have studied.

Throughout, the major theme is that all body systems are specialized and coordinated with each other to produce a dynamic and efficient organism. The systems are constantly adjusting to changes inside and outside the body. Although each system has its own specialized function, none operates without help from the others. As you will see, the structure of each system determines its particular function.

32

Protection, Support, and Movement

Concepts

1. The integumentary system consists of the outer body covering. Skin is the common vertebrate integument, as are its integumentary derivatives (glands, hair, and nails).
2. Animals possess a number of different kinds of skeletons: hydrostatic skeletons, endo- or exoskeletons. The vertebrate skeletal system consists of two main types of supportive connective tissue: cartilage and bone. Cartilage provides support and aids movement at the joints (articulations) of the body. Bone (osseous) tissue comprises the skeleton that supports the internal organs of the body. Bone tissue is the storehouse for calcium and phosphate, and bone marrow serves as the site for the manufacture of red blood cells and some white blood cells.
3. Muscles provide the ability for movement to many animal groups, from cnidarians to vertebrates. The latter use their skeleton in conjunction with muscles to accomplish movement. Although joints make a skeleton potentially movable, and bones provide a basic system of levers, they cannot move by themselves. The driving force behind movement is muscle tissue. Different types of muscles do different tasks. Skeletal muscles are involved in walking, smooth muscles in moving material through tubular organs and changing the size of tubular openings, and cardiac muscle in the beating of the heart.

Have You Ever Wondered:

[1] why mammals have been able to colonize terrestrial environments?
[2] how leather is made and where it comes from?
[3] what "goose bumps" are?
[4] what unique feature makes some people "double-jointed?"
[5] when the human heart has time to relax?
[6] how many skeletal muscles are in your body?

These and other useful questions will be answered in this chapter.

This chapter contains underlined evolutionary concepts.

In animals, structure and function have evolved together to accomplish the most advantageous results. Three such results are protection, support, and movement, which are represented by the integumentary, skeletal, and muscular systems, respectively.

Integumentary System: Protection

The outer covering of an animal, its **integument** (L. *integumentum,* cover) functions primarily in protection against mechanical and chemical injury, and invasion by microorganisms. Many other diverse functions of the integument have evolved in different animal groups. These include regulation of body temperature; excretion of waste materials; conversion of sunlight into vitamins; the reception of environmental stimuli, such as pain, temperature, and pressure; and the exchange of gases.

Some single-celled protists, such as protozoa, only have a plasma membrane for an external covering. Others, such as *Paramecium,* have a thick pellicle outside the plasma membrane that offers further environmental protection. Most multicellular invertebrates have various epidermal tis-

sue coverings; some have an added *cuticle* that is more protective than just an epidermis.

The epidermis in molluscs and corals is more advanced in that it contains mucous glands that secrete the calcium carbonate of the shell (a form of exoskeleton). Arthropods have the most complex of invertebrate integuments, in part because their integument also serves as a specialized exoskeleton (*see figure 22.4*). A number of invertebrate groups, such as molluscs and arthropods have specialized integumentary cells called *chromatophores* that allow the animal to change its color, depending on internal or external conditions.

Skin is the integument found in vertebrates. It is the largest organ of the vertebrate body.

Structure of the Skin

Skin has two main layers. The *epidermis* (Gr. *epi,* over + *derma,* skin) is the outermost layer of epithelial tissue, and the *dermis* is a thicker layer of connective tissue beneath the epidermis (figure 32.1). The skin is separated from the deeper tissues by a hypodermis ("below the skin"), or subcutaneous layer. The surface of human skin contains

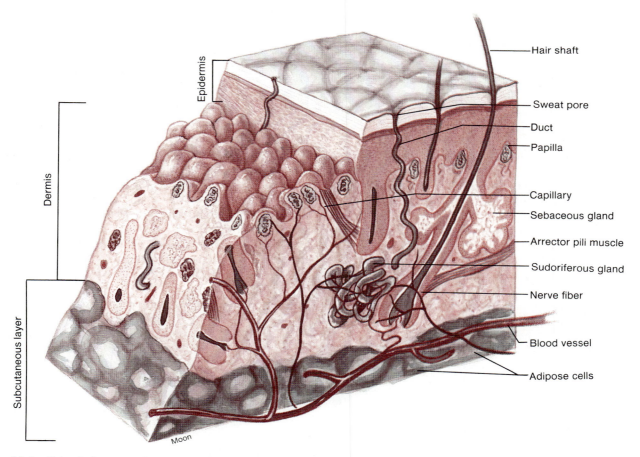

Figure 32.1 Skin. A diagram of the skin and underlying hypodermis. Skin is a two-layered organ with an outer epidermis and an inner dermis. Notice the various structures in the dermis.

grooves and ridges that form patterns unique to each person. These patterns are used in the identification procedure called *fingerprinting.*

The epidermis of mammalian skin is stratified squamous epithelium (colorplate 1e). Rapid cell divisions in the deepest layer of the epidermis push cells toward the surface of the skin. As cells reach the surface, they die and become *keratinized* (contain the protein keratin) or *cornified* (hardened) and make up the outer layer called the *stratum corneum.* Because keratin is virtually insoluble, the stratum corneum allows the skin to act as a barrier to prevent dehydration and act as the first line of defense against many toxic substances and microorganisms. [1] The prevention of dehydration is one of the evolutionary reasons mammals and other animals have been able to colonize terrestrial environments.

Most of the skin is composed of dermis, which is the strong, flexible connective-tissue meshwork of various types of fibers. Within the dermis, blood vessels, lymphatic vessels, nerve endings, hair follicles, and glands are present (figure 32.1). The dermis is composed of two layers that are not clearly separated. The deeper, thicker layer is called the *reticular layer.* Leather is made from the dermis of [2] mammalian skin by a tanning process that produces crosslinks among the proteins in the reticular layer.

The *hypodermis* lies underneath the integument and is composed of loose connective tissue, adipose tissue, and skeletal muscles. Adipose tissue (colorplate 1f) serves to store energy in the form of fat and provide insulation in cold environments. Skeletal muscles allow the skin above it to move somewhat independently of underlying tissues. Blood vessels thread from the hypodermis to the dermis and are absent in the epidermis.

Functions of the Skin

Through perspiration and the opening and closing of sweat pores, the skin is an effective regulator of body temperature. In fact, heat loss through radiation, conduction, convection, and evaporation accounts for approximately 95% of the body's heat loss. A small amount of waste urea is excreted through the skin. Up to 1 g of waste nitrogen may also be eliminated through the skin every hour.

The skin is an effective barrier for screening out excessive harmful ultraviolet rays from the sun, but it also lets in some necessary ultraviolet rays (box 32.1). This light is used to convert a chemical in the skin called 7–dehydrocholesterol into vitamin D_3. Vitamin D_3 is vital for the transport of calcium from the gut into the bloodstream and is required for the normal growth of bones and teeth.

The skin is also an important sensory organ, containing sensory receptors for heat, cold, touch, pressure, and pain. It helps protect an animal by means of its many nerve endings, which keep the animal responsive to factors in the environment that might harm it. The nerve endings also help an animal sense the environment so that adjustments can be made to maintain homeostasis.

Color of the Skin

Skin derives its color from three factors: the presence of **melanin** (Gr. *melas,* black), a dark pigment produced by specialized cells called *melanocytes;* the yellow pigment *carotene;* and the color of blood (oxyhemoglobin) reflected through the epidermis. The main function of melanin is to screen out excessive ultraviolet rays to protect the nucleus and its genetic material. Extra protection is provided when melanin is darkened by the sun and transferred to the outer skin layers, producing a suntan. A skin that is suntanned is less sensitive to sunlight than previously unexposed skin.

Glands of the Skin

The skin contains two types of glands: sudoriferous and sebaceous. **Sudoriferous glands** (L. *sudor,* sweat) are also called *sweat glands,* and in humans, are distributed over most of the body surface. They consist of a coiled portion embedded in the hypodermis and a duct leading up through the dermis to the surface of the epidermis (*see figure 32.1*). These glands secrete sweat by a process called *perspiration* (L. *per,* through + *spirare,* to breathe). Perspiration helps to regulate body temperature and maintain homeostasis, largely by the cooling effect of evaporation. In some animals, including humans, certain sweat glands also produce pheromones. A pheromone is a chemical secreted by an animal that communicates with other members of the same species to elicit certain behavioral responses.

Sebaceous (oil) glands (L. *sebum,* tallow or fat) are simple glands connected to hair follicles that are found in the dermis (*see figure 32.1*). Their main functions are lu-

---

Box 32.1

No Fun in the Sun

Why is excessive sunbathing potentially dangerous? The process of tanning is activated by the ultraviolet rays in sunlight. Overexposure to these rays can kill some skin cells, damage others so that normal secretions are stopped, increase the risk of skin cancer and mutations by affecting the genetic material in the nuclei of cells, and cause the immune system to become less effective. UV rays can also damage enzymes and plasma membranes, and interfere with cellular metabolism. If tissue destruction is excessive, cellular debris and toxic waste products enter the bloodstream and produce the fever associated with what is called sun poisoning.

brication and protection, which they accomplish by secreting substances. Secretions are produced by the breakdown of the interior cells, which become the oily secretion called sebum. **Sebum** is a semifluid composed almost entirely of lipid. Sebum serves as a permeability barrier, an emollient (skin-softening agent), and a protective agent against microorganisms. Sebum can also act as a pheromone.

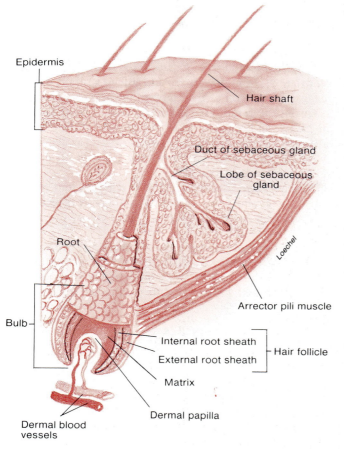

Epidermis

Hair shaft

Duct of sebaceous gland

Lobe of sebaceous gland

Root

Loechel

Arrector pili muscle

Bulb

Internal root sheath

External root sheath

Hair follicle

Matrix

Dermal papilla

Dermal blood vessels

Figure 32.2 Hair. The structure of hair and the hair follicle. Notice the relationship of the sebaceous gland to the hair root.

Hair

Hair is composed of cornified threads of cells that develop from the epidermis. Because hair arises from the skin, it is considered to be an appendage of the skin. The portion of hair that protrudes from the skin is the *hair shaft,* and the portion embedded beneath the skin is the *root* (figure 32.2). The lower portion of the root, located in the hypodermis, enlarges to form the *bulb.* An *arrector pili muscle* is attached to the connective tissue sheath of a *hair follicle* surrounding the bulb. When this muscle contracts, it pulls the follicle and its hair to an erect position. In humans, this is referred to as a "goose bump." In other mammals, this action helps warm the animal by producing an insulating layer of warm body air between the erect hair and skin. If the hair is erect as a result of the animal being frightened instead of being cold, the erect hair also makes the animal look larger and less vulnerable to attack.

[3]

Nails

Nails, like hair, are modifications of the epidermis. They are composed of flat, horny plates on the dorsal surface of the distal segment of the fingers and toes (figure 32.3). Nails appear pink because they are translucent, allowing the red color of the vascular tissue underneath to show through. The *lunula* is the half-moon-shaped, whitish area at the base of nail. It is white because the "red capillaries" in the underlying dermis do not show through. The developing nail is originally covered by thin layers of epidermis called the *eponychium* (Gr. *epi,* upon + *onyx,* nail), which remains at the base of the mature nail and is called the *cuticle.*

Stop and Ask Yourself

1. What is the major function of skin?
2. What are two types of glands found in the skin and what are several functions of each?
3. What determines the color of skin?

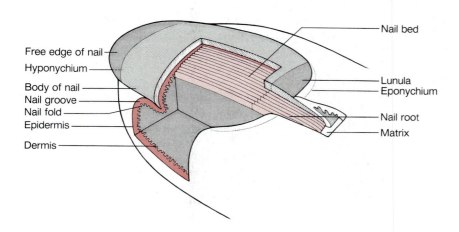

Free edge of nail

Hyponychium

Body of nail

Nail groove

Nail fold

Epidermis

Dermis

Nail bed

Lunula

Eponychium

Nail root

Matrix

Figure 32.3 Nails. A sagittal section of a fingertip illustrating a fingernail. Notice the associated anatomical structures.

Other Specializations

Many vertebrates show integumentary specializations for particular lifestyles. Because fish live in more stable environments than terrestrial vertebrates, they have a simpler skin structure. No keratinization occurs and only mucous glands are present. Fish have thin scales that develop in the skin from dermal-epidermal interactions. Some keratinization occurs in amphibian skin and mucous and poison glands are present. Heavily keratinized skin and horny scales evolved in the reptiles. Most species are able to shed their scales periodically. Apart from scent glands in some species, reptile skin is aglandular. Bird skin is characterized by the presence of feathers which protect the body surface, prevent dehydration, insulate, and form the flying surfaces. With the exception of a uropygial oil gland above the tail, bird skin is essentially aglandular. Claws, rather than nails, are found in most reptiles, birds, and mammals. Other keratinized derivatives of the skin are horns (not to be confused with bony antlers) and the baleen plates of the toothless whales.

Skeletal System: Support

Many invertebrates use their body fluids for internal support. For example, sea anemones, earthworms, and snails have a form of internal support called the *hydroskeleton* (colorplate 14a). The hydroskeleton is composed of a core of liquid (water or a body fluid such as blood) wrapped in a tension-resistant sheath. A hydroskeleton is similar to a balloon filled with water, because the force exerted against the incompressible fluid in one region can be transmitted to other regions. Contracting muscles can push against a hydroskeleton, and the transmitted force generates body movement. For example, a rapid surge of blood under high pressure can extend the hind leg spines of spiders.

As animals—particularly land animals—grew larger and thicker, they needed support for the body. The key to the success of insects and other arthropods is the *exoskeleton,* or external skeleton, which completely surrounds the animal (colorplate 14b). An exoskeleton provides protection, strong support, and counterforce for muscle movements. In arthropods, the epidermis of the exoskeleton secretes a thick, hard cuticle that makes the body covering waterproof. The cuticle also protects and supports the animal's soft internal organs. In crustaceans, such as crabs, lobsters, and shrimp, the exoskeleton also contains calcium carbonate crystals that make it a hard, inflexible armor. Besides providing shieldlike protection from enemies and resistance to general wear and tear, the exoskeleton also prevents internal tissues from drying out. This adaptation has been extremely important evolutionarily because it contributed to the successful colonization of land. The exoskeleton also serves as a site for attachment of muscles. Animals with exoskeletons, however, are limited in their growth by the space within the exoskeleton. They must either shed the skeleton periodically, as arthropods do, or increase the

skeleton's size, as molluscs do. The skeletal system of vertebrates is an internal skeleton (*endoskeleton*) consisting of two main types of supportive tissue: cartilage and bone.

Cartilage

Cartilage is a specialized type of connective tissue that provides support and aids movement at joints. Like other connective tissues, it consists of cells (*chondrocytes*), fibers, and ground substance. There are three types of cartilage, each with its distinctive matrix mixture of ground substance and fibers. The number and type of fibers in the matrix give each type of cartilage its characteristic strength and/or resiliency.

The most common type of cartilage is **hyaline cartilage** (colorplate 1i), which is strong and able to support weight. It forms the skeleton of the embryo, reinforces respiratory passages in the lungs, covers the portion of bone facing a joint cavity, and is essential to the growth of long bones. Hyaline cartilage is also found in the nose, and at the ventral ends of the ribs, where it allows the rib cage to expand and contract during breathing.

Fibrocartilage (colorplate 1k) is resilient and pliable because it contains many collagenous fibers. It is primarily found between the bodies of the vertebrae of the spinal column ("backbone").

The third type of cartilage, **elastic cartilage** (colorplate 1j) is richly supplied with elastic fibers that make it flexible and elastic. It is found in areas, such as the external ear, where lightweight support and flexibility are required.

Bones and Osseous Tissue

Bone (osseous) tissue is specialized connective tissue that supports the internal organs of many animals' bodies. It also functions as a storehouse and main supply of reserve calcium and phosphate, and red bone marrow serves as the site for the manufacture of red blood cells and some white blood cells. Bones aid movement by providing a point of attachment for muscles and transmitting the force of muscular contraction from one part of the body to another during movement. Bone is not dry, brittle, or dead, but is a living, changing, productive tissue that is continually resorbed (dissolved and assimilated), reformed, and replaced or renewed (box 32.2). Some bone regions are less solid than others and are known as *spongy* bone; others are more compact and are known as *compact* bone (figure 32.4).

Bone tissue is more rigid than other connective tissues because its homogeneous, organic ground substance also contains inorganic salts, mainly calcium phosphate and calcium carbonate. In the bones, these compounds and others form *hydroxyapatite crystals*. When the body needs the calcium or phosphate that is stored within the bones, the hydroxyapatite crystals ionize and release the required amounts.

As discussed in chapter 3, bone cells (osteocytes) are located in minute chambers called *lacunae* (s., lacuna), which are arranged in concentric rings around *osteonic*

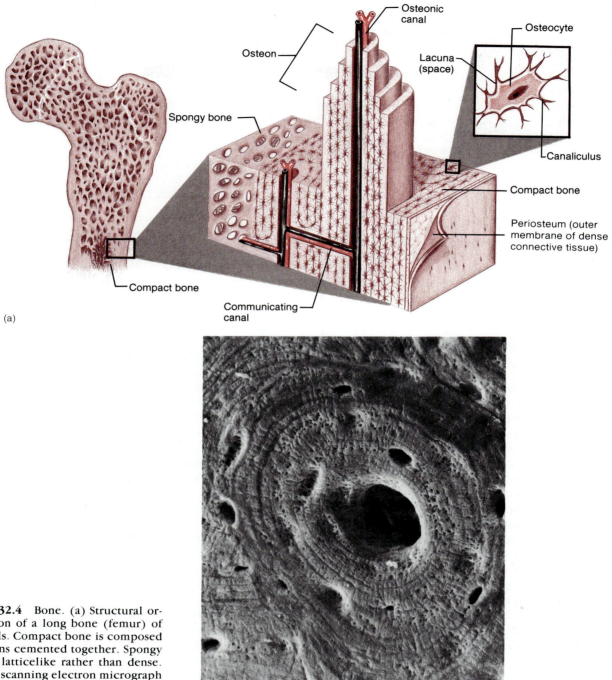

Figure 32.4 Bone. (a) Structural organization of a long bone (femur) of mammals. Compact bone is composed of osteons cemented together. Spongy bone is latticelike rather than dense. (b) The scanning electron micrograph is a single osteon in compact bone.

canals (figure 32.4). These cells communicate with nearby cells by means of cellular processes passing through small channels called *canaliculi* (s., canaliculus).

In compact bone, the osteocytes and associated layers of intracellular material cluster concentrically around an osteonic canal to form a cylinder-shaped unit called an *osteon*. A large number of these units cemented together make up compact bone. Each osteonic canal contains blood vessels and the canals are connected to each other by trans-verse communicating canals, through which blood vessels pass.

In spongy bone, the osteocytes are not arranged around osteonic canals and the space around the cells is filled with *red bone marrow*, which produces blood cells.

The Skeleton

Humans are born with as many as 300 bones, but some fuse during childhood to form the adult skeleton of 206 bones.

Box 32.2

Osteoporosis

Osteoporosis (Gr. *osteo*, bone + *poros*, passage + *osis*, disease process) is a disorder of the skeletal system in which there is an excessive loss of bone volume and mineral content. This disorder is most often associated with aging. The affected bones develop large spaces and canals that become filled with fibrous and fatty tissues. These bones are easily fractured and often break spontaneously.

Although osteoporosis may affect persons of either sex, it is most common in light-skinned females after menopause. Factors that increase the risk of osteoporosis include smoking, a lack of calcium in the diet, lack of physical exercise, and in females, the lack of the female sex hormone *estrogen*. (Estrogen is produced by the ovaries, which become inactive at menopause.)

Other than the bones that come in pairs, no two are alike. They may differ in size, shape, weight, and even composition. This diversity of form is directly related to the many structural or mechanical functions of the skeleton.

The most obvious function of the skeleton is to support the body and its internal organs. Besides support, the skeleton protects many of the internal organs. For example, the heart and lungs are safely enclosed within a rib cage that offers protection and freedom of movement, and the brain is cushioned within the cranium, a shock-absorbing bone case that evolved to protect the brain from the many bumps of everyday life.

The bones of the skeleton also act as a system of levers for the pulley action of muscles. This lever-pulley arrangement provides attachment sites on bones for muscles and tendons, and allows movement of the entire body, or just one finger or toe.

The skeleton is divided into two major parts: the axial skeleton and the appendicular skeleton (figure 32.5; table 32.1). The **axial skeleton** is so named because it forms the longitudinal axis of the body. It is made up of the skull, vertebral column, sternum, and ribs. The **appendicular skeleton** is composed of the appendages and the pectoral (shoulder) and pelvic girdles, which attach the upper and lower appendages to the axial skeleton.

Articulations and Levers

Bones give a vertebrate its structural framework, and muscles give it its power, but movable joints or **articulations** (L. *articulas,* small joint) provide the mechanisms that allow the animal to move. An articulation is the place where two adjacent bones and/or cartilages are joined, even if the joint does not allow movement.

Structurally, there are three types of joints: fibrous, cartilaginous, and synovial. *Fibrous joints* have no gap between the bones and are found primarily between the flat bones of the cranium. No movement can occur in this type of joint. *Cartilaginous joints* have no gap between the bones but do have fibrocartilage pads that enable some movement. The intervertebral discs between the bones of the vertebral column are an example of this type of joint.

The type of joint that allows the greatest range of movement is the *synovial joint.* Free movement is possible because the ends of the bones are covered with a smooth articular cartilage, the joint is lubricated by a thick fluid called synovial fluid, and the joint is enclosed in a flexible *articular capsule.* A synovial joint also has a synovial cavity (figure 32.6). The *synovial cavity* is the space between two articulating bones. The *synovial membrane* secretes the thick **synovial fluid** that lubricates the synovial cavity.

An *articular cartilage* caps the surface of the bones bordering the synovial cavity. Because of its thickness and elasticity, the articular cartilage acts as a shock absorber. In the knee joint, the articular cartilage is called the *medial* and *lateral meniscus* (pl., menisci). The menisci serve to cushion as well as guide the articulating bones. Many athletes, especially football players, tear these menisci in an injury commonly referred to as torn cartilage (box figure 32.3).

The *articular capsule* surrounds the synovial cavity in the noncartilaginous part of the joint. It extends from bone to bone across a joint, and reinforces the joint. However, this fibrous capsule is lax and pliable, permitting considerable movement. "Double-jointed" people have loose articular capsules, allowing an even greater range of movement.

Two other structures that are associated with joints, but not part of them, are bursae and tendon sheaths. **Bursae** (s., bursa) resemble flattened sacs and are filled with synovial fluid. Bursae are found wherever it is necessary to eliminate the friction that occurs when a muscle or tendon rubs against another muscle, tendon, or bone. A modification of a bursa is the **tendon sheath** surrounding long tendons that are subject to constant friction. Tendon sheaths are also filled with synovial fluid.

[4]

Stop and Ask Yourself

4. What are the three types of cartilage? What are the function and location of each type?
5. What are several functions of osseous tissue?
6. What are several functions of the skeleton?
7. How can a typical synovial joint be described?

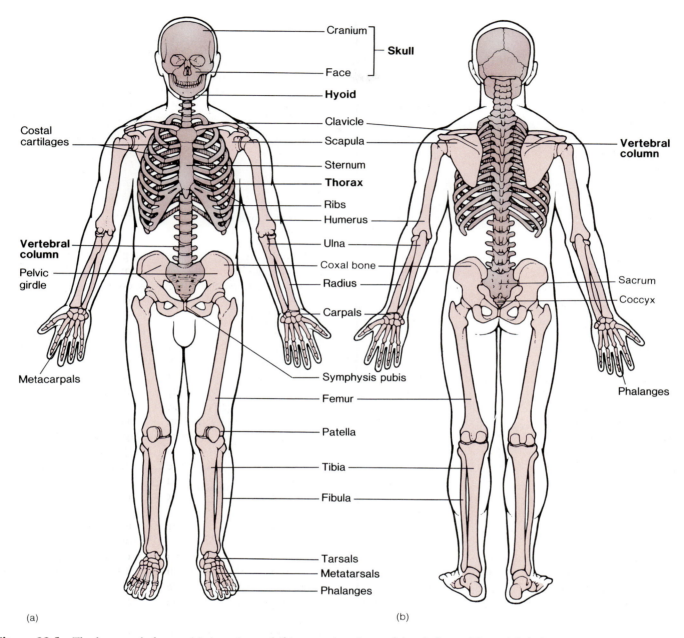

Figure 32.5 The human skeleton. (a) Anterior and (b) posterior views of the skeleton. The axial skeleton is shown in a different color to distinguish it from the appendicular skeleton.

Muscular System: Movement

Movement is a distinctive characteristic of certain protists and animals. Protists, such as *Amoeba proteus,* utilize *amoeboid movement (see figure 16.11).* Some invertebrates utilize cilia or flagella for movement. Muscles are found in various invertebrate groups from the primitive cnidarians to the more advanced arthropods (e.g., insect flight muscles). In more advanced animals, the muscles are attached to various skeletal systems to form a motor system, which allows for more complex movements. Joints make a skeleton potentially movable, and bones provide a basic system of levers, but bones and joints cannot move by themselves. The driving force, the power behind movement, is muscle tissue.

Muscle Tissue

The basic physiological property of muscle tissue is *contractility,* the ability to contract or shorten. In addition, muscle tissue has three other important physiological properties: *excitability* (or irritability) is the capacity to receive and respond to a stimulus; *extensibility* is the ability to stretch; and *elasticity* is the ability to return to its original shape after being stretched or contracted.

Table 32.1   Divisions of the Adult Human Skeleton (206 Bones)

Axial Skeleton (80 bones)		Appendicular Skeleton (126 bones)	
Skull (29 bones)		Upper extremities (64 bones)	
Cranium	8	Pectoral (shoulder) Girdle	4
Parietal (2)		Clavicle (2)	
Temporal (2)		Scapula (2)	
Frontal (1)		Arm	2
Ethmoid (1)		Humerus (2)	
Sphenoid (1)		Forearm	4
Occipital (1)		Ulna (2)	
Face	14	Radius (2)	
Maxillary (2)		Wrist	16
Zygomatic (2)		Carpals (16)	
Lacrimal (2)		Hands and fingers	38
Nasal (2)		Metacarpals (10)	
Inferior nasal concha (2)		Phalanges (28)	
Palatine (2)		Lower extremities (62 bones)	
Mandible (1)		Pelvic Girdle	2
Vomer (1)		Fused ilium, ischium, pubis	
Ossicles of the Ear	6	Thigh	4
Malleus (hammer) (2)		Femur (2)	
Incus (anvil) (2)		Patella (2)	
Stapes (stirrup) (2)		Leg	4
Hyoid (1)	1	Tibia (2)	
Vertebral Column (26 bones)		Fibula (2)	
Cervical vertebrae	7	Ankle	14
Thoracic vertebrae	12	Tarsals (14)	
Lumbar vertebrae	5	Foot and toes	38
Sacrum (5 fused bones)	1	Metatarsals (10)	
Coccyx (3 to 5 fused bones)	1	Phalanges (28)	
Thorax (25 bones)		Total (Appendicular)	126
Ribs	24		
Sternum	1		
Total (Axial)	80	Total (axial and appendicular)	206

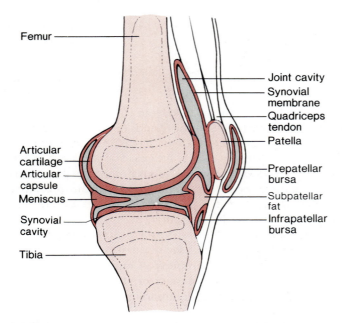

Femur — Joint cavity — Synovial membrane — Quadriceps tendon — Patella — Articular cartilage — Prepatellar bursa — Articular capsule — Meniscus — Subpatellar fat — Synovial cavity — Infrapatellar bursa — Tibia

Figure 32.6   Synovial joint. A typical synovial joint is represented by this illustration of a lateral view through the human knee.

When muscles contract and produce movement, important work is done. For example, food is passed along the digestive tract by a series of rhythmic waves of muscular contractions. Muscular contraction of the heart pumps blood from the heart to all parts of the body. As certain skeletal muscles in the body contract, legs move and an animal can run. Muscular contractions also help maintain posture, even when there is no obvious motion. Finally, the contraction of muscles produces much of the heat used in regulating body temperature.

Mammals have three types of muscle tissue: smooth, cardiac, and skeletal. The contractile cells of these tissues are called **muscle fibers**.

Smooth muscle is found in the walls of the gut and blood vessels, in the reproductive organs, and part of the eye. It is also called *involuntary* muscle because its contractions are not consciously controlled. Smooth-muscle fibers have a single nucleus, are spindle shaped, and are arranged in a parallel pattern to form sheets (colorplate 1p). Smooth muscle has the ability to maintain good tone even without nervous stimulation. A muscle has **tone** when a number of its fibers are contracted at any one time. Smooth muscle is slow to contract, but it can sustain prolonged contractions and does not fatigue easily.

Box 32.3

Knee Injuries—the "Achilles Heel" of Athletes

One of the most frequent sports injuries is commonly called torn cartilage (usually the medial meniscus) of the knee. When the cartilage is torn, it may become wedged between the articular surfaces of the femur and tibia, causing the joint to "lock." This type of injury is diagnosed by a surgical technique called **arthroscopy** ("looking into a joint"). During this surgery, the surgeon places a lighted scope about the size of a pencil into the joint capsule to view the structural damage. If the damage is local, and there do not seem to be complications, the surgeon makes another one-fourth-inch

incision and inserts microsurgical instruments to clear away damaged cartilage.

A more severe knee injury occurs from a tear in the anterior and posterior cruciate ligaments (box figure 32.3a). When these ligaments are torn, the knee joint becomes nonfunctional. To repair the cruciate ligaments, holes are drilled through the femur. Sutures are stitched to the damaged ligaments, and the sutures passed out through the holes (figure 32.3b). Six to eight weeks of healing is usually followed by about a week of rehabilitative therapy.

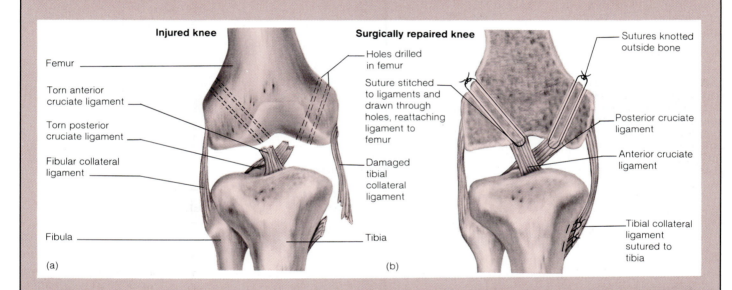

Box figure 32.3 Torn cartilage and arthroscopic surgery.

Cardiac muscle fibers are found in the heart, are *involuntary,* have a single nucleus, are *striated* (have dark and light bands), and are branched (colorplate 1q). This branching allows the fibers to interlock for greater strength. **Intercalated discs** are present where the individual fibers interlock. This arrangement of fibers permits contractions to spread quickly throughout the various heart chambers. [5] The heart does not fatigue because cardiac fibers relax completely between contractions.

Skeletal muscle, also called striated muscle, is *voluntary* muscle because its contractions are consciously controlled by the nervous system. Skeletal muscle fibers are multinucleated and striated (colorplate 1o). Skeletal muscle striations are due to the biochemical nature of the individual contractile elements, called myofibrils (box 32.4). Skeletal muscles are attached to the skeleton by ten-

dons, which are tough, fibrous bands or cords. When muscles contract, they shorten. Thus, muscles can only pull; they cannot push. Therefore, skeletal muscles work in *antagonostic pairs*. For example, one muscle of a pair bends a joint and brings the limb close to the body. The other member of the pair straightens the joint and extends the limb away from the body (figure 32.7).

In humans, over 600 individual muscles comprise the [6] skeletal muscular system. Some of the principal superficial muscles are shown in figure 32.8.

Skeletal Muscle Contraction

When observed with the light microscope, each skeletal muscle fiber has a pattern of alternate dark and light bands (colorplate 1o). This striation of whole fibers arises from

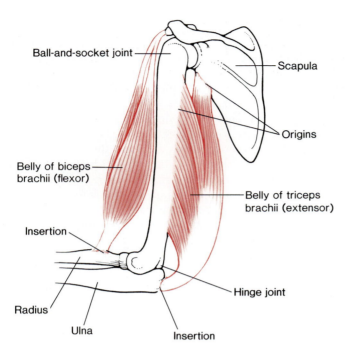

Figure 32.7 Antagonistic muscle pair. The biceps muscles is antagonistic to the triceps muscle, and the triceps is antagonistic to the biceps. When one muscle of an antagonistic pair contracts, the other one must relax, or movement does not occur. To bend the arm at the elbow joint and bring the lower arm closer to the upper body, the biceps muscle contracts and the triceps muscle relaxes. To extend the arm, the biceps relaxes, and the triceps contracts.

the alternating dark and light bands of the many smaller, threadlike **myofibrils** contained in each muscle fiber (figure 32.9a–c). Electron microscopy and biochemical analysis have shown that these bands are due to the placement of the muscle proteins **actin** and **myosin** within the myofibrils. Myosin occurs as thick filaments and actin as thin filaments. As figure 32.9c–e illustrates, the lightest region of a myofibril (the I band) contains only actin, whereas the darkest region (the A band) contains both actin and myosin.

The functional (contractile) unit of a myofibril is the **sarcomere**, each of which extends from one Z line to another Z line. Notice that the actin filaments are attached to the Z lines whereas myosin filaments are not (figure 32.9e). When a sarcomere contracts, the actin filaments slide past the myosin filaments as they approach one another. This process shortens the sarcomere. The combined decreases in length of the individual sarcomeres account for contraction of the whole muscle fiber, and in turn, the whole muscle. This movement of actin in relation to myosin is called the *sliding-filament model* of muscle contraction.

The actual contraction is accomplished by a ratchet mechanism acting between the two filament types. Myosin contains globular projections that attach to actin at specific active binding sites, forming attachments called *cross-bridges* (figure 32.10). Once cross-bridges have formed, they exert a force on the thin actin filament and cause it to move.

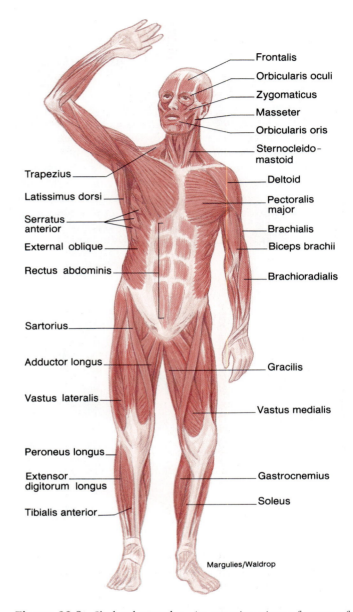

Margulies/Waldrop

Figure 32.8 Skeletal muscles. An anterior view of some of the superficial skeletal muscles of a human.

The energy for muscle contraction comes from the splitting of ATP by myosin. The hydrolysis of ATP by the myosin provides the energy for the power stroke and muscle contraction as follows:

$$ATP \quad ADP + P_i$$

$$actin + myosin \rightarrow actinomyosin$$

The ATP needed for muscle contraction is obtained from three sources:

1. The many mitochondria that are present in muscle cells generate ATP by aerobic respiration.

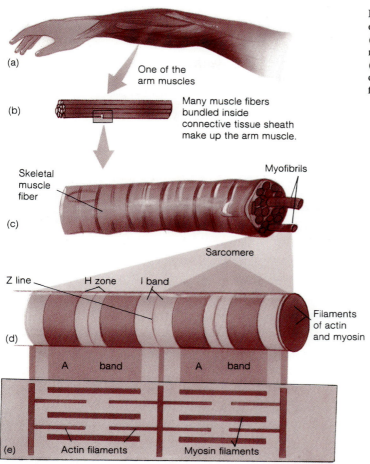

Figure 32.9 Structure of skeletal muscle tissue. (a) A skeletal muscle in the forearm consists of many muscle fibers (b) bundled inside a connective tissue sheath. (c) A skeletal muscle fiber contains many myofibrils, each consisting of (d) units called sarcomeres. (e) The characteristic striations of a sarcomere are due to the arrangement of actin and myosin filaments.

2. Muscle cells can store energy in the form of *creatine phosphate*. Creatine phosphate can be used to regenerate ATP as follows:

$$\text{creatine-P} + \text{ADP} \rightarrow \text{ATP} + \text{creatine}$$

3. When all of the creatine phosphate has been used up, a muscle cell can generate some ATP for a short period of time by means of anaerobic respiration.

Control of Muscle Contraction

When nerve impulses conducted by a motor nerve reach skeletal muscle fibers, the fibers are stimulated to contract via a motor unit. A **motor unit** consists of one motor nerve fiber and all the muscle fibers with which it communicates. A space separates the specialized end of the motor nerve fiber from the membrane (**sarcolemma**) of the muscle fiber. The motor end plate is the specialized portion of the sarcolemma of a muscle fiber surrounding the terminal end of the nerve. This arrangement of structures is called a **neuromuscular junction** or *cleft* (figure 32.11).

When nerve impulses reach the ends of the nerve fiber branches, a chemical called *acetylcholine* is released from

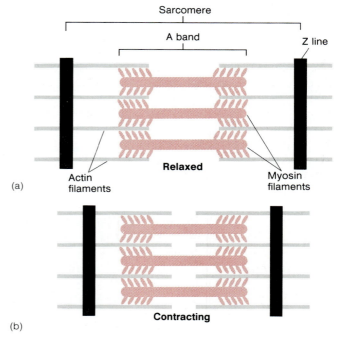

Figure 32.10 Sliding-filament model of muscle contraction. (a) A sarcomere in a relaxed position. (b) As the sarcomere contracts, the myosin filaments form attachments called crossbridges to the actin filaments and pull them so that they slide past the myosin filaments. Compare the size of the A band in (a) to that in (b).

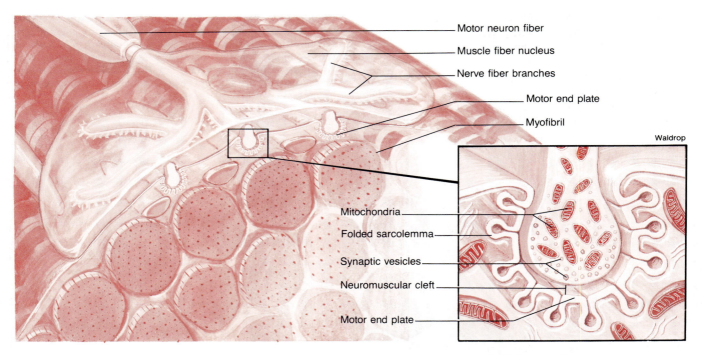

Waldrop

Motor neuron fiber
Muscle fiber nucleus
Nerve fiber branches
Motor end plate
Myofibril

Mitochondria
Folded sarcolemma
Synaptic vesicles
Neuromuscular cleft
Motor end plate

Figure 32.11   A nerve-muscle motor unit. A motor unit consists of one motor neuron and all the muscle fibers which it innervates. A neuromuscular junction, or cleft, is the site where the nerve fiber and muscle fiber meet.

Box 32.4

The Use and Disuse of Skeletal Muscles

Skeletal muscles vary in the speed with which they contract. Slow-contracting muscles are called *red muscles* because most of their fibers contain the red, oxygen-storing pigment, *myoglobin*. In addition, they have many blood vessels and contain many mitochondria to carry out aerobic respiration. For this reason, they can contract for long periods of time without undergoing fatigue. Fast-contracting muscles are called *white muscles* because their fibers contain less myoglobin, have a poorer blood supply, and fewer mitochondria than red muscle fibers. However, white muscle fibers can recycle calcium very rapidly; thus, they can contract more rapidly than red muscle fibers but fatigue faster as the substances needed to regenerate ATP are depleted.

A skeletal muscle is very responsive to use. For example, a muscle that is regularly used (exercised) tends to enlarge. This physiological response is called *hypertrophy*. Conversely, if a muscle is not used, it will decrease in size and strength (*atrophy*).

The way in which a muscle responds to use also depends upon the type of use. When a muscle contracts weakly, such as in swimming, slow, red muscle fibers are activated. Over time, these fibers will develop more mitochondria and blood vessels. This change will increase the muscle fiber's ability to resist fatigue during prolonged swimming. However, the size and strength of the fibers remain the same.

If one lifts weights, the fast, white muscle fibers are activated. Over time, these fibers develop new filaments of actin and myosin, their diameters increase, and the whole muscle enlarges (colorplate 14c). However, no new muscle fibers are produced.

Because the strength of a contraction is directly related to the diameter of the muscle fibers, an enlarged muscle can produce a stronger contraction. The ability to contract more strongly, however, does not increase the muscle's ability to resist fatigue during an activity such as swimming.

If the use of a muscle is discontinued, there will be a reduction in blood vessels, the number of mitochondria in the fibers, and the amount of actin and myosin. As a result, the entire muscle atrophies. Such atropy commonly occurs when limbs are immobilized by casts or when accidents or diseases interfere with motor impulses getting to the motor units.

storage *synaptic vesicles* in the nerve ending. Acetylcholine diffuses across the neuromuscular cleft between the nerve ending and the muscle-fiber sarcolemma and binds with acetylcholine receptors on the sarcolemma. The sarcolemma is normally polarized; the outside is positive and the inside is negative. When acetylcholine binds to the receptors, the polarity reverses—the outside becomes negative and the inside positive. This reversal of polarity flows in a wavelike progression into the muscle fiber by conducting paths called *transverse tubules*. Associated with the transverse tubules is the endoplasmic reticulum (*see figure 3.17*) of muscle cells, called *sarcoplasmic reticulum*. The electrical polarity causes calcium ions (Ca²⁺) to be released from the sarcoplasmic reticulum and diffuse into the cytoplasm. The calcium then binds with a regulatory protein called *troponin* that is on another protein called *tropomyosin*. This binding exposes the myosin binding sites on the actin molecule that had been blocked by tropomyosin (figure 32.12). Once the binding sites are open, the myosin filament can form cross-bridges, and the filament can slide, via a power stroke, resulting in muscular contraction.

Relaxation follows contraction. During relaxation, an active transport system pumps calcium back into the sarcoplasmic reticulum for storage. By controlling the nerve impulses that reach the sacroplasmic reticulum, the nervous system controls Ca²⁺ levels in skeletal muscle tissue, thereby exerting control over contraction.

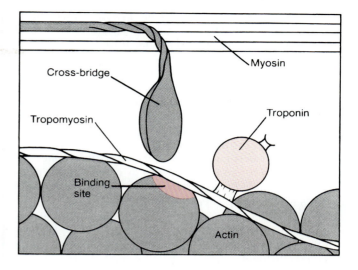

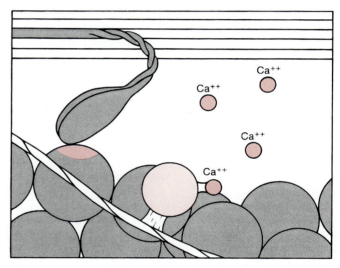

Figure 32.12 The attachment of Ca²⁺ to troponin causes movement of the troponin-tropomyosin complex, which exposes binding sites on the actin. The myosin cross-bridge can then attach to actin and undergo a power stroke.

Stop and Ask Yourself

8. What are several properties of muscle tissue?
9. What are several characteristics of smooth, cardiac, and skeletal muscle tissue?
10. What is a myofibril?
11. How does the sliding-filament theory explain muscle contraction?
12. How do nerves cause skeletal muscles to contract?

■ Summary

1. The integumentary system functions primarily in protection against mechanical injury and invasion by microorganisms. Secondary functions include temperature regulation, excretion of waste material, the conversion of a steroid by the sun rays into vitamins, and in the reception of stimuli such as pain, temperature, and pressure.

2. The skin is the largest organ of the mammalian body and consists of two main layers: the epidermis and dermis. Skin gets its color from three factors: the presence of the pigment melanin, the yellow pigment carotene, and the color of blood reflected through the epidermis. The skin contains two types of glands: sudoriferous (sweat) and sebaceous (oil). Skin structure varies considerably among vertebrates. Some of these variable structures include scales, hairs, feathers, claws, and nails.

3. Three types of skeletons are found in animals: hydrostatic, endo-, and exoskeletons. From a structural point of view, the skeleton of vertebrates is an internal skeleton (endoskeleton) and consists of two types of supportive tissue: cartilage and bone. Cartilage is a specialized type of connective tissue that provides support and aids movement at joints. Three types of cartilage are found in the mammalian body: hyaline, fibrocartilage, and elastic cartilage. Bone (osseous) tissue is the specialized connective tissue that

supports the internal organs of the body. The skelton of vertebrates is divided into two basic parts: the axial skeleton and the appendicular skeleton. Movement in protists and animals is accomplished by ameboid movement, cilia, flagella, muscles, and muscles attached to skeletons to form a motor system.

4. Movable joints called articulations allow movement. An articulation is the place where two adjacent bones and cartilages are joined. Structurally, there are three types of joints: fibrous, cartilaginous, and synovial. Synovial joints allow the greatest range of movement. Because bones are rigid structures and can be moved at certain joints in response to applied forces, they can be considered levers.

5. The power behind movement is muscle tissue. The three types of muscle tissue are smooth, cardiac, and skeletal. Muscle tissue exhibits contractility, excitability, extensibility, and elasticity. Skeletal muscles are attached to the skeleton by tendons and work in antagonistic pairs. In the human, over 600 muscles comprise the skeletal muscular system.

6. The functional (contractile) unit of a muscle myofibril is the sarcomere. The energy for muscle contraction comes from the splitting of ATP by myosin. Nerves control skeletal muscle contraction.

■ Key Terms

actin (p. 516)
appendicular skeleton (p. 512)
arthroscopy (p. 515)
articulations (p. 512)
axial skeleton (p. 512)
bone (p. 510)
bursae (p. 512)
cardiac muscle (p. 515)
cartilage (p. 510)
elastic cartilage (p. 510)
fibrocartilage (p. 510)
hyaline cartilage (p. 510)
integument (p. 507)
intercalated discs (p. 515)
melanin (p. 508)
motor unit (p. 517)
muscle fibers (p. 514)

myofibrils (p. 516)
myosin (p. 516)
neuromuscular junction (p. 517)
osteoporosis (p. 512)
sarcolemma (p. 517)
sarcomere (p. 516)
sebaceous (oil) glands (p. 508)
sebum (p. 509)
skeletal muscle (p. 515)
skin (p. 507)
smooth muscle (p. 514)
sudoriferous glands (p. 508)
synovial fluid (p. 512)
tendon sheath (p. 512)
tendons (p. 515)

■ Critical Thinking Questions

1. How does the structure of skin relate to its functions of protection, temperature control, waste removal, radiation protection, vitamin production, and environmental responsiveness?

2. What problems would an animal have if the spinal column lacked its normal curvature?

3. The statement, "Structures that encourage movement do not promote stability," is fundamental to understanding the structure and function of joints. What does this statement mean to you?

4. What is the role of ATP in muscle contraction and in muscle relaxation?

5. What is the sequence of events that allows the cross-bridges to attach to the thin filaments when a muscle is stimulated by a nerve?

■ Selected Reading

Books

Hancox, N. M. 1972. *Biology of Bone*. Cambridge: Cambridge University Press.

Tregear, R. 1966. *Physical Functions of the Skin*. New York: Academic Press.

Articles

Bonn, D. Hormones for healthy bones. *New Scientist* (February 19, 1987): 32–35.

Caplan, A. I. Cartilage. *Scientific American* October, 1981.

Carafoli, F. D., and Penniston, J. T. The calcium signal. *Scientific American* November, 1985.

Cohen, C. The protein switch of muscle contraction. *Scientific American* November, 1975.

Dunant, Y., and Israel, M. The release of acetylcholine. *Scientific American* April, 1985.

Hoyle, G. How muscle is turned on and off. *Scientific American* April, 1970.

Huxley, H. E. The mechanism of muscle contraction. *Scientific American* December, 1965.

Lester, H. A. The response to acetylcholine. *Scientific American* February, 1977.

Margaria, R. The sources of muscular energy. *Scientific American* March, 1972.

Merton, P. A. How we control the contraction of our muscles. *Scientific American* May, 1972.

Montagna, W. The skin. *Scientific American* February, 1965.

Murphy, R. A. 1988. Muscle cells of hollow organs. *News in Physiological Sciences* 3:124.

Murray, J. H., and Weber, A. The cooperative action of muscles. *Scientific American* February, 1974.

Napier, J. The antiquity of human walking. *Scientific American* April, 1967.

Oster, G. Muscle sounds. *Scientific American* March, 1984.

Porter, K. R., and Franzini-Armstrong, C. The sarcoplasmic reticulum. *Scientific American* March, 1968.

Ross, R., and Borstein, P. Elastic fibers in the body. *Scientific American* June, 1971.

Timmerman, M. Nerve and muscle: Bridging the gap. *New Scientist* (September 10, 1987):63–66.

Communication I: Nerves

33

Concepts

1. The nervous system works with the endocrine (hormonal) system to communicate, integrate, and coordinate the functions of the various organs and systems in the body.
2. The flow of information through the nervous system occurs in three main steps: the collection of information from outside and inside the body (sensory activities), the processing of this information in the nervous system, and the initiation of appropriate responses.
3. The cells that transmit information in the nervous system are neurons.
4. Most transmission between neurons is accomplished by means of chemicals called neurotransmitters.
5. The nervous system of vertebrates consists of the central nervous system, made up of the brain and spinal cord, and the peripheral nervous system, which is composed of the nerves in the rest of the body.
6. Nervous systems evolved through the gradual layering of additional nervous tissue over reflex pathways of more ancient origin. Increasingly complicated sensory organs and motor structures evolved along with other changes in the nervous systems in fish through mammals.

Have You Ever Wondered:

[1] what the most numerous cell type in the nervous system is?
[2] how nerves maintain an electrical charge?
[3] how flea sprays or powders work to kill fleas but not dogs or cats?
[4] how nerve pathways are analogous to a telephone cable?
[5] how the many spinal nerves are named?
[6] why a physician hits your knee with a hammer during a routine physical examination?
[7] what part of the brain is involved with language, both written and spoken?
[8] what causes us to fall asleep?
[9] what short-term memory is used for?
[10] when dreams occur?

These and other useful questions will be answered in this chapter.

This chapter contains underlined evolutionary concepts.

The two forms of communication that integrate body functions to maintain homeostasis are: (1) neurons, which transmit electrical signals that report information or initiate a quick response in a specific tissue; and (2) hormones, which are slower, chemical signals that initiate a widespread, prolonged response, often in a variety of tissues. This chapter focuses on the neuron and the functional organization of the nervous system. In the following chapter (chapter 34), the senses and the ways in which they collect information and transmit it along nerves to the central nervous system is presented. Chapter 35 will conclude the study of communication by presenting how hormones affect long-term changes in the body's performance.

Neurons: The Basic Units of the Nervous System

The functional unit of the nervous system is a highly specialized cell called the **neuron** (Gr. *nerve*). Neurons are specialized to produce signals that can be communicated over short to relatively long distances, from one part of the body to another. Neurons have two important properties: (1) *excitability,* the ability to respond to stimuli, and (2) *conductivity,* the ability to conduct a signal.

There are three functional types of neurons. **Sensory (receptor) neurons** either act as receptors of stimuli themselves or are activated by receptors. They are stimulated by changes in the internal or external environments and respond by sending signals to the brain and spinal cord, which are the major integrating centers of the body where information is processed. (2) **Interneurons** (also known as association or connecting neurons) are located in the integrating centers and receive signals from the sensory neurons and transmit them to motor neurons. Most neurons in the body, perhaps 90%, are interneurons. (3) **Motor (effector) neurons** send the processed information via a signal to the body's effectors (e.g., muscles), causing them to contract or to glands, causing them to secrete. The flow of information in the nervous system is called *integration* and is summarized in figure 33.1.

Neuron Structure: The Key to Function

Most neurons contain three principle parts: a cell body, dendrites, and an axon (figure 33.2). The **cell body** has a large, central nucleus, which contains a prominent nucleolus, and several organelles, such as endoplasmic reticula, mitochondria, microfilaments and microtubules, Golgi apparatus, and Nissl bodies. **Nissl bodies** are made of rough endoplasmic reticulum and free ribosomes and function in the production of proteins that are precursors of special molecules called neurotransmitters, such as acetylcholine. The motor neuron shown in figure 33.2a has many short, threadlike branches called **dendrites** (Gr. *dendron,* tree), which are actually extensions of the cell body and conduct signals toward the cell body. The **axon** is a relatively long, cylindrical process that conducts signals (information) away from the cell body. The conical tapering region of the axon, where it originates from the cell body, is the *axon hillock.* The axon may be covered with a laminated lipid sheath called **myelin,** which forms a thick pad of insulation called a *myelin sheath.* Some nerves have the myelin sheath rolled into layers called a **neurolemmocyte (Schwann cell).** In these nerves, the myelin sheath is segmented at regular intervals by gaps called **nodes of Ranvier.** The neurolemmocyte also assists in the regeneration of injured myelinated nerves.

Recall that the nervous system receives data (input stimulus), integrates it, and effects a change (output response) in the physiology of the organism. In a given neuron, the dendrites are the receptors, the cell body the integrator, and the ends of the axon are the effectors.

Although the neuron is the functional unit of the nervous system, most of the cells in the system are not neurons, but **neuroglial** (Gr. *glia,* glue) **cells.** Neuroglia are nonconducting cells that protect, support, repair, and nourish neurons. They are of particular interest because they are frequently the site of tumors in the nervous system. [1]

Neuron Communication

The language (signal) of a neuron is the electrical nerve impulse or action potential. The key to this nerve impulse is the axon's plasma membrane and its properties. Changes in membrane permeability and the subsequent movement of ions produce a nerve impulse that is conducted along the plasma membrane of the dendrites, cell body, and axon of each neuron.

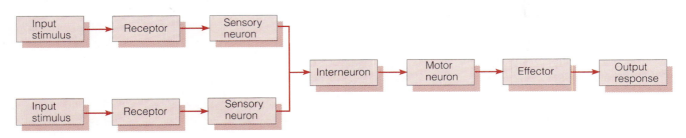

Figure 33.1 Integration. The stepwise flow of information in the nervous system.

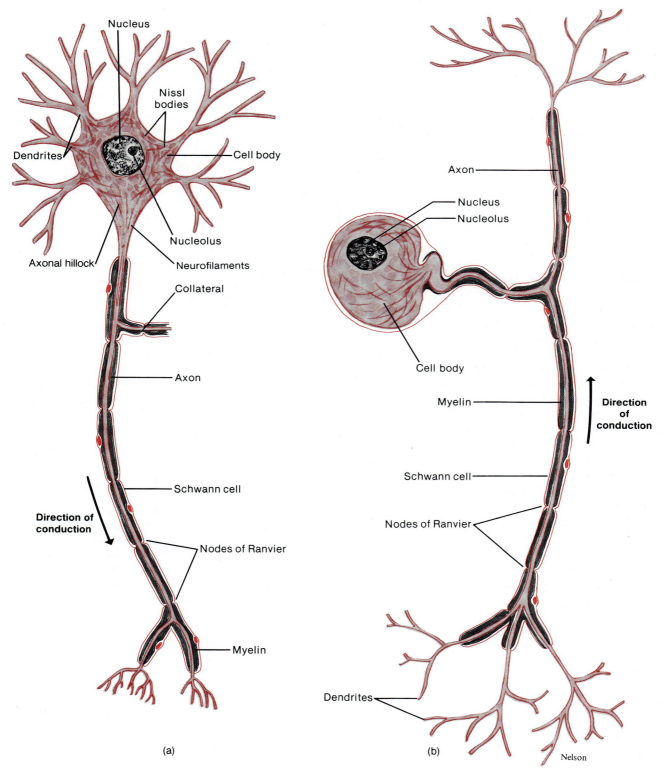

Nucleus

Nissl
bodies

Dendrites

Cell body

Nucleolus

Axonal hillock

Neurofilaments

Collateral

Axon

Schwann cell

**Direction of
conduction**

Nodes of Ranvier

Myelin

(a)

Axon

Nucleus

Nucleolus

Cell body

Myelin

**Direction
of
conduction**

Schwann cell

Nodes of Ranvier

Dendrites

(b)

Nelson

Figure 33.2 Neuron types. The structure of (a) motor and (b) sensory neurons.

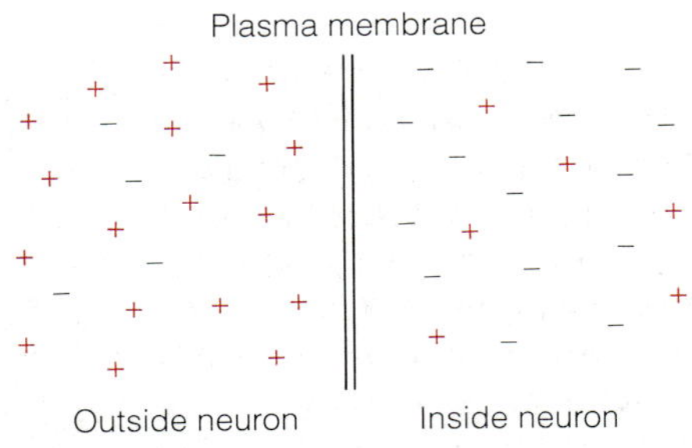

Plasma membrane

Outside neuron   Inside neuron

Figure 33.3 A resting membrane potential. Positively charged sodium ions are highly concentrated on the outside of the plasma membrane; some positive charged potassium ions, negatively charged protein molecules, and some negatively charged chloride ions are highly concentrated on the inside. When a neuron is in this resting condition, it is said to be polarized.

Resting Membrane Potential

A "resting" neuron is one that is not conducting a nerve impulse. (It may be compared to a battery that does not conduct a charge until it is switched on.) The plasma membrane of a resting neuron is polarized; the fluid on the inner side of the membrane is negatively charged with respect to the positively charged fluid outside the membrane (figure 33.3). The difference in electrical charge between the inside and the outside of the plasma membrane at any given point is due to the relative numbers of positive and negative ions on either side of the membrane, and the permeability of the plasma membrane to these ions. The difference in charge is called the *potential difference*, and it creates a potential for electrical activity along the membrane that is called the **resting membrane potential.**

The potential difference is measured in millivolts (mV). A millivolt is $1/1,000$ of a volt. Normally, the resting membrane potential is about -70mV due to the unequal distribution of various electrically charged ions and the selective permeability to the ions. Sodium (Na^+) ions are in a higher concentration in the fluid outside the plasma membrane, and potassium (K^+) and negative protein ions are in a higher concentration inside.

The Na^+ and K^+ ions are constantly diffusing and leaking through the plasma membrane, moving from regions of high concentrations to regions of lower concentration. (The large negative protein ions cannot move easily from the inside of the neuron to the outside.) However, the [2] concentration of Na^+ and K^+ ions on the inner and outer sides of the membrane remains constant due to the action of the **sodium-potassium pump,** which is powered by ATP (figure 33.4). The pump actively moves Na^+ to the outside of the cell and K^+ to the inside of the cell. Because it moves 3 Na^+ molecules out for each 2 K^+ that it moves in, the pump works to establish the potential difference across the membrane. Both ions leak back across the membrane—down their concentration gradients. K^+ however, moves more easily back to the outside, adding to the positive charge there and contributing to the membrane potential of -70mV.

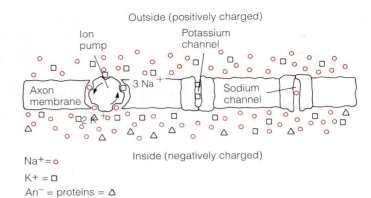

$Na^+ = \circ$
$K^+ = \square$
$An^- = proteins = \triangle$

Figure 33.4 Sodium-potassium pump and ion channels through which the ions diffuse. These mechanisms maintain a balance between the sodium ions (spheres) and potassium ions (squares) on both sides of the membrane and create a membrane potential. Note that the pump transports two K^+ for every three Na^+ ions, leading to a buildup of positive charge on the outside of the neuron; the inside thus becomes more negative relative to the outside.

Mechanism of Nerve Action: Changing the Resting Membrane Potential into the Action Potential (Nerve Impulse)

The change in the electrical potential across the plasma membrane is the key factor in the creation and subsequent conduction of a nerve impulse. A stimulus that is strong enough to initiate an impulse is called a *threshold stimulus.* When such a stimulus is applied to the resting plasma membrane, the permeability to Na^+ ions increases at the point of stimulation. The inflow of positive charge causes the membrane potential to go from -70mV toward 0. This loss in membrane polarity is called **depolarization** (figure 33.5). When depolarization reaches a certain level, special Na^+ channels (voltage-gated), that are sensitive to changes in membrane potential quickly open, and more Na^+ ions rush to the inside of the neuron. Shortly after the Na^+ ions move into the cell, the Na^+ gates close but now voltage-gated K^+ channels open, and K^+ ions rapidly diffuse outward. The movement of the K^+ ions out of the cell builds up the positive charge outside the cell again, and the membrane becomes **repolarized.** This series of membrane changes moves down the axon as an **action potential (nerve impulse).** Overall, the transmission of an action potential along the nerve plasma membrane may be visualized as a wave of depolarization and repolarization.

After each action potential, there is an interval of time where it is impossible for another action potential to occur because the membrane has become hyperpolarized (more negative than -70mV) due to the large number of K^+ ions that rushed out. This brief period is called the **refractory period.** During this period, the resting potential of the nerve membrane is being restored at the part of the membrane where the impulse has just passed. Afterwards, the fiber is "recharged" and is ready to transmit another impulse.

The Nervous System and the Senses

(a) Squid (*Sepioteuthis sepiodea*) have large neurons that are excellent for neurophysiological research.

(b) A rattlesnake (*Crotalus vergrandis*) has a pit organ beneath each eye that detects heat and allows the snake to locate warm prey in the dark.

(c) The head of a male cecropia moth contains long, feathery antennae. The antennae bear receptors by which the moth can detect minute quantities of the sex attractant released by the female of the species.

(d) The compound eye of a horsefly (*Tabanus*). Each eye is composed of densely packed photoreceptor units.

Hormones and Other Molecular Messengers

(a) A cecropia silk moth (*Hyalophora cecropia*) larva devouring a leaf. A surge of the hormone ecdysone causes the larva to form a pupa.

(b) The pupa forms a cocoon, an encasement of silk strands that enable the moth to overwinter.

(c) In the spring, the mature silk moth emerges. In the cocoon, a final surge of ecdysone, in the absence of juvenile hormone, causes the transformation into an adult moth.

(d) The owlish wings of the adult cecropia silk moth.

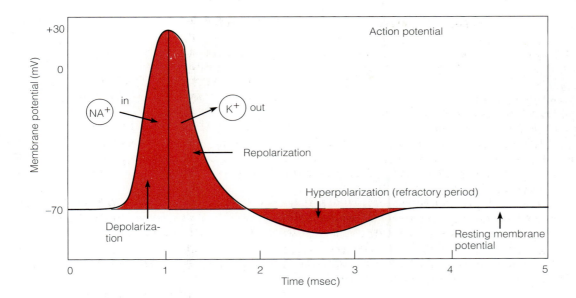

Figure 33.5 A nerve action potential as recorded on an instrument called an oscilloscope. During the depolarization phase of the action potential, Na⁺ ions rush to the inside of a neuron. The repolarization phase is characterized by a rapid increase of K⁺ ions on the outside of the neuron. The action potential is sometimes called a spike potential because of its shape on an oscilloscope screen.

A minimum stimulus (threshold) is necessary to initiate an action potential, but an increase in the intensity of the impulse does not increase the strength of the action potential. The principle that states that a neuron will fire at full power or not at all is known as the **all-or-none law.**

Saltatory Conduction

In a myelinated nerve, the action potential appears to "jump" from one node of Ranvier (*see figure 33.2*) to the next node. For this reason, conduction along myelinated fibers is known as **saltatory conduction** (L. *saltare,* to jump). It takes less time for an impulse to jump from node to node along a myelinated fiber than if it traveled smoothly along an unmyelinated fiber.

Transmission of the Action Potential Between Cells

After an action potential travels along an axon, it reaches the end of a branching axon terminal called the **end bulb (synaptic bouton).** The **synapse** is the junction between the axon of one neuron and the dendrite of another neuron. The space (junction) between the end bulb and the dendrite of the next neuron is called a **synaptic cleft** (figure 33.6). The nerve cell carrying the impulse toward a synapse is called the *presynaptic* ("before the synapse") neuron. It initiates a response in the receptive segment of a *post-synaptic* ("after the synapse") neuron leading away from the synapse. The presynaptic cell is always a neuron, but the postsynaptic cell can be a neuron, muscle cell, or gland cell.

In a synapse, two cells communicate by means of a chemical agent called a **neurotransmitter,** which is released by the presynaptic neuron. A neurotransmitter is capable of changing the resting potential in the plasma membrane of the receptive segment of the postsynaptic cell, allowing an action potential in that cell to continue the transmission of the impulse.

When a nerve impulse reaches an end bulb, it depolarizes the presynaptic membrane causing voltage-sensitive calcium channels to open. As a result, extracellular calcium ions (Ca²⁺) diffuse into the presynaptic end bulb. Calcium causes storage vesicles (containing the chemical neurotransmitter) to fuse with the plasma membrane. The neurotransmitter is released from the vesicles by exocytosis and empties into the synaptic cleft (figure 33.6c). One common neurotransmitter is the chemical *acetylcholine;* another is *norepinephrine.* (More than 50 other possible transmitters are known.)

Some of the released neurotransmitter binds with receptor protein sites in the postsynaptic membrane, causing changes in the membrane's permeability to certain ions. Due to the change in permeability in the postsynaptic membrane, Na⁺ and K⁺ ions rush through their open ion channels and permit a depolarization similar to that of the presynaptic cell. As a result, the nerve impulse is able to continue its path to an eventual effector. Once the neurotransmitter has crossed the synaptic cleft, it is quickly inactivated by being broken down into acetate and choline by the enzyme acetylcholinesterase. Without this breakdown, acetylcholine would remain and would continually stimulate the postsynaptic cell, leading to a diseased state.

Without knowing it, you have probably created a similar diseased state at the synapses of the fleas on your dog or cat. The active ingredient in most flea sprays and powders is parathion. It prevents the breakdown of acetylcholine in the fleas, as well as pets and people. However, because fleas are so small, the low dose that immobilizes the fleas does not effect the pets or humans.

[3]

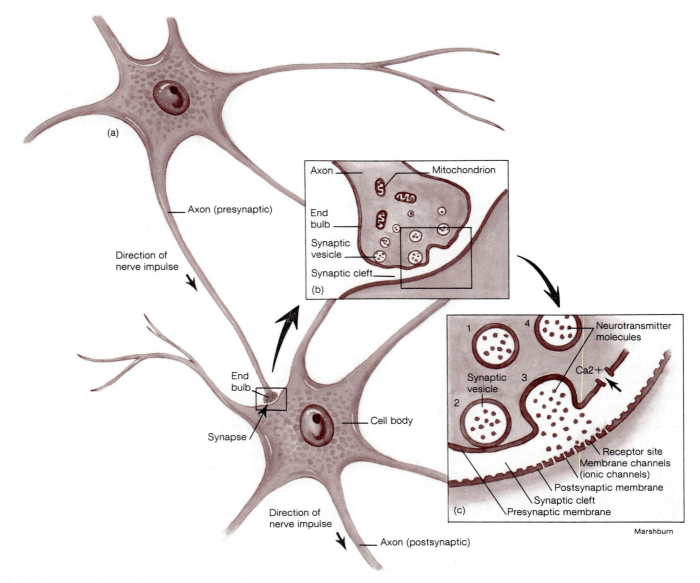

Figure 33.6 Transmission of a nerve impulse across a synapse. (a) Pre- and postsynaptic neurons with synaptic end bulb. (b) An enlarged view of the end bulb containing synaptic vesicles. (c) An enlargement of a portion of the end bulb showing exocytosis. The sequence of events in neurotransmitter release: (1) a synaptic vesicle containing neurotransmitter approaches the plasma membrane; (2) the vesicle fuses with the membrane; (3) exocytosis occurs; and (4) the vesicle reforms and begins to fill with more neurotransmitter.

Stop and Ask Yourself

1. What causes the resting membrane potential in a neuron?
2. How is a resting membrane potential maintained?
3. What are the different phases of a nerve action potential?
4. What is saltatory conduction?
5. How is the nerve action potential transmitted between nerve cells?

Invertebrate Nervous Systems

Neurons can be organized into systems to perform specialized tasks. In the previous chapters, we learned about some characteristics of nerves in particular animal groups. The trend toward more sophisticated nervous systems in invertebrate animals will be discussed now.

Among animals more complex than sponges, five general evolutionary trends in nervous system development are apparent. The first has been integrated throughout part 4 of this textbook: the more complex an animal, the more complicated its nervous system.

Of all animals, the cnidarians (hydras, jellyfishes, and sea anemones) have the simplest form of nervous organization. These animals have a nerve net (figure 33.7a). A *nerve net* is a latticework of nerve cells and fibers that permits the conduction of impulses from one area to another. Cnidarians lack brains and even local clusters of neurons. Instead, a nerve stimulus anywhere on the body initiates a nerve impulse that spreads across the nerve net to other regions of the body. In jellyfishes, this type of nervous system is involved in slow swimming movements and in

keeping the body right-side up. Despite the simplicity of such nerve nets, the neurons function at the cellular level in the way discussed earlier in this chapter.

Animals, such as flatworms and roundworms, that move in a forward direction have sense organs concentrated in the body region that first encounters new environmental stimuli. Thus, the second trend in nervous system evolution involves cephalization, which is a concentration of nervous tissue in the animal's anterior end. For example, a flatworm's nerve net contains *ganglia* (s., ganglion), which are distinct aggregations of nerve cell bodies in the head region. Ganglia function as a primitive "brain" (figure 33.7b). Distinct lateral nerve cords on either side of the body carry sensory information from the periphery to the head ganglia and carry motor impulses from the head ganglia back to muscles, allowing the animal to react to environmental stimuli.

These lateral nerve cords reveal that flatworms also exhibit the third trend in nervous system evolution: bilateral symmetry. Bilateral symmetry (a body plan with roughly equivalent right and left halves) could have led to paired nerves and muscles, paired sensory structures, and paired brain centers. This pairing facilitates coordinated ambulatory movements, such as climbing, crawling, flying, or walking.

In invertebrates, such as molluscs, segmented worms, and arthropods, the organization of the nervous system shows further advances. In these invertebrates, axons are joined into nerve cords, and, in addition to a small centralized brain, smaller peripheral ganglia help coordinate outlying regions of the animal's body. Ganglia can occur in each body segment or can be scattered throughout the body close to the organs they regulate (figure 33.7c). Regardless of the arrangement, these ganglia represent the fourth evolutionary trend: The more complex an animal, the more interneurons it will have. Because interneurons in ganglia do much of the integrating that takes place in nervous systems, the more interneurons, the more complex behavior patterns an animal can perform.

The fifth trend in the evolution of invertebrate nervous systems is a consequence of the increasing number of interneurons. The brain contains the largest number of neurons, and the more complex the animal, and the more complicated its behavior, the more neurons (especially interneurons) it will have concentrated into an anterior brain and bilaterally organized ganglia. Vertebrate brains, which will be discussed next, are the culmination of this trend.

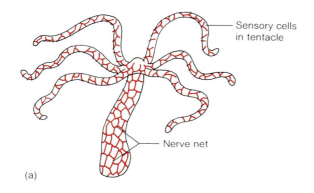

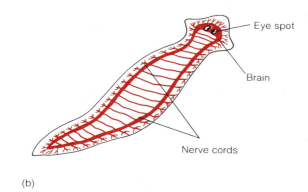

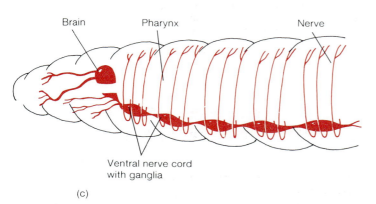

Figure 33.7 Simple invertebrate nervous systems. (a) The nerve net of *Hydra,* a cnidarian. (b) Brain and paired nerve cords of a planarian flatworm. (c) Brain, ventral nerve cord, ganglia, and the peripheral nerves of the earthworm, an annelid worm.

Vertebrate Nervous Systems

The evolution of vertebrate nervous systems is characterized by the modification of bilateral symmetry, the notochord, and the tubular nerve cord.

The *notochord* is a rod of nervous tissue encased in a firm sheath that lies ventral to the neural tube (*see figure*

10.7). It first appeared in marine chordates and is present in all vertebrate embryos, but is greatly reduced or absent in adults. In most vertebrate species, during embryological development, it is replaced by vertebrae serially arranged into a vertebral column. This vertebral column led to the development of strong muscles, allowing vertebrates to become fast-moving, predatory animals. Some of the other bones developed into powerful jaws, which facilitated the predatory nature of these animals.

A related character in vertebrate evolution was the development of a single, hollow nerve cord above the notocord. Early during evolution, the nerve cord underwent expansion and regional modification and specialization into a spinal cord and brain. Over time, the anterior end became variably thickened with nervous tissue and functionally divided into the hindbrain, midbrain, and forebrain. In the sensory world of the fast-moving and powerful vertebrates, the anterior sensory receptors became more complex and bilaterally symmetrical. For example, paired olfactory structures, such as eyes and ears, developed to better gather information from the outside environment.

The nervous system of vertebrates has two main divisions (figure 33.8). The *central nervous system* is composed of the brain and spinal cord, and is the site of information processing in the nervous system. The *peripheral nervous system* is composed of all the nerves of the body outside the brain and spinal cord. These nerves are commonly divided into two groups: *sensory* (*afferent*; L. *ad.,* toward +

ferre, to bring) nerves, which transmit information to the central nervous system; and *motor* (*efferent*; L. *ex,* away from) nerves, which carry commands away from the central nervous system. The motor nerves are in turn divided into the *voluntary* (*somatic*) nervous system, which relays commands to skeletal muscles, and the *involuntary* (*visceral* or *autonomic*) nervous system, which stimulates other muscles (smooth and cardiac) and glands of the body. The nerves of the autonomic nervous system fall into two categories called *sympathetic* and *parasympathetic* systems.

Nervous system pathways are composed of individual axons bundled together like the strands of a telephone cable. In the central nervous system, these bundles of nerve fibers are called *tracts*. In the peripheral nervous system, they are called *nerves*. The cell bodies from which the axons extend are often clustered into groups. These groups are called *nuclei* if they are in the central nervous system and *ganglia* if they are part of the peripheral nervous system. [4]

The Spinal Cord

The spinal cord serves two important functions; it is the connecting link between the brain and most of the body, and it is involved in spinal reflex actions. Thus, both voluntary and involuntary movement of the limbs, as well as certain organ functions, depend on this link.

The spinal cord is the part of the central nervous system that extends from the foramen magnum of the skull to the first lumbar vertebra (figure 33.9a). In cross section, there is a central canal that contains *cerebrospinal fluid* and a dark portion of H-shaped *gray matter* (figure 33.9b). The gray matter consists of nerve cell bodies and dendrites and is concerned mainly with reflex connections at various levels of the spinal cord. Extending from the spinal cord are the ventral and dorsal roots of the spinal nerves. These roots contain the main motor and sensory fibers, respectively, that contribute to the major spinal nerves. The white matter of the spinal cord gets its name from the myelin, which is a whitish color, that covers the nerve fibers. Many of the axons located in the white matter are bundled into sensory nerve tracts that ascend the cord and end at specific brain centers. Others are bundled into motor tracts that run in the opposite direction.

The spinal cord is surrounded by three layers of protective membranes called **meninges** (pl. of *menix,* membrane) (figure 33.9c). They are continous with similar layers that cover the brain. The outer layer, the *dura mater,* is a tough, fibrous membrane. The middle layer, the *arachnoid*, is delicate and connects to the innermost layer, the *pia mater*. The pia mater contains small blood vessels that nourish the spinal cord.

Spinal Nerves

In the human, 31 pairs of spinal nerves originate from the spinal cord. They provide a two-way communication sys-

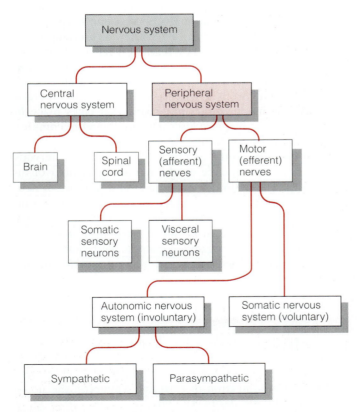

Figure 33.8 The divisions and nerves of the vertebrate nervous system.

tem between the brain and the arms, legs, neck, and trunk, via the spinal cord (figure 33.10). Although spinal nerves are not named individually, they are grouped according to the level of the vertebral column closest to where they arise [5] from the spinal cord, and each nerve is numbered in sequence. There are eight pairs of *cervical nerves* (numbered C1 to C8), twelve pairs of *thoracic nerves* (numbered T1 to T12), five pairs of *lumbar nerves* (numbered L1 to L5), five pairs of *sacral nerves* (numbered S1 to S5), and one pair of *coccygeal nerves*.

Spinal Reflexes

In addition to linking the brain and most of the body, the spinal cord also mediates reflex action. A **reflex** (L. *reflexus*, to bend back) is a predictable, involuntary response to a stimulus, such as quickly pulling your hand away from a hot stove. A reflex involving skeletal muscles is called a *somatic reflex*. A reflex involving responses of smooth muscle, cardiac muscle, or a gland cell is a *visceral* (autonomic) *reflex*. Visceral reflexes control the heart rate, respiratory rate, digestion, and many other bodily functions. Both types of reflex action allow the body to respond

quickly to external and internal changes in the environment to maintain homeostasis.

Most reflexes carried out by neurons connected to the spinal cord and not immediately involving the brain involve a receptor cell, sensory neuron, and motor neuron. Such a system is called a **reflex arc.** A few reflexes do not involve separate receptor cells or interneurons. (figure 33.11). In the *knee-jerk* or *patella reflex*, a tap of the patellar ligament suddenly stretches the quadriceps muscle, sending a nerve impulse from the receptor end of a sensory neuron to the spinal cord. Here the afferent neurons synapse directly with efferent motor neurons. The axon of the motor neuron carries the impulse rapidly to the motor end plates of the quadriceps femoris muscle group, stimulating it to contract, swinging the lower leg forward.

Part of a routine physical examination involves testing reflexes. The condition of the nervous system, particularly the functioning of the synapses, may be determined by examining reflexes. In case of injury to some portion of the nervous system, testing certain reflexes may indicate the location and extent of the injury. Also, an anesthesiologist may try to initiate a reflex to ascertain the effect of an anesthetic. [6]

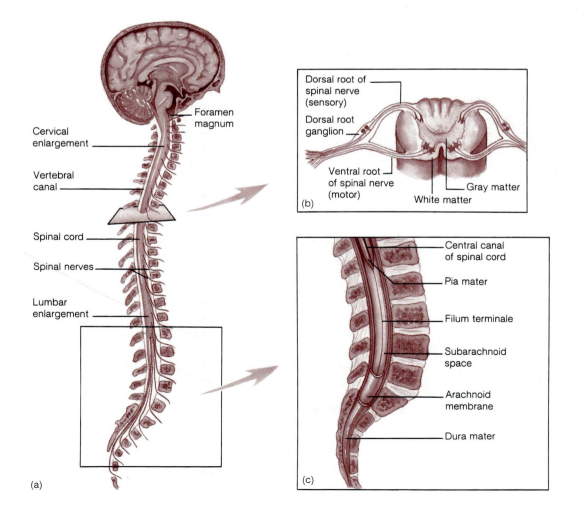

Cervical enlargement

Vertebral canal

Spinal cord

Spinal nerves

Lumbar enlargement

Foramen magnum

(a)

Dorsal root of spinal nerve (sensory)

Dorsal root ganglion

Ventral root of spinal nerve (motor)

White matter

Gray matter

(b)

Central canal of spinal cord

Pia mater

Filum terminale

Subarachnoid space

Arachnoid membrane

Dura mater

(c)

Figure 33.9 The human spinal cord. (a) The spinal cord begins at the foramen magnum. (b) A cross section through the spinal cord. (c) The three membranes (meninges) protecting the spinal cord.

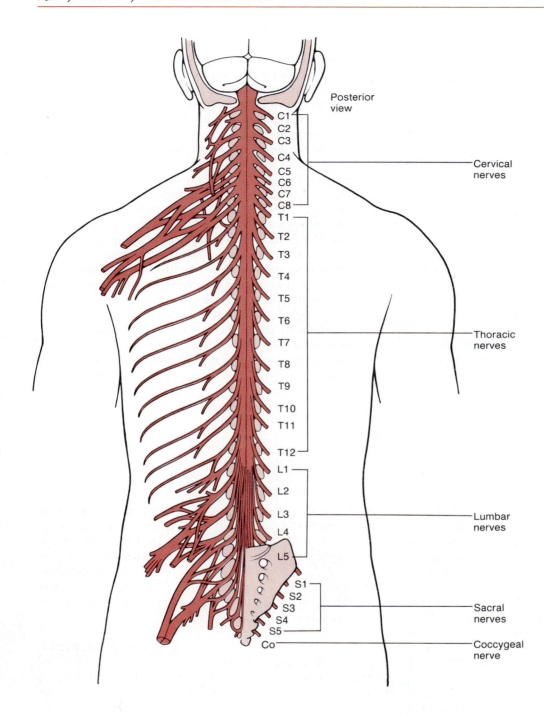

Posterior view

C1
C2
C3
C4
C5
C6
C7
C8

T1
T2
T3
T4
T5
T6
T7
T8
T9
T10
T11
T12

L1
L2
L3
L4
L5

S1
S2
S3
S4
S5

Co

Cervical nerves

Thoracic nerves

Lumbar nerves

Sacral nerves

Coccygeal nerve

Figure 33.10 An illustration of the spinal nerves. There are 31 pairs of spinal nerves in most vertebrates, such as the human shown here. They are numbered in sequence based on the part of the body where they leave the spinal cord: cervical region (C), thoracic region (T), lumbar region (L), or sacral (S) region.

The Brain

Anatomically, the brain begins as a continuation of the spinal cord. The protective meninges that wrap around the cord extend up and also wrap around the brain. The central canal of the spinal cord extends up into the brain and expands into large chambers called *ventricles*. The ventricles are filled with cerebrospinal fluid. The walls around the ventricles are composed of the actual brain tissue. During embryonic development, the brain undergoes regional ex-

pansion as a hollow tube of nervous tissue forms and develops into the primordial hindbrain, midbrain, and forebrain (figure 33.12).

Hindbrain   The hindbrain is continuous with the spinal cord and includes the medulla oblongata, cerebellum, and pons. The **medulla oblongata** is the enlargement where the spinal cord enters the brain (figure 33.12). It contains reflex centers for breathing, swallowing, cardiovascular function, and gastric secretion.

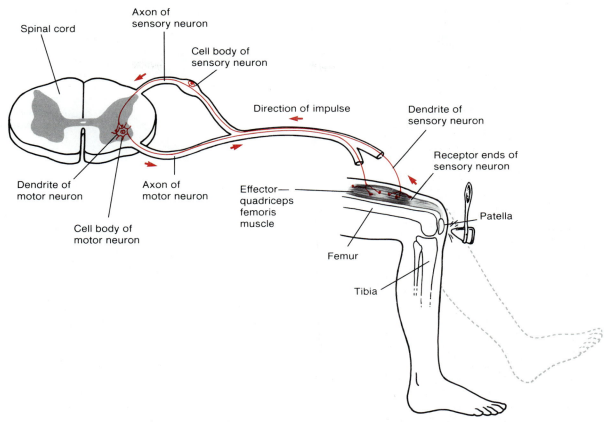

Figure 33.11 Reflex arcs. The knee-jerk reflex involves only two neurons: a sensory and motor neuron.

The **cerebellum** is an outgrowth of the medulla oblongata. It functions in the coordination of motor activity associated with limb movement, maintaining posture, and spatial orientation. It is also in this area of the brain that there is constant integration of sensory information from the eyes, muscles, tendons, skin, and inner ear. This integration serves to constantly monitor how the body and limbs are positioned, how much certain muscles are contracted, and in what direction the body limbs are moving. When the cerebellum receives commands from higher brain centers, it also uses this information to produce the most effective response possible.

The **pons** is a bridge of transverse nerve tracts from the cerebrum of the forebrain to both sides of the cerebellum. It also contains tracts that connect the forebrain and spinal cord.

During vertebrate evolution, the medulla and pons of the hindbrain have changed very little, whereas the cerebellum has grown noticeably (figure 33.13).

Midbrain The midbrain was originally a center for coordinating reflex responses to visual input. As the brain evolved, it took on added functions relating to tactile (touch) and auditory (hearing) input, but it did not change in size. The roof of the midbrain is called the *optic tectum*. It is a thickened region of gray matter where visual and auditory signals are integrated. In humans and mammals, most sensory information passes through the tectum on its way to higher brain centers (the forebrain) for processing. However, it still controls the reflexes of the iris of the eye and eyelids, and analyzes and relays information coming in from the ear via the auditory nerve. In fish and amphibians, the optic tectum is the largest part of the brain because it is the major integration center for their uniquely developed sensory abilities.

Forebrain The vertebrate forebrain has changed a great deal during vertebrate evolution. The forebrain has two main parts: the *diencephalon* and *telencephalon*(*see figure 33.12*). The diencephalon lies just in front of the midbrain and contains the pineal gland, pituitary gland, hypothalamus, and thalamus. In higher vertebrates, the *thalamus* relays all sensory information to higher brain centers. The *hypothalamus* lies below the thalamus. This area controls many functions, such as regulation of body temperature, sexual drive, carbohydrate metabolism, hunger, and thirst, via neural and endocrine (hormonal) action. The *pineal endocrine gland* controls some body rhythms. The *pituitary* is a major endocrine gland and will be discussed in detail in chapter 35.

In fish and amphibians, the diencephalon processes sensory (olfactory) information. In reptiles and birds, the most important part of the brain is the *corpus striatum*, which plays a role in their complex behavior patterns.

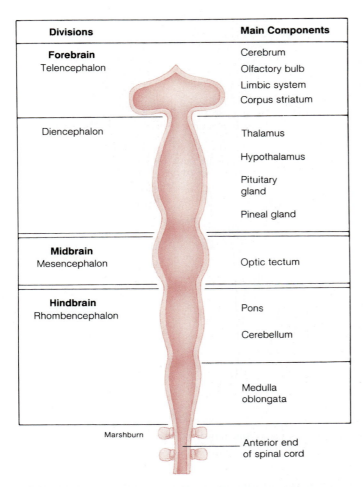

Divisions	Main Components
Forebrain Telencephalon	Cerebrum Olfactory bulb Limbic system Corpus striatum
Diencephalon	Thalamus Hypothalamus Pituitary gland Pineal gland
Midbrain Mesencephalon	Optic tectum
Hindbrain Rhombencephalon	Pons Cerebellum Medulla oblongata
	Anterior end of spinal cord

Marshburn

Figure 33.12 The human brain. Summary of the three major subdivisions of the human brain and some of the structures they contain.

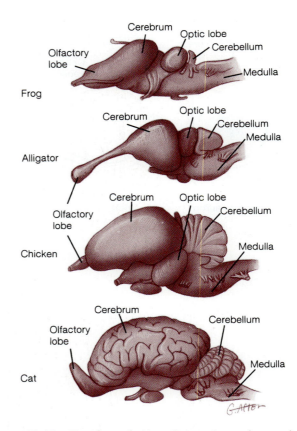

Figure 33.13 Vertebrate brains. Comparison of several vertebrate brains, as viewed from the side. The drawings are not drawn to the same scale. Notice the increase in size of the cerebrum from amphibians (frogs) to mammals (cat).

As the diencephalon slowly expanded during evolution to handle more and more sensory functions, the telencephalon (the front part of the forebrain) expanded rapidly in both size and complexity.

Just above the corpus striatum is the **cerebrum,** which is divided by a large fissure (groove) into *right* and *left cerebral hemispheres* (figure 33.14a). The parts of the brain related to sensory and motor integration changed greatly as vertebrates became more agile and inquisitive animals. Many functions shifted from the optic tectum to the expanding cerebral hemispheres. The increasing importance of the cerebrum affected many other brain regions, especially the thalamus and cerebellum. In mammals, there is a progressive increase in size and complexity of the outermost part of the cerebrum, called the *cerebral cortex.* This layer folds back on itself to a remarkable extent, suggesting that evolution of the mammalian cerebrum outpaced enlargement of the skullbones housing it. The original, deeper layers of the cerebrum, the *tegmentum* and *limbic structures,* help regulate emotional states and short-term memory (figure 33.14b). The cerebral cortex lies above these areas and is composed primarily of gray matter (unmyelinated neurons).

Different parts of the cerebrum have specific functions. For example, the cerebral cortex contains primary sensory areas and primary motor areas (figure 33.14c). Other areas of the cortex are involved in perception of visual or auditory signals from the environment, and in humans, the ability to use language—both written and spoken. [7]

In primitive mammals, each area of the cerebral cortex has specific sensory or motor functions. However, in mammals, such as primates (monkeys, apes, and humans), large areas of the cerebral hemispheres have no known function. Most neurobiologists believe that these areas are important in certain behavioral patterns and for the capacity of abstract thinking. One's "personality" may also be influenced by these areas.

Many tracts of white matter (myelinated neurons) are interspersed with the gray matter. These white-matter tracts connect various parts of the brain and communicate with the rest of the body. For example, the right and left cerebral hemispheres are connected by a large tract called the *corpus callosum* (*see figure 33.14b*). Its function is to communicate information from one side of the brain to the other. The two sides of the brain, however, function independently of one another.

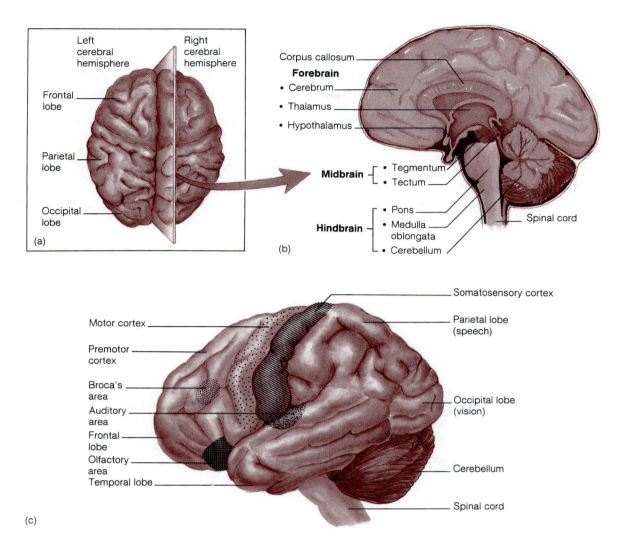

Figure 33.14 The human brain. (a) Left and right cerebral hemispheres. (b) Structures present in the major regions of a fully developed brain. The hindbrain is composed of the medulla oblongata, pons, and cerebellum. The optic tectum is found in the midbrain. The forebrain is composed of the hypothalmus, thalamus, and cerebrum. (c) Diagram of the cerebral cortex showing its four lobes and the primary projection areas of some of the senses.

Reticular Formation The **reticular formation** extends from the medulla into the thalmus. Its cells receive input from all types of sensory neurons entering the brain. Its output goes back down the spinal cord, where it amplifies or reduces incoming sensory signals. By this process, the nervous system's sensitivity to stimuli is adjusted. This part of the brain has changed little during vertebrate evolution.

The **reticular activating system** (**RAS**) extends from the reticular formation into the cerebral cortex. It acts as a filter that determines which sensory information reaches the level of consciousness in the cortex. Overall, the reticular formation and reticular activating system are responsible for general levels of consciousness, lethargy or liveliness, coma or total alertness. For example, sleep centers are located in the RAS. One center contains neurons that release a neurotransmitter called *serotonin*. Serotonin [8] has an inhibitory effect on the RAS; thus, high serotonin levels lead to drowsiness and sleep. Other neurotransmitters released from the RAS counteract the effects of serotonin, enabling the RAS to maintain the waking state.

Learning and Memory Learning requires memory, but what permits memory to occur is not yet known. Some believe that learning results from changes in nerve transmission at existing synapses rather than from the creation of new synapses.

Several types of learning exist. In humans, *short-term memory* (STM) is the ability to recall a telephone number long enough to dial the number. The capacity for STM appears to be limited to seven or eight bits of information. *Long-term memory* (LTM), such as the ability to recall the events of the day, has been shown to require the use of neurotransmitters in the limbic system, the use of the hippocampus, and protein synthesis. Memory appears to be stored in the association areas of the cerebrum. [9]

Brain Waves The electrical activity of the brain can be recorded in the form of an **electroencephalogram (EEG)**. When taking an EEG, electrodes are taped to different parts of the scalp, and an instrument called an electroencephalograph records the brain waves (figure 33.15).

When a human is awake, two types of waves are present: Alpha waves (α) have a frequency of about 6 to 13 per second and predominate when the eyes are closed. Beta waves (β) have a higher frequency of about 15 to 30 per second and predominate when the eyes are open.

During a normal 8-hour sleep pattern, there are usually five times when the brain waves become slower and have a larger amplitude than alpha waves. During each of these times, there are irregular flurries as the eyes move back and forth rapidly. When a human is awakened during the latter, called **REM** (*rapid eye movement*) **sleep**, he/she always reports that dreams were occurring. Breathing irregularities, faster heart rate, and twitching fingers accompany REM sleep.

[10]

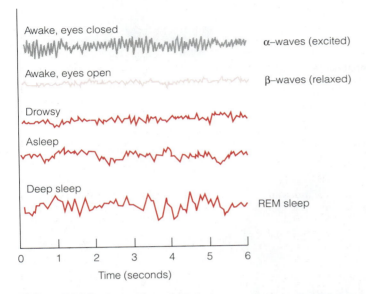

Figure 33.15 labels:
- Awake, eyes closed — α–waves (excited)
- Awake, eyes open — β–waves (relaxed)
- Drowsy
- Asleep
- Deep sleep — REM sleep
- Time (seconds): 0 1 2 3 4 5 6

Figure 33.15 Examples of EEG wave patterns. Recordings were made while the subject was excited, relaxed, and in various stages of sleep. During excitement, the brain waves are rapid, whereas in sleep they are much slower and of greater amplitude.

Stop and Ask Yourself

6. How is the brain of a vertebrate related to the spinal cord?
7. What are the three parts of the vertebrate hindbrain?
8. What is the function of the midbrain?
9. What are the different parts of the forebrain and their respective functions?

Cranial Nerves

In addition to the paired spinal nerves, the peripheral nervous system includes paired *cranial nerves* (figure 33.16 and table 33.1). Some of the nerves (e.g., optic nerve) contain only sensory axons, which carry signals to the brain. Others contain sensory and motor axons and are termed mixed nerves. For example, the vagus nerve has sensory axons leading to the brain as well as motor axons leading to the heart and smooth muscles of the visceral organs in the thorax and abdomen.

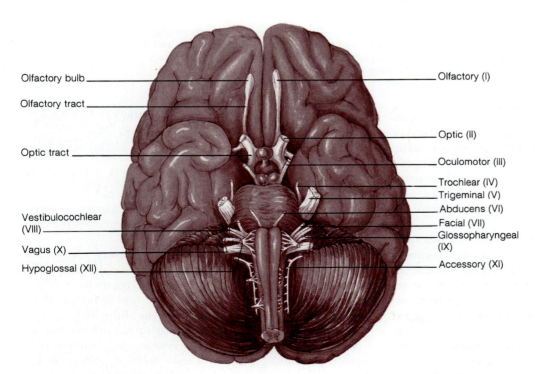

Figure 33.16 labels:
- Olfactory bulb
- Olfactory tract
- Optic tract
- Vestibulocochlear (VIII)
- Vagus (X)
- Hypoglossal (XII)
- Olfactory (I)
- Optic (II)
- Oculomotor (III)
- Trochlear (IV)
- Trigeminal (V)
- Abducens (VI)
- Facial (VII)
- Glossopharyngeal (IX)
- Accessory (XI)

Figure 33.16 Cranial nerves in the human brain. Except for the first pair, the cranial nerves arise from the brain stem. They are identified either by a Roman numeral indicating their order or by names describing their function or general distribution.

Table 33.1   Functions of the Cranial Nerves

Nerve	Type	Innervation
I Olfactory	Sensory	Smell
II Optic	Sensory	Vision
III Oculomotor	Primarily motor	Eyelids, eyes, adjust amount of light entering eyes, and focus lenses (motor)
		Condition of muscles (sensory)
IV Trochlear	Primarily motor	Eyes (motor)
		Condition of muscles (sensory)
V Trigeminal	Mixed	
Ophthalmic division		Eyes, tear glands, scalp, forehead, and upper eyelids (sensory)
Maxillary division		Upper teeth, upper gum, upper lip, lining of the palate, and skin of the face (sensory)
Mandibular division		Scalp, skin of the jaw, lower teeth, lower gum, and lower lip (sensory)
		Jaws, floor of the mouth (motor)
VI Abducens	Primarily motor	Eyes (motor)
		Condition of muscles (sensory)
VII Facial	Mixed	Taste receptors of the anterior tongue (sensory)
		Facial expression, tear glands, and salivary glands (motor)
VIII Vestibulocochlear	Sensory	
Vestibular branch		Equilibrium
Cochlear branch		Hearing
IX Glossopharyngeal	Mixed	Pharynx, tonsils, posterior tongue, and carotid arteries (sensory)
		Pharynx and salivary glands (motor)
X Vagus	Mixed	Speech and swallowing, heart, and visceral organs in the thorax and abdomen (motor)
		Pharynx, larynx, esophagus, and visceral organs of the thorax and abdomen (sensory)
XI Accessory	Motor	
Cranial branch		Soft palate, pharynx, and larynx
Spinal branch		Neck and back
XII Hypoglossal	Motor	Tongue

The Autonomic Nervous System

The vertebrate autonomic nervous system is composed of two divisions that act antagonistically (in opposition to each other) to control the body's involuntary muscles (smooth and cardiac) and glands (box 33.1). The **parasympathetic nervous system** comprises nerves that arise from the brain and sacral region of the spinal cord (figure 33.17). It consists of a network of long efferent nerve fibers that synapse at ganglia in the immediate vicinity of organs, and short efferent neurons that extend from the ganglia to the organs. The **sympathetic nervous system** comprises nerves that arise from the thoracic and lumbar regions of the spinal cord (figure 33.17). It is a network of short, efferent central-nervous-system fibers that extend to ganglia located near the spine, and long efferent neurons extending from the ganglia directly to each organ. Organs often receive input from both the parasympathetic and sympathetic systems. For example, parasympathetic input is responsible for stimulation of salivary gland secretions, stimulation of intestinal movements, contraction of pupillary muscles in the eyes, and relaxation of sphincter muscles. Sympathetic input controls antagonistic actions: inhibition of salivary gland secretions and intestinal movements, relaxation of pupillary muscles, and contraction of sphincters.

Stop and Ask Yourself

10. What three characteristics affected evolutionary development of the nervous system?
11. What part of the vertebrate forebrain became increasingly complex, especially in mammals?
12. How are cranial nerves named? Can you name the twelve pairs found in humans?
13. How would you describe the parasympathetic nervous system? The sympathetic system?

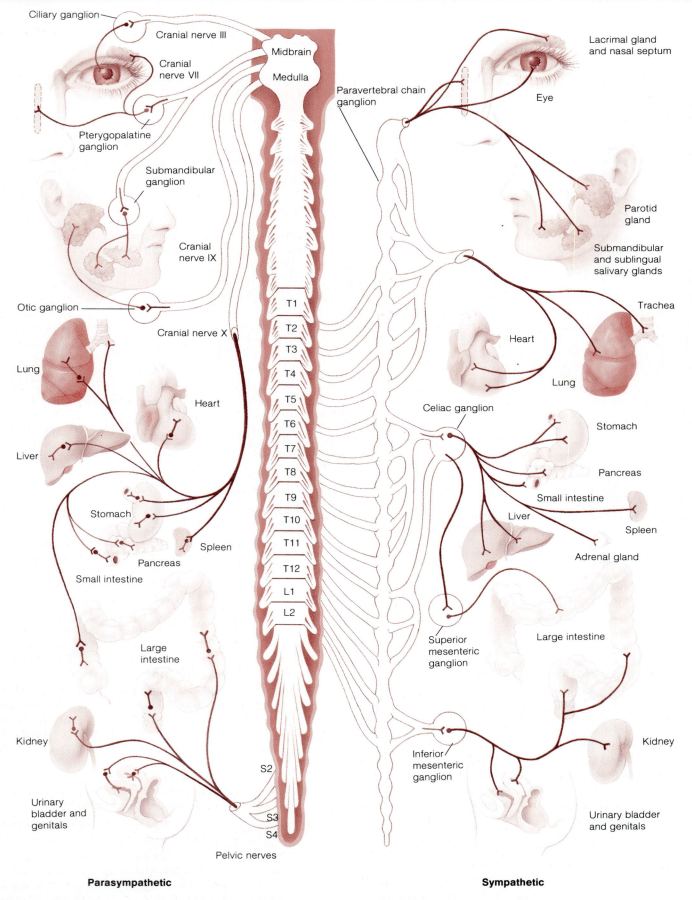

Figure 33.17 The autonomic nervous system. (Left) The parasympathetic system arises from the brain and sacral region of the spinal cord. (Right) The sympathetic system arises from the thoracic and lumbar regions of the spinal cord.

Box 33.1

The Autonomic Nervous System and Biofeedback

The activity of the autonomic nervous system in response to an ever-changing internal environment goes on without our conscious control or awareness. Some of these autonomic and involuntary functions involve the heart rate, blood pressure, breathing, blood-glucose levels, and body temperature. We can have some conscious control over these activities through **biofeedback**—a technique that allows a person to monitor and control his/her own bodily functions.

In biofeedback, a mechanical device is used to register and display signs of physiological responses to make the person aware of them and to enable him/her to monitor changes in them. In general, any physiological function that can be recorded, amplified by electronic means, and fed back to the person through any of the senses can be regulated to some extent by that person.

Biofeedback can work because every change in a physiological state is accompanied by a responsive change in a mental (or emotional) state. Conversely, every change in a mental state is accompanied by a change in the physiological state. The autonomic nervous system thus acts as a connecting link between the mental and physiological states.

Biofeedback control holds considerable promise as a means of treating some psychosomatic problems, (illnesses caused mainly by stress or psychological factors), such as ulcers, anxiety, and phobias. Biofeedback can also relieve muscle-tension headaches, reduce the pain of migraine headaches, achieve relaxation during childbirth, lower blood pressure, alleviate irregular heart rhythms, and control epileptic seizures. The greatest value of these biofeedback techniques is that they demonstrate that the autonomic nervous system is not entirely autonomous and automatic; some visceral responses can be controlled.

From Robert Carola, John P. Harley, and Charles Noback. 1990. *Human Anatomy and Physiology*. McGraw-Hill, Inc., New York.

■ Summary

1. The functional unit of the nervous system is the neuron. Neurons are specialized to produce signals that can be communicated from one part of the body to another. Neurons have two important properties: excitability and conductivity.

2. Neurons have three principle parts: a cell body, dendrites, and an axon.

3. The language (signals) of neurons is the electrical impulse or action potential.

4. The plasma membrane of a resting neuron is said to be polarized, meaning that the intracellular fluid on the inner side of the membrane is negatively charged with respect to the positively charged extracellular fluid outside the membrane. This state of affairs is maintained by the sodium-potassium pump and diffusion through membrane channels.

5. When a threshold stimulus is applied to a resting neuron, it causes it to depolarize and initiate an action potential. In myelinated neurons, the action potential jumps from one node of Ranvier to the next in a process known as saltatory conduction.

6. A nerve action potential is transmitted between cells at the synapse. This type of transmission is chemical because neurotransmitter molecules are involved.

7. Among animals more complex than sponges, five general evolutionary trends in the nervous system are apparent.

8. The vertebrate nervous system has two main divisions: the central nervous system is composed of the brain and spinal cord, and the peripheral nervous system is composed of all the nerves of the body outside the brain and spinal cord.

9. The spinal cord serves two important functions: it is the connecting link between the brain and most of the body, and it is involved in spinal reflex actions. A reflex is a predictable involuntary response to a stimulus.

10. The human body consists of 31 pairs of spinal nerves, which originate from the spinal cord.

11. The vertebrate brain can be divided into the hindbrain, midbrain, and forebrain. The hindbrain is continous with the spinal cord and includes the medulla oblongata, cerebellum, and pons.

12. The midbrain is a thickened region of gray matter where visual and auditory signals are integrated. The forebrain has two main parts: the diencephalon and telencephalon. The diencephalon contains the pineal gland, hypothalamus, and thalamus. The telencephalon expanded during evolution to give rise to the cerebrum.

13. The reticular formation and reticular activating system act as a filter to determine which sensory information reaches the level of consciousness in the cerebral cortex.

14. There are two types of learning: short-term and long-term.

15. The electrical activity of the brain can be recorded in the form of an electroencephalogram (EEG).

16. In addition to the paired spinal nerves, the peripheral nervous system includes 12 pairs of cranial nerves in reptiles, birds, and mammals. Fish only have the first 10 pairs.

17. The autonomic nervous system consists of two antagonistic parts: the sympathetic and parasympathetic divisions.

■ Key Terms

action potential (nerve impulse) (p. 524)
all-or-none law (p. 525)
axon (p. 522)
biofeedback (p. 535)
cell body (p. 522)
cerebellum (p. 530)
cerebrum (p. 532)
dendrites (p. 522)
depolarization (p. 524)
electroencephalogram (EEG) (p. 534)
end bulb (synaptic bouton) (p. 525)
interneurons (p. 522)
medulla oblongata (p. 530)
meninges (p. 528)
motor (effector) neurons (p. 522)
myelin (p. 522)
neuroglial cells (p. 522)
neurolemmocyte (Schwann cell) (p. 522)
neuron (p. 522)
neurotransmitter (p. 525)

Nissl bodies (p. 522)
nodes of Ranvier (p. 522)
parasympathetic nervous system (p. 537)
pons (p. 531)
reflex (p. 529)
reflex arc (p. 529)
refractory period (p. 524)
REM sleep (p. 534)
repolarized (p. 524)
resting membrane potential (p. 524)
reticular activating system (RAS) (p. 533)
reticular formation (p. 533)
saltatory conduction (p. 525)
sensory (receptor) neurons (p. 522)
sodium-potassium pump (p. 524)
sympathetic nervous system (p. 537)
synapse (p. 525)
synaptic cleft (p. 525)

■ Critical Thinking Questions

1. Would it not be more efficient to construct a reflex from one all-purpose neuron rather than interrupting the circuit with one or more synapses? Explain.
2. How can a neuron integrate information?
3. Surveying the functions of the evolutionarily oldest parts of the vertebrate brain gives us some idea of original functions of the brain. Explain this statement.
4. Many neurobiologists have stated that the hindbrain of a human is the most primitive element. How would you respond to this proposition?
5. What possible advantages and disadvantages are there in the evolutionary trend toward cephalization of the nervous system?

■ Suggested Readings

Books

Allport, S. 1986. *Explorers of the Black Box: The Search for the Cellular Basis of Memory*. New York: W.W. Norton, Co.

Articles

Bloom, F. E. Neuropeptides. *Scientific American* October, 1981.

Cave, L. J. 1983. Brain's unsung cells. *BioScience* 33:614–21.

Cowan, W. The development of the brain. *Scientific American* September, 1979.

Crick, F. H. C. Thinking about the brain. *Scientific American* September, 1979.

Dunant, Y., and Isrel, M. The release of acetylcholine. *Scientific American* April, 1985.

Gottlieb, D. I. GABAergic neurons. *Scientific American* February, 1988.

Hubel, D. H. Special issues on neurobiology. *Scientific American* September, 1979.

Keynes, R. Ion channels in the nerve-cell membrane. *Scientific American* March, 1987.

Kimelberg, H. K., and Norenberg, M. D. Astrocytes. *Scientific American* April, 1989.

Lester, H. The response of acetylcholine. *Scientific American* February, 1977.

Linas, R. Calcium in synaptic transmission. *Scientific American* October, 1982.

Luria, A. R. The functional organization of the brain. *Scientific American* March, 1970.

Morell, P., and Norton, W. Myelin. *Scientific American* May, 1980.

Wurtman, R. Nutrients that modify brain function. *Scientific American* August, 1982.

Communication II: Senses

Concepts

1. Sense organs permit an animal to detect changes in its body and objects and events in the world around it. Information collected by a sense organ is passed to the nervous system, which determines, evaluates, and initiates an appropriate response.
2. Sensory receptors initiate nerve impulses by opening channels in sensory neuron plasma membranes, depolarizing them, and causing a generator potential. Receptors differ from one another with respect to the nature of the environmental stimulus that triggers an eventual nervous impulse.
3. Many kinds of receptors have evolved among invertebrates and vertebrates, and each receptor is sensitive to a specific type of stimulus.
4. Each animal species has a unique perception of its body and environment due to the nature of its sensory receptors.

Have You Ever Wondered:

[1] if sounds occur when there is no one to hear them?
[2] which type of general sense receptor is most abundant in the human body?
[3] why metal feels colder on the skin than wood does?
[4] why a needle inserted into the skin produces great pain, but if the same needle is inserted directly into brain tissue, no pain is felt?
[5] why our sense of taste diminishes as we grow older?
[6] how a sturgeon can taste its food before it reaches its mouth?
[7] why a dog "sniffs" when it detects an odor?
[8] why dogs can hear the sound from a high-pitched dog whistle while humans cannot?
[9] what causes "sea-sickness?"
[10] why we sometimes see "spots" in front of our eyes?
[11] why some animals can see colors but others cannot?

These and other useful questions will be answered in this chapter.

This chapter contains underlined evolutionary concepts.

About 2,000 years ago, Aristotle identified five senses—sight, hearing, smell, taste, and touch—commonly referred to as the "five senses." Today we know that other senses exist in animals; these include a sense of equilibrium (balance), a sense of body movement, fine touch, touch-pressure (deep pressure), heat, cold, pain, and various tactile senses. In addition, receptors in the circulatory system register changes in blood pressure and blood levels of carbon dioxide and hydrogen ions, and receptors in the digestive system are involved in the perception of hunger and thirst. All of this information is used to help animals maintain homeostasis.

Overall, our impressions of the world are limited and defined by the senses. In fact, all knowledge and awareness depend on the reception and decoding of stimuli from the external environment and from within an animal's body. This chapter continues to examine how animals communicate externally and internally by discussing the senses.

Sensory Reception

[1] A favorite question is: Do sounds occur when there is no one to hear them? The answer is no. A "sound" is something that is received (*sensed*) by the ear and "heard" (*perceived*) by the brain. If there is no ear to receive sound waves, and no brain to translate those sound waves into what we consciously recognize as the "sound" of thunder, the thunder will send out sound waves, but there will be no perceived "sound." The same is true for the other senses.

Sensory receptors are structures made of cells that are capable of converting environmental information (*stimuli*) into nerve impulses. A *stimulus* is any form of energy an animal is able to detect with its receptors. All receptors are *transducers* ("to change over"), that is, they convert one form of energy into another. Because all nerve impulses are the same, different types of receptors convert different kinds of stimuli, such as light or heat, into a local electrical potential called a **generator potential.** If the generator potential reaches the sensory neuron's threshold, it will cause channels to open in its plasma membrane and create an action potential (*see figure 33.5*). The impulse then travels along the cell's axon toward a synaptic junction and becomes information going to the central nervous system.

As presented in chapter 33, all action potentials are alike. Furthermore, an action potential is an all-or-none phenomenon; it either occurs or it doesn't. How, then, does a common action potential give rise to different sensations, such as taste, color, or sound, or different degrees of sensation? Some nerve signals from specific receptors always end up in a specific part of the brain for interpretation, therefore, a stimulus that goes to the optic center will be interpreted as a visual stimulus. Another factor that characterizes a particular stimulus is the intensity of the stimulus. When the strength of the stimulus increases, the number of action potentials per unit of time also increases.

Thus, the passage of an all-or-none action potential can be used to transmit information about the intensity and type of stimulus by using the timing of the impulses and the "wiring" of neurons as additional sources of information for perception by the brain.

Sensory receptors have the following basic features:

1. They contain sensitive receptor cells or finely branched peripheral endings of sensory neurons that respond to a stimulus by causing a generator potential.
2. Their structure is designed to receive a specific stimulus.
3. Their primary receptor cells synapse with afferent nerve fibers that travel to the central nervous system along specific neural pathways.
4. In the brain, the nerve impulse is translated into a recognizable sensation, such as sound.

Sensory receptors can be classified by the stimuli to which they respond:

1. **Thermoreceptors** are widely distributed throughout the dermis of the skin and respond to temperature changes. They include infrared receptors.
2. **Nociceptors** or **free nerve endings** respond to potentially harmful stimuli that produce pain. Several million free nerve endings are distributed throughout the skin and internal tissues of vertebrates.
3. **Chemoreceptors** respond to chemical stimuli associated with taste and smell, and to changes in the level of carbon dioxide and hydrogen ions in the blood.
4. **Photoreceptors** respond to the energy of visible light waves.
5. **Mechanoreceptors** respond to and monitor physical stimuli associated with touch-pressure, muscle tension, joint position changes, air vibrations in the ear, and body movements and equilibrium. *Baroreceptors* (Gr. *baros*, weight) are a specialized type of mechanoreceptor that responds to changes in blood pressure.

Sight, hearing, equilibrium, smell, and taste are known as the *special senses* because they are found in restricted regions of the vertebrate body. The *general senses,* which include deep-pressure, heat, cold, pain, and body position, are widely distributed over the vertebrate body. Table 34.1 summarizes some of the vertebrate senses and their associated stimuli. Keep in mind, however, that the kind and number of receptors varies among different vertebrates. For example, humans do not have the infrared-sensitive receptors that a rattlesnake uses to detect warm-blooded prey in the dark (colorplate 15b). Unlike dogs, humans have more receptors that allow them to perceive color, whereas dogs have more olfactory receptors in their noses than do humans. Hence, humans and dogs sample the environment in different ways and have different perceptions of it.

Sensory nerve pathways from the different receptors lead to different parts of the cerebral cortex. Signals from receptors in the joints and skin travel to the somatosensory

Table 34.1	Some Vertebrate Receptors Associated with Different Sensory Modalities	
Receptors	**Sense**	**Stimulus**
Special		
Taste buds	Taste	Chemicals[a]
Olfactory cells	Smell	Chemicals[a]
Ear	Hearing	Sound waves[b]
	Equilibrium	Mechanical displacement[b]
Eye	Sight	Light[c]
General		
Temperature	Hot or cold	Heat flow[d]
Light touch	Touch	Touching skin[b]
Touch-pressure	Touch	Mechanical displacement of skin[b]
Free nerve endings	Pain	Tissue damage[a]
Proprioceptors	Limb placement	Mechanical displacement[b]

[a]Chemoreceptors; [b]Mechanoreceptors; [c]Photoreceptors; [d]Thermoreceptors.

Figure 34.1 A sensory homunculus map depicts the areas of the postcentral gyrus of the cerebral cortex devoted to sensation in different parts of the body. Notice that a disproportionately large area of the cortex is devoted to the fingers and tips.

cortex (*see figure 33.14*). Nervous tissue in this area is laid out like a map corresponding to the body surface. Some map regions are larger than others because they represent areas that are functionally more important and that have more receptors. For example, in humans, a large part of the somatosensory cortex receives stimuli from receptors located in the fingers, lips, and thumbs (figure 34.1).

Stop and Ask Yourself

1. What is a sensory receptor?
2. What are the basic characteristics of a sensory receptor?
3. What are the types of sensory receptors based on the stimuli to which they respond?

General Senses

Touch

Many animals rely on tactile stimuli to respond to contact with their environment. Insects are well endowed with hairs (setae) that are sensitive to both touch and vibration. The receptors provide them with information used in activities such as feeding, mating, and building nests. Sensory receptors in vertebrate skin detect stimuli that the brain interprets as light touch, touch-pressure, vibration, heat, cold, and pain.

Light touch is perceived when the skin is touched, but not deformed. Receptors of light touch include **free nerve endings** and **tactile (Meissner's) corpuscles** (figure

34.2). Free nerve endings are the most widely distributed receptors in the body, and are involved with pain and thermal stimuli, as well as light touch. The **bulbs of Krause** are mechanoreceptors, found in the dermis in certain parts of the body, that respond to and monitor some physical stimuli, such as position changes. [2]

The difference between light touch and touch-pressure (or deep pressure) on your skin can be shown by gently touching a pencil (light touch) and then squeezing the pencil as hard as you can (touch-pressure). Touch-pressure causes a deformation of the skin. Sensations of touch-pressure last longer than sensations of light touch, and are felt over a larger area. Receptors for touch-pressure are **Pacinian corpuscles** and the **organs of Ruffini**.

Temperature

Skin receptors for heat and cold (thermoreceptors) are the free nerve endings. Cold receptors respond to temperatures below skin temperature, and heat receptors respond to temperatures above skin temperature. So-called *cold spots* and *warm spots* are found over the surface of a mammalian body. A spot refers to a small area of several nerve endings

[3] that, when stimulated, yields a temperature sensation of warmth or cold. For example, when metal is placed on the skin, it absorbs heat and one feels a sense of coldness. If wood is placed on the skin, it absorbs less heat and therefore, feels warmer than the metal. The ability to detect changes in temperature has become well developed in a number of animals. Insects that require a meal of blood to complete their lifecycle (e.g., mosquitoes and bedbugs) can detect mammalian hosts by means of thermoreceptors. Some predatory snakes, such as rattlesnakes and pit vipers, have thermoreceptors in sensory pits on their heads that allow them to find their warm-blooded prey, especially at night, by sensing the infrared thermal radiation emitted by the prey (colorplate 15b).

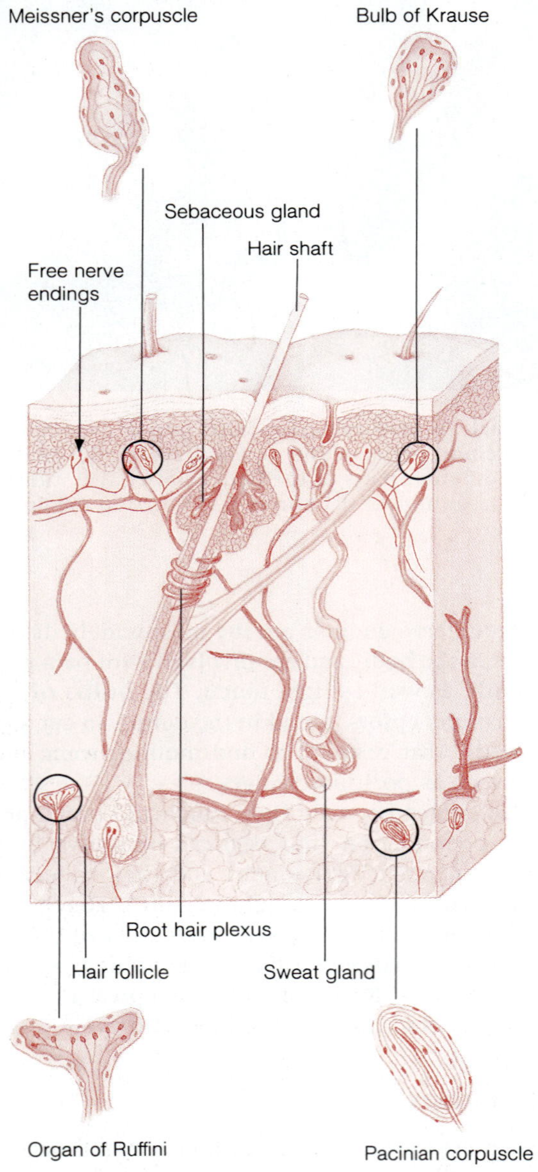

Pain

Pain receptors are specialized free nerve endings that are present throughout the body, except for the brain and intestines. These pain endings are also called **nociceptors**. There are about 3 million pain receptors distributed over the surface of the human body and the perception of pain seems to have an important emotional aspect mediated by the limbic system (box 34.1). [4]

Proprioception

Receptors in the muscles, tendons, and joints transmit impulses about the position of the body up the dorsal columns of the spinal cord. These impulses provide an animal with an awareness of the position of its body and its parts without actually seeing them. This sense of position is called **proprioception** (L. *proprius,* one's self + receptor) or the *kinesthetic sense.*

Figure 34.2 Different sensory receptors. Sensory receptors located in the skin for light-touch (Meissner's corpuscles), free nerve endings, touch-pressure (Ruffini's corpuscles and Pacinian corpuscles), position (bulbs of Krause), and pain (free nerve endings).

Stop and Ask Yourself

4. What are the receptors for light touch? Touch-pressure? Pain?
5. How do light touch and touch-pressure differ?
6. What is proprioception?

Special Senses

Taste

The receptors for taste, or **gustation** (L. *gustus,* taste) are chemoreceptors located in the tongue. The surface of the vertebrate tongue is covered with many small protuberances called *papillae* (s. *papilla*; L. *papula,* nipple). Papillae give the tongue its "bumpy" appearance (figure 34.3a).

Located in the crevices of the papillae are thousands of receptor organs called **taste buds** (figure 34.3b). They are barrel-shaped clusters of chemoreceptor cells and supporting cells arranged like alternating segments of an orange. Chemoreceptor cells have a life of only about 10 days, and are continually replaced. They are replaced with a decreasing frequency as humans get older, which explains why our sense of taste diminishes with age. Extending from each receptor cell are short *gustatory hairs* that project through a tiny opening called the *taste pore.* [5]

The four generally recognized basic taste sensations are sweet (sugars), sour (acids), bitter (alkaloids), and salty (electrolytes). The areas responsible for these basic tastes are located on specific parts of the tongue (figure 34.4). The exact mechanisms that stimulate a chemoreceptor taste cell are not known. One theory is that different types of gustatory stimuli cause proteins on the surface of the receptor-cell plasma membrane to change the permeability of the membrane, in effect, "opening and closing gates" to chemical stimuli and causing a generator potential.

The afferent pathway of taste receptors to the brain involves two cranial nerves. Taste buds on the posterior two-thirds of the tongue have a sensory pathway through the glossopharyngeal nerve (cranial nerve IX), and the anterior one-third of the tongue is served by the facial nerve (cranial nerve VII). These nerves convey the nerve impulse through the medulla and thalamus to the parietal lobe of the cerebral cortex, where they are interpreted.

Many invertebrate animals rely on specialized taste receptors, such as those found on snail antennae, insect mouthparts and legs, and octopus tentacles. In nonhuman vertebrates, taste buds may be found on other parts of the body besides the tongue. Taste buds are rare on the tongue of reptiles and birds. Instead, most are found in the pharynx. In fish and amphibians, they may also develop in the skin. For example, a sturgeon's taste buds are abundant on its projection called the rostrum. As the fish glides over the bottom, it can obtain a foretaste of potential food before the mouth reaches it. [6]

Smell

The sense of smell, or **olfaction** (L. *olere,* to smell + *facere,* to make) is due to *olfactory receptor cells* located in the roof of the vertebrate nasal cavity (figure 34.5). These cells, which are specialized endings of the fibers that make up the olfactory nerve, lie among supporting epithelial cells. They are densely packed; a dog has up to 40 million olfactory receptor cells/cm². Each olfactory cell ends in a tuft of cilia containing receptor sites for various chemicals.

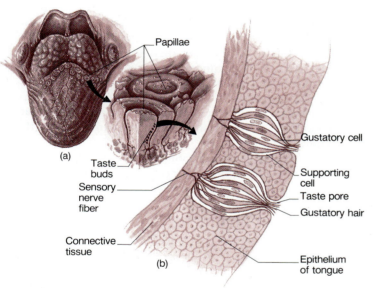

(a)

(b)

Papillae

Taste buds

Sensory nerve fiber

Connective tissue

Gustatory cell

Supporting cell

Taste pore

Gustatory hair

Epithelium of tongue

Figure 34.3 Taste. (a) Surface view of the human tongue showing the many papillae and the numerous taste buds in each papilla. (b) The gustatory cell and its associated gustatory hair is encapsulated by supporting cells.

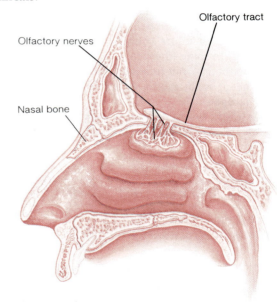

Olfactory tract

Olfactory nerves

Nasal bone

Figure 34.5 Smell. Position of olfactory receptors in a human nasal passageway. The receptor cells, which have cilia projecting into the nasal cavity, are supported by columnar epithelial cells. When these cells are stimulated by chemicals in the air, nerve impulses are conducted to the brain by the olfactory cranial nerve.

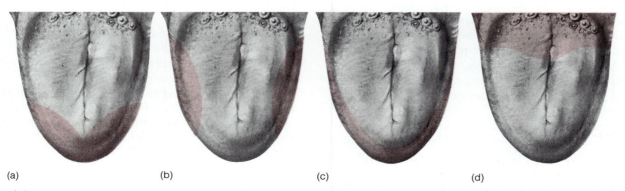

(a)   (b)   (c)   (d)

Figure 34.4   Patterns of taste receptor distribution on the dorsal part of the tongue. (a) Sweet receptors. (b) Sour receptors. (c) Salt receptors. (d) Bitter receptors.

Sensory impulses for olfaction are conveyed along the olfactory nerve (cranial nerve I) to the olfactory portion of the cerebral cortex where they are interpreted as odor and cause the sensation of smell.

Several theories have been proposed to explain how odors are perceived. The most likely one is that odor molecules have some kind of physical interaction with protein receptors located on the receptor-cell plasma membrane. Such an interaction somehow alters membrane permeability and leads to a generator potential. Recent research suggests that different types of odors are related to the shape of the molecules rather than to the atoms that make up the molecules. Ordinarily, odors are carried to the nose by air currents. When an animal sniffs, eddy currents are created that greatly increase the amount of air that reaches the olfactory receptors high in the nasal cavities. Sniffing is a semireflexive response and usually occurs when an odor attracts an animal's attention.

[7]

The detection of odors (olfaction) is one of the most ancient senses. In some invertebrates, specialized olfactory receptors respond to pheromones. For example, those on the antennae of male moths (*Cecropia*) can detect one bombykol molecule in 10^{15} molecules of air (colorplate 15c). Female silk moths secrete bombykol as a sex attractant, which enables a male to find a female even in the dark.

Stop and Ask Yourself

7. What are papillae?
8. What is the structure of a taste bud?

9. Which areas of the tongue respond to the four basic taste sensations?
10. What is one current theory to explain the mechanism of olfaction?

Hearing and Equilibrium

Hearing (audition) and equilibrium (balance) are considered together because both sensations are received in the same vertebrate organ—the ear. The ear has two functional units: the *auditory apparatus* is concerned with hearing, and the *vestibular apparatus* is concerned with posture and equilibrium.

The human ear has three divisions: the outer, middle, and inner ear. The *outer ear* consists of the *pinna* (auricle) and *external auditory canal* (figure 34.6a; table 34.2). The *middle ear* begins at the *tympanic membrane* (eardrum) and ends inside the skull, where two small membraneous openings, the *oval* and *round windows*, are located. Three small bones are located between the tympanic membrane and the oval window. Collectively called *ossicles*, they include the *malleus* (hammer), *incus* (anvil), and *stapes* (stirrup), because their shapes resemble these objects. The malleus adheres to the tympanic membrane and connects to the incus. The incus connects to the stapes, which adheres to the oval window. The *auditory (eustachian) tube* extends from the middle ear to the nasopharynx and permits equalization of air pressure between the middle ear and the throat.

The *inner ear* has three components. The first two, the *vestibule* and the *semicircular canals*, are concerned with equilibrium, and the third, the *cochlea*, is involved with

---

Box 34.1

The Body's Own Tranquilizers

Scientists have long searched for pain-killing drugs that are not as addicting as morphine and opium. One such line of drugs comes from the body itself.

In the 1970s, scientists discovered how opiate drugs function. Opiate molecules attach to special receptor sites of certain neurons in the central nervous system. This attachment slows down the firing of those neurons. Apparently, the decreased rate of firing decreases the sensation of pain. Many pain receptors are located in the spinal cord, where the pain impulse is first introduced into the central nervous system. It was also found that opiates had an especially strong effect on the thalamus, where pain is eventually processed into an actual perception.

Soon after, scientists isolated natural, short-chain neuropeptides from the brain and pituitary gland. They called these neuropeptides *endorphins*. Also discovered were the breakdown products of endorphins, smaller peptides that were named *enkephalins*.

Endorphins and enkephalans are morphinelike substances that occur naturally in the nervous system. They are the brain's own opiates, having the pain-killing effects of opiates, such as morphine. It is likely that endorphins and enkephalins work by binding to the same neuronal receptors that bind opiate drugs. These receptor sites in the central nervous system are associated with pain pathways.

In addition to moderating pain, enkephalins in the limbic system seem to counteract psychological depression by producing a state of euphoria similar to the feelings produced by opiate drugs.

Exercise also stimulates the release of endorphins. This release may explain, in part, why many people experience feelings of relaxation and contentment after a workout.

From Robert Carola, John P. Harley, and Charles Noback. 1990. *Human Anatomy and Physiology*. McGraw-Hill, Inc., New York.

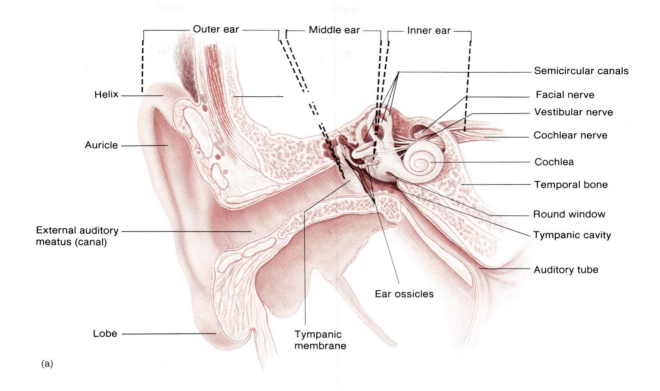

(a)

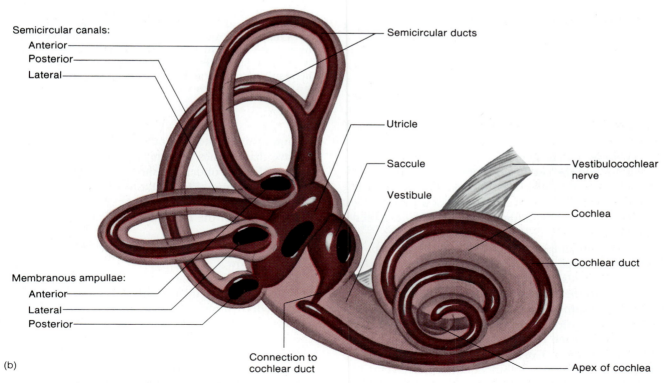

(b)

Figure 34.6 The ear. (a) Anatomy of the human ear. Note the outer, middle, and inner regions. (b) The inner ear, including the semicircular cannals, which are involved with equilibrium, and the cochlea, which is involved with hearing.

Table 34.2 The Human Ear: Structure and Function

Structure	Description
Outer ear	
Pinna (auricle)	The cartilaginous, exterior "flap" of the ear, designed to convey sound waves to the middle ear
External auditory canal	The canal leading from the floor of the outer ear to the tympanic membrane
Middle ear	
Tympanic membrane (eardrum)	The fibrous tissue extending across the deep inner end of the auditory canal and forming the partition between the external and middle ear
Auditory tube (eustachian tube)	The tube leading downward and inward from the middle ear to the nasopharynx
Ossicles	The malleus (hammer), incus (anvil), and stapes (stirrup), that form a connection between the tympanic membrane and the inner ear
Inner ear	
Vestibule	The central chamber of the inner ear; includes the utricle and saccule filled with fluid
Semicircular canals	The three small canals lying at right angles to each other; suspended in perilymph
Cochlea	The spiral structure containing the perilymph-filled vestibular and tympanic canals and cochlear duct
Organ of Corti	The organ of hearing composed of the basilar membrane, ciliated hair cells, and the tectorial membrane

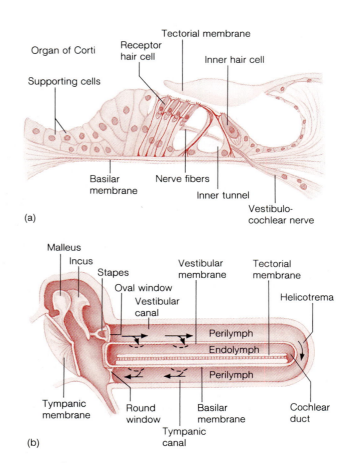

Figure 34.7 Hearing. (a) Enlarged cross-section through the organ of Corti, showing the receptor hair cells from the side. (b) The cochlea unwound, showing placement of the organ of Corti along its length. The arrows represent the pressure waves that move from the oval window to the round window. These cause the basilar membrane to vibrate, and the cilia of the receptor hair cells to bend against the tectorial membrane. The resulting nerve impulse travels along the vestibulocochlear nerve to the brain where sound is interpreted.

hearing (figure 34.6b). The semicircular canals are arranged so that there is one in each dimension of space. The base of each canal is called an *ampulla*. In the ampullae are hair cells whose cilia project into a fluid medium.

The *vestibule* is a chamber that lies between the semicircular canals and the cochlea. It contains two small sacs called the *utricle* and *saccule*, both of which contain hair cells whose cilia protrude into a gelatinous substance. Resting on this substance are calcium carbonate granules called *otoliths*.

The *cochlea* resembles a snail's shell because of its spiral shape. There are three areas in the cochlea: the *vestibular* and *tympanic canals* and the *cochlear duct*. The *basilar membrane* forms the lower wall of the cochlear duct (figure 34.7a). It contains ciliated cells that come into contact with a membrane called the *tectorial membrane*. The basilar membrane, ciliated cells, and the tectorial membrane are called the **organ of Corti.** When this organ sends out a nerve impulse along the vestibulocochlear nerve (cranial nerve VIII), it reaches the auditory cerebral cortex of the brain and is interpreted as sound. The process of hearing can be summarized as follows (figure 34.7b):

1. Sound waves enter the outer ear and create a pressure wave that reaches the tympanic membrane.
2. Air molecules under pressure cause the tympanic membrane to vibrate. The vibrations move the malleus, on the other side of the membrane.
3. The handle of the malleus strikes the incus, causing it to vibrate.
4. The vibrating incus moves the stapes into and out of the oval window.
5. The sound waves that reach the inner ear through the oval window set up pressure changes that vibrate the fluid (perilymph) in the inner ear. These vibrations are transmitted to the basilar membrane, causing it to ripple.
6. Receptor hair cells of the organ of Corti that are in contact with the overlying tectorial membrane are bent, causing a generator potential, which leads to an action potential that travels along the vestibulocochlear nerve to the brain for interpretation.

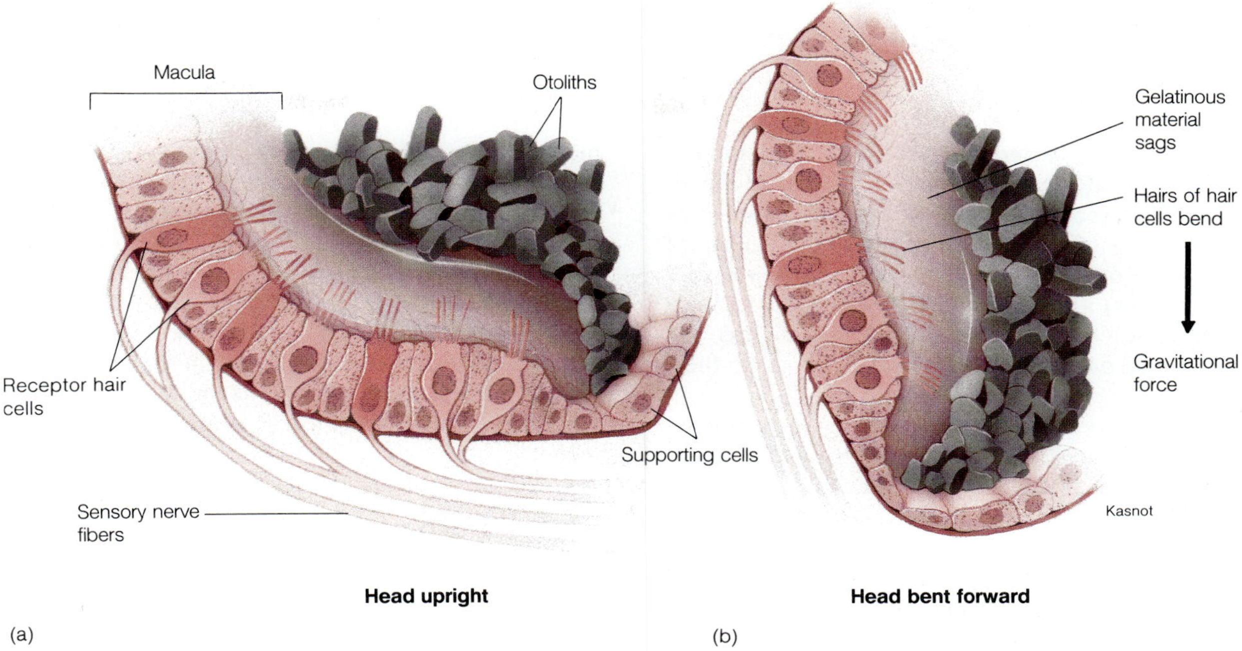

Figure 34.8 Static equilibrium (balance). Receptor hair cells in the utricle and saccule are involved in responsiveness to movement sideways or up or down. (a) When the head is upright, otoliths are balanced directly over the cilia of hair cells. (b) When the head is bent forward, the otoliths shift, and the hair of hair cells are bent. This bending of hairs causes a generator potential to be initiated.

7. Vibrations in the perilymph fluid dissipate from the cochlea through the round window into the middle ear.

Overall, the ability to hear depends on the flexibility of the basilar membrane. As mammals grow older, this flexibility decreases. Furthermore, humans are not able to hear low-pitched sounds, below 20 cycles/second, although some other vertebrates can (box 34.2). Young children can hear high-pitched sounds up to 20,000 cycles/second, but this ability decreases with age. Other vertebrates can hear sounds at much higher frequencies. For example, dogs can easily detect sounds of 40,000 cycles/second. Thus, dogs [8] can hear sounds from a high-pitched dog whistle that appears silent to a human.

The sense of equilibrium (balance) can be divided into two separate senses: *Static equilibrium* refers to the knowledge of movement in one plane (either vertical or horizontal), and *dynamic equilibrium* refers to the knowledge of angular and/or rotational movement.

When the body is still, the otoliths in the utricle and saccule rest on hair cells (figure 34.8a). When the head or body moves horizontally, or vertically, the granules in the utricle and saccule are displaced, causing the gelatinous material to sag (figure 34.8b). This displacement causes the hairs to bend slightly so that hair cells initiate a generator potential and then an action potential that travels by way of the vestibulocochlear nerve (cranial nerve VIII) to [9] the brain. Continuous movement of the fluid in the semicircular canals may cause motion sickness.

The important senses of hearing and balance are found in many animals living in many environments.

The sense of *balance* is highly developed in some invertebrates. For example, jellyfishes have an organ or equilibrium called a statocyst. Statoliths are dense crystals of sand grains or mineral salts and organic material found in the statocyst. When the body tilts, the statoliths also tilt. Neighboring hair cells bend and a receptor potential is generated.

Many arthropods are capable of *hearing*. For example, in the laboratory, mosquitoes and spiders are attracted to audible, high-frequency vibrations of a tuning fork. They possess a tympanum that vibrates and stimulates receptor cells attached to it. Certain nocturnal moths possess receptors in two tympanal organs that respond to the ultrasonic calls of bats (*see box 23.1*).

The early development of the vertebrate hearing apparatus can be traced to the lateral line and auditory system of fishes.

The lateral line organs of fish (*see chapter 27*) and in aquatic and larval amphibians contain hair-cell mechanoreceptors called neuromasts. When the water in the lateral line moves, it distorts the hair cells, causing a generator potential in the associated sensory neurons. Thus, the fish can detect the direction and force of water currents and the movement of other animals or prey in the water. In a number of other fish, parts of the lateral line organ have become modified to detect electrical currents in the surrounding water. Because most living organisms generate weak electrical fields, the ability to detect these fields may permit a fish to capture prey (electrolocation) or avoid predators. This is an especially valuable sense in deep, turbulent, or murky water, where vision is of little use. In fact, some fish actually generate electrical fields and then use their electroreceptors (electrocommunication) to detect how surrounding objects distort the field allowing the fish to navigate in murky or turbulent waters (*see figure 27.15*).

The lateral line and two inner ears, located behind the brain, allow fishes to hear by sending the information from mechanoreceptors in these two areas to the brain for interpretation. The front end of the lateral line organ evolved into the labryinth of structures found in the inner ear of vertebrates. The various hair cells became modified to detect sounds, gravity, and movement of the head.

In most reptiles and birds the ear has a tympanic membrane and a single auditory ossicle (the stapes or columella) that transmits sound vibrations across a tympanic cavity to the inner ear. Amphisbaenians (worm lizards) and snakes have secondarily lost the tympanic membrane, and the stapes simply abuts on a skull bone.

A frog's ear has a tympanic membrane that connects to the stapes and an operculum. By coupling and uncoupling the stapes and operculum, the frog can listen to either low-frequency or high-frequency sounds (*see figure 28.10*).

Stop and Ask Yourself

11. What is the organ of Corti?
12. How are sound waves converted into nerve impulses?
13. What is the function of the otoliths?
14. What are several functions of the lateral line organ of fish?

Vision

A large number of animals are sensitive to light and many have some kind of specialized photoreceptor for detecting light and transforming it to a nerve impulse.

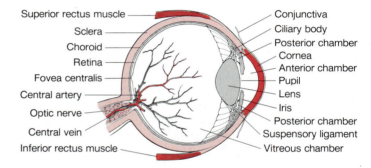

Figure 34.9 The internal anatomy of the human eyeball.

Many invertebrates respond to light-using receptors called *ocelli* or *eyespots*. These photo receptors can detect light but do not form an image. The ocelli are simply clusters of photosensitive cells arranged in a cuplike depression in the epidermis. Cnidarians are the simplest animals with image-forming eyes. Some molluscan eyes are closed, fluid-filled vesicles equipped with a transparent lens, a cornea, and a retina. A few annelids, crustaceans, and insects have compound eyes that contain closely packed photosensitive units. Some compound eyes have many thousands of these units (colorplate 15d). The compound eye has mediocre resolving power but is well adapted for detecting motion. In addition, most compound eyes can adapt to changes in light intensities, and some provide for color vision. Color vision is particularly important in day-flying, nectar-drinking insects, such as honeybees. Honeybees learn to recognize particular flowers by color, scent, and shape.

Vision is the primary sense used by vertebrates that live in a light-filled environment and consequently their photoreceptive structures are very well-developed. Most vertebrates have eyes (eyeballs) capable of forming visual images. As figure 34.9 and table 34.3 indicate, the eyeball has a *lens,* a *sclera* (the tough outer coat), a *choroid layer* (a thin middle layer), and an inner *retina* containing many light-sensitive receptors (photoreceptors). The transparent, light-focusing *cornea* is continuous with the sclera and

Table 34.3	The Human Eye: Structure and Function		
Structure	**Description**	**Structure**	**Description**
Supporting layer of eyeball		**Retinal layer of eyeball**	
Sclera	Opaque layer of connective tissue over posterior five-sixths of outer layer of eyeball; "white" of the eye; gives eyeball its shape, protects inner layers; perforated to allow optic nerve fiber to exit	Retina	Multilayered, light-sensitive membrane; innermost layer of eyeball; connected to brain by optic nerve; receives focused light waves, transduces them into nerve impulses that the brain converts into visual perceptions
Cornea	Transplant anterior portion of outer layer of eyeball; entry point for light	Rods	Specialized photoreceptor cells; not color sensitive; responsive in dim light and highly sensitive to light
Vascular layer of eyeball		Cones	Specialized color-sensitive photoreceptor cells; concentrated in fovea; not as light-sensitive as rods
Choroid	Thin membrane of blood vessels and connective tissue between sclera and retina	Fovea centralis	Depressed area in center of retina containing only cones; area of most acute image formation and color vision
Ciliary body	Thickened vascular layer in anterior portion of eyeball; contains muscles that help lens to focus by either increasing or decreasing tension on suspensory ligament of lens	**Cavities of eyeball**	
Ciliary processes	Inward extensions of ciliary body; help hold lens in place	Aqueous chambers	
Lens	Elastic, colorless, transparent body of epithelial cells behind iris; shape modified to focus on subjects at different distances (accommodation) through action of ciliary muscles	Anterior chamber	Anterior to iris, posterior to cornea (between cornea and iris); contains aqueous humor
Iris	Colored part of eye; thin, muscular layer; anterior extension of choroid; size of pupil, and thus amount of light entering eye, regulated by reflexive contracting and dilating of smooth muscle	Posterior chamber	Lies between iris and lens; contains aqueous humor
		Vitreous chamber	Largest cavity, fills entire space behind lens; contains vitreous humor
Pupil	Adjustable circular opening in iris; opens and closes reflexively relative to amount of light available		

covers the front of the eyeball. Choroid tissue also extends to the front of the eye to form the colored *iris, ciliary body,* and *suspensory ligaments.* The iris is heavily endowed with light-screening pigments and it has radial and circular smooth muscles for regulating the amount of light entering the pupil. A clear fluid (*aqueous humor*) fills the anterior and posterior chambers which lie between the lens and the cornea. The *lens* is located behind the iris, and a jellylike *vitreous body* fills the vitreous chamber behind the lens. The moist, mucous membrane that covers the eyeball is the *conjunctiva.*

Vertebrates can adjust the way in which light from either close-up or distant objects converges on the photoreceptors. Light rays entering the curved cornea are bent (refracted) towards a focal point. If the angle of bending is not enough, the focal point will be behind the retina. If it is too much, the focal point will be in front of the retina. Both conditions result in a blurry image. Movement of the lens, via the ciliary body, can adjust the path of the light rays. This process of focusing light rays precisely on the retina is called **accommodation.** Vertebrates rely on the coordinated stretching and relaxation of the eye muscles

and fibers (the ciliary body and suspensory ligament) that are attached to the lens for visual accommodation (figure 34.10).

In birds and mammals, the retina is well developed. Its innermost layer (basement layer) is composed of pigmented epithelium that covers the choroid layer. Nervous tissue that contains photoreceptors lies on this basement layer. The photoreceptors are called **rod** and **cone cells** because of their shape (figure 34.11). These specialized receptor cells apparently evolved from simple hair cells found in other sensory organs. Rods are sensitive to very dim light and are abundant in the periphery of the retina. Cones respond to high-intensity light and are involved in color perception. The cones of a human eyeball are densely packed in the *fovea centralis* (*see figure 34.9*), which is a funnel-shaped depression near the center of the retina. The fovea centralis is the most photosensitive portion of the retina. Vertebrates tend to move their eyes so that the image of the object they want to see most clearly falls on the fovea centralis. The retina is highly vascularized, and at times red blood cells escape from the capillaries. It is these cells that move easily through the vitreous humor of

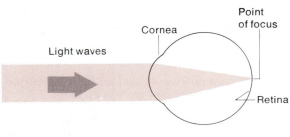

(a) **Normal eye**

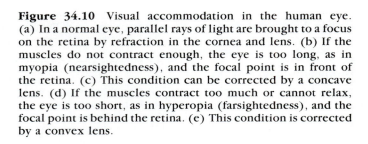

Figure 34.10 Visual accommodation in the human eye. (a) In a normal eye, parallel rays of light are brought to a focus on the retina by refraction in the cornea and lens. (b) If the muscles do not contract enough, the eye is too long, as in myopia (nearsightedness), and the focal point is in front of the retina. (c) This condition can be corrected by a concave lens. (d) If the muscles contract too much or cannot relax, the eye is too short, as in hyperopia (farsightedness), and the focal point is behind the retina. (e) This condition is corrected by a convex lens.

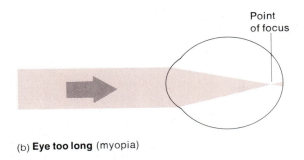

(b) **Eye too long** (myopia)

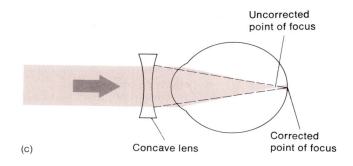

(c)

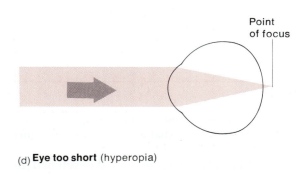

(d) **Eye too short** (hyperopia)

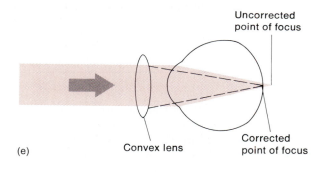

(e)

[10] the eye that we sometimes see as "spots" in front of our eyes. Their presence is normal and harmless.

Located on the top of each rod and cone is a stack of flattened membranes called *lamellae* (figure 34.11). Embedded in the lamellae of rods are rhodopsin molecules. Each **rhodopsin** molecule consists of a protein called *opsin* to which a side group, called *cis-retinal*, is attached. When the side group absorbs light energy, the molecule is converted to scotopsin plus retinene (figure 34.12a). The energy released from this reaction triggers the generator potential in an axon and then an action potential that leaves the eyeball via the optic nerve. From the optic nerve, signals travel to the thalamus, then to the primary visual cortex, and finally to the visual association cortex, where interpretation takes place. When the photoreceptor cells are not being stimulated (i.e., in the dark), energy from ATP

along with vitamin A converts retinene and scotopsin back to rhodopsin (figure 34.12b).

The 19th century poet Leigh Hunt said "colors are the smiles of Nature." Just how does a person distinguish one smile from another? The answer lies to a great extent in the three types of cone-shaped, color-sensitive cells in the retina of the eye. These three types of cone cells are found [11] among primates, birds, reptiles, and fishes. Each type responds differently to light reflected from a colored object, depending on whether the cells have within them red, green, or blue pigments. The pigments are light-absorbing proteins that are particularly sensitive to either the long-wave length (red), intermediate-wave length (green), or short-wave length (blue) region of the visible spectrum (figure 34.13). The relative amounts of light absorbed by each type of cone are translated into generator potentials

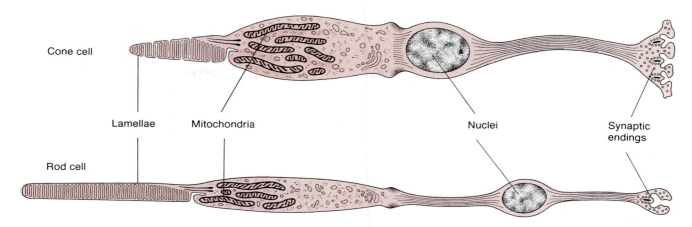

Figure 34.11 Rods and cones. Drawing of a longitudinal section of a rod and cone. The photosensitive pigment is located in the lamellae of the outer segment.

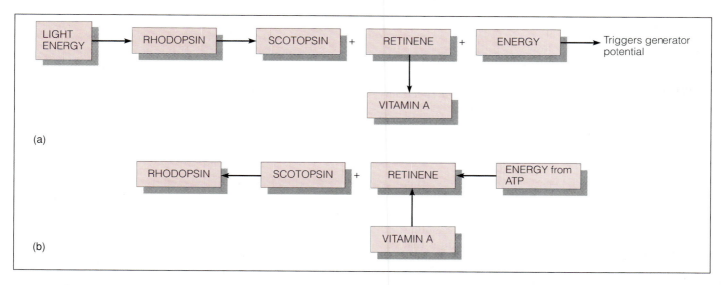

Figure 34.12 The rodopsin cycle. (a) In the presence of light, rhodopsin breaks down into the proteins scotopsin and retinene, and this reaction triggers a generator potential. Retinene becomes vitamin A. (b) In the dark, ATP energy and vitamin A are used to re-form rhodopsin.

by the retinal nerves and then transmitted as a nerve impulse to the brain, where the overall pattern evokes the sensation of a specific hue.

Most primitive fish, amphibians, and reptiles have a median eye complex, consisting of a pineal or parietal eye (or both), which monitors ambient light and appears to initiate physiological adjustments to light levels. This complex is represented by the pineal gland in birds and mammals.

Because light does not travel far in water, most fish have nearsighted eyes with a spherical lens located close to the cornea. As previously noted, a fish focuses on more distant objects by contracting a muscle that pulls the lens toward the retina. The eye of most tetrapods at rest is focused on more distant objects than a fish's eye. An amphibian accommodates for near objects by contracting a muscle that pulls the lens toward the cornea; except for snakes, amniotes do it by increasing the curvature of the lens.

Several other visual stimuli that some animals can respond to are polarized light (insects, squids, octopuses, amphibians, and sea turtles), ultrasound (bats), and magnetism (birds, *see chapter 30*).

Stop and Ask Yourself

15. How do some invertebrates "see"?
16. What is the function of the pupil?
17. What is accommodation?
18. What is the function of rhodopsin?
19. How do we perceive colors?

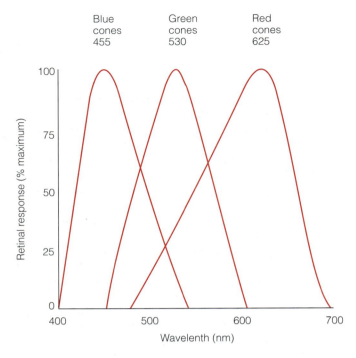

Figure 34.13 Spectral-activity curves. These show the sensitivity of the three eye color pigments to the visible spectrum of light.

Summary

1. A stimulus is any form of energy the body is able to detect by means of its receptors. Receptors are nerve endings of sensory neurons or specialized cells or structures that respond to stimuli, such as chemical energy, mechanical energy, photon energy, or radiant energy.

2. Receptors act as transducers of energy from one form to another. When a receptor is stimulated, a generator potential is initiated, causing an action potential to travel along a nerve pathway to a specific part of the cerebral cortex, where it is perceived.

3. Taste receptors are located in the tongue in taste buds. Olfaction occurs in specialized receptors located in the nose. Taste and olfaction are examples of chemical senses.

4. Mammalian hearing depends on the bending of receptor hair cells by changes in fluid pressure in the organ of Corti. The sense of balance also depends on information from hair cell receptors located in the vestibular apparatus (the semicircular canals and otoliths).

5. In vertebrate vision, light energy is used to change photopigment molecules (rhodopsin) in the rods and cones of the retina. Processing of this information thus begins in the retina and continues in the visual cortex of the brain, where images are perceived.

Key Terms

accommodation (p. 549)
bulbs of Krause (p. 541)
chemoreceptors (p. 540)
cone cells (p. 549)
free nerve endings (p. 541)
generator potential (p. 540)
gustation (p. 542)
mechanoreceptors (p. 540)
nociceptors (p. 540)
olfaction (p. 543)
organ of Corti (p. 546)

organs of Ruffini (p. 541)
Pacinian corpuscles
  (p. 541)
photoreceptors (p. 540)
proprioception (p. 542)
rhodopsin (p. 550)
rod cells (p. 549)
tactile (Meissner's)
  corpuscle (p. 541)
taste buds (p. 542)
thermoreceptors (p. 540)

Critical Thinking Questions

1. The comparison of the vertebrate eye to a camera is not very accurate. Why is this so?

2. Most animals lack cones in their retinas. How then does such an animal view the visual world?

3. Why is the sense of gravity considered a sense of equilibrium?

4. How does vitamin A deficiency result in night blindness?

5. What is the advantage to a wild mammal of having thermoreceptors on the tongue?

6. How would you expect your inner ear to behave in zero gravity?

Suggested Readings

Books

Barlow, H. B., and Mollon, J. D. 1982. *The Senses.* Cambridge: Cambridge University Press.

Kandel, E., and Schwartz, J. H. 1985. *Principles of Neural Science.* 2/e. New York: Elsevier.

Articles

Dowling, J. E. Night blindness. *Scientific American* October, 1966.

Finke, R. Mental imagery of the visual system. *Scientific American* May, 1986.

Geldard, F. A., and Sherrick, C. E. Space, time and touch. *Scientific American* July, 1986.

Gibbons, B. The intimate sense of smell. *National Geographic* September, 1986.

Gombrich, E. H. The visual immage. *Scientific American* September, 1972.

Hudspeth, A. J. The hair cells of the inner ear. *Scientific American* January, 1983.

Knudsen, E. I. The hearing of the barn owl. *Scientific American* December, 1981.

Livingston, M. S. Art, illusion, and the visual system. *Scientific American* January, 1988.

Loeb, G. E. The functional replacement of the ear. *Scientific American* February, 1985.

Macnichol, E. F. Three-pigment color vision. *Scientific American* December, 1964.

Masland, R. H. The functional architecture of the retina. *Scientific American* December, 1986.

Nassau, K. The causes of color. *Scientific American* October, 1980.

Nathans, J. 1989. The genes for color vision. *Scientific American* February, 1989.

Parker, D. E. The vestibular apparatus. *Scientific American* November, 1980.

Poggio, T., and Koch, C. Synapses that communicate motion. *Scientific American* May, 1987.

Schnapf, J. L., and Baylor, D. A. How photoreceptor cells respond to light. *Scientific American* April, 1987.

Stryer, L. The molecules of visual excitation. *Scientific American* July, 1987.

Vonrekesy, G. The ear. *Scientific American* August, 1957.

35

Communication III: Chemical Messengers

Outline

Concepts

1. Chemical messengers are involved in communication and maintaining homeostasis in an animal's body, and in the body's response to various stimuli. One type of chemical messenger is a hormone. Only those cells that have specific receptors for a hormone can respond to that hormone.
2. Hormones work along with nerves to communicate and integrate activities within the body of an animal.
3. The major endocrine glands of the vertebrate body include the pituitary, thyroid, parathyroids, adrenals, pineal, thymus, pancreas, and gonads. Various other tissues, however, have also been shown to secrete hormones. Examples include the kidneys, heart, digestive system, and placenta.

Have You Ever Wondered:

[1] how aspirin inhibits fever, pain, and inflammation?
[2] why a person who has a diseased liver or kidney experiences many diverse and bizarre side effects from the increased concentration of certain hormones?
[3] why you urinate more if you drink beer compared to an equal amount of water?
[4] what causes a person to be a dwarf? A giant?
[5] what causes a goiter?
[6] what causes rickets?
[7] what could cause a women to develop masculine characteristics, such as facial hair (a beard), a deep voice, and a reduction in breast size?
[8] why some animals (including humans) possess the "fight-or-flight" response?

These and other useful questions will be answered in this chapter.

This chapter contains underlined evolutionary concepts.

Chapters 33 and 34 discuss ways that the nervous and sensory systems work together to rapidly communicate information and maintain homeostasis in the body. In addition, many animals have a second, slower form of communication—the endocrine system with its chemical messengers.

Some biologists suggest that molecular messengers may initially have evolved in single-celled organisms to coordinate feeding or reproduction. As multicellularity arose, more complex organs evolved to govern the many individual coordination tasks, but control centers relied on the same kinds of molecular messengers that evolved in the simpler organisms. Some of the messengers worked fairly slowly but had long-lasting effects on distant cells; these became the modern hormones. Others worked more quickly, but influenced only adjacent cells for short periods; these became the neurotransmitters and local chemical messengers. Clearly, molecular messengers must have a very ancient origin and must have been conserved for hundreds of millions of years.

In evolution, new messengers are uncommon. Instead, "old" hormones are adapted to new purposes. For example, some protein hormones are very ancient and are found in species from bacteria to humans.

One key to the evolutionary success of any group of animals is proper timing of activity so that growth, maturation, and reproduction coincide with the times of year when climate and food supply favor survival. It seems likely that the chemical messengers regulating growth and reproduction were among the first to appear. These messengers were probably secretions of neurons. Later, specific hormones developed to play important regulatory roles in molting, growth, metamorphosis, and reproduction in various invertebrates (as has been presented in specific chapters in Part 3 of this textbook). Chemical messengers and their associated secretory structures became even more complex with the appearance of vertebrates.

Vertebrates possess two types of glands (figure 35.1). One type, *exocrine* (Gr. *exo,* outside + *krinein,* to separate) *glands,* secrete chemicals into ducts that, in turn, empty into body cavities or onto body surfaces (e.g., mammary, salivary, and sweat glands). The second type, *endocrine* (Gr. *endo,* within + *krinein,* to separate) *glands,* have no ducts and instead, secrete chemical messengers called hormones directly into the tissue space next to each endocrine cell. The hormones then diffuse into the bloodstream and are carried throughout the body to their target cells. By definition, a *target cell* is one having receptors to which chemical messengers selectively bind. The major vertebrate endocrine glands and organs are listed in figure 35.2 and their secretions are shown in table 35.1. The study of endocrine glands and their secretions is called **endocrinology.**

Besides the clearly defined hormones of the endocrine system, there exist other chemical messengers in animals that are less well characterized. In this chapter, these chemical messengers are discussed first, followed by the endocrine system and some specific hormones.

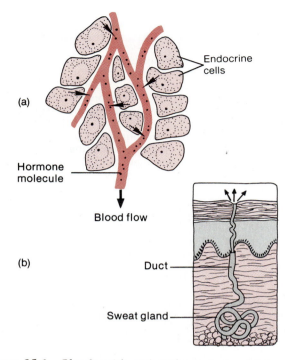

Figure 35.1 Glands with and without ducts. (a) An endocrine gland, such as the thyroid, secretes hormones into the extracellular fluid. From there, the hormones pass into blood vessels and travel to distinct sites in the body. (b) An exocrine gland, such as a sudoriferous (sweat) gland, secretes material (sweat) into a duct that leads to a body surface.

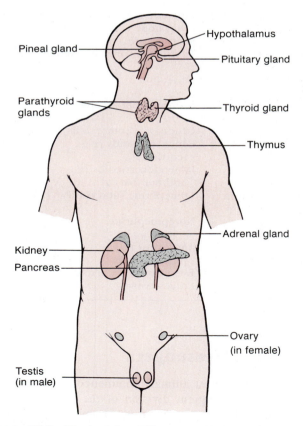

Figure 35.2 The location of human endocrine glands.

Table 35.1 Some Vertebrate Hormones		
Source	**Hormones**	**Target Cells and Principal Actions**
Anterior lobe of pituitary	Somatotropin (STH, or growth hormone [GH])	Growth of bone and muscle; promotes protein synthesis; affects lipid and carbohydrate metabolism; increases cell division
	Adrenocorticotropin (ACTH)	Stimulates secretion of adrenocortical steroids; involved in stress response
	Thyrotropin (TSH) or Thyroid-stimulating hormone	Stimulates thyroid gland to synthesize and release thyroid hormones concerned with growth, development, metabolic rate
	Gonadotropins:	
	Luteinizing or interstitial cell-stimulating hormone (LH or ICSH)	In ovary: formation of corpora lutea; secretion of progesterone; probably acts in conjuction with FSH
		In testis: stimulates the interstitial cells, thus promoting the secretion of testosterone
	Follicle-stimulating hormone (FSH)	In ovary: growth of follicles; functions with LH to cause estrogen secretion and ovulation
		In testis: acts on seminiferous tubules to promote spermatogenesis
	Prolactin (PRL)	Initiation of milk production by mammary glands; acts on crop sacs of some birds; stimulates maternal behavior in birds
Intermediate or posterior lobe of pituitary	Melanocyte-stimulating hormone (MSH)	Expansion of amphibian melanophores; contraction of iridophores and xanthophores; melanin synthesis; darkening of the skin; responds to external stimuli
Posterior lobe of pituitary	Anti-diuretic hormone (ADH or vasopressin)	Elevates blood pressure by acting on arterioles; promotes reabsorption of water by kidney tubules
	Oxytocin	Affects postpartum mammary gland, causing ejection of milk; promotes contraction of uterus; possible action in parturition and in sperm transport in female tract
Hypothalamus	Thyrotropin releasing hormone (TRH)	Stimulates release of TSH by anterior pituitary
	Adrenocorticotropin releasing hormone (CRH)	Stimulates release of ACTH by anterior pituitary
	Gonadotropin releasing hormone (GnRH)	Stimulates gonadotropin release by anterior pituitary
	Prolactin release inhibiting hormone (PIF)	Inhibits prolactin release by anterior pituitary
	Somatostatin	Inhibits release of STH by anterior pituitary
Thyroid gland	Thyroxine, triiodothyronine	Growth; amphibian metamorphosis; molting; metabolic rate in birds and mammals; growth, development
	Calcitonin	Lowers calcium level in blood by inhibiting calcium reabsorption from bone
Pancreas, islet cells	Insulin (from β cells)	Promotes glycogen synthesis and glucose utilization
	Glucagon (from α cells)	Raises blood glucose concentration
Adrenal cortex	Glucocorticoids (e.g., cortisol)	Promote synthesis of carbohydrate; protein breakdown; anti-inflammatory and antiallergic actions; mediates response to stress
	Mineralocorticoids (e.g., aldosterone)	Sodium retention and potassium loss through kidneys; water balance
Adrenal medulla	Epinephrine (adrenalin)	Mobilization of glucose; increased blood flow through skeletal muscle; increased oxygen consumption; heart rate increase
	Norepinephrine	Adrenergic neurotransmitter; elevation of blood pressure; constricts arterioles and venules
Testes	Androgens (e.g., testosterone)	Male sexual characteristics; spermatogenesis
Ovaries	Estrogens (e.g., estradiol)	Female sexual characteristics; oogenesis
Corpus luteum	Progesterone	Maintains pregnancy; stimulates development of mammary glands

Chemical Messengers

Development of most animals commences with fertilization and the subsequent division of the zygote. Further development is then dependent upon continued cell proliferation, growth, and differentiation. The integration of these events, as well as the communication and coordination of physiological processes, such as metabolism, respiration, excretion, movement, and reproduction, are dependent on chemical messengers—molecules synthesized and secreted by specialized cells. Chemical messengers can be categorized as follows:

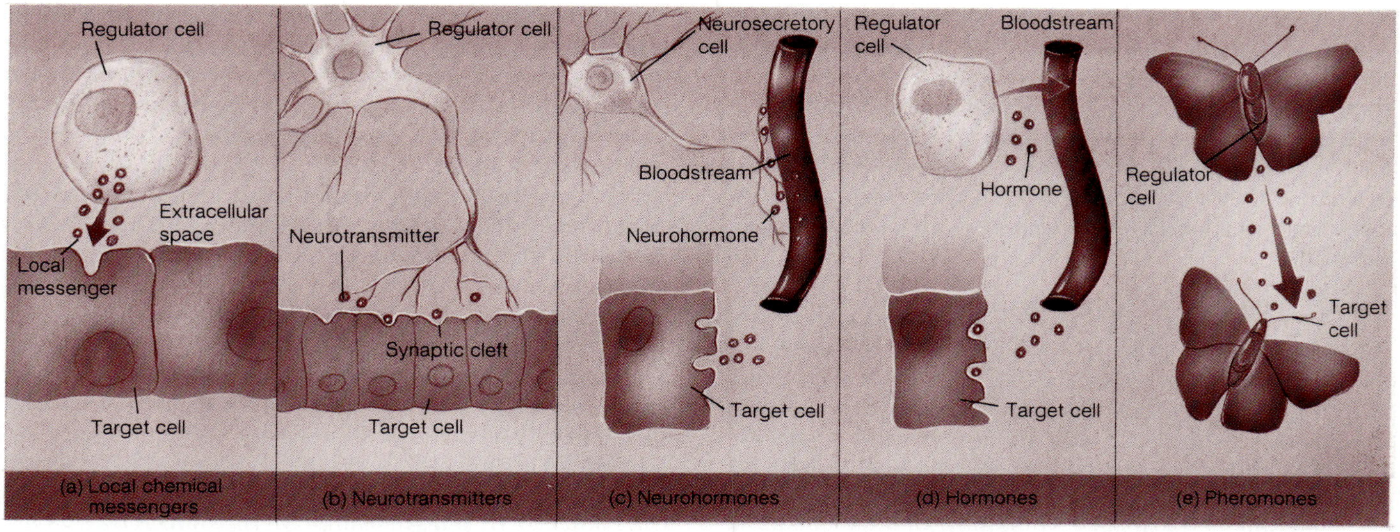

Figure 35.3 Molecular messengers: Targets and transport. (a) Short-distance local messengers act on an adjacent cell. (b) Individual nerve cells secrete neurotransmitters that cross the synaptic cleft to act on target cells. (c) Individual nerve cells can also secrete neurohormones that travel some distance in the bloodstream to reach a target cell. (d) Hormones are secreted by regulatory cells, usually in an endocrine gland, and enter the bloodstream, where they travel to target cells. (e) Pheromones are secreted by regulator cells in exocrine (ductless) glands. They leave the body and stimulate target cells in another animal.

1. **Local chemical messengers.** Many cells secrete chemicals that alter physiological conditions in the immediate vicinity (figure 35.3a). Most of these chemicals act on adjacent cells and do not accumulate in the bloodstream. Examples include some of the chemicals called *lumones* produced in the gut that help regulate digestion. In a wound, a substance called *histamine* is secreted by mast cells and participates in the inflammatory response. *Prostaglandins* are a group of molecules made from fatty acids that have a variety of physiological effects. For example, some prostaglandins stimulate smooth muscle to contract, and others cause smooth muscle to relax. Prostaglandins are involved with many aspects of reproduction, with menstrual cramps, allergic reactions to food, and the inflammatory response to infection. Aspirin inhibits the enzyme cyclooxygenase, which is needed for prostaglandin synthesis. This inhibition is thought to be why aspirin relieves fever, pain, and inflammation. Several other local chemical messengers include mitotic inhibitors called *chalones*; *growth factors* (mitogens) that induce mitosis; and *cell adhesion molecules* (CAMs) that signal tissue differentiation.

   [1]

2. **Neurotransmitters.** As presented in chapter 33, neurons secrete chemicals called *neurotransmitters* (e.g., acteylcholine) that act on immediately adjacent target cells (figure 35.3b). These chemical messengers reach high concentrations in the synaptic cleft, act quickly, and are actively degraded and recycled.

3. **Neurohormones.** Some specialized neurons (called neurosecretory cells) secrete *neurohormones*. Neurohormones enter the bloodstream and are transported to nonadjacent target cells where they exert their effects

(figure 35.3c). In mammals, for example, certain nerve cells in the hypothalamus release a neurohormone that causes the pituitary gland to release the hormone oxytocin, which induces powerful uterine contractions during labor.

4. **Hormones.** Endocrine glands or cells secrete *hormones* that are transported by the bloodstream to nonadjacent target cells (figure 35.3d). Many examples will be given in the next section.

5. **Pheromones.** *Pheromones* are chemical messengers released to the exterior of one animal that affect the behavior of another individual of the same species (figure 35.3e; *see table 23.2*).

6. **Neuromodulator.** *Neuromodulators* are hormones that modulate the response of neurons to neurotransmitters or other hormones.

Overall, scientists now recognize that the nervous and endocrine systems work together as an all-encompassing communicative and integrative network called the **neuroendocrine system.** In this system, chemical messengers achieve short- and long-term regulation of body function to maintain homeostasis by means of feedback systems.

Hormones and Their Feedback Systems

A **hormone** (Gr. *hormaein,* to set in motion or to spur on) is a specialized chemical messenger produced and secreted by an endocrine cell or tissue. It circulates through body fluids and affects the metabolic activity of a target cell or tissue in a specific way. Only rarely does a hormone operate independently. More typically, one hormone influences, depends upon, and balances another hormone in a controlled feedback network. Excess hormones that are not [2] used in the interactions with target cells are inactivated by the liver and kidneys. For this reason, a person with a liver or kidney disease may experience diverse and bizarre effects from the increasing concentrations of certain hormones.

Biochemistry of Hormones

Most hormones are proteins (polypeptides), derivatives of amino acids (amines), or steroids: a few are fatty acid derivatives. For example, hormones secreted by the pancreas are proteins. Those secreted by the thyroid gland are amines. Steroids are secreted by the ovaries, testes, and adrenal glands. The type of hormone depends on the embryonic origin of the gland. Glands derived from endoderm secrete proteins, ectodermal glands secrete amines, and mesodermal glands secrete steroids.

Hormones are effective in extremely small amounts. Only a few molecules of a hormone may be enough to produce a dramatic response in a target cell. In the target cell, hormones help control biochemical reactions in three different ways: (1) a hormone can increase the rate at which other substances enter or leave the cell; (2) it can stimulate a target cell to synthesize enzymes, proteins, or other substances; or (3) it can prompt a target cell to activate or suppress existing cellular enzymes. As is the case for enzymes, hormones are not changed by the reaction they regulate.

Feedback Control System of Hormone Secretion

Although hormones are always present in some amount in endocrine glands, they are not secreted continuously. In-

stead, the glands tend to secrete the amount of hormone that the body needs to maintain homeostasis. Regulation occurs through a *feedback control system,* where changes in the body or in the external environment are fed back to a central control unit (such as the brain), where the adjustment is made. A feedback system that produces a response that counteracts the initiating stimulus is called a *negative feedback system.* In contrast, a *positive feedback system* is one in which the initial stimulus is reinforced. Positive feedback systems are relatively rare in animals, because they usually lead to instability or pathological states.

Negative feedback systems monitor the amount of hormone secreted, altering the level of cellular activity as needed to maintain homeostasis. For example, suppose that the rate of chemical activity (metabolic rate) in body cells slows down (figure 35.4). The hypothalamus responds to this slow rate by releasing more *thyrotropin-releasing hormone* (TRH), which causes the pituitary gland to secrete more *thyrotropin,* or *thyroid-stimulating hormone* (TSH). This hormone, in turn, causes the thyroid gland to secrete a hormone called *thyroxine.* Thyroxine increases the metabolic rate, restoring homeostasis. Conversely, if the metabolic rate speeds up, the hypothalamus releases less TRH, the pituitary secretes less TSH, the thyroid secretes less thyroxine, and the metabolic rate slows down, once again restoring homeostasis.

Stop and Ask Yourself

1. What is a hormone?
2. What is a target cell?
3. What is the main difference between negative and positive feedback?

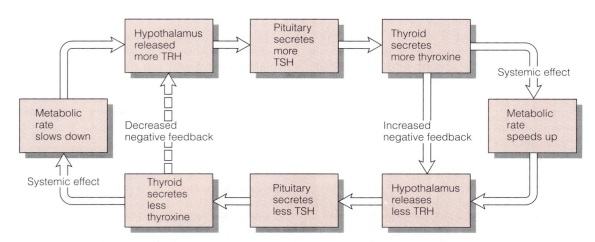

Figure 35.4 Hormonal feedback. A negative feedback system that helps control metabolic rate in higher vertebrates.

Mechanisms of Hormone Control

The function of a hormone is to modify the biochemical activity of a target cell or tissue. Two basic mechanisms are involved. The first, the fixed-membrane-receptor mechanism, applies to hormones that are proteins or amines. Because they are water-soluble and cannot diffuse across the plasma membrane, these hormones initiate their response by means of specialized receptors on the plasma membrane of the target cell. The mobile-receptor mechanism applies to steroid hormones. They are lipid-soluble and diffuse easily into the cytoplasm, where they initiate their response by binding to cytoplasmic receptors.

Fixed-Membrane-Receptor Mechanisms

According to one fixed-membrane-receptor mechanism, a water-soluble hormone is secreted by an endocrine cell and circulates through the bloodstream (figure 35.5a). At the cells of the target organ, the hormone acts as a "first or extracellular messenger," binding to a specific receptor site for that hormone on the plasma membrane (figure 35.5b). The hormone-receptor complex activates the enzyme adenylate cyclase in the membrane (figure 35.5c). The activated enzyme converts ATP into a nucleotide called cyclic AMP, which will become the "second (or intracellular) messenger." Cyclic AMP difuses throughout the cytoplasm and activates an enzyme called protein kinase, which causes the cell to respond with its distinctive phys-

iological activity (figure 35.5d). After inducing the target cell to perform its specific function, cyclic AMP is inactivated by the enzyme phosphodiesterase. In the meantime, the receptor on the plasma membrane has lost the first messenger and now becomes available for a new reaction.

Another fixed-membrane-receptor mechanism uses different second messengers to trigger cell responses to a variety of neurotransmitters, hormones, and growth factors. The plasma membrane of certain target cells contains small amounts of a phospholipid called phosphatidylinositol 4,5-biphosphate (PIP_2). When the hormone epinephrine, for example, binds to the receptor molecule, PIP_2 is cleaved into IP_3 and diacylglycerol (DG) (figure 35.6a,b). Both of these susbtances act as "second messengers." IP_3 instantly triggers the release of Ca^{2+} ions from storage sites (figure 35.6c). The Ca^{2+} ions act as a "third messenger," causing a number of physiological responses (e.g., activation of the actin-myosin system and enzymes). Meanwhile, DG activates protein kinase C, which phosphorylates a variety of proteins that lead to different physiological responses (figure 35.6d,e).

Mobile-Receptor Mechanism

Because steroid hormones pass easily through the plasma membrane, their receptors are inside the target cells. The mobile-receptor mechanism involves the stimulation of protein synthesis. After being released from a carrier protein in the bloodstream, the steroid hormone enters the

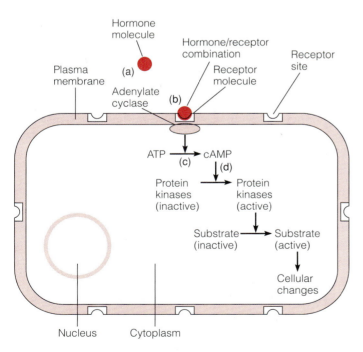

Figure 35.5 The steps (a-d) of the fixed-membrane-receptor mechanism of hormonal control. The response is mediated by a second messenger, cAMP, inside the cell. Other chemical messengers may be involved depending on the particular hormone and its target cell.

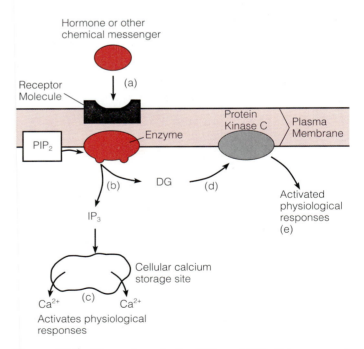

Figure 35.6 The actions (a-d) of IP_3 and DG: Other second messengers for hormone control. When a hormone or other chemical messenger binds to its receptor molecule, an enzyme cleaves PIP_2 into the second messengers IP_3 and DG. IP_3 liberates Ca^{2+} from storgae sites while DG activates protein kinase C. A cascade of physiological responses can follow such as cell division, secretion, or motility.

Figure 35.6 redrawn, with permission, from Norman K. Wessels and Janet L. Hopson, *Biology*. Copyright © 1988, McGraw-Hill Publishing Company, New York, New York.

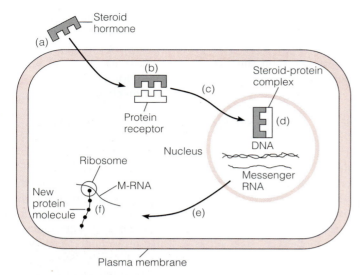

Figure 35.7 The mobile-receptor mechanism. In this case, steroid hormones enter the cell and initiate the steps leading to the synthesis of a protein.

target cell by diffusion and binds to a specific steroid receptor in the cytoplasm (figure 35.7a,b). This newly formed hormone-receptor complex acquires an affinity for DNA that causes it to enter the nucleus of the cell, where it binds to DNA and regulates the transcription of specific genes to form messenger RNA (figure 35.7c,d). The newly transcribed mRNA leaves the nucleus and moves to the rough endoplasmic reticulum, where it initiates protein synthesis (figure 35.7e,f). Some of the newly synthesized proteins may be enzymes whose effects on cellular metabolism constitute the cellular response attributable to the specific steroid hormone.

Stop and Ask Yourself

4. If a hormone is the "first messenger," what is the "second messenger?" Give specific examples.
5. What is the function of a receptor site on a plasma membrane?
6. What kinds of hormones stimulate protein synthesis in a cell?
7. How do IP$_3$ and DG function as "second messengers?"

The Vertebrate Endocrine System

Vertebrates have the most complex, but the best understood system of hormonal control. As the earliest vertebrates evolved, hormone-producing cells and tissues developed and came to be controlled in several ways. Some, such as the adrenal glands, are directed by sets of nerve cells in the brain; others are controlled by the pituitary;

and still others function independently of either nerves or the pituitary gland. Hormonal regulation in higher vertebrates became based primarily on a hierarchy involving the hypothalmus of the brain and the pituitary gland.

Pituitary Gland (Hypophysis)

The **pituitary gland** (also known as the **hypophysis**) is located directly below the hypothalamus (*see figure 35.2*). The pituitary has two distinct lobes: the anterior lobe (*adenohypophysis*) and the posterior lobe (*neurohypophysis*) (figure 35.8). There are several differences between the two lobes: (1) the adenohypophysis is larger than the neurohypophysis; (2) secretory cells called *pituicytes* are found in the adenohypophysis lobe, but not in the neurohypophysis; and (3) the neurohypophysis has a greater supply of nerve endings. Pituicytes produce and secrete hormones directly from the adenohypophysis, whereas the neurohypophysis obtains its hormones from the neurosecretory cells in the hypothalamus, stores them and secretes them when they are needed. These modified hypothalamic nerve cells project their axons down a stalk of nerve cells and blood vessels called the *infundibulum*, into the pituitary gland, creating a direct link between the nervous and the endocrine systems. Other hypothalamic neurosecretory cells secrete releasing hormones or factors (*see table 35.1*) into the portal blood vessels of the infundibular stalk. (A "portal" vessel is one that lies between two capillary beds.) These releasing hormones stimulate the pituicytes of the adenohypophysis to either secrete or inhibit their production of hormones.

The pituitary of many vertebrates (not humans) also has a functional *intermediate lobe* (*pars intermedia*) of mostly glandular tissue. Its secretions induce changes in the coloration of the body's surface of many animals by responding to external stimuli.

Hormones of the Neurohypophysis The neurohypophysis does not manufacture any hormones. Instead, the neurosecretory cells of the hypothalamus synthesize and secrete two hormones, antidiuretic hormone and oxytocin, which move down nerve axons into the neurohypophysis, where they are stored in the axon terminals until released. Action potentials cause release of these hormones, which are picked up by blood capillaries that lead to the general circulation. When these hormones contact cells with receptors specific for them, they exert their effects.

A *diuretic* is a substance that stimulates the excretion of urine, whereas an *antidiuretic* decreases urine secretion. When the body begins to lose water and become dehydrated, **antidiuretic hormone** (ADH, or **vasopressin**) is released and increases water absorption in the kidneys so that less urine is secreted. Because less urine is secreted, water is retained; this negative feedback system thus restores water homeostasis. Conversely, if you consume large quantities of beer, for example, you will stimulate urine secretion not only because you have increased the liquid

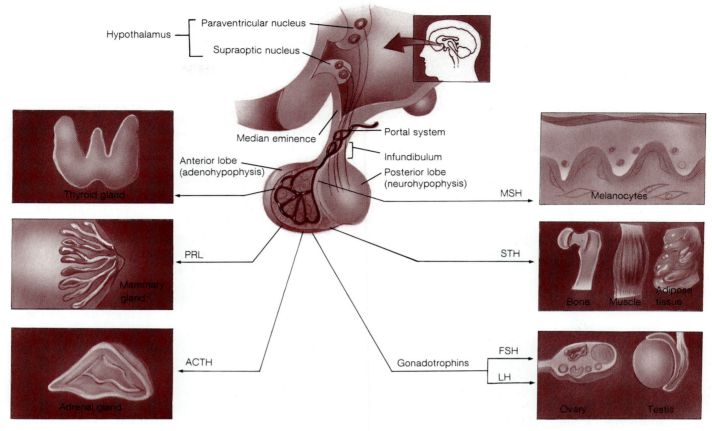

Figure 35.8 Functional links between the pituitary gland and the hypothalmus. Target areas for each hormone are shown in the relevant box. The blood vessels that make up the hypothalamic-hypophyseal portal system provide the functional link between the hypothalmus and the adenohypophysis, and the axons of the hypothalamic neurosecretory cells provide the link between the hypothalmus and neurohypophysis.

content of your body, which suppresses ADH secretion, but also because the alcohol in beer inhibits ADH secretion. ADH also causes smooth muscle in the walls of blood vessels to constrict, increasing blood pressure and the effective blood volume. These effects account for the hormone's other name: vasopressin.

Oxytocin plays a role in mammalian reproduction by its effect on smooth muscle. During labor, it stimulates contraction of the uterus to aid in expulsion of the offspring during childbirth, and promotes ejection of milk from the mammary glands to provide nourishment for the newborn during feeding.

Both ADH and oxytocin are thought to have evolved from a similar ancestral chemical messenger that helped control water loss and, indirectly, solute concentrations. For example, the neurohypophysis is notably larger in animals that live in arid parts of the world where water conservation is crucial. Also, the structure of the two hormones is similar except for a difference in two of the amino acids.

Hormones of the Adenohypophysis The true endocrine portion of the pituitary is the adenohypophysis, which synthesizes seven separate hormones (figure 35.8).

All of these hormones are polypeptides, and all but two are true *tropic hormones,* hormones whose primary target is another endocrine gland. The two nontropic hormones are growth hormone and prolactin.

Growth hormone (**GH**), or **somatotropin** (**STH**), does not influence a specific target organ; rather, it affects all parts of the body that are concerned with growth. It directly induces the cell division necessary for growth and protein synthesis in most types of cells by stimulating uptake of amino acids, RNA synthesis, and ribosome activity. STH indirectly promotes these activities by stimulating the release of somatomedins from the liver and other tissues. **Somatomedins** are growth factors that travel through the bloodstream to cartilage, bone, and other target tissues, where they trigger growth-related events. STH has its most dramatic effect on the growth rate of children and adolescents. For example, if a person is deficient in STH, the long bones stop growing and *pituitary dwarfism* results. (Recently, the supply of human STH has become widely available through genetic engineering. The gene that directs the synthesis of human STH has been introduced into bacteria, converting them into STH-synthesizing factories (*see figure 9.12*). Conversely, if the secretion of STH continues after adolescence, *gigantism* results (figure 35.9a).

[4]

When STH secretions are excessive during adulthood (when the long bones can no longer lengthen) acromegaly results. In acromegaly, the bones of the jaw, feet, and hands thicken, as do epithelial tissues of the nose, skin, eyelids, lips, and tongue (figure 35.9b).

Prolactin (PRL) has the widest range of actions of the adenohypophyseal hormones. It plays as essential role in many aspects of reproduction. For example, it stimulates reproductive migrations in many animals, including the movement of salamanders into water. It promotes nest building and maternal care in teleosts and birds. In some fishes, it also helps control water and salt balance and is thus essential for certain saltwater species to enter fresh water during their spawning runs.

Prolactin enhances mammary gland development and milk production in female mammals. (Recall that oxytocin stimulates milk ejection from the mammary glands, but not its production.)

Thyrotropin, or **thyroid-stimulating hormone (TSH)**, stimulates the synthesis and secretion of **thyroxine**, the main thyroid hormone. An excessive amount of TSH increases the blood flow into the thyroid gland. As a result, the cells grow excessively, producing an enlarged thyroid gland, called a *goiter* (figure 35.9c). (Goiters can also arise from other causes, as discussed below.) [5]

(a)

(b)

(c)

Figure 35.9 Some hormonal effects in humans. (a) Effect of somatotropin (growth hormone) on body growth. The people on the left are of average height, but the person on the right is affected by gigantism, which resulted from excessive STH production. (b) Acromegaly results from excessive STH production during adulthood. The effects of excessive STH can be seen in the facial structure as the affected individual ages. (c) Maria de Medici (1865) displays a mild case of goiter. During the late Renaissance, a rounded neck was considered a sign of beauty and occurred in regions of the world where iodine in the diet was insufficient for normal thyroid functioning.

Adrenocorticotropic hormone (ACTH) stimulates the adrenal gland to produce and secrete steroid hormones called *glucocorticoids*. Secretion of ACTH is regulated by the secretion of corticotropin-releasing factor from the hypothalamus, which in turn is regulated by a feedback system that involves such factors as stress, insulin, ADH, and other hormones.

Two *gonadotropins* (hormones that stimulate the gonads) are produced by the adenohypophysis: luteinizing hormone and follicle-stimulating hormone. **Luteinizing hormone (LH)** received its name from the corpeus luteum, a temporary endocrine tissue in the ovaries that secretes the female sex hormones estrogen and progesterone. In the female, an increase of LH in the blood stimulates ovulation, the monthly release of a mature egg(s) from an ovary. In the male, the target cells of LH are cells in the testes that secrete the male hormone testosterone. In the female, **follicle-stimulating hormone (FSH)** stimulates the follicle cells in the ovaries to develop into mature eggs and also stimulates them to produce estrogen. In the male, FSH stimulates the cells of the testes to produce sperm.

Although **melanocyte-stimulating hormone (MSH)** is present in human pituitary tissue, no direct physiological role for its normal secretion is known. In other vertebrates, this hormone is produced by the cells in the pars intermedia. It causes the skin of many fishes, amphibians, and reptiles to darken by promoting the dispersal of melanosomes (organelles that contain dark pigments) in the melanophores, which are cells in the skin that contain melanosomes. In the absence of this hormone, melanosomes become concentrated near the center of the cells and the skin becomes lighter in color. These physiological color changes help adapt the animal to changes in the color of its background. Temperature and emotional state can also affect the release of melanophore-stimulating hormone in some reptiles. Because skin pigment is deposited outside the pigment-producing cells in mammals, they do not show similar color changes.

In many fishes, amphibians, and reptiles a median pineal eye, parietal eye, or both are sensitive to light and produce a hormone called **melatonin.** Melatonin, like MSH, affects pigmentation in some animals. It also is involved in diurnal color changes, degree of exposure to sunlight, and certain seasonal rhythms. Only the pineal is found in birds and mammals, where it forms the **pineal gland** (*see figure 35.2*). (Cetaceans lack this gland.) Its distinctive cells have evolved from the photoreceptors of lower vertebrates, they synthesize melatonin and are most active in the dark. Light inhibits the enzymes needed for melatonin synthesis. Because of its cyclic production, melatonin has the potential of affecting many physiological processes and adjusting them to diurnal and seasonal cycles. Thus, melatonin may be a key molecule, attuned to photoperiodicity, which has been selected through evolution to effect adaptation to annual events. In humans, decreased melatonin secretion may help trigger the onset of puberty, the age at which reproductive structures start to mature.

> ## Stop and Ask Yourself
>
> 8. What hormones are secreted by the adenohypophysis?
> 9. What hormones are secreted by the neurosecretory cells of the hypothalamus?
> 10. What are some effects of prolactin in vertebrates?
> 11. What are the effects of gonadotropins in both the male and female?
> 12. What is the function of MSH? Of melatonin?

Thyroid Gland

The **thyroid gland** is located in the neck, anterior to the trachea (*see figure 35.2*). Two of its secretions are **thyroxine** and **triiodothyronine,** both of which influence overall growth, development, and metabolic rates (*see figure 35.4*). Another thyroid hormone, **calcitonin,** plays a role in controlling extracellular levels of calcium (Ca^{2+}) ions by promoting deposition of these ions into bone tissue when their levels rise. Once calcium returns to its homeostatic level, the thyroid cells decrease their secretion of calcitonin.

Because the thyroid hormones affect most cells of the body rather than specific target cells, the effects of malfunctions are far-reaching and can be serious. For example, overactivity of the thyroid gland due to a lack of iodine in the diet results in *hyperthyroidism,* which can also produce a goiter. Underactivity of the thyroid results in *hypothyroidism.* An underactive thyroid during prenatal development or infancy causes *cretinism,* which results in mental retardation and irregular bone development in humans.

Parathyroid Glands

The **parathyroid glands** are tiny, pea-sized glands embedded in the thyroid lobes, usually two glands in each lobe (*see figure 35.2*). The parathyroids secrete **parathormone (PTH)**, which regulates the levels of calcium (Ca^{2+}) and phosphate (HPO_2^{-4}) in the blood. An improper balance of calcium and phosphate ions in the blood can cause faulty transmission of nerve impulses, destruction of bone tissue, hampered bone growth, lack of gene activation, and muscle spasms.

When the level of calcium in the blood bathing the parathyroid glands is low, PTH secretion increases and has the following effects. It stimulates the activity of bone cells (osteoclasts) to break down bone tissue and release calcium ions into the blood. It also enhances the absorption of calcium from the small intestine into the blood. Adequate amounts of vitamin D are important for this process. (Vitamin D deficiency leads to *rickets,* a disorder arising from the lack of calcium.) Finally, PTH promotes the reabsorption of calcium by the kidney tubules, so that the amount of calcium excreted in the urine is decreased. The

[6]

negative feedback system for parathormone is shown in figure 35.10.

Adrenal Glands

The two **adrenal glands** rest on top of the kidneys (figure 35.11). Each gland is made up of two separate glandular tissues. The inner portion is called the *medulla*, and the outer portion, which surrounds the medulla, is the *cortex*.

Adrenal Cortex   The adrenal cortex secretes three classes of steroid hormones: *glucocorticoids* (hydrocortisone, cortisol), a single *mineralocorticoid* (aldosterone), and *sex hormones* (androgens, estrogens). The glucocorticoids, such as **cortisol,** help regulate overall metabolism and the level of blood sugar. They also function in defense responses to infection or tissue injury. **Aldosterone** helps maintain concentrations of solutes (such as sodium) in the extracellular fluid when either food intake or metabolic activity change the amount of solutes entering the bloodstream. Aldosterone also promotes sodium reabsorption in the kidneys and thus water reabsorption; hence it plays a major role in maintaining the homeostasis of extracellular fluid. Normally, the sex horomones secreted by the adrenal cortex have only a slight effect on the gonads of both the

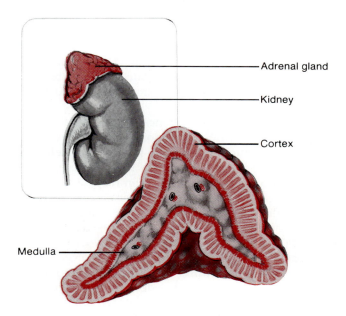

Figure 35.11   The adrenal gland. An adrenal gland consists of an outer cortex and an inner medulla.

male and female. They consist mainly of weak male hormones called *androgens* and lesser amounts of female hormones called *estrogens*. However, disorders of the adrenal gland can cause sex reversal changes. For example, a tumor of the adrenal cortex might cause a female to produce too many androgens, which will cause her to develop masculine characteristics, such as facial hair, a deep voice, and a reduction in breast size.   [7]

Adrenal Medulla   The adrenal medulla is under neural control. It contains neurosecretory cells that secrete **epinephrine (adrenalin)** and **norepinephrine (noradrenalin)**, both of which help control blood circulation and carbohydrate metabolism. Brain centers and the hypothalamus govern their secretion via sympathetic nerves.

During times of excitement, emergencies, or stress, the adrenal medulla contributes to the overall mobilization of the body through the sympathetic nervous system. In response to epinephrine and norepinephrine, the heart rate increases, blood flow increases to many vital organs, the airways in the lungs dilate, and more oxygen is delivered to all cells of the body. This group of events is sometimes called the "fight-or-flight" response and permits the body to react strongly and quickly to emergencies.   [8]

All vertebrates have groups of cells homologous to the mammalian adrenal cortical and chromaffin tissue, but they are not as intimately associated (figure 35.12). (Chromaffin tissue is a group of cells that actually produce and secrete hormones.) Many fishes have an elongated *interrenal body* located between the kidneys (comparable to the mammalian cortex), and a series of *islets of chromaffin cells* located alongside nerve ganglia. In amphibians, the cortical and chromaffin cells fuse to form a series of islets along the

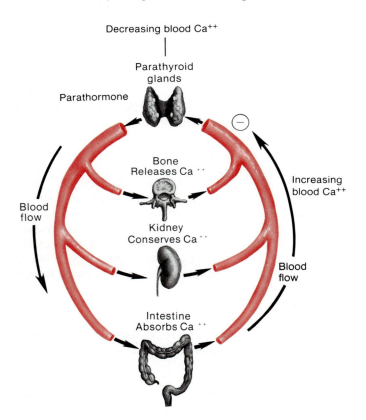

Figure 35.10   Hormonal feedback. The negative feedback mechanism of the parathyroid glands (parathormone). Parathormone stimulates the release of calcium from bone and the conservation of calcium by the kidneys. It indirectly stimulates the absorption of calcium by the intestine. The result increases blood calcium, which then inhibits the secretion of parathormone.

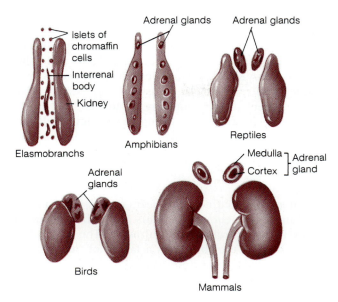

Figure 35.12 The adrenal glands of some respresentative vertebrates. The location of the chromaffin cells is shown in black and the cortical tissue is shaded.

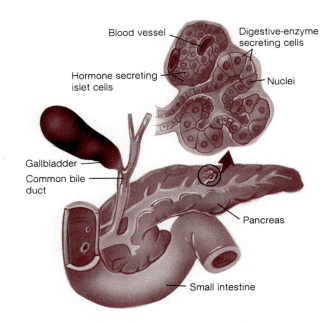

Figure 35.13 The pancreas. The hormone secreting cells of the pancreas are arranged in clusters or islets that are closely associated with blood vessels. Other pancreatic cells secrete digestive enzymes into ducts.

ventral surface of the kidneys. The association of cortical and chromaffin cells is more intimate in reptiles and in mammals, the chromaffin cells are completely within the cortex and form the medulla.

Pancreas

The **pancreas** is an elongated fleshy organ located posterior to the stomach (figure 35.13). It is a mixed gland

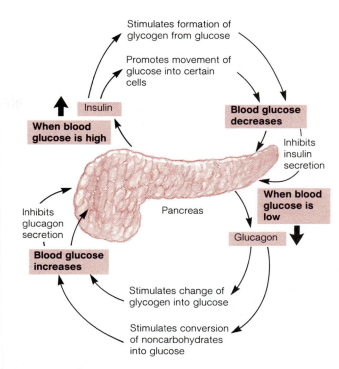

Figure 35.14 Insulin and glucagon. The negative feedback mechanism for regulating the secretion of glucagon and insulin. Insulin and glucagon function together to help maintain a relatively stable blood glucose concentration.

because it functions both as an exocrine (with ducts) gland to secrete digestive enzymes, and as an endocrine (ductless) gland. The endocrine portion of the pancreas makes up only about 1% of the gland. It is this portion that synthesizes, stores, and secretes hormones from clusters of cells called **pancreatic islets.**

The pancreas contains between 200,000 and 2,000,000 pancreatic islets scattered throughout the gland. Each islet contains four special groups of cells, called alpha (α), beta (β), delta (δ), and F cells. The α *cells* produce the hormone **glucagon,** and β *cells* produce **insulin.** The δ *cells* secrete **somatostatin,** the hypothalamic growth-hormone inhibiting factor that also inhibits the secretion of both glucagon and insulin. *F cells* secrete a pancreatic polypeptide of unknown function that is released into the bloodstream after a meal (box 35.1).

When the glucose levels in the blood are high, such as after a meal, β cells secrete insulin. Insulin promotes the uptake of glucose by the body's cells, including liver cells, where excess glucose can be converted to glycogen (a storage polysaccharide; *see figure 2.16*). Insulin and glucagon are crucial to the regulation of blood glucose levels. When glucose levels in the blood are low, the α cells secrete glucagon. Glucagon stimulates the breakdown of glycogen into glucose units, which can be released into the bloodstream to raise the level of blood glucose to restore the homeostatic level. Figure 35.14 illustrates the negative feedback system that regulates the secretion of glucagon and insulin and maintenance of appropriate blood glucose levels.

Stop and Ask Yourself

13. What effect does parathormone have on the levels of calcium in the blood?
14. What types of hormones are secreted by the adrenal cortex? By the three cell types in the pancreatic islets?
15. How do norepinephrine and epinephrine help the body respond to stress?
16. How does glucagon function in opposition to insulin?

Gonads

The gonads (ovaries and testes) secrete hormones that help regulate reproductive functions. In the male, **testosterone** is secreted by the testes and acts with luteinizing and follicle-stimulating hormones produced by the adenohypophysis to stimulate spermatogenesis. It is also necessary for the growth and maintenance of the male sex organs, it promotes the development and maintenance of sexual behavior, and stimulates the growth of facial and pubic hair, as well as enlargement of the larynx, which causes the voice to deepen.

Three major classes of ovarian hormones help to regulate female reproductive functions. **Estrogens** (estrin, estrone, and estradiol) help regulate the menstrual and estrus cycles and the development of the mammary glands and other female secondary sex characteristics. The **progestins** (progesterone) also regulate the menstrual and estrus cycles, the development of the mammary glands, and aid in the formation of the placenta during pregnancy. **Relaxin,** which is produced in small quantities, softens the opening of the uterus (cervix) at the time of delivery.

Thymus

The thymus is located near the heart (*see figure 35.2*). It is large and conspicuous in infants and children but diminishes in size throughout adulthood. The major hormonal product of the thymus is a family of peptide hormones, including **thymopoietin (TP)** and α_1 and β_4 **thymosin**, that appear to be essential for the normal development of the immune system.

Other Sources of Hormones

In addition to the major endocrine glands, other glands and organs carry on hormonal activity. Some of these are summarized in table 35.2.

Box 35.1

Diabetes

Several disorders classified under the term **diabetes** (Gr. *diabetes,* syphon) are associated with insulin deficiency or with the inability of cells to take up glucose. In *type I diabetes (juvenile diabetes)*, the afflicted person has little or no insulin because the β cells have been destroyed. This form of diabetes is relatively uncommon and usually occurs early in life. It can occur as a result of a heritable susceptibility, viral infection, or an autoimmune response mounted against the β cells. Without enough insulin, glucose accumulates in the blood but does not enter the cells. Because the cells cannot use the accumulated glucose, the body actually begins to starve. Brain, retinal, and gonadal tissues are particularly endangered because glucose is the only nutrient that can be utilized by these tissues. The use of fats (to replace glucose) for energy production in other tissues causes the accumulation of acetoacetic acid and keto acids in the blood. This buildup leads to acidosis, which can cause coma and death. Survival for type I diabetes is absolutely dependent upon regular injections of insulin. (Insulin is a small protein molecule and cannot be taken by mouth because it would be rapidly inactivated by protein-digesting enzymes in the digestive tract.) Without injections, the person would suffer severe metabolic and urinary disruptions.

In *type II (maturity-onset) diabetes,* insulin levels are nearly normal or even above normal, however, the target cells cannot respond to the hormone. Either the cells have an insufficient number of insulin receptors, or the receptors themselves are abnormal. Diet, exercise, and oral glycemic agents, which facilitate the uptake of glucose, are used to treat this type of diabetes.

Table 35.2   Other Sources of Vertebrate Hormones

Gland/Organ	Hormone	Function	Target Area
Placenta	Estrogens, progesterone, human chorionic gonadotropin (hCG)	Maintains pregnancy	Ovaries, mammary glands, uterus
Digestive tract	Secretin	Stimulates release of pancreatic juice to neutralize stomach acid	Cells of pancreas
	Gastrin	Produces digestive enzymes and HCL in stomach	Stomach mucosa
	Cholecystokinin (CCK)	Stimulates release of pancreatic enzymes and bile from gallbladder	Pancreas, gallbladder
Heart	Atriopeptin	Lowers blood pressure, maintains fluid balance	Blood vessels, kidneys
Kidneys	Erythropoietin	Stimulates red-blood-cell production	Bone marrow

■ Summary

1. For metabolic activity to proceed smoothly in the body, the chemical environment of each cell must be maintained within fairly narrow limits (homeostasis). This is accomplished using negative feedback systems that involve integrating, communicating, and coordinating molecules called messengers.

2. Chemical messengers are molecules secreted by specialized cells. These chemical messengers can be categorized as follows: local chemical messengers (lumones, prostaglandins, chalones, growth factors, cell adhesion molecules), neurotransmitters such as acetylcholine, neurohormones, hormones, pheromones (sex attractants), and neuromodulators (a hormone that modulates the response of a neuron).

3. A hormone is a specialized chemical messenger produced and secreted by an endocrine cell or tissue. Hormones are usually steroids, amines, proteins, or fatty acid derivatives. Secretion of hormones is often regulated by negative feedback systems.

4. The function of a hormone is to modify the biochemical activity of a target tissue, so called because it has receptors to which hormone molecules can bind. Hormonal activity may be accomplished by either the fixed-membrane-receptor mechanism (water soluble hormones) or the mobile-receptor mechanism (steroid hormones).

5. In all vertebrates, a neuroendocrine control center helps communicate and integrate activities for the entire body. This center consists of the hypothalamus and pituitary gland.

6. The vertebrate endocrine system consists of several major glands: the pituitary gland, thyroid gland, parathyroid glands, adrenal glands, pancreas, and gonads. In addition to these major glands, other glands and organs carry on hormonal activity. Examples include the placenta, thymus, digestive tract, heart, and kidneys.

■ Key Terms

adrenal glands (p. 564)
adrenocorticotropic hormone (ACTH) (p. 563)
aldosterone (p. 564)
antidiuretic hormone (vasopressin) (p. 560)
calcitonin (p. 563)
cortisol (p. 564)
diabetes (p. 566)

endocrinology (p. 555)
epinephrine (adrenalin) (p. 564)
estrogens (p. 566)
follicle-stimulating hormone (FSH) (p. 563)
glucagon (p. 565)
growth hormone (GH) (p. 561)
hormones (p. 557)

insulin (p. 565)
local chemical messengers (p. 557)
luteinizing hormone (LH) (p. 563)
melanocyte-stimulating hormone (MSH) (p. 563)
melatonin (p. 563)
neuroendocrine system (p. 557)

neurohormones (p. 557)
neuromodulator (p. 557)
neurotransmitters (p. 557)
norepinephrine (noradrenalin) (p. 564)
oxytocin (p. 561)
pancreas (p. 565)
pancreatic islets (p. 565)
parathormone (PTH) (p. 563)

Critical Thinking Questions

1. How is information encoded by hormones? How do cells "know what to do" in response to hormonal information?

2. Summarize your knowledge of how endocrine systems work by describing the "life" of a hormone molecule from the time it is secreted until it is degraded or used up.

3. All cells secrete or excrete molecules, and all cells respond to certain biochemical factors in their external environments. Could the origin of endocrine control systems lie in such ordinary cellular events? How might the earliest multicellular organisms have evolved some sort of endocrine coordination?

4. Mental states strongly affect the function of many endocrine glands. This link between the mind and the body occurs through the hypothalamus. Can you describe how thoughts are transformed into physiological responses in the hypothalamus?

5. Why is neurosecretion so important in adapting internal physiological processes to external stimuli?

6. Compared to enzymes and genes, hormones are remarkably small molecules. Wouldn't larger molecules be able to carry more information? Explain.

7. A patient with insulin-dependent diabetes will often eat a carbohydrtae snack or meal before taking a dose of insulin. How would this snacking help the insulin do its job for the diabetic?

Suggested Readings

Books

Crapo. L. 1985. *Hormones: The Messengers of Life*. New York: Freeman.

Hadley, M. 1988. *Endocrinology*. Englewood Cliffs: Prentice-Hall.

Articles

Am. Zoologist. 1983. Evolution of endocrine systems in lower vertebrates, a symposium honoring professor Aubrey Gorbman. pp 595–739.

Berridge, M. The molecular basis of communication within the cell. *Scientific American* October, 1985.

Binkley, S. A timekeeping enzyme in the pineal gland. *Scientific American,* April, 1979.

Bloom, F. E. Neuropeptides. *Scientific American* October, 1981.

Cantin, M., and Genest, J. The heart as an endocrine gland. *Scientific American* February, 1986.

Carmichael, S. W., and Winkler, H. The adrenal chromaffin cell. *Scientific American* August, 1985.

Crews, D. The hormonal control of behavior in a lizard. *Scientific American* August, 1979.

Guillemin, R., and Burgus, R. The hormones of the hypothalamus. *Scientific American* November, 1972.

Mcewen, B. Interactions between hormones and nerve tissue. *Scientific American* July, 1976.

Nathan, J., and Greengard, P. Second messengers in the brain. *Scientific American* August, 1977.

Notkins, A. The cause of diabetes. *Scientific American* November, 1979.

O'Malley, B., and Schrader, W. The receptors of steroid hormones. *Scientific American* February, 1976.

Orei, L., Yassalli, J. D., and Perrelet, A. The insulin factory. *Scientific American* September, 1988.

Pike, J. E. Prostaglandins. *Scientific American* November, 1971.

Circulation, Immunity, and Gas Exchange

36

Concepts

1. Animal circulatory systems transport substances from one part of the body to another, and between the animal's external environment and extracellular fluid.
2. The circulatory system of higher animals consists of a central pumping heart, blood vessels, blood, and an ancillary lymphytic system.
3. Vertebrates defend themselves against microorganisms, foreign matter, and cancer cells by both nonspecific and specific mechanisms.
4. In vertebrates, gas exchange occurs through the integument (cutaneous exchange), gills, or lungs. Gas movement occurs between the environment and cells of the body by diffusion down concentration gradients (from areas of higher to areas of lower concentration). In large and active animals, gas exchange is increased by ventilation—the active movement of air into and out of the respiratory system.
5. Some invertebrates depend solely on the diffusion of gases, nutrients, and wastes between body surfaces and individual body cells. Others have either open or closed circulatory systems for the transport of gases, wastes, and nutrients.

Have You Ever Wondered:

[1] why a mammalian red blood cell is shaped like a doughnut without a complete hole?
[2] if blood in our veins is bluish, why it appears red when we cut a vein and begin to bleed?
[3] why we sometimes have dark circles under our eyes?
[4] what a blood pressure reading of 120/80 mmHg means?
[5] why lymph nodes are removed during cancer operations?
[6] what causes infected tissue to be red, warm, painful, and puffy?
[7] why organ and tissue transplants are often rejected except between identical twins?
[8] why babies get more colds than do adults?
[9] how salamanders that do not have lungs breathe?

These and other useful questions will be answered in this chapter.

This chapter contains underlined evolutionary concepts.

All animals must maintain a homeostatic balance in their bodies, and this need requires that nutrients, wastes, and respiratory gases be circulated through the body. In very small organisms, this need can be met by simple diffusion processes. More complex body forms demand more sophisticated methods to ensure that the environment of all cells is maintained in a functional manner.

Because protozoa are small, with high surface-area-to-volume ratios (*see figure 3.3*), all they need for gas, nutrient, and waste exchange is simple diffusion. Sponges circulate the external environment through their bodies, instead of circulating an internal fluid (*see figure 17.5*). Some invertebrates, such as the cnidarian, *Hydra,* have a fluid-filled internal cavity called a *gastrovascular cavity* (figure 36.1a). This cavity supplies nutrients for all body cells lining the cavity, obtains oxygen from water in the cavity,

and releases carbon dioxide and other wastes into it. Simple body movement moves the fluid around.

The gastrovascular cavity of flatworms, such as a planarian, is more complex than that of *Hydra* because branches penetrate to all parts of the body (figure 36.1b). Because this branched gastrovascular cavity runs close to all body cells, diffusion distances for nutrients, gases, and wastes are not great. Body movements help distribute materials to various parts of the body.

Circulation

Beginning with the annelids, circulation functions become organized into a separate circulatory system. Various fluid-filled systems, called vascular systems, transport molecules throughout the bodies of many animals. More specifically, a *circulatory* or *cardiovascular system* (Gr. *kardia,* heart + L. *vascular,* vessel) is a specialized system that moves

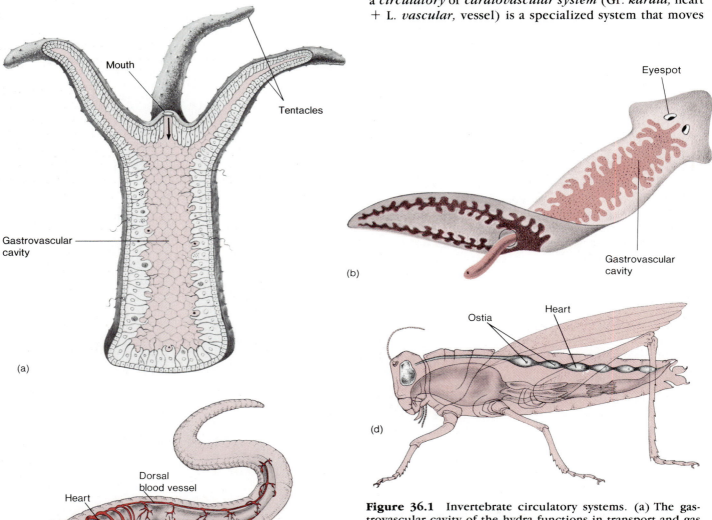

(a)

(b)

(c)

(d)

Figure 36.1 Invertebrate circulatory systems. (a) The gastrovascular cavity of the hydra functions in transport and gas exchange. Cells lining the cavity can exchange gases with the water in the cavity and release waste into it. (b) The planarian's gastrovascular cavity is branched, allowing for more effective distribution of materials. (c) The dorsal circulatory system of an earthworm. (d) The dorsal heart of a grasshopper pumps blood through an open circulatory system.

the fluid medium called blood or hemolymph in a specific direction determined by the presence of unidirectional blood vessels. Movement is accomplished by a muscular pumping heart.

There are two basic types of circulatory systems: open and closed. In an *open circulatory system*, the heart pumps blood out into the body cavity or at least through parts of the cavity, where blood bathes the cells, tissues, and organs. In a *closed circulatory system*, blood circulates in the confines of tubular vessels.

The annelids, such as the common earthworm, have a closed circulatory system in which blood travels through vessels delivering nutrients to cells and removing wastes (figure 36.1c). Earthworm hemoglobin functions in oxygen and carbon dioxide transport just as does the hemoglobin in vertebrate blood, but it is suspended in the fluid portion of blood rather than in red blood cells, as it is in vertebrates.

Most arthropods and molluscs have open circulatory systems, where the blood or hemolymph directly bathes the cells and tissues rather than being carried only in vessels (figure 36.1d). For example, an insect's heart pumps blood (hemolymph) through vessels that open into a body cavity (hemocoel).

Open circulatory systems are not very efficient transport systems, but the invertebrates that have these systems can lead active lives because their respiratory apparatus, the branching hollow tracheae, operate independently of the heart to bring oxygen close to each cell and to remove carbon dioxide.

All vertebrates have a closed circulatory system in which the walls of the heart and blood vessels are continuously connected, and the blood never leaves the blood vessels.

Blood generally is conducted from the heart, through arteries, arterioles, capillaries, venules, veins, and back to the heart. It is only at the capillary level that exchange occurs between the blood and extracellular fluid.

Characteristics of Blood

Because it is a liquid, blood is classified as a specialized type of connective tissue (colorplate 1m). As do other connective tissues, blood contains cellular elements called *formed elements,* and a fluid matrix, called *plasma*. Both the formed elements (red blood cells, white blood cells, and platelets) and plasma play important roles in homeostasis (table 36.1).

Overall, blood *transports* oxygen and carbon dioxide; *defends* the body against harmful microorganisms, cells, and agents; *destroys* foreign particles; is involved in *inflammation, coagulation,* (blood clotting), and the *immune response*; and helps *regulate* the pH of the body fluids. In birds and mammals, blood also helps *equalize body temperature* by carrying excess heat from areas of high metabolic activity (e.g., skeletal muscles) to the skin, where the heat can be dissipated from the body.

Plasma   Plasma is the straw-colored liquid part of blood. It is about 90% water and provides the solvent for dissolving and transporting nutrients. A group of proteins (albumins, fibrinogen, and globulins) comprise another 7% of the plasma. It is the concentration of these plasma proteins that influences the distribution of water between the blood and extracellular fluid. Because albumin respresents about 60% of the total plasma proteins, it plays the most

---

Table 36.1   Component Parts of Vertebrate Blood

	Function	Number Present per ml	Percent of Total
Formed Elements (40–50% of total volume)			
Red blood cells (erythrocytes)	Transport O_2, CO_2	4,500,000–5,500,000	
White blood cells (leukocytes)		5,000–10,000	
Granulocytes			
Neutrophils	Phagocytosis	3,000–6,000	
Eosinophils	Phagocytosis	100–400	
Basophils	Release anticoagulant (heparin) and histamine	25–100	
Agranulocytes			
Lymphocytes	Central to the immune response	1,000–3000	
Monocytes	Phagocytosis	150–200	
Platelets (thrombocytes)	Act in blood clotting to control blood loss	250,000–300,000	
Plasma Portion (50–60% of total volume)			
Water	Serves as solvent		90–92
Plasma proteins	Fight infection, blood clotting, lipid transport		7–8
Other solutes (e.g., electrolytes, lipids hormones, vitamins, gases)	Diverse roles in pH regulation, fluid regulation		1–2

important role with respect to water movement. Fibrinogen is necessary for blood coagulation (clotting), and the globulins (α and β) transport lipids and fat soluble vitamins. **Serum** is plasma from which the proteins involved in blood clotting have been removed. The gamma globulin portion functions in the immune response because it consists mostly of antibodies. The remaining 3% of plasma is composed of electrolytes, amino acids, glucose and other nutrients, various enzymes, hormones, metabolic wastes, and traces of many other inorganic and organic molecules. The lipids present in plasma include simple fatty acids, phospholipids, and cholesterol. Lipids that are being transported are generally bound with proteins to form lipoproteins.

Formed Elements   The formed-element fraction of blood consists of red blood cells, white blood cells, and platelets.

Red Blood Cells   **Red blood cells (erythrocytes)** (Gr. *erythros*, red + cells) make up about one-half the volume of mammalian blood. Each erythrocyte is shaped like a disk and is slightly concave on top and bottom, like a doughnut without the hole poked completely through [1] (figure 36.2a). This shape provides a larger surface area for gas diffusion than a flat disk or sphere. Also, when an erythrocyte is mature, it no longer has a nucleus or many organelles. Almost the entire weight of an erythrocyte consists of **hemoglobin** (Gr. *haima*, blood + L., *globulus*, little globe), an iron containing globular protein. The major function of erythrocytes is to pick up oxygen from the

environment, bind it to hemoglobin to form **oxyhemoglobin,** and transport it, via the blood vessels, to body tissues. Blood rich in oxyhemoglobin is bright red; as oxygen diffuses into the tissues, blood becomes darker and appears blue when observed through the blood vessel walls. However, when this deoxygenated blood is exposed [2] to oxygen (such as when a vein is cut and one begins to bleed), it instantaneously turns bright red. Hemoglobin also carries waste carbon dioxide (in the form of **carbaminohemoglobin**) from the tissues to the lungs (or gills) for removal from the body.

White Blood Cells   **White blood cells (leukocytes)** (Gr. *leukos*, white + cells) serve as scavengers that destroy microorganisms at infection sites, remove foreign chemicals, and remove debris that results from dead or injured cells. All leukocytes are derived from immature cells (called *stem cells*) in bone marrow by a process termed *hematopoiesis* (Gr. *hemato*, blood + *poiein*, to make). There are two categories of leukocytes: *granulocytes* and *agranulocytes* (*see table 36.1*).

The most numerous of the white blood cells are granulocytes, so named because they contain large numbers of granules in the cytoplasm. The three types of granulocytes are eosinophils, basophils, and neutrophils.

Eosinophils are phagocytic, and ingest foreign proteins and immune complexes rather than bacteria (figure 36.2b). Eosinophils also release chemicals that counteract the effect of certain inflammatory chemicals released during allergic reactions. **Basophils** are the least numerous of the white blood cells (figure 36.2c). Tissue basophils are

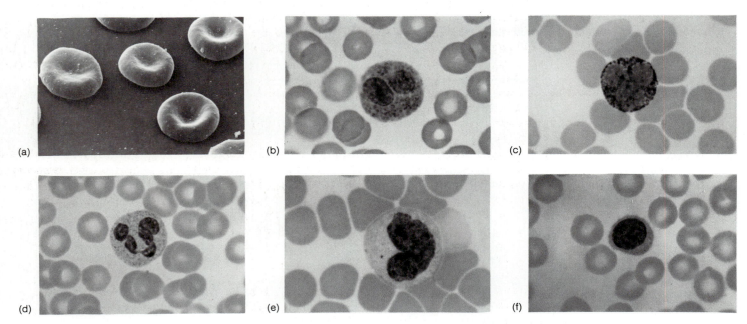

(a)   (b)   (c)

(d)   (e)   (f)

Figure 36.2   Blood cells. (a) Scanning electron micrograph showing the biconcave shape of erythrocytes. (b) An eosinophil is characterized by the presence of red-staining cytoplasmic granules. (c) A basophil is characterized by blue-staining granules. (d) A neutrophil is characterized by light-pink granules and a multilobed nucleus. Cells shown in b–d are also known as granulocytes. The agranulocytes consist of large monocytes (e) and lymphocytes (arrow) (f).

Immunity and Body Defenses

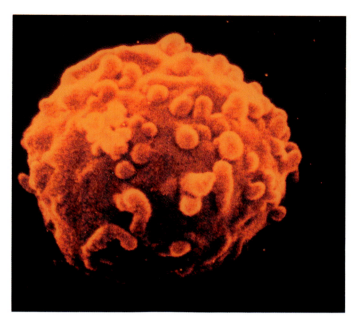

(a) A human lymphocyte, part of the body's protective armory of immune cells.

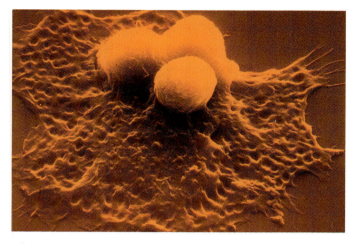

(b) A macrophage binding to three tumor cells.

(c) This bombardier beetle (*Brachinus*) defends itself by spraying a volatile irritant from glands.

(d) The porcupine (*Erethizon dorsatum*) has a spiny arsenal for protection.

Colorplate 18

Respiration and Gas Exchange

(a) The body of a marine flatworm (*Prostheceracus bellostriatus*) is so thin that oxygen and carbon dioxide diffuse into and out of each body cell from the surrounding seawater.

(b) This axolotl (*Ambystoma tigrinum*) has elaborate external gills that have a large surface area for exchanging gases with the water.

(c) Mosquito larvae. A mosquito larva's tracheae open through a spiracle at the end of a tube that is extended above the water surface.

(d) Llamas in the Peruvian Andes. These animals (*Lama guanicoe*) have three or four times more red blood cells than animals that live at sea level. The extra red blood cells enable the llama to live at high altitudes where oxygen tension is low.

called *mast cells*. Both basophils and mast cells bind to a particular antibody, causing the granules to release histamine and heparin. *Histamine* is a vasodilator that causes blood vessels to leak fluid at a site of inflammation, and *heparin* prevents blood clotting. Both chemicals enhance the migration of white blood cells to the inflammation site. **Neutrophils** are the most numerous of the white blood cells (figure 36.2d). They are chemically attracted to sites of inflammation and are active phagocytes.

The two types of agranulocytes are the **monocytes** and **lymphocytes** (figure 36.2e,f). Two distinct types of lymphocytes are recognized: B cells and T cells, both of which are central to the immune response. **B cells** originate in the bone marrow, and colonize the lymphoid tissue, where they mature. In contrast, **T cells** are associated with and influenced by the thymus gland before they colonize lymphoid tissue and play their role in the immune response. When B cells are activated, they divide and differentiate to produce **plasma cells**. Plasma cells have much more cytoplasm than do B cells, which enables them to accommodate the biochemical machinery necessary for the production of antibodies.

Platelets (thrombocytes) **Platelets** (so named because of their platelike flatness), or **thrombocytes** (Gr. *thrombus*, clot + cells), are disk-shaped cell fragments that function to initiate blood clotting. Platelets originate in bone marrow, where bits of cytoplasm are pinched off as "giant" cells called *megakaryocytes*. When a blood vessel is injured, platelets immediately move to the site and begin to clump together, attaching themselves to the damaged area, and begin the process of *coagulation*. Figure 36.3 summarizes the cellular components of vertebrate blood and the various cellular lineages.

Blood Vessels

Arteries are elastic blood vessels that carry blood away from the heart to the organs and tissues of the body. The central canal of an artery (and all blood vessels) is called a *lumen* (figure 36.4a). Surrounding the lumen of an artery is a thick wall composed of three layers, or *tunicae* (L. *tunica*, covering). The innermost covering is the *tunica intima*, followed by some connective and elastic tissue, then the *tunica media* (middle covering), and finally, the *tunica adventia* (outermost covering). The outer membrane that covers an artery is called the *serosa*.

Most **veins** are relatively inelastic, large vessels that carry blood from the body tissues to the heart. The wall of a vein contains the same three layers (tunicae) as arterial

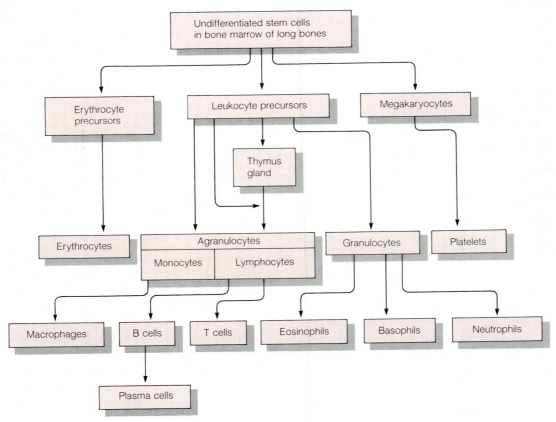

Figure 36.3 Cellular components of vertebrate blood. The process of blood cell production is called hematopoiesis. Notice that all blood cells initially begin their life in the bone marrow of long bones of the body.

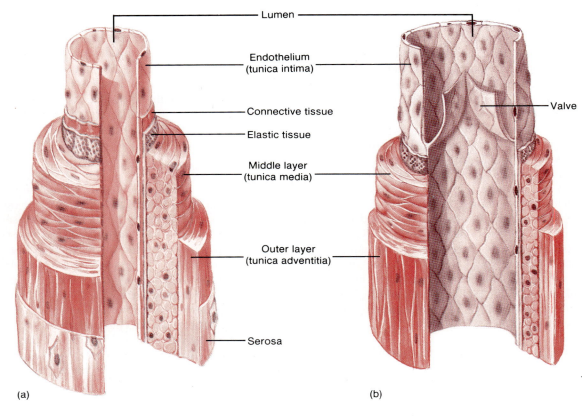

Lumen

Endothelium
(tunica intima)

Connective tissue

Elastic tissue

Middle layer
(tunica media)

Outer layer
(tunica adventitia)

Serosa

Valve

(a)

(b)

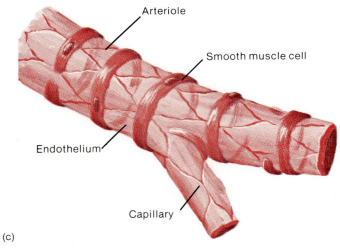

Arteriole

Smooth muscle cell

Endothelium

Capillary

(c)

Figure 36.4 Structure of blood vessels. (a) The wall of an artery, (b) vein, and (c) arteriole and capillary.

walls, but the tunica media is much thinner, and one or more valves is present (figure 36.4b). The valves permit blood to flow in only one direction, which is important in returning the blood to the heart. In specific locations, the blood flow slows in the veins, causing them to distend and become engorged (e.g., *varicose veins* in the legs and *hemorrhoids* in the walls of the rectum). In the skin below the eyes, the vessels are very visible, and engorged veins appear as dark circles.

[3]

Arteries lead to terminal *arterioles* (those closest to a capillary). The arterioles branch to form **capillaries** (L. *capillus,* hair), which are connected to venules and then to veins. Capillaries are generally composed of a single layer of endothelial cells (figure 36.4c) and are the most numerous blood vessels in the body. An abundance of capillaries makes an enormous surface area available for the exchange of gases, fluids, nutrients, and wastes between the blood and nearby cells.

Circulation of Blood

Blood circulates throughout the avian and mammalian body in two main circuits: the pulmonary and systemic circuits (figure 36.5). The *pulmonary circuit* supplies blood only

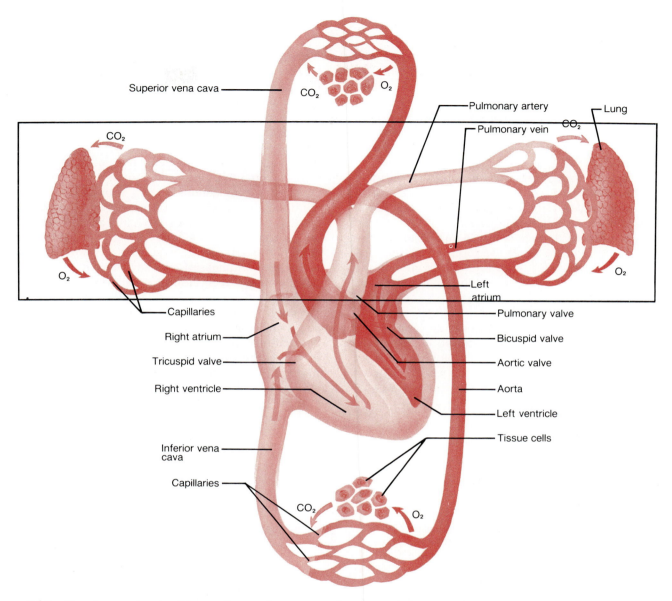

Figure 36.5 Circulatory circuits. The cardiovascular system of a mammal showing the main capillary beds and transport routes. The pulmonary circuit is shown in the box; the systemic circuit is outside the box.

to the lungs. It carries oxygen-poor (deoxygenated) blood from the heart to the lungs, where carbon dioxide is removed, and oxygen is added. It then returns the oxygen-rich (oxygenated) blood to the heart for distribution to the rest of the body. The *systemic circuit* supplies all the cells, tissues, and organs of the body with oxygen-rich blood and returns the oxygen-poor blood to the heart. Overall, the path of blood in the avian and mammalian body can be summarized as follows: right atrium → tricuspid valve → right ventricle → pulmonary valve → pulmonary arteries → lung capillaries → pulmonary veins → left atrium → bicuspid valve → left ventricle → aortic valve → aorta → arteries → arterioles → body capillaries → venules → veins → vena cavae → right atrium (box 36.1).

The Human Heart

The human heart is a hardworking pump that moves blood through the body. It pumps its entire blood volume (about 5 l) every minute; about 8,000 l of blood moves through 96,000 km (60,000 mi) of blood vessels every day. The heart of an average adult beats about 70 times per minute—more than 100,000 times per day. In a 70-year lifetime, the heart beats more than 2.6 billion times without tiring.

Most of the human heart is composed of cardiac muscle tissue (colorplate 1q) called *myocardium* (*myo,* muscle). The outer protective covering of the heart, however, is fibrous connective tissue called the *epicardium* (*epi,* upon). Connective tissue and endothelium form the inside

Box 36.1

Evolution of the Major Blood Vessels Leaving the Heart

As the vertebrate heart evolved, structural alterations also occurred in the *carotid artery* (supplying the head and brain), the *aorta* (supplying the major portion of the trunk), and the *pulmonary artery* (supplying the lungs)—the three major vessels by which blood leaves the heart.

Box figure 36.1a shows that the primitive heart of fishes pumps blood directly into the ventral aorta, which leads into six paired blood vessels called the *aortic arches*. The aortic arches are numbered I–VI. The ventral part of the first aortic arch is lost in gnathostomes, but the ventral part of the other five form the afferent brachial arteries, which lead into each of the gills. In the gills, the five pairs of aortic arches diverge into capillary beds, which lead into larger arteries that enter the dorsal aorta.

Those vertebrates that emerged onto the land, lacked gills. Thus, significant changes occurred in aortic arch development. During embryonic development, all six arches develop, but some degenerate, while others assume important anatomical roles. For example, arch III

(the carotid arch), develops into the carotid artery; arch IV (the systemic arch) develops into the aorta; and arch VI (the pulmonary arch) becomes the pulmonary artery (box figure 36.1b).

The continued simplification of the aortic arches and full separation of the systemic and pulmonary circuits were essential steps in the evolution of the diverse land vertebrates. For example, no bird or mammal could sustain its high rate of metabolism and activity without a high-pressure, fast-flowing internal transport system in which oxygenated and deoxygenated blood is never mixed. The anatomical trend toward this development is first seen in lungfish, and a key factor in its continual development is seen in the first amphibians that walked on land. In these amphibians, the lack of support from the water medium meant these animals had to expend considerable muscular energy to move about. This large demand for ATP required a more efficient delivery system so that adequate oxygen and nutrients could be supplied to the cells for their ATP production.

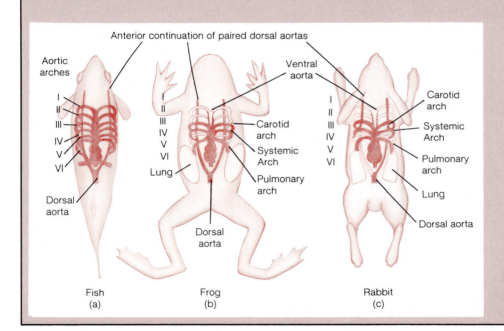

Box figure 36.1 Parts a-c show the evolution of the aortic arches as adaptations to terrestrial life.

of the heart, the *endocardium* (*endo*, inside). (Endothelium is a single layer of epithelial cells lining the chambers of the heart, as well as the lumen of blood vessels; *see figure 36.4*).

The left and right halves of the heart are two separate pumps, each containing two chambers (figure 36.6). In each half, blood first flows into a thin-walled *atrium* (L., antechamber; pl. atria), then into a thick-walled ventricle.

Valves are located between the upper (atria) and lower (ventricles) chambers. The *tricuspid valve* is located between the right atrium and right ventricle, and the *bicuspid valve* between the left atrium and left ventricle. (Collectively, these are referred to as the AV valves—*atrioventricular valves.*) The *pulmonary semilunar valve* is located at the exit of the right ventricle, and the *aortic semilunar valve* at the exit of the left ventricle. (Collectively, these

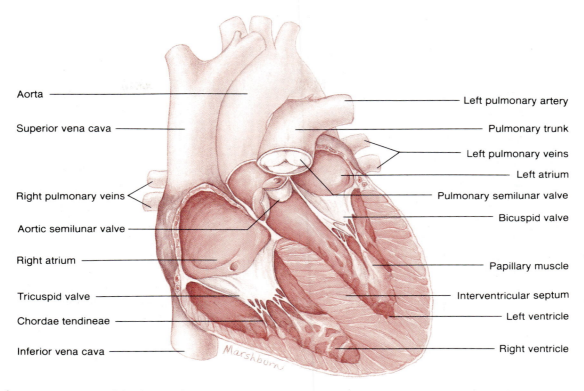

Aorta

Superior vena cava

Right pulmonary veins

Aortic semilunar valve

Right atrium

Tricuspid valve

Chordae tendineae

Inferior vena cava

Left pulmonary artery

Pulmonary trunk

Left pulmonary veins

Left atrium

Pulmonary semilunar valve

Bicuspid valve

Papillary muscle

Interventricular septum

Left ventricle

Right ventricle

Marshburn

Figure 36.6 The structures of the human heart.

are referred to as the *semilunar valves*.) All of these valves open and close due to blood pressure changes that are produced when the heart contracts during each heart beat. Just as do the valves in veins, heart valves keep blood moving in one direction, preventing backflow.

The heartbeat is an orchestrated sequence of muscle contractions and relaxations called the *cardiac cycle*. Each heartbeat is initiated by a "pacemaker," which is a small mass of tissue called the *sinoatrial node* (SA node) located at the entrance to the right atrium (figure 36.7). (Because the pacemaker is in the heart, nervous innervation is not necessary, which is why a heart transplant without nerves is possible.) The SA node initiates the cardiac cycle by producing an electrical impulse that spreads over both atria, causing them to contract simultaneously. The impulse then passes to the *atrioventricular node* (AV node), which is located on the inferior portion of the interatrial septum. From here, the impulse continues through the *atrioventricular bundle* (bundle of His), located at the tip of the interventricular septum. The atrioventricular bundle divides into right and left branches, which are continuous with the *Purkinje fibers* in the ventricular walls. Stimulation of these fibers causes the ventricles to contract simultaneously and eject blood into the pulmonary and systemic circulations.

The electrical impulse does not stop in the ventricles. It is conducted to the body surface, where it can be recorded as an *electrocardiogram* (ECG or EKG).

During each cycle, the atria and ventricles go through a phase of contraction called **systole** and a phase of relaxation called **diastole.** Specifically, while the atria are relaxing and filling with blood, the ventricles are also relaxed. As more and more blood accumulates in the atria, the blood pressure rises, and the atria contract, forcing the AV valves open and causing blood to rush into the ventricles. When the ventricles contract, the AV valves close, and the semilunar valves open, allowing blood to be pumped into the pulmonary arteries and aorta. After the blood has been ejected from the ventricles, they relax and start the cycle anew.

Blood Pressure

The fluid pressure that is generated by the ventricles contracting forces blood through the pulmonary and systemic circuits; the fluid pressure is called **blood pressure.** More specifically, blood pressure is the force exerted by the blood against the inner walls of blood vessels. Although such a force occurs throughout the vascular system, the term blood pressure is most commonly used to refer to systemic arterial blood pressure.

Arterial blood pressure rises and falls in a pattern corresponding to the phases of the cardiac cycle. When the ventricles contract (ventricular systole), their walls force the blood in them into the pulmonary arteries and the aorta. As a result, the pressure in these arteries rises sharply. The maximum pressure achieved during ventricular contraction is called the systolic pressure. When the ventricles relax (ventricular diastole), the arterial pressure drops, and the lowest pressure that remains in the arteries before the next ventricular contraction is called the diastolic pressure.

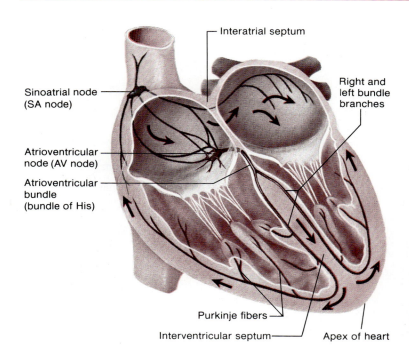

— Interatrial septum

Sinoatrial node
(SA node)

Right and
left bundle
branches

Atrioventricular
node (AV node)

Atrioventricular
bundle
(bundle of His)

Purkinje fibers

Interventricular septum

Apex of heart

Figure 36.7 The electrical conduction system of the human heart. The depolarization wave is initiated by the SA node and passes successively through the atrial myocardium to the AV node, the AV bundle, the right and left bundle branches, and the Purkinje fibers in the ventricular myocardium.

In humans, normal systolic pressure for a young adult is about 120 mmHg, which is the amount of pressure required to make a column of mercury [Hg] in a sphygmomanometer (sfig″mo-mah-nom′e-ter) rise a distance of 120 mm. Diastolic pressure is approximately 80 mmHg. These readings are by convention expressed as 120/80.

[4]

Other Vertebrates

The heart and blood vessels changed greatly as vertebrates moved from water to land and as endothermy evolved.

The fish heart consists of one atrium and one ventricle, through which blood flows in sequence (figure 36.8a). Blood returns to the heart through two major veins, flows through the chambers and leaves via the ventral aorta, which goes to the gills. In the gills, blood becomes oxygenated, loses carbon dioxide, and enters the dorsal aorta. The dorsal aorta distributes blood to all of the body organs and then returns to the heart via two major veins. Because blood only passes through the heart once, this system is called a *single circulation circuit*. It has the advantage of circulating oxygenated blood from the gills to all organs of the body almost simultaneoulsy. A disadvantage is that the narrow capillaries in the gills offer so much resistance that the blood is slowed down appreciably after it leaves the gills. This slow blood flow and rate of oxygen delivery to the cells limits the metabolic rate and activity that fish can sustain.

In amphibians and reptiles, the slow blood flow problem has been overcome by the evolution of a *double circulatory circuit,* in which blood passes through the heart twice during its circuit through the body. Amphibians and rep-

tiles have hearts that are not fully divided in two. In amphibians (figure 36.8b), there is a single ventricle that pumps blood both to the lungs and the rest of the body. However, because most amphibians absorb more oxygen through their skin than through their lungs or gills, blood returning from the skin also contributes oxygenated blood to the ventricle. The blood that is pumped out to the rest of the body is thus highly oxygenated and there is no need for two ventricles.

The heart of reptiles (figure 36.8c) is more advanced from an evolutionary perspective in that the ventricle is partially divided into a right and left side. Oxygenated blood from the lungs returns to the left side of the heart and is not mixed with deoxygenated blood in the right side of the heart. When the ventricles contract, blood is pumped out two aortae for distribution throughout the body, as well as to the lungs.

Both birds and mammals have complete double circulation systems in the heart (figure 36.8d) and two separate ventricles. This design is associated with two evolutionary advancements. (1) These animals have a high metabolic rate and require blood that is under considerable pressure. Because blood loses pressure as it passes through the lungs, returning the blood to the heart allows the blood pressure to be raised once again before blood is distributed to the rest of the body. A high blood pressure also allows for faster circulation of nutrients and the removal of wastes. (2) By keeping the deoxygenated and oxygenated blood separate in the right and left sides of the heart, blood that reaches the body tissues from the aorta contains as much oxygen as possible, allowing the high metabolic rate of these animals.

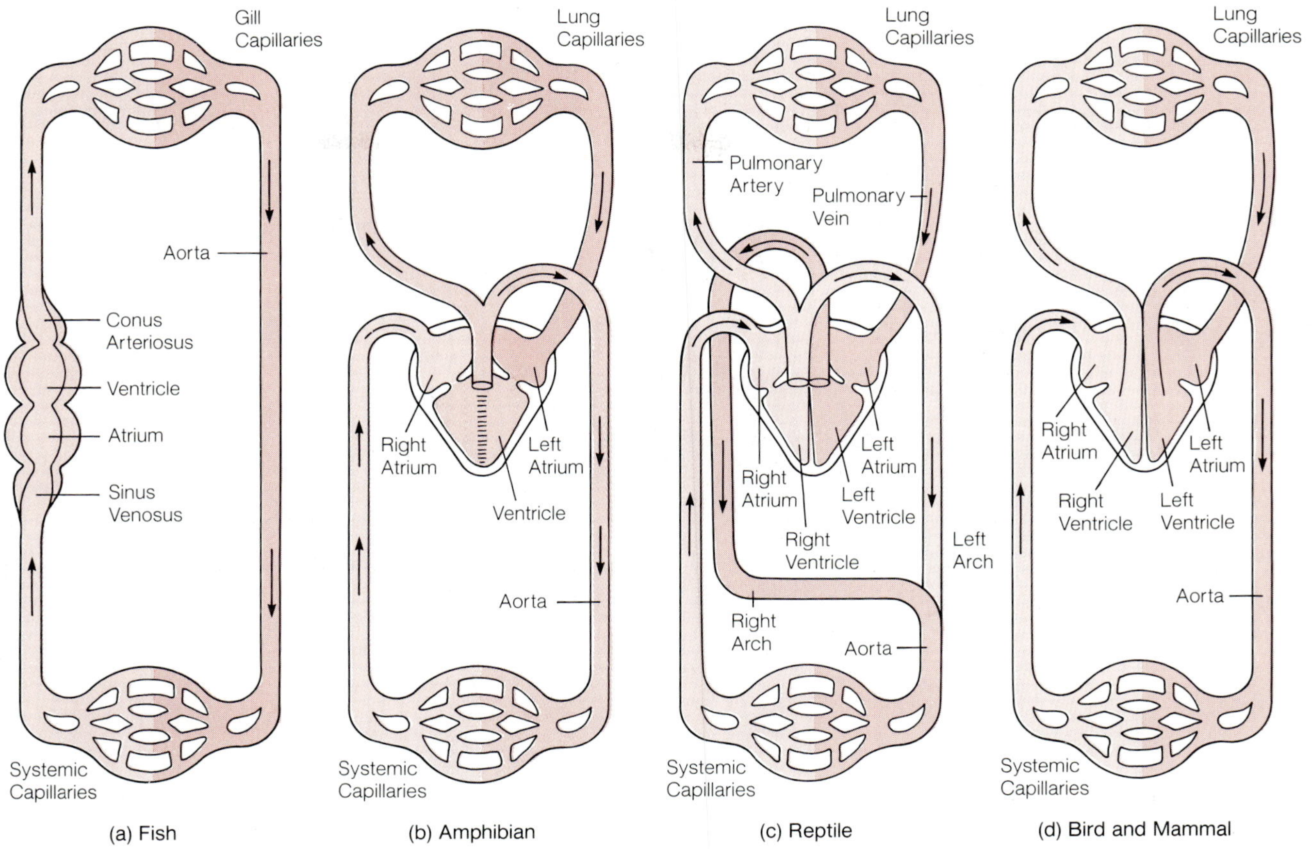

Figure 36.8 The heart and circulatory systems of various vertebrates. The arrows indicate the direction of blood flow. (a) The two-chambered fish heart in which the blood flows in sequence. (b) The amphibian heart has two atria and one ventricle. (c) In reptiles, the ventricle is partially divided, and oxygenated and deoxygenated blood are kept separate. (d) The heart of birds and mammals has two atria and two completely separate ventricles.

The Lymphatic System

The *lymphatic system* begins with very small vessels called lymphatic capillaries, which are in direct contact with the extracellular fluid surrounding tissues (figure 36.9). The system has four major functions: (1) to collect and drain most of the fluid that seeps from the bloodstream and accumulates in the extracellular fluid; (2) to return small amounts of proteins that have left the cells; (3) to transport lipids that have been absorbed from the small intestine; and (4) to transport foreign particles and cellular debris to disposal centers called *lymph nodes*. (It is worthy to note that lymph nodes near a cancer site may contain viable cancer cells. On entering the lymph nodes, cancer cells can multiply and establish secondary cancers by dispersing cells throughout the body via the lymphatic system. To [5] prevent the spread of cancerous cells, lymph nodes are usually removed during cancer operations.) The small lymphatic capillaries merge to form larger lymphatic vessels called lymphatics. *Lymphatics* are thin-walled vessels with valves that ensure the one-way flow of lymph. **Lymph** (L. *lympha,* clear water) is the extracellular fluid that accumulates in the lymph vessels. These vessels pass through the lymph nodes on their way back to the heart. Large concentrations of lymph nodes occur in several areas of the body and play an important role in the body's defense against disease.

In addition to the above parts, the lymphatic system consists of lymphoid organs, the spleen and thymus gland. The major components of the lymphatic system are summarized in table 36.2. The lymphatic system is also the bridge to the next section, on immunity, because it is vital to the body's defense against injury and attack.

Stop and Ask Yourself

1. What is the structure of an artery? A vein? A capillary?
2. How do the systemic and pulmonary circuits differ?
3. What is the cardiac cycle? What is blood pressure?
4. What is a single circulatory circuit? A double circulatory circuit?
5. What are the four major functions of the lymphatic system?

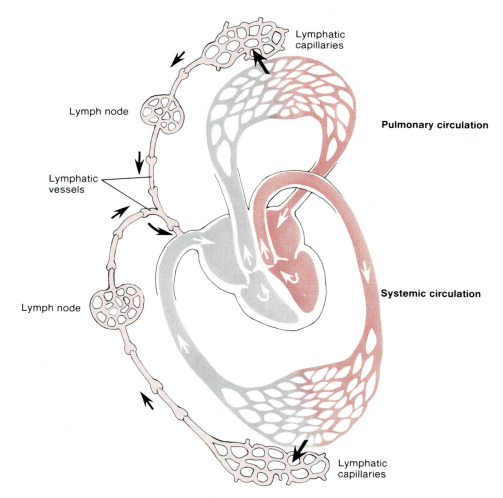

Figure 36.9 The lymphatic system. This system consists of one-way vessels that transport fluid from interstitial spaces back to the bloodstream.

Table 36.2	**Major Structural and Functional Components of the Lymphatic System in Vertebrates**		
Structure	**Function**	**Structure**	**Function**
Lymphatic capillaries	Collect excess extracellular fluid in tissues	Spleen	Filters foreign substances from blood, manufactures phagocytic lymphocytes, stores red blood cells, releases blood to the body when there is blood loss
Lymphatics	Carry lymph from lymphatic capillaries to veins in the neck where it is returned to the blood stream		
Lymph nodes	Destroy foreign substances at the upper entrances of the respiratory and digestive systems; play a role in antibody formation; exchange lymphocytes with lymph	Thymus gland (in mammals)	Forms antibodies in the newborn, is involved in the initial development of the immune system, site of T-cell differentiation
Tonsils	Destroy foreign substances at the upper entrances of the respiratory and digestive systems	Bursa of Fabricius (in birds)	A lymphoid organ found at the lower end of the alimentary canal in birds; the site of B-cell maturation

Immunity

To prevent unwanted cells, microorganisms, or agents from upsetting the body's homeostasis, animals have evolved defenses to keep such harmful materials from colonizing their bodies as well as to destroy internally caused dangers, such as cancer cells.

It is now generally agreed that invertebrates do not have immune systems with special cells that recognize and destroy very specific foreign agents. However, many do have innate, internal defense mechanisms. For example, molluscs and insects have granulocytes that are highly phagocytic to foreign agents and arthropods, primarily insects, are capable of encapsulating foreign agents. Arthropods also possess another type of internal defense mechanism called *melanization.* This process is characterized by the deposition of the dark pigment *melanin* around the foreign agent. *Nacrezation,* or pearl formation, is another innate defense mechanism found in molluscs.

Vertebrates are continuously exposed to microorganisms, foreign macromolecules, or cancer cells that can cause disease. Fortunately, they are equipped with an immune system that responds in defensive ways to this exposure. **Immunity** (L. *immunis,* free of burden) refers to the overall general ability of an animal to resist harmful attack. The *immune response* that results is a large and specific complex of defensive elements, widely distributed throughout the body, that help the animal defend against attack. The immune response can be divided into nonspecific and specific defenses (figure 36.10). **Immunology** is the branch of biology that deals with these immune defenses.

Nonspecific Defenses

Nonspecific defenses or *nonspecific immunity* refers to those general mechanisms that are inherited as part of the innate structure and function of each animal. These mechanisms act in concert as a first line of defense against intruders, before they can cause disease. Nonspecific defenses include biological barriers (inflammation and phagocytosis), chemical barriers (enzymatic action, interferons, complement), general barriers (fever), and physical barriers (skin, mucous membranes).

Biological Barriers The *inflammatory response* consists of a series of events that destroys invaders and restores the homeostasis of tissues. Specific chemicals and cells contribute to this response. For example, cells called *mast cells* are triggered to release histamine, which causes blood vessels to dilate. Blood flow to the capillaries also increases, which in turn increases their permeability. Fluid leaks out into the affected tissue, causing the warmth and redness of inflammation. Blood proteins that are part of the clotting mechanisms also enter the affected tissue and wall it off, causing swelling in an infected area. Phagocytes then [6]

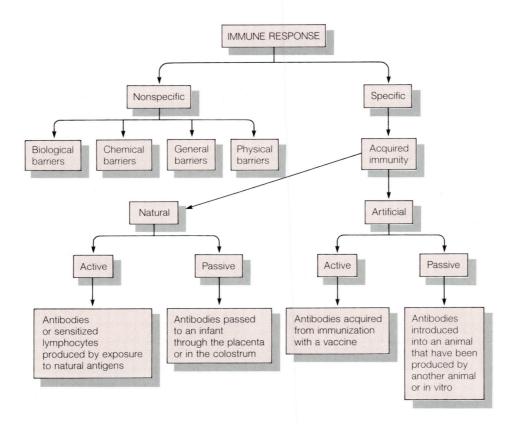

Figure 36.10 Various aspects of the immune response.

migrate toward the infected tissue and the invading cells are phagocytized. During the above steps, prostaglandins are released from injured cells and bind to free nerve endings, making them fire and start a pain impulse. After several days, the tissue is repaired and returned to normal.

Chemical Barriers Many chemicals, such as enzymes, are capable of destroying foreign invaders. For example, *lysozyme* is an enzyme found in tears, mucus, and sputum that helps degrade the walls of many bacteria and helps keep certain body sites free from infection. The skin of the toad *Xenopus* contains antimicrobial peptides called *magainins*. Magainins lyse bacterial cells by creating pores in their plasma membranes. When a cell becomes infected with a virus, a chemical called *interferon* is produced that helps protect healthy neigboring cells from the virus. Interferons stimulate these uninfected cells to produce substances that interfere with viral reproduction. Another group of chemicals that is found in blood plasma is the **complement system.** The complement system consists of about twenty plasma proteins that circulate in an inactive form. Once contact is made with an invader, the proteins are activated, one after another. These reactions can cause (1) the plasma membranes of invading cells to rupture, (2) phagocytic cells to move into the invaded area, or (3) the attachment of complement proteins to the surface of the invader, marking it for destruction by phagocytic cells.

General Barriers A *fever* (elevated body temperature) induced by an invader augments the defenses of an animal by stimulating white blood cells into action so that they can destroy the invader, decreasing available nutrients to the invader, and enhancing the specific chemical reactions that occur with antibodies and antigens.

Physical Barriers The skin is the first physical line of defense an animal has due to its dry and tough outer layer. Chemical defenses of the skin include acids in the oily secretions of the sebaceous glands and in sweat. The mucous membranes covering other body surfaces (e.g., the lining of the nose, mouth, and vagina) also produce protective secretions. In addition, all these body surfaces are home to millions of bacteria. These local residents produce substances that protect their territories—the skin and mucous membranes—from invasion by microorganisms that might cause disease.

Specific Defenses: The Immune Response

If the above nonspecific defenses are breached, the *specific immune response* is called upon for protection. This system consists of a number of immunological mechanisms in which certain white blood cells (lymphocytes) recognize the presence of particular foreign invaders or substances (antigens) and act to eliminate them. This elimination can occur by direct destruction of the antigens by the lymphocytes or by the formation of specialized proteins (antibod-

Table 36.3	Specific Defenses of the Immune System
Defense	**Function**
B cells	Lymphocytes that differentiate into plasma cells after stimulation by antigen
Plasma cells	Cells that are formed from stimulated B cells and secrete antibody
T cells	Lymphocytes that are produced in bone marrow and mature in the thymus gland
Helper T cells	Stimulate the rapid divisions of lymphocytes and along with them, mount an attack against foreign invaders
Natural Killer Cells	Destroy body cells infected with viruses or fungal parasites as well as cancer cells
Suppressor T Cells	Slow down or stop the immune response
Memory cells	A portion of the B and T cells produced during a first-time encounter with a specific invader that are held in reserve; they make possible a rapid response to subsequent encounters with the same type of invader
Macrophages	Highly phagocytic lymphoctytes; alert helper T cells to the presence of foreign intruders
Antibodies	Receptor molecules that bind specific foreign targets and tag them for destruction by phagocytes or the complement system
Lymphokines, Interleukins	Secretions by which white blood cells communicate with each other
Perforin-1	A protein secreted by certain T cells that punches a hole in target cells and kills them

ies) that either destroy the invader or target it for destruction by other cells (table 36.3).

As an overview, when B cells are stimulated by a foreign invader, they divide into plasma cells. Plasma cells respond to invaders by secreting specific proteins (antibodies) into the blood and lymph. These antibodies are specifically directed against the invader (antigen) that caused their formation. Because antibodies are soluble in blood and lymph, they provide **humoral** (L. *humor,* a liquid) **immunity.** The humoral response defends mostly against bacteria, bacterial toxins, and viruses that enter the animal's various body-fluid systems.

T cells do not secrete antibodies. Instead, they attack host cells that have been infected by microorganisms, tissue cells transplanted from one animal to another, or cancer cells. T cells (e.g., helper T cells) also influence the activity of other components of the immune system. Because

T cells must come into close physical contact with foreign cells or infected cells in order to destroy them, they are said to provide **cell-mediated immunity.**

Acquired Immunity **Acquired immunity** refers to the type of specific immunity that develops after exposure to an antigen or is produced after antibodies are transferred from one animal to another. It can be obtained by natural or artifical means, either actively or passively (*see figure 36.10*).

Naturally Acquired **Naturally acquired active immunity** occurs when an animal's immune system comes into contact with an appropriate antigenic stimulus during the course of daily activities. The immune system responds by producing antibodies and sensitized lymphocytes that inactivate the antigen. The immunity produced can be either lifelong (e.g., measles or chickenpox) or last for only a few years (e.g., tetanus).

Naturally acquired passive immunity involves the transfer of antibodies from one host to another. For example, in a pregnant mammal, some of the female's antibodies pass across the placenta to the offspring. Whatever immunity the female has will be transferred to the newborn. Certain other antibodies are able to pass from the female to her offspring in the first secretions (called colostrum) from the mammary glands. Unfortunately, naturally acquired passive immunity generally lasts only a short time (weeks or months at the most).

Artifically Acquired **Artifically acquired active immunity** results from immunizing an animal with a vaccine. A **vaccine** consists of a preparation of either killed microorganisms; living, weakened (attenuated) microorganisms; or inactivated bacterial toxins (called **toxoids**) that is administered to an animal to induce immunity. Such deliberate causing of antibody production is called **immunization.**

Artifically acquired passive immunity results from introducing into an animal antibodies that have been produced either in another animal or by specific in vitro methods. Although this type of immunity is immediate, it is short-lived (about 3 weeks). An example would be botulinum antitoxin produced in a horse and given to a human suffering from botulism food poisoning.

Antigens The mammalian immune system distinguishes between "self" and "nonself" by an elaborate recognition system. Prior to birth, the body somehow makes an inventory of the proteins and various other large molecules that are present in the body ("self") and removes most of the T cells specific for "self" molecules. Subsequently, the self-molecules can be distinguished from foreign, nonself substances, and lymphocytes can produce specific immunological reactions against the foreign material, leading to its removal.

All mammalian cells have "self" marker proteins on their plasma membrane surface. These markers result from the specific patterns the proteins form. The proteins themselves are encoded by genes called the **major histocompatibility complex** (MHC); hence the proteins are called *MHC markers.*

Lymphocytes can identify these MHC markers as being either "self" or "nonself." On their surfaces they bear specific recognition proteins that are only complementary to the "nonself" markers. Foreign ("nonself") substances (markers) to which lymphocytes respond are called **antigens** (*anti*body *gene*rator) or **immunogens.** Most antigens are large, complex molecules with a molecular weight generally greater than 10,000.

Antibodies **Antibodies** (immunoglobulins) are a group of recognition glycoproteins manufactured by plasma cells and present in the blood and tissue fluids of mammals. All antibody molecules have a basic Y structure composed of four chains of polypeptides (figure 36.11) connected to each other by disulfide bonds. The arms of the Y contain binding sites or fragments (Fab) for specific invaders (i.e., antigens). The tail of the Y can activate the complement system or bind to receptors on phagocytic cells (box 36.2).

Antibody-Mediated (Humoral) Immune Response B cells are of great importance in fighting invading organisms because they produce the antibodies that identify the antigens for destruction. The B cells carry some of their particular antibodies on their plasma membrane. When an antigen comes into contact with a B cell whose antibodies recognize the antigen, the B cell binds to the antigen. When stimulated by one kind of T cell, a helper T cell, the B cell divides many times, producing plasma cells that begin producing and secreting more of this particular antibody. These antibody molecules are carried through the circulation. If they encounter them, the anti-

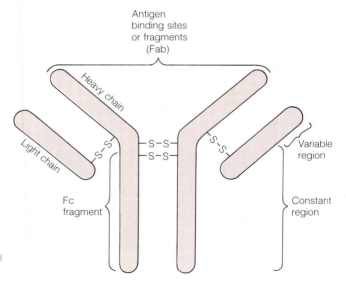

Figure 36.11 Antibody structure. The antibody molecule consists of two identical heavy chains and two light chains held together by disulfide bonds.

Box 36.2

Immunotoxins

One of the products of DNA research (colorplate 2) is the generation of **immunotoxins.** Immunotoxins are antibodies that have been attached to a specific toxin or toxic agent (*antibody + toxin = immunotoxin*). Immunotoxins are capable of killing specific target cells and no others, because the antibody will bind specifically to plasma membrane surface antigens found only on target cells. This approach is being used to treat certain types of cancer (box figure 36.2).

In this procedure, cancer cells from a person are injected into mice or rats to stimulate the production of specific antibodies against the antigens on the plasma membrane of the cancer cells. Using specific production techniques, **monoclonal antibodies** (MAbs) (cloned antibodies with a specific binding site for the antigen; i.e., the cancer cells) are produced, harvested, purified, and attached to an agent (drug) known to be toxic to the cancer cells.

When the immunotoxin is given to the person with the cancer, it circulates through the body and binds only to the cancer cells that have the specific surface receptors. After binding to the receptors, the immunotoxin is taken into the cancer cells by receptor-mediated endocytosis and released inside the cells. Once inside the cancer cells, the immunotoxin interferes with the metabolism of the target cells and kills them.

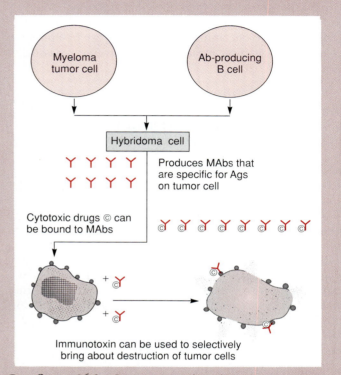

Box figure 36.2 Cancer research and immunotoxins.

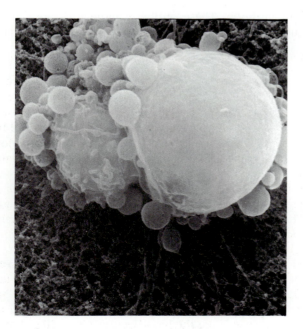

Figure 36.12 A natural killer (NK) cell destroying a cancer cell. Upon contact with a cancer cell, a natural killer cell releases toxic chemicals, such as perforin-1, which punch holes in the cancer cell's plasma membrane causing the cancer cell to die.

bodies bind to the antigen molecules, which are now marked for destruction by other parts of the immune system (e.g., macrophages, complement proteins).

Cell-Mediated Immune Response   Unlike B cells that produce the antibodies that subsequently mark antigens for destruction, T cells are directly involved with the destruction of invading cells, virus etc., as well as with the regulation of other parts of the immune system. T cell responses are called the cell-mediated immune response.

Natural killer (**NK**) **cells**, also called cytotoxic T cells, help in a mammal's defense by recognizing cell surface changes on cancer cells, virus-infected cells, fungi, bacteria, protozoa, or helminth parasites. These specialized T cells kill these cells or organisms as follows:

1. The NK cell is activated by interferons and/or interleukin-2 produced by helper T cells or infected cells.
2. The NK cell recognizes an MHC marker on the target cell and binds to it.
3. The NK cell inserts a pore-forming protein called *perforin-1* into the target cell's plasma membrane.
4. The NK cell injects secretions through the pore into the target cell.
5. The targeted cell is thus destroyed (figure 36.12).

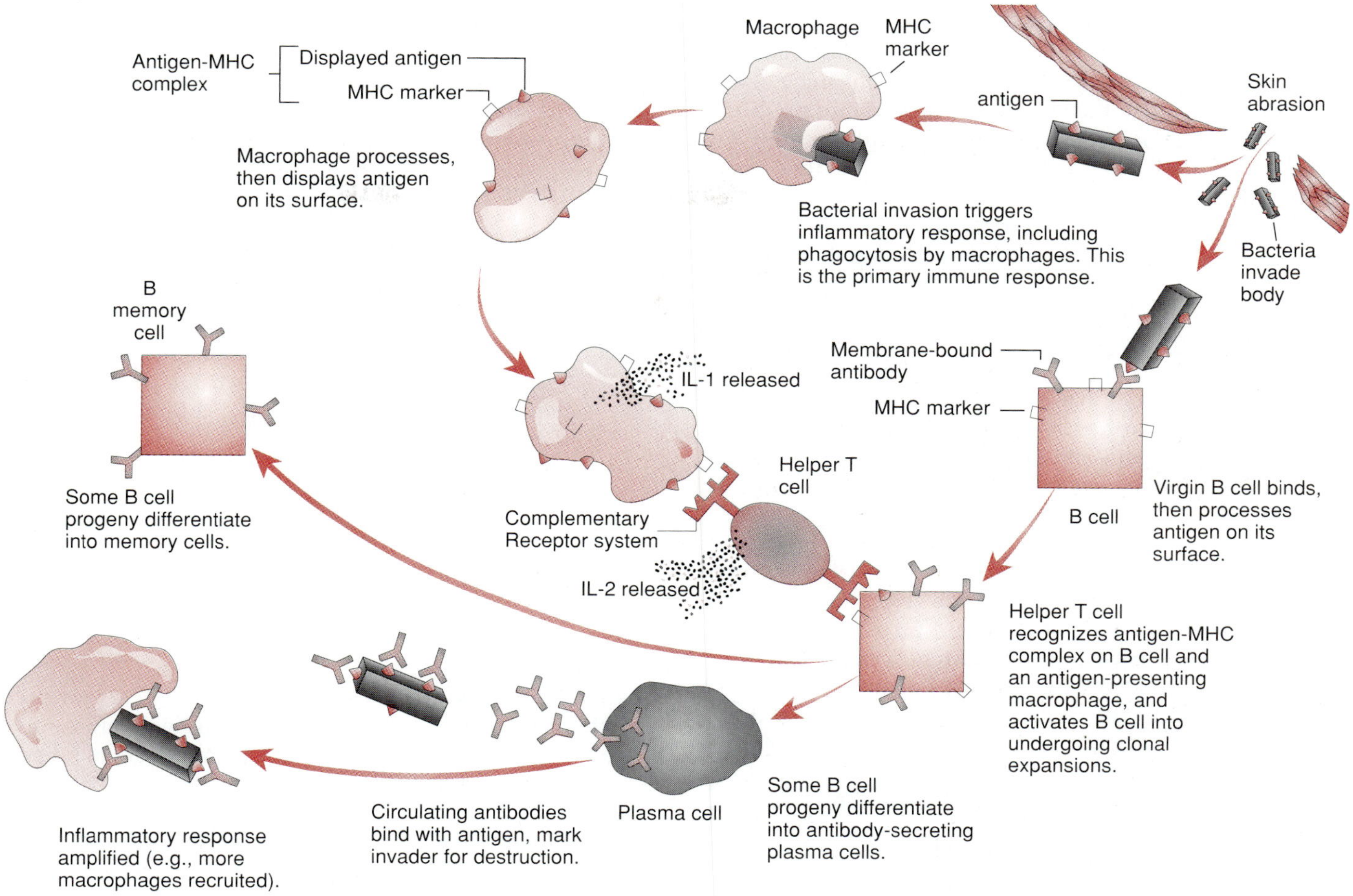

Figure 36.13 Cell-mediated immune response. Helper T cells activate B cells to differentiate into plasma cells that produce antibodies. Notice that different cells are involved: macrophages, helper T cells, and B cells. It is the interaction of these three cells that finally causes B cells to differentiate into antibody-producing plasma cells.

The above action of NK cells is also the reason why tissue and organ transplants are rejected in mammals. In the rejection mechanism, NK cells enter the transplanted tissue through the blood vessels, recognize it as foreign, attach to the tissue, and destroy it by the above mechanism. However, the body will tolerate tissue and organ transplants between identical twins who have identical sets of DNA and hence identical MHC regions.

T cells are also involved with regulating the activity of other parts of the immune system. For example, suppose several bacteria penetrate the first line of defense (the skin) through an abrasion or cut. Inflammation occurs, and the bacteria are phagocytized by macrophages (figure 36.13). The macrophage destroys most of the antigens (bacteria), but some of the antigens are moved to the surface of the plasma membrane of both macrophages and B cells, where they are displayed along side the MHC marker. It is this specific combination of MHC and antigen that is recognized by helper T cells.

When the macrophage reacts with a helper T cell, it releases interleukin-1 (IL-1). Interleukin-1 stimulates other helper T cells to secrete interleukin-2 (IL-2), which stimulates their growth and cell division. Some of the interleukin-2 produced by helper T cells acts on sensitized B cells. *Sensitized B cells* are ones that have recognized and processed the antigen onto their plasma membrane along side their MHC marker (figure 36.13). These stimulated B cells grow and divide into differentiated plasma cells. An antibody-mediated response continues as plasma cells secrete the specific antibody that binds with the antigen on the surface of the bacterium and targets it for destruction by macrophages or complement.

Primary and Secondary Immune Responses: Immunological Memory The **primary immune response** is the response an animal generates at its first encounter with an antigen. During this response, the antigen disappears from the blood because it is either bound to antibody and/or phagocytized by macrophages. Most of the B cells that are producing the antibodies also die. However, if the same antigen enters the body a second time, a **secondary immune response** is mounted that is faster

[7]

and more extensive than the primary response (figure 36.14). This rapid response occurs because the immune system has stored a "memory" of the antigen: some of the original B cells did not die but differentiated into **memory cells** (*see figure 36.13*) that will remain in the body for life. Memory cells from B-cell clones continually expose antibodies on their plasma-membrane surface. If the antigen invades the body again, these memory cells rapidly produce large numbers of antibody-secreting B cells.

[8] Overall, the body must build up a repertoire of memory clone cells for each antigen it encounters before it can produce secondary responses to invaders such as viruses. This process takes time (many months and years), which is why babies have so many colds and other microbial infections during their first two years of life. They must encounter many antigens and build up many clones of memory cells before they are immune to as many diseases as the average adult (box 36.3).

Evolution of the Vertibrate Immune System

The evolution of the immune system in vertebrates parallels that of the lymphatic system. Jawless vertebrates (cyclostomes), cartilaginous fishes (chondrichthyeans), and bony fishes (osteichthyeans) have networks of small, blind vessels that accompany the veins and help the capillaries drain the tissues. These vessels developed from veins and empty into them at frequent intervals. They represent the earliest stage in the evolution of the lymphatic system; however, no lymph nodes or lymphocytes are present.

Teleosts and tetrapods have independently evolved lymphatic systems in which lymphatic capillaries help drain most of the tissues of the body. Lymph nodes and lymphocytes occur in a few aquatic birds and are abundant in mammals. Additional aggregations of lymphatic tissue, T cells, and immunoglobulins evolved in birds and mammals, such as the distinct *bursa of Fabricius* in birds and

the *thymus* in mammals. Both of these organs are particularly large and active in late fetal and young individuals but gradually regress as the animal matures. They play a functional role in the processing of lymphocytes and the development of immunological maturity.

Stop and Ask Yourself

6. What is the relationship between the lymphatic and immune systems?
7. What is an MHC marker? How does it function?
8. How does an antibody differ from an antigen?
9. What is the difference between an antibody-mediated immune response and a cell-mediated immune response?
10. How do NK cells function?
11. How does immunological memory occur?

Gas Exchange

To take advantage of the rich source of energy represented by the organic matter on our planet, animals must solve two practical problems. They must break down and digest the organic matter so that it can enter the cells that are to metabolize it (this digestive process is described in the next chapter), and they must provide the cells with both an adequate supply of oxygen required for *aerobic respiration* and a way of eliminating the carbon dioxide produced by this type of respiration. This process of gas exchange with the environment, also called *external respiration,* is the subject of the rest of this chapter.

Respiratory Surfaces

Animals use four main types of respiratory surfaces to accomplish external respiration: tracheae, cutaneous (integument or body surface) exchange, gills, and lungs.

Invertebrate Respiration In single-celled protists, such as protozoa, diffusion is all that is needed to move gases into and out of the animal. Some multicellular invertebrates either have very flat bodies (e.g., flatworms) in which all body cells are relatively close to the body surface or are thin-walled and hollow (e.g., *Hydra*). With these designs, diffusion is sufficient to move gases into and out of the protozoan.

Other invertebrates, such as earthworms, that live in moist environments use integumentary exchange. Earthworms have capillary networks just under their skin surfaces, and they exchange gases with the air spaces among soil particles.

Most aquatic invertebrates carry out gas exchange with gills. The simplest gills are small, scattered projections of the skin, such as the gills of starfishes. Other aquatic invertebrates have their gas exchange structures in more re-

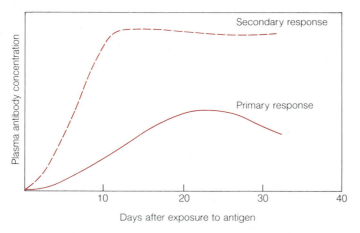

Figure 36.14 The antibody concentration increases more rapidly during a secondary immune response than during a primary response. Note that the secondary response peaks in 10 days, whereas the primary response does not peak for more than 20 days. The magnitude of the secondary response is also many times that of the primary response.

Box 36.3

AIDS: Acquired Immune Deficiency Syndrome

AIDS is a group of disorders that follow infection by the *human immunodeficiency virus* (HIV). Once the virus becomes established in the human body, it weakens the immune system, leaving the individual susceptible to opportunistic infections and a rare form of cancer. Currently, there is no vaccine against HIV, and there is no cure for those infected.

From 1981 to 1990, there were more than 90,000 cases of AIDS in the United States and more than 1 million carriers. The World Health Organization has estimated more than 10 million HIV carriers exist worldwide.

HIV is a *retrovirus*; its genetic material is RNA instead of DNA. Each viral particle consists of a protein core that surrounds the viral RNA and several copies of the enzyme reverse transcriptase (box figure 36.3). The outer lipid envelope is derived from the helper T cell (also called the T4 lymphocyte) that it infects. Once inside the helper T cell, the enzyme uses the viral DNA as a template for making DNA, which is then inserted into the helper T cell chromosome.

After an initial infection, antibodies to HIV can be detected in the blood but they do not target the infected helper T cells for destruction. When a secondary immune response is mounted, however, the infected cells are activated. During this activation, DNA is transcribed—including the viral DNA. Transcription yields copies of viral RNA, which are translated into viral proteins. The RNA and viral proteins are then assembled into new virus particles that bud from the helper T cell plasma membrane or are released when the cell undergoes lysis. The newly released viruses can now infect more helper T cells.

Eventually the helper T cell population is depleted, and the body cannot mount an immune response. During this time, some of the symptoms of AIDS occur, including flu-like symptoms, weight loss, fatigue, fever, and enlarged lymph nodes. The major opportunistic infection is a form of pneumonia caused by the protozoan, *Pneumocystis carinii*. Blue-violet spots that may appear on the skin are indicative of *Karposi's sarcoma,* a deadly form of skin cancer.

The HIV is transmitted via body fluids, primarily infected semen, blood, vaginal secretions, or breast milk. In the United States, transmission occurs most often among male homosexuals and intravenous drug users who share HIV-contaminated needles. Heterosexual transmission is on the rise in the United States. HIV can also be transmitted from infected mothers to their babies during pregnancy, birth, or breast feeding. Where blood is not screened for the HIV, contaminated blood can also spread the virus.

The drug AZT is being used to prolong the life of AIDS patients, although it is not a cure. At present, the prospects for developing an effective vaccine are not encouraging.

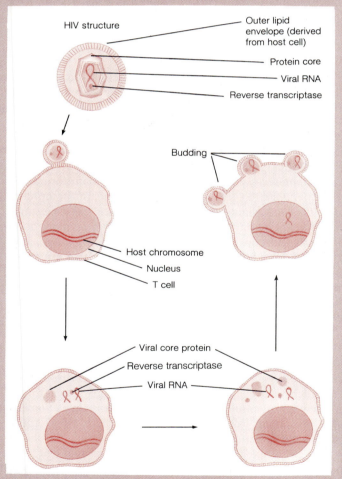

Box figure 36.3 HIV retrovirus activity.

stricted areas. Marine annelid worms have prominent lateral projections called *parapodia* (*see figure 21.3*) that are richly supplied with blood vessels and function as gills.

Many invertebrates (e.g., crustaceans, molluscs) have gills that are compact and protected with hard covering devices. Such gills are finely divided into highly branched structures to maximize area of gas exchange.

In some terrestrial invertebrates, such as insects, centipedes and some spiders, *tracheal systems* are present as networks of tubes that penetrate the tissues. These tubes, called tracheae, open to the outside through holes in the body surface called spiracles. Gases move into and out of the spiracles by diffusion.

Some aquatic insects also have a tracheal system that permits gas exchange in an aquatic environment. For example, some insect larvae hang head-down just below the surface of the water (colorplate 18c). They extend a posterior siphon tube into the air above the water surface and "snorkel" air into their tracheal system. Other aquatic insects have tracheal gills, which are expansions of the body surface that contain networks of tracheal system tubes.

Cutaneous Exchange Cutaneous exchange is used to supplement gas exchange by some vertebrates that have lungs or gills, such as some aquatic turtles, salamanders with lungs, snakes, fishes, and mammals. However, cutaneous exchange is most highly developed in frogs, toads, lungless salamanders, and newts.

Amphibian skin has the simplest structure of all the major vertebrate respiratory organs. In frogs, a uniform capillary network lies in a plane directly beneath the epidermis. This vascular arrangement facilitates gas exchange between the capillary bed and the environment by both diffusion and convection. To facilitate this exchange, amphibian skin is kept moist and is protected from injury by a slimy, mucous layer. Some amphibians obtain about 25% or more of their oxygen by this exchange, and the lungless [9] salamanders carry out all of their gas exchange through the skin and buccal-pharyngeal region.

Gills Gills are respiratory organs that have either a thin, moist, vascularized layer of epidermis to permit gas exchange across thin gill membranes, or a very thin layer of epidermis over highly vascularized dermis. Larval forms of a few fishes, amphibians, and some insects have *external gills* projecting from their body (colorplate 18b). Adult fishes have *internal gills*.

Gas exchange across gill surfaces is extremely efficient. It occurs as blood and water move in opposite directions on either side of the lamellar epithelium. For example, the water that passes over a gill first encounters vessels that are transporting blood with a low oxygen concentration *into* the body. Because the concentration (partial pressure) of the oxygen is lower in the blood than in the water, oxygen diffuses into the blood. Water then passes over the vessels carrying blood relatively high in oxygen from deep within the body. More oxygen diffuses inward, because this blood still has less oxygen than the surrounding water. Carbon dioxide also diffuses into the water because its concentra-

tion (pressure) is higher in the blood than in the water. This countercurrent mechanism provides very efficient gas exchange by maintaining a concentration gradient between the blood and water over the length of the capillary bed (*see figure 27.13*).

Lungs A **lung** is an internal sac-shaped respiratory organ. The typical lung of a terrestrial vertebrate comprises one or more internal blind pouches into which air is either drawn or forced. The respiratory epithelium of lungs is thin, well vascularized, and divided into a large number of small units, which greatly increase the surface area for gaseous exchange between the lung air and the blood. This blind-pouch construction, however, limits the efficiency with which oxygen and carbon dioxide are exchanged with the atmosphere because only a portion of the lung air is ever replaced with any one breath. Birds are an exception in that they have very efficient lungs (*see figure 30.16*). For example, a mammal removes approximately 25% of the oxygen from air with each breath, whereas a bird removes approximately 90%.

The evolution of the vertebrate lung is related to the swim bladder. The swim bladder is an air sac located dorsal to the digestive tract in the body of many modern fishes. Evidence indicates that both lungs and swim bladder evolved from a lunglike structure present in primitive fishes that were ancestors of both present-day fishes and tetrapods (amphibians, reptiles, birds, and mammals). These ancestral fishes probably had a ventral sac attached to their pharynx (*see figure 27.14*). This sac may have served as a supplementary gas-exchange organ when the fishes could not obtain enough oxygen through their gills (e.g, in stagnant or oxygen-depleted water). By swimming to the surface and gulping air into this sac, gas exchange could take place through its wall.

Further evolution of this blind sac off the pharynx proceeded in two different directions (*see figure 27.13*). The first was toward the arrangement seen in the majority of modern bony fishes, where the swim bladder lies dorsal to the digestive tract. The second evolutionary direction led to the lungs, which are ventral to the digestive tract. Ventral lungs are found in a few present-day fishes and in the tetrapods. The evolution of the structurally complex lung (figure 36.15) has paralleled the evolution of the large body sizes and high metabolic rates found in endothermic vertebrates (birds and mammals), which necessitated an increase in lung surface area for gas exchange, compared to the smaller body size and lower metabolic rates found in the ectothermic vertebrates.

Human Respiratory System

The model of external respiration in humans is typical of mammals. It is based on several physiological principles that apply to all air-breathing animals:

1. Air moves by bulk flow into and out of the lungs in the process called *ventilation*.
2. O_2 and CO_2 diffuse across the respiratory surface of the lung tissue.

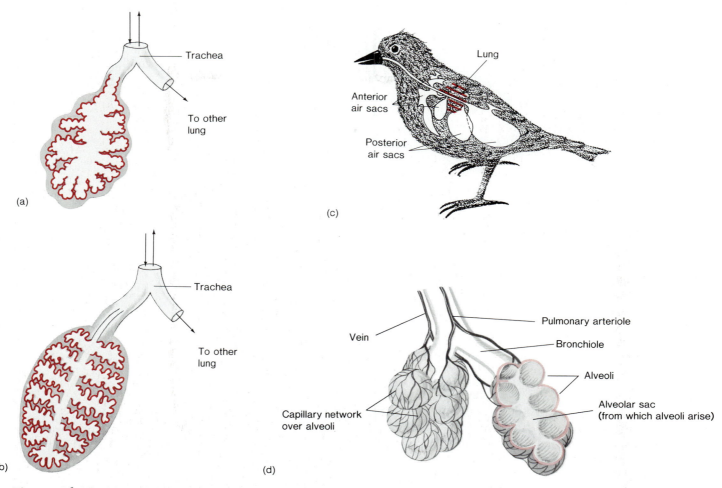

Figure 36.15 The evolution of the vertebrate lung showing the increased surface area from (a) amphibian, (b) reptiles, (c) birds, to (d) mammals. This evolution has paralleled the evolution of large body size and higher metabolic rates.

3. O_2 and CO_2 diffuse between the blood and interstitial fluid.
4. O_2 and CO_2 diffuse between the interstitial fluid and cells.

Air-Conducting Portion   The various organs of the human respiratory system are shown in figure 36.16a. Air normally enters and leaves this system through either *nasal* or *oral cavities.* From these cavities, air moves into the *pharynx,* which is a common area for both the respiratory and digestive tracts. The pharynx connects with the *larynx* (voice box) and with the esophagus that leads to the stomach. The *epiglottis* is a flap of cartilage that allows air to enter the *trachea* when breathing. It covers the trachea when swallowing to prevent food or water from entering.

During inhalation, air from the larynx moves into the *trachea* (windpipe), which branches into a right and left *bronchus* (pl. *bronchi*). After each bronchus enters the lungs, it branches into smaller tubes called *bronchioles,* then even smaller tubes called *terminal bronchioles,* and finally, *respiratory bronchioles* that connect to the gas exchange portion of the respiratory system.

Gas-Exchange Portion   The respiratory bronchioles have small tubes called *alveolar ducts* leading from them that connect to grapelike outpouching from their walls called **alveoli** (s., alveolus; L. *alveus,* hollow) (figure 36.16b). The alveoli are clustered together to form an *alveolar sac.* Surrounding the alveoli are many capillaries (figure 36.16c). Alveoli are referred to as the functional units of the lungs (gas exchange portion) because it is here that oxygen moves into the blood and carbon dioxide moves from the blood into the alveoli. Passive diffusion, driven by a partial pressure gradient, is enough to move oxygen across the respiratory surface and into the blood stream, and to move carbon dioxide from the blood stream into the alveoli (figure 36.17). Collectively, the alveoli provide a large surface area for gas exchange. To illustrate this important point, if the alveolar epithelium were removed from the lungs and put into a single layer of cells side by side, the cells would cover the area of a tennis court.

Ventilation   Breathing (also called *pulmonary ventilation*) has two phases: *inhalation,* the intake of air, and *exhalation,* the outflow of air. These air movements result

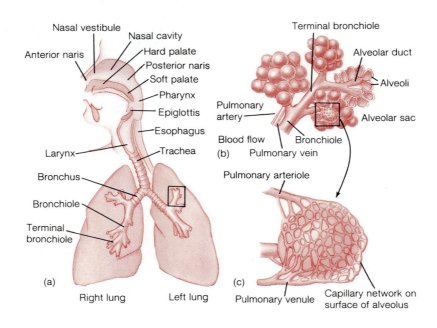

Figure 36.16 Organs of the human respiratory system. (a) The basic anatomy of the respiratory system. (b,c) The respiratory tubes end in minute alveoli, each of which is surrounded by an extensive capillary network.

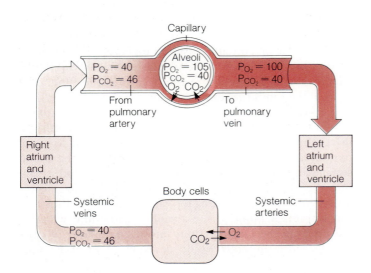

Figure 36.17 Gas exchange between the lungs and tissues. Gases diffuse according to partial pressure (P) differences, as indicated by the numbers and arrows.

from the rhythmic increases and decreases in the volume of the thoracic cavity. It is these changes in thoracic volume that lead to reversals in the pressure gradients between the lungs and the atmosphere; gases in the respiratory system follow these gradients. The mechanism of inhalation operates in the following way (figure 36.18):

1. Several sets of muscles contract, the main ones being the *diaphragm* and *intercostal muscles*. The intercostal muscles stretch from rib to rib, and when they contract, they pull the ribs closer together, causing them to rise upward and outward, enlarging the thoracic cavity.
2. The thoracic cavity is enlarged further when the diaphragm contracts and flattens.
3. The increased size of the thoracic cavity causes the pressure in the cavity to drop below the atmospheric pressure, allowing air to rush into the lungs, and they inflate.

During ordinary exhalation, the expulsion of air from the lungs occurs in the following way:

1. The intercostal muscles and the diaphragm relax, allowing the thoracic cavity to return to its original,

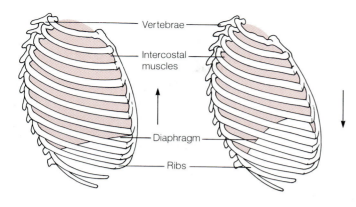

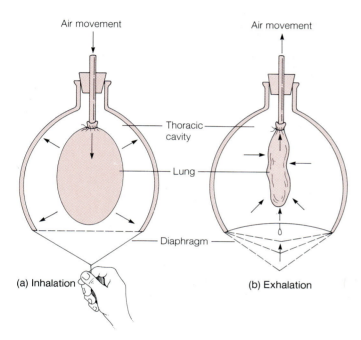

Figure 36.18 Ventilation of human lungs as an example of breathing in mammals. (a) During inhalation, muscle contractions lift the ribs up and out (upper diagram arrows) and lower the diaphragm. These movements increase the size of the thoracic cavity and decrease the pressure around the lungs. This negative pressure causes more air to enter the lungs. (b) Exhalation follows the relaxation of the rib cage and diaphragm muscles, as the increased pressure forces the air out of the lungs. The arrows indicate the direction pressure changes take in the thoracic (lower diagrams) cavity during inhalation and exhalation.

smaller size, and increasing the pressure in the thoracic cavity.

2. Abdominal muscles contract, pushing the abdominal organs against the diaphragm, further increasing the pressure within the thoracic cavity.

3. The above action causes the elastic lungs to contract and compress the air in the alveoli. With this compression, the alveolar pressure becomes greater than atmospheric pressure causing air to be expelled (exhaled) from the lungs.

Gas Exchange and Transport   As noted above, oxygen and carbon dioxide move down their concentration (pressure) gradients fast enough to allow adequate exchange across respiratory surfaces in the lungs. However, the amount of oxygen and carbon dioxide that can be carried dissolved in the blood is not enough to satisfy the need of a metabolically active animal. Several mechanisms have evolved to enhance gas exchange and transport.

When oxygen moves across the respiratory surface into the blood plasma, most of it immediately enters red blood cells, where it forms a weak, reversible bond with hemoglobin. Each hemoglobin molecule binds four oxygen molecules to form oxyhemoglobin (HbO_2). When the red blood cells enter tissues with a low oxygen partial pressure, the oxyhemoglobin releases its oxygen to satisfy the tissues' needs.

The partial pressure (concentration) of carbon dioxide is highest in the tissue cells and it diffuses from the cells into the bloodstream. In the blood stream, carbon dioxide is transported in three forms:

1. CO_2 dissolved in the blood plasma (about 7%).
2. CO_2 combined with hemoglobin to form carbaminohemoglobin ($HbCO_2$) (about 23%)
3. The bicarbonate ion (HCO_3^-) (about 70%).

Of these three, most of the carbon dioxide in the body is transported in the form of bicarbonate ion. When carbon dioxide enters the blood plasma, most of it quickly combines with water to form carbonic acid (H_2CO_3), which dissociates into the hydrogen (H^+) and bicarbonate (HCO_3^-) ions as follows:

$$CO_2 + H_2O \rightleftharpoons H_2CO_3 \rightleftharpoons H^+ + HCO_3^-$$

This same reaction also occurs within red blood cells but at a much faster rate (about 250 times faster) due to the presence of the enzyme carbonic anhydrase.

Because the partial pressure of carbon dioxide is lower in the alveoli of the lungs, the above three reactions proceed in the reverse direction. For example, H_2CO_3 dissociates to form CO_2 and H_2O, which diffuse down their concentration gradients and are exhaled from the body (box 36.4).

Stop and Ask Yourself

12. How do invertebrates accomplish gas exchange?
13. What different parts comprise the air-conducting portion of the human respiratory system?
14. Why are alveoli called the functional units of the respiratory system of humans?
15. How does ventilation occur in humans?
16. How is oxygen and carbon dioxide transported in the human body?
17. What is the relationship between the swim bladder of fishes and the lungs of mammals?

Box 36.4

How Deep-Diving Marine Mammals Adapt to Oxygen-Poor Environments

Deep-diving marine animals, such as the sperm whale, have evolved several mechanisms for remaining submerged for up to 1 hour. Most have a large number of red blood cells per ml of blood, large quantities of the oxygen-binding pigment *myoglobin* in their muscles, and hemoglobin that releases oxygen much more easily than other mammalian hemoglobin.

The **diving reflex** is an evolutionary adaptation that occurs in those mammals that dive (e.g., whales) downward in the sea. In this reflex, blood is shunted away from most regions of the body during a dive, and diverted mainly to the heart, brain, and skeletal muscles. Also, the skeletal muscles can incur an **oxygen debt**; they can continue to contract even after all oxygen is depleted building up lactate. The ATP energy for contraction comes from glycolysis, which forms lactic acid that accumulates in the tissues. When the whale surfaces, it quickly exhales its moist, oxygen-poor air, which condenses to form the "spout." A large amount of carbon dioxide is blown off in the "spout," and oxygen quickly combines with hemoglobin passing through the lungs.

The second part of the diving reflex is the slowing of the heart rate (called *bradycardia*). This slowing is an adaptation to ensure less rapid depletion of oxygen until the whale can return to the air for more oxygen.

■ Summary

1. A cardiovascular system allows the rapid exchange of materials throughout the body. This system consists of a heart, blood vessels, and blood.

2. Blood is a type of connective tissue made up of blood cells (red blood cells and white blood cells), plasma, and platelets.

3. The heart pumps blood through a series of vessels in the following order: arteries, arterioles, capillaries, venules, veins, and back to the heart.

4. The action of the heart consists of cyclic contraction (systole) and relaxation (diastole). It is the systolic contraction that generates blood pressure that forces blood through the closed system of vessels.

5. The lymphatic system consists of one-way vessels that help return fluid and protein to the circulatory system.

6. The vertebrate body uses nonspecific and specific immune responses to protect itself against the continuous onslaught of microorganisms, foreign invaders, and cancer cells.

7. The nonspecific response includes biological barriers (inflammation and phagocytosis), chemical barriers (enzymatic action, interferons, complement), general barriers (fever), and physical barriers (skin, mucous membranes).

8. The specific immune response is due to the coordinated activity of specialized cells (macrophages and T cells) and antibodies produced by B cells.

9. MHC markers on the surface of cells allow the lymphocytes to distinguish "nonself" from "self."

10. An immunization is the deliberate provocation of memory lymphocytes by a vaccine.

11. In those animals that respire aerobically, a constant supply of oxygen is needed. The process of acquiring oxygen and eliminating carbon dioxide is called external respiration.

12. The exchange of oxygen and carbon dioxide occurs across respiratory surfaces. Such surfaces include gills, cutaneous surfaces, and lungs.

13. The air-conducting portion of the respiratory system of air-breathing vertebrates moves air into (inhalation) and out off (exhalation) this system. The process of air movement is called ventilation.

14. Exchange of oxygen and carbon dioxide occurs by diffusion from areas of higher to areas of lower concentrations.

15. Once in the blood, oxygen diffuses into red blood cells and is bound to hemoglobin for transportation to the tissues. Carbon dioxide is transported in the form of the bicarbonate ion, carbonic acid, and bound to hemoglobin.

■ Key Terms

acquired immunity (p. 583)
alveoli (p. 589)
antibodies (p. 583)
antigens (immunogens) (p. 583)
arteries (p. 573)
artificially acquired active immunity (p. 583)
artificially acquired passive immunity (p. 583)
B cells (p. 573)
basophils (p. 572)
blood pressure (p. 577)
capillaries (p. 574)
cell-mediated immunity (p. 583)

complement system
(p. 582)
diastole (p. 577)
diving reflex (p. 592)
eosinophils (p. 572)
gills (p. 588)
hemoglobin (p. 572)
humoral immunity (p. 582)
immunity (p. 581)
immunization (p. 583)
immunology (p. 581)
immunotoxin (p. 584)
lung (p. 588)
lymph (p. 579)
lymphocytes (p. 573)
major histocompatibility
complex (MHC) (p. 583)
memory cells (p. 586)
monoclonal antibodies
(p. 584)
monocytes (p. 573)
natural killer cells (p. 584)
naturally acquired active
immunity (p. 583)

naturally acquired passive
immunity (p. 583)
neutrophils (p. 573)
oxygen debt (p. 592)
oxyhemoglobin (p. 572)
plasma cells (p. 573)
plasma (p. 571)
platelets (thrombocytes)
(p. 573)
primary immune response
(p. 585)
red blood cells
(erythrocytes) (p. 572)
secondary immune response
(p. 585)
serum (p. 572)
systole (p. 577)
T cells (p. 573)
toxoids (p. 583)
vaccine (p. 583)
veins (p. 573)
white blood cells
(leukocytes) (p. 572)

■ Critical Thinking Questions

1. Many invertebrates utilize the body cavity as a circulatory system. However, in humans the body cavity plays no role whatsoever in circulation. Why?

2. Describe the homeostatic functions of the vertebrate circulatory system. What functions are maintained at relative stability?

3. One of the most important goals of immunological research is developing techniques for transplanting skin between individuals. Why is this so important and why is it such a difficult problem?

4. How is the vertebrate immune system related to the circulatory system?

5. All antibodies are proteins. Is there some property of proteins that has favored the evolution of antibodies in this group of biomolecules rather than in others?

6. The immune system is capable of generating a million or more antibodies. How can such incredible diversity be programmed into a population of cells?

7. The area of an animal's respiratory surface is usually directly related to the body weight of the animal. What does this tell you about the mechanism of gas exchange?

8. How can seals and whales stay under water for long periods of time?

■ Suggested Readings

Books

Bruely, D. 1985. *Oxygen Transport to Tissues.* New York: Plenum.

Articles

Ada, G., and Nossal, G. The clonal-selection theory. *Scientific American* August, 1987.

Avery, M., Wang, N, and Taeusch, W. The lung of the newborn infant. *Scientific American* April, 1973.

Clements, J. A. Surface tension in the lungs. *Scientific American* December, 1962.

Comroe, J. H. The lung. *Scientific American* February, 1966.

Edelson, R., and Fink, J. The immunological function of the skin. *Scientific American* June, 1985.

Fedder, M., and Burggren, W. Skin breathing in verterbates. *Scientific American* November, 1985.

Fen, W. The mechanism of breathing. *Scientific American* June, 1960.

Golde, D. W., and Gasson, J. C. Hormones that stimulate the growth of blood cells. *Scientific American* January, 1988.

Perutz, M. F. The hemoglobin molecule. *Scientific American* November, 1964.

Robinson, T., Factor, S., and Sonnenblick, E. The heart as a suction pump. *Scientific American* June, 1986.

Scientific American. October 1988. Entire issue is devoted to articles on AIDS.

Tonegawa, S. The molecules of the immune system. *Scientific American* October, 1985.

Wiggers, C. The heart. *Scientific American* May, 1967.

Wood, W. White blood cells versus bacteria. *Scientific American* February, 1971.

Wood, J. E. The venous system. *Scientific American* January, 1968.

Young, J., and Cohn, Z. How killer cells kill. *Scientific American* January, 1988.

Zucker, M. The functioning of blood platelets. *Scientific American* June, 1980.

37

Nutrition and Digestion

Concepts

1. Animals are heterotrophic organisms that must obtain organic compounds from their environment.
2. In addition to organic nutrients needed for ATP energy and organic synthesis, animals require vitamins, minerals, and water.
3. Animal digestive tracts are specialized to transport, process, and absorb the nutrients required by all body cells.
4. The kind and degree of specialization in any animal's digestive tract is related to its metabolic requirements, body size, and the nature of its food supply.
5. Intracellular digestion occurs in some invertebrates; others utilize extracellular digestion, and some have evolved variations in extracellular digestion that allow them to exploit different food sources.
6. Vertebrate digestive systems have evolved into assembly lines where food is first broken down mechanically, and then chemically by digestive enzymes. The simple sugars, fats, triglycerides, amino acids, vitamins, and minerals that result are then taken into the circulatory and lymphatic systems for distribution throughout the animal's body and use in maintenance, growth, and energy production.

Have You Ever Wondered:

[1] why a light beer that is advertised as containing only 95 calories per 12 oz. actually contains 95,000 calories?
[2] why a high-fiber diet is good for you?
[3] what causes your mouth to be dry and have a bad taste in the morning—commonly known as "morning mouth?"
[4] why we belch, have a "noisy stomach," or "pass gas"?
[5] why vomiting causes a burning sensation in our esophagus and mouth?
[6] what causes "hunger pangs"?
[7] why most infants have a pudgy abdomen?
[8] why some snakes have salivary glands that produce venoms?
[9] why some mice and rabbits eat their own feces?

These and other useful questions will be answered in this chapter.

This chapter contains underlined evolutionary concepts.

Nutrition includes all of those processes by which

Nutrition includes all of those processes by which an animal takes in, digests, absorbs, stores, and uses food (*nutrients*) in order to meet its metabolic needs. **Digestion** (L. *digestio,* from + *dis,* apart + *gerere,* to carry) is the chemical and/or mechanical breakdown of food into particles that can be absorbed by the gut or individual cells.

This chapter discusses nutrition and the anatomy and physiology of digestive systems using the human as the primary example.

Digestion: Animal Strategies for Consuming and Using Food

Recall from chapters 16 and 17 that in some of the simplest forms of life (the protists and sponges), some cells take in whole food particles directly from the environment by diffusion, active transport, and/or endocytosis and break them down with enzymes to obtain nutrients. This strategy is called *intracellular* ("within the cell") *digestion* (figure 37.1a). Intracellular digestion circumvents the need for the mechanical breakdown of foods or for a gut or other cavity in which to chemically digest food. At the same time, however, intracellular digestion puts an upper limit on an animal's size and complexity—only very small pieces of food can be used.

To exploit the advantages of larger size, animals evolved structures and mechanisms necessary for *extracellular digestion:* the enzymatic breakdown of larger pieces of food into constituent molecules, usually in a special organ or cavity (figure 37.1b). Nutrients from the food then pass into body cells lining the organ or cavity and can take part in energy metabolism or biosynthesis.

In primitive, multicellular animals, such as cnidarians, the gut is a blind (closed) sac with only one opening that serves as both entrance and exit *(see figure 17.5)*. Some specialized cells in the cavity secrete digestive enzymes that begin the process of extracellular digestion (figure 37.1b). Other phagocytic cells that line the cavity engulf food material and continue intracellular digestion inside food vacules. Similar digestive patterns are seen in some flatworms *(see figure 18.3)*. Other parasitic flatworms, the tapeworms, are highly modified intestinal parasites that have no digestive system. They live in the host's intestines, surrounded by a nutrient broth of partially digested food. Tapeworms thus absorb and actively transport all of their nutrients across their body wall (tegument).

The development of the anus and complete digestive tract in the aschelminths *(see figure 19.6)* was an evolutionary breakthrough. A *complete digestive tract* permits the one-way flow of ingested material without mixing it with previously ingested material or waste. Complete digestive tracts also have the advantage of progressive digestive processing in specialized regions along the system. Food can be digested efficiently in a series of distinctly different steps. The many variations of the basic tubular plan of a complete digestive tract are correlated with different food-gathering mechanisms and diets. Many of these have been presented in the discussion of the many different animals in chapters 20 to 25.

Evolution of Nutrition

Nutrients in the food an animal eats provide the necessary chemicals for growth, maintenance, and energy production. Overall, the nutritional requirements of an animal are inversely related to its ability to synthesize molecules essential for life: the fewer such *biosynthetic abilities* an animal has, the more kinds of nutrients it must obtain from its environment. Green plants have the fewest such nutritional requirements because they can synthesize all their own complex molecules from simpler inorganic substances; they are called **autotrophs** (Gr. *auto,* self + *trophe,* nourishing). Animals that are called **heterotrophs** (Gr. *heteros,* another or different + *trophe,* nourishing)

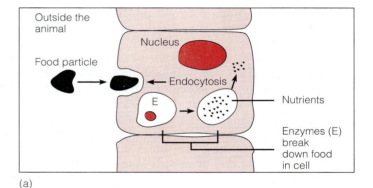

(a)

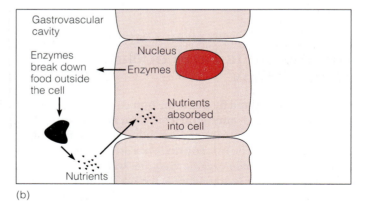

(b)

Figure 37.1 Intracellular and extracellular digestion. (a) A simple invertebrate such as a sponge has no gut and thus carries out intracellular digestion. Tiny food particles are taken into the body wall cells by endocytosis. Digestive enzymes in the vacuole then break the small particles into constituent molecules. (b) A *Hydra,* for example, has a gastrovascular cavity and so can take in and digest relatively large food particles. Cells lining the digestive cavity secrete enzymes into the cavity. There, the enzymes break down food materials into constituent nutrients, and the nearby cells absorb these nutrients.

Box 37.1

Types of Nutrition: A Modern View

One way of classifying organisms is according to their principal source of energy. Those that use light energy are termed *phototrophs,* and those that use chemically bound energy are termed *chemotrophs.*

Phototrophs can be subdivided according to whether they need only inorganic matter and carbon dioxide. If they can grow on such simple materials, they are termed *photolithotrophs* (G. *lithos,* stone). Most plants are photolithotrophs. If organic growth factors are required, as

in the case of some protozoa and algae, they are called *photoauxotrophs* (Gr. *auxo,* to increase).

Chemotrophs also can be subdivided according to whether their energy source is an inorganic substance, such as is the case with many bacteria; these organisms are classified as *chemolithotrophs.* If, as is the case with most animals, organic matter is their primary source of energy, the organisms are classified as *chemoorganotrophs.*

---

cannot synthesize many of their own organic molecules and must obtain them by eating other organisms or their products. Animals, such as rabbits, that subsist entirely on plant material are called **herbivores** (L. *herba,* plant + *vorare,* to eat), and those that eat only meat, such as eagles, are **carnivores** (L. *caro,* flesh). **Omnivores** (L. *omnius,* all), such as humans, eat both plant and animal matter, and **insectivores,** such as bats, eat primarily arthropods and soft-bodied invertebrates.

Much of animal evolution has been marked by losses in biosynthetic abilities. Once an animal routinely obtains essential, complex organic molecules in its diet, it can afford to lose the ability to synthesize those molecules. Moreover, the loss of this ability confers a selective advantage on the animal because the animal stops expending energy and resources to synthesize these molecules that are already in its diet. Thus, as the diet of animals became more varied, they tended to lose their abilities to synthesize such widely available molecules as some of the amino acids (box 37.1).

The Metabolic Fates of Nutrients in Heterotrophs

The nutrients ingested by a heterotroph can be divided into macronutrients and micronutrients. **Macronutrients** are needed in large quantities and include the carbohydrates, fats, and proteins. The **micronutrients** are needed in small quantities and include organic vitamins and inorganic minerals. Together, these nutrients make up the animal's *dietary requirements.* Besides these nutrients, animals require water.

Calories and Energy

The energy value of food is measured in terms of calories or Calories. A **calorie** (L. *calor,* heat) is the amount of energy required to raise the temperature of 1 g of water 1 °C. A calorie, with a small *c,* is also called a *gram calorie.* A **kilocalorie,** also known as a **Calorie** or *kilogram calorie*

(kcal), is equal to 1000 calories. In popular usuage, we talk about calories but actually mean Calories, because the larger unit is more useful for measuring the energy value of food. (If an advertisement says that a so-called light beer contains 95 calories per 12 oz., it really means 95,000 calories, 95 Calories, or 95 kcal.) [1]

Macronutrients

With a few notable exceptions, heterotrophs require organic molecules, such as carbohydrates, lipids, and proteins, in their diets. When these molecules are broken down by enzymes into their components, they can be used for energy production or as sources for the "building blocks" of life.

Carbohydrates: Carbon and Energy from Sugars and Starches   The major dietary source of energy for most heterotrophs is complex carbohydrates. With the exception of lactose, which comes from milk, and glycogen, which is synthesized by the liver, all carbohydrates originally come from plant sources (figure 37.2). This dietary need can be met by various polysaccharides, disaccharides, or any of a variety of simple sugars (monosaccharides). Carbohydrates also serve as a major carbon source. The animal body needs carbon atoms, usually in the form of methyl (CH_3-) or acetyl (CH_3-CH_2O-)

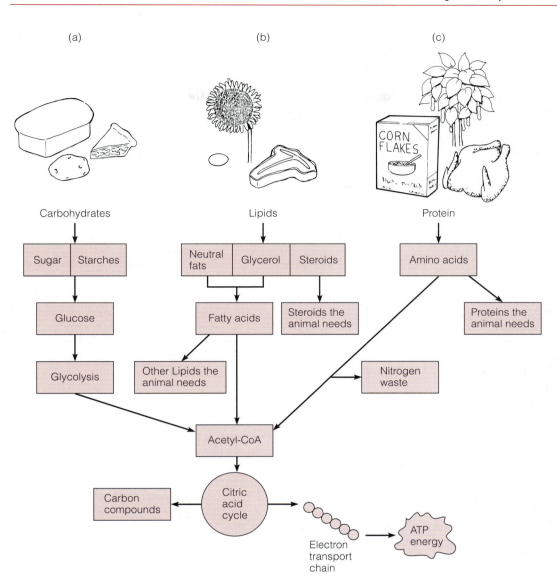

(a) (b) (c)

Carbohydrates

Lipids

Protein

Figure 37.2 Macronutrients in the diet: (a) carbohydrate foods are broken down to their constituent sugars and starches and ultimately into glucose. Individual cells use this sugar in glycoloysis and aerobic respiration to create new carbon compounds or ATP energy. (b) Lipids (fats) and oils in the diet are broken down to neutral fats, glycerol, and steroids. These molecules can be modified and incorporated into the lipids or steroids the animal needs for storing fat or generating hormones, or they can be converted to acetyl-CoA and enter the citric-acid cycle and electron-transport chain for ATP production. (c) Proteins are broken down to amino acids, which are incorporated into new proteins or modified to enter the citric acid cycle and electron-transport chain to produce ATP energy.

groups, for incorporation into important organic compounds. Many plants also supply cellulose, an indigestible polysaccharide. Cellulose is sometimes called dietary fiber. [2] It assists in the passage of food through the alimentary canal and may reduce the risk of cancer of the colon.

Lipids: Highly Compact Energy-Storage Nutrients Neutral lipids (fats) or triacylglycerols are contained in cooking fats and oils, butter, margarine, salad dressings and oils, meat and dairy products, nuts, chocolate, and avocados (figure 37.2). Lipids are the most concentrated source of food energy. They produce about 9 Calories (kcal) of usable energy per gram, more than twice as much energy available from an equal weight of carbohydrate or protein (table 37.1).

Many heterotrophs have an absolute dietary requirement for lipids, sometimes for specific types. For example, unsaturated fatty acids (e.g., linoleic, linolenic acid, and arachidonic acid) are required in human diets (*see figure 2.18*). Their most obvious function is to act as precursor

Table 37.1	The Average Caloric Values of Macronutrients
Macronutrient	**Calories per Gram**
Carbohydrates	4.1
Lipids	9.3
Proteins	4.4

molecules for the prostaglandins. Also, phospholipids and cholesterol are the primary constituents of biological membranes (*see figure 3.5*). Other lipids insulate the body and help maintain a constant temperature. Some lipids also supply or aid the absorption of the fat-soluble vitamins (A, D, E, K).

Proteins: Basic to the Structure and Function of Cells The animal sources of protein include meat, poultry, fish and other seafoods, eggs, milk and milk

Table 37.2	Physiological Roles of the Minerals Required in Large Amounts
Mineral	**Major Physiological Roles**
Calcium (Ca)	Component of bone and teeth; essential for normal blood clotting; needed for normal muscle, nerve, and cell function
Chlorine (Cl)	Principal negative ion in extracellular fluid; important in acid-base and fluid balance; needed to produce stomach HCl
Magnesium (Mg)	Component of many coenzymes; needed for normal nerve and muscle function, as well as carbohydrate and protein metabolism
Potassium (K)	Major constituent of bones, blood plasma; needed for energy metabolism
Phosphorus (P)	Major positive ion in cells; influences muscle contraction and nerve excitability; part of DNA, RNA, ATP, energy metabolism
Sodium (Na)	Principal positive ion in extracellular fluid; important in fluid balance; essential for conduction of nerve impulse, active transport
Sulfur (S)	Protein structure; detoxification reactions and other metabolic activity

Table 37.3	Some Physiological Roles of Trace Minerals
Mineral	**Physiological Roles**
Cobalt (Co)	Component of vitamin B_{12}; essential for red blood cell production
Copper (Cu)	Component of many enzymes; essential for melanin and hemoglobin synthesis; part of cytochromes
Fluorine (F)	Component of bone and teeth; prevents tooth decay
Iodine (I)	Component of thyroid hormones
Iron (Fe)	Component of hemoglobin, myoglobin, enzymes, and cytochromes
Manganese (Mn)	Activates many enzymes; an enzyme essential for urea formation and parts of the citric acid cycle
Molybdenum (Mo)	Constituent of some enzymes
Selenium (Se)	Needed in fat metabolism
Zinc (Zn)	Component of at least 70 enzymes; needed for wound healing and fertilization

products, including yogurt and cheese. The plant sources include beans and peas, nuts and peanut butter, cereal, bread, and pasta (figure 37.2). Proteins are needed for their amino acids, which heterotrophs use to build their own body proteins. Of the 20 naturally occurring amino acids—those found in proteins—about one-half cannot be synthesized and must be obtained from food in the diet; they are called the *essential amino acids*. The *nonessential amino acids* are also required to synthesize proteins, but these can be synthesized from other materials, such as other amino acids, and are not required in the diet. Proteins also form an important part of the blood. They are important regulators of osmotic pressure and water balance in the body, and help to maintain the pH of various body fluids. Excess protein is used as an energy source or converted into fat.

Micronutrients

Micronutrients are usually small ions and molecules that are used over and over for enzymatic reactions or as parts of certain proteins (e.g., iron in hemoglobin). Even though they are needed in small amounts, animals cannot synthesize them rapidly (if at all); thus, they must be obtained from the diet.

Minerals   Some minerals are needed in relatively large amounts and are called *essential minerals*, or *ma-*

crominerals. For example, sodium and potassium are vital to the functioning of every nerve and muscle in an animal's body. Large quantities of these minerals, especially sodium, are lost in the urine every day. In those mammals that sweat to help regulate body temperature, sodium loss also becomes important. A daily supply of calcium is needed for muscular activity and, with phosphorous, for bone formation. The function of the major essential minerals is given in table 37.2.

Other minerals are known as *trace minerals, trace elements,* or *microminerals.* Some of these are needed in only very small amounts for various enzymatic functions. Where known, the function of the major trace minerals is given in table 37.3.

Vitamins   Normal metabolic activity depends on very small amounts of more than a dozen organic substances called vitamins. A **vitamin** (L. *vita,* life) is the general term for a number of chemically unrelated, organic substances that occur in many foods in small amounts and are necessary for the normal metabolic functioning of the body. They may be *water soluble* or *fat soluble*. Most water-soluble vitamins are coenzymes needed in metabolism (table 37.4). The water-soluble vitamins are easily excreted by the kidneys. The fat-soluble vitamins have various functions (table 37.5).

The dietary need for vitamin C and the fat-soluble vitamins (A, D, E, and K) tends to be limited to the vertebrates. Even in closely related groups, vitamin requirements vary. For example, among vertebrates, humans and guinea pigs require vitamin C, but rabbits do not. Some birds require vitamin A; others do not.

Table 37.4 Water-Soluble Vitamins

Vitamin	Characteristics	Functions	Sources
Thiamine (Vitamin B$_1$)	Destroyed by heat and oxygen, especially in alkaline environment	Part of coenzyme needed for oxidation of carbohydrates, and coenzyme needed in synthesis of ribose	Lean meats, liver, eggs, whole-grain cereals, leafy green vegetables, legumes
Riboflavin (Vitamin B$_2$)	Stable to heat, acids, and oxidation; destroyed by alkalis and light	Part of enzymes and coenzymes needed for oxidation of glucose and fatty acids and for cellular growth	Meats, dairy products, leafy green vegetables, whole-grain cereals
Niacin (Nicotinic Acid)	Stable to heat, acids, and alkalis; converted to niacinamide by cells; synthesized from tryptophan	Part of coenzymes needed for oxidation of glucose and synthesis of proteins, fats, and nucleic acids.	Liver, lean meats, poultry, peanuts, legumes
Vitamin B$_6$	Group of three compounds; stable to heat and acids; destroyed by oxidation, alkalis, and ultraviolet light	Coenzyme needed for synthesis of proteins and various amino acids, for conversion of tryptophan to niacin, for production of antibodies, and for synthesis of nucleic acids	Liver, meats, fish, poultry, bananas, avocados, beans, peanuts, whole-grain cereals, egg yolk
Pantothenic Acid	Destroyed by heat, acids, and alkalis	Part of coenzyme needed for oxidation of carbohydrates and fats	Meats, fish, whole-grain cereals, legumes, milk, fruits, vegetables
Cyanocobalamin (Vitamin B$_{12}$)	Complex, cobalt-containing compound; stable to heat; inactivated by light, strong acids, and strong alkalis; absorption regulated by intrinsic factor from gastric glands; stored in liver	Part of coenzyme needed for synthesis of nucleic acids and for metabolism of carbohydrates; plays role in synthesis of myelin	Liver, meats, poultry, fish, milk, cheese, eggs
Folacin (Folic Acid)	Occurs in several forms; destroyed by oxidation in acid environment or by heat in alkaline environment; stored in liver where it is converted into folinic acid	Coenzyme needed for metabolism of certain amino acids and for synthesis of DNA; promotes production of normal red blood cells	Liver, leafy green vegetables, whole-grain cereals, legumes
Biotin	Stable to heat, acids, and light; destroyed by oxidation and alkalis	Coenzyme needed for metabolism of amino acids and fatty acids and for synthesis of nucleic acids	Liver, egg yolk, nuts, legumes, mushrooms
Ascorbic Acid (Vitamin C)	Closely related to monosaccharides; stable in acids, but destroyed by oxidation, heat, light, and alkalis	Needed for production of collagen, conversion of folacin to folinic acid, and metabolism of certain amino acids; promotes absorption of iron and synthesis of hormones from cholesterol	Citrus fruits, citrus juices, tomatoes, cabbage, potatoes, leafy green vegetables, fresh fruits

Source: Hole, John W. *Human Anatomy and Physiology.* © 1987 Wm. C. Brown Publishers, Dubuque, Iowa. All Rights Reserved. Reprinted by permission.

Stop and Ask Yourself

7. What are some functions of carbohydrates in vertebrates? Lipids? Proteins?
8. What is the difference between essential minerals and trace minerals?
9. What is a vitamin?
10. What are some water-soluble vitamins? Fat-soluble vitamins?

The Human Digestive System

Humans, bears, pigs, and many other vertebrates are omnivores. The digestive system of an omnivore has evolved the mechanical and chemical ability to process many kinds of foods. The following sections will examine the control of gastrointestinal motility, the major parts of the alimentary canal, and the accessory organs of digestion using the human as the representative example. The *alimentary*

Table 37.5 Fat-Soluble Vitamins

Vitamin	Characteristics	Functions	Sources
Vitamin A	Occurs in several forms; synthesized from carotenes; stored in liver; stable in heat, acids, and alkalis; unstable in light	Necessary for synthesis of visual pigments, mucoproteins, and mucopolysaccharides; for normal development of bones and teeth; and for maintenance of epithelial cells	Liver, fish, whole milk, butter, eggs, leafy green vegetables, and yellow and orange vegetables and fruits
Vitamin D	A group of sterols; resistant to heat, oxidation, acids, and alkalis; stored in liver, skin, brain, spleen, and bones	Promotes absorption of calcium and phosphorus; promotes development of teeth and bones	Produced in skin exposed to ultraviolet light; in milk, egg yolk, fish-liver oils, fortified foods
Vitamin E	A group of compounds; resistant to heat and visible light; unstable in presence of oxygen and ultraviolet light; stored in muscles and adipose tissue	An antioxidant; prevents oxidation of vitamin A and polyunsaturated fatty acids; may help maintain stability of cell membranes	Oils from cereal seeds, salad oils, margarine, shortenings, fruits, nuts, and vegetables
Vitamin K	Occurs in several forms; resistant to heat, but destroyed by acids, alkalis, and light; stored in liver	Needed for synthesis of prothrombin; needed for blood clotting	Leafy green vegetables, egg yolk, pork liver, soy oil, tomatoes, cauliflower

Source: Hole, John W., *Human Anatomy and Physiology.* © 1987 Wm. C. Brown Publishers, Dubuque, Iowa. All Rights Reserved. Reprinted by permission.

canal or *gastrointestinal tract* consists of the oral cavity (mouth), pharynx, stomach, esophagus, small intestine, large intestine (colon), rectum, and anus. The *accessory organs* of digestion include the salivary glands, pancreas, liver, and gallbladder (figure 37.3).

The process of digesting and absorbing nutrients includes:

1. *Ingestion*—eating.
2. *Peristalsis*—the involuntary, sequential muscular contractions that move ingested nutrients along the digestive tract.
3. *Secretion*—the release of hormones, enzymes, and specific ions and chemicals that take part in digestion.
4. *Digestion*—the conversion of large nutrient particles or molecules into small particles or molecules.
5. *Absorption*—the passage of usable nutrient molecules from the small intestine into the bloodstream and lymphatic system for the final passage to body cells.
6. *Defecation*—the elimination from the body of undigested and unabsorbed material as waste.

Gastrointestinal Motility and Its Control

As with any organ, the function of the gastrointestinal tract is determined by the type of tissues it contains. Most of the gastrointestinal tract has the same anatomical structure along its entire length (figure 37.4). From the outside inward, there is a thin layer of connective tissue called the *serosa.* (The serosa forms a moist epithelial sheet called the *peritoneum.* This peritoneum lines the entire abdominal cavity and all internal organs. The space it encompasses is the coelom.) Next is the *longitudinal smooth muscle layer* and *circular smooth muscle layer.* Underneath this muscle layer is the *submucosa.* The submucosa contains connective tissue that contains blood and lymphatic vessels. The *mucosa* faces the central opening called a *lumen.*

It is the coordinated contractions of the muscle layers of the gastrointestinal tract that mix the food material with various secretions, and move the food from the oral cavity to the rectum. The two types of movement involved are peristalsis and segmentation.

During **peristalsis** (Gr. *peri,* around + *stalsis,* contraction), food advances through the gastrointestinal tract when the rings of circular smooth muscle contract behind it and relax in front of it (figure 37.5a). Peristalsis is analogous to squeezing icing from a pastry tube. The small and large intestines also have rings of smooth muscles that repeatedly contract and relax, creating an oscillating back-and-forth movement in the same place, called **segmentation** (figure 37.5b). This movement mixes the bolus of food with digestive secretions.

Sphincters also influence the flow of material through the gastrointestinal tract and prevent backflow. Sphincters are rings of smooth or skeletal muscle located at the beginning or ends of specific regions of the gut tract. For example, there is a sphincter called the *cardiac sphincter* located between the esophagus and stomach, and a *pyloric sphincter* located between the stomach and small intestine.

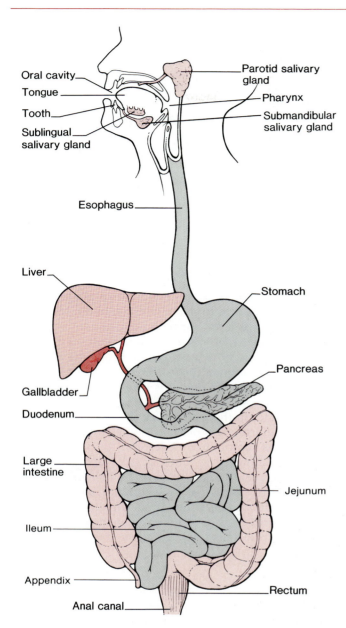

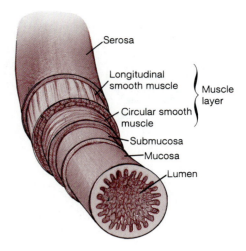

Figure 37.4 The gastrointestinal tract. A generalized illustration of the common structural layers of the gastrointestinal tract. The central lumen extends from the mouth to the anus.

Figure 37.3 Major organs and parts of the human digestive system. Food passes from the mouth through the pharynx and esophagus to the stomach. From the stomach, it passes to the small intestine, where nutrients are broken down and absorbed into the circulatory and lymphatic systems. Nutrients then move to the large intestine, where water is reabsorbed, and feces are formed. Feces exit the body via the anal canal.

The control of gastrointestinal activity is based on the volume and composition of food in the lumen of the gut. For example, ingested food distends the gut and stimulates mechanical receptors located in the gut wall. In addition, as carbohydrates, lipids, and proteins are digested, various chemical receptors in the gut wall are also stimulated. Signals from these mechanical and chemical stimuli travel through nerve plexuses in the gut wall to effect the muscular contraction that leads to peristalsis and segmentation, as well as the secretion of various substances (e.g., mucus,

enzymes) into the gut lumen. In addition to this local control, long-distance nerve pathways connect the receptors and effectors with the central nervous system. Either or both of these pathways function to maintain homeostasis in the gut.

The endocrine cells of the gastrointestinal tract produce hormones that help regulate secretion, digestion, and absorption. These hormones include secretin, gastrin, gastric inhibitory peptide, and cholecystokinin (table 37.6).

The Oral Cavity

The *oral cavity* (mouth) is protected by a pair of lips. The *lips* are highly vascularized, skeletal muscle tissue with an abundance of sensory nerve endings. Lips help retain food as it is being chewed and play a role in phonation (the modification of sound).

The oral cavity contains the teeth and tongue. As in most omnivores, there are four groups of adult teeth: *incisors, canines, premolars,* and *molars* (figure 37.6). Human teeth represent a mixture of the pointed teeth characteristic of carnivores and the chisel-shaped and flat teeth characteristic of herbivores. This versatility in omnivore teeth has led to the versatility of the omnivore diet. The mechanical processing of a wide range of foods is possible because the teeth are covered with enamel, the hardest material in the body, and because of the large force exerted by the jaws and teeth (box 37.2).

The *tongue* is a muscular organ that moves and manipulates food during chewing. It also monitors the texture and taste of foods.

The oral cavity is continuously bathed by **saliva,** a watery fluid secreted by three pairs of salivary glands (*parotid, submandibular,* and *sublingual; see figure 37.3*). Ducts of these exocrine glands empty into different parts of the oral cavity. Saliva humidifies the air passing through the

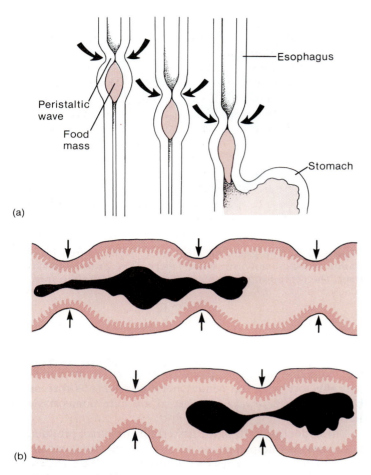

Figure 37.5 Peristalsis and segmentation. (a) Peristaltic waves move food through the esophagus to the stomach. (b) In segmentation, simultaneous muscular contractions of many sections of the intestine (arrows) help mix nutrients with digestive secretions.

mouth toward the lungs, moistens food and binds it together with *mucins* (glycoproteins), and forms the ingested food into a moist mass called a *bolus*. Saliva also contains bicarbonate ions (HCO_3^-) that buffer chemicals in the mouth, thiocyanate ions (SCN^-) and lysozyme that kill microorganisms, and it contributes an enzyme (*amylase*) necessary for the initiation of carbohydrate digestion.

Ordinarily, a constant flow of saliva flushes out the mouth and keeps the papillae on the tongue reasonably short (*see figure 34.3*). Because salivation is reduced during sleep, the papillae grow longer, trapping food and bacteria on a dry tongue. The bacteria digest the food and produce waste products that have a bad taste, leading to [3] the dryness and bad taste in "morning mouth."

The Pharynx and Esophagus

Chapter 36 presented the fact that both air and swallowed foods and liquids pass from the mouth into the *pharynx*—the common passageway for both the digestive and respiratory tracts. The epiglottis temporarily seals off the opening to the trachea so that swallowed food will not enter the trachea. Initiation of the swallowing reflex can be voluntary, but most of the time it is involuntary. When swallowing begins, sequential, involuntary contractions of smooth muscles in the walls of the *esophagus* propel the bolus or liquid to the stomach. Neither the pharynx nor the esophagus contribute to digestion.

During swallowing, as much as 0.5 l of air may enter the stomach. Some of it is released by *belching*. Air that remains in the stomach moves into the small intestine [4] where it produces a gurgling sound called *borborygmus*. Excess intestinal air is often removed by the passing of gas (*flatus*) from the anus long before the feces are ready to be expelled.

Table 37.6 Major Digestive Hormones			
Hormone	**Source**	**Stimulus for Secretion**	**Major Functions**
Cholecystokinin (CCK)	Small intestine	Amino acids and fatty acids	Stimulates gall bladder to release bile and the pancreas to secrete enzymes; inhibits gastrin-stimulated secretion of acid
Gastric-inhibitory peptide (GIP)	Small intestine	Acid or peptides in duodenum	Inhibits secretion of acid by stomach; helps regulate movement of chyme in duodenum by increasing gastric motility
Gastrin	Stomach	Distension of stomach; vagus nerve; protein in stomach	Stimulates secretion of acid, which stimulates secretion of pepsinogen by chief cells; increases motility
Secretin	Small intestine	Acid or peptides in duodenum	Stimulates pancreas and liver to secrete juices containing bicarbonate; inhibits gastrin-stimulated secretion of acid; inhibits motility of stomach and duodenum

The Stomach

The human *stomach* is a muscular distensible J-shaped sac having three main functions (figure 37.7a). It (1) stores and mixes the food bolus received from the esophagus, (2) secretes substances (enzymes, mucus, and HCl) that start the digestion of proteins, and (3) helps control the rate at which food moves into the small intestine via the pyloric sphincter.

The stomach is made up of an inner mucous membrane containing thousands of *gastric glands* (figure 37.7b). Three types of cells are found in these glands. **Parietal cells** secrete a solution containing HCl, and **chief cells** secrete *pepsinogen,* the precursor of the enzyme *pepsin.* Both of the cells are located in the pits of the gastric glands. The surface of the mucous membrane at the openings of the glands contains numerous **mucous cells** that secrete *mucus* that coats the surface of the stomach and protects it from the HCl and digestive enzymes. Each day, the stomach secretes about 2l of these substances, which constitute the *gastric fluid.* The surfaces of the upper gastrointestinal tract, esophagus and mouth, have a much thinner mucous-cell layer than the stomach, which is why vomiting can [5] cause a burning sensation in the esophagus or mouth. Endocrine cells in one part of the stomach mucosa release the hormone *gastrin,* which travels to target cells in the gastric glands, further stimulating them.

When the bolus of food enters the stomach, it causes the walls of the stomach to distend. This distention, as well as the act of eating, causes the gastric pits to secrete HCl (as H+ and Cl-) and pepsinogen. The H+ ions cause pepsinogen to be converted into the active enzyme pepsin. As pepsin, mucus, and HCl mix with and begin to break down proteins, smooth mucosal muscles contract and vigorously churn and mix the food bolus. During this mixing, some salts, sugars, and alcohol, if present, may be rapidly absorbed across the stomach wall. About 3 to 4 hours after a meal, the stomach contents have been sufficiently mixed and are a semiliquid mass called **chyme** (Gr. *chymos,* juice). The pyloric sphincter regulates the release of the chyme into the small intestine.

When the stomach is empty, peristalic waves cease; however, after about 10 hours of fasting, new waves may occur in upper region of the stomach. It is these waves that can [6] cause "hunger pangs" as sensory nerve fibers carry impulses to the brain.

The Small Intestine: Main Site of Digestion

Most of the food a human ingests is digested and absorbed in the *small intestine.* The human small intestine is about 4 cm in diameter and 7 to 8 m in length (*see figure 37.3*). It is intermediate in length between the small intestines of typical carnivores and herbivores of similar size, and reflects the human's omnivorous eating habits. The length of the small intestine is directly related to the total surface area available for absorbing nutrients, as determined by the many circular folds and minute projections of the inner gut surface (figure 37.8a; box 37.3). On the circular folds, thousands of fingerlike projections called **villi** (s. **villus**; L. "tuft of hair") project from each cm² of mucosa (figure 37.8b). Both the circular folds and villi are covered by simple columnar epithelial cells, each bearing numerous microvilli (figure 37.8d). These minute projections are so dense that the inner wall of the small intestine has a total surface area of approximately 300 m²—the size of a tennis court.

The first part of the small intestine is called the *duodenum,* and functions primarily in digestion. The next part is the *jejunum* and the last part is the *ileum:* both function in the absorption of nutrients.

The duodenum contains many digestive enzymes that are secreted by intestinal glands in the duodenual mucosa; other enzymes are secreted by the pancreas. It is here that digestion of carbohydrates and proteins is completed, and where the digestion of most lipids occurs. As chyme is acted upon by pancreatic and duodenal enzymes (as well as bile), it becomes a whitish, watery solution called **chyle** (L. *chylus,* juice), which moves into the jejunum and ileum. In these two regions, absorption of the end products of digestion (amino acids, simple sugars, fatty acids, glycerol, nucleic acids, water) take place. Much of this absorption involves active transport and the sodium-dependent ATPase

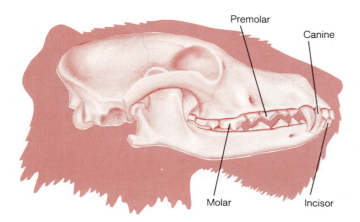

Figure 37.6 The teeth of an omnivorous mammal, such as this dog, are specialized for slicing, puncturing, tearing, and grinding, all of which are used to consume both animal flesh and plant material.

Premolar

Canine

Molar

Incisor

Box 37.2

Filter Feeding in Vertebrates

The baleen, or whalebone whales, include the humpback and blue whales (box figure 37.2a). These are the largest living animals on earth, and because they lack teeth, they have evolved a different approach to feeding. A full-grown blue whale exceeds 30 m in length and weighs as much as 120,000 kg. What food could support the life of such a large animal? The answer is krill. **Krill** (Norw., *kril,* fry or young fish) are small, shrimplike crustaceans.

Baleen whales are named for the fringes of *baleen,* or whale bone (not true bone) that hang down in "mustachelike" sheets inside their mouths (box figure 37.2b). Baleen is nothing more than hardened, shredded sheets of the gum epithelial layers. When these animals feed, they swim through dense schools of krill, straining huge amounts of water through the baleen meshwork. A large whale's expandable mouth holds 60 m³, or 60 tons, of water, which it filters in a few seconds. The filtering process allows the smaller plant life (phytoplankton) to escape with the seawater while the krill is retained to be swallowed at leisure. Tons of krill can be eaten by a whale in a short time.

Flamingos are also filter feeders. They have fringed filters hanging from the upper bill and a deep-sided, curved lower bill (box figure 37.2c,d). This bill represents a similar evolutionary solution to the same problem—removing small food particles from a dilute medium.

Flamingos are commonly misportrayed as denizens of lush tropical islands. In fact, they dwell in one of the world's harshest habitats—shallow, hypersaline lakes. Few animals can tolerate the unusual environments of these saline lakes. Those that can, thrive in the absence of competitors and build up their populations to enormous numbers. Hypersaline lakes provide these predators with ideal conditions for evolving a strategy of filter feeding on small molluscs, crustaceans, and insect larvae.

Flamingos pass water through their bills either by swinging their heads back and forth, permitting water to flow passively through, or by an active pump maintained by a large and powerful tongue. The tongue fills a large channel in the lower beak. It moves rapidly back and forth, up to five times per second, drawing water through the filters on the backwards pull and expelling it on the forward drive.

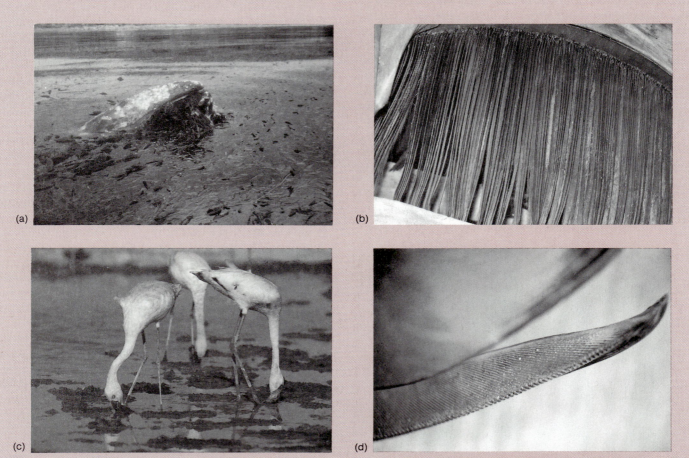

Box figure 37.2 Vertebrate filter feeders. (a) Gray whale feeding in a kelp bed. (b) Baleen. (c) Flamingos feeding. (d) Filters on the bill of a flamingo.

Nutrition and Digestion

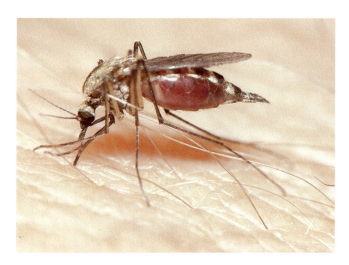

(a) The mosquito (*Culex*) is a fluid feeder.

(b)The giant acorn barnacle (*Balanus nubilus*) is a filter feeder, sweeping feathery appendages through the water to strain out food particles.

(c) The giant panda (*Ailuropoda melanoleuca*), an Asian mammal, survives on a diet of bamboo.

(d) The lion (*Panthera leo*) has long, sharp canine teeth and pointed incisors that allow it to tear flesh and slice through the tough hide of its prey.

Excretion and the Balancing of Water and Salt

(a) This oceanic island is covered with white guano, which was deposited by cormorants, gulls, and other nesting birds. Guano is excreted uric acid.

(b) This tropical lizard has a scaly hide that helps the animal retain its precious body water.

(c) The gray reef shark avoids losing water osmotically by maintaining a high concentration of urea in its body fluids so that they are osmotically equal to or higher than the seawater.

(d) These poison arrow frogs (*Dendrobates leucomelas*) are able to absorb needed salts and water through the belly skin that is in contact with the ground or another moist substrate.

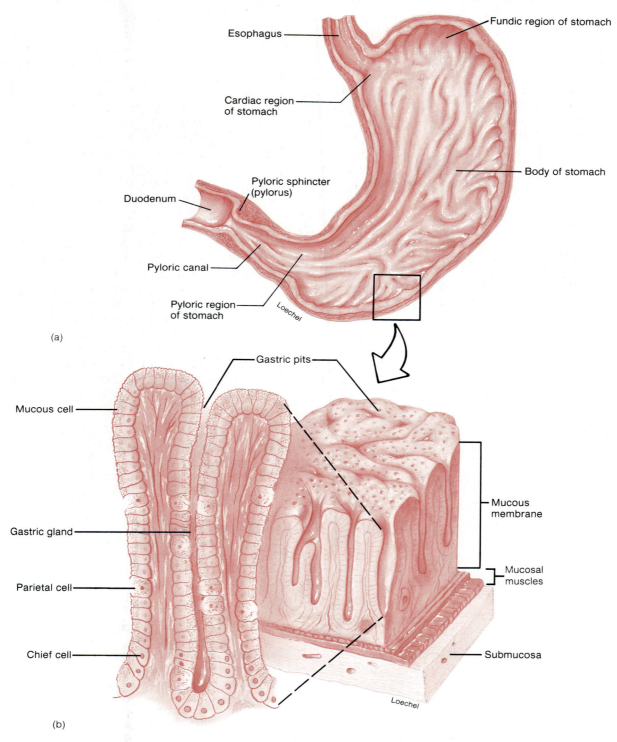

(a)

(b)

Figure 37.7 The stomach. (a) Food enters the stomach from the esophagus. (b) The mucosa of the stomach is covered with gastric glands. The gastric glands include mucous cells, parietal cells, and chief cells—each type producing a different secretion.

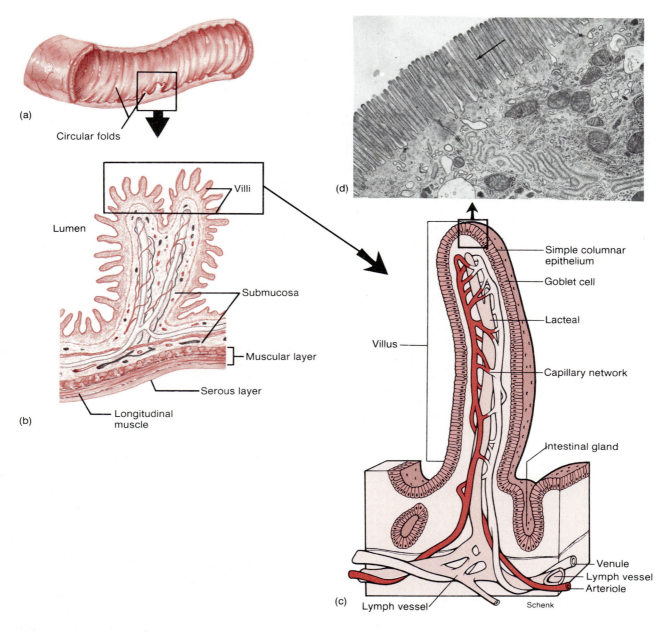

Figure 37.8 The small intestine. The small intestine absorbs food over a large surface area. (a) The lining of the intestine has many circular folds. (b,c) Fingerlike villi line the intestine. A single villus contains a central capillary network and a lymphatic lacteal, both of which transport nutrients absorbed from the lumen of the intestine. (d) The plasma membrane of the simple columnar epithelial cells covering the villi are folded into microvilli (arrow), which further increase the surface area facing the lumen.

pump. Sugars and amino acids are absorbed into the capillaries of the villi, whereas free fatty acids enter the epithelial cells of the villi and are recombined with glycerol to form triglycerides. The triglycerides are coated with proteins to form small droplets called *chylomicrons,* which enter the lacteals of the villi (figure 37.8c). From the lacteals, the chylomicrons move into the lymphatics and eventually into the bloodstream to be transported throughout the body.

Besides absorbing organic molecules, the small intestine absorbs water and dissolved mineral ions. About 9 l of water are absorbed per day in the small intestine, and the rest is absorbed in the large intestine.

The Large Intestine

The human *large intestine* has a diameter of about 6.5 cm and a length of 2 m. Unlike the small intestine, the large intestine has no circular folds, villi, or microvilli; thus, the

---

Box 37.3

The Short Lives of Intestinal Cells

The entire lumen of the gastrointestinal tract is lined with epithelium. These epithelial cells are continuously being lost and replaced at a very high rate, especially in the small intestine. Cell divisions produce new cells to replace the millions of cells shed from the tips of the villi into the gut lumen each day. On the average, cells on the surface of the villi in the small intestine live less than 48 hours, which is a very short life span compared to some other cells. Red blood cells live about 120 days, and nerve cells survive for the life of the animal. Why do these gut cells have such a short life span? There are several reasons.

The quick turnover of cells in the small intestine may be a mechanism that repairs the normal wear and tear caused by abrasion and the actions of digestive enzymes on the intestinal-lining cells, despite their mucous covering. It is also possible that only very young cells are able to carry out the vital absorptive processes in the intestine. Two-day-old cells may be so old that they can no longer function and are discarded.

---

surface area is much less. The small intestine joins the large intestine near a blind-ended sac, the *cecum* (L. *caecum,* blind gut) (*see figure 37.3*). The human cecum and its extension, the *appendix,* are nonfunctional storage sites and possibly represent evolutionary remains of a larger, functional cecum, such as is found in herbivores. The appendix contains an abundance of lymphoid tissue, and may function in the immune system.

The large intestine (also called the *colon*) ascends on the right side of the body (*ascending colon*), crosses over underneath the stomach (*transverse colon*), and descends on the left side of the body (*descending colon*). The descending colon assumes an "S" shape in the lower, left portion of the abdomen, where it is called the *sigmoid* (shaped like the letter S) *colon.* The sigmoid colon terminates in the *rectum,* which leads into the *anal canal,* the exit point of the gastrointestinal tract.

The major functions of the large intestine include the reabsorption of water and minerals, and the formation of feces. As chyle is moved along by peristalic waves, minerals diffuse or are actively transported from the chyle across the epithelial surface of the large intestine into the bloodstream. Water follows osmotically and is returned to the lymphatic system and bloodstream. When sufficient water reabsorption does not occur, *diarrhea* (Gr. *rhein,* to flow) results. If too much water is reabsorbed, fecal matter becomes too thick, and *constipation* results.

Many bacteria and fungi exist symbiotically in the large intestine. They feed on the chyle and further break down its organic molecules to waste products. In turn, they secrete amino acids and vitamin K, which are absorbed by the host's gut. What remains, feces, is a mixture of a large number of dead bacteria, fungi, undigested plant fiber, sloughed-off intestinal cells, and other waste products. The odor of feces is due to the breakdown products produced by the microorganisms (e.g., methane and hydrogen sulfide gas). Strong peristalic contractions of the muscle layer of the large intestine, combined with relaxation of the anal sphincter muscle, allow the expulsion of feces from the body.

Stop and Ask Yourself

15. What are three functions of the stomach?
16. How does the small intestine provide a large surface area for absorption of nutrients?
17. What are the major functions of the large intestine?
18. What is absorbed in the small intestine? The large intestine?

Role of the Pancreas in Digestion

The *pancreas* (Gr, *pan,* all + *kreas,* flesh) is an organ that lies just ventral to the stomach and has both endocrine and exocrine functions. Exocrine cells in the pancreas secrete digestive enzymes into the *pancreatic duct,* which merges with the hepatic duct from the liver to form a *common bile duct* that enters the duodenum. The pancreatic enzymes complete the digestion of carbohydrates and proteins and initiate the digestion of lipids. *Trypsin, carboxypeptidase,* and *chymotrypsin* digest proteins into small peptides and individual amino acids. *Pancreatic lipases* split triglycerides into smaller, absorbable glycerol and free fatty acids. *Pancreatic amylase* converts polysaccharides into disaccharides and monosaccharides. Table 37.7 summarizes the major glands, secretions, and enzymes of the digestive system.

The pancreas also secretes bicarbonate (HCO_3^-) ions that help neutralize the acidic chyme coming from the stomach. Bicarbonate raises the pH from 2 to 7 for optimal digestion to occur. If there were no such neutralization, the pancreatic enzymes could not function.

Role of the Liver and Gallbladder in Digestion

The human *liver* is the largest organ in the body (it weighs about 1.5 kg) and is located just under the diaphragm (*see figure 37.3*). In an infant, the liver occupies about 40% of the abdominal cavity, causing a "pudgy abdomen" and is

Table 37.7
Major Digestive Glands, Secretions, and Enzymes

Place of Digestion	Source	Secretion	Enzyme	Digestive Function
Mouth	Salivary glands	Saliva	Salivary amylase	Begins the digestion of carbohydrates; inactivated by stomach HCl
	Mucous glands	Mucus	—	Lubricates food bolus
Esophagus	Mucous glands	Mucus	—	Lubricates food bolus
Stomach	Gastric glands	Gastric juice	Lipase	Digests lipids into fatty acids and glycerol
			Pepsin	Digests proteins into polypeptides
	Gastric mucosa	HCl	—	Converts pepsinogen into active pepsin; kills microorganisms
	Mucous glands	Mucus	—	Lubricates
Small intestine	Liver	Bile	—	Emulsifies lipids; activates lipase
	Pancreas	Pancreatic juice	Amylase	Converts starch maltose
			Chymotrypsin	Digests proteins into peptides and amino acids
			Lipase	Digests lipids into fatty acids and glycerol (requires bile salts)
			Nuclease	Digests nucleic acids into mononucleotides
			Trypsin	Digests proteins into peptides and amino acids
	Intestinal glands	Intestinal juice	Enterokinase	Converts inactive trypsinogen into active trypsin
			Lactase	Converts lactose into glucose and galactose
			Maltase	Converts maltose into glucose
			Peptidase	Converts polypeptides into amino acids
			Sucrase	Converts sucrose into glucose and fructose
	Mucous glands	Mucus	—	Lubricates
Large intestine	Mucous glands	Mucus	—	Lubricates

responsible for about 5% of the total body weight. In an adult, the liver represents about 2% of the body weight.

In the liver are millions of specialized cells called *hepatocytes*. Hepatocytes take up nutrients absorbed from the intestines and release them into the bloodstream. Hepatocytes also maufacture the blood proteins prothrombin and albumin.

In addition to the above, some major metabolic functions of the liver include:

1. Removal of amino acids from organic compounds.
2. Formation of urea from worn-out tissue cells (proteins), and conversion of excess amino acids into urea to decrease body levels of ammonia.
3. Manufacturing most of the plasma proteins (albumins), forming fetal erythrocytes, destroying worn-out erythrocytes, and helping to synthesize the blood-clotting agents prothrombin and fibrinogen from amino acids.
4. Synthesis of nonessential amino acids.
5. Conversion of galactose and fructose to glucose.
6. Oxidation of fatty acids.
7. Formation of lipoproteins, cholesterol, and phospholipids (essential cell membrane components).
8. Conversion of carbohydrates and proteins into fat.

9. Modification of waste products, toxic drugs, and poisons (detoxification).
10. Synthesis of vitamin A from carotene, and along with the kidneys, participates in the activation of vitamin D.
11. Maintenance of a stable body temperature by raising the temperature of the blood passing through it. Its many metabolic activities make the liver the body's major heat producer.
12. The manufacture of bile salts, which are used in the small intestine for the emulsification and absorption of simple fats, cholesterol, phospholipids, and lipoproteins.
13. The liver is also the body's main storage center. It stores glucose in the form of glycogen, and with the help of insulin and enzymes, it converts glycogen back into glucose as it is needed by the body. The liver also stores fat-soluble vitamins (A, D, E, and K), and minerals, such as iron, from the diet. The liver can also store fats and amino acids and convert them into usable glucose as required.

The *gallbladder* (L. *galbinus,* greenish yellow) is a small organ located within the liver (*see figure 37.3*). The gallbladder stores the greenish fluid called bile that is continuously produced by the liver cells. Bile is very alkaline and contains pigments, cholesterol, lecithin, mucin, bilirubin, and bile salts that act as detergents to emulsify fats (form them into droplets suspended in water) and aid in fat digestion and absorption. (Recall that fats are insoluble in water.) Bile salts also combine with the end products of fat digestion to form micelles. **Micelles** are lipid aggregates with a surface coat of bile salts. Because they are so small, they are able to cross the microvilli of the intestinal epithelium.

Coordination of Ingestion and Digestion: A Question of Timing

Ingestion, digestion, absorption, and storage are highly efficient means of maintaining homeostatic nutrition for vertebrates. To maintain this efficiency, however, most of the organs of digestion must be controlled by feedback mechanisms.

Most animals, including humans, are conditioned to release saliva and gastric juices at the sight or smell of food. Their release is stimulated by nerve impulses from the brain and vagus nerve that stimulate the stomach to release *gastrin* (figure 37.9). Gastrin causes the gastric glands in the stomach to release pepsinogen and HCl. When food reaches the stomach and distends it, further secretion is initiated.

When chyme reaches the duodenum, the acid and fats cause the duodenum to release *secretin* and *cholecystokinin* (CCK). Both hormones slow the rate of peristalsis so that energy-rich fats can be completely digested and absorbed. Secretin also inhibits gastrin, which inhibits gastric-juice secretion and increases pancreatic-juice secretion. CCK stimulates the release of bile from the liver and gallbladder, and digestive enzymes and bicarbonate ions from the pancreas.

The presence of chyme in the duodenum triggers nerve signals that slow the further emptying of the stomach. This inhibition of the stomach is called the *enterogastric reflex,* and allows the small intestine more time to fully digest the chyme it contains before the stomach releases more.

Another feedback mechanism involves a hormone called *vasoactive intestinal peptide* (VIP). VIP is secreted by the duodenum when fats are present. VIP increases the secretion of pancreatic juice and inhibits the secretion of gastrin and in effect, signals the stomach that its job is finished, that the stomach can shut down its digestive processes.

Other basic feedback mechanisms ensure that an animal becomes hungry and eats when nutrients are required by the body cells, and that it stops eating when sufficient food is ingested. There is evidence to indicate that an animal senses hunger when cells in the hypothalamus or the liver respond to low levels of glucose, fats, and amino acids in the bloodstream. When these low levels are reached, the animal begins to graze, to hunt, or goes to a fast food restaurant. Once the levels of these nutrients reach homeostatic bloodstream levels, the animal stops eating.

The brain hormone oxytocin has been found to play a key role in stopping ingestion. When cholecystokinin reaches high levels in the bloodstream, it reaches the brain, which stimulates the pituitary gland to secrete oxytocin. When oxytocin is released, by some unknown mechanism, it causes the animal to stop eating.

Diversity in Digestive Structures

Because most animals spend the majority of their time acquiring food, feeding can be referred to as the universal pastime. The oral cavity (mouth), teeth, intestines, and

Figure 37.9 Feedback mechanism of enzyme secretion and gut motility. Most of the shown hormones have both positive (+) and negative (−) effects on different targets. For example, CCK stimulates the release of bile and pancreatic juice, but inhibits peristalsis.

Figure 37.9 redrawn, with permission, from Norman K. Wessels and Janet L. Hopson, *Biology.* Copyright © 1988, McGraw-Hill Publishing Company, New York, New York.

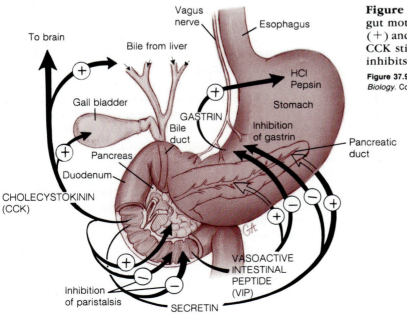

To brain

Vagus nerve

Esophagus

Bile from liver

HCl
Pepsin

Gall bladder

Stomach

GASTRIN

Bile duct

Inhibition of gastrin

Pancreatic duct

Pancreas

Duodenum

CHOLECYSTOKININ (CCK)

VASOACTIVE INTESTINAL PEPTIDE (VIP)

Inhibition of paristalsis

SECRETIN

(a)

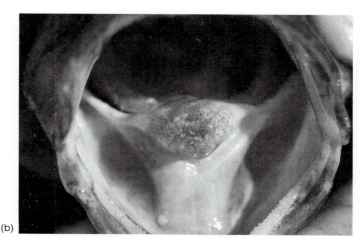

(b)

(c)

(d)

Figure 37.10   Tongues. (a) The rasping tongue and mouth of a lamprey. (b) A fish tongue. (c) The tongue of a chameleon catching an insect. (d) The tongue and supporting apparatus of a woodpecker.

other major digestive structures usually reflect the way an animal gathers food, the type of food it eats, and the way it digests that food.

A tongue or tonguelike structure develops in the floor of the oral cavity in many vertebrates. A lamprey has a protrusible tongue that bears horny teeth and is used to rasp its prey's flesh (figure 37.10a). In other fishes, a *primary tongue* may be present that bears teeth that help hold prey; however, this type of tongue is not muscular (figure 37.10b). Tetrapods have evolved mobile tongues that are used for gathering food. Frogs and salamanders and some lizards can rapidly project part of their tongue from the mouth to capture an insect (figure 37.10c). A woodpecker has a long and spiny tongue (figure 37.10d) that is used to gather insects and grubs. Ant- and termite-eating mammals also gather food with long, sticky tongues. Spiny papillae on the tongues of cats and other carnivores help these animals rasp flesh from a bone.

With the exception of birds, turtles, and baleen whales, most vertebrates contain teeth. (Birds lack teeth, probably to reduce body weight for flight.) Teeth are specialized, depending on whether an animal is feeding on plants or animals, and how it obtains its food. The teeth of snakes slope backward to aid in the retention of prey while swallowing (figure 37.11a), and the canine teeth of dogs are specialized for ripping food (37.11b). The teeth of herbivores, such as deer, have predominately grinding teeth, the front teeth of a beaver are used for chiseling trees and branches, and the elephant has two of its upper, front teeth specialized as weapons and for moving objects (figures 37.11c–e). Because humans are omnivores, they contain teeth that can perform a number of tasks—tearing, ripping, chiseling, and grinding (figure 37.11f).

Most fishes lack salivary glands in the head region. Lampreys are an exception because they have a pair of glands that secrete an anticoagulant that is needed to keep their prey's blood flowing as they feed. Terrestrial vertebrates

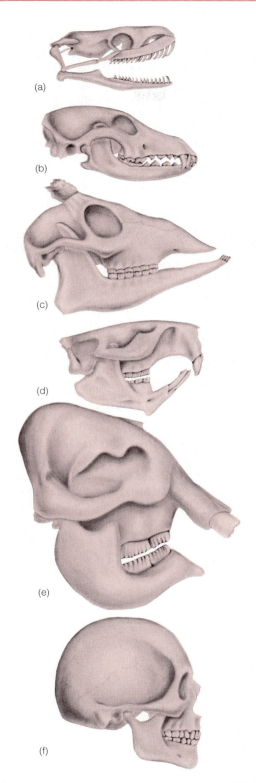

Figure 37.11 The arrangement of teeth in a variety of vertebrates: (a) Snake. (b) Dog. (c) Deer. (d) Beaver. (e) Elephant. (f) Human.

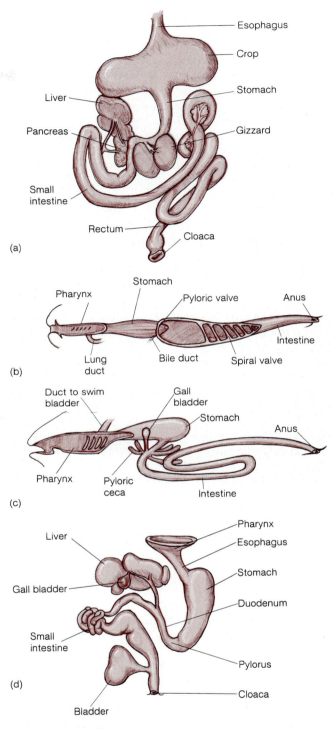

Figure 37.12 The arrangement of stomachs and intestines in a variety of vertebrates: (a) Pigeon. (b) Lungfish. (c) Teleost fish. (d) Frog.

have well-developed salivary glands. Modified salivary glands of some snakes produce venoms that are injected through their fangs to immobilize their prey. Because the secretion of oral digestive enzymes is not an important function in amphibians or reptiles, oral glands are absent.

The esophagus is short in fishes and amphibians but much longer in amniotes due to their longer necks. Grain- and seed-eating birds have a crop that develops from the caudal portion of the esophagus (figure 37.12a). Storing food in the crop ensures an almost continuous supply of food to the stomach and intestine for digestion. This design

[8]

allows these birds to have a high metabolic rate and reduces the frequency of feeding.

The stomach is an ancestral vertebrate structure that evolved as these animals began to feed on larger organisms that were caught at less frequent intervals and required storage. Some zoologists believe that the gastric glands and their HCl production evolved in the context of killing bacteria and helping preserve food. The synthesis of pepsinogen may have evolved later because the stomach is not essential for digestion. In fact, in those vertebrates that feed on very small particulate matter, the stomach may be lost, as probably occurred in some lungfishes, chimaeras, and teleosts (figure 37.12b).

Some fishes, some reptiles, and all birds have a gizzard (figure 37.12a) that is used to grind up food. The bird's gizzard develops from the posterior part of the stomach called the *ventriculus*.

Some of most unusual modifications of the stomach are seen in ruminant mammals—animals that "chew their cud." Examples include cows, sheep, and deer. This method of digestion has evolved in animals that need to eat large amounts of food relatively quickly, but can chew the food at a more comfortable or safer location. More important though, the ruminant stomach provides an opportunity for large numbers of microorganisms to digest the cellulose walls of grass and other vegetation. Cellulose contains a large amount of energy; however, animals generally lack the ability to produce the enzyme *cellulase*, which is needed to digest cellulose and obtain its energy. Because intestinal microorganisms can produce this enzyme, they have made the herbivorous life-style possible.

In ruminants, the upper portion of the stomach expands to form a large pouch, the *rumen*, and a smaller *reticulum*.

The lower portion of the stomach consists of a small antechamber, the *omasum*, with a "true" stomach, or *abomasum*, behind it (figure 37.13). Food first enters the rumen, where it encounters the microorganisms. Aided by copious secretions of fluid, body heat, and churning of the rumen, the food is partially digested and reduced to a pulpy mass. Later, the pulpy mass moves into the reticulum, from which mouthfulls are regurgitated as "cud" (L. *ruminare,* to chew the cud). At this time, food is thoroughly chewed for the first time. When reswallowed, the food enters the rumen where it becomes more liquid in consistency. When it is very liquid, the digested food material flows out of the reticulum and into the omasum and then the glandular region, the abomasum. Here the digestive enzymes are first encountered, and digestion continues in the usual mammalian fashion described above for humans.

The food of ruminants is attacked by microorganisms before gastric digestion, but in the typical nonruminant herbivore, microbial action on cellulose occurs after digestion. Rabbits, horses, and rats accomplish cellulose digestion by maintaining a population of microorganisms in their unusually large cecum, the blind pouch that extends from the colon (figure 37.14). Adding further to this efficiency, a few nonruminant herbivores, such as mice and rabbits, eat their own feces to process the remaining materials in them. [9]

Because of the importance of bile in fat digestion, the gallbladder is relatively large in carnivores and vertebrates in which fat is an important part of the diet. It is much reduced in bloodsuckers, such as the lamprey, and in animals that feed primarily on plant food (e.g., some teleosts, birds, rats, and other mammals).

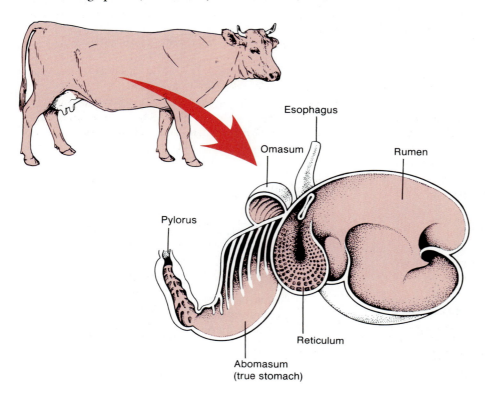

Figure 37.13 The four-chambered stomach of a ruminant.

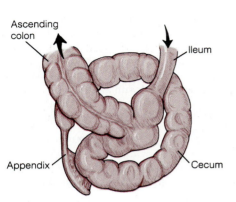

Figure 37.14 The extensive cecum of a nonruminant herbivore, such as a rabbit.

Every vertebrate has a pancreas; however, in lampreys and lungfishes it is embedded in the wall of the intestine and is not a visible organ. Both endocrine and exocrine tissues are present, but the cell composition varies.

The configuration and divisions of the small and large intestines vary greatly among vertebrates. They are closely related to the animal's type of food, body size, and levels of activity. For example, cyclostomes, chondrichthyean fishes, and primitive bony fishes have short, nearly straight intestines that extend from the stomach to the anus (*see figure 37.12b*). In more advanced bony fishes, the intestine increases in length and begins to coil (*see figure 37.12c*). The intestines are moderately long in most amphibians and reptiles (*see figure 37.12d*). In birds (*see figure 37.13a*) and mammals, the intestines are longer and have more surface area than other tetrapods. Birds typically have two ceca, and mammals a single cecum at the beginning of the large intestine. The large intestine is much longer in mammals than in birds, and it empties into the cloaca in most vertebrates.

Stop and Ask Yourself

19. What is the role of the pancreas in digestion?
20. What are some functions of the liver?
21. What is the function of secretin? Cholecystokinin?
22. How do vertebrate teeth reflect the eating habits of various animals?
23. How does the stomach of a ruminant function?

Summary

1. Nutrition describes all of those processes by which an animal takes in, digests, absorbs, stores, and uses food (nutrients) in order to meet the metabolic needs of the organism. Digestion is the mechanical and chemical breakdown of food into smaller particles that can be absorbed by the intestines.

2. Much of animal evolution has been marked by losses in biosynthetic abilities. This tendency has led to the evolution of the following nutritional types: insectivores, herbivores, carnivores, and omnivores.

3. The nutrients ingested by a heterotroph can be divided into macronutrients and micronutrients. Macronutrients are needed in large quantities and include the carbohydrates, lipids, and proteins. Micronutrients are needed in small quantities and include the vitamins and minerals.

4. The digestive system of vertebrates is one-way, leading from the mouth (oral cavity), to the pharynx, esophagus, stomach, small intestine, large intestine, rectum, and anus.

5. It is the coordinated contractions of the muscle layer of the gastrointestinal tract that mixes food material with various secretions and moves it from the oral cavity to the rectum. The two types of movement involved are segmentation and peristalsis.

6. Most digestion occurs in the duodenal portion of the small intestine. The products of digestion are absorbed in the walls of the jejunum and ileum. In the process of digestion, fats are made soluble by bile secreted from the liver. The liver has many diverse functions. It controls the fate of newly synthesized food molecules, stores excess glucose as glycogen, synthesizes many blood proteins, and converts nitrogenous and other wastes into a form that can be excreted by the kidneys.

7. The large intestine has little digestive or nutrient absorptive activity; it functions principally to absorb water, to compact the material that is left over from digestion, and to serve as a storehouse for microorganisms.

8. The evolution and structure of the digestive system in various invertebrates and vertebrates reflects their eating habits, their level of metabolism, and their body size.

Key Terms

autotrophs (p. 595)
calorie (p. 596)
Calorie (p. 596)
carnivores (p. 596)
chief cells (p. 603)
chyle (p. 603)
chyme (p. 603)
digestion (p. 595)
herbivores (p. 596)
heterotrophs (p. 595)

insectivores (p. 596)
kilocalorie (p. 596)
krill (p. 604)
macronutrients (p. 596)
micelles (p. 609)
micronutrients (p. 596)
mucous cells (p. 603)
nutrition (p. 595)
omnivores (p. 596)
parietal cells (p. 603)

peristalsis (p. 600) sphincters (p. 600)
saliva (p. 601) villi (villus) (p. 603)
segmentation (p. 600) vitamin (p. 598)

Critical Thinking Questions

1. "You are what you eat" is a common saying. Discuss the validity of this saying.

2. What advantages are there to digestion? Would it not be simpler for a vertebrate to simply absorb carbohydrates, lipids, and proteins from its food and use these molecules without breaking them down?

3. What might have been some evolutionary pressures acting on animals that led to the internalization of digestive systems?

4. Most people overrate the importance of the stomach in digestion. Why is this so?

5. Many digestive enzymes are produced in the pancreas and released into the duodenum. Why then, has the mammalian stomach evolved the ability to produce pepsinogen?

6. Trace the fate of a hamburger from the mouth to the anus, identifying sites and mechanisms of digestion and absorption.

7. Human vegetarians, unlike true herbivores, have no highly specialized fermentation chambers. Why is this so?

8. Why does it take longer to become hungry after a protein meal compared to a carbohydrate meal?

9. What are some negative feedback systems that help regulate the digestive process?

Suggested Readings

Books

Christian, J. L., and Gregor, J. L. 1985. *Nutrition for Living*. Menlo Park, Calif.: Benjamin Cummings.

Ruckenusch, Y, and Thivend, P. 1980. *Digestive Physiology and Metabolism in Ruminants*. Westport, CT: AVI Publsihing Company.

Articles

Baldwin, R. L. 1984. Digestion and metabolism of ruminants. *BioScience* 34(4):244–249.

Cohen, L. A. Diet and cancer. *Scientific American* November, 1987.

Davenport, H. W. Why the stomach does not digest itself. *Scientific American* January, 1972.

Degabriele, R. The physiology of the koala. *Scientific American* July, 1980.

Fernstrom, J. D., and Wurtman, R. J. Nutrition and the brain. *Scientific American* February, 1974.

Frisch, R. E. Fatness and fertility. *Scientific American* March, 1988.

Heinrich, B. The raven's feast. *Natural History* February, 1989.

Hume, I. Reading the entrails of evolution. *New Scientist* April, 1989.

Kappas, A., and Alveares, A. P. How the liver metabolizes foreign substances. *Scientific American* June, 1975

Kretchmer, N. Lactose and lactase. *Scientific American* October, 1972.

Moog, F. The lining of the small intestine. *Scientific American* November, 1981.

Scrimshaw, N. S., and Young, V. R. The requirements of human nutrition. *Scientific American* September, 1976.

Young, V. R., and Scrimshaw, N. S. The physiology of starvation. *Scientific American* October, 1971.

Temperature and Body Fluid Regulation

38

Concepts

1. Thermoregulation is a complex and important physiological process that maintains, to varying degrees, an animal's body temperature, despite variations in environmental temperature. Based on this regulation, animals can be categorized as endotherms or ectotherms, and homeotherms or heterotherms.
2. Some invertebrates have contractile vacuoles, flame-cell systems, nephridia, or Malpighian tubules for osmoregulation.
3. A vertebrate's urinary system functions in osmoregulation and excretion, both of which are necessary for internal homeostasis. Osmoregulation governs water and salt balance, and excretion eliminates metabolic wastes. In reptiles, birds, and mammals, the kidneys are the primary osmoregulatory structures.

Have You Ever Wondered:

[1] how sweating cools the body?
[2] why hummingbirds have to spend so much time feeding compared to larger birds?
[3] why bluefin tuna and mako sharks can swim faster than other fishes of a comparable size?
[4] why frogs have to live in warm, moist environments?
[5] why goose down was traditionally used as an insulator in outdoor vests and jackets?
[6] what blubber is?
[7] how kangaroo rats can live an entire lifetime drinking little or no water?
[8] why freshwater fishes never drink water?
[9] why saltwater fishes continually drink water?
[10] why a dog's nose is normally cold and wet?

These and other useful questions will be answered in this chapter.

This chapter contains underlined evolutionary concepts.

The earth's environments vary dramatically in temperature and amount of water present. In the polar regions, high mountain ranges, and deep in the oceans, the temperature remains near or below 0°C (32°F) throughout the year. Temperatures exceeding 40°C (103°F) are common in equatorial deserts. Between these two extremes, in the earth's temperate regions, wide fluctuations in temperature are common. The temperate regions have varying amounts of water, as well as varied habitats—fresh water, salt water, wetlands, mountains, and grasslands. Animals have successfully colonized these varied places on earth by evolving homeostatic mechanisms for maintaining a relatively constant internal environment, despite fluctuations in the external environment. This chapter covers three separate but related homeostatic systems that enable animals to survive the variations in temperature, water availability, and salinity (salt concentration) on the earth. The *thermoregulatory system* maintains an animal's body temperature and/or its responses to shifts in environmental temperature; the *osmoregulatory system* maintains the level and concentration of water and salts in the body; and the *urinary system* eliminates metabolic wastes from the body and functions in osmoregulation.

Homeostasis and Temperature Regulation

As presented in chapters 4 and 5, the temperature of a living cell affects the rate of its metabolic processes. An animal can grow faster and respond to the environment more rapidly if its cells are kept warm. In fact, the ability of some animals to maintain a constant (homeostatic), relatively high body temperature, is believed to be a major reason for their evolutionary success. This ability to control the temperature of the body is called **thermoregulation** ("heat control") and involves the nervous, endocrine, respiratory, and circulatory systems. It is no wonder that many zoologists regard thermoregulation as one of the most complex and highly integrated of the basic physiological processes.

The Impact of Temperature on Vertebrate Life

Every vertebrate's physiological functions are inexorably linked to temperature, because metabolism is very sensitive to changes in internal temperature. Thus, temperature has been a strong source of selective pressure on vertebrates. The rate of cellular respiration increases with temperature up to a certain point. When the temperature rises beyond the *temperature optima* at which enzymes most efficiently catalyze their chemical reactions, the rate declines as the enzymes begin to denature. The chemical interactions holding the enzymes together in their three-dimensional shape are disrupted (*see figure 4.5*). The results of enzyme evolution have frequently been enzymes with temperature

optima that reflect an animal's habitat. For example, a digestive enzyme in a trout might function optimally at 10°C, whereas another enzyme in the human body that catalyzes the same reaction functions best at 37°C (*see figure 4.6*). High temperatures cause the proteins in nucleic acids to denature and low temperatures may cause membranes to change from a fluid to a solid state, which can interfere with many cellular processes, such as active-transport pumps.

Animals can guard against these damaging effects of temperature fluctuations by balancing heat gains and heat losses with their environment.

Heat Gains and Losses

Animals produce heat as a by-product of metabolism, and either gain heat from, or lose it to, the environment. The total body temperature is a product of these factors and can be expressed as:

Body temperature = heat produced metabolically
+ heat gained from the environment
− heat lost to the environment

Vertebrates use four physical processes to exchange heat with the environment: conduction, convection, evaporation, and radiation (figure 38.1). **Conduction** is the direct transfer of thermal motion (heat) between molecules of the environment and those on the body surface of an animal. This transfer is always from an area of higher to one of lower temperature, because heat moves down thermal gradients. For example, when you sit on the cold ground you will lose heat, and when you sit on warm sand, you will gain heat.

Convection is the movement of air (or a liquid) over the surface of a body and contributes to heat loss (if the air is cooler than the body) or heat gain (if the air is warmer

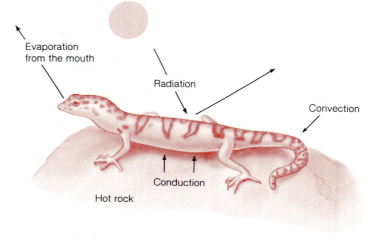

Figure 38.1 Heat gain and loss between an animal and its environment. Heat is either gained or lost by objects in direct contact with the animal (conduction), by air currents (convection), by electromagnetic waves (radiation), or by exhaled air (evaporation).

than the body). On a cool day, your body loses heat by convection because your skin temperature is higher than the surrounding air temperature.

Evaporation is loss of heat from the surface of a liquid as molecules escape in the form of a gas. Recall that humans and some other animals have sweat glands that actively move watery solutions through pores to the skin surface. When the skin temperature is high, water at the surface absorbs enough thermal energy to break the hydrogen bonds holding the individual water molecules together and they depart from the surface, carrying heat with them. As long as the environmental humidity is low enough to permit complete evaporation, sweating can rid the body of excess heat; however, the water must evaporate. Sweat dripping from one's forehead has no cooling effect at all.

[1]

Radiation is the emission of electromagnetic waves produced by objects, such as another animal's body or the sun. Radiation can transfer heat between objects that are not in direct contact with each other, as happens when an animal suns itself (figure 38.2).

Some Solutions to Temperature Fluctuations

Animals cope with temperature fluctuations in one of three basic ways: (1) They can occupy a place in the environment where the temperature remains constant and compatible with their physiological processes; (2) their physiological processes may have adapted to the range of temperatures that the animals are comfortable living in; or (3) they can generate heat internally to maintain a constant body temperature, despite fluctuations in the temperature of the external environment.

Figure 38.2 Radiation warms an animal. After a cold night in its den on the Kalahari desert, a meerkat stands at attention, allowing the large surface area of its body to absorb radiation from the sun.

Animals can be categorized as ectotherms or endotherms, based on whether their source of body heat is from internal processes or derived from the environment. **Ectotherms** (Gr. *ectos,* outside) derive most of their body heat from the environment rather than from their own metabolism. They have low rates of metabolism and are poorly insulated. In general, reptiles, fishes, amphibians, and invertebrates are ectotherms, although a few reptiles, insects and fishes can generate some internal heat. Ectotherms tend to move about the environment and find places where they will minimize heat or cold stress to their bodies. For example, to gain heat, they move out of the shade and onto a sunny rock (colorplate 21d). When the temperature begins to drop, they crawl into a crevice or under a rock where they will not cool as rapidly as the air.

Birds and mammals are called **endotherms** (Gr. *endos,* within) because they obtain their body heat from cellular processes. Having a constant source of internal heat allows them to maintain a nearly constant core temperature despite the surrounding environmental temperature. ("Core" refers to the body's internal temperature as opposed to the temperature near its surface.)

Most endotherms have bodies insulated by fur or feathers and a large amount of fat. This insulation enables heat to be retained more efficiently and a high core temperature to be maintained (colorplate 21a). Endothermy allows animals to stabilize their core temperature so that biochemical processes and nervous system functions can proceed at steady high levels of activity. Endothermy is an evolutionary advancement that has allowed some animals to colonize habitats denied to ectotherms.

Another way of categorizing animals is based on whether they maintain a constant or variable body temperature. Although most endotherms are **homeotherms** (maintain a constant body temperature), and most ectotherms are **heterotherms** (have a variable body temperature), there are many exceptions. Some endotherms vary their body temperatures seasonally (e.g., hibernation); others vary it on a daily basis.

For example, some birds (e.g., hummingbirds) and mammals (e.g., shrews) can only maintain a high body temperature for a short period because they usually weigh less than 10 g and have a body mass so small that not enough heat can be generated to compensate for the heat that is lost across their relatively large surface area. Humming- [2] birds must devote much of the day to locating and sipping nectar (a very high-calorie food source) as a constant energy source for metabolism. When not feeding, hummingbirds would rapidly run out of energy unless their metabolic rates decreased considerably. When resting at night, hummingbirds enter a sleeplike state, called **daily torpor,** and their body temperature approaches that of the cooler surroundings.

There are also some ectotherms that can maintain fairly constant body temperatures. Among these are a number of reptiles that can maintain fairly constant body temperatures by changing position and location during the day to equalize heat gain and loss.

In general, ectotherms prefer the tropics because they do not have to expend as much energy in maintaining body temperature, and more energy can be devoted to food gathering and reproduction. Indeed, in the tropics, amphibians are far more abundant than mammals. Conversely, in moderate to cool environments, endotherms have a selective advantage and are more abundant. Their high metabolic rates and insulation allow them to occupy even the polar regions (e.g., polar bears). In fact, the efficient circulatory systems of birds and mammals can be thought of as adaptations to endothermy and a high metabolic rate.

Stop and Ask Yourself

1. What is thermoregulation?
2. How do temperature extremes affect metabolic reactions?
3. What are three basic ways animals cope with temperature fluctuations?
4. How can animals be categorized based on how they obtain body heat?

Temperature Regulation in Fishes

The body temperature of most fishes is determined by their surrounding water temperature. Fishes that live in extremely cold water have "antifreeze" materials in their blood. Polyalcohols (e.g., sorbitol, glycerol) or water-soluble peptides and glycopeptides lower the freezing point of blood plasma and other body fluids. These fishes also have proteins or protein-sugar compounds that stunt the growth of ice crystals that begin to form. These adaptations enable these fishes to stay flexible and swim freely in a supercooled state (i.e., at a temperature below the normal freezing temperature of a solution).

Some very active fishes maintain a core temperature significantly above the temperature of the water. Bluefin tuna and the great white shark have their major blood vessels just under the skin. Branches deliver blood to the deeper, powerful, red swimming muscles, where smaller vessels are arranged in a countercurrent heat exchanger called the **rete mirabile** ("miraculous net") (figure 38.3a). The heat generated by these red muscles is not lost because it is transferred in the rete mirabile from venous blood passing outward to cold arterial blood passing inward from the body surface. This design enhances vigorous activity by keeping the swimming muscles several degrees warmer than the tissue near the surface of the fish (figure 38.3b). The evolutionary advantage of this system has been profound for these fishes. The power of their muscular contractions can be four times greater than those of similar muscles in fishes with cooler bodies. Thus, they can swim faster and range more widely through various depths in search of prey than can other predatory fish more limited to given water depths and temperatures.

[3]

Temperature Regulation in Amphibians and Reptiles

Animals, such as amphibians and reptiles, that have air rather than water as a surrounding medium are subjected to marked daily and seasonal temperature changes. Most of these animals are ectotherms, they derive heat from their environment, and their body temperatures vary with external temperatures.

Most amphibians have difficulty in controlling body heat because they produce little of it metabolically and rapidly lose most of it from their body surfaces. However, as previously noted, behavioral adaptations enable them to maintain their body temperature within a comfortable range most of the time. Amphibians have an additional thermoregulatory problem because they must exchange oxygen and carbon dioxide across their skin surface, and this mois-

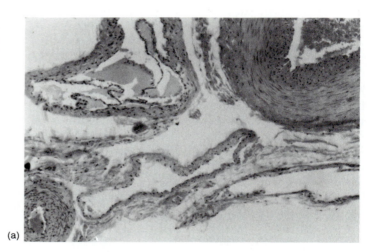

(a)

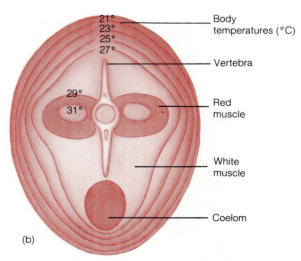

(b)

Figure 38.3 Thermogregulation in large, active fishes. (a) The rete mirabile of a bottlenose dolphin showing the thick-walled artery surrounded by thin-walled veins, creating a countercurrent exchange system that helps reduce the loss of metabolic heat. (b) Cross-section through the body of a bluefin tuna. The body temperature is highest around the red swimming muscles.

[4] ture layer acts as a natural evaporative cooling system. This problem of heat loss through evaporation limits the habitats and activities of amphibians to warm, moist areas. Some amphibians, such as bullfrogs, can vary the amount of mucus they secrete from their body surface—a physiological response that helps regulate evaporative cooling.

Reptiles have dry rather than moist skin, which reduces loss of body heat through the skin through evaporative cooling. They also have an expandable rib cage, which allows for more powerful and efficient ventilation. Reptiles are almost completely ectothermic. They have a low metabolic rate and warm themselves by behavioral adaptations (colorplate 21c). In addition, some of the more sophisticated regulatory mechanisms found in mammals are first found in reptiles. For example, in diving reptiles (e.g., sea turtles, sea snakes) body heat is conserved by routing blood through circulatory shunts into the center of the body. These animals can also increase heat production in response to the hormones thyroxine and epinephrine.

Temperature Regulation in Birds and Mammals

Birds and mammals are the most active and behaviorally complex vertebrates. They can live in habitats all over the earth because they are homeothermic endotherms; they can maintain constant body temperatures of 35 to 42°C with metabolic heat.

Various cooling mechanisms prevent excessive warming in birds. Because they have no sweat glands, birds pant to lose heat through evaporative cooling. Some species have a highly vascularized pouch (*gular pouch*) in their throat that they can flutter (a process called **gular flutter**) to increase evaporation from the respiratory system.

Some birds have evolved mechanisms for preventing heat loss. Feathers are excellent insulators for the body, especially downy-type feathers that trap a layer of air next to the body to reduce heat loss from the skin (figure 38.4a). [5] (This mechanism explains why goose down is such an excellent insulator and is used in outdoor vests and coats where protection from extreme cold is needed.) Aquatic species face the problem of heat loss from their legs and feet. To help solve this problem, they have peripheral countercurrent heat exchange vessels called a rete mirabile (figure 38.4b) in their legs to reduce heat loss. (Mammals that live in cold regions, such as the arctic fox and barren-ground caribou, also have these exchange vessels in their extremities [e.g., legs, tails, ears, nose]). Animals in hot climates, such as jackrabbits, have mechanisms to rid the body of excess heat (colorplate 21b).

Marine animals, such as seals and whales, maintain a body temperature around 36 to 38°C by having a thick layer of insulating fat called **blubber** just under the skin. [6] In the tail and flippers, where there is no blubber, a countercurrent heat exchanger occurs in the blood vessels.

Birds and mammals also use behavioral mechanisms to cope with external temperature changes. As do ectotherms, they can sun themselves or seek shade as the temperature fluctuates. Many animals huddle together to keep warm; others share burrows for protection from temperature extremes. Migration to warm climates and hibernation enable many different birds and mammals to survive through the harsh winter months.

Heat Production in Birds and Mammals

In endotherms, heat generation can warm the body as it dissipates throughout tissues and organs. Birds and mam-

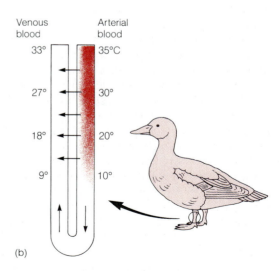

Figure 38.4 Insulation. (a) The thick layer of down feathers keeps these Chinstrap penguins warm. (b)The countercurrent heat exchanger in a bird foot. Some aquatic birds, such as this duck, possess countercurrent systems of arteries and veins (rete mirabile) in their legs that reduce heat loss. The arteries carry warm blood down the legs to warm the cooler blood in the veins, so that the heat is carried back to the body rather than lost through the feet that are in contact with a cold surface.

Figure 38.4b redrawn, with permission, from Neil A. Campbell, *Biology*. Copyright © 1987, Benjamin-Cummings Publishing Company, Menlo Park, CA

mals can generate heat (**thermogenesis**) by muscle contraction, ATPase pump enzymes, oxidation of fatty acids in brown fat, and metabolic processes.

Every time a muscle cell contracts, heat is generated by the actin and myosin filaments sliding over each other and the hydrolysis of ATP molecules. Both voluntary muscular work (e.g., running, flying, jumping) and involuntary muscular work (e.g., shivering) generate heat. The generation of heat by shivering is called **shivering thermogenesis.**

Birds and mammals have a unique capacity to generate heat by using specific enzymes of ancient evolutionary origin—the ATPase pump enzymes found in the plasma membranes of most cells. When the body cools, the hormone thyroxine is released from the thyroid gland. Thyroxine increases the permeability of many cells to Na$^+$ ions, which leak into the cells. These ions are then quickly pumped out by the ATPase pump. In the process, ATP is hydrolyzed, releasing heat energy. The hormonal triggering of heat production is called **nonshivering thermogenesis.**

Brown fat is a specialized type of fat found in newborn mammals, in mammals that live in cold climates, and in mammals that hibernate (figure 38.5). The brown color of this fat comes from the large number of mitochondria with their iron-containing cytochromes. Deposits of brown fat are found beneath the ribs and in the shoulders. A large amount of heat is produced when the brown fat cells oxidize fatty acids because little ATP is made. Blood flowing past brown fat is heated and contributes to warming the body.

The basal metabolic rate of birds and mammals is high and produces heat as an inadvertent but useful by-product (*see figure 4.14*).

In birds, amphibians, reptiles, and mammals, thermoregulation is controlled by specialized cells in the hypothalamus of the brain. There are two hypothalamic thermoregulatory areas: the heating center and the cooling center. The *heating center* controls vasoconstriction of superficial blood vessels, erection of hair and fur, and shivering or nonshivering thermogenesis. The *cooling center* controls vasodilation of blood vessels, sweating, or panting. Overall, body temperature is controlled by feedback mechanisms (with the hypothalamus acting as a thermostat) that trigger either the heating or cooling of the body (figure 38.6). Specialized nerve cells that sense temperature changes are located in the skin and other parts of the body. The warm cell receptors excite the cooling center and inhibit the heating center; the cold receptors have the opposite effects.

During the winter, various endotherms (e.g., skunks, woodchucks, chipmunks, ground squirrels) go into **hibernation** (L. *hiberna,* winter). When hibernating, the metabolic rate slows as do the heart and breathing rates. Mammals prepare for hibernation by building up fat reserves and growing long winter pelts. All hibernating animals have brown fat. Both increased fat deposition and fur growth are stimulated by decreasing day length. Another physiological state characterized by slow metabolism and inactivity is **aestivation,** which allows certain mammals to survive long periods of elevated temperature and diminished water supplies.

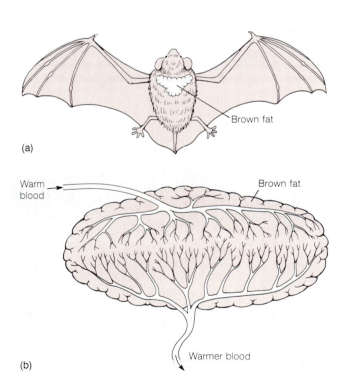

Figure 38.5 Brown fat. (a) Many mammals, such as this bat, have adipose tissue called brown fat located between the shoulder blades. (b) The area of brown fat is much warmer than the rest of the body. As blood flows through the brown fat, it is warmed.

Stop and Ask Yourself

5. What is the function of a rete mirabile?
6. How do amphibians maintain their body temperature?
7. How do birds and mammals generate heat?
8. What is the difference between hibernation and aestivation?

The Control of Water and Solutes

Animals also exchange water and solutes (e.g., salts and various ions) with their environment. Water and ion balance is a concern of all animals in all habitats.

Some simple marine invertebrates, such as protozoa and sponges, do not have specialized excretory structures because cell wastes simply diffuse into the surrounding isosmotic water. Many freshwater species, however, have contractile vacuoles that pump out excess water. Contractile vacuoles are energy-requiring devices that expel excess

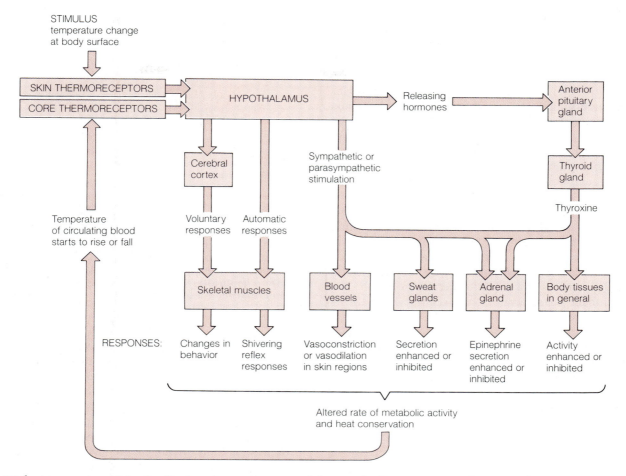

Figure 38.6 An overview of the feedback pathways that control the core body temperature of a mammal. The bold arrows are the major control pathways.

water from individual cells exposed to hypotonic environments (figure 38.7).

Cnidarians, echinoderms, and marine sponges lack special osmoregulatory and excretory organelles. Cells on the body surface regulate water and ions and eliminate wastes. In some freshwater species, cells on the body surface actively pump ions into the animal.

Many invertebrate excretory systems consist of tubules or collections of tubules known as *protonephridia*. Among the simplest of the protonephridia are *flame-cell systems*, such as those found in rotifers, some annelids, larval molluscs, and some flatworms (*see figure 18.5*). The protonephridial excretory system is composed of a network of excretory canals that open to the outside of the body through excretory pores. Bulblike *flame cells* are located along the excretory canals. Fluid filters into the flame cells from the surrounding interstitial fluid and is propelled by the beating cilia through the excretory canals and out of the body through the excretory pores. Flame-cell systems function primarily in eliminating excess water. Nitrogenous waste simply diffuses across the body surface into the surrounding water.

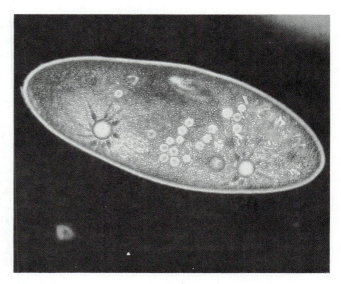

Figure 38.7 Contractile vacuoles in *Paramecium*. A photomicrograph showing the location of two contractile vacuoles in a stained *Paramecium*.

Most annelids (such as the common earthworm) and a variety of other invertebrates have a *metanephridial* excretory system. Recall that the earthworm's body is divided into segments, and that each segment has a pair of tubular excretory organs called *metanephridia* or simply *nephridia* (s., *nephridium*). Each nephridium (figure 38.8) begins with a ciliated funnel, the nephrostome, that opens from the body cavity of a segment into a coiled tubule. As the fluid is moved through the tubule by beating cilia, ions are reabsorbed and carried away by a network of capillaries surrounding the tubule. Each tubule leads to an enlarged bladder that empties to the outside of the body through an opening called the *nephridiopore*.

Some crustaceans have *antennal* or *green glands*; others have *maxillary glands* (*see figure 22.18*). Fluid filters into an antennal or maxillary gland from the hemocoel. Hemolymph pressure from the heart is the main driving force for filtration. Marine crustaceans have a short tubule and produce urine that is isosmotic to their hemolymph. The tubule is longer in freshwater crustaceans, which allows more surface area for transport of ions.

Insects have an excretory system made up of *Malpighian tubules* attached to the gut (*see figure 23.9*). The Malpighian tubules and the gut together serve as the excretory system. Excretion involves the active transport of potassium ions from the blood surrounding the tubules into the tubules and the osmotic movement of water follows. Nitrogenous waste (uric acid) also enters the tubules. As the fluid moves through the Malpighian tubules and into the gut, some of the water and certain ions are recovered. All of the uric acid passes out of the body.

The problems faced by invertebrates in controlling their water and ion balance are also faced by vertebrates. Generally, water losses are balanced precisely by water gains (table 38.1). Vertebrates gain water by absorption from liquids and solid foods in the small and large intestines, and metabolic reactions that yield water as an end product. Water is lost by evaporation from respiratory surfaces, evaporation from the integument, sweating, elimination in feces, and excretion by the urinary system.

Solute losses also must be balanced by solute gains. Solutes are taken in by absorption of minerals from the small and large intestines, through the integument or gills, from secretions of various glands or gills, and metabolism (e.g.,

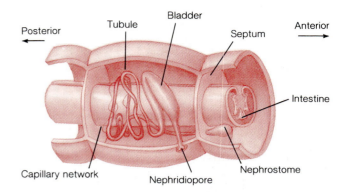

Figure 38.8 The earthworm nephridium. The nephridium opens by a ciliated nephrostome into the cavity of one segment, and the next segment contains the nephridiopore. The main tubular portion of the nephridium is coiled and is surrounded by a capillary network. Waste can be stored in a bladder before being expelled to the outside. Most segments contain two nephridia. Note the close relationship between blood vessels and the tubules.

the waste products of degradative reactions). Solutes are lost in sweat, feces, urine, gill secretions, and as metabolic wastes. The major metabolic wastes that must be eliminated are ammonia, urea, or uric acid.

Vertebrates live in salt water, fresh water, and on land; each of these environments presents different water and solute problems that have been solved in different ways. The next section of this chapter will present the various ways vertebrates employ to keep from losing or gaining too much water and in turn, maintain a homeostatic solute concentration in their body fluids. The disposal (excretion) of certain metabolic waste products is also coupled with osmotic balance and will be discussed by emphasizing the urinary system (box 38.1).

Osmoregulation

Vertebrates can be classified as osmoconformers or osmoregulators according to their osmoregulatory mechanisms. **Osmoconformers** maintain the osmotic concentration of their body fluids at about the same level as that of their environment. When the osmotic concentration of the en-

Table 38.1 Water Gain and Loss in a Human and Kangaroo Rat				
Vertebrate	**Water Gain (ml)**		**Water Loss (ml)**	
Human (daily)	Ingested in solid food	1200	Feces	100
	Ingested as liquids	1000	Urine	1500
	Metabolically produced	350	Skin and lungs	950
	Total	2550		2550
Kangaroo rat (over 4 weeks)	Ingested in solid food	6	Feces	3
	Ingested in liquids	0	Urine	13
	Metabolically produced	54	Skin and lungs	44
	Total	60		60

Box 38.1

Waste Products of Nitrogen Metabolism

Proteins and nucleic acids are the two main sources of nitrogenous wastes. Proteins are the source of over 90% of the total amount of excreted nitrogen; nucleic acids make up the remaining 5%. The major problem is to get rid of the ammonia (NH_3) that forms when amino groups ($-NH_2$) are split from amino acids derived from the metabolized protein or from excess dietary protein. Because ammonia is very toxic to most vertebrates, it is kept in low concentrations (0.0001 to 0.003 mg/100 ml) in the blood; higher concentrations can be lethal. For example, a mouse will die if the ammonia level in its blood reaches 5 mg/100 ml.

Most vertebrates convert ammonia to a less toxic form immediately after it is formed, thus allowing it to be retained until it can be excreted. Exceptions are teleost fishes, which excrete almost all of their nitrogen as ammonia through the gills and in the urine. The excretion of most of the nitrogen as ammonia is called **ammonotelic excretion.**

As noted in the text, most mammals need to drink water or eat food containing it to maintain a water balance. Mammals produce only a moderate amount of urine, not the quantities necessary to dilute the toxic ammonia. Most of the ammonia formed from amino acid metabolism is therefore converted to urea, which is less toxic than ammonia. The excretion of urea as the primary nitrogenous waste is called **ureotelic excretion.**

In a third category are animals that, because they inhabit dry habitats or for other reasons of water conservation, excrete a minimal amount of urine that contains little or no water. These animals (e.g., reptiles, some frogs, and birds) convert ammonia to uric acid or an-other highly insoluble substance. Because the substances are insoluble, only small amounts are retained in solution thus greatly limiting their toxicity. (For any substance to be toxic and achieve a biological effect, it must be in solution.) The excretion of nitrogen in the form of uric acid is called **uricotelic excretion.** Most uricotelic animals excrete their nitrogenous waste as a solid or semisolid urine, or uric acid crystals (colorplate 20a).

Ammonia, urea, and uric acid are the most common nitrogenous waste products, but not the only ones. Some sharks secrete trimethylamine oxide (TMO). Creatine and creatinine are also excreted in small quantities by a variety of animals. Some animals even wastefully excrete some of their excess amino acids.

Because urea, uric acid, and TMO are less toxic than ammonia, why don't more animals excrete most of their nitrogen in these forms? The answer is in the form of biological economics. The synthesis of these compounds from urea requires an expenditure of ATP energy. An animal using energy for an unnecessary function might be selected against during evolution. Therefore, an animal usually excretes its nitrogen in the form requiring the least expendature of energy, given the environment in which it lives.

A few animals excrete nitrogen that comes from the metabolism of purines (e.g., adenine and guanine). Purines can be broken down to ammonia only if the animal has the enzymes. Most animals excrete purine nitrogen as uric acid or as one or more of the intermediate products in this chain.

---

vironment changes, so does that of their body fluids. Obviously, this type of osmoregulation is not very efficient and has limited the distribution of those animals using it.

More efficient osmoregulatory mechanisms have evolved in other vertebrates so that they can control the osmotic concentration of their body fluids. These animals are called **osmoregulators** because they maintain an internal solute concentration that does not vary, regardless of the environment in which they live.

Those vertebrates that live in seawater have body fluids with an osmotic concentration that is about one-third less than that of the surrounding seawater, and water tends to continually leave their bodies. To compensate for this problem, they have evolved mechanisms to conserve water and prevent dehydration. Freshwater vertebrates have body fluids that are *hyperosmotic* with respect to their environment, and water tends to continually enter their bodies. To compensate for this problem, they have evolved mechanisms to excrete water and prevent fluid accumulation.

Land vertebrates have a higher concentration of water in their body fluids than in the surrounding air. They tend to lose water to the air through evaporation, and may use considerable amounts of water to dispose of wastes. Amphibians experience this problem when they are out of water, as do terrestrial reptiles, birds, and mammals, all of which must conserve water to prevent dehydration.

How Osmoregulation is Achieved

Vertebrates have evolved a variety of mechanisms to cope with the above osmoregulation problems, and most of them are adaptations of the urinary system. As presented in chapter 36, vertebrates have a closed circulatory system containing blood that is under pressure. This pressure forces blood through a membrane filter in a process called filtration. Proteins and large molecules are retained, and small molecules and ions pass through the filter. Some of these small molecules and ions, however, are important and must not leave the body in the filtrate. Thus, vertebrates have evolved mechanisms of selectively reabsorbing the needed molecules while at the same time secreting and/or allowing the unwanted ones to be excreted from the body in the filtered waste fluid. Selective absorption has allowed vertebrates to adapt to environments that have varying requirements for the reabsorption of particular molecules and ions. The primary organ involved in the filtration, reabsorption, and secretion of various materials is the kidney.

The Vertebrate Kidney

Vertebrates have two kidneys that are located in the back of the abdominal cavity, on either side of the aorta (figure 38.9). Each kidney has a coat of connective tissue called the *renal capsule* (L. *renes,* kidney). The inner portion of the kidney is called the *medulla*; the region between the capsule and medulla is the *cortex*.

Most vertebrate kidneys carry out the following functions:

1. Filtration, in which blood is passed through a filter that retains blood cells, proteins, and other large solutes but lets small molecules, urea, and ions pass through.
2. Reabsorption, in which selective ions and molecules are taken back into the bloodstream from the filtrate.
3. Secretion, whereby select ions and end products of metabolism (e.g., K^+, H^+, HN_4^+) that are in the blood are

added to the filtrate for removal from the body.
4. Excretion, in which the urine is voided from the body.

The filtration device of the mammalian kidney consists of over one million individual filtration secretion absorption structures called **nephrons**. At the beginning of the nephron is the filtration apparatus called the *glomerular capsule* (formerly Bowman's capsule), which looks rather like a tennis ball that has been punched in on one side (figure 38.10a). The capsules are located in the cortical

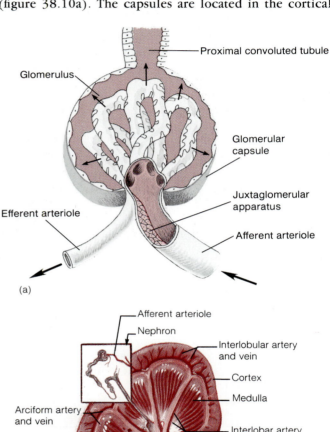

Glomerulus

Proximal convoluted tubule

Glomerular capsule

Juxtaglomerular apparatus

Afferent arteriole

Efferent arteriole

(a)

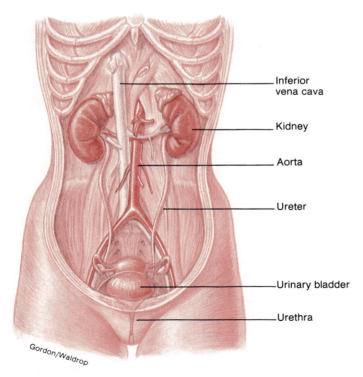

Inferior vena cava

Kidney

Aorta

Ureter

Urinary bladder

Urethra

Gordon/Waldrop

Figure 38.9 Component parts of the human urinary system. The position of the kidneys, ureters, urinary bladder, and urethra are shown.

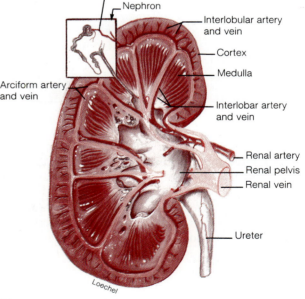

Afferent arteriole

Nephron

Interlobular artery and vein

Cortex

Medulla

Arciform artery and vein

Interlobar artery and vein

Renal artery

Renal pelvis

Renal vein

Ureter

Loechel

(b)

Figure 38.10 The filtration device of the mammalian kidney. (a) The glomerular capsule with arrows illustrating that high blood pressure forces water and ions through small perforations in the walls of the glomerulur capillaries to form the glomerular filtrate. (b) Interior of a kidney showing the positioning of the nephron and the blood supply to and from the kidney.

(outermost) region of the kidney (figure 38.10b). In each capsule, an afferent ("going to") arteriole enters and branches into a fine network of capillaries called the **glomerulus.** The walls of these glomerular capillaries contain small perforations that act as filters. Blood pressure forces fluid through these filters. The fluid is now known as *glomerular filtrate* and contains small molecules, such as glucose, ions (Ca^{2+}, PO_4^{3-}), and the primary nitrogenous waste product of metabolism, urea. Because the perforations are so small, large proteins and blood cells are retained in the blood and leave the glomerulus via the efferent ("outgoing") arteriole. The efferent arteriole then divides into a set of capillaries called the *peritubular capillaries* (figure 38.11) that wind profusely around the tubular portions of the nephron. Eventually they merge to form veins that carry blood out of the kidney.

Beyond the glomerular capsule are the proximal convoluted tubule, the intermediate tubule, and the distal convoluted tubule. At various places along these structures, selective reabsorption of the glomerular filtrate occurs, returning certain ions (e.g., Na^+, K^+, Cl^-) to the bloodstream. Both active (ATP-requiring) and passive processes are involved in the recovery of these substances. Potentially harmful compounds, such as hydrogen (H^+) and ammonium ($NH4^+$) ions, drugs, and various other foreign materials are secreted into the nephron. In the final portion of the nephron, called the collecting duct, water is reabsorbed so that the urine contains a salt concentration well above that of the blood. Thus the filtration, secretion, and reabsorption activities of the nephron do not simply remove wastes. They also maintain water and salt balance, and therein lies the importance of the homeostatic function of the kidney.

How the Mammalian Kidney Works

Mammalian, and to a lesser extent avian, kidneys can remove far more water from the glomerular filtrate than can the kidneys of reptiles and amphibians. Human urine is four times as concentrated as blood plasma, a camel's urine is eight times as concentrated, a gerbil's is 14 times as concentrated as the plasma, and some desert rats and mice have urine more than 20 times as concentrated as their plasma. This concentrated waste enables them to live in dry or desert environments, where there is little water available for them to drink. Most of their water is metabolically produced from the oxidation of carbohydrates, fats, and proteins in the seeds that they eat (*see table 38.1*). Mammals and birds achieve this remarkable degree of water conservation by a unique, yet simple, evolutionary advancement—the bending of the nephron tube in the form

[7]

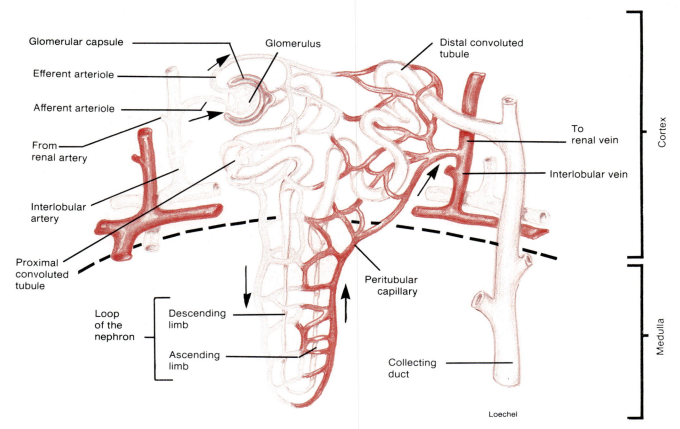

Glomerular capsule

Glomerulus

Distal convoluted tubule

Efferent arteriole

Afferent arteriole

From renal artery

To renal vein

Interlobular vein

Interlobular artery

Proximal convoluted tubule

Peritubular capillary

Cortex

Medulla

Loop of the nephron

Descending limb

Ascending limb

Collecting duct

Loechel

Figure 38.11 A vertebrate nephron. Glucose and some ions are reabsorbed in the proximal tubule; other ions and water are reabsorbed in the distal tubule. Final water resorption takes place in the collecting duct. Arrows indicate the direction of movement of materials in the nephron.

of a loop (figure 38.11). By bending, the nephron can greatly increase the salt concentration in the tissue through which the loop passes and use this gradient to draw large amounts of water out of the tube.

Countercurrent Exchange The kidney uses the *loop of the nephron* (formerly the loop of Henle) to set up a countercurrent flow similar to that in the gills of fishes or in the legs of birds, but with water being reabsorbed instead of oxygen or heat. Generally, the longer the loop of the nephron, the more water that can be reabsorbed. It follows that desert rodents (e.g., kangaroo rat) that form highly concentrated urine, have very long nephron loops (figure 38.12). Similarly, amphibians that are closely associated with aquatic habitats have nephrons that lack a loop.

The countercurrent flow mechanism used for concentrating urine is illustrated in figure 38.13. The process of reabsorption in the proximal convoluted tubule removes some salt (NaCl) and water from the glomerular filtrate and reduces its volume by approximately 25%. However, the concentrations of salt and urea are still isosmotic with the extracellular fluid.

As the filtrate moves to the descending limb of the loop of the nephron, it becomes further reduced in volume and more concentrated. Water moves out of the tubule by osmosis due to the high salt concentration (the "brine-bath") in the extracellular fluid. The walls of this portion of the tubule are impermeable to both salt and urea, therefore they cannot enter the tubule.

Notice in figure 38.13 that the highest urea-brine bath concentration is around the lower portion of the loop of the nephron. As the filtrate passes into the ascending limb, sodium (Na^+) ions are actively transported out of the filtrate into the extracellular fluid with chloride (Cl^-) ions follow-ing passively. Water cannot flow out of the ascending limb because the cells of the ascending limb are impermeable to water. Thus, the salt concentration of the extracellular fluid becomes very high. The salt flows passively into the descending loop, only to move out again in the ascending loop, creating a recycling of salt through the loop and the extracellular fluid. Because the flows in the descending and ascending limbs are in opposite directions, a countercurrent gradient in salt is set up. The osmotic pressure of the extracellular brine bath is made even higher because of the abundance of urea that moves out of the collecting ducts.

Finally, the distal convoluted tubule empties into the collecting duct, which is permeable to urea, and the concentrated urea in the filtrate diffuses out into the surrounding extracellular fluid. The high urea concentration in the extracellular fluid, coupled with the high concentration of salt, forms the urea-brine bath that causes water to move out of the filtrate by osmosis as it moves down the descending limb. The water is finally collected by the many peritubular capillaries surrounding each nephron and returned to the systemic circulation (*see figure 38.11*).

Control of Water Reabsorption The amount of water reabsorbed in the kidneys is determined by the permeability of the collecting ducts to water. When the collecting ducts are more permeable to water, more water is recovered and less eliminated; the urine is more concentrated. When the collecting ducts are less permeable to water, less water is recovered and more is eliminated. Urine becomes more dilute, and its volume increases.

A horomone called *antidiuretic hormone* (ADH) is an important regulator of water recovery. ADH opposes diuresis, which is the increased discharge of urine. Specifically, ADH increases the water permeability of the collecting duct walls. When the circulating level of ADH is high, more water is recovered, and the urine volume is decreased. Conversely, a decrease in ADH concentration decreases the permeability of the collecting ducts to water. The urine becomes more dilute, and its volume increases. ADH is produced by the hypothalamus and released into the blood from the posterior lobe of the pituitary gland. The rate of production and release of ADH is very responsive to changes in body water balance, which are detected by receptors that respond to blood pressure changes and osmotic changes in the blood. When the body loses too much water, more ADH is produced and released. When excess water accumulates in the body, less ADH is produced, and a larger volume of more dilute urine is excreted (figure 38.14a).

Control of Sodium Reabsorption Because sodium ions (Na^+) are by far the most numerous positively charged ions in the extracellular fluids, and because water moves osmotically in response to changes in osmotic gradient, control of the body's sodium content is closely related to the regulation of its water balance.

The sodium content of body fluids is regulated in several ways. Sodium reabsorption from the fluid passing through the kidney tubules is controlled by *aldosterone,* a hormone

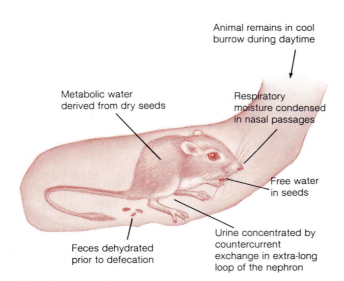

Animal remains in cool burrow during daytime

Metabolic water derived from dry seeds

Respiratory moisture condensed in nasal passages

Free water in seeds

Feces dehydrated prior to defecation

Urine concentrated by countercurrent exchange in extra-long loop of the nephron

Figure 38.12 The kangaroo rat is a master of water conservation. It has such efficient kidneys that it can concentrate urine 20 times that of its blood plasma. As a result, these kidneys as well as other adaptations prevent the unnecessary loss of water to the environment.

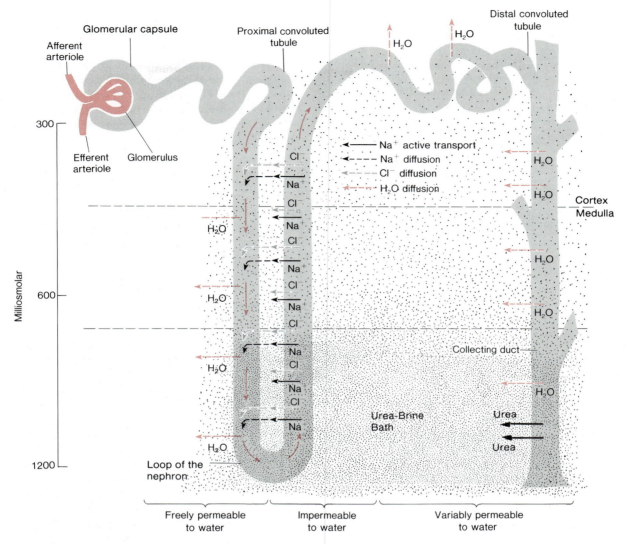

Figure 38.13 The movement of materials in the nephron and collecting duct. Active transport is shown with solid arrows; passive transport with dashed arrows. The shading at intervals along the tubules illustrates the relative concentration of the filtrate in milliosmoles.

released from the adrenal glands. Aldosterone promotes sodium reabsorption, and the rate of aldosterone secretion increases or decreases in response to the level of sodium ions in the blood.

Another sodium regulatory mechanism depends on responses of the kidney itself to blood pressure changes. A small blood pressure sensor called the *juxtaglomerular apparatus* is located along each afferent arteriole leading to the glomerular capsule. It responds to decreased blood pressure or decreased sodium concentration by releasing an enzyme called *renin* into the blood (figure 38.14b). Renin causes a plasma protein called *angiotensin I* to be converted to its active form, *angiotensin II*. Angiotensin II causes blood vessel constriction, which raises blood pressure. Angiotensin II also stimulates increased aldosterone secretion, promoting sodium and water reabsorption and increasing body fluid volume. Increased fluid volume in-

creases blood pressure. Thus, the kidney itself is involved in maintaining adequate blood pressure for glomerular filtration.

The renal pelvis of the kidney is continuous with a tube called the *ureter* that carries urine to a storage organ called the *urinary bladder* (*see figure 38.9*). Urine from two ureters (one from each kidney) accumulates in this organ. It leaves the body through a single tube, the *urethra,* which opens at the body surface at the end of the penis (in human males) or just in front of the vaginal entrance (in human females). As the urinary bladder fills with urine, tension increases in its smooth muscle walls. In response to this tension, a reflex response relaxes sphincter muscles at the entrance to the urethra. This involuntary reflex response is called *urination.* The two kidenys, two ureters, urinary bladder, and urethra constitute the *urinary system* of mammals.

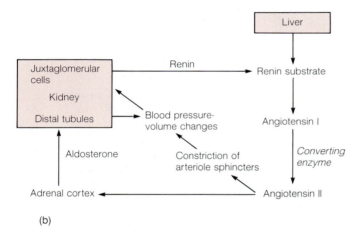

(a)

(b)

Figure 38.14 The regulation of body water and sodium. (a) The negative feedback effects of ADH. (b) A diagram of the role of the renin-angiotensin mechanism in blood pressure and blood sodium (Na⁺) regulation.

Kidney Variations

Other vertebrate kidneys have the same basic design presented above, although there have been some adaptations based on habitat characteristics. This chapter concludes with a presentation of how a few vertebrates maintain their water and solute levels in different habitats—in the seas, in fresh water, and on land (table 38.2).

Teleost Fishes   Kidneys evolved first in freshwater fishes. Because the body fluids of these fishes are hyperosmotic relative to fresh water, water tends to enter the fishes, causing excessive *hydration* or bloating. At the same time, body salts tend to move outward into the water. To solve this problem, freshwater fishes usually do not drink water; [8] their bodies are coated with mucus, which helps stem the inward movement of water; they transport salts by active transport across their gills; and they excrete a large volume of water as dilute urine.

Although most groups of animals probably evolved in the sea, marine bony fishes probably evolved from freshwater ancestors, as presented in chapter 27. In making the transition to the sea, they faced a new problem of water balance—their body fluids were hypoosmotic with respect to the seawater, and water tended to leave their bodies resulting in dehydration. To compensate, marine fishes [9] need to drink large quantities of water, and they excrete Na⁺, Cl⁻, and K⁺ ions through secretory cells in their gills. There are channels in plasma membranes of the kidneys

Table 38.2	How Various Vertebrates Maintain Water and Salt Balance			
Organism	**Environmental Concentration Relative to Body Fluids**	**Urine Concentration Relative to Blood**	**Major Nitrogenous Waste**	**Key Adaptation**
Freshwater fishes	Hypoosmotic	Hypoosmotic	Ammonia	Absorb salts through gills
Saltwater fishes	Hyperosmotic	Isosmotic	Ammonia	Secrete salts through gills
Sharks	Isosmotic	Isosmotic	Ammonia	Secrete salts through rectal gland
Amphibians	Hypoosmotic	Very hypoosmotic	Ammonia and urea	Absorb salts and urea through skin
Marine reptiles	Hyperosmotic	Isosmotic	Ammonia and urea	Secrete and urea salts through salt gland
Marine mammals	Hyperosmotic	Very hyperosmotic	Urea	Drink some water
Desert mammals	No comparison	Very hyperosmotic	Urea	Produce metabolic water
Marine birds	No comparison	Weakly hyperosmotic	Uric acid	Drink sea-water and uses salt glands
Terrestrial birds	No comparison	Weakly hyperosmotic	Uric acid	Drink fresh water

that actively transport the multivalent ions that are abundant in sea water (e.g., Ca^{2+}, Mg^{2+}, SO_4^{2-}, and PO_4^{3-}) out of the extracellular fluid and into the nephron tubes. The ions are then excreted in a concentrated urine. An organ that initially evolved to excrete water and reabsorb salts has adapted to the opposite demands.

Some fishes encounter both fresh and salt water during their life. Newborn Atlantic salmon swim downstream from the freshwater stream of their birth and enter the sea. Instead of continuing to pump salt in, as they have done in fresh water, the salmon must now rid their bodies of salt. To do so, they synthesize new ATPase enzyme-pump molecules. Years later, these same salmon migrate from the sea to their freshwater home to spawn. As they do, the enzyme-pumping mechanisms reverse themselves.

Sharks Sharks and their relatives (skates and rays) that live in the seas have solved their osmotic problem in ways different from the bony fishes. Instead of actively pumping ions out of their bodies through the kidneys, they have a *rectal gland* that secretes a highly concentrated salt solution. To reduce water loss, they use two organic molecules, urea and trimethylamine oxide (TMO), in their body fluids to raise the osmotic pressure to a level equal to or higher than that of the seawater (colorplate 20c).

Urea denatures proteins and inhibits enzymes, whereas TMO stabilizes proteins and activates enzymes. Together in the proper ratio, they counteract each other, raise the osmotic pressure, and do not interfere with enzymes or proteins. This reciprocity is termed the **counteracting osmolyte strategy.** A number of other fishes and invertebrates have evolved the same mechanism and employ pairs of counteracting osmolytes to raise the osmotic pressure of their body fluids.

Amphibians and Reptiles The amphibian kidney is identical to that of freshwater fishes, which is not surprising because amphibians spend a large portion of their time in fresh water, and when on land, they tend to seek out moist places. Amphibians take up water and salts in their food and drink, through the skin that is in contact with moist substrates, and through the urinary bladder (figure 38.15; colorplate 20d). This uptake counteracts what is lost through evaporation and prevents osmotic imbalance.

The urinary bladder of a frog, toad, or salamander is an important water and salt reservoir. For example, when the environment becomes dry, the bladder becomes larger for storing more urine. If an amphibian becomes dehydrated, a brain hormone causes water to leave the bladder and enter the body fluid.

Aquatic reptiles (crocodiles and alligators) occupy a habitat similar to that of freshwater fishes and amphibians;

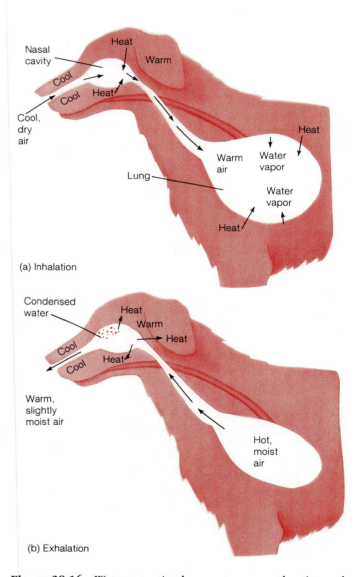

(a) Inhalation

(b) Exhalation

Figure 38.16 Water retention by countercurrent heating and cooling in a mammal. (a) When this animal inhales, the cool dry air passing through its nose is heated and humidified. At the same time, its nasal tissues are cooled. (b) When the animal exhales, it gives up heat to the previously cooled nasal tissue. The air carries less water vapor, and condensation occurs in the animal's nose.

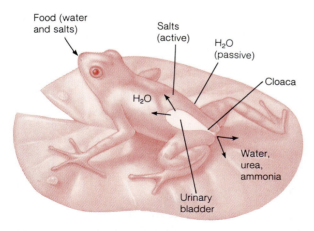

Figure 38.15 Water and salt uptake in an amphibian. Water and salts can enter the body in food, be absorbed through the skin, or from the urinary bladder.

therefore, they have similar kidneys. Marine reptiles (some crocodiles, sea turtles, sea snakes, and lizards) possess kidneys similar to those of their freshwater relatives. They eliminate excess salts through salt glands, which are often located near the eye or nose.

Terrestrial reptiles must conserve water to survive. They reabsorb much of the water in the glomerular filtrate before it leaves the kidney. Their urine is very concentrated and they secrete nitrogenous waste in the form of urea or uric acid, both of which require less water to excrete than does ammonia.

Birds and Mammals In most birds and mammals, far more water can be removed from the glomerular filtrate than in amphibians or reptiles, and the kidneys are the primary regulatory organs for controlling the osmotic balance of the body fluids. Some desert and marine birds often build up high salt concentrations in their bodies because they consume salty foods or seawater, and they lose water through evaporation and in their urine and feces. To rid themselves of this excess salt, these animals also have salt glands near the eye or in the tongue to remove excess salt from the blood and secrete it as tearlike droplets (box 38.2).

A major site of water loss in mammals is the lungs. To reduce this evaporative loss, many mammals have nasal cavities that act as countercurrent exchange systems to combat such loss (figure 38.16). When the animal inhales, air passes through the nasal cavities and is warmed by the surrounding tissues. In the process, the temperature of this tissue drops. When the air gets deep into the lungs, it is further warmed and humidified. During exhalation, as the warm moist air passes up the respiratory tree, it gives up its heat to the nasal cavity. As the air cools, much of the water condenses on the nasal surfaces and does not leave the body. This mechanism explains why a dog's nose is usually cold and moist. [10]

Stop and Ask Yourself

14. How do freshwater fishes osmoregulate? Saltwater fishes?
15. What is the function of the rectal gland of sharks?
16. How do amphibians conserve water?
17. What is a salt gland?
18. How do nasal cavities help conserve water?

Box 38.2

Extrarenal Secretion by Avian and Reptilian Salt Glands

Salt glands have been described in many species of birds and reptiles, including nearly all marine birds, ostriches, marine iguanas, sea snakes, sea turtles, crocodilians, and many terrestrial reptiles. In general, these animals are subjected to osmotic stress of a marine or desert environment.

The salt glands of some reptiles and birds occupy shallow depressions in the skull above the eye (box figure 38.2a,b). These glands consist of many lobes, each of which drains via branching secretory tubules, and a central canal into a collecting duct that empties into the nostril. Active secretion takes place across the epithelial cells of the secretory tubules. These cells have a large surface area and many mitochondria. As in other transport epithelia, adjacent cells are tied together by tight junctions (*see figure 3.7a*), which prevent the massive leakage of water or solutes past the cells, from one side of the epithelium to another.

The avian salt gland is organized as a countercurrent system that aids in concentrating the secreted salt solution. The capillaries are arranged so that the flow of blood is in the direction opposite to the flow of secretory fluid. This flow maintains a minimum concentration gradient from blood to tubular lumen along the entire length of the tubule.

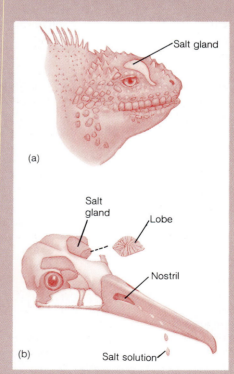

Box figure 38.2 Salt glands. (a) The reptilian salt gland is often located above the eyes. (b) The many-lobed salt glands of birds empty into the nostril.

Summary

1. Thermoregulation is a complex and important physiological process for maintaining heat homeostasis despite environmental changes.

2. Ectotherms generally obtain heat from the environment, whereas endotherms generate their own body heat from metabolic processes.

3. Homeotherms generally have constant core body temperatures, and heterotherms have a variable body temperature.

4. The high, constant body temperature of birds and mammals also depends on insulation, panting, sweating, specific behaviors, vasoconstriction or vasodilation of the peripheral blood vessels, and in some species, a rete mirabile system.

5. Thermogenesis involves mainly shivering, enzymatic activity, brown fat, and high cellular metabolism.

6. The hypothalmus is the temperature regulating center that functions as a thermostat with a fixed set point. This set point can either rise or fall during hibernation or torpor.

7. Some invertebrates have contractile vacuoles, flame-cell systems, nephridia, or Malpighian tubules for osmoregulation.

8. The osmoregulatory system of vertebrates governs water and salt levels; the excretory system eliminates metabolic wastes, water, and salts from the body.

9. In reptiles, birds, and mammals, the kidneys are important osmoregulatory structures. The functional unit of kidney is the nephron, composed of the glomerular capsule, proximal, intermediate, and distal tubules, and collecting duct. Mammalian and avian nephrons also have a loop of the nephron. The loop and the collecting duct are in the kidney's medulla; the other nephron parts lie in the kidney's cortex. Urine passes from the pelvis of the kidney to the bladder.

10. To make urine, kidneys produce a filtrate of the blood and reabsorb most of the water, glucose, and needed salts, while allowing wastes to pass from the body. Four physiological mechanisms are involved: filtration of the blood through the glomerulus, reabsorption of the useful substances, secretion of toxic substances, and concentration of the filtrate. In those animals with a loop of the nephron, salt and urea are concentrated in the extracellular fluid around the loop, allowing water to move by osmosis out of the loop and into the peritubular capillaries.

11. Freshwater animals tend to lose salts and take in water. To avoid hydration, freshwater fishes rarely drink water, have impermeable body surfaces covered with mucus, excrete a dilute urine, and take up salts through their gills.

12. Marine animals tend to take in salts from the seawater and to lose water. To avoid dehydration, they frequently drink water, have relatively permeable body surfaces, excrete a small volume of concentrated urine, and excrete salts from their gills.

13. Amphibians can absorb water across the skin and urinary bladder wall. Desert and marine reptiles and birds have salt glands to remove and secrete excess salt.

Key Terms

aestivation (p. 620)
ammonotelic excretion (p. 623)
blubber (p. 619)
brown fat (p. 620)
conduction (p. 616)
convection (p. 616)
counteracting osmolyte (p. 629)
daily torpor (p. 617)
ectotherms (p. 617)
endotherms (p. 617)
evaporation (p. 617)
glomerulus (p. 625)
gular flutter (p. 619)
heterotherms (p. 617)

hibernation (p. 620)
homeotherms (p. 617)
nephrons (p. 624)
nonshivering thermogenesis (p. 620)
osmoconformers (p. 622)
osmoregulators (p. 623)
radiation (p. 617)
rete mirabile (p. 618)
shivering thermogenesis (p. 620)
thermogenesis (p. 620)
thermoregulation (p. 616)
ureotelic excretion (p. 623)
uricotelic excretion (p. 623)

Critical Thinking Questions

1. Reptiles are said to be behavioral homeotherms. Explain what this means.

2. Why do very small birds and mammals go into a state of torpor at night?

3. How does the countercurrent mechanism help regulate heat loss?

4. In endotherms, what controls the balance between the amount of heat lost and the amount gained?

5. Why must marine vertebrates expend large amounts of ATP energy to keep their body fluids at a constant osmolarity?

6. With respect to their kidneys, how can whales and seals spend part of their life in both fresh and salt water?

7. If marooned on a desert isle, do not drink sea water; it is better to be thirsty. Why is this true?

Suggested Readings

Books

Hochachka, P., and Somero, G. 1984. *Biochemical Adaptations.* Princeton, N.J.: Princeton University Press.

Rankin, J. C., and Davenport, J. 1981. *Animal Osmoregulation.* New York: John Wiley and Sons.

Smith, H. 1961. *From Fish to Philospher.* New York: Doubleday.

Articles

Brock, T. D. 1985. Life at high temperatures. *Science* 230:132–135.

Carey, F. G. Fishes with warm bodies. *Scientific American* February, 1983.

Flam, F. Antifreezes in fish work quite similarly. *Science News* June 30, 1989.

French, A. R. 1988. The patterns of mammalian hiberbation. *American Scientist* 76(6):568–575.

Heatwole, H. 1978. Adaptations of marine snakes. *American Scientist* 66:594–604.

Irving, J. Adaptations to cold. *Scientific American* January, 1966.

Lee, R. E. 1989. Insect cold-hardiness: to freeze or not to freeze. *BioScience* 39:308–313.

Kolata, G. 1987. Managing the inland sea. *Science* 224:703–710.

Schmidt-Nielsen, K., and Schmidt-Nielsen, B. The desert rat. *Scientific American* July, 1953.

Schmidt-Nielsen, K. Countercurrent systems in animals. *Scientific American* May, 1981.

Schmidt-Nielsen, K. Salt glands. *Scientific American* January, 1959.

Silberner, J. Salt control. *Science News* January 25, 1986.

Smith, H. The kidney. *Scientific American* January, 1953.

Reproduction and Development

39

Concepts

1. All animals have the capacity for reproduction. The simplest form of reproduction is asexual. Asexual reproduction permits production of new individuals from one parent, but it does not produce new genetic combinations among offspring, as does sexual reproduction.
2. Many animals reproduce sexually. Sexual reproduction involves mechanisms that bring sperm and egg together for fertilization, and ensure that the fertilized egg has a suitable place to develop until the new animal is ready to function on its own.
3. Human fertilization and embryonic development are both internal.
4. Hormones coordinate the reproductive functions in both males and females. In females, hormones also maintain pregnancy, and after childbirth, stimulate the production and letdown of milk from the mammary glands.
5. The method of sex determination, the pattern of reproduction, and the structure of the gonads and reproductive passages vary greatly among vertebrates.
6. Sexual reproduction evolved in aquatic environments, and its modification for organisms living on dry land entailed evolutionary innovations to prevent the gametes and embryos from drying out.

Have You Ever Wondered:

[1] why the human testes are the only abdominal organs that lie outside of the abdominal cavity?
[2] what causes a common inguinal hernia?
[3] why some males are circumcised?
[4] what determines the size of a female's breasts?
[5] why urine is used to determine if a female is pregnant?
[6] when a female is considered pregnant?
[7] how a barber's pole got its red and white stripes?

These and other useful questions will be answered in this chapter.

This chapter contains underlined evolutionary concepts.

Reproduction is a basic attribute of all animals. Chapters 6 through 11 describe the general features of animal development and the control processes that allow a genotype to be translated into its phenotype. Although in modern zoology, development is "the center stage" in reproduction, the whole process includes the behavior, anatomy, and physiology of adult males and females, whether birds, bees, or humans. This chapter will focus on human reproduction, not only because of the subject's basic intrinsic interest to everyone, but because scientists know more about the biochemistry, hormones, anatomy, and physiology involved in human reproduction than they do for any other species. The chapter concludes with a comparative focus on different reproductive strategies in the invertebrates and the five major groups of vertebrates.

Reproduction in Vertebrates: Some Basic Strategies

Since the evolution of the first animals—from invertebrate ancestors to the first fishes, and on to birds, mammals, and finally humans, the basic use of male and female gametes has been preserved. Two evolutionary advantages of sexual reproduction emerged: (1) genes may recombine during crossing-over in meiosis (see figure 6.8), and (2) the random meeting of one sperm and one egg generates a new diploid set of genes. The variability derived from this new combination of genes is a crucial means of permitting groups of animals to adapt to changing environments—one of the basic tenets of evolution.

The reptiles and early mammals originated and shaped the anatomy and physiology that later insured successful human reproduction. The reptilian system perfected shelled, desiccation-resistant eggs. These eggs had the three basic embryonic membranes that still characterize the human embryo, as well as a flat embryo that developed and underwent gastrulation atop a huge yolk mass. This same process of gastrulation is still seen in human embryos, even though the massive yolk mass has been lost. It was in the early mammals that the mechanisms for maintaining the developing embryo within the female for long periods of time evolved. During this time, called **gestation** (L. *gestatio,* from + *gestare,* to bear), the embryo was nourished with nutrients and oxygen, yet it was protected from attack by the female's immune system. After birth, the first mammals nourished their young with milk from mammary glands, just as primates do today.

Vertebrate evolution has also given rise to the close link between reproductive biology and sexual behavior. The strong drive to reproduce dominates the lives of many vertebrates, as illustrated by the salmon's fateful spawning run or the rutting of bull elk. Females of most mammal species come into heat or **estrus** (Gr. *oistros,* a vehement desire; the period of sexual receptivity) about the same time each year. Estrus is usually timed so that the young will be born when environmental conditions make their survival most likely.

Individual female apes and monkeys are asynchronous: mating and births can take place over much of the year. Females mate only when in estrus, increasing the probability of fertilization. Human females show a less distinctive estrus phase and can reproduce throughout the year. They can also engage in sexual activity without reproductive purpose; no longer is sexual behavior precariously tied to ovulation. The source of this important reproductive innovation may be physiological or a result of concomitant evolution of the brain—a process that gave humans conscious control over their emotions and behaviors that are controlled by hormones, instincts, and the environment in other animals. This separation of sex from a purely reproductive function has evolved into the long-lasting pair ponds between human males and females (e.g., marriage) that further support the offspring. This type of behavior has also resulted in the transmission of culture—a key to the evolution and success of the human species.

With this background, the anatomy and physiology of the human male and female will be presented.

The Human Male Reproductive System

The reproductive role of the male is to produce sperm and deliver them to the vagina of the female. This function requires the following four structures:

1. The *testes* produce sperm and the male sex hormone, testosterone.
2. *Accessory glands* furnish a fluid for carrying the sperm to the penis. This fluid together with the sperm is called **semen** (L. *seminis,* seed).
3. *Accessory ducts* store and carry secretions from the testes and accessory glands to the penis.
4. The *penis* deposits semen into the vagina during sexual intercourse.

Production and Transport of Sperm

The paired **testes** (s., testis; L. *testis,* witness; the paired testes were believed to bear witness to a man's virility) are the male reproductive organs (*gonads*) that produce sperm (figure 39.1). Shortly after birth, the testes descend from the abdominal cavity into the **scrotum** (L. *scrautum,* a leather pouch for arrows), which hangs between the thighs. Because the testes hang outside the body, the temperature inside the scrotum is about 34°C compared to a 38°C core temperature. The lower temperature is necessary for active sperm production and survival. Muscles elevate or lower the testes depending on the outside air temperature.

[1]

Each testis is enclosed in a fibrous sac called the *tunica albuginea* (figure 39.2a). This sac extends into the testis as *septae* (s. *septum*), which divide it into compartments called *lobules*. Each testis contains over 800 tightly coiled **seminiferous tubules** (figure 39.2a-c), which produce

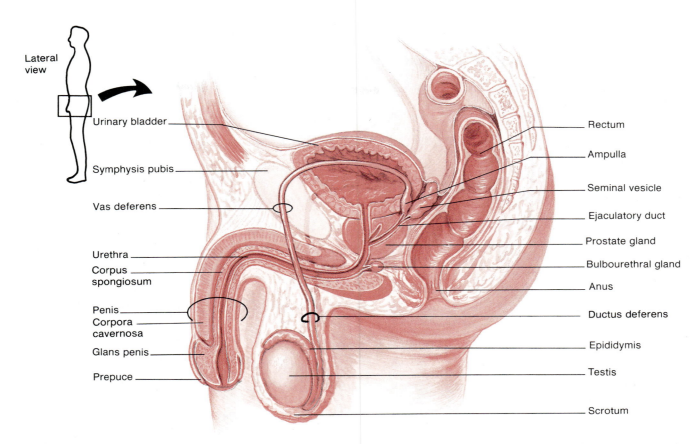

Figure 39.1   Sagittal view of the human male reproductive system. There are two each of the following structures: testis, epididymis, ductus deferens, seminal vesicle, ejaculatory duct, and bulbourethral gland.

thousands of sperm cells each second in healthy young men (*see figure 6.11*). The walls of the seminiferous tubules are lined with two types of cells: *spermatogenic cells,* which give rise to sperm, and *sustentacular cells,* which provide nourishment for the sperm as they are being formed and secrete a fluid (as well as the hormone *inhibin*) into the tubules to provide a liquid medium for the sperm. Between the seminiferous tubules are clusters of endocrine cells, called *interstitial endocrinocytes,* which secrete the male sex hormone *testosterone.*

The sperm produced in a testis are carried to the penis by a system of tubes. The seminiferous tubules merge into a network of tiny tubules called the *rete tesis* (L. *rete,* net), which merge into a coiled tube called the *epididymis.* The epididymis has three main functions: (1) it stores sperm until they are mature and ready to be ejaculated, (2) it contains smooth muscle that helps propel the sperm toward the penis by peristalic contractions, and (3) it serves as a duct system for the passage of sperm from the testis to the ductus deferens. The *ductus deferens* (formerly called the vas deferens or sperm duct) is the dilated continuation of the epididymis. Continuing upward after leaving the scrotum, the ductus deferens passes through the lower part of the abdominal wall via the inguinal canal. [2] If the abdominal wall weakens at the point where the ductus deferens passes through, an inguinal hernia may result. (In an inguinal hernia, the intestine may protrude down-

ward into the scrotum.) The ductus deferens then passes around the urinary bladder (*see figure 39.1*) and becomes enlarged to form the *ampulla.* The ampulla stores some sperm until they are ejaculated. Distal to the ampulla, the ductus deferens becomes the ejaculatory duct. The *urethra* is the final section of the reproductive duct system.

After the ductus deferens passes around the urinary bladder, several accessory glands add their secretions to the sperm as they are propelled through the ducts. These accessory glands are the seminal vesicles, prostate gland, and bulbourethral glands (*see figure 39.1*). The **seminal vesicles** secrete water, fructose, prostaglandins, and vitamin C. This secretion provides an energy source for the motile sperm and helps to neutralize the natural acidity of the vagina. (The pH of the vagina is about 3 to 4, but sperm motility and fertility are enhanced when it increases to about 6.) The **prostate gland** secretes water, acid phosphatase, cholesterol, buffering salts, and phospholipids. The **bulbourethral glands** secrete a clear alkaline fluid that acts as a lubricant in the urethra to facilitate the ejaculation of semen and to lubricate the penis prior to sexual intercourse. The fluid that results from the combination of sperm and glandular secretions is **semen.** The average human ejaculation produces about 3 to 4 ml of semen and contains 300 to 400 million sperm.

The penis has two functions. It carries urine through the urethra to the outside during urination, and it transports

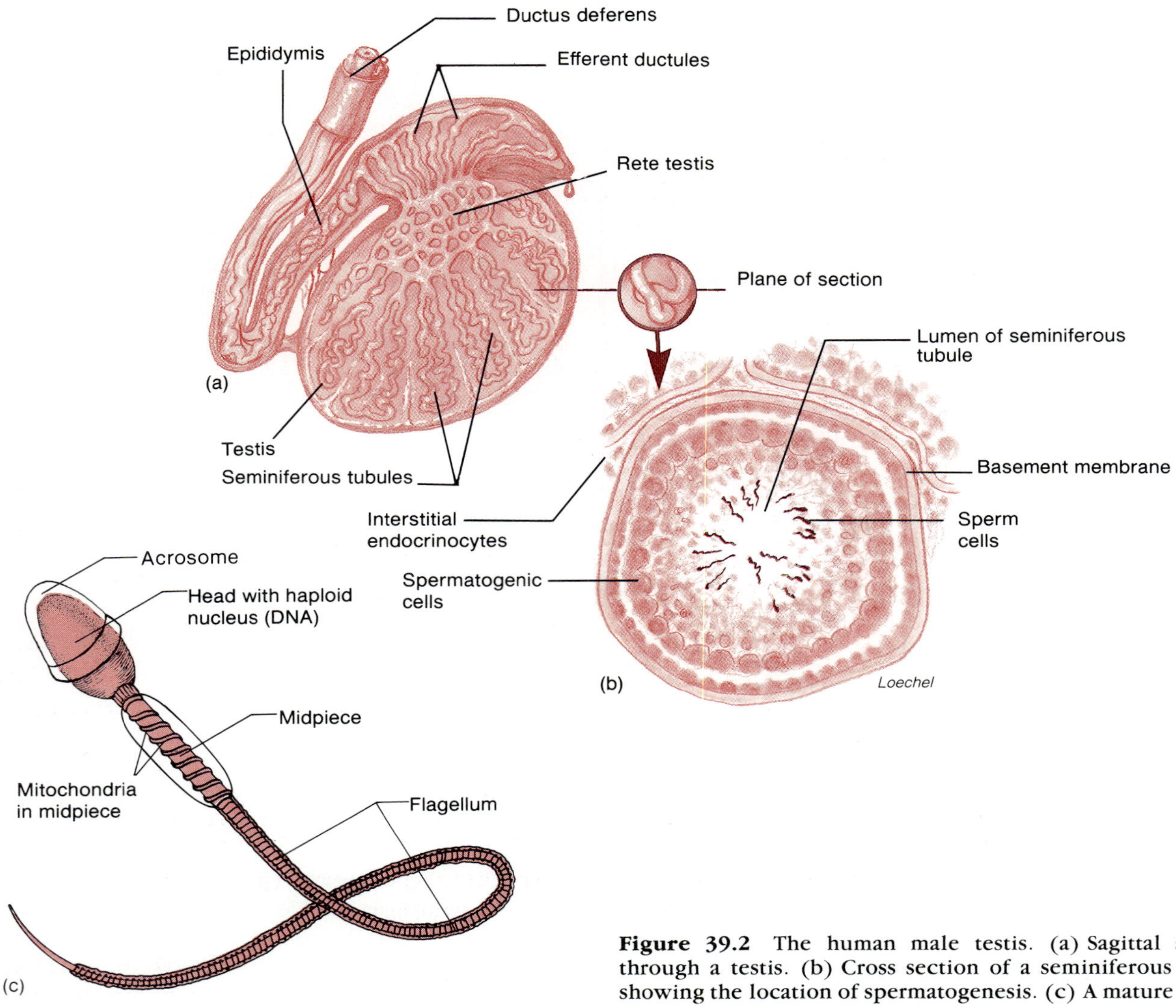

Figure 39.2 The human male testis. (a) Sagittal section through a testis. (b) Cross section of a seminiferous tubule showing the location of spermatogenesis. (c) A mature sperm.

semen through the urethra during ejaculation. In addition to the urethra, the penis contains three cylindrical strands of erectile tissue: two *corpora cavernosa* and the *corpus spongiosum* (*see figure 39.1*). The corpus spongiosum extends beyond the corpus cavernosa and becomes the expanded tip of the penis called the *glans penis*. The loosely fitting skin of the penis is folded forward over the glans to form the *prepuce* or *foreskin*. **Circumcision** is the removal of the prepuce for religious or health reasons. Today, many circumcisions are performed in the belief that the operation may decrease the occurrence of cancer of the penis.

[3]

A mature human sperm consists of a head, midpiece, and tail (figure 39.2c). The head contains the haploid nucleus, which is mostly DNA. The *acrosome,* a cap over most of the head, contains the enzyme hyaluronidase, that assists the sperm in penetrating the outer layer surrounding a secondary oocyte. The sperm tail contains an array of microtubules that bend to produce whiplike movements. The spiral mitochondria in the midpiece supply the ATP necessary for these movements.

Hormonal Control of Male Reproductive Function

Before a male can mature and function sexually, special regulatory hormones must come into play (table 39.1). Male sex hormones are collectively called **androgens** (Gr. *andros,* man + *gennan,* to produce). The hormones that travel from the brain and pituitary gland to the testes (and ovaries in the female) are called **gonadotropins.** As previously noted, the interstitial endocrinocytes produce the male sex hormone testosterone. Figure 39.3 shows the feedback mechanisms that regulate the production and secretion of testosterone, as well as its actions. When the level of testosterone in the blood decreases, the hypothalamus is stimulated to secrete GnRH (gonadotropin-releasing hormone). GnRH stimulates the secretion of FSH (follicle stimulating horomone) and LH (luteinizing hormone), also called ICSH (interstitial cell-stimulating hormone), into the bloodstream. (FSH and LH were first named for their functions in females, but their molecular

Table 39.1   Major Human Male Reproductive Hormones in An Adult

Hormone	Functions	Source
FSH (follicle-stimulating hormone)	Aids sperm maturation; increases testosterone production	Pituitary gland
GnRH (gonadotropin-releasing hormone)	Controls pituitary secretion	Hypothalamus
Inhibin	Inhibits FSH secretion	Sustentacular cells in testes
LH (luteinizing hormone) or ICSH (interstitial cell-stimulating hormone)	Stimulates testosterone secretion	Pituitary gland
Testosterone	Increases sperm production; stimulates development of male primary and secondary sex characteristics; inhibits LH secretion	Interstitial endocrinocytes in testes

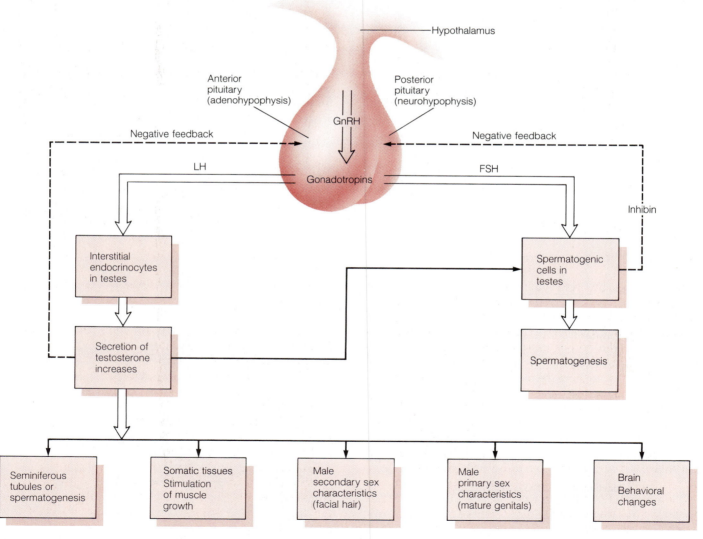

Figure 39.3   Hormonal control of reproductive function in adult human males. The feedback mechanisms by which the hypothalmus controls the maturation of sperm and the development of male secondary sexual characteristics are shown.

structure is exactly the same in males.) FSH causes the spermatogenic cells in the seminiferous tubules to initiate spermatogenesis, and LH stimulates the interstitial endocrinocytes to secrete testosterone. The cycle is completed when testosterone inhibits the secretion of LH, and another horomone, inhibin, is secreted. Inhibin inhibits the secretion of FSH from the anterior pituitary. This cycle maintains a constant rate (homeostasis) of spermatogenesis.

Stop and Ask Yourself

1. What is the advantage and disadvantage of estrus?
2. What are two evolutionary advantages to sexual reproduction?
3. What are the major constituents of semen?
4. How do LH and FSH function in males?

The Human Female Reproductive System

The reproductive role of females is more complex than that of males. Not only do females produce gametes (eggs, or ova), but after fertilization, they also nourish, carry, and protect the developing embryo. After the offspring is born, she may nurse it for a time. Another difference between the sexes is the monthly rhythmicity of the female reproductive system.

The female reproductive system consists of a number of structures with specialized functions (figure 39.4):

1. Two *ovaries* produce eggs and the female sex hormones called *estrogens,* and *progesterone.*
2. Two *uterine tubes,* one from each ovary, carry eggs from the ovary to the uterus. Fertilization usually occurs in the upper one-third of a uterine tube.
3. If fertilization occurs, the *uterus* receives the blastocyst and houses the developing embryo.
4. The *vagina* receives semen from the penis during sexual intercourse. It is the exit point for menstrual flow and is the canal through which the baby passes from the uterus during childbirth.
5. The *external genital organs* have protective functions and play a role in sexual arousal.
6. The *mammary glands,* contained in the paired breasts, produce milk for the newborn baby.

Production and Transport of the Egg

The female gonads are the paired **ovaries** (L. *ovum,* egg), which produce eggs and female hormones. The ovaries are located in the pelvic part of the abdomen, one on each

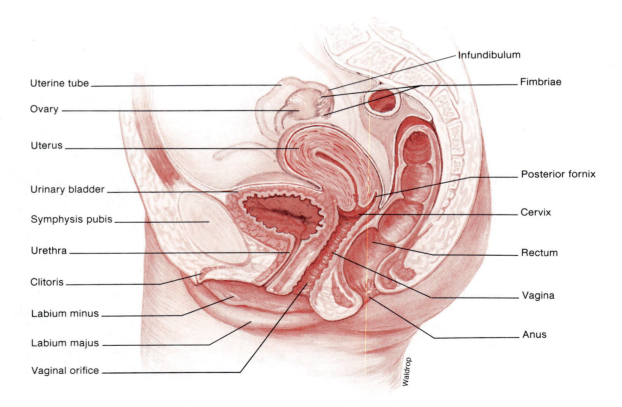

Figure 39.4 Sagittal view of the human female reproductive system. There are two uterine tubes that lead into the uterus and two ovaries.

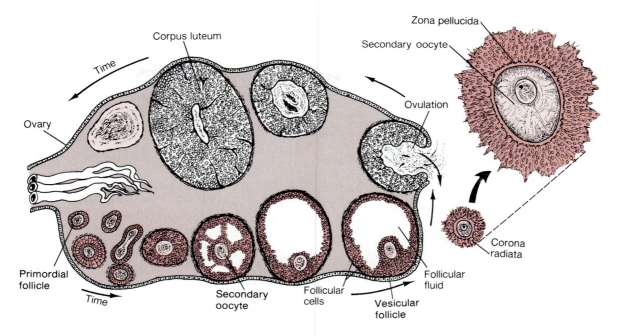

Figure 39.5   Cross section through a human ovary. As a follicle matures, the oocyte enlarges and becomes surrounded by a mantle of follicular cells and fluid. Eventually the mature follicle ruptures, and the secondary oocyte is released during ovulation.

side of the uterus. A cross section of an ovary reveals rounded vesicles called *follicles,* which are the actual centers of egg production (oogenesis) (figure 39.5). Each follicle contains an immature egg called a *primary oocyte,* and follicles are always present in several stages of development. Follicles are classified as either primordial or vesicular. *Primordial follicles* are not yet growing, and *vesicular follicles* are almost ready to release a secondary oocyte (commonly called an egg) in the process called **ovulation**. After ovulation, the lining of the follicle grows inward, forming the *corpus luteum* ("yellow body"), which serves as a temporary endocrine tissue and continues to secrete the female sex hormones estrogen and progesterone.

The paired tubes that receive the secondary oocyte from the ovary and convey it to the uterus are called *uterine tubes (see figure 39.4)*. The part of the uterine tube that encircles the ovary is called the *infundibulum*. The infundibulum is fringed with feathery *fimbriae*. Each month, as a secondary oocyte is released, it is swept by the motion of the fimbriae across a tiny space between the uterine tube and the ovary into the infundibulum.

Unlike sperm, the secondary oocyte is unable to move on its own. Instead, it is carried along the uterine tube toward the uterus by the peristalic contractions of the tube and the waving motions of the cilia in the mucous membrane of the tube (figure 39.6). Fertilization usually occurs in the first third of the uterine tube. A fertilized oocyte (zygote) continues its journey toward the uterus, where it will implant. The journey takes four to seven days. If fertilization does not occur, the secondary oocyte degenerates in the uterine tube.

The uterine tubes terminate in the **uterus**, a hollow, muscular organ located in front of the rectum and behind the urinary bladder (figure 39.7). The uterus terminates in a narrow portion called the *cervix,* which is the juncture between the uterus and the vagina. The uterus is made up of three layers of tissues. The outer *perimetrium* extends beyond the uterus to form the two *broad ligaments* that stretch from the uterus to the lateral walls of the pelvis. The middle, muscular layer is called the *myometrium* (Gr. *myo,* muscle + *metra,* womb) and makes up most of the

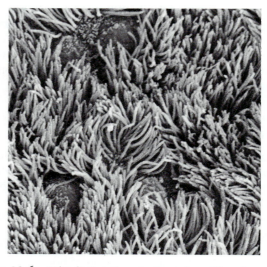

Figure 39.6   Cilia lining the uterine tubes. The electron micrograph shows the tiny, beating cilia on the surfaces of the uterine tube cell. They propel the secondary oocyte downward and perhaps the sperm upward.

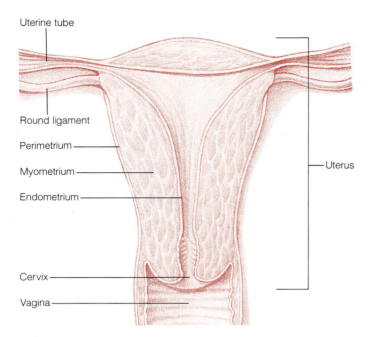

Figure 39.7 The layers of the human uterine wall.

uterine wall. The *endometrium* is the specialized mucous membrane that contains an abundance of blood vessels and simple glands.

The cervix leads to the **vagina,** a muscle-lined tube about 8 to 10 cm long. The wall of the vagina is composed mainly of smooth muscle and elastic tissue.

The external genital organs, or genitalia, include the *mons pubis, labia majora, labia minora, vestibular glands, clitoris,* and *vaginal orifice (see figure 39.4).* As a group, these organs are called the **vulva.** In most young women, the vaginal opening is partially covered by a thin membrane, the *hymen,* which may be ruptured during normal strenuous activities or may be stretched or broken during sexual activity.

The **mammary glands** (L. *mammae,* breasts) are modified sweat glands that produce and secrete milk. They contain varying amounts of adipose tissue. The amount of adipose tissue determines the size of the breasts, but the amount of mammary tissue does not vary widely from one women to another.

[4]

Hormonal Control of Female Reproductive Function

The male· is continuously fertile from puberty to old age, and throughout that period, sex hormones are secreted at a steady rate. The female, however, is fertile only during a few days each month, and the pattern of hormone secretion is intricately related to the cyclical release of a secondary oocyte from the ovary.

The development of a secondary oocyte in a follicle is controlled by the cyclical production of hormones (figure 39.8; table 39.2). *Gonadotropin-releasing hormone* (GnRH) from the hypothalamus acts on the anterior pituitary gland, which releases *follicle-stimulating hormone*

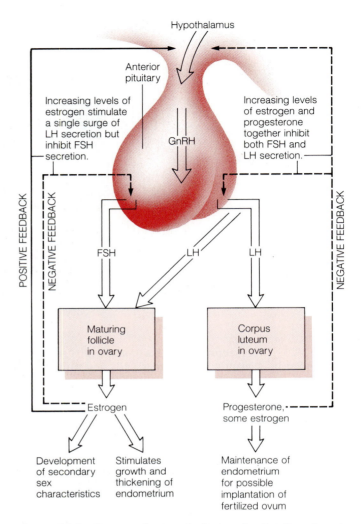

Figure 39.8 Hormonal control of reproductive functions in an adult human female. Feedback loops involving the hypothalmus, anterior pituitary, and ovaries are included. GnRH stimulates the release of both FSH and LH. Two negative feedback systems and a positive feedback system (one of the few in the human body) control the ovarian cycle.

(FSH) and *luteinizing hormone* (LH) to bring about the oocyte's maturation and release from the ovary. These hormones regulate the **menstrual cycle,** which is the cyclic preparation of the uterus to receive a fertilized egg, and the **ovarian cycle,** during which the oocyte matures, and ovulation occurs. This monthly preparation of the uterine lining for the fertilized egg normally begins at puberty with **menarche,** the first menstruation. When a female reaches about 45 to 55 years of age, the ovaries lose their sensitivity to FSH and LH, they stop making normal amounts of progesterone and estrogen, and the monthly menstrual cycle ceases in what is called the **menopause.**

One way to understand the hormonal pattern in the normal monthly cycle is to follow the development of the oocyte and the physical events in the menstrual cycle (figure 39.9; table 39.3). On the average, it takes about 28 days to complete one menstrual cycle, although the range

Table 39.2 Major Human Female Reproductive Hormones

Hormone	Functions	Source
Estrogen	Stimulates thickening of uterine wall; stimulates maturation of oocyte; stimulates development of female sex characteristics; inhibits FSH secretion; increases LH secretion	Ovarian follicle, corpus luteum
FSH (follicle stimulating hormone)	Causes immature oocyte and follicle to develop; increases estrogen secretion; stimulates new gamete formation and development of uterine wall after menstruation	Pituitary gland
GnRH (gonadotropin-releasing hormone)	Controls pituitary secretion	Hypothalamus
hCG (human chorionic gonadotropin)	Prevents corpus luteum from disintegrating; stimulates estrogen and progesterone secretion by corpeus luteum	Embryonic membranes and placenta
LH (luteinizing hormone)	Stimulates further development of oocyte and follicle; stimulates ovulation; increases progesterone secretion; aids development of corpus luteum	Pituitary gland
Oxytocin	Stimulates uterine contractions during labor; milk release during nursing	Pituitary gland
Prolactin	Promotes milk secretion by mammary glands after childbirth	Pituitary gland
Progesterone	Stimulates thickening of uterine wall	Corpus luteum

may be from 22 to 45 days. During this time, the following events take place.

1. The controlling center for ovulation and menstruation is the hypothalamus. It releases, on a regular cycle, GnRH, which stimulates the anterior pituitary to secrete FSH and LH (*see figure 39.8*).
2. FSH promotes the development of the oocyte in one of the immature ovarian follicles.
3. Estrogen is produced by the follicles, causing a buildup and enrichment of the endometrium, as well as the inhibition of FSH production.
4. The elevated estrogen level about midway in the cycle triggers the anterior pituitary to secrete LH, which causes the mature follicle to enlarge rapidly and release the secondary oocyte (ovulation). LH also causes the collapsed follicle to become another endocrine tissue, the *corpus luteum.*
5. The corpus luteum secretes estrogen and progesterone, which act to complete the development of the endometrium and maintain it for 10 to 14 days.
6. If the oocyte is not fertilized, the corpus luteum disintegrates into a *corpus albicans,* and estrogen and progesterone secretion cease.
7. Without estrogen and progesterone, the endometrium breaks down, and **menstruation** occurs. The menstrual flow, which is composed mainly of sloughed-off endometrial cells, mucus, and blood, remains in a liquid

form as it is drained from the body, because the dying cells of the endometrium release the enzyme *fibrinolysin* that prevents clotting.

8. As the progesterone and estrogen levels decrease further, the pituitary renews active secretion of FSH, which stimulates the development of another follicle, and the monthly cycle begins again.

Hormonal Regulation in the Pregnant Female

Pregnancy sets a new series of physiological events into motion. The ovaries are directly affected, because as the embryo develops, the cells of the embryo and placenta release the hormone *human chorionic gonatotropin* (hCG), which keeps the corpus luteum from disintegrating. The progesterone that it secretes is necessary to maintain the uterine lining. After a time, the placenta takes over the production of progesterone, and the corpus luteum degenerates. By the end of two weeks following implantation, the concentration of hCG is so high in the female's blood, and therefore in her urine as well, that an hCG immunological test is used to test for pregnancy. As the embryo develops, other hormones are secreted. For example, *prolactin* and *oxytocin* induce the mammary glands to secrete and eject milk after childbirth. Oxytocin and prostaglandins also stimulates the uterine contractions that expel the baby from the uterus during childbirth.

[5]

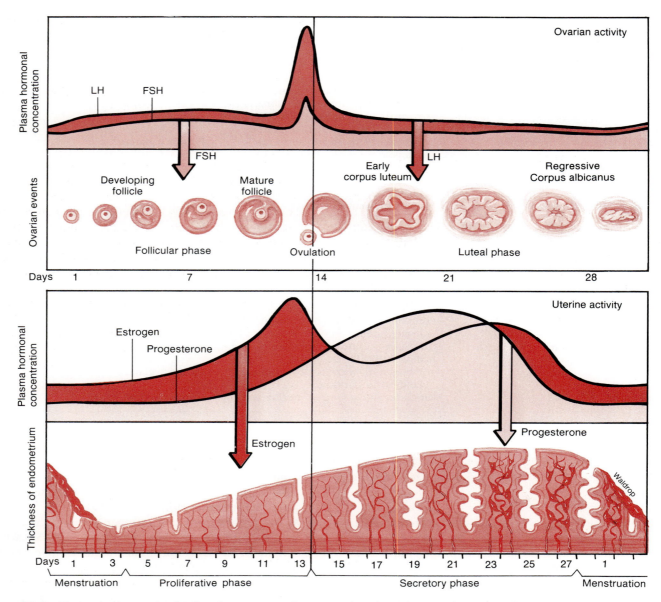

Figure 39.9 The major events in the female ovarian and menstrual cycles. The two charts correlate the gonadotropins, ovarian hormones, follicle development, ovulation, and changes in uterine anatomy during the cycles.

Table 39.3	Summary of the Menstrual Cycle Events	
Phase	**Events**	**Duration in Days***
Follicular	Follicle matures in the ovary; menstruation (endometrium breaks down); endometrium rebuilds	1–5
Ovulation	Secondary oocyte released from ovary	6–14
Luteal	Corpus luteum forms; endometrium thickens and becomes glandular	15–28

*Using a 28–day menstrual cycle as an example.

Stop and Ask Yourself

5. What is the anatomy of a human uterus?
6. What are the external genital organs in a human female?
7. What events take place in the menstrual cycle?

Pregnancy and Prenatal Development in the Human

This section covers the main event in reproduction—the nine-month pregnancy period, during which time the female's body carries, nourishes, and protects the embryo as it grows to a full-term baby.

Events of Prenatal Development: From Zygote to Newborn

The development of a human being may be divided into *prenatal* ("before birth") and *postnatal* ("after birth") periods. During the prenatal period, the developing individual begins life as a zygote, then becomes a ball of cells called a morula, and eventually becomes a blastocyst that implants in the endometrium. From 2 weeks after fertilization until the end of the eighth week of its existence, the individual is called an *embryo*. From nine weeks until birth it is a *fetus*, and is called a *newborn,* or *baby* when it is completely outside its mother's body.

Pregnancy is arbitrarily divided into *trimesters,* periods of three months each. The first trimester begins at fertil-

ization and is the time when most of the organs are formed. The next two trimesters are mainly periods of growth for the fetus.

The First Trimester After fertilization, usually in the first one-third of the uterine tube, the zygote goes through several cleavages as it is transported down the tube (figure 39.10). It eventually becomes a solid ball of cells called a *morula* and by the fourth day, it develops into a 50 to 120 cell blastula stage called a **blastocyst.** The mammalian blastocyst is characterized by an outer layer of cells called the *trophoblast* and an *inner cell mass.*

The next stage of development occurs when the blastocyst adheres to the uterine wall and implants. During implantation, the trophoblast cells invade the endometrium. Implantation is usually completed 11 to 12 days after fertilization; from then on the female is considered to be pregnant. [6]

One of the unique features of mammalian development is that most of the cells of the early embryo make no contribution to the embryo's body, giving rise instead to membranes. Only the inner cell mass gives rise to the embryonic body. Eventually, these cells become arranged in a flat sheet

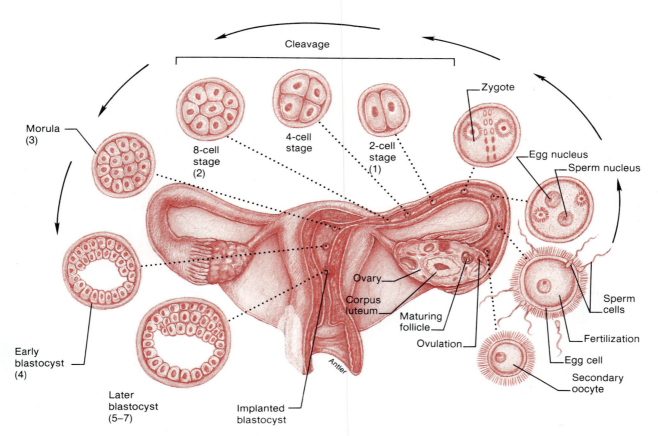

Figure 39.10 The early stages in human development. The numbers indicate the days after fertilization. The secondary oocyte is fertilized in the upper one-third of the uterine tube, undergoes cleavage while traveling down the tube, and finally implants in the endometrium of the uterus.

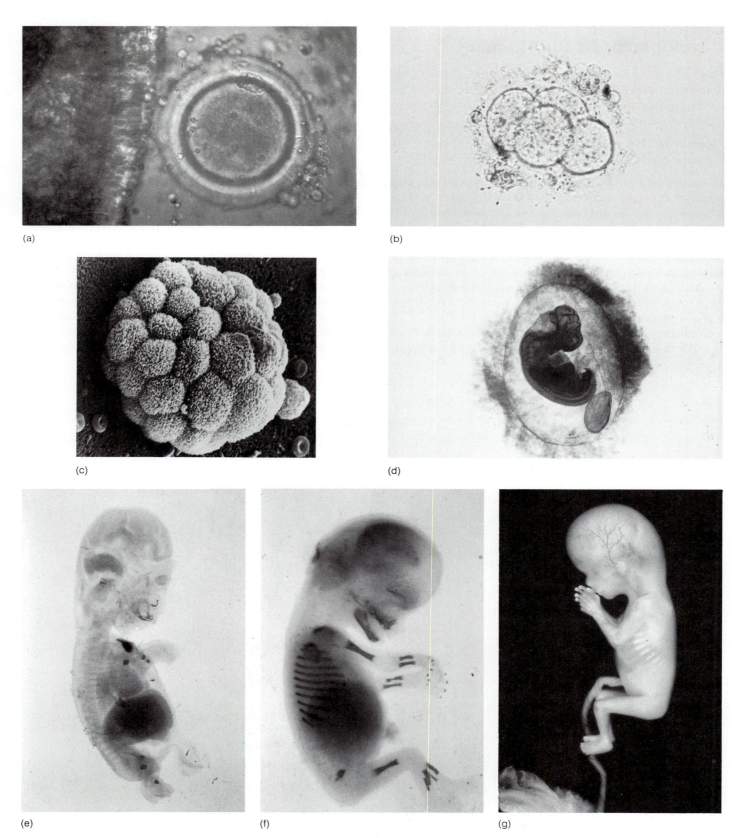

Figure 39.11 Human embryonic development. (a) A fertilized zygote. (b) The 4-cell stage. (c) A blastula. (d) A 5-week-old embryo. (e) The embryo at 8 weeks. (f) The fetus at 10 weeks. (g) The fetus at about 3 months.

that undergoes a gastrulation similar to that of reptiles and birds (*see figure 10.12*).

Once gastrulation is completed, the rest of the first trimester is devoted to organogenesis and growth (figure 39.11a-g). Regulatory events and inductive-tissue interactions occur that shape most of the organ systems.

The Second Trimester   By the start of the second trimester (fourth month), all the major systems have formed, and growth is spectacular. By now, the pregnant mother is quite aware of fetal movements. The heartbeat can be heard with a stethoscope. During the sixth month, the upper and lower eyelids separate, and the eyelashes form. During the seventh month, the eyes open. It is during this period that the bones begin to ossify.

The Third Trimester   The third trimester extends from the seventh month until birth. It is during this time that the fetus has developed sufficiently (with respect to the circulatory and respiratory systems) to potentially survive if born prematurely. During the last month, the weight of the fetus *doubles.*

The Placenta: Exchange Site and Hormone Producer

The lengthy pregnancy characteristic of mammals is possible in part because of the embryonic membranes that originated in the reptiles: the amnion, yolk sac, chorion, and allantois (*see figure 10.13*). The latter two gave rise to the embryonic parts of the placenta. The **placenta** is the organ that sustains the embryo and fetus throughout the pregnancy and through which an exchange of gases, nutrients, and wastes takes place between the maternal and fetal systems (figure 39.12). The tiny, fingerlike projections that were sent out from the blastocyst during implantation develop into numerous *chorionic villi,* which contain embryonic blood vessels. These blood vessels do not merge with the mother's; the two bloodstreams remain separate throughout the pregnancy. The placenta remains connected to the abdomen of the fetus by the **umbilical cord,** in which the fetus's umbilical arteries and veins spiral about each other. (These spiral vessels are the source of the red-and-white-striped design of the barber pole, the medieval advertisement for a surgeon—barbers of the day.)   [7]

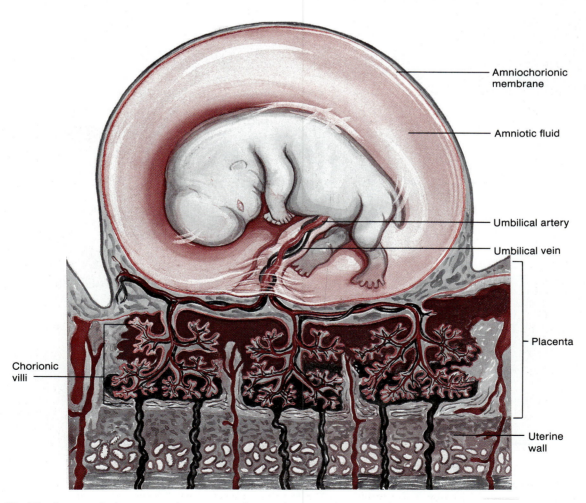

Amniochorionic membrane

Amniotic fluid

Umbilical artery

Umbilical vein

Placenta

Uterine wall

Chorionic villi

Figure 39.12   The fetus and placenta as they appear during the seventh week of development.

Birth: An End and A Beginning

About 266 days after fertilization, or 280 days from the beginning of the last menstrual period, the human infant is born. The birth process is called **parturition** (L. *parturire,* to be in labor). During parturition, the mother's uterine muscles begin to contract, and the cervix begins to dilate, or open. The hormone *relaxin,* produced by the ovaries and placenta, acts to cause the mother's pelvic bones to slightly separate so that the baby can pass through the birth canal with relative ease.

Changing hormone levels initiate parturition (box 39.1). When it is time for the baby to be born, its pituitary gland secretes *adrenocorticotropic hormone* (ACTH), which stimulates the adrenal glands to secrete steroids. These steroids stimulate the placenta to produce prostaglandins that, along with the hormone oxytocin from the mother's pituitary, cause the uterus to begin powerful muscular contractions. The contractions build in strength and increase in frequency over a period that usually lasts anywhere from 2 to 18 hours. During that time, the cervix becomes fully dilated, and the amniotic sac ruptures. Usually within an hour of these events, the baby is expelled from the uterus (figure 39.13a–c). After the baby emerges, uterine contractions continue, and the **afterbirth** is expelled (figure 39.13d). The umbilical cord is severed, and the newborn embarks on its nurtured existence in the outside world. (In mammals other than humans, the female bites through the cord to sever it.)

Milk Production and Lactation

Lactation (L. *lactare,* to suckle) includes both milk secretion (production) by the mammary glands and milk release from the breasts. (Mammary glands, which are an evolutionary innovation of mammals evolved from sweat glands in the skin.) During pregnancy, the breasts enlarge

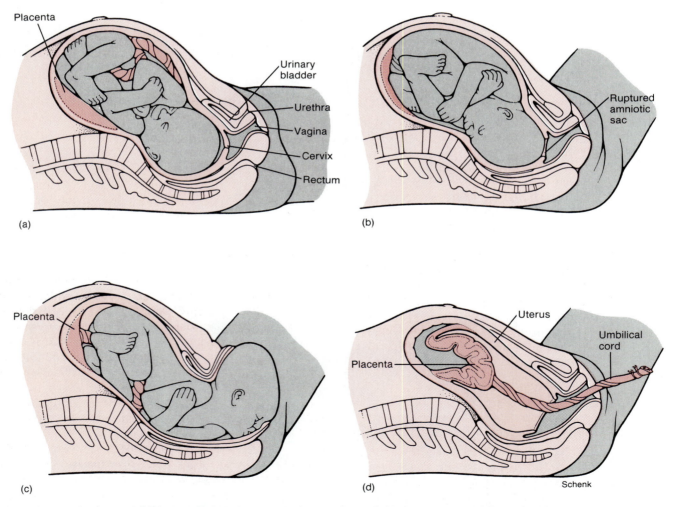

(a)

(b)

(c)

(d)

Schenk

Figure 39.13 The stages of labor and parturition. (a) The position of the fetus prior to labor. (b) The ruptured amnionic sac and early dilation of the cervix. (c) The expulsion stage of parturition. (d) Passage of the afterbirth.

Box 39.1

The Function of the Fetus During Childbirth

Until recently it was thought that the fetus played no active role in its own birth process. There is now evidence that the fetus triggers the release of "stress" hormones during parturition that help it survive the arduous process of birth and adjust to life outside the mother's uterus.

In addition to the pressure the fetus feels while passing though the birth canal, it is also periodically deprived of oxygen when powerful uterine contractions compress the umbilical cord and placenta. During these periods of stress, the fetus produces very high levels of epinephrine and norepinephrine, both of which are classified as *catecholamines*. Catecholamines are generally produced to help a person react favorably to incidents of extreme stress. The secretion of catecholamines allows the fetus to counteract the period of low oxygen and other potentially harmful situations throughout most of parturition.

The above stress situations are actually beneficial, because the presence of unusually high levels of cate-

cholamines permits the newborn to adjust to new conditions outside the mother's uterus immediately after birth. Postnatal adjustments include the necessity to breathe, the breakdown of fat and glycogen into usable fuel for cells, the acceleration of the heart rate and cardiac output, and the increase of blood flow to the brain, heart, and skeletal muscles.

The surge of catecholamines during parturition also causes the newborn's pupils to dilate, even when strong light is present. This alertness may help the infant form an early bond with its mother.

The production of fetal catecholamines is a direct result of the adrenal glands' response to stress. In adults, however, the secretion of catecholamines begins with the stimulation of the sympathetic nervous system. It is also of interest that the adrenal glands of the fetus are proportionately larger than those of the adult.

From Robert Carola, John P. Harley, and Charles R. Noback. 1990. *Human Anatomy and Physiology*. McGraw-Hill, Inc., New York.

---

in response to increasing levels of the hormone *prolactin*. Before birth, placental secretions of estrogen and progesterone inhibit the secretion of milk from the breasts. After the placenta has been expelled from the uterus, the concentrations of estrogen and progesterone drop, and the breasts begin to produce copious amounts of milk.

The actual release of milk from the mother's breasts does not occur until 1 to 3 days after the baby is born. During these first days, the suckling baby receives **colostrum**, a high-protein fluid that is present in the breast at birth. Colostrum contains an abundance of maternal antibodies and thus helps strengthen the baby's immune system. It also functions as a laxative, removing fetal wastes called **meconium** from the intestines. After about 3 days, the prolactin secreted from the pituitary stimulates milk production. When the newborn suckles, the pituitary is stimulated into releasing oxytocin as well as prolactin. Oxytocin triggers the release of milk from the mammary glands.

Stop and Ask Yourself

8. What five stages comprise the prenatal period?
9. What are the main events of the first trimester of pregnancy? The second and third trimesters?
10. What is the function of the placenta?
11. What causes parturition to occur?
12. How does lactation occur?

Reproductive Diversity

The individuals of every species, whether single cells or animals, must reproduce to pass on their genetic information to a new generation. This last section focuses on the basic reproductive strategies found in various animal groups.

Invertebrates

The most common and direct form of *asexual reproduction* is mitotic cell division, the means by which unicellular organisms multiply. Reproduction by cell division confers a type of immortality on unicellular organisms that has no counterpart among animals. Although it is true that the life of a particular unicellular organism ends when it divides, the organism does not die, as animals eventually do. One individual simply divides to produce two new individuals. Genetically, each of these new individuals is identical to the original individual (cell) and is referred to as a *clone*.

In animals, asexual reproduction is common in cnidarians, bryozoans, and a few other invertebrates (flatworms, echinoderms, annelids). In some of these animals, asexual reproduction has economic applications. For example, sponge producers chop sponges into small pieces and place them where each piece can grow into a new sponge. In some parasites, such as the flukes, enormous numbers of offspring are produced at several stages in their life cycles, which greatly enhances their total reproductive potential. This trait is adaptive for these parasites, because many individuals are lost during transfer from one host to another.

Asexual reproduction is rare among the higher invertebrates. In the lower invertebrates, the forms of asexual reproduction are binary fission, budding (both external and internal), fragmentation, and multiple fission.

Binary fission is common among some protozoa, cnidarians, and annelids. In this process, the body of the parent divides into two approximately equal parts, and each part then grows into a new individual (*see figure 16.2*).

Budding is an unequal division of the organism. The new individual arises as an outgrowth (bud) from the parent. In external budding, such as that found in the cnidarians, the bud eventually detaches and grows into a new individual. In internal budding, as in the freshwater sponges, gemmules, which are collections of many cells surrounded by a body wall, are produced. When the body of the parent dies and degenerates, each gemmule gives rise to a new individual.

Fragmentation is asexual reproduction in which an animal breaks into two or more parts, each capable of becoming a new animal. Fragmentation occurs in the Platyhelminthes, Rhynchocoela, and Echinodermata.

Multiple fission occurs in some protozoa (*see figure 16.16*). In these protists, the nucleus divides repeatedly before division of the cytoplasm. This process gives rise simultaneously to many daughter cells.

Some invertebrates (e.g., rotifers, nematodes, crustaceans, and insects) reproduce by *parthenogenesis*. As presented in chapter 6, parthenogenesis is the spontaneous activation of a mature egg followed by normal egg divisions and subsequent embryonic development. This form of reproduction, however, is relatively uncommon.

Hermaphroditism occurs when an animal has both functional male and female reproductive systems (e.g., annelids, some hydroids, some crustaceans, and sea slugs). This form of reproduction is especially beneficial to sessile animals, such as barnacles.

Vertebrates

Almost all vertebrates reproduce sexually; only a few lizards and fish normally reproduce parthenogenetically. Sexual reproduction evolved among aquatic animals and then spread to the land as animals became terrestrial. Many aquatic animals simply release their gametes into the water. Fertilization is accomplished by the release of sperm by the males into the same area containing the eggs. Although simple and effective, this pattern of *external fertilization* posed a problem for many animals. As soon as eggs and sperm were released into the sea water, they became greatly diluted, and the chance of a sperm meeting an egg greatly decreased. To overcome this problem, many animals evolved the synchronous release of sperm and eggs during a few brief, well-defined periods.

All fishes reproduce in aquatic environments. In bony fishes, fertilization is usually external, and eggs are produced that contain only enough yolk to sustain the developing fish for a short time. After this yolk is consumed, the growing fish must seek food. Although many thousands of eggs are produced and fertilized, very few survive and grow to maturity. Some succumb to fungal and bacterial infections, others to siltation, and still others to predation. Thus, to assure reproductive success, the development of the fertilized egg is very rapid, and the young that result achieve maturity within a short period of time.

The invasion of land by vertebrates meant facing for the first time the danger of drying out or desiccating—especially the small and vulnerable gametes. Obviously, the gametes could not simply be released near one another on the land, because the gametes would quickly desiccate.

The amphibians were the first vertebrates to invade the land. They have not, however, become adapted to a completely terrestrial environment; their life cycle is still inextricably linked to water. Among most amphibians, fertilization is still external, just as it is among the fishes. Among the frogs and toads, the male grasps the female and discharges fluid containing sperm onto the eggs as she releases them into the water (figure 39.14a).

The developmental period is much longer in amphibians than in fishes, although the eggs do not contain appreciably more yolk. An evolutionary advancement of amphibians is the presence of two periods of development: larval and adult stages. The development of the aquatic larval stage is rapid, and the organism spends a lot of time eating and growing. When a sufficient size has been reached, the larval form undergoes a developmental transition called metamorphosis into the adult (often terrestrial) form.

The reptiles were the first group of vertebrates to completely abandon the aquatic habitat because they solved the problem of sexual reproduction on land. A crucial advancement first found in reptiles is the evolution of internal fertilization (figure 39.14b). With internal fertilization, the gametes are protected from drying out, freeing the animals from returning to the water to breed.

Many reptiles are *oviparous* (L. *ovum*, egg + *parere*, to bring forth), and the eggs are deposited outside the body of the female. Others are *ovoviviparous* (L. *ovum*, egg + *vivere*, to live, + *parere*, to bring forth). They form eggs that hatch in the body of the female, and the young are born alive.

The shelled egg and extraembryonic membranes, also first seen in reptiles, constitute two other important evolutionary adaptations to life on land. By enclosing the developing organism in a shell and having an egg that contains a large amount of yolk encased in a membranous covering, the eggs can be laid in dry places without danger of desiccation. As the embryo develops, the extraembryonic chorion and amnion help protect it, the latter by creating a fluid-filled sac for the embryo. The allantois permits gas exchange and stores excretory products. Complete development can occur within the egg, and when the egg

Figure 39.14 Vertebrate reproductive strategies. (a) A male wood frog (*Rana sylvatica*) clasping the female in amplexus, a form of external fertilization. As the female releases eggs into the water, the male releases sperm over them. (b) Reptiles, such as these turtles, were the first terrestrial vertebrates to develop internal fertilization. (c) Birds are ovivaparous animals. Their shelled eggs have large yolk reserves, and the young develop and hatch outside the mother's body. Birds may show advanced parental care. (d) A placental mammal, this female dog is nursing her puppies.

hatches, the newly emerged animal is well equipped to face the environment.

Birds have retained the important adaptations for life on land that evolved in the early reptiles. With the exception of most waterfowl, birds lack a penis. Internal fertilization is achieved by males simply depositing semen against the cloaca. Sperm then migrate up the cloaca and fertilize the eggs before their hard shells are formed. This method of mating occurs more quickly than the internal fertilization practiced by reptiles. All birds are oviparous, and the eggshells are much thicker than those of reptiles. Thicker shells permit birds to sit on their eggs and warm them. This brooding, or incubation, hastens the development of the embryo. When many birds hatch from their eggs, they are incapable of surviving on their own. Extensive parental

care and feeding of young are more common among birds than fishes, amphibians or reptiles (figure 39.14c).

The most primitive mammals, the monotremes (e.g., the duckbilled platypus and spiny anteater) lay eggs, as did the reptiles from which they evolved (figure 39.14d). All other mammals are viviparous.

Mammalian viviparity was another major evolutionary adaptation, and it has taken two forms. The marsupials developed the ability to nourish their young in a pouch after a short gestation inside the female. The other, much larger group, the placentals, retains the young inside the female, where they are nourished by the mother by means of a placenta. Even after birth, mammals continue to nourish their young. Mammary glands are a unique mammalian adaptation that permit the female to nourish the young with milk that she produces. Some mammals nurture their young until adulthood, when they are able to mate and fend for themselves. In humans, the long-lasting pair bond between one male and one female has evolved into marriage, which further supports the highly dependent human offspring. As noted at the beginning of this chapter, mammalian reproductive behavior also contributes to the transmission and evolution of culture that is the key to the evolution of our species.

Stop and Ask Yourself

13. What reproductive strategy developed in bony fishes to help assure survival? In amphibians? In reptiles?
14. What is the difference between a bird egg and an amphibian egg?
15. What reproductive adaptation evolved in humans that is not present in other mammals?

■ Summary

1. The reptiles and first mammals shaped the anatomy and physiology that later ensured successful human reproduction.

2. The reproductive role of the human male is to produce sperm and deliver them to the vagina of the female. This function requires four different structures. The testes produce sperm and the male sex hormone, testosterone. Accessory glands furnish a fluid, called semen, for carrying the sperm to the penis. Accessory ducts store and carry secretions from the testes and accessory glands to the penis. The penis deposits semen into the vagina during sexual intercourse.

3. Before a male can mature and function sexually, special regulatory hormones (FSH, GnRH, inhibin, LH, and testosterone) must function.

4. The reproductive roles of the human female are more complex than those of the male. Not only do females produce eggs, but after fertilization, they also nourish, carry, and protect the developing embryo. They may also nourish the infant for a time after it is born. The female reproductive system consists of two ovaries, two uterine tubes, a uterus, vagina, and external genitalia. The mammary glands contained in the paired breasts produce milk for the newborn baby.

5. The human female is fertile only during a few days each month, and the pattern of hormone secretion is intricately related to the cyclical release of a secondary oocyte from the ovary. Various hormones regulate the menstrual and ovarian cycles.

6. Pregnancy sets a new series of physiological events into motion that are directed to housing, protecting, and nourishing the embryo.

7. The development of a human may be divided into prenatal and postnatal periods. Pregnancy is arbitrarily divided into trimesters.

8. The placenta is the organ that sustains the embryo and fetus throughout the pregnancy. The birth process is called parturition and occurs about 266 days after fertilization.

9. Lactation includes both milk secretion (production) by the mammary glands and milk release from the breasts.

10. Sexual reproduction first evolved in aquatic animals. The invasion of land meant facing the danger of the gametes and embryos desiccating. The five major groups of vertebrates have evolved reproductive strategies that reflect the environment in which they reproduce.

■ Key Terms

afterbirth (p. 646)
androgens (p. 636)
blastocyst (p. 643)
bulbourethral glands (p. 635)
circumcision (p. 636)
colostrum (p. 647)
estrus (p. 634)
gestation (p. 634)
gonadotropins (p. 636)
lactation (p. 646)
mammary glands (p. 640)
meconium (p. 647)
menarche (p. 640)
menopause (p. 640)
menstrual cycle (p. 640)
menstruation (p. 641)

ovarian cycle (p. 640)
ovaries (p. 638)
ovulation (p. 639)
parturition (p. 646)
placenta (p. 645)
prostate gland (p. 635)
scrotum (p. 634)
semen (p. 635)
seminal vesicles (p. 635)
seminiferous tubules (p. 634)
testes (p. 634)
umbilical cord (p. 645)
uterus (p. 639)
vagina (p. 640)
vulva (p. 640)

■ Critical Thinking Questions

1. Is the fertility of a women affected by the length of a given menstrual cycle or whether the cycles are regular or irregular? Explain.

2. Looking at a variety of animals, what are the advantages to restricting reproduction to a limited time period? Why do so many animals have a sharply defined reproductive season during the year?

3. In most sexual species, males produce far more gametes than do females. Why does this occur when in most cases, only one male gamete can fertilize one female gamete?

5. Why are the accessory glands of the male so important in reproduction?
6. Why do human females have a menstrual cycle?
7. Why is the placenta so important to an embryo?

Suggested Readings

Books

Jones, R. E. 1984. *Human Reproduction and Sexual Behavior.* Englewood Cliffs, N.J.: Prentice-Hall.

Masters, W. H., and Johnson, V. E. 1966. *Human Sexual Response.* Boston: Little, Brown.

Margulis, L., and Sagan, D. 1986. *Origins of Sex: Three Billion Years of Genetic Recombination.* New Haven, Conn.: Yale University Press.

Nilsson, L. 1986. *Behold Man: A Photographic Journey of Discovery Inside the Body.* Boston: Little, Brown.

Nilson, L. 1977. *A Child is Born.* New York: Delacorte Press.

Singer, P., and Wells, D. 1985. *Making Babies: The New Science and Ethics of Conception.* New York: Charles Scribner and Sons.

Articles

Beaconsfield, P., Birdwood, G., and Beaconsfield, R. The placenta. *Scientific American* August, 1980.

Epel, D. The program of fertilization. *Scientific American* November, 1977.

Frisch, R. Fatness and fertility. *Scientific American* March, 1988.

Grobstein, C. External human fertilization. *Scientific American* June, 1979.

Hrdy, S. Daughters or sons: Can parents influence the sex of their offspring? *Natural History* April, 1988.

Lagercrantz, H., and Slotkin, A. The stress of being born. *Scientific American* April, 1986.

Segal, S. J. The physiology of human reproduction. *Scientific American* March, 1974.

Todd, J. T. The perception of human growth. *Scientific American* February, 1980.

Ulmann, A., Teutsch, G., and Philibert, D. RU 486. *Scientific American* June, 1990.

Wassarman, P. M. Fertilization in mammals. *Scientific American* December, 1988.

Behavior and Ecology

Part VI

C hapter 1 of this textbook stressed unity and diversity. All life is unified at molecular, cellular, and evolutionary levels. A fundamental unity also exists at an environmental level. All living organisms are partners in the use of the resources of the earth. In particular, humans have a responsibility to see that the resources of this planet are preserved for all organisms in all generations. It is only through studying the interactions of organisms with one another and with the physical attributes of their environment, that this goal will ever be realized. Even in our "enlightened" generation, waste and misuse is unbridled. In many ways, our technologies have permitted abusive practices unheard of by previous generations. Scientists, theologians, and philosophers must lead the way in protecting our fragile ecosystems.

Part VI (chapters 40 through 42) present some behavioral interactions among organisms and some of the other interactions between organisms and their environment. Many complex relationships exist, and we can fully understand living things only in the context of populations, communities, and ecosystems.

40

Animal Behavior

Concepts

1. Animal behavior refers to the varied activities that an animal performs during its lifetime. Internal motivation, environmental stimuli, and stimulus filtering mechanisms influence specific behavioral responses.
2. Ultimately, the behavior of an animal is influenced and shaped by natural selection. Certain animals are selected for their behavioral traits that allow them to survive and reproduce.
3. Some behavior patterns are genetically controlled stereotyped actions that are not affected by experience. These patterns allow an animal to respond immediately to a specific environmental stimulus. Within genetic limits, other behaviors are readily modified through learning and are more adapted to changing environmental conditions.
4. Behavior is affected by the internal state of an animal. Internal motivation is affected by physiological factors such as hormones and nerve impulses. These factors may be influenced by environmental factors, such as day length, season, or temperature.

Have You Ever Wondered:

[1] if an earthworm placed on a fishhook feels any pain?
[2] why a greylag goose always rolls her egg back to the nest?
[3] if animals are motivated?
[4] how tadpoles know how to swim?
[5] how a bird sings the right song?
[6] how baby chicks know how to follow their mother?
[7] why some birds are not afraid of scarecrows?
[8] how a trainer teaches an animal to do tricks?
[9] how an animal knows about its surroundings?

These and other useful questions will be answered in this chapter.

This chapter contains underlined evolutionary concepts.

Animal behavior can be defined as the activities animals perform during their lifetime. These activities include locomotion, feeding, breeding, capture of prey, avoidance of predators, and social behavior. Animals also send signals, respond to signals or stimuli, carry out maintenance behaviors, make choices, and interact with one another. This chapter examines some of these aspects of animal behavior.

Four Approaches to Animal Behavior

Observations of animal behavior have been made by theologians and philosophers for centuries. Only in the last century, however, has there been significant progress in understanding behavior.

One approach to studying behavior is that of the **experimental psychologists** who study learning in laboratory animals, such as the white rat. A rat is placed in a box or maze in which it receives a reward if a lever is pressed or if it runs a maze successfully.

Ethologists study behavior in a different way than do psychologists. **Ethology** (Gr. *ethologica,* depicting character) is the study of animal behavior in which evolution and the natural environment are important considerations. The leaders of this approach have been Konrad Lorenz, Niko Tinbergen, and Karl von Frisch who were awarded the Nobel Prize in Physiology or Medicine in 1973. Ethologists observe the behavior of a variety of animals in their natural environments. They study the behavior of closely related species in order to consider the evolution and origin of certain behavior patterns. Unlike the psychologists, ethologists rarely deal with learning and are interested instead in questions of animal communication, mating behavior, and social behavior.

Behavioral ecology emphasizes the ecological aspects of animal behavior. Predator-prey interactions, foraging strategies, reproductive strategies, habitat selection, intraspecific and interspecific competition, and social behavior are topics of interest to behavioral ecologists.

Sociobiology is the study of the genetic basis of social behavior. It combines many aspects of ethology and behavioral ecology. Sociobiologists emphasize the importance of natural selection on individuals living in groups. Examples of sociobiological concepts are altruism and kin selection.

Ultimate and Proximate Causes

"Why do animals do what they do?" is a frequently asked question by behavioral scientists. The more immediate causes of behavior that are ecological and physiological in nature, such as eating to satisfy hunger, are called *proximate causes*. Another level of causation in behavior that occurs on the evolutionary time scale is that of *ultimate causes*. A display serves not only to attract a mate but to successfully reproduce and pass on genetic information to the next generation.

Anthropomorphism

Anthropomorphism (Gr. *anthropos,* man + *morphe,* form) is the application of human characteristics to anything not human. It is important when observing animals to avoid assigning human feelings to their behavior. Humans like to explain behavior in terms of their own experiences, but it is not likely to be accurate with regard to other animals, especially those lower on the phylogenetic scale. Consider the example of placing an earthworm on a fishhook. Does the fishhook "hurt" the earthworm causing it to writhe in "pain?" Both of the descriptive words, "hurt" and "pain," are based on human experience and conscious awareness. A better explanation of the above example that reduces the anthropomorphic interpretation is that when the earthworm is placed on the hook, certain receptors are stimulated, and nerve impulses are generated and travel along reflex neural circuits. The muscles are stimulated and the worm begins to wriggle in an attempt to escape from the hook. This explanation more closely describes what has been observed and does not attempt to suggest what the earthworm "feels." [1]

Instincts

There has been considerable controversy over the use of the term instinct in animal behavior. An **instinct** may be defined as a genetically controlled activity that occurs automatically in response to a stimulus, without benefit of learning or experience. Some zoologists do not consider it to be a precise term and do not use it. However, the term is so ingrained in the scientific vocabulary that it is difficult to avoid. Certainly, animals have specific and identifiable innate behavior patterns that are characteristic for a particular species.

Konrad Lorenz proposed that instincts are composed of *fixed action patterns*, which are movements that are innate and characteristic for the species. The classic example of an instinct is the egg-rolling behavior of a greylag goose, in which the female rolls her egg back to the nest (figure 40.1). The fixed action pattern consists of chin-tucking movements by the female to push or pull the egg toward the nest. Once the chin-tucking movements begin, they continue to completion, even if the egg is removed. [2]

Another example of an instinct is that of a toad feeding on a mealworm. The fixed action pattern is carried out in a stereotyped manner with little variation. Once under way, the behavior pattern is continued to completion. The importance of instinctive behavior and fixed action patterns

Figure 40.1 Egg-rolling behavior of a greylag goose (*Anser anser*). This fixed action pattern is a stereotyped movement in which an egg is retrieved from outside the nest. Many ground-nesting birds show this behavior. In fact, these birds will retrieve anything "egglike" near the nest, including baseballs and tin cans.

is that responses of the animal to important stimuli in its normal environment may occur immediately.

Certain instinctive behavior patterns have three stages. The first, *appetitive behavior*, is a searching phase in which the animal behaves so as to increase the probability of encountering the stimulus that will release the particular instinctive activity. The second stage is known as *consummatory behavior*, and is a highly stereotyped, fixed action pattern that occurs in response to a releasing stimulus. An example would be the act of eating, once food, which is the releasing stimulus, is located. The third stage is a *refractory period* during which the animal does not respond to releasing stimuli and does not exhibit appetitive behavior.

Stop and Ask Yourself

1. What are four ways to study animal behavior?
2. Why is anthropomorphism a problem when describing the behavior of animals?
3. What are some characteristics of an instinct?
4. What are some examples of proximate and ultimate causes of animal behavior?

Motivation and Spontaneous Behavior

Animals do not always respond in the same way to a given stimulus. The threshold of response may vary with changes in the animal's internal state. The tendency to perform a specific behavior is referred to as **motivation.** The animal is said to be motivated to perform a given behavior or possess a drive to perform that behavior. Examples would be a thirst drive, hunger drive, or mating drive. The use of the term "drive" has been criticized in that it provides no information about the mechanism involved.

The discipline of neurophysiology offers some possible explanations for motivation. It is known that neurons can be spontaneously active and discharge without any input. This activity may explain some of the spontaneous behavior [3] that occurs in animals. For example, a dog wakes and begins looking for food. Konrad Lorenz suggested that there was a build-up of *action specific energy* in the central nervous system of the dog over a period of time. As the energy accumulates, there is a continual lowering of the threshold of response to the stimulus that will release the behavior.

Also, during certain seasons of the year, increasing day length activates a reproductive drive or behavior in many species of animals. Increased levels of hormones involved in reproduction are found in the blood of these animals and contribute greatly to the increase in internal readiness or motivation of an animal to reproduce.

Sign Stimulus

For a given instinctive behavior to occur, a signal is usually presented by one individual that elicits the behavior in another individual. Such a signal is called a **sign stimulus**. A sign stimulus is a releasing stimulus that is important in communication within species. A sign stimulus is usually simple and obvious, but may be part of a large stimulus configuration. A classic example is the red belly of the male stickleback fish during reproduction. Males have a distinctive red belly when in breeding condition, and females have a swollen belly when they are ripe with eggs. The red belly serves as a territorial signal to other males. Figure 40.2 illustrates some of the models used in an experiment to determine what the specific signal is. Crude models with red undersides stimulate aggression by males. A more accurate model that lacked red did not elicit aggression. The conclusion is that the red belly is the sign stimulus in this species of fish. Other characteristics of the sign stimulus are also important. If a male or model is in a head-down

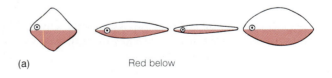

(a)   Red below

(b)   No red below

Figure 40.2 Models used to test aggression in male sticklebacks. The crude models (a) with colored area below released more aggressive attacks than the more realistic model (b) that lacked the color.

or threat posture, an even greater aggressive response is elicited from a territorial male. These two components of a stimulus configuration are apparently additive in their effects.

Stimulus Filtering

Because the nervous system of an animal is continually receiving information or stimuli from a variety of sources, a mechanism must exist to filter or remove unimportant stimuli. This **stimulus filtering** allows an animal to respond only to relevant stimuli.

Although the sense organs act as filters for many stimuli, most incoming stimuli are filtered through the central nervous system. This ability was recognized very early by ethologists and the term **innate releasing mechanism (IRM)** was proposed as a stimulus filtering mechanism that releases a particular behavior pattern in response to a specific sign stimulus. The nature of an IRM is exemplified by the work of Gordon Burghardt on the feeding behavior of garter snakes (*Thamnophis*) and water snakes (*Nerodia*). These snakes detect their prey by flicking the tongue out of the mouth, picking up chemicals from the air or contact with the prey, and then placing the tip of the tongue into pits in the roof of the mouth called the Jacobson's organ, which contains chemoreceptors for taste and smell. The chemical odors of some prey species release feeding behavior, whereas others do not.

Burghardt discovered that when different species of garter snakes were presented with prey odors of fish, frogs, salamanders, or worms, the snakes preferred the prey type on which they usually fed in their normal environment. These preferences were also observed in newborn snakes that had never fed, and they could not be altered by force feeding the young snakes with artificial food.

However, some snakes may change their prey preference. Burghardt found that water snakes obtained near a fish hatchery where goldfish were abundant preferred goldfish to the native minnows on which they would normally feed.

There appears to be a genetically controlled (inherited) response or bias in the filtering mechanism toward certain prey, but it is important to note that they can be modified by experience.

Communication

Communication involves the transfer of information from one animal to another. Communication requires a sender and receiver that are mutually adapted to each other. The animal acting as the sender must send a clear signal to the receiving animal. Communication can occur within species (*intraspecific*) or between species (*interspecific*). Intraspecific communication in animals is especially important

for reproductive success. Examples of interspecific communication include warning signals, such as the rattle of a rattlesnake's tail and the presentation of its hindquarters and tail by a skunk.

Animals use a variety of signals for communication, including visual, auditory, tactile, and chemical signals. The characteristics of a signal system are shaped by natural selection. It is important to note that animals have evolved combinations of signals that may be more effective than any single signal.

Visual Communication

Visual communication is important to many animals because a large amount of information can be conveyed in a short period of time. Most animals (e.g., cephalopod molluscs, arthropods, and most vertebrates other than mammals) that have well-developed eyes have color vision. Many fishes, reptiles and birds exhibit brilliant color patterns that usually have a signaling function. Most mammals have plain, darker colors and lack color vision because they are nocturnal, as were their probable ancestors—nocturnal insectivores. A notable exception is primates in which color vision has appeared again along with colorful displays.

A visual signal may be present at all times, as are the bright facial markings of a male mandrill (colorplate 131). The signal may be hidden, or located on a less exposed part of an animal's body and then suddenly presented. Some lizards, such as green anoles, can actually change their color through activities of pigment cells in the skin.

Visual signals have some disadvantages in that the line of sight may be blocked by various objects in the environment and/or may be difficult to see over a long distance. Also, they are usually not effective at night and may be detected by predators.

Acoustic Communication

Acoustic or sound communication is commonly used by arthropods and vertebrates. These animals must expend energy to produce sounds, but sounds can be used during night or day. Sound waves also have the advantage of traveling around objects and may be produced or received while an animal is in the open or concealed. Sounds may carry a large amount of information because of the many possible variations in frequency, duration, volume, or tone.

Communication systems are closely adapted to the environmental conditions in which they are used and the function of the signal. For example, tropical forest birds produce low-frequency calls that pass easily through dense vegetation. Many primates in tropical forests produce sounds that travel over long distances. Other examples include the calls of territorial birds that sit at a high point to deliver the signal more effectively, and the alarm calls of many small species of birds. Some of the more complex acoustic signals that have been studied are birdsong and human speech.

Sound waves are known to travel much farther in water than in the air. Some fishes, whales, and other aquatic animals utilize sound communication very effectively. The "songs" of humpback whales can travel for hundreds of miles.

Tactile Communication

Tactile communication refers to the communication between animals in physical contact with each other. The antennae of many invertebrates and the touch receptors in the skin of vertebrates function in tactile communication (*see figure 34.2*). Some examples of tactile communication are birds preening the feathers of other birds and primates grooming each other (figure 40.3).

Chemical Communication

Chemical communication is another common mode of communication. Unicellular organisms with chemoreceptors can recognize members of their own species. Chemical signals are well-developed in insects, fishes, salamanders, and mammals. Some advantages of chemical signals are that they: (1) usually provide a simple message that can last for hours or days; (2) are effective night or day; (3) can pass around objects; (4) may be transported over long distances; and (5) take relatively little energy to produce. Disadvantages of chemical signals are that they cannot be changed quickly and are slow to act.

Chemicals that are synthesized by one organism and affect the behavior of another organism are called *phero-mones*. Chemical signals are usually detected by olfactory receptors in the receiving animal. Many animals mark their territories by depositing odors that act as chemical signals to other animals of the same species. For example, many male mammals mark specific points in their territories with pheromones that serve to warn other males of their occupancy in the area. The same pheromones may also attract females that are in breeding condition.

Differences in the chemical structure of pheromones may be directly related to their function. Pheromones used for marking territories and attracting mates usually last longer because of their higher molecular weights. Airborne signals have lower molecular weights and disperse easily. For example, the sex attractant pheromones of female moths are airborne and can be detected by males a few kilometers away. Some chemicals released by ants to signal alarm are known to fade away within one minute.

Figure 40.3 Subordinant male chacma baboon grooming a dominant male.

Development of Behavior

For a normal behavior pattern to develop, the genes that code for the formation of the structures and organs involved in the behavior must be present. For example, in vertebrates, normal locomotion movements will not occur without proper development and growth of the limbs. This process requires some interaction with the animal's environment because proper nourishment, water balance, and other factors must be maintained for normal development to occur.

Maturation

Some behavior patterns appear only after a specific developmental stage or time is reached. During **maturation**, there is improvement in the performance of the behavior pattern as parts of the nervous system and other structures complete development. A classic example is tail movement [4] in frog embryos that are near hatching. While still in the egg membranes, they start moving their tails as they would if they were swimming, and the coordination of movements improves with time. These improved movements are due to maturation, not practice or experience.

Instinct/Learning Interactions

In recent years, many behavioral scientists have concluded that both instinct and learning are important in the behavior of animals. Interaction of inherited and learned components shapes a number of behavior patterns. For example, young bobcats raised in isolation without the chance to catch live prey did not attack a white rat placed with them, unless the rat tried to escape. At first, their attacks were not efficient, but after some experience, they were seizing prey by the neck and rapidly killing them. Apparently, inherited components of this behavior pattern are refined by learning. Under normal conditions, the learning or experiences occur during play with litter mates.

Another example involving instinctive and learned components to behavior may be seen in the nut-cracking behavior of squirrels. Squirrels gnaw and pry to open a nut. Inexperienced squirrels are not efficient; they gnaw and pry at random on the nut. Experienced squirrels, however, are very efficient. They gnaw a furrow on the broad side, then wedge their lower incisors into the furrow, and crack the nut open (figure 40.4).

[5] Many species of birds produce relatively simple vocalizations that are considered innate or instinctive. The perching birds of the order Passeriformes are known for their rich, complex vocalizations often called *songs*. The songs may function as reproductive signals and they typically result from an interaction of inherited and learned components of a song. Some birds inherit a predisposition to learn the songs characteristic of their species.

Imprinting

During **imprinting**, a young animal develops an attachment toward an animal or object (figure 40.5a). The attachment usually forms only during a specific sensitive period called the *critical period* that occurs soon after hatching or birth and is not reversible (figure 40.5b). Imprinting is a rapid learning process that apparently occurs without reinforcement.

[6] Some pioneering work on imprinting was done by Konrad Lorenz. He did experiments with geese in which he allowed the geese to imprint on him (figure 40.5c). The

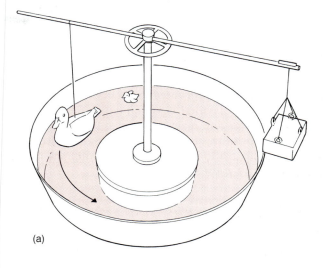

(a)

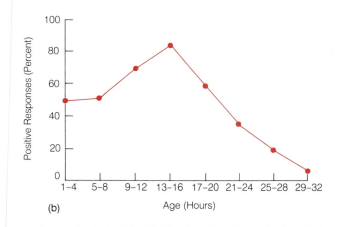

(b)

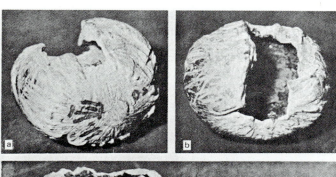

Figure 40.4 Nut-cracking behavior in squirrels involves instinctive and learned components. (a,b) An inexperienced squirrel has opened this hazelnut as evidenced by the many gnawing marks made at random. (c,d) An experienced squirrel has opened this hazelnut efficiently by gnawing and prying on the broad side of the nut to crack it open.

(c)

Figure 40.5 Imprinting in young birds. (a) A circular arena apparatus is used to imprint the young duckling on its "mother" (a stuffed model). (b) The critical period for imprinting is most likely to occur 13 to 16 hours after birth. (c) Imprinted geese following Konrad Lorenz.

goslings followed him as though he was their mother. In nature, many species of birds in which the young follow the parent soon after birth use imprinting so the young can identify with or recognize their parent(s). They can then be led successfully to the nest or to water. Visual and/or auditory cues may also be important in imprinting systems.

Imprinting has been reported to influence social development in some mammals. Dogs have a sensitive period between 3 and 10 weeks of age, during which social contacts are formed. If prevented from having normal social contacts, their behavioral development is abnormal. Mate selection in birds and mammals is also known to be influenced by imprinting. Imprinting is not limited to young animals. Some adult female mammals imprint on their young, often by auditory cues.

Learning

Learning produces changes in the behavior of an individual that are due to experience. Learning is considered adaptive because it allows an animal to respond quickly to changes in its environment. Once an animal learns something, its behavioral choices are increased.

An animal's ability to learn may be correlated with the predictability of certain characteristics of its environment. Where certain changes in the habitat occur regularly and are predictable, the animal may respond to a stimulus with an instinctive behavior very rapidly and without modification. An animal would not necessarily benefit from learning in this situation. However, where certain environmental changes are unpredictable and cannot be anticipated, an animal may modify its behavioral responses through learning or experience. This modification is adaptive because it allows an animal to not only change its response to fit a given situation but also to improve its response to subsequent, similar environmental changes.

Several different categories of learning have been identified, ranging from habituation (the simplest form of learning) to insight learning (the most complex form) that involves cognitive processes.

Habituation

Habituation is the simplest and perhaps most common type of behavior occurring in many different animals. Habituation involves a waning or decrease in response to repeated or continuous stimulation. Simply, an animal learns not to respond to stimuli in its environment that are constant and probably relatively unimportant. By habituating to unimportant stimuli an animal conserves energy and time which might be better spent on other important functions. For example, after time, birds learn to ignore scarecrows that previously caused them to flee. Squirrels in a city park adjust to the movements of humans and automobiles. If the stimulus should be withheld, then the response returns rapidly. Habituation is considered to be innate and does not involve any conditioning. Habituation is believed to be controlled through the central nervous system, and should be distinguished from sensory adaptation. *Sensory adaptation* involves repeated stimulation of receptors until they stop responding. For example, if one enters a room where there is an unusual odor, the olfactory sense organs stop responding to these odors, after some time. [7]

Classical Conditioning

Classical conditioning is a type of learning documented by the Russian physiologist, Ivan Pavlov in 1927. In his classic experiment on the salivary reflex in dogs, he presented food right after the sound of a bell (figure 40.6). After a number of such presentations, the dogs were conditioned—they associated the sound of the bell with food. It is then possible to bring about the dog's usual response to food—salivation—with just the sound of the bell. The food is a positive reinforcement for salivating behavior, but responses can also be conditioned using negative reinforcement.

Classical conditioning is very common in the animal kingdom. Birds learn to avoid certain brightly colored caterpillars that have a noxious taste. Because birds may associate the color pattern with the bad taste, animals with a similar color pattern may also be avoided.

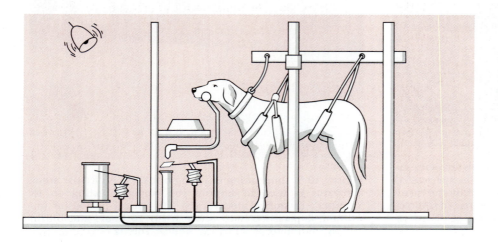

Figure 40.6 The apparatus used by Pavlov to demonstrate classical conditioning. If a ringing bell is presented just before a dog is given food, over several repetitions the dog will begin to associate the stimulus with the food. The dog becomes conditioned to salivate with the ringing of the bell alone.

Instrumental Conditioning

In **instrumental conditioning** (also known as trial-and-error learning), the animal shows appetitive behavior during which it carries out certain searching actions, such as walking and moving about. If the animal should, for example, find food during these activities, the food acts as reinforcement for the behavior and is the reward associated with the behavior. If this association is repeated several times, the animal has learned that the behavior leads to reinforcement.

A classic example of instrumental learning is that of a white rat in a *Skinner box,* developed by B. F. Skinner, a prominent psychologist (figure 40.7). When placed in the box, the rat will begin to explore. It moves all about the box and, by accident, eventually presses a lever and is rewarded with a food pellet. Because food rewards are provided with each pressing of the lever, the reward is associated with the behavior. Through repetition, the rat learns to press the lever right away to receive the reward. In this type of learning, the animal is said to be instrumental in providing its own reinforcement.

[8]   In instrumental conditioning, a ''shaping'' of the behavior can be manipulated by providing the reinforcement (food) whenever the animal comes close to the lever and continuing to supply reinforcement when the animal touches the lever. Finally, the animal learns to press the lever to obtain food.

The learning of new motor patterns by young animals often involves instrumental conditioning. A young bird learning to fly or a young mammal at play may improve coordination of certain movements or behavior patterns by practice during these activities.

Figure 40.7 Instrumental conditioning in a Skinner box. The white rat through random movements accidentally pressed the lever and received a reward. Through repetition the behavior of pressing the lever comes to be associated with the reward.

Latent Learning

Latent Learning, sometimes called exploratory learning, involves making associations without immediate reinforcement or reward. The reward is not obvious. There is, however, apparently some motivation for an animal to learn about its surroundings. For example, if a rat is placed in a maze that has no food or reward, it will explore the maze, although rather slowly. If food or other reward is provided, the rat quickly runs the maze. Apparently previous learning of the maze had occurred, but remained latent or hidden until an obvious reinforcement was provided. Latent learn-   [9] ing allows an animal to learn about its surroundings as it explores. Knowledge about an animal's home area may be important for its survival, perhaps enabling it to escape from a predator or capture prey.

Insight Learning

In **insight learning** the animal uses cognitive or mental processes to associate experiences and solve problems. The classic example is the work of Wolfgang Kohler in 1929 on chimpanzees that were trained to get food rewards by using tools (figure 40.8a). One chimpanzee was given some bamboo poles that could be joined together to make a longer pole. Some bananas were supported from the ceiling. Once a longer pole was formed, the chimp went right away to the fruit and used the long pole to knock the bananas to the cage floor. Kolher believed that the animal used insight learning to get the bananas.

In addition, Jane van Lawick-Goodall has observed chimpanzees in the wild using tools to accomplish various tasks. For example, they use a twig or stem that has been stripped of its leaves to remove termites from their tunnels (figure 40.8b).

Stop and Ask Yourself

9. What happens during the maturation of a behavior pattern?
10. What is meant by ''an inherited disposition to learn''?
11. What are some functions of imprinting in animals?
12. What does learning do for an animal that instinct does not?

Social Behavior

Social behavior typically refers to any interactions among members of the same species, but it has been applied to animals of different species, excluding predator-prey interactions.

Living In Groups

Animal populations are often organized into groups. A group of animals may form an *aggregation* for some simple purpose, such as feeding, drinking, or mating. Several *Drosophila* flies on a piece of rotting fruit is an example of an aggregation. A true animal *society* is a stable group of individuals of the same species that maintain a cooperative social relationship. This association typically extends beyond the level of mating and taking care of young. Social behavior has evolved independently in many species of animals; complex social organizations may be found in invertebrates as well as vertebrates.

One major benefit of belonging to a group may be that it offers protection against predators (figure 40.9). There is safety in numbers, and the detection of predators may be enhanced by having several individuals on alert to warn against an intruder. In a group, some individuals can be on alert while others are free to be involved in other activities. Also, cooperative hunting and capture of prey would increase the feeding efficiency of predators.

A disadvantage of group living may be competition for resources. The value of group living depends on the species and behaviors involved.

Agonistic Behavior and Territories

In a society of animals, there is usually some maintenance of social structure and spacing of group members. This pattern is often accomplished through *agonistic behavior* in which one animal is aggressive or attacks another animal that responds by either returning the aggression or submitting. In rare cases, agonistic behavior is lethal, but usually the animals are not killed or even severely injured. In many species, males vent much of their aggression in the form of *threat displays*. Displays typically involve signals that serve to warn other males of an intention to defend an area or territory (figure 40.10). Although agonistic behav-

Figure 40.8 Insight learning in chimpanzees. (a) This chimpanzee put two sticks together and stood on stacked boxes to reach the bananas that had been suspended from the ceiling. (b) This chimpanzee is using a tool to obtain ants for food.

Figure 40.9 A flock of starlings reacts to a bird of prey by forming a tight group, reducing the chances of the predator singling out any particular individual.

Figure 40.9 redrawn, with permission, from Manning, *An Introduction to Animal Behavior*, **3rd edition**. Copyright © 1979, Benjamin/Cummings Publishing Company, Menlo Park, California. Reprinted by permission.

ior may seem antisocial, it is important for maintaining the social order. Agonistic behavior is especially important in the maintenance of territories and dominance hierarchies.

A territorial animal uses agonistic behavior to defend a site or area against certain other individuals. The site is known as its **territory,** and competing individuals are excluded from it. Many male birds and mammals occupy a breeding territory for part of the year. A male will actively defend his area against other males so that he can attract a female and court her without interference from other males. Territories may offer certain advantages to the occupants in addition to being a location for attracting a mate and rearing young. Territories may contain a food supply or provide shelter to avoid predators and unfavorable climate.

Dominance Hierarchies

In **dominance hierarchies** a group of animals are organized in such a way that some members of the group have greater access to resources, such as food or mates, than do others. Those near the top of the order have first choice of resources, whereas those near the bottom go last and may do without if resources are in short supply. An example of a dominance hierarchy is the "peck-order" of chickens in a pen. When a small group of chickens are placed together, fights will occur among them until a linear hierarchy of dominance is established. Higher-ranked chickens are among the first to eat and may peck lower-ranked chickens. Once the hierarchy is set, peaceful coexistence is possible. Occasional fights will occur if a bird tries to move up in the order.

Dominance hierarchies exist in many vertebrate groups, the most common being in the form of linear relationships,

Figure 40.10 A baboon exhibiting a threat display toward other members of the troop.

although triangular relationships may form. In baboons, the strongest male is usually highest in the rank order. But, sometimes, older males may band together to subdue a stronger male and lead the troop.

Altruism

In **altruism,** an individual gives up or sacrifices its own reproductive potential to benefit another individual. For example, one individual of a group of crows gives an alarm call to warn other individuals of the group of an approaching predator, even though the call may attract the predator to the sender of the signal. How is it possible that such behavior evolved? Are normal natural selection processes at work here?

To be successful in a biological sense, an animal must produce as many young as possible, thereby passing its genes to succeeding generations. However, passing on genes can be accomplished by aiding a relative and its young because they probably share some genes. In terms of reproductive potential or output, it is theoretically possible that an individual may pass more genes to the next generation by aiding the survival of relatives than it would rearing its own young.

A well-known example of altruism occurs in societies of hymenopteran insects, such as honeybees. The male drones are haploid, and the female workers and queen are diploid, resulting in a genetic asymmetry. Diploid workers share, on the average, 3/4 of their genes with their full sisters and only 1/2 with their own offspring. Thus, female honeybees may have more genes in common with their sisters than with their own offspring. The workers may pass more genes to the next generation by helping their mother produce more full sisters, some of whom may become reproductive queens, than if they produce their own young.

William Hamilton proposed the idea of *kin selection* to explain how selection acting on related animals can affect the fitness of an individual. In this way, a gene carried by a particular individual may pass to the next generation through a related animal. An individual's fitness is therefore based on both the genes it passes on, as well as those common genes passed on by relatives. An altruistic gene could therefore be passed on by the individual carrying it or by a relative who also carries it. Obviously, for kin selection to work, individuals of a group must be able to identify relatives, as occurs in small groups of primates and in social insects.

Stop and Ask Yourself

13. What are some benefits of being a member of a group?
14. Why is agonistic behavior so common in animal societies?
15. What are some benefits of having a territory?
16. How does a dominance hierarchy become established in a group of animals?

Summary

1. Animal behavior includes the many activities of an animal during its lifetime. There are several approaches to the study of animal behavior: experimental psychology, ethology, behavioral ecology, and sociobiology.

2. Natural selection influences animal behavior just as it does other characteristics of an animal. Certain behavioral traits that allow them to survive and reproduce are favored.

3. An instinct is a genetically-controlled behavior that consists of stereotyped movements called fixed action patterns. Instincts allow an animal to respond to a specific environmental stimulus and may not be affected by experience. Instincts are characteristic for a species.

4. The motivation or internal readiness of an animal to perform a given behavior is influenced by various physiological and environmental factors. Certain environmental factors, such as increasing day length, may greatly affect the internal physiological state of an animal by changing hormonal levels or increasing activity of the central nervous system.

5. Stimulus filtering mechanisms in the nervous system serve to remove unimportant or irrelevant stimuli so that the animal responds only to certain stimuli that release specific behaviors. Stimulus filtering mechanisms may be modified by experience.

6. Communication in animals requires the use of clear signals by one animal and their reception by another. Visual, acoustic, tactile, and chemical signals are important channels in communication systems.

7. Certain behavior patterns require time for maturation, during which an improvement of the behavior occurs as parts of the nervous system and other structures complete development.

8. Many behavior patterns require an interaction of instinctive and learned components for efficient performance. In some instances, an animal may inherit a disposition to learn a specific behavior. Also, an animal may learn certain behavior patterns only during a specific sensitive period early in life.

9. Through learning, an animal can adjust quickly to changes in its environment. Learning is adaptive for animals in an environment where changes are not predictable. The types of learning known to occur in animals include habituation, classical conditioning, instrumental conditioning, latent learning, and insight learning.

10. Many animal species live in groups in which various benefits are provided. Groups range from simple aggregations to more complex social organizations, or societies.

11. Agonistic behavior in the form of attacks or threat displays is important in spacing the members of a species or establishing and defending a territory. In some instances, a dominance hierarchy may exist in which the members may be ranked in order from the most dominant individual to the most subordinate individual. Once the hierarchy is established, agonistic behavior is reduced in the group.

12. In some societies, behavior in which one individual sacrifices its reproductive potential to help another individual occurs. Altruism may be a result of kin selection, in which aiding one's relatives enhances the spreading of these genes that are shared with relatives.

Key Terms

altruism (p. 663)
animal behavior (p. 655)
anthropomorphism (p. 655)
behavioral ecology (p. 655)
classical conditioning (p. 660)
communication (p. 657)
dominance hierarchies (p. 663)
ethologists (p. 655)
ethology (p. 655)
experimental psychologists (p. 655)
habituation (p. 660)
imprinting (p. 659)

innate releasing mechanism (IRM) (p. 657)
insight learning (p. 661)
instinct (p. 655)
instrumental conditioning (p. 661)
latent learning (p. 661)
learning (p. 660)
maturation (p. 658)
motivation (p. 656)
sign stimulus (p. 656)
sociobiology (p. 655)
stimulus filtering (p. 657)
territory (p. 663)

Critical Thinking Questions

1. What are several approaches to the study of animal behavior?

2. What roles do motivation and sign stimuli play in the instinctive behaviors?

3. What are some characteristics of imprinting? What are some examples of imprinting?

4. How can habituation be related to an animal's everyday life?

5. How can you distinguish between classical and instrumental conditioning? What are some examples of each that might occur in nature?

6. How is kin selection related to natural selection?

7. Will a sea gull incubate an infertile chicken egg placed in her nest? Explain your answer.

8. An advertisement for Boone Trail Nugget Company features a hiker eating a bowl of cereal on a mountaintop. What type of learning does this type of advertisement rely on? Explain your answer.

9. How do ethologists distinguish between proximate and the ultimate cause of behavior?

Suggested Readings

Books

Alcock, J. 1989. *Animal Behavior*. Sunderland, MA: Sinauer Associates, Inc.

Drickamer, L. C., and Vessey, S.H. 1986. *Animal Behavior: Concepts, Processes, and Methods*. Boston: Prindle, Weber, and Schmidt.

Gould, J. L. 1982. *Ethology: Mechanisms and Evolution of Behavior*. New York: Norton.

Manning, A. 1979. *An Introduction to Animal Behavior*. Reading, MA: Addison-Wesley.

Wallace, R. A. 1979. *Animal Behavior: Its Development, Ecology, and Evolution*. Glenview, IL: Scott, Foresman and Co.

Wilson, E. O. 1975. *Sociobiology: The New Synthesis*. Cambridge, Mass.: Belknap Press.

Articles

Axelrod, R., and Hamilton, W. D. 1981. The evolution of co-operation. *Science* 211:1390–1396.

Gould, J., and Marler, P. Learning by instinct. *Scientific American* January, 1987.

Hess, E. H. Imprinting in a natural laboratory. *Scientific American* February, 1972.

Page, R. E. 1989. Genetic specialists, kin recognition, and nepotism in honey bee colonies. *Nature* 338:576–579.

Partridge, B. L. The structure and function of fish schools. *Scientific American* June, 1982.

Scheller, R. H., and Axel, R. How genes control an innate behavior. *Scientific American* March, 1984.

Trivers, R. L. 1971. The evolution of reciprocal altruism. *Quarterly Review Biology* 46:35.

41

Ecology I: Individuals and Populations

Concepts

1. Ecology is the study of the relationships of organisms to their environment, and to other organisms. In part, it involves the study of how abiotic factors, such as energy, temperature, moisture, and light influence individuals.
2. Ecology also involves the study of populations. Populations grow, and growth is regulated by population density, the carrying capacity of the environment, and interactions between members of the same population.
3. Ecology also involves the study of individuals interacting with members of their own and other species. Herbivory, predator-prey interactions, competition for resources, and other kinds of interactions influence the makeup of animal populations.

Have You Ever Wondered:

[1] whether or not some animals can tolerate colder winter temperatures than others in the same species?
[2] why deer hunters in northern climates take larger animals than hunters in the south?
[3] whether or not bears really hibernate?
[4] why some animals never need to drink water?
[5] how an animal can tell that winter is coming, and that it is time to accumulate food reserves?
[6] what information life insurance companies use for establishing insurance rates?
[7] how long a bird, like a robin, can expect to live?
[8] why mosquitoes appear so quickly after a few days of rain?
[9] why predators usually do not hunt their prey to extinction?
[10] why a parasite usually does not kill its host?

These and other useful questions will be answered in this chapter.

This chapter contains underlined evolutionary concepts.

All animals have certain requirements for life. In searching out these requirements, animals come into contact with members of their own species and other species. The interactions that result mold the lives of the animals involved. Sometimes these interactions are obvious and even violent, such as interactions that occur between predator and prey. Other interactions may occur without the participants seeing each other. The multitude of interactions of animals with their environment and other animals are the basis for chapter 41.

Animals and Their Abiotic Environment

An animal's **habitat** (environment) includes all living (biotic) and nonliving (abiotic) characteristics of the area in which the animal lives. Abiotic characteristic of a habitat include factors such as availability of oxygen and inorganic ions, light, temperature and current or wind velocity. *Physiological ecologists* study abiotic influences and have found that animals live within a certain range of values for any environmental factor. This range is called the **tolerance range** for that factor. At either limit of the tolerance range, one or more essential functions cease. A certain range of values within the tolerance range, called the **range of optimum,** defines the conditions under which an animal is most successful (figure 41.1).

Combinations of abiotic factors are necessary for an animal to survive and reproduce. When one of these is out of the range of tolerance for an animal, that factor becomes a **limiting factor.** For example, even though a stream insect may have the proper substrate for shelter, adequate current to bring in food and aid in dispersal, and the proper ions to ensure growth and development, inadequate supplies of oxygen will make life impossible.

Often, the response of an animal to any of these factors is to orient itself with respect to it; such orientation is called **taxis.** For example, a response to light is called *phototaxis.* If an animal favors well-lighted environments, and moves toward a light source, it is said to display positive phototaxis. If it prefers low light intensities, it is said to display negative phototaxis.

Physiological ecologists have found that tolerance ranges often change in response to altered environmental conditions (e.g., changes in season). Catfish maintained at 25°C have a tolerance range for temperature that is 2 to 3°C lower than catfish maintained at 30°C. Such changes in tolerance ranges are called **acclimation**. [1]

Energy

Energy is the ability to do work. For animals, work includes everything from foraging for food to moving molecules around within cells. To supply their energy needs, animals ingest other organisms. That is, animals are **heterotrophic** (Gr. *hetero,* other + *tropho,* feeder). **Autotrophic** (Gr. *autos,* self + *tropho,* feeder) organisms carry on photosynthesis or other carbon fixing activities that supply their food source. An accounting of the total energy intake of an animal and a description of how that energy is used and lost is referred to as an **energy budget** (figure 41.2).

The total energy contained in food eaten by an animal is called the *gross energy intake.* Some food is undigested, and some food that is digested may not be absorbed into the bloodstream. Energy in unabsorbed, and undigested food is lost in the feces. Other food may be absorbed, but later lost to excretion in urine or perspiration. Energy lost in feces and through excretion is called *excretory energy.*

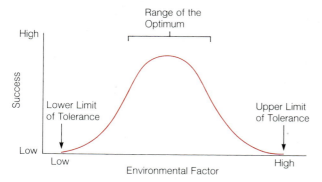

Figure 41.1 The tolerance range of an animal can be depicted by plotting changes in an environmental parameter versus some index of success. (Perhaps egg production, longevity, or growth.) The graphs that result are often, though not always, bell-shaped. The range of the optimum is the range of values of the parameter within which success is greatest.

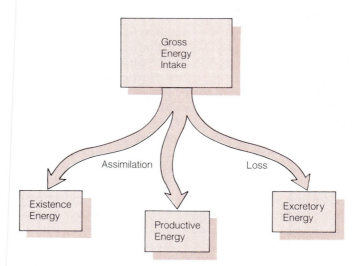

Figure 41.2 Energy budgets of animals. The gross energy intake of an animal is the sum of energy lost in excretory pathways, plus energy assimilated for existence and productive functions. The relative sizes of the boxes in this diagram are not necessarily proportional to the amount of energy devoted to each function. The total energy in an energy budget of an individual, and the amount of energy devoted to productive functions, depends upon various internal and external factors (e.g., time of year, reproductive status).

After excretory energy is accounted for, the remainder of gross energy intake is absorbed and is available for cell work. Some of this energy supports minimal maintenance activities, such as pumping blood; exchanging gases; supporting repair processes; and maintaining nervous, sensory, and endocrine functions. These base-level energy expenditures are called *standard metabolism* or *basal metabolism*. (These terms are not exactly equivalent. The former is used in reference to animals, such as lizards, that tend to assume a body temperature close to that of their surroundings, and the latter is used in reference to birds and mammals, who regulate body temperatures over a wide range of environmental temperatures. See the discussion of temperature classifications in chapter 38.) In addition to minimal maintenance activities, animals must support necessary activities, such as catching food, escaping predators, and finding shelter. The energy devoted to the sum of these activities is called *existence energy*.

After existence and excretory functions are accounted for, any energy that is left can be devoted to growth, mating, nesting, and caring for young. The portion of gross energy intake that exceeds excretory and existence energies is called *productive energy*. Survival requires that individuals acquire enough energy to supply these productive functions. Favorable energy budgets are sometimes difficult to attain, especially in temperate regions where winter often makes food supplies scarce.

Temperature

Part of an animal's existence energy is expended in regulating body temperature. Temperature influences the rates of chemical reactions, both in chemistry laboratories and in animal cells. Rates of chemical reactions in animal cells (metabolic rate) affect the overall activity of the animal. As discussed in chapter 38, the body temperature of an animal seldom remains constant because of an inequality between heat loss and heat gain. Heat energy can be lost to objects in an animal's surroundings as infrared and heat radiation; to the air around the animal through convection; and as evaporative heat loss. On the other hand, heat is gained from solar radiation, infrared and heat radiation from objects in the environment, and metabolic activities. Thermoregulatory needs influence many habitat requirements, such as the availability of food, water, and shelter.

Body size and metabolic rates are interrelated. In a particular taxonomic category, smaller animals have higher rates of metabolism than do larger animals. For example, the metabolic rate of a mouse is in excess of 20 times higher per gram of tissue than the metabolic rate of an elephant. The reason for this relationship between size and metabolic rate is not entirely clear. A traditional explanation is that small animals have a higher surface area to volume ratio, and thus as body size increases, there is relatively less body surface over which heat is lost (figure 41.3). This explanation, however, is less than satisfactory. Ectotherms (animals that use external heat sources for regulation of body temperature), whose rate of heat loss is less influenced by body size, show the same relationship of metabolic rate to body mass as endotherms.

The surface-area explanation, however, probably does explain another observation. Members of a species living in cold climates tend to be larger than members of the same species in warm climates. The smaller surface area per unit of body mass of large animals reduces heat loss compared to smaller animals of the same species. [2]

Animal Inactivity   When food becomes scarce, or when they are not feeding for other reasons, animals are subject to starvation. This problem is especially severe for small endotherms, whose metabolic rates are high and whose energy reserves are small. Under these circumstances, metabolic activities may decrease dramatically.

Torpor is a time of decreased metabolism and lowered body temperature that occurs in daily activity cycles. Hummingbirds and bats must feed almost constantly when they are active. To survive daily periods of inactivity, their body temperature and metabolic rate drop dramatically. A lower metabolic rate conserves energy and allows hummingbirds and bats to survive periods when they do not feed. At the end of a period of torpor, metabolic activity and body temperature rise quickly.

Hibernation is a time of decreased metabolism and lowered body temperature that may last for weeks or months. True hibernation occurs in small mammals, such as rodents, shrews, and bats. The set-point of a hibernator's thermoregulatory center drops to about 20°C, but thermoregulation is not suspended. This maintenance of thermoregulation distinguishes hibernation from the winter inactivity of ectotherms. Snakes, for example, overwinter in retreats that are unlikely to experience freezing temperatures. In an unusually cold winter, however, animals

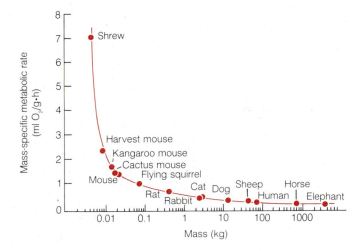

Figure 41.3 The effects of body size on metabolic rate for some mammals. Metabolic rate is expressed in ml O_2 consumed/g of body weight/hour. Note that the x-axis has a logarithmic scale to compress the difference between the weights of mouse and elephant. Metabolic rates and body size are inversely related.

Temperature Regulation

(b) This antelope jackrabbit (*Lepus alleni*) must get rid of excess body heat. Its huge, thin, highly vascularized ears have a large surface area for heat exchange.

(a) This Canadian muskox (*Ovibos moschatus*) must conserve body heat in order to survive in a cold environment. Its thick hair provides an excellent insulating blanket.

(c) This gaping alligator (*Alligator mississippiensis*) can keep its body temperature high and its brain temperature low by locking its jaws open as it rests.

(d) After a cold night, this eastern collared lizard (*Crotaphytus collaris*) basks in the Texas sunshine to raise its body temperature.

Animal Adaptations

(a) The arctic snowshoe hare (*Lepus americanus*) in its winter and (b) summer pelage. For most of the year this species is hidden by its cryptic coloration.

(c) The color pattern of this tiger (*Panthera tigris*) provides effective camouflage that helps when stalking prey.

(d) This decorator crab (*Loxorhynchus crispatus*) camouflages itself to gain protection from predators.

(e) Mullerian mimicry. These six species of *Heliconius* are all distasteful to bird predators. A bird that consumes any member of the six species will be likely to avoid all six species in the future.

(f) A mutualistic relationship between a clownfish (*Amphiprion perideraion*) and a sea anemone. The clownfish provides the anemone with scraps of its food and the anemone provides protection for the clownfish. The clownfish is apparently covered by mucus that does not elicit discharge of the anemone's stinging organelles called nematocysts.

(g) Flowers and their insect pollinators have coevolved for millions of years. Typical hummingbird-pollinated flowers are red and they produce large amounts of nectar. As a hummingbird feeds on the nectar with its long, thin beak, its head contacts the flower parts involved in pollination.

(Continued)

(h) Cleaning symbiosis is a common mutualistic relationship between marine animals. This Nassau grouper is being cleaned by two gobies.

(i) A cleaner shrimp is shown on the tentacles of a sea anemone.

(j) A mutualistic relationship between *Acacia* and the ant *Psuedomyrex*. The *Acacia* provides ants with nectar, protein, and nest sites. The ants defend the *Acacia* from herbivores and entangling vegetation.

(k) Another important mutualistic relationship involves algae called zooxanthellae and coral. In return for a place to live, the zooxanthellae provide this cactus coral (*Mycetophilia*) with food from algal photosynthesis.

may freeze to death because they have no internal thermoregulatory abilities.

[3] *Winter sleep* occurs in some larger mammals. These mammals are sustained through periods of winter inactivity by large energy reserves. Body temperatures do not drop substantially, and sleeping animals can wake and become active very quickly—a fact quickly learned by any rookie zoologist probing around the den of a sleeping bear!

Moisture

All life's processes occur in the watery environment of the cell. Water loss is inevitable, and it must be replaced if life is to continue. Water is lost in urine, feces, respiration, and through evaporation from body surfaces. Animals display numerous adaptations to lower or reduce the impact of water loss. Some animals reduce water loss by producing essentially dry feces. Others produce a concentrated urine and utilize excretory products that can be excreted in semisolid form. Controlling evaporative water loss involves a difficult set of compromises between water conservation and evaporative cooling. Often environments that require water conservation also require cooling. The nocturnal (night active) life-style of many desert animals is apparently an adaptation to reduce the need for evaporative cooling in a hot environment. Some animals, such as camels, can tolerate water losses of up to 25 to 40% of their total body weight.

[4] Water is acquired from food, drink, and metabolism. *Metabolic water* is water produced as a by-product of cellular metabolism. Numerous arthropods, and some mammals have such efficient water conservation mechanisms that metabolic water is sufficient for replacing water loss.

Aquatic animals also face water-regulation problems. Maintaining proper salt and water balances in freshwater environments involves conserving salts that tend to be lost to the environment and countering the osmotic influx of water into animal tissues. The problem in marine environments is just the opposite. Sea water is hypertonic to the tissues of many marine animals. As on land, water tends to be lost to the environment, and salts tend to accumulate in body tissues.

Light

Animals are also influenced by light conditions. *Photoperiod* is the length of the light period in a 24-hour day and is an accurate index of seasonal change. As spring approaches and daylight hours lengthen, birds enter breeding condition. Lengthening daylight hours indicate that con-
[5] ditions will soon be suitable for producing and rearing offspring. Similarly, changing photoperiod in autumn indicates that food supplies will soon be dwindling. This change may be an important cue to stockpile food reserves, or to find a suitable wintering site.

Light also influences cycles of daily activity, called **circadian** (L. *circa,* about + *diem,* day) **rhythms**. Many people have experienced air travel to different time zones and

the accompanying upset in timing of daily activities. This upset, commonly known as "jet lag," is a result of our innate (biological) clock telling us it is time for one activity (e.g., sleep) and external cues (e.g., daylight) telling us it is time for another (daytime) activity. Similar problems result from a change between "day shift" and "night shift" at a job. Recent research indicates that photoperiod is responsible for resetting our biological clocks so the new time of activity can seem normal.

Many animals have circadian rhythms. Freshwater and marine invertebrates have daily patterns of migration between the upper and lower levels of a water column. Many stream invertebrates have periods of activity at dawn and dusk. Terrestrial animals are either diurnal or nocturnal. These cycles of activity are controlled by biological clocks that are probably influenced by photoperiod.

Many animals have marked phototactic responses that help them find appropriate habitats. For example, many invertebrates living in the substrates of lakes and streams are negatively phototactic. Their daily activities are thus directed away from light sources and into sheltering substrates.

Stop and Ask Yourself

1. The tolerance range of a certain fish for dissolved oxygen is between 5 and 14 parts per million oxygen. What does this mean?
2. What is a shift in the tolerance range of an organism for a particular environmental parameter called?
3. What are three avenues of energy expenditure that contribute to the dissipation of gross energy intake?
4. Why is photoperiod a more reliable indicator of seasonal change than is temperature?

Populations

Populations are groups of individuals of the same species that occupy a given area at the same time, and have unique attributes. These attributes include growth rate, age structure, sex ratio, and mortality rate.

Population Growth

Animal populations change over time as a result of birth, death, and dispersal. One way to characterize a population with regard to the birth and death of individuals is with a *life table.* A group of individuals born into a population in a given time interval is called a *cohort.* For each cohort:

x is the age of individuals at the beginning of a time interval;

I_x is the number of individuals alive at the beginning of an interval, the *survivorship;*

d_x is the number of individuals dying within an interval, the *mortality;*

q_x is the *mortality rate, d_x/I_x*; and

e_x is the the average time left for individuals at the beginning of an interval, the *life expectancy.*

[6] Life tables, such as that in table 41.1, are used by life insurance companies to determine risks for any age group and to establish insurance policy rates.

Similar information can be derived from *survivorship curves* (figure 41.4). The y-axis of a survivorship graph is a logarithmic plot of numbers of survivors, and the x-axis of the graph is a linear plot of age. Note that there are three kinds of survivorship curves. Individuals in type I (convex) populations survive to an old age, then die rapidly. Environmental factors are relatively unimportant in influencing mortality, and most individuals live their potential lifespan. Some human populations approach type I survivorship. Individuals in type II (diagonal) populations have a constant probability of death throughout their lives. The environment has an important influence on death, and the environment is no harsher on the young than on the old. Populations of birds and rodents often have type II survivorship curves. Individuals in type III (concave) populations experience very high juvenile mortality. Those reaching adulthood, however, experience a much lower mortality rate. Fish and many invertebrates display type III survivorship curves.

Mortality in natural populations is often high. In birds, for example, it is not uncommon for each cohort to ex-

Table 41.1	A Life Table for Males and Females in the United States (1981)			
Age (years)	Survivorship (number alive at beginning of age interval)	Mortality (number dying during age interval)	Mortality Rate	Life Expectation (average number of years left to an individual at beginning of age interval)
x	l_x	d_x	q_x	e_x
Male				
0–1	100,000	1,824	0.0182	68.3
1–5	98,176	323	.0033	68.5
5–10	97,853	220	.0023	64.7
10–15	97,633	238	.0024	59.9
15–20	97,395	735	.0075	55.0
20–25	99,660	927	.0096	50.4
25–30	95,733	799	.0083	45.9
30–35	94,934	845	.0089	41.3
35–40	94,089	1,154	.0123	36.6
40–45	92,935	1,840	.0198	32.0
45–50	91,095	3,013	.0331	27.6
50–55	88,082	4,542	.0516	23.5
55–60	83,540	6,913	.0827	19.6
60–65	76,627	9,618	.1255	16.1
65–70	67,009	12,014	.1793	13.1
70–75	54,995	14,160	.2575	10.4
75–80	40,835	14,798	.3624	8.1
80–85	26,037	12,354	.4745	6.3
85 and over	13,683	13,683	1.0000	4.7
Female				
0–1	100,000	1,370	0.0137	75.9
1–5	98,630	252	.0025	75.9
5–10	98,378	164	.0017	72.1
10–15	98,214	141	.0014	67.2
15–20	98,073	285	.0029	62.3
20–25	97,788	314	.0032	57.5
25–30	97,474	339	.0035	52.7
30–35	97,135	461	.0047	47.9
35–40	96,674	687	.0071	43.1
40–45	95,987	1,093	.0114	38.4
45–50	94,894	1,710	.0180	33.8
50–55	93,184	2,425	.0260	29.3
55–60	90,759	3,649	.0402	25.1
60–65	87,110	5,096	.0585	21.0
65–70	82,014	7,370	.0899	17.1
70–75	74,644	10,784	.1445	13.6
75–80	63,860	15,078	.2361	10.4
80–85	48,782	17,237	.3534	7.8
85 and over	31,545	31,545	1.0000	5.7

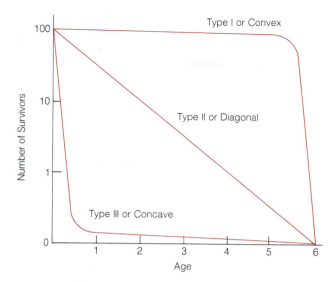

Figure 41.4 Survivorship curves are a plot of number of survivors (a logarithmic plot) versus age. Type I curves apply to populations in which individuals are likely to live out their potential life span. Type II curves apply to populations in which mortality rates are constant throughout age classes. Type III curves apply to populations in which mortality rates are highest for the youngest cohorts.

[7] perience greater than 50% mortality each year. The potential life-span of the American robin, for example, is 11 years; however, the average robin probably lives only 1.4 years.

Exponential Growth   The potential for population growth can be demonstrated in the laboratory under conditions that provide abundant resources and space. For example, placing a few fruit flies in a large culture jar containing abundant food allows the fruit flies to reproduce very rapidly. One female fruit fly will lay in excess of 50 eggs. Reproductive adults develop in about 14 days, and there are approximately equal numbers of male and female offspring. For each female that began the population, 50 flies would be expected 2 weeks later. For each female in the second generation, 50 more flies would be produced after 2 more weeks, and so on. In other words, the population is experiencing **exponential growth.** Exponential growth is shown in figure 41.5 in arithmetic and logarithmic plots. Rather than increasing by adding a constant number of individuals to the population in every generation, the population increases multiplicatively.

All populations do not display the same capacity for growth. Factors such as the number of offspring produced, the likelihood of survival to reproductive age, the duration of the reproductive period, and the length of time it takes to reach maturity all influence reproductive potential. The capacity of a population to increase maximally is called its **biotic potential** or **intrinsic rate of growth.** It is symbolized by *r,* and defined as the rate of population increase under optimal conditions.

Logistic Population Growth   It should be obvious that exponential growth cannot occur indefinitely. In some years, for example, American robin populations are larger than in other years; on average, however, robin populations are relatively constant. The constraints placed on a population by climate, food, space, and other environmental factors is called **environmental resistance**. The population size that a particular environment can support is called the environment's **carrying capacity,** and is symbolized by *K* (box 41.1). In these situations, growth curves assume a sigmoid, or flattened-S shape (figure 41.6a).

The effects of environmental resistance are often not instantaneous. For example, a field of grain may support a large population of young field mice, but as mice mature and acquire larger appetites, resources may dwindle. Populations on the increase, therefore, may exceed the carrying capacity. Eventually, increased death rate will cause the population to decrease to, or go below, *K*. Growth curves may thus cycle around *K* or become J shaped (figure 41.6b,c).

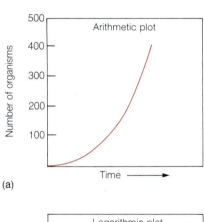

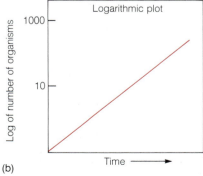

Figure 41.5 Exponential growth. (a) The number of organisms (y axis) is plotted on an arithmetic scale and results in a curve with increasing slope. (b) The number of organisms is plotted on a logarithmic scale and results in a graph with a straight line.

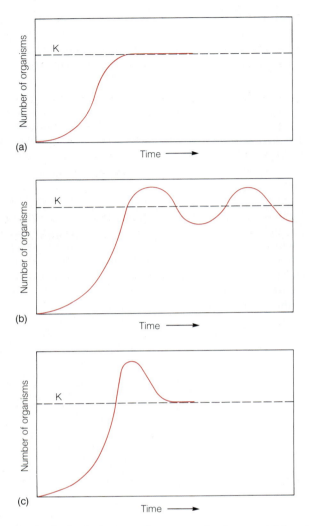

(a)

Time ⟶

Number of organisms

K

(b)

Time ⟶

Number of organisms

K

(c)

Time ⟶

Number of organisms

K

Figure 41.6 Logistic growth curves take into consideration the fact that limited resources place an upper limit on population size. (a) When carrying capacity (*K*) is reached, population growth levels off, creating an S-shaped curve. (b-c) During its exponential growth phase, a population may overshoot carrying capacity because demand on resources may lag behind population growth. When that happens, numbers may cycle on either side of *K*, or form a J-shaped curve.

Population Regulation

The conditions that must be met for an animal to survive are unique for every species. What many species have in common, however, is that population density and competition impact populations in predictable ways. In addition, unstable environmental conditions may prevent animal populations from ever reaching the carrying capacity of the environment.

Population Density Some factors influence the number of animals in a population without regard to the number of individuals per unit space (density). For example, weather conditions often limit populations. An ex-

tremely cold winter with little snow cover may devastate a population of lizards sequestered beneath the litter of the forest floor. Regardless of the size of the population, a certain percentage of individuals will freeze to death. Activities of humans, such as construction and deforestation, often affect animal populations in a similar fashion. Such factors are said to regulate populations in a *density-independent* fashion.

Other regulatory factors are more severe when population density is high (or sometimes very low) than they are at other densities. Such regulatory factors are said to be *density dependent*. Animals often use territorial behavior, song, and scent marking to tell others to look elsewhere for reproductive space. These actions become more pronounced as population density increases, and are thus density dependent. Other density dependent factors include competition for resources, disease, predation, and parasitism.

Intraspecific Competition *Competition* occurs when animals utilize similar resources and, in some way, interfere with each other's procurement of those resources. Competition between members of the same species is called **intraspecific competition** and is often intense, because the resource requirements of individuals are nearly identical. Intraspecific competition may occur without individuals coming into direct contact. (The "early-bird that gets the worm" doesn't necessarily see later arrivals.) In other instances, the actions of one individual directly affect another. Territorial behavior and the actions of socially dominant individuals are examples of direct interference.

K and r Selection In some instances, the logistic-growth model accurately characterizes a population. Populations whose size tends to evolve toward their carrying capacity are said to be **K-selected.** They are regulated by density-dependent factors, and are found in relatively predictable environments. K-selected populations tend to be made up of long-lived, slowly maturing species. Reproductive rates are relatively low, and the young may receive substantial parental care. Most populations of large mammals are K-selected.

In other instances, climatic (and other) factors may prevent populations from achieving their carrying capacity. When population density is maintained substantially below the carrying capacity, ecologists say that the population is **r-selected.** Populations that are r-selected tend to be regulated by density-independent factors. Individuals may occupy marginally adequate habitats and reproduce very quickly when conditions permit. One female produces many young, and developmental stages mature quickly with little parental care. Efficient dispersal systems distribute eggs, young, or adults. In other words, the biotic potential of r-selected populations has been maximized during evolution. A mosquito population that increases rapidly a few days after a rainy period is a familiar example of r-selection.

[8]

Box 41.1

Exponential and Logistic Growth—
The Mathematical Models

If you go on in science, you will learn that mathematics is very useful in helping scientists to describe and analyze the natural world. The descriptions of exponential and logistic growth are good examples.

Exponential growth can be represented by the following differential equation:

$$dN/dt = rN$$

In this equation, dN/dt is the change in the number of individuals (N) in a population per time interval (t), when that time interval is very small. rN is the instantaneous growth rate, expressed as the number of individuals added to a population per unit time. Biotic potential (r) is the increase in numbers of individuals per reproductive individual per time interval, and (N) is the number of individuals in the population. Substituting a few numbers into the equation will demonstrate the impact of r and N on population growth rates:

r	N	rN
0.2	50	10
0.4	50	20
0.2	100	20
0.4	100	40

The formula for logistic growth takes carrying capacity of the environment (K) into consideration:

$$dN/dt = rN([K - N]/K)$$

In this formula, $[K - N]/K$ is a measure of environmental resistance. When the population (N) is small, $[K - N]/K$ approaches one, and population growth nearly achieves its full potential. When the population is large, however, $[K - N]/K$ approaches zero, and population growth is suppressed. This relationship is illustrated with some simple calculations:

r	N	K	$rN([K - N]/K)$
0.5	10	100	45
0.5	90	100	5

Stop and Ask Yourself

5. What does the survivorship curve of a type III population look like?
6. Are environmental factors, such as weather, very important, moderately important, or unimportant in determining survivorship of juvenile stages in r-selected species?
7. What factor(s) are taken into consideration in logistic growth calculations that place a limit on exponential growth?
8. What is the difference between density-independent and density-dependent population regulation? Give examples of each.

Interspecific Interactions

Members of other species can affect all characteristics of a population. Interspecific interactions include herbivory, predation, competition, and symbiosis. Animals, however, are rarely limited by artificial categories that zoologists cre-

ate to help organize life's complexity. As you study the following material, realize that animals often do not interact with other animals in only one way. The nature of interspecific interactions may change as an animal matures, as seasons change, or as the environment changes.

Herbivory

Animals that feed on plants by cropping portions of the plant, but usually not killing the plant, are called *herbivores*. Some are *grazers* (e.g., cattle and bison), which means they feed on grasses and other herbaceous vegetation. *Browsers* (e.g., deer and rabbits), feed on the leaves and twigs of woody plants. Other forms of herbivory include *frugivory* (feeding on fruits), seed eating, nectar and sap feeding, and pollen eating.

Predation

Predators are animals that feed by killing and eating other organisms. Predators influence the distribution and abundance of prey, and conversely, the distribution and abundance of prey influences predator populations. For example, the lake trout (*Salvelinus namaycush*) is the only

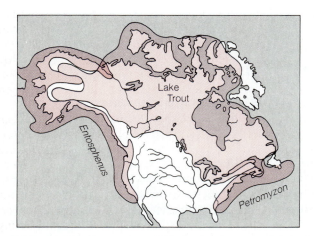

Figure 41.7 The distribution of lake trout (*Salvelinus namaycush*) in North America before the construction of the Welland Canal (light shading). Heavily shaded areas show the distribution of sea lampreys.

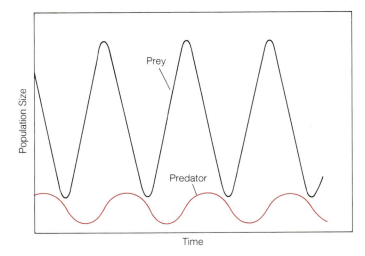

Figure 41.8 Mathematical predictions of predator and prey abundance. Plotting population size versus time produces cycles in which fluctuations in numbers of predators lag behind fluctuations in numbers of prey. Note that predator populations are usually considerably smaller than prey populations.

freshwater fish whose native distribution included virtually all of North America but never crossed the narrow Bering Strait into Siberia. In contrast, lake trout have crossed larger expanses of ocean to the freshwater systems of islands in northern Canada. Predation by the Pacific sea lamprey (*Entosphenus*) has probably prevented the movement of lake trout to Siberian freshwater systems (figure 41.7). This explanation is supported by the decline of lake trout in the Great Lakes after the construction of the Welland Canal allowed the Atlantic sea lamprey (*Petromyzon*) to move into the Great Lakes (*see box 27.1*).

Mathematical models have been generated that predict that numbers of predators and prey should cycle over time. Changes in the predator population are predicted to lag behind changes in prey population (figure 41.8). There have been numerous laboratory and field investigations of these models, and the predictions have generally not been supported. In some instances, the prey tend to become extinct; if no other food source is available, the predators will also die. In other instances, cycles exist, but they are not in the predicted phase or they occur in the absence of predation (figure 41.9).

[9] Interactions between predators and prey are complex and are affected by many characteristics of the environment. Coexistence of predator and prey apparently requires a habitat that allows prey to take refuge when their numbers are small. Under these conditions, predator populations decline, but some survive by finding other sources of food.

Interspecific Competition

When competition for resources occurs between members of different species, one species may be forced to move or become extinct, or the two species may share the resource and coexist.

Competitive Exclusion The **competitive exclusion principle** has influenced the thinking of ecologists

for over 50 years. It states that two species with exactly the same requirements for food, habitat, nest sites, and other conditions of life cannot coexist.

Numerous investigations in laboratories and observations in nature, however, have failed to document the universality of the competitive exclusion principle. Part of the reason for this failure is the fact that two species rarely have identical requirements for food, space, and other conditions of life.

Although it is an exception to the results of most investigations, competitive exclusion apparently did occur in the citrus groves of southern California. California red scale is an insect pest of citrus trees. A wasp parasite of scale insects, called the golden-naveled scalesucker, lays its eggs under the scalelike citrus pest, and larvae feed on and eventually kill the scale insect. In hopes of more efficient pest control, a second scalesucker, the Lingnan scalesucker, was introduced to southern California from China. The Lingnan scalesucker was able to outcompete the golden-naveled scalesucker, and within 2 years, the Lingnan species completely replaced the native species at some plots. Some laboratory studies, such as those of Thomas Park also suggest that competitive exclusion may occur in some cases of interspecific competition (figure 41.10).

Coexistence Most studies have shown that competing species can coexist. Coexistence can occur when species utilize resources in slightly different ways, and when the effects of interspecific competition are less severe than the effects of intraspecific competition. These ideas can be illustrated by the studies of Robert MacArthur on five species of warblers that all used the same caterpillar prey. Warblers partitioned their spruce-tree habitats by dividing a tree into preferred regions for foraging. Although

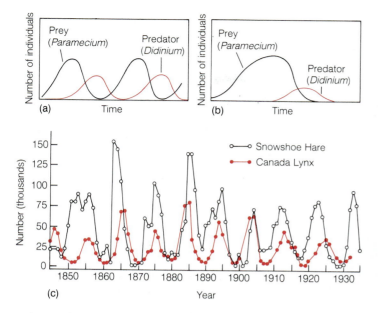

Figure 41.9 A test of the mathematical models of predator and prey abundance with two ciliated protozoa, *Didinium* (predator) and *Paramecium* (prey). (a) The populations were maintained, and cycled only by periodically introducing new individuals. (b) If new individuals were not introduced, the prey was fed on until extinction occurred. With its food source exhausted, *Didinium* also went extinct. (c) Cycling in populations of Canada lynx and snowshoe hare. The data are based upon numbers of pelts handled by the Hudson Bay Company. At first glance, the cycles resemble the mathematical predictions. Note, however, that cycles between predator and prey are almost exactly in phase. Other populations of snowshoe hare show similar cycles in the absence of predation by lynx.

there was some overlap of foraging regions, competition was limited, and the five species coexisted (figure 41.11).

Resources can be divided among competing species in many ways. Partitioning, however, cannot occur indefinitely because the total quantity of any resource is limited. If individuals in a population cannot obtain enough of a resource to support reproduction, then similarities between species become important limiting factors, and competitive exclusion may occur.

Coevolution

The evolution of ecologically related species is sometimes coordinated such that each species exerts a strong selective influence on the other. The adaptations that result are fascinating. This kind of evolution is called **coevolution**.

Coevolution may occur when species are competing for the same resource, or during predator-prey interactions. In the evolution of predator-prey relationships, for example, natural selection should favor the development of protective characteristics in prey species. Similarly, selection favors characteristics in predators that allow them to become better at catching and immobilizing prey. Coevolution in predator-prey relationships occurs when a change toward greater predator efficiency is countered by increased elu-

siveness of prey. If evolutionary rates are balanced, neither species would be expected to win this evolutionary "arms race." If not balanced, then one species could become extinct, and its role would likely be taken over by another species. Although there are many examples of predator-prey relationships, the evidence documenting their development through coevolution is scanty.

Coevolution is obvious in the relationships between some flowering plants and their animal pollinators (color-plate 22g). Early land plants were dependent on wind and rain to achieve pollination. Cross-pollination became more likely when pollen began to be carried from one flower to another by insects, birds, or bats. Coevolution is obvious in plants that display adaptations for attracting specific pollinators, or when animals display adaptations for extracting pollen and nectar from specific plants.

Flowers attract pollinators with a variety of elaborate olfactory and visual adaptations. Insect-pollinated flowers are usually yellow or blue, because insects best see these wavelengths of light. In addition, petal arrangements often provide perches for pollinating insects. Flowers pollinated by hummingbirds, on the other hand, are often tubular and red. Hummingbirds have a poor sense of smell, but see red very well. The long beak of hummingbirds is an adaptation that allows them to reach far into tubular flowers. Their hovering ability means they have no need of a perch. One particularly striking example of coevolution between a

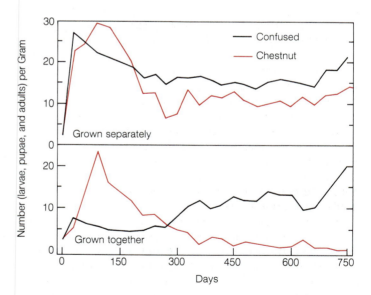

Figure 41.10 Thomas Park and his coworkers investigated interspecific competition between two species of flour beetles Confused and Chestnut. When grown separately, under the same conditions, each species achieved the carrying capacity of the culture container. When grown together, however, one of the two species eventually became extinct. Conditions of temperature and humidity also influenced which of the two species became extinct. Highly simplified environments constructed for laboratory experiments probably do not reflect the conditions experienced by the same organisms in natural environments. Therefore, one must always be careful in interpreting results of experiments such as these.

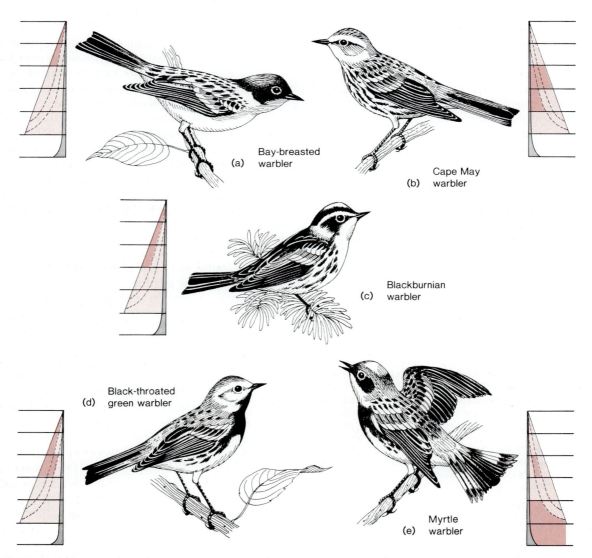

(a) Bay-breasted warbler

(b) Cape May warbler

(c) Blackburnian warbler

(d) Black-throated green warbler

(e) Myrtle warbler

Figure 41.11 Robert MacArthur found that five species of warblers (a–e) coexisted by partitioning resources. Partitioning occurred by dividing up spruce trees into preferred foraging regions.

hummingbird and a flowering plant occurs in the tropics. *Ensifera ensifera* is a hummingbird with an extremely long bill (in excess of 8 cm). It feeds almost exclusively on passion flowers, which store their nectar deep within tubular flowers. The structure of the flower excludes virtually all other nectar-feeding animals, limiting competition for the hummingbird and reducing the chances of cross-pollination with other plant species.

Symbiosis

Some of the best examples of adaptations arising through coevolution are seen when two different species live in continuing, intimate associations. These relationships are called **symbiosis** (Gr. *sym*, together + *bio*, life). Such interspecific interactions influence the species involved in dramatically different ways. In some instances, one member of the association benefits, and the other is harmed. In other cases, life without the partner would be impossible for both.

Parasitism **Parasitism** is a form of symbiosis in which one organism lives in or on a second organism, called a *host.* The host usually survives, at least long enough for the parasite to complete one or more life cycles. Parasitism is a very common way of life in the animal kingdom. Its impact on humans is greatest in the less-developed regions of the world, but even affluent societies must contend with infections of pinworms, mites, ticks, and lice.

The relationships between a parasite and its host(s) are often very complex. Some parasites have life histories involving multiple hosts. The *definitive* or *final host* is the host that harbors the sexual stages of the parasite. A fertile female in a definitive host may produce and release hundreds of thousands of eggs in its lifetime. Each egg that is released gives rise to an immature stage that may be a

[10]

parasite of a second host. This second host is called an *intermediate host,* and asexual reproduction occurs in this host. More than one intermediate host and more than one immature stage may be involved in some life cycles. For the life cycle to be completed, the final immature stage must have access to a definitive host. The likelihood that a particular egg will mature into a reproductive adult is usually remote. Producing large numbers of eggs is one of many parasitic adaptations that help ensure that a few individuals will eventually complete the life cycle. In any symbiotic relationship, however, one-half of the story is missed by focusing only on the parasite. Hosts may evolve mechanisms that help them resist parasitic infections. Many examples of coevolutionary interactions between host and parasite are cited in Part 4 of this textbook.

Parasitoids are animals that deposit eggs or other developmental stages on another animal. As immatures develop, they live off host tissues and eventually kill the host. Because the death of the host is the usual outcome, parasitoids combine features of parasite and predator.

Commensalism **Commensalism** is a symbiotic relationship in which one member of the relationship benefits, and the second is neither helped nor harmed. The distinction between parasitism and commensalism is somewhat difficult to apply in natural situations. Whether or not the host is harmed often depends on factors such as the nutritional state of the host. Thus, symbiotic relationships may be commensal in some situations and parasitic in others. One example of a commensal relationship involves *Entamoeba gingivalis,* a common ameba that lives in the mouth of humans. Although it is often associated with decay, it does not cause tooth or gum decay. It is passed from individual to individual by direct contact. Commensals may be provided with food, transportation (a relationship called *phoresis*), or shelter by their hosts.

Mutualism **Mutualism** is a symbiotic relationship in which both members of the relationship benefit. Examples of mutualism abound in the animal kingdom. Clownfish and sea anemones live together in close association. The clownfish gain protection from the anemones and in turn provide some food to the anemones (colorplate 22f). Another example of mutualism is the cleaning of marine fish by other organisms. In cleaning mutualism, small fish or invertebrates may swim into the mouth of larger fish to remove ectoparasites. Characteristic markings and behavior identify cleaner species, which enjoy relative safety in the mouths of predatory species, such as barracudas and groupers.

Coevolution may occur in any symbiotic relationship. A host species that has a long history of association with a particular parasite may evolve greater resistance to the parasite. Similarly, parasites that are extremely virulent may kill their host before the parasite has a chance to reproduce. Thus, lower pathogenicity on the part of the parasite is advantageous for most parasites. Long-standing symbiotic associations, therefore, often tend to be commensalistic or mutualistic in nature.

Commensalism or mutualism, however, are not guaranteed outcomes for all symbiotic relationships. It may be advantageous for a parasite to kill or weaken its host if perpetuation of a parasite species requires that a host be eaten. For example, certain acanthocephalans are larval parasites of amphipods and adult parasites of mallard ducks. Mallards acquire infections by eating freshwater crustaceans, called amphipods. When infected by acanthocephalan larvae, the behavior of amphipods changes dramatically. Their usual negative phototaxis is reversed so that they leave the bottom substrates of a lake or river, and swim near the surface where they are easy prey for mallard ducks.

Other Interspecific Adaptations

Many other characteristics of animals have been shaped by interspecific interactions. Predator/prey systems have led to some unusual adaptations, (e.g., coloration patterns) in prey species.

Camouflage Cryptic coloration (L. *crypticus,* hidden) occurs when an animal takes on color patterns in its environment. There are many examples of animals that blend in with their surroundings. In some cases, these are prey animals that are camouflaged to prevent them from being seen by predators. For example, the Arctic hare acquires a white coat at its fall molt to blend in with its snowy background (colorplate 22a). In the spring, however, when the snow melts and bare ground is exposed, another molt produces a brown coat (colorplate 22b).

Countershading is common in eggs of frogs and toads. These eggs are darkly pigmented on top and lightly pigmented on the bottom. When viewed by a bird or other predator from above, the dark of the top side hides the eggs from detection against the darkness below. On the other hand, when viewed from below by a fish, the light undersurface blends with the bright air/water interface. Similar countershading occurs in many species of fish, and decorator crabs mask themselves with algae and other concealing organisms to help themselves blend in with their environment (colorplate 22d).

Predators use camouflage to avoid detection by prey. The rufous-red coat overlaid with black stripes that breaks up the outline of a tiger in the forests of India, China, and Persia successfully hides the tiger as it stalks its prey (colorplate 22c).

Aposematic Coloration and Mimicry Some animals that protect themselves by being dangerous or distasteful to predators advertise their condition by conspicuous coloration. This coloration is a clear signal to predators to stay away. Certain dendrobatid frogs of Central and South America have toxic secretions in their skin, and are brightly colored to warn potential predators (*see box 28.1*). The sharply contrasting white stripe of a skunk and bright colors of poisonous snakes give similar messages. These color patterns are examples of warning or **aposematic coloration** (Gr. *apo*, away from + *sematic*, sign).

Resembling conspicuous animals may also be advantageous. **Mimicry** (L. *mimus*, to imitate) occurs when a species resembles one, or sometimes more than one, other species and gains protection by the resemblance. **Batesian mimicry** occurs when one species, called the *mimic*, resembles a second species, called the *model*, that is protected by aposematic coloration. An example of this form of mimicry occurs with monarch and viceroy butterflies (Colorplate 23i,j). Monarch butterflies lay their eggs on milkweed plants, and as caterpillars grow, they incorporate toxins from the milkweed into their tissues. Jays or other birds that eat insects apparently learn about monarchs through experience. Jays that have experienced the taste of a monarch will not only refuse to eat monarchs, but they will also refuse to eat the viceroys that mimic the monarchs.

A second form of mimicry, called **Mullerian mimicry**, occurs when the model and the mimic are both distasteful. A predator does not have to taste both of the species involved to encounter the distasteful form (colorplate 22e).

In this chapter, you have learned that animals spend their lives searching for combinations of physical require-ments that each must have to survive and reproduce. In these activities, animals interact with other organisms, and these interactions may either promote or interfere with the animal providing for its own needs. An even more complete understanding of how an animal lives comes from studying interactions that occur when many species share a common habitat. These interactions occur in communities and eco-systems and are covered in chapter 42.

Stop and Ask Yourself

9. What is competitive exclusion? Under what circum-stances is coexistence of competing species possi-ble?
10. What is parasitism? How does it differ from com-mensalism and mutualism? What is coevolution?
11. How might coevolution of parasites and their hosts lead to mutualism?
12. Under what circumstances might it be advantageous for parasites to kill their hosts?
13. What is cryptic coloration? Give an example.
14. What is mimicry? What are two forms it may take?

■ Summary

1. Many abiotic factors influence where an animal may live. Animals have a tolerance range and a range of optimum for environmental factors.
2. Energy for animal life comes from consuming autotrophs or other heterotrophs. It is expended in excretory, exist-ence, and productive functions.
3. Temperature, water, and light are important environmental factors that influence animal life-styles.
4. Animal populations change in size over time. Change can be characterized using life tables and survivorship curves.
5. Animal populations grow exponentially until the carrying capacity of the environment is achieved, at which point environmental resistance restricts population growth.
6. Population regulation may occur by density-independent or density-dependent mechanisms. Population control mechanisms may prevent some populations from achieving their carrying capacity.
7. Interspecific interactions influence animal populations. These interactions include herbivory, predator/prey inter-actions, interspecific competition, coevolution mimicry, and symbiosis.

■ Key Terms

acclimation (p. 667)
aposematic coloration (p. 678)
autotrophic (p. 667)
Batesian mimicry (p. 678)
biotic potential or intrinsic rate of growth (p. 671)
carrying capacity (p. 671)
circadian rhythms (p. 669)
coevolution (p. 675)
commensalism (p. 677)
competetive exclusion principle (p. 674)
countershading (p. 677)
cryptic coloration (p. 677)
energy budget (p. 667)
environmental resistance (p. 671)
exponential growth (p. 671)
habitat (p. 667)
heterotrophic (p. 667)
intraspecific competition (p. 671)
K-selected (p. 671)
limiting factor (p. 667)
mimicry (p. 678)
Mullerian mimicry (p. 678)
mutualism (p. 677)
parasitism (p. 676)
parasitoids (p. 677)
r-selected (p. 671)
range of optimum (p. 667)
symbiosis (p. 676)
taxis (p. 667)
tolerance range (p. 667)

■ Critical Thinking Questions

1. Explain how a 5 part per million level of dissolved oxygen might be a limiting factor for a fish in the summer, but may be completely suitable for that same fish in the winter.

2. Compare the survivorship curves for human populations of developed and third-world countries.

3. Assuming a starting population of 10 individuals, a doubling time of 1 month, and no mortality, how many generations would it take a hypothetical population to achieve 10,000 individuals?

4. Why do you think that winter inactivity of many small mammals takes the form of hibernation, whereas winter inactivity in larger mammals is in the form of winter sleep?

5. Explain the compromises between cooling and water conservation that are made by small desert mammals.

6. Are most parasites K-selected or r-selected? Explain.

■ Suggested Readings

Books

Brewer, R. 1988. *The Science of Ecology.* Philadelphia: Saunders College Publishing.

Ehrlich, P. R., and Roughgarden, J. 1987. *The Science of Ecology.* New York: Macmillan Publishing Company.

Krebs, C. J. 1985. *Ecology: The Experimental Analysis of Distribution and Abundance.* New York: Harper and Row.

Schmidt-Nielsen, K. 1984. Scaling: *Why is Animal Size So Important?* Boston: Cambridge University Press.

Articles

Ehrlich, A. H. 1985. The human population: size and dynamics. *American Zoologist* 25:395.

Fingeman, M. (ed.). 1988. Energetics and Animal Behavior (a symposium). *American Zoologist* 28(3):813–938.

Lewin, R. 1988. Food scarcity hones competitive edge. *Science* 241:165.

May, R. M., and Seger, J. 1986. Ideas in ecology. *American Scientist* 74:256.

Peterson, I. 1986. Ecological energy: bigger is better. *Science News* 130:341.

Wallace, A. June 16, 1988. Mechanisms of coexistence. *Nature* 333:597.

42

Ecology II: Communities and Ecosystems

Concepts

1. All populations living in an area make up a community. Communities have unique attributes that can be characterized by ecologists.
2. Communities and their physical surroundings are called ecosystems. Energy flow through an ecosystem is one way. Energy that comes into an ecosystem must support all organisms living there before it is lost as heat. Matter in the ecosystem, on the other hand, is recycled.
3. The earth can be divided into ecosystem types according to their characteristic plants, animals, and physical factors.
4. Concepts of community and ecosystem ecology provide a basis for understanding many of our ecological problems including: human population growth, pollution, and resource depletion.

Have You Ever Wondered:

[1] why we have an energy crisis?
[2] why predators, such as hawks and eagles, are relatively uncommon, but other kinds of animals, such as robins and rabbits, are often so numerous?
[3] where many of the atoms that make up your body's tissues were in the years before you were born?
[4] what was the earliest natural application of solar power?
[5] whether or not the earth's current human population can be supported for many years to come?
[6] why predators are often the first to feel the effects of poisons released into the environment?

These and other useful questions will be answered in this chapter.

This chapter contains underlined evolutionary concepts.

All populations living in an area make up a **community**. When we take off our shoes to explore the animals living in streams we enter a foreign environment. However, we may have some ideas about what we will find because ecologists have found that community interactions, even in very different communities, have common themes.

Communities cannot be considered apart from their physical environment. Communities and their physical environment are called **ecosystems**, and by studying ecosystems, ecologists discover how communities are influenced by their physical surroundings. A stream community, for example, depends on supplies of carbon, phosphorus, nitrogen, water, and energy. At the same time, however, populations alter their physical environment. Stream animals reshape the stream by digging into its banks or by moving substrate. Even in the act of dying, a stream animal changes the characteristics of its environment by contributing organic matter to the stream bed.

The purpose of this chapter is to examine the interactions occurring in communities and ecosystems. The principles that explain these interactions can help us understand many of today's environmental problems.

Community Structure and Diversity

Stream communities are not just random mixtures of species; instead, they have a unique organization. Most communities have certain members that are of overriding importance in determining community characteristics. For example, a stream may have a large population of rainbow trout that helps determine the makeup of certain invertebrate populations on which the trout feed. Species that are responsible for establishing community characteristics are called **dominant** or **keystone species**.

Communities are also characterized by cycles of activity. Stream animals are most active around sunset and sunrise. In other communities, some animals are *nocturnal* (night active). Nocturnal animals may avoid predators and the hot dry conditions of the daylight hours. Most birds (except owls), on the other hand, are *diurnal* (day active) and are dependent on keen vision for gathering food and avoiding predators. Other patterns of activity may follow the seasons, or conditions of temperature or moisture. Thus, communities are said to have a *temporal structure*.

Many communities also have a *spatial structure*, which is most obvious in the layering that occurs in a forest or a lake. Most photosynthesis occurs in the upper layers of forests and lakes. Animals feeding on this plant life must live in or visit these upper regions. Eventually, organic matter falls to the forest floor, or lake bottom, where it supports other organisms.

Community structure is also reflected in the variety of animals present in a community, called the **community (species) diversity** or richness. Forces of nature and human activities influence community diversity. Factors that promote high diversity include: diverse resources, high productivity, climatic stability, moderate levels of predation, and moderate levels of disturbance from outside the community. Pollution often reduces the species diversity of ecosystems.

The Ecological Niche

An important concept of community structure is the *ecological niche*. The niche of any species includes all the attributes of an animal's life-style: where it looks for food, what it eats, where it nests, and what conditions of temperature and moisture it requires. By measuring and plotting the tolerance ranges of an animal for environmental factors, a multidimensional image can be constructed that describes how an animal functions in a community (figure 42.1). Theoretically, competition results when the niches of two species overlap. Although niche overlap occurs in many communities, it tends to be minimized as communities mature. Niche overlap may be reduced if a common resource is divided in very specific ways. This mechanism is called *niche diversification*.

Although the niche concept is very difficult to quantify, it is valuable in helping us perceive community structure. It illustrates that members of a community tend to complement each other in resource use. They tend to partition

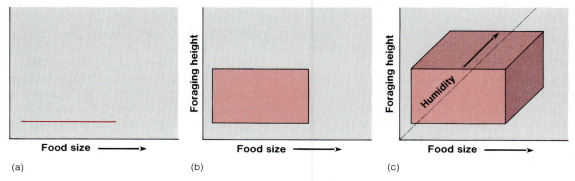

Figure 42.1 The niche as a multidimensional space. Each characteristic of an animal's niche can be plotted on a separate axis of a graph. A visualization of a three dimensional niche involving (a) food size, (b) foraging height, and (c) humidity is shown here. Theoretically, additional dimensions could be added.

resources rather than compete for them. It also helps us visualize the role of an animal in the environment.

Community Stability

As with individuals, communities are born and they die. Between those events is a time of continual change. Some changes in the life of a community are not permanent changes. These include the replacement of old individuals by those of the next generation and seasonal fluctuations in the abundance of some animals. Other changes, however, are permanent and directional. Some permanent changes are the result of climatic or geological events. Others are brought about by the members of the community. It is the latter that are important for understanding community structure and diversity.

The dominant members of a community often change a community in predictable ways in a process called **succession** (L. *successio,* to follow). Communities may begin in areas nearly devoid of life. Geological events, such as the retreat of a glacier, may form a deep, nutrient-poor lake. This lake, however, will not remain in its early postglacial condition forever. Nutrients that can support new life wash into the lake from surrounding hillsides. Death, decay, and additional nutrients from the shore provide the raw materials that further enrich the lake. Animals (e.g., lake trout) that require deep, cold water can no longer survive, and are replaced by species that live in warmer, shallower waters. Over thousands of years, the lake is filled with organic matter and inorganic sediments. Eventually, what was once a lake is transformed into a marsh, then a field, and finally a forest. Each successional stage is called a *seral stage,* and the entire successional sequence is a **sere** (ME. *seer,* to wither). Succession occurs because the dominant life forms of a sere gradually make the area less favorable for themselves, but more favorable for organisms of the next successional stage.

The final community (the forest in our example) is called the **climax community.** It is different from the seral stages that preceded it because it can tolerate its own reactions. The accumulation of the products of life and death no longer make the area unfit for the individuals living there. Climax communities usually have complex structure and high diversity. Species of climax communities tend to be K selected. The stability of a climax community generally depends on a stable environment, although moderate, short-lived disturbances are tolerated.

If succession begins in an area that did not previously support a community, it is called *primary succession.* Examples include the establishment of a community on a newly exposed stretch of sandy beach or a new volcanic island. Succession also may occur in areas that previously supported a community, but where some outside disturbance (e.g., fire) upset the previous community. This kind of succession, called *secondary succession,* occurs more rapidly than primary succession because some organisms are already living there (figure 42.2).

Stop and Ask Yourself

1. What are dominant or keystone species?
2. What is community diversity, and what factors promote high diversity?
3. In what way is succession brought about by the dominant members of a sere? How is the climax community different from a sere?
4. What is the difference between primary and secondary succession?

Ecosystems

In this section you will learn two important facts. First, energy on this planet is not recyclable. Although energy is [1] constantly coming to the earth from the sun, when we use energy stored in the earth's energy reserves it can never be reused. Second, everything else that we use comes from our planet. Eventually, materials we use will be returned to the earth, and will be reused by future generations.

Trophic Structure

There must be enough energy in an ecosystem to support the activities of all organisms living there. This energy enters an ecosystem as sunlight and is incorporated into the chemical bonds of living and decaying tissues (*see figure 4.1*). The sum of all living and decaying tissues in an ecosystem is called the **biomass** (Gr. *bios,* life + ME. *masse,* to knead). Sooner or later, the energy in the biomass is lost from an ecosystem as heat.

The sequence of organisms through which energy moves in an ecosystem is called a **food chain.** One relatively simple food chain might look like the following:

$$\text{grass} \rightarrow \text{grazing insects} \rightarrow \text{shrews} \rightarrow \text{owls}$$

It is more realistic to envision complexly interconnected food chains, called **food webs,** that involve many kinds of organisms (figure 42.3). Because food webs can become very complex, it is convenient to group organisms based on the form of energy used. These groupings are called **trophic levels.**

Producers Producers (autotrophs) obtain food (complex organic compounds) from inorganic materials and an energy source. They form the first trophic level of an ecosystem. The producers that are most familiar to us are green plants. They are called **photoautotrophs** (Gr. *photos,* light + *autos,* self + *trophe,* to feed), their source of energy is the sun, and they convert energy to food through the reactions of photosynthesis. Less than 1% of the sunlight reaching the earth's atmosphere is transformed by photosynthesis. The rest is reflected back into space by the atmosphere (30%), absorbed by the atmosphere (20%),

(a)

(b)

Figure 42.2 Succession. (a) Primary succession on a sand dune. Beach grass is the first species to become established on a sand dune. It stabilizes the dune so that shrubs, and eventually trees, can become established. (b,c) Secondary succession after fires in Yellowstone National Park. The lodgepole pine shown here is a fire-dependent species. The maintenance of a healthy forest requires periodic burning. Investigations reveal that fires reoccurring in 60- to 100-year intervals maintain these forests.

(c)

or absorbed by the earth (50%). The total amount of energy fixed by producers is called **gross primary production.** A portion of gross primary production is used for maintenance needs of the producers. The rest, called **net primary production,** is available to other trophic levels. Primary production is measured in kilocalories, or grams, of organic matter produced/square meter of earth/year (figure 42.4).

Few ecosystems obtain energy from sources other than the sun. Producers in ecosystems that get energy other than from the sun are called **chemolithoautotrophs** (Gr. *chemeai,* to alloy metals + *lithos,* stone). They obtain energy by oxidizing inorganic compounds.

The largest communities known to rely on chemolithoautotrophy are the hydrothermal vent communities that occur 2,000 to 3,000 m below the surface of the Pacific Ocean. Hydrothermal vents are hot springs associated with regions where the sea floor is spreading. Usually such depths are sparsely inhabited because energy input from surface waters is very low. Vent communities, however, have thriving populations of clams, polychaetes, and many other marine animals. The energy for these systems is derived from geothermal sources.

Consumers Other trophic levels are made up of consumers (heterotrophs). Consumers obtain their energy by eating other organisms. Herbivores (primary consumers) eat producers. Some carnivores (secondary consumers) eat herbivores, and other carnivores (tertiary consumers) eat the carnivores that ate the herbivores. Consumers also include scavengers that feed on large

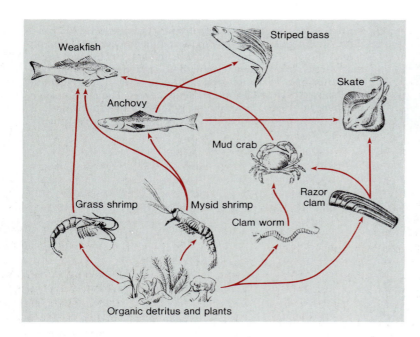

Figure 42.3 A simplified food web of a shallow saltwater ecosystem. Primary production by rooted aquatic vegetation and phytoplankton (not shown) provide food for shrimp, razor clams, and clam worms. Mud crabs, anchovies, weakfish, and skates are found at the secondary consumer trophic level (i.e., they eat the primary consumers). Skates and weakfish also function, along with the striped bass, at the tertiary consumer level.

chunks of dead and decaying organic matter. Scavengers include earthworms and vultures.

Decomposers Feeding at any consumer level is never 100% efficient. If herbivores crop 20% of net primary production, 80% is left behind. Eventually this primary production dies. Thanks to the decomposers, however, leftovers do not accumulate. Decomposers break down dead organisms and feces. Bacteria and fungi perform most of

the decomposition by digesting organic matter extracellularly and absorbing the products of digestion. (They are said to be *saprophytic*.)

Efficiencies The efficiency with which the animals of a trophic level convert food into new biomass depends upon the nature of the food (figure 42.5). Much of the production of a forest is tied up in inedible, or undiges-

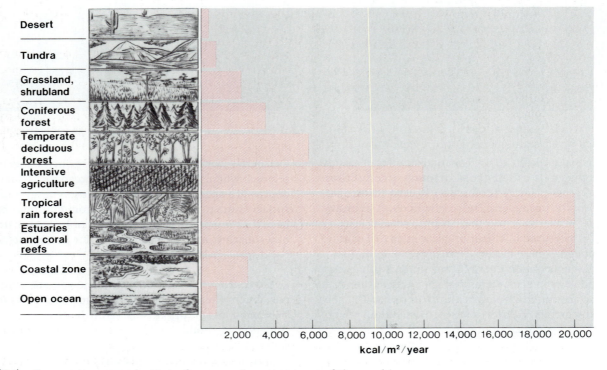

Figure 42.4 Gross primary production of some major ecosystems of the world.

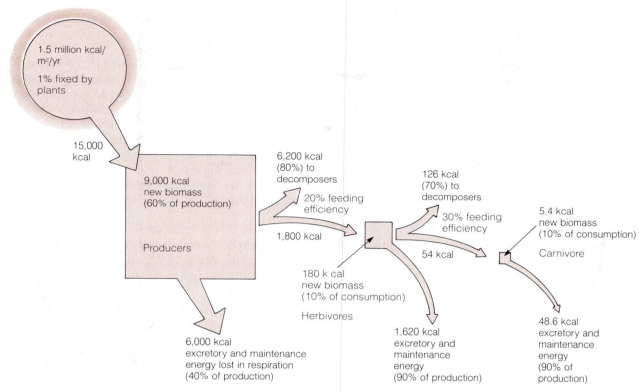

Figure 42.5 Energy flow through ecosystems. Approximately 1.5 million kilocalories of radient energy strike a square meter of the earth's surface each year. Less than 1% (15,000 kcal/m²/yr) is converted into chemical energy by plants. Of this, approximately 60% is converted into new biomass, and 40% is lost in respiration. The herbivore trophic level harvests approximately 20% of net primary production, and decomposers get the rest. Of the 1,800 kcal moving into the herbivore trophic level, 10% (180 kcal) is converted to new biomass, and 90% (1,620 kcal) is lost in respiration. Carnivores harvest about 30% of the herbivore biomass, and 10% of that is converted to carnivore biomass. At subsequent trophic levels, harvesting efficiencies of about 30% and new biomass production of about 10% can be assumed. All of these percentages are approximations. Absolute values depend on the nature of the primary production (e.g., forest vs. grassland) and characteristics of the herbivores and carnivores (e.g., ectothermic vs. endothermic).

tible, cellulose. On the other hand, a greater proportion of a grassland is available for herbivores. Carnivores are more efficient at assimilation than herbivores because most of an herbivore's biomass is digestible. Further, the efficiency with which an animal converts food into new biomass depends upon maintenance requirements of the animals involved.

Ecological Pyramids Ecological pyramids express the relationships between trophic levels in a graphic form. A **pyramid of numbers** shows the numbers of individuals at each trophic level of an ecosystem, a **pyramid of biomass** represents the biomass at each trophic level, and a **pyramid of energy** represents the amount of energy tied up in each trophic level (figure 42.6). All of these

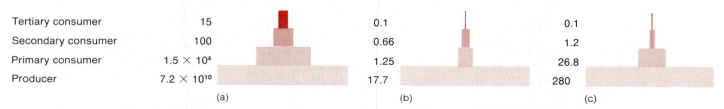

		(a)	(b)	(c)
Tertiary consumer	15		0.1	0.1
Secondary consumer	100		0.66	1.2
Primary consumer	1.5×10^4		1.25	26.8
Producer	7.2×10^{10}		17.7	280

Figure 42.6 Ecological pyramids for data from an experimental pond. (a) Pyramid of numbers (individuals/m²). (b) Pyramid of biomass (dry g/m²). This figure represents a relatively long time interval. In open-water systems, instantaneous producer biomass is often less than herbivore biomass, because producers are small and reproduce rapidly, and consumers are large and longer lived. (c) Pyramid of energy (dry mg/m²).

pyramids have the same shape. The area devoted to each level of the pyramid generally decreases from the producer to the highest consumer levels. Thus, predators (e.g., [2] hawks and eagles) are much less common than animals such as robins and rabbits.

The trophic level concept is useful to help us visualize what happens to energy in ecosystems. In real ecosystems, however, trophic levels are difficult to describe. While the producer level is uniform, most animals function at more than one trophic level. Herbivores often consume insects when they graze, and carnivores move back and forth between secondary and tertiary consumer levels.

Cycling

Did you ever wonder where the calcium atoms in your bones were 100 or even 100 million years ago? Perhaps they were in the bones of an ancient reptile, or in the sediments of prehistoric seas. Unlike energy, all matter is [3] cycled from nonliving reservoirs to living systems, and back to nonliving reservoirs. Matter moves through ecosystems in **biogeochemical cycles**.

Nutrient Cycling   A nutrient is any element essential for life. Approximately 97% of living matter is made of oxygen, carbon, nitrogen, and hydrogen. *Gaseous cycles* involving these elements utilize the atmosphere or oceans as a reservoir. Elements such as sulfur, phosphorus, and calcium are less abundant in living tissues than those with gaseous cycles, but they are no less important in sustaining life. The nonliving reservoir for these nutrients is the earth, and the cycles involving these elements are called *sedimentary cycles*.

Carbon Cycle   Carbon is very plentiful on our planet and is rarely a limiting factor. The reservoir for carbon is carbon dioxide (CO_2) in the atmosphere or water. Carbon enters the reservoir when organic matter is oxidized to CO_2. CO_2 is released to the atmosphere or water where autotrophs incorporate it into organic compounds. In aquatic

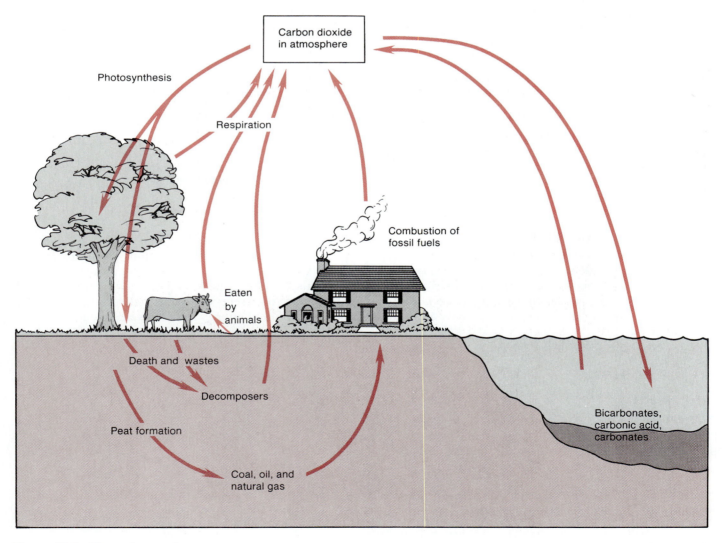

Figure 42.7   The carbon cycle.

systems, some of the CO_2 combines with water to form carbonic acid. Because this reaction is reversible, carbonic acid can supply CO_2 to aquatic plants for photosynthesis when CO_2 levels in the water decrease. Carbonic acid can also release CO_2 to the atmosphere.

Some of the carbon in aquatic systems is tied up as calcium carbonate in the shells of molluscs and skeletons of echinoderms. Accumulations of mollusc shells and echinoderm skeletons has resulted in limestone formations that are the bedrock of much of the United States. Geological uplift, volcanic activities, and weathering returns much of this carbon to the earth's surface and the atmosphere.

A large supply of carbon is also tied up in *fossil fuels*. During the 65 million years of the Carboniferous period (which ended about 280 million years ago), organic matter accumulated in peat bogs. Incomplete decomposition and conditions of high temperature and pressure converted this carbon to coal, natural gas, and oil. Since the beginning of the Industrial Revolution, humans have been burning fossil fuels and returning large quantities of this carbon to the atmosphere as CO_2 (figure 42.7).

Nitrogen Cycle   Nitrogen is an essential component of amino acids and the proteins that they form. The res-ervoir for this nitrogen is the atmosphere, which is about 78% nitrogen (figure 42.8). Gaseous nitrogen, however, is not directly available to most plants. Instead, plants use nitrogen as ammonia (NH_3) or nitrate (NO_3^-). In these forms, nitrogen can be removed from the soil or water and incorporated into protein. Animals, of course, get their nitrogen by eating plants or other animals. When proteins and amino acids are metabolized, nitrogen is excreted as ammonia, urea, or uric acid. The latter two are readily converted to the ammonium ion (NH_4^+). Ammonia and the ammonium ion, in turn, are converted into nitrate and nitrite (NO_2) by bacteria in an energy-yielding reaction called nitrification. Nitrites and nitrates are then taken up by plants and incorporated into proteins.

In another part of the nitrogen cycle, the nitrogen in nitrates is returned to the air as gaseous nitrogen (N_2). *Denitrifying bacteria* live in anaerobic environments and use nitrates as a source of oxygen for cellular respiration. Denitrifying bacteria are offset by *nitrogen fixation*. Nitrogen-fixing bacteria (in association with leguminous plants such as soybeans) and cyanobacteria (blue-green algae), convert gaseous nitrogen to nitrates.

Other components of the nitrogen cycle include the industrial production of nitrates, the conversion by light-

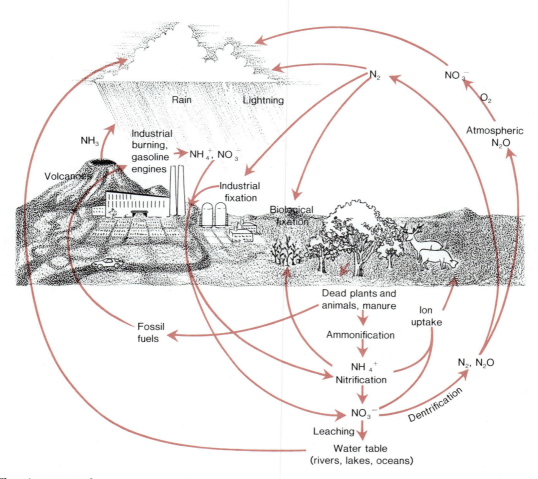

Figure 42.8   The nitrogen cycle.

ening of N_2 to nitrates, and volcanic activities that release nitrates from the earth's crust.

Other Nutrient Cycles Phosphorus is an essential component of DNA, RNA, ATP, and vertebrate skeletons. Its cycle involves a sedimentary reservoir that contains about 0.000003% phosphorus. Animals acquire phosphorus in their diet, and phosphorus is returned to the soil in excretion and decay. Mining operations remove phosphates from the soil for the production of fertilizers and detergents. Overuse of these can promote algal growth in our rivers and lakes and can cause the early death of our waterways.

Sulfur is found in elemental form in the earth's crust, and as sulfate (SO_2) in the atmosphere. It is an essential component of amino acids but is rarely limiting. Animals acquire sulfur from plants, which extract it from the soil. Decaying plants and animals produce hydrogen sulfide and sulfate and returns sulfur to the soil. The burning of high-sulfur coal also sends sulfate into the atmosphere.

Calcium is essential for muscle contraction, bone formation, nerve impulse conduction, and many other animal functions. The reservoir for calcium is the earth's crust (limestone beds) and our oceans. Animals acquire calcium in food and drink and return it to the soil in death, decay, and excretion.

Biogeochemical cycles have been disrupted by human activities. For example, the mining of phosphates for fertilizers increases the rate at which phosphates are lost through terrestrial runoff to ocean depths. Activities such as these can lead to serious shortages of some nutrients.

Water Cycling The cycling of water is one of the most basic of all cycles. Water moves between land, oceans, and the atmosphere. Oceans contain 97% of the earth's water, and water that evaporates from oceans eventually falls as rain, snow, ice, or fog. Some water that falls on land may return to the atmosphere by evaporation from soil, or from the leaves of plants. Other water is temporarily absorbed in the soil. Rain that does not evaporate or contribute to groundwater runs off into lakes and streams. Eventually, streams make their way to rivers, and rivers to oceans. In completing its cycle, water acts as a powerful force. It sculpts our land, transports nutrients to the sea, and powers electric generators. The water cycle is called the **hydrological cycle** and is the earliest form of solar power because it is powered entirely by sun-induced evaporation from our oceans.

Stop and Ask Yourself

5. What is the source of energy for individuals at the producer trophic level? At the secondary consumer trophic level?
6. Approximately what portion of the energy intake at one trophic level is converted into biomass at that trophic level? What happens to the rest of the energy?
7. What is the main reservoir for carbon in the carbon cycle?
8. What do nitrogen-fixing bacteria do?

Ecosystems of the Earth

The earth can be divided into ecosystem types according to characteristic plants, animals, and physical factors. *Terrestrial ecosystems* are divided into distinctive associations of plant and animal populations called **biomes** (figure 42.9). Biomes have certain geographical boundaries, but different parts of the world with similar climatic and geographical features have similar biome types. *Aquatic ecosystems* are subdivided into freshwater and marine ecosystems, each of which is further subdivided.

Terrestrial Ecosystems

Tundra exists in two forms in North America. Arctic tundra occurs in the far northern reaches of North America (figure 42.10). For most of the year, it is covered with snow and ice. A very short summer provides only about 60 days during which mosses, lichens, grasses, and shrubs can grow. No trees occur here. Precipitation is a scanty 25 cm per year, yet the summer soils are wet and boggy. Just below the surface of the ground is a permanently frozen subsoil called permafrost. Productivity is low. The animal life of arctic tundra consists of rodents, musk oxen, caribou, owls, foxes, and weasels.

Alpine tundra occurs in mountains, above the altitude at which trees grow. Widely fluctuating temperatures, moderate precipitation, and grasses and shrubs characterize this ecosystem.

Northern coniferous forests are characterized by cool summers, cold winters, short (130 day) growing seasons, and moderate precipitation (40 to 100 cm per year). Most of the precipitation comes in the form of heavy winter snows. Soils are acidic and relatively infertile. Plants characteristic of the northern coniferous forests are spruce, fir, and pine. Because coniferous forests are "evergreen," some limited primary production can occur on warmer winter days. Pine needles that accumulate on the forest floor are very slow to decompose, and fire is important in promoting seed germination and release of nutrients locked in the pine needles. The diversity of animals in northern coniferous forests is not great, though population sizes can be quite large. Insects abound in the summer months, and other residents include snowshow hares, lynx, wolves, caribou, and moose (figure 42.11).

Temperate deciduous forests occur in regions with moderate climate and well-defined summer and winter seasons. Precipitation ranges from 75 to 150 cm per year, and soils are relatively fertile. Deciduous trees, which lose their leaves in fall, dominate these forests. Decomposition of fallen leaves enriches forest soils. Vertical stratification of temperate deciduous forests is well developed, with most production occurring in the tops of the trees (the forest canopy). Insects are the most common herbivores, and white-tailed deer are the largest herbivores (figure 42.12). Carnivores of presettlement temperate deciduous forests included wolves and mountain lions.

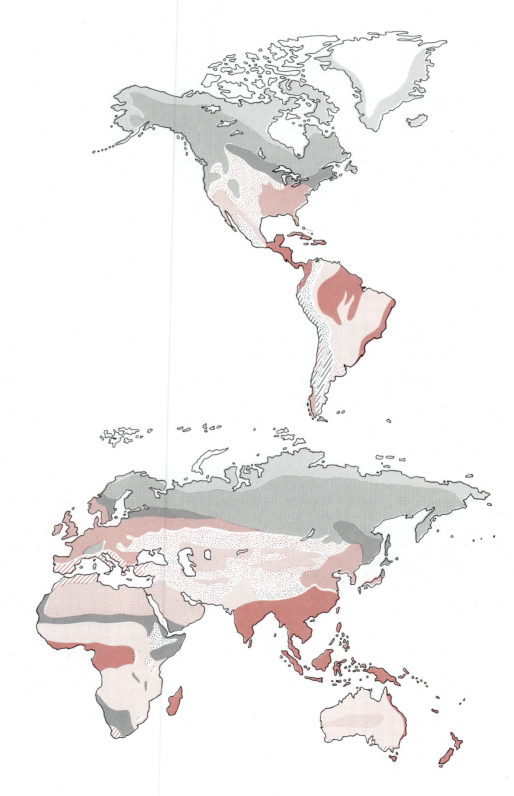

Figure 42.9 Biomes of the world.

Figure 42.10 Arctic tundra is dominated by mosses, lichens, grasses, and shrubs. Animal life includes rodents, caribou (shown here), and arctic foxes.

Figure 42.11 Northern coniferous forests are dominated by spruce and fir. Animal life includes snowshoe hares, wolves, and moose (shown here).

Figure 42.12 Temperate deciduous forests provide beautiful landscapes in the fall. Animal life includes white-tailed deer.

The *grasslands* of North America are called prairies. Similar ecosystems in South America and Africa are called *tropical savannahs*. A few scattered trees may be present in grasslands, but grasses, such as big and little bluestem, predominate. The nature of the vegetation of a grassland depends on the amount of moisture. Moist prairies in North America are called tall-grass prairies, and drier prairies are called short-grass prairies. As in northern coniferous forests, fire plays a role in maintaining grasslands by killing trees and shrubs that recover from disturbance very slowly. Because much of the primary production of a grassland dies each year, organic matter accumulates quickly in grassland soils. The fertile soils of the prairies of North America have been exploited for agricultural purposes, and few tracts of virgin prairie remain. Because most of their primary production is edible, grasslands support large populations of herbivores, such as numerous insects and bison (figure 42.13).

Chaparral occurs in relatively dry (30 to 75 cm of rain per year) areas of the world, including the southwestern United States. Most precipitation comes during the winter, and temperatures fluctuate between 13 and 23°C. Plants of the chaparral are low and shrubby and have tough, waxy leaves. Animals include insects, rodents, rabbits, lizards, snakes, and mule deer (figure 42.14). The chaparral is another ecosystem in which fire is an important regulator of plant and animal populations.

Deserts occur around the world between the latitudes of 30°N and 30°S. Rainfall is less than 25 cm per year. Temperatures fluctuate widely; they are very high during daylight hours, and low at nighttime. Plants of very hot deserts are predominately cacti and desert shrubs. Cooler deserts are characterized by sagebrush. Animals of the desert include birds, rodents, numerous insects, and reptiles (figure 42.15). The evolution of desert animals has involved delicate compromises between the needs for evaporative cooling and water-based excretion, and the need to conserve water.

Tropical rain forests are found, among other places, in the equatorial regions of Central America, and northern South America. Their temperatures fluctuate between 20 and 25°C, and rainfall approaches 200 cm per year. Because of the high moisture, high temperatures, and the 12-month growing season, nutrient cycling occurs rapidly. Therefore, soils of tropical rain forests are relatively poor in nutrients because most of the nutrients are tied up in living organisms. Life forms of tropical rain forests are very diverse. Trees are broad-leaved and nondeciduous, and the forest is highly stratified. A nearly complete forest canopy creates a densely shaded forest floor. The rich diversity of plant life creates diverse food and habitat resources for animals. The stable temperatures and high humidity create ideal conditions for both soft- and hard-bodied ectothermic animals; so populations of annelids, arthropods, amphibians, and reptiles abound (figure 42.16).

Figure 42.13 Grasslands of North America are dominated by grasses such as bluestem. They once provided abundant food for grazing animals, such as bison.

Figure 42.14 Chaparral occurs in the southwestern United States. It is characterized by low, shrubby vegetation, and animals such as mule deer (shown here), rodents, rabbits, and many reptiles.

Figure 42.15 Deserts are dominated by cacti and desert shrubs. Animal life of deserts includes birds, rodents, numerous species of insects and reptiles, such as this scaly lizard.

Figure 42.16 Tropical rain forests provide the earth with a great variety of plant and animal species. The interior of the forest is dark and moist and is the last refuge on earth for many animals, such as this pygmy marmoset.

Aquatic Ecosystems

Aquatic environments are divided into freshwater and marine ecosystems, both of which have diverse habitats.

Freshwater Ecosystems Freshwater ecosystems are classified according to water flow. *Lotic* (flowing water) ecosystems include brooks, streams, and rivers. Some of the primary production of streams is provided by attached aquatic plants and algae. However, most of the primary production for lotic ecosystems comes from surrounding terrestrial habitats as runoff during rains and autumn leaf falls. Because most lotic ecosystems are relatively shallow and have long shorelines, terrestrial production is more than adequate to supply energy needs. Numerous aquatic insects and other invertebrates divide resources. Some shred larger pieces of plant matter. Others, such as blackfly larvae and caddis fly larvae, filter floating microscopic debris from the water (figure 42.17). Predators include numerous species of fish, as well as insects, such as stonefly larvae and dobsonfly larvae.

Current is probably the most important physical factor in lotic habitats. Some animals are strong swimmers, capable of negotiating rapids and waterfalls. Others retreat from the current by living on or in the substrate. The latter are called **benthic** (Gr. *benthos,* depth of the sea) animals. Many stream insects display adaptations for living in moving water. Adaptations include streamlining, flattening, and ballasting themselves with pebbles (figure 42.18).

Lentic habitats, including lakes and ponds, have standing water. Vertical stratification of lakes is common, especially in temperate regions. In the summer, the upper regions

(the *epilimnion*) of a lake warm quickly. Because the deeper water (the *hypolimnion*) is cool, and because cool water is denser than warm water, mixing between upper and lower levels of the water column is prevented. Photosynthesis (and thus oxygen production) occurs only in the upper levels of a lake because light is filtered by water with increasing depth. The main reactions occurring in the depths of a lake are those of decomposition. The deepest parts of a lake are often depleted of oxygen in the summer. In the fall, however, the surface waters cool, and winds cause mixing of the entire water column in what is called *turnover*. In the winter, the water again becomes layered under the ice, but winds and temperature changes cause another turnover in the spring.

A much greater share of the production of lentic ecosystems is from production occurring in the body of water. Most of this production is from microscopic algae floating near the surface. Organisms that swim or float in surface waters and are at the mercy of wind, waves, and current are a part of the **planktonic** (Gr. *planktos,* to wander) community. Deep, cold lakes that are nutrient poor and relatively unproductive are said to be *oligotrophic*. Shallow, warm, nutrient-rich lakes are highly productive, and are said to be *eutrophic*. Succession in a lake, called *eutrophication,* was described earlier. The rate of eutrophication can be greatly increased by adding nutrients to a lake from fertilizers and other sources.

Other freshwater ecosystems include wet grassland areas, called marshes; wet woodland areas, called swamps; and peat bogs. They occur where the water table is near the surface of the land. Freshwater wetlands are important as hatcheries and rearing grounds for wildlife.

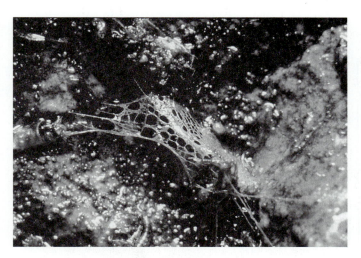

Figure 42.17 Some caddis fly larvae spin nets to catch microscopic debris floating in the current.

Figure 42.18 Stream insects display a great variety of adaptations for living in moving water. This mayfly nymph shows streamlining and flattening, which allows the insect to maintain itself in strong currents.

Marine Ecosystems The most important physical characteristic that distinguishes marine ecosystems from freshwater ecosystems is the salinity of the water. Marine ecosystems have water containing a mixture of salts, which make up 3.5% of the mass of a quantity of seawater. This section briefly examines five marine ecosystems: estuaries, littoral zones, neritic zones, oceanic zones, and coral reefs.

Estuaries occur where fresh water meets salt water, and the salinity of seawater is diluted. Estuaries are very productive because freshwater streams carry nutrients into the estuary, and tidal currents bring nutrients from the ocean. The primary producers in estuaries include plankton, larger attached aquatic vegetation, and emergent vegetation, such as cordgrass. Animals and decomposers feed on this production, and find refuge in it. One of the very important functions of estuarine ecosystems is to serve as nursery grounds for many marine fishes, molluscs, and crustaceans. Estuaries are also important feeding grounds for wading and swimming birds.

Estuaries have suffered from human activities. They have been drained and filled to develop subdivisions and condominiums, and have been excavated to harbor pleasure boats. They have also been polluted by oil explorations and drilling. Their productivity is threatened by acid rain. In recent years, these activities have been slowed, but have not stopped.

Littoral zones extend from the splash zone of ocean waves to the low tide marks (figure 42.19). These regions are subject to the actions of both waves and tides. Some littoral animals make their living in splash zones. These animals, such as periwinkle snails, may live on algae attached to the rocky shoreline.

Other littoral animals live in the *intertidal zone*, which is a region submerged at high tide and exposed at low tide. Animals use a variety of strategies to survive during low tide. More motile animals, such as fish and some crabs, can move in and out of intertidal zones with the tides. Other animals, such as amphipods, isopods, ghost crabs, and numerous polychaetes, burrow beneath the sand and mud of the intertidal areas. A shovel full of intertidal sand and mud will reveal thousands of hidden residents. Yet other inhabitants of the littoral zone are found in tidal pools. Small basins in the rock may be completely submerged at high tide, but exposed at low tide. Water trapped in tidepools by ebbing tides supports a rich animal life including numerous echinoderms, barnacles, limpets, and chitons. Attached animals withstand periods of exposure by closing their shells to avoid drying.

Neritic ecosystems consist of relatively shallow water that extends from the littoral zone to the edge of the continental shelves. Oceanic currents bring nutrient-rich bottom waters to the surface, where they stimulate photosynthesis by planktonic algae (*phytoplankton*). In some communities, attached algae, such as the vast kelp beds of the north Pacific, are locally important in contributing to primary production. Animal members of the plankton (*zooplankton*), include the numerous crustaceans and larval stages of barnacles, echinoderms, and polychaetes. Neritic zones support some of the largest vertebrate predators and provide food for human consumption.

Oceanic ecosystems extend from the continental shelves into the relatively unproductive open ocean. Virtually all primary production in the open ocean occurs in the upper waters, a region called the *photic zone*. Phytoplankton again forms the base of the food web. There is continual darkness in the *aphotic (abyssal) zone* below 200 m. Life at these depths is dependent on production from the surface waters drifting to great depths. (The vent communities of the ocean floor are an exception that was described earlier.) Animals found at these great depths may have luminescent organs that serve as searchlights, lures for prey, and sexual attractants (*see box figure 4.1*).

Coral reefs are one of the most highly productive ecosystems in the world. They are associations of stony corals

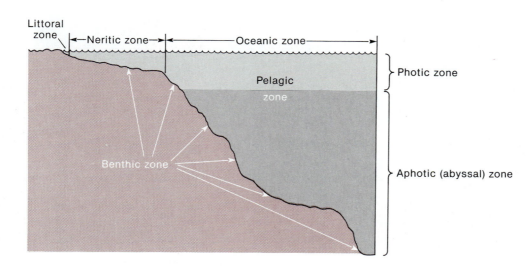

Figure 42.19 Diagrammatic section of an ocean, showing the distribution of major marine ecosystems.

(phylum Cnidaria, class Anthozoa) and algae (called coralline algae and zooxanthellae). Reefs form over thousands of years in warm, clear, relatively shallow waters. The greatest depth at which they can grow is between 50 and 60 m. Below this depth, light is insufficient to support the growth of the dinoflagellates necessary for reef formation.

Stop and Ask Yourself

9. What are the physical characteristics, dominant plants, and dominant animals of nothern coniferous forests?
10. What are the physical characteristics, dominant plants, and dominant animals of tropical rain forests?
11. What is the source of most nutrients entering lotic ecosystems?
12. What is a definition for the following terms: benthos, plankton, littoral zone, neritic zone, and oceanic zone?

Ecological Problems

In the last few hundred years of our history, humans have attempted to provide for the needs and wants of a growing human population. In our search for longer and better lives, however, humans have lost a sense of being a part of our world's ecosystems. Now that you have studied some general ecological principles, it should be easier to understand many of our ecological problems.

Human Population Growth

An expanding human population is the root of most of our other environmental problems. Humans, like other animals, have a tendency to undergo exponential population growth. The earth, like any ecosystem on it, has a carrying capacity and a limited supply of resources. When human populations achieve that carrying capacity, populations should stabilize. If they do not stabilize in a fashion that limits human misery, then war, famine, and/or disease is sure to be the vehicle that accomplishes our ecological destiny.

What is our planet's carrying capacity? The answer to that question is not simple. In part, it depends on the standard of living that we desire for ourselves and whether or not we expect resources to be distributed equally among all populations. Currently, the earth's population stands at over 5 billion people. Virtually all environmentalists agree that number is too high if all people are to achieve the affluence of developed countries. [5]

Unless intense efforts are made to curb population growth, world populations could double in the next 50 years. Looking at the age characteristics of world populations helps to explain why human populations will grow rapidly. The **age structure** of a population is the proportion of a population that is in prereproductive, reproductive, and postreproductive classes. Age structure is often represented by an *age pyramid*. Figure 42.20 shows an age pyramid for a developed country and for a lesser developed country. In less-developed countries, the age pyramid has a very broad base, indicating high birth rates. As in many natural populations, these high birth rates are offset by high infant mortality. However, what happens when less-developed countries begin accumulating technologies that reduce prereproductive mortality and prolong the life of the elderly? Unless reproductive practices change, a population explosion occurs. Unfortunately, cultural practices change very slowly.

In developed countries, population growth tends to be slower, and the proportion of the population in each reproductive class is balanced. Birth rates in the United States have decreased in recent years, probably because of an increased emphasis on careers for both men and women. Discounting immigration, and assuming a birth rate of two

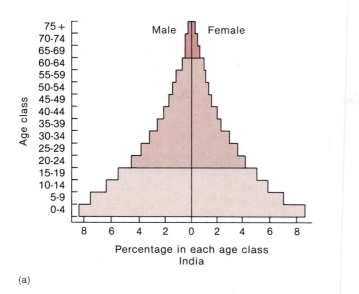

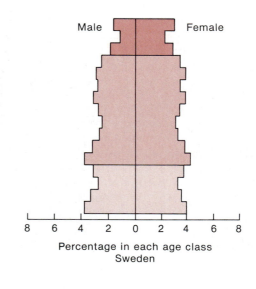

Figure 42.20 Human age pyramids. (a) In less-developed countries, a greater proportion of the population is in the prereproductive age classes. High mortality in that age class compensates for high birth rates. As technologies reduce infant mortality and prolong the life span of the elderly, populations increase rapidly. (b) In developed countries, the age structure is more rectangular because of reduced mortality in all age classes.

children per family, the population of the United States could stabilize at about 250 million in the 21st century. It is unrealistic, however, to assume that immigration to the United States will cease. If one assumes 2 million immigrants per year, and that most immigrants will bring along the reproductive practices of less-developed countries, then the U.S. population may increase to 500 million by 2050. Given our current standard of living, even 200 million people living in the United States is too many. Problems of homelessness, hunger, resource depletion, and pollution all stem from trying to support too many people at our current standard of living.

The problems faced by developed countries are compounded by the fact that individuals in less-developed countries are beginning to demand their "fair share" of world resources. As described in the introduction of this text, we have no choice but to face the problems of population growth with education, strong family-planning programs, and a reexamination of the values of our "bigger and better" technologies.

Pollution

Pollution is any detrimental change in an ecosystem. Most kinds of pollution are the results of human activities. When human populations are large and affluence demands more and more goods and services, pollution problems are compounded. Because you have studied sections of this chapter on ecosystem productivity and nutrient cycling, you should now be able to understand why these problems exist.

Pollution of our waters takes many forms. Industries generate toxic wastes, heat, and plastics, some of which will persist for centuries. Every household in the world gen-

erates human wastes that must be disposed of. All too often, industrial and human wastes find their ways into groundwater or into streams, lakes, and oceans. When they do, water becomes unfit for human consumption and unfit for wildlife.

Air pollution also presents serious problems. Burning fossil fuels releases sulfur dioxide and nitrogen oxides into the atmosphere. Sulfur dioxide and water combine to produce sulfuric acid. When *acid rain* falls, it lowers the pH of lakes, often many miles from the site of sulfur dioxide production. Carbon dioxide released in burning fuels has accumulated in the atmosphere, causing the *greenhouse effect*. Carbon dioxide reflects solar radiation back to the earth, causing an increase in world temperature, polar ice caps to melt, and ocean levels to rise. The release of chlorinated fluorocarbons from areosol cans, air conditioners, and refrigerators contributes to the depletion of the earth's ultraviolet filter—the atmospheric ozone layer. As a result, the incidence of skin cancer is likely to increase.

When wastes and poisons enter food webs, organisms at the highest trophic levels usually suffer the most. Very tiny amounts of a toxin incorporated into primary production can quickly build up as carnivores feed on herbivores that have concentrated toxins in their tissues. This problem is especially severe when the material is not biodegradable (not broken down by biological processes). The accumulation of matter in food webs is called **biological magnification.** [6]

Resource Depletion

Other environmental problems arise because humans have been too slow to realize that an ecosystem's energy is used

Box 42.1

Preserving A Global Resource—Biodiversity

Regular news reports of whales beached on our shores, leopards exploited for their hides, and rhinos killed for their horns pluck at the heart strings of Americans almost weekly. These unfortunate circumstances, however, are unique only in the ability of these animals to draw our nation's sympathy.

No one knows for sure how many species there are in the world. About 1.4 million species have been described. Taxonomists estimate that there are 4 to 30 million more. Much of this unseen, or unnoticed, biodiversity is unappreciated for the free services it performs. Forests hold back flood waters and recycle CO_2 and nutrients. Insects pollinate crops and control insect pests, and subterranean organisms promote soil fertility through decomposition. Many of these undescribed species would, upon study, provide new food crops, petroleum substitutes, new fibers, and pharmecuticals. All of these functions require not just remnant groups but large healthy populations. Large populations promote the genetic diversity required for surviving environmental changes. When genetic diversity is lost, it is lost forever. Our heroic attempts to save endangered species come far too late. Even when they succeed, they salvage only a tiny portion of an original gene pool.

The biological diversity (biodiversity) of all natural areas of the world is threatened. Acid rain, pollution, urban development, and agriculture know no geographic or national boundaries. The main threats to biological diversity arise from habitat destruction by expanding human populations. Humans are either directly or indirectly exploiting about 40% of the earth's net primary production. Often this involves converting natural areas to agricultural uses, frequently substituting less efficient crop plants for native species. Habitat loss displaces thousands of native plants and animals.

Some of the most important threatened natural areas include tropical rain forests, coastal wetlands, and coral reefs. Of these, tropical rain forests have been an important focus of attention. Tropical rain forests cover only 7% of the earth's land surface, but they contain more than 50% of the world's species. Tropical rain forests are being destroyed rapidly, mostly for agricultural production. About 76,000 km² (an area greater than the area of the country of Costa Rica) are being cleared each year (*see figure 1.8b*). At current rates of destruction, most tropical rain forests will be gone in the 21st century. According to some estimates, we are losing 17,500 rain forest species/year. Clearing of tropical rain forests achieves little, because the thin, nutrient-poor soils of tropical rain forests are exhausted within 2 years (box figure 42.1). Sadly, rain forests are a nonrenewable resource. Seeds of rain forest plants germinate rapidly, but seedlings are unprotected on sterile, open soils. Even if a forest were able to become reestablished, it would take many centuries to return to climax rain forest.

There are solutions to the problems of threatened biodiversity. None of the solutions, however, is quick and easy. First, more money needs to be appropriated for training taxonomists and ecologists, and for supporting their work. Second, all countries of the world need to realize that biodiversity, when preserved and managed properly, is a source of economic wealth. Third, we need a system of international ethics that values natural diversity for the beauty it brings to our lives. Anything short of these steps will surely lead to severe climatic changes and mass starvation.

Box figure 42.1 Severe erosion quickly followed the removal of the tropical rain forest on these slopes in Madagascar.

only once. When a quantity of energy is lost to outer space as heat, it is gone from the earth forever. As with energy, other resources are also being squandered by human populations. Overgrazing and deforestation has led to the spread of our world's deserts. Exploitation of tropical rain forests has contributed to the extinction of many plant and animal species (box 42.1).

Stop and Ask Yourself

13. What is an age pyramid? How do age pyramids help predict future population trends?
14. What is acid rain? What kinds of pollutants contribute to its formation?
15. Why are chlorinated flurocarbons thought to contribute to problems with skin cancer?
16. Why are higher trophic levels often most severely affected by poisons in the food web?

Epilogue

It is fitting to end this textbook as it began—by looking at the relationship of zoology to our world. Zoology is a discipline that transcends disciplinary and societal boundaries. Though we zoologists sometimes get caught up in the narrow focus of our disciplinary "niche," we need to be willing to step back and look at the larger implications of what we do. Many of the lessons from our field of study have direct practical applications in our world. If your career choice is in one of the fields of zoology, you will participate in discovering, teaching, and applying these lessons. If not, you can still apply some of the lessons learned in studying this textbook. As we have just seen, not all of our society's values are perfect. As a factory worker, a lawyer, a farmer, a theologian, or a business executive, you can help reshape those values by committing yourself to living in harmony with, and working to preserve, the animal kingdom and its environment.

■ Summary

1. All populations living in an area make up a community.
2. Communities can be characterized by dominant members, cycles of activity, and the variety of organisms present.
3. Organisms have roles in their community. The niche concept helps ecologists visualize those roles.
4. Communities often change in predictable ways. Successional changes often lead to a stable climax community.
5. Energy that supports ecosystem reactions is not recyclable. Energy that is fixed by producers is eventually lost as heat.
6. Unlike energy, nutrients and water are cycled through ecosystems. Cycles involve movements of materials from nonliving reservoirs in the atmosphere or the earth to biological systems and back to the reservoirs again.
7. The earth is divided into ecosystem types according to characteristic plants, animals, and physical factors. Terrestrial ecosystems are called biomes. Aquatic ecosystems are subdivided into freshwater and marine ecosystems.
8. Human population growth is the root of most of our environmental problems. Trying to support too many people at the standard of living present in developed countries has resulted in air and water pollution and resource depletion.

■ Key Terms

age structure (p. 694)
benthic (p. 692)
biogeochemical cycles (p. 686)
biological magnification (p. 695)
biomass (p. 682)
biomes (p. 688)
chemolithoautotrophs (p. 683)
climax community (p. 682)
community (p. 681)
community (species) diversity (p. 681)
dominant or keystone species (p. 681)
ecosystems (p. 681)
food chain (p. 682)
food webs (p. 682)
gross primary production (p. 683)
hydrological cycles (p. 688)
net primary production (p. 683)
photoautotrophs (p. 682)
planktonic (p. 692)
pyramid of biomass (p. 682)
pyramid of energy (p. 682)
pyramid of numbers (p. 682)
sere (p. 682)
succession (p. 682)
trophic levels (p. 682)

■ Critical Thinking Questions

1. What is the usefulness of the concept of an ecological niche?
2. Which of the following would be a more energetically efficient strategy for supplying animal protein for human diets? Explain your answers.
   (a) Feeding people beef raised in feedlots on grain. Or, feeding people beef that has been raised in pastures.
   (b) Feeding people sardines and herrings. Or, processing sardines and herrings into fishmeal that is subsequently used to raise poultry, which is used to feed people.
3. Explain why the biomass present at one trophic level of an ecosystem decreases at higher trophic levels.

4. What is the role of fire in grassland and coniferous forest ecosystems? Should fires be allowed to burn themselves out in land managed by the U.S. Park Service?

5. Why are wetlands and estuaries important ecosystems?

■ Suggested Reading

Books

Hunt, C. E. 1988. *Down By The River: The Impact of Federal Water Projects and Policies on Biological Diversity.* Washington, D.C.: Island Press.

Kennedy, I. R. 1988. *Acid Soil and Acid Rain.* Champaign: Research Studies Press.

Norton, B. G. 1988. *The Preservation of Species: The Value of Biological Diversity.* Princeton: Princeton University Press.

Wilson, E. O. (ed.) 1988. *Biodiversity.* Washington, D.C.: National Academy Press.

Articles

Ehrlich, P. R. 1987. Population biology, conservation biology, and the future of humanity. *BioScience* 37:757–763.

Harden, G. 1986. Cultural carrying capacity: a biological approach to human problems. *BioScience* 36:599–607.

Managing planet earth. *Scientific American* September, 1989 (special issue).

Moore, J. A. 1985. Science as a way of knowing: Human ecology. *American Zoologist* 25:483.

Pearce, F. 1988. Gaia: a revolution comes of age. *New Scientist* 117:32–33.

Repetto, R. Deforestation in the tropics. *Scientific American* April, 1990.

Romme, W. H., and Despain, D. G. The Yellowstone fires. *Scientific American* November, 1989.

Schindler, D. W. 1988. Effects of acid rain on freshwater ecosystems. *Science* 239:149–157.

Scott, J. M., Csuti, B., Jacobi, J. D., and Estes, J. E. 1987. Species richness: a geographic approach to protecting future biological diversity. *BioScience* 37:782–788.

Glossary

A

aboral (ab-or'al) The end of a radially symmetrical animal opposite the mouth.

acanthella (a-kan'thel-a) Developing acanthocephalan larva between an acanthor and a cystcanth, in which the definitive organ systems are developed. 302

acanthor (a-kan'thor) Acanthocephalan larva that hatches from the egg. 302

acclimation (ak'li-ma'shen) The change in tolerance of an animal for a condition in its environment. 667

accommodation (ah-kom"o-da'shun) The adjustment of the eye for various distances. 549

acetabulum (as"e-tab'u-lum) Sucker; the ventral sucker of a fluke; a sucker on the scolex of a tapeworm. 281

Acheulean industry (ah shoo'le-an in'dah-stre) A prehistoric tool-making technology characterized by small stone flakes that served as knives; larger, pointed tools used as hand axes; and truncated tools used as cleavers. The Acheulean industry is associated with *Homo erectus.* 215

acid (as'id) A substance that ionizes in water to release hydrogen ions. 21

acoelomate (a-se'lah-mat) Without a body cavity. 233

acquired immunity (a-kwir-ed i-mu'ni-te) The type of specific immunity that develops after exposure to a suitable antigen or is produced after antibodies are transferred from one individual to another. 583

acrosome (ak-ro-som') The enzyme-filled cap on the head of a sperm. Used in egg penetration. 147

actin (ak'tin) A protein in a muscle fiber that, together with myosin, is responsible for contraction and relaxation. 516

action potential (ak'shun po-ten-shal) The sequence of electrical changes occurring when a nerve cell membrane is exposed to a stimulus that exceeds its threshold. 524

activator (ak'te-va'tor) A chemical that speeds up an enzyme's reaction rate. 64

active transport (ak'tiv trans'port) A process that requires an expenditure of energy to move molecules across a cell membrane; usually moved against the concentration gradient. 44

adaptation (ad'ap-ta"shen) Structures or processes that increase an organism's potential to successfully reproduce in a specified environment. 185

adaptive radiation (a-dap-tiv ra'de-a"shen) Evolutionary change that results in the formation of a number of new characteristics from an ancestral form. 197

adenosine diphosphate (ah-den'o-sen di-phos'phate) A nucleoside composed of the pentose sugar D-ribose, adenine, and two phosphates. 67

adenosine monophosphate (ah-den'o-sen mon-o-fos'fat) AMP molecule; created when the terminal phosphate is lost from a molecule of adenosine diphosphate; AMP. 67

adenosine triphosphate (ah-den'o-sen tri-fos'fat) ATP molecule; stores energy and releases energy for use in cellular processes; ATP. 65

adhesive gland (ad-he'siv gland) Attachment glands. 276

adipose tissue (ad'i-pos tish'u) Fat-storing tissue. 54

adrenal glands (ah-dre'nal glandz) The endocrine glands located on the tops of the kidneys. 564

adrenocorticotropic hormone (ah-dre"no-kor'te-ko-trop'ik hor'mon) Hormone secreted by the anterior lobe of the pituitary gland that stimulates activity in the adrenal cortex; ACTH. 563

aerobic (a"er-ob'ik) An oxygen-dependent form of respiration. 71

aestivation (es'te-va'shun) The condition of dormancy or torpidity during the hot summer months. 620

afterbirth (af'ter-berth") The placental and fetal membranes expelled from the uterus after childbirth. 646

age structure (aj struk'cher) The proportion of a population that is in prereproductive,

reproductive, and postreproductive classes. 694

agricultural revolution (ag're-kul'cher-al rev'ah-loo"shen) A cultural revolution that was marked by a shift toward an agricultural economy; it began about 10,000 years ago. 220

airfoil (ar'foil) A surface, such as a wing, that provides lift by using currents of air it moves through. 475

aldosterone (al-dos'ter-on) A hormone secreted by the adrenal cortex that functions in regulating sodium and potassium concentrations and water balance. 564

all-or-none law (al'or-nun' la) The phenomenon in which a muscle fiber contracts completely when it is exposed to a stimulus of threshold strength. 525

allantois (ah-lan'tois) One of the extraembryonic membranes formed in the embryo of an amniote; forms as a ventral outgrowth of the gut, enlarges during development, and functions in waste (uric acid) storage and gas exchange. 156

alleles (al-els') Alternate forms of a gene that occur at the same locus of a chromosome. 104

allopatric speciation (al'o-pat'rik spe'se-a'shen) Speciation that occurs in populations separated by geographic barriers. 197

allosteric change (al"o-ster'ik chanj) Pertaining to an effect produced on the biological function of a protein by a compound not directly involved in that function. 64

altricial (al-trish'al) An animal that is helpless at hatching or birth. 484

altruism (al'troo-iz'em) The principle or practice of unselfish concern for or devotion to the welfare of others. 663

alveolus (al-ve'o-lus) An air sac of a lung; a saclike structure. 589

ambulacral groove (am'byul-ac"ral groov) The groove along the length of the oral surface of a sea star arm. Ambulacral grooves contain tube feet. 397

ametabolous metamorphosis (a'me-tab'a-lus met'ah-mor'fe-sis) Development in which

699

the number of molts is variable; matures resemble adults, and molting continues into adulthood. 373

amictic (e-mik'tic) Pertaining to female rotifers producing only diploid eggs that cannot be fertilized; or to the eggs produced by these females. 294

amino acid (ah-me'no as'id) A relatively small organic compound that contains an amino group (-NH₂) and a carboxyl group (-COOH); the structural unit of a protein molecule. 29

amnion (am'ne-on) The innermost of the protective extraembryonic membranes surrounding reptilian, avian, and mammalian embryos; also called the amniotic membrane. 156

amniote lineage (am'ne-ot lin'e-ij) The line of vertebrate ancestry leading from ancient amphibians to modern reptiles, birds, and mammals. 439

amphid (am"fe-d) A sensory organ on each side of the "head" of nematodes. 296

amplexus (am-plek'sus) The clasping of a female amphibian by a male during external fertilization of eggs. 449

anabolism (an"ah-bol-izm) The metabolic process by which larger molecules are formed from smaller molecules. 58

anaerobic (an-a"er-ob'ik) Characterized by the absence of oxygen. 71

analogous (a-nal'ah-gus) A similarity between two structures that is not the result of a common evolutionary origin (e.g., bird and insect wings).

anaphase (an'ah-faz) Stage in mitosis during which duplicate chromosomes move to opposite poles of the cell. 89

anapsid lineage (an-ap'sid lin'e-ij) The evolutionary pathway leading to modern turtles. 454

anatomy (ah-nat'o-me) Branch of science dealing with the form and structure of body parts. 16

androgen (an'dro-jen) A male sex hormone, such as testosterone. 636

aneuploidy (an'yoo-ploi-de) The addition or deletion of one or more chromosomes in a cell. 125

animal behavior (an'e-mal be'hav-or) The science that studies the behavior of animals. 655

animal pole (an'ah-mal pol) The upper pole of an egg that contains less yolk than the opposite (vegetal) pole and is, therefore, less dense. 147

Animalia (an'ah-mal'e-ah) The animal kingdom; it includes organisms that are eukaryotic and multicellular, lack cell walls, and are nourished by ingestion. 227

Anthropoidea (an'thra-poi-dea) The suborder of primates that includes monkeys, apes, and humans. 205

anthropomorphism (an"thro-po-mor'fizm) The attribution of human characteristics to nonhuman beings and objects. 655

antibody (an'ti-bod"e) A specific substance produced by cells in response to the presence of an antigen; it reacts with the antigen; also known as immunoglobulin (Ig). 583

anticodon (an'ti-ko"don) A sequence of three bases on transfer RNA that pairs with codons of messenger RNA to position amino acids during protein synthesis. 134

antidiuretic hormone (an"ti-di-u-ret'ik hor'mon) Hormone released from the posterior lobe of the pituitary gland that enhances the conservation of water by the kidneys; ADH. 560

antigen (an'ti-jen) A foreign (nonself) substance (such as a protein, nucleoprotein, polysaccharide, and some glycolipids) to which lymphocytes respond; also known as an immunogen because it induces the immune response. 583

antiparallel (an'ti-par"ah-lel') Refers to opposing strands of DNA that are oriented in opposite directions. 131

aposematic coloration (ah'pos-mat'ik kul'e-ra"shen) Sharply contrasting colors of an animal that warn other animals of unpleasant or dangerous effects. 678

archenteron (ar-ken"te-ron') The embryonic digestive tract that is formed during gastrulation. 149

Aristotle's lantern (ar"i-stot'els lan'tern) The series of ossicles making up the jawlike structure of echinoid echinoderms. 401

artery (ar'ter-e) A vessel that transports blood away from the heart. 573

arthroscopy (ar-thros'ko-pe) Examination of the interior of a joint with an arthroscope. 515

articulation (ar-tik'u-la'shun) The joining together of parts at a joint. 512

artificially acquired active immunity The type of immunity that results from immunizing an animal with a vaccine. 583

artificially acquired passive immunity The type of immunity that results from introducing antibodies that have been produced either in another animal or by specific in vitro methods into an animal. 583

aschelminth (ask'hel-minth) Any worm of the phylum Aschelminthes. 291

Ascidiacea (as-id'e-as"e-ah) A class of urochordates whose members are sessile as adults, and solitary or colonial.

ascon (as'kon) The simplest of the three sponge body forms. Asconoid sponges are vaselike, with choanocytes directly lining the spongocoel. 258

aster (as'ter) A structure seen in a cell during the prophase of mitosis; composed of a system of microtubules arranged in astral rays around the centrosome. 89

asymmetry (a-sim'i-tre) Without a balanced arrangement of similar parts on either side of a point or axis. 231

atom (at'om) Smallest particle of an element that has the same properties. 16

atomic mass (ah-tom-ik mass) A mass unit determined by arbitrarily assigning the carbon-12 isotope a mass of 12 atomic mass units. 17

atomic number (ah-tom-ik num'ber) A value equal to the number of protons of an element. 17

Australopithecus (au-stra'lo-pe'the-kus) A genus of early hominoids. 211

autosomes (au"te-somz') Chromosomes other than sex chromosomes. 115

autotroph (aw'to-trof) An organism that uses carbon dioxide as its sole or principle source of carbon. 595

autotrophic (au'te-trof'ic) Having the ability to synthesize food from inorganic compounds and an energy source. 667

axial filament (ak'si-al fil'ah-ment) The axoneme, which consists of nine pairs of microtubules arranged in a circle around two central tubules, this is called a 9 + 2 pattern of microtubules. 51

axial skeleton (ak'se-al skel'e-ton) Portion of the skeleton that supports and protects the organs of the head, neck, and trunk. 512

axon (ak'son) A nerve fiber that conducts a nerve impulse away from a neuron cell body. 522

axoneme (ak'so-nem) The axial thread of the chromosome where the axial combination of genes are located. 51

B

B cell (B sel) A type of lymphocyte derived from bone marrow stem cells that matures into an immunologically competent cell under the influence of the bursa of Fabricus in the chicken, and the bone marrow in nonavian species; following interaction with antigen, it becomes a plasma cell, which synthesizes and secretes antibody molecules involved in humoral immunity; B lymphocyte. 573

balanced polymorphism (bal'enst pol'e-morf-ism) Occurs when different phenotypic expressions are maintained at a relatively stable frequency in a population. 195

Barr body (bar bod'e) A heterochromatic X chromosome usually found in the nucleus of female mammals. 119

basal body (basal bod'e) A centriole that has given rise to the microtubular system of a cilium or flagellum and that remains attached at the base of the meiotic structure, just beneath the plasma membrane. 52

base (bas) A substance that ionizes in water to release hydroxyl ions (OH⁻) or other ions that combine with hydrogen ions.   21

basophil (ba′so-fil) White blood cell characterized by the presence of cytoplasmic granules that become stained by basophilic dye.   572

Batesian mimicry (bats′e-an mim′ik-re) Occurs when one species, called the mimic, resembles a second species, the model, that is protected by aposematic coloration.   678

behavioral ecology (be-hav′yer-al e-kol′o-je) The scientific study of all aspects of animal behavior as related to the environment.   655

benign tumor (be-nin too′mor) Not malignant; not recurrent; favorable for recovery.   172

benthic (ben′thik) Refers to the bottom substrate of an ocean, lake, stream, or other body of water.   692

bilateral symmetry (bi-lat′er-al sim′i-tre) A form of symmetry in which only the midsagittal plane will divide an organism into mirror images; bilateral symmetry is characteristic of actively moving organisms that have definite anterior (head) and posterior (tail) ends.   231

binary fission (bi′ne-re fish′en) Asexual reproduction in protists in which mitosis is followed by cytoplasmic division, producing two new organisms.   238

binomial nomenclature (bi-no″me-al no″men-kla′cher) A system for naming in which each kind of organism (a species) has a name of two parts, the genus and the species epithet.

biofeedback (bi″o-fed′bak) Furnishing information to an individual, usually in an auditory or visual mode, on the state of one or more physiological variables such as heart rate, blood pressure, or skin temperature.   535

biogeochemical cycles (bi′o-je′o-kem′i-kal si′kels) The cycling of elements between reservoirs of inorganic compounds and living matter in an ecosystem.   686

biogeography (bi′o-je-og′re-fe) The study of the distribution of life on Earth.

biological chemistry or **biochemistry** (bi″o-kem′is-tre) Branch of science dealing with the chemistry of living organisms.   16

biological magnification (bi′o-loj′i-kal mag-ni-fi-ka′shen) The concentration of substances in animal tissues as the substances are passed through ecosystem food webs.   695

biology (bi-ol′o-je) The study of life.

biomass (bi′o-mas) The part of an ecosystem consisting of living matter.   682

biomes (bi′omz) Distinctive associations of plant and animal populations; characterized by certain geographical boundaries and specific climatic and geographical features.   688

biotic potential (bi-ot′ik pa-ten′shal) The capacity of a population to increase maximally.   671

bipedal (bi′ped-al) Walking on two, rather than four, appendages.   208

biramous appendages (bi-ra′mus ah-pen′dij-ez) Appendages having two distal processes connected to the body by a single proximal process.   346

bladder worm (blad′er worm) The unilocular hydatid cyst of a tapeworm.   285

blastocoel (blas′to-sel) The fluid-filled cavity of the blastula.   149

blastocyst (blas′to-sist) An early stage of embryonic development consisting of a hollow ball of cells.   643

blastoderm (blas′to-derm) A small disk of cells at the animal end of the embryo of a reptile or bird that results from early cleavages.   155

blastomeres (blas′to-merz) Any of the cells produced by cleavage of a zygote.   148

blastopore (blas′to-por) The point at which cells on the surface of the blastula move to the interior of the embryo during gastrulation.   149

blastula (blas′tu-lah) An early stage in the development of an embryo; it consists of a sphere of cells enclosing a fluid-filled cavity (blastocoel).   149

blood (blud) The fluid that circulates through the heart, arteries, capillaries, and veins.   54

blood pressure (blud presh′ur) The force (energy) with which blood is pushed against the walls of blood vessels and circulated throughout the body when the heart contracts.   577

blubber (blub′er) The fat found between the skin and muscle of whales and other cetaceans, from which oil is made.   619

bone (bon) The hard, rigid form of connective tissue constituting most of the skeleton of vertebrates; composed chiefly of calcium salts.   510

bone (osseous) tissue (bon tish′u) A type of connective tissue.   54

book lungs (book lunz) Modifications of the arthropod exoskeleton into a series of internal plates that provide surfaces for exchange of gases between the blood and air.   348

bothria (both-re-ah) Dorsal or ventral grooves, which may be variously modified, on the scolex of a cestode.

bottleneck effect (bot′el-nek′ i-fect) Changes in gene frequency that result when numbers in a population are drastically reduced, and genetic variablity is reduced as a result of the population being built up again from relatively few surviving individuals.   193

brown fat Mitochondria-rich, heat generating adipose tissue of endothermic invertebrates.   620

buccal pump (buk′el pump) The mechanism by which lung ventilation occurs in amphibians; muscles of the mouth and pharynx create positive pressure to force air into the lungs.   445

buccopharyngeal respiration (buk′o-fah-rin′je-al res′pah-ra′shen) The diffusion of gases across moist linings of the mouth and pharynx of amphibians.   445

budding (bud′ing) The process of forming new individuals asexually in many different invertebrates.   97

buffer (buf′er) A substance that can react with a strong acid or base, to form a weaker acid or base and thus resist a change in pH.   22

bulb of Krause (bulb ov Krause) A sensory receptor in the skin believed to be the sensor for touch-pressure; also called bulbous corpuscle.   541

bulbourethral glands (bul″bo-u-re′thral glandz) Glands that secrete a viscous fluid into the male urethra during sexual excitement.   635

bursa (bur′sah) A saclike, fluid-filled structure lined with synovial membrane that occurs near a joint.   512

C

calcitonin (kal″si-to′nin) A thyroid hormone that lowers calcium and phosphate levels in the blood; also called thyrocalcitonin.   563

calorie (kal′o-re) A unit used in the measurement of heat energy and the energy value of foods.   596

Calorie (kal′o-re) Amount of heat.   59

calyx (ka′liks) A boatshaped or cuplike central body of an entoproct or crinoid.   304

capillary (kap′i-ler″e) A small blood vessel that connects an arteriole and a venule.   23, 574

carbaminohemoglobin (kar″bah-me′no-he″mo-glo′bin) Compound formed by the union of carbon dioxide and hemoglobin.

carbohydrate (kar″bo-hi′drat) An organic compound that contains carbon, hydrogen, and oxygen, with a 2:1 ratio of hydrogen to oxygen atoms.   25

carcinogen (kar-sin′o-jen) A substance that causes cancer.   172

cardiac muscle (kar′de-ak mus′el) Specialized type of muscle tissue found only in the heart.   515

carnivore (kar′ne-vor) One of the flesh-eating animals of the order Carnivora; also, any organism that eats flesh.   596

carrying capacity (kar′e-ing ka-pas′i-te) The maximum population size that an environment can support.   671

cartilage (kar′ti-lij) Type of connective tissue in which cells are located within lacunae and are separated by a semisolid matrix. 54, 510

caste (kast) One of the distinct forms of polymorphous social insects. 374

catabolism (kat″ah-bol′ism) Metabolic process by which large molecules are broken down into smaller ones; **catabolic metabolism.** 58

catalysis (kah-tal′i-sis) An increase in the velocity of a chemical reaction or process produced by the presence of a substance that is not consumed in the net chemical reaction or process. 60

catalyst (kat-ah-list) A substance that increases the rate of a chemical reaction, but is not permanently altered by the reaction. 60

cecum (se′kum) 1. Each arm of the blind ending, Y-shaped digestive tract of trematodes (phylum Platyhelminthes). 2. A region of the vertebrate digestive tract where fermentation can occur. It is located at the proximal end of the large intestine. 495

cell adhesion molecule (sel ad-he′zhen mol′e-kul) Specific plasma membrane glycoprotein that cells use to maintain physical contact with each other. 162

cell body (sel bod′e) Portion of a nerve cell that includes a cytoplasmic mass and a nucleolus, and from which the nerve fibers extend. 522

cell cycle (sel si′kl) The regular sequence during which a cell grows, prepares for division, and divides to form two daughter cells. 87

cell-mediated immunity (sel me-de-a-tid i-mu′ne-te) Immunity resulting from T cells coming into close physical contact with foreign cells or infected cells to destroy them; it can be transferred to a nonimmune individual by the transfer of cells. 583

cellular respiration (sel′u-lar res″pi-ra′shun) Process by which energy is released from organic compounds within cells. 68

centriole (sen′tre-ol) A cellular organelle that functions in the organization of the mitotic spindle during mitosis. 52

centromere (sen′tro-mer) Portion of a chromosome to which the spindle fiber attaches during mitosis. 89

cephalization (sef′al-iz-a″shen) The development of a head with an accumulation of nervous tissue into a brain. 231

cercaria (se-kar-e-a) Juvenile digenetic trematode, produced by asexual reproduction within a sporocyst or redia. 282

cerebellum (ser″e-bel′um) Portion of the brain that coordinates skeletal muscle movement. 530

Cestoidea (ses-toid-e-ah) A class of flatworms; tapeworms.

chalone (kal′on) A group of tissue-specific water soluble proteins that are produced within the tissue, and that inhibit mitosis of

cells of that tissue and whose action is reversible. 91

chelicerae (ke-lis′er-ae) One of the two pairs of anterior appendages of arachnids, may be pincerlike or modified for piercing and sucking or other functions. 346

chemistry (kem′is-tre) The science dealing with the elements and atomic relations of matter, and various elemental compounds. 16

chemolithoautotroph (kem′o-lith-o-au′te-trof) Organisms that synthesize organic matter by oxidizing inorganic compounds. 683

chemoreceptor (ke″mo-re-sep′tor) A receptor that is stimulated by the presence of certain chemical substances. 540

chiasma (ki-as′mah) A decussation of X-shaped crossing; the places where pairs of homologous chromatids remain in contact during late prophase to anaphase of the first meiotic division, indicating where an exchange of homologous segments has taken place between nonsister chromatids by crossing over. 93

chief cell (chef sel) Cell of gastric gland that secretes various digestive enzymes, including pepsinogen. 603

chitin (ki′tin) The polysaccharide found in the exoskeleton of arthropods. 345

chloragogen tissue (klor′ah-gog′en tish′u) Cells covering the dorsal blood vessel and digestive tract of annelids; function in glycogen and fat synthesis, and urea formation. 336

choanocytes (ko-an′o-sitz) Cells of sponges that create water currents and filter food. 257

chordamesoderm (kor′dah-mez″o-derm) Tissue in the amphibian gastrula that forms between ectoderm and endoderm in the dorsal lip region of the blastopore; develops into the mesoderm and notochord. 153

chorion (kor′e-on) The outermost extraembryonic membrane of the embryo of an amniote; becomes highly vascular and aids in gas exchange. 156

chromatid (kro′mah-tid) A member of a duplicate pair of chromosomes. 89

chromatin (kro-mah-tin) Nuclear material that gives rise to chromosomes during mitosis. 53, 115

chromatophores (kro-mah-tah-forz) Cells containing pigment that, through contraction and expansion, produce temporary color changes. 320

chromosome (kro″mo-som) Rodlike structure that appears in the nucleus of a cell during mitosis; contains the genes responsible for heredity. 53

chromosome diminution (kro′mo-som dim′a-noo′sun) The loss of whole chromosomes or parts of chromosomes (genes) early in embryonic development. 168

chromosome maps (kro″mah-som maps) Representations of chromosomes indicating the position of gene loci. 118

chrysalis (kris′ah-lis) The pupal case of a butterfly that frms from the exoskeleton of the last larval instar. 373

chyle (kil) The milky fluid taken up by the lacteals from the food in the intestine during digestion; consists of lymph and droplets of triglyceride fat (chylomicrons) in a stable emulsion. 603

chyme (kim) Semifluid mass of food materials that pass from the stomach to the small intestine. 603

cilia (sil′e-ah) Microscopic, hairlike processes on the exposed surfaces of certain epithelial cells. 51

circadian rhythms (sur′ke-de′an rith′em) Daily cycles of activity. 669

circumcision (ser″kum-sizh′un) The removal of all or part of the prepuce or foreskin. 636

citric acid cycle (sit-rik as′id si′kl) A series of chemical reactions in the mitochondrion by which various molecules are oxidized and energy is released from them; Kreb's cycle. 75

cladograms (klad′o-gramz) Diagrams depicting the evolutionary history of taxa, which are derived from phylogenetic systematics (cladistics). 229

class (klas) A level of classification between phylum and order. 225

cleavage (kle′vij) The early mitotic and cytoplasmic divisions of an embryo. 148

climax community (kli′maks ka-myoo′ni-te) A final, relatively stable stage in an ecological succession. 682

clitellum (klit′el-um) The region of an annelid responsible for secreting mucus around two worms in copula and for secreting a cocoon to protect developmental stages. 337

cloaca (klo-a-kah) A common opening for excretory, digestive, and reproductive systems. 434

closed circulatory system (klosd sir′ku-lah-tory sis′tem) A circulatory system in an animal in which blood is confined to vessels throughout its circuit. 319

clouds of electronegativity (klouds ov e-lek″tro-neg″ah-tiv′i-te) The distribution of electrons around the nucleus of an atom. 17

cnidocytes (ni-do-sitz) The cell that produces and discharges the stinging organelles (nematocysts) in members of the phylum Cnidaria. 261

cocoon (ka-koon) The protective covering of a resting or developmental stage; sometimes refers to both the covering and the contents. 279, 373

codominance (ko-dom′ah-nens) An interaction of two alleles such that both alleles are expressed in a phenotype. 109

codon (ko'don) A sequence of three bases on messenger RNA that specifies the position of an amino acid in a protein. 133

coelom (se'lom) The fluid-filled body cavity of many animals; this term is often used to designate the eucoelom, a body cavity lined by mesoderm. 233

coenzyme (ko-en'zim) An organic nonprotein molecule, frequently a phosphorylated derivitive of a water-soluble vitamin, that binds with the protein molecule (apoenzyme) to form the active enzyme (holoenzyme). 65

coevolution (ko-ev'ah-loo"shen) The evolution of ecologically related species such that each species exerts a strong selective influence on the other. 675

cofactor (ko'fak-tor) A coenzyme, with which another must unite in order to function. 65

colloblasts (kol'ah-blasts) Adhesive cells on the tentacles of ctenophorans used to capture prey. 270

colonial hypothesis (kah-lo-ne-al hi-poth'e-sis) A hypothesis formulated to explain the origin of multicellularity from protist ancestors; animals may have been derived when protists associated together and cells became specialized and interdependent. 254

colony (kol'a-ne) An aggregation of organisms. 97

colostrum (ko-los'trum) The first secretion of the mammary glands following the birth of an infant. 647

comb rows (kom roz) Rows of cilia that serve as the locomotory organs of ctenophorans. 270

combination reaction (kom'be-na'sun re-ak'shun) The type of reaction that occurs when two or more atoms, ions, or molecules combine to form a more complex substance. 20

commensalism (kah-men'sal-izm) Living within or on an individual of another species without harm. 677

communication (ko-mu'ni-ka-shun) Act on the part of one organism (or cell) that alters the probability pattern of behavior in another organism (or cell) in an adaptive fashion. 657

community (kah-myoo'ni-te) The different kinds of organisms living in an area. 681

comparative anatomy (kom-par'ah-tiv ah-nat'ah-me) The study of animal structure in an attempt to deduce evolutionary pathways in particular animal groups.

comparative embryology (kom-par'ah-tiv em'bre-ol"o-je) The study of animal development in an attempt to deduce evolutionary pathways in particular animal groups. 308

Competitive Exclusion Principle (kom-pet'i-tive ik-skloo'zhen prin'se-pel) The idea

that two species with identical niches cannot coexist. 674

competitive inhibition (kom-pet'i-tiv in'i-bish'en) A type of enzyme control where the active site is occupied by a molecule other than the normal substrate, preventing binding of the substrate. 64

complement system (kom'ple-ment sis'tem) A group of circulating plasma proteins that plays a major role in an animal's defensive immune response. 582

complete linkage (kom-plet link'aj) Two genes positioned so close to one another on the same chromosome that recombination between them does not occur. 118

compound (kom'pownd) A substance composed of two or more elements joined by chemical bonds.

condensation reaction (kon'den-sa'shun re-ak'shun) The type of reaction that occurs when smaller molecules unite to form larger molecules and the production of one or more molecules of water. 21

conduction (kon-duk'shun) The conveyance of energy, such as heat, sound, or electricity. 616

cone cell (kon sel) A color sensitive photoreceptor cell concentrated in the retina. 549

conjugation (kon'ju-ga"shen) A form of sexual union used by ciliates involving a mutual exchange of haploid micronuclei. 250

connective tissue (ko-nek-tiv tish'u) A basic type of tissue that includes bone, cartilage, and various fibrous tissues. 53

continental drift (kon'ti-nen"tel drift) The break up and movement of land masses of the earth. The earth had a single land mass about 250 million years ago. This mass broke apart into continents, which have moved slowly to their present positions. 187

contractile vacuoles (kon-trak'til vak"u-ol') An organelle that collects and discharges water in protists. 238

convection (kon-vek'shun) The act of conveying or transmission. 616

convergent evolution (kon-ver'jent ev'o-loo"shen) Evolutionary changes that result in members of one species resembling members of a second unrelated (or distantly related) species.

coracidium (kor"ah-sid'e-um) Larva with a ciliated epithelium hatching from the egg of certain cestodes; a ciliated oncosphere. 286

coralline algae (kor'ah-lin al'je) Any red alga that is impregnated with calcium carbonate. Coralline algae often contribute to coral reefs. 272

corona (ko-ro'nah) A crown; an encircling structure. 293

cortisol (kor'ti-sol) A glucocorticoid secreted by the adrenal cortex. 564

counteracting osmolyte (kown"ter-acting os-mo-lyt) An osmolyte (ion) that counteracts another ion. 629

countercurrent mechanism (koun'ter kur'ent mek'ah-niz'em) The passive exchange of something between fluids moving in opposite directions past each other. 430

countershading (koun'ter-shad-ing) Contrasting coloration that helps conceal the animal (e.g., the darkly pigmented top and lightly pigmented bottom of frog embryos). 677

covalent bond (ko'va-lent bond) Chemical bond created by the sharing of electrons between atoms. 18

coxal glands (koks'el glands) An organ of excretion found in some arthropods. 348

crossing-over (kros'ing o'ver) The exchange of material between homologous chromosomes, during the first meiotic division, resulting in a new combination of genes. 93

cryptic coloration (kript'ik kul'e-ra"shen) Occurs when an animal takes on color patterns of its environment. 677

crystaline style (kris'telin stil) A proteinaceous, rodlike structure in the digestive tract of a bivalve (Mollusca) that rotates against a gastric shield and releases digestive enzymes. 316

cultural evolution (kul'cher-al ev'ah-loo-shen) The change in cultures over time. 219

culture (kul'cher) The ability to change behavior through learning. 219

cutaneous respiration (kyoo-ta'ne-us res'pah-ra'shen) Exchange of gases across thin, moist surfaces of the skin. 445

cuticle (ku-tikel) A noncellular, protective, organic layer secreted by the external epithelium (hypodermis) of many invertebrates, refers to the epidermis or skin in higher animals. 293, 344

cystacanth (sis"ta-kant) Juvenile acanthocephalan that is infective to its definitive host. 302

cysticercosis (sis"ti-ser-ko'sis) Infection with the larval forms (*Cysticercus cellulosae*) of *Taenia solium*. 286

cysticercus (sis"ti-ser'kus) Metacestode developing from the oncosphere in most Cyclophyllidae; usually has a tail and a well-formed scolex; **cysticercoid.** 285

cytochrome (si-te-krom) Several iron-containing pigments that serve as electron carriers in aerobic respiration. 78

cytokinesis (si'ta-kin-e'sis) The division of the cytoplasm of a cell. 87

cytopharynx (si'to-far'inks) A region of the plasma membrane and cytoplasm of some ciliated and flagellated protists specialized for endocytosis. 238

cytoplasm (si'to-plazm) The contents of a cell surrounding the nucleus. 34

cytoplasmic inclusion (si'to-plaz'mic in-kloo'zhun) The basic food material or stored product of a cell's metabolic activities.   52

cytoplasmic organization (si'to-plaz'mik or'ga-ni-za"shen) *see* unicellular organization.

cytopyge (si'to-pij) A region of the plasma membrane and cytoplasm of some ciliated protists specialized for exocytosis of undigested wastes.   238

cytoskeleton (si"to-skel'e-ton) In the cytoplasm of eukaryotic cells, an internal framework of microtubules, microfilaments, and other fine strands by which organelles and other structures are anchored, organized, and moved about.   50

D

daily torpor (daily tor'por) Daily sluggishness that some animals experience.   617

deamination reaction (de-am"i-na'shun re-ak'shun) A reaction in which an amino group, -NH₂, is removed from a compound.   80

decomposer (de-kem-poz-ir) Mostly heterotrophic bacteria and fungi that obtain organic nutrients by breaking down the remains or products of other organic compounds; their activities help cycle the simple compounds back to the autotrophs.   58

decomposition reaction (de-kom"po-zish'un re-ak'shun) The type of reaction that occurs as a result of the breakage of chemical bonds to form two or more simple products (atoms, ions, or molecules).   20

degeneracy (de-gen'er-ah-cy) The genetic code is said to be degenerate because more than one three base sequence in DNA can code for one amino acid.   133

delayed fertilization (di-lad' fur'teli-za'shen) Occurs when fertilization of an egg does not occur immediately following coitus, but may be delayed for weeks or months.   502

deletions (di-le'shenz) The loss of a portion of a chromosome.   127

deme (deem) A small, local subpopulation.   198

denaturation (de-na"chur-a'shun) Disruption of bonds holding a protein in its three-dimensional form, such that its polypeptide chain(s) unfolds partially or completely.   63

dendrite (den'drit) Nerve fiber that transmits impulses toward a neuron cell body.   522

dental formula (den'tel for'myu-lah) A notation that indicates the number of incisors, canines, premolars, and molars in the upper and lower jaw of a mammal.   494

deoxyribonucleic acid (de'oks-e-ri'bo-nuk"la-ik as'id) A polymer of deoxyribonucleotides that is in the form of a double helix; DNA is the genetic molecule of life in that it

codes for the sequence of amino acids in proteins.   29, 130

depolarization (de-po"lar-i-za'shun) The loss of an electrical charge on the surface of a membrane.   524

dermal branchiae (der'mal branch'e-ae) Thin folds of the body wall of a sea star that extend between ossicles and function in gas exchange and other exchange processes.   396

determination (de-ter"mi-na'shun) The loss of pluripotentiality in any embryonic part and its start on the way to an unalterable fate.   160

deuterostomes (du'te-ro-stoms") Animals in which the anus forms from, or in the region of, the blastopore; often characterized by enterocoelous coelom formation, radial cleavage, and the presence of a dipleurulalike larval stage.

diabetes (di"ah-be'tez) Condition characterized by a high blood glucose level and the appearance of glucose in the urine due to a deficiency of insulin; diabetes mellitus.   566

diaphragm (di"ah-fram') The domed respiratory muscle between thoracic and abdominal compartments of mammals.   497

diapsid lineage (di-ap'sid lin'e-ij) The vertebrate lineage leading from early amniotes to modern reptiles and birds.   454

diastole (di-as'to-le) Phase of the cardiac cycle during which a heart chamber wall is relaxed.   577

differential gene activation (dif'e-ren'shal jen ak'ta-va-shun) The process whereby some genes function and others do not during the synthesis and processing of DNA.   168

differentiation (dif'ah-ren'she-a'shen) The development of embryonic structures from a nondescript form in the early embryo to their form in the adult.   149, 162

digestion (di-jest'yun) The process by which larger molecules of food substances are broken down into smaller molecules that can be taken up by the digestive system; hydrolysis.   595

dihybrid cross (di-hi'brid kros) A mating between individuals heterozygous for two traits.   107

diploblastic (dip"lo-blas'tik) Animals whose body parts are organized into layers that are derived embryologically from two tissue layers--ectoderm and endoderm. Animals in the phyla Cnidaria and Ctenophora are diploblastic.   233

direct flight (di-rekt' flit) Insect flight that is accomplished by flight muscles acting on wing bases and in which a single nerve impulse results in a single wing cycle; also called synchronous flight. *see* indirect [asynchronous] flight.   366

directional selection (di-rek'shen-al si-lek'shen) Natural selection that occurs when individuals at one phenotypic extreme have

an advantage over individuals with more common phenotypes.   193

disaccharide (di-sak'ah-rid) A sugar produced by the union of two monosaccharide molecules.   26

disruptive selection (dis-rup'tive si-lek'shen) Natural selection that occurs when individuals at the most common phenotypes are at a disadvantage; produces contrasting subpopulations.   194

dissociation (dis-so"se-a'shun) The ability of some molecules to break up into charged ions in water.   21

diving reflex (div-ing re'flex) The reflex certain animals have to stay under water for prolonged periods of time.   572

dominance hierarchy (dom'i-nans hi'e-rar'ke) The physical domination of some members of a group by other members, in relatively orderly and long-lasting patterns.   663

dominant (dom'ah-nent) A gene that masks one or more of its alleles. *see* recessive.   104

dominant species (dom'ah-nent spe'shez) A species that exerts an overriding influence in determining the characteristics of an ecosystem; also called a keystone species.   681

duplications (doo"ple-ka"shens) The presence of two copies of one or more loci in a chromosome.   127

E

ecdysis (ek-dis'is) The shedding of the arthropod exoskeleton to accommodate increased body size or a change in morphology (as may occur in molting from immature to adult); to molt; may also refer to the shedding of the outer epidermis of the skin of reptiles.   345

ectoderm (ek'ta-durm") The outer embryological tissue layer; gives rise to skin epidermis and glands, also hair and nervous tissues in some animals.   149

ectoplasm (ek'to-plaz-em) The outer, viscous cytoplasm of a protist.   238

ectotherm (ek'to-therm) Having a variable body temperature derived from heat acquired from the environment.   617

egestion vacuole (e-jes'chen vak'u-ol) A membrane-bound vesicle within the cytoplasm of a protist that functions in expelling wastes from a protist.   238

elastic cartilage (e-las'tik kar'ti-lij) A specialized fibrous connective tissue found in adults.   510

electroencephalogram (e-lek"tro-en-sef'ah-lo-gram) A recording of the potentials of the skull generated by currents emanating spontaneously from nerve cells in the brain; EEG.   534

electrolyte (e-lek'tro-lit) A substance that ionizes in a water solution.   21

electron (e-lek′tron) A small, negatively charged particle that revolves around the nucleus of an atom. 16

electron transport chain In a cell membrane, electron carriers and enzymes positioned in an organized array that enhance oxidation-reduction reactions; such systems function in the release of energy that is used in ATP formation and other reactions. 77

electroreception (i-lek′tro-re-sep′chen) The ability to detect weak electrical fields in the environment. 43

element (el′e-ment) A basic chemical substance. 16

elephantiasis (el″e-fan-ti′ah-sis) A chronic filarial disease most commonly occurring in the tropics due to infection of the lymphatic vessels with the nematode *Wucherceria* spp. 299

embryology (em″bre-ol′a-je) The study of development from the egg to the point that all major organ systems have formed. 145

embryonic diapause (em′bre-on′ik di′ah-pauz′) The arresting of early development to allow young to hatch, or be born, when environmental conditions favor survival. 502

end bulb (end bulb) A tiny swelling on the terminal end of telodendria at the distal end of an axon; also called the synaptic bouton. 525

end product inhibition (end pro-duct in′i-bish′en) The shut down of a biochemical pathway by inhibiting the activity of an enzyme. 64

endergonic (end″er-gon-ik) Characterized by the absorption of energy; said of chemical reactions that require energy to proceed. 60

endocrinology (en″do-kri-nol′o-je) The study of the endocrine system and its role in the physiology of the body. 555

endocytic vacuoles (en′do-sit′ik vak′u-olz) Membrane bound vacuoles formed by invagination of large areas of the plasma membrane; also called food vacuoles. 238

endocytosis (en″do-si-to′sis) Physiological process by which substances may move through a cell membrane. 45

endoderm (en″da-durm′) The innermost embryological tissue layer; gives rise to the inner lining of the gut tract. 149

endoplasm (en′do-plaz-em) The inner, fluid cytoplasm of a protist. 238

endoplasmic reticulum (en-do-plaz′mic re-tik′u-lum) Cytoplasmic organelle composed of a system of interconnected membranous tubules and vesicles; ER; *rough* ER has ribosomes attached to the side of the membrane facing the cytoplasm while *smooth* ER does not. 47

endostyle (en″do-stil′) A ciliated tract within the pharynx of some chordates that is used in forming mucus for filter feeding. 413

endosymbiont hypothesis (en′do-sim′bi-ont hi-poth′e-sis) The idea whereby the evolution of the eukaryotic cell might have occurred when a large anaerobic ameboid prokaryote ingested small aerobic bacteria and stabilized them instead of digesting them. 35

endotherm (en′do-therm) Having a body temperature determined by heat derived from the animal's own metabolism; contrasts with **ectotherm.** 617

energy (en′er-je) An ability to cause something to move, and thus, to do work. 59

energy budget (en′er-je buj′it) An accounting of the way in which energy coming into an ecosystem from the sun is lost or processed by organisms of the ecosystem. 667

energy-level shell (en′er-je lev′al shel) The distribution of atoms around the nucleus of an atom. 17

entropy (en′tro-pe) A measure of the degree of disorganization of a system; how much energy in a system has become so dispersed (usually as heat) that it is no longer available to do work. 59

environmental resistance (en-vi′ren-ment′al ri-zis′tens) The constraints placed on a population by climate, food, space, and other environmental factors. 671

enzyme (en′zim) A protein that is synthesized by a cell and acts as a catalyst in a specific cellular reaction. 60

enzyme-substrate complex (en′zim sub-strat com-plex) The binding of a substrate molecule to the active site of an enzyme. 61

eosinophil (e″o-sin′o-fil) White blood cells characterized by the presence of cytoplasmic granules that become stained by an acid dye. 572

epiblast (ep′i-blast) An outer layer of cells in the embryo of an amniote that forms from the proliferation and movement of cells of the blastoderm. 155

epiboly (ep-ib′ol-e) A spreading and thinning of ectoderm from the animal pole of an amphibian gastrula toward the vegetal pole. 153

epidermis (ep′i-durm″is) A sheet of cells covering the surface of an animal's body. 260

epigenesis (ep′i-jen′i-sis) The mistaken belief that the egg contains all the materials from which the embryo is constructed. 145

epinephrine (ep″i-nef′rin) A hormone secreted by the adrenal medulla during times of stress; adrenalin. 564

epistasis (ah-pis′tah-sis) An interaction between genes in which one gene prevents the expression of a gene at a second locus. 110

epithelial tissue (ep″i-the′le-al tish′u) The cellular covering of internal and external surfaces of the body; consists of cells joined by small amounts of cementing substances. Ep-

ithelium is classified into types based on the number of layers deep and the shape of the superficial cells. 53

epitoky (ep′i-to′ke) The formation of a reproductive individual (epitoke) that differs from the nonreproductive (atoke) form of that species. 333

estrogen (es′tro-jen) Hormone that stimulates the development of female secondary sex characteristics. 566

estrous cycle (es′tres si′kel) A recurrent series of changes in the reproductive physiology of female mammals other than primates; females are receptive, physiologically and behaviorally, to the male only at certain times in this cycle. 502

estrus (es′trus) The recurrent, restricted period of sexual receptivity in female mammals other than human females marked by intense sexual urges. 634

ethologist (e-thol′o-jist) A person who studies the whole patterns of animal behavior in natural environments, stressing the analysis of adaptation and the evolution of the patterns. 655

ethology (e-thol′o-je) The study of whole patterns of animal behavior in natural environments, stressing the analysis of adaptation and the evolution of the patterns. 655

euchromatic (u-chrom′a-tic) To have active genes. 115

eucoelom (u-se′lom) A body cavity completely lined by mesodermally derived tissues. 234

eukaryote (u-kar′e-ot) Having a true nucleus; a cell that has membranous organelles, most notably the nucleus. 34

eutely (u′te-le) Condition where the body is composed of a constant number of cells or nuclei in all adult members of a species (e.g., rotifers, some nematodes, and acanthocephalans). 291

evaporation (i-vap′e-ra′shun) The act or process of evaporating. 617

evolutionary systematics (ev′ah-loo″she-ner-e sys′tah-mat″iks) The study of the classification of, and evolutionary relationships among, animals; evolutionary systematists attempt to reconstruct evolutionary pathways based upon resemblances between animals that result from common ancestry. 229

exergonic (ek″ser-gon′ik) Characterized or accompanied by the release of energy; said of chemical reactions that release energy, so that the products have a lower free energy than the reactants. 59

exocytosis (eks′o-si-to′sis) The process by which substances are moved out of a cell; the substances are transported in the cytoplasmic vesicles, the surrounding membrane of which merges with the plasma membrane in such a way that the substances are dumped outside. 45

exons (eks'onz) The coding base sequence of a gene of a eukaryotic organism; exons are interrupted by noncoding sequences called introns. 133

exoskeleton (eks'o-skel''ah-ton) A skeleton that is formed on the outside of the body (as in the exoskeleton of arthropods). 344

experimental psychologist (ek-sper'i-men-tal si-kol'o-jest) A person who studies the mind and mental operations by the use of experimental methods. 655

exponential growth (ek'spo-nen''shal groth) Population growth in which the number of individuals double in each generation. 671

F

facilitated diffusion (fah-sil'i-tat''id di-fu'zhun) Diffusion in which substances are moved through membranes from a region of higher concentration to a region of lower concentration by carrier molecules. 42

family (fam'ah-le) The level of classification between order and genus. 225

feedback inhibition Control mechanism whereby an increase in some substance or activity inhibits the very process leading to the increase. 81

fermentation (fer''men-ta'shun) Degradative pathway that begins with glycolysis and ends with the electrons being transferred back to one of the breakdown products or intermediates. 73

fertilization membrane (fer''ti-li-za'shen mem'bran) A membrane that raises off the surface of an egg after sperm penetration; prevents multiple fertilization. 147

fibrocartilage (fi''bro-kar'ti-lij) The type of cartilage made up of parallel, thick, compact bundles, separated by narrow clefts containing typical cartilage cells (chondrocytes). 510

fibrous connective tissue (fi'brus ko-nek'tiv tish'u) The tissue that is made up of fibers that are very densely packed (e.g., tendons and ligaments). 53

filtration (fil-tra-shun) Movement of material through a membrane as a result of hydrostatic pressure. 43

first law of thermodynamics The total amount of energy in the universe remains constant, more energy cannot be created and existing energy cannot be destroyed; energy can only undergo conversion from one form to another. 59

fission (fish'un) Asexual reproduction in which the cell divides into two (binary fission) or more (multiple fission) daughter parts, each of which becomes an individual organism. 97

flagella (flah-jel'ah) Relatively long motile processes that extend out from the surface of a cell. 51

flame cell Specialized hollow excretory or osmoregulatory structure consisting of one to several cells containing a tuft of cilia (the ''flame'') and located at the end of a minute tubule; flame bulb. 277

flavin adenine dinucleotide (FAD) (fla'vin ad'e-nen di''nuc'leo- tid) A coenzyme that is a condensation product of riboflavin phosphate and adenylic acid; it forms the prosthetic group of certain enzymes. 75

fluke (flook) Any trematode worm; a member of the class Trematoda or class Monogenea. 279

follicle-stimulating hormone (fol'i-kl stim'u-la''ting hor'mon) A hormone secreted by the anterior pituitary gland that stimulates the development of an ovarian follicle in a female or the production of sperm in a male; FSH. 563

food chain (food chan) A linear sequence of organisms through which energy is transferred in an ecosystem. 58, 682

food webs (food webz) A sequence of organisms through which energy is transferred in an ecosystem; rather than being a linear series, food webs have highly branched energy pathways. 682

fossils (fos'elz) Any remains, impressions, or traces of organisms of a former geological age.

founder effect (found'er i-fekt) Changes in gene frequency that occur when a few individuals from a parental population colonize new habitats; the change is a result of founding individuals not having a representative sample of the parental population's genes. 193

fragmentation (frag''men-ta'shun) Division into smaller units. 97

free nerve ending (free nerv ending) A free nerve ending that acts as a pain receptor. 541

Fungi (fun'ji) The kingdom of life whose members are characterized by being eukaryotic, multicellular, and saprophytic (mushrooms, molds). 227

G

Galapagos Islands (gah-lah''pe-gos' i'landz) An archipelago on the equator in the Pacific Ocean about 1,000 km west of Ecuador. Charles Darwin's observations of the plant and animal life of these islands were important in the formulation of the theory of evolution by natural selection. 180

gamete (gam'et) Mature haploid cell (sperm or egg) that functions in sexual reproduction. 91

gametogenesis (gam''e-to-jen'e-sis) The formation of gametes by way of meiosis. 94

gastric shield (gas'trik sheld) A chitinized plate in the stomach of a bivalve (phylum Mollusca) upon which the crystalline style is rotated. 316

gastrodermis (gas-tro-derm'is) The endodermally derived lining of the gastrovascular cavity of Cnidaria. 260

gastrovascular cavity (gas'tro-vas'ku-lar kav'i-te) The large central cavity of cnidarians that serves as a chamber for receiving and digesting food. 261

gastrozooid (gas'tro-zo'oid) A feeding polyp in a colonial hydrozoan (phylum Cnidaria). 263

gastrulation (gast'ru-la''shen) The embryological processes that result in the formation of the gastrula; results in the formation of the embryonic gut, ectoderm, and endoderm. 149

gemmules (jem'yoolz) Resistant, overwintering capsules formed by freshwater, and some marine, sponges that contain masses of mesenchyme cells; ameboid mesenchyme cells are released and organize themselves into a sponge. 259

gene amplification (jen am'ple-fa-ka'shun) The selective synthesis of additional DNA. 167

gene flow (jen flow) Changes in gene frequency in a population that result from emigration or immigration. 193

gene insertion (jen in-sur'shen) The process by which one or more genes from one organism are incorporated into the genetic make-up of a second individual. 142

gene pool (jen pool) The sum of all genes in a population. 190

gene rearrangement (jen re-a'rang-ment) The unique mechanism in the immune system whereby pieces of genes are moved around in cells to produce a variety of combined genes. 167

generator potential (jen'e-ra''tor po-ten'shal) A graded potential that travels only a short distance along the plasma membrane of a sensory cell. 540

genes (jenz) A heritable unit in a chromosome; a series of nucleotide bases on the DNA molecule that codes for a single polypeptide. 104

genetic drift (je-net'ik drift) Occurs when chance events influence evolution; also called neutral selection. 192

genetic recombination (je-net'ik re-kombe-na'shun) Crossing-over; a major source of genetic variation in a population or a given species. 93

genetics (je-net'iks) The study of the mechanisms of transmission of genes from parents to offspring. 103

genotype (je'no-tip) The specific gene combinations that characterize a cell or an individual. 104

genus (je'nus) The level of classification between species and family. 225

gerontology (jer″on-tol′o-je) The scientific problems of aging in all their aspects, including clinical, biological, and sociological. 291

gestation (jes-ta′shun) Period of development of the young in viviparous animals, from the beginning of fertilization of the ovum until birth. 634

gestation period (jest-a′shun per′e-ed) The time between fertilization and birth in viviparous animals. 502

gill (gil) An aquatic respiratory organ for obtaining oxygen and getting rid of carbon dioxide.

gill arches (gil arch′ez) Bony or cartilaginous gill supports of some vertebrates; also called visceral arches. 430

gill lamellae (gil la-mel′a) Thin plates of tissue on gill filaments that contain the capillary beds across which gases are exchanged. 430, 588

glochidium (glo-kid′e-um) A larval stage of freshwater bivalves in the family Unionidae; it lives as a parasite on the gills or fins of fishes. 317

glomerulus (glo-mer′u-lus) A capillary tuft located within the capsule (Bowman's) of a nephron. 625

glucagon (gloo′kah-gon) Hormone secreted by the pancreatic islets of Langerhans that causes the release of glucose from glycogen. 565

glycocalyx (gli″ko-kal′iks) The glycoprotein and polysaccharide covering that surrounds many cells. 37

glycolysis (gli-kol′i-sis) The conversion of glucose to pyruvic acid with the release of some energy in the form of ATP. 71

Golgi apparatus (gol′je ap″ah-ra′tus) A cytoplasmic organelle that functions in preparing cellular products for secretion. 48

gonadotropin (go-nad″o-trop′in) A hormone that stimulates activity in the gonads. 636

gonozooid (gon′o-zo″id) A polyp of a hydrozoan cnidarian that produces medusae. 263

gray crescent (gra kres′ent) A dark arching band that forms on the surface of the amphibian zygote opposite the point of sperm penetration; forms in the region where gastrulation will occur. 147

gross primary production (gros pri′mer-e pro-duk′shen) The total energy fixed by autotrophs in an ecosystem. 683

growth hormone (groth hor′mon) A hormone released by the anterior lobe of the pituitary gland that promotes the growth of the organism. 561

gular flutter (gu-lar flut′er) The type of breathing experienced by some birds. 619

gustation (gus-ta′shun) The act of tasting or the sense of taste 542

H

habitat (hab′i-tat) The native environment of an organism. 667

habituation (hah-bich″u-a′shun) The gradual adaptation to a stimulus or to the environment. 660

hairworm (har-worm) A free-living, threadlike Nematomorph. 302

half-life (haf′lif) 1. The time it takes for one-half of the radioactivity of an isotope to be released. 2. The time required for half of the atoms in a given amount of radioactive matter to decay. 18

haploid (hap′loid) Having one member of each pair of homologous chromosomes; haploid cells are the product of meiosis and are often gametes. 117

Hardy-Weinberg equilibrium (har′de win′berg e′kwe-lib′re-em) The condition in which the frequency of genes in a population does not change from one generation to another; the conditions defined by Hardy-Weinberg equilibrium define the conditions under which evolution does not occur. 191

head-foot (hed foot) The body region of a mollusc that contains the head and is responsible for locomotion as well as retracting the visceral mass into the shell. 310

hectocotylus (hek′to-kot″i-lus) A modified arm of some male cephalopods that is used in sperm transfer. 321

hemimetabolous metamorphosis (hem′i-met-ab″ol-us met-ah-morf′a-sis) A type of insect metamorphosis in which immature insects are different in form and habitats from the adult. It is different from holometabolous metamorphosis in that there is a gradual series of changes in form during the transition from immature to adult. 373

hemizygous (hem′i-zi′ge) An individual having one member of a pair of genes. 119

hemocoel (hem′o-sel) Large tissue spaces within arthropods that contain blood; derived from the blastocoel of the embryo. 348

hemoglobin (he″mo-glo′bin) Pigment of red blood cells responsible for the transport of oxygen and carbon dioxide. 572

Henson's node (hen′sonz nod) A depression that forms at the anterior margin of the primitive streak of an amniote embryo; the site of inward migration of epiblast cells. 155

herbivore (her′bi-vor) A plant-eating animal. 596

hermaphroditism (her-maf′ro-di-tizm) A state characterized by the presence of both male and female organs in the same animal. 99

heterochromatic (het′er-o-chrom″a-tik) Having inactive genes. 115

heterodont (het′e-ro-dont) Having a series of teeth specialized for different functions. 494

heterotherm (hed′e-ro-therm) An animal whose body temperature fluctuates markedly. 617

heterotroph (het″er-o-traf) An organism that obtains both inorganic and organic raw materials from the environment in order to live; animals, fungi, many protistans, and most bacteria are heterotrophs. 595

heterotrophic (het′er-o-trof′ik) Organisms that obtain energy by feeding on other organisms. 667

heterozygous (het′er-ozi″ges) Having different expressions of a gene on homologous chromosomes. 105

hibernation (hi″ber-na′shun) Condition of mammals that involves passing the winter in a torpid state in which the body temperature drops to nearly freezing and the metabolism drops close to zero. 620

Higgins larva The larval stage of the phylum Loricifera; has an introvert, thorax, and abdomen. 303

holoblastic (hol′o-blas″tik) Division of a zygote that results in separate blastomeres. 148

holometabolous metamorphosis (hol′o-met-ab″ol-us met-ah-morf′a-sis) A type of insect metamorphosis in which immatures, called larvae, are different in form and habitats from the adult; the last larval molt results in the formation of a pupa; radical cellular changes in the pupal stage end in adult emergence. 373

homeobox (ho′me-o-box) Describes the conserved sequence that is part of the coding region of homeotic genes. 166

homeostasis (ho″me-o-sta′sis) A state of equilibrium in which the internal environment of the body remains constant. 40

homeotherm (ho′me-o-therm) Having nearly uniform body temperature, regulated independently of the environmental temperature. 617

homeotic gene (ho″me-o-tic jen) A gene that helps determine body shape by controlling the developmental fate of groups of cells. 166

homodont (ho′mo-dont) Having a series of similar, unspecialized teeth. 494

Homo erectus (ho′mo a-rekt′us) An early hominid that live 1.6 million to 200,000 years ago. 214

Homo habilis (ho′mo hab′il-is) The first hominid. 214

homologous (ho-mol′o-ges) Structures that have a common evolutionary origin; the wing of a bat and the arm of a human are homologous, each can be traced back to a common ancestral appendage.

homologous chromosomes (ho-mol′o-gus kro′mo-som) Chromosomes that carry genes for the same traits. 93

Homo sapiens neanderthalensis (ho′mo sape′enz ne-an′der-thol-en-sis) The first subspecies of modern humans; present about 100,000 years ago.   215

Homo sapiens sapiens (ho'mo sape'enz sape'enz) Modern humans.   221

homozygous (homo-zi'ges) Having the same expression of a gene on homologous chromosomes.   104

hormone (hor'mon) A chemical secreted by an endocrine gland that is transmitted by the blood stream or body fluids.   171, 557

humoral immunity (hu'mor-al i-mu'no-te) The type of immunity that results from the presence of soluble antibodies being soluble in blood and lymph.   582

hunters and gatherers (hun'terz and gath'er-erz) A way of life in which 25 to 30 individuals exist in a community by hunting game and gathering plant products rather than depending upon agriculture or industry. *Homo erectus* and *Homo habilis* were hunters and gatherers.   219

hyaline cartilage (hi'ah-lin kar'ti-lij) The type of cartilage with a glassy, translucent appearance.   510

hydration shell (hi-dra'shun shel) The orientation of water molecules around ions.   23

hydraulic skeleton (hi-dro'lik skel'e-ton) The use of body fluids in open circulatory systems to give support and facilitate movement; muscles contracting in one part of the body force body fluids into some distant tissue space, thus causing a part of the body to extend or become turgid.   313

hydrocarbon (hi"dro-kar'bon) An organic molecule that contains only carbon and hydrogen, and has its carbons bonded in a linear fashion.   25

hydrogen bond A weak to moderate attractive force between a hydrogen atom bonded to an electronegative atom and one pair of electrons of another electronegative atom.   19

hydrological cycles (hi'dro-loj'i-kal si'kelz) The cycling of water between reservoirs in oceans, lakes, and groundwater, and the atomosphere.   688

hydrolysis (hi-drol'i-sis) The splitting of a molecule into smaller portions by the action of a water molecule.   20

hydrostatic skeleton (hi'dro-stat"ik skel'e-ton) The use of body cavity fluids, confined by the body wall, to give support (e.g., the hydrostatic skeleton of nematodes and annelids).   261

hypoblast (hi'po-blast) An inner layer of cells that results from the proliferation and movement of cells in the blastoderm of an avian or reptilian embryo.   155

hypothesis (hi-poth'e-sis) A tentative explanation of a question; an explanation based upon careful observations.

I

immunity (i-mu'ni-te) Refers to the overall general ability of a host to resist a particular disease; the condition of being immune.   581

immunization (im"u-ni-za'shun) The process of rendering a subject immune, or of becoming immune.   583

immunology (im"u-nol'o-je) The science concerned with the response of the organism to antigenic challenge, the recognition of nonself, and the chemical aspects of the immune phenomena.   581

immunotoxin (im"u-no-tok"sin) A monoclonal antibody that has been attached to a specific toxin or toxic agent (antibody + toxin = immunotoxin) and can kill specific target cells.   584

imprinting (im'print-ing) A rapid kind of learning of certain species-specific behavior patterns that occurs with the exposure to the proper stimulus at a critical stage of early life.   659

incomplete dominance (in'kem-plet' dom'i-nens) An interaction between two alleles in which the heterozyous state results in a phenotype intermediate between either homozygous state.   109

incomplete linkage (in'kem-plet' link'ij) The condition when two loci are far enough from one another on the same chromosome that recombination can occur between them.   118

indirect flight (in'dah-rekt flit) The insect flight mechanism in which flight muscles move the wings by acting on the thorax rather than directly on the wing bases; indirect flight is asynchronous because a single nerve impulse to a flight muscle results in many wing cycles *see* direct flight   366

induced fit (in-duced fet) The phenomenon of change in an enzyme's shape following binding of substrate.   61

industrial revolution (in-dus'tre-al rev'ah-loo"shen) The social and economic changes that resulted from the mechanization; began in the 18th century.   220

inhibitor (in-hib'i-tor) Any substance that interferes with a chemical reaction, growth, or other biological activity.   64

inorganic molecule (in"or-gan'ik mol'e-kul) A molecule that lacks a carbon atom.   24

insectivore (in-sek'ti-vor) Any organism that eats insects.   596

instars (in'starz) Any of the developmental stages of an arthropod.   373

instinct (in'stinkt) Behavior that is highly stereotyped, more complex than the simplest reflexes, and usually directed at particular objects in the environment.   655

insulin (in'su-lin) A hormone secreted by the pancreatic islets that functions in the control of carbohydrate metabolism.   565

integument (in-teg'u-ment) A covering (e.g., the skin).   507

intercalated disc (in-ter"kah-lat'ed disk) Membranous boundary between adjacent cardiac muscle cells.   515

intercellular junction (in"ter-sel'u-lar jungk'shun) The type of connection between individual cells.   40

intermediate filament (in"ter-me'de-at fil'ah-ment) The chemically heterogeneous group of protein fibers, the specific proteins of which can vary with cell type.   49

interneuron (in"ter-nu'ron) A neuron located between a sensory neuron and a motor neuron.   522

interphase (in'ter-faz) Period between two cell divisions when a cell is carrying on its normal functions.   87

intraspecific competition (in'tra-spi-sif'ik kom'pi-tish'en) The interaction that occurs when two individuals of the same species seek the same resources.   671

introns (in'tronz) Noncoding base sequences in the genes of eukaryotic organisms.   133

introvert (in'tro-vert) The anterior narrow portion that can be withdrawn (introverted) into the trunk of a sipunculid worm.   303

inversions (in-ver'zhenz) A rearrangement in the structure of a chromosome in which two breaks occur and a segment of the chromosome is flip-flopped; genes in that segment occur in a reverse order.   127

ion (i'on) An atom or group of atoms with an electrical charge.   19

ionic bond An association between ions of opposite charges.   20

isomer (i'so-mer) Any compound exhibiting, or capable of exhibiting, isomerism; compounds of the same molecular formula but with different arrangements of the atoms.   26

isotope (i'so-top) Atoms of the same element having different masses.   17

J

Jacobson's organs (ja'keb-sonz or'ganz) Olfactory receptors present in most reptiles; blind ending sacs that open through the secondary palate into the mouth cavity; they are used to sample airborne chemicals.   465

Johnston's organs (jon'stonz or'ganz) Mechanoreceptors (auditory receptors) found at the base of the antennae of male mosquitoes and midges.   369

joule (jool) A standard unit of energy.   59

K

K-selected (ka si-lekt'ed) Organisms whose populations are maintained near the carrying capacity of the environment.   671

karyotyping (kar″i-o-tip′ing) The determination of the number and structure of chromosomes in an individual.   124

keystone species (kee-ston spe′shez) *see* dominant species.

kilocalorie (kil′o-kal″o-re) A unit of heat equal to 1,000 calories.   59, 596

kinetic energy The energy associated with a body by virtue of its motion.   59

kinetochore (ki-ne′to-kor) A centromere; serves as an attachment site for the microtubules of the mitotic apparatus.   89

kingdom (king′dom) The highest level of classification of life; the most widely accepted classification system includes five kingdoms: Monera, Protista, Fungi, Plantae, and Animalia.   225

krill (kril) Any of the small, pelagic, shrimp-like crustaceans.   604

L

labial palps (la′be-al palps) 1. Chemosensory appendages found on the labium of insects (Insecta, Arthropoda). 2. Flaplike lobes surrounding the mouth of bivalve molluscs that direct food toward the mouth.   316

lactation (lak-ta′shun) The production of milk by the mammary glands.   646

larva (lar′vah) 1. The immature, feeding stage of an insect that undergoes holometabolous metamorphosis. 2. The immature stage of any animal species in which adults and immatures are different in body form and habitat.   259

lateral line system (lat′er-al lin sis′tem) 1. A line of sensory receptors along the side of some fish and amphibians used to detect water movement (phylum Chordata). 2. The external manifestation of a lateral excretory canal of nematodes (phylum Nematoda).

leucon (lu′kon) The sponge body form that has an extensively branched canal system; the canals lead to chambers lined by choanocytes.   258

light microscope (lit mi′kro-skop) The type of microscope in which the specimen is viewed under ordinary illumination.   38

limiting factor (lim′i-ting fak′tor) A nutrient or other component of an organism's environment that is in relatively short supply and, therefore, restricts the organism's ability to reproduce successfully.   667

linkage group (ling′kij groop) Genes linked to the same chromosome that tend to be inherited together.   118

linkage map (ling′kij map) A representation of a chromosome on which the position of various loci are shown; determined based upon the frequency of recombination between loci.   118

linked genes (lingkt genz) Genes located on the same chromosome.   118

lipid (lip′id) A fat, oil, or fatlike compound that usually has fatty acids in its molecular structure.   27

local chemical messenger (local kem′i-kal messenger) A chemical that acts on nearby cells.   557

locus (lo′kus) The position of a gene in a chromosome.   109

loose connective tissue (loos ko-nek′tiv tish′u) The type of tissue in which the matrix contains strong, flexible fibers of the protein collagen that are interwoven with fine, elastic and reticular fibers.   53

lophophore (lof′a-for) Tentacle-bearing ridge or arm within which is an extension of the coelomic cavity in lophophorate animals (e.g., brachiopods, ectoprocts, and phoronids).   381

lung (lung) An organ of the respiratory system.   588

luteinizing hormone (lu-te-in-iz″ing hor′mon) A hormone secreted by the anterior pituitary gland that controls the formation of corpus luteum in females and the secretion of testosterone in males.   563

lymph (limf) Fluid transported by the lymphatic vessels.   579

lymphocyte (lim′fo-sit) A type of white blood cell that functions to provide protection to an animal.   573

lysosome (li′so-som) Cytoplasmic organelle that contains digestive and hydrolytic enzymes.   48

M

macronucleus (mak′ro-nuk″le-us) A large nucleus found within the Ciliata (Protista) that regulates cellular metabolism.   249

macronutrient (mak″ro-noo′tre-ent) An essential nutrient that has a large minimal daily requirement (greater than 100 mg) (e.g., calcium, phosphorous, magnesium, potassium, sodium, and chloride).   596

major histocompatibility complex A large set of cell surface antigens in each individual, encoded by a family of genes, that serve as a unique biochemical marker of individual identity; it can trigger T-cell responses that may lead to rejection of transplanted organs. MHC antigens are also involved in the regulation of the immune response and the interactions between immune cells; MHC.   583

malignant (mah-lig′nant) The power to threaten life; cancerous.

malignant tumor (mah-lig′nant too′mor) A tumor that tends to become progressively worse and to result in death.   172

malpighian gland (mal-pig′e-an gland) Blind tubules opening into the hindgut of almost all insects and some myriapods and arachnids, and that function as excretory organs.   392

malpighian tubules (mal-pig′e-an tu′bulz) The blind-ending excretory tubules that join the midgut of insects and some other arthropods.   348

mammary gland (mam′ar-e gland) The breast.   640

mandibles (man′dib-elz) 1. The lower jaw of vertebrates. 2. The paired, grinding and tearing mouthparts of arthropods, which were derived from anterior head appendages.   354

mantle (man′tel) The outer fleshy tissue of molluscs that secretes the shell.   310

mantle cavity (man′tel kav′i-te) The space between the mantle and the visceral mass of molluscs.   311

mass (mas′) The quantity of matter in a material.   16

mastax (mas′tax) The pharyngeal apparatus of rotifers.   294

matter (mat′er) Anything that has weight and occupies space.   16

mechanoreceptor (mek″ah-no-re-sep′tor) A sensory receptor that is sensitive to mechanical stimulation such as changes in pressure or tension.   540

meconium (me-ko′ne-um) A dark green mucilaginous material in the intestine of the full term fetus, being a mixture of the secretions of the intestinal glands and some amniotic fluid.   647

median eye (me′de-an i) A photoreceptor located middorsally on the head of some vertebrates; it is associated with the vertebrate epithalamus.   465

medulla oblongata (me-dul′ah ob″longah′tah) Portion of the brain stem located between the pons and the spinal cord.   530

medusa (me-du′sah) The sexual stage in the life cycle of cnidarians; the jellyfish body form.   261

meiosis (mi-o′sis) Process of cell division by which egg and sperm cells are formed.   91

melanin (mel′ah-nin) Dark pigment normally found in skin and hair.   508

melatonin (mel″ah-to′nin) A hormone secreted by the pineal gland.   563

memory cell A lymphocyte capable of initiating the antibody-mediated immune response upon detection of a specific antigen molecule for which it is genetically programmed. It circulates freely in the blood and lymph, and may live for years.   586

menarche (me-nar′ke) The first menstrual period.   640

meninges (me-nin′jez) A group of three membranes that covers the brain and spinal cord.   528

menopause (men′o-pawz) Termination of the menstrual cycle.   640

menstrual cycle (men′stroo-al si′k′l) The period of the regularly recurring physiologic changes in the endometrium that culminate in its shedding (menstruation).

menstruation (men″stroo-a′shun) Loss of blood and tissue from the uterus at the end of a female reproductive cycle. 640, 641

meroblastic (mer′ah-blas″tik) The division of a zygote in which cleavages do not completely divide the embryo. 148

mesenchyme (mez′en-kim) Undifferentiated mesoderm. 149

mesoderm (mez′ah-durm) The embryonic tissue that gives rise to tissues located between the ectoderm and endoderm (e.g., muscle, skeletal tissues, and excretory structures). 149

mesoglea (mez-o-gle′ah) A gel-like matrix found between the epidermis and gastrodermis of cnidarians. 261

mesothorax (mez′o-thor″aks) The middle of the three thoracic segments of an insect; usually contains the second pair of legs and the first pair of wings. 365

messenger RNA (mes′en-jer r-n-a) A single-stranded polyribonucleotide; formed in the nucleus from a DNA template and carries the transcribed genetic code to the ribosome where the genetic code is translated into protein. 132

metabolism (me-tab′o-lizm) All of the chemical changes that occur within cells. 58

metacercaria (me′ta-ser-ka′re-ah) Stage between the cercaria and adult in the life cycle of most digenetic trematodes; usually encysted and quiescent. 282

metamerism (me-tam″a-riz′em) A segmental organization of body parts. 327

metamorphosis (met″ah-mor′fo-sis) Change of shape or structure, particularly a transition from one developmental stage to another as from larva to adult form. 163

metanephridium (met′ah-ne-frid′e-um) An excretory organ found in many invertebrates; it consists of a tubule that has one end opening at the body wall and the opposite end in the form of a funnel-like structure that opens to the body cavity. 332

metaphase (met-ah-faz′) Stage in mitosis when chromosomes become aligned in the middle of the spindle. 89

metastasis (me-tas′tah-sis) The spread of cancer from one part of the body to another part. 172

metathorax (met′ah-thor″aks) The posterior of the three segments of an insect thorax; it usually contains the third pair of walking legs and the second pair of wings (Arthropoda). 365

micelle (mi-sel′) A supermolecular colloid particle, most often a packet of chain molecules in parallel arrangement. 609

microfilament Component of the cytoskeleton; involved in cell shape, motion, and growth. 50

micronucleus (mi′kro-nuk″le-us) A small body of DNA that contains the hereditary information of ciliates (Protista); exchanged between protists during conjugation. 249

micronutrient (mi″kro-nu′tre-ent) A dietary element essential in only small quantities. 596

microscopy (mi-kros′ko-pe) Examination with a microscope. 38

microtubule (mi″kro-tu′bul) A hollow cylinder of tubulin subunits; involved in cell shape, motion, and growth; functional unit of cilia and flagella. 49

mimicry (mim′ik-re) When one species resembles one or more other species; often protection is afforded the mimic species. 678

miracidium (mi-rah-sid′e-um) The first stage larva of a trematode that undergoes further development in the body of a snail. 282

mitochondrion (mi″to-kon′dre-on) Eukaryotic organelle that specializes in aerobic respiration. 49

mitosis (mi-to′sis) Nuclear division in which the parental number of chromosomes is maintained from one cell generation to the next. Basis of reproduction of single-cell eukaryotes; basis of physical growth (through cell divisions) in multicellular eukaryotes. 87

mitotic apparatus (mi-to′tic ap′a-rat′es) Collectively, the asters, spindle, centrioles, and microtubules of a dividing cell. 89

modifier genes (mod′ah-fi′er genz) Genes that alter the expression of genes at a second locus. 110

molecular biology (mo-lek′yah-ler bi-ol′ah-je) The study of the biochemical structure and function of organisms.

molecular genetics (mo-lek′yah-ler je-net′iks) The study of the biochemical structure and function of DNA. 130

molecule (mol′e-kul) A particle composed of two or more atoms bonded together. 18

molting (molt′ing) *see* ecdysis. 473

Monera (mon′er-ah) The kingdom of life whose members are characterized by having cells that lack a membrane-bound nucleus, as well as other internal, membrane-bound organelles (they are prokaryotic); bacteria and cyanobacteria. 227

monoclonal antibody (mon″o-klon′al an′ti-bod″e) An antibody of a single type that is produced by a population of genetically identical plasma cells (a clone); a monoclonal antibody is typically produced from a cell culture derived from the fusion product of a cancer cell and an antibody-producing cell. 584

monocyte (mon′o-sit) A type of white blood cell that functions as a phagocyte. 573

monoecious (ma-ne-shos) Having both male and female gonads in the same organism; hermaphroditic. 99

monogamous (mah-nog′ah-mus) Having one mate at a time. 482

monohybrid cross (mono-hi′brid kros) A mating between two individuals heterozygous for one particular trait. 104

monosaccharide (mon″o-sak′ah-rid) A simple sugar, such as glucose or fructose, that represents the structural unit of a carbohydrate. 25

morphogenesis (mor″fo-gen′e-sis) The evolution and development of form, as the development of the shape of a particular organ or part of the body. 151, 163

morula (mor′yah-lah) A stage in the embryological development of some animals that consists of a solid ball of cells. 149

mosaic evolution (mo-za-ik ev′ah-loo″shen) A change in a portion of an organism (e.g., a bird wing) while the basic form of the organism is retained. 200

motor neuron (mo′tor nu′ron) A neuron that transmits impulses from the central nervous system to an effector. 522

motor unit (mo′tor unit) A motor neuron and the muscle fibers associated with it. 517

mucous cell (mu′kus sel) A glandular cell that secretes mucus. 603

Muller's larva A free-swimming ciliated larva that resembles a modified ctenophore, characteristic of many marine polyclad turbellarians. 279

Mullerian mimicry (mul′er-e-an mim′ik-re) Occurs when two similar species are both distasteful to predators. 678

multiple alleles (mul′te-pel al-els′) The presence of more than two alleles in a population. 109

multiple fission (mul′te-pel fish′on) Asexual reproduction by the splitting of a cell or organism into many cells or organisms. 238

muscle fiber (mus′l fi-ber) The contractile unit of a muscle. 514

muscle tissue (mus′l tish′u) The type of tissue that allows movement. The three kinds are skeletal, smooth, and cardiac. 54

mutation pressure (myoo-ta′shen presh′er) A measure of the tendency for gene frequencies to change through mutation. 193

mutualism (myoo′choo-ah-liz-em) A relationship between two species in which both members of the relationship benefit. 677

myelin (mi′e-lin) Fatty material that forms a sheathlike covering around some nerve fibers. 522

myofibril (mi″o-fi′bril) Contractile fibers found within muscle cells. 516

myosin (mi′o-sin) A protein that, together with actin, is responsible for muscular contraction and relaxation. 516

N

naiads (na'ad) The aquatic immature stage of any hemimetabolous insect.   373

natural killer cell A non-T, non-B lymphocyte present in nonimmunized individuals that exhibits MHC-independent cytolytic activity against tumor cells.   584

natural selection (nach'er-al si-lek'shen) A theory, conceived by Charles Darwin and Alfred Wallace of how some evolutionary changes occur.   184

naturally acquired active immunity The type of immunity that develops when an individual's immunologic system comes into contact with an appropriate antigenic stimulus during the course of normal activities; it usually arises as the result of recovering from an infection.   583

naturally acquired passive immunity The type of immunity that involves the transfer of antibodies from one individual to another.   583

nematocyst (ni-mat'ah-sist) An organelle characteristic of the Cnidaria, that is used in defense, food gathering, and attachment.   261

neoteny (ne-ot'e-ne) The tendency to remain in the larval stage, although gaining sexual maturity.   275

nephridiopore (ne-frit-i-o'por) The opening to the outside of a nephridium.   278

nephron (nef'ron) The functional unit of a kidney, consisting of a renal corpuscle and a renal tubule.   624

nervous tissue (ner'vus tish'u) The type of tissue composed of individual cells called neurons.   54

net primary production (net pri'mer-e produk'shen) The total energy converted into biomass by autotrophs. Net primary production is equal to gross primary production less energy lost in maintenance functions of autotrophs.   683

neuroendocrine system (nu"ro-en'do-krin sys'tem) The combination of the nervous and endocrine systems.   557

neuroglial cell (nu-rog'le-al sel) A nonconducting cell of the central nervous system that protects, nutures, and supports the nervous system; also called a glial cell.   522

neurohormone (nu"ro-hor'mon) A hormone produced by nervous tissue.   557

neurolemmocyte (nu-ro-lem-o-sit) The cell that surrounds a fiber of a peripheral nerve and forms the neurolemmal sheath and myelin; **Schwann cell.**   522

neuromuscular junction (nu"ro-mus'ku-lar jungk'shun) The junction between nerve and muscle; **myoneural junction.**   517

neuron (nu'ron) A nerve cell that consists of a cell body and its processes.   522

neurotransmitter (nu"ro-trans-mit'er) Chemical substance secreted by the terminal end of an axon that stimulates a muscle fiber contraction or an impulse in another neuron.   525, 557

neurulation (noor'yah-la"shen) External changes along the upper surface of a chordate embryo that result in the formation of the neural tube.   154

neutral selection (nu'tral si-lek'shen) *see* genetic drift.   192

neutron (nu'tron) A neutral particle with a mass approximately the same as a proton that exists in atomic nuclei.   16

neutrophil (nu'tro-fil) A type of phagocytic white blood cell.   573

New World monkeys (nu wurld monk'ez) Monkeys of the Americas, characterized by the presence of a prehensile tail.   205

nicotinamide adenine dinucleotide A local electron carrier that transfers hydrogen atoms and electrons within metabolic pathways; a free-moving carrier, not membrane bound in a transport system; NAD⁺.   65

nictitating membrane (nik'ti-tat-ing mem'bran) The thin, transparent lower eyelid of amphibians and reptiles.   446

Nissl bodies (nis'l bod'ez) Membranous sacs that occur within the cytoplasm of nerve cells and have ribosomes attached to their surfaces.   522

nociceptor (no"se-sep'tor) A sensory receptor responding to potentially harmful stimuli that produce pain.   540

node of Ranvier (nod ov Ran-ve-a) A constriction of myelinated nerve fibers at regular intervals at which the myelin sheath is absent and the axon is enclosed only by Schwann cell processes; also known as a neurofibril node.   522

nonamniote lineage (non-am'ne-ot lin'e-ij) The vertebrate lineage leading to modern amphibians.   439

nondisjunction (non'dis-junk"shen) The failure of homologous chromosomes to separate during meiosis.   125

nonpolar covalent bond (non-po-lar cova'lent bond) The type of bond that is formed when electrons spend as much time orbiting one nucleus as the other; thus, the distribution of charges is symmetrical.   18

norepinephrine (nor"ep-i-nef'rin) A neurotransmitter released from the axon ends of some nerve fibers; **noradrenalin.**   564

notochord (no"ta-kord') A rodlike, supportive structure that runs along the dorsal midline of all larval chordates and many adult chordates.   411

nuclear envelope Double membrane forming the surface boundry of a eukaryotic nucleus.   52

nuclear transplantation (nu'kle-ar trans"plan-ta'shun) The transplantation of a nucleus from one cell to another cell.   161

nucleic acid (nu-kle-ik as'id) A substance composed of nucleotides bonded together; RNA, DNA.   29

nucleolus (nu-kle'o-lus) A small structure that occurs within the nucleus of a cell and contains RNA.   53

nucleoplasm (nu-kle-o-plazm") The fluid parts of the nucleus of a cell.   34

nucleosome (noo-kle'ah-som) An association of DNA and histone proteins that makes up chromatin.   115

nucleotide (nu'kle-o-tid) A component of a nucleic acid molecule consisting of a sugar, a nitrogenous base, and a phosphate group.   30

nucleus (nu'kle-us) Cell nucleus; a spheroid body within a cell, contained in a double membrane, the nuclear envelope, and containing chromosomes and one or more nucleoli.   34

numerical taxonomy (noo'mer'i-kal takson'ah-me) A system of classification in which there is no attempt to distinguish true and false similarities.   229

nutrition (nu-trish'un) The study of the sources, actions, and interactions of nutrients.   595

nymphs (nimfs) The immature stages of a paurometabolous insect; resemble the adult but are sexually immature and lack wings (Arthropoda).   373

O

ocellus (o-sel-as) A simple eye or eyespot in many inverbarbetes.   278

odontophore (o-dont"o-for') The cartilaginous structure that supports the radula of molluscs.   311

Old World monkeys (old wurld monk'ez) Monkeys of Africa and Asia; lack prehensile tails.   205

Oldowan industry (old'ah-wan in'de-stre) Tool making style used by *Homo habilis*; characterized by relatively primitive stone tools.   214

olfaction (ol-fak'shun) The act of smelling; the sense of smell.   543

ommatidia (om'ah-tid"e-ah) The sensory units of the arthropod compound eye.   370

omnivore (om-niv'or) Subsisting upon both plants and animals.   596

onchosphere (ong'ko-sfer) The larva of a tapeworm contained within the external embryonic envelope and armed with six hooks.   285

oncogene (ong'ko-jen) A gene found in the chromosome of tumor cells whose activation is associated with the initial and continual conversion of normal cells into cancerous cells.   172

oncomiracidium (an'ko-mir-a-sid'e-um) Ciliated larva of a mongenetic trematode. 281

oogenesis (o"o-jen'e-sis) The process by which an egg cell forms from an oocyte. 94

open circulatory system (o'pen sur'ku-le-tor''e sis'tem) A circulatory system in which blood is not confined to vessels in a part of its circuit within an animal; blood bathes tissues in blood sinuses. 313

operculum (o-pur'ku-lum) A cover. 1. The cover of a gill chamber of a bony fish (Chordata). 2. The cover of the genital pores of a horseshore crab (Meristomata, Arthropoda). 3. The cover of the aperature of a snail shell (Gastropoda, Mollusca). 281, 311, 426

opisthaptor (a'pis-thap'ter) Posterior attachment organ of a mongenetic trematode. 280

oral sucker (o'ral suk'er) The sucker on the anterior end of a tapeworm. 281

order (or'der) The level of classification between class and family. 225

organ (or'gan) A structure consisting of a group of specialized tissues that performs a specialized function. 54

organ of Corti (or'gan ov Cor-te) The organ of hearing; also called the spiral organ. 546

organ of Ruffini (or'gan ov Ruffini) Sensory receptor in the skin believed to be a sensor for touch-pressure, position sense of a body part, and movement; also known as corpuscle of Ruffini. 541

organelle (or"gah-nel) A part of a cell that performs a specific function. 34

organic evolution (or-gan'ik ev'ah-loo"shen) The change in an organism over time; a change in the sum of all genes in a population. 177

organic molecule (or-gan'ik mol'e-kul) A molecule that contains one or more carbon atoms. 24

osmoconformer (oz-mo'con-form-er) An organism whose body fluids have the same or very similar osmotic pressure as that of its aquatic environment; a marine organism that does not utilize energy in osmoregulation.

osmoregulator (oz-mo'reg-u'lat-er) An organism which regulates its internal osmotic pressure.

osmosis (oz-mo'sis) Diffusion of water through a selectively permeable membrane due to the existence of a concentration gradient. 42

osmotic pressure (oz-mot'ik presh'ur) The amount of pressure needed to stop osmosis; the potential pressure of a solution due to the presence of nondiffusible solute particles in the solution. 42

osteoporosis (os'te-o-po-ro'sis) Reduction in the amount of bone mass. 512

ovarian (o-va're-an) Pertaining to the ovary. 640

ovary (o'var-e) The primary reproductive organ of a female; where eggs (ova) are produced. 638

oviparous (o-vip'er-us) Organisms that lay eggs that develop outside the body of the female. 350

ovoviviparous (o'vo-vi-vip'er-us) Organisms with eggs that develop within the reproductive tract of the female and are nourished by food stored in the egg. 350

ovulation (o"vu-la'shun) The release of an egg shell from a mature ovarian follicle. 639

oxidation (ok"si-da'shun) Process by which oxygen is combined with a chemical substance; the removal of hydrogen or the loss of electrons; the opposite of reduction.

oxidation reaction (ok"si-da'shun re-ak'shun) The type of reaction that occurs when an atom or molecule loses electrons or hydrogen atoms. 21

oxidative respiration (ok"si-da-tiv res"pi-ra'shun) The electron-stripping process that occurs in the mitochondrion that requires molecular oxygen. 68

oxygen debt (ok'si-jen det) The amount of oxygen that must be supplied following physical exercise to convert accumulated lactic acid to glucose. 592

oxyhemoglobin (ok"si-he"mo-glo'bin) Compound formed when oxygen combines with hemoglobin. 572

oxytocin (ok"si-to'sin) Hormone released by the posterior lobe of the pituitary gland that causes smooth muscles in the uterus and mammary glands to contract. 561

P

Pacinian corpuscle (Pa-cin-ian kor'pus'l) A sensory receptor in skin, muscles, body joints, body organs, and tendons that is involved with the vibratory sense and firm pressure on the skin; also called a lamellated corpuscle. 541

paleontology (pa'le-on-tol'o-je) The study of early life forms on earth.

pancreas (pan'kre-as) Glandular organ in the abdominal cavity that secretes hormones and digestive enzymes. 565

pancreatic islet (pan"kre-at'ik i'let) An island of special tissue in the pancreas. 565

parapatric speciation (par'ah-pat'rik spe'she-a'shen) Speciation that occurs in small, local populations, called demes. 198

parapodia (par'ah-pod"e-ah) Paired lateral extensions on each segment of polychaetes (Annelida); may be used in swimming, crawling, and burrowing. 329

parasitism (par'ah-si'tiz-em) A relationship between two species in which one member lives at the expense of the the second. 676

parasitoids (par'ah-si'toids) Animals that deposit eggs or other developmental stages on another animal; matures feed on host tissues and eventually kill the host. 677

parasympathetic nervous system (par"ah-sim"pah-thet'ik ner-vus sys-tem) Portion of the autonomic nervous system that arises from the brain and sacral region of the spinal cord. 537

parathromone (par"ah-thor'mon) Hormone secreted by the parathyroid glands that helps to regulate the level of blood calcium and phosphate; parathyroid hormone or PTH. 563

parathyroid gland (par"ah-thi'roid gland) One of the small glands located within a lobe of the thyroid gland. 563

parenchyma (pa"ren'ka-ma) A spongy mass of mesenchyme cells filling spaces around viscera, muscles, or epithelia in lower animals. 275

parietal cell (pah-ri'e-tal sel) Cell of a gastric gland that secretes hydrochloric acid and intrinsic factor. 603

parthenogenesis (par'the-no-jen'e-sis) A modified form of sexual reproduction by the development of a gamete without fertilization, as occurs in some bees, wasps, and certain lizards. 97

parturition (par"tu-rish'un) The process of childbirth. 646

pattern formation (pat'ern for-ma'shun) The process by which the pattern of an organ unfolds during development. 165

paurometabolous metamorphosis (por'o-me-tab'a-lus met-ah-morf'a-sis) A form of insect development in which immatures resemble parents and molting is restricted to the immature stages. 373

pedicellariae (ped'e-sel-ar"i-ae) Two or three clawed pincerlike structures found on the body wall of many echinoderms. They are used in cleaning and defense. 397

pellicle (pel-ik-el) A thin, frequently non-cellular covering of an animal (e.g., the protective and supportive pellicle of protists occurs just below the plasma membrane). 238

pentaradial symmetry (pen'tah-ra'de-al sim'i-tre) A form of radial symmetry found in the echinoderms in which body parts are arranged in five's around an oral/aboral axis. 395

peptide bond (pep'tid bond) The covalent bond that joins individual amino acids together. 29

peristalsis (per"i-stal'sis) Rhythmic waves of muscular contraction that occur in the walls of various tubular organs. 600

peristomium (per"i-stom'e-um) The segment of the body of an annelid that surrounds the mouth. 329

peroxisome (pe-roks'i-som) Membranous cytoplasmic vesicle that contains enzymes responsible for the production and decomposition of hydrogen peroxide. 49

pH scale (p-h skal) The numerical scale that measures acidity and alkalinity. 21

phagocytosis (fag"o-si-to'sis) Process by which a cell engulfs and digests substances. 45

phagolysosome (fag"o-li'so-som) The organelle that is formed when a lysome combines with a vesicle. 45

pharyngeal gill slits (far-in'je-al gil slitz) A series of openings between the digestive tract, in the pharyngeal region, and the outside of the body. One of the four unique characteristics of the chordates. 411

phasmid (faz-mid) Sensory pit on each side near the end of the tail of nematodes of the class Phasmidea. 296

phenotype (fe'no-tip) The expression that results from an interaction of one or more gene pairs and the environment. 104

photoautotrophs (fo'to-au'te-trofs) Organisms that synthesize organic matter using the energy of light. 682

photoreceptor (fo"to-re-sep'tor) A nerve ending that is sensitive to light energy. 540

phyletic gradualism (fi-let'ik graj'oo-el-izm) The idea that evolutionary change occurs showly over millions of years. 198

phylogenetic systematics (fi-lo-je-net'ik sis-tem'at-iks) The study of the phylogenetic relationships among organisms in which true and false similarites are differentiated. 229

phylum (fi'lum) The level of classification between kingdom and class; members are considered a monophyletic assemblage derived from a single ancestor). 225

physiology (fiz"e-ol'oje) The branch of science that deals with function. 16

pilidium larva (pi-lid-e-um lar'va) The free-swimming, hat-shaped larva of nemertine worms. 287

pinacocytes (pin'ah-ko'sitz) Thin, flat cells covering the outer surface, and some of the inner surface, of poriferans. 45, 256

pineal gland (pin'e-al gland) A small structure located in the central part of the brain.

pinocytosis (pin"o-si-to'sis) Cell-drinking; the engulfment of liquid particles. 45

pit organs (pit or-ganz) Receptors of infrared radiation (heat) on the head of some snakes (pit vipers). 465

pituitary gland (pi-tu'i-tar"e gland) Endocrine gland that is attached to the base of the brain and consists of anterior and posterior lobes; hypophysis.

placenta (plah-sen'tah) Structure by which an unborn child is attached to its mother's uterine wall and through which it is nourished. 645

placid (plac-id) Plates on Kinorhyncha. 295

planktonic (plangk'ton-ik) Small organisms that passively float or drift in a body of water. 692

Plantae (plant'a) One of the five kingdoms of life; characterized by being eukaryotic and multicellular, and having rigid cell walls and chloroplasts. 227

plasma (plaz'mah) The fluid or liquid portion of circulating blood. 571

plasma cell (plaz'mah sel) A mature, differentiated B lymphocyte chiefly accupied with antibody synthesis and secretion; a plasma cell lives for only five to seven days. 573

plasma membrane (plaz'mah mem'brane) Outermost membrane of a cell; its surface has molecular regions that detect changes in external conditions. 34

platelet (plat'let) Cytoplasmic fragment formed in the bone marrow that functions in blood coagulation; thrombocyte. 573

pleiotropy (ple'o-trop"e) A condition in which a gene has two or more phenotypic effects. 110

pleurocercoid larva (ploor"o-ser-coid lar'va) Metacestode that develops from a procercoid larva; it usually shows little differentiation. 286

pneumatic sacs (noo-mat'ik saks) Gas chambers that develop as outgrowths of the digestive tract of fish; function in gas exchange or buoyancy regulation. 430

polar covalent bond (po-lar co-va'lent bond) The type of bond that is formed by asymmetrical moving electrons. 19

polyandrous (pol"e-an'drous) Having more than one male mate. 482

polygenes (pol'e-genz) Genes at multiple loci that influence a trait in a quantitative fashion. 110

polygynous (pa-llij'a-nus) Having more than one female mate. 482

polyp (pol'ip) The attached, usually asexual, stage of a cnidarian. 261

polyphyletic (pol'e-fi-let'ik) An assemblage of organisms that includes muliple evolutionary lineages. 237

polyploid (pol-e-ploid) Having more than two sets of chromosomes. 117, 128

polysaccharide (pol"e-sak'ah-rid) A carbohydrate composed of many monosaccharide molecules joined together. 26

polyspermy (pol-e-sper'me) The fertilization of an egg by more than one sperm. 147

polytene chromosome (pol'e-tene kro'mosom) A multistranded chromosome that is formed by repeated replication without mitosis. 168

pons (ponz) A portion of the brain stem above the medulla oblongata and below the midbrain. 531

population genetics (pop'ya-la"shen je-net'iks) The study of events occurring in gene pools. 190

porocytes (por'o-sitz) Tubular cells found in a sponge body wall that create a water channel to an interior chamber. 256

postanal tail (post-an'al tal) A tail that extends posterior to the anus; one of the four unique characteristics of chordates. 412

postmating isolation (post-mat'ing i'sah-la'shen) Isolation that occurs when fertilization is prevented even though mating has occurred. 196

potential energy The energy an object has by virtue of its position. 59

precocial (pre-ko'shel) Having developed to a high degree of independence at the time of hatching or birth. 484

preformation (pre-for-ma'shen) The erroneous idea that gametes contain miniaturized versions of all of the elements present in an adult. 145

premating isolation (pre-mat'ing i'sah-la'shen) When behaviors or other factors prevent animals from mating. 196

primary consumer (pri-mer-e kon-su'mar) A plant that captures the energy from the sun. 58

primary immune response (pri-mer-e i-mun response) The initial immune response following antigen exposure. 585

primary producer An autotrophic organism; able to build its own complex organic molecules from simple inorganic substances in the environment. 58

primary transcript (pri'mer-e tran'skript) The messenger RNA molecule produced from DNA. It often includes base sequences that are removed before translation at the ribosome. 133

primates (pri'matz) The order of mammals whose members include humans, monkeys, apes, lemurs, and tarsiers. 204

primitive streak (prim'i-tiv strek) A medial thickening along the dorsal margin of an amniote embryo that forms during the migration of endodermal and mesodermal cells into the interior of the embryo. 155

Principle of Independent Assortment (prin'ce-pel ov in'di-pen'dent ah-sort'ment) One of Mendel's observations on the behavior of hereditary units during gamete formation. A modern interpretation of this principle is that genes carried in one chromosome are distributed to gametes without regard to the distribution of genes in nonhomologous chromosomes. 107

Principle of Segregation (prin'ce-pel ov seg're-ga'shen) One of Mendel's observations on the behavior of hereditary units during gamete formation. A modern interpretation of the Principle of Segregation is that genes exist in pairs, and during gamete formation, members of a pair of genes are distributed into separate gametes. 105

procercoid larva (pro-ser′koid lar′va) Cestode developing from a corcidium in some orders; it usually has a posterior cercomer. 286

proglottid (pro-glot′id) One set of reproductive organs in a tapeworm strobila; usually corresponds to a segment. 284

prokaryote (pro-kar′e-ot) Single-celled organism that has no membrane-bound nucleus or other internal organelles (e.g., bacteria). 34

prolactin (pro-lak′tin) Hormone secreted by the anterior pituitary gland that stimulates the production of milk from the mammary glands.

prophase (pro′faz) The stage of mitosis during which the chromosomes become visible. 89

Prosimii (pro′sime-i) A suborder of primates that includes lemurs and tarsiers. 205

prostate gland (pros″tat gland) Gland located around the male urethra below the urinary bladder that adds its secretions to seminal fluid during ejaculation. 635

protandry (pro-tan′dre) The condition in a monoecious organism in which male gonads mature before female gametes; prevents self-fertilization. 269

protein (pro′te-in) Nitrogen-containing organic compounds composed of amino acid molecules joined together. 29

prothorax (pro′thor′aks) The first of the three thoracic segments of an insect; usually contains the first pair of walking appendages. 365

Protista (pro-tist′ah) The kingdom whose members are characterized by being eukaryotic and unicellular or colonial. 227

proton (pro′ton) A positively charged particle found in the atomic nucleus. 16

protonephridium (pro′to-ne-frid′e-um) Primitive osmoregulatory or excretory organ composed of a tubule terminating internally with a flame bulb or solenocyte; the unit of a flame bulb system. 277

protostomes (pro′to-stoms″) Animals in which the embryonic blastopore becomes the mouth; often possess a trochophore larva, schizocoelous coelom formation, and spiral embryonic cleavage. 308

protostyle (pro′to-stil″) A rotating mucoid mass into which food is incorporated in the gut of a gastropod (phylum Mollusca). 313

pseudocoelom (soo′do-se′lom) A body cavity that is not entirely lined by mesoderm. 234

pseudopodia (soo′dah-po′de-ah) Temporary cytoplasmic extensions of amebas that are used in feeding and locomotion. 243

punctuated equilibrium model (pungk′choo-at′ed e′kwe-lib′riam mod′el)

The idea that evolutionary change can occur rapidly over periods of thousands of years and that these periods of rapid change are interrupted by periods of constancy (stasis). 199

pupal stage (pu′pal staj) The stage in holometabolous metamorphosis of an insect during which radical cellular changes result in a change from the larval to the adult body form. 373

puparium (pu-par′e-um) A pupal case formed from the last larval exoskeleton. *see* pupa stage. 373

purine (pyoor′en) A nitrogen-containing organic compound that contributes to the structure of a DNA or RNA nucleotide; uric acid is also derived from purines. 130

pyramid of biomass (pir′ah-mid ov bi′omas) A representation showing the total mass of living matter in each trophic level of an ecosystem. 682

pyramid of energy (pir′ah-mid ov en′er-je) A representation showing the total energy in each trophic level of an ecosystem. 682

pyramid of numbers (pir′ah-mid ov num′berz) A representation showing the number of individuals in each trophic level of an ecosystem. 682

pyrimidine (pi-rim′i-den) A nitrogen-containing organic compound that is a component of the nucleotides making up DNA and RNA. 130

Q

quantitative traits (kwon′ti-ta′tiv trats) Genetic traits that are determined by multiple interacting loci, and for which there is a range of phenotypes between phenotypic extremes. 110

R

r-selected (ar si-lekt′ed) Populations that are maintained below the carrying capacity of the environment. 671

radial symmetry (ra′de-al sim′i-tre) A form of symmetry in which any plane passing through the oral/aboral axis divides an organism into mirror images. 231

radiation (ra′de-a′shun) A form of energy that includes visible light, ultraviolet light, and X rays; means by which body heat is lost in the form of infrared rays.

radioactive (ra″de-o-ak′tiv) Having the property of radioactivity. 17

radioactive dating (ra″de-o-ak′tiv da-ting) A technique in which isotopes can serve as "radioactive clocks" for measuring the passage of time. 18

radioisotope (ra″de-o-i′so-top) An isotope that is radioactive; i.e., one having an unstable nucleus, which gives it the property of

decay by one or more of several processes. 17

radula (raj′oo-lah) The rasping, tongue-like structure of most molluscs that is used for scraping food; composed of minute chitinous teeth that move over a cartilaginous odontophore. 311

ram ventilation (ram ven′te-la′shen) The movement of water across gills as a fish swims through the water with its mouth open. 429

range of optimum (ranj ov op′te-mum) The range of values for of a condition in the environment that is best able to support survival and reproduction of an organism. 667

recepetor-mediated endocytosis (re-sep′tor me′de-at-ed en″do-si-to′sis) The type of endocytosis that involves a specific receptor on the plasma membrane that recognizes an extracellular molecule and binds with it. 45

recessive (ri-ses′iv) A gene whose expression is masked when it is present in combination with a dominant allele. 104

recombinant DNA (re′kom-be-nant d-n-a) The incorporation of DNA from one organism into that of another organism (usually a bacteria) so that the second organism produces a desired protein. 140

recombination nodule (re′kom-be-na″shen noj′ool) Proteins that are involved in the cutting and rejoining of homologous chromosomes during crossing over. 118

red blood cell (erythrocyte) (red blood sel) The type of blood cell that contains hemoglobin and no nucleus. 572

redia (re′de-ah) A larval, digenetic trematode, produced by asexual reproduction within a miracidium, sporocyst, or mother redia. 289

reduction A half reaction in which an atom is reduced in oxidation number and thereby gains electrons. 21

reflex (re′fleks) A rapid, automatic response to a stimulus. 529

reflex arc (re′fleks ark) A nerve pathway consisting of a sensory neuron, interneuron, and motor neuron, that forms the structural and functional bases for a reflex. 529

refractory period (re-frak′to-re pe′re-od) Time period following stimulation during which a neuron or muscle fiber will not respond to a stimulus. 524

regulator genes (reg′ya-la′ter genz) Genes that control the activity of genes coding for enzymes or structural proteins. 137

relaxin (re-lak′sin) A hormone from the corpus luteum that inhibits uterine contractions during pregnancy.

REM sleep (r-e-m sleep) Rapid eye movement; a phase of sleep associated with dreaming and characterized by rapid movements of the eyes. 534

renette (re′net) An excretory structure found in some worms. 297

repolarized (re″po-lar-i-zed) The reestablishment of polarity, especially the return of cell membrane potential to resting potential after depolarization. 524

reproductive isolation (re′prah-duk″tiv i-sah-la′shen) Occurs when individuals are prevented from mating even though they may occupy overlapping ranges. *see* premating and postmating isolation. 196

resting membrane potential (resting mem′bran po-ten′shal) The potential for electrical activity along the plasma membrane of a neuron that is in a state of polarization. 524

rete mirabile (re′te ma-rab′a-le) A newtwork of small blood vessels arranged so that the incoming blood runs countercurrent to the outgoing blood and thus, makes possible efficient exchange of gases between the two bloodstreams.

reticular activating system (re-tik′u-lar ak-ta-vat-ing sys-tem) A network of branched nerve cells in the brainstem; involved with the adjustment of many behavioral activities, including the sleep-wake cycle, awareness, levels of sensory perception, emotions, and motivation; also called the arousal system. 533

reticular formation (re-tik′u-lar for-ma′shun) A complex network of nerve fibers within the brain stem that functions in arousing the cerebrum. 533

rhabdite (rab′dit) A rodlike structure in the cells of the epidermis or underlying parenchyma in certain tubellarians that are discharged in mucous secretions. 276

rhodopsin (ro-dop′sin) Light-sensitive substance that occurs in the rods of the retina; visual purple. 550

rhynchocoel (ring′ko-sel) In nemerteans, the dorsal cavity that contains the inverted proboscis. 287

ribonucleic acid (ri′bo-nuk″la-ik as′id) A single stranded polymer of ribonucleotides; RNA is formed from DNA in the nucleus and carries a code for proteins to ribosomes where the proteins are synthesized. 29, 130

ribosomal RNA (ri′bo-som″al r-n-a) A form of ribonucleic acid that makes up a portion of ribosomes. 132

ribosome (ri-bo-som) Cytoplasmic organelle that consists of protein and RNA, and functions in protein synthesis. 46

rod cell (rod sel) A type of light receptor that is responsible for color vision. 549

Rotifera (ro-tif′e-ra) A phylum of pseudocoelomate animals characterized by a ciliated crown or corona.

S

saliva (sah-li′vah) The enzyme-containing secretion of the salivary glands. 601

salt (sawlt) A compound produced by a reaction between an acid and a base. 21

saltatory conduction (sal′tah-tor-e kon-duk′shun) A type of nerve impulse conduction in which the impulse seems to jump from one node to the next. 525

sarcolemma (sar′ko-lem′ah) The cell membrane of a muscle fiber. 517

sarcomere (sar′ko-mer) The contractile unit of a myofibril. The repeating units, delimited by the Z bands along the length of the myofibril. 516

satellite (sat′elit) A portion of a chromosome, near the tip of a chromatid, that is set off from the rest of the chromatid by a constriction. 115

scalid (sca-lid) A set of complex spines found on the Kinorhyncha. 295

scanning electron microscope (skan-ing e-lek′tron mi″kro-skop) The type of microscope in which an electron beam, instead of light, forms an image for viewing, allowing much greater magnification and resolution. 38

scanning tunneling microscope (skan-ing tun′l-ing mi″kro-skop) The type of microscope that uses a needle probe and electrons to determine the surface features of specimens. 38

scientific method (si-en-tif′ik meth′od) A system of thought and procedure that reflects how a scientist goes about defining and studying a problem.

scolex (sko′leks) The attachment or holdfast organ of a tapeworm, generally considered the anterior end. 284

scrotum (skro′tum) A pouch of skin that encloses the testes. 634

sebaceous gland (se-ba′shus gland) Gland of the skin that secretes sebum; oil gland. 508

sebum (se′bum) Oily secretion from the sebaceous gland. 509

second law of thermodynamics Physical and chemical processes proceed in such a way that the entropy of the universe (the system and its surroundings) increases to the maximum possible. 59

secondary consumer (sek′en-der′e kon-su′mar) An animal that preys upon a primary consumer. 58

secondary immune response (sek-en′der′e i-mun response) The immune response that follows a second exposure to a specific antigen. 585

segmentation (seg″men-ta′shun) In many animal species, a series of body units that may be externally similar to or quite different from one another. 600

selection pressure (si-lek′shen presh′er) The tendency for natural selection to occur; natural selection occurs whenever some genotypes are more fit than other genotypes. 193

selective permeability (si-lek′tiv per″me-ah-bil′i-te) The ability of the cell membrane to let some things in and keep others out. 40

sensory (receptor) neuron (sen′so-re nu′ron) A nerve cell that conveys nerve impulses from sensory receptors in the body to the central nervous system; also called an afferent neuron. 522

sequential hermaphroditism (si-kwen′shal her-maf′ro-di-tizm) The type of hermaphroditism that occurs when an animal is one sex during one phase of its life cycle and an opposite sex during another phase. 99

sere (ser) A successional stage in an ecosystem. 682

serially homologous (ser′e-al-e ho-mol′o-ges) Metameric structures that have evolved from a common form; the biramous appendages of crustaceans are serially homologous. 355

serum (se′rum) The fluid portion of coagulated blood; the protein component of plasma. 572

sex chromosomes (seks kro′mo-somz) A chromosome that carries genes determining the genetic sex of an individual. 115

sex-influenced traits (seks in′floo-enst trats) Traits that behave as if they are dominant in one sex but recessive in the other sex. 121

simple diffusion (sim′pel di-fu′zhun) The process of molecules spreading out randomly until they are evenly distributed. 41

siphon (si′fon) A tubular structure through which fluid flows; siphons of some molluscs allow water to enter and leave the mantle cavity. 313

skeletal muscle (skel′e-tal mus′l) Type of muscle tissue found in muscles attached to skeletal parts. 515

smooth muscle (smooth muscle) Type of muscle tissue found in the walls of the hollow organs; visceral muscle. 514

sodium-potassium pump The active transport mechanism that functions to concentrate sodium ions on the outside of a plasma membrane and potassium ions on the inside of the membrane. 524

somatic cell (so-mat′ik sel) Ordinary body cell; pertaining to or characteristic of a body cell. 91

somatomedin (so″mah-to-me′din) Any of a group of peptides formed in the liver and other tissues and found in the plasma that mediate the effect of growth hormone on cartilage.

somatostatin (so″mah-to-stat′in) A hormone produced by the hypothalamus and by the delta cells of the pancreas; inhibits release of growth hormone.

somatotrophin (so″mah-to-tro′pin) A growth hormone.

somites (so'mitz) Segmental thickenings along the side of a vertebrate embryo that result from the development of mesoderm. 153

speciation (spe'she-a'shen) The process by which two or more species are formed from a single ancestral stock. 196

species (spe'shez) A group of populations in which genes are actually, or potentially, exchanged through multiple generations; numerous problems with this definition make it difficult to apply in all circumstances. 196, 225

species diversity (spe'shez di-vur'si-te) *see* community diversity.

spermatogenesis (sper"mah-to-jen'e-sis) The production of sperm cells. 94

spermatophores (sper-mat"ah-fors) Encapsulated sperm that can be deposited on a substrate by a male and picked up by a female, or transferred directly to a female by a male. 321

sphincter (sfingk'ter) A ringlike band of muscle fibers that constricts a passage or closes a natural orifice. 600

spicules (spik'ulz) Skeletal elements secreted by some mesenchyme cells of a sponge body wall; may be made of calcium carbonate or silica. 257

spiracle (spi'rah-kel) An opening for ventilation. The opening(s) of the tracheal system of an arthropod or an opening posterior to the eye of a shark, skate, or ray. 348

spongin (spun'jin) A fibrous protein that makes up the supportive framework of some sponges. 257

sporocyst (spor'o-sist) Stage of development of a sporozoan protozoan, usually with an enclosing membrane, the oocyst; also an asexual stage of development in some trematodes. 282

stabilizing selection (sta'be-liz'ing si-lek'shen) Natural selection that results in the decline of both extremes in a phenotypic range; results in a narrowing of the phenotypic range. 194

statoblast (stad'o-blast) A biconvex capsule containing germinative cells and produced by most freshwater ectoprocts by asexual budding. 384

statocysts (stat"o-sists) An organ of equilibrium and balance found in many invertebrates. Statocysts usually consist of a fluid-filled cavity containing sensory hairs and a mineral mass called a statolith. The statolith stimulates the sensory hairs, which helps orient the animal with regard to the pull of gravity. 264

steroid (ste'roid) An organic substance whose molecules include complex rings of carbon and hydrogen atoms. 28

strobila (stro-bi'lah) The chain of proglottids constituting the bulk of the body of adult tapeworms; includes the head, neck, and proglottids. 284

structural genes (struk'cher-al genz) Genes that code for enzymes or structural proteins rather than proteins that control the function of other genes. 137

substrate (sub'strat) The substance upon which an enzyme acts. 61

substrate level phosphorylation (sub'strat level fos"for-i-la'shun) The generation of ATP by coupling strongly exergonic reactions with the synthesis of ATP from ADP and phosphate. 67

succession (sek-sesh'en) A sequence of community types that occurs during the maturation of an ecosystem. 682

sudoriferous gland (su"do-rif'er-us gland) A sweat gland. 508

surface tension (ser'fas ten'shun) Force that tends to hold moist membranes together due to an attraction between water molecules. 23

swim bladder (swim blad'er) A gas-filled sac, which is usually located along the dorsal body wall of bony fish; it is an outgrowth of the digestive tract and regulates buoyancy of a fish. 427

sycon (si'kon) A sponge body form characterized by choanocytes lining radial canals. 258

symbiosis (sim"bi-o'sis) The biological association of two individuals or populations of different species, classified as mutualism, commensalism, or parasitism, depending on the advantage or disadvantage derived from the relationship. 35

symbiosis (sim"bi-o'sis) 1. The living together or close association of two dissimilar organisms, each of these organisms being known as a symbiont. 2. Members of separate species living in continuing, intimate associations. 676

sympathetic nervous system (sim"path-thet'ik ner'vus sis'tem) Portion of the autonomic nervous system that arises from the thoracic and lumbar regions of the spinal cord. 537

sympatric speciation (sim'pat'rik spe'she-a'shen) Speciation that occurs in populations that have overlapping ranges. 198

synapse (sin-aps) The junction between the axon end of one neuron and the dendrite or cell body of another neuron. 525

synapsid lineage (sin-ap'sid lin'e-ij) The evolutionary lineage leading from early reptiles to the mammals. 454

synapsis (si-nap'sis) The place in reduction division when the pairs of homologous chromosomes lie alongside each other in the first meiotic division. 93

synaptic cleft (si-nap'tik kleft) The narrow space between the terminal ending of a neuron and the receptor site of the postsynaptic cell. 525

synaptonemal complex (si-nap'to-nem'al kom-plex) The network that holds the sister chromatids in a precise union so that each gene is located directly across from its sister gene on a homologous chromosome. 93

syncytial hypothesis (sin-sit'e-al hi-poth'e-sis) The idea that multicellular organisms could have arisen by the formation of cell boundaries within a large multinucleate protist. 255

syngamy (sin'ga-me) The fertilization of one gamete with another individual gamete to form a zygote; found in most animals that have sexual reproduction. 91

synovial fluid (si-no've-al floo'id) Fluid secreted by the synovial membrane in a joint. 512

system (sis-tem) A group of organs that act together to carry on a specialized function. 54

systematics (sis'tah-mat"iks) The study of the classification and phylogeny of organisms. 225

systole (sis'to-le) Phase of the cardiac cycle during which a heart chamber wall is contracted. 577

T

T cell A type of lymphocyte derived from bone marrow stem cells that matures into an immunologically competent cell under the influence of the thymus. T cells are involved in a variety of cell-mediated immune reactions; also known as a T lymphocyte. 573

tactile (Meissmer's) corpuscle (tak-tl kor'pus'l) A sensory receptor in the skin that detects light pressure. 541

tagmatization (tag'mah-ti-za"shen) The specialization of body regions of a metameric animal for specific functions. The head of an arthropod is specialized for feeding and sensory functions, the thorax is specialized for locomotion, and the abdomen is specialized for visceral functions. 328

taste bud (tast bud) Organ containing the receptors associated with the sense of taste. 542

taxis (tak'sis) The movement of an organism in a particular direction in response to an environmental stimulus. 667

taxon (tak'son) A group of organisms that are genetically (evolutionarily) related. 225

taxonomy (tak'son'ah-me) The study of the naming and classification of organisms. 225

tegument (teg'u-ment) The external covering in cestodes and trematodes; once called a cuticle. 280

telophase (tel'o-faz) Stage in mitosis during which daughter cells become separate structures. 89

tendon (ten′don) A cord or bandlike mass of white fibrous connective tissue that connects a muscle to a bone. 515

tendon sheath (ten-don sheath) The sheath surrounding a tendon. 512

test cross (test kros) A cross between an individual of an unknown genotype and an individual that is homozygous recessive for the trait in question in order to determine the unknown genotype. 103

testis (tes′tis) Primary reproductive organ of a male; a sperm-cell producing organ.

testosterone (tes-tos′te-ron) Male sex hormone secreted by the interstitial cells of the testes.

tetrapods (te′trah-podz) A nontaxonomic designation used to refer to amphibians, reptiles, birds, and mammals. 439

theory (the′o-re) A generalized statement that accounts for a body of facts.

theory of evolution by natural selection (the′o-re of ev′ah-loo″shen by nach′er-al si-lek′shen) A theory conceived by Charles Darwin and Alfred Russell Wallace of how some evolutionary changes occur. 177

theory of inheritance of acquired characteristics (the′o-re of in-her′i-tenz of ah-kwird kar′ik-te-ris″tiks) The mistaken idea that organisms develop new organs, or modify existing organs as environmental problems present themselves, and that these traits are passed on to offspring. 177

thermodynamics (ther″mo-di-nam′iks) The branch of science that deals with heat, energy, and the interconversion of these. 59

thermogenesis (ther″mo-jen′a-sis) The generation of heat. 620

thermoreceptor (ther″mo-re-sep′tor) A sensory receptor that is sensitive to changes in temperature; a heat receptor. 540

thermoregulation (ther″mo-reg″u-la′shun) Heat regulation. 616

tissue (tish′u) A group of similar cells that performs a specialized function. 53

tolerance range (tol′er-ens ranj) The range of variation in an environmental parameter that is compatible with the life of an organism. 667

tonicity (to-nis′i-te) The state of tissue tone or tension; in body fluid physiology, the effective osmotic pressure equivalent. 42

torsion (tor′shen) A developmental twisting of the visceral mass of a gastropod mollusc that results in an anterior opening of the mantle cavity and a twisting of nerve cords and the digestive tract. 311

totipotent (to-tip′o-tent) Characterized by the ability to develop in any direction. 160

toxoid (tok′soid) A bacterial exotoxin that has been modified so that it is no longer toxic, but will still stimulate antitoxin formation when injected into a person or animal. 583

tracheae (tra′che-e) The small tubes that carry air from spiracles through the body cavity of an arthropod; arthropod tracheae are modifications of the exoskeleton. 348

transcription (tran-skrip′shen) The formation of a messenger RNA molecule that carries the genetic code from the cytoplasm to the nucleus of a cell. 132

transdetermination (trans″de-term-in-ation) When a cell or tissue shifts from one determined state to another. 161

transfer RNA (trans′fer r-n-a) A single-stranded polyribonucleotide that carries amino acids to a ribosome and positions those amino acids by matching the tRNA anticodon with the messenger RNA codon.

translation (trans-la′shen) The production of a protein based upon the code in messenger RNA. 126

translocation (trans-lo-ka′shen) The movement of a segment of one chromosome to a nonhomologous chromosome.

transmission electron microscope (trans-mis′en e-lek′tron mi′kro-skop) The type of microscope that produces highly magnified images of ultrathin tissue sections or other specimens. 38

transposons (trans-poz′ons) Genes whose function involves their movement between chromosomes. 138

trichinosis (trik″i-no′sis) A disease resulting from infection by *Trichinella spiralis* larvae upon eating undercooked meat; characterized by muscular pain, fever, edema, and other symptoms. 299

trichocysts (trik′o-sists) An anchoring structure present in the ectoplasm of some ciliates. 249

triploblastic (trip′lo-blas″tik) Animals whose body parts are organized into layers that are derived embryologically from three tissue layers--ectoderm, mesoderm, and endoderm. Platyhelminthes and all coelomate animals are triploblastic. 233

trochophore larva (trok″o-for lar′va) A larval stage characteristic of many molluscs, annelids, and some other protostomate animals. 308

trophic levels (trof′ik lev′elz) The feeding level of an organism in an ecosystem; green plants and other autotrophs function at producer trophic levels; animals function at the consumer trophic levels. 682

tube feet (toob feet) Muscular projections from the water vascular system of echinoderms that are used in locomotion, gas exchange, feeding, and attachment. 396

tubular nerve cord (too′byah-ler nerv kord) A hollow nerve cord that runs middorsally along the back of chordates; one of four unique chordate characteristics; also called the neural tube and, in vertebrates, the spinal cord. 412

tumor (too′mor) A growth of tissue resulting from abnormal new growth and reproduction (neoplasia). 172

tympanal organs (tim-pan′al or′ganz) Auditory receptors present on the abdomen or legs of some insects. 369

U

umbilical cord (um-bil′i-kal kord) Cordlike structure that connects the fetus to the placenta. 645

umbo (um′bo) The rounded prominence at the anterior margin of the hinge of a bivalve (Mollusca) shell; it is the oldest part of the shell. 315

unicellular organization (yoo′ne-cel″yah-lar or′gah-ni- za″shen) The life form in which all functions are carried out within the confines of a single plasma membrane; members of the kingdom Protista display unicellular oraganization; also called cytoplasmic organization. 237

uniformitarianism (yoo′nah-for′mi-tar′e-an-ism) The idea that today the earth is shaped by forces of wind, rain, rivers, volcanoes, and geological uplift, just as it was formed in the past. 180

uterus (u′ter-us) Hollow muscular organ located within the female pelvis in which a fetus develops. 639

V

vaccine (vak′sen) A preparation of either killed microorganisms; living, weakened (attenuated) microorganisms; or inactivated bacterial toxins (toxoids); administered to induce development of the immune response and protect the individual against a pathogen or toxin. 583

vagina (vah-ji′nah) Tubular organ that leads from the uterus to the vestibule of the female reproductive tract. 640

valves (valvz) 1. Devices that permit a one-way flow of fluids through a vessel or chamber. 2. The halves of a bivalve (Mollusca) shell. 315

vegetal pole (vej″e-tal pol) The lower pole of an egg; usually more dense than the animal pole because it contains more yolk. 147

vein (van) A vessel that carries blood toward the heart. 573

veliger larva (vel′i-jer lar′va) The second free-swimming larval stage of many molluscs; develops from the trochophore and forms rudiments of the shell, visceral mass, and head-foot before settling to the substrate and undergoing metamorphosis. 314

vestigial structures (ve-stij′e-al struck′cherz) Visible evidence of a structure that was present at an earlier stage in the evolution of an organism. One of the sources of evidence for evolution.

villus (vil'us) Tiny, fingerlike projection that extends outward from the inner lining of the small intestine.   603

visceral mass (vis'er-al mas) The region of a mollusc's body that contains visceral organs.   310

viscosity (vis-kos'i-te) The tendency for a fluid to resist flowing due to the internal friction of its molecules.   24

vitamin (vi'tah-min) An organic substance other than a carbohydrate, lipid, or protein that is needed for normal metabolism but cannot be synthesized in adequate amounts by the body.   598

viviparous (vi-vip'er-us) Having eggs that develop within the female reproductive tract and are nourished by the female.   350

vulva (vul'vah) The external genital organs in the female.   640

W

water vascular system (wah'ter vas'ku-lar sis'tem) A series of water-filled canals and muscular tube feet present in echinoderms; provides the basis for locomotion, food gathering, and attachment.   395

white blood cell (leukocyte) (white blood sel) A type of blood cell, usually of the scavenger type that injests foreign material in the bloodstream and tissues.   572

winter sleep (win'ter sleep) A period of inactivity in which a mammal's body temperature remains near normal and the mammal is easily aroused.

wobble hypothesis (wob'el hi-poth'e-sis) The observation that the third base in the tRNA anticodon does not always pair with a specific base in the mRNA codon. The wobble hypothesis explains why there are fewer different tRNA molecules than there are different mRNA codons.   135

work (wurk) The exertion or effort that is needed to accomplish something.   59

Y

yolk plug (yok plug) The large yolk-filled cells that protrude from beneath the ecto-derm at the blastopore of the amphibian gastrula.   153

yolk sac (yok sac) The stored food reserve (yolk) and its surrounding membranes, which is found in embryonic reptiles, birds, and mammals.   156

Z

zonite (zo-nit) The individual body unit of a Kinorhyncha.   295

zooid (zo-oid) An individual member of a colony of animals, such as colonial cnidarians and ectoprocts.   279, 383

zoology (zo-ol'o-je) The study of animals.

zooxanthellae (zo'o-zan-thel"e) A group of dinoflagellates that live in mutualistic relationships with some cnidarians. They promote high rates of calcium carbonate deposition in coral reefs.   272

zygote (zi'got) Cell produced by the fusion of an egg and sperm; fertilized egg cell.   91

Credits

Photographs

Part Openers

Part One, Two: © David M. Phillips/ Visuals Unlimited; **Part Three:** © Len Rue, Jr./Visuals Unlimited; **Part Four:** Science VU-Homosassa Springs/Visuals Unlimited; **Part Five:** © John Gerlach/Visuals Unlimited; **Part Six:** © Leonard Lee Rue III/Visuals Unlimited.

Chapter 1

1.1: Ealing Corporation; **1.2:** Science VU/ Visuals Unlimited; **1.4:** © A. Graffham/ Visuals Unlimited; **1.7a:** © Hank Andrews/ Visuals Unlimited; **1.7b:** © Noah Poritz/ Visuals Unlimited; **1.8a:** © Wm. Grenfell/ Visuals Unlimited; **1.8b:** © G. Prance/ Visuals Unlimited.

Chapter 2

Box 2.1: © John D. Cunningham/Visuals Unlimited.

Chapter 3

3.12a-c: © S.J. Singer; **3.16:** © W.J. Johnson/Visuals Unlimited; **3.20a:** © Keith Porter; **3.21a, 3.27:** © K.G. Murti/Visuals Unlimited; **3.21c:** © M. Schliwa/Visuals Unlimited; **3.25b, Box 3.2b-d:** © David M. Phillips/Visuals Unlimited; **3.26a:** © E.G. Pollack.

Chapter 4

4.5b-c: IBM UK Scientific Centre and Open University; **Box 4.1:** © Robert F. Myers/ Visuals Unlimited.

Chapter 5

5.10a: © H. Fernandez-Moran.

Chapter 6

6.5a-d, 6.12b,d: © John D. Cunningham/ Visuals Unlimited; **6.8a:** © James Kezar; **6.12a:** © Carolina Biological/Visuals Unlimited; **6.12c:** © Mike Milligan/Visuals Unlimited; **6.12e:** © Daniel W. Gotshall/ Visuals Unlimited.

Chapter 7

7.1: American Museum of Natural History.

Chapter 8

8.1a: From Leland G. Johnson: BIOLOGY © 1987 Wm. C. Brown Publishers, Dubuque, Iowa. All rights reserved. Reprinted by permission.

Chapter 9

9.12a-c: Potter, Huntington and Dressler, David: LIFE Magazine/ © Time, Inc.

Chapter 10

10.2a: © William Byrd; **10.2b:** © Gerald Schatten; **10.2c-d:** Tegner, M.J. and Epel, D.: Science, Vol. 179, pp.685–688, 16 February 1973/ © AAAS; **10.5a-1, 10.8b:** © Carolina Biological/Visuals Unlimited; **10.8a:** © Bill Beatty/Visuals Unlimited; **10.10a-c:** © Robert Waterman; **10.10d-f:** Kessel and Shih: SCANNING ELECTRON MICROSCOPY IN BIOLOGY/ © Springer-Verlag, Berlin.

Chapter 11

11.10b: Kathryn Tosney, from N.K. Wessels: TISSUE INTERACTIONS AND DEVELOPMENT/ © 1977 Benjamin-Cummings Publishing Company, Redwood City, California; **11.11:** © Carolina Biological/Visuals Unlimited; **11.13:** © John D. Cunningham/Visuals Unlimited; **11.14b:** © Bo Lambert.

Chapter 12

12.3, 12.9: The Bettmann Archive.

Chapter 13

13.5: Science VU/Visuals Unlimited; **13.6:** © Arthur Morris/Visuals Unlimited; **13.8:** © William J. Weber/Visuals Unlimited.

Chapter 14

14.2, 14.6b: © Milton H. Tierney, Jr./ Visuals Unlimited; **14.3a, 14.6a, 14.22:** © John D. Cunningham/Visuals Unlimited; **14.3b:** © Ron Garrison, San Diego Zoo/Photo Researchers, Inc.; **14.4:** © Michael Fogden; **14.5a:** © Leonard Lee Rue III/Visuals Unlimited; **14.5b:** © Len Rue, Jr./Visuals Unlimited; **14.6c:** Science VU/Visuals Unlimited; **14.14a-f:** © Alan Walker/Visuals Unlimited.

Chapter 15

15.3: © David M. Phillips/Visuals Unlimited; **15.4, 15.10:** © John D. Cunningham/Visuals Unlimited; **15.5:** © Louis Borie/Visuals Unlimited; **15.6:** © Howard A. Miller, Sr./Visuals Unlimited; **15.7:** © Leonard Lee Rue III/ Visuals Unlimited; **15.9:** © Daniel W. Gotshall/Visuals Unlimited.

Chapter 16

16.3a-d: © Dennis Diener/Visuals Unlimited; **16.5a:** © David M. Phillips/ Visuals Unlimited; **16.6:** © Terry Hazen/ Visuals Unlimited; **16.8a:** © Carolina Biological/Visuals Unlimited; **16.12:** © K.W. Jeon/Visuals Unlimited; **16.14, 16.15b:** © John D. Cunningham/ Visuals Unlimited; **16.15a:** © M. Schliwa/ Visuals Unlimited; **16.17, 16.19:** © Michael Abbey/Visuals Unlimited; **16.18a:** © Karl Aufderheide/Visuals Unlimited; **16.18b:** © K.G. Murti/Visuals Unlimited; **16.18c:** © T.E. Adams/Visuals Unlimited; **16.20:** © Biophoto Associates/Visuals Unlimited; **16.21:** © D.J. Patterson/Oxford Scientific Films.

Chapter 17

17.2c: © Stan Elems/Visuals Unlimited; **17.4a:** © Nancy Sefton/Photo Researchers, Inc.; **17.4b, 17.20a, 17.23a-b:** © Daniel W. Gotshall/Visuals Unlimited; **17.4c:** © Rick Wallace/Visuals Unlimited; **17.6, 17.13a, 17.14:** © Carolina Biological/Visuals Unlimited; **17.16a, Box 17.1:** © Edward Hodgson/Visuals Unlimited; **17.16b:** © N.G. Daniel/Tom Stack & Associates; **17.19:** © Neville Coleman/ Visuals Unlimited; **17.20b:** © Milton H. Tierney, Jr./Visuals Unlimited; **17.23c:** © Valorie Hodgson/Visuals Unlimited; **17.23d:** © W. Ober/Visuals Unlimited; **17.24a:** © Bob DeGoursey/ Visuals Unlimited.

Chapter 18

18.1: © Gary R. Robinson/Visuals Unlimited; **18.16b:** © Kessel & Shih/Peter Arnold, Inc.

Chapter 19

19.2a: © Peter Parks/Oxford Scientific Films; **19.3a:** © Carolina Biological/Visuals Unlimited; **19.13:** Science VU-AFIP/Visuals Unlimited; **19.15:** © R. Calentine/Visuals Unlimited; **19.16b:** © Lauritz Jensen/Visuals Unlimited.

Chapter 20

20.4b: Science VU-Polaroid/Visuals Unlimited; **20.6a:** © William J. Weber/ Visuals Unlimited; **20.7a:** © C.H. King/ Visuals Unlimited; **20.7b:** © A. Kerstitch/ Visuals Unlimited; **20.8a:** © Oxford Scientific Films/Animals Animals; **20.8b:** © Daniel W. Gotshall/Visuals Unlimited; **20.8c:** © Valorie Hodgson/ Visuals Unlimited; **20.14a:** © Bruce Iverson/Visuals Unlimited; **20.14b:** © Oxford Scientific Films; **20.14c, 20.16:** © R. Calentine/Visuals Unlimited; **20.17a:** © William Ferguson; **20.17b:** © John D. Cunningham/Visuals Unlimited; **20.17c:** © R.J. Siezen/Visuals

Unlimited; **20.17d:** © Michael DiSpezio; **20.21:** © Kjell B. Sandved/Photo Researchers, Inc.; **20.23:** © James Culter/Visuals Unlimited; **20.24a:** © Robert A. Ross; **Box 20.1:** © John Forsythe/Visuals Unlimited.

Chapter 21
21.3: © R. DeGoursey/Visuals Unlimited; **21.5:** © William H. Hughes/Visuals Unlimited; **21.15a:** © John D. Cunningham/Visuals Unlimited; **Box 21.2:** © C.P. Hickman/Visuals Unlimited.

Chapter 22
22.1: © Roger Klocek/Visuals Unlimited; **22.5:** © A. Graffham/Visuals Unlimited; **22.7a, 22.15, 22.19a:** © John D. Cunningham/Visuals Unlimited; **22.11a:** © Michael Fogden/Animals Animals; **22.13a:** © David Scharf/Peter Arnold, Inc.; **22.13b:** © R.F. Ashley/Visuals Unlimited; **22.14a-b:** Science VU-W. Burgdorfer/Visuals Unlimited; **22.14c:** © R. Calentine/Visuals Unlimited; **22.19b:** © William C. Jorgensen/Visuals Unlimited.

Chapter 23
23.1a, 23.12a: © John D. Cunningham/Visuals Unlimited; **23.1b:** Science VU-ESA, Ries/Visuals Unlimited; **23.1c:** © Carolina Biological/Visuals Unlimited; **23.1d:** © Dan Kline/Visuals Unlimited; **23.8c:** © Stanley L. Flegler/Visuals Unlimited; **Box 23.3:** © Richard Thom/Visuals Unlimited.

Chapter 24
24.2a, 22.4b, 24.9a: © John D. Cunningham/Visuals Unlimited; **24.2b:** © A.J. Copley/Visuals Unlimited; **24.4a,c:** © Daniel W. Gotshall/Visuals Unlimited; **24.6a, 24.11a:** © Stan Elems/Visuals Unlimited; **24.7a:** © William C. Jorgensen/Visuals Unlimited; **24.8a:** Science VU-D. Foster, WHOI/Visuals Unlimited; **24.10a:** © R.F. Ashley/Visuals Unlimited; **24.12a:** © Carolina Biological/Visuals Unlimited; **Box 24.2b:** Science VU-WHOI/Visuals Unlimited.

Chapter 25
25.1a: © Stan Elems/Visuals Unlimited; **25.7a-b:** © John D. Cunningham/Visuals Unlimited; **25.9:** Science VU-Polaroid/Visuals Unlimited; **25.13:** © Alan N. Baker; **Box 25.1:** © Daniel W. Gotshall/Visuals Unlimited.

Chapter 26
26.1a: © John D. Cunningham/Visuals Unlimited; **26.4a:** © William C. Jorgensen/Visuals Unlimited; **26.4b:** © Daniel W. Gotshall/Visuals Unlimited; **26.7a:** © Runk/Schoenberger/Grant Heilman; **Box 26.1:** © Gary R. Robinson/Visuals Unlimited.

Chapter 27
27.6a: Field Museum of Natural History; **27.7a, 27.13b:** © Fred Hossler/Visuals Unlimited; **27.8:** © John D. Cunningham/Visuals Unlimited; **27.9a:** Steinhart Aquarium, California Academy of Sciences; **27.17:** © Daniel W. Gotshall/Visuals Unlimited; **27.18a-b:** American Museum of Natural History; **Box 27.2:** © Roger Klocek/Visuals Unlimited.

Chapter 28
28.11: © Nada Pecnik/Visuals Unlimited; **28.12a-b:** © E.S. Ross; **28.13b:** © Dan Kline/Visuals Unlimited; **28.14a:** Science VU/Visuals Unlimited; **28.14b:** © M.J. Tyler/Australasian Nature Transparencies; **28.15a-d:** © Jane Burton/Bruce Coleman, Inc.; **Box 28.1:** © Dennis Paulson/Visuals Unlimited.

Chapter 29
29.4: © Biophoto Associates/Photo Researchers, Inc.; **29.5:** © Nathan W. Cohen/Visuals Unlimited; **29.6:** © John D. Cunningham/Visuals Unlimited; **29.8e:** © Bayard Brattstrom/Visuals Unlimited; **29.9a:** © Joe McDonald/Visuals Unlimited; **29.11:** © Carolina Biological/Visuals Unlimited.

Chapter 30
30.1: © George Herben/Visuals Unlimited; **30.2a, 30.15, 30.16b:** © John D. Cunningham/Visuals Unlimited; **30.2b:** American Museum of Natural History; **30.7d, 30.8b:** © S. Maslowski/Visuals Unlimited; **30.8a:** © Gary R. Robinson/Visuals Unlimited; **30.11c:** © Hans-Rainer Duncker; **30.12, 30.16a:** © Karl Maslowski/Visuals Unlimited; **Box 30.1a:** © Tim Hauf/Visuals Unlimited; **Box 30.1b:** © Scott Johnson/Animals Animals.

Chapter 31
31.3a: © Tom McHugh/Photo Researchers, Inc.; **31.3b:** © J. Alcock/Visuals Unlimited; **31.3c:** © Bruce Berg/Visuals Unlimited; **31.3d:** © Wm. Grenfell/Visuals Unlimited; **31.4a, 31.14:** © Walt Anderson/Visuals Unlimited; **31.4b:** © William J. Weber/Visuals Unlimited; **31.13a:** © Dennis W. Schmidt/Valan Photos; **31.15:** © Michael S. Quinton/Visuals Unlimited; **Box 31.1b:** © Tom J. Ulrich/Visuals Unlimited; **Box 31.2:** © Bruce Cushing/Visuals Unlimited.

Chapter 32
32.4b: From R.G. Kessel and R.H. Kardon: TISSUES AND ORGANS: A TEXT-ATLAS OF SCANNING ELECTRON MICROSCOPY/ © 1979 W.H. Freeman, New York.

Chapter 34
34.1: British Museum (Natural History), London.

Chapter 35
35.9a: © Bettina Cirone/Photo Researchers, Inc.; **35-9b(1-4):** Department of Illustrations, Washington University School of Medicine, and American Journal of Medicine, 20 (Jan. 1956), p.133; **35.9c:** The Bettmann Archive.

Chapter 36
36.2a: © Stanley L. Flegler/Visuals Unlimited; **36.2b-f:** © John D. Cunningham/Visuals Unlimited; **36.12:** © Andrejs Liepins/Science Photo Library/Photo Researchers, Inc.

Chapter 37
37.8d: © K.R. Porter; **37.10a, Box 37.1b,d:** © John D. Cunningham/Visuals Unlimited; **37.10b:** © Daniel W. Gotshall/Visuals Unlimited; **37.10c:** © Dwight Kuhn; **37.10d:** Science VU/Visuals Unlimited; **Box 37.2a:** © Gary R. Robinson/Visuals Unlimited; **Box 37.1c:** © Joe McDonald/Visuals Unlimited.

Chapter 38
38.2: © Joe McDonald/Visuals Unlimited; **38.3a:** © A. Rawson/Visuals Unlimited; **38.4a:** © G. Prance/Visuals Unlimited; **38.7:** © Carolina Biological/Visuals Unlimited.

Chapter 39
39.6: © David M. Phillips/Visuals Unlimited; **39.11a,c:** © Petit Format/Nestle/Photo Researchers, Inc.; **39.11b:** © Alexander Tsiaras/Science Source/Photo Researchers, Inc.; **39.11d-f:** © Science VU/Visuals Unlimited; **39.11g:** © Carolina Biological/Visuals Unlimited; **39.14a:** © John Serrao/Visuals Unlimited; **39.14b:** © A. Gurmankin/Visuals Unlimited; **39.14c:** © Karl Maslowski/Visuals Unlimited; **39.14d:** © SIU/Visuals Unlimited.

Chapter 40
40.4a-d: Reprinted with permission of Holt, Rinehart and Winston from ETHOLOGY by Irenaus Eibl-Eibesfeldt/ © 1970 Holt, Rinehart and Winston, Orlando, Florida; **40.3, 40.10:** © Leonard Lee Rue III/Visuals Unlimited; **40.5c:** Nine Leen: LIFE Magazine/ © Time, Inc.; **40.7:** © Will Rapport/Photo Researchers, Inc.; **40.8b:** © S. Halperin/Animals Animals.

Chapter 42
42.2a, 42.15: © John D. Cunningham/Visuals Unlimited; **42.2b:** © Tom J. Ulrich/Visuals Unlimited; **42.2c:** © Peter K. Ziminski/Visuals Unlimited; **42.10, 42.14:** © Leonard Lee Rue III/Visuals Unlimited; **42.11:** © Len Rue, Jr./Visuals Unlimited; **42.12:** © Charlie Heidecker/Visuals Unlimited; **42.13:** © Marilyn Maring/Visuals Unlimited; **42.16:** © Leonard Lee Rue IV/Visuals Unlimited; **42.17:** © Glenn Oliver/Visuals Unlimited; **42.18:** © Dwight Kuhn; **Box 42.1:** © Walt Anderson/Visuals Unlimited.

Colorplate 1
1a-b,d,f-j,m,o-p: © Edwin A. Reschke; **1c,n,q:** © Manfred Kage/Peter Arnold, Inc.; **1e:** © Fred Hossler/Visuals Unlimited; **1k:** © John D. Cunningham/Visuals Unlimited; **1l:** © Victor B. Eichler.

Colorplate 2
2a: © Carolina Biological/Visuals Unlimited; **b-e:** © Dan McCoy/Rainbow.

Colorplate 3
3a-b: Genentech; **3c:** Monsanto Agricultural; **3d:** Science VU-Jackson Laboratory/Visuals Unlimited.

Colorplate 4
4a-d: © Carolina Biological/Visuals Unlimited; **4e:** © Nathan Cohen/Visuals Unlimited; **4f:** © Joe McDonald/Visuals Unlimited.

Colorplate 6
6a,c: © Daniel W. Gotshall/Visuals Unlimited; **6b:** © James R. McCullagh/Visuals Unlimited; **6d:** © John D. Cunningham/Visuals Unlimited.

Colorplate 7
7a: © S. Maslowski/Visuals Unlimited; **7b:** © Richard Walters/Visuals Unlimited; **7c:** © BioMedia Associates; **7d:** © John D. Cunningham/Visuals Unlimited; **7e:** © Zig

Lezcynski/Animals Animals; **7f:** © Tom Stack/Tom Stack & Associates.

Colorplate 8
8a: © P. Starborn/Visuals Unlimited; **8b:** © John D. Cunningham/Visuals Unlimited; **8c:** © T.E. Adams/Visuals Unlimited; **8d:** © Glenn Oliver/Visuals Unlimited; **8e-f:** © Robert and Linda Mitchell.

Colorplate 9
9a: © Michael DiSpezio; **9b:** © Carolina Biological/Visuals Unlimited; **9c:** © Robert Dunne/Photo Researchers, Inc.; **9d:** © H. Wes Pratt/BPS; **9e:** © Bruce Iverson/Visuals Unlimited; **9f:** © C. McDaniel/Visuals Unlimited; **9g:** © Daniel W. Gotshall/Visuals Unlimited; **9h:** © R. DeGoursey/Visuals Unlimited; **9i:** © David L. Meyer; **9j:** © Carl Roessler/Tom Stack & Associates.

Colorplate 10
10a: © Michael DiSpezio; **10b:** © Russ Kinne/Photo Researchers, Inc.; **10c:** © Ed Robinson/Tom Stack & Associates; **10d-f:** © Daniel W. Gotshall/Visuals Unlimited; **10g:** © John D. Cunningham/Visuals Unlimited; **10h-i,m:** © Patrice Ceisel/Visuals Unlimited; **10j:** © Albert Copley/Visuals Unlimited; **10k:** © R. DeGoursey/Visuals Unlimited; **10l:** © Fred Rohde/Visuals Unlimited.

Colorplate 11
11a-b: © Dwight Kuhn; **11c:** © Joel Arrington/Visuals Unlimited; **11d:** © Tom McHugh/Photo Researchers, Inc.; **11e:** © W. Banaszewski/Visuals Unlimited; **11f:** © John D. Cunningham/Visuals Unlimited; **11g-h:** © Joe McDonald/Visuals Unlimited; **11i:** © John Serrao/Visuals Unlimited.

Colorplate 12
12a: © Tim Hauf/Visuals Unlimited; **12b:** © Valorie Hodgson/Visuals Unlimited; **12c:** © Stephen Dalton/Animals Animals; **12d:** © Joe McDonald/Visuals Unlimited; **12e:** © Tom J. Ulrich/Visuals Unlimited; **12f:** © Thomas Gula/Visuals Unlimited; **12g:** © Dale Jackson/Visuals Unlimited; **12h:** © Walt Anderson/Visuals Unlimited.

Colorplate 13
13a-b: © John D. Cunningham/Visuals Unlimited; **13c:** © Leonard Lee Rue III/Visuals Unlimited; **13d:** © G. Prance/Visuals Unlimited; **13e:** © Steve McCutcheon/Visuals Unlimited; **13f:** Science VU/Visuals Unlimited; **13g:** © Bruce Cushing/Visuals Unlimited; **13h:** © David S. Addison/Visuals Unlimited; **13i:** © Glenn Oliver/Visuals Unlimited; **13j:** © S. Maslowski/Visuals Unlimited; **13k:** © M. Long/Visuals Unlimited; **13l:** © Christine Case/Visuals Unlimited.

Colorplate 14
14a: © Gary R. Robinson/Visuals Unlimited; **14b:** © Richard Walters/Visuals Unlimited; **14c:** © Bruce Berg/Visuals Unlimited; **14d:** © Runk/Schoenberger/Grant Heilman.

Colorplate 15
15a: © Daniel W. Gotshall/Visuals Unlimited; **15b:** © Milton H. Tierney, Jr./Visuals Unlimited; **15c:** © Richard C. Johnson/Visuals Unlimited; **15d:** © Richard Walters/Visuals Unlimited.

Colorplate 16
16a: © Dick Poe/Visuals Unlimited; **16b:** © John Serrao/Visuals Unlimited; **16c:** © W. Kirkland/Animals Animals; **16d:** © Richard C. Johnson/Visuals Unlimited.

Colorplate 17
17a: © David M. Phillips/Visuals Unlimited; **17b:** © W. Johnson/Visuals Unlimited; **17c:** © Thomas Eisner and Daniel Aneshansley; **17d:** © William J. Weber/Visuals Unlimited.

Colorplate 18
18a: © Stan Elems/Visuals Unlimited; **18b:** © Victor H. Hutchison/Visuals Unlimited; **18c:** © Martin Rotker/Taurus Photos; **18d:** © David L. Pearson/Visuals Unlimited.

Colorplate 19
19a: © John Shaw/Tom Stack & Associates; **19b:** © Daniel W. Gotshall/Visuals Unlimited; **19c:** © Ann B. Swengel/Visuals Unlimited; **19d:** © Joe McDonald/Visuals Unlimited.

Colorplate 20
20a: © John D. Cunningham/Visuals Unlimited; **20b:** © R. DeGoursey/Visuals Unlimited; **20c:** © Daniel W. Gotshall/Visuals Unlimited; **20d:** © Milton H. Tierney, Jr./Visuals Unlimited.

Colorplate 21
21a: © Tom J. Ulrich/Visuals Unlimited; **21b:** © Dan Kline/Visuals Unlimited; **21c:** © Milton H. Tierney, Jr./Visuals Unlimited; **21d:** © Robert and Linda Mitchell.

Colorplate 22
22a: © Len Rue, Jr./Visuals Unlimited; **22b:** © Tom J. Ulrich/Visuals Unlimited; **22c:** © E.R. Degginger; **22d,f:** © Daniel W. Gotshall/Visuals Unlimited; **22e:** © Paul Opler/Visuals Unlimited; **22g:** © Anthony Mercieca/Photo Researchers, Inc.; **22h:** © Marty Snyderman/Visuals Unlimited; **22i:** © Valorie Hodgson/Visuals Unlimited; **22g:** © Daniel Jantzen; **22k:** © William H. Hughes/Visuals Unlimited.

Line Art

Chapter 1
Table 1.1: From Leland G. Johnson, *Biology*, 2nd edition. Copyright © 1987, Wm. C. Brown Publishers, Dubuque, Iowa. Reprinted by permission. **Figure 1.3:** From Leland G. Johnson, *Biology*, 2nd edition. Copyright © 1987, Wm. C. Brown Publishers, Dubuque, Iowa. Reprinted by permission.

Chapter 2
Figure 2.4: From John W. Hole, Jr., *Human Anatomy and Physiology*, 5th edition. Copyright © 1987, Wm. C. Brown Publishers, Dubuque, Iowa. Reprinted by permission. **Figure 2.8:** From Leland G. Johnson, *Biology*, 2nd edition. Copyright © 1987, Wm. C. Brown Publishers, Dubuque, Iowa. Reprinted by permission. **Figure 2.9:** From Kent Van De Graaf and Stuart Ira Fox, *Concepts of Human Anatomy and Physiology*, 2nd edition. Copyright © 1989, Wm. C. Brown Publishers, Dubuque, Iowa. Reprinted by permission. **Figure 2.16:** From Kent Van De Graaf and Stuart Ira Fox, *Concepts of Human Anatomy and Physiology*, 2nd edition. Copyright © 1989, Wm. C. Brown Publishers, Dubuque, Iowa. Reprinted by permission. **Figure 2.17:** From John W. Hole, Jr., *Human Anatomy and Physiology*, 5th edition. Copyright © 1987, Wm. C. Brown Publishers, Dubuque, Iowa. Reprinted by permission.

Chapter 3
Figure 3.3: From Leland G. Johnson, *Biology*, 2nd edition. Copyright © 1987, Wm. C. Brown Publishers, Dubuque, Iowa. Reprinted by permission. **Figure 3.4:** From John W. Hole, Jr., *Human Anatomy and Physiology*, 5th edition. Copyright © 1987, Wm. C. Brown Publishers, Dubuque, Iowa. Reprinted by permission. **Figure 3.6a:** From Leland G. Johnson, *Biology*, 2nd edition. Copyright © 1987, Wm. C. Brown Publishers, Dubuque, Iowa. Reprinted by permission. **Figure 3.8:** From John W. Hole, Jr., *Human Anatomy and Physiology*, 5th edition. Copyright © 1987, Wm. C. Brown Publishers, Dubuque, Iowa. Reprinted by permission. **Figure 3.11:** From John W. Hole, Jr., *Human Anatomy and Physiology*, 5th edition. Copyright © 1987, Wm. C. Brown Publishers, Dubuque, Iowa. Reprinted by permission. **Figure 3.13:** From John W. Hole, Jr., *Human Anatomy and Physiology*, 5th edition. Copyright © 1987, Wm. C. Brown Publishers, Dubuque, Iowa. Reprinted by permission. **Figure 3.17a,b:** From Leland G. Johnson, *Biology*, 2nd edition. Copyright © 1987, Wm. C. Brown Publishers, Dubuque, Iowa. Reprinted by permission. **Figure 3.18b:** From Leland G. Johnson, *Biology*, 2nd edition. Copyright © 1987, Wm. C. Brown Publishers, Dubuque, Iowa. Reprinted by permission. **Figure 3.19:** From Neil A. Campbell, *Biology*. Copyright © 1987, Benjamin-Cummings Publishing Company, Inc., Menlo Park, California. Reprinted by permission. **Figure 3.20:** From Leland G. Johnson, *Biology*, 2nd edition. Copyright © 1987, Wm. C. Brown Publishers, Dubuque, Iowa. Reprinted by permission. **Figure 3.23:** From George V. Kelvin, Science Graphics. Reprinted by permission. **Figure 3.24:** From Norman K. Wessels and Janet L. Hopson, *Biology*. Copyright © 1988, McGraw-Hill Publishing Company, New York, New York. Reprinted by permission. **Figure 3.25a,b:** From Leland G. Johnson, *Biology*, 2nd edition. Copyright © 1987, Wm. C. Brown Publishers, Dubuque, Iowa. Reprinted by permission. **Box figure 3.1a,b:** From T. Uzzell and C. Spolsky, "Origin of the Eukaryotic Cell" in *American Scientist*, 62:334-343, 1974. Reprinted by permission. **Box figure 3.2c,d:** From Thomas D. Brock, *Biology of Microorganisms*, 3rd edition. Copyright © 1979, Prentice-Hall, Inc., Englewood Cliffs, New Jersey. Reprinted by permission.

Chapter 4
Figure 4.5a: From Leland G. Johnson, *Biology*, 2nd edition. Copyright © 1987, Wm. C. Brown Publishers, Dubuque, Iowa. Reprinted by permission. **Figure 4.6a,b:** From Kent Van De Graaf and Stuart

Ira Fox, *Concepts of Human Anatomy and Physiology*, 2nd edition. Copyright © 1989, Wm. C. Brown Publishers, Dubuque, Iowa. Reprinted by permission. **Figure 4.9:** From Kent Van De Graaf and Stuart Ira Fox, *Concepts of Human Anatomy and Physiology*, 2nd edition. Copyright © 1989, Wm. C. Brown Publishers, Dubuque, Iowa. Reprinted by permission. **Figure 4.13:** From John W. Hole, Jr., *Human Anatomy and Physiology*, 5th edition. Copyright © 1987, Wm. C. Brown Publishers, Dubuque, Iowa. Reprinted by permission.

Chapter 5

Figure 5.2: From Leland G. Johnson, *Biology*, 2nd edition. Copyright © 1987, Wm. C. Brown Publishers, Dubuque, Iowa. Reprinted by permission. **Figure 5.3a,b:** From Leland G. Johnson, *Biology*, 2nd edition. Copyright © 1987, Wm. C. Brown Publishers, Dubuque, Iowa. Reprinted by permission. **Figure 5.12:** From Neil A. Campbell, *Biology*. Copyright © 1987, Benjamin-Cummings Publishing Company, Inc., Menlo Park, California. Reprinted by permission.

Chapter 6

Figure 6.1: From Stuart Ira Fox, *Human Physiology*, 3rd edition. Copyright © 1990, Wm. C. Brown Publishers, Dubuque, Iowa. Reprinted by permission. **Figure 6.2:** From John W. Hole, Jr., *Human Anatomy and Physiology*, 5th edition. Copyright © 1987, Wm. C. Brown Publishers, Dubuque, Iowa. Reprinted by permission. **Figure 6.3:** From Stuart Ira Fox, *Human Physiology*, 3rd edition. Copyright © 1990, Wm. C. Brown Publishers, Dubuque, Iowa. Reprinted by permission. **Figure 6.5a-d:** From John W. Hole, Jr., *Human Anatomy and Physiology*, 5th edition. Copyright © 1987, Wm. C. Brown Publishers, Dubuque, Iowa. Reprinted by permission. **Figure 6.6:** From Leland G. Johnson, *Biology*, 2nd edition. Copyright © 1987, Wm. C. Brown Publishers, Dubuque, Iowa. Reprinted by permission. **Figure 6.7:** From John W. Hole, Jr., *Human Anatomy and Physiology*, 5th edition. Copyright © 1987, Wm. C. Brown Publishers, Dubuque, Iowa. Reprinted by permission. **Figure 6.8:** From Leland G. Johnson, *Biology*, 2nd edition. Copyright © 1987, Wm. C. Brown Publishers, Dubuque, Iowa. Reprinted by permission. **Figure 6.10:** From Leland G. Johnson, *Biology*, 2nd edition. Copyright © 1987, Wm. C. Brown Publishers, Dubuque, Iowa. Reprinted by permission. **Figure 6.11a,b:** From John W. Hole, Jr., *Human Anatomy and Physiology*, 5th edition. Copyright © 1987, Wm. C. Brown Publishers, Dubuque, Iowa. Reprinted by permission.

Chapter 7

Figure 7.6: From Leland G. Johnson, *Biology*, 2nd edition. Copyright © 1987, Wm. C. Brown Publishers, Dubuque, Iowa. Reprinted by permission.

Chapter 8

Figure 8.1: From W. M. Becker, *The World of the Cell*. Copyright © 1986, Benjamin/Cummings Publishing Company, Menlo Park, California. Reprinted by permission. **Figure 8.6:** From Leland G. Johnson, *Biology*, 2nd edition. Copyright © 1987,

Wm. C. Brown Publishers, Dubuque, Iowa. Reprinted by permission. **Figure 8.8:** From Leland G. Johnson, *Biology*, 2nd edition. Copyright © 1987, Wm. C. Brown Publishers, Dubuque, Iowa. Reprinted by permission. **Figure 8.11:** From Leland G. Johnson, *Biology*, 2nd edition. Copyright © 1987, Wm. C. Brown Publishers, Dubuque, Iowa. Reprinted by permission. **Figure 8.13:** From Leland G. Johnson, *Biology*, 2nd edition. Copyright © 1987, Wm. C. Brown Publishers, Dubuque, Iowa. Reprinted by permission.

Chapter 9

Figure 9.2: From Leland G. Johnson, *Biology*, 2nd edition. Copyright © 1987, Wm. C. Brown Publishers, Dubuque, Iowa. Reprinted by permission. **Figure 9.3:** From Leland G. Johnson, *Biology*, 2nd edition. Copyright © 1987, Wm. C. Brown Publishers, Dubuque, Iowa. Reprinted by permission. **Figure 9.5:** From Leland G. Johnson, *Biology*, 2nd edition. Copyright © 1987, Wm. C. Brown Publishers, Dubuque, Iowa. Reprinted by permission. **Figure 9.6:** From W. K. Purves and G. H. Orians, *Life: The Science of Biology*. Copyright © 1983, Sinauer Associates, Sunderland, Massachusetts. Reprinted by permission. **Figure 9.8a:** From Leland G. Johnson, *Biology*, 2nd edition. Copyright © 1987, Wm. C. Brown Publishers, Dubuque, Iowa. Reprinted by permission. **Figure 9.9:** From M. W. Strickberger, *Genetics*, 2nd edition. Copyright © 1976, Macmillan Publishing Company, New York, New York. Reprinted by permission. **Figure 9.10:** From Stuart Ira Fox, *Human Physiology*, 3rd edition. Copyright © 1990, Wm. C. Brown Publishers, Dubuque, Iowa. Reprinted by permission. **Figure 9.12:** From Leland G. Johnson, *Biology*, 2nd edition. Copyright © 1987, Wm. C. Brown Publishers, Dubuque, Iowa. Reprinted by permission.

Chapter 10

Figure 10.3: From Leland G. Johnson, *Biology*, 2nd edition. Copyright © 1987, Wm. C. Brown Publishers, Dubuque, Iowa. Reprinted by permission. **Figure 10.11:** From Leland G. Johnson, *Biology*, 2nd edition. Copyright © 1987, Wm. C. Brown Publishers, Dubuque, Iowa. Reprinted by permission. **Figure 10.12:** From Neil A. Campbell, *Biology*. Copyright © 1987, Benjamin-Cummings Publishing Company, Inc., Menlo Park, California. Reprinted by permission. **Figure 10.13:** From Neil A. Campbell, *Biology*. Copyright © 1987, Benjamin-Cummings Publishing Company, Inc., Menlo Park, California. Reprinted by permission.

Chapter 11

Figure 11.1: From Walbodt and Holder, *Developmental Biology*. Copyright © 1987, McGraw-Hill Publishing Company, New York, New York. Reprinted by permission. **Figure 11.2:** From Norman K. Wessels and Janet L. Hopson, *Biology*. Copyright © 1988, McGraw-Hill Publishing Company, New York, New York. Reprinted by permission. **Figure 11.4:** From Leland G. Johnson, *Biology*, 2nd edition. Copyright © 1987, Wm. C. Brown Publishers, Dubuque, Iowa. Reprinted by permission. **Figure 11.5:** From Thomas D. Brock, *Biology of Microorganisms*, 3rd edition. Copyright © 1979, Prentice-Hall, Inc.,

Englewood Cliffs, New Jersey. Reprinted by permission. **Figure 11.6:** From Bonner-Fraser, *Journal of Cell Biology*, 1984, 98:1947-1960. **Figure 11.7:** From Norman K. Wessels and Janet L. Hopson, *Biology*. Copyright © 1988, McGraw-Hill Publishing Company, New York, New York. Reprinted by permission. **Figure 11.8a,b:** From Walbodt and Holder, *Developmental Biology*. Copyright © 1987, McGraw-Hill Publishing Company, New York, New York. Reprinted by permission. **Figure 11.10:** From B. I. Balinksy, *An Introduction to Embryology*, 5th edition. Copyright © 1981, Saunders College Publishers, a division of Holt, Rinehart, and Winston, Inc., Orlando, Florida. Reprinted by permission. **Figure 11.14:** From Leland G. Johnson, *Biology*, 2nd edition. Copyright © 1987, Wm. C. Brown Publishers, Dubuque, Iowa. Reprinted by permission. **Figure 11.15:** From Neil A. Campbell, *Biology*. Copyright © 1987, Benjamin-Cummings Publishing Company, Inc., Menlo Park, California. Reprinted by permission. **Figure 11.16:** From Neil A. Campbell, *Biology*. Copyright © 1987, Benjamin-Cummings Publishing Company, Inc., Menlo Park, California. Reprinted by permission. **Figure 11.17:** From Leland G. Johnson, *Biology*, 2nd edition. Copyright © 1987, Wm. C. Brown Publishers, Dubuque, Iowa. Reprinted by permission. **Figure 11.18:** From Norman K. Wessels and Janet L. Hopson, *Biology*. Copyright © 1988, McGraw-Hill Publishing Company, New York, New York. Reprinted by permission. **Box figure 11.1:** From Walbodt and Holder, *Developmental Biology*. Copyright © 1987, McGraw-Hill Publishing Company, New York, New York. Reprinted by permission.

Chapter 12

Figure 12.2: From Leland G. Johnson, *Biology*, 2nd edition. Copyright © 1987, Wm. C. Brown Publishers, Dubuque, Iowa. Reprinted by permission. **Figure 12.4:** From E. P. Volpe, *Understanding Evolution*, 5th edition. Copyright © 1985. Wm. C. Brown Publishers, Dubuque, Iowa. Reprinted by permission. **Figure 12.5:** From Leland G. Johnson, *Biology*, 2nd edition. Copyright © 1987, Wm. C. Brown Publishers, Dubuque, Iowa. Reprinted by permission.

Chapter 13

Table 13.2: From Leland G. Johnson, *Biology*, 2nd edition. Copyright © 1987, Wm. C. Brown Publishers, Dubuque, Iowa. Reprinted by permission. **Figure 13.3:** From Norman K. Wessels and Janet L. Hopson, *Biology*. Copyright © 1988, McGraw-Hill Publishing Company, New York, New York. Reprinted by permission. **Figure 13.4:** From Leland G. Johnson, *Biology*, 2nd edition. Copyright © 1987, Wm. C. Brown Publishers, Dubuque, Iowa. Reprinted by permission. **Figure 13.7:** From Leland G. Johnson, *Biology*, 2nd edition. Copyright © 1987, Wm. C. Brown Publishers, Dubuque, Iowa. Reprinted by permission. **Figure 13.10:** From Leland G. Johnson, *Biology*, 2nd edition. Copyright © 1987, Wm. C. Brown Publishers, Dubuque, Iowa. Reprinted by permission.

Chapter 14

Figure 14.11: From N. Toth, "The First Technology" in *Scientific American*, April 1987:114. Scientific American, New York,

New York. Reprinted by permission. **Figure 14.13:** From Leland G. Johnson, *Biology*, 2nd edition. Copyright © 1987, Wm. C. Brown Publishers, Dubuque, Iowa. Reprinted by permission. **Figure 14.15:** From R. Lewin, *Human Evolution: An Illustrated Introduction.* Copyright © 1984. W.H. Freeman and Company Publishers, New York, New York. Reprinted by permission. **Figure 14.23:** From R. Lewin, *Human Evolution: An Illustrated Introduction.* Copyright © 1984. W.H. Freeman and Company Publishers, New York, New York. Reprinted by permission. **Box figure 14.1:** From N. Toth, "The First Technology" in *Scientific American*, April 1987:115. Scientific American, New York, New York. Reprinted by permission.

Chapter 15
Table 15.1: From Leland G. Johnson, *Biology*, 2nd edition. Copyright © 1987, Wm. C. Brown Publishers, Dubuque, Iowa. Reprinted by permission. **Figure 15.1:** From Leland G. Johnson, *Biology*, 2nd edition. Copyright © 1987, Wm. C. Brown Publishers, Dubuque, Iowa. Reprinted by permission. **Figure 15.8:** From Pough et al., *Vertebrate Life*, 3rd edition. Copyright © 1989, Macmillan Publishing Company, New York, New York. Reprinted by permission. **Figure 15.11:** From C. F. Lytle and J. E. Wodsedalek, *General Zoology Laboratory Manual, Complete Version*, 11th edition. Copyright © 1991, Wm. C. Brown Publishers, Dubuque, Iowa. Reprinted by permission. **Figure 15.12:** From Leland G. Johnson, *Biology*, 2nd edition. Copyright © 1987, Wm. C. Brown Publishers, Dubuque, Iowa. Reprinted by permission. **Figure 15.13:** From Leland G. Johnson, *Biology*, 2nd edition. Copyright © 1987, Wm. C. Brown Publishers, Dubuque, Iowa. Reprinted by permission.

Chapter 16
Figure 16.1: From W. D. Russell-Hunter, *A Life of Invertebrates.* Copyright © 1979. Macmillan Publishing Company, New York, New York. Reprinted by permission. **Figure 16.2:** From Vicki Pearse, John Pearse, Mildred Buchsbaum, and Ralph Buchsbaum, *Living Invertebrates.* Copyright © 1987, Boxwood Press, Pacific Grove, California. Reprinted by permission. **Figure 16.8:** From C. F. Lytle and J. E. Wodsedalek, *General Zoology Laboratory Manual, Complete Version*, 11th edition. Copyright © 1991, Wm. C. Brown Publishers, Dubuque, Iowa. Reprinted by permission. **Figure 16.9:** From Vicki Pearse, John Pearse, Mildred Buchsbaum, and Ralph Buchsbaum, *Living Invertebrates.* Copyright © 1987, Boxwood Press, Pacific Grove, California. Reprinted by permission. **Figure 16.13:** From Robert D. Barnes, *Invertebrate Zoology*, 3rd edition. Copyright © 1987, Saunders College Publishers, a division of Holt, Rinehart, and Winston, Inc., Orlando, Florida. Reprinted by permission.

Chapter 17
Figure 17.3: From K. G. Grell, *Zeitschrift für Morphologie der Tiere*, 73:297-314, 1972. University of Tubingen, Germany. Reprinted by permission. **Figure 17.7:** From S. Miller, *General Zoology Laboratory Manual.* Copyright © 1991, Wm. C. Brown Publishers, Dubuque, Iowa. Reprinted by permission. **Figure 17.8:** From Vicki Pearse,

John Pearse, Mildred Buchsbaum, and Ralph Buchsbaum, *Living Invertebrates.* Copyright © 1987, Boxwood Press, Pacific Grove, California. Reprinted by permission. **Figure 17.9:** From J. A. Pechenik, *Biology of the Invertebrate*, 2nd edition. Copyright © 1991, Wm. C. Brown Publishers, Dubuque, Iowa. Reprinted by permission. **Figure 17.10:** From Leland G. Johnson, *Biology*, 2nd edition. Copyright © 1987, Wm. C. Brown Publishers, Dubuque, Iowa. Reprinted by permission. **Figure 17.12:** From Leland G. Johnson, *Biology*, 2nd edition. Copyright © 1987, Wm. C. Brown Publishers, Dubuque, Iowa. Reprinted by permission. **Figure 17.13:** From Vicki Pearse, John Pearse, Mildred Buchsbaum, and Ralph Buchsbaum, *Living Invertebrates.* Copyright © 1987, Boxwood Press, Pacific Grove, California. Reprinted by permission. **Figure 17.17:** From C. F. Lytle and J. E. Wodsedalek, *General Zoology Laboratory Manual, Complete Version*, 11th edition. Copyright © 1991, Wm. C. Brown Publishers, Dubuque, Iowa. Reprinted by permission. **Figure 17.18:** From Robert D. Barnes, *Invertebrate Zoology*, 3rd edition. Copyright © 1987, Saunders College Publishers, a division of Holt, Rinehart, and Winston, Inc., Orlando, Florida. Reprinted by permission. **Figure 17.21:** From C. F. Lytle and J. E. Wodsedalek, *General Zoology Laboratory Manual, Complete Version*, 11th edition. Copyright © 1991, Wm. C. Brown Publishers, Dubuque, Iowa. Reprinted by permission. **Figure 17.22:** From T. F. Goreau, N. I. Goreau, T. J. Goreau, "Corals and Coral Reefs" in *Scientific American*, August 1979:132. Scientific American, New York, New York. Reprinted by permission. **Figure 17.24:** From Vicki Pearse, John Pearse, Mildred Buchsbaum, and Ralph Buchsbaum, *Living Invertebrates.* Copyright © 1987, Boxwood Press, Pacific Grove, California. Reprinted by permission.

Chapter 18
Figure 18.3: From Vicki Pearse, John Pearse, Mildred Buchsbaum, and Ralph Buchsbaum, *Living Invertebrates.* Copyright © 1987, Boxwood Press, Pacific Grove, California. Reprinted by permission. **Figure 18.4:** From Leland G. Johnson, *Biology*, 2nd edition. Copyright © 1987, Wm. C. Brown Publishers, Dubuque, Iowa. Reprinted by permission. **Figure 18.5:** From Vicki Pearse, John Pearse, Mildred Buchsbaum, and Ralph Buchsbaum, *Living Invertebrates.* Copyright © 1987, Boxwood Press, Pacific Grove, California. Reprinted by permission. **Figure 18.6:** From Vicki Pearse, John Pearse, Mildred Buchsbaum, and Ralph Buchsbaum, *Living Invertebrates.* Copyright © 1987, Boxwood Press, Pacific Grove, California. Reprinted by permission. **Figure 18.8:** From Vicki Pearse, John Pearse, Mildred Buchsbaum, and Ralph Buchsbaum, *Living Invertebrates.* Copyright © 1987, Boxwood Press, Pacific Grove, California. Reprinted by permission. **Figure 18.9:** From P. E. Lutz, *Invertebrate Zoology.* Copyright © 1986, Benjamin/Cummings Publishing Company, Menlo Park, California. Reprinted by permission. **Figure 18.11:** From P. E. Lutz, *Invertebrate Zoology.* Copyright © 1986, Benjamin/Cummings Publishing Company, Menlo Park, California. Reprinted

by permission. **Figure 18.12:** From P. E. Lutz, *Invertebrate Zoology.* Copyright © 1986, Benjamin/Cummings Publishing Company, Menlo Park, California. Reprinted by permission. **Figure 18.13:** From P. E. Lutz, *Invertebrate Zoology.* Copyright © 1986, Benjamin/Cummings Publishing Company, Menlo Park, California. Reprinted by permission. **Figure 18.14:** From L. M. Prescott, J. Harley, and D. Klein, *Microbiology.* Copyright © 1990, Wm. C. Brown Publishers, Dubuque, Iowa. Reprinted by permission. **Figure 18.15:** From L. M. Prescott, J. Harley, and D. Klein, *Microbiology.* Copyright © 1990, Wm. C. Brown Publishers, Dubuque, Iowa. Reprinted by permission. **Figure 18.16a:** From I. Sherman and V. Sherman, *The Invertebrates: Function and Form.* Copyright © 1970, Macmillan Publishing Company, New York, New York. Reprinted by permission. **Figure 18.17:** From V. Zaman, *Atlas of Medical Parasitology.* Copyright © 1979, Lea and Febiger, Philadelphia, Pennsylvania. Reprinted by permission. **Figure 18.18:** From V. Zaman, *Atlas of Medical Parasitology.* Copyright © 1979, Lea and Febiger, Philadelphia, Pennsylvania. Reprinted by permission. **Figure 18.19:** From J. A. Pechenik, *Biology of the Invertebrate*, 2nd edition. Copyright © 1991, Wm. C. Brown Publishers, Dubuque, Iowa. Reprinted by permission. **Figure 18.20a:** From Sterrer, *Systematic Zoology*, 21:151, 1972.

Chapter 19
Figure 19.1: From Robert D. Barnes, *Invertebrate Zoology*, 3rd edition. Copyright © 1987, Saunders College Publishers, a division of Holt, Rinehart, and Winston, Inc., Orlando, Florida. Reprinted by permission. **Figure 19.2b:** From Robert D. Barnes, *Invertebrate Zoology*, 3rd edition. Copyright © 1987, Saunders College Publishers, a division of Holt, Rinehart, and Winston, Inc., Orlando, Florida. Reprinted by permission. **Figure 19.3b:** From P. E. Lutz, *Invertebrate Zoology.* Copyright © 1986, Benjamin/Cummings Publishing Company, Menlo Park, California. Reprinted by permission. **Figure 19.4:** From J. A. Pechenik, *Biology of the Invertebrate*, 2nd edition. Copyright © 1991, Wm. C. Brown Publishers, Dubuque, Iowa. Reprinted by permission. **Figure 19.5:** From P. E. Lutz, *Invertebrate Zoology.* Copyright © 1986, Benjamin/Cummings Publishing Company, Menlo Park, California. Reprinted by permission. **Figure 19.6:** From P. E. Lutz, *Invertebrate Zoology.* Copyright © 1986, Benjamin/Cummings Publishing Company, Menlo Park, California. Reprinted by permission. **Figure 19.7:** From P. E. Lutz, *Invertebrate Zoology.* Copyright © 1986, Benjamin/Cummings Publishing Company, Menlo Park, California. Reprinted by permission. **Figure 19.8:** From P. E. Lutz, *Invertebrate Zoology.* Copyright © 1986, Benjamin/Cummings Publishing Company, Menlo Park, California. Reprinted by permission. **Figure 19.9:** From V. Zaman, *Atlas of Medical Parasitology.* Copyright © 1979, Lea and Febiger, Philadelphia, Pennsylvania. Reprinted by permission. **Figure 19.10:** From V. Zaman, *Atlas of Medical Parasitology.* Copyright © 1979, Lea and Febiger, Philadelphia, Pennsylvania. Reprinted by permission. **Figure**

19.11: From V. Zaman, *Atlas of Medical Parasitology*. Copyright © 1979, Lea and Febiger, Philadelphia, Pennsylvania. Reprinted by permission. **Figure 19.12:** From V. Zaman, *Atlas of Medical Parasitology*. Copyright © 1979, Lea and Febiger, Philadelphia, Pennsawvania. Reprinted by permission. **Figure 19.14:** From V. Zaman, *Atlas of Medical Parasitology*. Copyright © 1979, Lea and Febiger, Philadelphia, Pennsylvania. Reprinted by permission. **Figure 19.16a:** From J. A. Pechenik, *Biology of the Invertebrate*, 2nd edition. Copyright © 1991, Wm. C. Brown Publishers, Dubuque, Iowa. Reprinted by permission. **Figure 19.17:** From Robert D. Barnes, *Invertebrate Zoology*, 3rd edition. Copyright © 1987, Saunders College Publishers, a division of Holt, Rinehart, and Winston, Inc., Orlando, Florida. Reprinted by permission. **Figure 19.18:** From Robert D. Barnes, *Invertebrate Zoology*, 3rd edition. Copyright © 1987, Saunders College Publishers, a division of Holt, Rinehart, and Winston, Inc., Orlando, Florida. Reprinted by permission. **Figure 19.19:** From Vicki Pearse, John Pearse, Mildred Buchsbaum, and Ralph Buchsbaum, *Living Invertebrates*. Copyright © 1987, Boxwood Press, Pacific Grove, California. Reprinted by permission.

Chapter 20

Figure 20.2: From Leland G. Johnson, *Biology*, 2nd edition. Copyright © 1987, Wm. C. Brown Publishers, Dubuque, Iowa. Reprinted by permission. **Figure 20.3:** From Vicki Pearse, John Pearse, Mildred Buchsbaum, and Ralph Buchsbaum, *Living Invertebrates*. Copyright © 1987, Boxwood Press, Pacific Grove, California. Reprinted by permission. **Figure 20.4:** From W. D. Russell-Hunter, *A Life of Invertebrates*. Copyright © 1979. Macmillan Publishing Company, New York, New York. Reprinted by permission. **Figure 20.5:** From L. Hyman, *The Invertebrates, Vol. VI*. Copyright © 1967, McGraw-Hill Publishing Company, New York, New York. Reprinted by permission. **Figure 20.6:** From J. A. Pechenik, *Biology of the Invertebrate*, 2nd edition. Copyright © 1991, Wm. C. Brown Publishers, Dubuque, Iowa. Reprinted by permission. **Figure 20.10:** From Vicki Pearse, John Pearse, Mildred Buchsbaum, and Ralph Buchsbaum, *Living Invertebrates*. Copyright © 1987, Boxwood Press, Pacific Grove, California. Reprinted by permission. **Figure 20.11:** From Vicki Pearse, John Pearse, Mildred Buchsbaum, and Ralph Buchsbaum, *Living Invertebrates*. Copyright © 1987, Boxwood Press, Pacific Grove, California. Reprinted by permission. **Figure 20.12:** From Leland G. Johnson, *Biology*, 2nd edition. Copyright © 1987, Wm. C. Brown Publishers, Dubuque, Iowa. Reprinted by permission. **Figure 20.13:** From Vicki Pearse, John Pearse, Mildred Buchsbaum, and Ralph Buchsbaum, *Living Invertebrates*. Copyright © 1987, Boxwood Press, Pacific Grove, California. Reprinted by permission. **Figure 20.15:** From W. D. Russell-Hunter, *A Life of Invertebrates*. Copyright © 1979, Macmillan Publishing Company, New York, New York. Reprinted by permission. **Figure 20.18:** From Vicki Pearse, John Pearse, Mildred Buchsbaum, and Ralph Buchsbaum, *Living Invertebrates*. Copyright © 1987, Boxwood Press, Pacific Grove, California. Reprinted by permission. **Figure 20.19:** From Vicki Pearse, John Pearse, Mildred Buchsbaum, and Ralph Buchsbaum, *Living Invertebrates*. Copyright © 1987, Boxwood Press, Pacific Grove, California. Reprinted by permission. **Figure 20.20:** From Robert D. Barnes, *Invertebrate Zoology*, 3rd edition. Copyright © 1987, Saunders College Publishers, a division of Holt, Rinehart, and Winston, Inc., Orlando, Florida. Reprinted by permission. **Figure 20.22:** From Leland G. Johnson, *Biology*, 2nd edition. Copyright © 1987, Wm. C. Brown Publishers, Dubuque, Iowa. Reprinted by permission. **Figure 20.24a,b,c:** From Beck and Braithwaite, *Invertebrate Zoology Laboratory Workbook*, 3rd edition. Copyright © 1968, Macmillan Publishing Company, New York, New York. Reprinted by permission.

Chapter 21

Figure 21.2: From Snodgrass, *Principles of Insect Morphology*. Copyright © 1935, McGraw-Hill Publishing Company, New York, New York. Reprinted by permission. **Figure 21.4:** From W. D. Russell-Hunter, *A Life of Invertebrates*. Copyright © 1979, Macmillan Publishing Company, New York, New York. Reprinted by permission. **Figure 21.7:** From Vicki Pearse, John Pearse, Mildred Buchsbaum, and Ralph Buchsbaum, *Living Invertebrates*. Copyright © 1987, Boxwood Press, Pacific Grove, California. Reprinted by permission. **Figure 21.8:** From W. D. Russell-Hunter, *A Life of Invertebrates*. Copyright © 1979, Macmillan Publishing Company, New York, New York. Reprinted by permission. **Figure 21.9:** From Robert D. Barnes, *Invertebrate Zoology*, 3rd edition. Copyright © 1987, Saunders College Publishers, a division of Holt, Rinehart, and Winston, Inc., Orlando, Florida. Reprinted by permission. **Figure 21.10:** From W. D. Russell-Hunter, *A Life of Invertebrates*. Copyright © 1979, Macmillan Publishing Company, New York, New York. Reprinted by permission. **Figure 21.12:** From W. D. Russell-Hunter, *A Life of Invertebrates*. Copyright © 1979, Macmillan Publishing Company, New York, New York. Reprinted by permission. **Figure 21.13:** From Leland G. Johnson, *Biology*, 2nd edition. Copyright © 1987, Wm. C. Brown Publishers, Dubuque, Iowa. Reprinted by permission. **Figure 21.14:** From Vicki Pearse, John Pearse, Mildred Buchsbaum, and Ralph Buchsbaum, *Living Invertebrates*. Copyright © 1987, Boxwood Press, Pacific Grove, California. Reprinted by permission. **Figure 21.16:** From J. A. Pechenik, *Biology of the Invertebrate*, 2nd edition. Copyright © 1991, Wm. C. Brown Publishers, Dubuque, Iowa. Reprinted by permission. **Figure 21.17:** From W. D. Russell-Hunter, *A Life of Invertebrates*. Copyright © 1979, Macmillan Publishing Company, New York, New York. Reprinted by permission.

Chapter 22

Figure 22.3: From W. D. Russell-Hunter, *A Life of Invertebrates*. Copyright © 1979, Macmillan Publishing Company, New York, New York. Reprinted by permission. **Figure 22.6:** From Robert D. Barnes, *Invertebrate Zoology*, 3rd edition. Copyright © 1987, Saunders College Publishers, a division of Holt, Rinehart, and Winston, Inc., Orlando, Florida. Reprinted by permission. **Figure 22.7:** From Robert D. Barnes, *Invertebrate Zoology*, 3rd edition. Copyright © 1987, Saunders College Publishers, a division of Holt, Rinehart, and Winston, Inc., Orlando, Florida. Reprinted by permission. **Figure 22.8:** From Robert D. Barnes, *Invertebrate Zoology*, 3rd edition. Copyright © 1987, Saunders College Publishers, a division of Holt, Rinehart, and Winston, Inc., Orlando, Florida. Reprinted by permission. **Figure 22.9:** From Vicki Pearse, John Pearse, Mildred Buchsbaum, and Ralph Buchsbaum, *Living Invertebrates*. Copyright © 1987, Boxwood Press, Pacific Grove, California. Reprinted by permission. **Figure 22.10a:** From Robert D. Barnes, *Invertebrate Zoology*, 3rd edition. Copyright © 1987, Saunders College Publishers, a division of Holt, Rinehart, and Winston, Inc., Orlando, Florida. Reprinted by permission. **Figure 22.11b:** From Vicki Pearse, John Pearse, Mildred Buchsbaum, and Ralph Buchsbaum, *Living Invertebrates*. Copyright © 1987, Boxwood Press, Pacific Grove, California. Reprinted by permission. **Figure 22.12:** From J. A. Pechenik, *Biology of the Invertebrate*, 2nd edition. Copyright © 1991, Wm. C. Brown Publishers, Dubuque, Iowa. Reprinted by permission. **Figure 22.16:** From Vicki Pearse, John Pearse, Mildred Buchsbaum, and Ralph Buchsbaum, *Living Invertebrates*. Copyright © 1987, Boxwood Press, Pacific Grove, California. Reprinted by permission. **Figure 22.17:** From Vicki Pearse, John Pearse, Mildred Buchsbaum, and Ralph Buchsbaum, *Living Invertebrates*. Copyright © 1987, Boxwood Press, Pacific Grove, California. Reprinted by permission. **Figure 22.18:** From Vicki Pearse, John Pearse, Mildred Buchsbaum, and Ralph Buchsbaum, *Living Invertebrates*. Copyright © 1987, Boxwood Press, Pacific Grove, California. Reprinted by permission. **Box figure 22.1:** From Herbert W. Levy, "Orb-Weaving Spiders and Their Webs" in *American Scientist* 66:734-742. Scientific Research Society, Research Triangle Park, North Carolina. Reprinted by permission.

Chapter 23

Figure 23.2: From Vicki Pearse, John Pearse, Mildred Buchsbaum, and Ralph Buchsbaum, *Living Invertebrates*. Copyright © 1987, Boxwood Press, Pacific Grove, California. Reprinted by permission. **Figure 23.3:** From C. Gillott, *Entomology*. Copyright © 1980, Plenum Publishing Corp., New York, New York. Reprinted by permission. **Figure 23.4:** From M. D. Atkins, *Insects in Perspective*. Copyright © 1978, Macmillan Publishing Company, New York, New York. Reprinted by permission. **Figure 23.5:** From Vicki Pearse, John Pearse, Mildred Buchsbaum, and Ralph Buchsbaum, *Living Invertebrates*. Copyright © 1987, Boxwood Press, Pacific Grove, California. Reprinted by permission. **Figure 23.6:** From Vicki Pearse, John Pearse, Mildred Buchsbaum, and Ralph Buchsbaum, *Living Invertebrates*. Copyright © 1987, Boxwood Press, Pacific Grove, California. Reprinted by permission. **Figure 23.7:** From Robert D. Barnes, *Invertebrate Zoology*, 3rd edition. Copyright © 1987, Saunders College Publishers, a division of Holt, Rinehart, and Winston, Inc., Orlando, Florida. Reprinted by permission. **Figure 23.8:** From Vicki Pearse, John Pearse, Mildred Buchsbaum, and Ralph Buchsbaum, *Living Invertebrates*. Copyright © 1987, Boxwood Press, Pacific Grove, California. Reprinted by permission. **Figure 23.10:** From Carolina Biological Supply Co., Burlington, North Carolina. Reprinted by permission. **Figure 23.11:** From Carolina Biological Supply Co., Burlington, North Carolina. Reprinted by permission. **Figure 23.12:** From Vicki Pearse, John Pearse, Mildred Buchsbaum, and Ralph Buchsbaum, *Living Invertebrates*. Copyright © 1987, Boxwood Press, Pacific Grove, California. Reprinted by permission. **Box figure 23.2:** From W. D. Russell-Hunter, *A Life of Invertebrates*. Copyright © 1979, Macmillan Publishing Company, New York, New York. Reprinted by permission.

Chapter 24

Figure 24.1: From Vicki Pearse, John Pearse, Mildred Buchsbaum, and Ralph Buchsbaum, *Living Invertebrates*. Copyright © 1987, Boxwood Press, Pacific Grove, California. Reprinted by permission. **Figure 24.3:** From Vicki Pearse, John Pearse, Mildred Buchsbaum, and Ralph Buchsbaum, *Living Invertebrates*. Copyright © 1987, Boxwood Press, Pacific Grove, California. Reprinted by permission. **Figure 24.5:** After L. Hyman, *The Invertebrates, Vol. VI*. Copyright © 1967, McGraw-Hill Publishing Company, New York, New York. **Figure 24.6:** From Vicki Pearse, John Pearse, Mildred Buchsbaum, and Ralph Buchsbaum, *Living Invertebrates*. Copyright © 1987, Boxwood Press, Pacific Grove, California. Reprinted by permission. **Figure 24.7:** From Vicki Pearse, John Pearse, Mildred Buchsbaum, and Ralph Buchsbaum, *Living Invertebrates*. Copyright © 1987, Boxwood Press, Pacific Grove, California. Reprinted by permission. **Figure 24.8:** From P. E. Lutz, *Invertebrate Zoology*. Copyright © 1986, Benjamin/Cummings Publishing Company, Menlo Park, California. Reprinted by permission. **Figure 24.11c:** From Robert D. Barnes, *Invertebrate Zoology*, 3rd edition. Copyright © 1987, Saunders College Publishers, a division of Holt, Rinehart, and Winston, Inc., Orlando, Florida. Reprinted by permission. **Figure 24.12:** From Robert D. Barnes, *Invertebrate Zoology*, 3rd edition. Copyright © 1987, Saunders College Publishers, a division of Holt, Rinehart, and Winston, Inc., Orlando, Florida. Reprinted by permission. **Box figure 24.1:** From P. E. Lutz, *Invertebrate Zoology*. Copyright © 1986, Benjamin/Cummings Publishing Company, Menlo Park, California. Reprinted by permission.

Chapter 25

Figure 25.3: From Vicki Pearse, John Pearse, Mildred Buchsbaum, and Ralph Buchsbaum, *Living Invertebrates*. Copyright © 1987, Boxwood Press, Pacific Grove, California. Reprinted by permission. **Figure 25.5:** From Vicki Pearse, John Pearse, Mildred Buchsbaum, and Ralph Buchsbaum, *Living Invertebrates*. Copyright © 1987, Boxwood Press, Pacific Grove, California. Reprinted by permission. **Figure 25.6:** From J. A. Pechenik, *Biology of the Invertebrate*, 2nd edition. Copyright © 1991, Wm. C. Brown Publishers, Dubuque, Iowa. Reprinted by permission.

Figure 25.10: From Leland G. Johnson, *Biology*, 2nd edition. Copyright © 1987, Wm. C. Brown Publishers, Dubuque, Iowa. Reprinted by permission. **Figure 25.11:** From Robert D. Barnes, *Invertebrate Zoology*, 3rd edition. Copyright © 1987, Saunders College Publishers, a division of Holt, Rinehart, and Winston, Inc., Orlando, Florida. Reprinted by permission. **Figure 25.12:** From Robert D. Barnes, *Invertebrate Zoology*, 3rd edition. Copyright © 1987, Saunders College Publishers, a division of Holt, Rinehart, and Winston, Inc., Orlando, Florida. Reprinted by permission.

Chapter 26

Figure 26.2: From Vicki Pearse, John Pearse, Mildred Buchsbaum, and Ralph Buchsbaum, *Living Invertebrates*. Copyright © 1987, Boxwood Press, Pacific Grove, California. Reprinted by permission. **Figure 26.3:** From Vicki Pearse, John Pearse, Mildred Buchsbaum, and Ralph Buchsbaum, *Living Invertebrates*. Copyright © 1987, Boxwood Press, Pacific Grove, California. Reprinted by permission. **Figure 26.5:** From C. F. Lytle and J. E. Wodsedalek, *General Zoology Laboratory Manual, Complete Version*, 11th edition. Copyright © 1991, Wm. C. Brown Publishers, Dubuque, Iowa. Reprinted by permission. **Figure 26.6:** From Vicki Pearse, John Pearse, Mildred Buchsbaum, and Ralph Buchsbaum, *Living Invertebrates*. Copyright © 1987, Boxwood Press, Pacific Grove, California. Reprinted by permission. **Figure 26.8:** From A. S. Romer and T. S. Parsons, *The Vertebrate Body*, 6th edition. Copyright © 1986, Saunders College Publishers, a division of Holt, Rinehart, and Winston, Inc., Orlando, Florida. Reprinted by permission.

Chapter 27

Figure 27.7b: From R. Steele, *Sharks of the World*. Copyright © 1985, Cassel PLC, London, England. Reprinted by permission. **Figure 27.10:** From K. F. Lagler, *Ichthyology*. Copyright © 1977, John Wiley & Sons, Inc., New York, New York. Reprinted by permission. **Figure 27.12:** From R. Eckert and D. Randall, *Animal Physiology*, 2nd edition. Copyright © 1983, W.H. Freeman and Company Publishers, New York, New York. Reprinted by permission. **Figure 27.13a,b:** From R. Eckert and D. Randall, *Animal Physiology*, 2nd edition. Copyright © 1983, W.H. Freeman and Company Publishers, New York, New York. Reprinted by permission. **Figure 27.14:** From A. S. Romer and T. S. Parsons, *The Vertebrate Body*, 6th edition. Copyright © 1986, Saunders College Publishers, a division of Holt, Rinehart, and Winston, Inc., Orlando, Florida. Reprinted by permission.

Chapter 28

Figure 28.1: From Duellman and Treub, *Biology of Amphibians*. Copyright © 1986, McGraw-Hill Publishing Company, New York, New York. Reprinted by permission. **Figure 28.3a,b:** From A. S. Romer and T. S. Parsons, *The Vertebrate Body*, 6th edition. Copyright © 1986, Saunders College Publishers, a division of Holt, Rinehart, and Winston, Inc., Orlando, Florida. Reprinted by permission. **Figure 28.3c:** From Pough, et al., *Vertebrate Life*, 3rd edition. Copyright © 1989, Macmillan Publishing Company, New York, New York.

Reprinted by permission. **Figure 28.5:** From A. S. Romer and T. S. Parsons, *The Vertebrate Body*, 6th edition. Copyright © 1986, Saunders College Publishers, a division of Holt, Rinehart, and Winston, Inc., Orlando, Florida. Reprinted by permission. **Figure 28.6:** From Pough, et al., *Vertebrate Life*, 3rd edition. Copyright © 1989, Macmillan Publishing Company, New York, New York. Reprinted by permission. **Figure 28.8:** From C. Gans et al., "Bullfrog (Rana Calesbeana) Ventilation: How Does the Frog Breathe?" in *Science*, 163:1223-1224, 1969. Copyright © 1969, American Association for the Advancement of Science. Reprinted by permission. **Figure 28.13a,b:** From Duellman and Treub, *Biology of Amphibians*. Copyright © 1986, McGraw-Hill Publishing Company, New York, New York. Reprinted by permission.

Chapter 29

Figure 29.1: From W. F. Walker, *Functional Anatomy of the Vertebrates: An Evolutionary Perspective*. Copyright ©. Saunders College Publishers, a division of Holt, Rinehart, and Winston, Inc., Orlando, Florida. Reprinted by permission. **Figure 29.7:** From W. F. Walker, *Functional Anatomy of the Vertebrates: An Evolutionary Perspective*. Copyright ©. Saunders College Publishers, a division of Holt, Rinehart, and Winston, Inc., Orlando, Florida. Reprinted by permission. **Figure 29.8:** From Pough, et al., *Vertebrate Life*, 3rd edition. Copyright © 1989, Macmillan Publishing Company, New York, New York. Reprinted by permission. **Figure 29.9:** From Porter, *Herpetology*, 3rd edition. Copyright ©. Saunders College Publishers, a division of Holt, Rinehart, and Winston, Inc., Orlando, Florida. Reprinted by permission. **Figure 29.10:** From Pough, et al., *Vertebrate Life*, 3rd edition. Copyright © 1989, Macmillan Publishing Company, New York, New York. Reprinted by permission. **Box figure 29.1:** From W. F. Walker, *Functional Anatomy of the Vertebrates: An Evolutionary Perspective*. Copyright ©. Saunders College Publishers, a division of Holt, Rinehart, and Winston, Inc., Orlando, Florida. Reprinted by permission.

Chapter 30

Figure 30.1: From Wallace and Mahan, *Introduction to Ornithology*, 3rd edition. Copyright ©. Macmillan Publishing Company, New York, New York. Reprinted by permission. **Figure 30.2b:** From Wallace and Mahan, *Introduction to Ornithology*, 3rd edition. Copyright ©. Macmillan Publishing Company, New York, New York. Reprinted by permission. **Figure 30.3:** From Wallace and Mahan, *Introduction to Ornithology*, 3rd edition. Copyright ©. Macmillan Publishing Company, New York, New York. Reprinted by permission. **Figure 30.4:** From Wallace and Mahan, *Introduction to Ornithology*, 3rd edition. Copyright ©. Macmillan Publishing Company, New York, New York. Reprinted by permission. **Figure 30.5:** From Wallace and Mahan, *Introduction to Ornithology*, 2nd edition. Copyright ©. Macmillan Publishing Company, New York, New York. Reprinted by permission. **Figure 30.7:** From R. Eckert and D. Randall, *Animal Physiology*, 2nd edition. Copyright

© 1983, W.H. Freeman and Company Publishers, New York, New York. Reprinted by permission. **Figure 30.9:** From F. B. Gill, *Ornithology*. Copyright © 1990, W.H. Freeman and Company Publishers, New York, New York. Reprinted by permission. **Figure 30.13:** From Wallace and Mahan, *Introduction to Ornithology*, 3rd edition. Copyright ©. Macmillan Publishing Company, New York, New York. Reprinted by permission. **Figure 30.14:** From Wallace and Mahan, *Introduction to Ornithology*, 3rd edition. Copyright ©. Macmillan Publishing Company, New York, New York. Reprinted by permission.

Chapter 31

Figure 31.5: From A. S. Romer and T. S. Parsons, *The Vertebrate Body*, 6th edition. Copyright © 1986, Saunders College Publishers, a division of Holt, Rinehart, and Winston, Inc., Orlando, Florida. Reprinted by permission. **Figure 31.6:** From M. Hildebrand, *Analysis of Vertebrate Structure*. Copyright ©. John Wiley & Sons, New York, New York. Reprinted by permission. **Figure 31.11:** From W. F. Walker, *A Functional Anatomy of Vertebrates: An Evolutionary Perspective*. Copyright ©. Saunders College Publishers, a division of Holt, Rinehart, and Winston, Inc., Orlando, Florida. Reprinted by permission.

Chapter 32

Figure 32.1: From Kent M. Van De Graaff, *Human Anatomy*, 2nd edition. Copyright © 1988, Wm. C. Brown Publishers, Dubuque, Iowa. Reprinted by permission. **Figure 32.2:** From Kent M. Van De Graaff, *Human Anatomy*, 2nd edition. Copyright © 1988, Wm. C. Brown Publishers, Dubuque, Iowa. Reprinted by permission. **Figure 32.3:** From Kent M. Van De Graaff, *Human Anatomy*, 2nd edition. Copyright © 1988, Wm. C. Brown Publishers, Dubuque, Iowa. Reprinted by permission. **Figure 32.4:** From John W. Hole, Jr., *Human Anatomy and Physiology*, 5th edition. Copyright © 1987, Wm. C. Brown Publishers, Dubuque, Iowa. Reprinted by permission. **Figure 32.5:** From Kent M. Van De Graaff, *Human Anatomy*, 2nd edition. Copyright © 1988, Wm. C. Brown Publishers, Dubuque, Iowa. Reprinted by permission. **Figure 32.6:** From Kent M. Van De Graaff, *Human Anatomy*, 2nd edition. Copyright © 1988, Wm. C. Brown Publishers, Dubuque, Iowa. Reprinted by permission. **Figure 32.7:** From Kent M. Van De Graaff, *Human Anatomy*, 2nd edition. Copyright © 1988, Wm. C. Brown Publishers, Dubuque, Iowa. Reprinted by permission. **Figure 32.8:** From Kent Van De Graaf and Stuart Ira Fox, *Concepts of Human Anatomy and Physiology*, 2nd edition. Copyright © 1989, Wm. C. Brown Publishers, Dubuque, Iowa. Reprinted by permission. **Figure 32.9c-e:** From John W. Hole, Jr., *Human Anatomy and Physiology*, 5th edition. Copyright © 1987, Wm. C. Brown Publishers, Dubuque, Iowa. Reprinted by permission. **Figure 32.10:** From John W. Hole, Jr., *Human Anatomy and Physiology*, 5th edition. Copyright © 1987, Wm. C. Brown Publishers, Dubuque, Iowa. Reprinted by permission. **Figure 32.11:** From Stuart Ira Fox, *Human Physiology*, 3rd edition. Copyright © 1990, Wm. C. Brown

Publishers, Dubuque, Iowa. Reprinted by permission. **Figure 32.12:** From Stuart Ira Fox, *Human Physiology*, 3rd edition. Copyright © 1990, Wm. C. Brown Publishers, Dubuque, Iowa. Reprinted by permission. **Box figure 32.3:** From Carola et al., *Human Anatomy and Physiology*. Copyright © 1990, McGraw-Hill Publishing Company, New York, New York. Reprinted by permission.

Chapter 33

Figure 33.2: From Kent M. Van De Graaff, *Human Anatomy*, 2nd edition. Copyright © 1988, Wm. C. Brown Publishers, Dubuque, Iowa. Reprinted by permission. **Figure 33.4:** From Carola et al., *Human Anatomy and Physiology*. Copyright © 1990, McGraw-Hill Publishing Company, New York, New York. Reprinted by permission. **Figure 33.6:** From Carola et al., *Human Anatomy and Physiology*. Copyright © 1990, McGraw-Hill Publishing Company, New York, New York. Reprinted by permission. **Figure 33.7:** From Leland G. Johnson, *Biology*, 2nd edition. Copyright © 1987, Wm. C. Brown Publishers, Dubuque, Iowa. Reprinted by permission. **Figure 33.8:** From Neil A. Campbell, *Biology*. Copyright © 1987, Benjamin-Cummings Publishing Company, Inc., Menlo Park, California. Reprinted by permission. **Figure 33.9a:** From John W. Hole, Jr., *Human Anatomy and Physiology*, 5th edition. Copyright © 1987, Wm. C. Brown Publishers, Dubuque, Iowa. Reprinted by permission. **Figure 33.10:** From John W. Hole, Jr., *Human Anatomy and Physiology*, 5th edition. Copyright © 1987, Wm. C. Brown Publishers, Dubuque, Iowa. Reprinted by permission. **Figure 33.12:** From C. Starr and R. Taggert, *Biology: The Unity and Diversity of Life*, 4th edition. Copyright © 1987, Wadsworth, Inc., Belmont, California. Reprinted by permission. **Figure 33.13:** From Kent M. Van De Graaff, *Human Anatomy*, 2nd edition. Copyright © 1988, Wm. C. Brown Publishers, Dubuque, Iowa. Reprinted by permission. **Figure 33.14b:** From Leland G. Johnson, *Biology*, 2nd edition. Copyright © 1987, Wm. C. Brown Publishers, Dubuque, Iowa. Reprinted by permission. **Figure 33.14c:** From John W. Hole, Jr., *Human Anatomy and Physiology*, 5th edition. Copyright © 1987, Wm. C. Brown Publishers, Dubuque, Iowa. Reprinted by permission. **Figure 33.16:** From John W. Hole, Jr., *Human Anatomy and Physiology*, 5th edition. Copyright © 1987, Wm. C. Brown Publishers, Dubuque, Iowa. Reprinted by permission. **Figure 33.17:** From John W. Hole, Jr., *Human Anatomy and Physiology*, 5th edition. Copyright © 1987, Wm. C. Brown Publishers, Dubuque, Iowa. Reprinted by permission.

Chapter 34

Table 34.3: From Carola et al., *Human Anatomy and Physiology*. Copyright © 1990, McGraw-Hill Publishing Company, New York, New York. Reprinted by permission. **Figure 34.1:** From T. L. Peele, *Neuroanatomical Basis for Clinical Neurology*, 2nd edition. Copyright © 1961, McGraw-Hill Publishing Company, New York, New York. Reprinted by permission. **Figure 34.2:** From Kent M.

Van De Graaff, *Human Anatomy*, 2nd edition. Copyright © 1988, Wm. C. Brown Publishers, Dubuque, Iowa. Reprinted by permission. **Figure 34.3:** From Kent M. Van De Graaff, *Human Anatomy*, 2nd edition. Copyright © 1988, Wm. C. Brown Publishers, Dubuque, Iowa. Reprinted by permission. **Figure 34.4:** From Kent M. Van De Graaff, *Human Anatomy*, 2nd edition. Copyright © 1988, Wm. C. Brown Publishers, Dubuque, Iowa. Reprinted by permission. **Figure 34.5:** From Kent M. Van De Graaff, *Human Anatomy*, 2nd edition. Copyright © 1988, Wm. C. Brown Publishers, Dubuque, Iowa. Reprinted by permission. **Figure 34.7:** From J. R. McClintic, *Physiology of the Human Body*, 2nd edition. Copyright ©. John Wiley & Sons, Inc., New York, New York. Reprinted by permission. **Figure 34.8:** From Kent M. Van De Graaff, *Human Anatomy*, 2nd edition. Copyright © 1988, Wm. C. Brown Publishers, Dubuque, Iowa. Reprinted by permission. **Figure 34.9:** From Kent M. Van De Graaff, *Human Anatomy*, 2nd edition. Copyright © 1988, Wm. C. Brown Publishers, Dubuque, Iowa. Reprinted by permission. **Figure 34.11:** From Kent M. Van De Graaff, *Human Anatomy*, 2nd edition. Copyright © 1988, Wm. C. Brown Publishers, Dubuque, Iowa. Reprinted by permission. **Figure 34.13:** From Kent M. Van De Graaff, *Human Anatomy*, 2nd edition. Copyright © 1988, Wm. C. Brown Publishers, Dubuque, Iowa. Reprinted by permission.

Chapter 35

Table 35.1: From J. Barrett et al., *Biology*. Copyright © 1986. Prentice-Hall, Englewood Cliffs, New Jersey. **Figure 35.1:** From John W. Hole, Jr., *Human Anatomy and Physiology*, 5th edition. Copyright © 1987, Wm. C. Brown Publishers, Dubuque, Iowa. Reprinted by permission. **Figure 35.2:** From John W. Hole, Jr., *Human Anatomy and Physiology*, 5th edition. Copyright © 1987, Wm. C. Brown Publishers, Dubuque, Iowa. Reprinted by permission. **Figure 35.4:** From Carola et al., *Human Anatomy and Physiology*. Copyright © 1990, McGraw-Hill Publishing Company, New York, New York. Reprinted by permission. **Figure 35.5:** From John W. Hole, Jr., *Human Anatomy and Physiology*, 5th edition. Copyright © 1987, Wm. C. Brown Publishers, Dubuque, Iowa. Reprinted by permission. **Figure 35.6:** From Norman K. Wessels and Janet L. Hopson, *Biology*. Copyright © 1988, McGraw-Hill Publishing Company, New York, New York. Reprinted by permission. **Figure 35.7:** From John W. Hole, Jr., *Human Anatomy and Physiology*, 5th edition. Copyright © 1987, Wm. C. Brown Publishers, Dubuque, Iowa. Reprinted by permission. **Figure 35.8:** From Kent Van De Graaf and Stuart Ira Fox, *Concepts of Human Anatomy and Physiology*, 2nd edition. Copyright © 1989, Wm. C. Brown Publishers, Dubuque, Iowa. Reprinted by permission. **Figure 35.10:** From John W. Hole, Jr., *Human Anatomy and Physiology*, 5th edition. Copyright © 1987, Wm. C. Brown Publishers, Dubuque, Iowa. Reprinted by permission. **Figure 35.11:** From John W. Hole, Jr., *Human Anatomy and Physiology*, 5th

Copyright © 1975, Holt, Rinehart, and Winston, Orlando Florida. Reprinted by permission. **Figure 40.9:** From Manning, *An Introduction to Animal Behavior*, 3rd edition. Copyright © 1979, Benjamin/Cummings Publishing Company, Menlo Park, California. Reprinted by permission.

Chapter 41

Figure 41.1: From R. Brewer, *The Science of Ecology*. Copyright ©. Saunders College Publishers, a division of Holt, Rinehart, and Winston, Inc., Orlando, Florida. Reprinted by permission. **Figure 41.3:** From K. Schmidt-Nielsen, *Animal Physiology*, 3rd edition. Copyright © 1983, Cambridge University Press, Cambridge, England. Reprinted by permission. **Figure 41.4:** From R. Brewer, *The Science of Ecology*. Copyright ©. Saunders College Publishers, a division of Holt, Rinehart, and Winston, Inc., Orlando, Florida. Reprinted by permission. **Figure 41.5:** From Leland G. Johnson, *Biology*, 2nd edition. Copyright © 1987, Wm. C. Brown Publishers, Dubuque, Iowa. Reprinted by permission. **Figure 41.7:** From C. C. Lindsey, ''Problems in Zoology of the Lake Trout, *Salvelinus namaycush*'' in *Journal of Fisheries Research Board of Canada*, 21:977-944, 1964. Reprinted by permission. **Figure 41.8:** From R. Brewer, *The Science of Ecology*. Copyright ©. Saunders College Publishers, a division of Holt, Rinehart, and Winston, Inc., Orlando, Florida. Reprinted by permission. **Figure 41.9c:** From S. Charles Kendeigh, *Ecology with Special Reference to Animals and Man*. Copyright © 1974, Prentice-Hall, Englewood Cliffs, New Jersey. Reprinted by permission. **Figure 41.10:** From Andrewartha and Birch, *The Distribution and Abundance of Animals*. Copyright © 1954, The University of Chicago Press, Chicago, Illinois. Reprinted by permission. **Figure 41.11:** From R. H. MacArthur, ''Population Ecology of Some Warblers in Northeastern Coniferous Forests'' in *Ecology*, 39:599-619. Reprinted by permission.

Chapter 42

Figure 42.1: From Leland G. Johnson, *Biology*, 2nd edition. Copyright © 1987, Wm. C. Brown Publishers, Dubuque, Iowa. Reprinted by permission. **Figure 42.3:** From Leland G. Johnson, *Biology*, 2nd edition. Copyright © 1987, Wm. C. Brown Publishers, Dubuque, Iowa. Reprinted by permission. **Figure 42.4:** From Leland G. Johnson, *Biology*, 2nd edition. Copyright © 1987, Wm. C. Brown Publishers, Dubuque, Iowa. Reprinted by permission. **Figure 42.6:** From R. H. Whittaker,

Communities and Ecosystems. Copyright © 1975, Macmillan Publishing Company, New York, New York. Reprinted by permission. **Figure 42.7:** From Leland G. Johnson, *Biology*, 2nd edition. Copyright © 1987, Wm. C. Brown Publishers, Dubuque, Iowa. Reprinted by permission. **Figure 42.8:** From Leland G. Johnson, *Biology*, 2nd edition. Copyright © 1987, Wm. C. Brown Publishers, Dubuque, Iowa. Reprinted by permission. **Figure 42.9:** From Leland G. Johnson, *Biology*, 2nd edition. Copyright © 1987, Wm. C. Brown Publishers, Dubuque, Iowa. Reprinted by permission. **Figure 42.19:** From Leland G. Johnson, *Biology*, 2nd edition. Copyright © 1987, Wm. C. Brown Publishers, Dubuque, Iowa. Reprinted by permission. **Figure 42.20a,b:** From Robert Leo Smith, *The Ecology of Man*, 2nd edition. Copyright © 1976, Harper and Row Publishing. Reprinted by permission.

Illustrators

Gwen Afton:
10.1, 10.4, 10.6, 10.7, 10.9, 10.12, 10.13, 11.3, 11.4, 11.5, 11.6, 11.7, 11.8, 11.9, 11.10, 11.17a, 11.18, Box figure 11.1, 14.9, 22.2, 22.3, 22.4, 22.6, 22.7, 22.8, 22.9, 22.10, 22.11, 22.16, 22.17, 22.18, 22.20, 33.13, 37.4, 37.9, 37.12, 37.14.

Hans and Cassady, Inc.:
2.10, 2.11, 2.12, 2.13, 2.14, 2.15, 2.16, 2.17, 2.18, 2.19, 2.20, 2.22, Box figure 2.1, Box figure 2.2, 4.12, 4.13, 6.2, 6.4, 6.5, 6.6, 6.7, 6.8, 6.9, 6.10, 6.11, 7.2, 7.3, 7.4, 7.5, 7.6, 7.7, Box figure 7.1, Box figure 7.2, Box figure 7.3, 8.2, 8.3, 8.4, 8.5, 8.6, 8.7, 8.12, 8.13, 8.14, Box figure 8.1, 9.1, 9.2, 11.1, 11.2, 11.15, 11.16, 11.17, 23.10, 23.11, 23.12, Box figure 23.2, 36.13, Box figure 36.2.

Ruth Krabach:
1.5, 1.6, 14.7, 14.8, 14.10, 14.12a, 14.16a, 14.18a, 14.20a, 14.21a, 15.2, 15.8, 18.2, 18.3, 18.5, 18.6, 18.7, 18.8, Box figure 32.3, 37.6, 37.11.

Carlyn Iverson:
Colorplate 5.

Marsha Hartsock/Kessler Hartsock Associates:
26.1, 26.2, 26.3, 26.5, 26.6, 26.7, 26.8, 27.1, 27.2, 27.3, 27.4, 27.5, 27.6, 27.7, 27.9, 27.10, 27.11, 27.14, 27.15, 27.16, 28.1, 28.2, 28.3, 28.4, 28.5, 28.6, 28.7, 28.8, 28.9, 28.10, 29.1, 29.2, 29.3, 29.7,

29.8, 29.9, 29.10, Box figure 29.2, 30.3, 30.4, 30.5, 30.6, 30.9, 30.10, 30.14.

Marlene Hill-Werner:
23.2, 23.3, 23.4, 23.5, 23.6, 23.7, 23.8, 23.9, 24.1, 24.3, 24.5, 24.6b, 24.7b, 24.8b, 24.9b, 24.10b, 24.11b, 24.12b, Box figure 24.1, 25.2, 25.3, 25.4, 25.5, 25.8, 25.10, 25.11, 25.12, 36.16, Box figure 36.3, 38.1, 38.8, 38.12, 38.15, 38.16, Box figure 38.22, 38.18.

Marjorie C. Leggitt:
12.1, 12.6, 12.7, 12.8.

John McGee:
14.12b, 14.16b, 14.18b, 14.20b, 14.24b.

Nancy Marshburn:
33.6, 33.9, 33.12, 33.14, 33.16, 34.2, 34.7.

Iris Nichols:
31.5, 31.6, 31.7, 31.9, 31.10, 31.11, 31.12, 31.13, Box figure 31.1a, c.

Rolin Graphics:
1.9, 2.1, 2.2, 2.3, 2.4, 2.5, 2.6, 2.7, 2.8, 2.9, 2.21, 3.1, 3.3, 3.5, 3.6, 3.8, 3.11, Box figure 3.1, Box figure 3.2, 4.1, 4.2, 4.3, 4.4, 4.5, 4.6, 4.7, 4.8, 4.9, 4.10, 4.11, 4.14, 5.1, 5.2, 5.3, 5.4, 5.6, 5.7, 5.8, 5.11, 5.12, Box figure 5.1, 6.1, 8.9, 8.10, 8.11, 9.4, 9.8, 9.10, 9.11, 9.12, 11.12, 13.1, 13.2, 13.3, 13.4, 13.9, 13.10, 13.11, 14.1, 14.11, 14.15, 14.23, Box figure 14.1, 19.4, Box figure 21.1, Box figure 21.2, Box figure 24.2, 25.1, 27.12, 27.13, 28.13, 30.1, 30.7, 30.11, 30.13, 31.1, 31.2, 33.1, 33.3, 33.4, 33.5, 33.8, 33.15, 34.12, 34.13, 35.4, 35.5, 35.6, 35.7, 36.3, 36.8, 36.10, 36.11, 36.14, 36.17, 36.18, 37.1, 37.2, 38.4b, 38.5, 38.6, 38.14, 39.3, 39.7, 39.8, 40.1, 40.2, 40.6, 40.9, 40.10, 41.1, 41.2, 41.3, 41.4, 41.5, 41.6, 41.7, 41.8, 41.9, 41.10, 42.5, 42.9.

Nadine Sokol:
3.2, 3.7, 3.9, 3.10, 3.12, 3.14, 3.15, 3.17, 3.18, 3.19, 3.20, 3.21, 3.22, 3.24, 3.25, 3.26, 5.5, 5.9, 5.10, 6.3, 8.1, 9.7, 15.12, 15.13, 16.1, 16.2, 16.4, 16.5, 16.7, 16.9, 16.10, 16.11, 16.13, 16.16, 16.22, 17.1, 17.2, 17.3, 17.5, 17.7, 17.8, 17.10, 17.11, 17.12, 17.13, 17.18, 17.21, 17.22, 17.24, 18.9, 18.10, 18.11, 18.12, 18.13, 18.14, 18.15, 18.17, 18.18, 18.20, 19.1, 19.2, 19.3, 19.5, 19.6, 19.7, 19.8, 19.9, 19.10, 19.11, 19.12, 19.14, 19.17, 19.18, 19.19, 32.9, 35.3, 35.8, 35.12, 35.13.

Kevin Somerville:
20.1, 20.2, 20.3, 20.4, 20.5, 20.9, 20.10, 20.11, 20.12, 20.13, 20.15, 20.18, 20.19, 20.20, 20.22, 20.24, 21.1, 21.2, 21.4, 21.6, 21.7, 21.8, 21.9, 21.10, 21.11, 21.12, 21.13, 21.14, 21.15, 21.17.

Index

Correlation Chart

Chapters in Miller/Harley, *Zoology*, correlated with laboratory exercises
in some zoology laboratory manuals.

Chapters in Miller/Harley *Zoology*, 1991	Exercises in Miller *General Zoology Laboratory Manual*, 2nd edition, 1991	Exercises in Lytle and Wodsedalek *General Zoology Laboratory Guide*, 11th edition, 1991 (complete version)	Exercises in Lytle and Wodsedalek *General Zoology Laboratory Guide*, 11th edition, 1991 (short version)
Chapters			
1. Zoology in Science and History	1		
3. Cells, Tissues, Organs, and Systems	1, 2	1, 2	1, 2
4. Energy and Enzymes	3		
5. Harvesting Energy Stored in Nutrients	3		
6. The Principles of Cell Division and Reproduction	3	3	3
7. Inheritance Patterns	4		
10. Descriptive Embryology	5	4	4
15. Animal Classification, Organization, and Phylogeny	6		
16. Animallike Protists	7	5	5
17. Multicellular and Tissue Levels of Organization	8, 9	6, 7	6, 7
18. The Triploblastic, Acoelomate Body Plan	10	8	8
19. The Pseudocoelomate Body Plan: Aschelminths	11	10	10
20. Molluscan Success	12	11	11
21. Annelida: The Metameric Design	13	12	12
22. The Arthropods: Blueprint for Success	14	13	13
23. The Insects and Myriapods	14	13	13
25. The Echinoderms	15	14	14
26. Hemichordata and Invertebrate Chordates	16	15	15
27. Fishes: Vertebrate Success in Water	16, 17–23	16, 17	
28. Amphibians: The First Terrestrial Vertebrates	16, 17–23	18	16
29. Reptiles: The First Amniotes	16, 17–23		
30. Birds: Feathers, Flight, and Endothermy	16, 17–23		
31. Mammals: Endothermy, Hair, and Viviparity	16, 17–23	19, 20	17
32. Protection, Support, and Movement	17	16–20	16, 17
33. Communication I: Nerves	18	16–20	16, 17
36. Circulation, Immunity, and Gas Exchange	19	16–20	16, 17
37. Nutrition and Digestion	21	16–20	16, 17
38. Temperature and Body Fluid Regulation	22	16–20	16, 17
39. Reproduction and Development	23	16–20	16, 17